2014 10th Conference on Ph.D. Research in Microelectronics and Electronics

(PRIME 2014)

Grenoble, France
30 June – 3 July 2014

IEEE Catalog Number:	CFP14622-POD
ISBN:	978-1-4799-4993-9

Copyright © 2014 by the Institute of Electrical and Electronic Engineers, Inc
All Rights Reserved

Copyright and Reprint Permissions: Abstracting is permitted with credit to the source. Libraries are permitted to photocopy beyond the limit of U.S. copyright law for private use of patrons those articles in this volume that carry a code at the bottom of the first page, provided the per-copy fee indicated in the code is paid through Copyright Clearance Center, 222 Rosewood Drive, Danvers, MA 01923.

For other copying, reprint or republication permission, write to IEEE Copyrights Manager, IEEE Service Center, 445 Hoes Lane, Piscataway, NJ 08854. All rights reserved.

***This publication is a representation of what appears in the IEEE Digital Libraries. Some format issues inherent in the e-media version may also appear in this print version.**

IEEE Catalog Number: CFP14622-POD
ISBN 13:   978-1-4799-4993-9

Additional Copies of This Publication Are Available From:

Curran Associates, Inc
57 Morehouse Lane
Red Hook, NY 12571 USA
Phone: (845) 758-0400
Fax: (845) 758-2633
E-mail: curran@proceedings.com
Web: www.proceedings.com

2014 10th Conference on Ph.D. Research in Microelectronics and Electronics (PRIME 2014)

Grenoble, France
30 June - 3 July 2014

IEEE Catalog Number: CFP14622-POD
ISBN: 978-1-47994-993-9

10th Conference on Ph.D. Research In Microelectronics and Electronics

PRIME 2014

CONFERENCE PROGRAM

Monday, June 30th

Welcome - Workshop on Green Electronics I
Chair: Emmanuel Pistono, *Univ. de Grenoble, IMEP-LAHC*
Monday, June 30th, 10:20 – 10:40
Room: Petit Salon

Session 1: Workshop on Green Electronics I
Monday, June 30th, 10:40 – 12:20
Room: Petit Salon

1A1 – Invited - IP rights, and particularly patents - How and what for?   N/A
10:40 Sigrid Thomas[1], Jeremy Harrison[2]
       [1]*CEA*
11:40 [2]*Cabinet Beaumont*

1A2 – Invited – Mixed-signal verification challenges   1
11:40 Nicolas Delorme
12:20 *Asyn*

12:20-13:20 - Lunch Break

Session 2: Workshop on Green Electronics II

Monday, June 30th, 13:20 – 15:00
Room: Petit Salon

13:20 14:10	**2A1 - Invited - Electronics for Energy Management** Bernard Courtois *CMP*	N/A
14:10 15:00	**2A2 - Invited - Reduction of IPs Energy Consumption with Ultra-Low Voltage Supply** Fady Abouzeid *STMicroelectronics*	2

15:00-15:20 - Coffee Break

Session 3: Workshop on Green Electronics III

Monday, June 30th, 15:20 – 17:00
Room: Petit Salon

15:20 16:10	**3A1 - Invited - Power Management of Ultra Low Power Radio and Microcontroller Communication** Frédéric Hasbani *STMicroelectronics*	N/A
16:10 17:00	**3A2 - Invited - RF Power Gating Techniques for Ultra Low Power Communication Systems** Sylvain Bourdel[1], I. Ben Amor[2], J.F. Pons[2], N Dehaese[2], R. Vauché[2], J. Gaubert[2] *[1]Univ. Grenoble Alpes, IMEP-LAHC* *[2]IM2NP*	3

Tuesday July 1st

Session 4: Opening and Plenary

Chairs: Dominique Morche (CEA-Leti), Gilles Sicard (Univ. Grenoble, TIMA), Yannis
Le Guennec (Univ. Grenoble, IMEP-LAHC)
Tuesday, July 1st, 09:00 – 09:30
Room: Petit Salon

Session 5A: Simulation Approaches of Analog and Digital Systems

Tuesday, July 1st, 09:30 – 10:10
Room: B221
Chair: Marie-Minerve Louërat, *Université Pierre et Marie Curie, LIP6, Paris*

09:30 **5A1 -Accurate Modeling of Ultra Low-Power Sigma-Delta Modulator** 4
09:50 Luca Giuffredi, Giorgio Pietrini and Andrea Boni
University of Parma

09:50 **5A2 - System-on-Chip Verification: TLM-to-RTL Assertions Transformation** 8
10:10 Zeineb Bel Hadj Amor, Laurence Pierre and Dominique Borrione
Univ. Grenoble Alpes TIMA

Session 5B: ADC/DAC/Mixed I
Tuesday, July 1st, 09:30 – 10:10
Room: Petit Salon
Chair: Marcello De Matteis, *University of Milano Bicocca, University of Salento*

5B1 - A 1.2V low-power high-resolution noise-shaping ADC using multistage time 12
09:30 **encoding converters for Biomedical Applications**
09:50 Francisco Javier Pérez Sanjurjo, Enrique Prefasi Sen and Luis Hernandez Corporales
Universidad Carlos III Madrid

09:50 **5B2 - A Low Power Second Order Current Mode Continuous Time Sigma Delta ADC** 16
10:10 **with 98 dB SNDR**
Sina Parsnejad, Feyyaz Melih Akcakaya and Gunhan Dundar
Bogazici University

Session 5C: Signal Processing
Tuesday, July 1st, 09:30 – 10:10
Room: B225
Chairs: *Emil Novakov, Univ. Grenoble Alpes IMEP-LAHC*

5C1 - Continuous Time Analog to Digital Conversion in Interferer Resistant Wake Up 20
09:30 **Radios**
09:50 Alin Ratiu[1,2,3], Dominique Morche[2], Bruno Allard[3], Xuefand Lin-Shi[3], Jacques Verdier[3]
[1]Univ. Grenoble Alpes, [2]CEA-Leti, [3]Univ. de Lyon

5C2 - Green improvements of IEEE 802.11 directional multi-gigabit physical layer 24
09:50 **specifications**
10:10 Marc-Antoine Bouzigues[1], Isabelle Siaud[1], Maryline Hélard[2] and Anne-Marie Ulmer-Moll[1]
[1]Orange, [2]IETR - UMR CNRS 6164

10:10-10:40 - Coffee Break

Session 6A: Simulation Approaches of Analog and Digital Systems

Tuesday, July 1st, 10:40 – 12:20
Room: B221
Chair: Marie-Minerve Louërat, *Université Pierre et Marie Curie, LIP6, Paris*

6A1 - An Efficient Simulation Methodology for Electrical Energy Systems   28
10:40 Alessandro Sassone, Sara Vinco, Massimo Poncino and Enrico Macii
11:00 *Politecnico di Torino*

6A2 - Macromodel-based Signal and Power Integrity simulations of an LP-DDR2   32
11:00 **interface in mSiP**
11:20 Gianni Signorini[1,2], Stefano Grivet-Talocia[3], Igor Simone Stievano[3] and Luca Fanucci[2]
[1]Intel Mobile Communications GmbH, [2]University of Pisa, [2]Politecnico di Torino

11:20 **6A3 - A SystemC Bluetooth Network Simulator**   36
11:40 Cristiano Scavongelli and Massimo Conti
Università Politecnica delle Marche

11:40 **6A4 - A new algorithm for convergence verification in circuit level simulations**   40
12:00 Francesco Lannutti[1], Francesco Menichelli[1], Paolo Nenzi[2] and Mauro Olivieri[1]
[1]Sapienza University of Rome, [2]ENEA

6A5 -Simulation methodology for Large-Bandwidth Track-and-Hold microwave   44
12:00 **circuit**
12:20 Arnaud Meyer[1,2], Patricia Desgreys[1], Hervé Petit[1], Bruno Louis[2], Vincent Petit[2]
[1]Telecom Paristech, [2]Thales Systèmes aéroportés

Session 6B: ADC/DAC/Mixed I

Tuesday, July 1st, 10:40 – 12:20
Room: Petit Salon
Chair: Marcello De Matteis, *University of Milano Bicocca, University of Salento*

6B1 - Invited - Design of Class-D Amplifier for Audio Portable Solutions   48
10:40 Angelo Nagari
11:20 *STMicroelectronics, Grenoble*

11:20 **6B2 - Time Interleaved Current Steering DAC for Ultra-High Conversion Rate**   49
11:40 Da Feng[1], Sai-Weng Sin[1], Edoardo Bonizzoni[2] and Franco Maloberti[2]
[1]University of Macau, [2]University of Pavia

11:40 **6B3 - Calibrated Switched Capacitor Integrators based on Current Conveyors and its**   53
12:00 **application to Delta Sigma ADC**
Harish Balasubramaniam and Klaus Hofmann
Technical University of Darmstadt

	6B4 - Statistical Analysis of Harmonic Distortion in a Differential Bootstrapped	57
12:00	**Sample and Hold Circuit**	
12:20	Gaël Kamdem De Teyou[1,2], Hervé Petit[1], Patrick Loumeau[1] and Hussein Fakhoury[1]	
	[1]Télécom ParisTech, [2]Renesas Mobile	

Session 6C: Signal Processing

Tuesday, July 1st, 10:40 – 12:20
Room: B225
Chairs: Luca Fanucci, *University of Pisa / Emil Novakov, Univ. Grenoble Alpes IMEP-LAHC*

	6C1 - Multiple-Event Direct to Histogram TDC in 65nm FPGA Technology	61
10:40	Neale Dutton[1,2], Johannes Vergote[1,2] and Salvatore Gnecchi[1,2], Lindsay Grant[2], David	
11:00	Lee[2], Sara Pellegrini[2], Bruce Rae[2], Robert Henderson[1]	
	[1]University of Edinburgh, [2]STMicroelectronics United Kingdom	

11:00	**6C2 - High-Speed Serial Interface with a Full Digital Delay-Loop**	66
11:20	Christian Harder, Bastian Mohr and Stefan Heinen	
	RWTH Aachen University	

11:40	**6C4 - A vector implementation of a fast Fourier transform on DSP and NVIDIA CUDA**	70
12:00	**platforms**	
	Bartosz Pikacz and Jacek Gambrych	
	Warsaw University of Technology	

12:00	**6C5 - Experimental validation of a new power line communication system for battery**	74
12:20	**management**	
	Jérémie Jousse[1,2], Nicolas Ginot[2], Christophe Batard[2] and Elisabeth Lemaire[1]	
	[1]CEA LITEN, [2] PRES LUNAM - UMR CNRS 6164	

12:20-13:20 - Lunch Break

Session 7A: NEMS, MEMS, Sensors

Tuesday, July 1st, 13:20 – 15:00
Room: B221
Chair: Catherine Dehollain, *Swiss Federal Institute of Technology (EPFL)*

	7A1 - Carbon Nanotube Based Temperature Sensors Fabricated by Large-Scale Spray	78
13:20	**Deposition**	
13:40	Engin Cagatay, Aniello Falco, Alaa Abdellah and Paolo Lugli	
	Technische Universität München	

13:40	**7A2 - Methodology Modeling of MaE-fabricated Porous Silicon Nanowires**	82
14:00	Aleandro Antidormi[1,2], Diego Chiabrando[1,2], Maria Grazia Graziano[1], Luca Boarino[2] and	
	Gianluca Piccinini[1]	

[1]Politecnico di Torino, [2]I.N.Ri.M, Istituto Nazionale Ricerca Metrologica

7A3 - Analysis and Modeling of Four-Folded Vertical Hall Devices in Current Domain   86
14:00 Hadi Heidari, Edoardo Bonizzoni, Umberto Gatti and Franco Maloberti
14:20 *University of Pavia*

7A4 - 3-Terminal Tungsten CMOS-NEM Relay   90
14:20 Martín Riverola, Gabriel Vidal-Álvarez, Francesc Torres and Núria Barniol
14:40 *Universitat Autònoma de Barcelona*

7A5 - Tunable Transimpedance Sustaining-Amplifier for High Impedance CMOS-   94
14:40 **MEMS Resonators**
15:00 Guillermo Sobreviela, Arantxa Uranga and Nuria Barniol
Universitat Autònoma de Barcelona

Session 7B: Reliability Analysis of Analog and Digital Systems
Tuesday, July 1st, 13:20 – 15:00
Room: Petit Salon
Chair: Ian O'Connor, *Lyon Institute of Nanotechnology (INL) University of Lyon*

7B1 - Invited - AUTOMICS: A Novel CAD Framework for Substrate Modeling   98
13:20 Ramy Iskander
14:00 *Université Pierre et Marie Curie, LIP6, Paris*

7B2 - Sensitivity based Methodologies for Process Variation Aware Analog IC   102
14:00 **Optmization**
14:20 Engin Afacan, Gönenç Berkol, Faik Başkaya and Günhan Dündar
Bogazici University

14:20 **7B3 - Impact of enhanced contact doping on minority carriers diffusion currents**   106
14:40 Camillo Stefanucci, Pietro Buccella, Maher Kayal and Jean Michel Sallese
École polytechnique fédérale de Lausanne

14:40 **7B4 - Reliability Analysis of Logic Circuits Using Probablistic Techniques**   110
15:00 Satish Grandhi, Christian Spagnol and Emanuel Popovici
University College Cork, Ireland

Session 7C: Energy Harvesting
Tuesday, July 1st, 13:20 – 15:00
Room: B225
Chair: Jean-Marc Duchamp, *Univ. de Grenoble IMEP-LAHC.*

7C1 - A 40mV Start up Voltage DC–DC Converter for Thermoelectric Energy Harvesting Applications 114

13:20

13:40 Carlo Veri[1], Mirko Pasca[1], Stefano D'Amico[1], L. Francioso[2]
[1]University of Salento, [2] CNR- IMM

7C2 - Wire-bonds Used as Matching Inductor in RF Energy Harvesting Applications 118

13:40

14:00 Dino Michelon[1,2], Emmanuel Bergeret[1], Mathieu Egels[1] and Antonio Di Giacomo[2]
[1]IM2NP, [2]STMicroelectronics

7C3 - FEM modeling of vertically integrated nanogenerators in compression and flexion modes 122

14:00

14:20 Ran Tao, Ronan Hinchet, Gustavo Ardila Rodreguez, Laurent Montes and Mireille Mouis
Univ. Grenoble Alpes IMEP-LAHC

7C4 - Design of a low power wireless sensor network node for distributed active vibration control system 126

14:20

14:40 Mateusz Zielinski[1], Fabien Mieyeville[1], David Navarro[1] and Olivier Bareille[2]
[1]Lyon Institute of Nanotechnology (INL) University of Lyon,[2]LTDS

7C5 - Co-design of Dual-band GSM Filtenna based on Printed-IFA for Energy Harvesting 130

14:40

15:00 Manh Ha Hoang[1,2], Van Hieu Nguyen[2], Thi Quynh Van Hoang[2] and Tan Phu Vuong[1]
[1]Univ. Grenoble Alpes IMEP-LAHC, [2]Ho Chi Minh City of Technology

15:00-15:20 - Coffee Break

Session 8A: Power Amplifier and Detector

Tuesday, July 1st, 15:20 – 17:00
Room: B221
Chair: Estelle Lauga-Larroze, *Univ. Grenoble Alpes IMEP-LAHC*

8A1 - Sub-Threshold Based Power Detector for Low-Cost Millimeter-Wave Applications 134

15:20

15:40 Ayssar Serhan, Estelle Lauga-Larroze and Jean-Michel Fournier
Univ. Grenoble Alpes IMEP-LAHC

8A2 - Structured Design to Optimize the Output Power of Stacked Power Amplifiers 138

15:40

16:00 Elena Sobotta, Robert Wolf, David Fritsche and Frank Ellinger
TU Dresden

8A3 - 66-87 GHz Power Amplifier with 20dBm 1-dB compression point and 35% peak PAE in a 55nm SiGe technology 142

16:00

16:20 David Del Rio[1,2], Roc Berenguer[1,2], Ainhoa Rezola[1,2] and Juan Francisco Sevillano[1,2]
[1]CEIT, [2] Technological Campus of University of Navarra (TECNUN)

8A4 - A Linear Model of Efficiency for Switched-Capacitor RF Power-Amplifiers 146

16:20 Antonio Passamani[1,2], Davide Ponton[2], Gerhard Knoblinger[2] and Andrea Bevilacqua
16:40 [1]University of Padova Italy, [2]Intel Mobile Communications Austria

8A5 - Analysis and Design of a High Power, High Gain SiGe BiCMOS Output Stage for 150
16:40 **Use in a Millimeter-Wave Power Amplifier**
17:00 Ramses Pierco, Timothy De Keulenaer, Guy Torfs and Johan Bauwelinck
INTEC/IMEC - Ghent University

Session 8B: CMOS Sensor Design

Tuesday, July 1st, 15:20 – 17:00
Room: Petit Salon
Chair: Gianluca Piccinini, *Politecnico di Torino*

8B1 - Invited -Sensor Interfaces: Keys to Success of Integrated Sensor Systems N/A
15:20 Franco Maloberti
16:00 *University of* Pavia

8B2 - Base-Station Design for Passive UHF RFID Tags with Pulse-Width Modulated 154
16:00 **Backscattering**
16:20 Kerem Kapucu and Catherine Dehollain
École polytechnique fédérale de Lausanne

8B3 - Backside Illuminated Wafer-to-Wafer Bonding Single Photon Avalanche Diode 158
16:20 **Array**
16:40 Yu Zou[1], Danilo Bronzi[1], Federica Villa[1] and Sascha Weyers[2]
[1]Politecnico di Milan, [2]Fraunhofer IMS

8B4 - 5x5 SPAD Matrices for the Study of the Trade-offs between Fill Factor, Dark 162
Count Rate and Crosstalk in the Design of CMOS Image Sensors
16:40 Manuel Moreno Garcia, Rocio Del Rio Fernandez, Oscar Guerra Vinuesa and Angel
17:00 Rodriguez Vazquez
Instituto de Microelectronica de Sevilla

Session 8C: Material and Process Challenges

Tuesday, July 1st, 15:20 – 17:00
Room: B225
Chair: Irina Ionica, *Univ. Grenoble Alpes IMEP-LAHC*

8C1 - Structural, magnetic and dielectric properties of nanocomposites for RF 166
applications
15:20 Hélène Takacs[1,2], Bernard Viala[1], Jean-Hervé Tortai[2], Juvenal Alarcon Ramos[1], Marie
15:40 Bousquet[1], Florence Duclairoir[3], Cécile Gourgon[2]
[1]CEA Leti, [2]LTM –CNRS- UJF, CEA INAC

8C2 - Fabrication and characterization of ECM memories based on a solid electrolyte Ge2Sb2Te5　　170

15:40
16:00 Bocquet Marc, Ouled-Khachroum Toufik, Putero Magali, Deleruyelle Damien and Rebora Charles
IM2NP - Aix Marseille University

8C3 -Role of Nanowire Length in Morphological and Electrical Properties of Silicon Nanonets　　174

16:00
16:20 Pauline Serre[1], Pierre Chapron[2,3], Quentin Durlin[3], Anaïs Francheteau[3], Arthur Lantreibecq[3] and Céline Ternon[3]
[1]LTM, [2] Univ. Grenoble Alpes IMEP-LAHC, [3]Univ. Grenoble Alpes LMGP

8C4 - Wavy channel thin film transistor for area efficient high performance and low power applications　　178

16:20
16:40 Amir Hanna, Mohamed Ghoneim, Galo Sevilla and Muhammad Hussain
King Abdullah University Of Science and Technology

8C5 - Investigation of the optics system carbonaceous contamination induced by chemically amplified resist outgassing under e-beam radiation　　182

16:40
17:00 Armel-Petit Mebiene-Engohang[1], Marie-Line Pourteau[2], Jean-Christophe Marusic[2], Laurent Pain[2], Sylvain David[3], Sebastien Labau[3] and Jumana Boussey[3]
[1]STMicroelectronics, [2]CEA-Leti, [3]LTM-CNRS

Wednesday July 2nd

Session 9A: Power Electronics: Integration, Modeling and Applications

Wednesday, July 2nd, 08:50 – 10:10
Room: B221
Chair: Nicolas Rouger, *Univ. Grenoble Alpes G2ELab, CNRS G2Elab*

9A1 - A Suitable Inductor Modeling for DC-DC Converters　　186

08:50
09:10 Andrea Mocci, Alessandro Serpi, Ignazio Marongiu and Gianluca Gatto
DIEE, University of Cagliari

9A2 - Monolithically Integrated Voltage Level Shifter for Wide Bandgap Devices-Based Converters　　190

09:10
09:30 Romain Grezaud[1,2], François Ayel[1], Nicolas Rouger[2] and Jean-Christophe Crebier[2]
[1]CEA Leti, [2] Univ. Grenoble Alpes G2Elab, CNRS G2ELab

9A3 - Extensive Electro-Thermal Simulation Methodology for Automotive High Power Circuits   194

09:30
09:50
Adrian-Gabriel Bajenaru[1], Cristian Mihai Boianceanu[1], Fabio Ballarin[2] and Gheorghe Brezeanu[3]
[1]Infineon Technologies Romania, [2]Infineon Technologies Italia, [3]Politehnica University of Bucharest

9A4 - Two-Dimensional Optical Beam Induced Current measurements in 4H-SiC bipolar diodes   198

09:50
10:10
Hassan Hamad, Pascal Bevilacqua, Christophe Raynaud and Dominique Planson
INSA de Lyon - Ampère Laboratory

Session 9B: Emerging Technologies for Digital Circuits

Wednesday, July 2nd, 08:50 – 10:10
Room: Petit Salon
Chair: Edith Beigné, *CEA-Leti*

9B1 - Invited - Towards the Use of Functionality-Enhanced Devices : A Transversal Design Approach   202

08:50
09:30
Pierre-Emmanuel Gaillardon
Swiss Federal Institute of Technology (EPFL)

9B2 - Safe Operation Region Characterization for Quantifying the Reliability of CMOS Logic Affected by Process Variations   203

09:30
09:50
Usman Khalid, Antonio Mastrandrea and Mauro Olivieri
Sapienza University of Rome

09:50
10:10
9B3 - High Performance Electronics on Flexible Silicon for Brain Computing   207

Galo Torres Sevilla, Jhonathan Rojas and Muhammad Mustafa Hussain
King Abdullah University of Science and Technology

10:10-10:40 - Coffee Break

Session 10A: Digital Techniques I

Wednesday, July 2nd, 10:40 – 12:20
Room: B221
Chair: Edith Beigné, *CEA-Leti*

10A1 - Towards Formal Verification of Reset Sequence in Fully Asynchronous Digital Circuits   211

10:40
11:00
Oleksandr Melnychenko[1] and Hans-Peter Kreuter[2]
[1]Vienna University of Technology, [2]Infineon Technologies Austria AG

10A2 - A New Circuit Topology for Floating High Voltage Level Shifters   215

11:00 Dawei Liu, Simon Hollis and Bernard Stark
11:20 *University of Bristol*

10A3 - Probabilistic Saboteur-based Simulated Fault Injection Techniques for Low   219
11:20 **Supply Voltage Interconnects**
11:40 Sergiu Nimara, Alexandru Amaricai, Oana Boncalo and Mircea Popa
Politehnica University Timisoara

11:40 **10A4 - Design of a secure architecture for scalar multiplication on elliptic curves**   223
12:00 **Simon Pontié and Paolo Maistri**
Univ. Grenoble Alpes TIMA

10A5 - ASIC design of a Phoneme Recogniser based on Discrete Wavelet Transforms   227
12:00 **and Support Vector Machines**
12:20 Michelle Cutajar, Edward Gatt, Ivan Grech and Owen Casha
University of Malta

Session 10B: Millimeter Wave Circuits

Wednesday, July 2nd, 10:40 – 12:20
Room: Petit Salon
Chair: José Luis Gonzalez-Jimenez, *CEA-Leti*

10:40 **10B1 - Invited - ESD Co-Design Methodologies for RF and mmW Circuits**   231
11:20 Roc Berenguer
CEIT

10B2 - Filterless millimetre-wave optical generation using optical phase modulators   232
11:20 **without DC bias**
11:40 Rabiaa Guemri[1], Frédéric Lucarz[1], Daniel Bourreau[1], Camilla Kärnfelt[1], Jean-Louis de
Bougrenet de la Tocnaye[1], Trevor Hall[2]
[1]Télécom Bretagne, [2]University of Ottawa

11:40 **10B3 - A Digitally Controlled Threshold Adjustment Circuit in a 0.13um SiGe BiCMOS**   236
12:00 **Technology for Receiving Multilevel Signals up to 80Gbps**
Timothy De Keulenaer, Guy Torfs, Ramses Pierco and Johan Bauwelinck
INTEC/IMEC - Ghent University

12:00 **10B4 -A 60 GHz down-conversion mixer using a novel topology in 65 nm CMOS**   240
12:20 Chong Wang, Zhiqun Li, Qin Li, Yang Liu, Jia Cao and Zhigong Wang
Southeast University, China

12:20-13:20 - Lunch Break

Session 11A: Signal Generation Circuits

Wednesday, July 2nd, 13:20 – 15:00
Room: Petit Salon
Chair: Roc Berenguer, *CEIT*

13:20
13:40 **11A1 - A High Conversion Gain Millimeter-Wave Frequency Doubler in 65nm CMOS**   244
Yang Liu, Zhiqun Li, Qin Li, Chong Wang and Zhigong Wang
Southeast University, China

11A2 - Integrated Multi-band Fractional-N PLL for FMCW Radar Systems at 2.4 and   248
13:40 **5.8 GHz**
14:00 Niko Joram, Bastian Lindner, Jens Wagner and Frank Ellinger
TU Dresden

14:00 **11A3 - A Low Power, Small Area, Fully Integrated 5.5GHz CMOS LC-VCO**   252
14:20 Shaahin Haddadinejad[1], Achim Noculak[2], Michael Hinz[1] and Bernd Meinerzhagen[1]
[1]BST TU Braunschweig, [2]Institut fuer Theoretische Elektrotechnik der RWTH Aachen

11A4 - Comparative Analyses of Phase Noise in Differential Oscillator Topologies in   256
14:20 **28 nm CMOS Technology**
14:40 Ilias Chlis[1,2], Domenico Pepe[1] and Domenico Zito[1,2]
[1]Tyndall National Institute, [2]University College Cork

14:40 **11A5 - Pulsed oscillations generator based on initial conditioned and switched cross**   260
15:00 **coupled MOS oscillator. Application to the synchronization of the pulsed oscillations**
Clement Jany[1], Alexandre Siligaris[1], Philippe Ferrari[2] and Pierre Vincent[1]
[1]CEA-Leti, [2]Univ. Grenoble Alpes IMEP-LAHC

15:00-15:20 - Coffee Break

Session 12A: Amplifiers

Wednesday, July 2nd, 15:20 – 17:00
Room: Petit Salon
Chair: Piero Malcovati, *University of Pavia*

12A1 - High Precision Bidirectional Chopper Instrumentation Amplifier With   264
15:20 **Negative and Positive Input Common Mode Range**
15:40 Matei Nicolae Stan[1,2], Laurentiu Creosteanu[1] and Gheorghe Brezeanu[2]
[1]On Semiconductor, [2]University Politehnica of Bucharest

12A2 - Comparative Study of a Fully Differential Op Amp in FinFET and Planar   268
Technologies
15:40 Sébastien Morrison[1,2], Bertrand Parvais[2], Gerd Vandersteen[1], Kenichi Miyaguchi[1],
16:00 Abdelkarim Mercha[2] and Piet Wambacq[1]
[1]VUB, [2]imec

12A3 - An novel architecture for current-feedback instrumentation amplifiers with rail-to-rail input range　　　272

16:00
16:20　Francesco Del Cesta[1], Aurelio Nunzio Longhitano[1], Paolo Bruschi[1] and Massimo Piotto[2]
[1]University of Pisa, [2]IEIIT Pisa

12A4 - A Bootstrap Transimpedance Amplifier for High Speed Optical Transcutaneous Wireless Links　　　276

16:20
16:40　Tianyi Liu, Zhicheng Cai, Jens Anders and Maurits Ortmanns
University of Ulm

12A5 - High Accuracy Current Sense Amplifier With Extended Input Common Mode Range　　　280

16:40
17:00　Razvan Puscasu[1,2], Pavel Brinzoi[1], Laurentiu Creosteanu[1] and Gheorghe Brezeanu[2]
[1]ON Semiconductor, [2]University Politehnica of Bucharest

Session 11B - 12B: Company Fair
Wednesday, July 2nd, 13:20 – 17:00
Room: Grand Salon

Thursday July 3rd

Session 13A: Digital Techniques II
Thursday, July 3rd, 08:50 – 10:10
Room: B221
Chair: Jean-Frederic Christmann, *CEA-Leti*

13A1 - Test and Diagnosis of FPGA Cluster Using Partial Reconfiguration　　　284

08:50　Rehman Saif-Ur, Mounir Benabdenbi and Lorena Anghel
09:10　*Univ. Grenoble Alpes, TIMA*

09:10　**13A2 - A New Hardware Implementation of The Advanced Encryption Standard**　　　288
09:30　**Algorithm for Automotive Application**
　　　Riccardo Cassettari, Luca Fanucci and Giorgio Boccini
　　　University of Pisa

13A3 - FPGA Design for the Decoding Functions of the Physical Layer Adaptation Subsystem of XG-PON Optical Network Unit/Terminal　　　292

09:30
09:50　Georgios Georgis[1], Charalambos Tzeranis[1], George Synnefakis[2] and Dionysios Reisis[1]
[1]National and Kapodistrian University of Athens, [2]inAccess S.A.

13A4 - Fast Register Criticality Evaluation in a SPARC Microprocessor　　　296

09:50　Kais Chibani, Michele Portolan and Régis Leveugle
10:10　*Univ. Grenoble Alpes, TIMA*

Session 13B: Power Converter and Integrated Control

Thursday, July 3rd, 08:50 – 10:10
Room: Petit Salon
Chair: Pierre-Olivier Jeannin, *Univ. Grenoble Alpes G2Elab*

08:50
09:30
13B1 – Invited - Challenges and Benefits of Microelectronics for Power Electronics: from Integrated Optical Driving to Optimized Power Semiconductor Switches   **300**
Nicolas Rouger
Univ. Grenoble Alpes / CNRS / G2Elab

09:30
09:50
13B2 - An Improved DC-Link Voltage Equalization for Three-Level Neutral-Point Clamped Converters   **301**
Mario Porru, Alessandro Serpi, Ignazio Marongiu and Alfonso Damiano
University of Cagliari

09:50
10:10
13B3 - Simplified Review of DCDC Switching Noise and Spectrum Contents   **305**
Adnan Fares[1], Sami Ajram[1] and Guy Cathébras[2]
[1]SL3J SYSTEMS, [2]LIRMM

Session 13C: Modeling and Characterization for Emerging Devices

Thursday, July 3rd, 08:50 – 10:10
Room: B225
Chair: Marie-Minerve Louërat, *Université Pierre et Marie Curie, LIP6, Paris*

08:50
09:10
13C1 - Quantifying the Figures of Merit of Graphene-Based Adiabatic Pass-XNOR Logic (PXL) Circuits   **309**
Valerio Tenace and Andrea Calimera
Politecnico di Torino

09:10
09:30
13C2 - 3D Modeling of CNT Networks for sensing applications   **313**
Simone Colasanti, Vijay Deep Bhatt and Paolo Lugli
Technical University Munich (TUM)

09:30
09:50
13C3 - A Quantitative Approach to Testing in Quantum dot Cellular Automata: NanoMagnet Logic Case   **317**
Giovanna Turvani, Fabrizio Riente, Mariagrazia Graziano and Maurizio Zamboni
Politecnico di Torino

09:50
10:10
13C4 - A Compact Model for Phase Change Memory Cells   **321**
Erika Covi[1,2], Athanasios Kiouseloglou[1,3], Alessandro Cabrini[1] and Guido Torelli[1]
[1]University of Pavia, [2]Laboratorio MDM, IMM – CNR, [3]CEA – Leti

10:10-10:40 - Coffee Break

Session 14A: ADC/DAC/Mixed II

Thursday, July 3rd, 10:40 – 12:20
Room: B221
Chair: Marc Sabut, *STMicroeletronics*

10:40 11:00	**14A1 - High Resolution Current-Mode CCO-Based Continuous Time Delta-Sigma Modulators for Sensor-Array Applications** Anouar Laifi, Mohammed Adib Al Abaji and Roland Thewes *TU Berlin*	325

11:00 11:20	**14A2 - A 32-Channel 12-bits 65nm Wilkinson ADC for CMS Central Tracker** Tommaso Vergine[1,2], Marcello De Matteis[1,3], Andrea Baschirotto[2] and Alessandro Marchioro[4] *[1]University of Pavia, [2]University of Milano Bicocca, [3]University of Salento, [4]CERN*	329

11:20 11:40	**14A3 - Design of a Low-Power Calibratable Charge-Redistribution SAR ADC** Soheil Aghaie, Jan Henning Mueller, Ralf Wunderlich and Stefan Heinen *RWTH Aachen University*	333

11:40 12:00	**14A4 - A 10 bit 12.8 MS/s SAR Analog-to-Digital Converter in a 250 nm SiGe BiCMOS Technology** Johannes Digel, Markus Grözing and Manfred Berroth *University of Stuttgart*	337

Session 14B: Voltage and Current References

Thursday, July 3rd, 10:40 – 12:20
Room: Petit Salon
Chair: Piero Malcovati, *University of Pavia*

10:40 11:20	**14B1 - Invited - Continuous Time Analog Filters Design in Nanometer-Scale CMOS Technologies** Marcelo De Matteis[1,2], Andrea Baschirotto[3] *[1]University of Pavia, [2]University of Salento, [3]University of Milano Bicocca*	341

11:20 11:40	**14B2 - A compact low-noise fully differential bandgap voltage reference with intrinsic noise filtering** Aurelio Longhitano[1], Francesco Del Cesta[1], Paolo Bruschi[1] and Roberto Simmarano[2] *[1]University of Pisa, [2]sensichips*	345

11:40 12:00	**14B3 - A 65nm CMOS Technology Radiation-Hard Bandgap Reference Circuit** Tommaso Vergine[1,2], Stefano Michelis[3], Marcello De Matteis[2,4] and Andrea Baschirotto[2] *[1]University of Pavia, [2]University of Milano Bicocca, [3]CERN, [4] University of Salento*	349

12:00 **14B4 - A modified CMOS nano-power resistorless current reference circuit**   353
12:20 Shailesh Singh Chouhan and Kari Halonen
Aalto University

Session 14C: RF/mmW Measurement & Modeling Techniques
Thursday, July 3rd, 10:40 – 12:20
Room: B225
Chair: Roc Berenguer, *CEIT*

14C1 - Towards the determination of GaN HEMT large signal model parameters by   357
10:40 **Time Domain Reflectometry method**
11:00 Marian Bernát, Alexander Šatka, Aleš Chvála, Jaroslav Kováč, Ľubomír Sládek and
Daniel Donoval
Slovak University of Technology in Bratislava

11:00 **14C2 - A fast and functional technique for the noise figure measurement of**   361
11:20 **differential amplifiers**
Yogadissen Andee[1], Jérôme Prouvée[1], François Graux[2] and François Danneville[3]
[1]*CEA-Leti,* [2]*Rohde&Schwarz,* [3]*IEMN*

11:20 **14C3 - Half-Thru De-embedding Method for Millimeter- Wave and Sub-Millimeter-**   365
11:40 **Wave Integrated Circuits**
Vipin Velayudhan, Emmanuel Pistono and Jean-Daniel Arnould
Univ. Grenoble Alpes, IMEP-LAHC

11:40 **14C4 - Design of passive filters using dual-mode embedded dielectric resonator**   369
12:00 Úrsula Martínez-Iranzo, Bahareh Moradi and Joan Garcia-Garcia
Universidad Autónoma de Barcelona

12:00 **14C5 - The Impact of the Q-Factor of the Parasitic Capacitances of RF Transistors on**   373
12:20 **their Load Modulation Capabilities**
David Seebacher[1], Wolfgang Bösch[1], Peter Singerl[2] and Christian Schuberth[2]
[1]*TU Graz,* [2]*Infineon Technologies Austria AG*

12:20-13:20 - Lunch Break

Session 15A: Device Technical Trends
Thursday, July 3rd, 13:20 – 15:00
Room: B221
Chair: Quentin Rafhay, *Univ. Grenoble Alpes IMEP-LAHC*

13:20 **15A1 - Comprehensive Analysis of traps in InGaP/GaAs HBT by GR noise**   377
13:40 Ahmad Al Hajjar, Jean-Christophe Nallatamby and Michel Prigent
Xlim

15A2 - Optimization of Low-Resistance State Performance in Ge-rich GST Phase Change Memory   381

13:40 Athanasios Kiouseloglou[1,2], Gabriele Navarro[1], Alessandro Cabrini[2], Guido Torelli[2] and
14:00 Luca Perniola[1]
[1]CEA-Leti, [2]University of Pavia

15A3 - Design Considerations for Monolithically Integrated Fully-Depleted CMOS   385
14:00 **Image Sensors**
14:20 Jean-Baptiste Lincelles[1], Olivier Marcelot[1], Pierre Magnan[1] and Olivier Saint-Pé[2]
[1]ISAE, [2]Airbus Defense and Space

14:20 **15A4 - TIA optimization for on-package multi-core optical network receivers**   389
14:40 Robert Polster[1], Jose-Luis Gonzalez Jimenez[1] and Eric Cassan[2]
[1]CEA-Leti, [2]EF UMR 8622, University Paris Sud

15A5 - Characterization and modeling of low frequency noise in 0.13 µm BiCMOS   393
14:40 **SiGe :C heterojunction bipolar trasnsistors**
15:00 Marcelino Seif[1], Fabien Pascal[1], Bruno Sagnes[1] and Sebastien Haendler[2]
[1]IES - Université Montpellier 2, [2]STMicroelectronics

Session 15B: Sensors on Flexible Substrate

Thursday, July 3rd, 13:20 – 15:00
Room: Petit Salon
Chair: Catherine Dehollain, *Swiss Federal Institute of Technology (EPFL)*

13:20 **15B1 - Invited - Towards Flexible and Conformable Electronics**   397
14:00 Ravinder S. Dahiya
University of Glasgow

14:00 **15B2 - Integrated Low-Noise Current Amplifier for Glass-Based Nanopore Sensing**   399
14:20 Pietro Ciccarella[1], Marco Carminati[1], Raquel Fraccari[2], Azadeh Bahrami[2] and Giorgio
Ferrari[1]
[1]Politecnico di Milano, [2]Imperial College London

14:20 **15B3 - Bendable Piezoresistive Sensors by Screen Printing MWCNT/PDMS**   403
14:40 **Composites on Flexible Substrates**
Saleem Khan[1,2], Leandro Lorenzelli[2] and Ravinder Singh Dahiya[3]
[1]University of Trento, [2]Fondazione Bruno Kessler, [3]University of Glasgow

14:40 **15B4 - Thickness effects of ZnO thin films on flexible ozone sensors**   407
15:00 Mónica Acuautla[1], Sandrine Bernardini[1], Marc Bendahan[1] and Emmanuelle Pietri[2]
[1]Aix – Marseille University - CNRS, IM2NP, [2]GENES'INK

Session 15C: Analog Techniques

Thursday, July 3rd, 13:20 – 15:00
Room: B225
Chair: Lionel Geynet, *Oridao*

15C1 - Temperature Study of High-Drive Capability Buffer for Phase Change Memories
13:20
13:40 Athanasios Kiouseloglou[1,2], Erika Covi[2,3], Gabriele Navarro[1], Alessandro Cabrini[2], Luca Perniola[1] and Guido Torelli[2]
[1]CEA-Leti, [2]University of Pavia, [3]Laboratorio MDM, IMM – CNR

411

13:40 **15C2 - Large Bandwidth Tunable Analog Equalizers Based on an InP DHBT**
14:00 **Differential Pair Amplifier Cell for 100-GBaud Communication Systems**
Ronan Mettetal[1,2], Jean-Yves Dupuy[2], Achour Ouslimani[1] and Jean Godin[2]
[1]ECS-Lab/ENSEA, [2]III-V Lab

415

14:00 **15C3 - Low Power Inductor-less CML Latch and Frequency Divider for Full-Rate 20**
14:20 **Gbps in 28-nm CMOS**
Laszlo Szilagyi, Guido Belfiore, Ronny Henker and Frank Ellinger
Dresden University of Technology

419

14:20 **15C4 - A 2.4 GHz Fast Settling Wake-Up Receiver Frontend**
14:40 Christoph Tzschoppe, Robert Kostack and Frank Ellinger
Dresden University of Technology

423

15C5 -Design of a CMOS Image Sensor with a 10-bit Two-Step Single-Slope A/D
14:40 **Converter and a Hybrid Correlated Double Sampling**
15:00 Yeonseong Hwang, Seongjoo Lee and Minkyu Song
Dongguk University

427

Closing Ceremony – Leaf Awards

Thursday, July 3rd, 15:00 – 16:00
Room: Petit Salon

WELCOME PRIME 2014

It's our pleasure to welcome you to the 10th conference on Ph. D. Research in Microelectronics and Electronics (PRIME 2014). This event, dedicated to Ph. D. Students in Microelectronics takes place in Grenoble, France, from June 30th to July 3rd. Grenoble is the capital of the French Alps and fifth World's Most Inventive City (Forbes).

PRIME 2014 is organized by the CEA-LETI, the IMEP-LAHC laboratory and the TIMA Laboratory with the sponsorship of CNRS, CEA, GIANT, Grenoble INP, METRO, Région Rhones-Alpes, Grenoble, TIMA Laboratory and the technical co-sponsorship of the IEEE CAS Section.

The technical program covers all aspects of the Microelectronics design from digital aspects to analog, RF and Sensor designs, to current technical trends. The technical program committee, with the help of 8 track chairs and expert reviewers had selected 105 papers. These papers are organized in 27 oral sessions, including 8 invited papers.

All the presented papers at PRIME 2014 will be published in the IEEExplore database, which provides a large audience and an excellent dissemination of the event. As all previous PRIME conferences, the traditional PRIME Leaf Certificates will reward the best papers of this event during the closing session on Thursday July 3rd.

PRIME 2014 is an opportunity for Ph. D. Students to present their research activity and meet other people in the research community. This year, there will be two short courses on IP's and start-up story, and four short courses on Green Electronics topic on June 30th. Moreover, an afternoon (on July 02d) will be dedicated to a Company Fair for allowing Ph. D. Students to get in touch with Microelectronics Companies.

We hope that PRIME will be also an opportunity for you to discover Grenoble. Thanks to the "Mairie de Grenoble" you will be able to visit a part of the "Musée de Grenoble" where the welcome event will take place. For the gala dinner, you will visit another place very well-known and appreciated by the Grenoble inhabitants: "La Bastille".

We would like to thank colleagues who voluntarily worked to make this PRIME 2014 edition possible: the expert reviewers; the members of the Technical Program and Steering Committees; The invited speakers; and last but not least, the local colleagues who offered their skill, time, and extensive knowledge to make PRIME 2014 a memorable event.

Dominique Morche and Gilles Sicard
General Chairs

Program Committee

Hassan Aboushady	UPMC
Eduard Alarcón	Technical University of Catalunya
Federico Alimenti	UNIPG
Aytac Atac	RWTH-Aachen
Elmar Bach	Infineon
Diego Barrettino	SUPSI
Andrea Baschirotto	UNIMIB
Edith Beigné	CEA-Leti
Didier Belot	ST Microelectronics
Roc Berenguer	CEIT
Andoni Beriain	CEIT and Tecnun (University of Navarra)
Sylvain Blayac	CMPGC, Ecole des Mines
Edoardo Bonizzoni	UNIPV
Vincent Bourguet	UFRN
Ralf Brederlow	Texas Instruments Deutschland GmbH
Alessandro Cabrini	UNIPV
Jean-Frederic Christmann	CEA LETI
Massimo Conti	UNIVPM
Jan Craninckx	imec
Stefano D'Amico	UNISALENTO
Ravinder Dahiya	University of Glasgow
Gian-Franco Dalla Betta	UNITN
Marcello De Matteis	University of Milan-Bicocca
Carl James Debono	University of Malta
Wojciech Debski	Silicon Radar GmbH
Catherine Dehollain	Swiss Federal Institute of Technology (EPFL)
Manuel Delgado-Restituto	IMSE
Gregory Denbeaux	Albany Univ.
Jean-Marc Duchamp	Université de Grenoble Alpes IMEP-LAHC
Günhan Dundar	Boğaziçi University
Ahmed Elwakil	University of Sharjah
Emmanuelle Encrenaz	UPMC
Luca Fanucci	University of Pisa
Vittorio Ferrari	UNIBS
Giuseppe Ferri	UNIVAQ
Dimitri Galayko	UPMC
Tzeno Galchev	University of Freiburg
Didier Gigmes	ICR, Aix-Marseille Univ.
José Luis Gonzalez-Jimenez	CEA-LETI
Massimo Gottardi	FBK
Marco Grassi	University of Pavia
Stefan Heinen	RWTH-Aachen
Robert Henderson	University of Edinburgh
Frank Henkel	IMST
Ramy Iskander	UPMC
David Johns	University of Toronto

PRIME 2014   Program Committee

Peter Kennedy	University College Cork
Juha Kostamovaara	University of Oulu
Estelle Lauga-Larroze	IMEP-LAHC Université de Grenoble Alpes
Yannis Le Guennec	Université de Grenoble Alpes IMEP-LAHC
Marie-Minerve LouËrat	LIP6
Paolo Madoglio	INTEL
Torsten Maehne	LIP6
Akram Malak	LIP6
Piero Malcovati	University of Pavia
Franco Maloberti	UNIPV
Rui Paulo Martins	University of Macau
Nicola Massari	FBK
Peter Mole	Intersil
Dominique Morche	CEA-Leti
Christophe Muller	IM2NP
Angelo Nagari	ST Microelectronics
Tobias Noll	RWTH-Aachen
Pierluigi Nuzzo	University of California, Berkeley
Ian O'Connor	INL
Tom O'Dwyer	Analog Devices
Gaetano Palumbo	UNICT
Daniele Passeri	UNIPG
Roberto Passerone	UNITN
Emmanuel Pistono	Université de Grenoble Alpes IMEP-LAHC
Jean-Michel Portal	Im2np, Aix-Marseille University
Wolfgang Pribyl	Graz University of Technology
Rüdiger Quay	Fraunhofer IAF
Valerio Re	UNIBG
Patrick Reynaert	KU Leuven
Angel Rodriguez-Vazquez	IMSE-CNM
Nicolas Rouger	G2elab
Stefan Rusu	INTEL
Sergio Saponara	Università di Pisa
Naohiko Shimizu	Tokai University
Gilles Sicard	TIMA
Gilles Sicard	Université de Grenoble Alpes TIMA
Hector Solar	CEIT
David Stoppa	FBK
Thierry Taris	IMS, Bordeaux Univ.
Himanshu Thapliyal	University of South Florida
Roland Thewes	TU Berlin
Guido Torelli	UNIPV
Alberto Tosi	Politecnico di Milano
Maurizio Valle	UNIGE
Michel Vasilevski	UFRN
Udo Weimar	University of Tübingen

Mixed-signal verification challenges

Nicolas Delorme

Asygn, Montbonnot-Saint-Martin, France

The verification of today's complex mixed-signal SoCs poses major difficulties: image sensors with several million pixels, RF tranceivers with GHz signals carrying thousands of symbols, heterogeneous systems with scattered time constants often suffer from some form of verification coverage reduction, resulting in design robustness degradation, unnecessary overdesign or even functional failures. This talk addresses application examples where the benefit of appropriate tools and methodologies for improved verification coverage is described.

Reduction of IPs Energy Consumption with Ultra-Low Voltage Supply

Fady Abouzeid

STMicroelectronics, Crolles, France

In the near future, a number of systems will be powered using energy constrained devices or scavenging technologies, enabling new applications such as medical monitoring, sensors and next-generation portable video gadgets. This will require the electronic circuits to operate with utmost energy efficiency while performing the required functionality. A major opportunity to reduce the energy consumption of digital circuits is to scale supply voltages below 0.4V driving them to sub- (or near-) threshold operation. This workshop presents the expected energy gain and challenges in the ultra-low-voltage (ULV) regime. Our recent ULV achievements will be then described in relation to specific standard cell libraries, adapted CAD margins and Sign-off and ULV digital demonstrators operating down to 0.3V. Last, the enhanced ULV performances offered by the FDSOI28 technology will be highlighted, in relation to an improved ION/IOFF ratio and sub-threshold slope, with promising Silicon results showcased near 0.35V.

RF Power Gating Techniques for Ultra Low Power Communication

Sylvain Bourdel[1], I. Ben Amor[2], J.F. Pons[2], N Dehaese[2], R. Vauché[2], J. Gaubert[2]

[1]Univ. Grenoble Alpes, IMEP-LAHC

[2]IM2NP

sylvain.bourdel@phelma.grenoble-inp.fr

Actually, in the mobile application area, the reduction of power consumption appears to be a bottleneck when designing transceiver architectures for those systems. This is especially true when considering two large parts of the actual electronics market: Wireless Sensors Networks (WSN) and Internet of Things (IoT). Both of them require wireless communications and a large autonomy while being battery-powered.

In order to decrease power consumption, four main complementary approaches are possible at several levels: i) at the technological level with technologies such as SOI or FD-SOI, ii) at the circuit level where sub-threshold or current reuse topologies can be implemented, iii) at the system level where zero-IF or non-coherent architectures enable substantial power reduction and, iv) at the application level with accurate power management or power efficient protocols.

Power gating is an often used technique and is implemented at the system or application level where the system is duty-cycled between two consecutive frames. Although power gating or clock gating are common techniques in digital area, power gating is rarely used at the symbol time scale in a RF design. Since the RF part consumes a large amount of the DC power, RF gating is a promising approach for Ultra Low Power systems.

In this workshop we will present several approaches to implement RF power gating at the bit scale. We will specially present solutions in the case of the UWB Impulse Radio with their possible applications. The feasibility of RF gating for narrow band systems will also be discussed.

Accurate Modeling of Ultra Low-Power $\Sigma\Delta$ Modulator

Luca Giuffredi
Dip. di Ingegneria dell'Informazione
University of Parma
I-43124, Parma, Italy
Email: luca.giuffredi@studenti.unipr.it

Giorgio Pietrini
Dip. di Ingegneria dell'Informazione
University of Parma
I-43124, Parma, Italy

Andrea Boni
Dip. di Ingegneria dell'Informazione
University of Parma
I-43124, Parma, Italy

Abstract—**This paper presents a behavioural model suitable for the simulation of low-power Sigma-Delta Modulators. Second-order effects affecting the settling behaviour of the switched-capacitor integrator was included, leading to improved accuracy. Due to the oversampling mode of the converter, transistor-level simulations are extremely time consuming. Accurate behavioural models are thus mandatory in the first design phase of the modulator, in particular when the involved analog blocks must be optimized for minimum power consumption at some converter resolution.**

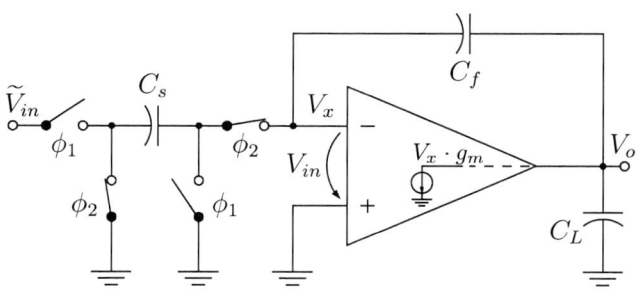

Fig. 1.   Switched-capacitor integrator (single-ended version).

I. Introduction

The increasing market demand for portable electronic devices equipped with multiple sensors is forcing a relevant research interest in high resolution A/D converters (ADCs) featuring a very low power consumption. Considering the signal bandwidth (below 1 kHz) and an effective resolution higher than 15 bit, Sigma-Delta ($\Sigma\Delta$) oversampling ADC are the best choice. The basic block of the $\Sigma\Delta$ modulator implements the function of integration: a simplified schematic is reported in Fig. 1. In deep sub-μm technology the $\Sigma\Delta$ modulator is the real bottleneck if power consumption must be minimized, due to the presence of operational amplifiers being usually power hungry. Indeed, the dynamic performance (bandwidth and slew-rate limiting) of such amplifiers have a relevant impact on the signal-to-noise and distortion ratio (SNDR) of the ADC. Unfortunately, transistor-level simulations of a $\Sigma\Delta$ modulator are extremely time consuming due to oversampling mode of this converter type. For this reason high-level behavioural models of the modulator are mandatory for the designer in order to enable fast simulations and to identify the op-amp's specifications (DC-gain, bandwidth and slew-rate) leading to the required converter performance (linearity and harmonic distortion). For this reason there has been an intensive research effort in system level macro-models (based on MATLAB, Simulink or similar tools) [1], [2], [3], [4]. Although $\Sigma\Delta$ modulator is a discrete-time system, the SC integrator is usually modeled with two continuous-time circuits corresponding to the sampling and integration configuration. State-of-the-art behavioural models reported in literature are based on a few approximations. In particular, the op-amp in the integration phase ($\phi_2 = 1$ in Fig. 1) is assumed to work in the slew-rate limited regime for only a small fraction of the time available for this phase. Furthermore the transconductance (g_m) of the op-amp input stage is usually assumed to be time-

independent, thus leading to an invariable time constant in the linear settling regime. Unfortunately both approximations fail when the op-amp is optimized for minimum power consumption, but compatible with the required ADC specification.

Indeed, very low power modulator designs require that the involved op-amps work for a large fraction of available time under slew-rate limiting and with a narrow small-signal bandwidth, leading to incomplete output settling [5], [6]. Furthermore, since the time interval where the amplifier exhibits a quasi-linear behaviour is quite short, the approximation of an invariable time constant (as in the case of a constant g_m) is no more acceptable. Therefore, state-of-the-art models requires some modifications and improvements in order to enable their usage under ultra low-power design constraints.

Starting from [3] and [4] an improved behavioural model of the SC integrator is proposed. A good agreement with transistor-level simulation were obtained by improving the model accuracy under both slew-rate limiting and linear settling regimes. The model was validated in the design of a third order $\Sigma\Delta$ modulator in a 65 nm technology. The simulated spectra of the modulator output are in extremely good agreement with the results returned by transistor-level simulations, thus proving the effectiveness of the added features, being mandatory in a very low power design.

II. First order model

In conventional behavioural models of $\Sigma\Delta$ modulators a first order transfer function is usually assumed for the op-amp in Fig. 1.

$$A(s) = \frac{v_o}{v_{in}} = \frac{A_o}{1 + \dfrac{s}{\omega_{p1}}} \simeq \frac{A_o\omega_{p1}}{s} = \frac{\omega_{ta}}{s} \qquad (1)$$

$$\omega_{ta} = \frac{g_m}{C_L + \dfrac{C_f C_s}{C_f + C_s}} = \frac{g_m}{C_L + \beta \cdot C_s} = \frac{g_m}{C_{Lop}} \quad (2)$$

where A_o, ω_{p1} and ω_{ta} are respectively: the DC-gain, the pole frequency and unity gain transition angular frequency of op-amp in open loop condition. C_{Lop} is its capacitive load in integration phase, while we define β as the feedback factor. Closed loop transfer function $A_{CL}(s)$ can be calculated from the op-amp transfer function:

$$A_{CL}(s) = \frac{v_o}{\widetilde{v}_{in}} = \frac{A(s)}{1 + \beta \cdot A(s)} = \frac{\dfrac{1}{\beta}}{1 + \dfrac{s}{\beta \omega_{ta}}} \quad (3)$$

As shown in (3) the unity gain transition frequency ω_t of the loop gain ($\omega_t = \beta \omega_{ta}$) determines the time constant of the response and must be calculated considering the equivalent load capacitance of the op-amp, C_{Leff}:

$$\omega_t = \beta \cdot \frac{g_m}{C_L + \beta \cdot C_s} = \frac{g_m}{C_s + C_L \left(1 + \dfrac{C_s}{C_f}\right)} \quad (4)$$

$$C_{Leff} = C_s + C_L \left(1 + \frac{C_s}{C_f}\right) \quad (5)$$

III. DISCONTINUOUS INITIAL CONDITIONS

Considering the single-ended integrator in Fig. 1, some charge transfer takes place just after the switching from sampling to integration, occurring at $t = t_0 \equiv nT_{CK} - \frac{T_{CK}}{2}$, and before the op-amp starts the slew-rate limiting phase [7]. Due to this effect the voltage at nodes V_x and V_o are discontinuous and jump at $t = t_{0+}$ to the opposite direction with respect to the settling values (Fig. 2). Computing exactly $V_o(t_{0+})$ and $V_x(t_{0+})$ is mandatory for accurately estimating the length of the slew-limited region and thus the final settling error, Fig. 2. Since the op-amp output impedance at $t = t_{0+}$ is high and thus no charge is transferred to the capacitors, the charge conservation must be considered for both the negative input and output terminals of the op-amp:

$$V_x(t_{0+}) = \beta \left[(V_o(t_{0+}) - V_o(n-1)) - \widetilde{V}_{in}(n-1)\frac{C_s}{C_f} \right] \quad (6)$$

$$V_o(t_{0+}) = V_o(n-1) + \frac{C_f}{C_f + C_L} V_x(t_{0+}) \quad (7)$$

where $V(n-1)$ notation, corresponding to $z^{-1}V[z]$ in discrete-time domain, stands for $V[(n-1)T_{CK}]$.

Solving charge conservation equations (6) and (7), the values of V_o and V_x at $t = t_{0+}$ are found:

$$\begin{cases} V_o(t_{0+}) = V_o(n-1) - \widetilde{V}_{in}(n-1)\dfrac{C_s}{C_{Leff}} \\ V_x(t_{0+}) = -\dfrac{C_s}{C_{Leff}}\left(1 + \dfrac{C_L}{C_f}\right)\widetilde{V}_{in}(n-1) \end{cases} \quad (8)$$

where the relevant contribution due to the output capacitive load C_L, including the op-amp output parasitic capacitance and input capacitance of following stages, is highlighted.

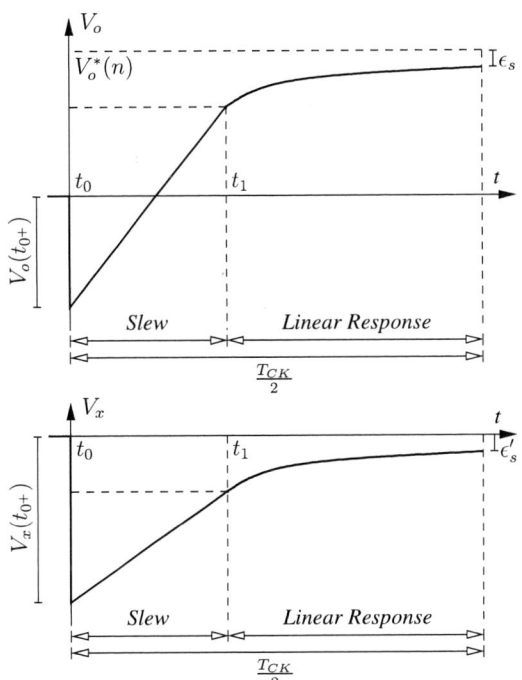

Fig. 2. Example of V_o and V_x transient in ϕ_2 (integration) phase. ϵ_s and c'_s are the settling errors with respect to asymptotic values (i.e. $V_o^*(n)$ and 0 V for V_o and V_x respectively).

IV. SECOND ORDER MODEL

For improved accuracy a second order model transfer function was considered for the op-amp. Indeed, the contribution of the second pole (ω_{p2}) cannot be neglected since it is usually close to the op-amp unity gain angular frequency due to low power constraints. Considering an input frequency well above the first pole (i.e. ω_{p1}), the open and closed loop transfer functions can be approximated as [8]:

$$A(s) = \frac{A_o}{\left(1 + \dfrac{s}{\omega_{p1}}\right)\left(1 + \dfrac{s}{\omega_{p2}}\right)} \simeq \frac{\overbrace{A_o \cdot \omega_{p1}}^{\omega_{ta}}}{s\left(1 + \dfrac{s}{\omega_{p2}}\right)} \quad (9)$$

$$A_{CL}(s) = \frac{A(s)}{1 + \beta \cdot A(s)} = \frac{1}{\beta} \cdot \frac{\overbrace{\beta \omega_{ta}\omega_{p2}}^{\omega_o^2}}{s^2 + \underbrace{\omega_{p2}}_{2\xi\omega_o} \cdot s + \underbrace{\beta \omega_{ta}\omega_{p2}}_{\omega_o^2}} \quad (10)$$

ω_o and ξ defined in (10) are respectively the natural frequency and damping ratio. Recalling ω_t (unity gain transition frequency) and phase margin (PM) definitions:

$$\begin{cases} \beta \omega_{ta} = \omega_t \sqrt{1 + \tan^2(90° - PM)} \\ \omega_{p2} = \dfrac{\omega_t}{\tan(90° - PM)} \end{cases} \quad (11)$$

Taking integrator's equations in time domain from [7] and including (11) in (10), the $V_o(t)$ in linear region is obtained:

$$V_o(t) = V_o(n-1) + \frac{C_s}{C_f} \cdot \tilde{V}_{in}(n-1) + \left(1 + \frac{C_s}{C_f}\right) V_x(t_{0+}) \cdot$$

$$\cdot e^{-\frac{\omega_t \cdot t}{2 \cdot \tan(90° - PM)}} \cdot 2\cos(\omega_o \sqrt{1-\xi^2} \cdot t) \qquad (12)$$

where ω_t depends on ω_{ta} (being a function of g_m) as in (11). The exponential envelope term in (12) dominates the sinusoidal one, so the second can be neglected. For this reason it was not expressed in terms of ω_t and PM.

Therefore an accurate equation for $V_o(t)$ is obtained with only two circuit-dependent parameter, i.e. PM and ω_t which can be easily obtained by small-signal transistor-level simulation performed on the stand-alone op-amp.

V. TRANSCONDUCTANCE VARIATION

The input transistors of the op-amp in Fig. 1 exhibit a transconductance which depends on the differential voltage, $V_{in}(t)$. The impact of the g_m variation on the charge transfer error in SC amplifier and filters has been already reported ([7], [9]). Such effect causes a non-negligible non-linearity in the ADC input-output characteristic and cannot be neglected in high-resolution $\Sigma\Delta$ modulators. To this aim, the time interval where the op-amp works in linear settling mode (i.e. after the slew-rate limited phase) can be divided in several time steps $(t_{k+1} - t_k)$, with a specific value of $g_m(t_k)$. In each step the input transconductance g_m is calculated by an iterative algorithm [9].

$$g_m(t_k) = g_{m0} \cdot \left(\sqrt{1 - \alpha V_{in}^2(t_k)} - \frac{\alpha V_{in}^2(t_k)}{\sqrt{1 - \alpha V_{in}^2(t_k)}} \right) \qquad (13)$$

$$\alpha = \frac{\mu c_{ox}(W/L)}{4I_H} \qquad (14)$$

α parameter can be estimated by means of simple DC circuit simulations on the unity gain buffer op-amp's configuration. I_H is the bias current of op-amp's source-coupled input pair (Fig. 5), μ is the carrier electrical mobility, c_{ox} is the gate oxide capacitance per unit area, W/L is the aspect ratio of $MP1$ and $MP2$ and $g_{m0} = \sqrt{\mu c_{ox}(W/L)I_H}$ (assuming strong-inversion bias).
From (13) the unity loop gain transition frequency, $\omega_t(t_k)$ is updated at each time step:

$$\omega_t(t_k) = \frac{g_m(t_k)}{C_{Leff} \sqrt{1 + \tan^2(90° - PM)}} \qquad (15)$$

From the equations in linear region and with the value of ω_t at $t = t_k$, V_{in} at $t = t_{k+1}$ is calculated and used for updating the value of input transconductance at the next time step.
Physical consistency of (13) requires $g_m(t_k)$ be real and strictly positive, leading to a couple of mandatory constraints for the iterative algorithm:

$$1 - \alpha V_{in}^2(t_k) \geqslant 0 \implies V_{in}^2(t_k) \leqslant \frac{1}{\alpha} \qquad (16)$$

$$g_m(t_k) > 0 \implies V_{in}^2(t_k) < \frac{1}{2\alpha} \qquad (17)$$

The second condition is clearly stronger, thus prevails:

$$|V_{in}(t_k)| < \sqrt{\frac{1}{2\alpha}} = \sqrt{2} \cdot \sqrt{\frac{I_H}{\mu c_{ox}(W/L)}} = \sqrt{2} \cdot V_{ov} \qquad (18)$$

where V_{ov} represents the overdrive voltage.

It is worth to be noticed that (18) corresponds to the constraints on the op-amp differential pair ($MP1$ and $MP2$ in Fig. 5) leading to the quasi-linear response [10]. Since the iterative algorithm taking into account the g_m dependency on V_{in} is used only in the linear settling interval (i.e. $t \geq t_1$ in Fig. 2) such constraints is satisfied only if t_1 is accurately estimated.

VI. SIMULATION RESULTS

The improved model described in this paper was validated by means of two design cases in STM 65 nm digital CMOS technology. Op-amp's architecture selected in this design, depicted in Fig. 5, exploits current cancellation through local positive feedback [11]. The first design was optimized without internal compensation, with benefits in terms of bandwidth and power consumption. In the second design Miller compensation was introduced (C_c) leading to larger stability margins. Both cases refers to the same $\Sigma\Delta$ modulator's architecture: single-loop 3rd order (Fig. 4) fully-differential CIFF (Cascade of Integrators in Feed-Forward configuration) modulator with 16 bits of target resolution.

Results reported in Tab. I and in Fig. 3 highlights the good performance of the proposed model. Indeed, after digital filtering with 250 Hz filter bandwidth, the returned effective number of bit (ENOB) and the amplitude of main in-band harmonics are in good agreement with the transistor-level simulation performed with commercial simulators (i.e. Spectre and Eldo). Both simulations were performed using a 36 Hz input sine-wave with 0.9 V amplitude. Fourier Transform (FFT) spectra are calculated over 2^{17} samples with Hann windowing. Op-amp parameters (unity loop gain transition frequency, phase margin, slew-rate, α, C_L) provided to the proposed Simulink model are obtained by simple AC, TRAN and DC transistor-level simulation of the op-amp in integration phase.

It is worth noticing that using state-of-the-art models for $\Sigma\Delta$ modulators ([3], [4]) on the same circuits, the obtained results are really too optimistic and far from the Spectre simulation results: ENOB = 18.58 bits; 3rd harmonic amplitude = -121.6 dB and 5th harmonic amplitude = -118 dB. This therefore confirms the importance of the improvements carried out on the original models in the case of a low power modulator implementation.

TABLE I.   RESULTS COMPARISON

Metric	Two stage op-amp		Miller op-amp		Unit
	Spectre	This model	Spectre	This model	
ENOB	15.65	15.54	16.36	16.31	bit
3rd harmonic	-98.2	-98	-101.7	-101.9	dB
5th harmonic	-102.8	-101	-110.9	-108	dB

VII. CONCLUSION

An improved Simulink model focused on the simulation of harmonic distortion in $\Sigma\Delta$ modulators was proposed. The

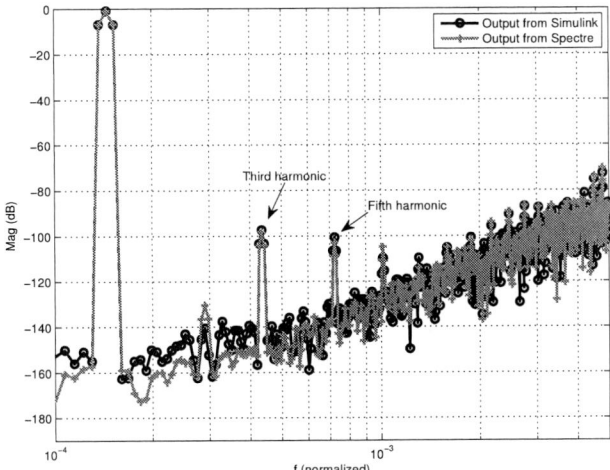

(a) Normal two stage op-amp integrator spectrum

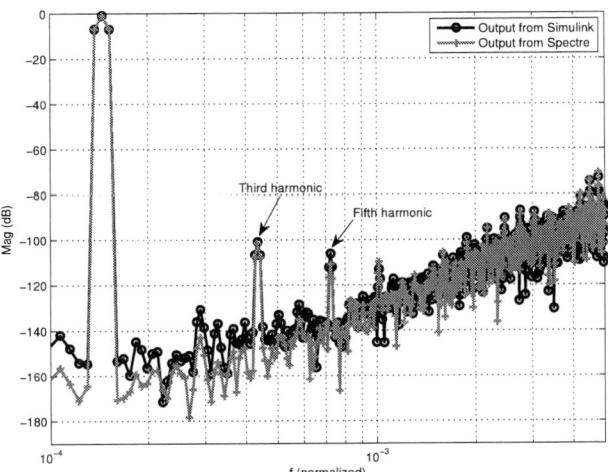

(b) Miller compensated two stage op-amp integrator spectrum

Fig. 3. Third order $\Sigma\Delta$ modulator spectra: black line (o) as returned from the proposed model; gray line (+) as returned from transistor-level simulator.

TABLE II.   MODULATOR METRICS

Parameters	Symbol	Value	Units
Sampling Frequency	f_s	250	kHz
Signal Bandwidth	f_B	250	Hz
Oversampling Ratio	OSR	512	
Effective Num. of Bits	ENOB	16	bits
Signal-to-Noise-plus-Distortion Ratio	SNDR	99	dB
Input Voltage Range		[-1, +1]	V
Supply Voltage	V_{dd}	1.2	V

simulated output spectra are in strict agreement with the results obtained with time consuming transistor-level simulation. The proposed model allows to optimize the op-amp specifications in a power-limited modulator design. Effectiveness of the model was proved in two design cases, thus showing that this model is not strictly op-amp's circuit dependent.

ACKNOWLEDGMENT

The authors wish to thank Dr. Marco Ronchi and Dr. Elio Guidetti of STM for their useful suggestions and support to the paper.

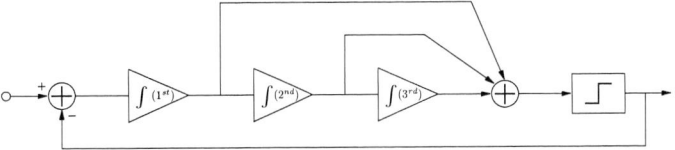

Fig. 4.   Third order $\Sigma\Delta$ modulator structure.

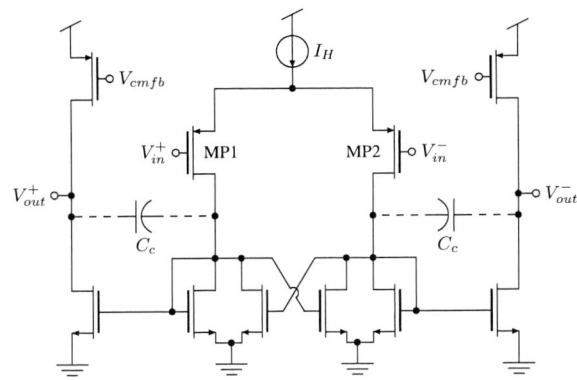

Fig. 5.   Simplified full-differential op-amp schematic. C_c is the compensation capacitor and V_{cmfb} is the output common mode feedback voltage.

REFERENCES

[1] F. Medeiro, B. Perez-Verdu, A. Rodriguez-Vazquez, and J. Huertas, "Modeling opamp-induced harmonic distortion for switched-capacitor sigma delta modulator design," in *Circuits and Systems, 1994. ISCAS '94., 1994 IEEE International Symposium on*, vol. 5, May 1994, pp. 445–448 vol.5.

[2] F. Medeiro, A. Pérez-Verdú, B. Rez-Verd, and A. Rodríguez-Vázquez, *Top-Down Design of High-Performance Sigma-Delta Modulators*, ser. Kluwer international series in engineering and computer science: Analog circuits and signal processing.   Springer, 1999.

[3] S. Brigati, F. Francesconi, P. Malcovati, D. Tonietto, A. Baschirotto, and F. Maloberti, "Modeling sigma-delta modulator non-idealities in simulink(r)," in *Circuits and Systems, 1999. ISCAS '99. Proc. IEEE International Symposium on*, vol. 2, Jul 1999, pp. 384–387 vol.2.

[4] P. Malcovati, S. Brigati, F. Francesconi, F. Maloberti, P. Cusinato, and A. Baschirotto, "Behavioral modeling of switched-capacitor sigma-delta modulators," *IEEE Trans. Circuits Syst. I*, vol. 50, no. 3, pp. 352–364, Mar 2003.

[5] R. Naiknaware and T. Fiez, "Power optimization of delta sigma analog-to-digital converters based on slewing and partial settling considerations," in *Circuits and Systems, 1998. ISCAS '98. Proc. IEEE International Symposium on*, vol. 1, May 1998, pp. 360–364 vol.1.

[6] I. Williams, L.A. and B. Wooley, "A third-order sigma-delta modulator with extended dynamic range," *IEEE J. Solid-State Circuits*, vol. 29, no. 3, pp. 193–202, Mar 1994.

[7] W. Sansen, H. Qiuting, and K. A. I. Halonen, "Transient analysis of charge-transfer in sc filters-gain error and distortion," *IEEE J. Solid-State Circuits*, vol. 22, no. 2, pp. 268–276, Apr 1987.

[8] D. A. Johns and K. Martin, *Analog Integrated Circuit Design*.   New York: John Wiley and Sons, Inc, 1997.

[9] C. Shi, X. Li, and M. Ismail, "A novel nonlinear settling model of ota for high-resolution switched-capacitor sigma-delta modulator design," in *Microelectronics, 1999. ICM '99. The Eleventh International Conference on*, Nov 2000, pp. 47–50.

[10] P. R. Gray and R. G. Meyer, *Analysis and Design of Analog Integrated Circuits*.   John Wiley and Sons, Inc, 1995.

[11] J. Roh, S. Byun, Y. Choi, H. Roh, Y.-G. Kim, and J.-K. Kwon, "A 0.9-v 60- μw 1-bit fourth-order delta-sigma modulator with 83-db dynamic range," *IEEE J. Solid-State Circuits*, vol. 43, no. 2, pp. 361–370, Feb 2008.

978-1-4799-4993-9/14 $31.00 © 2014 IEEE

System-on-Chip Verification:
TLM-to-RTL Assertions Transformation

Zeineb Bel Hadj Amor, Laurence Pierre, Dominique Borrione
TIMA Laboratory (CNRS-INPG-UJF)
46 Avenue Felix Viallet, 38031 Grenoble France
Email: Zeineb.Bel-hadj-amor@imag.fr, Laurence.Pierre@imag.fr, Dominique.Borrione@imag.fr

Abstract—The Electronic System Level design flow aims to manage the great complexity of today's Systems-on-Chip: design and verification methodologies start from abstraction levels higher than RTL (Register Transfer Level), referred to as Transaction Level Modeling (TLM). At this level, virtual prototypes are used for early validation, software development, and as golden reference for the derived RTL designs. Assertion-based verification, a verification technique widely adopted for RTL designs, started to prove its efficiency at TLM. Verification is the bottleneck of the ESL design flow. Hence, there is a real need for a complete verification flow covering all the abstraction levels of the design flow.

In this paper, we describe the implementation of an approach for temporal assertions refinement from TLM to RTL, using a set of transformation rules. The reuse of TLM assertions is the basis of an Assertion-based verification flow.

I. INTRODUCTION

The complexity of today's Systems-on-Chip has increased drastically, making the use of RTL (Register Transfer Level) design methodologies time consuming and error prone. Two basic approaches could help to address these issues: the first one increases the design and verification level to a higher level above RTL. The second one introduces the reuse methodology by means of IP (Intellectual Property) components. Electronic System Level (ESL) methodology [1] combines the benefits of these two approaches. ESL flow is a generic term for a set of abstraction levels starting from the system level (SL) to the physical implementation. It complies with the need for HW/SW co-design and early functional verification.

Figure 1 shows a typical ESL design flow as described by [2], [3]. Starting with a textual specification, the first step consists in obtaining a specification of the entire system written in an executable language capturing the basic functionalities and requirements regardless architectural and timing concerns. Based upon this step, preliminary analysis steps are made (performance estimation, power consumption, area) to guide architectural decisions leading to Hardware/Software partitioning. This step results in an architectural specification model. At this level, designers can use SystemC transaction level modeling (SystemC-TLM) library to reach a threshold of simulation performance that enables HW/SW co-design [4]. The system level architectural system is the virtual prototype for the SW development path and the golden reference for the HW design and verification activities. Hence, it has to be verified with respect to the original specification and requirements.

The system level architectural model will be incrementally refined, giving rise to the RTL model for the HW. Various refinement procedures may be applied e.g., high level synthesis for computational components, or replacement by an actual bus model for the abstract communication channels.

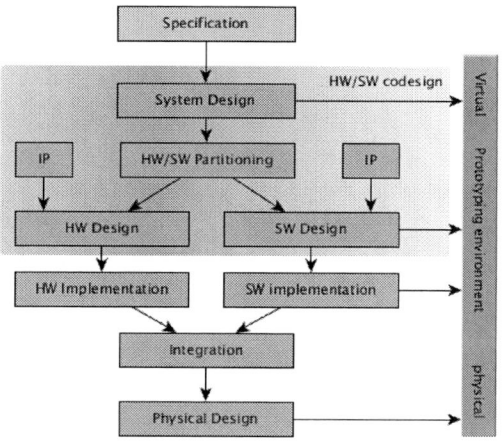

Fig. 1. Idealized ESL design flow

The resulting implementation has to be verified with respect to the golden model. There are increasing demands on effective verification flows that can help to reduce the verification cost. In this context, *Assertion-Based-Verification* (ABV), a well known verification methodology for RTL models [5], has increasingly started to be applied on Transaction Level models. The validation of the TLM design using ABV has several benefits since it significantly reduces debug time [6], offers a formal representation of the system requirements and enhances the use of the transactional model as golden model. We assume here that assertions are specified using the PSL (Property Specification Language) IEEE standard [7] and are *used in a simulation-based context* thanks to the ISIS tool described in [8], to avoid the state explosion issues related to complex SoCs formal verification. We focus on the *reuse of TLM assertions for RT models* as the basis of the verification flow. It will ensure the functional compliance between golden and refined models and minimize the verification effort.

The contribution described here is the implementation of a tool for the automatic refinement of TLM assertions into RTL assertions, as specified in [9]. Minimal user guidance is required, to specify the actual protocols and timing that are introduced in the TLM-to-RTL design flow.

978-1-4799-4993-9/14 $31.00 © 2014 IEEE

II. ASSERTION-BASED VERIFICATION

An assertion is a specification of an expected behavior. The result of compilation of the assertion is a verification IP called "monitor" or "checker". Since its approval by the IEEE, PSL (IEEE 1850) has met a huge success in the hardware verification community. The main part of PSL is the "Temporal Layer". The Foundation Language (FL) of this layer includes temporal operators that allow the user to describe relations over time, and Sequential Extended Regular Expressions (SEREs) to portray sequences of events.

A. PSL assertions on RTL models

PSL assertions are evaluated over execution traces, obtained by sampling the simulation results. RTL designs are pin-accurate and cycle-accurate. At this level, the communication between the RTL components is made through signals and the synchronization is often made through clock signals. Hence, the evaluation points of the execution traces are obtained by sampling on clock ticks.

Example: The property below is related to a HDLC controller IP [10]. It states that, every time the IP receives the abort sequence (which is the sequence: zero followed by seven consecutive ones), then the signal *AbortFound* shall be set high before the arrival of a new frame.

```
default clock is (DataEn'event and DataEn='1');
assert always({not Data ; Data[*7]}
              |-> AbortFound before! StartOFFrame);
```

As we can see, the synchronization signal is *DataEn* and the DUV observed signals are *Data*, *AbortFound* and *StartOFFrame*. This assertion involves several PSL operators, among them ";" that is the SERE concatenation operator, "[*7]" that uses the SERE consecutive repetition operator to express that 7 consecutive ones are expected on *Data*, "|->" that is the SERE implication operator, and the "before!" temporal operator.

At this level, the Horus tool [11] can be used for the automatic generation of RTL checkers from such PSL assertions.

B. PSL assertions on TLM platforms

TLM is functionally accurate, it uses function calls (*transactions*) rather than signals or wires to communicate between modules. TLM is widely used with SystemC [12] thanks to the SystemC-TLM OSCI (Open SystemC Initiative) library. TLM classes define the content of transactions and how they are controlled within the system. At this level, the communications are modeled as function calls and the synchronization mechanism is based on events. Transactional PSL assertions are defined to express temporal requirements on the interactions in the SoC. Thus, for the TLM monitors, simulation traces must be sampled according to communication events involved in the assertions, and the Boolean expressions are the communication actions and conditions. For these reasons, we use an ad doc observation mechanism on the communication functions calls and their parameters to trigger the monitors accordingly [8].

Example: Let us illustrate these notions with a property related to an image processing platform that performs spectral compression on image packets, see Figure 2 [9]. The *leon_a* processor configures the *DMA_a* component to get the image from the *IO* module to the memory *Mem_a* where raw data

are sub-sampled. The packet is then put on the second memory *Mem_b* where a 2D-FFT is applied to obtain the corresponding spectrum. The resulting packet is then compressed by the *leon_b* processor which configures *DMA_b* to send the output packet back to the *IO* module.

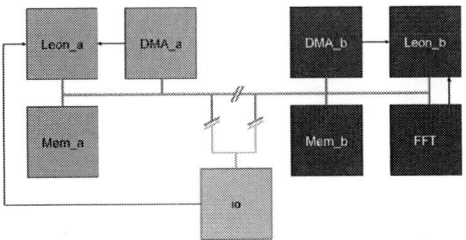

Fig. 2. Image processing platform (Astrium)

An example of property of interest is: "*each time the leon_a configures DMA_a for a new transfer (i.e., writes a "start" in its control register), the DMA_a end-of-transfer interrupt must be generated before the next configuration*".

The ISIS monitor of this property observes two communication functions: the *leon_a* "write(address,data,size)" method (and its parameters "address" and "data") for the configuration of the *DMA_a*, and the *DMA_a* "generate_irq()" method for the interrupt generation.

C. PSL assertions refinement

Since the underlying synchronization paradigms of RTL (signals, clock) and TLM (transactional events) are different, a direct reuse of TLM assertions at the RT level is not possible. A refinement process must be established. This refinement has to map the concurrent data bundled together in the transaction into a set of clocked sequences of signals representing the implemented communication protocol, while ensuring that a property with the same meaning is checked.

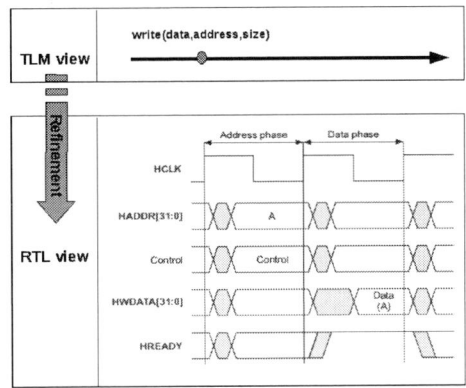

Fig. 3. The refinement of the simple write operation (AHB)

As an example, let us consider that an abstract bus description is refined into a AMBA High-performance Bus (AHB) [13] in the RTL model. As shown in Figure 3, a simple write operation is made of two phases: a one-cycle address phase in which control signals and address are sent, followed by a one-cycle data phase for data transfer. The TLM atomic "write" function call is therefore transformed on the RTL view into a sequence of signals and conditions over time.

A set of PSL transformation rules was proposed in [9]. A transformation rule $L \xrightarrow{C} R$ maps a TLM temporal expression L into the corresponding RTL or mixed expression R, according to given time constraints C. The role of a constraint is to specify delays introduced by the actual communication protocol. For example, for the image processing platform with an AHB bus, the constraint related to the "write" function specifies that there is one-cycle delay between sending the address and sending data on the next clock tick.

III. THE REFINEMENT TOOL

A first specification of this tool was briefly sketched in [9]. This specification has been slightly improved, fully completed, and implemented as the tool described here. Its principles are depicted in Figure 4. Some information such as the implemented communication protocol cannot be deduced automatically from the RTL design. However, the tool tries to minimize the user's effort by pre-filling files as much as possible. Moreover, the user's inputs are verified in order to avoid errors and inconsistency.

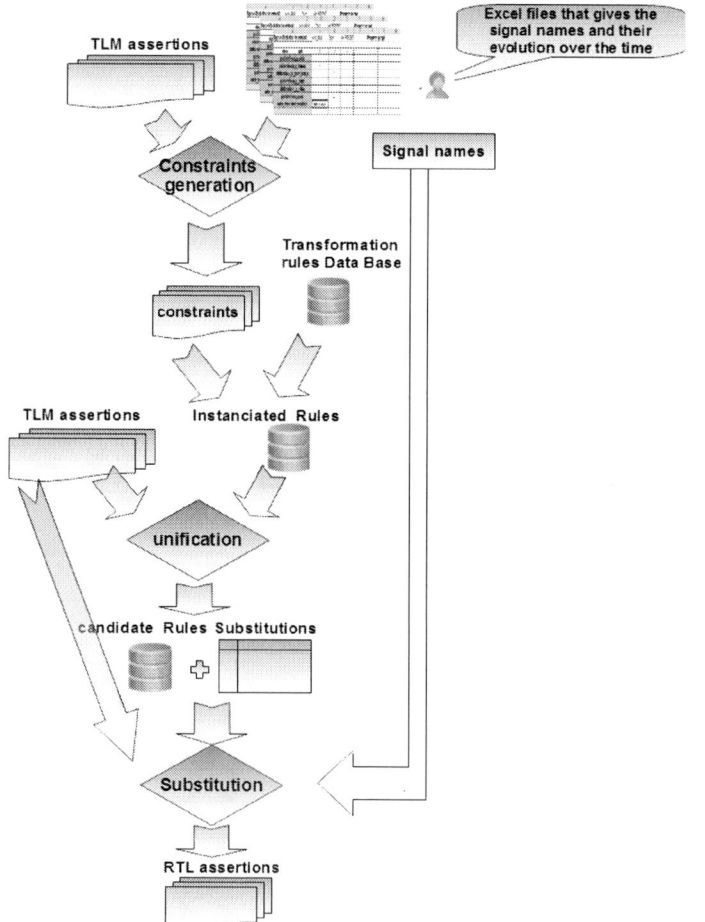

Fig. 4. The flow chart of the transformation tool

Given a transactional PSL property, the tool pre-processes the assertion in order to detect the communication functions observed by the TLM monitors, and their parameters. Then it generates a pre-filled Excel file for each observed component. In the first Excel sheet of the file, the used communication

functions and their parameters are displayed. The user is asked to supply the signals associated with these function calls and to their parameters in the RTL design. Figure 5 shows an example of a generated Excel file for the assertion of section II-B, when the abstract bus description has been refined into an AHB implementation: here the user essentially has to give the name of the clock signal, and the identifiers of the AHB signals involved in the "write" operation. The tool verifies the user inputs according to the given RTL platform description.

Once checked, the tool pre-fills a new sheet for each communication function. In each one, the signals corresponding to the function and their parameters are displayed in the first column. The user simply has to specify the protocol implemented by each communication function by filling the chronogram for each signal. To this aim, he uses the chronograms given by the reference manual of the bus he chooses to introduce as concrete communication channel (e.g., the one shown on Figure 3). Figure 6(a) shows an example for the "write" operation of Figure 3 (including the bus arbitration phase). As illustrated by Figure 6(b), the assertion transformation constraints C are automatically generated from this information. The instantiated rules are the outcome of the injection of the constraints into the rules data base.

This facilitates the forthcoming unification procedure. This process aims to find the set of transformation rules for which the left hand side L can match an expression in the PSL transactional assertion. The output of the unification procedure is a set of candidate rules and substitutions for each corresponding rule. If no structural match is found the substitutions set is empty. The set of candidate transformation rules $T = \{T_1, T_2, ..T_n\}$ that will be applied to the property correspond to the set of rules having a non empty set of substitutions. Then the substitutions are applied on the right side R of each transformation rule T_i. Finally, the resulting instantiated rules $L \xrightarrow{C} R$ are applied to the TLM PSL property and the structural interface details are injected, which results in the refined RTL assertion.

This transformation process has been implemented as an extension of the ISIS tool. Hence, the user can use the same environment for TLM checkers generation, assertions refinement and hybrid level checkers generation.
Experimental results are reported in [9] for the image processing case study. The complexity of the resulting RTL assertions is about two times the complexity of the original assertions in terms of temporal operators, and four times in terms of atomic expressions. The original simulation time at the TLM level is 15.87 seconds for processing 1000 images. When the simulation is instrumented to check the property of section II-B, the simulation time is 16.57 seconds (4.4% overhead, for 11000 monitor activations). After RTL refinement (i.e., the bus is replaced by a RTL description of a AHB bus), the simulation time is 127.50 seconds for processing 100 images. The CPU time for the instrumented simulation becomes 132.31 seconds (3.8% overhead, for 5833434 monitor activations).

As far as we know, few results have been presented so far for the refinement of assertions from TLM to RTL. In [14], the assertions under consideration are only Boolean expressions (no temporal operator), they express computational invariants that relate primary outputs to primary inputs for a given

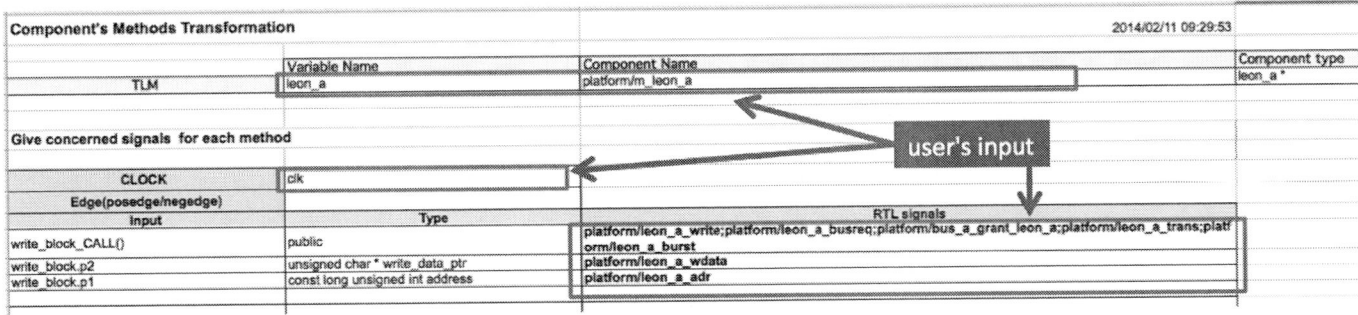

Fig. 5. The generated Excel file for the "write" method of the assertion of section II-B

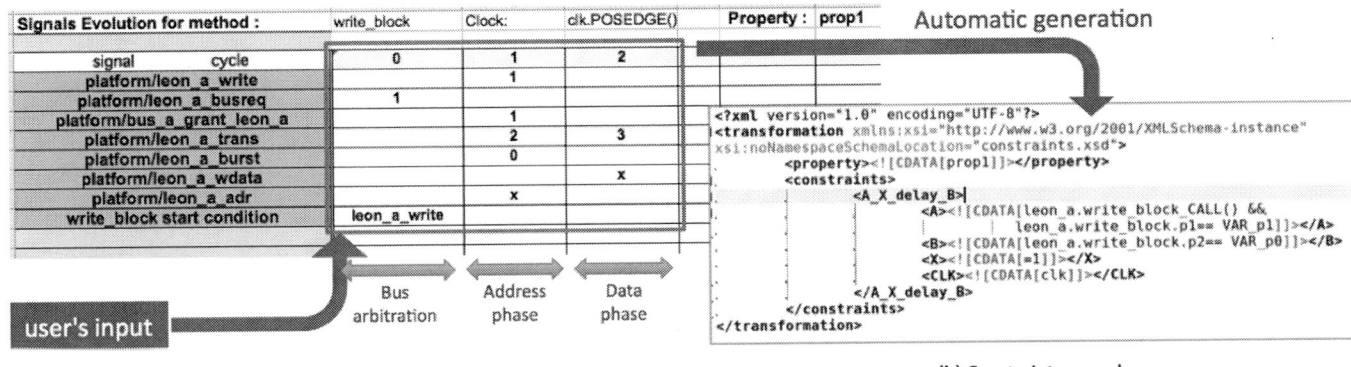

Fig. 6. The "write" method signals evolution and the corresponding constraint

platform component. The work of [15] proposes an interesting discussion about the major concerns for the refinement of temporal assertions. However, the associated tool imposes the preservation of the syntactic structure of the properties (temporal templates must be unchanged), which is a restrictive solution.

IV. CONCLUSION

Today's RTL-based validation technology is inadequate to the task of validating the industry's highly complex platform oriented architectures. In order to address verification issues at the RT level, the ESL design and verification flow has moved the design strategy from "design first, then verify" to "design and verify" [16].

Transaction level assertions refinement is the key for an efficient verification flow. A tool performing the automatic refinement of TLM assertions toward mixed and RT levels was presented. More transformation rules are currently under development to support more sophisticated communication protocols such as burst transfer. A support for SEREs operators is also under development, to enhance the checkers generation at the signal level.

REFERENCES

[1] G. Martin and B. Bailey, *ESL Models and their Application.* Springer US, 2010.

[2] F. Rogin and R. Drechsler, *Debugging at the Electronic System Level.* Springer Netherlands, 2010.

[3] A. Gerstlauer, "System design methodology," Univ. of Texas, 2010. [Online]. Available: http://www.cerc.utexas.edu/~jaa/soc/lectures/8-2. pdf

[4] S. Rigo, R. Azevedo, and L. Santos, Eds., *Electronic System Level Design, An open-source approach.* Springer, 2009.

[5] H. Foster, "Applied Assertion-Based Verification: An Industry Perspective," *Foundations and Trends in Electronic Design Automation*, vol. 3, 2009.

[6] *Assertion Based Design.* Kluwer Academic Publishers group, 2004.

[7] *IEEE Std 1850-2005, IEEE Standard for Property Specification Language (PSL)*, IEEE Std. IEEE Std 1850-2005, October 2005.

[8] L. Ferro and L. Pierre, "ISIS: Runtime Verification of TLM Platforms," in *Proc. Forum on specification & Design Languages (FDL'09)*, September 2009.

[9] L. Pierre and Z. Bel Hadj Amor, "Automatic Refinement of Requirements for Verification throughout the SoC Design Flow," in *Proc. International Conference on Hardware/Software Codesign and System Synthesis (CODES+ISSS'13)*, October 2013.

[10] L. Pierre, F. Pancher, R. Suescun, and J. Quévremont, "On the Effectiveness of Assertion-Based Verification in an Industrial Context," in *Proc. 18th International Workshop on Formal Methods for Industrial Critical Systems (FMICS)*, September 2013.

[11] K. Morin-Allory, Y. Oddos, and D. Borrione, "Horus: A tool for Assertion-Based Verification and on-line testing," in *Proc. MEMOCODE'08*, June 2008.

[12] IEEE, *IEEE Std 1666-2005, IEEE Standard SystemC Language Reference Manual.* IEEE, 2005.

[13] ARM, "AMBA Specification," www.arm.com, 1999.

[14] N. Bombieri, F. Fummi, G. Pravadelli, and A. Fedeli, "Hybrid, Incremental Assertion-Based Verification for TLM Design Flows," *IEEE Design & Test of Computers*, vol. 24, no. 2, 2007.

[15] T. Steininger, "Automated Assertion Transformation Across Multiple Abstraction Levels," Ph.D. dissertation, Technische Universität München, 2009.

[16] M. Vardi, "Formal Techniques for SystemC Verification," in *Proc. DAC'07*, June 2007.

A 1.2V low-power high-resolution noise-shaping ADC using multistage time encoding converters for Biomedical Applications

J. P. Sanjurjo, E. Prefasi, L. Hernandez
Universidad Carlos III de Madrid
Madrid, Spain
eprefasi@ing.uc3m.es

Abstract—The high-resolution and low power consumption ADCs demand in read-out circuits for biopotential systems has been increased in the last few years. This paper presents a new architecture to implement this kind of ADC′s using multistage time encoding converters. Due to the low voltage supply and low power demanded on this type of applications, the proposed ADC is formed by a second-order multibit noise-shaping converter using a time domain integrating quantizer as first stage and a Differential Gated-Ring Oscillator (DGRO) as second stage of the multistage architecture (MASH). The first-order noise shaping behavior of the DGRO allows to obtain a total third order noise shaping performance in the final ADC output. Moreover, using the arrangement proposed in this work, the low power requirements demanded in biopotential read-out circuits can be achieved. This because the multi-bit flash quantizer used in standard noise-shaping ADCs has been replaced by a time domain integrating quantizer that uses a one bit comparator and a PWM DAC. In addition the second stage of the MASH structure is used to quantize the width of a digital pulse with the benefit of first order noise shaping. Hence, the combination of a GRO with an integrating quantizer may produce a hardware-efficient multistage ADC (MASH) due to the digital nature of the GRO. As an example, the transistor level performance of a MASH 2-1 ADC with the proposed architecture has been evaluated. The transistor level simulations show that the ADC can achieve an ENOB = 15bits in a signal bandwidth of 16kHz using a 0.18μm CMOS technology at 1.2V.

Keywords— Sigma Delta Modulation, Time Encoding, Integrating Quantizer, Gated-Ring Oscillator.

I. INTRODUCTION

The SoCs that cope with the monitoring of biopotential signals (such as ECG and EEG), routinely used for medical diagnosis, have a ultra-low power consumption and miniature size as the most important design requirements. The need for long-term recordings that minimizes patient discomfort has increased the demand for high-resolution with low-power low-voltage portable biopotential acquisition systems [1]. In addition current biomedical ICs still require further improvement of power efficiency as their analog back ends consume significant power. To relax the requirements of amplification and filtering stages in the typical processing and digitization chain, oversampled high resolution ADCs (>12

bits) are increasingly used as biopotential readout solutions [2],[3]. In such applications, the active anti-aliasing filter can be replaced by a simple single-pole passive filter. Moreover, high resolution ADCs help cope better with motion artifacts. However, the architecture proposed in [3] is composed by three operational amplifiers, a flash comparator of 4 bits and a Data Weighted Averaging (DWA). The most demanding blocks in term of power of this approach are the loop filter opamps and the comparators in the flash quantizer. In addition, the amplitude domain multi-bit conversion implies a limitation in the reduction of the power supply due to the flash quantizer.

In the last years multi-bit noise-shaping ADCs using time domain integrating quantizers instead of classical amplitude domain approaches have been successfully implemented in low voltage CMOS technologies, both with continuous time integrators [4] and with switched capacitor circuits [5]. These two approaches meet the requirement of low power and low voltage demanded in biopotential read-out circuits. This because the multi-bit flash quantizer used in standard noise-shaping ADCs [3] has been replaced by a time domain integrating quantizer that uses a one bit comparator and a PWM DAC. Still they have some lacks in terms of power efficiency. The first implementation [4] would need to increase the clock of the system to undesired values to obtain higher resolutions. The second implementation [5] needs high power demanding OpAmps in its discrete time filter to do third order noise shaping. In the other hand also in the recent literature solutions to reduce the power supply of A/D converters have been presented by means of voltage-to-time conversion. Examples of this are the Gated Ring Oscillators (GRO) [6].

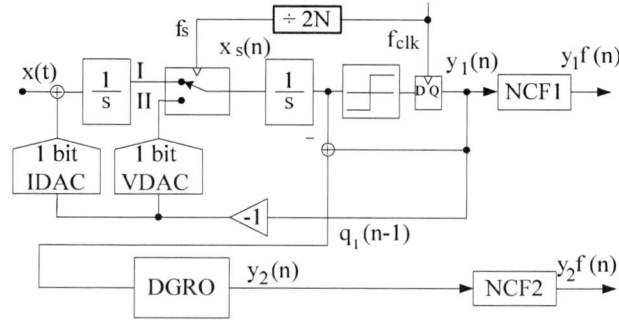

Fig. 1. Simplify block diagram of the MASH 2-1 ADC

This work has been funded by projects FP7-2013-IAPP-610484 and TEC201016330 of CICYT, Spain.

978-1-4799-4993-9/14 $31.00 © 2014 IEEE

In this work a multi-stage MASH 2-1 third order noise-shaping ADC is presented to alleviate the problems mentioned before. Figure 1 shows a simplify block diagram of the proposed architecture. The first stage of the MASH structure is a second-order noise shaping ADC using a time domain integrating quantizer [4]. The second stage is a differential Gated Ring Oscillator (DGRO) that behaves as a first order noise-shaping converter [6]. Using this arrangement third order noise-shaping can be achieved using the proper Noise Cancellation Filters (NCFs). In this way some of the issues mentioned before are alleviated. Compare with [4] the need of increasing the clock to reach higher resolutions is trade-off by the multistage noise-shaping increase. Moreover, compare with [5] high power consuming OpAmps are no longer needed to reach higher resolutions as the second stage is a digital voltage-to-time converter. To obtain the proper behavior of the MASH architecture the NCFs need to be implemented in the digital domain. These filters can be implemented in the processing part of the data-acquisition chain together with the decimation filter to lower the total power consumption.

The paper is organized as follows. Section II discusses the system of Fig. 1. Section III shows the transistor level simulations of the circuit of Fig. 1 and analyzes the results compare with standard high resolution third-order noise-shaping converters. Finally, Section IV presents the conclusions of this work.

II. System Architecture

In order to reduce the power supply of the ADC this paper presents time-encoding quantification and voltage-to-time conversion. To avoid the power consumption of a third opamp and to increase the stability of the chip, a MASH architecture has been implemented. As it is already said, the first stage is a second order noise-shaping converter with a time-domain integrating quantizer, and the second stage is a voltage-to-time converter using DGRO architecture.

A. 2nd order noise-shaping ADC using integrating quantizer

The first block of the system depicted in Fig. 1 is similar to a second order $\Sigma\Delta$ modulator with Dual-slope quantizer [7] in seizing that the integrator holds the quantization error after each sample.

But the integrator quantizer starts into an oscillation until the end of the period after the quantization has taken place [7]. In a second order noise-shaping modulator the equation for first stage of [7] has been affected and the final equation for $y_1[n]$ will be:

$$Y_1(z) = X_s(z) + Q_1(z)\cdot(1\text{-}z^{-1})^2 = X(z)\cdot(1\text{-}z^{-1})^{-1} + Q_1(z)\cdot(1\text{-}z^{-1})^2 \quad (1)$$

As it can be seen in (1) the first stage of the MASH performs as a second-order noise-shaping ADC.

This second block quantizes the width of a digital pulse (PWM error signal from first stage) with the benefit of first order noise shaping. A DGRO composed by two GRO is selected. Each GRO is composed by a chain of seventeen inverters. The selection of each GRO depends on the sign of the first output $y_1[n]$.

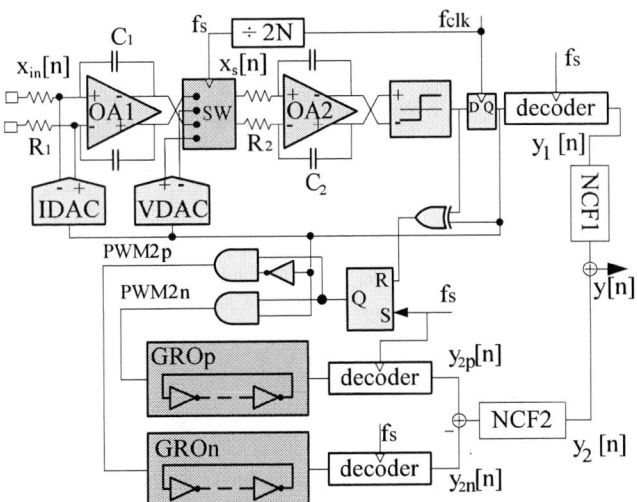

Fig. 2. Simplified schematic of MASH 2-1 architecture

B. Differential Gated-Ring Oscillator (DGRO)

With this amount of inverters and using the differential architecture this block makes a quantization with more than 5 bits. The equations of the behavior of the system are already explained in [6], for this reason it will not be analyzed in detail. But basically it can be seen that this DGRO realizes first order noise shaping without the need of any feedback signal.

C. MASH 2-1 noise-shaping ADC

To connect the two stages of the MASH an extra digital logic has to be implemented. Fig. 2 shows the logic that connects both stages. This logic generates a PWM signal representing the error of the first stage [6]. Then, this error signal is fed to the second stage. The DGRO then quantize this error signal with first order noise shaping.

Finally, to perform third-order noise-shaping the NCFs have to be selected. With a second order modulator on the first stage and a DGRO on the second one the equations of the NCFs and the final output Y(z) will be:

$$NCF1 = z^{-1} \quad (2)$$

$$NCF2 = K^{-1}\cdot(1\text{-}z^{-1})^2 \quad (3)$$

$$Y(z) = X(z) + Q_2(z)\cdot(1\text{-}z^{-1})^3 \quad (4)$$

III. Circuit Architecture

A 0.18μm CMOS technology has been chosen to implement the circuit presented in this work. The simplified schematic of the proposed 3rd order noise-shaping multi-stage ADC is shown in Fig. 2. The first stage is a 2nd-order noise shaping converter composed by two loops. The inner loop is a time domain quantizer composed by an active integrating quantizer, a one bit comparator and a one bit PWM DAC. The outer loop is a continuous time filter composed by an active integrator and one bit current IDAC. The second stage is first order noise shaping DGRO. For this design we have selected a power supply of 1.2V and a clock (CLK in Fig. 2) of f_{CLK}=18.432MHz. This clock frequency corresponds to a sampling frequency f_S=1.152 MHz which implies an 8-level integrating quantizer.

978-1-4799-4993-9/14 $31.00 © 2014 IEEE

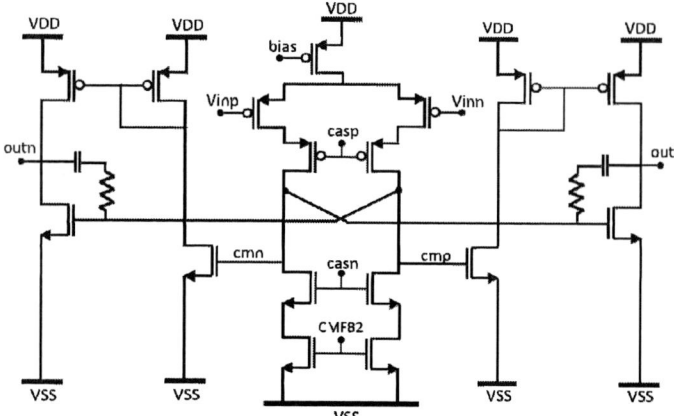

Fig. 3. Operational amplifier OA1 used in the input integrator

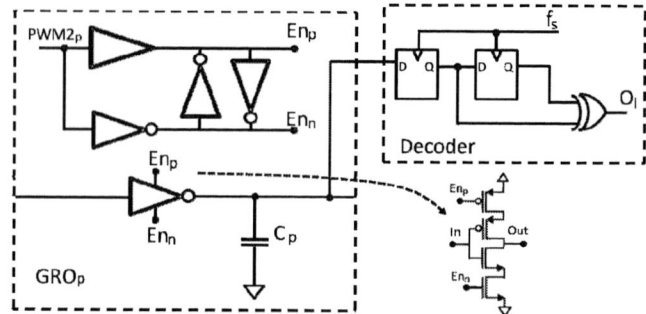

Fig. 4. Simplified schematic of one stage of the GRO

A. Operational amplifiers

The input of the first opamp (OA1) determines the linearity and the power consumption of the whole ADC. The opamp OA1 adopts the architecture proposed in [3] also used in [4].

OA1 uses a two stage class A/AB topology and it is Miller-compensated. In comparison to folded cascode and two stage class A topologies this architecture allows a better power tradeoff. But the A/AB topology requires a separate common mode feedback circuit. This is a consequence of current mirrors used in the second stage: the common mode output voltage of the first stage affects the bias condition of the second stage but does not affect the second stage output voltage. In the other hand, the opamp used in the integrating quantizer of the first stage (OA2) has lower specifications in terms of performance, as its distortion is shaped by the noise-shaping effect of the ADC. OA2 uses a one stage folded cascode topology with continuous time common mode feedback circuit. The designed GxBW for OA1 is 10 MHz and its DC gain equal to 80dB. For the OpAmp used in the integrating quantizer (OA2) the GxBW can be lowered up to 5MHz and its DC gain up to 70dB without a lost in overall performance.

B. Feedback DACs and Switch

The two feedback loops are composed by a one-bit voltage DAC (VDAC) for the inner loop and a one-bit current DAC (IDAC) for the outer loop. The total current drawn by IDAC is equal to 4uA as a trade-off between thermal and quantization noise at the input of the ADC. The two DACs are controlled by the sampling signal fs, working only on the second phase of Ts. The inputs of the DACs determine their current polarity. The switches of the integrating quantizer are implemented with transmission gates. They do not need to be driven by special circuits, since their distortion is shaped by the noise-shaping effect.

C. Gated Ring Oscillator

In order to obtain a higher resolution on the output of the system the two GROs inside of the DGRO are made with 17 inverters. A simplified scheme of one module of the GRO is shown in Figure 4. Each module is composed by two separated blocks.

The first one (GROp) has a calibration block to synchronize the enable signal (PWM2p) of the GRO to activate the pMOS and nMOS gated transistors (ENp/ENn) of the starved inverters of the GRO chain at the same time. The physical values of the pMOS transistors (L=180nm, W=900nm) and the nMOS transistors (L=180nm, W=300nm) are set to perform an output with symmetric behavior in order to reduce the mismatch of the GROs. The frequency of the oscillation is controlled by the programmable capacitor Cp.

The second block is composed by two flip-flops to detect a variation on the phase of the inverter through a XOR gate. The output of the XOR gate (Oi) is added to the other 16 phases of the GRO at the sample rate fs. The output of this adder is signal Y2p[n] in Fig. 2.

IV. CIRCUIT SIMULATIONS

A complete transistor level model of the system of Fig. 2 has been simulated using Cadence/Virtuoso. A summary of the performance of the ADC is shown in Table I. The value of the parameters selected for the simulation of the ADC have been chosen in order to optimizes the power consumption and reach the proposed resolution for this paper (ENOB=15 bits). Figure 5 depicts the simulated power spectral density (PSD) of the MASH digital output (signal y[n] in Fig. 2). The PSD in Fig. 5 shows the expected third order noise-shaping behavior predicted in section II. The calculated Signal-to-Noise-plus-Distortion Ratio (SNDR) in Fig. 5 is equal to 84dB with an input level of -2dBFull-Scale in a signal bandwidth of 16kHz. The clock frequency for this simulation has been set to f_{CLK}=18.432MHz corresponding to a sampling frequency f_S=1.152MHz and an Oversampling Ratio of OSR=36. The oscillation frequency of the GRO is set to fs. Using a power supply of 1.2V the simulated power consumption of the ADC can be estimated to be 70uW and 20uW for the first and second stage respectively. Figure 6 shows the simulated SNDR versus the input voltage of the complete MASH ADC. Analyzing the results of Fig. 6 a Dynamic Range of DR=90dB can be estimated.

V. CONCLUSIONS

This work shows an architecture that seizes the efficiency of an amplitude-to-time conversion circuit implemented with an integrating quantizer combined with a DGRO based time digitizer. The result is a multistage MASH 2-1 converter with a second order noise shaping single-bit continuous time first stage which produces a pulse width modulated error signal.

978-1-4799-4993-9/14 $31.00 © 2014 IEEE

The pulse width modulated signal permits the use of a mostly digital first order noise shaping second stage without analogue integrators, allowing a total third order noise shaping behavior of the MASH ADC. This makes the ADC especially suited for the high-resolution and low power consumption ADCs demand in read-out circuits for biopotential systems.

As a proof of concept a transistor level simulation of the complete MASH 2-1 ADC has been realized. The transistor level simulation of the ADC shows an ENOB=15bits in a signal bandwidth of 16 kHz using a 0.18μm CMOS technology at 1.2V.

If we compare the proposed circuit with a standard noise-shaping modulator of the same resolution, the proposed ADC only uses a 1 bit comparator instead of a FLASH quantizer. In addition the ADC does not need any calibration logic (i.e. DWA) due the 1 bit feedback loop.

Compared to a standard noise-shaping integrating ADC, the ADC proposed in this work can achieve the same resolution but using a lower OSR. In addition the noise shaping behavior can be achieved using less opamps, as the multistage architecture proposed in this work used a second opamp-less digital DGRO stage.

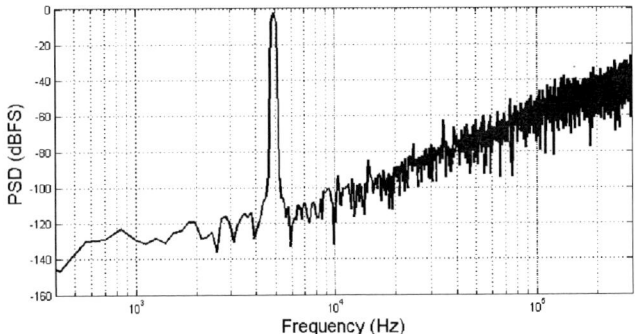

Fig. 5. PSD of the proposed third order noise-shaping ADC

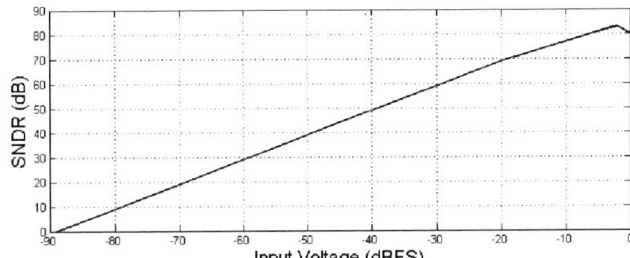

Fig. 6. Dynamic Range of the proposed MASH ADC

TABLE I. SUMMARY OF THE MASH ADC PERFORMANCE

Parameters	Value
Technology	0.18um CMOS
Number of Bits in int. quantizer(N)	3 (8)
Order of the ADC	2+1
OSR/ sampling Frequency (fs) (MHz)	36/1.152
Clock Frequency (fclk) (MHz)	18.432
Input Signal Bandwidth (kHz)	16
Input Voltage Full-Scale (mVp diff)	260
Power Supply (V)	1.2
ADC performance	
SNDR(max) dB	84
Dynamic Range (dB)	90
ENOB	15
Estimated Power consumption	
1st stage (uW)	70
2nd stage (uW)	20

REFERENCES

[1] R. F. Yazicioglu et al., "200 μW eight-channel EEG acquistion ASIC for ambulatory EEG systems," IEEE Journal of. Solid-State Circuits, vol. 43, no. 12, pp. 3025–3038, Dec. 2008.

[2] Texas Instruments. ADS1298 8-Channel, 24-bit Analog-to-Digital Converter with Integrated ECG Front End Datasheet. 2011.

[3] S. Pavan, N. Krishnapura, R. Pandarinathan, P. Sankar, "A Power Optimized Continuous-Time ΔΣ ADC for Audio Applications," IEEE Journal of Solid-State Circuits, vol.43, no.2, pp.351-360, Feb. 2008.

[4] F. Cannillo, E. Prefasi, L. Hernandez, E. Pun, F. Yazicioglu, C. Van Hoof, "1.4V 13μW 83dB DR CT-ΣΔ modulator with Dual-Slope quantizer and PWM DAC for biopotential signal acquisition," in IEEE ESSCIRC Proc., Sept. 2011, pp.267-270.

[5] N. Maghari, U. Moon, "A third-order DT ΔΣ modulator using noise-shaped bidirectional single-slope quantizer," in IEEE Int. Solid-State Circuits Conference Dig. Tech. Papers, Feb. 2011, pp.474-4

[6] L. Hernandez, E. Prefasi "Multistage ADC based on integrating quantizer and gated ring oscillator." Electronics Letters, vol 49, no 8, pp. 526-527, April 2013

[7] E. Prefasi, E. Pun, L. Hernandez and S. Paton, "Second-order multi-bit ΣΔ ADC using a Pulse-Width Modulated DAC and an integrating quantizer," Proc. 16th IEEE International Conference on Electronics, Circuits and Systems, pp. 37–40, Dec. 2009.

A Low Power Second Order Current Mode Continuous Time Sigma Delta ADC with 98 dB SNDR

Sina Parsnejad, Melih Akcakaya, Gunhan Dundar
Department of Electrical and Electronics Engineering
Boğaziçi University
Istanbul, Turkey
Email: sina.parsnejad@boun.edu.tr

Abstract—A second order CT $\Sigma\Delta$ modulator for audio frequency sensory systems in 180μm TSMC CMOS process is presented. The overall power consumption is distributed evenly among segments of the loop to attain adequate number of bits without the need to sacrifice power. The design incorporates a C-gm based current mode structure with 2nd order noise shaping, a 25 kHz bandwidth, a sampling frequency of 12.8 MHz marking an OSR of 256 and a total power consumption of 4.3μW. Consequently the proposed loop achieves a FOM of 1fJ/conv.

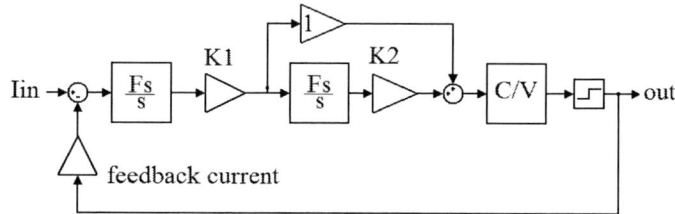

Fig. 1: Schematic of the proposed ADC architecture.

I. Introduction

The dawn of post PC computation has surfaced new research paradigms in battery powered biomedical monitoring and implant systems. There is a growing demand for mobile sensory systems in biomedical systems, lab on a chip devices, self powered structure health monitoring etc. To increase the limited battery life and avoid inflicting damage to environment, the system needs to be designed for low supply voltages and low power consumption while establishing a minimum precision especially in the case of sensor systems.

The presented work is a data converter block for audio frequency sensory systems. A CT $\Sigma\Delta$ modulator is an attractive choice of ADC implementation for it possesses an inherent anti aliasing filter and relaxed requirements on integrators, thus eliminating the need for excess filtering and sampling circuitry and mitigating power consumption. They also do not require complex switching and clocking schemes, thus paving the way for very high OSR.

The paper is organized as follows; in section II the architecture of the modulator and the elements incorporated in the design are introduced and explained. Section III concentrates on simulation results of the circuit including the dynamic range and maximum recorded SNDR on schematic and layout. Lastly, a comparison to the state of art similar $\Sigma\Delta$ modulators is conducted followed by a final conclusion.

II. Architecture of the Modulator

The system architecture is depicted in Fig .1. The pie chart in Fig .2 demonstrates the distribution of power consumption between components. Due to single node/pole architecture of C-gm integrators, they are traditionally used in high frequency low power systems [1]. However, it is concluded that such an

architecture is a perfect candidate for bandwidths as low as 25 kHz. The limited required BW and the single pole nature of integrators make it possible to incorporate a sampling frequency on 12.8 MHz on a single bit quantizer. Consequently, the required area for the integrator is extensively reduced. A 2nd order filter was selected to reduce the power consumption and overall complexity. The low bandwidth requirements of the system opens the possibility of using a large OSR while optimizing the ratio of integrator and latch power consumption to acquire an optimal operating frequency with minimal number of integrators and maximum SNDR. This trade off proved to be accurate since it acquired almost the maximum achievable SNDR with a second order system which is adequate for the designated purpose. A feed forward architecture was selected and optimized to have minimal voltage variations on internal nodes thus limiting the consumption of integrators. Moreover, a feed forward path does not require additional summing circuitry due to current mode nature of design. A single bit latch is selected because it provides our design with a rather simple feedback circuitry since it simplifies the DAC section thus saving power. A single bit ADC also reduces the linearity requirements of the gain stage and occupies much less area which meets design density requirements. Due to current mode operation of the circuit, no summing circuitry is required to add the feed forward paths which leads to a much simpler and thus, a power efficient architecture. However, a current to voltage converter is necessary since the quantizer is in voltage mode and a large gain is required for efficient operation of the comparator. This circuit also acts as an amplifier and transducer to the voltage mode quantizer. Two distinct supply voltages were used in the design of this circuit. The latch and the feedback segment incorporate a 1.2V supply, ensuring adequate accuracy. The rest of circuit is run under a 0.8V

978-1-4799-4993-9/14 $31.00 © 2014 IEEE

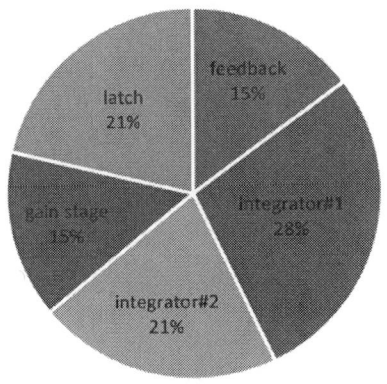

Fig. 2: Power consumption distribution of loop.

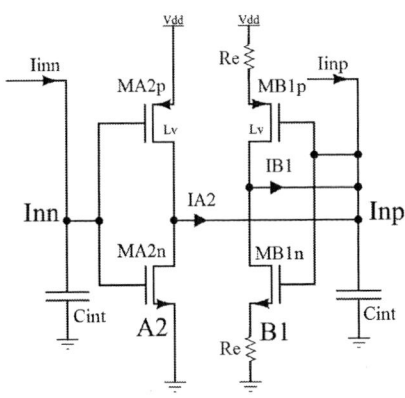

Fig. 4: Simplified integrator unit.

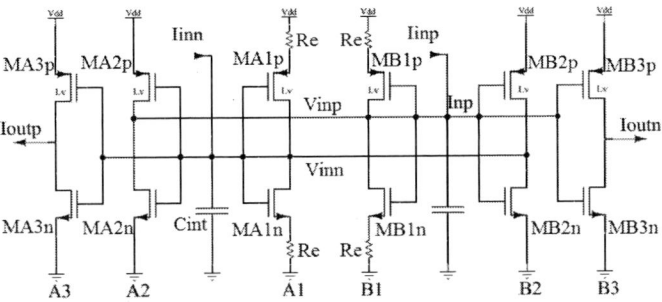

Fig. 3: Architecture of proposed single integrator unit.

supply voltage, minimizing the consumption of these stages.

A. Integrators

Fig. 3 depicts the architecture of the integrator units. It is fundamentally composed of two cross coupled inverter pairs [1]. A gain boosting method [2] has been implemented through the enhancement resistors R_e. Unlike the previous approaches, inverters are used instead of current mirror circuits. This modification results in a wide linear region of operation which together with minimal current and voltage swing due to the choice of structure results in minimal 3rd harmonic in SNDR performance. The enhancement resistors R_e are considerably small. Thus, it is safe to assume that there is only one internal node in each differential path to which a large capacitance is attached. Hence, any extra wiring and gate capacitance are only added to the total capacitance. Consequently, the circuit has only one pole which make it a perfect candidate for integrating circuitry. Moreover, the circuit does not require any additional biasing stages which adds to the overall simplicity of design. In order to analyze circuit performance, the model represented in Fig. 4 is used. The process of integrating input current into a voltage signal on the input node can be described as follows:

$$V_{inp} = (I_{inp} - G_{B1}V_{inp} - G_{A2}V_{inn}) \times \frac{1}{Cs + g_0} \quad (1)$$

$$V_{inn} = (I_{inn} - G_{A1}V_{inn} - G_{B2}V_{inp}) \times \frac{1}{Cs + g_0} \quad (2)$$

$$V_{inp} = \frac{Cs + g_0 + G_{A1} + G_{A2}}{(Cs + g_0 + G_{B1})(Cs + g_0 + G_{A1}) - G_{A2}G_{B2}} \quad (3)$$

where G represents the overall transconductance of each individual inverter and g_0 represents the overall input transconductance seen on the input node. Having a matched circuit with $G_{A1} = G_{B1}$ and $G_{A2} = G_{B2}$, it is concluded that:

$$V_{inp} = I_{inp} \frac{1}{Cs + g_0 + G_{A1} - G_{A2}} \quad (4)$$

$$I_{outp} = I_{inp} \frac{G_{A3}(s)}{Cs + g_0 + G_{A1} - G_{A2}} \quad (5)$$

if the G_{A1} is slightly less than G_{A2} due to the enhancement resistor R_e, it would cause the value of g_0 to diminish. Consequently, the 3 dB frequency of circuit pole would get below 1 kHz and the overall DC gain of circuit will decrease, causing more aggressive noise shaping. However, it should be noted that g_0 is not to be completely eradicated for it would make the DC phase prone to a change of 180 degrees and make the loop unstable at very low frequencies. Long transistors with small width were selected for implementation in the integrators. The aim is to increase the resistance seen at the internal nodes, decreasing the value of g_0. Having such long lengths also causes the transconductance of inverters to decrease and as a result the value of integrating capacitance can diminish as well, saving chip area. It should be noted that the amount of threshold voltage on both NMOS and LvPMOS devices is close to 0.3V. Consequently, the DC operating point of integrators close to 0.4V. For optimum linearity, the transconductance of PMOS and NMOS transistors on each inverter are to be closely matched. The term G_{A3}(s) can be represented as follows [3]:

$$G_{A3}(s) = 2g_m(1 - \frac{s}{z_1}) \quad (6)$$

$$z_1 = \frac{g_m - g_{ds}}{2C_{gd}} \quad (7)$$

The unwanted zero is located at approximately 1GHz frequency which means that the maximum operating range of

978-1-4799-4993-9/14 $31.00 © 2014 IEEE   17

TABLE I: Corner analysis results for an input of -4.8 db

Corners	TT	FF	SS	FnSp	SnFp
SNDR	91.4dB	90.7dB	87.4dB	90.2dB	91.4dB

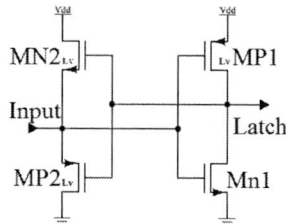

Fig. 5: Gain stage architecture.

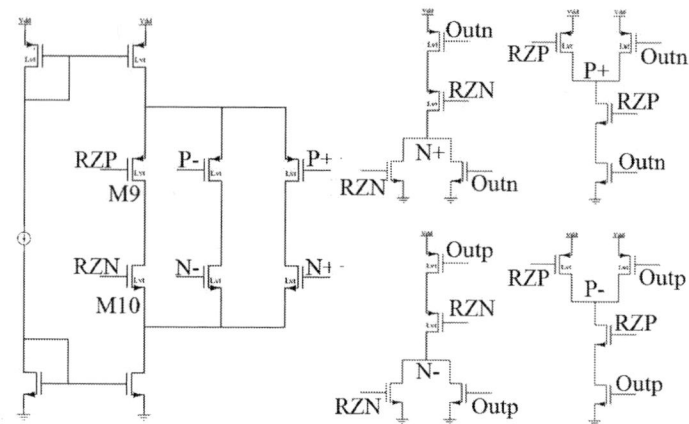

Fig. 6: Feedback architecture.

this integrator is 100 MHz. It is obvious from equation (7) that this current mode integrator is sensitive to g_m matching. As a result, the transistors inside the integrator are to be matched perfectly. It is also the case that any process variations between PMOS and NMOS transistors would alter the operating voltage of the internal nodes. However, the effects of process variation may only cause the bias voltage of input to change 30 mV for the most extreme corners of simulation. Table 2 demonstrates the circuit performance under corner analysis simulation. It is worth noting that it is possible to have as many feed forward paths as desired since adding any additional feed forward path is the equivalent of having an extra pair of inverters in the circuit. However, each extra feed forward results in extra power consumption and the value of the path is only an ordinal number. Thus, it is best to use a single feed forward path with a gain of 1 and design the loop accordingly.

B. Gain stage

An interface between the current mode blocks and the quantizer was deemed necessary. Such circuit is to convert a nano-scale current into voltage with adequate gain and the least amount of phase shift while having a very low input resistance. A modified version of [4] was used in the designs. The proposed circuit is depicted in Fig. 5. The proposed structure scales the input resistance by $(1 + G)$ where G is the voltage gain of inverters [5]. The inverter is to have a low phase shift at the clock frequency of 12.8 MHz while maintaining a reasonable gain of at least 20dB. Thus, a power budget of 1uW was provided for the converter to meet the required characteristics.

C. Comparator

The quantizer proposed in [6] is used for the design. It is of crucial importance to have a minimum amount of delay and a precision of sub millivolt since the design incorporates only one latch and any mismatch or offset on this design would compromise the SNDR of the second order system. Consequently, a huge portion of power consumption is devoted to optimizing this segment. In contrast to loop elements, this segment is run on a supply voltage of 1.2v to boost its accuracy and speed. The input transistors are low threshold so that the

latch segment can interact with the low Vdd segments of the circuit.

D. Feedback

The signal imposed delay to the comparator for small inputs could have impairing effects on performance the same way clock jitter affects circuit [1]. It is also the case that rise time and fall time of latch output are considerably large and unequal. Besides, the crossover point of the falling and rising latch signals is unpredictable and asymmetric regarding feedback transistor bias points. Such non-idealities would introduce uncontrolled and undecided feedback current to the circuit for short burst of time with pernicious effects on SNDR. Hence, it is best if a Return to Zero (RZ) feedback signal with sharp changing edges is used [1]. In this work a clock signal of 12.8 MHz with a pulse width of 3ns is implemented. The phase during which the clock value is high, is used as the return to zero period. Fig. 6 and 7 depict the feedback circuit and its corresponding control signals. It should be noted that the latch circuit is not affected by the short clock pulse width since it demonstrates a maximum delay of 1.3 ns. This switching architecture is data independent and so is the feedback charge injection. Consequently, no excess noise is produced. During the RZ phase, the negative and positive current paths are connected together through the dummy path composed of M9 and M10. As a result, the current flow of current sources is never interrupted and the voltage on the mirrors is kept constant, thus eliminating the excess time required to charge these nodes after the RZ phase. The digital circuitry and the Boolean functions driving the feedback circuit are shown in Fig. 6. It should be noted that for compatibility with latch circuit, the feedback circuit operates under 1.2V.

III. SIMULATION RESULTS

The proposed 2nd order CT sigma delta data converter is designed in 0.18μm with dual supply voltages of 0.8V and 1.2V. The simulation is done using 262144 points in Eldo environment and the post processing was performed in Matlab. As depicted in Fig. 8, the peak SNDR was calculated to be 98dB. The dynamic range plot is demonstrated in Fig. 9. Table 2 compares the proposed circuit with a number of similar

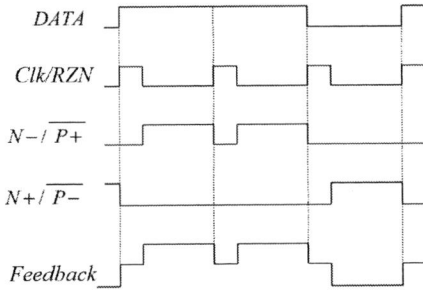

Fig. 7: Feedback wave forms.

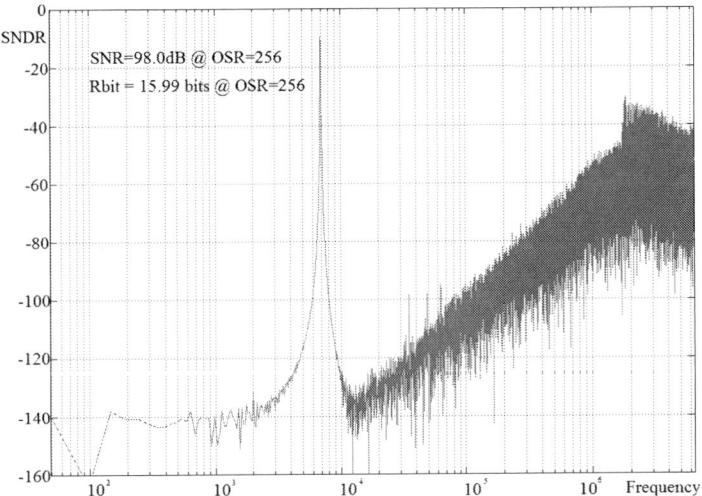

Fig. 8: Output signal FFT.

state of the art designs. The total power consumption of the modulator is $4.3\mu W$ which in turn produces a FOM of 1 fJ/conv based on the following formula:

$$FOM = \frac{Total power}{2 \times BW \times 2^{nbits}} \quad (8)$$

IV. CONCLUSION

A low power consuming current mode continuous time $\Sigma\Delta$ data converter is presented. A high SNDR performance

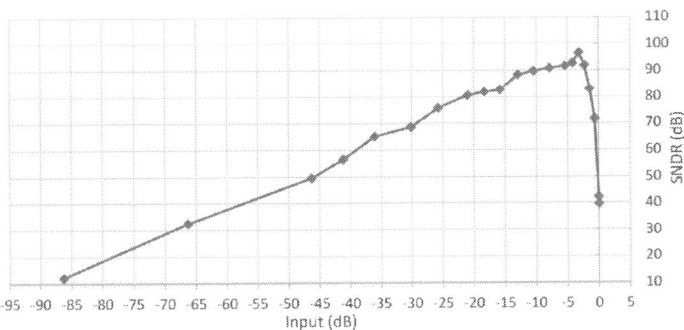

Fig. 9: Dynamic Range.

TABLE II: performance summary and comparison results of this work

Ref	Tech	Power	BW	OSR	SNDR	FOM(fJ/conv)
[7]	180nm	$90\mu W$	24KHz	64	93.5dB	54
[8]	180nm	$90\mu W$	24KHz	64	93.5dB	49
[9]	130nm	$28.6\mu W$	20KHz	64	79.1dB	97
[10]	130nm	$42.6\mu W$	20KHz	64	97.3dB	17
[11]	180nm	$38\mu W$	8KHz	128	92dB	73
This work	180nm	$4.3\mu W$	25KHz	256	98db	1

is achieved which in turn results in a very good FOM. Due lack of proper current mode latching systems, a voltage mode latch and a high gain transconductance is used. Because of the simple structure of the system, no auxiliary DAC is incorporated and the biasing system is self sufficient. For optimum FOM performance, an analog 0.8V and a digital 1.2V supply are utilized. Simulation results indicate a maximum SNDR of 98db with an OSR of 256 and power consumption of $4.3\mu W$, resulting in a minimum FOM of 1fJ/conv.

REFERENCES

[1] Aboushady, *Design for reuse of current-mode continuous-time sigma-delta analog-to-digital converters*, Ph.D. thesis, Ph. D. thesis, University of Paris VI, Department of Electronics, Communications and Computer Science, 2002.

[2] U. Yazkurt and G. Dundar, "Dc-gain enhancement technique for differential current-mode integrators," *Electronics Letters*, vol. 46, no. 11, pp. 750–752, May 2010.

[3] Rajesh H Zele and David J Allstot, "Low-power cmos continuous-time filters," *Solid-State Circuits, IEEE Journal of*, vol. 31, no. 2, pp. 157–168, 1996.

[4] Träff, "Novel approach to high speed cmos current comparators," *Electronics Letters*, vol. 28, no. 3, pp. 310–311, 1992.

[5] ATK Tang and C Toumazou, "High performance cmos current comparator," *Electronics Letters*, vol. 30, no. 1, pp. 5–6, 1994.

[6] Libin Yao, Michiel Steyaert, and Willy MC Sansen, *Low-power low-voltage sigma-delta modulators in nanometer CMOS*, vol. 868, Springer, 2006.

[7] Shanthi Pavan, Nagendra Krishnapura, Ramalingam Pandarinathan, and Prabu Sankar, "A power optimized continuous-time $\sigma\delta$ adc for audio applications," *Solid-State Circuits, IEEE Journal of*, vol. 43, no. 2, pp. 351–360, 2008.

[8] S. Pavan, N. Krishnapura, R. Pandarinathan, and P. Sankar, "A 90μw 15-bit $\sigma\delta$ adc for digital audio.," in *Solid State Circuits Conference, 2007. ESSCIRC 2007. 33rd European*, Sept 2007, pp. 198–201.

[9] Jinghua Zhang, Yong Lian, Libin Yao, and Bo Shi, "A 0.6-v 82-db 28.6-w continuous-time audio delta-sigma modulator," *Solid-State Circuits, IEEE Journal of*, vol. 46, no. 10, pp. 2326–2335, 2011.

[10] Jinghua Zhang, Libin Yao, Bo Shi, and Yong Lian, "A 1-v 42.6μw 1.5-bit continuous-time delta-sigma modulator for audio applications," in *Microelectronics and Electronics (PrimeAsia), 2010 Asia Pacific Conference on Postgraduate Research in*, Sept 2010, pp. 73–76.

[11] Da Qi, Yuan wen Li, Long Cheng, Jun Xu, Fan Ye, and Jun-Yan Ren, "An ultra low power sigma-delta modulator for hearing aid with double-sampling," in *ASIC, 2009. ASICON '09. IEEE 8th International Conference on*, Oct 2009, pp. 1141–1144.

Continuous Time Analog to Digital Conversion in Interferer Resistant Wake Up Radios

Alin Ratiu[*†], Dominique Morche[*], Bruno Allard[†], Xuefang Lin-Shi[†], Jacques Verdier[†]

[*]Univ. Grenoble Alpes, F-38000 Grenoble, France

CEA, LETI, MINATEC Campus, F-38054 Grenoble, France

Email: {alin.ratiu, dominique.morche}@cea.fr

[†]Univ. de Lyon, Ampère, INSA de Lyon, Lyon, France

Email: {bruno.allard, xuefang.shi, jacques.verdier}@insa-lyon.fr

Abstract—Continuous time digital filtering offers the possibility of implementing low power, tunable, low frequency filters required for interferer rejection in ultra low power radios. The bottleneck of such architectures lies in the limited linearity of the continuous time analog to digital conversion (CT-ADC) stage. This paper presents an interferer resistant wake-up radio architecture based on continuous time digital filters and discusses different ways of improving the linearity of the analog to digital conversion stage without sacrificing its power consumption.

I. INTRODUCTION

The specifications of the next generation of wireless sensor networks (WSN) push the required autonomy of WSN nodes to durations of 5 to 10 years. Recent advances in deep sub-micron CMOS technology have reduced the power consumption of the sensing/processing part of the node leaving the wireless receiver as a major consumer of power [1]. Receiver duty cycling solves the WSN node power issue, but introduces other problems like network synchronization, clock distribution and network latency.

An alternative solution, complementary to the one previously presented has been proposed during the past years and focuses on the use of a second, ultra low power co-receiver called wake-up receiver (WU-RX) which remains switched on all the time. Whenever the WU-RX intercepts a transmission request, it wakes up the main receiver which handles the reception of the bulk of the data. The sensitivity of the WU-RX has to match that of the main receiver in order to avoid missing any transmission requests. A fundamental difference between the two receivers lies in the expected communication speed: the WU-RX can have a data rate as low as 100kbps since the transmission requests are usually very short bit sequences. Recent implementations of WU-RX have pushed the sensitivity to levels very close to those of standard receivers [2] and [3]. The main drawback of the existing architectures is that they cannot distinguish between desired signals and the undesired ones, called interferers. The presence of such signals can cause the receiving node to completely miss transmission requests, forcing the transmitting node to spend more power resending the communication requests. The existing solutions are often limited to rejecting only certain interferer configurations [4].

As a consequence, there is a strong motivation to provide wake-up radios with better interferer rejection. The tight power requirements make it difficult to conceive receivers with a sufficiently high linearity, allowing subsequent processing for interferer rejection. This paper presents a new WU-RX architecture with improved interferer robustness, based on the utilization of a continuous time digital filtering stage. The focus of this article lies in the exploration of different design trade-offs for the most non-linear part of the system, the CT-ADC.

The rest of this paper is organised as follows: part II presents the WU-RX architecture and the motivation behind the choice of the filtering scheme, while part III presents the scheme chosen for the CT-ADC. Different CT-ADC circuit specifications and their impact on the system performances are presented in part IV. Part V concludes the paper.

II. WAKE-UP RECEIVER ARCHITECTURE

Standard WU-RX architectures usually use an input RF filter which limits the effective receiver bandwidth to several MHz, acting as a basic interferer rejection filter. Improving upon this level of selectivity requires the use of intermediate frequency (IF) filtering in order to reduce the filter roll-off requirements. Given the tight power budget required for a wake-up application, it is difficult to generate a precise frequency for the local oscillator. As a consequence, this paper focuses on a well-known "uncertain IF" wake up receiver architecture [3] with a slight modification of the IF stage which enables interferer rejection (Fig. 1). This system uses a local oscillator with an uncertain frequency (thus drastically reducing its power consumption) along with an energy detector for demodulation. The main drawback of this architecture is that the frequency of the received signal at IF is unknown, hence filtering it requires a tunable filter capable of changing its transfer function according to the position of the useful signal on the frequency axis.

For the implementation of such a filter several solutions are compared: *a*) analog filters: active GM-C implementations demand a large amount of power for the operation of the required op-amps [5]; *b*) digital, discrete time filters: while this solution offers a very good programmability, the main drawback lies in the fact that it requires the use of a clock signal (similar to the one required for the local oscillator),

978-1-4799-4993-9/14 $31.00 © 2014 IEEE

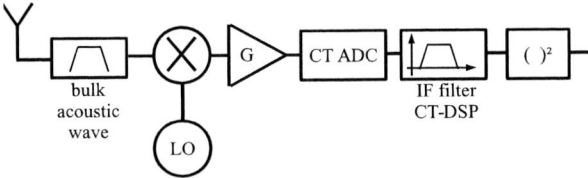

Fig. 1. Proposed, interferer resistant, wake up receiver architecture

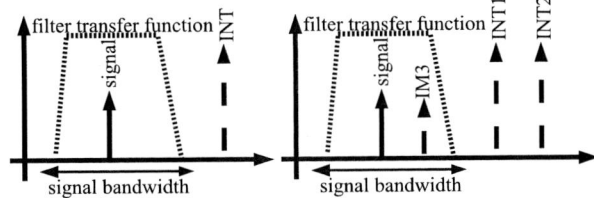

Fig. 2. Possible interferer configurations: left - interferer rejection limited by the filter characteristic; right - interferer rejection limited by the system linearity

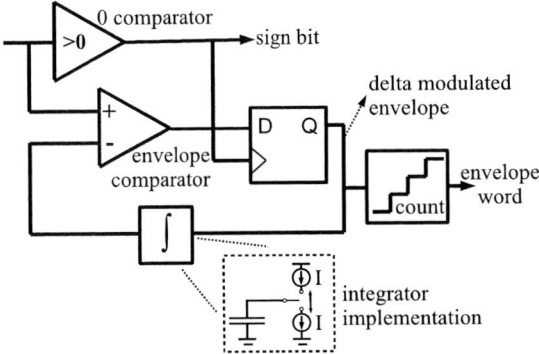

Fig. 3. Architecture of the CT-ADC

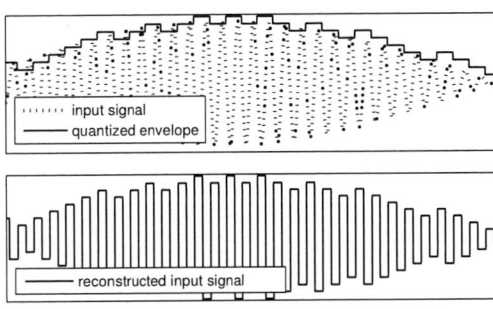

Fig. 4. Time domain output of the chosen quantization scheme

hence negatively impacting the power budget of the receiver; *c*) switched capacitor filters: while offering a very good programmability, this solution requires a system clock as well as the use of power consuming op-amps [6]; and *d*) digital, continuous time filters: this new paradigm in signal processing, presented in [7], offers the same degree of programmability as the discrete time case, without requiring any system clock. Furthermore, it offers the advantage of naturally scaling its power consumption based on the input signal characteristics: when no signal is present, the power consumption drops to very low levels. Consequently, the chosen implementation of the IF filtering stage is based on continuous time digital signal processing (CT-DSP).

III. CT ADC / DSP

In a single interferer scenario Fig. 2 (left) the selectivity of the receiver can be completely characterized by the transfer function of the IF filter. However, in a two interferer situation Fig. 2 (right), the selectivity may be limited by the linearity of the CT-ADC conversion: since the filtering occurs after the analog to digital conversion, in-band third order intermodulation products remain unfiltered thereby decreasing the in-band signal to noise and distortion ratio (SNDR) and limiting the sensitivity of the receiver.

As it has been shown in [8], the linearity of any CT-ADC depends on the number of output quantization levels. Increasing the number of quantization levels increases the output activity of the CT-ADC and the power consumption in the CT-DSP. For the purposes of our application, this represents a key trade-off point as the filter requires a maximal linearity for a minimum of power consumption.

CT-ADCs rely on conveying information in the sequence of digital output levels as well as in the precise transition timing between these levels. Any transistor level implementation suffers from non-ideal behaviour of the respective circuit, such as non zero comparison time at the comparator level. Since

these non-idealities have a negative impact on the linearity of the CT-ADC, they constitute a second trade-off point between the amount of tolerable non ideal behaviour and the power spent on improving this behaviour. This paper focuses on finding the optimal trade-off point between activity and power consumption as well as on studying the system performance in the presence of different non ideal behaviors of the involved CT-ADC blocks. Results presented here facilitate transistor level design of the CT-ADC with a strong focus on minimizing the power consumption.

The CT-ADC architecture of choice [9] is presented in Fig. 3. The quantization scheme employed relies on separating the information from the input signal in a sign bit and an envelope word. During each fundamental period of the input signal its envelope is quantized in a delta-like manner - Fig. 4. The multiplication of the sign bit with the digital envelope word represents the quantized version of the input signal. Results presented in the original paper show that the power consumption of the system for 2GHz input signals is 550μW. Since the power consumption of this system scales linearly with activity, it is expected that a redesign of the architecture, for lower frequency signals - 100MHz, consumes around 20μW which is within the power budget of the WU-RX.

This quantization scheme can be viewed as a hybrid mix between a high linearity conversion - due to the use of several quantization levels for the envelope word, and a very low power, low performance conversion - similar to the single bit conversion, as an output event is generated only once per fundamental period. This characteristic behaviour provides a good starting point for the output activity/power trade-off. The drawback of this quantization scheme is the fact that its linearity is not only dependant of the average output activity, given by the fundamental frequency, but also on the effective input signal bandwidth. As it has been shown in [9], the total in-band harmonic distortion (THD) is a direct function of the ratio between the central signal frequency and its bandwidth (1), herein called bandwidth ratio (BR). Although this equation shows how the total distortion depends on the BR, it does not give any insight on how the distortion is distributed among different intermodulation products. This question is answered through simulation in the next part of this paper.

$$ THD_{quant} = \log_2 \left(\frac{\sqrt{12}}{\pi} \frac{F_c}{BW} \right) = \log_2 \left(\frac{\sqrt{12}}{\pi} BR \right) \quad (1) $$

IV. Architecture Optimization

The principal performance metric of the CT-ADC is its spurious free dynamic range (SFDR) in a two-tone configuration. Optimizing the architecture for achieving a very low power consumption can be done by sacrificing the performance of unessential circuit blocks while emphasizing those which have the greatest impact on the SFDR. Consequently, an extensive simulation plan is used to study the link between different implementation non-idealities and the overall performance of the system. Even though all simulations have been done with noiseless input signals, results can be easily extrapolated to noisy situations given the fact that the CT-ADC has a noise transfer function of 1. By defining a minimum output in-band SNR required for reception (SNR_{min}) the maximum input SNR can be computed given the SFDR, using (2). The thermal noise of the CT-ADC is expected to have a minimal impact on the performance of the system, since this block is used at the end of the reception chain.

$$ SNR_{in} = SNR_{min} - 10 \log_{10} \left(1 - 10^{\frac{SNR_{min} - SFDR}{10}} \right) \quad (2) $$

A. Intermediate Frequency Selection

A characteristic of the chosen CT-ADC architecture is the direct relation which exists between the linearity and the BR, as shown in Fig. 5. It is important to note that, at this point, the actual signal frequency is not important, as the presented results are valid for any central signal frequency and bandwidth that meet the BR requirements.

Supposing the input BAW filter limits the receiver bandwidth to 10MHz, the maximum interferer spacing which creates in-band IM3 is 5MHz. The starting point for the required SFDR is set to 25dB, thus the minimum BR respecting this specification is around 20, giving us an IF of 100MHz. Even though the IF is variable due to the architecture choice, this paper focuses on the results obtained in a typical scenario.

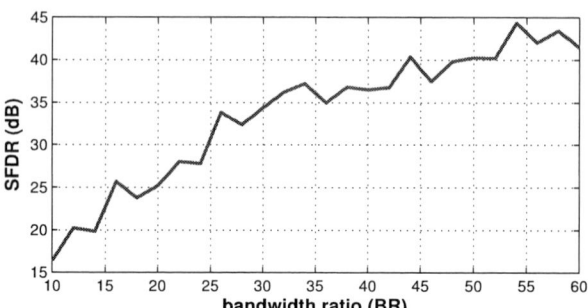

Fig. 5.   Linearity of the CT - ADC versus the bandwidth ratio

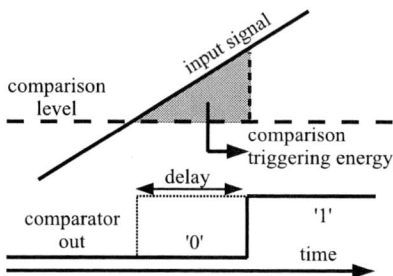

Fig. 6.   Illustration of the supposed comparator behaviour, requiring a certain amount of energy to trigger

B. Comparator Design Margins

The system linearity degradation is studied in the presence of a non-ideal comparator behaviour. The chosen comparator model supposes that a fixed amount of energy is required for any triggering to take place, as shown in Fig. 6. The comparison delay, the time difference between the actual level crossing and the comparator triggering, depends on the slope of the signal in the neighbourhood of the level crossing (i.e. slow signals incur a larger delay while fast signals - a small one). This model corresponds to a standard non-latched and non-clocked comparator used in CT systems [10]. Transistor level simulations of the previously proposed comparator design, in 28nm FDSOI, confirm the constant tripping energy behaviour.

Results showing the degradation of the SFDR in the presence of a non-zero comparator tripping energy are presented in Fig. 7. As expected, the envelope comparator has much more relaxed delay requirements, since its output is processed only during the falling edges of the sign bit: as long as the comparison has correctly taken place by that time, the effective time delay is not important. On the other hand, the 0 comparator has much more stringent energy requirements. Considering an input swing of 300mV and a central frequency of 100MHz, the comparison time delay must be lower than 1.5ns (corresponding to 75pJ of energy).

C. Integrator Design Margins

In this section we study the matching requirements for the two current sources presented in Fig. 3. Any mismatch in

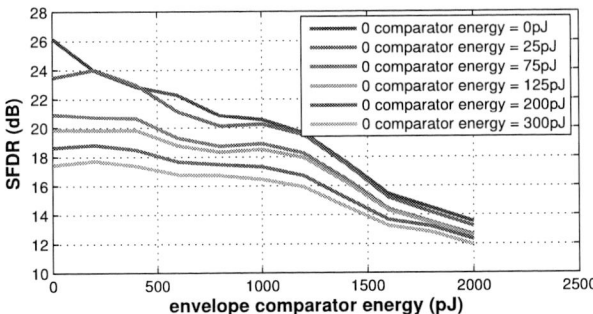

Fig. 7. CT-ADC linearity versus the envelope comparator triggering energy for different 0 comparator triggering energies and a bandwidth ratio of 20

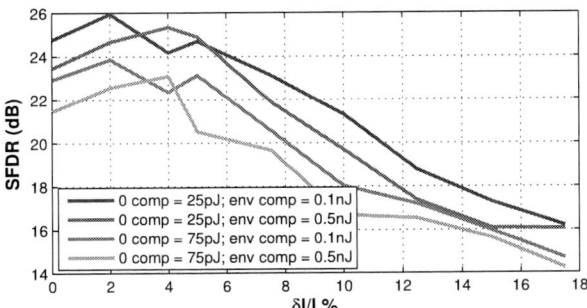

Fig. 8. CT-ADC linearity versus mismatch in the current generated by the two current sources for a bandwidth ratio of 20

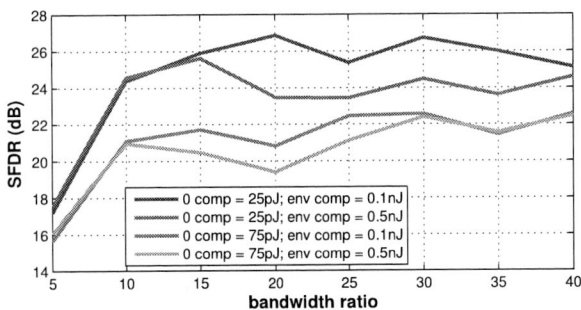

Fig. 9. CT - ADC linearity versus the bandwidth ratio for several comparator design configurations and an integrator current source mismatch of 8%

impacting the linearity of the system. As stated earlier, the single interferer performance is not limited by the linearity of the CT-ADC, but rather by the filter transfer function.

V. CONCLUSION

In this paper we have explored the possibility of using a CT-ADC-DSP system as an ultra low power interferer rejection filter for wake up radios. The main challenge for such a system is the design of a very energy efficient CT-ADC as the linearity of this stage limits the interferer rejection performance of the system. The results in this paper also present different trade-off points that help facilitate design choices for an ultra low power implementation of the circuit.

Preliminary comparator transistor level implementations in FDSOI 28nm technology node, show promising results, as a 18pJ zero comparator requires only $4\mu W$ of average power.

REFERENCES

[1] N.M. Pletcher, "Ultra-low power wake-up receivers for wireless wensor wetworks", Ph.D. dissertation, University of California, Berkley, 2008

[2] S. Drago et al., "A 2.4GHz 830pJ/bit duty-cycled wake-up receiver with -82dBm sensitivity for crystal-less wireless sensor nodes", in IEEE ISSCC Dig. Tech. Papers, Feb 2010, pp. 224-225

[3] N.M. Pletcher, S. Gambini, J.M. Rabaey, "A 52 uW wake-up receiver with -72dBm sensitivity using an uncertain-IF architecture", in IEEE ISSCC Dig. Tech. Papers, Feb 2009, pp. 269-280

[4] X. Huang et al., "A 915MHz ultra low power 2-tone transciever with enhanced interference resilience", IEEE JSSC, Dec 2012, vol. 47, no. 12, pp. 3197-3207

[5] S. Hori et al., "Low-Power Widely Tunable Gm-C Filter Employing an Adaptive DC-blocking, Triode-Based MOSFET Transconductor", in IEEE Transactions on Circuits and Systems, vol. 61, no.1, Jan 2014, pp. 37-47

[6] K.W.H. Ng, W.S.-L. Cheung, H.C. Luong, "A 44-MHz Wideband Switched-Capacitor Bandpass Filter Using Double Sampling Pseudo-Two-Path Techniques", in IEEE JSSC, vol. 40, no. 3, Mar 2005, pp. 781-784

[7] Y. Tvidis, "Event-driven, continuous-time ADCs and DSPs for adapting power dissipation to signal activity", in IEEE International Symposium on Circuits and Systems (ISCAS), May 2010, pp. 3581-3584

[8] B. Schell, Y. Tsividis, "Analysis and simulation of continuous-time digital signal processors", in Signal Processing, Volume 89, Issue 10, Oct 2009, pp. 2013-2026

[9] D. Lachartre, "A 550μW inductorless bandpass quantizer in 65 nm CMOS for 1.4-to-3GHz digial RF receivers", in VLSIC Dig. Tech. Papers, Jun 2011, pp. 166-167

[10] M. Kurchuk et al. "GHz-range continuous-time programmable digital FIR with power dissipation that automatically adapts to signal activity", in IEEE ISSCC Dig. Tech. Papers, Feb 2011, pp. 232-234

the value of the two currents contributes to the non-linear behaviour of the CT-ADC, since that mismatch is not be reflected in the digital reconstruction of the envelope, where the counter supposes a constant quantization step. The chosen model for this circuit imperfection supposes that one of the current sources has a value of $I_1 = I + \delta I$ while the other is equal to $I_2 = I - \delta I$, with δI a Gaussian random variable. Fig. 8 presents the dependence of the output SFDR versus the standard deviation of $\delta I / I$ for four different comparator tripping energy configurations. Results confirm that matching is a critical parameter. However, 8% mismatch can be tolerated in the target system which does not require device oversizing.

D. System Performance

The output SFDR versus the BR for different design configurations is plotted in Fig. 9. This allows us to determine the worst case interferer performance for the CT-ADC-DSP block. Depending of the frequency difference between the two interferers, interferer rejection of over 15dBs is expected (eq. 2), considering a 12dB margin is kept for the output SNR (corresponding to a bit error rate of 1e-3 for OOK modulation). Emphasis should be put on the design of a fast, power efficient 0 comparator, as the comparison energy of this block has the greatest impact on the overall performance of the system. Mismatch in the two integrator current sources should be kept under 8%, while sacrificing the envelope comparator tripping energy would enable a low power design without drastically

Green improvements of IEEE 802.11 directional multi-gigabit physical layer specifications

Marc-Antoine Bouzigues, Isabelle Siaud,
and Anne-Marie Ulmer-Moll
Orange Labs
35512 Cesson-Sévigné, France
{marcantoine.bouzigues, isabelle.siaud,
annemarie.ulmermoll}@orange.com

Maryline Hélard
Institute for Electronics and Telecommunications
of Rennes (IETR) –UMR CNRS 6164
35708 Rennes, France
maryline.helard@insa-rennes.fr

Abstract—**In the context of green communications, millimeter wave bands are investigated to ensure mobile access and backhauling. Such transmissions can be performed with the IEEE 802.11 Directional Multi-Gigabit (DMG) modes, designed for very high throughput in the 60 GHz band. However, the DMG Orthogonal Frequency-Division Multiplexing (OFDM) physical (PHY) layer specifications corresponding to the highest data rates are not energy-efficient. Therefore, in this article, we propose easily implementable green improvements for IEEE 802.11 DMG OFDM PHY: a change of error correcting code and a new modulation and coding scheme. Binary error rate performance evaluations are made to show the green benefits of our improvements.**

Keywords—Millimeter waves; orthogonal frequency-division multiplexing; green communications; multi-gigabit transmissions.

I. INTRODUCTION

Over the last years, due to the increase of Information and Communication Technologies (ICTs) carbon footprint, interest for green communications grows among the ICTs actors, resulting in the creation of consortium GreenTouch [1].

In this context, millimeter (mm) wave bands are investigated to ensure mobile access and backhaul transmissions. Mid-2013, Millimeter-Wave Evolution for Backhaul and Access (MiWEBA) project, resulting from an international cooperation between Europe and Japan, has been launched and aim to design heterogeneous networks (HetNets) with mm-wave cells [2].

Mm-wave transmission can be performed with the popular IEEE 802.11 standard which describes Directional Multi-Gigabit (DMG) modes allowing very-high throughput in the 60 GHz band [3]. However, Orthogonal Frequency-Division Multiplexing (OFDM) modes corresponding to the highest data rates have not been designed to maximize energy-efficiency and can be improved under certain propagation conditions.

Within the framework of MiWEBA challenges, this paper describes green improvements of IEEE 802.11 DMG OFDM physical (PHY) layer. The first improvement is a change of forward error correction (FEC) technique allowing a reduction of complexity with similar Binary Error Rate (BER) performance. The second improvement is a new modulation and coding scheme allowing an increase of BER performance under certain propagation conditions. Our improvements are evaluated in different propagation scenarios: Line Of Sight (LOS) and Obstructed/Non Line of Sight (OLOS/NLOS) transmissions.

The paper is organized as follows. Section II describes the original physical layer and channel models. The first improvement is described and evaluated in section III and the second in section IV. Section V concludes this paper.

II. SYSTEM DESCRIPTION

In this section, we present the original IEEE 802.11 DMG PHY layer, particularly the OFDM modes. Then, we describe the channel models used to perform the simulations.

A. Directional multi-gigabit physical layer

The DMG PHY [3] is an enhancement of standard IEEE 802.11 [4] that enables very high throughputs in the 60 GHz band. It supports 32 Modulation and Coding Schemes (MCS) divided into three modulation methods: control modulation (MCS 0), single-carrier (SC) modulation (MCS 1-12 and 25-31) and OFDM modulation (MCS 13-24). Highest data rates (up to 6756.75 Mbps) are obtained with OFDM modulation whose main timing-related parameters are presented in Table I.

TABLE I.   MAIN TIMING-RELATED PARAMETERS

Parameter	Value
Number of data subcarriers	336
Number of pilot subcarriers	16
Number of DC subcarriers	3
DFT size (N_{DFT})	512
Bandwidth size (W)	2640 MHz
Subcarrier frequency spacing	5.16 MHz = W/N_{DFT}

At the transmitter, before the OFDM modulation, MCS 13 to 24 are generated using a systematic Low-Density Parity-Check (LDPC) encoder using four different parity-check matrix corresponding to 1/2, 5/8, 3/4 and 13/16 code rates. Coded bits are then mapped to complex constellations: Spread Quadrature Phase Shift Keying (SQPSK), Quadrature Phase Shift Keying (QPSK), 16-Quadrature Amplitude Modulation (QAM) or 64-QAM.

At the receiver, a zero-forcing equalization is performed before the demodulation and coded bits are soft-decoded using the min-sum algorithm [5] with 10 iterations.

B. Channel models

In order to evaluate the system performance for different propagation environments, we use a propagation channel model denoted CEPD (Canal Enregistré de Propagation Déterministe) which combines multi-rate filtering theory with a statistical analysis of multipath signature of the propagation channel to generate multipath models in accordance with system parameters [6].

Two realizations of the model are used to simulate LOS and OLOS/NLOS transmissions in multipaths residential environments. With LOS propagation, the transmission is realized thanks to the direct path between the transmitter and the receiver, channel delay spread is 1.73 ns. With OLOS/NLOS propagation the transmission is realized thanks to an indirect path, channel delay spread is 12.40 ns.

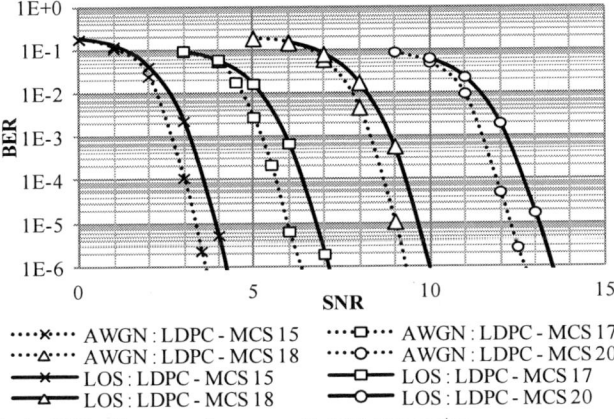

Fig. 1. BER reference performance with LOS propagation

····×···· AWGN : LDPC - MCS 15   ····□···· AWGN : LDPC - MCS 17
····△···· AWGN : LDPC - MCS 18   ····○···· AWGN : LDPC - MCS 20
——×—— LOS : LDPC - MCS 15   ——□—— LOS : LDPC - MCS 17
——△—— LOS : LDPC - MCS 18   ——○—— LOS : LDPC - MCS 20

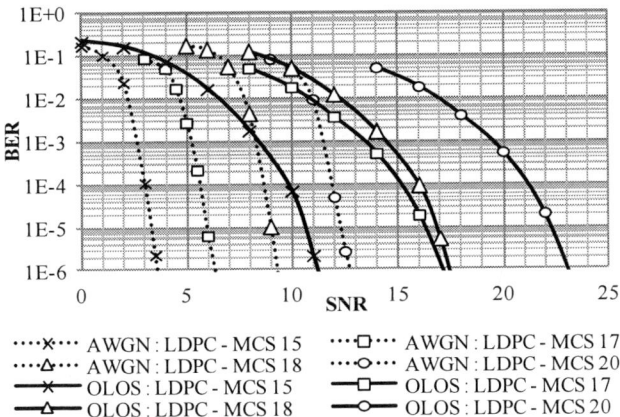

Fig. 2. BER reference performance with OLOS/NLOS propagation

····×···· AWGN : LDPC - MCS 15   ····□···· AWGN : LDPC - MCS 17
····△···· AWGN : LDPC - MCS 18   ····○···· AWGN : LDPC - MCS 20
——×—— OLOS : LDPC - MCS 15   ——□—— OLOS : LDPC - MCS 17
——△—— OLOS : LDPC - MCS 18   ——○—— OLOS : LDPC - MCS 20

Fig. 1 and Fig. 2 show the BER reference performance with LOS and OLOS/NLOS propagation of MCS 15, 17, 18 and 20 whose main characteristics are given in Table II. In both cases, BER performance is compared with the ideal propagation scenario with only Additive White Gaussian Noise (AWGN).

TABLE II.   MODULATION AND CODING SCHEMES USED IN THE STUDY

Index	Modulation	Code Rate	Data Rate (Mbps)
15	QPSK	1/2	1386.00
17	QPSK	3/4	2079.00
18	16-QAM	1/2	2772.00
20	16-QAM	3/4	4158.00

III.   ERROR CORRECTING CODE MODIFICATION

LDPC are advanced iterative codes whose major drawback is complexity. Thus, most systems integrating LDPC use suboptimal decoding algorithms to reduce latency and complexity [7]. High complexity and iterative process also mean high hardware consumption. Thus, if we manage to obtain the same BER performance with a correcting code with lower complexity, we improve the system energy-efficiency.

Standard IEEE 802.11 also describes Very High Throughput (VHT) OFDM modes for bands below 6 GHz [4][8]. Those modes can be use with either LDPC or Binary Convolutive Codes (BCC) with puncturing in order to generate 1/2, 2/3, 3/4 and 5/6 code rates.

BCC have less complexity than LDPC and can be use to perform the same code rates thanks to puncturing. Their recommended soft-decoding process is performed using Viterbi algorithm [4] and is not iterative, meaning lower complexity and lower latency.

In analogy to the VHT OFDM PHY specifications [8], we include BCC codes to the DMG OFDM PHY [3] in order to evaluate the performance difference between the different FEC techniques. We use the VHT OFDM PHY specifications convolutional encoder and puncturing patterns and we adapt the first data interleaving operation to the DMG OFDM timing-related parameters.

A. Convolutional encoder and puncturing

The convolutional encoder of coding rate 1/2 uses the standard generator polynomials g_0=133 and g_1=171 (in octal). Decoding is performed with the Viterbi algorithm [9]. The 3/4 code rate is obtained by puncturing the encoded data. The stolen bits are replaced at the receiver by dummy bits as described in IEEE 802.11 standard [4].

B. Data interleaving

The first interleaving operation ensures that successive coded bits are mapped onto non-successive data subcarriers. It is realized with a matrix interleaver with 28 rows and $12 \cdot m$ columns with m the number of coded bit per complex symbol. The second interleaving operation ensures that successive bits are mapped alternatively onto less and more significant bits of the constellation. It is realized as described in IEEE 802.11 standard [4] with a block interleaver.

C. BER performance

BER performance for LOS and OLOS/NLOS propagation of MCS 15, 17, 18 and 20 with BCC is represented respectively on Fig. 3 and Fig. 4 and compared with the ideal propagation scenario with only AWGN.

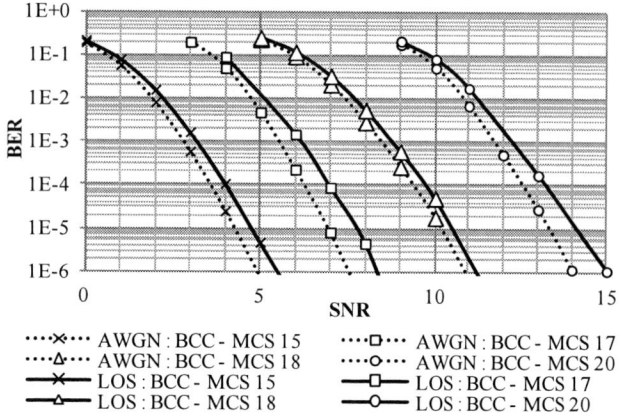

Fig. 3. BER BCC performance with LOS propagation

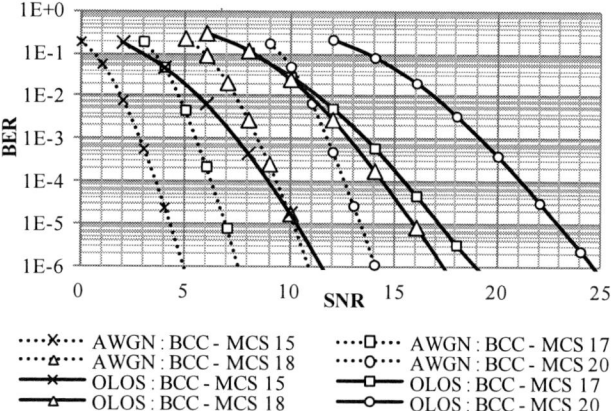

Fig. 4. BER BCC performance with OLOS/NLOS propagation

We call multipath degradation the difference (in dB) between the system BER performance in multipath situation and the system BER performance with only AWGN. In OLOS/NLOS situation, a comparison between Fig. 2 and Fig.4 shows that BCC codes are less subject to multipath degradation than LDPC codes and can challenge their performance as shown on Fig 5.

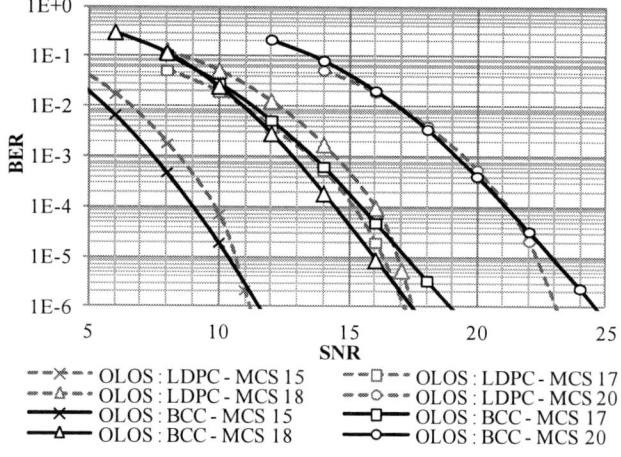

Fig. 5. BER BCC and LDPC performance comparison with OLOS/NLOS propagation.

In NLOS situation, it is worth using BCC for MCS without puncturing when the targeted BER is over 10^{-6} because they have better performance. With puncturing, it is worth using BCC if the targeted BER is over 10^{-5} because they have similar performance with lower complexity.

However, as the power amplifier is the hardware component responsible for the greatest part of the transmitter power consumption [10], in LOS situation LDPC codes are better, because with BCC it is necessary to increase the transmission power to obtain the same performance.

IV. A NEW MODULATION AND CODING SCHEME

As DMG OFDM PHY supports 64-QAM modulation and 1/2 coding rate, it is able to perform MCS 64-QAM 1/2. But, it corresponds to the same data rate as MCS 20 (16-QAM 3/4) and is not implemented in systems because classical rate adaptation algorithm identify MCS given their corresponding data rate [11].

However, previous results of this paper show that code rate 1/2 suffers less from multipath degradation than code rate 3/4, especially with OLOS/NLOS propagation. Thus, given the propagation conditions, it is worth increasing the modulation order rather than increasing the code rate.

Fortunately, novel algorithms have been developed in recent years which are able to select the MCS based on the propagation conditions [12][13]. It is now possible to have several MCS corresponding to the same data rate in order to increase the number of choices for the selection algorithm.

Then, we propose to add the combination of 64-QAM modulation and 1/2 code rate to DMG OFDM PHY specifications. This new MCS is referred as MCS 20' in the remaining of this article.

Fig. 6 shows BER performance of MCS 20 and MCS 20' with LDPC. LOS, OLOS/NLOS and AWGN situations are compared.

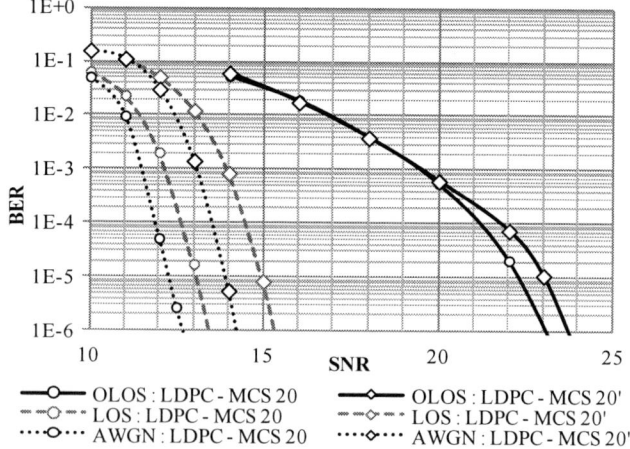

Fig. 6. BER LDPC performance of MCS 20 and 20'

Even if MCS 20' suffers less from multipath degradation than MCS 20 it is not worth using it with the classical LDCP scheme because MCS 20 performance is better.

978-1-4799-4993-9/14 $31.00 © 2014 IEEE

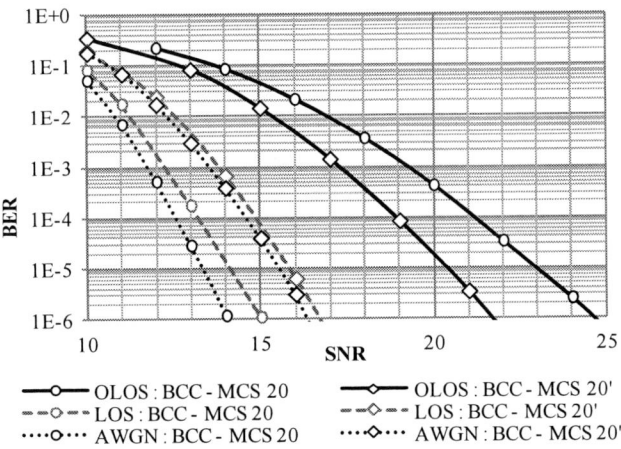

Fig. 7. BER BCC performance of MCS 20 and 20'

However, as shown by Fig. 7 when BCC are used, MCS 20' has better performance than MCS 20 in OLOS/NLOS situation. Then, it is worth using MCS 20' instead of MCS 20 in OLOS/NLOS situation, when BCC are used, because it has low multipath degradation.

Moreover, by using MCS 20' with BCC instead of MCS 20 with LDPC in OLOS/NLOS situation, we increase the BER performance by more than 2 dB while reducing the system complexity.

V. CONCLUSION

In this paper, we propose two easily implementable green improvements for IEEE 802.11 DMG OFDM PHY specifications and we evaluate their BER performance under LOS and OLOS/NLOS propagation scenarios.

The first improvement is the replacing of LDPC codes by BCC codes in order to lower the system complexity. Results show that with OLOS/NLOS propagation it is worth using BCC when the targeted BER is higher than 10^{-5} because BCC have similar performance than LDPC with less hardware complexity. Moreover, with code rate 1/2, BCC has better performance than LDPC when the targeted BER is higher than 10^{-6}. With LOS propagation, it is better to keep LDPC codes in order not to increase the power amplifier consumption.

The second improvement is the adding of a new MCS, 64-QAM 1/2, which corresponds to the same data rate as MCS 20, 16-QAM 3/4. Results show that with OLOS/NLOS propagation, the new MCS suffers less from multipath degradation than MCS 20 and has better performance when BCC codes are used.

Generally, we show that even if current systems are mainly design to improve the achievable maximum data rate at the expense of energy-efficiency, it is easy to modify such systems in order to improve their energy-efficiency without sacrificing their data rate. Evolved algorithms such as iterative error

coding can have similar performance than less complex techniques under certain propagation conditions.

Moreover, recent green improvements such as novel link adaptation techniques allow for a smarter selection of transmission modes given the propagation conditions. Thus, increasing the number of available transmission modes with the adding of several MCS corresponding to the same data rate or allowing for the selection of the error correction code will increase the system flexibility and performance if the system is able to identify the propagation conditions.

Further work should study the hardware implementation complexity difference between classical schemes and ours in terms of gate density and power consumption.

ACKNOWLEDGMENT

This work has been supported by MiWEBA project under international cooperation program of ICT-2013 EU-Japan supported by FP7 in EU and MIC in Japan.

REFERENCES

[1] GreenTouch Technical Committee, "Green Meter Research Study: Reducing the Net Energy Consumption in Communications Network by up to 90% by 2020", A GreenTouch White Paper, June 2013.

[2] http://www.miweba.eu/project.html

[3] IEEE Computer Society, "Wireless LAN Medium Access Control (MAC) and Physical Layer (PHY) Specifications: Enhancements for Very High Throughput in the 60 GHz Band," IEEE Std 802.11ad-2012, Dec. 2012.

[4] IEEE Computer Society, "Wireless LAN Medium Access Control (MAC) and Physical Layer (PHY) Specifications," IEEE Std 802.11-2012, Mar. 2012.

[5] M. Fossorier, M. Mihaljevic, and H. Imai, "Reduced complexity iterative decoding of low-density parity check codes based on belief propagation," IEEE Transactions on Communications, vol. 47, May 1999.

[6] I. Siaud, A.M. Ulmer-Moll, N. Malhouroux-Gaffet, and V. Guillet, "Short-Range Wireless Communications," ISBN-13: 978-0-470-69995-9 - John Wiley & Sons Ed., Chap. 18, Feb. 2009.

[7] J.-B. Doré, "Optimisation conjointe de codes LDPC et de leurs achitectures de décodage et mise en oeuvre sur FPGA," PhD Thesis, INSA of Rennes, May 2008.

[8] IEEE Computer Society, "Wireless LAN Medium Access Control (MAC) and Physical Layer (PHY) Specifications: Enhancements for Very High Throughput in Bands below 6 GHz," IEEE Std 802.11ac-2013, Dec. 2013.

[9] A. J. Viterbi, "Error bounds for convolutional codes and an asymptotically optimum decoding algorithm," IEEE Transactions on Information Theory, Apr. 1967.

[10] T. Bohn et al., "Most Promising Tracks on Green Radio Technologies," INFSO-ICT-247733 EARTH, WP4-Green Radio, Deliverable D4.1, Dec. 2010.

[11] S. Kant and T. Lindstrøm Jensen, "Fast Link Adaptation for IEEE 802.11n," Master's Thesis, Aalborg University, Aug. 2007.

[12] A.-M. Ulmer-Moll, and I. Siaud, "Procédé de sélection d'une interface de transmission parmi plusieurs pour un même dispositif de télécommunication et dispositif correspondant.," WO2011/083238A1 PCT/FR2010/052727, July 14, 2011.

[13] D. Qiao, S. Choi, A. Jain, and K. G. Shin, "MiSer: An Optimal Low-Energy Transmission Strategy for IEEE 802.11a/h," MobiCom'03, pp. 161-175, Septembre 2003.

978-1-4799-4993-9/14 $31.00 © 2014 IEEE

An Efficient Simulation Methodology for Electrical Energy Systems

Alessandro Sassone, Sara Vinco, Massimo Poncino, Enrico Macii

Dipartimento di Automatica e Informatica, Politecnico di Torino, Torino, Italy

Abstract—**Electrical energy systems (EESs) represent a wide class of systems involving consumption, generation, distribution and storage of energy. Examples of such systems can be found at various scales, ranging from smart systems-on-chip to smart grids. The conventional design methodology uses the model-based approach provided by commercial platforms such as Matlab/Simulink, and relying on built-in model libraries. This paper presents a modeling and simulation methodology for EESs based on the SystemC standard (and its Analog and Mixed-Signal extension SystemC-AMS). Simulations show that the proposed approach provides accuracy comparable to Matlab/Simulink results, with higher modularity and an average speedup of 36x.**

I. INTRODUCTION

An Electrical Energy System (EES) is a system where electrical energy is generated, distributed, stored and consumed, ranging from micro-scale smart systems-on-chip to large energy distribution smart grids. Despite of the different contexts and scales, all EESs incorporate components of similar nature, such as energy/power sources, energy storage devices, power converters/transformers, power transfer interconnects, and loads. Moreover, they also exhibit similarities in functional properties and optimization objectives.

Designing an EES is a complex task traditionally supported by model-based commercial tools such as Matlab/Simulink. This approach allows the exploration of various design alternatives and it guarantees reliability, as each tool provides a library of built-in models, that can be easily instantiated. However, commercial tools rely on closed and proprietary backbones, that are not extensible and that may change with new software releases, with no guarantee of across-version compatibility. Furthermore, such tools have a critical gap in the efficient co-simulation of physical dynamics (i.e., continuous-time) and of computational elements (i.e., discrete-time) of the system. This clearly limits the possibility of designing EESs following a systematic approach guided by user-defined optimizations.

Recently several solutions proposed in the literature addressed these two limitations applying methods borrowed from the domain of electronic systems design [1]–[6]. In all these solutions the different EES components are pre-characterized with models at a given level of abstraction and using a given semantics. Therefore the issue of the *modularity* of the models, (i.e., the possibility of replacing a model of a component with a different one) is not generally addressed [2].

In this paper we introduce a modeling and simulation methodology that relies on the open-source tool SystemC with its Analog/Mixed-Signal (AMS) extension. Similarly to [3], our solution uses standard modeling and simulation languages, while having no limitations in the target EES as in [1]. With respect to those solutions, we define the model interfaces for energy information of EES components, that can thus be seamlessly plugged into the SystemC simulation environment, regardless of their level of detail and semantics. This allows to perform multi-level system simulations in a single open platform, where components can be simulated at different levels of detail.

II. BACKGROUND

A. EESs Simulation

Several approaches for modeling and simulating EESs at various scales have been proposed in the last years [1]–[6]. They address different application contexts: general-purpose [1], [3] or focused on some type of EES, e.g., a smart grid [4], [5] or a fuel-cell hybrid electric vehicle [6].

In terms of simulation approach, these solutions can be arranged in two groups. *Pure software simulation* falls back on simulation engines implemented in a variety of languages, ranging from Matlab [5] to SystemC [3] and to ad-hoc C/C++-based simulators [1], [4]. It is the most cost-effective approach, but the accuracy depends on the types of models used. *Hardware-in-the-loop* combines one or more real devices with software simulated models through sensors and actuators [6]. The resulting accuracy is higher w.r.t. software simulation, but its application is restricted to small- and mid-scale EESs. All these solutions rely on a set of pre-characterized models of the various EES components. Therefore they have been implemented assuming a given level of abstraction and a given semantics of the used models and hence they do not allow to replace a model of a component with a different one.

B. SystemC and its AMS extension

SystemC extends C++ with libraries for the hardware description and an event-based simulation kernel [7]. It is widely used in digital design for early-stage analysis, design-space exploration, and functional verification.

SystemC-AMS is the extension of SystemC standardized for modeling and simulating analog/mixed-signal subsystems [8]. SystemC-AMS provides three abstract models of computation to cover a wide variety of domains: Electrical Linear Network (ELN), Linear Signal Flow (LSF), and Timed Data Flow (TDF). In this paper the TDF and the ELN are considered as they are the only ones adopted in the proposed approach. ELN models electrical networks through the instantiation of linear electrical primitives (e.g., resistors or capacitors) where each primitive is associated with a corresponding electrical

---

This work was supported by the EC co-funded CONTREX (Design of embedded mixed-criticality CONTRol systems under consideration of EXtra-functional properties) project Grant Agreement FP7-ICT-611146.

978-1-4799-4993-9/14 $31.00 © 2014 IEEE

equation. The SystemC-AMS solver analyzes the ELN network to derive the equations modeling system behavior, that will be solved to determine the system state at any simulation time. It is important to note that ELN is conservative, in the sense that energy conservation laws are satisfied by the equation system. TDF features the modeling of discrete time processes, where signals are considered as uniformly sampled directed signals.

III. EES Modeling and Simulation

The proposed modeling and simulation methodology defines an architectural template of an EES system where the components are classified according to the their role in the power domain (Section III-A). Each class of components is characterized by a specific interface for sharing the appropriate information, consisting of both power information and environmental parameters affecting the component behavior. Finally, all components are instantiated in a SystemC(-AMS) simulation environment (Section III-B), where each component behavior is modeled with library models or user-defined ones.

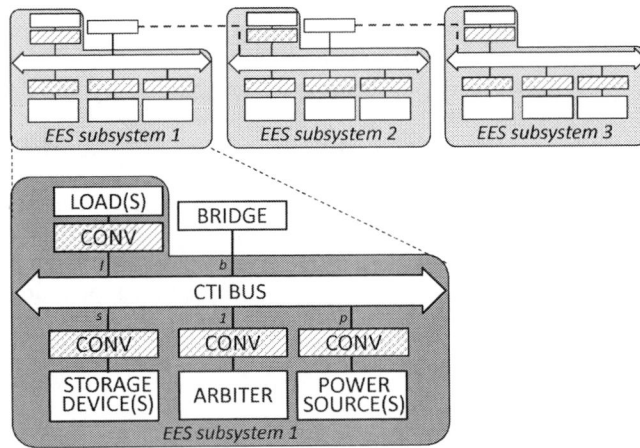

Figure 1. Template of the reference architecture, including the CTI bus and the connected components.

A. Component classification

Figure 1 depicts the EES template including its main components and their typical number in a system. The system features a certain number ($l \geq 1$) of *loads*, i.e., components that require a given amount of power to implement a certain functionality (e.g., digital cores, MEMS, analog and RF devices). As a result, load components must share information about the required load current (I) and their operating voltage (V).

Power is provided by either energy storage devices or power sources. *Energy storage devices* (ESDs) can be of different natures, such as batteries to supercapacitors and fuel cells. ESDs share their voltage (V) whereas they are provided with the current load (I) of the system. Furthermore, they must communicate their state of charge (SOC) and their nominal capacity (E), so that they can be activated through an enable signal (En) depending on the actual energy capability. *Power sources* are almost infinite sources of power, such as photovoltaic cells or thermoelectric energy generators, used for either satisfying the loads power demand or charging

ESDs. Their interface includes the supplied current (I) and voltage (V), and the activation signal (En) provided by the arbiter. An EES must contain at least one ESD or one power source ($s + p \geq 1$). Additional environmental parameters, such as temperature (T) and solar radiation (G), can be included depending on the kind of ESD or power source adopted.

An *arbiter* must be added to manage ESDs and power sources. It monitors the state of charge of ESDs, to determine whether they can provide energy or they have to be charged. Furthermore, the arbiter determines which ESD or power source to use, based on the loads request for power. Therefore its interface includes a couple of ports (SOC, E) for each ESD and an activation signal (En) for each ESD and power source.

All components are connected through a *charge transfer interconnect bus* (CTI bus). The CTI bus allows (either as an ideal conductor or with some power loss) for the energy to combine and to propagate within the system. As a result, its interface consists of an I and V ports for each EES component.

Each component is connected to the CTI bus through a *converter* module, necessary to maintain compatibility of voltage levels between EES components. The converter interface includes a couple of I and V ports for the input and output current and voltages, respectively.

An EES may involve several subsystems interconnected each other, as showed in Figure 1. Such interconnections are managed by *bridges*, which are used to connect the CTI bus to another CTI bus through a couple of I and V ports.

B. Modeling and Simulation

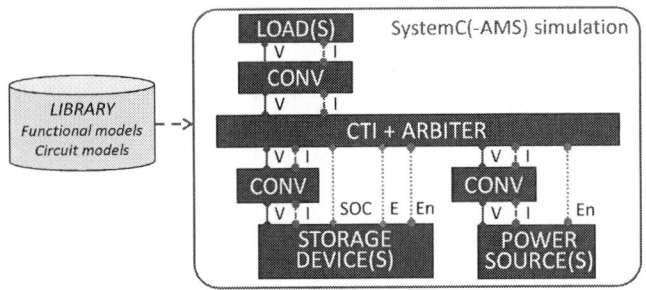

Figure 2. Construction of the SystemC(-AMS) system from a library of possible implementations of the EES components.

The definition of the EES template and of the component classification constitute a skeleton of the simulation environment, as they specify the interface of each component. This Section shows how to complete the simulation environment by instantiating the behavior of each EES component, as showed in Figure 2.

a) Interface: Each EES component is instantiated with a SystemC module declaration (`SC_MODULE`). Ports are declared as SystemC TDF ports (`sca_tdf::sca_in` or `sca_tdf::sca_out`) of type double, as adopting a TDF interface is crucial to speed up simulation. Figure 3 provides an example of ESD SystemC(-AMS) interface declaration.

b) Implementation: The SystemC modules must be populated with an implementation of the components behavior. This paper aims at presenting how to reproduce consolidated approaches within the proposed methodology, rather than providing new ones. Most of the models proposed in literature

978-1-4799-4993-9/14 $31.00 © 2014 IEEE

can be divided into two main categories: functional models and circuit models.

- *Functional models* use a function to implement component evolution (e.g., modeling an equation or even a simple waveform) [9]–[11]. Such models can thus be implemented as TDF processes that reproduce the evolution in C++. The left part of Figure 3 provides an example of battery model realized with the Peukert's law.

- *Circuit models* emulate the component behavior using an equivalent electrical circuit. A vast amount of circuit models exists in the literature, e.g., targeting batteries [12], converters [13] and power sources [14]. Such models can be implemented by describing the equivalent circuit as a network of SystemC-AMS ELN components, which reproduces the elements and the topology of circuit specification. The ELN subsystem is then wrapped by the TDF interface, with the adoption of converters, to preserve efficient synchronization with the rest of the system. The right part of Figure 3 provides an example of battery modeled with a circuit model.

The proposed approach has an high degree of modularity, thus enabling a design space exploration with a specific trade off between accuracy and simulation speed. Components of interest can be described with detailed "low-level" models for high accuracy, whereas other components can have simplified models for high speed. As showed in Figure 3, both the battery models have the same interface, compliant with the standard ESD interface described in Section III-A. Then, SystemC(-AMS) allows to easily replace the module implementation, to evaluate different configurations or alternative models at different levels of detail.

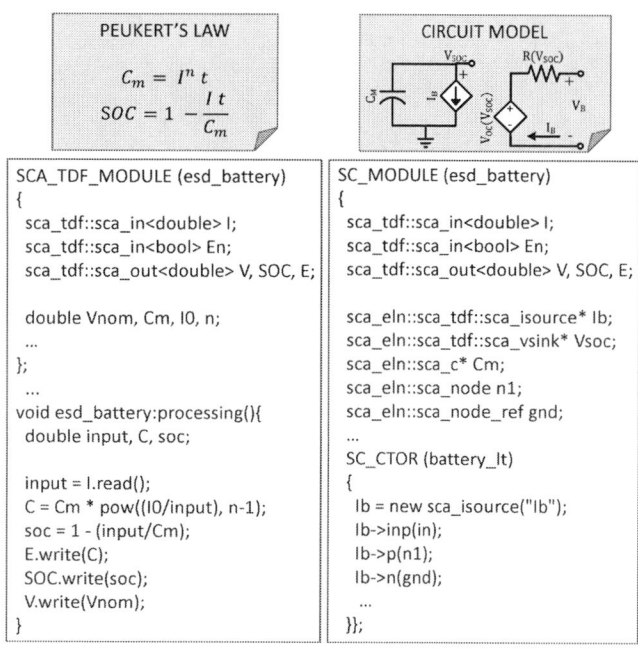

Figure 3. Application of the methodology to a battery. Interface is the one defined for ESD components while the adopted implementation is either Puekert's law [9] or a circuital model [12].

IV. SIMULATION RESULTS

In this section we demonstrate the effectiveness of the proposed simulation methodology w.r.t. state-of-the-art tools (Matlab/Simulink) in terms of accuracy and simulation performance. All simulations have been run by using Matlab/Simulink R2013a, SystemC 2.3 and SystemC-AMS 2.0. Simulation times are calculated as an average over a number of executions.

The experiment reproduces a simple charge allocation policy applied to the EES depicted in Figure 4 and containing:

- a Li-ion rechargeable battery by Qinetiq (capacity of 5.8Ah and nominal voltage of 3.69V) modeled with the simplified equivalent circuit model depicted on the right hand side of Figure 3 and presented in [12];

- a power source represented as a photovoltaic (PV) panel composed of 5 Sunpower A300 PV cells connected to a module performing maximum power transfer tracking (MPTT) [15] and AC-DC conversion. Daily solar irradiation profile is implemented as in [10];

- two load devices corresponding to the power state profiles of two commercial cores running two different applications; power information is represented as a pair of current and voltage waveform over time.

- three DC-DC converters represented with the model borrowed from [10], where the conversion efficiency is function of the input current, input voltage, and output voltage;

- a CTI bus modeled as an ideal current conductor from the energy providers (battery and PV panel) to the energy consumers (load devices) and having a constant reference voltage of 3.0V.

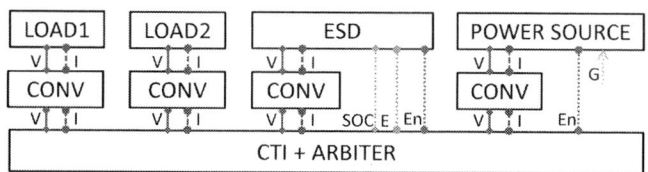

Figure 4. EES used for the charge allocation policy experiment.

The charge allocation policy is implemented in the CTI arbiter. As long as the the power drawn from the PV panel satisfies the power demand of the loads, loads are supplied by the power source. Otherwise, the loads are supplied by the battery, until its SOC reaches the value of 10%. Finally, when the PV panel is able to provide power and the power demanded by the loads is 0, the battery is charged by the power source. Although this policy is quite simplistic, the purpose here is not to develop sophisticated policies but rather to show the efficiency and the flexibility of the system.

The system has first been modeled in Matlab/Simulink in a top down approach and using the Simscape library. Then, it was described in SystemC(-AMS), by following the methodology proposed in this work. Table I compares simulation accuracy and performance of four code versions: SystemC-AMS and Matlab/Simulink both with time steps 0.1s and 1s. The table shows, for each version, the number of samples (column *Samples (#)*), the estimated system runtime (column *System RT (s)*) and simulation time (column *Sim. time (s)*). Furthermore,

978-1-4799-4993-9/14 $31.00 © 2014 IEEE

Table I. ACCURACY AND EFFECTIVENESS OF THE PROPOSED APPROACH W.R.T. MATLAB/SIMULINK.

	Time step (s)	Samples (#)	System RT (s)	Error (avg. %)	Sim. time (s)
Matlab/	1.0	31,158	31,147.1	-	27.75
Simulink	0.1	311,451	31,144.7	-	288.21
SystemC	1.0	31,146	31,145.1	0.35	0.78
(-AMS)	0.1	311,420	31,141.8	0.34	7.69

the table shows the average error (column *Error (avg %)*) in the estimation of the battery voltage for the SystemC-AMS versions using the Matlab/Simulink version with time step 0.1s as a reference.

The SystemC versions exhibits *very high accuracy*, and it is also almost insensitive to the size of time step: the average error is 0.35% with time step 1s (maximum error 0.58%) and 0.34% with time step 0.1s (maximum error 0.73%).

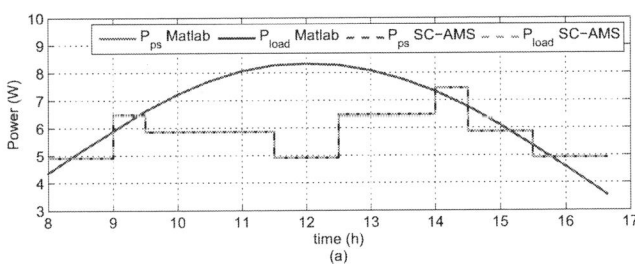

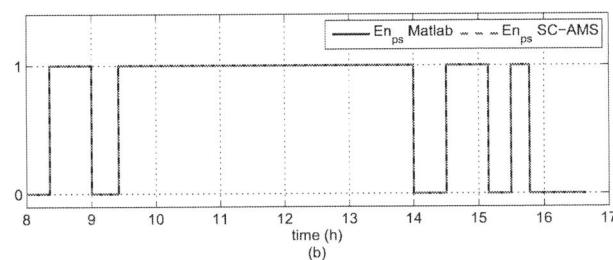

Figure 5. Comparison of Matlab/Simulink and SystemC(-AMS) simulations with time step 1s: power drawn from the power source and power demand of the loads (a); activation signal of the power source (b).

The high level of accuracy is highlighted by Figure 5, that shows the power drawn from the PV panel P_{ps} and the power demand of the loads P_{load} (a), and the activation signal En_{ps} of the PV panel during the simulation (b). The simulated operating time is concentrated on the peak power period of the PV panel (8am-5pm) and ends when the system is not able to satisfy the power demand. The curves of the Matlab/Simulink and SystemC-AMS implementations are almost completely overlapped, denoting the high accuracy of the SystemC(-AMS) implementation. This is a an evident result of the correct processing of the simple charging allocation policy implemented. Moreover, this experiment demonstrates the effectiveness of the simulation methodology in considering environmental parameters such as the solar irradiation.

Concerning *simulation speed*, SystemC-AMS proved to be much faster than Matlab/Simulink. The speedup w.r.t. Matlab/Simulink version is about 35x with time step 1s and 37x with time step 0.1s. The reason of such a slow execution is

due to the Matlab/Simulink internal solver, whose computation is far heavier than the efficient TDF and ELN SystemC implementations.

These considerations prove that SystemC(-AMS) can be a very efficient alternative to Matlab/Simulink. Indeed, besides being open source and extensible, it also is accurate and fast w.r.t. the corresponding Matlab/Simulink implementations.

V. CONCLUSIONS

This paper proposed a methodology for simulating EESs based on an open source tool typical of the digital domain, i.e., SystemC(-AMS). Experimental results show that the proposed approach provides a faster simulation w.r.t. Matlab/Simulink (avg. 36x), still preserving a high level of accuracy (avg. error is always lower than 0.35%). We envision that this solution will allow effective simulation of more complex systems, thus enhancing design of EESs. Furthermore, the adoption of a C++ based simulation environment will enable the simultaneous simulation of the power and functional domains, thus enhancing the accuracy of the proposed models.

REFERENCES

[1] S. Yue, D. Zhu, Y. Wang, M. Pedram, Y. Kim, and N. Chang, "SIMES: A Simulator for Hybrid Electrical Energy Storage Systems," in *Proc. of ACM/IEEE ISLPED*, 2013, pp. 33–38.

[2] Y. Kim, D. Shin, M. Petricca, S. Park, M. Poncino, and N. Chang, "Computer-Aided Design of Electrical Energy Systems," in *Proc. of ACM/IEEE ICCAD*, 2013, pp. 194–201.

[3] J. Molina, X. Pan, C. Grimm, and M. Damm, "A Framework for Model-Based Design of Embedded Systems for Energy Management," in *Proc. of IEEE MSCPES*, 2013, pp. 1–6.

[4] S. Nassif, G.-J. Nam, J. Hayes, and S. Fakhouri, "Applying VLSI EDA to Energy Distribution System Design," in *Proc. of IEEE ASPDAC*, Jan. 2014, pp. 91–96.

[5] M. A. Al Faruque and F. Ahourai, "A Model-Based Design of Cyber-Physical Energy Systems," in *Proc. of IEEE ASPDAC*, Jan. 2014, pp. 97–105.

[6] L. Gauchia and J. Sanz, "A Per-Unit Hardware-in-the-Loop Simulation of a Fuel Cell/Battery Hybrid Energy System," *IEEE TIE*, vol. 57, no. 4, pp. 1186–1194, 2010.

[7] Open SystemC Initiative, *SystemC*, http://www.systemc.org.

[8] Accellera Systems Initiative, *SystemC-AMS and Design of Embedded Mixed-Signal Systems*, http://www.systemc-ams.org/.

[9] W. Peukert, "Über die Abhängigkeit der Kapazität von der Entladestromstärke bei Bleiakkumulatoren," in *Elektrotechnische Zeitschrift*, 1897, p. 20.

[10] S. Park, Y. Wang, Y. Kim, N. Chang, and M. Pedram, "Battery Management for Grid-connected PV Systems with a Battery," in *Proc. of ACM/IEEE ISLPED*, 2012, pp. 115–120.

[11] M. Knauff, C. Dafis, D. Niebur, H. Kwatny, and C. Nwankpa, "Simulink Model for Hybrid Power System Test-bed," in *Proc. of IEEE ESTS*, 2007, pp. 421–427.

[12] M. Petricca, D. Shin, A. Bocca, A. Macii, E. Macii, and M. Poncino, "An automated framework for generating variable-accuracy battery models from datasheet information," in *Proc. of ACM/IEEE ISLPED*, 2013, pp. 365–370.

[13] Y. Choi, N. Chang, and T. Kim, "DC-DC Converter-Aware Power Management for Low-Power Embedded Systems," *IEEE TCAD*, vol. 26, no. 8, pp. 1367–1381, Aug. 2007.

[14] A. Bauer, J. Hanisch, and E. Ahlswede, "An Effective Single Solar Cell Equivalent Circuit Model for Two or More Solar Cells Connected in Series," *IEEE PHOT*, vol. 4, no. 1, pp. 340–347, Jan. 2014.

[15] Y. Kim, N. Chang, Y. Wang, and M. Pedram, "Maximum power transfer tracking for a photovoltaic-supercapacitor energy system," in *Proc. of ACM/IEEE ISLPED*, 2010, pp. 307–312.

Macromodel-based Signal and Power Integrity simulations of an LP-DDR2 interface in mSiP

Gianni Signorini[*†], Stefano Grivet-Talocia[‡], Igor Simone Stievano[‡] and Luca Fanucci[†]

[*]Intel Mobile Communications GmbH
Am Campeon 10-12, Neubiberg, Germany 85579
email: gianni.signorini@intel.com
[†]Department of Information Engineering, University of Pisa
Via G.Caruso 16, Pisa, Italy 56122
[‡]Department of Electronics and Telecommunications, Polytechnic University of Turin
Corso Duca degli Abruzzi 24, Torino, Italy 10129

Abstract—Signal and Power Integrity (SI/PI) analyses assume a paramount importance to ensure a secure integration of high-speed communication interfaces in low-cost highly-integrated System-in-Package(s) (SiP) for mobile applications. In an iterative fashion, design and time-domain SI/PI verifications are alternated to assess and optimize system functionality. The resulting complexity of the analysis limits simulation coverage and requires extremely long runtimes (hours, days). In order to ensure post-silicon correlation, electrical macromodels of Package/PCB parasitics and high-speed I/Os can be generated and included in the testbenches to expedite simulations. Using as example an LP-DDR2 memory interface to support the operations of a mobile digital base-band processor, we have developed and applied a macromodelling flow to demonstrate simulation run-time speed-up factors (x1200+), and enable interface-level analyses to study the effects of Package/PCB parasitics on signals and PDNs, as well as the corresponding degradation in the timing budget.

I. INTRODUCTION

The design of state-of-the-art mobile platforms is becoming even more and more challenging: to accomodate the demand for advance computing abilities and ubiquitus connectivity, an ever increasing number of data-processing architectures, multi-mode multi-band wireless communication interfaces and leading-edge technology peripherals are integrated in shrunk form factor electronic PCBs (Printed Circuit Boards). Furthermore, the trends and the competition in mobile-market dictate a low-cost nature of the devices, the minimization of the power consumption, an optimal battery efficiency and short design periods [1], [2], [3]. Relevant progresses on multi-layer PCB cross-sections and packaging technologies have made possible the integration of multiple silicon dies inside a single package, satisfying the request for small feature-size and high-performance devices.

In the context of modern digital base-band processors for mobile applications, the most common package structures adopted in commercial devices are reported in figure 1, namely single-die fcBGA, multiple-dies SiP (System-in-Package), PoP (Package-on-Package) and PiP (Package-in-Package). For these devices, packages do not only provide a mechanical support and a first-level of interconnection towards the outer world, but realize also internal chip-to-chip

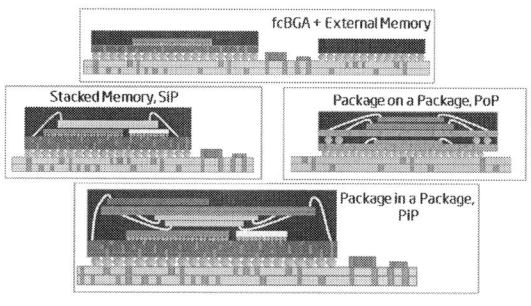

Fig. 1. Common packaging structures of Processor-Memory links: fcBGA + external memory, stacked-dies SiP, PoP, PiP

communication links [4]. The routing of signals and power distribution networks (PDNs) is highly dense, and the risk of performance degradation is increased due to potential mutual interferences between different portions of the system. To guarantee system reliability and assess perfomance prior to tapeout, a tremendous effort is required to perform reliable time-domain simulations, including input-output circuit (I/O) descriptions, as well as the effects of package/PCB parasitics on PDNs and interconnections [4]; unfortunately, such system-level simulations are extremely critical to be performed, and often even prevented due to the resulting complexity.

In this paper, we will present a fast and reliable full macromodel-based flow, aimed at supporting the design challenges for a secure integration of an LP-DDR2 (Low-Power Double Data-Rate) interface in a mSiP (SiP for mobile applications); in particular, we will consider a structure composed of a digital base-band processor and memory device, placed in a stacked configuration. Focusing on Signal and Power Integrity (SI/PI) aspects, section II discusses the major implications of low-cost, small-area and high-integration constraints on the design procedures of packages and PCBs. Section III presents a macromodel-based approach to expedite and increase the reliability of complex SI/PI time-domain verifications; the benefits of such an approach will be presented in section IV.

II. LP-DDR2 INTERFACE IN A MSIP

The LP-DDR2 memory interface is characterized by a 16/32-bit parallel bus supporting high-speed datarates

978-1-4799-4993-9/14 $31.00 © 2014 IEEE

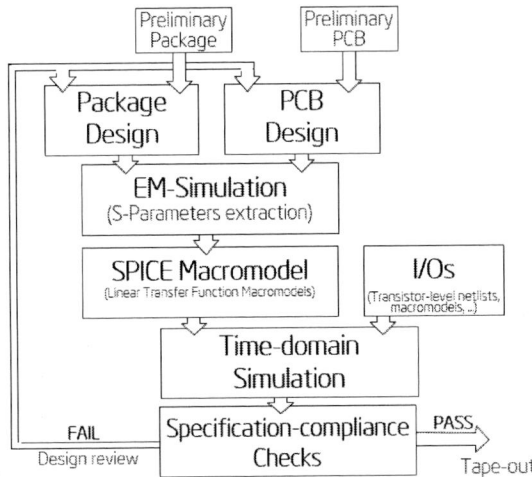

Fig. 2. Flow-chart diagram that describes the iterative "design/optimization" process for package and PCB layouts

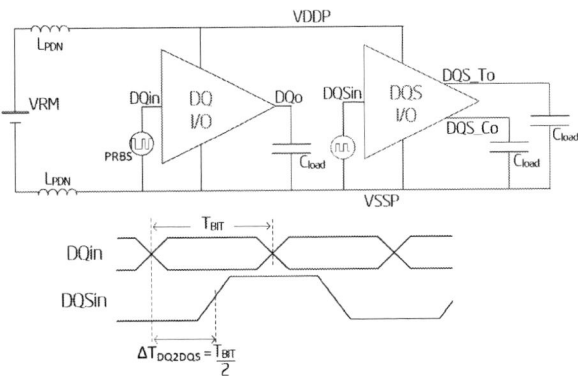

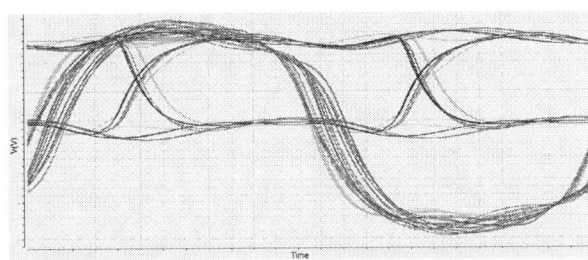

Fig. 4. Structure of the test-bench to determine the impact of inductive parasitic components of PDN on the DQ and DQS output eye-diagrams.

Fig. 5. Eye-diagrams of DQ and DQS lines with an L_{PDN} of 500pH (green), 1.5nH (blue) and 2nH (red)

(200...1066 Mb/s) with a considerably large voltage swing (1.2V for unterminated applications) [5]. For the case of mSiP in a stacked-dies configuration, the processor-memory interconnections are completely routed inside the package, leading to a relevant area occupation, intevitably implying critical crossings of memory signals with other communication interfaces and/or sensitive portions of the system. Furthermore, the package distributes the supply to both the memory interface on processor side (core-logic, timing interfaces and I/Os) and the memory device itself. Low-cost constraints reduce to the minimum the number of available metal layers in the package stackup; small-feature size also limit the number solderballs and solderbumps reserved for power/ground connections. Furthermore, the number of available bypass capacitors on the PCB and their placement depend on product requirements, and often cannot be optimal. Because of all these constraints, common good-design practices (target-impedance for PDN, avoidance of return-path discontinuities (RPDs), etc.) cannot be completely implemented and require careful trade-offs with costs and area implications. As depicted in figure 2, following an iterative approach and starting from preliminary layouts, the impact of package and PCB parasitics on system performances are accurately assessed through time-domain simulations, aimed at studying the compliance with target operating specifications. Based on simulation results, designs are reworked and optimized, or delivered for the final physical implementation [4], [6].

SSN-induced jitter in LP-DDR2 interface

One of the most critical effects to be optimized in mSiP is the so-called Simultaneous Switching Noise (SSN), graphically illustrated in figure 3.
The short rise and fall times required by the communication interface, together with the large voltage swing (1.2V), imply large pulses of current to flow through the supply rails of each switching I/O (3-a). This dynamic current, in combination with parasitic components of PDN and a weak current return-path (3-b), produces relevant voltage fluctuations around the nominal supply value (3-c), inevitably affecting all the other I/Os that share the same supply-domain. SSN effects have to be

carefully analyzed: the injected critical voltage fluctuations on signals and supplies introduce a data-dependent jitter on memory lines. As an example, considering the 'write'-operation of the LP-DDR2 interface (i.e., data are sent from the processor to the memory), in the testbench of figure 4 we assume a pure-inductive parasitic component (L_{PDN}) on the supplies and we investigate its dependence on the jitter at the output of a DQ and a DQS I/Os (respectively, $t_{j,DQ}$ and $t_{j,DQS}$); corresponding simulation results are depicted in figure 5. The current required to perform the logic-state transitions at the output of the DQ pad (i_{DD}) couples with the inductor, and triggers a supply variation $\Delta V_{DD}(t)$ that follows $\Delta V_{DD}(t) = L_{PDN} \cdot \frac{\partial}{\partial t} i_{DD}(t)$. The input transitions on the DQS pad occur after a time ΔT_{DQ2DQS}, ideally set to be half bit period, and the residual bouncing on its supply implies a jitter $t_{j,DQS}$ on the switching event. $t_{j,DQS}$ rapidly increases with L_{PDN}, reducing the residual timing margins of a factor Δt_{setup} and Δt_{hold}; a graphical representation of setup/hold time degradation due to SSN effects is reported in figure 6.

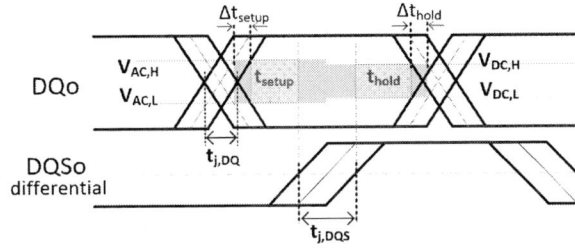

Fig. 6. Reduction of setup and hold time (Δt_{setup}, Δt_{hold}) due to jitter on DQ and DQS line ($t_{j,DQ}$, $t_{j,DQS}$).

978-1-4799-4993-9/14 $31.00 © 2014 IEEE

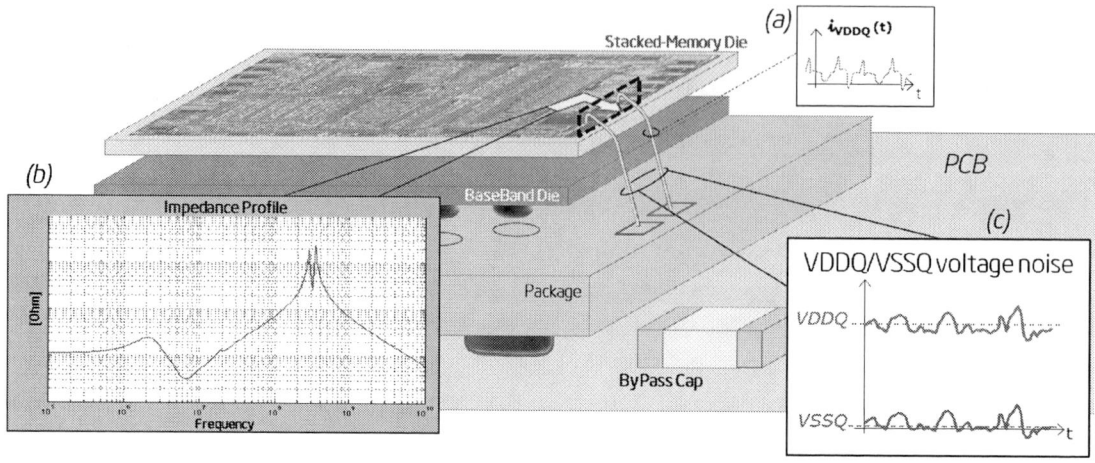

Fig. 3. Power Distribution Network components, supply-current peaks, frequency-dependent impedance profile and resulting supply-voltage noise (SSN effects)

III. MACROMODELS FOR SIGNAL AND POWER INTEGRITY SIMULATIONS

In order to assess the quality of the memory interface after its integration in a mSiP, complex time-domain simulations shall be performed including the netlists for all the transmitting and receiveing ends, and the complete set of parasitics that affect the signal routing and the power distribution networks. In this paper, we propose a full macromodel-based simulation flow, which is able to cast each system component as a compact SPICE-compatible behavioural netlist.

A. LTFM Macromodels of Package and PCB

Real PDN structures are much more complicated than the simple inductive model of Figure 4. In fact, only full-wave electromagnetic characterizations are able to represent all parasitics with sufficient accuracy. In this paper, a commercial hybrid 2.5D full-wave solver is adopted for the extraction of the scattering parameter matrix (S-parameters) of signal interconnections and PDN of the complete memory interface. Due to the width of the parallel bus (16 or 32 bit) and the number of involved power/ground terminals (relevant because of the number of I/Os), the ports included in the EM-extraction can easily exceed 100+. For this reason, we adopt a rational curve fitting methodology with passivity enforcement [8] to cast the linear structure as a lumped Linear Transfer Function Model (LTFM), which is converted to a state-space system and realized as a SPICE-compatible behavioral netlist. This procedure is standard and not further commented here [7].

B. The need for I/O Macromodels

To improve the correlation of SI/PI simulation results with post-silicon measurements, post-layout extracted views containing on-chip parasitics of devices and interconnections should be used for each I/O. Unfortunately, this results in a tremendous explosion of netlist complexity, and prevents the execution of interface-level transient simulations. However, black-box macromodelling methodologies can be applied to the same netlists, and equivalent representations can be extracted and used in SI/PI testbenches. In this context, in order to guarantee confidence in the results and enable complex analyses, macromodels shall be:

- *compact*, to expedite analysis and extend the simulation coverage;

- *accurate*, offering a superior accuracy of currents and voltages compared to corresponding netlists, both at output and at the supply terminals.

C. Mpilog Macromodels of Drivers and Receivers

In this paper, Mpilog macromodels [9] have been developed and used to represent the behaviour of LP-DDR2 I/Os. The generation of the macromodels is based on a DC sweep and a transient simulation, stimulating the device-under-modelling (DUM) with suitable voltage stimuli at input, output and supply terminals. Post-processing the resulting current and voltage waveforms at the same pins, Mpilog tunes a non-linear parametric mathematical model, in order to reproduce both the static and dynamic $i - v$ characteristics of the DUM for low and high logic-state. Time-domain weighting functions multiply these non-linear functions to implement the dynamics of logic-state transitions, both at output and supply-terminals. The structure of Mpilog macromodels and the generation procedures are well described in [9], [10]. For a single-ended driver structure [10], model equations are:

$$
\begin{aligned}
i_O =\ & w_H k_H(v_{\mathrm{dd}}) f_{sH}(v_O) & i_{\mathrm{dd}} =\ & w_H k_H(v_{\mathrm{dd}}) f_{sH}(v_O) \\
& + w_H f_H(v_O, v_{\mathrm{dd}}, \partial/\partial t) & & + w_H f_H(v_O, v_{\mathrm{dd}}, \partial/\partial t) \\
& + w_L k_L(v_{\mathrm{dd}}) f_{sL}(v_O) & & + w_{dH} f_{dH}(v_{\mathrm{dd}}, v_O, \partial/\partial t) \\
& + w_L f_L(v_O, \partial/\partial t)) & & + w_{dL} f_{dL}(v_{\mathrm{dd}}, v_O, \partial/\partial t) \\
& + w_L f_{cL}(v_{\mathrm{dd}}, \partial/\partial t) & & + \delta_i(t).
\end{aligned}
$$

f_{sH} and f_{sL} represent the i-v static output characteristics when the driver is kept, respectively, in a fixed high and low logic-state, while f_H, f_L, f_{dH} and f_{dL} are discrete-time Local-Linear State-Space (LLSS) models that account for the non-linear dynamics of the buffer; w_H, w_L, w_{dH} and w_{dL} are time-varying weighting function to reproduce logic-state evolutions; k_H and k_L accounts for the effect of supply fluctuations on the static characteristics; δ_i reproduces the current drawn from the pre-driver stages.
These equations are then synthesized as electrical representation, enabling the simulation in any SPICE circuit solver.

978-1-4799-4993-9/14 $31.00 © 2014 IEEE

IV. RESULTS

Table 1 reports the time required to perform a transient simulation of a 40-bit PRBS pattern using several different electrical representations of a DQ transmitting pad (schematic, post-layout RC-extracted netlist and an Mpilog model). Neglecting layout-induced parasitics, the I/O schematic netlist can only deliver approximate results and potentially lead to inaccurate predictions about system performances. Post-layout RC-extracted netlist ensures post-silicon correlation, but its complexity implies a significant increase in simulation runtime (x6); however, an Mpilog macromodel can be generated from this netlist and be used to represent the I/O with the same accuracy while offering a tremendous simulation speed-up (x1293). This is proven in figure 7, depicting the output voltage and the supply currents of the model and the corresponding RC-extracted netlist for a 1-0-1-0 logic-state transition.

TABLE I. TRANSIENT SIMULATION RUNTIME FOR A 40-BIT PRBS INPUT PATTERN USING DIFFERENT I/O ELECTRICAL REPRESENTATIONS

	Schematic	RC-full	Mpilog
Runtime [sec]	122.4	732.1	0.566

A macromodel-based SI/PI testbench has been developed to analyze the 'write'-mode operation of the complete 32bit LP-DDR2 interface. In figure 8, the eye-diagrams of the 32 DQ lines and the corresponding 4 DQS strobes are superimposed. Such an analysis would not have been possible using the detailed RC-extracted netlists (due to convergence issue), while it takes only 1h 42m 38s using the macromodel-based approach for the simulation of a 200-bit PRBS pattern.

V. CONCLUSIONS

Low-cost, small feature-size and area constraints in the design of SiP for mobile applications, require accurate SI/PI time-domain simulations to support the layout of packages and PCBs, ensuring performance compliance with target operating specifications. A macromodel-based approach for SI/PI analysis enables interface-level simulations to study the impact of

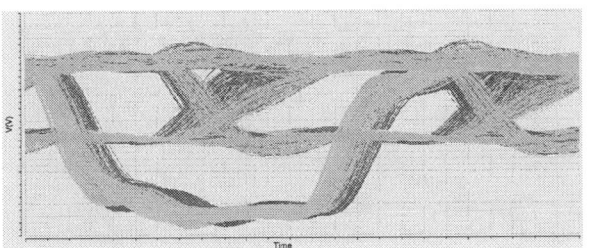

Fig. 8. Eye-diagram of DQ<31:0> lines and the corresponding DQS<3:0> for a *Write*-operation

package/PCB parasitics on system functionality, hardly achievable using transistor-level descriptions due to an excessive complexity of the resulting netlist. Mpilog has been used to generate I/O behavioural models, offering outstanding runtime speedup (x1200+), superior accuracy and easy integration in SPICE simulation environments. Their use in combination with LTFM equivalent netlists of Package/PCB parasitics ensures reliable performance predictions, expedite simulations and improves the effectiveness of optimization processes for the design of the overal system.

ACKNOWLEDGMENT

The authors would like to thank Intel Mobile Communications GmbH for the support offered to this work: special thanks go to Vincenzo Costa, Venkatesh Kasturirangan, Dr.Alexander Olbrich, Dr. Christoph Heer, Pietro Brenner, Kay Schiller and Alexander Ruehl for the valuable discussions.

REFERENCES

[1] S. K. Pienimaa; N. I. Martin, *"High-Density Packaging for Mobile Terminals,"* IEEE Transactions on Advanced Packaging, Vol. 27, No. 3, Aug. 2004, pp. 467-475.

[2] Leung, L.L.-W.; Sham, M.L.; Ma, W.; Chen, Y.C.; Lin, J.R.; Chung, T., *"System-in-Package (SiP) Design: Issues, Approaches and Solutions,"* Electronic Materials and Packaging, 2006. EMAP 2006. International Conference on , vol., no., pp.1,5, 11-14 Dec. 2006

[3] Romero, C.; Seungwook Park; Youngdo Kweon; Mijin Park, *"Advanced high density interconnection substrate for mobile platform application,"* Microsystems, Packaging, Assembly and Circuits Technology Conference (IMPACT), 2011 6th International , vol., no., pp.214,217, 19-21 Oct. 2011

[4] Pulici, P.; Vanalli, G.P.; Dellutri, M.A.; Guarnaccia, D.; Lo Iacono, F.; Campardo, G.; Ripamonti, Giancarlo, *"Signal Integrity Flow for System-in-Package and Package-on-Package Devices,"* Proceedings of the IEEE, vol.97, no.1, pp.84,95, Jan. 2009

[5] JEDEC, *"LOW POWER DOUBLE DATA RATE 2 (LPDDR2) - Specification Document,"* JESD209-2F, Jun 2013

[6] Swaminathan, M.; Daehyun Chung; Grivet-Talocia, S.; Bharath, K.; Laddha, V.; Jianyong Xie, *"Designing and Modeling for Power Integrity,"* Electromagnetic Compatibility, IEEE Transactions on, vol.52, no.2, pp.288,310, May 2010

[7] IdEM 9. [Online]. Available: www.idemworks.com

[8] Chinea, A.; Grivet-Talocia, S.; Olivadese, S.B.; Gobbato, L., *"High-Performance Passive Macromodeling Algorithms for Parallel Computing Platforms,"* Components, Packaging and Manufacturing Technology, IEEE Transactions on, vol.3, no.7, pp.1188,1203, July 2013

[9] I.S. Stievano, I.A. Maio, F.G. Canavero, *"M[pi]log, Macromodeling via Parametric Identification of Logic Gates,"*, IEEE Transactions on Advanced Packaging, pp. 15-23, vol. 27, n. 2, February, 2004

[10] MOCHA - *MOdelling and CHAracterization for SiP Signal and Power Integrity Analysis* - WP2 Report, Deliverable 2.1. [Online]. Available: www.mocha.polito.it

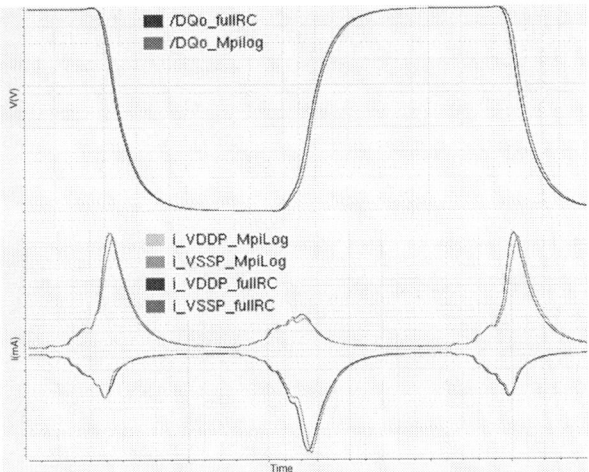

Fig. 7. Comparison of output voltage and supply-currents for post-layout RC-extracted netlists and the corresponding Mpilog SPICE macromodel

A SystemC Bluetooth Network Simulator

Cristiano Scavongelli
Department of Information Engineering
Università Politecnica delle Marche
Ancona, Italy
c.scavongelli@univpm.it

Massimo Conti
Department of Information Engineering
Università Politecnica delle Marche
Ancona, Italy
m.conti@univpm.it

Abstract—As the wireless networks utilization grows up and the technology becomes more powerful, the complexity of the possible applications increases. With the current hardware potentiality, it's unavoidable to make use of simulation tools to develop, improve and optimize the designed networks. Many network simulators are available, but they mainly concentrate on the protocol aspects of the design, i.e. they work on a very high level of abstraction. Rather, in many applications it may be useful, important or even critical to observe and manage at the same time high-level and low-level parameters like the power in the channel, the packet loss, or more specific network configuration details. In this paper we present a Bluetooth network simulator that let the user to work both on application level and on link management level. The Bluetooth simulator has been implemented using SystemC and it has been developed with the aim of modeling a real microcontroller-based device.

Keywords—SystemC, Bluetooth, Wireless networks

I. INTRODUCTION

In recent years, wireless networks have become an ubiquitous, affordable and well-accepted reality. This is particularly true for networks such as Bluetooth and Wi-Fi, which today are almost everywhere and are used for the most different applications. The main reasons behind this widespread availability are perhaps the low costs and the high reliability of the technology, and these same reasons allow the designer to use them in many different scenarios.

In many cases, when using an off-the-shelf chip, the designer can limit himself to work on the highest levels of the protocol stack and to simply implement the application layer. In other cases, the designer must go down through the stack and work on a lower level where he can see and control some network parameters that he cannot access from the application layer. Let's think for example to a real-time, multi-to-one audio transmission on a moderate-to-low data rate network. Due to the real-time requirement, the transmission and the reconstruction of the audio signal at the receiver end must not suffer from the delays introduced by the higher protocol layers, and hence must be managed really close to the transmission/reception hardware. The same holds if the data rate becomes low or if there are a lot of nodes in the network: in these cases the overhead introduced by the protocol stack must be reduced or even avoided at all.

Moreover, in a real-time application there might not be corrupted packets retransmissions, and the designer might want to keep the corrupted packets anyway, rather than flush them away. For instance, in the real-time audio example, the corrupted packet might be used "as it is", instead of introducing an empty frame. The corrupted packets are another example of data the application layer doesn't see.

The simulations can help in the earlier phases of the design to choose high-level parameters such as the network topology, the number of slaves, the payload size, and so on. A simulation can help to optimize lower-level details such as the transmission delays and times, the noise protection, the bandwidth utilization, and so on. For example, in [1] and [2] the authors use high-level simulations to estimate the power consumption of Bluetooth nodes. The careful choice of these aspects before the real implementation can help to reduce the costs and the design time-to-market. Unfortunately, the network simulators nowadays available allow the designer to work only on the highest levels of the protocol stack, and only with models very far from the final implementation. For example, the system described in [3] allows the user to simulate a complete wireless or wired, TCP/IP-based network, but only from above the TCP/IP layer. This simulator has been adapted for Bluetooth networks by the authors of [4]. The same holds true for the simulator presented in [5], which includes a very abstract modeling paradigm for the protocol layer of Wi-Fi and WiMax. Even in this second case, the designer must re-implement his own layers if he wants to work on a more detailed representation of the network. The same philosophy lies beneath the simulator described in [6], which introduces an abstract representation of the upper layers of Bluetooth stack and a simplified version of the lower ones, and the designer works above the TCP/IP layer. An interesting comparative study of general-purpose simulators of this kind and with similar problems is presented in [7].

Talking about the Bluetooth, there are another two interesting simulators. The system presented in [8] implements the layered structure of the Bluetooth stack from the application layer to the baseband, and allow the designer to use it through a graphical interface. The system presented in [9], instead, uses the protocol behavior model given by the IEEE to implement the simulator. The authors use it to evaluate the performances of Web traffic over Bluetooth.

All these simulators use high-level descriptions of the system. In this work, we present a Bluetooth simulator based on SystemC [10]. SystemC let us to mix in the same design cycle-accurate descriptions with higher-level descriptions. Therefore it let us to create a simulator with hardware-like blocks that model the actual hardware used to transmit/receive the data, and application-level blocks where the designer can

978-1-4799-4993-9/14 $31.00 © 2014 IEEE

insert its own link-management (LM) or application routines. Such architecture can be used either to simulate application code, or to optimize the link policy with the deep control of every Bluetooth parameter. Moreover, the simulator can be used also as a starting point for the design of customized Bluetooth systems, i.e. for the design of Bluetooth chips with dedicated features. The specifications followed by the model are those of the 2.0 version of the standard [11]. The simulator for now doesn't implement the SCO links, the adaptive frequency hopping, and the low-power modes (sniff, park and hold), which are currently under development.

The paper is organized as follows. In Section III we'll describe the architecture of our simulator, and, as a simple case of study, in Section III we'll present the analysis of the network creation phase, obtained with our simulator. Finally, in Section IV we'll draw down some conclusion and an outline of future works.

II. THE SIMULATOR STRUCTURE

The structure we decided to use for our simulator recalls the structure of a real Bluetooth system, and it's shown in Fig. 1. An USB-based Bluetooth key contains a microcontroller that manages the real transmission on the wireless channel, and an USB bus. On the other side, there is a computer, which can be seen as a very powerful microcontroller, and another USB bus. Abstracting this model, our simulator defines a BT controller, a simple bus, and an application-layer controller. The bus in the middle is a simple 8-lines parallel bus, but it can be modeled as a different bus without any problem. In a common MCU, the user code manages the integrated peripherals writing some memory register, and/or managing the interrupts issued by the hardware. In our simulator, the two microcontrollers can be "programmed" writing the code inside the main loop or the code inside the interrupt service routines, and can communicate with the hardware peripherals (the intermediate bus controller and the baseband hardware) writing some "registers" defined in a shared memory. Moreover, in many MCUs that contain packet-based peripherals (such as the USB), the main memory is used also to pass to the peripheral the data to be transmitted. We used the same approach in our simulator. First of all, hence, we'll describe in detail the memory management scheme and all a designer needs to work at application or LM layer. Then we'll describe the implementation of the baseband (BB) layer.

A. Working on application or LM layer

As we said, the two MCU models define three objects: a main loop, an interrupt service routine (ISR), and a memory map. For simplicity, the memory map is the same in the two MCUs, even though the application layer doesn't have a BB controller. In this architecture, the memory is simply an array which pointer is shared among all the modules that need to access it, and a "registry" is a reserved location in it with a special meaning. An ISR, instead, is a SystemC process triggered by the interrupt line, an input port for the MCU module.

Let's consider a particularly important example. Fig. 2 shows the control register used to launch interrupts from the

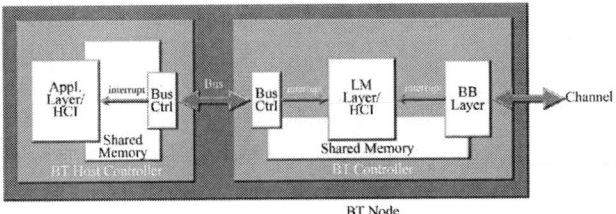

Fig. 1 - Simulator architecture.

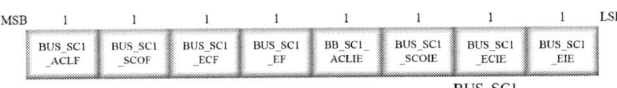

Fig. 2 - A control register.

bus. It contains four control bits (BUS_SC1_EIE, BUS_SC1_ECIE, BUS_SC1_SCOIE, BUS_SC1_ACLIE) that the designer can use to enable a particular interrupt source. "EIE", for example, stands for "Event Interrupt Enable": setting this bit, the ISR will be activated when the bus completed the reception of an Host-Controller Interface (HCI) event packet, i.e. of a response to an HCI command. Similarly, "ECIE" stands for "End Command Interrupt Enable", and it is set to enable the bus to issue an interrupt when it has completed the reception of a new HCI command. (Obviously, the event is received only by the application MCU, and the command only by the LM MCU.) The register contains also four status flags that are set by the bus controller upon the reception of a new HCI packet and that can be checked by the code inside the ISR in order to identify the event that issued the interrupt. The reading or the writing of a single bit is extremely simple: for each register there are a set of defines that allow the direct manipulation of the various registers fields. In our example, to enable the four bus interrupt sources, we have simply to write:

```
BUS_SC1_EIE      = 1 ;
BUS_SC1_ECIE     = 1 ;
BUS_SC1_SCOIE    = 1 ;
BUS_SC1_ACLIE    = 1 ;
```

in any part of the application program, as we'd do in a real MCU. A first advantage of this approach is that it is extremely simple, for the programmer point of view, to control the hardware connected to the MCU. Moreover, this way is rather simple to add to the MCU other hardware: just add the new register definitions to the memory map and in the hardware. Third, if the goal of the project is the development of a new protocol or application, this approach let a designer to use the simulator without knowing at all the SystemC.

More complex is the approach used to implement the HCI packet exchange through the intermediate bus, and it has been conceived in order to maximize the transmission performances. The idea, basically, is to implement in the shared memory a packets queue and a couple of control registers that the various hardware can use to process them.

978-1-4799-4993-9/14 $31.00 © 2014 IEEE

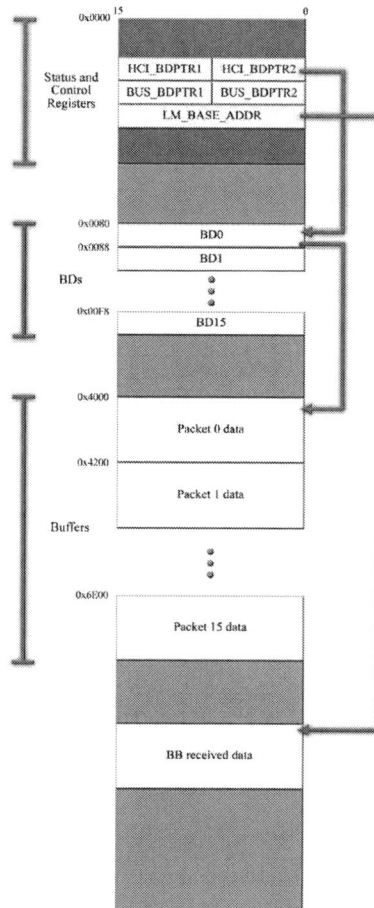

Fig. 3 - The HCI packets management system.

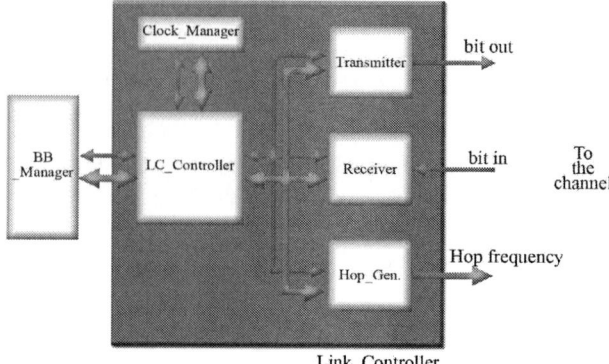

Fig. 4 - Block diagram of the baseband layer.

Fig. 3 shows the main idea behind this memory management. Among the registers, there are some that are used as pointers to the effective memory location where the HCI packets are located. In one case (for the LM_BASE_ADDR register), this addressing is direct, i.e. the content of the 16-bits register is the effective base address of the packet. In the other cases, the addressing is indirect, and the register points to a memory location where is located a buffer descriptor (BD) for the searched packet. This BD, in turn, contains the base address of the packet.

Working with the shared memory and the registers present the same difficulties of working with a new microcontroller: the designer needs to study the documentation and learn what he has to touch to do something. Anyway, the BD-based structure of the simulator is wrapped in a set of primitives that can be called to parse a received packet or to send a particular command. We just implemented a series of commands needed to create the network, and more will be added in future works, but the addition of a new command or event is simply a matter of have a look at the standard definitions in [11] and then copy-and-paste the commands already defined.

B. Working on BaseBand layer

As we said before, the BB layer is a hardware-like model that implements (and hides) the details of the Bluetooth communication over the wireless channel. The model developed is not completely synthesizable, yet, but that's the style of coding. A high level structure of this layer is shown is Fig. 4. The BB_Manager module is a state machine that rules the main steps of the various operations the node can perform: for example, it's the BB_Manager that decides when the node can go from inquiry scan mode to page scan mode. The Link_Controller then will send the appropriate packets following the timings defined by the standard. Internally, the Link_Controller contains the transmission and reception modules, the hop frequency generator, a clock generator for the timings, and a control state machine. This control state machine enables and disables the transmitter and the receiver, following the commands from the BB_Manager and following the time slots division of the time axis. The transmitter and the receiver will then perform the encodings/decodings, the CRCs generation/check, and the actual transmission/read of the channel.

From the style of coding point of view, the main difference between the BB layer and the LM or application layer is that at the BB layer data and trigger/flag signals are carried by separate lines, mimicking a hardware implementation. On LM or application layer, instead, the communications are carried almost exclusively using the shared memory and the registers. Conceptually, it's almost the same thing, but using two different paradigms helps to match more closely a real world implementation and to allow different kind of designers to work in a way they are more used to.

III. A CASE OF STUDY

Figs. 5 and 6 present the signals of particularly important wires in the simulator. There are the input and output data signals to/from the master, where we can observe the data packets exchanged, and the lines that the LC_Controller uses to enable the receiver and the transmitter, both for the master and the slave. These lines can be used to identify loss of synchronization among the devices, as well as a possibly incorrect scheduling of the packets. Also shown are the channels used for the frequency hopping.

978-1-4799-4993-9/14 $31.00 © 2014 IEEE

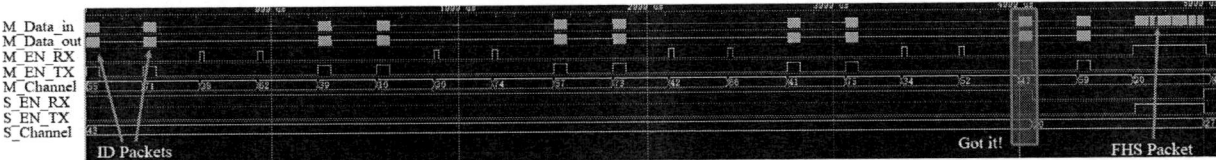

Fig. 5 - Signal tracings during the inquiry phase.

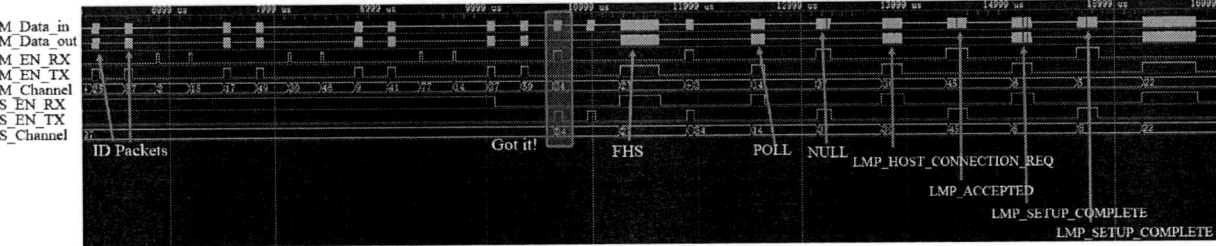

Fig. 6 - Signal tracings during the page phase.

Fig. 5 shows what happens during the inquiry phase of the network creation. The master sends continuously the ID packet, changing each time the transmission channel and sending two packets for each time slot, as the standard requires. The slave, on the contrary, continuously listens to the same channel. When the master transmits on a channel different from that the slave is listening to, obviously in the reception half time slot there won't be any answer, and the receiver enable line will be active only for the duration of the pre-defined reception window. When the master happens to transmit in the slave channel, the slave will catch the master packet and will reply in the (master) reception time slot. The activity on the channel keeps the receiver active for the whole transmission, and the packet is successfully transferred to the upper layers. The receiver enable line can also help to find noise-corrupted packets: when an error is found on the header CRC, or during the decoding of the payload (for the DM packets), the receiver will stop listening the channel, in order to save power.

Fig. 6 shows what happens after a successful inquiry, i.e. during the paging phase. As with the inquiry, the master starts sending repeatedly the ID packet, changing channel for each one, while the slave listens always to the same channel. When there is a match, the master sends a FHS packet, and with the first POLL packet the master and the slave are synchronized. Hence, can now begin the LM packet exchange that allows the slave higher layers to confirm or reject the just established connection. In the case shown, the slave accepts the request and the data transfer can start.

IV. CONCLUSIONS

In this paper we presented a novel network simulator for the Bluetooth protocol. While the other simulators currently available are very abstract models that mainly only let the designer to work above the TCP/IP layer, our simulator mimics a more real-world MCU-based implementation. Our simulator let the designer to work at application or link manager level using a software-like approach, and at baseband level using a hardware-like approach. To support the coexistence of these two opposite approaches, the simulator has been developed using SystemC, and in such a way that the designer doesn't need to know anything about SystemC in order to work at

application or LM level. The simulator supports the 2.0 version of the Bluetooth standard, except for the adaptive frequency-hopping, for the sniff, hold and park modes, and for SCO logical links, that are currently under development.

REFERENCES

[1] D. Macci, A. Zanotto, F. Leonardi and D. Petri, "A high-level model for estimating power consumption of Bluetooth devices," in *IEEE International Instrumentation and Measurement Technology Conference*, Vancouver Island, Canada, 2008.

[2] D. Macii, R. Corradi and D. Petri, "A measurement-based power consumption simulator for Bluetooth modules," *IEEE Transactions on Instrumentation and Measurement*, vol. 58, no. 5, pp. 1592-1601, May 2009.

[3] B. Kock, C. Wijting, M. Kuipers and R. Prasad, "WIP-Sim: a novel object-oriented event-driven IP network simulator," in *IEEE 54th Vehicular Technology Conference*, 2001.

[4] G. Kuijpers, S. Gameiro, N. Sousa, B. Kuipers and R. Prasad, "Bluetooth implementation in the WING-IP simulator," in *The 13th IEEE International Symposium on Personal, Indoor and Mobile Radio Communications*, 2002.

[5] D. Bultmann, M. Muhleisen, K. Klagges and M. Schinnenburg, "openWNS - open Wireless Network Simulator," in *European Wireless*, 2009.

[6] C. Lee and A. Helal, "NS-based Bluetooth LAP simulator," in *Proceedings of the 26th Annual IEEE Conference on Local Computer Networks*, 2001.

[7] A. Kumar, S. Kaushik, R. Sharma and P. Ray, "Simulators for wireless networks: a comparative study," in *International Conference on Computing Sciences*, 2012.

[8] X. Zhang and G. Riley, "Bluetooth Simulations for wireless sensor networks using GTNetS," in *Proceedings of the IEEE Computer Society's 12th Annual International Symposium on Modeling, Analysis and Simulation of Computer and Telecommunications Systems*, 2004.

[9] P. Latkoski and L. Gavrilovska, "Web traffic over Bluetooth: modeling, analysis and performance evaluation," in *Proceedings, of the 3rd International Conference on Wireless and Mobile Communications*, 2007.

[10] T. Grotker, S. Liao, G. Martin and S. Swan, System Design with SystemC, Springer, 2002.

[11] "802.15.1: Wireless medium access controlo (MAC) and physical layer (PHY) specifications for wireless personal area networks (WPANs)," IEEE Standard for Information technology, 2005.

A new algorithm for convergence verification in circuit level simulations

Francesco Lannutti, Francesco Menichelli, Paolo Nenzi, Mauro Olivieri

Abstract—**In this paper we present a new algorithm, based on Kirchhoff's Current Law (KCL) verification, for assessing convergence in circuit-level simulation and its implementation in the NGSPICE open source circuit simulator. We start from the analysis of false convergence problems appeared in the analysis of test circuits that we found to be related to the convergence checks that NGSPICE simulator inherits from SPICE3F5. These checks are inaccurate in some cases, so we propose to replace them with a new, efficient, algorithm based on KCL verification. The methodology to verify KCL of the circuit is detailed in this paper, along with the extraction flow, required to compute every KCL contribution. We finally present a case study of a circuit showing non converging solution, while the same circuit can be solved using the proposed algorithm.**

Index Terms—**KCL, Kirchhoff, NGSPICE, SPICE, Circuit Simulation.**

I. INTRODUCTION

CIRCUIT simulation has been a horsework for integrated circuit design since its introduction in 1972 at U.C. Berkeley [4], [10], and has been the basis for subsequent commercial as well as public domain releases. NGSPICE [6] is a widely used open-source implementation of the SPICE engine based upon the latest version released by U.C. Berkeley. As a public domain tool, NGSPICE has been used for circuit analysis both in research works as well as in the industry [1], [2], [3], [8], [9], as is being subject of continuous improvements [7], [11].

However, the convergence check criteria implemented in NGSPICE (and generically in any Berkeley SPICE engine) do not inherently guarantee the satisfaction of the Kirchhoff Current Law (KCL), leading to possible false convergence problems. The contribution of this article is in the mathematical identification of the problem and the development of a nover convergence check algorithm based on KCL verification. We report the methodology to verify KCL, along with the extraction flow required to compute every KCL contribution. The proposed algorithm has been implemented in the NGSPICE software code. We furthermore present a final circuit case study showing a non-converging solution when analyzed with a conventional SPICE engine, while showing a correct solution when the proposed convergence check algorithm is applied.

F. Lannutti, F. Menichelli and M.Olivieri are with the Department of Information Engineering, Electronics and Telecommunications, Sapienza University, Rome, RM, 00184 Italy - e-mail: {lannutti,menichelli,olivieri}@diet.uniroma1.it

P.Nenzi is with ENEA, Frascati, RM, Italy - e-mail: paolo.nenzi@enea.it

II. BACKGROUND

Any SPICE simulator uses the Modified Nodal Analysis (MNA) to assemble the circuit matrix, which extends the Nodal Analysis to include voltage sources and inductors.

In the Nodal Analysis, every matrix row represents the Kirchhoff Current Law (KCL) of the considered node and the circuit matrix A is the conductances matrix G.

In the Modified Nodal Analysis, current nodes are added to take care of the direct unknown current branches, due to voltage sources and inductors mentioned above, so it has a more complicated structure, which can be represented in Eq. 1:

$$A = \begin{bmatrix} G & B \\ C & D \end{bmatrix} \tag{1}$$

and it's composed by 4 contributions:

1) G is the conductances matrix, equivalent to the one in the Nodal Analysis;
2) B is the current branches matrix, due to every direct current branch unknown; it contains only ones and zeroes;
3) C is another current branches matrix and it's due to the MNA voltage constraints when at least a current branch is present in B; it contains ones and zeroes as well;
4) D is the controlled sources matrix and it contains also the inductors entries; its elements have resistance dimension.

The linear system which SPICE solves is represented in Eq. 2:

$$\begin{bmatrix} G & B \\ C & D \end{bmatrix} * \begin{bmatrix} V \\ I \end{bmatrix} = \begin{bmatrix} I_s \\ V_s \end{bmatrix} \tag{2}$$

If the circuit has non-linear devices, as usual in SPICE simulations, G is the **Jacobian Matrix**, represented as J, resulting from the linearization through Taylor's series expansion and solving a sequence of linear systems until convergence is reached.

A. The Newton-Raphson Method

The Newton-Raphson Method is based upon the Eq. 3:

$$\vec{U}\left(\vec{F}(\vec{V}), \vec{I}\right) = 0 \Rightarrow \tag{3a}$$

$$\Rightarrow \vec{F}(\vec{V}) + \vec{I} = 0 \Rightarrow \tag{3b}$$

$$\Rightarrow \vec{F}(\vec{V_0}) + J(\vec{V_0}) * (\vec{V_1} - \vec{V_0}) + h.o.t + \vec{I} = 0, \tag{3c}$$

where $F(V_k)$ is the non-linear function, J is the Jacobian Matrix and h.o.t stands for higher order terms, which are dropped by the linearization process.

978-1-4799-4993-9/14 $31.00 © 2014 IEEE

The Newton-Raphson Method is iterative, so it needs a termination criterion; valid options are:

1) A maximum iteration limit
2) The limitation of the difference between two adjacent solutions

SPICE implements both of them, leveraging on the second one for every iteration until convergence is satisfied or it's necessary to start an homotopy method, leaving the first one as worst case.

Considering what has been said, it's possible to write the following linear system at every Newton-Raphson iteration:

$$\vec{F}(\vec{V_k}) + J(\vec{V_k}) * (\vec{V_{k+1}} - \vec{V_k}) + \vec{I_{k+1}} = 0, \quad (4)$$

where k is the iteration step.

Both $\vec{V_{k+1}}$ and $\vec{I_{k+1}}$ are direct unknowns.

Rearranging the Eq. 4, it becomes:

$$J(\vec{V_k}) * \vec{V_{k+1}} + \vec{I_{k+1}} = J(\vec{V_k}) * \vec{V_k} - \vec{F}(\vec{V_k}). \quad (5)$$

Assembling this development back into the matrix notation, it becomes:

$$\begin{bmatrix} J(\vec{V_k}) & B \\ C & D \end{bmatrix} * \begin{bmatrix} \vec{V_{k+1}} \\ \vec{I_{k+1}} \end{bmatrix} =$$
$$= \begin{bmatrix} J(\vec{V_k}) * \vec{V_k} & \vec{F}(\vec{V_k}) \mid \vec{I_s} \\ & \vec{V_s} \end{bmatrix}, \quad (6)$$

which is exactly what SPICE solves as $A\vec{x} = \vec{b}$.

It's known that the Newton-Raphson Method cannot guarantee global convergence when it's applied on a non-monotonic function.

B. State of the art of convergence algorithms for circuit simulation

As already mentioned in [5], the circuit simulator criteria to check if a Newton-Raphson iteration is terminated correctly comprehends three convergence tests:

1) Delta-V is satisfied
2) $f(\vec{V_k})$ is satisfied
3) Delta-I is satisfied

In order to verify Delta-V, the Eq. 7 is applied, which tests the difference between the current value of V and the previous one against a fixed threshold:

$$\left| \vec{V_{k+1}} - \vec{V_k} \right| < RELTOL*$$
$$MAX \left| \vec{V_{k+1}}, \vec{V_k} \right| + VNTOL. \quad (7)$$

A similar check is performed after Delta-V to test $f(\vec{V_k})$, using the Eq. 8, which tests if every indirect current, known by every device model in the given circuit, is limited:

$$\left| f(\vec{V_{k+1}}) - f(\vec{V_k}) \right| < RELTOL*$$
$$MAX \left| f(\vec{V_{k+1}}), f(\vec{V_k}) \right| + ABSTOL. \quad (8)$$

Since direct currents are also unknowns, a similar test has to be performed for them; this task is accomplished by the Delta-I convergence test:

$$\left| \vec{I_{k+1}} - \vec{I_k} \right| < RELTOL*$$
$$MAX \left| \vec{I_{k+1}}, \vec{I_k} \right| + ABSTOL. \quad (9)$$

C. The False Convergence Phenomenon

During the simulation, there are some curves which satisfy Delta-V, but not $f(\vec{V_k})$, and other curves which satisfy the opposite criterias. The False Convergence Phenomenon appears when the iteration is so slow to pass $f(\vec{V_k})$ conergence test, without assuring that the KCL is satisfied. A good circuit simulator should verify the KCL, instead of using $f(\vec{V_k})$ convergence test [5].

III. KCL VERIFICATION

In order to avoid the False Convergence Phenomenon, $f(V_k)$ convergence test has to be replaced by the KCL Verification for the entire circuit, after that Delta-V is satisfied for every voltage node.

The KCL formulation in Eq. 10 is derived by the previous matrix notation considering $\vec{V_{k+1}} = \vec{V_k}$, or, equally, extracting $\vec{F}(\vec{V_k}), \vec{I_{k+1}}, \vec{I_s}$ from every device model and writing the KCL. In addition, when convergence in satisfied, $\vec{I_{k+1}} = \vec{I_k}$.

$$\left| F^n(\vec{V_k}) + I_k^n - I_s^n \right| < RELTOL*$$
$$MAX \left| F^{nj}(\vec{V_k}), I_k^{nj}, I_s^{nj} \right| + ABSTOL, \quad (10)$$

where n is the voltage node index, j is the current branch index of the n_{th} voltage node, F^{nj} are the indirect unknown current branches at every voltage node and F^n is the total sum of all the F^{nj} at every voltage node. I_k^n is the bunch of direct unknown current branches and I_s^{nj} is the same thing of F^{nj} but for the current sources.

IV. $\vec{F}(\vec{V_k})$ IN SPICE

The left hand side part of the Eq. 10 can be calculated by changing every device model to directly output every contribution of $F^{nj}, I_k^{nj}, I_s^{nj}$ independently and then assembling them inside the same device routine, which calculates the Jacobian entries of that device.

This approach isn't straightforward, because it requires changing every model, but it's efficient, because $\vec{b}$ is already calculated inside every device model as $J(\vec{V_k}) * \vec{V_k} - \vec{F}(\vec{V_k}) + \vec{I_s}$, which exactly includes $\vec{F}(\vec{V_k})$ and $\vec{I_s}$ for current sources. So it's enough to compute the same intermediate terms in another way to make $\vec{F}(\vec{V_k})$ directly visible alone and let the compiler optimize everything.

In devices which have $\vec{F}(\vec{V_k}) \neq 0$, $\vec{I_k} = 0$ and $\vec{I_s} = 0$, $\vec{F}(\vec{V_k})$ represents the device total current between two voltage nodes, either internal or external ones, and can be composed by three different terms:

1) A *Purely Linear and Static* Device Model (Resistor) always has $\vec{b} = J(\vec{V_k}) * \vec{V_k} - \vec{F}(\vec{V_k}) = 0$, so $\vec{F}(\vec{V_k}) = J(\vec{V_k}) * \vec{V_k} = G * \vec{V_k}$.

2) A *Purely Linear and Dynamic* Device Model (Capacitor) always has $\vec{b} = J(\vec{V_k}) * \vec{V_k} - \vec{F}(\vec{V_k}) = G_{eq} * \vec{V_k} - \vec{I_{tot}}$, so $\vec{F}(\vec{V_k}) = \vec{I_{tot}}$; $\vec{I_{tot}}$ depends on the chosen integration method and includes the impressed current, which is the current at the previous timestep.

3) A *Non-Linear* Device Model (Diode is the simplest one) always has two parts: *Static* and *Dynamic*. The Static part serves to determine the Operating Point, while the Dynamic part is needed during the Transient Analysis. In this kind of devices, $\vec{b} = J(\vec{V_k}) * \vec{V_k} - \vec{F}(\vec{V_k})$, where $J(\vec{V_k}) * \vec{V_k}$ is the linearized part of the Non-Linear device. Here it's possible to extract $\vec{F}(\vec{V_k})$ summing every current between two voltage nodes, considering the Non-Linear current, when present, instead of its linear approximation.

A "special case" appears when $\vec{F}(\vec{V_k}) = G(\vec{V_k}) * \vec{V_k}$. In this case:

$$\vec{b} = J_f(\vec{V_k}) * \vec{V_k} - \vec{F}(\vec{V_k}) =$$
$$= \left[J_g(\vec{V_k}) * \vec{V_k} + G(\vec{V_k}) \right] * \vec{V_k} - G(\vec{V_k}) * \vec{V_k} =$$
$$= J_g(\vec{V_k}) * \vec{V_k}^2 + G(\vec{V_k}) * \vec{V_k} - G(\vec{V_k}) * \vec{V_k} =$$
$$= J_g(\vec{V_k}) * \vec{V_k}^2, \quad (11)$$

so $\vec{b}$ doesn't contain $\vec{F}(\vec{V_k})$. This means that, in this case, $\vec{F}(\vec{V_k})$ has to be calculated in the same manner it's calculated for a linear static device, but remembering that actually it's non-linear, so it changes at every Newton-Raphson iteration.

In devices which have $\vec{F}(\vec{V_k}) = 0$, $\vec{I_k} \neq 0$ and $\vec{I_s} = 0$, like independent voltage sources and inductors, only the direct unknown current branches have to be considered for the KCL Verification.

The remaining devices have $\vec{F}(\vec{V_k}) = 0$, $\vec{I_k} = 0$ and $\vec{I_s} \neq 0$ and they are independent current sources. Here the contribution for the KCL Verification is simply its own current.

A. Nodes Classification

Since the KCL Verification is really needed on non-linear nodes, a proper node classification is mandatory to know which nodes are non-linear.

Definition 1: a node is non-linear if it connects to at least one non-linear branch.

Definition 2: a node is linear if it connects only linear branches.

By knowing this classification, it's possible to skip the KCL Calculation and Verification when it isn't needed.

V. GMIN STEPPING INCOERENCIES

During the KCL Verification, all the netlists which need GMIN stepping algorithm to reach convergence, never verified the KCL. After a thorough investigation, we were able to attribute this wrong behavior to an incoerent implementation of the GMIN stepping algorithm in NGSPICE.

There are two different kind of incoerencies:

1) The GMIN is applied on the main diagonal which appears as is after that the circuit matrix has been

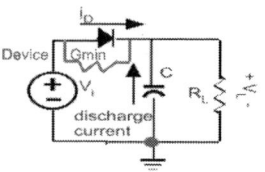

Fig. 1. Device GMIN Circuit

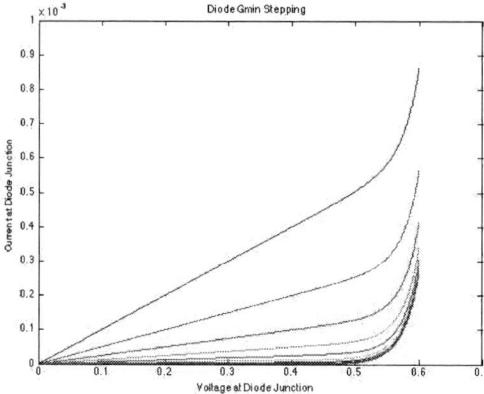

Fig. 2. Device GMIN Stepping

reordered to remove zeroes from it, but these diagonal elements aren't the circuit diagonal elements anymore.

2) The GMIN is applied on every row of the circuit matrix, disregarding if it's a voltage or current node; in the latter case, what is actually summed on current nodes it's a resistance with a GMIN value, not a conductance.

So GMIN Algorithm has been restored to the original SPICE2 formulation and this helped to reach the real convergence we are looking for, by using the KCL Verification.

Moreover, we implemented a new homotopy method to smooth the exponential non-linearity near the curve knee. So, considering the circuit in Fig. 1, we have exploited the already existing Device GMIN in parallel with exponential components, like diodes.

Starting stepping from a good value of GMIN (in our case 1e-3 - Fig. 2), the non-linear component is actually linear and this is an optimal starting point for the N-R iteration.

VI. IMPLEMENTATION RESULTS

The presented algorithm has been implemented in NGSPICE and verified on the ISCAS85 benchmark suite [13]. ISCAS85 suite has been implemented with a standard cell library, which uses BSIM4v7 devices along with resistors and capacitors. Both standard and back-annotated netlists have been tested.

In our previous work [7] on this benchmark suite, we found that not all netlists converged to a valid operating point and we had to arise the target GMIN value.

Just inserting KCL Verification and increasing the GMIN target value to at least 1e-18, we still faced the False Convergence of *c7552_ann* netlist, but we were able to make *c6288_ann* converge with a such small precision (red line).

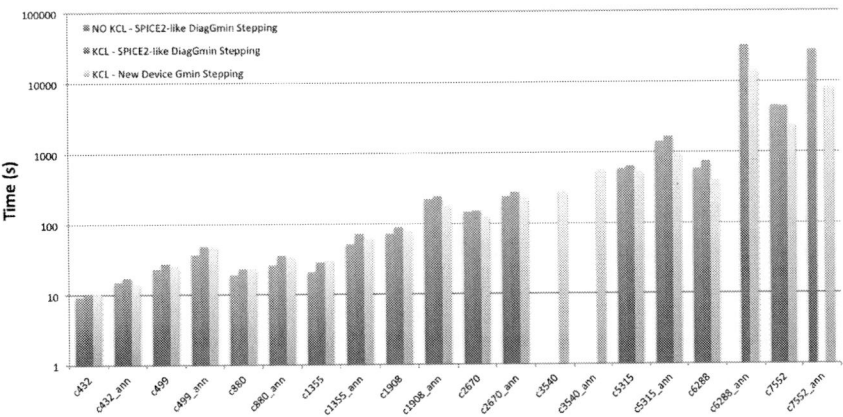

Fig. 3. Execution Time Comparison between NGSPICE without KCL and NGSPICE with KCL

By inserting also the new Device GMIN Stepping Algorithm, we were able to make all the netlists converge with such small precision (green line).

Regarding speed, KCL Verification has introduced a lot of calculation due to the "special" nodes mentioned above, but the slowdown has been contained in about 15% of time loss overall (Fig. 3). This good behavior is due to the efficient implementation previously discussed.

Introducing the Device GMIN Stepping Algorithm, we managed to speedup all the simulations by an averge of 1.31X and let all the netlist converge with the same accuracy, too. This happens because this new homotopy method doesn't let the simulation switch to Source Stepping Algorithm and reduces the number of fill-ins, as showed in Table I.

VII. CONCLUSION

The KCL Verification methodology has been presented and implemented in NGSPICE.

This new convergence test avoids the False Convergence Phenomenon, using an efficient implementation, and it substitutes the $f(\vec{V_k})$ convergence test, which is present in every non-linear device model, under the model developer responsibility.

This new algorithm highlighted an incoherent implementation of the GMIN Stepping Algorithm, which prevented convergence in netlists that trigger it.

By using this new homotopy algorithm and the KCL Verification, every netlist of the ISCAS85 benchmark suite converged with a small precision, showing also an average speedup of 1.31X.

Future work will be devoted to implementing KCL Verification in all device models already encoded in NGSPICE.

TABLE I
GMIN RESULT COMPARISON

Algorithm	Time (s)	Fill-ins	DC Iterations
Old GMIN	1264.530	124242	533
New GMIN and KCL	860,160	119450	443

REFERENCES

[1] Abbas, Z., Olivieri, M., Impact of technology scaling on leakage power in nano-scale bulk CMOS digital standard cells, Microelectronics Journal, 2014. Elsevier. ISSN: 0026-2692. Volume 45, Issue 2, February 2014, Pages 179-195.

[2] Abbas, Z.; Mastrandrea, A.; Olivieri, M., "A Voltage-Based Leakage Current Calculation Scheme and its Application to Nanoscale MOSFET and FinFET Standard-Cell Designs," IEEE Transactions on Very Large Scale Integration (VLSI) Systems, 2014 [Online]. http://dx.doi.org/10.1109/TVLSI.2013.2294550

[3] Abbas Z., Genua, V., Olivieri, M., A novel logic level calculation model for leakage currents in digital nano-CMOS circuits, 7th Conference on Ph.D. Research in Microelectronics and Electronics (PRIME), Trento, Italy, 3-7 July 2011, pp. 221 224. IEEE.

[4] Nagel, L. W., and Pederson, D. O., SPICE (Simulation Program with Integrated Circuit Emphasis), Memorandum No. ERL-M382, University of California, Berkeley, Apr. 1973

[5] K. Kundert, I. Clifford, Achieving Accurate Results With a Circuit Simulator, Cadence Design Systems (Analog Division), San Jose, CA.

[6] P. Nenzi, H. Vogt, Ngspice Users Manual, [Online]. At http://ngspice.sourceforge.net/docs/ngspice-manual.pdf

[7] F. Lannutti, P. Nenzi, M. Olivieri, KLU sparse direct linear solver implementation into NGSPICE, Mixed Design of Integrated Circuits and Systems (MIXDES), 2012 Proceedings of the 19th International Conference , vol., no., pp.69,73, 24-26 May 2012

[8] Mastrandrea A., Olivieri, M., Menichelli, F., A delay model allowing nano-CMOS standard cells statistical simulation at the logic level, 7th Conference on Ph.D. Research in Microelectronics and Electronics (PRIME), Trento, Italy, 3-7 July 2011, pp. 217 220. IEEE.

[9] Olivieri, M., Mastrandrea, A., Logic Drivers: A Propagation Delay Modeling Paradigm for Statistical Simulation of Standard Cell Designs, IEEE Transactions on Very Large Scale Integration (VLSI) Systems, vol.22, no.6, pp.1429,1440, June 2014.

[10] Pescovitz, D. "1972: The release of SPICE, still the industry standard tool for integrated circuit design". Lab Notes: Research from the Berkeley College of Engineering, 2002.[Online]. At http://coe.berkeley.edu/labnotes/0502/history.html.

[11] F. Ramundo, P. Nenzi, M. Olivieri, First integration of MOSFET Band-To-Band-Tunneling current in BSIM4, Microelectronics Journal, Volume 44, Issue 1, Jan. 2013, Pages 26-32. Elsevier.

[12] Sanyal, A., Rastogi, A., Chen, W., Roy, K., Kundu, S., An Efficient Technique for Leakage Current Estimation in Nanoscaled CMOS Circuits Incorporating Self-Loading Effects, IEEE Transaction on Computers, vol. 59, no. 7, pp. 922-932, July. 2010.

[13] J. Xu, Perform the SPICE Simulation of ISCAS85 Benchmark Circuits for Research, [Online]. At http://www.ece.uic.edu/ masud/iscas2spice.htm

Simulation methodology for Large-Bandwidth Track-and-Hold microwave circuit

Arnaud Meyer[*][†], Patricia Desgreys [*], Hervé Petit [*], Bruno Louis [†], Vincent Petit [†]

[*]Télécom ParisTech, France, 46 rue Barrault, 75013, Paris, Email: armeyer@enst.fr

[†]Thales Systèmes aeroportés, France, 2 av Gay Lussac, 78990, Élancourt

Abstract—A step-by-step simulation methodology for large-bandwidth Track-and-hold (T/H) microwave circuit is proposed. A T/H circuit is characterized accurately under the Cadence© environment. With the consideration of a specific windowing function, linearity simulation analysis could be done effectively. Moreover, the use of an input frequency generation function, in accordance with theoretical calculation, allows to treat the entire input bandwidth (BW). A simulated switched emitter follower (SEF) structure in a 0.13-μm SiGe BiCMOS technology with 24 GHz effective bandwidth and a 4 GS/s clock illustrates our methodology.

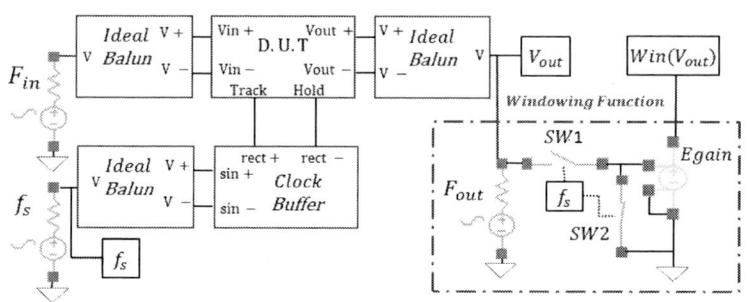

Fig. 1.   The used environment for T/H circuit simulation

I. INTRODUCTION

A classical microwave receiver is composed of a low noise amplifier (LNA), a mixer and a Track-and-hold (T/H) circuit in front of the analog-to-digital-converter (ADC) [1]. Usually, frequency mixing requires space and consumes power. Therefore, a T/H circuit with sub-sampling action, could replace positively the mixer and the LNA if the target specifications in terms of noise are compatible. Such circuits present different facets which require different analysis in track/hold mode and in sampling mode. Until now, some articles cover the subject in terms of specific analyses [2],[3]. The aim of this paper, is to propose a step-by-step methodology to cover principal aspects of large bandwidth (BW) T/H circuits and to overcome analysis problem. In addition, the use of a simple windowing function, which allows to process the signal when linearity analysis must be done, is introduced. Moreover, a process able to generate adequate input frequencies in sub-sampling mode is detailed. The purpose analysed architecture is based on the most common used structure the switched emitter follower (SEF) [4],[5] and principal specifications are demonstrated.

In the next section, the considered simulation environment is described. Section III presents the possible simulation scenario, and section IV concludes the paper.

II. SPECIFIC ENVIRONMENT SIMULATION

For our analysis, we considered the schematic of Fig. 1 under Cadence© design environment. The schematic is based on a differential approach, but it could be used for single-ended structure. For sake of simplicity, we do not consider in our analysis the output buffer and we assume that an ideal quantizer is used. Regarding the microwave clock, we considered a sinusoidal one. The device under test (DUT) is based on the SEF structure of Fig. 2.

A. Simulated Switched emitter follower

In track mode, $Q3$, $Q4$, $Q5$ and $Q8$ are ON, thus capacitor C_{Hold} is loaded. In hold mode, $Q7$, $Q6$ are ON, hence added currents in Rl creates drop voltages and $Q3$, $Q4$ are blocked. The added capacitors $C_F \simeq C_{be}$ enhance the isolation against input signal feed-through [6].

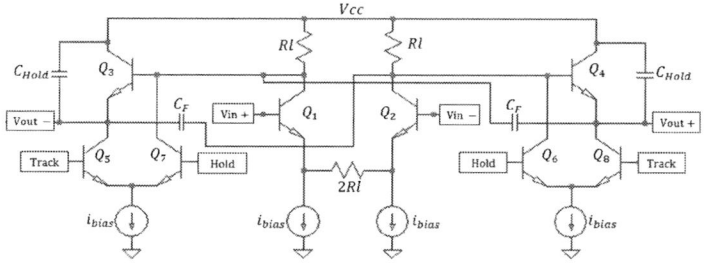

Fig. 2.   The simulated large bandwidth T/H structure

B. Rectangular windowing function

In order to simulate this structure periodic steady state (PSS) analysis combined with periodic X frequency (PXF) analysis could be used with the special sampled option. However, the linearity simulation with Quasi Periodic Steady State (QPSS) analysis will be impacted by the zero-hold nature of the DUT [2]. Our solution consists in treating the hold part of the signal with a windowing-rectangular function to approximate a discretized sample as closely as possible. Considering $V_{out}[t]$ the output signal, the equivalent response of windowing function in frequency domain with $0 < \alpha \leq 0.25$ is:

$$W_{in}(\omega) = \int_{-Te \cdot \alpha}^{Te \cdot \alpha} V_{out}[t] \exp^{-j\omega t} dt, \qquad (1)$$

$$Egain * W_{in}(f) = Te \cdot Sinc[\alpha 2\pi F_{out} Te]. \qquad (2)$$

978-1-4799-4993-9/14 $31.00 © 2014 IEEE

To be efficient, the chosen value of duty cycle α should be equal to the hold time divided by two or more. With the sampling clock $f_s = 1/Te$, assuming that the input signal is maintained during $Te/2$, $\alpha \leq 0.25$ and $Egain = 1/\alpha$.

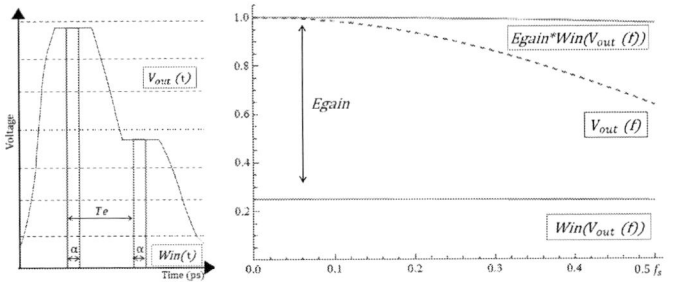

Fig. 3. (a) Illustration of the windowing function on the signal and (b) the equivalent frequency response for $\alpha = 0.25$,we can observe how the linear gain compensate the attenuation

Such function could be done with the implementation of an ideal VerilogA T/H but, because of hidden-state, it will increase simulation time and complexity [7]. In consequence of this limitation we use a couple of complementary switches $SW1/SW2$ piloted by the input clock signal (Fig. 1). The clock synchronization is used for QPSS simulation convergence, therefore it will also be used to control the switches. We combine it with an ideal buffer gain $Egain$ to compensate the linear attenuation coming from the function (2) as shown in Fig. 3. The complementary switches must be closed/opened during the determined windowing time.

C. Input frequency generation function

One of the interesting applications of a large bandwidth T/H is the capability to translate numerous frequencies located in different Nyquist bands into baseband. In order to use QPSS results to calculate total harmonic distortion (THD), it is useful to know the frequencies values created by non-linearity. We assume that major non-linear contributions come from third and second harmonics, therefore (THD) can be express:

$$THD = \frac{\sqrt{P_{H2}^2 + P_{H3}^2}}{P_{Fund}}. \tag{3}$$

Where P_{Hn} is the power of the corresponding harmonic n. The following Fig. 4 shows how to choose correctly the input frequencies (F1,F2..Fn) in order to produce, in baseband, the same fundamental (Fund), the same harmonic 2 (H2) and the same harmonic 3 (H3).

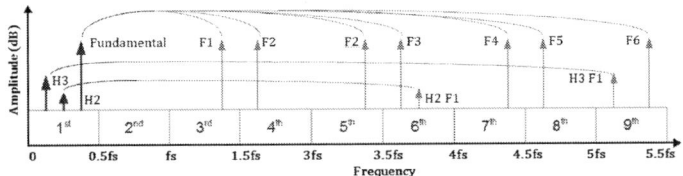

Fig. 4. Input harmonics generation used to analyze the full input spectrum, numbers indicated nyquist bands which is considered under sub-sampling modes

Using right input frequencies it is easier to simulate large BW T/H and generated frequencies can be reuse further to

characterize the structure in measurements. In the case that we consider a bandpass signal with a frequency range f_l (lower frequency) to f_u (upper frequency), then we have to respect Nyquist condition sampling frequency $f_s > 2B$ with the band $B = f_u - f_l$. When considering sub-sampling, the band is located at an integer number of bandwidth [8]. Considering the location k of the output frequency $Fout$ in the band B, with $k \in \mathbb{Q}$, the relation can be written :

$$B = \frac{fs}{2}, \quad k = \frac{Fout}{B}. \tag{4}$$

If we consider the track input bandwidth (BW) and the clock frequency (f_s) we can calculate the total number of Nyquist Band $N_B \in \mathbb{Q}$ and the corresponding Nyquist band number $n \in \mathbb{N}^*$:

$$N_B = \frac{BW}{fs}, \quad 1 < n < N_B \tag{5}$$

The Nyquist location of a sub-sampling frequency depends on its location into the band and can be expressed as:

$$Fin = B(n_{odd} - 1 + k), \quad Fin = B(n_{even} - k) \tag{6}$$

Combining the equation (4), with equation (5) and considering equation (6) we found the relation between the output and the input. Equivalent equation can be written in function of two coefficients $a \in \mathbb{N}^*$ and $b \in \mathbb{N}^*$ depending on the Nyquist band location n:

$$Fin = \frac{fs}{2}(a + \frac{2}{fs}b \cdot Fout). \tag{7}$$

Depending on a and b the input frequency is defined by the following modulo ($MOD(x,\mathbb{N})$) operation for a fixed $Fout$:

$$a = n - MOD(n,2) , \quad b = 2 * MOD(n,2)\text{-}1. \tag{8}$$

Using the previous equation (7), it is straightforward to calculate harmonic 2 and harmonic 3 values in order to calculate THD (3):

$$n_{H2} = floor(\frac{2Fin}{0.5fs}), \quad n_{H3} = floor(\frac{3Fin}{0.5fs}), \tag{9}$$

$$Fout_{H2} = \frac{1}{b}(2Fin\text{-}\frac{fs}{2}a), \quad Fout_{H3} = \frac{1}{b}(3Fin\text{-}\frac{fs}{2}a). \tag{10}$$

For example, with $f_s = 4$ GHz ,$BW = 24$ GHz and $Fout = 1.2$ GHz we can calculate corresponding input frequencies witch fall in the 11th Nyquist band. With B=2 GHz, $k = \frac{Fout}{fs} = \frac{3}{5}$, $N_B = 12$. For $n_{odd} = 11$ we calculate :

$$Fin_{11} = B(n_{odd} - k) = 2 \cdot 10^9 \cdot (\frac{52}{5}) = 20.8 \text{ GHz}. \tag{11}$$

Using equation (10) and equation (7) in an Ocean© script, combined with QPSS analysis , the input frequencies can be generated depending on the desired output frequency as showed in Fig. 4. After implementation of the described function we can follow up with the different analysis.

III. DEDICATED SIMULATIONS FOR T/H CIRCUITS

Track and hold circuits could be simulated in track mode or in sampling mode. Depending on performance the designer want to simulate, the dedicated environment in Table I allows to treat all the simulation scenarios. Considering a microwave circuit, scattering parameters (SP) small signal analysis is commonly used. The following table resumes the principal simulations executed by order and detailed in this paper.

TABLE I
SIMULATIONS DONE WITH SPECTRERF© EXECUTED BY ORDER

Simulation Analysis	Track mode	Step	Sampling mode	Step
Bandwidth	SP	1st	PSS-PXF	2nd
Noise Figure	SP-Noise	3rd	PSS-Pnoise	4th
THD/P-1dB	PSS	5th	QPSS	6th

A. Bandwidth

One of the main interest of our circuit is to obtain a high input bandwidth (BW) therefore we start by the 1st analysis to obtain the input bandwidth. In Fig. 5 the scattering parameter $(S_{[2,1]})$ in function of F_{in} is plotted showing a value of 28 GHz at -3 dB which corresponds to the effective BW. After the consideration of the input bandwidth in track mode, we now concentrate on the sampling mode.

The 2nd analysis is done in frequency-domain through a PSS analysis with $f_s = 4$ GHz. This one allows to compute the impulse response of the system. Secondly, PXF analysis is chosen with sampling option. This analysis will compute all the transfer functions of the inputs with regard to one selected output. Considering the right sample, we can obtain conversion loss of the T/H and compare track bandwidth to sampling-bandwidth impacted by conversion loss.

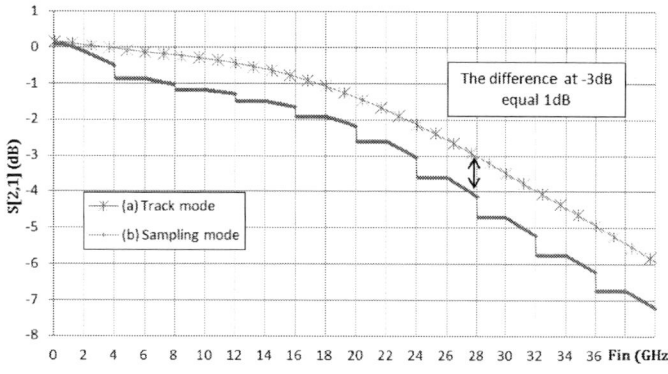

Fig. 5. Obtained scattering parameter S(2,1) in respect to input frequency in (a) track mode and (b) in sampling mode sliced by f_s

The comparison of both curves in Fig. 5, shows that the input track gain at -3 dB decreases by 1 dB with conversion losses. Therefore, the final effective input BW is equal to 24 GHz. In parallel, it is important to size the output isolation during the hold-mode to quantify the feed-through:

$$Isolation_{dB} = S_{[2,1]dB|Hold} - S_{[2,1]dB|Track}. \tag{12}$$

We shall pay a particular attention to the considered direct bias current (dc) in this mode. When a load is applied at the

output of the circuit, a non-realistic isolation could-be obtained as illustrated in Fig. 6 (b). The problem comes from the initial conditions (ic). Because of a leaking current in the output load, the considerate ic is wrong and corresponding dc will never exist in sampled mode. In order to overcome this problem we may force the capacitor voltage to a value equivalent to a possible held value Fig. 6 curve (a).

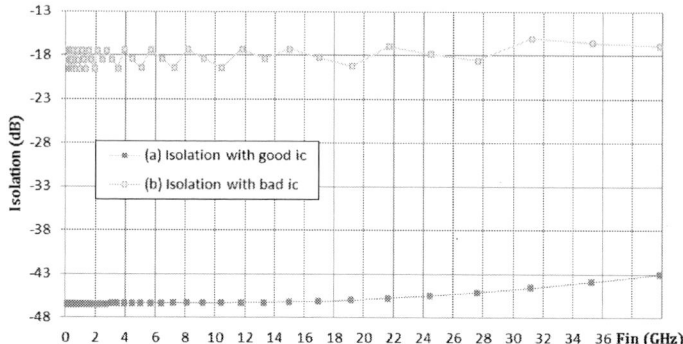

Fig. 6. Obtained output isolation (12) from the input in Hold mode (a) with good initial condition (b) with bad initial condition

B. Noise Figure

Commonly, thermal noise in sampled circuits is the main noise source [3]. This noise is linked to an equivalent resistive element in the circuit which is uniformly distributed. Its spectral density is equal $4kTR$, were k is the Boltzman constant, T is the temperature in kelvin and R is the equivalent resistor. We sample the stationary input noise every $1/f_s$, therefore the resulting noise is cyclostationary [9] and aliased in baseband. Theoretical calculation of an aliasing noise P_{als} with a first order transfer function for the input bandwidth (BW) under a single-side-band consideration (SSB) [2], has to be taken:

$$P_{als|dB} = 10 \cdot Log(\frac{\pi}{2}BW \cdot \frac{1}{0.5f_s}). \tag{13}$$

From the previous calculation the noise factor NF_{Sample} in sampling mode can be approximate by:

$$NF_{Sample} = P_{in|dBm} - P_{kT|dBm} - P_{BW|dB} - P_{als} - SNR_{out}. \tag{14}$$

With P_{in} is the input power of the signal, P_{kT} is a noisy 50 Ω resistor, P_{BW} is the equivalent noise input bandwidth and SNR_{out} is the signal to noise ratio at the output. In track mode the noise factor NF_{Track} is expressed by:

$$NF_{Track} = SNR_{in} - SNR_{out}, \quad SNR_{in} = P_{in} - P_{kT} - P_{BW}. \tag{15}$$

In accordance with theoretical result, noise analysis simulation can be done. To perform a good check of results, we start by evaluating the noise in the track mode.

The 3rd step is carried out using noise source option in SP parameter analysis and plotted in Fig. 7 . The obtained results show an average value of $NF_{Track} \simeq 13$ dB.

The 4th simulation step concerns the cyclostationary noise, it requires a specific analysis through a PSS, followed by a specific Periodic noise (P-noise) in sampling mode. In this

978-1-4799-4993-9/14 $31.00 © 2014 IEEE

paper all the simulations are done using fs=4 GHz and a power input signal of P_{in}=-6 dBm. The last result shows a average value of $NF_{Sample} \simeq 28$ dB.

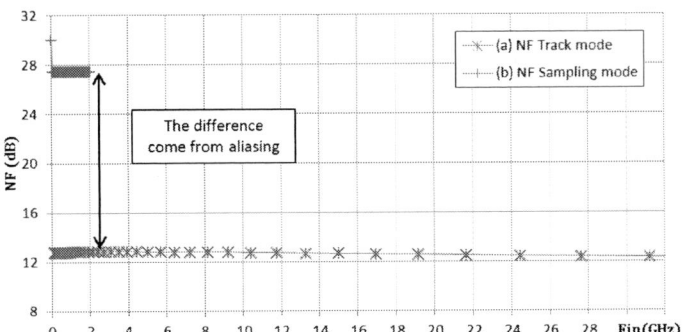

Fig. 7. Obtained Noise figure in (a) track mode and (b) sampling mode, we can observe how the noise increase by aliasing effect

The observed difference in Fig. 7 results from the contribution of aliasing noise into baseband. The comparison of the theoretical calculation $P_{als|cal} \simeq 14$ dB (13) with the plotted simulated results $P_{als|sim} \simeq 13$ dB Fig. 7, shows a small difference of 1dB. This difference comes from the consideration of an approximate value for the input bandwidth BW=24 GHz in the $\Gamma_{als|cal}$ calculation. After achieving verification of the obtained $NF_{sampling}$ value we can calculate the SNR_{out}. Under a SSB consideration the SNR_{out} value equal to 47 dB. Once the final SNR_{out} is evaluated, the linearity specification could be analyzed.

C. Linearity

According to [10], we start by calculating the compression point at 1dB (P_{1dB}) from specifications. Here, for example we calculate a minimal $P_{1dB} \simeq 2$ dBm for one tone excitation. Using the compression function in PSS analysis, we obtain the following curve in track mode with one tone excitation :

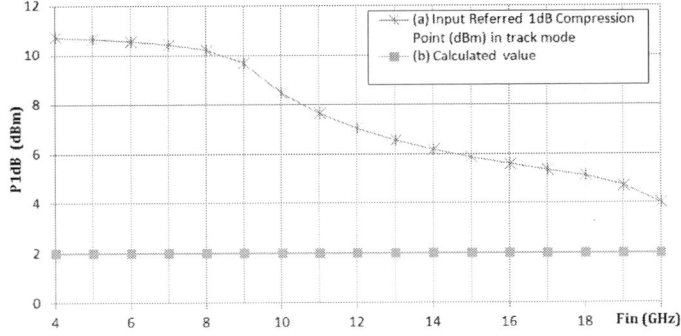

Fig. 8. Obtained $P1$ dB from extrapolation at -20 dBm input power P_{in} in (a) track mode and (b) calculated value

The obtained simulation results indicates that our targeted P_{1dB} is not reached. Based on this calculation we continue with the **5th** analysis in order to obtain THD (3) in track mode Fig. 9. We are now interested in the linearity of the T/H in sampling mode, hence we consider the **6th** analysis. By using windowing function, it is possible to directly obtain the

magnitude of the different harmonics. Using the generation function described previously to simulate the THD, we obtain the following result:

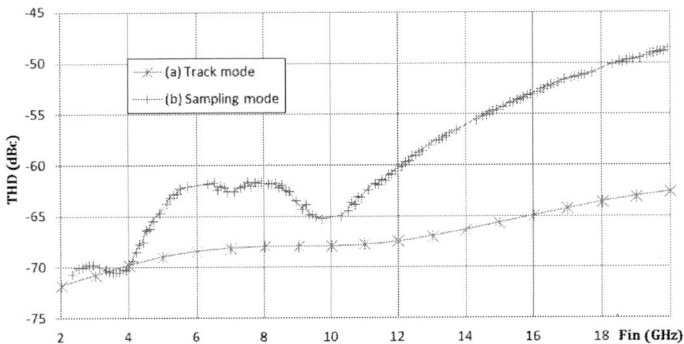

Fig. 9. Obtained THD in (a) track mode and (b) sampling mode

Considering the ideal current sources used and the Harmonic 2 reduction by differential operation, the worst average $THD_{sampling} \simeq$ -48 dB. In track mode the worst average $THD_{Track} \simeq$ -63 dB is significantly lower. The difference between the two curves comes from the non-linearity created by mixing products between the clock and input signal.

IV. CONCLUSION

We have detailed how to simulate a large bandwidth T/H circuit through a step-by-step methodology. Based on analytical calculation we demonstrated the validity of the results obtained by simulation. In addition, the use of a specific generation frequency function combined with a windowing function demonstrates how to analyze the linearity of the structure over a large input spectrum. Regarding simulated results, we concluded that high performance T/H microwave circuit can be efficiently analyzed under the Cadence© environment with our methodology.

REFERENCES

[1] D. Jakonis, K. Folkesson, J. Dbrowski, P. Eriksson, and C. Svensson. A 2.4-ghz rf sampling receiver front-end in 0.18μm cmos. *Solid-State Circuits, IEEE Journal of*, 40(6):1265–1277, June 2005.
[2] K. Kundert. Simulating switched-capacitor filters with spectre rf. *http://www.designers-guide.org/*, (Version 6c), 2006.
[3] B. Murmann. Thermal noise in track-and-hold circuits: analysis and simulation techniques. *Solid-State Circuits Magazine, IEEE*, June 2012.
[4] A. Moriyama, S. Taniyama, and T. Waho. A low-distortion switched-source-follower track-and-hold circuit, 2012 19th ieee international conference on. In *Electronics, Circuits and Systems (ICECS)*, Dec 2012.
[5] Yevgen Borokhovych and J. Christoph Scheytt. 10 gs/s 8-bit bipolar tha in sige technology. In *NORCHIP, 2011*, pages 1–4, Nov 2011.
[6] S. Shahramian, A.C. Carusone, and S.P. Voinigescu. Design methodology for a 40-gsamples/s track and hold amplifier in 0.18- μm sige bicmos technology. *Solid-State Circuits, IEEE Journal of*, 41, Oct 2006.
[7] K. Kundert. Hidden state in spectre rf application note. *http://www.designers-guide.org/*, (5.0), 2003.
[8] R.G. Vaughan, N.L. Scott, and D.R. White. The theory of bandpass sampling. *Signal Processing, IEEE Transactions on*, Sep 1991.
[9] J. Phillips and K. Kundert. Noise in mixers, oscillators, samplers, and logic an introduction to cyclostationary noise. In *Custom Integrated Circuits Conference, 2000. CICC.*, 2000.
[10] B. Razavi. *RF Microelectronics*, chapter 2.2 Effects of non-linearity. Prentice HALL, 2010.

Design of Class-D Amplifier for Audio Portable Solutions

Angelo Nagari

ST-Microelectronics, Grenoble, France

Audio Amplifier in portable systems (i.e. Mobile Phones) or in thermally close environment (i.e. car radio) faced the historical problem of low efficiency while using linear amplifier classical approach such as Class A/B/AB. In latest year the adoption of Class-D switching amplifier has been widely accepted for their higher efficiency (~90%) thus removing the problem of wasting power and thermal dissipation. Moreover, latest smart-phone business creates new multimedia experience requiring more and more loudness with small speaker. This leaded to an explosion of R&D activity to linearize the Class-D amplifier's behavior and improve mixed-signal solution with a clear benefit of sub-nm lithography. In this lecture the latest design development for portable system is described showing all design solution for improving audio quality, speaker's safety and loudness.

Time Interleaved Current Steering DAC for Ultra-High Conversion Rate

Da Feng, Sai-Weng Sin
State-Key Lab. of Analog and Mixed-Signal VLSI
University of Macau
Macao, CHINA
E-mails: fengda@gmail.com, terryssw@umac.mo

Edoardo Bonizzoni, Franco Maloberti
Dep. of Electrical, Computer and Biomedical Engineering
University of Pavia
Via Ferrata 1, Pavia, ITALY
E-mails: [edoard.bonizzoni, franco.maloberti]@unipv.it

Abstract—**A four-path time interleaved current steering DAC is presented. It requires the same number of unity current generators of the plain counterpart, thanks to the use of a digital $\Sigma\Delta$ modulator, thus leading to a lower number of unity current switchings. The benefit is that the non-linearity caused by clock feedthrough is attenuated. Behavioral level simulation results show that the SFDR of a 12-bit DAC operating at 12 GS/s can be 60 dB.**

I. Introduction

Digital-to-analog converters (DACs) with medium-high resolution and conversion rate in the several GS/s range are more and more necessary for applications in the communication field. The DAC must provide very high sampling rates because new standards employ ultra wide bandwidths, [1]–[4]. The extremely high cutoff frequency, f_T, of the CMOS transistors designed with deep sub-micron technologies enables the switching of currents at extremely high rates. Thus, the current steering DAC architecture is the most adequate solution for ultra-high speed applications.

The current-steering architecture has been widely used in circuits and experimentally demonstrated very good performances in the hundreds MS/s range, [5]–[7]. The current steering scheme gives at the output a current injected into a resistive load. At very high frequency, the current flows through a coaxial cable and the load must have the cable matching value, typically $50\,\Omega$.

The classical current steering architecture uses 2^N unity current generators switched toward the output(s) (for the single ended or the differential implementations) by differential switches made by MOS transistors. The matching between unity current generators and the switching strategy (i.e. the use of binary, unary, segmented selection, possibly using the random walk), [6]–[10], are well studied. All those issues are not discussed here because they are assumed known and properly applied to the design of current steering DACs.

What this paper considers is the limit that becomes significant when the sampling rate is in the multi-GS/s range: the clock feedthrough. When a MOS transistor switches on or off, a given amount of charge, corresponding to the charge in the channel, is injected at the two sides of the channel.

Moreover, there is an injection of charge caused by the capacitive coupling between gate and output node. If the injection is non linear, harmonic distortion results. The limit is not very important at relatively low speeds, but is relevant at very high speed because the signal charge given by unity current multiplied by the clock period becomes low.

This paper first analyzes the problem associated to the clock feedthrough and then proposes a time interleaved scheme capable to reduce the clock feedthrough non-linearity by the interleaving factor.

II. Clock Feedthrough

The circuit of Fig. 1 represents the unity cell of an N-bit current steering DAC switching its unity current, I_u, on the load resistances R_1 or R_2 ($R_1 = R_2 = R_L$), under the control of the driving signals V_1 and V_2. The current generators kI_u and $(2^N - 1 - k)I_u$ are for the other unity elements. They determine the output voltages for a given input code, k.

The switching of that single cell causes injection of charge into the output loads. The transistor turning on needs charge to create the channel; the one turning off disperses the channel. The charge in the channel is given by

$$Q_{ch} = (C_{ox}WL)(V_{GS} - V_{th}) \tag{1}$$

where C_{ox} is the oxide capacitance per unit gate area.

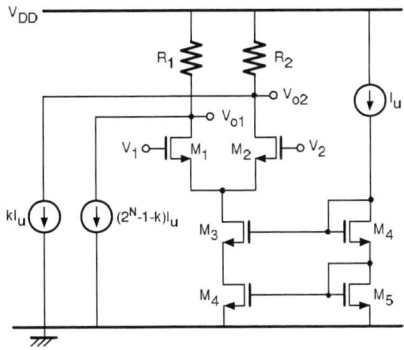

Fig. 1.   Unity current steering cell schematic diagram.

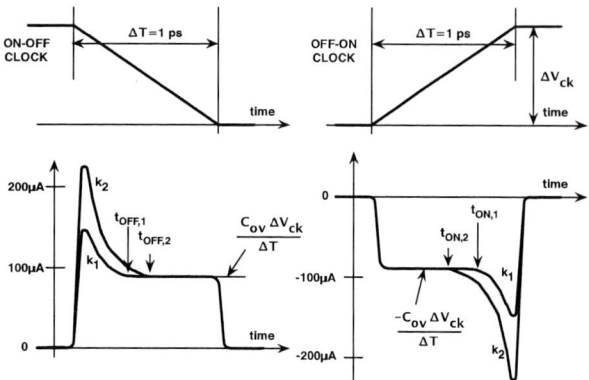

Fig. 2.   Current waveforms at clock transitions.

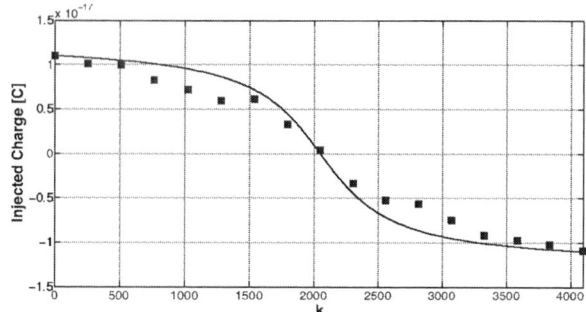

Fig. 4.   Total charge injected with differential RZ driving at clock transitions.

If the conversion speed is several GS/s, the clock period is hundreds of ps and the rise and fall time of the phases driving the switches must be extremely small. Since the switching is almost instantaneous, in first approximation, we can assume that there is no difference between the injections on the two switch sides: half of the channel charge goes to the source and half to the drain. Moreover, there is the C_{ov} coupling, being C_{ov} the overlap capacitance per unit width. Therefore, the clock feedthrough charge is approximated by

$$Q_c = \frac{1}{2}(C_{ox}WL)(V_{GS} - V_{th}) + Wx_{ov}C_{ov}\Delta V_{ck} \qquad (2)$$

where x_{ov} is the overlap extent.

Since deep sub-micron technologies use oxides with high dielectric constant, C_{ox} is relatively large. The estimation of Q_c for a minimum area transistor realized with a 65 nm technology ranges from 80 aC to 150 aC. With $1 V_{FS}$, 12-bit and $R_L = 50 \, \Omega$, the unity current is 2.929 μA that, for 3GS/s and 50% return to zero, leads to 498 aC, only 3-6 times larger than the one due to the clock feedthrough.

Computer simulations give rise to a quantitative estimation of the clock feedthrough. The circuit of Fig. 1 implemented with a 65-nm CMOS technology, minimum transistor sizes,

$R_L = 50 \, \Omega$, $N = 12$, and $I_u = 2.929$ μA is the test vehicle. The rising and falling edge of the driving signals $V_{1,2}$ is 1 ps.

The current injected into the output node depends on the overdrive voltage determined by the value of the number of unity cells which are switched on. Fig. 2 shows typical waveforms. The figure outlines the time at which the transistor goes off and on. The off-to-on and the on-to-off waveforms seem symmetrical but in reality there is a slight difference. Because of the very fast switching, the peak of the glitch of current is much larger than the switched current itself.

Fig. 3 shows the charge injected for the off-to-on and on-to-off transitions as a function of the input code k, where the dots and the lines (solid and dashed) denote behavioral level simulation results and polynomial fitting curves, respectively. The result is not linear because the charge on the channel depends on the overdrive voltage. The charge due to the capacitive coupling is almost output voltage independent. Moreover, the transition on-to-off differs from the inverse of the one of the complementary case. The two curves can be used to estimate the injected charge for non-return-to zero (NRZ), differential NRZ, return-to zero (RZ) and differential RZ. The NRZ cases give rise to very poor results. Even the single ended RZ is source of significant non-linearity. As expected, the differential RZ is the optimal solution.

With a differential RZ mode, the positive injection partially balances the negative injection; the resulting residual charge, $Q_{c,d,RZ}$ gives rise to the diagram of Fig. 4. There is a linear

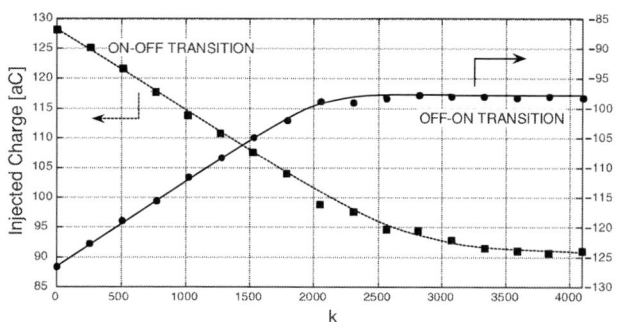

Fig. 3.   Charge injected as a function of the input code k.

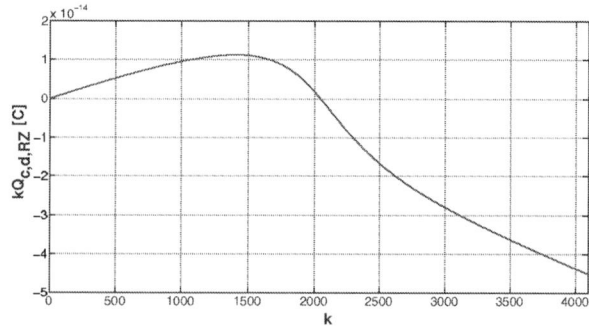

Fig. 5.   Distortion current as a function of the input code k.

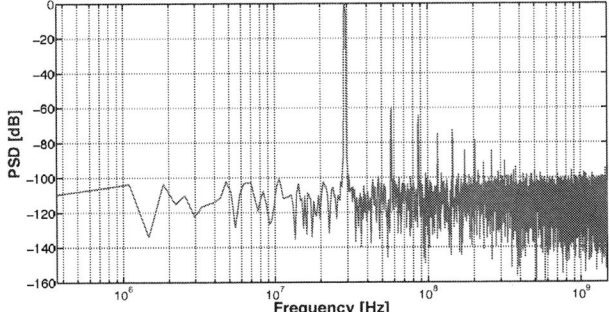

Fig. 6. Simulated output spectrum of a 12-bit 3-GS/s current steering DAC.

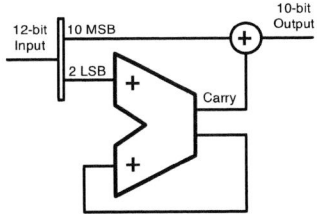

Fig. 8. Digital $\Sigma\Delta$ used before the time interleaved DAC to reduce the resolution from 12-bit to 10-bit.

component, whose effect is a gain error, and a non linear component that causes harmonic distortion. When k unity current sources switch on and off, the total injected charge is the one shown in Fig. 5. For sine wave symmetrical with respect to the mid point ($k = 2048$), there is a remarkable distortion when the amplitude is larger than 0.1 full scale peak to peak.

The distortion current, $I_d = Q_{c,d,RZ}/T$, added to the average unity current and multiplied by the number k of current sources performing the on-to-off transition gives the output voltage

$$V_{out}(k) = k\left[\bar{I}_u + \frac{Q_{c,d,RZ}(k)}{T}\right]R_L \qquad (3)$$

When the conversion rate is high, the distortion current becomes several $\bar{I}_u$ and the non-linearity gives rise to significant harmonic distortion.

The accuracy of the simulation results has been indirectly verified with the experimental results given in [2]. The paper describes a 12-bit 2.9 GS/s current steering DAC with 2.5 V_{FS}. It achieves 80 dB SNR. A behavioral simulation of a 12-bit 3-GS/s DAC which accounts for the $kQ_{c,d,RZ}$ behavior of Fig. 5 obtains the output spectrum shown in Fig. 6. Harmonic tones are well visible. The SFDR is dominated by the second harmonic tone. It is at -60 dB$_c$, matching the experimental result published in [2].

III. Time Interleaved Architecture

Possible methods for reducing the clock-feedthrough non-linearities are: use low switching rate, use a low number of switching elements, use a large full scale voltage to increase the unity current. The first method is not possible for ultra-high speed. The second is a limit to resolution. The last one is possible within limits.

The use of a time-interleave scheme increases the conversion speed. Fig. 7 shows a 4-path architecture with RZ. Every channel samples the digital input every 4 clock periods and the outputs have a 50% duty cycle for the required return-to-zero. The architecture gives rise to a $(1+z^{-1})$ transfer function. There is a zero at Nyquist ($f_N = f_{ck}/2$) and a non negligible attenuation as the frequency increases. However, the simple digital IIR filter ($1/(1 + 0.875z^{-1})$) compensates for the loss until more than half of f_N.

The time interleaved solution augments the conversion rate, but does not help in reducing the number of unity current generators that switch on and off. A possible solution is to use a single pair of switches for the parallel connection of a binary power of unity current sources. The disadvantage, however, is a mismatch in the on-resistance of the switches and this is source of non-linearity. The solution proposed here is to exploit the oversampling, always used in DACs. It grants a gray region between the band of interest and its replica used by the reconstruction filter.

Suppose that the used oversampling is 8. The use of a digital $\Sigma\Delta$ can reduce the number of bits at the input of each DAC by 2 so that the number of unity elements switched in each path is divided by 4 and the number of total switched elements remains unchanged. The digital $\Sigma\Delta$, as Fig. 8 shows, processes the 2-LSB of the digital word. It is a simple 2-bit accumulator whose carry-out is added to the 10 MSB of the input. The error caused by the truncation from 12 to 10 bit passes through a noise shaping function ($1 - z^{-1}$), enough to limit the loss of the SNR to a fraction of bit. The digital $\Sigma\Delta$ operates at the full speed of the converter. That is challenging but possible with deep sub micron technologies. Moreover, parallel processing reduces the computation speed.

IV. Simulation Results

The above described method has been validated at the behavioral level using the Matlab-Simulink^{TM} environment. The scheme of Fig. 7 with 4 paths made by 12-bit DACs with return-to-zero output gives rise to the output spectrum

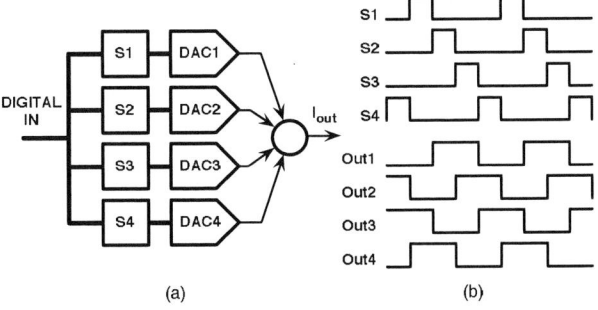

Fig. 7. a) Four paths time-interleaved RZ current steering DAC. b) S_i are the sampling periods, Out$_i$ the phases controlling the four outputs.

978-1-4799-4993-9/14 $31.00 © 2014 IEEE

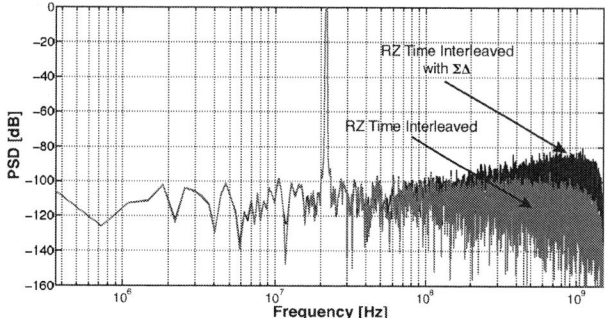

Fig. 9. Simulated output spectrum of a time interleaved RZ DAC with and without $\Sigma\Delta$.

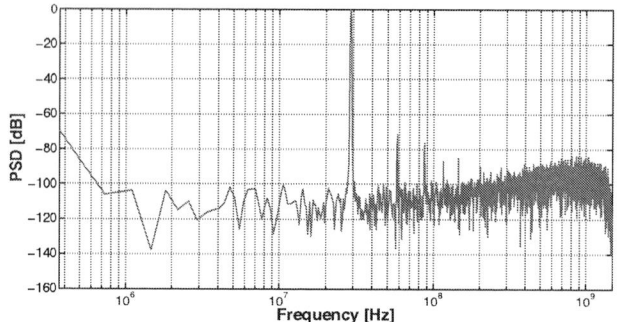

Fig. 11. Simulated output spectrum of the proposed DAC when considering clock feedthrough effect.

of Fig. 9. There is, as expected, a zero at Nyquist due to the $(1 + z^{-1})$ transfer function determined by the time-interleaved RZ. The quantization noise power from zero to $f_N/8$ leads to a SNR of equal to 83 dB corresponding to 13.5-bit. The use of the digital $\Sigma\Delta$ which reduces the required number of unity current sources from 2^{12} to 2^{10} gives rise to the second spectrum of Fig. 9. The SNR is 80.7 dB equivalent to 13.1-bit. The 0.4-bit cost is affordable when considering the benefit on the SFDR shown below.

The non-linear injection of charge caused by clock-feedthrough gives rise to the spur signal shown in Fig. 10. It comprises a component at the input frequency and high order harmonics, as shown by the spectrum of Fig. 11. The second harmonic tone is at -72 dB_c. Compared to the spectrum of Fig. 6, there is a 12 dB improvement in the achieved SFDR.

The design can trade the 12-dB benefit granted by the $\Sigma\Delta$ time interleaved architecture with other features. It can reduce the full scale voltage across the 50 Ω loads or can increase the conversion speed. For the latter option, the method permits to increase by a factor 4 the conversion speed. The considered case can extend its operation up to 12 GS/s, while achieving a SFDR of 60 dB.

V. Conclusion

The non-linearity of the clock feedthrough is the main limit to SFDR for high conversion rate current steering DACs. This

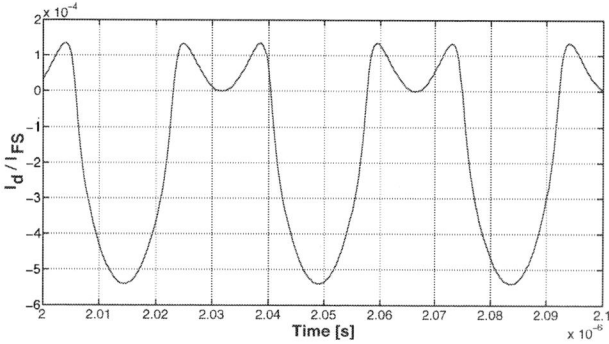

Fig. 10. Transient simulation of the distortion current.

study, after quantifying the limit, shows that a 65-nm CMOS time interleaved architecture with four 10-bit DACs driven by a digital $\Sigma\Delta$ modulator processing the 12-bit input obtains 60-dB SFDR at 12 GS/s.

Acknowledgment

This work has been partially supported by Research Grants from University of Macau and Macao Science and Technology Development Fund (FDCT).

References

[1] W.-H. Tseng, C.-W. Fan, and J.-T. Wu, "A 12-Bit 1.25-GS/s DAC in 90 nm CMOS With > 70 dB SFDR up to 500 MHz," *IEEE Journal of Solid-State Circuits*, vol. 46, no.12, pp. 2845-2856, Dec. 2011.

[2] C.-H. Lin, F.M.L. van der Goes, J.R. Westra, J. Mulder, Y. Lin, E. Arslan, E. Ayranci, X. Liu, and K. Bult, "A 12 bit 2.9 GS/s DAC With IM3 < -60 dBc Beyond 1 GHz in 65 nm CMOS", *IEEE Journal of Solid-State Circuits*, vol. 44, no.12, pp. 3285-3293, Dec. 2009.

[3] G. Engel, S. Kuo, and S. Rose ,"A 14b 3/6GHz Current-Steering RF DAC in 0.18μm CMOS with 66dB ACLR at 2.9GHz", *IEEE ISSCC Dig. Tech. Papers*, pp. 458-460, Feb. 2012.

[4] B. Schafferer and R. Adams, "A 3V CMOS 400mW 14b 1.4GS/s DAC for Multi-Carrier Applications", *IEEE ISSCC Dig. Tech. Papers*, pp. 458-460, Feb. 2004.

[5] C.-H. Lin and K.Bult, "A 10-b, 500-MSample/s CMOS DAC in 0.6 mm^2", *IEEE Journal of Solid-State Circuits*, vol. 33, no.12, pp. 1948-1958, Dec. 1998.

[6] J. Bastos, A.M. Marques, M.S.J. Steyaert, and W. Sansen, "A 12-Bit Intrinsic Accuracy High-Speed CMOS DAC", *IEEE Journal of Solid-State Circuits*, vol. 33, no.12, pp. 1959-1969, Dec. 1998.

[7] J. Deveugele and M. S. J. Steyaert, "A 10-bit 250-MS/s Binary-Weighted Current-Steering DAC", *IEEE Journal of Solid-State Circuits*, vol. 41, no. 2, pp. 320-329, Feb. 2006.

[8] A.Van Den Bosch, M.A.F. Borremans, M.S.J. Steyaert, and W. Sansen, "A 10-bit 1-GSample/s Nyquist Current-Steering CMOS D/A Converter", *IEEE Journal of Solid-State Circuits*, vol. 36, no. 3, pp. 315-324, Mar. 2001.

[9] P. Palmers and M.S.J. Steyaert, "A 10-Bit 1.6-GS/s 27-mW Current-Steering D/A Converter With 550-MHz 54-dB SFDR Bandwidth in 130-nm CMOS", *IEEE Journal of Solid-State Circuits*, vol. 57, no. 11, pp. 2870-2879, Nov. 2010.

[10] G.A.M. Van der Plas, J. Vandenbussche, W. Sansen, M.S.J. Steyaert, and G.G.E.Gielen, "A 14-bit Intrinsic Accuracy Q^2 Random Walk CMOS DAC", *IEEE Journal of Solid-State Circuits*, vol. 34, no. 12, pp. 1708-1718, Dec. 1999.

978-1-4799-4993-9/14 $31.00 © 2014 IEEE

Calibrated Switched Capacitor Integrators based on Current Conveyors and its application to Delta Sigma ADC

Harish Balasubramaniam
Integrated Electronic Systems Lab
TU Darmstadt
Merckstr.25, 64283 Darmstadt
harish.balasubramaniam@ies.tu-darmstadt.de

Klaus Hofmann
Integrated Electronic Systems Lab
TU Darmstadt
Merckstr.25, 64283 Darmstadt
klaus.hofmann@ies.tu-darmstadt.de

Abstract— **In this paper a fully differential current conveyor based switched capacitor (SC) integrator is presented. To operate the current conveyor (CCII) integrator with high linearity, calibration is used to correct for its non idealities. Simulation results are presented to show the effectiveness of the calibration technique. The presented integrator is applied to design of a 2nd order Delta Sigma ADC. A comparison of the calibrated and uncalibrated ADC outputs reveals that the calibration technique improves the dynamic range and SNDR significantly. The oversampled ADC gives more than 85 dB DR and 76 dB SNDR for bandwidth of 0.5 MHz and clock frequency of 80 MHz respectively.**

Keywords— *Sigma Delta ADC, Oversampling ADC, Calibration, Current Conveyor Integrator*

I. INTRODUCTION

Switched capacitor (SC) circuits are widely used for design of various kinds of ADCs such as Ramp, Dual Slope and Delta Sigma converters. The important block in these ADCs is the SC integrator which needs to be ideal and lossless. Any non ideality in the integrator leads to gain errors, non linearity errors which degrade the ADC performance. Traditional SC integrators use operational amplifiers (opamps) as the main building block. Depending on the technology constraints, designers have relied invariably on different opamp topologies such as cascade, folded cascode and cascode for designing ADCs. However in deep submicron technologies challenges in designing an opamp to operate at lower supply voltages with high performance, increases drastically. Hence alternative SC circuits consisting of inverters, open loop buffers, comparators were proposed by the circuit designers for low voltage regimes. Second generation current conveyors (CCII) are similar to opamps with the added advantage of providing also current mode of operation in addition to voltage mode. Many CCII topologies are operated in open loop mode thus negating the need for compensation techniques commonly found in opamps. Operating in open loop mode also brings the added advantage of increased operating bandwidth and the ability to provide gain independent bandwidth. CCIIs have thus found wide acceptance in design of filter networks [1][2][3].

Designers also actively explored the use of CCIIs for SC integrators [4][5]. These integrators were useful in older technologies where the capacitors used for sampling and integration were orders of magnitude larger than the non idealities present in the CCII. The effect of CCII non idealities

on the operation of integrator was thus less pronounced. In deep submicron technologies, however the focus is always on reducing area and using low power which necessitates the use of lower capacitance values for sampling and integration purposes. The use of low value capacitors coupled with the increase in the CCII non idealities make the SC integrators based on current conveyors unviable for ADC applications. In this paper an improved CCII integrator architecture which cancels the CCII non idealities is proposed. A calibration technique is introduced which helps improve the overall performance of the integrator making it a viable option for implementing ADCs. Section II describes the calibration technique and integrator operation. Section III presents the application example of a delta sigma modulator and finally section IV shows the simulation results to validate the concepts.

II. PROPOSED CALIBRATED CCII INTEGRATOR

Fig.1 shows the proposed CCII integrator architecture where the non ideal CCII characteristics are given in Equation (1) [6]. Compared to the solution given in [5] which analyzed an integrator with ideal CCII properties, this architecture considers the effect of non idealities present in the CCII. The proposed architecture offers many improvements compared to the one in [5]. The voltage on the integration capacitor *Cint* is isolated by a separate unity gain buffer. The use of separate unity gain buffers to drive higher output loads allows use of CCII with low current requirements. Low current in the CCII requires only smaller transistor sizes which in turn also reduces the parasitic at the three ports. The X and Y ports of the two CCIIs are cross coupled to generate an effect similar to virtual ground effect of the opamps. The cross coupling make the inputs settle to their stable DC points allowing the charge transfer between *Cin* and *Cint* to take place quickly.

The integration capacitor *Cint* is made variable and an additional variable capacitor *Ccomp* is introduced both of which are controlled by a calibration algorithm to compensate for the CCII non ideality as explained below. Equation (2) shows the transfer function of an ideal delaying CCII integrator, under assumption of unity gains for the voltage and current gain of the CCII and negligible parasitics at the CCII ports. This derivation is largely invalid in new CMOS technologies since the presence of parasitics at the X, Y and Z ports and non ideal voltage and current gains significantly alter the transfer function of the integrator.

978-1-4799-4993-9/14 $31.00 © 2014 IEEE

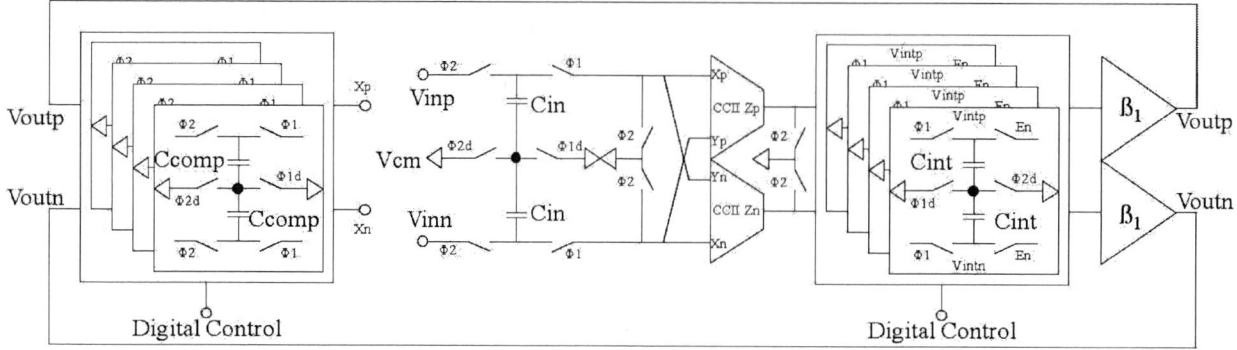

Fig.1 Integrator Architecture

The parasitics at Z port play a more important role in comparison to the parasitic at the X and Y port. This is because the parasitics at X and Y port are usually negligible or small in advanced technologies when compared to the values of the Cin and $Cint$ capacitors. Hence in the following analysis we consider the effect of Z port parasitics and the frequency dependent gains of CCII. The integrator operates in two clock phases along with delayed version of the clocks to prevent signal dependent charge injection effects. $\Phi 1$ is the integration phase or the charge transfer phase. $\Phi 2$ is the sample and hold phase. Each phase of the integrator introduces errors. During the integration or the charge transfer phase the voltage sampled on the Cin capacitor is transferred to the $Cint$ capacitor. This charge transfer phase introduces gain errors because of two reasons. 1) The parasitic at Z port takes away a part of the charge from $Cint$ capacitor and 2) The presence of frequency dependent voltage and current gains $\beta(s)$ and $\alpha(s)$. This error can be removed by adjusting the value of the $Cint$ capacitor until the required integration coefficient is reached.

$$\begin{pmatrix} Vx \\ Iz \\ Iy \end{pmatrix} = \begin{pmatrix} \beta(s) & Zx = \{Rx + sLx\} & 0 \\ 0 & \alpha(s) & 1/(Zz = \{Rz \| Cz\}) \\ 1/(Zy = \{Ry \| Cy\}) & 0 & 0 \end{pmatrix} \begin{pmatrix} Vy \\ Ix \\ Vz \end{pmatrix} \quad (1)$$

$$V_{out}(z) = \frac{Cin}{Cint}\left(\frac{z^{-1}*V_{in}(z)}{1 - z^{-1}}\right) \quad (2)$$

$$V_{out}(z) = \frac{\alpha(s)*\beta_o(s)*Cin}{(Cint \| Zz)}\left(\frac{z^{-1}*V_{in}(z)}{1 - z^{-1}\left(1 - K + \left(\frac{\alpha(s)*\beta_o(s)*Ccomp}{Cint \| Zz}\right)\right)}\right) \quad (3)$$

During the sample and hold phase the input is sampled onto the Cin capacitor and the voltage on the $Cint$ capacitor is transferred to the output using the voltage buffer. To prevent the $Cint$ capacitor from losing charge through the parasitic Rz resistor at the Z port it is disconnected during the hold phase and the Z port is autozeroed. This autozeroing however also removes the charge from the parasitics Cz capacitance at the Z port resulting in charge loss errors. The addition of a compensation capacitor $Ccomp$ cancels out this charge loss by sampling the lost charge in the form of voltage and then feeding it back to the $Cint$ capacitor during subsequent charge transfer phases. Additionally the unity voltage buffer also introduces frequency dependent gain errors defined by the gain factor $\beta_1(s)$. Equation (3) shows the resulting transfer function of the integrator where $\beta_o(s) = \beta(s)*\beta_1(s)$.

The parameter K represents the charge loss resulting from autozeroing of Z port and the value of $Ccomp$ is chosen to cancel out this loss factor. The calibration routine sets the values of $Cint$ and $Ccomp$ so that an ideal integrator results as shown by the transfer function in (2). The calibration logic works in two steps by finding the correct value for the $Cint$ capacitor first followed by the $Ccomp$ capacitor. In first step, all the capacitors in the integrators are zeroed to remove any charge, Ccomp is made 0 and the integrator is given a known input signal $Vin=Vtest$ for one clock cycle. The output voltage of the integrator is compared with a reference voltage $Vrefint$ which is the ideal output voltage the integrator should give. In the presence of CCII non idealities as described above the voltage $Vout$ does not reach the desired $Vrefint$ voltage. The calibration routine adjusts the values of the $Cint$ capacitor by repeating the first step until the desired output voltage $Vout=Vrefint$ is reached. The ideal integrator coefficient which is defined as $Vrefint/Vtest$ is thus accurately achieved using the first step.

The calibration logic then finds the value of the $Ccomp$ capacitor which can cancel out the loss factor K in the transfer function. It does this by zeroing all capacitors initially, setting $Cint$ to the value from first step and and applying $Vtest$ voltage for two clock cycles. The output voltage is then compared to the $2*Vrefint$ voltage. The output voltage does not reach $2*Vrefint$ in the absence of $Ccomp$ capacitor at the end of second clock cycle. The algorithm adjusts the value of the $Ccomp$ capacitor until the required output voltage is reached. At this point the $Ccomp$ value cancels out the charge loss factor K resulting in an ideal integrator which can then be used to design various kinds of ADCs. The variable Cint and Ccomp capacitors have a programmability of 8 bits each to adjust its value across a wide range.

III. INTEGRATOR & MODULATOR DESIGN

In order to validate this architecture an integrator was designed consisting of the CCII and unity buffers as shown in Fig.2. This CCII configuration is made of translinear loops and uses cascoding at Z port to improve the Z impedance. The main properties of the CCII are given Table I. The class AB type buffer gives an approximate nominal gain of 0.904 at low frequencies and has a 3dB bandwidth of 320MHz. The buffer draws a static power of about 400μW. The proposed integrator was used in the design of a 2nd order modulator as shown in Fig.3.

978-1-4799-4993-9/14 $31.00 © 2014 IEEE

The modulator uses low distortion topology based on unity STF stages to improve its performance [7]. The integrator coefficients are chosen $\{c1,c2,a1,a2\} = \{0.25,1,8,4\}$ such that they reduce the output voltage swing requirements of the integrator stages. A high resolution low offset comparator based on four preamplifiers and dynamic latch is used during the calibration process to correctly adjust the capacitor values (see Fig.4). The preamplifier is based on differential NMOS stages with output offset cancellation technique to achieve low offsets. Each preamplifier uses internal resistive common mode feedback to keep track of the output voltage of preamps. A fully differential dynamic clocked comparator consisting of cross coupled inverters and differential NMOS pair with switched current sources forms the final stage to generate the full scale comparator outputs. Fig.5 shows the circuits used in the calibration comparator. The comparator outputs are captured by a RS latch made of digital AND gates. The calibration comparator consumes about 800µA of current and is used only during the calibration process.

IV. SIMULATION RESULTS

Fig.6 shows the output waveforms for the integrator with and without the calibration. The integrator is clocked at 160MHz frequency and is given a square wave input of 2mV amplitude at 5MHz frequency. For an integrator coefficient of 1, it can be seen that the waveforms for uncalibrated case exhibit loss and gain errors whereas the calibrated integrator generates an ideal triangular output with frequency of 5MHz. By applying the same input square wave but with 1MHz frequency, the integration process is allowed to continue for longer time period as shown in Fig.7. The uncalibrated integrator performs inaccurately due to the accumulation of the charge loss errors and gain errors. It starts saturating due to loss accumulation and generates a square wave instead of a triangular wave.

Although the integrator is capable of reaching frequencies up to 160MHz, the delta sigma modulator topology was simulated using the proposed integrator at lower frequency of 80MHz. Fig.8 and Fig.9 show the SNDR plot of the modulator with and without calibration for an input signal of -36dB at 253.9 KHz. The plot show that the gain and leakage are

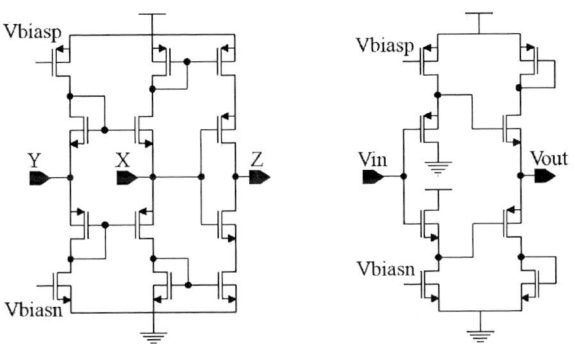

Fig.2 (a) CCII Topology   (b) Class AB Buffer

TABLE I   CCII Properties

Tech.(nm) / Supply(V)	90 / 1
Power(µW)	125
Voltage Gain / Current Gain	0.93 / 0.96
3dB Voltage BW / 3dB Current BW (GHz)	1.91 / 0.86
Z port Parasitic $R_Z(k\Omega)$ / $C_Z(fF)$	360 / 26.4
X port Parasitic $R_X(\Omega)$	504

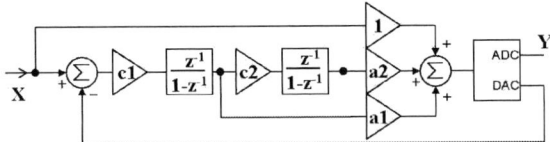

Fig.3 Modulator Architecture

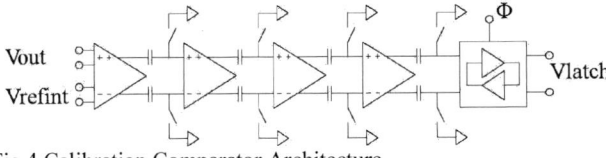

Fig.4 Calibration Comparator Architecture

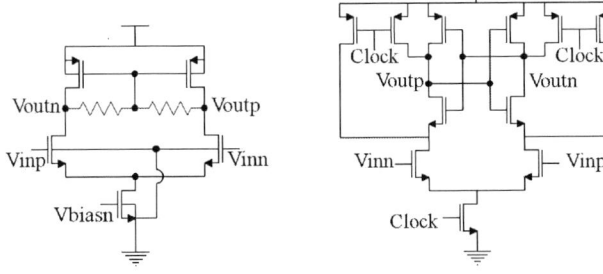

Fig.5 (a) Preamplifier   (b) Dynamic Comparator

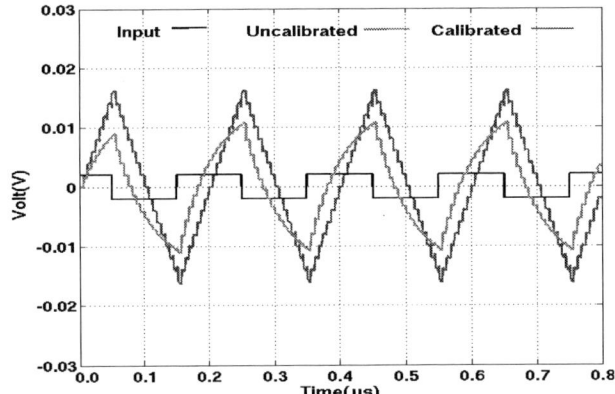

Fig.6 Integrator Output Plot1

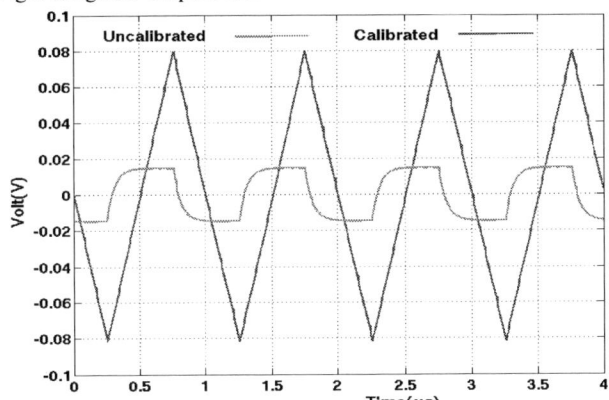

Fig.7 Integrator Output Plot2

978-1-4799-4993-9/14 $31.00 © 2014 IEEE   55

completely minimized resulting in an integrator useful for ADC applications. Fig.10 shows the dynamic range of the modulator plotted for various input signal amplitudes. For signal amplitudes above -12dBFS the SNDR decreases. A reason might be the peak signal currents flowing through the CCII approach full scale resulting in it adding more nonlinearity. The total power consumption of the modulator is under 7mW. The power estimation excludes the calibration logic (calibration comparator and the algorithm) since they are utilized only during the calibration process. Using a successive approximation type algorithm the total number of clock cycles required for calibration of one integrator is approximately 160-200 clock cycles. Hence for calibration of two integrators of the modulator, the total clock cycles required would be approximately 320-400 clock cycles. Table II reveals the complete performance of the modulator. All reference voltages used by the modulator including the calibration comparator are assumed to be generated externally. The proposed solution provides comparable performance to the modulators having similar specifications but with the added advantage of not requiring high accuracy or complex analog blocks.

TABLE II   Modulator Performance

	This Work	[8]	[9]
Tech.(nm) / Supply(V)	90 / 1	90/1.2	90/1.2
Sampling Rate (MS/s)	80	80	90
fb Bandwidth(MHz)	0.5	0.5	0.5
OSR	80	80	90
Power (mW)	6.98	5.35	3.7
SNDR (dB)	76.5	68	76
DR (dB)	86	70	77
$FOM = \dfrac{Power * 10^{12}}{2^{ENOB} * 2 * f_b}$	0.426	2.1	0.64

V. CONCLUSION

An improved current conveyor based integrator architecture has been presented and its performance verified through simulations. The presented integrator was used in the design of delta sigma modulators whose performance improvement over the uncalibrated version was verified. The major non idealities in the integrator are corrected by the calibration allowing a relaxed design requirement for the analog blocks in the integrator.

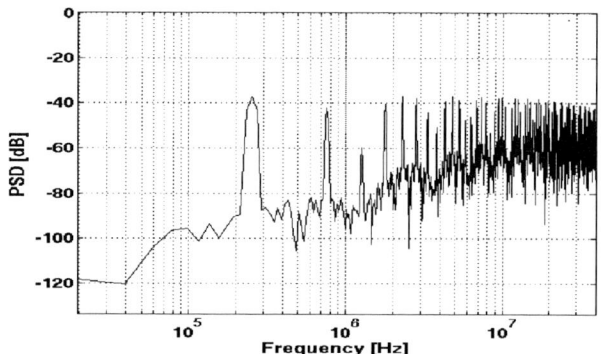

Fig.8 Uncalibrated FFT Plot

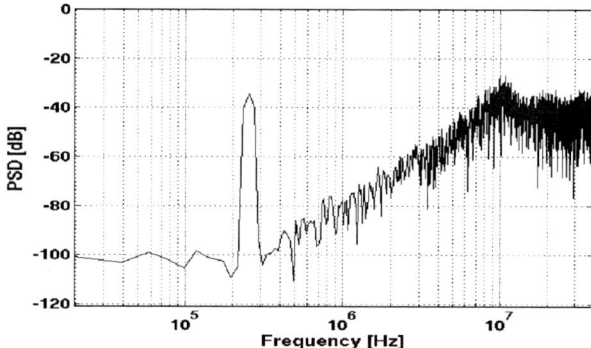

Fig.9 Calibrated FFT Plot

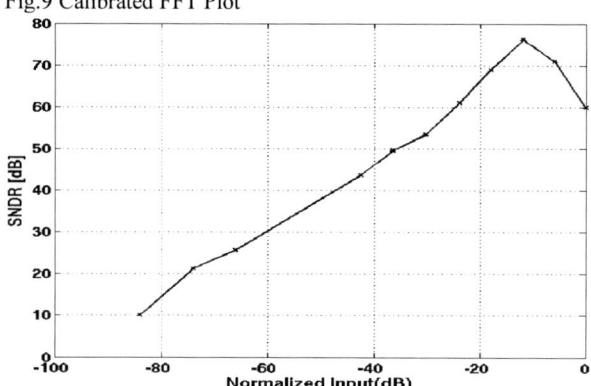

Fig.10 Dynamic Range

REFERENCES

[1] Popovic, J. Pavasovic, A. Vasiljevic, D., "CMOS voltage-controlled oscillator based on current conveyor," 21st International Conference on Microelectronics, 1997. Proceedings., 1997, vol.2, pp.755,758, 14-17 Sep 1997

[2] Chao-Chyun Chen, Sheng-Chou Lee and Shen-Iuan Liu, "A Capacitor Multiplication Technique Using a Second-generation Current Conveyor in the Loop Filter of the Phase-locked Loops", International Journal of Electrical Engineering, vol. 14, pp.239-245, June 2007.

[3] Gaudet, V.C. Gulak, P.G., "CMOS implementation of a current conveyor-based field-programmable analog array," Conference Record of the Thirty-First Asilomar Conference on Signals, Systems & Computers, 1997, vol.2, pp.1156,1159 vol.2, 2-5 Nov. 1997

[4] Tutyshkin, A.A. Korotkov, A.S., "Current conveyor based switched-capacitor integrator with reduced parasitic sensitivity," 1st IEEE International Conference on Circuits and Systems for Communications, 2002, pp.78,81, 2002

[5] Kaabi, H. Motlagh, M.-R.J. Ayatollahi, A., "A novel current-conveyor-based switched-capacitor integrator,". IEEE International Symposium on Circuits and Systems, 2005, pp.1406,1408 Vol. 2, 23-26 May 2005

[6] F. Khateb, N. Khatib, D. Kubanek, " Low voltage Ultra Low Power Current Conveyor Based on Quasi-Floating Gate Transistors", Radioengineering, CZ, vol.21: no.2. pp. 725-735.

[7] J. Silva, U. Moon, J. Steensgaard and G.C. Temes: "Wideband low-distortion delta sigma ADC topology", Electronic Letters, Vol. 37, pp. 737-738, June 2001.

[8] Morgado et al., "A 100kHz–10MHz BW, 78-to-52dB DR, 4.6-to-11mW flexible SC ΣΔ modulator in 1.2-V 90-nm CMOS," Proceedings of the ESSCIRC, 2010, pp.418,421, 14-16 Sept. 2010

[9] L. Bos et al., "Multirate cascaded discrete-time low-pass ΔΣ modulator for GSM/bluetooth/UMTS,"IEEE J. Solid-State Circuits, vol. 45, pp. 1198–1208, Jun. 2010

978-1-4799-4993-9/14 $31.00 © 2014 IEEE

Statistical Analysis of Harmonic Distortion in a Differential Bootstrapped Sample and Hold Circuit

Gaël Kamdem De Teyou[*][†], Hervé Petit [*] Patrick Loumeau [*] and Hussein Fakhoury [*]

[*]Télécom ParisTech, France, 46 rue Barrault, 75013, Paris, France

[†]Renesas Mobile, 5 rue de la Chataigneraie, 35510 Cesson-Sévigné, France

Abstract—The bootstrap technique is known to increase the linearity of Sample and Hold (S/H) circuit by reducing the input signal dependency of the transistor-switch resistance. But some nonlinearities remain due to parasitic capacitances, mobility degradation and back gate effect resulting in a second order harmonic spurious which can be reduced with a differential architecture. However mismatch between channels limits this technique. In this paper, we provide a general framework to analyze the residual nonlinearity in bootstrapped S/H. Statistical laws are also provided converting harmonic distortion specifications into matching requirements for differential sampling and therefore provide key rules for S/H designers.

Index terms : Bootstrap, Differential architecture, Sample and Hold, ADCs, Linearity, HD, Mismatch

INTRODUCTION

Sample and Hold (S/H) circuits are essential elements in data acquisition systems such as Analog-to-Digital Converters (ADC). They are difficult to design because they must achieve simultaneously high linearity, high speed, large voltage swings and low power consumption. The linearity is all the more desired in applications such as wireless communications where an interfering signal can generate harmonics that may completely block the in-channel signal.

Unfortunately, S/H circuits are nonlinear because their transistor-switch resistance depends of the input signal [1]. To remove this nonlinearity, there are some analogue techniques such as the use of a PMOS switch in parallel with an NMOS switch. But the most popular is the clock bootstrapping [2] which removes a significant portion of nonlinearities by making the value of the transistor-switch gate-source voltage as independent as possible of the input signal. But a residual dependency remains due to parasitic capacitances[6], mobility degradation and back gate effect. It results in a second-order harmonic.

A lot of work has be done on the analog design side [2] [3] [4], but little work has been done in creating a general model of S/H. Based on Volterra series, [5] proposes a deterministic model for S/H and a digital calibration

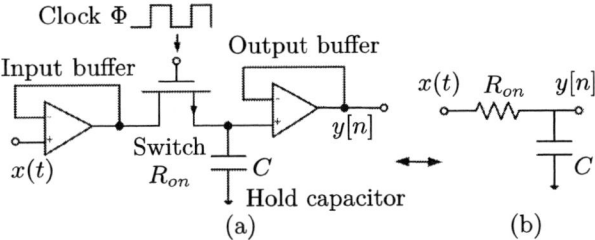

Figure 1.   S/H diagram (a) and its equivalent first order model (b).

scheme but this model is complex and holds only for single ended architecture. The novelty of this paper consists in:

- A deterministic model in time and in frequency domain of bootstrapped S/H both for single ended and differential architectures.
- A statistical analysis of bootstrapped S/H for differential signal in function of mismatches
- Determining the probability for the Harmonic Distortion (HD) to be lower than a critical value for any mismatch dispersion.

Therefore, for a level of performance determined by a minimum HD and its probability of achievement we can specify the required mismatch dispersion. This practical information becomes of relevant importance to establish robust design with safe margins.

I. ANALYSIS OF BOOTSTRAP CIRCUIT

Fig. 1(a) shows a basic S/H circuit. In the sampling mode, the transistor is ON and the circuit is equivalent to a RC low-pass filter, with R the ON resistance of the transistor-switch given by:

$$R = \frac{1}{\mu C_{ox} \frac{W}{L} \left(v_{gs} - v_{th} - \frac{v_{ds}}{2} \right)} \qquad (1)$$

Since the gate-source voltage v_{gs} of the transistor-switch can be written as $v_{gs} = v_{dd} - x$, R depends of the value of the input signal $x(t)$. This will cause significant distortion in the sampled voltage on the hold capacitor. Therefore it is important to modify the circuit and make R independent as possible of $x(t)$. The bootstrap circuit [2] achieves it and its logical structure is shown in Fig. 2.

978-1-4799-4993-9/14 $31.00 © 2014 IEEE

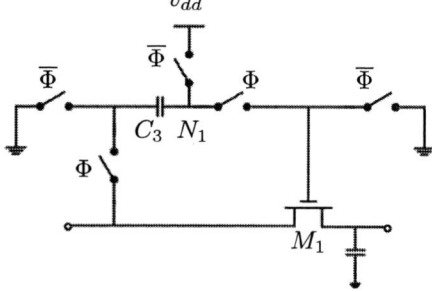

Figure 2. Logical structure of bootstrap circuit.

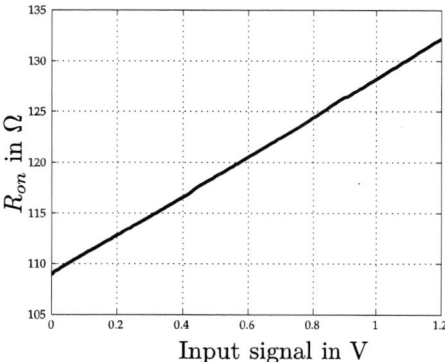

Figure 3. Simulations result of the ON resistance of a bootstrapped S/H as a function of the input signal in 65 nm CMOS process with supply voltage of $v_{dd} = 1.2$ V.

When the sampling switch M_1 is OFF, $\Phi = 0$ and C_3 is precharged to v_{dd}. When $\Phi = 1$, a constant voltage equal to v_{dd} is established between the gate and the source of M_1. Ideally, R_{on} is now independent of the input signal. But in practice, parasitic capacitance at node N_1, mobility degradation and back gate effect limit the linearity that can be achieved. At a first order we can consider that R varies linearly with the input signal:

$$R(x) = b_{on} + a_{on}x \tag{2}$$

Simulations results show that (2) is realistic as we can see on Fig. 3 where a bootstrapped switch is simulated in 65 nm CMOS technology.

Therefore throughout this paper, without loss of generality we will consider that the time constant τ varies linearly with the input signal as in (3). Furthermore montecarlo simulation of a bootstrapped circuit in 65 nm CMOS process shows that b can be model as a random variable normally distributed and that a and b are correlated. A correlation coefficient $\rho(a, b) = 0.9$ was obtained between a and b. So we consider that a can be expressed linearly in function of b as in (3). For this simulation we obtain $\alpha = -39.5$ ns V^{-1} and $\beta = 0.5$ V^{-1}:

$$\tau(x) = RC = b + ax$$
$$a = \alpha + \beta b \tag{3}$$

II. SINGLE ENDED ARCHITECTURE

During the sampling phase, the S/H of Fig. 1(a) behaves like a first order lowpass filter as in Fig. 1(b) and can be described by a first order differential equation with non constant coefficients :

$$\dot{y} = -\frac{1}{\tau(x)}\Big[y - x\Big] \tag{4}$$

To resolve (4), we make the following assumptions:

- The sampling duration T_s/n is several times bigger than the time constant τ. As a consequence $\exp(-Ts/n\tau) \ll 1$. For example with ADCs of concern, $C \sim 1$ pF, $R_{on} \sim 120\,\Omega$, $f_s \sim 300$ MHz and $n = 2$. It comes that $\exp(-Ts/n\tau) = 9.3\,e{-}7$
- The S/H is weakly nonlinear i.e $ax \ll b$ in (3)
- To avoid memory effect, the charge of the hold capacitor is set to zero after each sample

By resolving (4) by the variation of constants method, the nth sample at the S/H output can be splitted into a desired component and an undesired component :

$$y[n] = \overbrace{(h \star x)(nT_s)}^{desired\ component}$$
$$- \frac{a}{b}\underbrace{\Big\{(h \star x^2)(nT_s) - \Big(h \star \big[x.(h \star x)\big]\Big)(nT_s)\Big\}}_{undesired\ component} \tag{5}$$

With h the impulse response of the linear S/H which is obtained by taking the static component b of the time constant $\tau(x)$:

$$h(t) = \frac{1}{b}\exp(-\frac{t}{b})u(t) \tag{6}$$

$u(t)$ in (6) is the Heaviside step function. The desired part is what should be obtained if the ON resistance were totally constant while the undesired component comes from the variations of the ON resistance with the input signal $x(t)$. As $a \to 0$, the resistance becomes constant and the undesired component decreases. (5) is far more simple than what has been obtained in [5].

Taking the particular case of a sinusoidal input $x(t) = A\sin(2\pi f_0 t)$, the Discrete Time Fourier Transform (DTFT) $Y(f)$ of $y[n]$ is:

$$Y(f) \simeq \overbrace{\frac{aA^2}{4b}\Big[H(f_0) + H(-f_0) - 2\Big]}^{Offset} + \overbrace{\frac{1}{2}AH(f_0)\delta(f - f_0)}^{desired\ part}$$
$$+ \underbrace{\frac{aA^2}{4b}H(2f_0)K(f_0)\delta(f - 2f_0)}_{second\ harmonic} \tag{7}$$

$H(f)$ in (7) is the transfer function of the S/H and $K(f) = H(f) - 1$.

We notice that firstly the S/H exhibits an offset which can easily be removed digitally. Secondly the nonlinearity is mostly characterized by the presence of a dominant

Frequency	According to (7)	By simulation
HD_2 in dB	61.25	61.20

Table I

SIMULATION RESULTS OF A S/H WITH $\tau(x) = (0\,125e{-}9 + 0\,023e{-}9x$. THE INPUT SIGNAL IS $x(t) = 0.6\sin(2\pi f_o t)$, $f_o = 20\,\text{MHz}$, $f_s = 300\,\text{MHz}$.

Frequency	$\frac{\Delta b}{b} = 0.062$	$\frac{\Delta b}{b} = -0.041$
Simulated HD_2	77.09	80.23
HD_2 with (10)	76.96	80.52

Table II

SIMULATION RESULTS OF A DIFFERENTIAL BOOTSTRAPPED S/H WITH $\overline{\beta} = 0.125\,\text{ns}$. THE INPUT SIGNAL IS $x(t) = 0.6\sin(2\pi f_o t)$, $f_o = 20\,\text{MHz}$ AND $f_s = 300\,\text{MHz}$.

harmonic of second order which will degrade the dynamic performances such as SFDR and SNDR.

Fig. 4 presents the simulation of the output spectrum of a S/H with a time constant varying linearly with the input signal and Tab. I compares the simulation results with (7). Considering a sinusoidal input, the worst spurious is effectively at $2f_0$. The Second Harmonic Distortion (HD_2), defined as the ratio of the power of fundamental signal to the power of the second harmonic matches very well with (7).

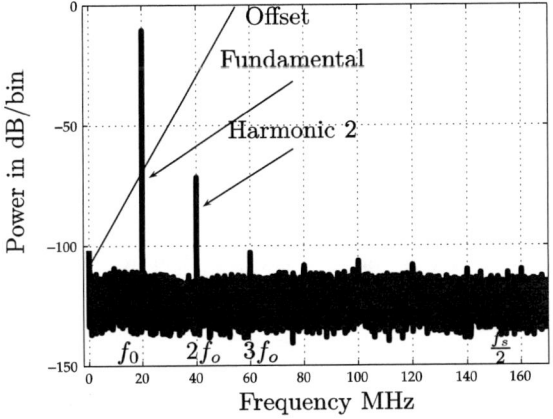

Figure 4. Output spectrum of a S/H with $\tau(x) = (0\,125e{-}9 + 0\,023e{-}9x$. The input signal is $x(t) = 0.6\sin(2\pi f_o t)$, $f_o = 20\,\text{MHz}$, $f_s = 300\,\text{MHz}$ and the number of fft points is 16384

III. DIFFERENTIAL ARCHITECTURE

In the previous section, we saw that the worst spurious is the one which occurs at frequency $2f_0$. Using the differential architecture, this spurious can be mitigated. To do it, for an input signal $x(t)$, we send the signal $x_1(t) = \frac{1}{2}x(t)$ on the first channel and on the second channel we send the signal $x_2(t) = -\frac{1}{2}x(t)$.

From (7), the second harmonic of the $i^{th}_{1,2}$ channel is obtained by dividing the amplitude by two :

$$H_{2i} = \frac{a_i A^2}{16 b_i} H_i(2f_0) K_i(f_0) \qquad (8)$$

Where $H_i(f) = \frac{1}{1+j2\pi f b_i}$ is the transfer function of the ith channel and $K_i(f) = H_i(f) - 1$. The time constant of channel $i_{1,2}$ is $\tau_i = b_i + a_i x_i(t)$. The second harmonic of the whole S/H is obtained by making the difference of the second harmonic of channels 1 and 2:

$$H_2 = H_{21} - H_{22}$$
$$\simeq \frac{\lambda A^2}{16} \overline{H}(2f_0)\overline{K}(f_0)\left[\frac{\theta}{\lambda} + \overline{K}(2f_0) + \overline{H}(f_0)\right]\frac{\Delta b}{\overline{b}} \qquad (9)$$

$\overline{b}$ is the average value of b, $\overline{H} = \frac{1}{1+j2\pi f \overline{b}}$, the average transfer function, $\overline{K} = \overline{H} - 1$, $\lambda = \frac{\alpha}{\overline{b}} + \beta$ and $\theta = \frac{\alpha}{\overline{b}}$. α and β are defined in (3)

The HD_2 can be derived from (9) and (7) as :

$$HD_2 = 10\log_{10}\left[|c|^2 / \left(\frac{\Delta b}{\overline{b}}\right)^2\right] \qquad (10)$$

With c the complex number given by:

$$c = \frac{8\overline{H}(f_0)}{\lambda A \overline{H}(2f_0)\overline{K}(f_0)\left[\frac{\theta}{\lambda} + \overline{K}(2f_0) + \overline{H}(f_0)\right]} \qquad (11)$$

Table. II compares some simulation results with (10). As we can see this analytical model match very well and the HD_2 is considerably better than that was obtained in section II.

IV. PROBABILITY DENSITY FUNCTION OF HD_2

Device mismatches are inherent to any manufacturing processes whatever the technology is (CMOS, SiGe, ...) and are commonly described by random variables normally distributed. Since these mismatches are responsible for channel mismatch errors, it is convenient to rely on a probabilistic characterization of the time constant mismatches. As a consequence, the harmonic distortion is modeled by a statistical distribution related to the standard deviation σ of time constant mismatch as in [8].

Readily, the relative mismatch on b can be modeled as a gaussian variable: $\frac{\Delta b}{\overline{b}} = \sigma\epsilon$, with $\epsilon \sim \mathcal{N}(0,1)$. From (10) the HD_2 can be rewritten as:

$$HD_2 = 10\log_{10}\left(\frac{|c|^2}{\sigma^2\epsilon^2}\right) = F(\epsilon^2) \qquad (12)$$

As the HD_2 is function of the random variable ϵ^2, it is also a random variable whose dispersion depends on standard deviation of mismatches σ. The random variable ϵ^2 follows a chi-squared distribution with one degree of freedom χ_2^1 and the function F is strictly monotone. Therefore the Probability Density Function (PDF) p of

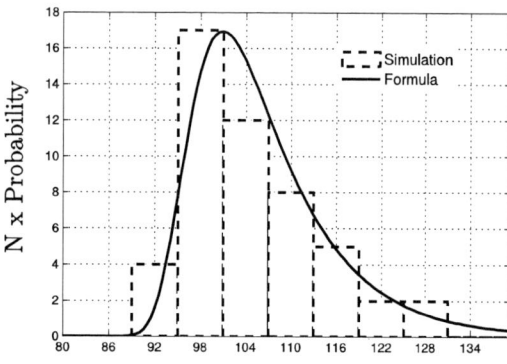

HD_2 in dB

Figure 5. Statistical distribution of second harmonic distortion of a differential bootstrapped S/H in 65 nm CMOS process with an input signal $x(t) = 0.6 \sin(2\pi f_o t)$, $f_o = 20$ MHz, $f_s = 300$ MHz, a relative mismatch of 1.2 % and N = 50 points.

the HD_2 can be obtained from that of ϵ^2 with a change of variables. All calculations done, we find that the PDF of the HD_2 is:

$$
\begin{aligned}
p(HD_2) &= \frac{1}{F'\left[F^{-1}(HD_2)\right]} \chi_2^1\left(F^{-1}(HD_2)\right) \\
&= \sqrt{\frac{C}{2\pi}} \frac{\log(10)}{10\sigma} 10^{-\frac{HD_2}{20}} \exp^{-\left[\frac{C}{2\sigma^2} 10^{-\frac{HD_2}{10}}\right]}
\end{aligned}
\tag{13}
$$

Fig. 5 shows the statistical distribution of the HD_2 obtained by Monte-Carlo simulation of a differential bootstrapped S/H in 65 nm. This distribution matches very well with the analytical model of (13).

V. Reliability of HD_2

Using an analogy to the usual yield, we introduce the robustness criterion $HD_2(\eta)$, that states the HD_2 remains higher than this threshold value with a probability $1 - \eta$.

$$
\begin{aligned}
\eta &= p\left(HD < HD(\eta)\right) \\
&= 1 - p\left(\epsilon^2 < F^{-1}(HD(\eta))\right) \\
&= 1 - \mathrm{erf}\left(\sqrt{\frac{\left[F^{-1}(HD(\eta))\right]^2}{2}}\right)
\end{aligned}
\tag{14}
$$

erf is the Gauss error function.
Inverting (14) with respect to $HD(\eta)$ gives :

$$
HD_2(\eta) = -10\log_{10}\frac{\sigma^2}{|c|^2} - 10\log_{10} K(\eta)
\tag{15}
$$

With $K(\eta) \simeq \frac{\pi}{2}(1 - \eta)^2$. (15) states the HD_2 remains higher than the threshold value $HD_2(\eta)$ with the probability $1 - \eta$ allowing to control the reliability of any mismatch calibration process like in [8].
Fig. 6 shows the harmonic distortion law as a function of the standard deviation of mismatches for a frequency of 20 MHz. To obtain a HD_2 of 100 dB with a yield of 0.999, the mismatch should be less or equal than 1 %.

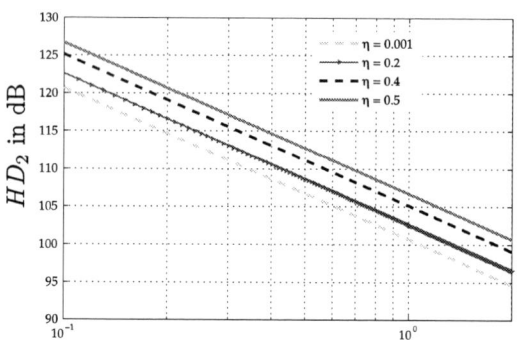

Standard deviation σ of mismatch in %

Figure 6. Harmonic distortion law.

VI. Conclusion

In this paper we have derived a deterministic model describing boostrapped S/H circuits both for single ended and differential architecture. For single ended architecture, the second harmonic is dominant and it can be mitigated with a differential architecture. A statistical law linking the Second Harmonic Distortion ratio (HD_2) to the standard deviation of mismatches was also provided for a differential architecture. Numerical results show that mismatches must be reduced significantly under 1% to obtain substantial HD_2 in the order of 100 dB.

A perspective of this work is to consider higher order harmonics and so to derive others dynamic performances such as SFDR, THD and SNDR.

Acknowledgment

The authors wish to thank specially Yann Le Guillou and Stéphane Paquelet to have initiated this work.

References

[1] J. M. Rabaey et al, "Embedding mixed-signal design in systems-on-a-chip" *Proc. IEEE*, vol. 94, no. 6, pp. 1070–1088, Jun. 2006.

[2] A. M. Abo and P.R Gray "A 1.5 V, 10bit, 14.3 MS/s CMOS Pipeline Analog-to-Digital Converter". *IEEE Journal of Solid-State Circuits*, vol. 34, pp. 599-606, 1999

[3] D. Aksin, M. Al-Shyoukh and F. Maloberti "Switch Bootstrapping for Precise Sampling Beyond Supply Voltage". *IEEE Journal of Solid-State Circuits*, vol. 41, pp. 1938-1943, 2006

[4] C. J. B. Fayomi, G. W. Roberts and M. Sawan "Low-voltage CMOS analog bootstrapped switch for sample-and-hold circuit: design and chip characterization". *IEEE International Symposium on Circuits and Systems*, May 2005

[5] P. Satarzadeh et al "Digital Calibration of a Nonlinear S/H" *IEEE Journal Of Selected Topics In Signal Processing.*, vol. 3, NO. 3, pp. 454-471, June 2009

[6] Parastoo Nikaeen "Digital Compensation Of Dynamic Acquisition Errors At The Front-End Of ADCs" *PhD thesis, University Of Stanford*, September 2008

[7] Patrick Satarzadeh et al. "Adaptive Semiblind Calibration of Bandwidth Mismatch for Two-Channel Time-Interleaved ADCs". *IEEE Transactions on Circuits and Systems*, vol. 56, pp. 2075 - 2088, 2009

[8] S. Paquelet, G. Kamdem De Teyou and Y. Le Guillou "TIADC SFDR Requirements Analysis". *IEEE NEWCAS Conference*, 2013

978-1-4799-4993-9/14 $31.00 © 2014 IEEE

Multiple-Event Direct to Histogram TDC in 65nm FPGA Technology

Neale Dutton[*†] & Johannes Vergote[*†] & Salvatore Gnecchi[*†]
Lindsay Grant[*], David Lee[*], Sara Pellegrini[*], Bruce Rae[*], Robert Henderson[†]
[*]ST Microelectronics Imaging Division, Edinburgh, United Kingdom
[†]The University of Edinburgh, Edinburgh, United Kingdom

Abstract—A novel multiple-event Time to Digital Converter (TDC) with direct to histogram output is implemented in a 65nm Xilinx Virtex 5 FPGA. The delay-line based architecture achieves 16.3 ps temporal accuracy over a 2.86ns dynamic range. The measured maximum conversion rate of 6.17 Gsamples/s and the sampling rate of 61.7 Gsamples/s are the highest published in the literature. The system achieves a linearity of -0.9/+3 LSB DNL and -1.5/+5 LSB INL. The TDC is demonstrated in a direct time of flight optical ranging application with 12mm error over a 350mm range.

I. INTRODUCTION

Time to Digital Converters (TDC) have myriad applications including Time of Flight (TOF) optical ranging [1], flow cytometry [2], advanced microscopy [3] and positron emission tomography (PET) [4], yet few TDC designs are fully optimised for high throughput in these applications [5]. Furthermore, recent advances in optical sensors with large arrays of CMOS Single Photon Avalanche Diodes (SPADs) have presented the problem of processing the large amount of temporal information that these detectors produce. In order to meet these high throughput demands, the system conversion rate can be increased by improving both the TDC and the signal processing.

Realising the TDC and processing on a Field Programmable Gate Array (FPGA) is a cost effective and a practical method for many applications [6]. Delay-line TDCs have been implemented successfully in FPGA taking advantage of the regular structure and optimized clock routing [7][8][9]. As Figure 1(a) demonstrates these converters process only one event per clock cycle, losing all the subsequent photons in the same clock cycle. To improve this, multiple time-interleaved TDC channels are used as shown in Figure 1(b) [1][2][5]. This approach reduces the loss of incoming events at the converter stage, known as converter pile-up. This provides both quicker measurements and less distortion in the output data [10]. The number of channels dictates the maximum number of photons that can be processed in one clock cycle. However, delay-line TDCs are inherently area intensive, using 2^N delay cells for N-bit temporal dynamic range. Therefore implementing multiple time-interleaved delay-line TDCs to achieve higher throughput significantly reduces the available logic for other functions or requires a larger FPGA (with increased costs). To overcome this, a parallel approach to the single channel TDC design is realised on FPGA. Figure 1(c) shows the proposed architecture alongside the single channel and time-interleaved designs

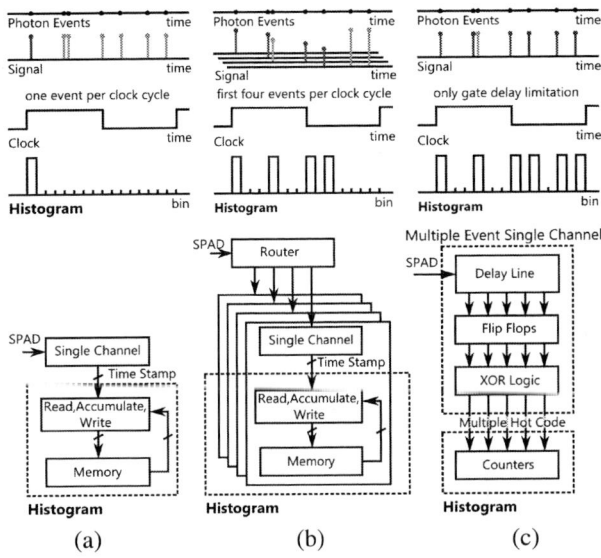

Figure 1. TDC architecture comparison. (a) Single channel: throughput limited to one event per clock cycle (b) Multi channel: throughput limited by the number of channels. Example with 4 channels (c) Proposed multiple-event approach: higher throughput limited only by the gate delay of the chosen technology

highlighting the different limitations in the three systems. The effect of the converter pile-up is reduced without increasing the number of channels by higher throughput. This limits the conversion to a rate equal to the reciprocal of the gate delay, which in DSM CMOS technology can reach tens of Gsamples/s.

This work presents an optimisation of the TDC and signal processing. The architecture of the delay-line TDC is re-designed, with parallelised outputs, allowing multiple events to be recorded per clock cycle. A temporal histogram is created in parallel achieving an order of magnitude increase in conversion rate of any published single channel TDC to date.

II. MULTIPLE EVENT DIRECT TO HISTOGRAM TDC

A. Multiple Hot Thermometer Coding Scheme

All TDC designs output a binary time-stamp natively or via a binary conversion step such as a Thermometer to Binary converter (T2B). This conversion process is a primary limiting factor in increasing converter throughput and in lowering converter dead-time. A 16.9ps temporal resolution delay-line TDC was demonstrated on a Xilinx

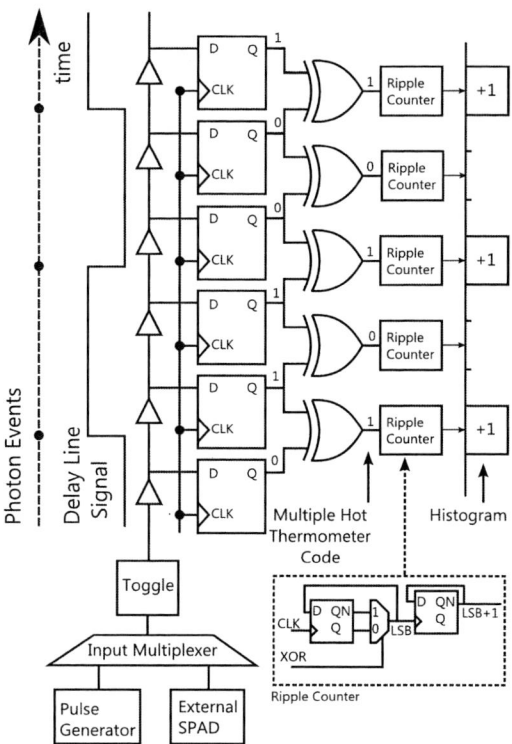

Figure 2. The Multiple Event TDC Concept: XOR gates replace conventional thermometer decoding scheme

Virtex 5 FPGA with the highest single channel converter throughput demonstrated in FPGA logic of 300Msamples/s [7]. That work doubled the maximum conversion rate of delay-line TDCs by implementing a toggle flip-flop, triggering on both rising and falling edges of the input pulse rather than using only the rising edge and clearing the delay-line after each event. Building on this Dual-Data Rate (DDR) approach, if the inter-arrival time of two input events is lowered below the converter sampling period then both the positive and negative edges on the delay-line will be sampled. In this situation, a conventional T2B converter with error correction encodes only the first rising edge event on a delay-line, assuming the subsequent falling edge event is an error and correcting for it [11]. This work presents the new 'multiple-event' approach, shown in Figure 2, replacing the T2B logic with an Exclusive OR (XOR) operation encoding the position of many events in a Multiple-Hot Thermometer Coding (MHTC) scheme.

B. Direct to Histogram Output

In time-domain imaging a temporal histogram is conventionally created by incrementing a memory location determined by the TDC output code [1]. Like the T2B converter, this step is parallelised: a synchronous ripple counter is directly connected to each MHTC XOR output. Each ripple counter then forms one bin of the temporal histogram. The TDC and ripple counters are enabled for an exposure period to create the histogram. Programmable protection logic is implemented to avoid ripple counter overflow. Figure 2 gives an overview of the complete TDC system implemented on the FPGA.

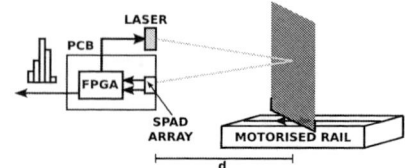

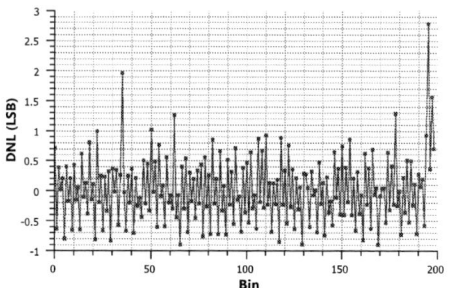

Figure 3. Optical setup for Time of Flight measurements

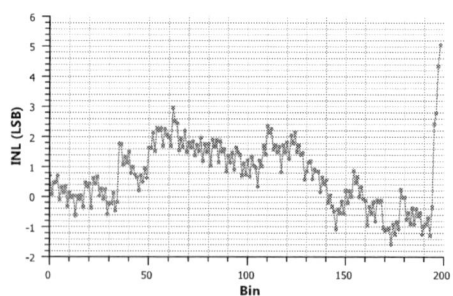

Figure 4. Differential Non Linearity (DNL) of the TDC

III. MEASUREMENT RESULTS

The TDC design is implemented on a Xilinx Virtex 5 LX50. A 200 delay-element delay-line is created with adder carry lines, as in previous works using the CARRY4 primitives [7][12]. Place and route constraints were created with the Xilinx FPGA Editor tool to ensure the delay-line and output flip-flops successfully mapped onto the FPGA logic slices. Figure 3 demonstrates the TOF optical setup. The FPGA was mounted on a custom PCB along with a 3×3 SPAD array module. The module contains nine $8\mu m$ diameter PW-NISO SPADs detailed in [13]. The FPGA is connected as timing-master to a PicoQuant PDL 800-D laser driver and LHD-D-C-470 470nm 63ps FHWM pulsed laser head. The PCB and laser were attached to a fixed end point of a motorised rail. A white card target ($\sim 85\%$ reflectivity) was attached to the moving section of the rail. The experiment was run in a dark room at low ambient light and $\sim 25°C$ temperature.

The clock frequency was set at the maximum allowable frequency from the static timing analysis at 316MHz. This translates to a clock period of 3.20ns and an unambiguous distance range of 446mm before the TOF signal wraps and the histogram contains two peaks (from 446 to 474mm). The laser trigger signal, generated by divide-by-8 logic, runs at a repetition rate of 39.5MHz to be compatible with

Figure 5. Integral Non Linearity (INL) of the TDC

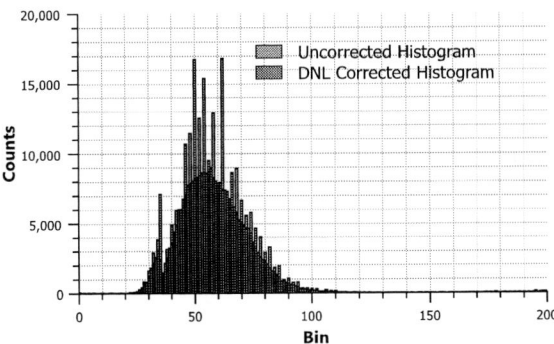

Figure 6. Implemented DNL correction: a comparison between uncorrected and corrected histograms

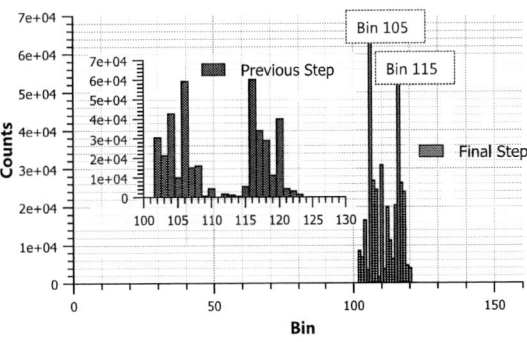

Figure 7. Converter resolution measurement - showing the minimum separation between two distinguishable peaks. Inset histogram for comparison with previous measurement step

the laser input.

The statistical code density test was employed to characterise the linearity of the TDC [14]. The test was performed activating a single SPAD under ambient light conditions for 50 exposures with the TDC recording 3.1×10^6 SPAD counts for each exposure (on average 15,500 counts per bin). Figure 4 demonstrates the Differential Non Linearity (DNL) of the converter. The DNL is a combination of the CARRY4 logic and the clock routing. A repeating pattern from the CARRY4 is noted. The large DNL spikes are caused at the boundary region between clock regions. For delay-lines longer than 80 delay elements in a Xilinx Virtex 5 FPGA, the crossing of clock regions is unavoidable. The problem can be mitigated by trialling multiple placements of the delay-line logic selecting the best-case linearity [15]. However, this iterative process is a time consuming solution.

In Figure 5, a gain error is evident by the slope of the Integral Non Linearity (INL) curve. The spike at bin 196 is from the clock boundary. In a TOF measurement system, the gain error can be calibrated out with a look-up table [14].

Post-processed scaling based on DNL is applied to the output histogram. For each ith bin, the value is multiplied by a correction factor $CF_i = 1/(1 + DNL_i)$. Figure 6 demonstrates the effectiveness of this approach comparing the output histogram with the DNL corrected histogram.

An accuracy of 16.3ps is measured permitting a sampling rate of 61.3Gsamples/s (TDC equivalent of ADC sampling rate). In reality the maximum achievable conversion rate is instead the reciprocal of the temporal resolution, defined as the minimum time difference between two distinguishable events in a two peak histogram (TDC equivalent of ADC bandwidth). The temporal resolution is measured by inputting two pulses from the pulse generator with known temporal separation down the delay-line. In a series of experiments, the pulse separation was decremented using an FPGA programmable delay cell. The second peak of the histogram approached the first and eventually amalgamated to a single pulse. The minimum separation at which one pulse could be discerned from the other is the temporal resolution. The maximum achievable conversion rate is calculated as the reciprocal of this value. The main graph in Figure 7 demonstrates the output

histogram of the final experimental run with two distinct peaks with a 10bin (=162ps) separation demonstrating a maximum achievable conversion rate of 6.17Gsamples/s. The inset graph shows complete peak separation in the previous experimental run for comparison.

The optical TOF measurements were performed with a single SPAD input. The TOF distance measurement was calibrated to account for the system offsets:

- FPGA to laser synchronisation pulse
- Laser synchronisation pulse to laser output
- SPAD to front end of the delay-line.

The calibration was performed by setting the physical target at 250mm from the PCB and laser, aligning the histogram peak on the first bin. The experiment was repeated moving the target from the calibration point (250mm) in 5mm steps up to 600mm (350mm of relative distance). The mean of the corrected histogram (μ) provides the centroid of the histogram peak and was used to calculate the target distance (d):

$$ d = \frac{c \cdot \text{TOF}}{2} = \frac{c \cdot \mu \cdot \tau_{\text{res}}}{2} \qquad (1) $$

$$ c = 3 \times 10^8 ms^{-1}, \tau_{\text{res}} = 16.3 \times 10^{-12} s $$

Figure 8 demonstrates the TOF ranging results and calculated error over the measured 350mm range. In the upper range between 350mm and 446mm the measurements were discarded due to the signal wrapping causing ambiguity in the calculated distance. Table I compares this work against a range of recent TDC implementations. A drawback of the proposed design is the lower dynamic range in comparison to the previous works. The extension of the dynamic range would occupy a larger area in the FPGA in the form of a longer delay line and more counters. The throughput is improved by an order of magnitude compared to a similar FPGA implementation with the same accuracy while a more advanced FPGA would improve the TDC accuracy.

IV. CONCLUSIONS

A re-designed parallelised-output architecture of the delay-line TDC is presented. Multiple events are sampled per clock cycle and a temporal histogram is created in parallel, in real-time, achieving an order of magnitude increase in conversion rate of any published single channel

Table I
PERFORMANCE COMPARISON

Parameter	Unit	Niclass et al. [1]	Yousif et al. [5]	Favi et al. [7]	Menninga et al. [15]	This Work
Technology	nm	180	130	65	40	65
Implementation		ASIC	ASIC	Xilinx Virtex 5 FPGA	Xilinx Virtex 6 FPGA	Xilinx Virtex 5 FPGA
Accuracy	ps	208	31	16.9	9.8	16.3
Distance Resolution	mm	31.2	4.65*	2.54*	1.47*	2.44
DNL	LSB	-0.52	< 1.25	[−1, +3.55]	[−1, +1.5]	[−0.9, +3]
INL	LSB	0.73	< 1.45	[−2.99, +2.58]	[−2.25, +1.61]	[−1.5, +5]
TDC Clock Freq.	MHz	600	500	300	600	316
Single Channel Conversion Rate (Theoretical)	Msamples/s	25	500	300	300	61 700
Single Channel Conversion Rate (Measured)	Msamples/s	25	500	300	300	6 170
TDC Dynamic Range	ns	853	2	53.33	53.33	2.86
Distance Range (Measured)	cm	10 000	N/A	N/A	N/A	35.0
Unambiguous Distance Range	cm	12 800	30*	800*	800*	44.6
Max Error	cm	2.3	>0.67*	0.90*	0.33*	1.22

Calculated value for comparison

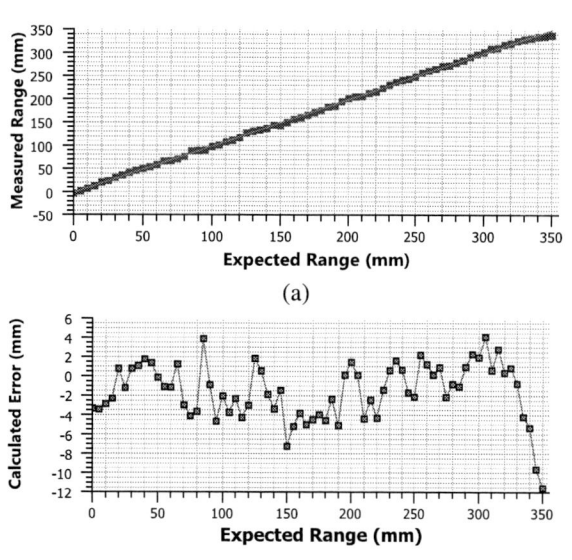

Figure 8. Range measurements: a) measurement to 350mm b) error calculated as the deviation from the expected range value

TDC. The system was successfully demonstrated in a short range optical TOF experiment. Applying the multiple-event concept and parallelised direct-to-histogram output to other TDC architectures may yield significant increases in conversion rate for a wide range of applications.

REFERENCES

[1] C. Niclass, M. Soga, H. Matsubara et al., "A 100m-Range 10-Frame/s 340 × 96-Pixel Time-of-Flight Depth Sensor in 0.18- μm CMOS," IEEE J. Solid-State Circuits, vol. 48, no. 2, pp. 559–572, Feb 2013.

[2] D. Tyndall, B. R. Rae, D. D.-U. Li et al., "A high-throughput time-resolved mini-silicon photomultiplier with embedded fluorescence

lifetime estimation in 0.13 μm CMOS," IEEE Trans. Biomed. Circuits Syst., vol. 6, no. 6, pp. 562–70, Dec. 2012.

[3] D. D.-U. Li, S. Ameer-Beg, J. Arlt et al., "Time-domain fluorescence lifetime imaging techniques suitable for solid-state imaging sensor arrays." Sensors (Basel, Switzerland), vol. 12, no. 5, pp. 5650–69, Jan. 2012.

[4] L. H. C. Braga, L. Gasparini, L. Grant et al., "A Fully Digital 8×16 SiPM Array for PET Applications With Per-Pixel TDCs and Real-Time Energy Output," IEEE J. Solid-State Circuits, vol. 49, no. 1, pp. 301–314, 2014.

[5] A. S. Yousif and J. W. Haslett, "A Fine Resolution TDC Architecture for Next Generation PET Imaging," IEEE Trans. Nucl. Sci., vol. 54, no. 5, pp. 1574–1582, Oct. 2007.

[6] M. Bogdan, H. Frisch, M. Heintz et al., "A 96-channel FPGA-based Time-to-Digital Converter (TDC) and fast trigger processor module with multi-hit capability and pipeline," Nuclear Instruments and Methods in Physics Research Section A: Accelerators, Spectrometers, Detectors and Associated Equipment, vol. 554, no. 1-3, pp. 444–457, Dec. 2005.

[7] C. Favi and E. Charbon, "A 17ps time-to-digital converter implemented in 65nm FPGA technology," Proceeding of the ACM/SIGDA International Symposium on Field Programmable Gate Arrays, p. 113, 2009.

[8] S. S. Junnarkar, P. O'Connor, and R. Fontaine, "FPGA based self calibrating 40 picosecond resolution, wide range Time to Digital Converter," 2008 IEEE Nuclear Science Symposium Conference Record, pp. 3434–3439, Oct. 2008.

[9] S. Rogacki and T. H. Zurbuchen, "A time digitizer for space instrumentation using a field programmable gate array," Review of Scientific Instruments, vol. 84, no. 8, p. 083107, Aug. 2013.

[10] J. Arlt, D. Tyndall, B. R. Rae et al., "A study of pile-up in integrated time-correlated single photon counting systems," Review of Scientific Instruments, vol. 84, no. 10, pp. 103 105–103 105–10, Oct. 2013.

[11] N. Dutton, "Error correction in thermometer codes," Apr. 30 2013, US Patent 8,432,304. [Online]. Available: http://www.google.com/patents/US8432304

[12] L. Zhao, X. Hu, S. Liu et al., "The Design of a 16-Channel 15 ps TDC Implemented in a 65 nm FPGA," IEEE Trans. Nucl. Sci., vol. 60, no. 5, pp. 3532–3536, Oct 2013.

[13] J. A. Richardson, E. A. G. Webster, L. A. Grant et al., "Scaleable Single-Photon Avalanche Diode Structures in Nanometer CMOS Technology," IEEE Trans. Electron Devices, vol. 58, no. 7, pp. 2028–2035, Jul. 2011.

[14] J. Kalisz, "Review of methods for time interval measurements with

picosecond resolution," *Metrologia*, vol. 41, no. 1, pp. 17–32, Feb. 2004.

[15] H. Menninga, C. Favi, M. Fishburn *et al.*, "A multi-channel, 10ps resolution, FPGA-based TDC with 300MS/s throughput for open-source PET applications," in *Nuclear Science Symposium and Medical Imaging Conference (NSS/MIC), 2011 IEEE*, Oct 2011, pp. 1515–1522.

High-Speed Serial Interface with a Full Digital Delay-Loop

Christian Harder, Bastian Mohr and Stefan Heinen
Integrated Analog Circuits and RF Systems
RWTH Aachen University, D-52056 Aachen, Germany
Email: charder@ias.rwth-aachen.de

Abstract—This paper presents a high-speed serial interface with a PLL-less clock and data recovery circuit in 65 nm CMOS. A digitally controlled delay line combined with a sample&hold register is used as self-calibrating time-to-digital converter, measuring the phase offset between data and clock. The same line is then utilized to delay the clock appropriately to allow error-free sampling of the data. In contrast to analog DLLs, initial lock can be achieved after transmission of four bits, requiring minimal protocol overhead for synchronization in burst-mode transmissions. The CDR circuit is optimized for 1.228 Gbit/s and consumes 1.9 mA from 1.2 V supply, with a maximum jitter of 1.8 ps. The overall power consumption is 10.7 mW.

I. INTRODUCTION

Modern mobile communication transmitter systems have to cover a broad range of wireless standards, such as IEEE 802.11 (WLAN), WCDMA based UMTS, LTE and also legacy standards like GSM. For modern standards like WLAN and LTE, a powerful interface for high data rate communication between baseband processor and radio frequency IC (RFIC) is needed. At the same time, the interface should have both, no impact on the RF front end and minimal sensitivity to distortions caused by the transmitter. The interface presented in [1] provides a sufficient data rate when using two lanes, but is limited by the driving strength of the CMOS serializer. To overcome this limitation and reduce pin count, this paper presents a CML based interface which only needs a single data lane. Furthermore, the impact of the clock-data alignment is optimized to fulfill DigRF requirements as specified in [2].

This paper is organized as follows: In Section II, the requirements are derived from the planned application. In Section III, the interface is discussed on a system level and the procedure of self-calibration of the delay loop is shown. Section IV is focused on the circuit implementation of critical components. Simulation results are presented in Section V.

II. INTERFACES REQUIREMENTS

The interface is designed to fit in a digital centric mobile communications transmitter involving radio frequency digital-to-analog conversion, based on the design presented in [3]. To improve the resolution, the data width is increased to 15 bit for inphase and quadrature data (I and Q) each. Another 1 bit is added to each block to debug the on-chip signal flow. Considering 8b10b encoding [4], it results in 20 bit sized

TABLE I
INTERFACE DATA RATES

Mode	BB Clock / MHz	PLL Multiplier	Data Rate / Mbit	IQ Word Length / Bits
Sync WLAN	20	20	800	12+4
Sync 3GPP	30.72	20	1228.8	12+4
HS1P	26	24	1248	12
HS1S	26	28	1456	12

blocks, which are each equivalent to two symbol intervals (SI). As I and Q data are transmitted using only one channel, the interface has to be capable of at least $R_{min} = 1228.8$ Mbit/s as calculated in (1). Hereby, 30.72 MHz is the baseband frequency of Long Term Evolution (LTE).

$$R_{min} = 30.72\,\text{MHz} \cdot 20\,\text{bit} \cdot 2 = 1228.8\,\text{Mbit/s}. \quad (1)$$

Table I shows the different data rates, which are supported by the interface. For testing, the interface should support a synchronous mode, where the interface's parallel clock is two times the signals baseband sample clock *bbclk*. To increase compatibility, the interface should fulfill the requirements of the DigRF v4 standard, which defines protocols for communication between ICs that support next generation mobile broadband technologies, such as LTE and Mobile WiMAX [2]. This standard suggests to spend at maximum five SI for clock synchronization of an interface.

III. SYSTEM DESCRIPTION

The block diagram of the interface is depicted in Fig. 1. There is one dedicated differential current mode data channel for each, receive (RX) and send (TX) data, as seen from the FPGA. Another channel transmits the half rate clock from the FPGA to the transmitter. After amplification of the input signals, a digital delay-loop provides the serial clock to serializer and deserializer, which are realized using current mode logic to minimize ground bouncing and V_{DD}-sag [5] and enhance the serializers performance compared to an earlier CMOS implementation [1].

A. Bit Clock Alignment

To realize a fast synchronization of incoming data and clock signals, a combination of a fixed delay line and a

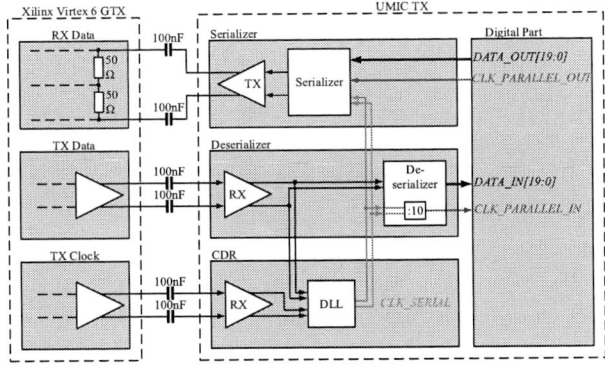

Fig. 1. Interface block diagram.

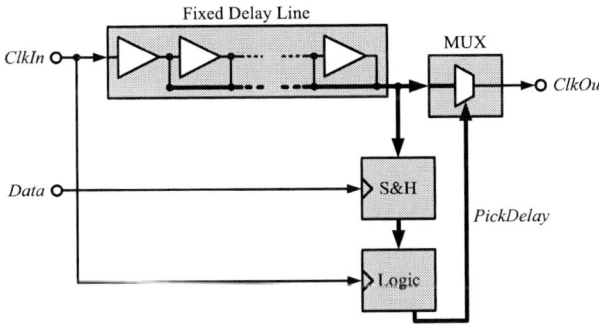

Fig. 2. Simplified block diagram of the digital delay-loop.

sampling&hold register is proposed to form a time-to-digital converter as shown in Fig. 2 [6]. Assuming every buffer within the line to have a delay of ΔT_{buffer}, the total delay is $n \cdot \Delta T_{buffer}$, with n the number of buffers within the line. To permit correct phase measurement, the total delay has to be greater than $\frac{1}{2}T$, whereas T is the period of the serial clock. For simplification $n = 8$ and (2) is assumed:

$$\frac{2}{3}T = n \cdot \Delta T_{buffer} > \frac{1}{2}T \qquad (2)$$

If a rising edge of the data signal triggers the S&H register, digital logic gets an input like for example "00000111" (phase shift: 0°) or "11111100" (phase shift: 180°) as each bit equals $\frac{1}{12}T$.

All buffers within the fixed delay line are also directly connected to a MUX; the buffer having a phase shift of 90° relative to the data is then selected as clock output. If both, the S&H register and the delay align logic (DAL) are assumed to be infinitely fast, the first bit of a data-package can be sampled correctly by following circuitry.

B. Deserializer

The deserializer consists out of a differential $100\,\Omega$ terminated CML receiver very similar to the implementation inside the FPGA [7, p. 184] and a small amplifier for signal regeneration. The shift-register with serial dual data rate (DDR)

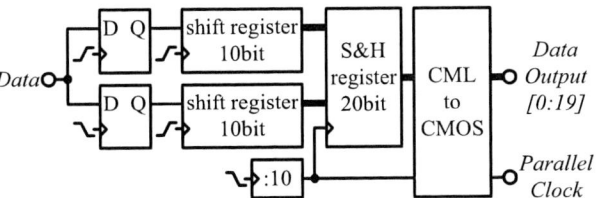

Fig. 3. Deserializer block diagram.

input and parallel output storing 20 bit, is build using only custom CML cells. For easier readability, Fig. 3 shows only non-differential signals. To ease timing restrictions, the data is shifted on the clocks rising edge, while the parallel clock is synchronized to the falling edge. Hence, registers have as much time to settle as half a clock period. The parallel clock is recovered from the serial clock using a divider-by-10. Both, the clock and the 20 bit data bus are converted to single-ended CMOS afterwards for the use in the standard cell digital part. Word recovery and further decoding is then done with the method presented in [1].

In total, the deserializer consists of 52 registers draining $20\,\mu A$ each and 21 CML-to-CMOS converters draining $21{\cdot}36\,\mu A$. Including the input amplifier draining $90\,\mu A$, the deserializer's total current consumption is about $1.9\,mA$ at $1.2\,V$ supply.

C. Serializer

The serializer, shown in Fig. 4, is also completely build out of CML logic to introduce minimal distortions to the rest of the chip. While the half-rate clock design results in relaxed timing constrains in general, the trigger signals for initializing the parallel read-in of new data into the shift registers every 10 clock cycles have to be sufficiently fast. Both registers are triggered on opposite edges to allow the maximum possible time for settling of the outputs. This is achieved by simply swapping the clock lines, as current mode logic is inherently differential. A fast MUX then combines alternately even and odd numbered bits into a full rate serial stream which is then amplified for transmission.

As the FPGA has its own clock and data recovery circuit, there is no need to align the serial data to any clock. The serializer consists of 45 registers and several buffers for increasing the driving strength of some signals. The total current adds up to approximately $1.4\,mA$, not including the final output driver consuming $3.7\,mA$ from $1.2\,V$ supply. This driver has to supply $3.5\,mA$ output current as required for LVDS operation [8].

IV. CIRCUIT IMPLEMENTATION

The diagram presented in Fig. 5 depicts the implementation of the digital delay-loop including the calibration. The MUX selects one of four different signals as *SamplingClk* to clock the calibration (Section IV-A). The DAL is implemented in a hardware description language and then synthesized, therefore all input signals are converted to CMOS.

978-1-4799-4993-9/14 $31.00 © 2014 IEEE

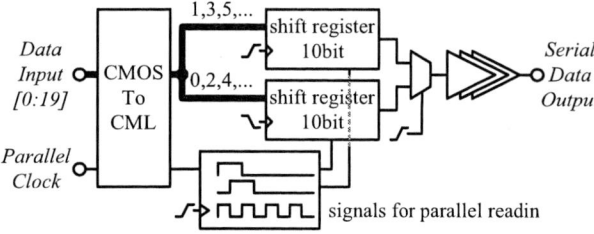

Fig. 4. Serializer block diagram.

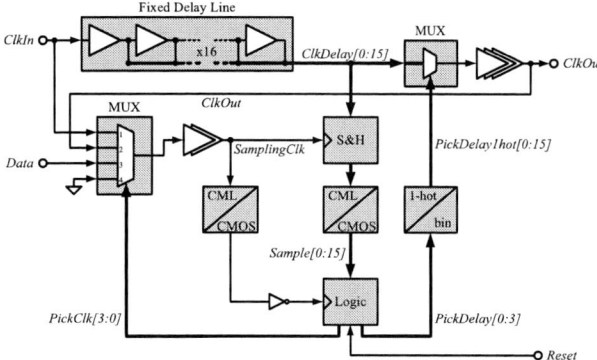

Fig. 5. Implemented digital delay-loop with calibration.

To provide sufficient settling time for the output of the S&H register *Sample*, the DAL is clocked on the falling edge of *SamplingClk* while the register itself is triggered on the rising one. It is reasonable to not constantly use *ClkIn* as internal clock, as the DAL only needs to react on changes of *Sample*, occurring only at the rising edge of *SamplingClk*. The DAL consumes an area of about $60 \times 60\,\mu m^2$ after synthesis.

A. Calibration circuit

For calibration, a MUX selects one out of four different signals as *SamplingClk* to trigger both the S&H circuit and DAL. As the delay lines total delay $n \cdot \Delta T_{buffer}$ varies with process, temperature and supply voltage, the length of half a clock period $\frac{T}{2}$ has to be measured in multiples of ΔT_{buffer} firstly. Therefore, *ClkIn* is chosen to trigger a measurement to calculate how many buffer delays equal $\frac{T}{4}$. This value is the optimum phase difference between *ClkOut* and *Data* for sampling. This calibration step also enables the usage of the circuit for a larger frequency range without redesign. Of course, requirement (2) still represents the lower frequency boundary, while the resolution ΔT_{buffer} and timing constrains within the DAL limit the maximum operating frequency.

The second calibration step is necessary due to the huge increase of ΔT_{buffer}, when a certain buffer is selected as *ClkOut*, due to the capacitive loading of the MUX. Instead of increasing the driving strength of each buffer, which would lead to much greater power consumption, the difference in delay of loaded and unloaded state is measured once and subtracted later. It should be mentioned, that the number of

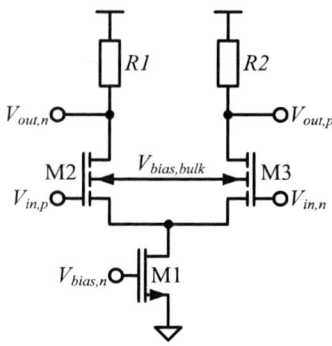

Fig. 6. CML buffer with resistive load and bulk biasing.

measurements done for calibration has be to be kept as low as possible, as the time-to-digital converters limited resolution can cause huge errors when adding up to many imprecise values. After initial calibration, the phase shift between data and clock is measured as described in Section III.

As the speed of the DAL limits the frequency of *Sampling-Clk* for an increasing number of buffers within the line, it is also possible to operate the logic at a lower frequency by the cost of increased calibration and lock time. Still the locking is very fast compared to analog solutions.

If the calibration algorithm does not use the full range of the delay chain, e.g. the clock period is determined between 7 and 13 buffer delays by corner simulation, its area can be downsized. Reducing the measurement range reduces the bitwidth of the DAL and hence increases the logics maximum clock frequency, enabling fastest initial lock.

B. Power Save Mode

After the synchronization process is completed DAL, S&H and CML-to-CMOS conversion can be turned off to save power and reduce noise. This of course only leads to valid results when clock and data stay aligned. The higher level logic of the interface turns the synchronization logic back on after a certain number of transmitted bursts. This measure saves in total $900\,\mu A$ during transmitter operation.

C. Delay Buffers

Fig. 6 shows the CML buffer used in the delay line. To keep parasitic capacitances low, the design of the delay buffers is kept simple, consisting out of a standard CML buffer [9] which has a current source M1 providing $30\,\mu A$, a differential n-MOS input pair (M2, M3) and a resistive load of $20\,k\Omega$. The input pair is built within its own deep n-well isolation to be able to put its bulk as high as $\frac{3}{4}V_{DD}$. As the single ended output voltage swing $\Delta V_{single} = V_{out,max} - V_{out,min}$ and the threshold voltage V_{th} fulfills equation (3) the switching speed could be maximized and rise and fall times equalized by putting the bulk potential in between these two values.

$$V_{th} \approx \Delta V_{single} = 30\,\mu A \cdot 20\,k\Omega \qquad (3)$$

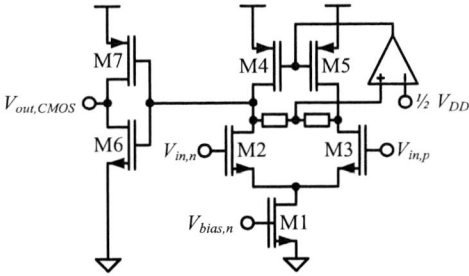

Fig. 7. Schematic of CML2CMOS converter.

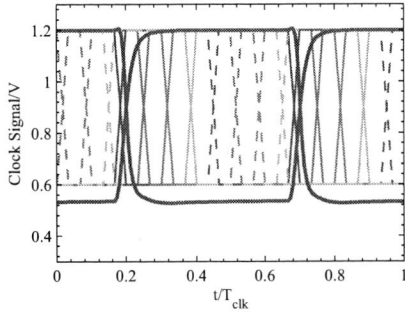

Fig. 8. Clock input with different phase offsets.

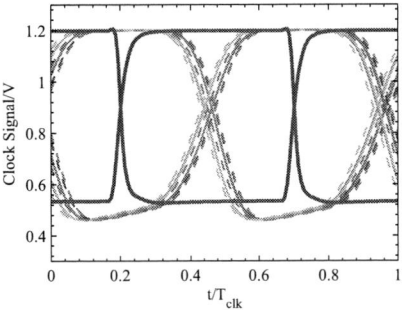

Fig. 9. Clock output after synchronization.

D. CML-to-CMOS converter

The schematic of the CML-to-CMOS converter is depicted in Fig. 7. Clock signals with precise 50 % duty cycle relax the integration of the DSP of the transmitter. To achieve symmetric switching, the differential stages output voltages are stabilized by a common mode feedback amplifier, which keeps the DC level at the inverter at precisely $\frac{1}{2}V_{DD}$.

V. SIMULATION RESULTS

The design was implemented in a 65 nm mixed-signal CMOS technology. Simulation results are obtained using parasitic extraction from the finished layout. They proof reliable operation at the designed data rate of 1228.8 Mbit/s. Fig. 8 shows the input clock with different phases related to the data signal. Fig. 9 shows how the delay loop aligns the clock to the data. It is observable, that the proposed digital delay loop sets a sufficient phase offset for error-free data sampling, only limited by the resolution of the time-to-digital converter. Due to the fixed delay line and the regular shape of the signal in lock state, a maximum jitter of 1.8 ps is achieved. This low jitter relaxes the timing constraints for the whole DSP block of the transmitter.

The effect of increased delay through capacitive loading, mentioned in Section IV-A, is typically about $7 \cdot \Delta T_{buffer}$, while $\frac{T}{2} = 11 \cdot \Delta T_{buffer}$. Still, the calibration compensates this effect and allows the delay line to operate with very low currents by at the same time keeping the resolution constant compared to the unloaded case.

Referring to the DigRF standard, one 10 bit symbol with "1010101010"-pattern (1 SI) is used for synchronization, which is well below the specified limit of 5 SI. The system actually needs only 4 bit to align the delay, but waiting for 1 SI gives the S&H circuit more time to settle and further increases the margin between the clock and the data.

This combination of fast lock time and low jitter shows the supremacy of this design, compared to delay loops for clock recovery with an analog phase detector. Higher data rates of up to 2 Gbit/s can be achieved at the expense of a higher CML bias current.

VI. CONCLUSION

This paper presents a high-speed low-noise serial interface for mobile communication transmitters. The interface enables a data rate of 1228.8 Mbit/s per lane and bidirectional transmission between a baseband processor and the RFIC. It is build out of custom CML logic cells and a digital delay loop for clock alignment featuring self-calibration. The systems total power consumption is 10.7 mW without clock buffer, the locking is faster than 1 SI and the maximum jitter 1.8 ps.

REFERENCES

[1] B. Mohr, J. Mueller, Y. Zhang, R. Leys, S. Schenk, U. Bruening, and S. Heinen, "High speed interface for digital centric transmitters," in *Ph.D. Research in Microelectronics and Electronics (PRIME), 2013 9th Conference on*, June 2013, pp. 233–236.

[2] "MIPI alliance specification for DigRF^{SM} v4," MIPI Alliance, January 2009.

[3] B. Mohr, N. Zimmermann, B. T. Thiel, J. H. Mueller, Y. Wang, Y. Zhang, F. Lemke, R. Leys, S. Schenk, U. Bruening, R. Negra, and S. Heinen, "An rfdac based reconfigurable multistandard transmitter in 65 nm cmos," in *Radio Frequency Integrated Circuits Symposium (RFIC), 2012 IEEE*, june 2012, pp. 109 –112.

[4] A. X. Widmer and P. A. Franaszek, "A dc-balanced, partitioned-block, 8b/10b transmission code," *IBM Journal of Research and Development*, vol. 27, no. 5, pp. 440–451, Sept. 1983.

[5] K. Chabrak, F. Bachmann, G. Hueber, K. Seemann, L. Maurer, Z. Boos, and R. Weigel, "Design of a high speed digital interface for multi-standard mobile transceiver rfics in 0.13 mu:m cmos," in *Microwave Conference, 2005 European*, vol. 3, oct. 2005, p. 4 pp.

[6] S.-H. Han, J.-H. Lee, and H.-J. Yoo, "Fast lock-on time mixed mode dll with 10 ps jitter," *Electronics Letters*, vol. 35, no. 20, pp. 1700–1701, Sep 1999.

[7] *Virtex-6 FPGA GTX Transceivers User Guide*, Ug366 (v2.6) ed., Xilinx, July 2011.

[8] *LVDS Owner's Manual*, National Semiconductor Std., 2008.

[9] G. P. Massimo Alioto, *Model and Design of Bipolar and MOS Current-Mode Logic*. Kluwer Academic Publishers, 2005.

Vector implementation of the fast Fourier transform on DSP and NVIDIA CUDA platforms

Bartosz Pikacz
Warsaw University of Technology
Nowowiejska 15/19, 00-665 Warszawa, Poland
Email: B.Pikacz@ise.pw.edu.pl

Jacek Gambrych
Warsaw University of Technology
Nowowiejska 15/19, 00-665 Warszawa, Poland
Email: J.Gambrych@stud.pw.edu.pl

Abstract—This paper presents two implementations of the fast Fourier transform decomposed into vector operations. This approach is appropriate for cases where data to be transformed is stored in an unorthodox order, and as such is well suited for radar and sonar data processing. The described procedures performing a vector FFT were implemented for TigerSHARC DSP and NVIDIA CUDA platforms. Their performance was measured and compared with highly optimized library procedures.

I. Introduction

One of the most frequently implemented DSP algorithms is the fast Fourier transform (FFT). The most basic, but the most popular version of this algorithm is the radix-2 type algorithm [1]. The majority of DSP platforms can execute the instructions specific to this algorithm (e.g. reverse bit addressing, simultaneous addition and subtraction of a pair of variables) and have the FFT algorithm available in dedicated libraries. These libraries are highly optimized, but require some specific arrangement of input data (the natural or bit-reversed order of samples in continuous memory space). In some applications the natural order of input data is different and the rearrangement of data proves problematic. In such cases, specialized functions may be a better choice (for the sake of computation time) than library functions preceded by data rearrangement.

In radar and sonar systems with a coherent pulse transmission, Doppler analysis is applied to samples collected from every single range resolution cell, but the acquired samples are naturally stored in a memory according to the time of their collection (Fig. 1). This forms a complex $R \times N$ input data matrix, where R is the number of range cells and N is the number of pulses transmitted coherently in a batch called the coherent processing interval. The matrix is written column-wise (N data vectors corresponding to the pulses, each containing R complex samples), but read as row vectors (R N-point transforms). This data order is thus incompatible with the majority of FFT procedures, because samples for a single transform are scattered in memory with stride R ($2R$, if a single cell contains only a real or an imaginary part of the sample).

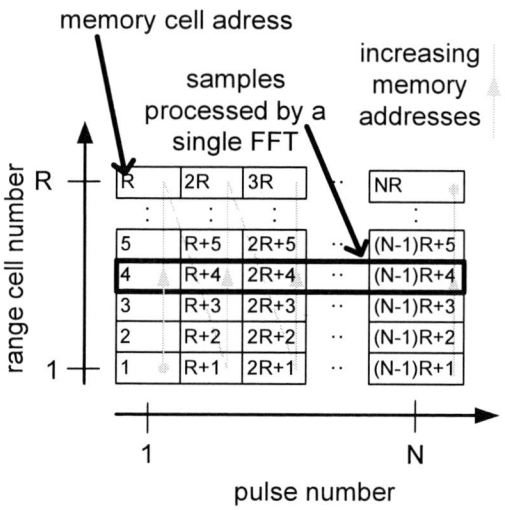

Fig. 1. Data arrangement in memory

II. Comparison of the library and the proposed FFT procedure for the TigerSHARC DSP

A complex N-point radix-2 FFT consists of $\frac{N}{2}\log_2 N$ complex butterflies. An ADSP-TS201 TigerSharc processor can execute up to two floating-point multiplications in a single core clock cycle [2],[3], thus a single butterfly may be executed during two cycles and a complete transform requires at least $N\log_2 N$ cycles. This number may by reduced by a simple trick: the first two stages of the transform use twiddle factors equal to $-j$ and -1, so a complex *multiplication* by them does not require an actual multiply instruction. Simplified butterflies from the first two stages may be executed during a single cycle while a complete transform requires at least $N\log_2 \frac{N}{2}$ cycles.

The relative performance of the FFT procedure implementation may be defined as a quotient of the theoretical execution time ($N\log_2 \frac{N}{2}$ cycles) and the measured time. The relative performance of a library FFT procedure [4] as a function of N is presented in Figure 2 (red thick curve). As expected, the measured relative performance is worse than the theoretical one. For small N, the low performance is caused by the overhead and stalls from procedure execution, loop initializa-

978-1-4799-4993-9/14 $31.00 © 2014 IEEE

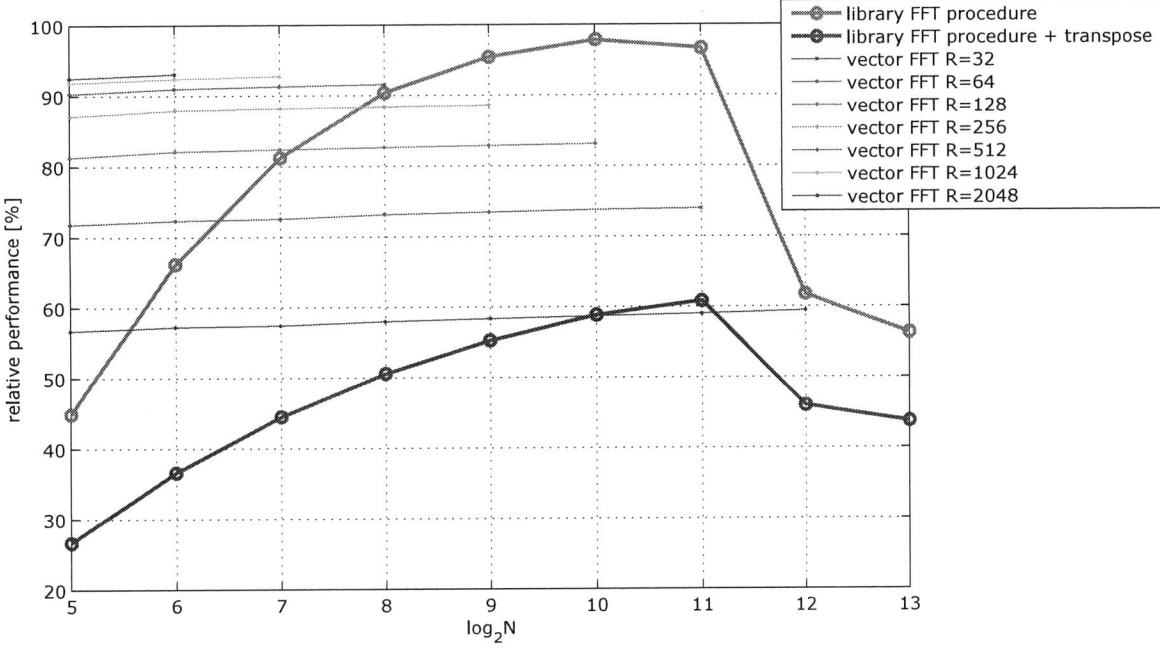

Fig. 2. Relative performance of the library and the proposed FFTs for ADSP-TS201

tion and non-sequential access to an uncached memory during the first stage of the transform. For larger N, these overhead and stalls are equal or even superior, but their time is much shorter than the time of the complete transform. Therefore, the performance reaches almost 100% for $N = 2^9 \dots 2^{11}$. The performance drastically drops for $N = 2^{12} \dots 2^{13}$ because of a limited cache memory: for $N = 2^{12}$ only half of samples are cached, while for $N = 2^{13}$ – just a quarter of them.

The library FFT procedure requires input samples to be stored in a natural order in a continuous memory space. As a result, it cannot directly process the data acquired from a radar and stored in a memory in the order presented in Figure 1. The input matrix must be transposed to deliver properly arranged data prior to calling the library procedure. The transpose operation requires non-sequential access (read or write) to a memory, which is inefficient: reading every R-th word in a memory may be even three times slower than reading a continuous block of the same size. If we add the measured execution time of R N-point library FFTs to the time of $R \times N$ matrix transpose, the resulting relative performance of a single transform is significantly worse (almost twice slower) than previously (thick blue curve in Figure 2). A very good performance of the library procedures was wasted by the need for data rearrangement. Therefore, the library FFT procedure is hardly usable for data arranged in this manner and needs to be substituted by the more efficient procedure proposed below.

The proposed implementation of the FFT procedure for the TigerSHARC processor is based on the assumption that the access to memory should be sequential in order to directly process the data arranged in the way presented in Fig. 1. This

can be achieved using a vector approach to input data: every operation that is normally carried out for individual samples (e.g., radix-2 butterfly has 2 input and 2 output samples) should be carried out for the whole column of data matrix (Fig. 3). This solution has the following advantages increasing overall performance:

- sequential memory access eliminates stalls (opening and closing memory pages), simplifies and reduces pointer arithmetic,
- simple vector operations of equations (1) and (2) can be efficiently coded as a loop with multiple streams,

$$\mathbf{X}'_m = \mathbf{X}_m + w\mathbf{X}_n \qquad (1)$$

$$\mathbf{X}'_n = \mathbf{X}_m - w\mathbf{X}_n \qquad (2)$$

- a single procedure performing R transforms at a time reduces the overhead as compared to the case of R procedures performing a single transform,
- in-place computations save memory,
- computations may be executed in parallel by processors connected in one cluster [5].

The relative performance of the vector FFT procedure has been tested for various problem sizes (R, N) and presented in Figure 2. These sizes were limited during the experiments by the size of two internal memory blocks of TigerSHARC ($R \cdot N \leq 2^{17}$) – hence the curves are broken off for some N. For a fixed R the measured performance is virtually independent of N. This fact is caused by a constant relation of vector operations execution time (equations (1) and (2))

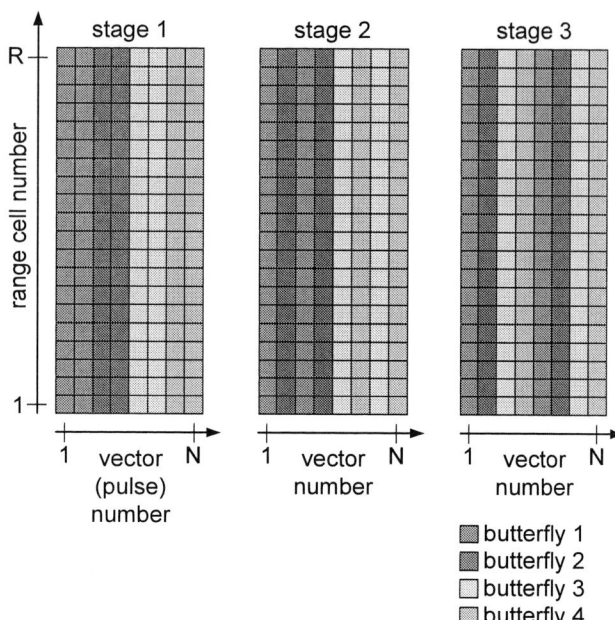

Fig. 3. Data processed by vector operations during all three (for $N = 8$) stages of the vector radix-2 FFT

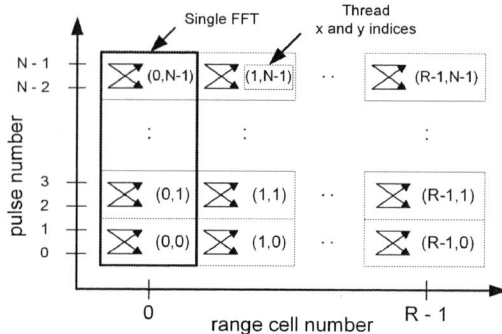

Fig. 4. FFT kernel function thread organization

to their initialization time (i. e. calculation of twiddle factors w and addresses of vectors in memory) – regardless of the number of executed butterflies. On the other hand, the relative performance increases with R: the time of vector operations is longer, while the time of their initialization is constant. As may be noticed, for almost all cases the performance of the proposed vector FFT is better than the performance of the library procedure preceded by transposition. Moreover, for input matrix sizes typical for the radar Doppler processing ($N \leq 256$, $R \geq 256$), the vector procedure is comparable or more efficient than the library procedure without data shuffling.

III. LIBRARY AND THE PROPOSED FFT PROCEDURE FOR CUDA PLATFORM

CUDA (Compute Unified Device Architecture) is a graphic device architecture and software infrastructure, which enables using GPUs for general-purpose scientific and engineering computing, created by NVIDIA in 2007 year. Because of a high computational power of GPU chips and a simple programming model based on C language, CUDA is increasingly used in digital signal processing. NVIDIA develops and distributes several libraries dedicated to CUDA platform. The libraries involve e.g., CUBLAS (the GPU version of the standard BLAS library) and cuFFT – the CUDA Fast Fourier Transform library. This makes it possible to perform computations with simple user interface.

One of the DSP issues that has recently become more popular in the context of the CUDA technology is radar signal processing e.g. [6], [7]. One of the most fundamental operations on radar signal is Doppler processing – MTD

(Moving Target Detection). The simplest and fastest realization of MTD is the FFT analysis of appropriately aggregated samples from the same range cell and from consecutive pulses. What is important, the cuFFT library enables to perform FFT on column-major matrix without the necessity of its transposition[1].

However, such data organisation has more advantages which library function does not take into account, e.g. there is no need to perform a bit reverse order permutation, because it can be done previously, at the data collection stage. This approach allows to simplify address arithmetic and reduces the number of fixed-point operations and data read operations necessary to obtain bit reverse order. This is one of the reasons to write the vector FFT kernel function solely dedicated to MTD filtration and compare its performance with the library function.

There is a number of principles that were taken into account while writing the kernel function that performs vector FFT computation. First of all, the global memory access pattern should be coalesced [9]. In fact, because of the data organisation, this condition is easy to fulfill. In 2D array of threads, threads within the same row process the same butterflies for successive range cells and threads within one column perform all butterflies for one FFT vector (Fig. 4).

Secondly, we chose radix-4 DIT FFT rather than radix-2 to reduce the number of floating point operations and shared memory read/write operations. Thirdly, the time spent on the kernel function calls should be minimized. It would then be advisable to develop one function that performs all stages of FFT. Such an approach has another advantage – global memory reads/writes are performed only once and only the shared memory is used for communication between the threads. On the other hand, this implementation causes some problems with matrices which have more than 128 rows. In CUDA graphic processors with compute capabilities 2.x, 3.0 and 3.5 the amount of shared memory per multiprocessor is limited to 48 kB [10]. In the simplest and most efficient version of the proposed algorithm, one block of threads computes N-point FFTs of multiple columns. The minimum

---

[1]Such a computation may be performed using a plan created by the function *cufftPlanMany* [8].

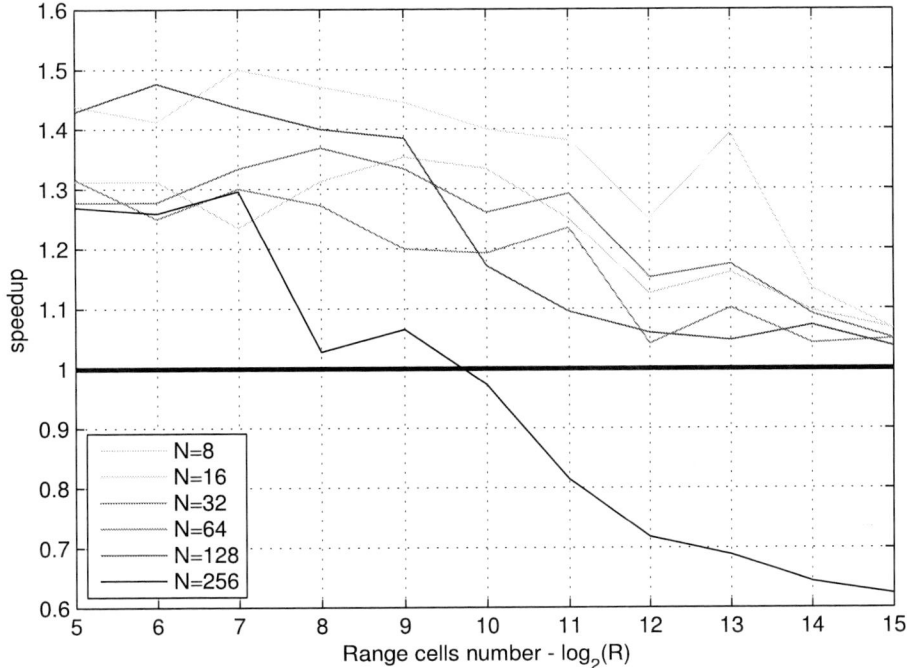

Fig. 5. Speedup of the proposed algorithm against cuFFT library versus the number of range cells.

x-dimension of block size is 16 because of the half-warp size[2]. One single-precision complex sample requires 8 B. Therefore, the minimum declared amount of the shared memory is $(N \times 16 \times 8)$ B. Thus, for N greater than 128, the amount of shared memory is the major limitation on the number of active warps and thread blocks per multiprocessor. It can be seen on diagram 5, which shows the ratio of the algorithm execution time and the library function execution time versus the number of range cells.

The calculations were performed on NVIDIA Tesla K20c GPU. As can be seen, if the number of pulses is greater than 128, then the low occupancy of multiprocessor causes a significant slowdown of the algorithm. For the number of pulses less than or equal to 128, the speedup never falls below 1 and for the majority of real data sizes commonly encountered in radar applications, the acceleration is greater than 1.2. As the numerical errors of both procedures are on the same level, the implementation of the dedicated vector FFT function may be considered satisfactory.

IV. CONCLUSIONS

Both presented implementations have proven to be more efficient than highly optimized library procedures. Depending on the platform and the data size, the speedup reached from 10% to 200%. For example for the tracking radar emitting coherently 128 pulses and processing 512 range cells, the speedup was about 40% for the CUDA platform and about 100% for the TigerSHARC processor. This improvement is

[2]It is important because of coalesced global memory access.

significant, as radars are always real-time systems and the Doppler filtration is generally the most computationally complex stage of data processing. This fact shows the advantages of using specialised procedures optimised to the input data arrangement.

REFERENCES

[1] A. V. Oppenheim, R. W. Schafer, and J. R. Buck, *Discrete-time signal processing.* Upper Saddle River: Prentice-Hall, Inc., 1999.

[2] *ADSP-TS201 TigerSHARC Processor Programming Reference*, http://www.analog.com/static/imported-files/processor_manuals/34605284816196ts201_pgr.pdf, Analog Devices, Inc., Norwood, 2005.

[3] *ADSP-TS201 TigerSHARC Processor Hardware Reference*, http://www.analog.com/static/imported-files/processor_manuals/396096833ts201_hwr.pdf, Analog Devices, Inc., Norwood, 2004.

[4] B. Lerner, *Engineer-to-Engineer Note, EE-218, Writing Efficient Floating-Point FFTs for ADSP-TS201 TigerSHARC Processors*, http://www.analog.com/static/imported-files/application_notes/EE-218.pdf, Analog Devices, Inc., 2004.

[5] B. Franke and M. F. P. O'Boyle, "Compiler parallelization of C programs for multi-core DSPs with multiple address spaces," *First IEEE/ACM/IFIP International Conference on Hardware/ Software Codesign and Systems Synthesis*, 2003.

[6] K. Szumski, M. Malanowski, J. Kulpa, W. Porczyk, and K. Kulpa, "Real-time software implementation of passive radar," *Radar Conference, 2009. EuRAD 2009*, pp. 33–36, 2009.

[7] M. Bernaschi, A. Di Lallo, A. Farina, R. Fulcoli, E. Gallo, and L. Timmoneri, "Use of a graphics processing unit for passive radar signal and data processing," *Aerospace and Electronic Systems Magazine*, pp. 52–59, 2012.

[8] *CUFFT Library User's Guide*, NVIDIA Corporation, 2013.

[9] S. Cook, *CUDA Programming: A Developer's Guide to Parallel Computing with GPUs.* Morgan Kaufmann, 2013.

[10] *CUDA C Programming Guide*, NVIDIA Corporation, 2013.

Experimental validation of a new power line communication system for battery management

Jérémie JOUSSE[*‡§], Nicolas GINOT[‡], Christophe BATARD[‡] and Elisabeth LEMAIRE[*]

[*]CEA, LITEN, Laboratory for Electrochemical Storage (LSEC),
National Institute for Solar Energy (INES),
50 avenue du Lac Léman
73375 Le Bourget-du-Lac, France.
Emails: jeremie.jousse@cea.fr, elisabeth.lemaire@cea.fr

[‡]PRES LUNAM - Université de Nantes - UMR CNRS 6164,
Rennes Electronic and Telecommunications Institute (IETR),
La Chantrerie - rue Christian Pauc
BP 50609 - 44306 Nantes Cedex 3, France.
Emails: nicolas.ginot@univ-nantes.fr, christophe.batard@univ-nantes.fr

[§]Novéa Énergies,
3 rue Fourier
49070 Beaucouzé, France.

Abstract—A new Power Line Communication (PLC) solution over a DC powerline is presented for remote management of batteries. This solution rely on the well known Controller Area Network (CAN) protocol and is designed to be directly compatible with existing CAN controllers. The presented PLC-CAN system is a low complexity and low cost solution suitable for short and low power DC bus used in small scale autonomous systems. Simulations of the communication channel over the frequency range [0.5 - 10] MHz are conducted to validate the concept's feasibility. Experimental measurements on prototypes are then presented and show achievable data rates of 116 kbit/s.

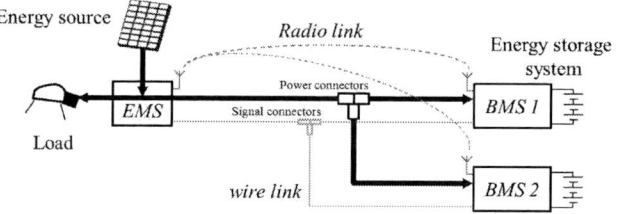

Fig. 1. System architecture with radio communication link and wire link.

I. INTRODUCTION

Since the first commercial release of lithium-ion batteries in 1991, this technology has been adopted in a wide variety of applications thanks to its high energy density and lifespan. However, its high susceptibility to abusive conditions such as overcharge requires constant monitoring of the cells by a Battery Management System (BMS) to ensure safe and reliable operation [1]. In most applications the battery and its BMS are physically separated from the load and from the Energy Management System (EMS) which controls it. Since these two systems must exchange information, a data link must be implemented. Today this connectivity involves the implementation of a separate medium for data as shown on figure 1. The latter can be dedicated wires, optical fibers or even a radio link, depending on specific needs such as isolation, speed, energy consumption or price. However these solutions present respectively mechanical robustness and cost issues.

This work focuses on small scale autonomous photovoltaic systems, applications range from active road-signs to photovoltaic street lighting or parking metering. These applications require a modular energy storage solution to tightly adjust the capacity for reliable operation throughout the year while keeping the overall system price low. Moreover, maintenance costs are high due to a lot of geographically dispersed units.

The potentialities of Power Line Communication (PLC) are thus studied to use the existing DC power line as a communication channel instead of additional cables. Unlike traditional PLC solutions used on the AC grid, this work focuses on a simple and innovative solution directly compatible with the Controller Area Network (CAN) protocol which is already widely used in the industry. As shown on figure 2, this solution rely on a single carrier generator in order to implement a bidirectional communication bus between the EMS and several BMS.

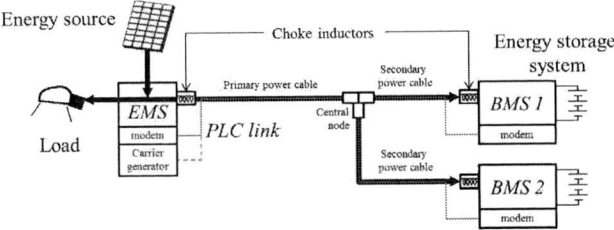

Fig. 2.   System architecture with Power Line Communication.

II.   SYSTEM DESCRIPTION

In this work we propose to use the DC power link between EMS and BMS to implement a PLC channel. Such technology is already broadly used in consumer electronic products allowing extending an ethernet Local Area Network through the power network of a building. Such systems implements electrical isolation from the AC grid and complex modulation schemes to cope with the unpredictable noises of the power network while allowing high communication rates. Some studies seek to adapt this technology to industrial power networks in order to allow remote management of electrical drives [2]. However these features also lead to bulky, expensive and energy consuming modems which are unsuitable for the studied applications.

The innovative communication solution presented consists in the design of a new PLC modem allowing communication between 5 V or 3.3 V powered microcontrollers through a 20 V to 30 V DC power network. Because of the broad adoption and availability of the CAN protocol in industrial applications, it has been chosen to design this modem to be directly compatible with the CAN controllers embedded in a large range of commercially available integrated circuits. The original CAN definition does not specify the physical layer of the OSI model, however some have been standardized and the two-wire differential bus depicted in the ISO 11898-1 standard is the most used. The proposed solution thus replaces the traditional CAN transceiver connected to a dedicated differential bus with a PLC modem connected to the DC power bus.

A similar solution is already proposed by the company Yamar and have been assessed in [3], however the proposed modem uses a complex DQPSK (Differential Quadrature Phase-Shift Keying) modulation scheme and needs to emulate part of the natural CAN operation to ensure compatibility with existing CAN controllers.

The basic principle of operation of the CAN network for transmitting messages and competing for access to the bus rely on the ability to force dominant and recessive levels on the physical bus to encode binary data. This is the core feature which made it a CSMA/CD (Carrier Sense Multiple Access / Collision Detection) protocol. It allows random multiple accesses to the bus with an intrinsic and nondestructive prioritization of the messages.

The presented solution translates this principle of operation in the high frequency domain by using a single carrier generator superimposed on the DC power signal. Each modem then have the ability to shift the impedance it exhibits at the carrier frequency in order to modulate the carrier signal. when a node on the bus wishes to transmit data, the CAN controller signals directly drive the modulation of the carrier signal. When the modem exhibits a low impedance on the bus, the carrier signal amplitude is noticeably reduced for all the nodes on the bus, this corresponds to the dominant level. When all the modems connected to bus exhibit a high impedance, the carrier signal amplitude is maximized, this corresponds to the recessive level. Only the physical medium is modified and the proposed solution rely on the original CAN protocol for bus arbitration and higher level management.

In order to validate this implementation, the ability to differentiate these recessive and dominant levels all over the bus must be assessed. Thus modeling of the various parts and simulation of the whole system are needed.

III.   SYSTEM MODELING

The proposed solution involves constant alterations of the communication channel characteristics in order to modulate a common carrier signal. Thus, transfer functions from the carrier generator input to the various modems outputs must be calculated to assess the behavior of the communication channel for each combination of modem states. The complete system was thus modeled within the MATLAB software and thorough simulations were conducted at various frequencies to assess the carrier signal amplitude at each modem output in both recessive and dominant configurations.

The first obstacle in imposing a high frequency signal over the DC power bus is the low impedance presented by the energy generators and consumers, i.e. power converters and batteries. In order to delimit a protected high impedance channel, choke inductors are placed at each end of the bus and fulfill the same function as the termination resistors of the standard CAN differential pair. These termination inductors are chosen to exhibit a high impedance at the carrier frequency. Otherwise too much power would be required to apply a high amplitude carrier signal through the impedance of the battery and converters. However the DC impedance of these chokes must also be as low as possible to avoid power losses on the DC power bus. Finally cost and size of these components should also be taken into consideration when dimensioning the system.

The length of the communication channel is also a limiting factor since it behaves as a transmission line. Reflexions of an emitted signal cause spatial fluctuations of the channel impedance at distances multiples of $\lambda/4$ from the load. Thus the communication channel length must be kept an order of magnitude smaller than the carrier frequency wavelength. Otherwise the distinction between dominant and recessive levels cannot be guaranteed in every location of the bus.

The minimum carrier frequency is thus limited by the impedance of the choke inductors, which must be high enough to isolate the carrier signal from the DC power. On the other hand the maximum frequency is limited by the bus length which must stay small compared to the wavelength.

As shown on figure 3, each modem is composed of a variable impedance decoupled from the DC power bus by a capacitor. This impedance can either be in a low impedance

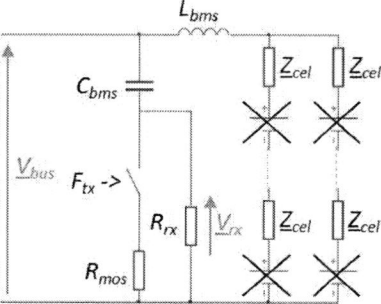

Fig. 3. Electrical model of a BMS modem, in the high frequency domain the DC sources are suppressed and the cells impedances can be neglected compared to the inductor impedance.

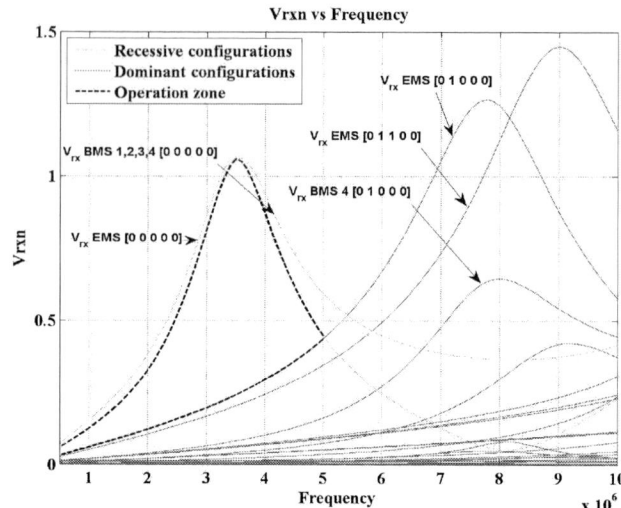

Fig. 4. Simulation of the modems outputs for a carrier signal amplitude of 3 V over the frequency range [0.5 - 10] MHz. The labels on some curves indicate the modem designation and the driving signals F_{tx} from all the modems on the bus : [EMS BMS1 BMS2 BMS3 BMS4]. A 0 indicates that the modem's switch is open and a 1 indicates that the modem's switch is closed (dominant configuration).

state corresponding to the serial resistance R_{mos} of the MOS-FET switch or in a high impedance state corresponding to the resistor R_{rx}. The latter is needed to protect the switch by dampening the oscillations generated when switching, however it also limits the number of modems on the bus since each new modem is added in parallel on the bus and lowers the global impedance of the recessive configuration. Here again a compromise must be found and time domain simulations are needed to determine the optimal value. The MOSFET switch is modeled by an ideal switch driven by the input signal F_{tx} and a serial resistor R_{mos}. The V_{rx} voltage between the switch terminals is the raw output voltage of the modem which needs to be processed before it can be fed into a CAN controller. inductors and capacitors are also modeled by ideal components while the power cables are approximated by a transmission line whose parameters were measured on a real application power cables with a network analyzer.

If we consider a cable of length ℓ loaded at one end by a short circuit or an open circuit condition, the transmission line theory states that the impedance seen at the other end of the cable is identical for both conditions at a signal frequency f_0 such as:

$$f_0 = \frac{1}{8\ell}\frac{1}{\sqrt{LC}} \qquad (1)$$

With L and C being the characteristic inductance and capacitance of the cable per unit length. This f_0 frequency is thus the maximum theoretical carrier frequency allowing to differentiate recessive and dominant configurations. We consider a configuration similar to the one depicted on figure 2 where 4 BMS are linked to a central node with cables of length ℓ_1 and then to an EMS with a single cable of length ℓ_2. If only one BMS is on a low impedance state we show that the frequency f_0 is close to $f_0 = f_1/2$ with $f_1 = \omega_1/2\pi$ and ω_1 solution from equation 2:

$$-tan(\omega_1\sqrt{LC}\ell_1).tan(\omega_1\sqrt{LC}\ell_2)+1-3tan^2(\omega_1\sqrt{LC}\ell_1) = 0 \qquad (2)$$

We thus find a frequency f_0 equal to 4.9 MHz.

The simulation results shown on figure 4 have been computed for that same configuration. The red curves represent the modems outputs for the various dominant configurations where at least one modem present a low impedance at the carrier signal frequency. The blue ones represent the modems outputs for the only recessive configuration where all the modem are in a high impedance state. Only two blue curves can be distinguished because of the symmetrical layout of the four BMS. It can be seen that the recessive and dominant configurations can be differentiated up to a frequency of 5 MHz. Above this frequency the wavelength of the carrier signal become of the same magnitude than the length of the communication channel. The reflexion phenomena then lead to an inversion of the dominant and recessive levels at some modems outputs.

IV. EXPERIMENTAL RESULTS

In order to assess the sizing and modeling detailed in section III, each function of the system has been implemented on a printed circuit board. Each of these circuits has been tested independently and then the whole system was assessed. As shown on figure 5, a Marconi 2024 function generator was used to generate the carrier signal at various frequencies. For this experiment an ISG500 function generator was used to drive the BMS 4 modem at a frequency of 170 kHz.

Figures 6(a) and 6(b) shows the evolution of the carrier signal modulation for frequencies of 3 MHz and 8 MHz. As stated in section III and shown on figure 4, the gap between recessive and dominant levels amplitudes reaches a maximum around 3 MHz. Some differences can be seen between simulation predictions and experimental data because of the impact of parasitic parameters which were neglected in the model such as MOSFET parasitic gate capacitance or inductors serial resistance. Figure 6(b) shows clearly the impact of a too high carrier frequency when the wavelength becomes close to the bus length. The carrier signal is modu-

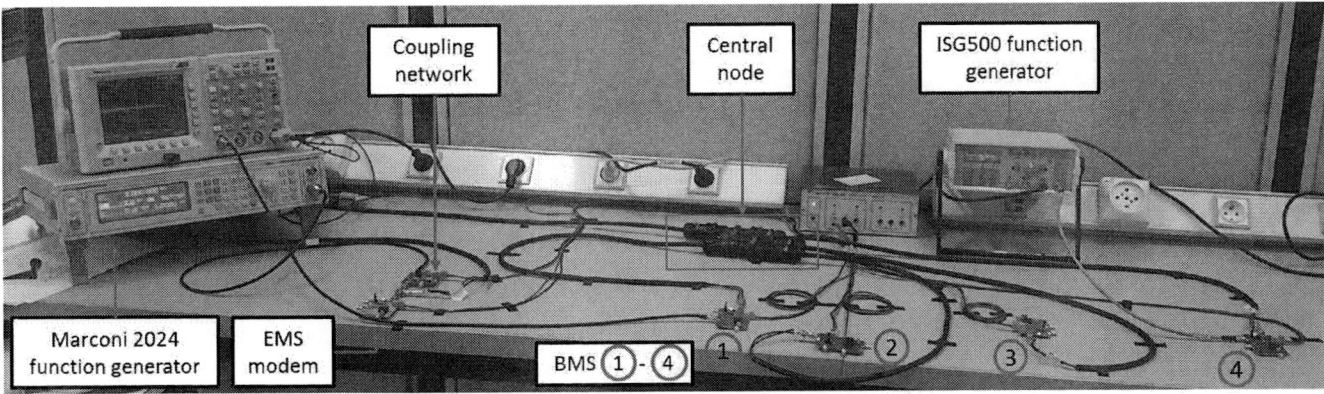

Fig. 5. Experimental setup used for the assessment of the PLC-CAN solution.

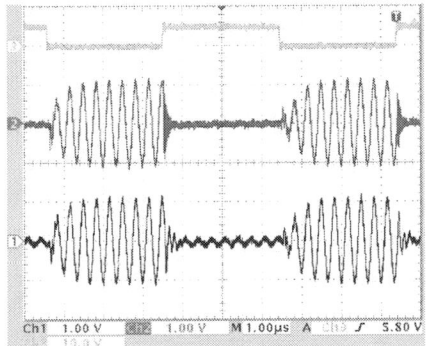

(a) 3 MHz carrier signal.

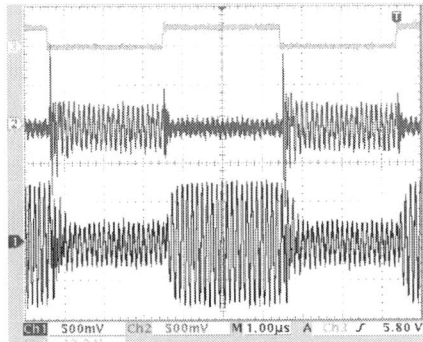

(b) 8 MHz carrier signal.

Fig. 6. Evolution of the carrier modulation for various frequencies. On each figure channel 3 is the F_{tx} driving signal of BMS 4 (green curve), channel 2 is the output voltage V_{rx} from BMS 4 (red curve) and channel 1 is the output voltage V_{rx} from the EMS (blue curve).

lated by the BMS 4 modem and whereas the carrier amplitude at the BMS 4 modem output behave as intended, the signal received by the EMS modem is reversed. At some locations on the bus the amplitude of the carrier frequency may even be invariant, which means there is no physical difference between recessive and dominant configurations, and thus no possible communication.

These results show that raw data rates of 200 kbit/s can easily be achieved. The CAN protocol specifies a standard data frame of 110 bits which comprises up to 64 bits of data [4]. Considering the use of the proposed communication scheme through a CAN controller leads us to the following relation :

$$\text{useful\_data\_rate} = \text{raw\_data\_rate} \cdot \frac{\text{useful\_data\_length}}{\text{data\_frame\_length}} \quad (3)$$

Useful data rates of 116 kbits/s are thus achievable with the proposed PLC-CAN solution.

V. CONCLUSION

An innovative PLC-CAN communication scheme is presented for remote management of batteries equipped with a digital BMS. The described solution uses a single carrier generator and a new kind of CAN transceiver to transpose the original CAN implementation to a PLC medium. This solution is thus directly compatible with the CAN controllers embedded in many commercially available integrated circuits. Simulation of the communication channel shows the concepts feasibility and its various limits in terms of frequency and number of nodes on the bus. Experimental measurements then confirm the simulation results and allow estimation of the achievable data rates. Further work will include complete validation of the communication system through implementation of the digitization stage of the modems and connection to actual CAN controllers.

REFERENCES

[1] L. Lu, X. Han, J. Li, J. Hua, and M. Ouyang, "A review on the key issues for lithium-ion battery management in electric vehicles," *Journal of Power Sources*, vol. 226, no. null, pp. 272–288, Mar. 2013. [Online]. Available: http://dx.doi.org/10.1016/j.jpowsour.2012.10.060

[2] N. Ginot, M. A. Mannah, C. Batard, and M. Machmoum, "Application of Power Line Communication for Data Transmission Over PWM Network," *IEEE Transactions on Smart Grid*, vol. 1, no. 2, pp. 178–185, 2010. [Online]. Available: http://ieeexplore.ieee.org/lpdocs/epic03/wrapper.htm?arnumber=5530396

[3] F. Grassi, S. A. Pignari, and J. Wolf, "Assessment of CAN performance for Powerline Communications in dc differential buses," in *2009 IEEE International Conference on Microwaves, Communications, Antennas and Electronics Systems*. IEEE, Nov. 2009, pp. 1–6. [Online]. Available: http://ieeexplore.ieee.org/xpls/abs_all.jsp?arnumber=5385959

[4] A. RACHID and F. COLLET, "Bus CAN," *Techniques de l'ingénieur Systèmes d'information et de communication*, vol. base docum, no. ref. article : s8140, 2014.

978-1-4799-4993-9/14 $31.00 © 2014 IEEE

Carbon Nanotube Based Temperature Sensors Fabricated by Large-Scale Spray Deposition

Engin Cagatay*, Aniello Falco, Alaa Abdellah, and Paolo Lugli

Institute for Nanoelectronics

Technische Universität München

Arcisstrasse 21 80333 Munich, Germany

*Email: engin.cagatay@nano.ei.tum.de

Abstract—We present high-performance temperature sensors which utilize spray deposited carbon nanotube (CNT) films as the active sensing material. In order to evaluate device performance, the change in device resistance with respect to change in temperature is monitored. The fabricated sensors show very good electrical response to the change in temperature. Relative change in resistance as high as 25% for an increase of temperature from 0°C to 80°C can be achieved. Also in this work, a comparative experiment is carried out in order to investigate the effect of CNT film thickness on the characteristics of CNT-based temperature sensors. Results show that films with higher thickness exhibit a lower temperature coefficient of resistance (TCR) as compared to those with lower thickness, while demonstrating higher stability and reproducibility. Finally, another important feature of the devices presented is that the change in resistances with respect to temperature change is nearly linear.

I. Introduction

With the advance of technology and demand for higher device performance across different application areas, the need for new materials with better properties becomes pronounced. Carbon nanotubes (CNT) are good candidates for many device applications, because since their discovery in 1991, they have shown novel electrical and optical properties [1]. As a result, they have been utilized for different applications in thin-film transistors [2], [3], chemical sensors [4], [5] and as transparent conductive electrodes [6], [7]. CNTs are also known to have good electrical response to change in temperature, which makes them very promising candidates for temperature sensing applications [8], [9].

In this work, CNT-based temperature sensors are fabricated with spray deposition technique and characterized. This technique enables high-throughput, low-cost processes for the fabrication of high-quality CNT films under ambient conditions onto a wide range of substrate materials having nearly arbitrary geometry [10]. For evaluation of device performance, the change in device resistance with respect to change in temperature is observed. Moreover, a comparative experiment is carried out in order to investigate the effect of CNT film thickness on the temperature sensor characteristics. The paper is organized as follows: In Section II, the materials utilized and methods used in fabrication of the sensors are presented. Results are shown in Section III first with a device having 5

This work was partly supported by the European Commission under grant agreement PITN-GA-2012-317488-CONTEST and by the TUM graduate school.

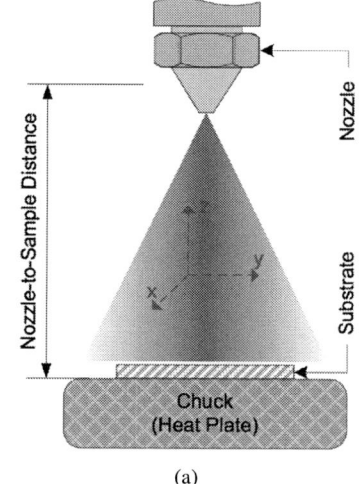

(a)

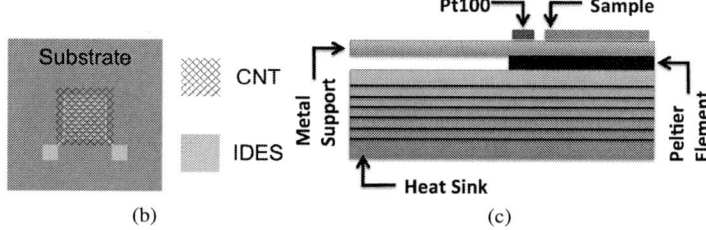

(b)　　　　　　　(c)

Fig. 1: a) Schematic representation of the spray valve and placement of the sample, b) The architecture of the fabricated device, c) Schematic representation of the measurement setup.

layer of CNT film thickness and then they are compared with those of two other devices having thicker CNT layers. Finally, the results are concluded in Section IV with final remarks.

II. Materials and Methods

Spray deposition method has been utilized in this work to deposit the CNT films. For this purpose an automated spray system, which consists of an air atomizing spray valve and an overhead motion platform, has been used. Once the process parameters are optimized, it provides good thickness control during deposition resulting in high-uniformity films and high reproducibility. The process parameters of interest are material flow rate, atomizing gas (N_2) pressure, nozzle-to-sample distance, substrate temperature, and motion speed. The CNT films are sprayed layer by layer following the method presented in [11] in order to end up with high-quality films, where 1 layer corresponds to CNT density of 5 CNTs/μm^2. Figure 1(a) shows the schematic representation of the spray valve and placement of the sample in the automated spray system from side view.

A. Solution Preparation

Due to Van der Waals forces between CNTs, they tend to stick each other and create bundles. In order to prevent this and achieve high-uniformity films during spraying, CNTs are first dispersed in an aqueous solution in spray deposition method. For this purpose a high molecular weight cellulose derivative, sodium carboxymethyl cellulose (CMC), is used. The first step of the solution preparation is to add sufficient amount of CMC into DI water in order to get a 0.5 wt% aqueous solution. After stirring the solution for at least 12 hours at room temperature, the SWNTs (P3-SWNT, Carbon Solutions Inc.) are added in order to achieve a concentration of 0.03 wt% of SWNTs in the final solution. The next step consists of the sonication of the solution for 15 minutes using a horn sonicator (Branson Sonifier S-450D). This step ensures that the individual CNTs are dispersed to a high extent in the solution. Finally, the solution is centrifuged at 15000 rpm for 90 minutes and the spray solution is obtained by decanting the top 80 % of the supernatants.

B. CNT Temperature Sensor Fabrication

During fabrication, silicon wafers with 200 nm of thermally grown silicon dioxide are used as substrates. First, a photolithography process is carried out to define an interdigitated electrode structure (IDES) on the samples (Figure 1(b)). Then, a 5nm/40nm Cr/Au layer is evaporated in order to realize the device metallization with IDES layout. The spacing between the electrodes of the structure is defined to be 200 μm for all samples, while the total active area of each device is 3x3 mm^2. After lift-off, the CNT film with a thickness of 5 layers is spray deposited on the IDES using the automated spray system mentioned above. Before the samples can be used for testing, it is necessary to remove the CMC dispersant from the sprayed CNT layer. Otherwise, because of the insulating nature of the CMC dispersant, the CNTs cannot touch each other and form a percolating network. Hence, a chemical treatment is carried out in HNO$_3$ for at least 12 hours at room temperature. As a result, the CMC dispersant is completely removed. After rinsing the samples in DI water and drying, they are encapsulated. The need for encapsulation arises from the fact that, as reported in [12], the electronic properties of CNTs are extremely sensitive to the presence of the oxygen in the environment. Therefore, following a heat treatment process on hot plate at 100°C for 2 hours, samples are encapsulated using a UV curable epoxy (OP-4-20632, Dymax) and glass. After curing the epoxy under UV light for 15 minutes, the samples become ready to be characterized.

C. CNT Temperature Sensor Characterization

The setup used for characterization of the devices in this work is composed of a heat sink, a Peltier heating element positioned under the sample for temperature control and a

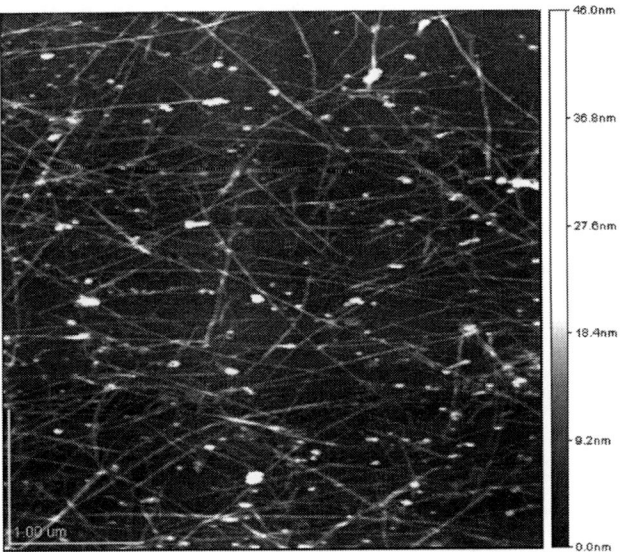

Fig. 2: AFM image of CNT film spray deposited on Device-A.

Pt100 commercial temperature sensor as the reference (Figure 1(c)). The main role of the heat sink is to enable cooling down to 0°C. Moreover, the reference sensor is positioned as close to the sample as possible to measure the temperature of the latter more accurately. In order to study the performance of the sensors, measurements with different temperature protocols have been conducted. The temperature protocols followed are 25°C-30°C 5 loops and 0°C-80°C 1 loop, where the temperature of the sensor is changed by steps of 5°C. At the beginning of each protocol, the temperature is kept constant at the first temperature step for a certain amount of time for the purpose of temperature stabilization. Then, it is changed by steps of 5°C and kept constant for a certain amount of time at each step. In order to plot the resistance versus temperature graph, the resistance values in each step were averaged for the last 30 seconds of that step.

III. RESULTS AND DISCUSSION

A. CNT Temperature Sensor Operation

The first step of the device charecterization is the inspection of the deposited CNT film by atomic force microscopy (AFM) (Figure 2). From the AFM image it is clearly seen that the deposited CNT film is uniform and does not contain bundles. Temperature sensor characterization is carried out by applying a sensing current and measuring the voltage drop across the two terminals of the device while the temperature of the sensor is changed. As a result, the change in device resistance with respect to change in temperature is monitored and recorded using a LabView program.

In order to observe the behaviour of the device over a wide range of temperatures, 1 loop of measurement is first carried out between 0°C-100°C (Figure 3). Figure 3(a) shows the temperature data measured by the reference temperature sensor during the measurement. In Figure 3(b), where the resistance data from Device-A is shown, decreasing resistance with increasing temperature indicates that the CNT film in this device has a semiconducting nature. Moreover, one of

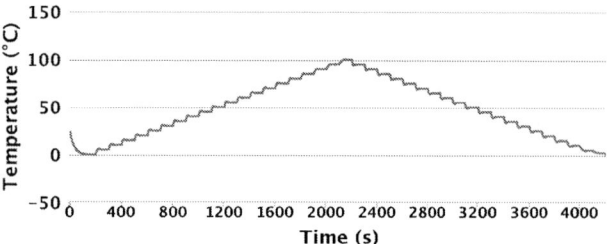

(a) Temperature data measured by the reference temperature sensor

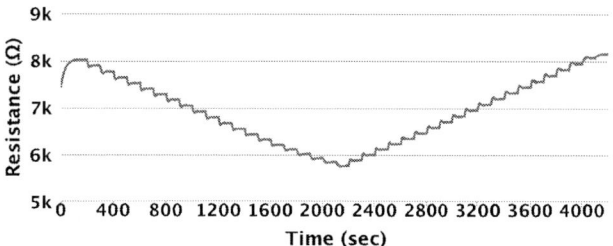

(b) Resistance data of Device-A during the measurement

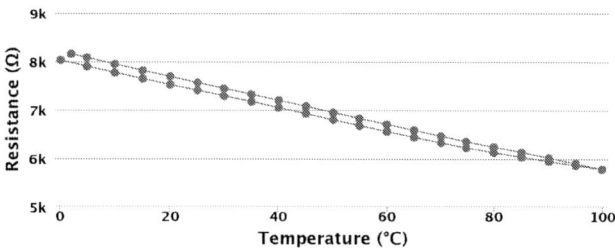

(c) Resistance vs temperature graph obtained by plotting the resistance data of Device-A and temperature data from reference sensor

Fig. 3: 0°C-100°C 1 loop measurement results for Device-A: a) Temperature values recorded by Pt100 reference sensor, b) Resistance values of the sample and c) change in resistance in response to change in temperature.

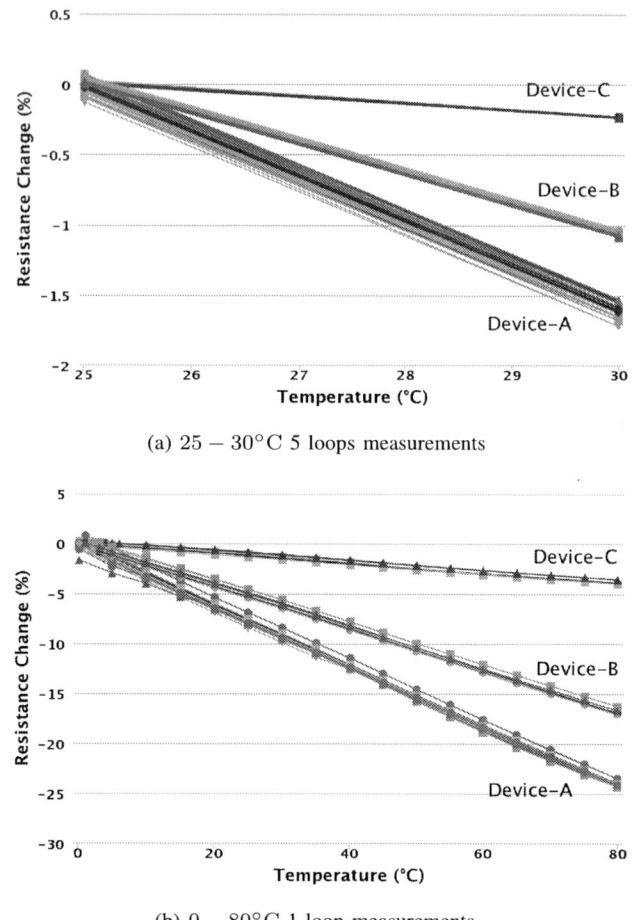

(a) $25 - 30°C$ 5 loops measurements

(b) $0 - 80°C$ 1 loop measurements

Fig. 4: Results obtained from a) $25 - 30°C$ 5 loops and b) $0 - 80°C$ 1 loop measurements of Device-A, Device-B and Device-C.

the most noticible features observed in the sensor response is the immediate change in resistance due to the change in temperature. Starting from $85°C$ the resistance starts to increase slightly during each temperature step, causing a small hysteresis in resistance which is also observed on Figure 3(c). Despite this small hysteresis, the device response between $0°C$ and $100°C$ can be said to be nearly linear.

Plotting relative change in resistance versus temperature data provides a better view of the device performance. Therefore, the rest of the figures show relative change in resistance versus temperature graphs of the measured data, which are calculated using the formula:

$$Rel.\ Change = \frac{R_f - R_i}{R_i} \cdot 100 (\%) \qquad (1)$$

where R_f is the average resistance at each temperature step and R_i is resistance at $0°C$, respectively. For instance, Figure 4 demonstrates the two different type of measurements done using Device-A. Figure 4(a) shows the $25°C$-$30°C$ 5 loops measurements, in which the temperature of the sample is varied

between $25°C$ and $30°C$ five times. This way, reproducibility of the measurement is studied. From the figure, it is clear that the resistance values are quite overlapping. Moreover, while resistance values seem to differ slightly for different measurements, they do not deviate too much. Another type of measurement is $0°C$-$80°C$ 1 loop, in which the temperature of the sample is increased from $0°C$ to $80°C$ and decreased back to $0°C$ by $5°C$ steps for one time. As shown in Figure 4(b), the measurements are quite reproducible in different runs. This superior device stability has been possible due to two reasons: i) the fixed oxygen impurity content of the CNT film throughout the measurements thanks to the epoxy and glass encapsulation of the device, ii) high CNT film quality as a result of the spray deposition method used.

The average resistance value at $25°C$ throughout all measurements with Device-A is measured to be $8451.60\ \Omega$. Additionally, resistance values recorded during these measurements overlap to a great extent. This means that the measurements are quite reproducible using the CNTs deposited by spray deposition technique and encapsulated with epoxy and glass. Temperature coefficient of resistivity (TCR) is generally used as a figure of merit for temperature sensors, which is defined as:

978-1-4799-4993-9/14 $31.00 © 2014 IEEE

TABLE I: An Overview of the Fabricated Devices

Device	Num. of layers	Resistance (Ω)	TCR (°C^{-1})	Rel std dev @25°C	Rel std dev @30°C
A	5	8451.60	-0.002954	0.280%	0.518%
B	13	1264.63	-0.002087	0.208%	0.330%
C	25	115.19	-0.000368	0.106%	0.126%

$$TCR = \frac{R_{100} - R_0}{100 \cdot R_0} \qquad (2)$$

where R_{100} is the resistance at 100°C and R_0 is resistance at 0°C. The TCR value calculated for this device is -0.002954 °C^{-1}. The negative TCR value indicates again a semiconductive behaviour of the CNT film used.

B. Effect of CNT Film Thickness

In order to observe the effect of CNT film thickness on temperature sensing performance and reproducibility, two additional devices with higher number of CNT film layers have been fabricated, namely Device-B and Device-C with 13 layers and 25 layers of CNT film, respectively (See Table I). The fabrication procedure and the rest of the parameters for these two devices, such as IDES spacing and encapsulation, are the same as Device-A. Figure 4 shows the results obtained with the same type of measurements carried out with Device-B and Device-C. 25°C-30°C 5 loops and 0°C-80°C 1 loop measurements show that with increased CNT film thickness the spread in device resistance gets smaller.

Device performances are compared further using additional parameters, such as average resistance at 25°C, TCR and relative standard deviation of resistances calculated at 25°C and 30°C (See Table I). Increasing TCR towards more positive values suggest that increasing thickness of CNT film leads to more metallic behaviour. Because, as reported in [13], CNTs with metallic behaviour have a positive TCR, whereas semiconducting ones have negative. This is also supported by the fact that with increasing film thickness, the average resistance decreases, i.e. the film becomes for conductive. Moreover, by comparing the relative standard deviation of device resistances, the spread in the measurements can be observed. Corresponding values in Table I suggest that with increasing thickness the measurements become more stable and reproducible. As a final comparison, the plotted data in Figure 4 suggest that Device-A yields the highest change in resistance. Moreover, for Device-C, the data from different runs seem to overlap more. This suggests that increase in CNT thickness results in more metallic CNT film behaviour and more reproducible measurements.

IV. CONCLUSION

We report a high performance temperature sensor, in which spray deposited CNT films are used as the active sensing material. Presented results show that the sensor has good electrical response to changes in temperature. For instance, Device-A with 5 layers of CNT film shows a relative resistance change of 25% for an increase of temperature from 0°C to 80°C. Another important feature of the reported device is that the change in resistance with temperature is nearly linear for all measurement protocols.

Also, in order to study the effect of CNT film thickness on temperature sensing performance and reproducibility, additional devices with higher film thickness have been fabricated. Parameters like average resistance at 25°C, TCR, relative standard deviation and the relative change in device resistances were examined to compare these devices. Results obtained show that devices with a thicker CNT film exhibit lower resistance and TCR. They also have lower relative standard deviation of resistance suggesting that they are more stable and reproducible in terms of device performance.

REFERENCES

[1] L. Hu, D. S. Hecht, and G. Grner, "Carbon nanotube thin films: Fabrication, properties, and applications," *Chemical Reviews*, vol. 110, no. 10, pp. 5790–5844, 2010.

[2] M. Ha, Y. Xia, A. A. Green, W. Zhang, M. J. Renn, C. H. Kim, M. C. Hersam, and C. D. Frisbie, "Printed, sub-3v digital circuits on plastic from aqueous carbon nanotube inks," *ACS Nano*, vol. 4, no. 8, p. 43884395, August 2010.

[3] C. Wang, J. Zhang, , and C. Zhou, "Macroelectronic integrated circuits using high-performance separated carbon nanotube thin-film transistors," *ACS Nano*, vol. 4, no. 12, p. 71237132, November 2010.

[4] D. Zhang, K. Ryu, X. Liu, E. Polikarpov, J. Ly, M. E. Tompson, and C. Zhou, "Transparent, conductive, and flexible carbon nanotube films and their application in organic light-emitting diodes," *Nano Lett.*, vol. 6, no. 9, p. 18801886, October 2006.

[5] A. K. K. Kyaw, H. Tantangand, T. Wu, L. Ke, J. Wei, H. V. Demir, Q. Zhang, and X. W. Sun, "Dye-sensitized solar cell with a pair of carbon-based electrodes," *J. Phys. D, Appl. Phys.*, vol. 45, no. 16, p. 165103, April 2012.

[6] J. Li, Y. Lu, Q. Ye, M. Cinke, J. Han, and M. Meyyappan, "Carbon nanotube sensors for gas and organic vapor detection," *Nano Lett.*, vol. 3, no. 7, p. 929933, July 2003.

[7] M. Penza, R. Rossi, M. Alvisi, G. Cassano, and E. Serra, "Functional characterization of carbon nanotube networked films functionalized with tuned loading of au nanoclusters for gas sensing applications," *Sens. Actuators B, Chem.*, vol. 140, no. 1, p. 176184, June 2009.

[8] A. D. Bartolomeo, M. Sarno, F. Giubileo, C. Altavilla, L. Iemmo, S. Piano, F. Bobba, M. Longobardi, A. Scarfato, D. Sannino, A. M. Cucolo, and P. Ciambelli, "Multiwalled carbon nanotube films as small-sized temperature sensors," *Journal of Applied Physics*, vol. 105, no. 6, 2009.

[9] M. De Volder and D. Reynaerts and C. Van Hoof and S. Tawfick and A. J. Hart, "A temperature sensor from a self-assembled carbon nanotube microbridge," *Sensors, 2010 IEEE*, pp. 2369–2372, November 2010.

[10] A. Abdellah, A. Abdelhalim, M. Horn, G. Scarpa, and P. Lugli, "Scalable spray deposition process for high-performance carbon nanotube gas sensors," *Nanotechnology, IEEE Transactions on*, vol. 12, no. 2, pp. 174–181, March 2013.

[11] A. Abdellah, A. Abdelhalim, F. Loghin, P. Kohler, Z. Ahmad, G. Scarpa, and P. Lugli, "Flexible carbon nanotube based gas sensors fabricated by large-scale spray deposition," *Sensors Journal, IEEE*, vol. 13, no. 10, pp. 4014–4021, October 2013.

[12] P. G. Collins, K. Bradley, M. Ishigami, and A. Zettl, "Extreme oxygen sensitivity of electronic properties of carbon nanotubes," *Science*, vol. 287, no. 5459, pp. 1801–1804, 2000.

[13] K. Cheng-Yung, C. L. Chan, G. Chie, L. Chien-Wei, S. Shiuan-Hua, and J. Ting, "Nano temperature sensor using selective lateral growth of carbon nanotube between electrodes," *Nanotechnology, IEEE Transactions on*, vol. 6, no. 1, pp. 63–69, January 2007.

978-1-4799-4993-9/14 $31.00 © 2014 IEEE

Methodology Modeling of MaE-fabricated Porous Silicon Nanowires

Aleandro Antidormi*†, Diego Chiabrando*†, Maria Grazia Graziano*, Luca Boarino† and Gianluca Piccinini*

*Department of Electronics and Telecommunications, Politecnico di Torino, Torino, Italy

Email: {aleandro.antidormi, diego.chiabrando, mariagrazia.graziano, gianluca.piccinini}@polito.it

†Nanofacility Piemonte, Electromagnetic Division, I.N.Ri.M., Torino, Italy

Abstract—**Porous Silicon Nanowires (PS-NWs) represent very promising electronic devices with a wide range of applications, all taking advantage of the irregular microscopic structure of PS. The strong variability of the electrical properties of the material with technological process makes computer simulation of PS-nanowires a cumbersome task. Here we present a simple model of PS-NWs which can be implemented for simulations in physics-based software TCAD Atlas thus giving a contribution to the investigation of the effects of the peculiar structure on the material resistivity. Extensive simulations have been performed on PS-Nws with microscopic characteristics deduced from our experimental fabricated devices. We investigated the dependence of current from applied voltage and channel doping also in relation with the electrical field inside the device and with the carriers' mobility.**

Finally we suggest an electrical model for PS-NWs easily implemented in commercial circuit simulators (Eldo in this case). It will allow to exploit the actual simulation tools also for the analysis and design of circuits where PS-based devices are present.

I. INTRODUCTION

Silicon Nanowires (Si-NWs) are largely being studied in the emerging scenario of (nano)electronics for their versatile properties, their obvious integrability with standard Silicon based technologies, and for the fabrication techniques in several cases exploiting techniques based on self-assembly not requiring lithographic processes. A major distinction should be done between pure-cristalline Si-NWs, mainly used for computation applications [1]–[3] and NWs presenting an irregular, non cristalline, i.e. porous, structure. Porous Silicon (PS) is an interesting material with several possibilities of application, in particular in gas sensing [4]- [5]. It is composed of a network of small Si crystals with a typical size of a few nanometers. This structure makes transport properties of PS rather unique. On one hand, PS cannot be considered a bulk wide-bandgap semiconductor because of its granular structure in the nanometer scale. On the other hand, it is not a insulator, since its crystallites retain the lattice structure of the bulk material. Moreover, the electrical properties of this material largely depend from the geometrical dimensions of the pores as well as their concentration. The consequence is a complete absence of numerical models in commercial device simulators apt to describe the behaviour of PS, as well as of any other kind of models. This is a crucial missing point, especially from the electronic applications point of view, since the design of PS sensors systems cannot be verified and engineered as it would be necessary especially in critical applications, as gas sensing might be.

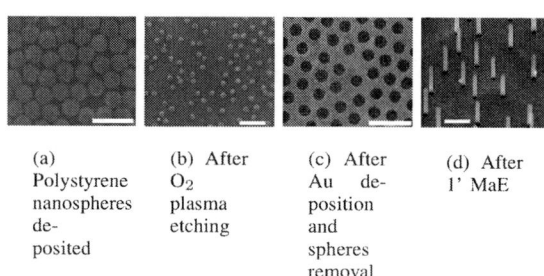

| (a) Polystyrene nanospheres deposited | (b) After O_2 plasma etching | (c) After Au deposition and spheres removal | (d) After 1' MaE |

Fig. 1. SEM images: Si NWs fabrication steps (marker: 500 nm)

In this paper we aim at filling the gap in two steps. First we introduce a simplified method to model a PS nanowire in a physical level simulator (Silvaco Atlas) using as a reference real geometrical data on PS nanowires (PS-NWs) we fabricated. The goal of this first step, which is the main focus of this paper, is to get an solid insight on the behaviour of this simplified structure for varying silicon doping and considering realistic distribution of pores in the NW volume. Second we suggest an equivalent electrical model of PS-NWs that, starting from the physical level results, could be simulated using spice-like simulators (Eldo from Mentor Graphics in our case).

The structure of the paper is as follows: in the first section we present a short description of the technological process to fabricate our PS-NWs and images showing the resulting devices. The description of the model for PS-NW is treated in section II with the results of some parametric analysis we performed. Finally, an electrical model of PS-NWs is introduced for future developments.

II. TECHNOLOGICAL PROCESS

A. Fabrication of Si NWs

Metal-assisted etching (MaE) is the technique used to fabricate our PS-NWs [7]. This technique consists in an electroless etching of silicon in HF/H_2O_2 solution. A thin film noble metal catalyzer ($Ag, Au, Pt, AuPd$) is deposited on the surface of the silicon sample. Depending on the geometrical properties of the deposited metal film, different structures can be obtained (nanowires, pores or even more complicated objects). In our case, we chose highly p-type doped silicon (10-20mΩ· cm <100>). Polystyrene (PS) nanospheres (diameter: 210nm ± 5%) were spread on the silicon sample by spin coating in order to obtain a self-assembled hexagonal-packed

978-1-4799-4993-9/14 $31.00 © 2014 IEEE

monolayer (fig. 1-a). Their diameter has been later reduced of a factor 2 via oxigen plasma etching (fig. 1-b). At this point, a 20 nm thick Au film was deposited on the surface using e-gun evaporator. This machine allows to achieve a continuous film also in case of very reduced thickness. Finally, PS nanospheres have been removed in ultrasonic bath obtaining an "antidot" metal pattern (fig. 1-c).

We performed MaE dipping the sample in a solution of ((H_2O:H_2O_2:HF 1:1:3)) at a temperature of $60°C$ for 1 minute. Fig. 1-d clearly shows NWs whose diameter is the same of that of the reduced nanospheres.

B. TEM analisys

Transmission Electron Microscopy (TEM) was performed after having scratched the fabricated NWs on a TEM grid. Hence, their structural properties could be investigated and extrapolated data have been used as a starting point for subsequent simulations. Fig. 2-a shows the internal structure of a NW: darker dots are nanocrystals of silicon immersed in a porous structure. Using ImageJ software it has been possible to locate the nanocrystals exactly and, consequently to statistically evaluate mean and distributions of their geometrical properties. Starting from these images, a binary (b/w) copy is generated (fig. 2-b) and a graph of the area distribution can be plotted, like in fig. 2-c. According to the results in the picture nanocrystals have an average area of 4.1 nm.

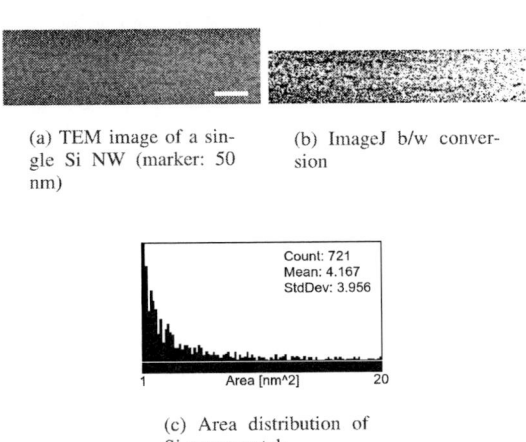

(a) TEM image of a single Si NW (marker: 50 nm)

(b) ImageJ b/w conversion

(c) Area distribution of Si nanocrystals

Fig. 2. TEM image analisys

III. PS-NW MODELING AND SIMULATIONS

In this section we describe the procedure we followed to construct a simple model for a porous silicon nanowire; this model is suitable for numerical analysis in the physics-based software TCAD Atlas [9]. Electrical characterization of a PS-NW can then be performed and data on the main electrical quantities of the material can be obtained together with their dependence on physical and geometrical parameters.

As it is evident from Fig. 2-b, representing the results of elaboration of an image of the fabricated nanowires, silicon nanocrystals are immersed among non-crystalline regions forming an irregularly networked pore structure. At the best of authors' knowledge, no models are present in physical simulators for porous materials, nor physical data are present to describe their electrical properties (mobility, energy gap, dielectric constant,...); the reason for this being the unavoidable dependence of the electrical parameters of the materials from the dimensions of the pores and from their effective shape. Our research for a simulative model of PS-NWs is also driven by a second reason. The uniform distribution of the pores along the channel and the statistics of their geometrical properties make an analytical treatment of the electrical problem rather impractical. Consequently, the simulative approach seems to be the most promising way to tackle the problem and, in order for it to be effective, a physical model of these devices is strongly required.

A. PS-NW model

We modelled a PS-NW as a 3D-wire with square section composed of p-doped crystalline silicon ($N_A = 10^{15}$ atoms/cm^3). Its lateral surfaces are drilled through slit pore segments randomly arranged and filled with air. These pores have a squared section; their number, depth and position are randomly chosen according to the geometrical properties of our fabricated devices in a way which will be explained in the following. An image of a silicon nanowire with this structure is shown in Fig. 3-a.

This is the simplest model of a pore in a PS-NW: air has been inserted in the volume created from the slits. However, in general, non-crystalline silicon and/or gas fill that space, thus changing the electrical conductivity of the whole device. To keep things treatable by Silvaco Atlas software, air is actually the proper choice. Secondarily, this allowed us to focus attention on the dependence of the nanowire resistivity from the physical parameters of the device. It is worthy to be mentioned that the peculiar geometry chosen for the pores does not restrict the range of devices to be simulated and subsequently does not limit the validity of the obtained results. In fact, we left the pores the possibility to spatially overlap and intersect each other so that ramified porous entities can arise thus making the pores' geometry rather arbitrary.

The contacts have been chosen to be non-rectifying metal contacts in order not to include the effects of the Schottky contacts in the transport process.

We have described the structure of the PS-NW in Atlas software by means of a rectangular mesh, denser in correspondence of the interfaces between different materials (silicon-air/metal) and sparser elsewhere. This guarantees a faster convergence of the numerical method implemented in the nodes of the grid. The image of a mesh as described in Atlas is depicted in fig.3-b.

To compute the relevant electrical quantities in the PS-NW we adopted a Boltzmann distribution of the carriers in the channel. Moreover, their mobility has been chosen to depend on the local electrical field according to the model FLDMOB implemented in Silvaco Atlas. It takes into account the effect of velocity saturation of the carriers through a reduction in the effective mobility according to the Caughey and Thomas expression which provides a smooth transition between the low-field and high-field regimes ([9]). As it will become clear, non-uniformities of the electrical field in the channel are responsible for a gradual reduction of the current with the applied voltage. This mobility model coherently describes the current saturation effect in presence of non-homogeneous

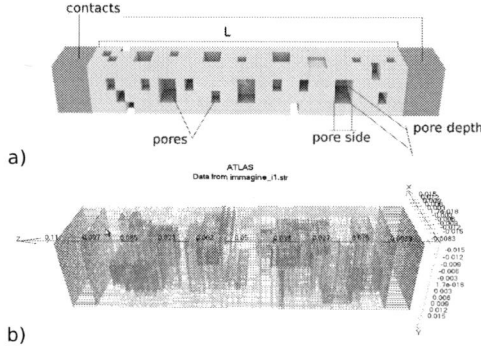

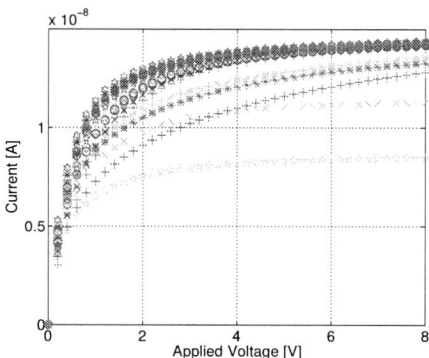

Fig. 3. a) Scheme of the model for Porous Silicon Nanowire; b) Atlas mesh of a PS NW (pores are coloured in purple).

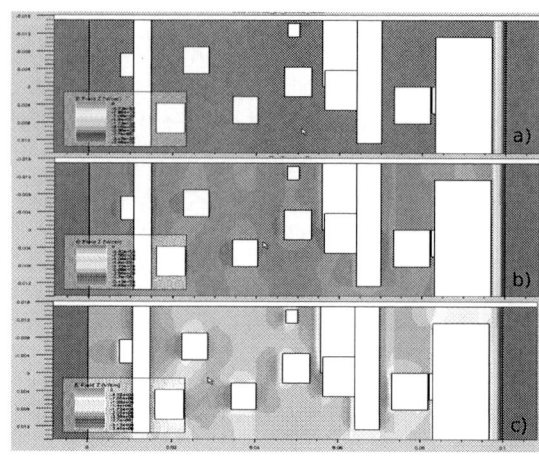

Fig. 5. Electric field distribution in a section of a PS-NW with $L = 100nm$ and section $900nm^2$ for a) $V = 0.2V$, b) $V = 2.5V$ and c) $V = 8V$. The uniform electric field for low voltages becomes non-uniform for higher voltages provoking the velocity saturation of carriers in some channel regions. Higher voltages increases the extensions of saturated regions and total current consequently plateaus.

Fig. 4. Voltage-Current characteristics of some PS-NWs having different geometrical parameters (position, depth, side). Their position along the wire is uniformly distributed, their depth and side are normally distributed.

electrical fields.

The nanowire we analysed through simulations has a channel length of $100nm$ and a squared section of $900nm^2$. In order to obtain valuable information about the fabricated nanowires, we simulated many devices whose pores present geometrical characteristics similar to those shown in fig. 2-b. According to the results of a statistical analysis on the actual pores, their distribution along the simulated wire is uniform while the pore depths and sides are normally distributed with average values respectively $\mu_{depth} = 19.2nm$, $\mu_{side} = 6.3nm$. The corresponding standard deviations are $\sigma_{depth} = 11.4nm$, $\sigma_{side} = 3.2nm$.

B. Simulation Results

Fig. 4 shows the current-voltage characteristics for some actual realizations of silicon nanowires: each curve corresponds to a particular device with specific distribution and properties of pores. As it is clear, the current behaviour is strongly non-linear, each curve reaching a saturation value for sufficiently high voltage applied. Specifically, in the regime of low voltage, the current in each nanowire increases linearly with a slope which is influenced by the concentration and dimensions of the pores. A larger amount of silicon removed by the pores produces a greater differential resistivity of the device since less conductive paths are available for current. As the voltage is increased, the current rate gradually decreases. This behaviour can be justified by considering the distribution

of the electrical field in the channel (Fig. 5): for sufficiently high voltage the electric field ceases to be uniform, becoming more intense in certain particular regions (lighter regions in the figures). There, the electric field will soon reach a value such that the mobility saturation will occur according to the transport mobility chosen. The current flow in that region will consequently be limited by the mobility of the carriers thus reducing the overall charge flow along the device. Current will continue to increase with voltage until more regions will be interested by a strong electrical field. Then the current will plateau. The voltage at which the current deviates from linearity and the saturation current are actually dependent on the specific configuration of pores along the channel.

Fig. 6 shows the differential resistivity of a PS-NW averaged over the ensemble of devices which have been analysed. The vertical bars in the figure represent the standard deviation. The resistivity naturally grows with voltage, while the increment of the corresponding standard deviation can be easily explained: for high voltages, a small difference in the variation of current for two devices can result in a large difference in the differential resistivities: these consequently present a larger deviation from the average value.

We finally discuss the results of the simulation of the ensemble of different nanowires for different doping channels. Higher doping levels produce lower resistivity as shown in Fig. 7 where the average resistivity and its standard deviation computed at $V = 1.0V$ are depicted. Moreover, larger deviations from the average values are found for increasing doping level, due to the fact that, for fixed applied voltage, electrical field in more doped channels is higher with respect to the electrical fields in channels with lower doping. In more doped devices this reduces the carriers' mobility (increasing resistivity) counteracting the effects of the augmented concentration of carriers through doping. Doping can then be used to trim the sensitivity of the sensor, as well as applied voltage has to be carefully defined at the PS-NW sensor engineering phase. In other words, detailed data on what can be obtained are essential during a sensor design and simulation phase, thus

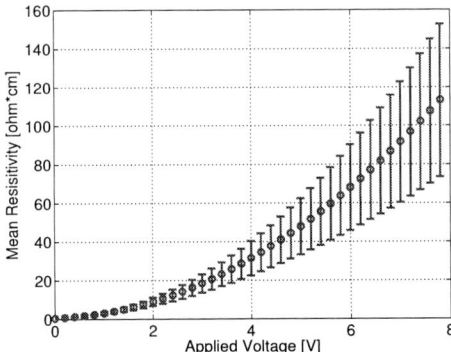

Fig. 6. Average differential resistivity of an ensemble of PS-NWs with $L = 100nm$ and section $900nm^2$ as a function of the applied voltage. The vertical bars represent the standard deviation.

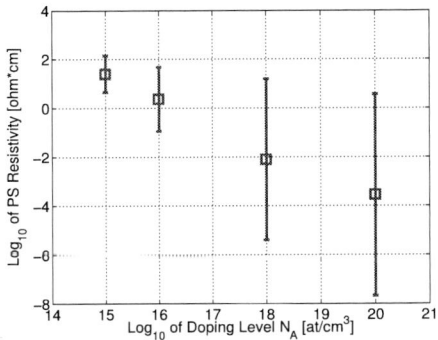

Fig. 7. Differential Resistivity computed at $V = 1.0V$ averaged over a large number of nanowires with different channel doping.

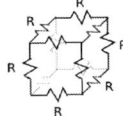

Fig. 8. Elementary block of the circuit model of a PS-NW.

confirming the importance of the modelling attempt presented in this work.

IV. A POSSIBLE ELECTRICAL MODEL FOR PS-NW

We conclude the paper with the suggestion of a possible electrical model which could be implemented in commercial circuit simulators, like Eldo. This would allow to exploit the existing simulation tools for the analysis and design of circuits where PS-NWs are present. Once the physical parameters of the device have been explored through a physics-based simulation (Sect.III) and the electrical model has been defined, the analysis and design of complex circuits by means of a commercial simulator is dramatically simplified and time-efficient.

The whole geometrical structure of our nanowire is subdivided in many elementary blocks of cubic shape. Each cube is substituted by the block depicted in fig.8, a 3D 4-terminal component made of simple resistors representing the main possible conductive paths for carriers in a piece of material ([5]). The values of the resistances included in each block

is actually variable, depending on the applied voltage, the physical dimension chosen for the elementary cube and from the doping level of silicon. In general they can be obtained from the electrical field distribution computed via a physical simulation. By connecting more of these blocks together, the equivalent circuit model for a PS-nanowire is obtained. The validity of the model can be extended to the deep nanometer regime, when quantum confinement effects start to affect the transport properties. In our simple model these effects can be included by adding some more conductive paths (resistances) in the elementary blocks. For example, resistances of proper value could be inserted across the pores to describe the tunnelling current contribution while quantum resistances can be added if ballistic regime is to be included in the computations. We actually developed the elementary block in Eldo and built a whole model for a PS-NW. However, a complete analysis of the values of the resistances and a validation of the model is yet to come.

V. CONCLUSION

In this paper we described the state of our research in fabrication and modelling of Porous Silicon NWs. Starting from a statistical analysis of the fabricated NWs, we devised a simple model for a PS NW which can be implemented in the physics-based simulator TCAD Atlas. We used it to investigate the current-voltage characteristics of an ensemble of PS-NWs with different pores' distribution and doping: we revealed the relation between the current non-linear behaviour in terms of the electrical field in the channels. Finally, a simple modular electrical circuit for a PS NW has been suggested, capable of being implemented in commercial circuit simulators.

In the future, electrical characterization of the fabricated nanowires could give more information on the electrical properties of Porous Silicon and good agreement between numerical and experimental data will be sought.

REFERENCES

[1] S. Frache, M. Graziano, M. Zamboni, "A Flexible Simulation Methodology and Tool for Nanoarray-based Architectures". In: IEEE International Conference on Computer Design, Amsterdam, 3-6 October. 2010, pp. 60-67

[2] S. Frache, M. Graziano, M. Zamboni, "Nanoarray Architectures Multilevel Simulation", ACM J. on Emerging Technologies in Computing Systems, Vol 10, N.1, pp. 1-20, 2014

[3] S. Frache, D. Chiabrando, M. Graziano, E. Enrico, L. Boarino, M. Zamboni, "Silicon nanoarray circuits design, modeling, simulation and fabrication". In: IEEE International conference on Nanotechnology (IEEE-NANO), Birmingham, UK, 20-23 August. 2012, pp. 1-5

[4] L. Boarino, F. Geobaldo, S. Borini, A.M. Rossi, P. Rivolo, M. Rocchia, E. Garrone and G. Amato, "Local environment of boron impurities in porous silicon and their interaction with NO 2 molecules", *Phys. Rev. B* 64, 205308, 2001.

[5] Z. Gaburro, P. Bettotti, M. Saiani, L. Pavesi et. al., "Role of microstructure in porous silicon gas sensors for NO_2", *Appl.Phys.Lett.* 85, n. 4, pp. 555-557, Jul. 2004.

[6] Ben-Chorin, M. and Möller, F. and Koch, "Nonlinear electrical transport in porous silicon", Phys. Rev. B vol. 49, n.4, pp. 2981-2984, 1994.

[7] Z. Huang, N. Geyer, P. Werner, J. de Boor and U. Gösele, "Metal-Assisted Chemical Etching of Silicon: A Review", *Advanced Materials*, Vol.23, n.2, pp. 285-308, Jan. 11, 2011.

[8] M.J. Sailor, *Properties of Porous Silicon*, London, U.K.:L. Canham, IEE Inspec, 1997.

[9] Silvaco Int, ATLAS User's Manual A 2D-3D Numerical Device Simulator <http://www.silvaco.com>

Analysis and Modeling of Four-Folded Vertical Hall Devices in Current Domain

Hadi Heidari, Edoardo Bonizzoni, Umberto Gatti, and Franco Maloberti
Department of Electrical, Computer, and Biomedical Engineering
University of Pavia, Pavia, Italy
E-mails: {hadi.heidari, edoardo.bonizzoni, franco.maloberti}@unipv.it, gattiu@alice.it

Abstract—This paper presents a four-folded current-mode vertical Hall device. The current spinning technique is applied to a vertical Hall sensor driven in current mode to eliminate the offset and to increase the sensitivity. Different geometries have been studied and simulated by using a simulator based on finite element method. A four-folded three contacts vertical Hall device model displayed the lowest residual offset and the best sensitivity. Simulations results, obtained in two different environments, are compared and discussed. COMSOL results are validated with respect to the electrical behavior of an 8-resistor Verilog-A model implemented in Cadence environment. Simulations show that the achieved sensitivity can be better than 160 mT⁻¹, a remarkable performance for vertical Hall sensors.

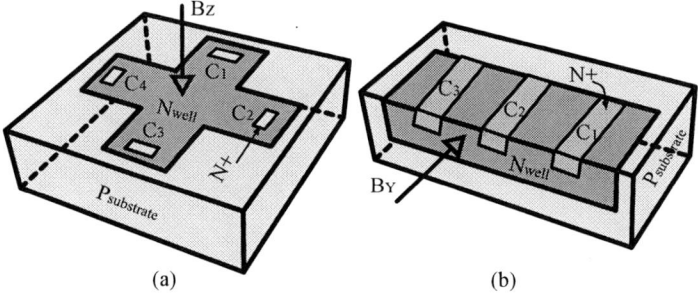

Fig. 1. (a) Horizontal Hall effect device and (b) three contacts vertical Hall device cross-sections in CMOS technology.

I. Introduction

Hall magnetic sensors have widely been used as they are fully compatible with integrated CMOS technologies.

The horizontal Hall effect device, also referred to as Hall plate, has an horizontal plate toward the semiconductor substrate and measures the magnetic field, B_Z, perpendicular to the sensor plate surface, as shown in Fig. 1(a). The use of a symmetric structure permits to cancel the offset by applying the current spinning method.

Recently, vertical Hall sensors (VHS) have been proposed. They are sensitive to the in-plane component of the magnetic field, B_Y, and detect the magnetic field in the plane of the sensor plate, as shown in Fig. 1(b). Several geometries have been studied in [1], [2], [3] and [4]. Among them, a four-folded geometric structure that uses the current-spinning technique, provides high resolution and low offset performance [5]. However, the sensitivity of vertical Hall devices is low. Values published in the open literature are 1.2 %T⁻¹, 1.16 %T⁻¹ and 4 %T⁻¹, as reported in [4], [5] and [6], respectively. This paper shows that the use of current as output quantity instead of voltage increases the sensitivity. Moreover, integrating a current into a capacitor for a given time slot provides relatively large signals and averages the noise.

The outline of the paper is as follows: in the next Section, the current-mode technique for four-folded vertical Hall devices and its operation during current spinning four phases are discussed. In Section III, simulations in 2D COMSOL and implementation of Verilog-A model of the four-folded vertical Hall device are presented. Comparison of model simulation

results achieved in COMSOL and in Cadence environment are also discussed. Section IV draws some conclusions.

II. Current-Mode Vertical Hall Sensor

Typically, the output of Hall devices is voltage. Voltage-mode Hall sensors can be biased in two ways: voltage biasing and current biasing. The current biasing consists in injecting a current, I_{bias}, into one terminal, for instance C_1 in Fig. 2(a), and drawing it from another non-adjacent terminal (for instance, C_3). When biasing in the voltage mode, a voltage is applied to contact C_1 with contact C_3 grounded. In both configurations, a magnetic field gives rise to the Hall voltage (V_{Hall}) across the other two terminals. In the current biased voltage-mode, the current-related sensitivity is calculated as

$$S_{V_I} = \left| \frac{V_{Hall}}{I_{bias} \times B} \right| \tag{1}$$

where the units are V·A⁻¹·T⁻¹. In the voltage biased voltage-mode, the voltage-related sensitivity is defined as

$$S_{V_V} = \left| \frac{V_{Hall}}{V_{bias} \times B} \right| \tag{2}$$

where the units are Tesla inverse (V/(V·T)=T⁻¹) [4].

On the contrary, a current-mode Hall sensor uses current as output signal. This technique has been already described in [7] for horizontal Hall sensors.

Since different structures have been proposed in the literature, this study preliminarily compares them, when used in

978-1-4799-4993-9/14 $31.00 © 2014 IEEE

the current-mode using 2D COMSOL. The architectures with five contacts, [2] and [3], and eight symmetrical contacts, [1], demonstrate limitations for offset performance and complexity of readout. The current-mode four-folded vertical Hall device model turned out to be the optimum selection. Fig. 2(a) shows its cross-section. Four lined up N-wells with three n-plus contacts each make the device. Metal connections create a ring of external contacts. The device has four terminals, C_1, C_2, C_3 and C_4. If a current enters one of the terminals, say C_1, and exits from terminal C_3, supposing to have equal structures, the current through C_2 and C_4 is zero with zero magnetic field. A magnetic field parallel to the surface in Y-direction alters the current flow, as shown in Fig. 2. The solid-lines represent the current trajectory in the zero magnetic field case and the dashed-lines show the current trajectory at nonzero magnetic field and under the influence of Lorenz force. A differential current, I_{Hall}, at the two output terminals results. The sensitivity of the device is

$$S_I = \left| \frac{I_{Hall}}{I_{bias} \times B_Y} \right| \tag{3}$$

In order to compare the operation of the vertical Hall sensor in voltage-mode and current-mode let draw it as shown in Fig. 2(b) and Fig. 2(c), respectively. The effect of the magnetic field in the four sections of the sensor is depicted in the figures. For the voltage-mode operation, the current path is shortened in S_4 and lengthened in S_2. However, because of the symmetry, the voltage of the output terminal is in the middle for both S_2 and S_4 (V_{O1} and V_{O2}). Thus, they are ineffective for the magnetic sensitivity. As shown shortly, we can use equivalent resistances from pairs of the four terminals. They, as shown in the figure, are $(R-\Delta R_V)$, $(R+\Delta R_V)$ on the top section, and $(R+\Delta R_V)$, $(R-\Delta R_V)$ in the bottom section. V_{O1} and V_{O2} can be expressed as

$$V_{O1} = \frac{V_{bias}}{2R}(R - \Delta R_V) \tag{4}$$

$$V_{O2} = \frac{V_{bias}}{2R}(R + \Delta R_V) \tag{5}$$

The differential Hall voltage is given by

$$V_{HV} = V_{bias} \frac{\Delta R_V}{R} \tag{6}$$

The operation in the current-mode, shown in Fig. 2(c), connects the output terminals to ground. The equations describing the equivalent model are

$$I_1 = \frac{V_x}{(R - \Delta R_I)} \tag{7}$$

$$I_2 = \frac{V_x}{(R + \Delta R_I)} \tag{8}$$

that yield the differential Hall current as

$$I_{HI} = I_{bias} \cdot 2 \cdot \frac{\Delta R_I}{R} \tag{9}$$

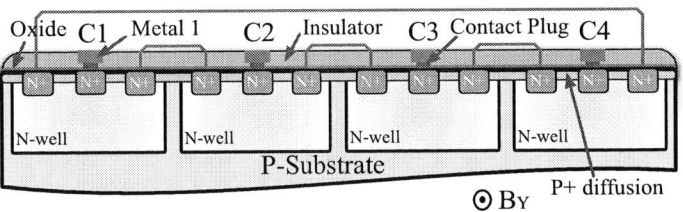

(a)

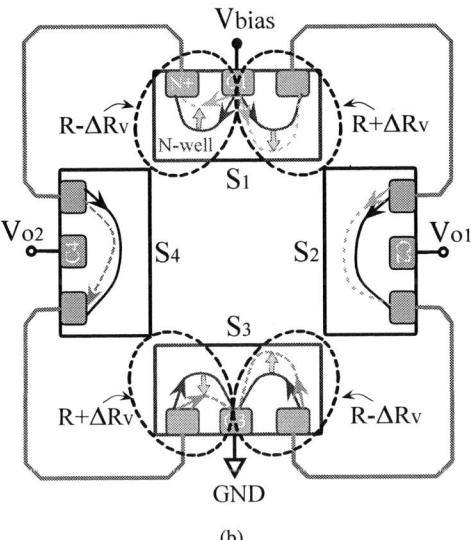

(b)

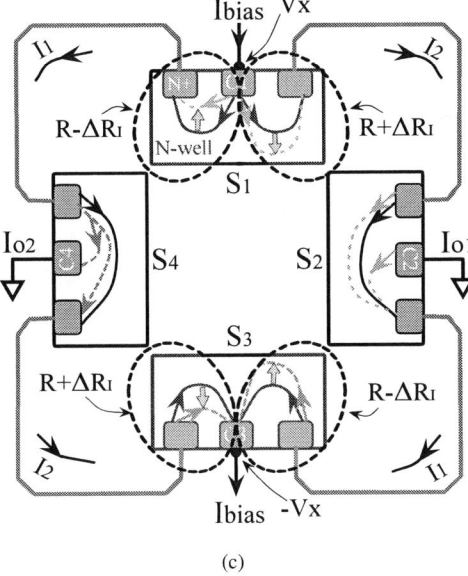

(c)

Fig. 2. (a) Three contacts four-folded vertical Hall device cross-section, (b) vertical Hall device in voltage-mode, (c) vertical Hall device in current-mode configuration: biasing conditions and currents flows.

The last expression gives a sensitivity two times bigger than the one obtained in the voltage-mode, supposing $\Delta R_V = \Delta R_I$. However, since in the current mode even S_2 and S_4 are active, we have $\Delta R_V < \Delta R_I$.

This work analyzes the influence of current injection and

current output signals on vertical Hall effect sensors performances, including noise and offset, with the aid of COMSOL simulations. Moreover, an accurate 8-resistor network model for the four-folded vertical Hall device is described in Verilog-A and tested in a Cadence environment. Simulation results obtained in COMSOL and in Cadence show excellent matching and system potentiality.

III. Models Description

This Section presents the symmetrical models of the current-mode vertical Hall device. First, the COMSOL model is illustrated and simulation results are presented. Later on, the 8-resistor model described in Verilog-A and Cadence-based simulation results follow.

A. COMSOL Model and Simulations

A two-dimensional model of the current-mode vertical Hall device has been implemented and simulated in COMSOL with the model parameters summarized in Table I. Fig. 3(a) shows the four configurations of the model geometry and the surface

TABLE I
Model Parameters

Symbol	Value	Parameter
N_D [cm^{-3}]	$7.78 \cdot 10^{16}$	Doping
$R_{\square}$ [Ω/sq]	1028	N-well Sheet Resistance
σ_n [S/m]	1040	N-well Conductivity
σ_p [S/m]	10	P-Sub Conductivity
B_Y [T]	$0 \sim 0.5$	Magnetic Field
t [m]	$6 \cdot 10^{-6}$	Silicon Thickness
I_0 [μA]	100	Bias Current

electrical distribution of the three contacts four-folded vertical Hall device when a magnetic field of 0.5 T is applied. The simulation uses a nominal bias current of 100 μA.

In order to compensate for the mismatches due to possible masks misalignment during fabrication, e.g. non-equal distance between contacts, the sensor uses the current spinning method. Simulations include a mismatch in the terminal C_4 of

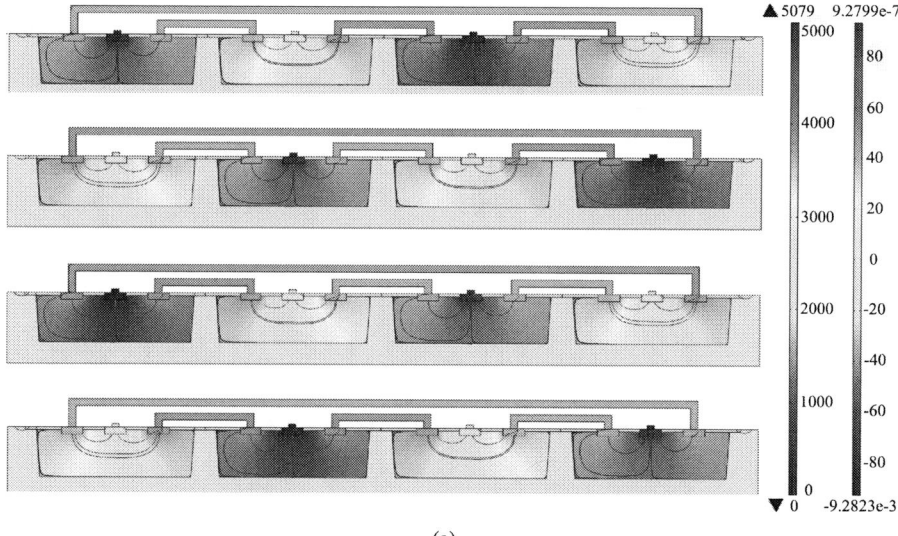

(a)

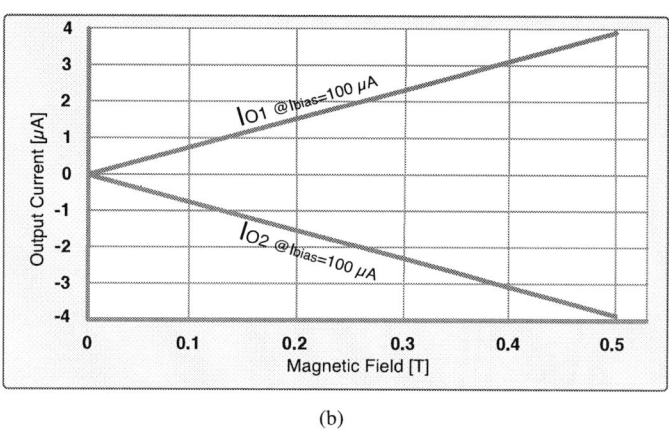

(b)

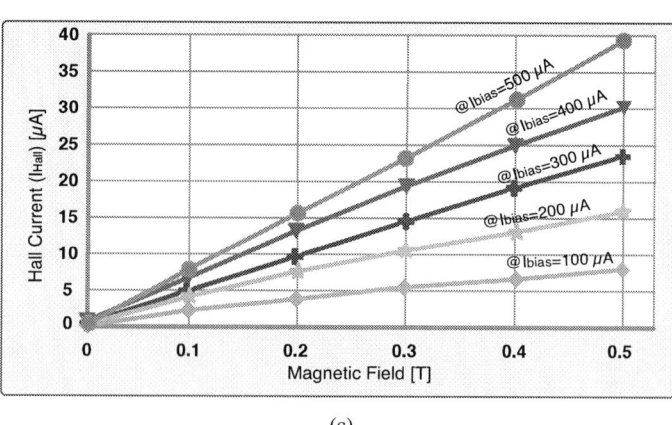

(c)

Fig. 3. (a) Simulation of three contact vertical Hall device in COMSOL environment and current streamline of four modes of operating in the current-mode, (b) Simulated sensor output currents as a function of the magnetic field at 100 μA bias current after four phases, and (c) Simulated sensor Hall current as a function of the different biasing current at different magnetic field.

978-1-4799-4993-9/14 $31.00 © 2014 IEEE

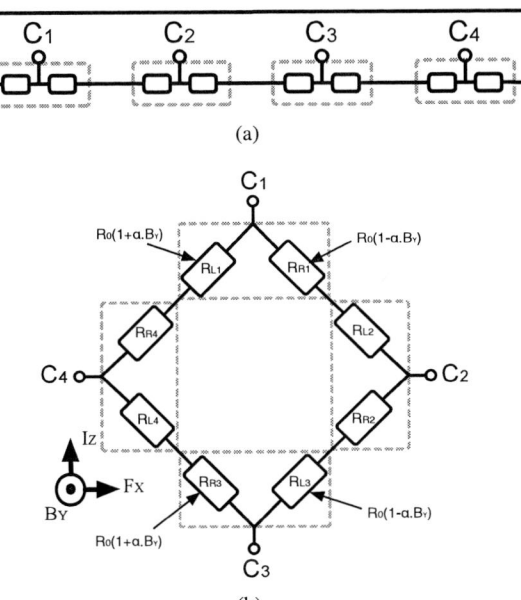

(a)

(b)

Fig. 4. (a) The equivalent model implemented in Verilog-A using eight resistors, (b) A same model in a Wheatstone bridge depiction.

the sensor. Fig. 3(b) shows the simulated average output currents of the vertical Hall device, when changing the magnetic field within the 0 to 0.5 T range after the four current spinning phases. The offset is zero and the maximum differential output current (Hall current) is almost 8 μA for a magnetic field equal to 0.5 T. These current levels can be transformed into suitable voltages by integrating the current signal over a given period of time.

Fig. 3(c) draws the simulated Hall currents when the bias current (I_{bias}) is ranging from 100 μA to 500 μA with a step of 100 μA and the magnetic field ranges from 0 to 0.5 T with steps of 0.1 T.

B. Verilog-A Model

The vertical Hall device is simplified into a series of eight resistors, as schematically shown in Fig. 4(a). Each contact terminal has two resistors, one on the left (R_L) and one on the right (R_R). The initial values of all resistors at zero magnetic field are equal (R_0). Fig. 4(b) shows the same model arranged in a Wheatstone bridge, which has been modeled and described using Verilog-A language so that it can be simulated in the Cadence environment. It includes four electrical terminals (C_1, C_2, C_3 and C_4) and eight resistors (four R_L and four R_R). The values of these resistors are controlled by three parameters: the external magnetic field, the initial value of resistors, R_0, and a magnetic resistance coefficient, α:

$$R_L = R_R = R_0(1 \pm \alpha.B_Y) \qquad (10)$$

The magnetic resistance coefficient, α, is defined as the average of the initial values of resistors, R_0, in the presence and absence of the magnetic field.

Simulations have been performed in Cadence using 100 μA input bias current and considering the magnetic field in the

range from 0 to 0.5 T. Results are summarized and compared with COMSOL simulations in the following sub-Section.

C. Simulations Comparison

In order to show the accuracy of the Verilog-A model, the simulation results have been compared with the ones achieved with the COMSOL model, as summarized in Table II. The results are in excellent agreement and the difference is less than 0.1%.

TABLE II
NUMERICAL COMPARISON BETWEEN THE SIMULATIONS OF THE VERTICAL HALL PLATE IN
COMSOL AND VERILOG-A

Magnetic Field	COMSOL		VERILOG-A	
$B_Y[T]$	$I_{Hp}[\mu A]$	$I_{Hn}[\mu A]$	$I_{Hp}[\mu A]$	$I_{Hn}[\mu A]$
0	0	0	0	0
0.1	0.59772	-0.59772	0.59798	-0.59798
0.2	1.43061	-1.43061	1.43021	-1.43021
0.3	2.26121	-2.26121	2.26102	-2.26102
0.4	3.08317	-3.08317	3.08351	-3.08351
0.5	3.90042	-3.90042	3.90018	-3.90018

IV. CONCLUSION

This study shows the effectiveness of a current-mode Hall sensor. The symmetric four-folded model has been simulated using COMSOL and modeled with an equivalent circuit using Verilog-A for behavioral simulations in the Cadence environment. Simulation and modeling results show that the vertical Hall sensor obtains a sensitivity better than 16 %T^{-1}, which reveals that the proposed technique enables superior performance of magnetic field sensitivity, in terms of signal to noise ratio compared to sensitivity performance of voltage-mode Hall sensors.

As future work, the analysis and the design of a complete Hall sensor microsystem and peripheral readout circuits will be continued, all using a conventional CMOS technology.

REFERENCES

[1] M. Banjevic, "High bandwidth cmos magnetic sensors based on the miniaturized circular vertical hall device," Ph.D. dissertation, EPFL, 2005. [Online]. Available: http://infoscience.epfl.ch/record/166133

[2] T. Kaufmann, M. Vecchi, P. Ruther, and O. Paul, "A computationally efficient numerical model of the offset of cmos-integrated vertical hall devices," *Sensors and Actuators A: Physical*, vol. 178, pp. 1–9, 2012.

[3] O. Paul, R. Raz, and T. Kaufmann, "Analysis of the offset of semiconductor vertical hall devices," *Sensors and Actuators A: Physical*, vol. 174, pp. 24–32, 2012.

[4] G.-M. Sung and C.-P. Yu, "2-d differential folded vertical hall device fabricated on a p-type substrate using cmos technology," *IEEE Sensors Journal*, vol. 13, no. 6, pp. 2253–2262, June 2013.

[5] C. Sander, R. Raz, P. Ruther, O. Paul, T. Kaufmann, M. Cornils, and M. Vecchi, "Fully symmetric vertical hall devices in cmos technology," in *Proc. of IEEE Sensors*, Nov 2013, pp. 1–4.

[6] E. Schurig, "Highly sensitive vertical hall sensors in cmos technology," Ph.D. dissertation, EPFL, 2005. [Online]. Available: http://infoscience.epfl.ch/record/33603

[7] H. Heidari, U. Gatti, E. Bonizzoni, and F. Maloberti, "Low-noise low-offset current-mode hall sensors," in *Proc. of 9th Conference on Ph.D. Research in Microelectronics and Electronics (PRIME)*, June 2013, pp. 325–328.

978-1-4799-4993-9/14 $31.00 © 2014 IEEE

3-Terminal Tungsten CMOS-NEM Relay

Martín Riverola, Gabriel Vidal-Álvarez, Francesc Torres and Núria Barniol
Dept. d'Enginyeria Electrònica
Universitat Autònoma de Barcelona
08193, Cerdanyola del Vallés, Spain
martin.riverola@uab.es

Abstract—**The present work describes the design, fabrication and experimental results of a 3-terminal laterally actuated tungsten nanoelectromechanical (NEM) relay which is monolithically integrated in a 0.35 μm commercial standard CMOS technology. The movable structure is released by means of a simple one-step maskless wet etching. The switch shows an abrupt switching with less than 5 mV/decade and a good on-off current ratio of $\sim 10^4$ although it exhibits an on-state contact resistance R_{ON} around 500 MΩ. Also, the relay is cycled up to 1500 times in ambient conditions showing great endurance but variability in its contact.**

Keywords—**Switches; relays; tungsten; CMOS technology.**

I. INTRODUCTION

Nowadays we live in a digital era surrounded by plenty of electronic devices that have become a pervasive part of our daily lives, available everywhere and anytime. This fact has been possible thanks to the rapid growth in integrated circuit (IC) technology over the last decades. As we know, the main block of these electronic devices is the IC chip which is generally built using the technology so-called complementary metal oxide semiconductor (CMOS) that uses symmetric pairs of p-type and n-type metal oxide semiconductor field effect transistors (MOSFETs) to achieve complementary switching behaviour.

Basically, the rapid CMOS technology advancement was achieved following simple scaling rules to reduce the dimensions of the metal oxide semiconductor (MOS) transistors, thus achieving better performance and more integration density, reaching more functionality per chip and reduced cost per area [1]. However, in the deep-submicron regime is increasingly difficult to continue this scaling-down approach: a power crisis has emerged due to a fundamental issue inherent in the MOSFET operating principle [2, 3].

The CMOS circuit power dissipation can be expressed as a sum of two power dissipation mechanisms [4]: active power dissipation P_A and standby power dissipation P_S which are given by

$$P_A = KCV_{DD}^2 f \tag{1}$$
$$P_S = I_{OFF}V_{DD} \tag{2}$$

where in (1), K is the switching factor, C is the total capacitance and f is the operating clock frequency. In (2), I_{OFF} is the total leakage current. V_{DD} is the applied voltage in both terms.

As CMOS is scaled to smaller dimensions and the supply voltage V_{DD} is reduced to maintain reliability and reasonable active power dissipation, the threshold voltage V_{TH} needs to

be scaled down at the same rate as V_{DD} in order to achieve the desired circuit switching speed [4]. However, the supply voltage is not expected to be scaled down further in smaller technology nodes due to the thermal limit k_BT/q [5], where k_B is the Boltzmann constant, T is the temperature and q is the elementary charge. As a result, the P_S is increasing rapidly becoming the dominant factor in total power dissipation at small channel lengths.

Nanoelectromechanical (NEM) relays are proposed to solve the standby power dissipation problem because they have zero leakage current and abrupt ON/OFF transition, since the device can be considered as an ideal switch. In fact, the use of mechanical switches for computing is not a new idea, but dates back to the 1930's [6]. Due to impressive advances in manufacturing techniques and photolithography of bulk-planar processing technology over the past few decades, the interest in the development of surface micromachining processes for micro- and nanoelectromechanical systems (M/NEMS) has been renewed for several applications in mechanical computing such as ultra-low power IC applications [7], static random access memories (SRAM) [8], NEM relays integrated in CMOS circuitry [9] and operating in extremely hard conditions [10].

This paper presents the results on the design, fabrication and electrical testing of electrostatically actuated tungsten (W) NEM relays which are defined using the commercial standard CMOS technology AMS 0.35 μm. Section II outlines the design flow for fabricating the relays. Section III shows the theoretical analysis. Section III reports the results of the electrical testing, including current-voltage (I-V) curves and cycling data. Finally, section IV presents the conclusions.

II. CMOS-NEM RELAY DESIGN AND FABRICATION

A. NEM Relay Design

The designed relay is based on electrostatic in-plane actuation of a cantilever beam which is defined using a three terminal (3-T) topology. These 3T's are denoted as source (S), gate (G) and drain (D) like in a MOSFET switch device as shown in Fig. 1. In this configuration, the actuated structure is electrically connected to the S electrode and its position is determined by the voltage difference between the G and S terminals V_{GS}, which can be high without affecting the drain current I_{DS}. In fact, I_{DS} is determined by the drain-to-source voltage V_{DS}, which can be small to prevent excessive Joule heating in the contact electrode. In Fig. 1, the relay geometry is depicted showing their main design parameters. In Table I, the design parameter values are summarized.

978-1-4799-4993-9/14 $31.00 © 2014 IEEE

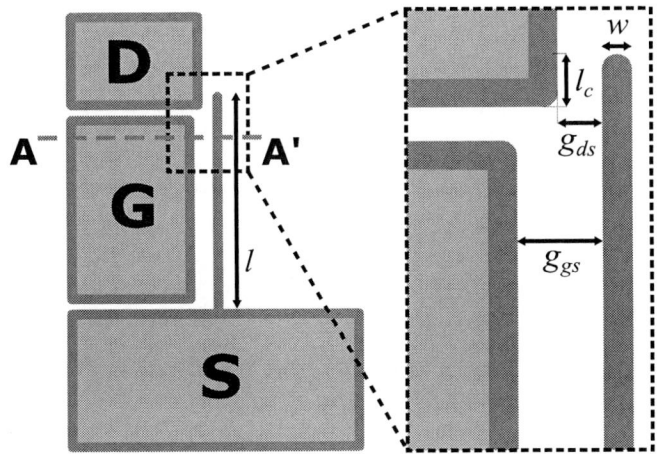

Fig. 1. Schematic showing the 3-T relay geometry and the main design parameters which consists of the suspended beam fixed by one end (S), the actuation electrode (G) and the contact electrode (D).

TABLE I. CANTILEVER DESIGN PARAMETERS AND THEIR VALUES

Design Parameters		Value (μm)
length	l	20, 30, 40, 50
width	w	0.5
thickness	t	1
gate-to-source gap	g_{gs}	0.55
drain-to-source gap	g_{ds}	0.45
contact length	l_c	1.5

B. CMOS-M/NEMS Fabrication Approach

The fabrication process is performed using the commercial standard CMOS technology AMS 0.35 μm which is based on a p-substrate, 2 polysilicon layers with a minimum feature size of 0.35 μm and 4 stack metal layers of TiN-Al-TiN. As we know, in CMOS technology the VIA layers are used to interconnect different metal layers. In this work, a cantilever using VIA3 layer is defined violating the design rules of the technology (fixed size of the VIA layer). VIA3 is chosen because is made of tungsten (W), a hard material which provides us great endurance against mechanical stress [11].

Fig. 2(a) shows the cross-section view along A-A' in Fig. 1 of the designed relay using the above mentioned technology. As we can see, VIA3 is used to define the beam as well as the wall/limit of the electrodes (see Fig. 1 and 2). The minimum separation between VIA3 layers defined by the design rules is 0.45 μm. Note that the electrodes are made by a sandwich of MET4 and MET3 with VIA3 between them. Note also that the design is complemented by a grounded n-well to avoid parasitic electrical signal in the MEMS performance.

In order to complete the whole fabrication sequence is necessary to release the movable structures because they are buried in SiO$_2$ (Fig. 2(a)). To do that, after the CMOS process is completed, one-step maskless wet etching process is done by means of a buffered solution of hydrofluoric acid (BHF). In this way, the structural layer is released as cross-sectional Fig. 2(b) view represents. A scanning electron microscope (SEM) image of a fabricated and released 3-T relay 30 μm long is shown in Fig. 3.

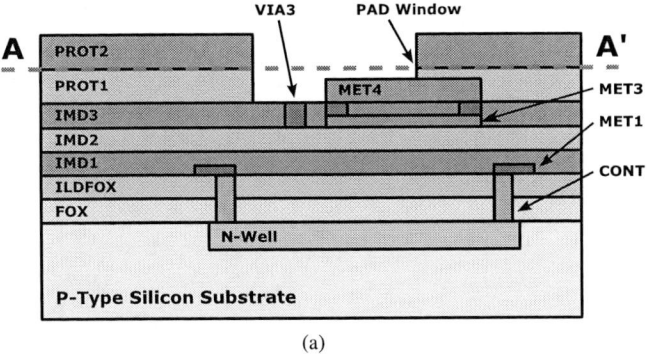

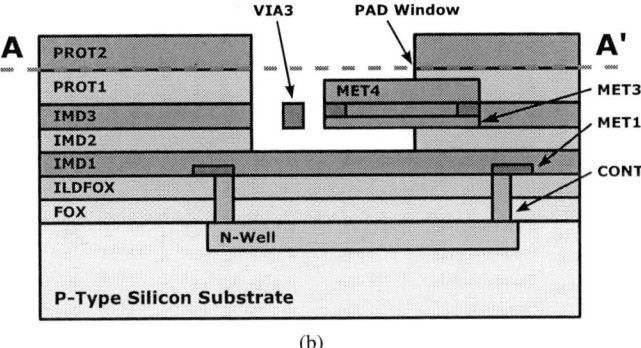

Fig. 2. Cross-section (taken along A-A' in Fig. 1) of a designed relay (a) before and (b) after post-CMOS one-step maskless wet etching process done by means of a BHF solution.

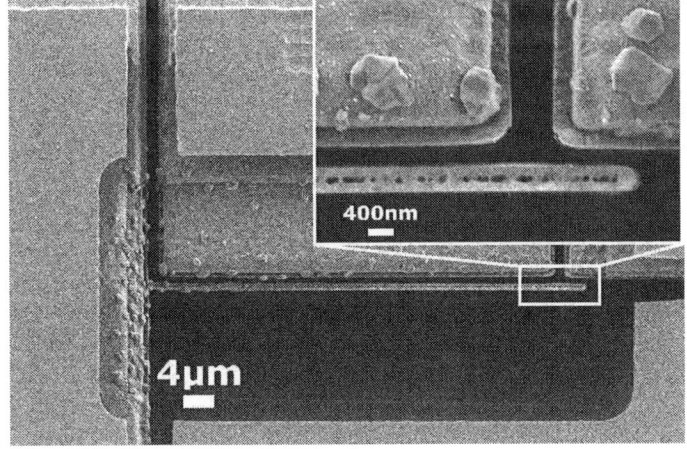

Fig. 3. SEM image of a released 3-T CMOS-NEM relay 30 μm long.

III. THEORETICAL ANALYSIS

Simulations are performed using ARCHITECT parametric libraries provided by the Coventor MEMS design software. In Fig. 4(a), a transient simulation of a 20 μm long cantilever's tip displacement as a function of the applied voltage is depicted. The transient simulation is used here in a pseudo-static manner to get the full contact which is so-called pull-in voltage V_{PI}. The pulse source is a ramp from 0 to 70 V in 10 s. Note that the cantilever tip touches the contact electrode at 40 V (Fig. 4(b)). As the actuation voltage increases, the tip of the cantilever flattens onto the actuation pad until it touches the actuation electrode and the end of the tip is lifted up (Fig. 4(c)).

978-1-4799-4993-9/14 $31.00 © 2014 IEEE

A way to reduce the V_{PI} is increasing the beam length, since the longer it is the softer will behave. The beam length is parametrized because it is the only degree of freedom that the CMOS technology allow us to modify. The other parameter dimensions are set by the design rules of the chosen technology.

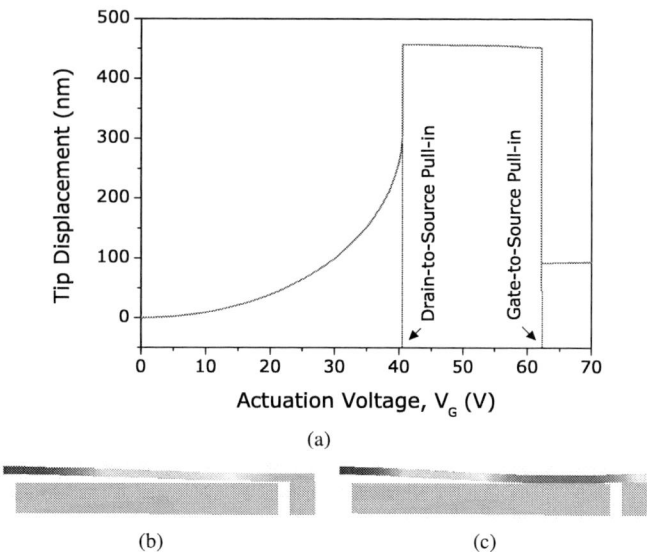

(a)

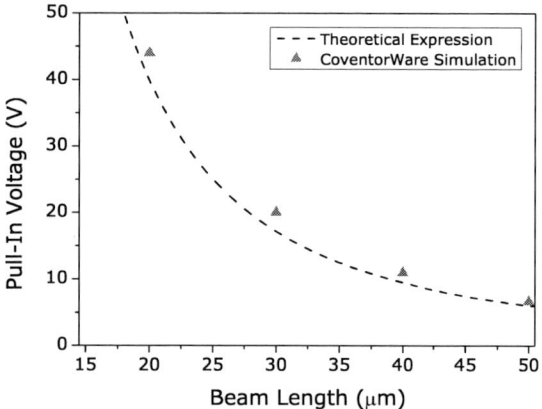

(b)   (c)

Fig. 4. Simulation (a) of a cantilever with l=20 μm applying a voltage ramp from 0 to 70 V in 10 s. Cantilever (b) makes pull-in at 40 V with the drain and (c) at 62 V with the gate electrode.

Fig. 5. Pull-in comparison between theoretical pull-in expression [12] and CoventorWare FEM solver simulations of cantilevers with lengths of 20, 30, 40 and 50 μm.

An easy manner to theoretically calculate the V_{PI} is assuming that the cantilever sustains a linear deformation shape deflection [12] which follows as

$$V_{PI} = \sqrt{\frac{0.88kg^3}{\epsilon A_C}} \quad (3)$$

where k is the spring constant of the cantilever, g is the gap separation between the movable beam and the actuation electrode and A_C is the actuation area. In Fig. 5, the V_{PI} calculated through (3) is compared with the simulation performed using the CoventorWare FEM solver for each designed beam length. As we can see, the analytical expression has quite a good fitting compared with the FEM simulation.

IV. EXPERIMENTAL RESULTS

A. I-V Curves

To evaluate the basic performance of the VIA3 NEM relay, we first examine 5 individual switching events in a cantilever of length 20 μm. Fig. 6 demonstrates the I_{DS}-V_{GS} characteristic of a typical hysteretic behaviour of a NEM relay. To perform this experiment we used the semiconductor analyzer Agilent B1500A sweeping V_G from 0 to 45 V and then back to 0 V, fixing V_{DS} to 10 V. As we can see, the I_{ON}/I_{OFF} ratio is approximately 10^4 exhibiting an abrupt switching less than 5 mV/decade, the V_{PI} is 40 V and the release voltage V_{RE} around 35 V. The maximum I_{DS} current is approximately 20 nA, thus the contact resistance R_{ON} is around 500 MΩ.

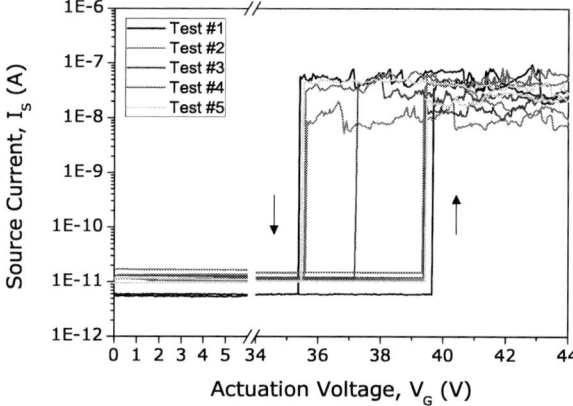

Fig. 6. Hysteresis I-V curves of 5 switching cycles at room ambient. Currents are plotted in logarithmic scale. The measured cantilever beam is 20 μm long.

B. Stress Test

Using the set-up measurement proposed in Fig. 7, we performed interrupted cycling tests of 20 s of duration until reaching a total of 1500 s with a pulse signal of 20 Vpp and 1 Hz over a cantilever of length 30 μm. These measurements were taken fixing V_{DS} to 10 V and fixing the compliance current at 1 μA (using a resistor of 10 MΩ).

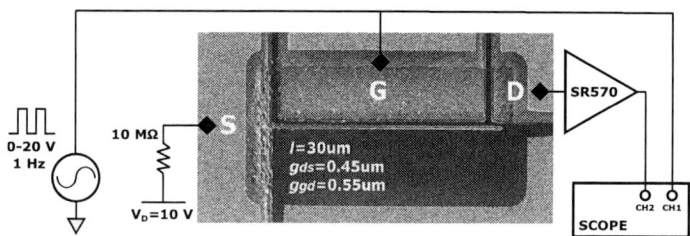

Fig. 7. Measurement scheme for stress test of the device. The output is amplified by a current preamplifier (SR570) with a sensibility of 1 nA/V.

An I-V curve was performed in order to obtain the V_{PI} voltage to set the pulse levels of the waveform as shown in Fig. 8. The first 20 s and the last 20 s recorded of the cycling test are shown in Fig. 9(a) and Fig. 9(b) respectively. Finally, another I-V curve was recorded trying to see what changes were produced in the electrical response due to the stress cycling test (Fig. 8). After 1500 cycles, the maximum current level in the R_{ON} is increased while V_{PI} is still happening at 16 V, proving the yield of the relay.

978-1-4799-4993-9/14 $31.00 © 2014 IEEE

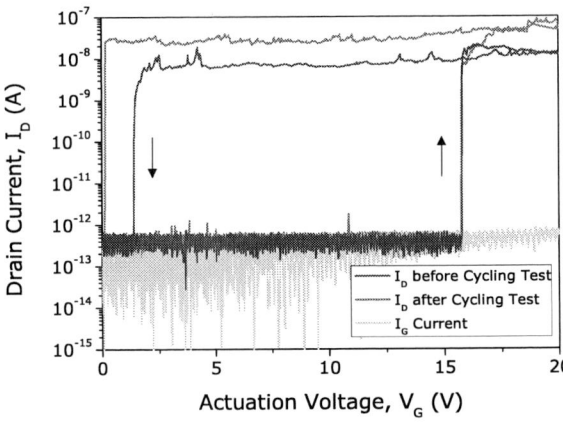

Fig. 8. I-V curves recorded before and after the cycling stress test. The gate current I_G at the noise floor level shows no contact on the gate electrode.

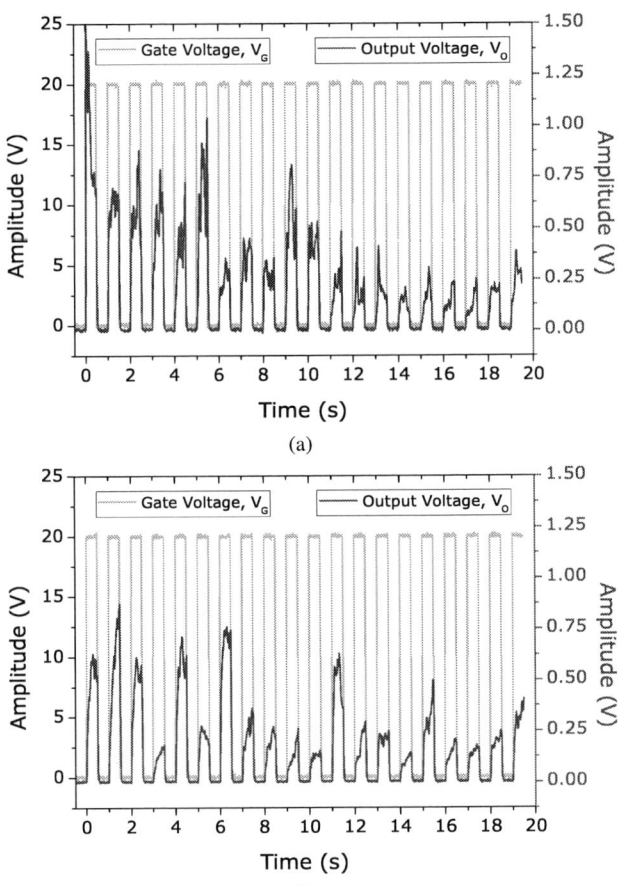

Fig. 9. Time domain evolution of (a) first and (b) last 20 switching events.

Note that the change in amplitude of the output voltage over these long cycles indicates that the on-state contact resistance R_{ON} is varying in each cycle. We think that these amplitude changes along with the non-expected high R_{ON} value are due to how the cantilever comes into contact with the electrode. To elucidate this hypothesis we took SEM images of the contact cross-section as shown in Fig. 10. It seems that the cantilever makes contact firstly on the TiN layer which protrudes more than the rest of layers reducing in this way the expected effective contact area.

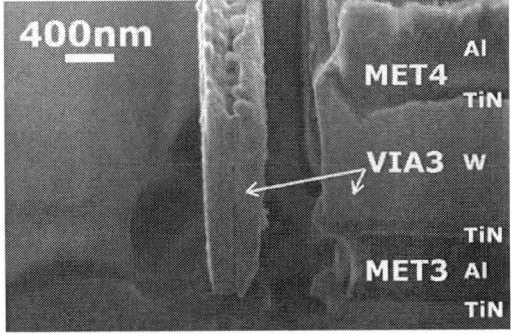

Fig. 10. SEM image of the contact cross-section of a released relay.

V. CONCLUSION

In summary, a 3-T NEM relay was successfully fabricated using a commercial standard CMOS technology and its characteristics were investigated. It showed abrupt on/off switching and an excellent on/off current ratio of 10^4. Furthermore, the relay showed repetitive switching operation up to 1500 cycles in ambient conditions and is still operative. However, it exhibits a high (> 500 MΩ) and a varying contact resistance in every cycle which has to be improved with new design features.

ACKNOWLEDGMENT

This work has been partially funded by the Spanish Government and the European Union FEDER program under project TEC2012-32677 (NEMS-in-CMOS).

REFERENCES

[1] Y. Taur, and T. H. Ning, *Fundamentals of Modern VLSI Devices*, Cambridge University Press, 1998.

[2] E. J. Nowak, *Maintaining the benefits of CMOS scaling when scaling bogs down*, IBM Jou. of Res. and Develop., vol. 46, pp. 169-186, 2002.

[3] ITRS. Available at *http://www.itrs.net*

[4] B. Davari, R. H. Dennard, and G. G. Shahidi, *CMOS Scaling for High Performance and Low Power – The Next Ten Years*, Proc. of the IEEE, vol. 83, no. 4, pp. 595-606, 1995.

[5] H. Kam, T.-J. King Liu, E. Alon, and M. Horowitz, *Circuit-Level Requirements for MOSFET-Replacement Devices*, IEDM, pp. 1, 2008.

[6] V. Pott, H. Kam, R. Nathanael, J. Jeon, E. Alon, and T.-J. King Liu, *Mechanical Computing Redux: Relays for Integrated Circuit Applications*. Proc. of the IEEE, vol. 98. no. 12, pp. 2076-2094, 2010.

[7] J. O. Lee, Y.-H. Song, M.-W. Kim, M.-H. Kang, J.-S. Oh, H.-H Yang, and J.-B Yoon, *A Sub-1-Volt Nanoelectromechanical Switching Device*, Nature Nanotechnology, vol. 8, no. 1, pp. 36-40, 2012.

[8] S. Chong, K. Akarvardar, R. Parsa, J.-B. Yoon, R. T. Howe, S. Mitra, and H.-S. Philip Wong, *Nanoelectromechanical Relays Integrated with CMOS SRAM for Improved Stability and Low Leakage*, IEEE/ACM Int. Conf. on Comp.-Aid. Design – Dig. of Tech. Papers, pp. 478-484, 2009.

[9] S. Chong, B. Lee, S. Mitra, R. T. Howe, and H.-S. Philip Wong, *Integration of NEM Relays with Silicon CMOS with Functional CMOS-NEM Circuit*, IEDM, pp. 30.5.1 - 30.5.4, 2011.

[10] T. He, R. Yang, V. Ranganathan, S. Rajgopal, M. A. Tupta, S. Bhunia, M. Mehregany, and Philip X.-L. Feng, *Silicon Carbide (SiC) Nanoelectromechanical Switches and Logic Gates with Long Cycles and Robust Performance in Ambient Air and at High Temperature*, IEEE Int. Elec. Dev. Meeting, pp. 4.6.1-4.6.4, 2013.

[11] Y. Chen, R. Nathanael, J. Jeon, J. Yaung, L. Hutin, and T.-J King Liu, *Characterization of Contact Resistance Stability in MEM Relays with Tungsten Electrodes*, J. Microelectromech. Syst., vol. 99, pp. 1-3, 2012.

[12] G. Abadal, Z. J. Davis, B. Helbo, X. Borrisé, R. Ruiz, A. Boisen, F. Campabadal, J. Esteve, E. Figueras, F. Pérez-Murano, and N. Barniol, *Electromechanical Model of a Resonating Nano-Cantilever-based Sensor for High-Resolution and High-Sensitivity Mass Detection*, Nanotechnology, vol. 12, pp. 100-104, 2001.

Tunable Transimpedance Sustaining-Amplifier for High Impedance CMOS-MEMS Resonators

G. Sobreviela, A. Uranga and N. Barniol.
Dept. of Electronic Engineering
Universitat Autònoma de Barcelona
08193, Cerdanyola del Vallés, Spain
guillermo.sobreviela@uab.es

Abstract— **This paper presents a full-custom tunable gain transimpedance amplifier (TIA) designed to work as a sustaining amplifier for a monolithic CMOS-MEMS oscillator. Based on a capacitive detection of the MEMS current, the implemented differential structure allows to compensate a wide range of motional resistances (between 1.4MΩ to 20MΩ) exhibiting an exceptional low input current noise of 32fA/√Hz @ 20 MHz.**

Keywords— *CMOS-MEMS; MEMS-based Oscillatorr; Radio-Frequency MEMS.*

I.   INTRODUCTION

The importance of the quartz crystals during the last decades in the field of frequency references has been due to their extremely high quality factor Q, their excellent stability against temperature variations and aging mechanisms and excellent phase noise performance. However, quartz crystals remain discrete elements, being not integrable in CMOS technologies increasing size and cost.

One of the solutions proposed to achieve fully integrated oscillators is based on the use of microelectromechanical resonators which can be implemented on-chip with the CMOS technology as it has been reported [1-3].This leads to microelectromechanical reference oscillators that try to achieve the quality standards of the currently commercialized quartz oscillators.

The design of the CMOS-MEMS oscillator can be divided into three stages: the first stage consists on the design of the resonator, the second is the selection of the detection scheme that is used to transduce the movement of the resonator and finally, the development of the CMOS sustaining amplifier for oscillation.

The design of the CMOS MEMS is limited by the layers available in the technology and the design rules. The technology layers fix the thickness of the layers and the mechanical properties of the material (mass density and young modulus). The design rules define the final dimensions of the structure, in terms of distances between the different layers. As a direct consequence, low quality factors and high motional resistances are found, compared with the quartz crystals counterpart.

According to the implemented MEMS, the design of the

This work has been supported by the Spanish Government and European Union FEDER program under project NEMS-in-CMOS TEC-2012-32677.

circuitry must be adapted to the operating frequency of the resonator and the motional resistance, so as to force the fulfilment of the conditions of phase-loop and gain in order to achieve oscillation.

In this article we develop a tunable gain (123 to 146 dBΩ @ 20MHz) sustaining-amplifier. The implemented amplifier allows sensing the movement of different CMOS-MEMS and feed it back to achieve a monolithic CMOS MEMS oscillator. The structure of the paper is as follows: Section II describes both the MEMS fabrication and electrical behavior. In section III the possible different design strategies are analyzed and a detailed description of the implemented sustained amplifier is presented. Section IV shows the simulated oscillator performance. Finally, conclusions are exposed in Section V.

II.   MEMS RESONATORS

The implemented resonators consist basically of two electrodes (excitation and read-out electrodes), and a movable structure [2]. The devices have been implemented in a commercial technology (0.35 μm AMS). The structural layer along with the driver and read-out electrodes are defined by using the poly, metal or even a stack of the metal layers available in the technology, while the sacrificial layer is always the silicon oxide. To electrically characterize the movement of the resonator, an electrostatic excitation and capacitive read-out is performed. In particular, a DC voltage (V_{DC}) is applied to the suspended structure, while an AC excitation voltage (V_{AC}) is applied to the excitation electrode. The resulting electrostatic force induces the movement of the suspended structure at the excitation frequency. This oscillation generates a change in the capacitance constituted by the read-out electrode and the suspended structure that can be quantified by measuring the induced output current (assuming $V_{DC} \gg V_{AC}$):

$$I \cong V_{DC} \frac{\partial C_2}{\partial t} + C_0 \frac{\partial V_{AC}}{\partial t} \qquad (1)$$

where C_2 is the variable capacitance between the suspended structure and the read-out electrode, and C_0 is the static capacitance between the excitation and read-out electrodes. The first term corresponds to the motional current, whereas the second term corresponds to the feedthrough or parasitic current.

The behaviour of the MEMS configured as a two port (input and output port) scheme can be modelled by an RLC equivalent circuit that models motional current. An additional capacitance C_0 is added between the input and output ports to model the feedthrough current. For our resonators, motional resistances of the order of 10^6-10^7 Ω are expected, with resonance frequency around 20 MHz. These large values force the output current of the MEMS to be in the range of several nA

III. CMOS-Oscillator design

The design of the oscillator circuit needs of a transimpedance sustaining amplifier in charge of amplifying the motional current and generate a voltage that is used to excite the MEMS. In order to start oscillation Barkhausen's criteria must be met at the oscillation frequency: the loop gain must be higher than unity and the phase shift must be equal to 0°. A differential Pierce configuration for the overall oscillator has been chosen as depicted in Fig.1.

A. Design Approaches

Basically three different types of MEMS sensing architectures can be found in the literature: common base detection, resistive detection and capacitive detection [4]. Note that sampled techniques are not applicable due to the relatively high resonance frequencies of the MEMS resonators (@20 MHz). According to the range of our expected input currents, the reduction of the input noise is a must in order to choose the transimpedance architecture.

In the common base detection the generated motional current is connected to the drain of a CMOS device. Basically the amplifier structure corresponds to a diode-connected topology (current mirror) that presents as a main advantage a high bandwidth. When using a diode-connected topology amplifier, the input noise is proportional to the trasconductance of the input MOS device. As a consequence, a high gain implies always a high input current noise.

In the resistive detection, the motional current is sensed by a resistance. In this way the current is converted into a voltage that can be further amplified. In this case the amplifier noise presents as a main source of noise the contribution of the resistance. A reduction of the noise can be achieved increasing the value of the resistance which reduces the bandwidth.

Finally, the capacitive detection integrates the motional current into a capacitance and generates a voltage as output. In this case the input noise is attributed to the needed bias resistance, which value can be very larger without implying a bandwidth reduction.

Since the capacitive detection is the least noisy structure we have implemented a capacitive topology. The capacitive detection integrates the motional current into a capacitance and generates a voltage as output.

While the sustaining amplifier topology has been chosen in terms of noise, the fulfillment of the Barkhausen's criteria has moved us to the implementation of a differential

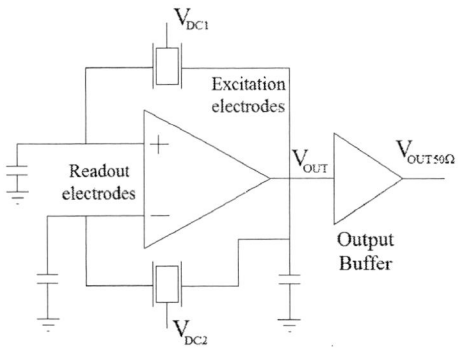

Fig. 1: Block diagram of differential configuration formed by two identical MEMS, a differential amplifier and an output

configuration [2]. As it has been exposed previously, the electrical MEMS behavior corresponds to a RLC circuit with a parallel capacitance that origins a parallel resonant frequency.

The effect of this static capacitance contributes to the degradation of the phase shift and magnitude, since close to the parallel resonant frequency the phase achieves a value of 90°. If the parallel resonant frequency value is close to the series resonant frequency the shift phase is reduced. The differential approach is based on the subtraction of the parasitic currents provided by two identical MEMS resonators (i.e. with the same physical dimensions) using a differential amplifier. Fig. 1 shows, at block level, the different parts that compose the full amplifier: two identical MEMS and a differential amplifier. This architecture allows applying independent DC voltages on each beam (V_{DC1} and V_{DC2}). By looking at equation 1, it is seen how only the resonator with applied V_{DC} contributes with motional and parasitic current, while the resonator with $V_{DC}=0$ only contributes with the parasitic current. Therefore, this technique eliminates current component caused by all the static capacitances in parallel with the RLC circuit, making the MEMS resonator behave like a pure RLC.

B. Differential Transimpedance amplifier circuit design

As it has been explained, a capacitive detection has been preferred as a design approach to reduce the noise of the circuit. In this way the generated motional current will be integrated into a capacitance in order to convert it to a voltage.

Since the MEMS current is very small, the generated voltage needs to be amplified to be fed back to the MEMS and obtain a self-oscillating system. The chosen architecture is based on a differential voltage to voltage sustaining amplifier. The gain of the full system will depend on the sustaining amplifier gain (A_V) and the total capacitance at the input node (C_{in}). Note that this capacitance is formed by: the input capacitance of the circuit, the capacitance of the connection metals and the capacitance formed between the readout electrode and the n-well that supports the MEMS. In particular:

$$G = \frac{1}{2\pi f C_{in}} A_V \quad (2)$$

According to expression 2, the input capacitance should be as small as possible to allow a high gain. The sustaining amplifier is formed by three stages (see fig. 2):

- The first stage is based on a differential amplifier with a current mirror as a load. In order to polarize the input nodes, two PMOS devices working as a pseudo resistor (sub-threshold operation) have been used due to their high resistance, in the range of 10^{12} Ω. The voltage gain of this stage will depend on the differential pair transconductance (g_m) and the output resistance of this first stage. In particular, the output resistance will be fixed by the input resistance of the second stage.

- The second stage is a source follower with a shunt-shunt feedback implemented with a PMOS device working in the linear region. The gate voltage is controlled by an external voltage (V_C) which can take values between 0-1.5 V. This PMOS plays the role of variable resistance, providing tunable input resistance in terms of the DC voltage applied to the gate. This variable input resistance affects the output resistance of the first stage and in consequence its gain.

- The third stage consists of a common source configuration with a current mirror load. This stage is self-biased by a PMOS transistor working as pseudo-resistor. The purpose of this stage is to raise the global gain of the sustaining amplifier by 33 dB. The output voltage is centered in 1.2 V, and the output swing voltage goes from 0.4 V to 2.2 V. An output buffer is employed for testing purposes to load the 50Ω impedance of the measurement instrumentation.

The sustaining amplifier designed with a tunable gain allows achieving the gain and phase conditions of Barkhausen stability criteria for different MEMS resonators. To ensure the phase loop condition, the sustain-amplifier feedback should be given by the positive input.

IV. SIMULATION RESULTS

In order to simulate the AC transimpedance gain of the amplifier, 20fF capacitance at the input and output of the MEMS resonator are assumed. These capacitances correspond to the capacitance formed between the driver (excitation or read-out) and the n-well, where the MEMS is embedded. This capacitance is added in parallel to the input capacitance of the amplifier, whose value is approximately of 30fF.

Fig. 3 shows the transimpedance gain of the sustaining-amplifier. A slope of 20 dB / dec is generated due to the capacitive integration of the input current. According to the design, a tunable gain is achieved modifying the voltage of $V_{control}$. In particular, a variable gain between 123 to 146 dBΩ (1,4MΩ - 20MΩ) is achieved for a 20 MHz frequency. At this frequency, the simulated input noise of the amplifier is 32fA/$\sqrt{Hz}$.

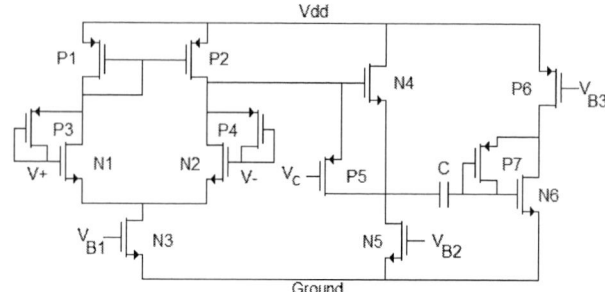

Fig. 2: Circuit diagram of the variable gain differential amplifier formed by three stages.

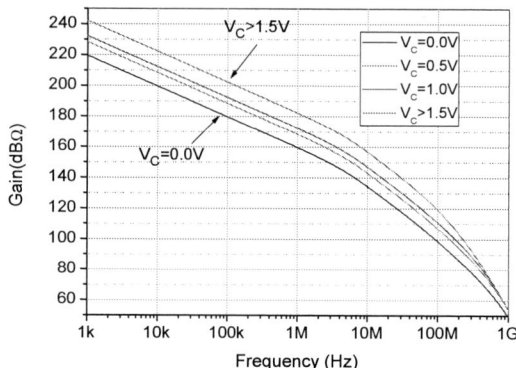

Fig. 3: Simulated transimpedance gain for different V_C values included within the effective range of values (0 – 1.5V)

A periodic steady static analysis has been performed on the oscillator system for the MEMS resonator with the following equivalent RLC$\parallel$C$_0$ values: R=5,6MΩ, C=5aF, L=9,58H, C$_0$=120aF. This value has been extracted from a C-C-Beam resonator previously designed by the research group [3]. The amplifier gain has been set to 146dBΩ using a V_C value of 1,5V in order to achieve the oscillation criterion for the gain and the phase. Under these conditions the output signal is a periodic function of time with a period of 43.13 ns and a peak-to-peak amplitude of 1,9V as it is shown in Fig 4.

The oscillator output signal has been analyzed in order to obtain the phase noise. Figure 5 shows the simulation results. In this analysis a noise floor of -112 dBc/Hz is obtained.

In order to assess the ability of self-sustaining-oscillator the system has been tested for different theoretical resonators. Different motional resistances values have been used to simulate the MEMS. Fig. 6 represents the maximum open loop gain for a fixed resonant frequency versus the motional resistance R_m of the resonator. If the open loop gain is higher than zero dB, the system can reach the oscillation state. Fig. 6 proves that the system can oscillate for any MEMS presenting a high motional resistance R_m of the order of MΩ for frequencies between 1MHz and 50MHz.

The values of gain and output capacitance of the resonator have been extracted from the layout. From a Monte Carlo simulation (process and mismatch), the high stability of the gain has been proven.

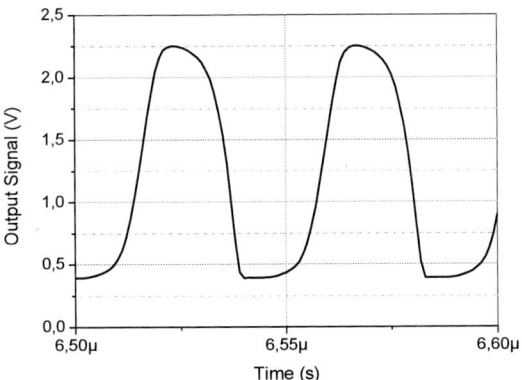

Fig. 4: Periodic steady state simulation output signal.

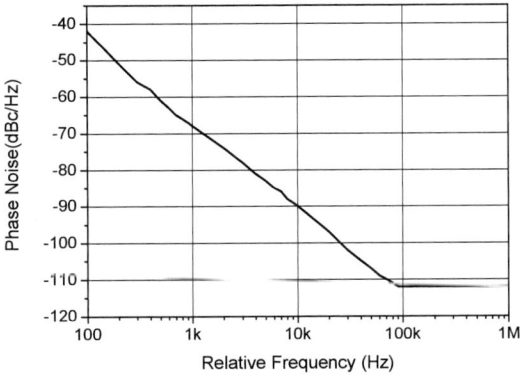

Fig. 5: Simulated oscillator phase noise for a carrier frequency of 23.18 MHz.

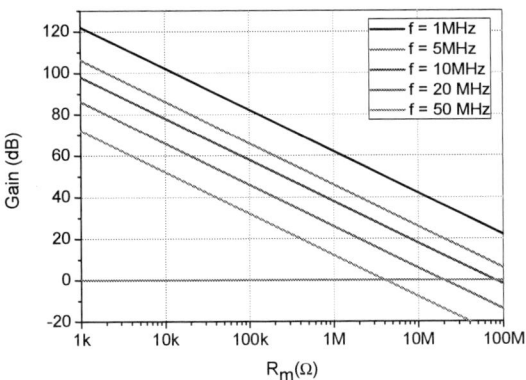

Fig 6: Simulations of the behavior of the open-loop gain for different conditions of frequency and motional resistance.

V. CONCLUSIONS

A tunable transimpedance amplifier has been designed with the capability to self-sustain the oscillation of the system for oscillation resonators presenting a motional resistance in the range of MΩ. The amplifier presents excellent parameters compared with the state of the art.

We want to point out its reduced input noise 32fA/√Hz @ 20MHz, really competitive value in the field of MEMS-CMOS oscillators. Its variable gain controlled by an external voltage provides a range of gains between 1,4MΩ - 20MΩ @ 20MHz. Table II contains data for the most recent TIA amplifiers [6, 7, 8, 2] designed to perform the function of sustaining-amplifiers published in recent years. As can be seen, the proposed oscillator presents advantages both in terms of input noise, area, power consumption and variable loop gain.

Currently, the design has been sent to fabrication. Authors expect to present the experimental measurements results at the conference.

REFERENCES

[1] C. Nguyen, "Integrated Micromechanical Radio Front-Ends" (Invited Plenary Talk), VLSI-TSA '08, 2008, pp. 3-4J. Clerk Maxwell, A Treatise on Electricity and Magnetism, 3rd ed., vol. 2. Oxford: Clarendon, 1892, pp.68-73.

[2] J. Verd, A. Uranga, J. Segura and N. Barniol "CMOS-MEMS Oscillator with Bias Voltage Bellow 3V". The 17th International Conference on Solid State Sensors, Actuators and Microsystems. Transducers 2013.

[3] E. Marigó, J. Verd, J. L. López, A. Uranga and N. Barniol. "Packaged CMOS-MEMS free-free beam oscillator". J. Micromechanics and Microengineering. Vol. 23 Number 11. 2013.

[4] T. A. Roessig, R. T. Howe, A. P. Pisano, and J. H. Smith, "Surface-micromachined 1MHz oscillator with low-noise Pierce configuration". 1998 Solid State sensor and actuators workshop.

[5] A. Uranga , J . Verd , JL .López, J. Teva, F. Torres, J. Giner, G Murillo, G Abadal, N Barniol. "Electrically enhanced read-out system for a high frequency CMOS-MEMS resonator". ETRI Journal.Vol. 31 pp. 478-480. 2010.

[6] Ming-Huang Li, Cheng-Syun Li, Li-Jen Hou, Yu-Chia Liu, and Sheng-Shian Li "A 1.57mW 99dBΩ CMOS transimpedance amplifier for VHF micromechanical reference oscillators". Circuits and Systems (ISCAS), 2012 IEEE International Symposium on.Pp. 209-212

[7] S. Seth, S. Wang, T. Kenny, B. Murmann."A -131 – dBc/Hz, 20-MHz MEMS Oscillator with a 6.9 mW, 69 kΩ, Gain Tunable CMOS TIA". ESSCIRC, 2012.

[8] Tung-Tsun Chen, Jui-Cheng Huang; Yung-Chow Peng; Chia-Hua Chu; Chung-Hsien Lin; Chun-Wen Cheng; Cheng-Syun Li; Sheng-Shian Li. "A 17.6 MHz 2.5V Ultra-low polarization voltage mems oscillator using an innovative high gain-bandwidth fully differential trans-impedance voltage amplifier" Micro Electro Mechanical Systems (MEMS), 2013 IEEE 26th International Conference on. Pp 741-744.

Table II: Specifications @24MHz of the reported sustaining amplifier circuit in comparison with some of the last reported TIAS.					
TIA	[6]	[7]	[8]	[2]	This work
Integration	Hybrid	Hybrid	Hybrid	Monolithic	Monolithic
Technology CMOS	0.35 µm	0.35 µm	0.18 µm	AMS 0.35 µm	AMS 0.35 µm
Input configuration	Singled	Differential	Differential	Differential	Differential
Transimpedance Gain	25kΩ-89kΩ @ 280 MHz	12 kΩ-69kΩ. @ 20 MHz	316 kΩ @ 60MHz	4.9 MΩ @24MHz	1.4MΩ-20MΩ @20MHz
Input-referred current noise	-	-	< 2.5 pA/√Hz	80fA/√Hz @24MHz	32fA/√Hz @20MHz
Power	1.57 mW	6.9 mW	5.9 mW	1.5 mW	1.2 mW
Layout area	0.024 mm²	0.15 mm²	0.048 mm²	0.003 mm²	0.0045 mm²

A novel CAD framework for substrate modeling

(Invited Paper)

Hao Zou*, Yasser Moursy*, Ramy Iskander*, Marie-Minerve Louërat* and Jean-Paul Chaput*
*Laboratoire LIP6, Université Pierre et Marie CURIE (UPMC), Paris, France

Email: (first name).(last name)@lip6.fr

Abstract—This paper presents a novel Computer-Aided-Design (CAD) framework for 3D extraction of the substrate electrical network. The proposed CAD tool (framework) models efficiently the minority carrier propagation inside substrate network especially for smart power ICs. Today, the minority carrier propagation into the substrate is ignored in existing SPICE simulators. It can be simulated using finite element methods in TCAD. Generally, TCAD simulations are accurates but take long time. Thus, they become of limited help for large scale ICs involving hundreds of transistors. In the context of the FP7 AUTOMICS project, the extraction tool will take into consideration the minority carriers effects. It will allow the designer to predict the minority carrier propagation through the substrate. This can be useful in evaluation the efficiency of ESD protection and latchup faults due to this leackage current in the substrate specially in HV/HT applications. With the proposed substrate network, the three-dimensional layout parasitics are constructed and substrate noise is simulated before first silicon fabrication. A simple diode example is illustrated to demonstrate the principal idea of the extraction tool.

I. Introduction

Smart Power ICs merge high-voltage and low-voltage devices on the same silicon chip [1]. This makes them highly demanded in automotive embedded systems, at a competitive cost. Merging standard low voltage architectures with high power devices on a single chip is a difficult task since these devices operate at different voltage levels. Switching of the power stage for many automotive high voltage applications (e.g. driving motor) cause the minority carrier injection to propagate into the substrate. The disturbances caused by minority carrier propagation may dramatically affect the functionality of low voltage devices existing on the same chip, even at long distances. [2].

To evaluate the noise coupling into the substrate, numerous investigations for substrate model extraction have been studied recently [3] [4] [7]. The boundary-element method and the interpolation method [5] [6] face the choice between a relatively large amount of memory use and less accurate extraction and computation, even when neglecting the minority carriers diffusion. A physically based TCAD simulators are available. They are accurate and considered as powerful tools to model the minority carriers diffusion [8]. For large-scale IC involving hundreds of high voltage devices, TCAD becomes very limited due to the amount of components, the time consuming FEM simulation, and the complexity of layout and placement of devices. All these limitations impact significantly the estimation of the circuit's ESD and latch-up behavior for a

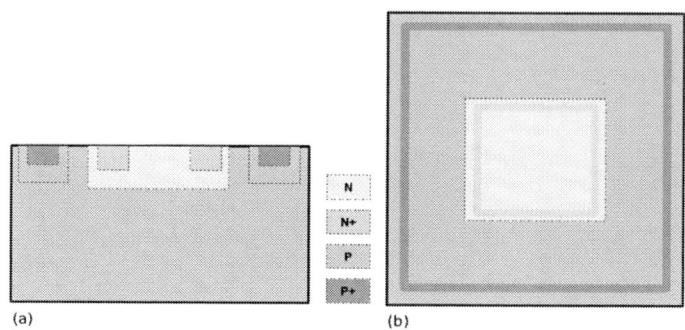

Fig. 1. Simple diode layout: (a) diode cross section view, (b) top side view.

successful design in CMOS technology. Therefore, an accurate and even simple to use substrate extraction CAD tool is highly demanded.

In this paper, a fundamental extraction flow is illustrated. We present the modeling of the substrate by a three-dimensional extraction of enhanced resistors and diodes that has been presented in [12], [13]. Pattern based extraction strategies with intelligent computation algorithms are employed, for the purpose of delivering highly accurate, efficient and simple tool. This will facilitate the usage and dramatically reduce the cycle of development and verification of substrate noise before first-silicon fabrication.

This paper is organized as follows: in section II, we introduce a simple design diode as a case study, in section III, a general extraction flow is illustrated. Finally, the concluding remarks are in section IV.

II. Case study

We study the simple diode structure presented in figure 1. The diode consists of an N-well in the middle of the P-doped substrate and is surrounded by P+ doped diffusion area. All those layers are presented as two-dimensional shapes (rectangles and/or multi-paths) in the layout database in our design environment.

III. Proposed extraction flow

The diagram representing the proposed extraction flow is depicted in figure 2. The extraction tool is developed using OpenAccess [14] which is a common and open C++ database for integrated design tools and design methodologies. OpenAccess give access to shapes, geometries and the coordinates

978-1-4799-4993-9/14 $31.00 © 2014 IEEE

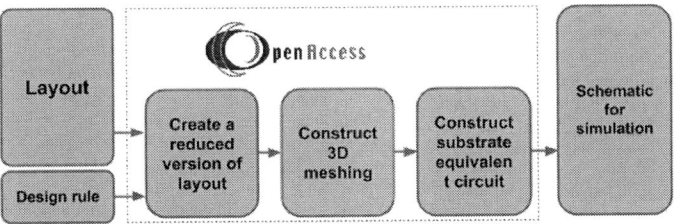

Fig. 2. The overall substrate extraction flow proposed.

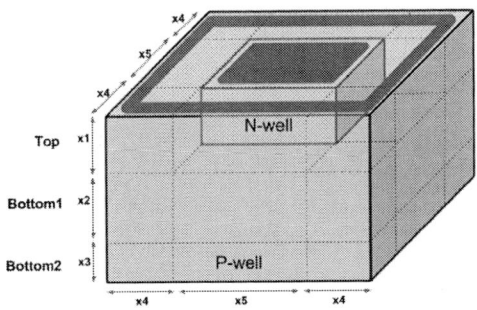

Fig. 3. Three-dimension view of reduced layers: the bigger light-blue cube representing the P-type area, the small gray cube representing the N-type area, and green shapes on the top for the diffusion area

of the physical layers. With this layout geometrical database, we are able to extract the substrate network and produce an OpenAccess schematic view that is ready for simulation.

Four main steps are involved in this extraction flow:

1) Design rules definition to define extraction rules to apply.
2) A version of layout which comprises only the layers contributing in the substrate parasitics (more details discussed in section III-B);
3) Three-dimensional meshing construction by using the developed reduced layout (more details in section III-C);
4) An equivalent three-dimensional substrate network extraction that produces the corresponding circuit netlist (more detail in section III-D).

A. Design rule definition

During the preprocessing stage, the substrate extraction rules are declared in a standard Extensible Markup Language (XML) database file. An example of extraction rules is presented below.

```
1  <Wells>
2    <well name="Top" depth="x1e-6"/>
3    <well name="Bottom1" depth="x2e-6"/>
4    <well name="Bottom2" depth="x3e-6"/>
5  </Wells>
6  <Bypass>
7    <layer name="MET1" preserve="true"/>
8    <layer name="VIA1"/>
9  </Bypass>
10 <Operations>
11   <merge name="DNTUB"/>
12   <contact diffusion="DIFF"/>
13   <setFilePath path="netlist/"/>
14 </Operations>
```

Three sections implementing extraction rules are shown:

1) Substrate well depths are given in the section with keyword "Wells", between line 1 to 5;
2) "Bypass" section automatically removes the unused layer data, from line 6 to 9 (c.f. details given in the next subsection);
3) Customized extraction operations are used to customize the extraction results, with section keyword "Operations", between line 10 to 14.

B. Reduced layout extraction

Traditionally, for a given layout, many layers are used and stored in layout database. Practically, only few of them will be reused for extraction strategy of substrate network. Metal Layers, such as METAL layers and VIA layers, etc will be ignored since they do not contribute to substrate noise. To define unneeded layers, we declare them inside the section "Bypass" in extraction rules file. Consequently, the extraction tool will keep only the desired layers data inside the layout database during the preprocessing stage. For example, lines 7 and 8 inform the extraction tool to ignore METAL1 and VIA1 layers during extraction.

Unfortunately, the obtained reduced layers may not completely satisfy the requirements of the meshing construction process, due to the complexity or repetitiveness of the drawn shapes. As an example from figure 1, the highly-doped N+ multi-part path shape forming the contacts is contained in the doped N rectangle shape. A simple "merge" operation applies for N doped layer could inform the extraction tool to ignore the path shape since it has been contained in a bounding rectangular shape. Moreover, other operations are defined such as:

- "remove" to remove the target layer,
- "overlaps" to merge overlapped shapes,
- "setPath" to set produced file path

As a result of layout reduction, only diffusion, N-well and P-substrate shapes will be considered as shown in figure 3.

C. Three-dimensional Meshing construction

Commercially existing layout design tools draw the layers by using two-dimensional geometrical shapes, such as a rectangle, a path segment, a path, a polygon, etc. Firstly, we studied these two-dimensional geometrical shapes. Secondly, we implemented the extraction strategy by constructing a two-dimensional rectilinear meshing strategy. Construction strategies aims at constructing a highly accurate two-dimensional meshing network that fully emulates meshing in TCAD simulations. An example of a generated two-dimensional meshing matrix is shown in figure 4(b). Based on the original reduced layer in figure 4(a), it produce an 5x5 mesh cells. Each of these mesh cells represent four boundaries where minority carrier injected currents can be evaluated.

Applying the substrate extraction rules from the "Wells" section of the XML rules file, the substrate is subdivided into

978-1-4799-4993-9/14 $31.00 © 2014 IEEE

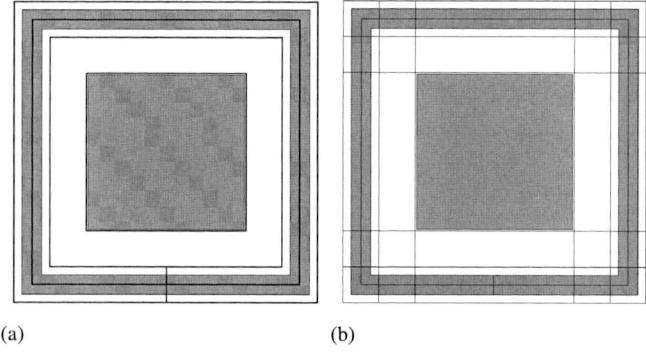

(a)   (b)

Fig. 4. Two-dimension view of: (a) reduced version of layers, (b) meshing network with diffusion (DIFF) layer.

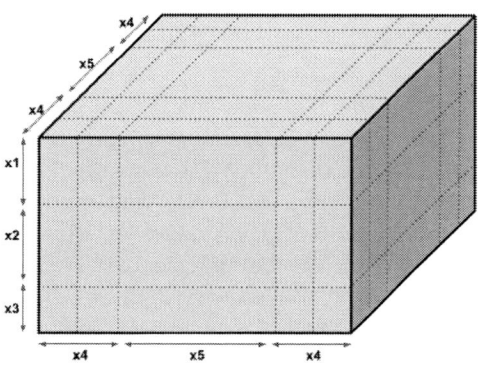

Fig. 5. Constructed 3D meshing network with 5x5x3 cuboids.

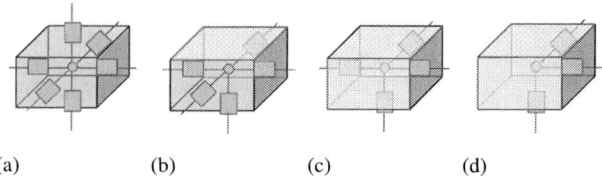

(a)   (b)   (c)   (d)

Fig. 6. Components extraction strategy for cuboid: 6(a) inside the substrate, 6(b) in the face, 6(c) in the edge, 6(d) in the vertex.

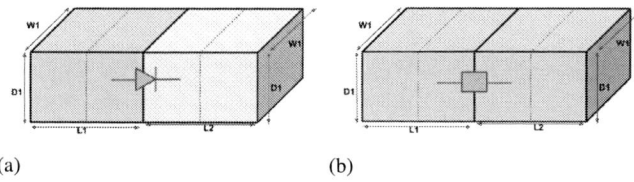

(a)   (b)

Fig. 7. Component extraction strategy at neighbouring cells: (a) a diode is extracted if two neighbouring cells are doping type different, (b) a resistor is extracted if two neighbouring cells are the same doping type.

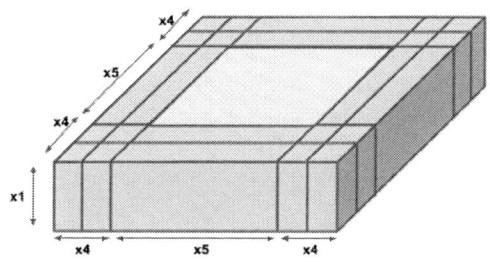

Fig. 8. Top layer of the substrate.

three conducting wells with different depths in the Z-direction. In our example, 5x5x3 cuboids shown in figure 5 will be generated using the depth value of each well. Each cuboid inside the 3D representation, stores *technological* information such as doping type, *geometrical* information (width, length and depth) and *algorithmic* information such as coordinate in the three-dimensional space.

D. Three-dimensional schematic extraction

Basically, the obtained three-dimensional meshing represents the whole substrate network. Therefore, the tasks for schematic extraction is to deal with these cuboids *technological*, *geometrical* and *algorithmic* information mentioned in the previous section.

Algorithmically, each cuboid in the three-dimensional matrix directly defines the number of components to extract, as shown in figure 6. This figure illustrates a cuboid inside the 3D matrix and how it will generate substrate components in different cases:

(a) A cell in the middle of matrix will generate six components, four in horizontal plane and two in the vertical plane.

(b) A cell in the face of the 3D matrix will generate five components because no neighbouring cuboid exists on top of it.

(c) A cell in the edge of the 3D matrix will generate only four components,

(d) A cell in the vertex of 3D matrix will generate only three of them.

Technologically, only three parasitic components can be extracted: the *EPFL enhanced diode* [12], [13], the *EPFL enhanced resistors* [12], [13] and the *homo-junction contacts* [9]. the type of component to extract is determined by the doping types of two neighbouring cuboids, as shown in figure 7, either to be a resistor (same doping type) or to be a diode (different doping type). Homo-junction contacts will be discussed below.

Geometrically, the size of each component is computed by examining neighbor cuboids in three-dimension, e.g. length(resistor)=L1/2+L2/2 for the resistor and length_AnodeP(diode)=L1/2, length_CathodeN(diode)=L2/2 for the diode P doping and N doping length respectively, both device areas equal to the cross-sectional area D1xW1 of the cuboid. Figure 7 illustrates these calculations.

E. Example of schematic extraction for one layer

According to the extraction method presented above, we illustrate the extraction results of the upper topmost layer of the diode in figures 9 and 10. A total number of 5x5 cuboids including one N-type cuboid in the middle of P-type cuboids, with the depth of "Top" layer equals X1 as shown in figure 8. After scanning all these cuboids, the extraction tool produces

978-1-4799-4993-9/14 $31.00 © 2014 IEEE

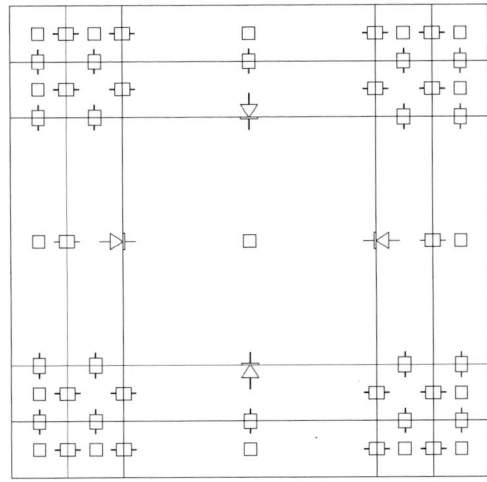

Fig. 9. Extracted view for the Top Layer.

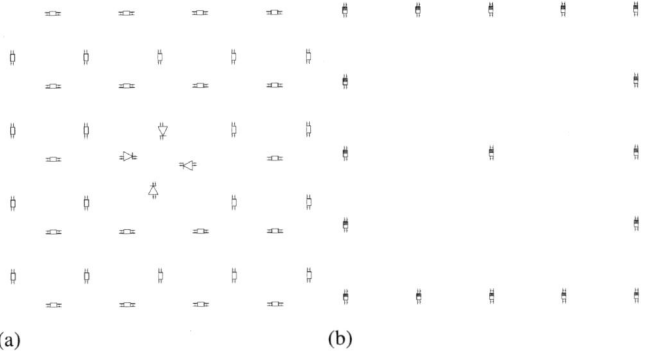

(a)   (b)

Fig. 10. Extracted netlist for Top layer: (a) horizontal contribution, (b) vertical contribution.

the substrate netlists into two separate files to distinguish the horizontal and vertical contributions to the substrate network. These contributions are shown in figures 10(a) and 10(b) respectively. In addition, each component is drawn in the individual layout view containing the meshing lines as shown in figure 9. We note that for the vertical contribution in figure 10(b), sixteen vertical homo-junction contacts [9] are generated: fifteen components generated by the cuboids in the edge of the substrate, plus one component generated at the center which is reduced into one center contact of N-contact.

IV. CONCLUSION

A novel Computer-Aided-Design (CAD) framework for substrate modeling is presented. The extraction steps of a simple diode illustrate the principal layout-to-netlist extraction flow. The substrate parasitic circuit can be directly back-annotated to the corresponding circuit for the purpose of simulation. In this way, the designer will be able to simulate substrate noise coupling before first-silicon fabrication. We have to mention that the simulation speed is shorter than that of the TCAD with an acceptable error. The next enhancement is to extend the extraction tool to deal with complete chips containing hundreds of devices. Consequently, the extraction

tool will allow the designer to simulate and predict the circuit's ESD protection performance and latch-up immunity for automotive applications.

ACKNOWLEDGMENT

This work has been sponsored by the European commission under the European FP7 AUTOMICS project.

REFERENCES

[1] B. Murari and F. Bertotti and G. Vignola, *Smart Power ICs*, 2rd ed. pp.218-220   Springer-Verlag, Berlin, 2002.

[2] M. Schenkel, *Substrate current effects in smart power ICs*, Ph.D. dissertation, Diss., Technische Wissenschaften ETH Zürich, Nr. 14925, 2003.

[3] Salman, E.; Jakushokas, R.; Friedman, E.G.; Secareanu, R.M.; Hartin, O.L., "Input port reduction for efficient substrate extraction in large scale ICs," Circuits and Systems, 2008. ISCAS 2008. IEEE International Symposium on , vol., no., pp.376,379, 18-21 May 2008

[4] Silva, J. M S; Silveria, L.M., "Issues in parallelizing multigrid-based substrate model extraction and analysis," Integrated Circuits and Systems Design, 2004. SBCCI 2004. 17th Symposium on , vol., no., pp.123,128, 7-11 Sept. 2004

[5] Smedes, T.; Van Der Meijs, N.P.; van Genderen, AJ.; Elias, P. J H; Vanoppen, R. R J, "Layout Extraction of 3D Models for Interconnect and Substrate Parasitics," Solid State Device Research Conference, 1995. ESSDERC '95. Proceedings of the 25th European , vol., no., pp.397,400, 25-27 Sept. 1995.

[6] Smedes, T.; Van Der Meijs, N.P.; van Genderen, AJ., "Extraction of circuit models for substrate cross-talk," Computer-Aided Design, 1995. ICCAD-95. Digest of Technical Papers., 1995 IEEE/ACM International Conference on , vol., no., pp.199,206, 5-9 Nov. 1995

[7] Verghese, N.K.; Allstot, D.J.; Wolfe, M.A, "Fast parasitic extraction for substrate coupling in mixed-signal ICs," Custom Integrated Circuits Conference, 1995., Proceedings of the IEEE 1995 , vol., no., pp.121,124, 1-4 May 1995

[8] C. Stefanucci and P. Buccella, *Optimization Strategy of Numerical Simulations Applied to EPFL Substrate Model*, accepted at MIXDES 2014.

[9] C. Stefanucci and P. Buccella, *Impact of enhanced contact doping on minority carriers diffusion currents*, accepted at PRIME 2014.

[10] P. Buccella and C. Stefanucci, *Spice Simulation of Substrate Minority Carriers Propagation with Equivalent Electrical Circuit*, accepted at MIXDES 2014

[11] Y. Moursy and al. *AUTOMICS: A novel approach for substrate modeling for automotive applications*, 18th IEEE European Test Symposium, May, 2013

[12] F. Lo Conte, J.-M. Sallese, and M. Kayal, *Circuit level modeling methodology of parasitic substrate current injection from a high-voltage h-bridge at high temperature*, IEEE Transactions on Power Electronics, vol. 26, no. 10, pp. 2788-2793, 2011.

[13] F. Lo Conte, J.-M. Sallese, M. Pastre, F. Krummenacher, and M. Kayal, *Global modeling strategy of parasitic coupled currents induced by minority-carrier propagation in semiconductor substrates*, IEEE Transactions on Electron Devices, vol. 57, no. 1, pp. 263-272, 2010.

[14] OpenAccess: www.si2.org

Sensitivity based Methodologies for Process Variation Aware Analog IC Optimization

Engin Afacan, Gönenç Berkol, Faik Başkaya, and Günhan Dündar

Department of Electrical and Electronics Engineering, Bogazici University, Bebek, Istanbul, Turkey 34342

engin.afacan@boun.edu.tr

Abstract—With the continuous downscaling of CMOS technology over the last two decades, reliability of CMOS circuits has become a more critical design issue due to the worsening effects of process variations. Conventionally, variability analysis has been performed following the design process after which the design is modified based on the variation effects, if necessary. However, increased variation and mismatch problems have enforced designers to consider robustness as a design objective that should be maximized. Besides the variability problem, increased non-idealities with more advanced technologies have complicated circuit analysis, and caused unacceptably long design times. Therefore, design automation tools for analog circuits have become crucial to keep the synthesis time within acceptable limits even if variability analysis is included. In this paper, two different methodologies are proposed and discussed for variation aware design automation of analog circuits: *Over-Design* approach that is based on guard-banding the circuit performance and *Variation-Aware Design* approach depending on looking for more robust solutions in the dedicated search space. The superiorities of the proposed approaches are discussed via the synthesis results of a two stage OTA example.

I. Introduction

Process tolerances have deteriorated with the aggressive downscaling of feature size. As a result, it is highly difficult to handle variations in different fabrication steps such as line-edge roughness (LER) that is induced by gate etching and the lithography process [1], oxide thickness fluctuations (OTF) that causes the fluctuation of the voltage drop across the oxide layer, further changing V_{th}, and random dopant fluctuations (RDF) that significantly affect V_{th}, [2]. Thus, if a circuit was designed to achieve specific nominal values of performances, a dispersion occurs between the expected and the actual performances in a population of manufactured ICs [3]. In addition to the variability phenomenon, comprehensive analyses and complicated trade-offs among various aspects of performances are required for analog circuit design in advanced technologies [4]. The high-order non-idealities complicate the analysis of analog circuits, resulting in drastically increased design times.

Analog circuit design can be decomposed into two levels: circuit and layout levels. At first, the circuit topology and design objectives are determined. Then, the design space is explored via consecutive simulations on the SPICE platform. When a solution satisfying all design constraints is found, the layout is drawn and post-layout simulations are performed considering layout parasitics. If the solution still satisfies the constraints, then the design is fabricated. If there are any unsatisfied constraint, the layout and the

design are revised. However, in practice, some additional steps should be added to this procedure to capture process variation effects and guarantee a certain circuit performance after the fabrication. In order to observe the process variation effects, some additional analyses (Monte Carlo, Sensitivity, and Corner etc.) are performed for the candidate solution. Then, the design is revised if the yield of the candidate design does not meet the yield objective. Considering complexity of analog design and increased variation problems, design automation tools have become vital to reduce the design time. In the literature, analog design automation approaches can be roughly classified into three categories: knowledge-based approaches, simulation-based approaches, and equation-based approaches. In knowledge based analogue synthesis tools, optimization depends on designer's experience. On the other hand, equation based optimization tools are also used for analog circuit sizing, which utilizes analytical equations to evaluate the circuit performance. Even though such tools provide fast synthesis, the accuracy is still a problem due to the simplified models. On the other hand, using complicated models reduces the optimizer efficiency in terms of CPU time. In simulation-based approaches, the circuit performances are evaluated via a SPICE-class simulator utilizing qualified models, hence accuracy is excellent for such tools. Another advantage of simulation based approaches manifests itself during variability simulations. Knowledge based approaches require many additional simulations to construct a database for variability analysis, which require excessive times. Considering equation based approaches, variation effects should also be modelled analytically as well as circuit performances. Again, the efficiency and accuracy balance should be revised during variation aware circuit synthesis, which results in either worsening accuracy by using simplified models or increased synthesis time by using complicated models. In addition to this trade-off, the modelling task of variability is quite hard to handle. On the other hand, simulation based approaches utilize again SPICE platform for variability analysis which provides easier, more accurate, and quite efficient variation aware circuit synthesis.

In this study, two different methodologies are proposed for robust automation of analog circuits: *Over-Design* and *Variation-Aware Design*. The paper is organized as follows; a summary on the process variation problem and sensitivity based variation analysis are given in Section-II. Analog circuit optimizer is described in Section-III. Proposed variation aware circuit design approaches are explained in Section IV. Variation aware circuit synthesis algorithms and synthesis results for a two stage CMOS OTA are presented in Section

978-1-4799-4993-9/14 $31.00 © 2014 IEEE

V. Finally, Section VI concludes this study.

II. PROCESS VARIATION AND SENSITIVITY BASED VARIABILITY ANALYSIS

Pelgrom model [5] given in Equation 1 has been widely used to model the mismatch phenomenon.

$$\frac{\sigma^2 \Delta I}{I^2} = \frac{\sigma^2 \Delta \beta}{\beta^2} + 4 \frac{\sigma^2 \Delta V_{th}}{(V_{GS} - V_{TH})^2} \quad (1)$$

However, variations on the wafer stem from two different sources of process non-uniformity. These are stochastic (local) and systematic (global) variations. The Pelgrom model assumes that both components of variation are random. However, considering the entire die, global variations must contain a systematic component instead of being fully random [6]. On the other hand, physical parameters, such as oxide thickness, substrate doping etc., which are directly related to electrical parameters, can also be used to model the mismatch effect. The advantage of the physical models is that the correlation between different parameters are automatically inserted into the variation analysis while electrical models need additional correlation parameters [7].

There are several ways to perform a variability analysis such as Monte Carlo (MC) Analysis, Design of Experiment (DoE), and Sensitivity based Estimation (SbE). Conventionally, Monte Carlo analysis is commonly used because of excellent accuracy [4]. However, excessively long estimation time limits the usage of this approach in automatic sizing tools, in which the variability analysis would be run many times. Another estimation method is Design of Experiments that requires many initial simulation tasks to construct a database for each circuit. This case also degrades the efficiency. Therefore, in this study, Sensitivity based Estimation approach is utilized to decrease the time budget used by the variation analysis. Sensitivity based approach depends on the construction of the sensitivity matrix including the first derivatives of performance metrics with respect to the uncertain parameters (Equation 2) [8].

$$\sigma^2_y = \sum_i \left(\frac{\delta y}{\delta x_i}\right)^2 \sigma^2_{x_i} \quad (2)$$

σ^2_y is the variance of the performance, $\delta y / \delta x$ is the sensitivity, and $\sigma^2_{x_i}$ is the variance of the ith uncertain parameter. Using sensitivity analysis, the effect of each uncertain parameter on each performance metric can be obtained. Considering the physical parameters, sensitivity equation takes form as shown in Equation 3.

$$\begin{pmatrix} \sigma^2_{Gain} \\ \sigma^2_{BW} \\ \sigma^2_{Rout} \\ \sigma^2_{Power} \\ \vdots \\ \vdots \\ \sigma^2_{Area} \end{pmatrix} = \begin{pmatrix} \frac{\delta I_1}{\delta \Delta W} & \frac{\delta I_1}{\delta t_{ox}} & \frac{\delta I_1}{\delta V_{fb}} & \frac{\delta I_1}{\delta \mu_0} & \frac{\delta I_1}{\delta \Delta L} \\ \frac{\delta I_2}{\delta \Delta W} & \frac{\delta I_2}{\delta t_{ox}} & \frac{\delta I_2}{\delta V_{fb}} & \frac{\delta I_2}{\delta \mu_0} & \frac{\delta I_2}{\delta \Delta L} \\ \frac{\delta I_3}{\delta \Delta W} & \frac{\delta I_3}{\delta t_{ox}} & \frac{\delta I_3}{\delta V_{fb}} & \frac{\delta I_3}{\delta \mu_0} & \frac{\delta I_3}{\delta \Delta L} \\ \frac{\delta I_4}{\delta \Delta W} & \frac{\delta I_4}{\delta t_{ox}} & \frac{\delta I_4}{\delta V_{fb}} & \frac{\delta I_4}{\delta \mu_0} & \frac{\delta I_4}{\delta \Delta L} \\ \vdots & \vdots & \vdots & \vdots & \vdots \\ \frac{\delta I_n}{\delta \Delta W} & \frac{\delta I_n}{\delta t_{ox}} & \frac{\delta I_n}{\delta V_{fb}} & \frac{\delta I_n}{\delta \mu_0} & \frac{\delta I_n}{\delta \Delta L} \end{pmatrix} \times \begin{pmatrix} \sigma^2_{\Delta W} \\ \frac{\sigma^2_{t_{ox}}}{WL} \\ \frac{\sigma^2_{V_{fb}}}{WL} \\ \frac{\sigma^2_{\mu_0}}{WL} \\ \vdots \\ \frac{\sigma^2_{N_{sub}}}{WL} \end{pmatrix} \quad (3)$$

Furthermore, including correlation of dependent parameters improves the estimation accuracy. Considering the correlation between parameters, the variance of the output takes the form;

$$\sigma^2_y = \sum_i \left(\frac{\delta y}{\delta x_i}\right)^2 \sigma^2_{x_i} + \sum_i \sum_j \left(\frac{\delta y}{\delta x_i}\right) \sigma_{x_i} \rho_{ij} \left(\frac{\delta y}{\delta x_j}\right) \sigma_{x_j}, i \neq j \quad (4)$$

where ρ_{ij} is the correlation of two uncertain parameters.

III. ANALOG CIRCUIT OPTIMIZER

A modified version of the optimizer developed in [9] is used in this study. The optimizer utilizes an Evolutionary Strategies (ES) algorithm and Simulating Annealing (SA) in search and selection parts, respectively. HPSICE software is used as the performance estimator. In Figure 1, the optimizer flow and ES algorithm is summarized, where μ and λ denote parents and individuals. P, g, and T variables are population, generation, and temperature, respectively.

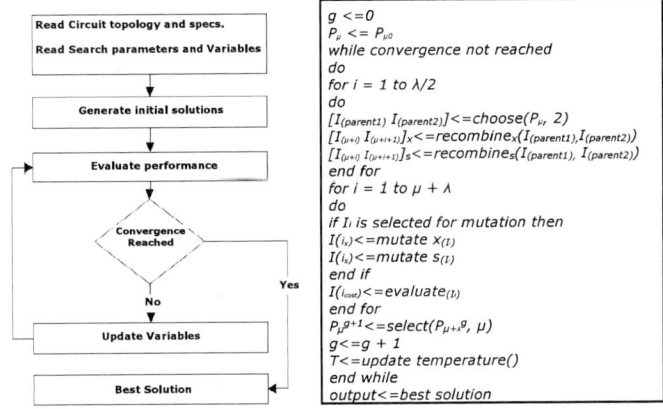

Figure 1.   Optimizer flow diagram and Pseudocode of the ES algorithm.

Optimization starts with a randomly generated population. All individuals are separately evaluated via SPICE simulations and cost calculation is performed. Then, individuals having lower cost values than the average cost value of the population are chosen as the new parents of the next generation and the population evolves until convergence is reached. The cost is calculated as;

$$Cost = Cost_{perf} + Cost_{penalty}$$
$$Cost_{perf} = \sum_i^n w_i . perf_i^2 \quad (5)$$
$$perf_i = \frac{upper_i - f_i}{upper_i - lower_i}, perf_{i,min} = 0$$

where n, $perf_i$, $upper_i$, and $lower_i$ are the number of performance specifications, i^{th} instant, upper, and lower values of performance metrics, respectively. $C_{penalty}$ is calculated according to the operation point of all transistors, where triode and cut-off regions are penalized. In addition to the search parameters, cost function weights (w_i) are also automatically evolved. Thus, if any performance metric is not satisfied, the corresponding cost function weight is automatically increased and vice versa.

978-1-4799-4993-9/14 $31.00 © 2014 IEEE

IV. VARIATION AWARE CIRCUIT DESIGN STRATEGIES

A. Over-Design Approach

Over-Design approach is the well-known method to make a circuit more robust, in which the design is guardbanded anticipating the worst case performance degradation. Thus, even if the transistor parameters are degraded, the circuit performance can still remain within the design margins as depicted in Figure 2(a). However, dramatic increase in power consumption restricts the use of larger devices in low power applications. Furthermore, some circuits are very sensitive to transistor dimensions and over-designing is not applicable for such circuits. *Over-Design* approach is still valid and used for specific applications due to its ease of use.

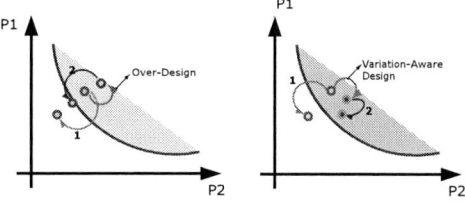

Figure 2. (a) *Over-Design* approach provides robustness by increasing the constraints. (b) *Variation-Aware Design* provides a local search around the solution to get more robust circuits.

B. Variation-Aware Design Approach

This approach is based on inserting the reliability specification of the circuit into the design constraints, thus the solution should satisfy not only electrical constraints, but also the reliability requirements. It is possible to find more robust circuits against process variations via a little sacrifice from some electrical specifications (power, area, etc.) The disadvantage of this method is that manual search for the solution that satisfies both electrical and reliability constraints may take longer time. Therefore, this method becomes efficient only by using automatic sizing algorithms that search a wide region of the whole solution space.

V. VARIATION AWARE ANALOG CIRCUIT OPTIMIZATION

The flow diagram of *Over-Design* method that is implemented into the optimizer is shown in Figure 3. The synthesis continues until the optimizer narrows the search space down to a certain point. When the decision mechanism determines that the solution is sufficiently close to the optimal point, reliability simulation is started to be performed for the current population. Then, design constraints are expanded with respect to the degradation data coming from the reliability simulations and a new cost value is calculated with this new constraints. If the degradation in any circuit performance metric is more than the expected, the optimizer automatically increases the related cost function weight. Therefore, the optimizer is enforced to move towards solution that adequately meets the initial design constraints.

Another method to design robust analog circuits via optimization is *Variation-Aware Design* approach that is

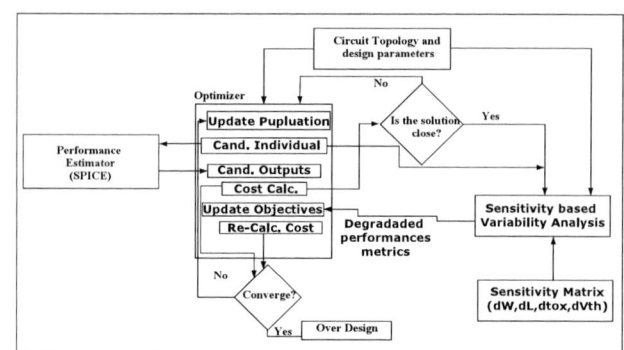

Figure 3. *Over-Design* Flow Diagram

shown in Figure 4. In this approach, reliability is defined as a new constraint in addition to electrical constraints such as gain, bandwidth, power etc. This new constraint has its own evolutionary strategy parameters, and evolves through the generations as well as the other constraints. Thanks to the automatic update mechanism of the cost function weights through generations, addition of this new constraint does not cause any problem during cost function calculation and selection process. Again, the reliability is not taken into account until the circuit optimization has proceeded enough so that reliability simulations can be performed. Similar to the previous approach, when the optimizer narrows the design space, the cost value comes closer to zero. Meanwhile, reliability simulations start to be performed. Then, reliability parameters of the current population are updated with respect to the variation amount in the design constraints. The cost value is also updated considering the reliability term. If the performance of the circuit varies more than the reliability constraint, the optimizer automatically eliminates the solution and increases the weight of the reliability constraint to find a more reliable solution.

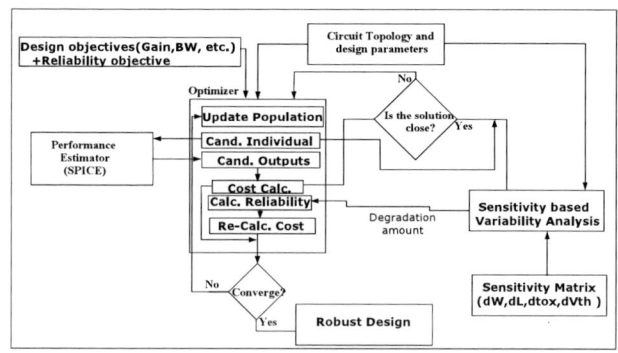

Figure 4. *Variation-Aware Design* Flow Diagram.

A two stage basic OTA circuit, shown in Figure 5, is used as the implementation example by using TSMC 180nm technology. Electrical specifications, gain, bandwidth, output resistance, power consumption, and chip area are determined as design constraints. Transistor dimensions (W, L), bias current, and compensation capacitance are the design parameters (22 design parameters). A first order sensitivity matrix was constructed for the design considering variations

Table I. Synthesis Results

	*Over-Design* approach Synthesis Results for 3 independent run					
Run	New Boundaries- Synthesis Results	$BW(12-10)[kHz]$	$Gain(65-60)[dB]$	$R_{out}(6-8)[k\Omega]$	$Power(5e^{-4}-10e^{-4})[W]$	$Area(1e^{-9}-10e^{-9})[m^2]$
1	New Upper-New Lower	13.5-11.5	70.2-65.2	5.4-7.4	4.99-9.99	1-10
	Synthesis Result	13	68.7	7327	7e-4	3.54
2	New Upper-New Lower	12.283-10.283	71.6-66.6	4.8-6.8	4.99-9.99	1-10
	Synthesis Result	11.229	75.23	4007	3.57	8.67
3	New Upper-New Lower	13.767-11.767	67.58-62.58	5.3-7.3	4.99-9.99	1-10
	Synthesis Result	11.889	68.2	6000	3.29e-4	6.55
	Variation-Aware Design approach Synthesis Results for 3 independent run					
Run	Synthesis Results	$BW(13-11)[kHz]$	$Gain(75-70)[dB]$	$R_{out}(4-6)[k\Omega]$	$Power(5e^{-4}-10e^{-4})[W]$	$Area(1e^{-9}-10e^{-9})[m^2]$
1	Normal Synthesis Result-variation	12-2.8	69.52-3.58	2.9-0.566	2.6-0	4.54-0
	Variation Aware Synthesis result-variation	11.8-0.69	67.77-1.5	3.9-0.192	2.19-0	5.94-0
2	Normal Synthesis Result-variation	11-2.2	69.61-2.59	6.9-0.98	1.51-0	4.64-0
	Variation Aware Synthesis result-variation	10.7-0.42	66.77-1.62	4.9-0.3	2.03-0	8.29-0
3	Normal Synthesis Result-variation	11.3-2.9	69.05-2.52	5.58-0.67	2.11-0	4.89-0
	Variation Aware Synthesis result-variation	11.6-0.6	66.7-1.82	6.12-0.245	5.86-0	5.86-0

in V_{th}, t_{ox}, W, and L for each transistor (48 uncertain parameters). Voltage and temperature variations are not considered in this study, but they can also be added to these adaptive algorithms. The optimizer and sensitivity analysis were implemented on the MATLAB platform. An Intel I7 chipset with 2.80GHz clock frequency was utilized during the synthesis process. Population size was kept 50 (30 parents + 20 offspring's) and the number of generations was determined as 20 for all runs. Average synthesis times are around 150 sec. and 900 sec. for the normal synthesis and variation aware synthesis, respectively. Synthesis results are given in Table I. In the first approach, design constraints were slightly relaxed to make possible to overdesign. As seen from the results, the optimizer is quite successful to satisfy the updated constraints that were expanded with respect to the variation data coming from the sensitivity analysis. On the other hand, considering the bottom table results, solutions having less variations can be achieved thanks to the *Variation-Aware Design* approach by sacrificing some other design specifications. Obviously, the chip area worsens through the optimizer looking for more robust solution. However, the overdesign approach may be unsuccessful to expand the design margins when more aggressive specifications are required.

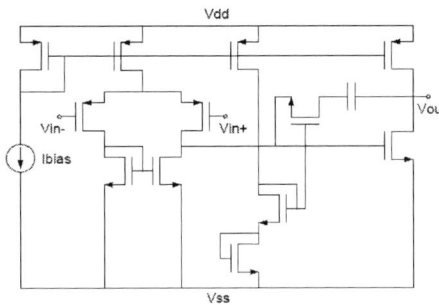

Figure 5.   A two stage CMOS OTA

VI.   Conclusion

Variability has become a crucial reliability problem for modern CMOS technologies as a result of increased fabrication errors due to the scaling difference between device dimensions and process tolerances. Therefore, the designer should increase the focus on the variability analysis after the regular design process. Moreover, increased non-idealities

of advanced technologies complicates the design process. Considering combination of these two phenomena, design automation tools are needed to keep the design within acceptable limits. In this sense, two different variation aware design optimization approaches (*Overdesign* and *Variation-Aware Design*) are proposed in this paper. In *Over-Design* approach, slightly relaxed constraints are expanded considering variations to guard-band the design. On the other hand, in the robust-design approach, the optimizer researches the dedicated design space considering variation effects.

Acknowledgment

This study is supported by the research grant of the Scientific and Technological Research Council of Turkey (TUBITAK) project under the project number 112E005.

References

[1] A. Asenov, S. Kaya, and A. R. Brown, "Intrinsic parameter fluctuations in decananometer mosfets introduced by gate line edge roughness," *Electron Devices, IEEE Transactions on*, vol. 50, no. 5, pp. 1254–1260, 2003.

[2] Y. Ye, S. Gummalla, C.-C. Wang, C. Chakrabarti, and Y. Cao, "Random variability modeling and its impact on scaled cmos circuits," *Journal of computational electronics*, vol. 9, no. 3-4, pp. 108–113, 2010.

[3] M. Conti, P. Crippa, S. Orcioni, and C. Turchetti, "Parametric yield formulation of mos ic's affected by mismatch effect," *Computer-Aided Design of Integrated Circuits and Systems, IEEE Transactions on*, vol. 18, no. 5, pp. 582–596, 1999.

[4] B. Liu, "Computational intelligence techniques for automated design of analog and high-frequency circuits," Ph.D. dissertation, KATHOLIEKE UNIVERSITEIT LEUVEN, 2012.

[5] M. J. Pelgrom, A. C. Duinmaijer, A. P. Welbers *et al.*, "Matching properties of mos transistors," *IEEE Journal of solid-state circuits*, vol. 24, no. 5, pp. 1433–1439, 1989.

[6] G. Tulunay, G. Dundar, and A. Ataman, "A new approach to modeling statistical variations in mos transistors," in *Circuits and Systems, 2002. ISCAS 2002. IEEE International Symposium on*, vol. 1. IEEE, 2002, pp. I–757.

[7] P. G. Drennan, C. C. McAndrew, and J. Bates, "A comprehensive vertical bjt mismatch model," in *Bipolar/BiCMOS Circuits and Technology Meeting, 1998. Proceedings of the 1998*. IEEE, 1998, pp. 83–86.

[8] C. C. McAndrew, J. Bates, R. T. Ida, and P. Drennan, "Efficient statistical bjt modeling, why β is more than i c/i b," in *Bipolar/BiCMOS Circuits and Technology Meeting, 1997. Proceedings of the.* IEEE, 1997, pp. 28–31.

[9] Ö. S. Sönmez and G. Dündar, "Simulation-based analog and rf circuit synthesis using a modified evolutionary strategies algorithm," *INTEGRATION, the VLSI journal*, vol. 44, no. 2, pp. 144–154, 2011.

Impact of enhanced contact doping on minority carriers diffusion currents

Camillo Stefanucci, Pietro Buccella, Maher Kayal and Jean Michel Sallese
Swiss Federal Institute of Technology (EPFL), Lausanne, Switzerland

Email: camillo.stefanucci@epfl.ch

Abstract—**Minority carriers diffusion currents are particularly important in parasitic substrate couplings of Smart Power ICs. In CMOS technologies the P-substrate potential is imposed by P+ contacts and N-wells by N+ highly doped implantations. The doping concentration discontinuity of these contact regions can have a big impact on parasitic diffusion currents of minority carriers. This work gives a description of these effects by device physical simulations of PN junctions under different injection levels of minority carriers. The perturbation of boundary conditions for electrons diffusion is also studied inside the substrate bulk in case a highly-doped substrate is used for high-voltage technologies.**

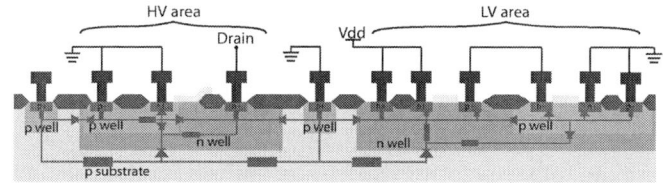

Fig. 1. Cross section of high-voltage NDMOS and low-voltage NMOS devices in high-voltage Smart Power IC technology with equivalent parasitic network using diodes and diffusion resistors to model minority carriers propagation through the substrate.

I. INTRODUCTION

Smart Power ICs integrate on the same silicon chip high-voltage and low-voltage devices [1]. In these systems power transistors are used to drive actuator loads, but during switching operation they inject minority carriers in the substrate. These carriers diffuse and are collected by nearby devices creating malfunctioning in the chip sometimes destructive for the presence of triggered latch up [2].

To simulate these substrate coupling mechanisms in circuit simulators a new modeling methodology has been developed in recent years [3]. It simulates the substrate with an equivalent distributed parasitic network that can be automatically extracted from IC layout [4] as schematically shown in Fig. 1. This network is composed only by diodes and resistors that model the diffusion processes of minority carriers in the substrate with the mean of equivalent voltages and currents [5].

The minority carriers diffusion is solved imposing boundary conditions to the contacts assumed to be ohmic. For minority carriers this means that fully recombination is assumed along metallization. However, in technology realization high-doped implant wells P+ or N+ are used for contact deposition and annealing [6]. These high-doped regions can change the surface recombination mechanisms of minority carriers leading to uncorrect current simulation from the substrate parasitic network if neglected.

For this reason, this paper focuses on the description of the physics mechanisms of minority carriers at high-low doping junctions. In Section II the changes of boundary conditions at the doping discontinuities are analysed for electrons and holes. The simple PN junction diode in forward bias is considered in Section III as case study to show the effects of contact doping on diffusion currents. The analysis of these

effects for low-injection and high-injection regimes is then presented. Finally, in Section IV the investigation results are used to describe effects of doping discontinuity in the substrate coupling currents of Smart Power ICs where highly doped P+ substrate is normally used [7]. Conclusions are summarized in Section V.

II. HOMOJUNCTION BARRIER FOR MINORITY CARRIERS

Devices like diodes and bipolar junction transistors (BJT) are dominated by diffusion currents of minority carriers. This is also the case of substrate parasitic currents in Smart Power ICs. To model analytically these currents, diffusion equation is solved applying as boundary condition to the metal contact complete recombination (ohmic contact).

However, in order to technologically realize not-rectifying contacts, N+ or P+ implantation regions are normally used. It is well known that in PP+ or NN+ homojunctions a space charge region is formed at the interface and a built in potential depending on doping profile appears [8].

On the other side, it is often neglected that the resulting bending of the band diagram dramatically changes the boundary conditions used to solve continuity equations of minority carriers and to compute diffusion currents. By extension this error is also present in the analytical formulations of ideal 1D bipolar transistor analysis. To overcome this issue the high-low doped homojunction interface can be modeled with a non-total recombination of minority carriers.

In Fig. 2 the equilibrium band diagram of a PN junction with P+ and N+ contacts at the extremities is reported. We can notice that at NN+ and PP+ junctions barriers for holes (in N side) and for electrons (in P side) appear which disturb the carriers diffusion towards the contacts. This phenomenon is

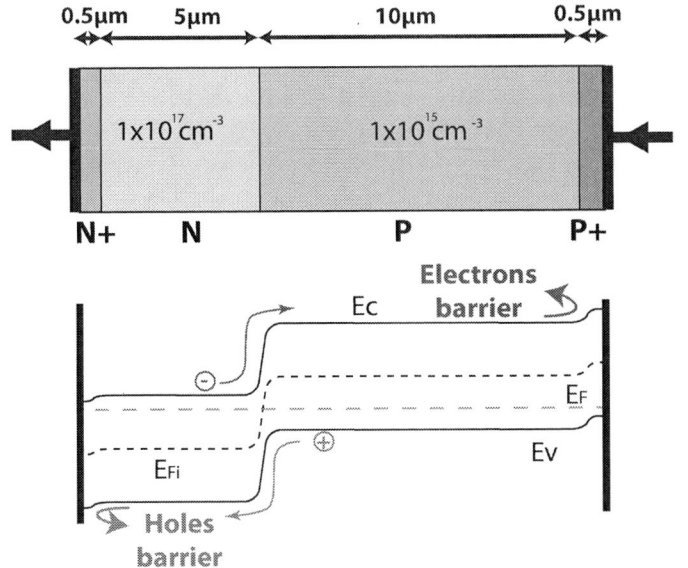

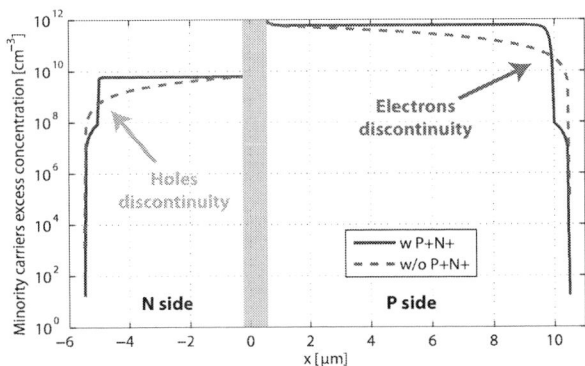

Fig. 3. Simulated discontinuity of minority carriers at high-low junction interfaces in the case of P+ and N+ contact regions (see Fig. 2) compared to the same PN junction biased at 0.4 V without homojunctions.

Fig. 2. PN junction with P+ and N+ end contacts and relative band diagram illustrating the creation of barriers for minority carriers at homojunctions. E_C: conduction band; E_V: valence band; E_F: Fermi level; E_{Fi}: intrinsic Fermi level.

at the base of minority carriers mirror (MCM) used to confine minority carriers in photovoltaic devices [9].

As a matter of fact at homojunctions there is an accumulation of minority carriers in one side and a drop of concentration in the more doped side. This discontinuity of excess concentration is due to the Boltzmann statistics. Out of equilibrium the mass action law imposes that the product between minority and majority carriers concentration is the same and it is dependent on the quasi-fermi potentials splitting. Assuming quasi-neutrality we can analytically express the relation between minority carriers concentration at the discontinuity applying the mass-action law. In case of PP+ homojunction with N_P and N_{P+} the doping of P and P+ side respectively, the electrons excess concentrations $\hat{n}$ on the two sides are related by the equation:

$$\hat{n}_{P+} \simeq \frac{N_{P+}}{2}\left(\sqrt{1 + \frac{4\hat{n}_P(N_P + \hat{n}_P)}{N_{P+}^2}} - 1\right) \quad (1)$$

We notice that in low-injection condition (i.e. the excess of minority carriers is much lower than the doping concentration $\hat{n}_P \ll N_P$) we have a concentration jump depending only on doping ratio between the two sides ($\hat{n}_{P+} \simeq \hat{n}_P N_P/N_{P+}$). In very high-injection conditions $\hat{n}_P \gg N_P$ this discontinuity tends instead to vanish as a consequence of the barrier lowering ($\hat{n}_{P+} \simeq \hat{n}_P$).

III. CONTACT DOPING EFFECTS IN DIODES

In order to investigate the effect of homojunction barriers for minority carriers, physical simulations of a PN junction have been performed.

Technology computer aided design (TCAD) Sentaurus Device [10] software is used including mobility (Arora's model) and lifetime variation with doping concentration.

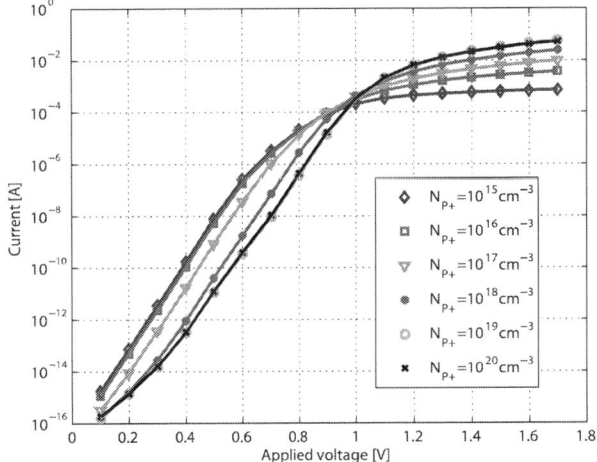

Fig. 4. Simulated IV characteristic of diode in Fig. 2 for different doping concentrations of P+ contact region (other doping concentrations unchanged).

In Fig. 3 the simulated discontinuity of excess carriers concentration of the N+NPP+ diode of Fig. 2 at $0.4V$ forward bias is reported compared with the case of the same PN junction without P+ and N+ doped contacts. It can be seen that the boundary conditions of minority carriers drastically change as expected by Eq. 1. For N+, P+ doping concentration of $10^{19} cm^{-3}$ we have 2 decades drop on the N-side ($N_N/N_{N+} = 10^2$) and 4 decades drop on the P-side ($N_P/N_{P+} = 10^4$) for holes and electrons respectively. The diffusion current proportional to the gradient of minority carriers is then reduced and the total current passes from $167pA$ to $322fA$ in the presence of P+, N+ (for a cross section of $5\mu m^2$).

To deeply study the current lowering, the doping concentration of the P+ contact has been varied from 10^{15} to 10^{20} cm^{-3}. The simulated currents for forward bias conditions are reported in the IV characteristics of Fig. 4. We recognize different behaviors in low-injection regime ($V_P < 1$ V) and high-injection regime ($V_P > 1$V).

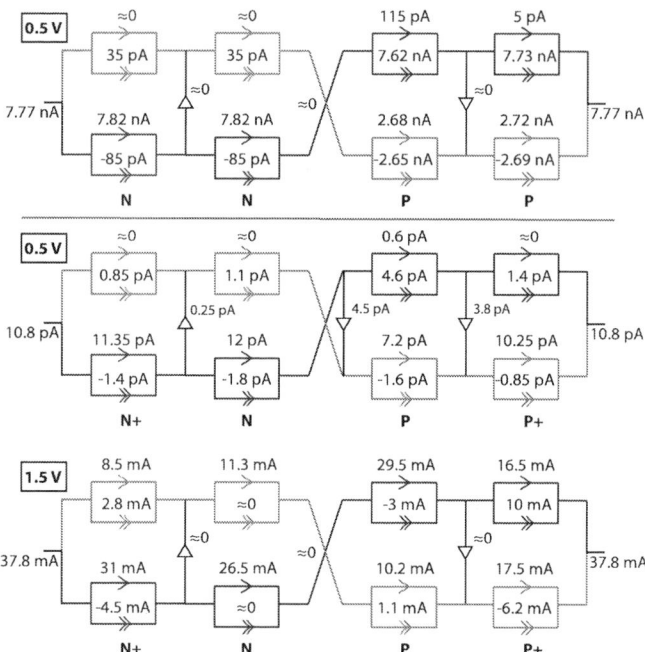

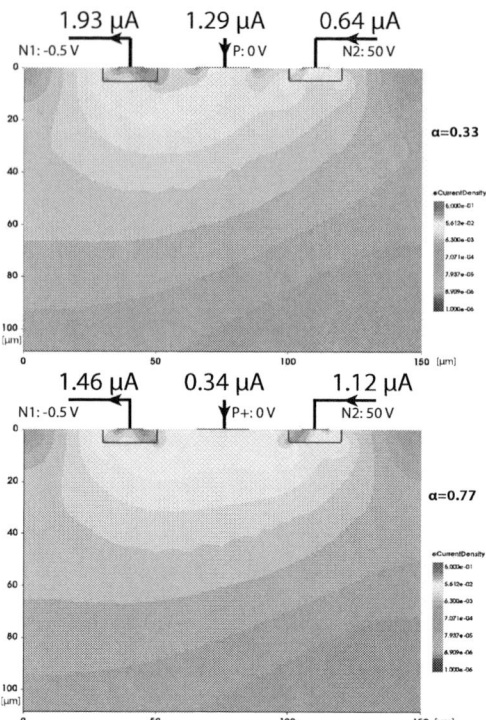

Fig. 5. Flow diagram of simulated currents for electrons (blue branch) and holes (orange branch) in the middle of N+,N,P,P+ regions in low-injection (0.5 V) and high-injection (1.5 V). Drift currents are represented with single arrow, diffusion currents with double arrow and recombination currents with white empty arrow.

Fig. 6. Electrons current density simulation (color) and simulated total currents for a low-doped P substrate ($10^{15}cm^{-3}$) in the case of P contact (top) and P+ contact (bottom).

A. Low injection case: effects on diffusion currents

In low-injection the effect of diffusion current lowering by highly doped contacts is verified and it is only proportional to doping variation. This means that changing the P+ contact width has no effect on the discontinuity of electrons at PP+ interface and then the simulated currents do not have significative changes.

However, the current lowering is not linear with the doping N_{P+} since also recombination currents play a role when P+ and N+ contacts are added. To better visualize this aspect, the flow diagram of electrons and hole drift-diffusion currents is presented in Fig. 5. Looking at the minority carriers top branch of the diagram, for 0.5V the currents are diffusion-dominated. In case of P+, N+ contacts these diffusion currents are strongly reduced by the discontinuity of carriers at the barriers. This lowering is higher in P side because of the higher diffusivity of electrons.

B. High injection case: effects on drift currents

In high-injection the behavior is completely opposite. This is due to the change of currents from diffusion to drift also for minority carriers as reported in the flow diagram of Fig. 5. For 1.5V forward bias the minority carriers currents are now drift-dominated.

As a consequence, in high injection the resistive effect is more important and we get higher currents for higher doped semiconductors where the resistance is lower as expected.

IV. DOPING EFFECTS ON SUBSTRATE COUPLING CURRENTS OF SMART POWER ICs

The same effects reported for diodes can be extended to parasitic bipolar transistor couplings in Smart Power ICs. To investigate this, the coupling between the typical aggressor-victim configuration with two N wells is considered. The 2D geometry under study is composed of two N wells of 200 μm^2 doped $10^{17}cm^{-3}$ at a distance of $50\mu m$ in a low-doped P substrate of 500 μm thickness. The simulations with Sentaurus Device TCAD were performed forward biasing the N1 well at -0.5V (low injection) keeping the substrate at 0V and the N2 well reverse biased at 50V (typical supply for high-voltage technology) activating the lateral parasitic bipolar transistor.

The electrons (minority carriers) simulated current densities and total currents are reported in Figs. 6 for two cases: a P substrate ohmic contact obtained only by metal deposition and a contact with a P+ implantation of 500 nm thickness and $10^{19}cm^{-3}$ acceptor dopant concentration. Two main differences can be detected: the injected current is decreased while the coupling evaluated in terms of transport factor $\alpha = I_{N2}/I_{N1}$ is increased.

These results can be explained using the previous analyses carried out on PN junction. The reduction of emitter current from $1.93\mu A$ to $1.46\mu A$ is indeed related to the different boundary conditions of minority carriers at the P+ base contact.

On the other side, for the coupling α we can distinguish 3 main mechanisms of minority carriers propagation (see Fig.

978-1-4799-4993-9/14 $31.00 © 2014 IEEE

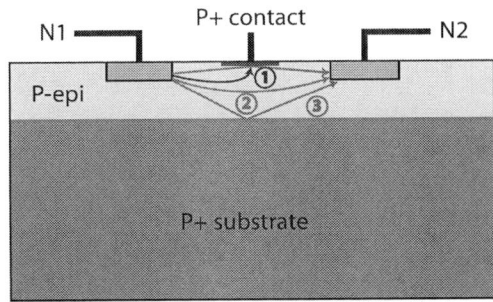

Fig. 7. Propagation mechanisms of minority carriers: direct well-contact path ①, well-well coupling ② and MCM reflection ③.

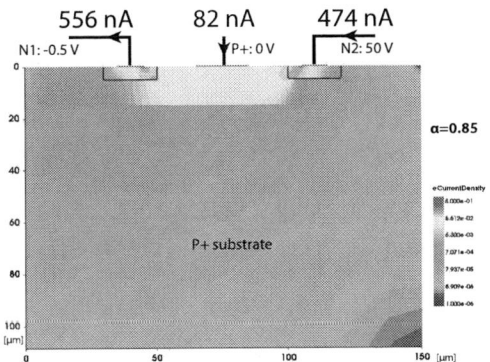

Fig. 8. Electrons current density simulation (color) and simulated total currents for a substrate with low-doped epi-layer ($10^{15} cm^{-3}$) and P+ substrate ($10^{17} cm^{-3}$).

7): the direct propagation from the injecting well to a P contact, the direct propagation between two N wells and the propagation aided by the presence of MCM. In the presented case the P+ MCM partially reflects back electrons towards the collector increasing the coupling.

Another typical situation of doping discontinuity in high-voltage technologies where this effect is evident is in the substrate bulk. High doped substrate are often used after a low-doped P epi layer to recombine more the minority carriers and to reduce the substrate noise injection [7]. This create a homojunction PP+ barrier in the substrate that changes the boundary conditions for diffusion of electrons. The PP+ interface constitutes a MCM that confines the minority carriers on the top layer.

From simulation results reported in Fig. 8, we notice that the presence of a P+ substrate reduces once again by a factor of 3 the injection of minority carriers. However, the presence of the PP+ MCM helps the propagation of electrons towards the N2 well, so the emitter efficiency α is higher than the case without P+ substrate. This guided confinement of minority carriers can be eventually used to develop protection strategies to collect injected electrons [11].

Nevertheless, the size of contacts and wells, their relative position and the epi-layer thickness can play a significant role in the presented coupling mechanisms and their impact must be further analyzed.

V. CONCLUSION

High-low homojunctions creates a barriers for minority carriers propagation. This is the typical case of P+ or N+ contacts. In high injection the contact resistive effect behaves as expected, while in low injection diffusion currents are strongly affected by different boundary conditions of minority carriers. This is particularly evident in low doped regions and in P substrate where electrons have higher diffusivity coefficient. The resulting discontinuity of minority carriers is only dependent on doping discontinuity between PP+ or NN+ homojunctions. In PN junctions this leads to a lower diffusion current.

By extension, TCAD simulation of parasitic bipolar in Smart Power ICs showed that the presence of P+ regions reduce the injected electrons current in low-doped P substrate. However, the coupling increases as a result of the minority carrier mirror effect propagation. As a consequence in Smart Power ICs substrate modeling, the presence of homojunctions cannot be neglected and the minority carrier discontinuity must be included.

ACKNOWLEDGMENT

This work has been sponsored by the European commission under European FP7 AUTOMICS project.

REFERENCES

[1] B. Murari, F. Bertotti and G. Vignola, *Smart Power ICs*, 2nd ed., pp. 218-220,Springer-Verlag, Berlin, 2002.

[2] R. Zhu, V. Khemka, A. Bose, T. Roggenbauer, "Substrate Majority Carrier-Induced NLDMOSFET Failure and Its Prevention in Advanced Smart Power IC Technologies", IEEE Transactions on Device and Materials Reliability, vol.6, no.3, pp.386-392, September, 2006

[3] F. Lo Conte, J. M. Sallese, M. Pastre, F. Krummenacher, M. Kayal, "Global Modeling Strategy of Parasitic Coupled Currents Induced by Minority-Carrier Propagation in Semiconductor Substrates", IEEE Transactions on Electron Devices, vol.57, no.1, pp.263-272, January, 2010.

[4] Y. Moursy et al., "AUTOMICS: A novel approach for substrate modeling", 18th IEEE European Test Symposium, May, 2013.

[5] F. Lo Conte, J. M. Sallese, M. Pastre, F. Krummenacher and M. Kayal, "A Circuit-Level Substrate Current Model for Smart-Power ICs", IEEE Transactions on Power Electronics, vol. 25, n. 9, September, 2010.

[6] S. Li and Y. Fu, "Smart Power Technology and Power Semiconductor Devices", in "3D TCAD Simulation for Semiconductor Processes, Devices and Optoelectronics", Springer, New York, 2012.

[7] A. Moscatelli, A. Merlini, G. Croce, P. Galbiati and C. Contiero, "LDMOS Implementation in a 0.35 μm BCD Technology (BCD6)", International Symposium on Power Semiconductor Devices (ISPSD), Toulouse, France, 2000

[8] J. Oehmen, L. Hedrich, M. Olbrich and E. Barke, "A Methodology for Modeling Lateral Parasitic Transistors in Smart Power ICs", Proceedings of IEEE International Behavioral Modeling and Simulation Workshop (BMAS), 2005.

[9] H. L. Chuang, M. E. Klausmeier-Brown, M. R. Melloch and M. S. Lundstrom, "Effective minority-carrier hole confinement of Si-doped, n+n GaAs homojunction barriers", Journal of Applied Physics, vol. 66, no. 1, 1989.

[10] Synopsys Sentaurus Device, http://www.synopsys.com/Tools/TCAD/DeviceSimulation

[11] C. Y. Huang, M. J. Cher and C. Y. Wu, "Low-temperature characteristics of well-type guard rings in epitaxial CMOS", IEEE Transactions on Electron Devices, vol. 43, no. 12, 1996.

Reliability Analysis of Logic Circuits Using Probabilistic Techniques

Satish Grandhi, Christian Spagnol, Emanuel Popovici
Department of Electrical and Electronic Engineering,
University College Cork
Cork, Ireland
sagrand@ue.ucc.ie, christian.spagnol@ue.ucc.ie, e.popovici@ucc.ie

Abstract—**The low reliability of advanced CMOS devices has become a critical issue that can potentially supersede the benefits of the technology shrinking process. This is making the design time reliability assessment and optimization a mandatory step in the IC design flow. As part of our ongoing research, we describe an algorithm based on probability analysis and logic principles for computing the impact of gate failures on the circuit output. We also propose a Bound and Propagate based methodology to handle the reconvergent fanout issue. A reliability evaluator has been developed around the open source logic synthesis tool 'abc' to allow integration and evaluation of our method in the context of an IC design flow. This approach had tremendously reduced the computation time while maintaining adequate precision. Simulation results for several benchmark circuits demonstrate the accuracy and the simulation time advantages when compared to MonteCarlo simulations.**

Keywords:Reliability, Reconvergent Fanout, AIG, abc

I. INTRODUCTION

As CMOS technology enters the nanometer era they force the design tolerance limits to their lowest possible levels. Nanotechnology specific issues, e.g., V_{dd} reduction, higher impact of process variation and temperature related aging resulted in increased device failure rates, making CMOS ICs less reliable [1], [2]. Therefore, design time reliability assessment and optimization are becoming a mandatory step in the IC design flow. As part of our ongoing research, we are developing a reliability aware logic synthesis tool. The first step in this process is to develop an efficient algorithm that computes circuit reliability. Reliability analysis of logic circuits deals with computing the impact that the gate level have on the circuit Primary Outputs(PO). Traditional approach to reliability analysis begins with elementary SPICE simulations to estimate the circuit error probability. Several analytical approaches for computing reliability have been previously reported [3], [4]. As we represent the circuit in the AIG format, a novel algorithm based on probability principles is proposed, with the prime focus being AND & Inverter gates. The algorithm uses the dynamic weighted average algorithm (DWAA) [5] approach to estimate the impact of reconvergent fanout on Static Probabilities. Further, Bound and Propagate

This work was supported by the Seventh Framework Programme of the European Union, under Grant Agreement number 309129 (i-RISC project).

methodology is introduced to account for the statistical dependence on error probabilities due to reconvergent fanout.

This paper is organised as follows. Section II, presents the data structure used in this work. Section III discusses the mathematics behind the error model. Section IV demonstrates the implementation, along with the design flow incorporating the tool. Section V introduces the error bounds. Section VI concludes and outlines our ongoing research effort.

II. AND INVERTER GRAPH

Structural representation, logic synthesis and technology mapping of a Boolean function are important issues in the design of digital circuits. One of the major decision in designing an EDA tool capable of resolving these issues is the selection of the right data structure as it determines the speed and efficiency of the tool. Binary Decision Diagrams (BDDs) has been used extensively but they have reached the limit of their scalability due to the always increasing complexity of modern circuitry. To overcome these limitations, And Inverter Graph (AIG) [6] are used to represent the circuits in this work.

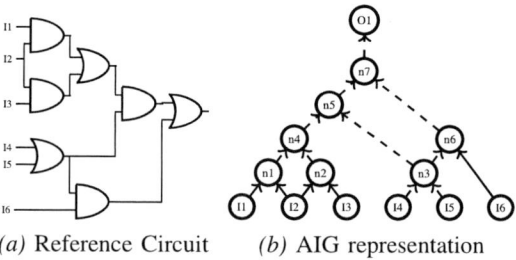

(a) Reference Circuit *(b)* AIG representation

Fig. 1. And Inverter representation of Combinational Circuit

AIG is a Boolean network composed of two-input-ANDs and inverters. Fig. 1 depicts a simple combinational circuit and its corresponding AIG representation. The circle represent the 2-Input AND gates and edges with a dash line indicate negation i.e. inversion of that input. AIG unifies equivalence checking, synthesis and technology mapping and offer better performance [7], [8], [9]. In the next section, we present the probabilistic gate error models for AND & inverter logic gates. This concept can be extended to any other generic gate as well. Some of the common conventions used through out the paper are listed below:

978-1-4799-4993-9/14 $31.00 © 2014 IEEE

- $P_1(Z)$ - Probability of node 'Z' to be 1
- $P_0(Z)$ - Probability of node 'Z' to be 0
- $P_\epsilon(Z)$ - Probability of error on node 'Z'
- $P_f(y)$ - Failure probability of Logic Gate 'y'
- $R(y)$ - Reliability of Logic Gate 'y'

III. Gate Error Models

An unreliable AND gate can be modeled as an ideal (error free) AND gate followed by a faulty buffer that represents the stochastic behavior of the errors. This model moves the entire error statistic on the output and so it implicitly assumes a symmetrical error behavior in relation to the inputs. The two nodes Z* and Z are named as internal and the external output node. Next, the analysis of the output error due to two possible reasons: (i) propagation of the errors onto the gate input nodes and (ii) intrinsic errors within the gate, is presented.

Reliable Condtions			Unreliable Condtions			Error Probability
A	B	Z*	A	B	Z*	
0	0	0	c/0	c/0	c/0	
			c/0	ϵ/1	c/0	
			ϵ/1	c/0	c/0	
			ϵ/1	ϵ/1	ϵ/1	$\mathbf{P[A=0,B=0,A_c,B_\epsilon]}$
0	1	0	c/0	ϵ/0	c/0	
			c/0	c/1	c/0	
			ϵ/1	ϵ/0	c/0	
			ϵ/1	c/1	ϵ/1	$\mathbf{P[A=0,B=1,A_c,B_c]}$
1	0	0	ϵ/0	c/0	c/0	
			ϵ/0	ϵ/1	c/0	
			c/1	c/0	c/0	
			c/1	ϵ/1	ϵ/1	$\mathbf{P[A=1,B=0,A_c,B_\epsilon]}$
1	1	1	ϵ/0	ϵ/0	ϵ/0	$\mathbf{P[A=1,B=1,A_\epsilon,B_\epsilon]}$
			ϵ/0	c/1	ϵ/0	$\mathbf{P[A=1,B=1,A_\epsilon,B_c]}$
			c/1	ϵ/0	ϵ/0	$\mathbf{P[A=1,B=1,A_c,B_\epsilon]}$
			c/1	c/1	c/1	

TABLE I
IDEAL AND GATE WITH UNRELIABLE INPUTS

A. 2-Input Ideal AND Gate

As AIG's comprises of only 2-input AND gates, we restrict our analysis to 2 input AND gates and note that its extension to multi-input AND gates is straightforward. The static probability values of the internal output node Z* can be expressed as:

$$P_{z^*}(1) = P[A = 1, B = 1] \qquad (1)$$

We note that error on the inputs need not necessarily result in a wrong output value. Consider error free '0' on one of the input pins. This would mask the error on the other input node from being propagated onto the output. Similarly, consider the scenario where the input pins are set to '0' & '1' and both are in error. This event of double error mutually negates each other and will still result in a correct output state. Hence, due to the masking and double error events, there are no simple rules to predict the state of the output. Tab. I presents an exhaustive enumeration on all the possible cases with the associated output status. To explain the table, consider the case of input A=0 and B=0. Then, the internal output Z* evaluates

to '0'. Now, each of the inputs can be in error (ϵ) or correct (c) state. The state of the inputs determines if Z*=0 is correct or not. It is clear that, for these inputs values, the internal output node is in error if and only if both the inputs are in error. It is evident from Tab. I that only 6 of the possible 16 cases result in error on the internal output. The probabilities for each of these events to occur are presented in the last column.

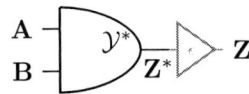

Fig. 2. Unreliable AND Gate Model

In order to arrive at a close form representation of AND gate output node error probability, we assume (an alternative approach is presented later) that the inputs of the gates are independent. This allows for the use of simple formulas to compute reliability and reduce the overall algorithm running time. The internal node error probability is the sum of all the six terms in Tab. I, thanks to the previous assumption each of these can be factorized, this results in an erroneous output and evaluates to :

$$
\begin{aligned}
P_\epsilon(\mathbf{Z^*}) = & P_\epsilon(\mathbf{A})P_\epsilon(\mathbf{B})P_\mathbf{A}(0)P_\mathbf{B}(0) + \\
& P_\epsilon(\mathbf{A})(1 - P_\epsilon(\mathbf{B}))P_\mathbf{A}(0)P_\mathbf{B}(1) + \\
& (1 - P_\epsilon(\mathbf{A}))P_\epsilon(\mathbf{B})P_\mathbf{A}(1)P_\mathbf{B}(0) + \\
& [P_\epsilon(\mathbf{A}) + P_\epsilon(\mathbf{B}) - P_\epsilon(\mathbf{A})P_\epsilon(\mathbf{B})]P_\mathbf{A}(1)P_\mathbf{B}(1)
\end{aligned}
\qquad (2)
$$

Z*		Gate	Z		Error Probability
Ideal	State	Fault	Value	State	
0	c	c	0	c	
	c	f	1	ϵ	$\mathbf{P[Z^*=0,Z_c,P_f]}$
	ϵ	c	0	ϵ	$\mathbf{P[Z^*=0,Z_\epsilon,P_c]}$
	ϵ	f	1	c	
1	c	c	1	c	
	c	f	1	ϵ	$\mathbf{P[Z^*=1,Z_c,P_f]}$
	ϵ	c	1	ϵ	$\mathbf{P[Z^*=1,Z_\epsilon,P_c]}$
	ϵ	f	0	c	

TABLE II
FAULTY AND GATE WITH UNRELIABLE INPUTS

B. Intrinsic Gate Error Effects

Tab. II presents the analysis of the possible configurations of the faulty buffer. Column 1 represents the state of the internal node Z*. Column 2 defines the state of the gate whether it is faulty or not. Based on these two conditions, the value of the output is computed in column 3. The current of the output, whether it is erroneous or correct, is also defined. Column 4 defines the probabilistic state when the output is in error.

We employ the Binary Symmetric Channel (BSC) model to represent the erroneous buffer behavior. The gate output static and error probabilities can be defined as:

$$
\begin{aligned}
P_z(1) &= P_z^*(1)(1 - P_F) + P_Z^*(0)P_F \\
P_z(0) &= P_z^*(0)(1 - P_F) + P_Z^*(1)P_F \\
P_\epsilon(Z) &= P_F + P_\epsilon(Z^*) - 2 * P_F * P_\epsilon(Z^*)
\end{aligned}
\qquad (3)
$$

Benchmark	Gate Count		Avg Error Deviation on all outputs %			Runtime{s}	
	AND	Inverter	$\epsilon = 0.001$	$\epsilon = 0.01$	$\epsilon = 0.05$	MC_Sims	Tool
cu	55	29	5.75	2.78	9.48	7051.89	**0.393**
x2	55	32	6.06	7.24	9.24	5356.85	**0.924**
parity	45	61	3.51	6.38	7.55	10215.41	**1.042**
cm150a	61	71	3.36	2.81	9.43	16477.93	**1.558**
cordic	82	84	1.57	2.24	9.57	22749.73	**0.966**
mux	85	92	2.32	3.13	6.25	22019.47	**0.295**
b9	104	78	3.79	3.38	6.57	43578.04	**0.827**
count	128	130	5.34	7.84	9.84	52890.57	**4.691**

TABLE III

MCNC BENCHMARK CIRCUITS BASED ACCURACY & PERFORMANCE EVALUATION FOR DIFFERENT GATE ERRORS (ϵ)

The modeling of unreliable Inverter follows similar line of flow. It is not presented here due to space constraints.

IV. CAD Tool: Reliability Evaluator

In this section we present the algorithm and the experimental setup used to demonstrate the accuracy of the proposed approach against MonteCarlo simulations. Further, we report the accuracy and simulation time savings compared to MonteCarlo.

A. Computation Algorithm

Alg. 1 presents the methodology employed within the tool to compute circuit reliability. It accounts for the error introduced by both Inverter & AND gates. Using Eq. 2 and 3, the error due to the AND gate is computed both on the internal and external output nodes. This flow has been integrated into the open source tool 'abc' [10] and automates the error probability computation.

Algorithm 1 Generic Method for Reliability Evaluation

INPUT:N, total number of nodes in the AIG network, Error Probability of Individual Gates and Switching activity P_{SA} on the primary input nodes (PI's)
OUTPUT: Output error probability
  for all nodes I= 1 to N **do**
    if Input Nodes are inverted **then**
      Account for the inverter error
    end if
    Compute Internal node error probability using Eq. 2
    Compute Output node error probability using Eq. 3
  end for

B. Experimental Flow

Fig. 3 depicts the complete experimental setup developed to compare the algorithm results with fault inserted gate level simulations. The sample size used for reliability analysis is 100k, 50k and 10k for 0.001, 0.01 and 0.05 error scenario's respectively. We assume that the primary inputs are independent and set the switching activity on all the pins to 0.5. The switching activity numbers are flexible and can be set to any other value. In Tab. III, columns 1 and 2 give the name and number of gates in the benchmark circuit. Column 3 captures the accuracy of the method when compared with MonteCarlo

Simulations while Column 4 highlights the significant time savings the proposed algorithms achieves when compared with MonteCarlo simulations. From the table, it is clear that the proposed algorithm maintains accuracy within 10% while significantly speeding up on the computation time. This error is justified because the current algorithm does not takes the impact of reconvergence on the error probabilities into account. As we represent the logic only in terms of AND and Inverter gates, we have not performed comparative study with the previous published analytical methodologies. In the next section, we introduce a novel methodology, which is still under research, to overcome this limitation.

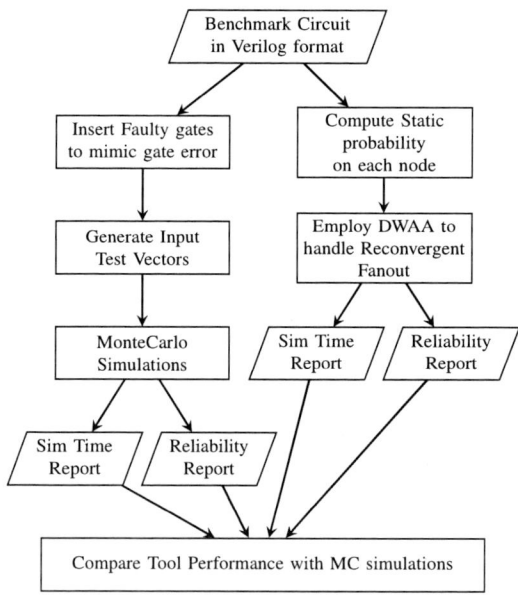

Fig. 3. Design Flow for Reliability Computation

V. Reconvergent Fanout : Bounding Approach

The methodology developed in the previous section does not accounts for the impact of the reconvergent fanout gate error on the final reliability of the output node. The statistical dependence among signals in a combinational circuit is possible due to reconvergent fanout assuming that the primary inputs are independent. When reconvergent fanout gates are present in a circuit, the signals at the inputs of reconvergent gates are correlated. Ignoring these correlations

978-1-4799-4993-9/14 $31.00 © 2014 IEEE

can produce erroneous results as seen in Tab. III. Computing the node error probability in the presence of reconvergent fanout is complex because Eq. 2 does not hold true. This is because each of the terms in Eq. 2 cannot be factorized due to dependencies between the four probabilities. There is no closed form solution to solve the terms in Eq. 2 in an exact manner. Iterative approaches do exist but their complexity grows exponentially for each of the 6 terms and their running time is not acceptable for application in synthesis tools. This section presents a novel methodology to deal with the impact of reconvergent fanout gate error on the overall circuit error probability.

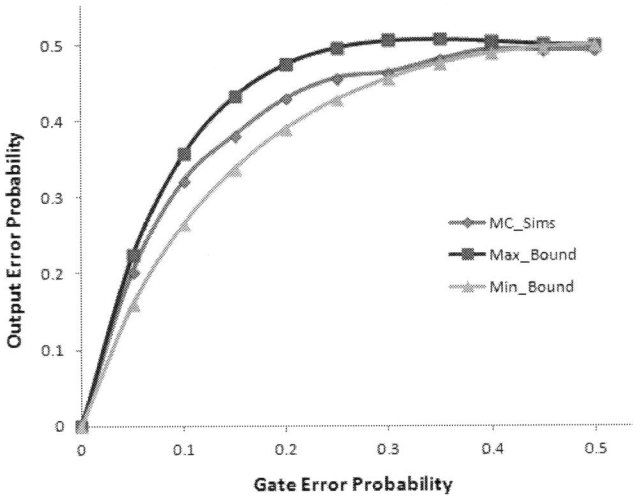

Fig. 4. Bounding Error

A. Bounding Node Error Probability

The methodology of Bound and Propagate is now introduced. It accounts for the reconvergent fanout gate error while not increasing the overall simulation time. The algorithm computes the upper and lower bounds for the error probability on all the gates with the reconvergent fanout. Eq. 2 accounts for the six possible scenarios that would result in an output error. Bounding each of these terms singularly results in loose bounds that would quickly converges to the upper/lower bound of 1/0, respectively. To avoid this scenario, only the total gate output error probability is bounded. As depicted in Fig. 4, an error on any of the input nodes can be propagated onto the output iff the other input node is set to '1'. If the other node is at '0', the error would be masked. Now to obtain the bounds, two cases are presented; a pessimistic and an optimistic scenario. The plot in Fig. 5 illustrates the accuracy of the error bounds when applied to MCNC benchmark circuit 'B9'.

Pessimistic Rule: The maximum error bound on the output node of the 'AND' gate is defined as the summation of the error probabilities on each of the input nodes when the other node is set to '1'. This is represented by Eq. 4. From Tab. I, it is clear that this rule is pessimistic since many cases exist where a single error get "masked".

$$P_\epsilon(Z) \leq P_\epsilon^{Max}(A)P_1(B) + P_\epsilon^{Max}(B)P_1(A) \triangleq P_\epsilon^{Max}(Z) \tag{4}$$

Optimistic Rule: The minimum error bound on the output node of the 'AND' gate is defined as the maximum of the product of the error on the input node and the probability of other node set to '1'. This is represented by Eq. 5.

$$P_\epsilon(Z) \geq Max\{P_\epsilon(A)P_1(B), P_\epsilon(B)P_1(A)\} \tag{5}$$

VI. Conclusion

As part of our ongoing research, we are developing novel reliability aware synthesis algorithms to improve circuit reliability. In this process, reliability evaluation is the most primitive and important step towards developing a reliability aware logic synthesizer. This paper described a simple technique to study the gate failure rate on the overall circuits. The reliability numbers obtained with the proposed algorithm are within 10% margin compared to the MC simulations. This error is within acceptable limits as the main focus of this approach is to quickly compare hundreds of logically equivalent realizations and select higher reliability circuit configuration. The next step is to study the impact of reconvergent fanout on reliability estimation in greater detail to develop tighter bounds and eventually integrate it into the synthesis tool.

References

[1] S. Borkar. Designing reliable systems from unreliable components: the challenges of transistor variability and degradation. *Micro, IEEE*, 25(6):10–16, 2005.

[2] C. Constantinescu. Trends and challenges in vlsi circuit reliability. *Micro, IEEE*, 23(4):14–19, 2003.

[3] M.R. Choudhury and K. Mohanram. Reliability analysis of logic circuits. *Computer-Aided Design of Integrated Circuits and Systems, IEEE Transactions on*, 28(3):392–405, 2009.

[4] Jie Han, Hao Chen, Erin Boykin, and Jos Fortes. Reliability evaluation of logic circuits using probabilistic gate models. *Microelectronics Reliability*, 51(2):468 – 476, 2011.

[5] S. Ercolani, M. Favalli, M. Damiani, P. Olivo, and B. Ricco. Estimate of signal probability in combinational logic networks. In *European Test Conference, 1989., Proceedings of the 1st*, pages 132–138, Apr 1989.

[6] A Biere. The aiger and-inverter graph (aig) format, version 20070427. *Johannes Kepler University*.

[7] Robert Brayton and Alan Mishchenko. Abc: An academic industrial-strength verification tool. In *Computer Aided Verification*, pages 24–40. Springer, 2010.

[8] Rashmi Mehrotra, Tom English, Michel Schellekens, Steve Hollands, and Emanuel Popovici. Timing-driven power optimisation and power-driven timing optimisation of combinational circuits. *Journal of Low Power Electronics*, 7(3):364–380, 2011.

[9] Thiago Figueiro, Renato P Ribas, and André Inácio Reis. Constructive aig optimization considering input weights. In *Quality Electronic Design (ISQED), 2011 12th International Symposium on*, pages 1–8. IEEE, 2011.

[10] A Mishchenko et al. Abc: A system for sequential synthesis and verification. *URL http://www. eecs. berkeley. edu/~ alanmi/abc*, 2007.

Fig. 5. B9 Benchmark Circuit Error Bounds

A 40mV Start up Voltage DC – DC Converter For Thermoelectric Energy Harvesting Applications

C. Veri[1], M. Pasca[1], S.D'Amico[1,2]
[1]Department of Innovation Engineering
University of Salento
Lecce, Italy
carlo.veri@studenti.unisalento.it

L. Francioso[2]
[2]Institute of Microelectronics and Microsystems
CNR – IMM
Lecce, Italy

Abstract— **A low start up voltage DC – DC converter for thermoelectric energy harvesting is presented in this paper. Output voltage of a thermoelectric energy generator (TEG) provides an output voltage from 40mV to 400mV, depending of thermal gradient. In order to increase input voltage, a boost converter is used. The proposed DC – DC converter is composed of two main sections. One provides a high duty cycle pulse width modulation for a forward control. It is used when the voltage coming from the TEG, is in the 40mV to 150mV range. As the TEG voltage it is higher than 150mV, a feedback circuit is switched on. It provides a more accurate control of the output voltage. Entire DC – DC converter is implemented in a 65nm bulk CMOS technology.**

Keywords—component; DC – DC Converter, low startup voltage, energy harvesting, thermoelectric generator.)

I. INTRODUCTION

During the last years, the interest of scientific community and caregivers towards low cost solutions for energy autonomous and wearable biometric monitoring sensors has grown very fast. In order to extend the life-time of traditional batteries, intensive research is currently focused on the development of portable power generators able to harvest energy from different environmental sources and convert it into electricity. To this aim, a thermoelectric energy generator is used. It is able to generate an output voltage of 40mV with a thermal gradient of 6°K. Nominal voltage supplies are higher than 40mV. Therefore a DC – DC converter is required in order to increase this value. State of art is rich of interesting architectures: in [1] a boost converter is shown. In that work, authors describes a low start – up voltage interface circuit, but since the converter's control circuitry derives its power from its own output, the output capacitor must be pre-charged to approximately 650mV before the converter can operate. A batteryless low voltage interface circuit is shown in [2]. It uses a mechanically assisted step-up process that needs vibration at start up. Other solution is represented of [3]. With 95mV start up voltage, its maximum efficiency reaches 72%. A more interesting work is [4]. A 3 – stage stepping up architecture provides to increase minimum input voltage with a peak efficiency of 73%. However, the conversion efficiency is generally low for an input voltage below 100mV, making it less attractive for applications with small temperature differences. In this work, a batteryless DC – DC converter is presented. Start up voltage is 40mV, enabling operation at a lower thermal gradient. Its maximum efficiency is 85%. The paper is organized as follows. Section II introduces the architecture and operation principles of the proposed converter. Detailed circuit implementations and simulation results are presented in Sections III and IV, respectively. A concluding remark is given in Section V.

II. PROPOSED DC – DC CONVERTER

In design of voltage multipliers, more powerful technique is characterized by various steps of conversion. At first, the low voltage DC is converted in AC therefore, by means suitable circuits, it returns in DC with a higher value than before. Through this process, it is possible to improve conversion efficiencies for very low start up voltage. Figure 1 shows the architecture of the proposed DC – DC converter. It consists of three functional blocks: a feedforward low start up control circuit that provides a high duty cycle PWM signal. This circuit operates for input voltages (V_{IN}) from 40mV to 150mV. A feedback network presents in output a variable duty cycle clock signal. It is supplied by the output voltage of boost converter, therefore it is possible to avoid the presence of external pre – charged battery. It works for voltage range from 150mV to 400mV. Through the use of this PWM, a boost converter increases the input voltage. Its output is fixed at 1V independently of V_{IN}. As the input voltage exceeds 40mV, local oscillator (LO) becomes operational, providing two differential outputs. In order to facilitate the start up at extremely low voltage, feedback network is initially off. Afterwards, feedforward control circuit switches on, and its PWM signal changes three different values of duty cycle, depends on V_{IN}. It operates with varies V_{IN} range, in order to achieve right duty cycle PWM signal. Charge pump wired to the LO and TEG, provides an output DC voltage greater than 10 times of V_{IN}. The presence of diodes D1 and D2 prevents exceeding the value of 1.2V, which could break down MOS M7 and M8. Duty cycle variation circuit accepts DC voltage from charge pump and transforms it into a high duty cycle PWM. Boost converter is composed of an external inductance of 1mH, an external capacitor, and two MOS switches M7 and M8 that are driven of control circuitry.

After C_{OUT} is charged to 1V and V_{IN} is higher than 150mV, output comparator enables feedback network. Main task of the

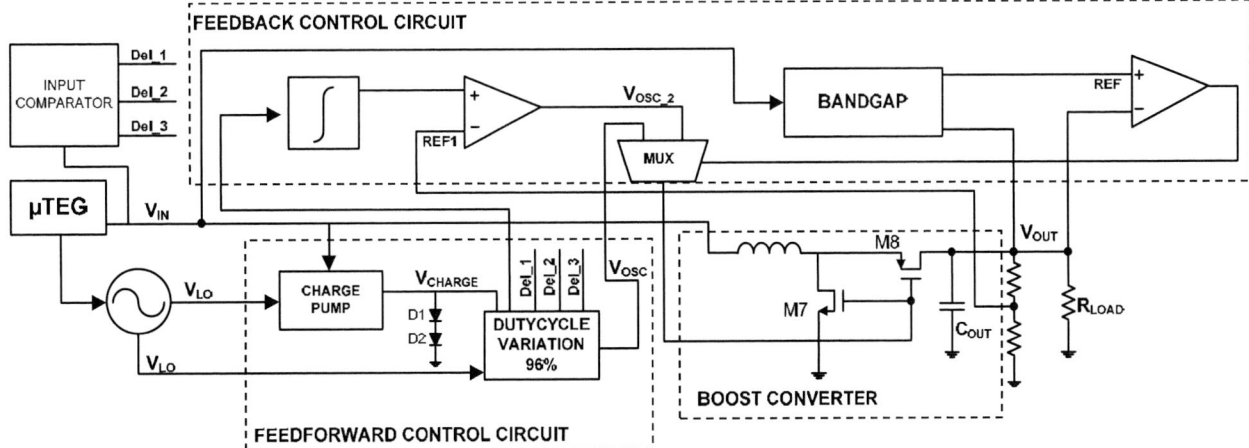

Figure 1 – Schematic of DC – DC Converter

latter is to control the duty cycle of PWM, in order to maintain output voltage to 1V, as Vin changes due to temperature variation. During operation of feedback circuit, the feedforward control circuit is disabled.

III.   CIRCUIT IMPLEMENTATIONS

A. Local Oscillator

This is the most critical component to be design in order to operate at low voltage. It has been designed as an NMOS cross coupled with source degeneration architecture. Figure 2 is shows the oscillator schematic:

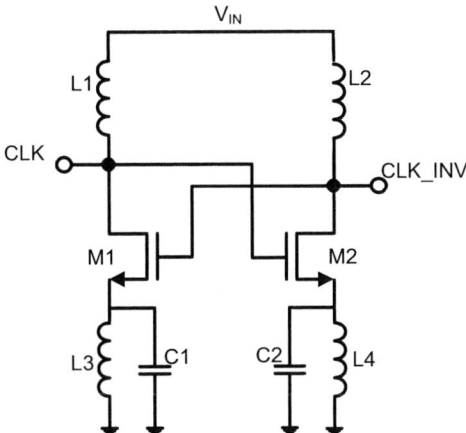

Figure 2 – Schematic of Local Oscillator

Through source degeneration, oscillator starts to work from very low voltage. The oscillation frequency (232 kHz) depends on L1 and L2 (offchip) values and the input capacitance of the charge pump, where oscillator is connected. This makes the oscillation frequency stable over all V_{IN}. Figure 3 shows output voltage of the oscillator, when V_{IN} is 40mV.

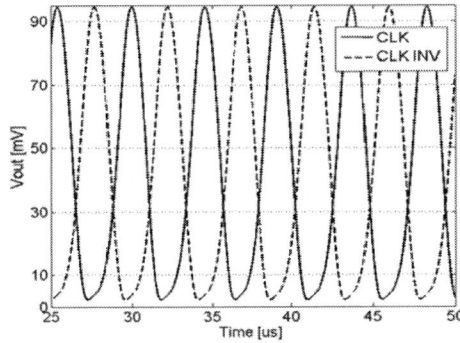

Figure
3 – Oscillator Output Voltage at 40mV V_{IN}

B. Boost Converter

Boost Converter is the circuit that raises input voltage provided by the TEG. Figure 4 shows architecture:

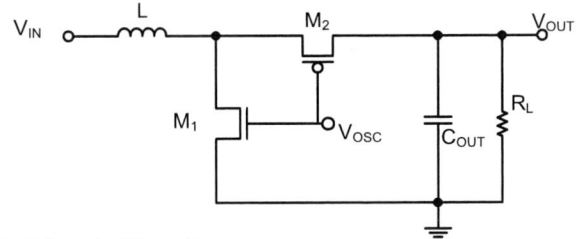

Figure 4 – Schematic of Boost Converter

Two MOS M1 and M2 are switched by a PWM signal. When M1 is on, current in inductor L increases linearly. Next, turning off M1 (and hence turning on M2) transfers the current flowing through the inductor, to the output capacitor and load. The steady state output voltage of the boost regulator is:

$$V_{OUT} = \frac{V_{IN}}{1 - D} \qquad (1)$$

Where D is the duty cycle of the PWM signal.

As equation 1 shows, voltage gain depends on PWM signal. Therefore, lower V_{IN} needs higher duty cycle value to get the same output voltage.

C. Feedforward control circuit

DC – DC converter has a section that drives boost converter when V_{IN} is in the range from 40mV to 150mV. Its task is to provide a suitable PWM signal with a high duty cycle, that enables increasing input voltage from TEG up to 1V. The PWM control signal is changed according to V_{IN} as table I shows:

VIN range	Duty Cycle
40mV ÷ 60mV	96%
60mV ÷ 90mV	90%
90mV ÷ 150mV	86%

Table 1 – V_{IN} range vs Duty Cycle

Duty Cycle is changed by regulating delay introduced of delay circuit. This is got by properly adding capacitances that are activated by input comparator, supplied by output voltage of boost converter. A charge pump is used to get a voltage higher than V_{IN}. Structure of charge pump is a series of 25 Pelliconi's single cell. As describe in [5], this architecture has many advantages such as eliminating body effect by connecting each device substrate to its source. It also uses very simple clocking scheme and it has no diode – connected MOS. Therefore higher gain is achieved. Each stage is driven by a simple two-phase clocking scheme; the final stage is the same as the others, therefore no specific output stage is needed. Figure 5 shows structure of single cell.

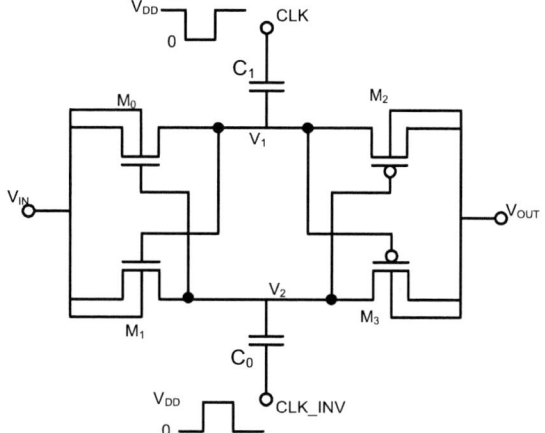

Figure 5 – Pelliconi Single Cell

The voltage at the output of the charge pump is converted into a high duty cycle PWM signal, which drives MOS in boost converter. Therefore, the PWM amplitude must be higher than the NMOS threshold V_{TH}.
High duty cycle circuit is shown to next picture:

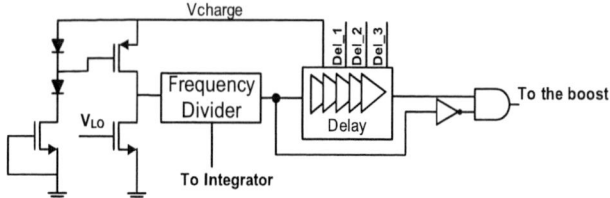

Figure 6 – Schematic of high duty control circuit

Output voltage of local oscillator is directly connected to the circuit. Oscillator frequency is divided 64 times, in order to prevent strong reduction of total efficiency at low input voltage. After division, delay circuit shifts the clock signal at its inputs. NAND gate with the clock and its delayed copy at its input enables a high duty cycle PWM signal. Output PWM signal at 40mV V_{IN} waveform is shown in figure 7:

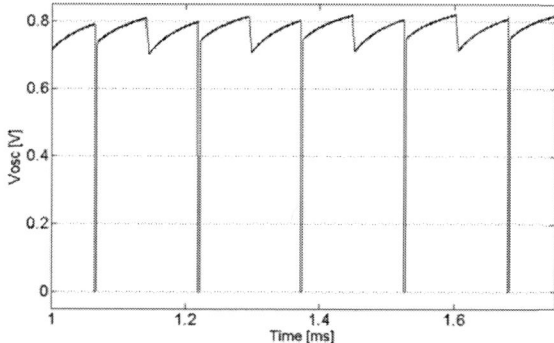

Figure 7 – Clock signal with 96% duty cycle

D. Feedback control circuit

The feedback circuit that achieves an automatic control of duty cycle PWM signal, is enabled as V_{IN} is larger than 150mV and the output voltage is higher than 1V. Every circuit of feedback stage is supplied by the output voltage of boost converter. This section, during feedforward circuit works, is off in order to improve globally efficiency. It starts to operate when V_{IN} is in the range from 150mV to 400mV and an output comparator, driving a multiplexer, provides to select clock signal, according to input voltage from TEG. As shown in figure 8, integrator circuit has in input a square waveform from frequency divider, in high duty cycle circuit. In output of this, a triangle wave is compared with a variable REF1, depending of V_{OUT} to boost converter.

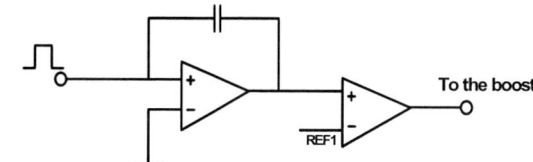

Figure 8 – Schematic of Feedback Circuit

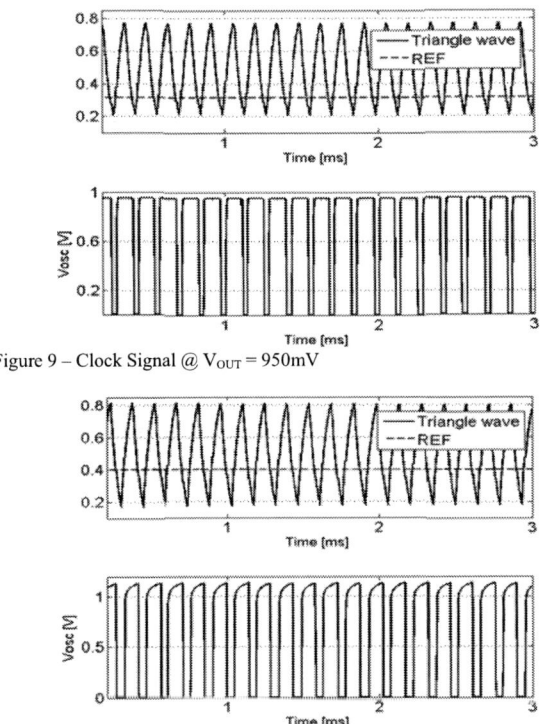

Figure 9 – Clock Signal @ V_{OUT} = 950mV

Figure 10 – Clock Signal @ V_{OUT} = 1.2V

Figures 9 and 10 shows normal operation of feedback control circuit. As V_{OUT} decreases under 1V, a lower value of REF1 is compared with triangle wave and so the output of comparator is a waveform with high duty cycle. In this case, according with (1), V_{OUT} increases and reaches an approximate constant value of 1V. On the other hand, if V_{OUT} gets over 1V, REF1 increases and so in output of comparator will be a clock signal with lower duty cycle. In this case, V_{OUT} decreases. A bandgap circuit provides an accurate references voltages value to switch between feedforward and feedback control.

IV. SIMULATION RESULTS

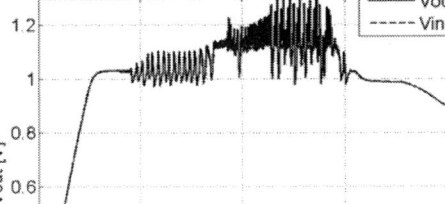

Figure 11 – V_{OUT} vs V_{IN}

The proposed ultra – low input voltage DC – DC converter for thermoelectric energy harvesting applications is designed in a standard 65nm CMOS technology. To characterize electrical properties, the input voltage source is modeled by an ideal voltage source ranging from 40mV to 400mV. In figure 11, output voltage of boost and input voltage from TEG are shown. Operation has been simulated with a variable V_{IN}. At first, from 0 to 25ms, with an input voltage of 40mV feedforward circuit works and V_{OUT} reaches 1V. After this time, input voltage raises and feedback circuit starts (and hence feedforward circuit turns off). As it can see, varying V_{IN}, feedback control circuit is able to maintain a constant value of V_{OUT} centered around 1.07V. Difference between maximum and minimum output voltage is 100mV. With an input voltage of 400mV, maximum efficiency reaches 85%.

V. CONCLUSIONS

In this paper, a low startup DC – DC converter is presented for batteryless operation. With a feedforward and feedback control circuits, providing a regulated 1V output voltage with a minimum input voltage of 40 mV and maximum efficiency of 85%. It is well suited for thermoelectric energy harvesting from body heat.

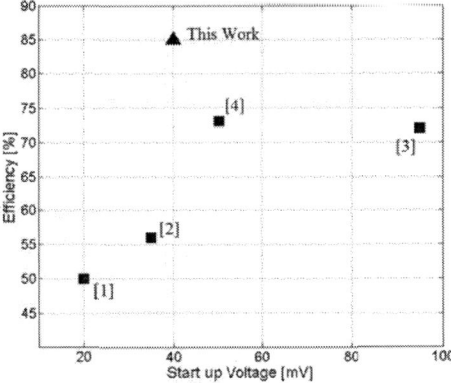

Figure 12 – State of Art Comparison

REFERENCES

[1] E. Carlson, K. Strunz, B. Otis, "20mV Input Boost Converter for Thermoelectric Energy Harvesting" *IEEE JSSC, Avr., 2010.*

[2] Y. K. Ramadass, A. P. Chandrakasan, "A Batteryless Thermoelectric Energy – Harvesting Interface Circuit with 35mV start up Voltage", *IEEE ISSCC Dig. Tech. Papers , Feb 2010*

[3] H. Chen, K. Ishida, K. Ilkeuchi, X. Zhang, K. Honda, Y. Okuma, Y. Ryu, M. Takamiya, T. Sakurai, "A 95mV start up step up converter with V_{TH} tuned oscillator by fixed – charge programming and capacitor pass – on scheme", in *IEEE ISSCC Dig Tech. Papers, 2011,*

[4] P. S. Weng, H. Y. Tang, P. C. Ku and L. H. Lu, "50mV Input Batteryless boost converter for thermal energy harvesting" IEEE JSSC, A standard CMOS technology", *IEEE JSSC,Avr 2013.*

[5] R. Pelliconi, D. Iezzi, A. Baroni, M. Pasotti, and P. Rolandi,"Power efficient charge pump in deep sub – micron standard CMOS technology", *IEEE JSSC, 2003*

Wire-bonds Used as Matching Inductor in RF Energy Harvesting Applications

Dino Michelon[*†], Emmanuel Bergeret[*], Matthieu Egels[*] and Antonio Di Giacomo[†]

[*]IM2NP (UMR 7334), Marseille, France

[†]STMicroelectronics, Rousset, France

Email: dino.michelon@st.com

Abstract—**This document presents an alternative antenna matching strategy for RF Harvesting circuits, based on the use of wire-bonds as external matching inductors. In the following, the electrical characteristics of the bond-wires are evaluated in terms of inductance, quality factor and UHF behavior. Pad de-embedding techniques are applied to reduce parasitic influence. A HFSS model is built as well in order to provide a predictive tool for further improvements. The measurements and simulations are compared and a real-case application is presented.**

Keywords—*bond wires, energy harvesting, impedance match, electromagnetic simulation, pad de-embedding.*

I. INTRODUCTION

During the last years, there is a rising interest toward the Wireless Sensors Networks (WSN): an easy to deploy population of smart sensor nodes capable of receive and send data to each other. One of the main challenges of WSNs is to provide energy for every communicating node. The standard option is to provide a battery to every device but its short life is quite limiting in terms of costs, flexibility, reliability and environmental concerns. A lot of interest is gathering around an alternative solution called Energy Harvesting (EH), which consist of harvest the required energy directly from node surroundings. Possible sources are thermal, vibrational or electromagnetic energy [1]. In this paper the focus will be on the latter. More precisely the RF harvesting in the UHF band (900 MHz and 2.4 GHz) which provides long ranges for smaller antennas.

In general an RF harvester is composed of 3 main blocks (Fig. 1): the RF front end composed of the antenna and an impedance matching circuit, the RF rectifier which converts the received signal to a DC voltage. The last block, sometimes optional, is the power management unit that boosts the DC voltage according to load specifications (for example MCUs and sensors).

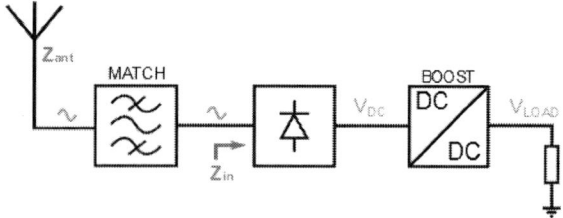

Fig. 1. RF Energy Harvesting System

The harvested energy is usually low (tens of μW) therefore maximum efficiency is required in every sub-block of the system. Moreover particular attention must be paid to correctly match all the blocks in order to provide the best power transfer between antenna and load.

The silicon chip is usually mounted inside a rugged plastic package. The electrical contact with the outside world is provided by gold wire links between the metal pins of the package and the chip pads. This step is called wire bonding. From an electrical point of view this wire provides a parasitic inductance that is usually estimated through the rule of thumb of 1 nano-henry for millimeter of wire. The proposed solution is to exploit this unwanted characteristic in order to create inductors of known value. These components will be used to bring to resonance the input capacitance of the IC, without the need of integrated coil or external discrete components.

II. INPUT IMPEDANCE AND MATCHING

In general the input impedance of the harvester IC is capacitive. For low harvested energy it can be approximated by the sum of parasitic capacitance introduced by pads and ESD circuits. If the energy increases, a nonlinear contribution of impedance due to active devices is present as well. The latter varies with bias point and load [2]. An ad-hoc inductive antenna is often used to resonate out the input reactance in order to meet the conditions specified by the maximum power transfer theorem [3]. Moreover, under the hypothesis of a perfect match, it can be shown that increasing the Q factor of the input impedance of the IC (Q_{IC}) increases the voltage feed to the circuit (at fundamental frequency):

$$V_{rf} = \sqrt{2 \cdot Re(Z_{IC}) \cdot P_{rf} \cdot (1 + Q_{IC}^2)} \qquad (1)$$

It is not always easy to develop a new antenna therefore an on-the-shelf antenna is preferred in some circumstances. Common antennas however are available only for standard impedances (namely 50 Ω or 75 Ω). In this case an additional matching circuit must be provided [4], [5]. In the case under study, the input capacitance of the harvester is around 1 pF so the resonant inductor must be quite small (some nH). In this case is impossible to use a discrete external inductor and the only viable solution are Silicon-integrated coils, which are costly and offer low Q-factors (they suffer of resistive and substrate coupling losses). A possible alternative is to use wire-bond inductors and it will be presented in the following.

978-1-4799-4993-9/14 $31.00 © 2014 IEEE

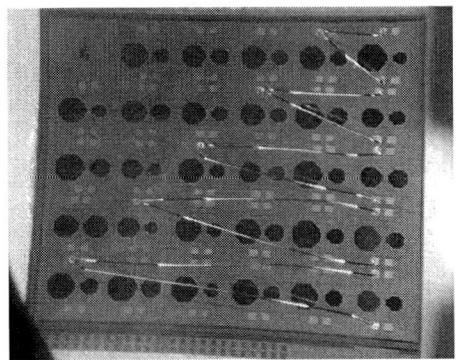

Fig. 2. Test die with 5 (double) wire bonds of different length.

III. WIRE-BONDS

A. Preliminary tests

A set of preliminary studies preceded the actual implementation. The aim is to confirm the rule of thumb of $1 \, nH \cdot mm^{-1}$, its linearity with bond length and to test the repeatability of this solution. Five different bonding patterns are used in order to evaluate the evolution of the parasitic inductance with wire length (Fig. 2). The study starts from 1.6 mm (shortest bond) to about 8 mm (longest bond). Every test inductor is formed by two distinct in-series wire bonds in order to increase the final equivalent length. The wire has a diameter of 25 μm.

After the measurements (S-matrix with Vector Network Analyzer), the data is processed with Agilent ADS. The aim is to delete the influence of the parasitics (the pads capacitance above all) in order to determine the exact impedance of the bond-wire. This procedure is called pad de-embedding. The measured S-matrix is converted to the equivalent Z (or Y-matrix) and the entire structure is modelled, respectively, as a T or Pi two-port network (Fig. 3) [6]. The inductance can be calculated, depending on the model, as:

$$L_Z = \frac{Im(Z_{11} - Z_{12})}{2\pi f} + \frac{Im(Z_{22} - Z_{21})}{2\pi f} \quad (2)$$

$$L_Y = \frac{1/Im(-Y_{12})}{2\pi f} \quad (3)$$

Pi network offers some advantages when dealing with pad de-embedding. In fact the equivalent Y-matrix of the pads (also called OPEN test structure) has Y_{11} equals to port1 pad admittance and Y_{22} equals to port2 pad admittance (Y_{12} and Y_{21} are 0). The new Y-matrix (without pads) is straightforward to derive: $Y = Y_{meas} - Y_{pads}$.

The result of the de-embedding is presented in Fig. 4. As can be seen, the inductor value becomes more stable in the whole frequency range and the self-resonance peak disappears.

The test structure has been commissioned to two different manufactures in order to compare the variability of the process. Afterwards the same measuring and de-embedding procedure has been carried out. The final values of impedance are plotted in Fig. 6. Measurements confirm the linear relation between bond-wire inductance and distance between pads. The mismatch between the two manufactures is very small and

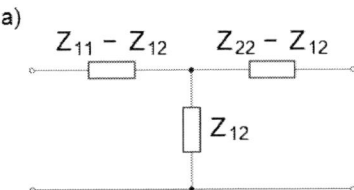

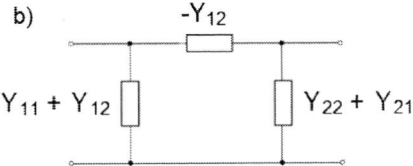

Fig. 3. T (a) and Pi (b) representations of a 2-port network with Z and Y matrix components.

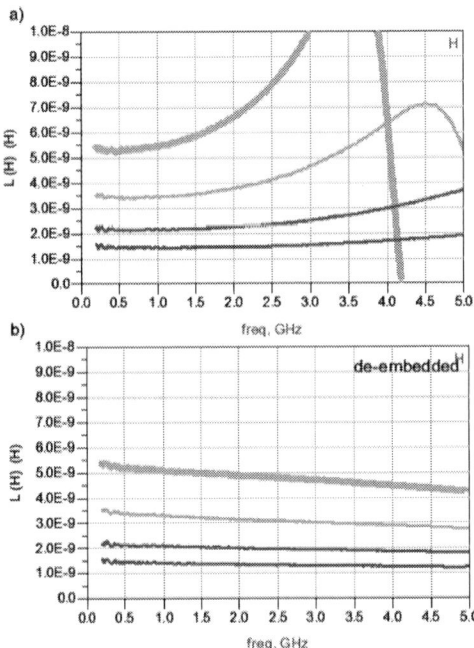

Fig. 4. Inductor frequency response for raw measures (a) and with pad de-embedding (b).

increases only for long distances. The linear relation of the inductance, inferred from data, is about $0.6 \, nH \cdot mm^{-1} + 0.4 \, nH$.

B. Simulations

The bonding procedure is relatively time consuming and, now that the inductive behavior of the bond-wires is confirmed, it would be useful to develop a predictive mechanism that will help us to understand the parameters involved in the final performances (pad size, pad height, wire diameter, etc.). To do so, a HFSS electromagnetic simulation has been created to simulate the bond-wires and the environment (pads, silicon substrate, oxides).

Thanks to simulations, different solutions can be easily compared. The developed parametric model (Fig. 5) makes possible to understand the influence of every physical parameter. Thus the final quality of the inductors can be predicted

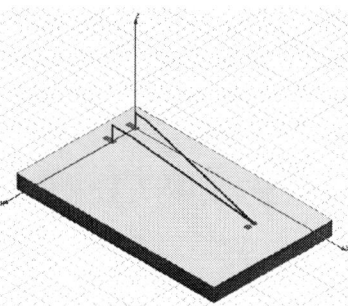

Fig. 5.   HFSS model of a double wire-bond based on manufactured samples.

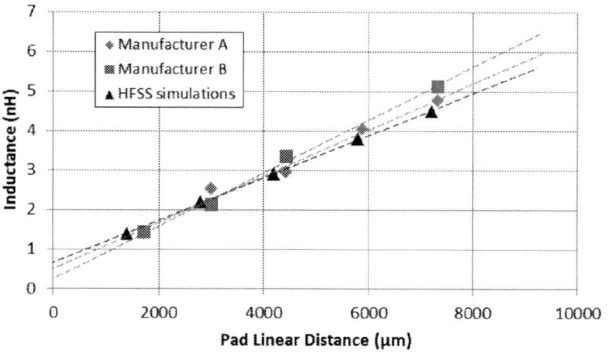

Fig. 6.   Comparison of measured and simulated inductance of wire bonds for different lengths.

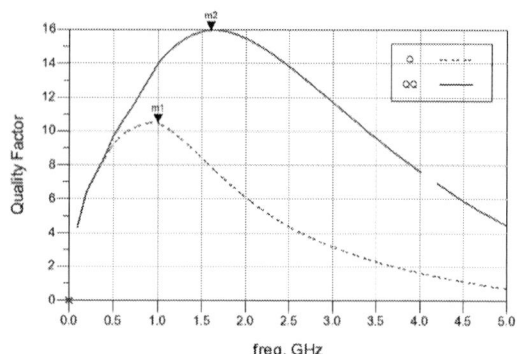

Fig. 7.   Simulated Q-factor of a bonding wire with (solid line) and without (dotted line) pad capacitance influence.

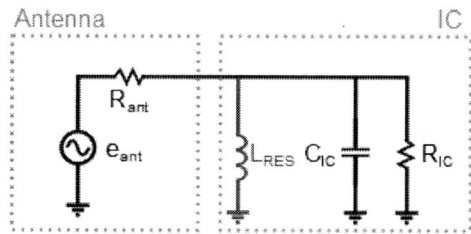

Fig. 8.   Schematic of the matching strategy adopted for tests: the additional inductor L_{RES} is mounted in parallel to IC input capacitance.

and fine-tuned. This procedure is cheaper and faster than an actual physical test (that requires a wafer, a bonding process and long and destructive measures).

The model has been used for the following simulations:

- *Standard:* It has been done to compare the results between the model and test-chip measurements.

- *Pad improvement 1 (Small pads, 60 μm x 60 μm):* Smaller pads mean lower parasitic capacitance. In fact the final inductance remains unchanged but the quality factor and self-resonant frequency increase.

- *Pad improvement 2 (Top metals pad):* In this case as well, the final capacitance is reduced by using an RF pad structure (bottom-most metal layer as ground plane and top-most layers as RF pads). Q-factor increases as in the previous configuration.

As outlined in Fig. 6, the simulation matches the measures very well. The simulated Q-factor reaches a value of 16 for the longest bond (Fig. 7). This is a worst-case value because for the smaller inductors the series resistance decreases (shorter wire) and Q reaches 23. This is an encouraging result, considering that Silicon-integrated coils hardly do better than 15.

IV.   APPLICATION TO REAL CIRCUITS

The final test is the matching of real RF harvesting circuits. A lot of different matching circuits can be found in literature; each one offers some particular characteristics like wide bandwidth, impedance transformation ratio or low component count. For this first set of tests a simple solution (only one

inductor) has been preferred to more complex circuits. It allows a fast and painless implementation with aforementioned wire-bond inductors. Moreover it does not required additional Silicon-integrated components so it can be used with existing ICs.

The input capacitance (C_{IN}) is brought to resonance through an external inductor L_{RES} mounted in parallel to it (Fig. 8). The reactive parts of L and C cancel-out each other and the IC presents an input impedance that is purely resistive. In this case the perfect power transfer between antenna and IC is achieved when R_{ant} equals R_{IC}. The input capacitance is known (about 1 pF, depending on the circuit topology) and the required resonant inductor is found with Eq. 4.

$$L_{RES} = \frac{1}{C_{IC} \cdot (2\pi f_{res})^2} \tag{4}$$

The working frequency influences the final value of L therefore a solution to limit the size of the inductor is to increase the frequency. For that reason the circuits that exhibit low input capacitance have a target frequency of 2400 MHz; working at 900 MHz would require an inductor difficult to manufacture with wire-bonds. On the other side, circuits with higher C_{IC} require lower resonant frequency to avoid extremely small inductors. The required inductors and the resulting length of the bond-wires (calculated according to resonant frequency and input capacitance) for 4 different RF harvesters are shown in Table I. The value spaces between 9 mm and 32 mm.

The side of silicon chips under analysis is less than 5 mm. In order to achieve a wire-bond of more than 10 mm it is necessary to take advantage of the IC package. An example is

978-1-4799-4993-9/14 $31.00 © 2014 IEEE

TABLE I.   REQUIRED WIRE-BOND LENGTH FOR VARIOUS CIRCUITS

	Frequency	C_{IC}	L_{RES}	Wire length
Circuit A	900 MHz	1.6 pF	19.6 nH	31.9 mm
Circuit B	900 MHz	3.45 pF	9.1 nH	14.5 mm
Circuit C	2400 MHz	0.75 pF	5.9 nH	9.1 mm
Circuit D	2400 MHz	0.45 pF	9.8 nH	15.6 mm

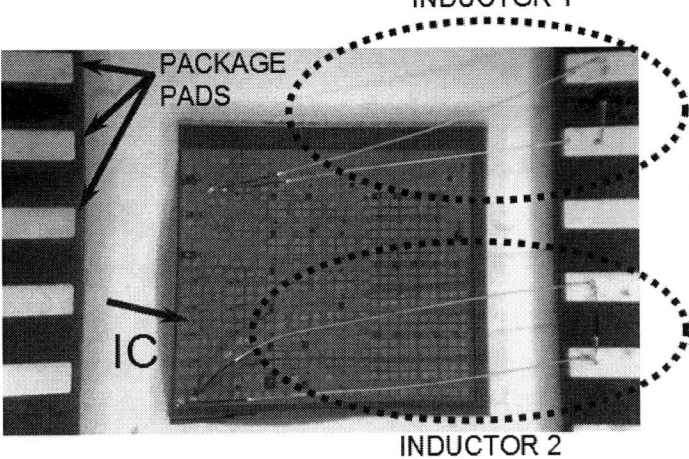

Fig. 9.   A silicon die with a couple of bond-wires used as inductors (about 9 nH each).

presented in Fig. 9. In this case the inductor is formed by 3 different electrically-connected wire-bonds. Two wires go from the IC to 2 (not used) package pads; the latters are connected by a third smaller wire. A total wire length of about 15 mm is achieved.

Fig. 10 show the response of one of the circuits designed to resonate at 900 MHz. The measures confirm the expected result: at about 924 MHz the input impedance has almost lost its imaginary part (181 Ω - 0.7j Ω). However the frequency of resonance is 24 MHz higher than expected. This is due to the limited precision of the bonding process which does not guarantee a complete control over the final length of the wire.

V. CONCLUSIONS

The use of external matching circuit can be a solution but Silicon integrated passive devices do not always meet required performances. The proposed wire-bond inductor, on the other side, can be a better option since it shows good electrical characteristics. Moreover it is easy to size because it shows good linearity between wire length and final inductance. Additionally the results of its simulated model match measurements; this is useful in order to evaluate some optimizations like pad size reduction. Thus bond-wire inductor represents a valuable option whenever a high-quality discrete external inductor or integrated coil would be too expensive and present bad electrical performances for the specific application. Finally, this paper demonstrates how a bond-wire inductor can be used to bring to resonance the capacitive input impedance of a family of UHF harvesters. The input reactive part is successfully eliminated but an unwanted frequency shift highlights that attention must be paid in terms of bonding precision and process repeatability.

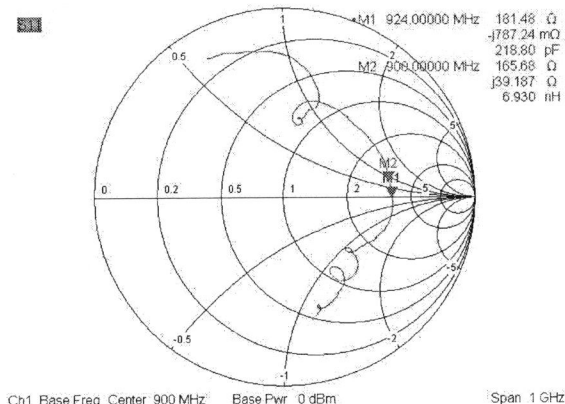

Fig. 10.   Smith-plot of the input impedance.

REFERENCES

[1] S. Sudevalayam and P. Kulkarni, "Energy Harvesting Sensor Nodes: Survey and Implications," *IEEE Communications Surveys & Tutorials*, vol. 13, no. 3, pp. 443–461, 2011.

[2] E. Bergeret, J. Gaubert, P. Pannier, and J. M. Gaultier, "Modeling and Design of CMOS UHF Voltage Multiplier for RFID in an EEPROM Compatible Process," *IEEE Transactions on Circuits and Systems II: Express Briefs*, vol. 54, no. 10, pp. 833–837, Oct. 2007.

[3] P. Nikitin, K. Rao, S. Lam, V. Pillai, R. Martinez, and H. Heinrich, "Power reflection coefficient analysis for complex impedances in RFID tag design," *IEEE Transactions on Microwave Theory and Techniques*, vol. 53, no. 9, pp. 2721–2725, Sep. 2005.

[4] G. Papotto, F. Carrara, and G. Palmisano, "A 90-nm CMOS Threshold-Compensated RF Energy Harvester," *IEEE Journal of Solid-State Circuits*, vol. 46, no. 9, pp. 1985–1997, Sep. 2011.

[5] J. Curty, N. Joehl, C. Dehollain, and M. Declercq, "A 2.45 GHz remotely powered RFID system," in *Research in Microelectronics and Electronics, 2005 PhD*, vol. 1, no. Id. IEEE, 2005, pp. 149–152.

[6] D. M. Pozar, *Microwave Engineering*.   3rd ed. John Wiley and Sons, 2004.

FEM modeling of vertically integrated nanogenerators in compression and flexion modes

R. Tao, R. Hinchet, G. Ardila, L. Montès and M. Mouis

IMEP-LAHC, Joint Research Unit, CNRS / Grenoble-INP / Université Joseph Fourier / Université de Savoie
Grenoble-INP/Minatec, 3 parvis Louis Néel
Grenoble, France
ran.tao@imep.grenoble-inp.fr

Abstract—**This paper analyzes the working principle and structure strength of vertically integrated piezoelectric nanowires into active devices for sensing or energy harvesting applications. Finite element method simulations have been used to evaluate the performances of the devices working in flexion and compression modes under varying input pressure. The geometric influence on the energy generation is also analyzed.**

Keywords—*nanogenerators; piezoelectric nanowires; flexible device; energy harvesting; sensing*

I. INTRODUCTION

Emerging energy transducers provide the harvesting of energy from the environment. Solar, thermal and mechanical energies [1] are converted into electrical energy to power electronic devices. Among them, mechanical energy is relatively stable and ubiquitous, ergo a reliable energy source. Piezoelectric energy harvesters at sub-micron-scale have been subjects of active research and development for harvesting large-span mechanical energy existing in our living environment. Vertically integrated nanogenerators (VING) based on insulator-surrounded piezoelectric ZnO nanowires (NWs) occupy an important place in the research of electromechanical energy transfer due to the high piezoelectric coefficients and easy fabrication of ZnO NWs [2]. In this paper, the performance of VING in both flexion and compression modes is evaluated using FEM (Finite Element Method) modeling. Other than energy harvesting, the energy transduction can also be used for sensing applications.

II. DEVICE STRUCTURE

A. Device Discription

Fig. 1a presents the cross section of the reference plate device consisting of the VING structure between a top thin metallic electrode and a flexible aluminum plate at the bottom (25 μm). ZnO NWs with a diameter of 50 nm and a length of 600 nm [3] are encapsulated by PMMA insulating layer. The device has a surface area of 1 cm² with a NW density varying around 10^{10} cm⁻². Although the active part of the device is made of billions of NWs, this study is only dealing with the unit cell at the central zone of the device (*Fig. 1b-c*) with appropriate boundary conditions for fast FEM modeling. The density of NWs is described by the geometry ratio R between the NW diameter and the cell size (*Fig. 1b*). A density of 10^{10}

cm⁻² corresponds to R=0.5. The device can work in flexion and compression modes. In flexion mode, when the device is bent as a doubly clamped plate, the individual NW has been demonstrated to be in compression or extension (depending on the position of the NW on the plate) and to be elongated or shortened in the c-axis direction [4]. The input strain on the NG cell is calculated from a 2D model of the plate bent by a hydrostatic pressure (*Fig. 1b*). When the device works in compression mode, the individual NW will also be compressed along its c-axis. For the NG cell (*Fig. 1c*) under compressive pressure, two assumptions are considered in the model: (1) the cell is constrained in x and y directions due to the symmetric boundary condition, (2) the compressive pressure on the top surface of the cell is approximately equal to the one on the plate surface. In both cases, the piezoelectric effect converts the time variation of nanowire deformation into a displacement current between the top and bottom electrodes.

a)

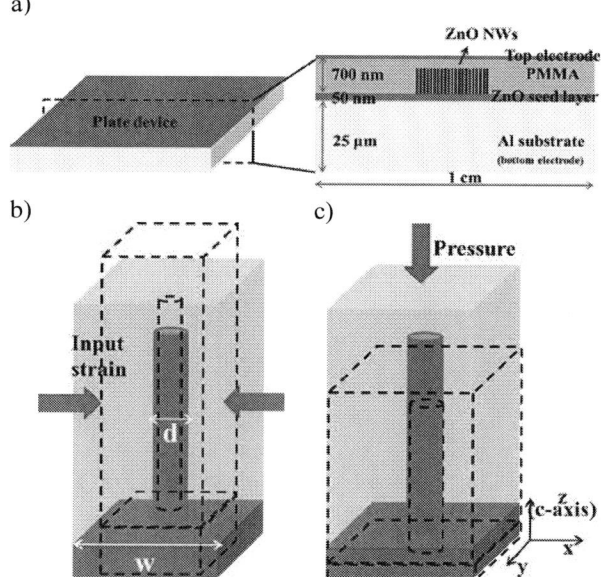

Fig. 1 a) Schematic of the cross section of the VING device; b) deformation of the NG cell in flexion mode, the geometry ratio is defined as R = d / w; c) deformation of the NG cell in compression mode.

B. Discuss on the Structure Strength

In the model, the aluminum substrate has a Young's modulus of 70 GPa, which corresponds to a typical flexural strength of 55 MPa [5], while the flexural strength of the general PMMA is generally 50-70 MPa [6]. According to the FEM simulation using solid mechanical properties, the stress and strain are proportional to the input pressure when the plate device has an elastic deformation. For the device working in flexion mode with an input pressure of 100 Pa, the maximum stress in the aluminum layer is 7.5 MPa and the maximum stress in the PMMA layer is 2 MPa. Thus the maximum input pressure is decided by aluminum substrate, which can sustain 700 Pa hydrostatic bending pressure. Meanwhile, through conservative estimates in the tensile test, the maximum elastic strain of a ZnO NW is 2% [7]. The transversal strain limitation can be calculated using the tensile strain, in this case, 9.8×10^{-3}. When the membrane is bent by 700 Pa pressure, the transversal strain of the NW is only 3.4×10^{-4}, much lower than the elastic limitation.

The situation is very different for the device working in compression mode, previous work has shown that VING can sustain as high as 1 MPa compressive pressure [8].

III. NG CELLS IN FLEXION AND COMPRESSION MODES

A. Influence of the Input Pressure on the Device

Fig. 2 to *Fig. 5* show the simulation results of the NG cells in flexion and compression modes under varying input pressure applied on the device. The geometry ratio of the cells is 0.5, corresponding to a NW density of 10^{10} cm^{-2}. The input pressure is varied from 100 Pa to 700 Pa for the flexion mode and from 0.1 MPa to 1 MPa for the compression mode. These pressure ranges are chosen to work on the elastic regime of the materials used and to obtain comparable values of voltage on both modes. In both cases, the potential generated by the unit cell increased proportionally to the input pressure (*Fig. 2*). The potential/pressure sensitivity of the NG cell is 0.9 V/kPa in the flexion mode, while this value is 0.2 V/MPa in the compression mode.

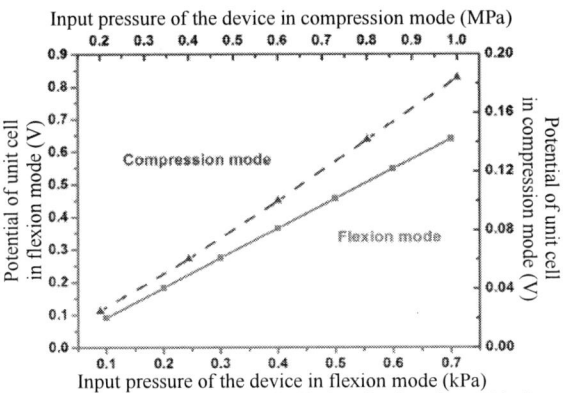

Fig. 2 The potential of an individual cell changing with input pressure in flexion and compression modes.

Fig. 3 and *Fig. 4* present the strain energy at the NG cell and electric energy produced by the NG cell changing with the input pressure applied on the device. The strain energy increases proportionally to the square of the input pressure in both modes, as well as the increasing trend of the electric energy in the flexion mode. This explains the constant cell energy conversion efficiency (1.1%) for the device under flexion (see *Fig. 5*). On the other hand, as shown in *Fig. 5*, the electric energy generated by compressing the cell is improved with higher input pressure, resulting in an increasing cell energy conversion efficiency (5% at 1 MPa), which is 3.5 times higher than the one in flexion mode.

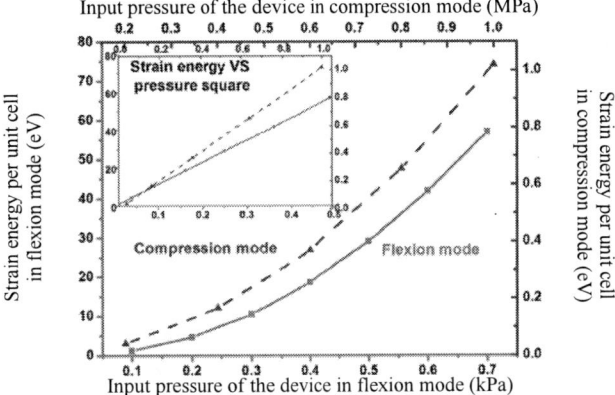

Fig. 3 The strain energy of an individual cell changing with input pressure in flexion and compression modes. Inset: the strain energy varying with the square of the pressure.

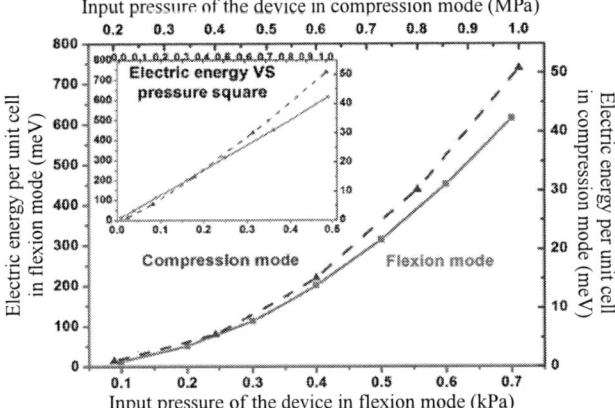

Fig. 4 Electric energy produced by an individual cell changing with input pressure applied on the device in flexion and compression modes. Inset: Electric energy varying with the square of the input pressure.

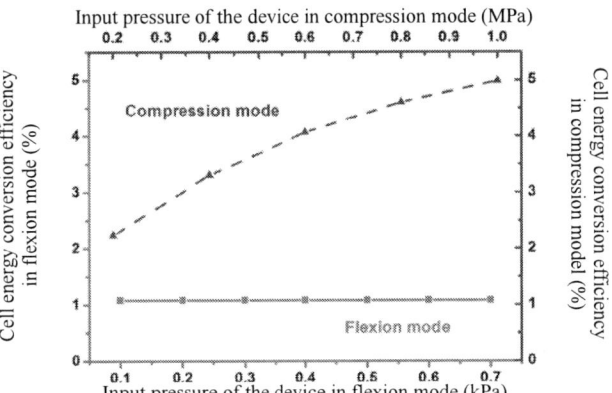

Fig. 5 The cell energy conversion efficiency changing with input pressure of the device in flexion and compression modes.

With a high sensitivity to the input pressure, the flexion mode can be used in pressure sensors, but also in mechanical energy harvesters under small forces. The compression mode has a high energy conversion efficiency which is even superior to conventional ZnO thin film based generators [7].

B. Influence of the NW Density of the Device

In a single NG cell, the density of ZnO NW is calculated by the geometry ratio R. The geometry ratio varies from 0.1 to 1, corresponding to the ZnO volume ratio from 0.6% to 62.8% and the NW density from 4×10^8 cm^{-2} to 4×10^{10} cm^{-2}. In flexion mode, the input pressure is 100 Pa, while in compression mode, the pressure is 0.1 MPa. To explain the different behaviors of NG cell in flexion and compression modes (*Fig. 6* to *Fig. 9*), mechanical and electrical energy transfer mechanism is considered. In compression mode, the lower the NWs density, the higher is the mechanical energy stored in the device for a given pressure. This is due to the fact that the Young's modulus is much lower in PMMA than in ZnO. With a lower NW density, smaller fraction of ZnO results a softer NG layer that stores more strain energy. In the same way, NG working in flexion mode also has a smaller equivalent Young's modulus with a lower NW density. However, as the thickness of the NG layer is negligible compared to the substrate, the core cell achieves an almost constant input strain for a given pressure. Thus the strain energy that is stored in the cell increases with the NW density (*Fig. 6*).

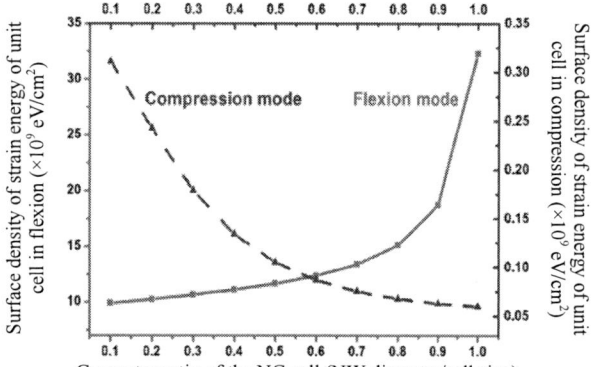

Fig. 6 Surface density of the strain energy of an individual cell changing with the geometry ratio in flexion and compression modes.

At the same time, the equivalent capacitance increases with the NW density due to large relative permittivity of ZnO. A higher NW density improves the capacity to store electrical energy in the NG in both modes. As a result, piezoelectric potential and electric energy go through a maximum that results from a trade-off between the two mechanisms with contrary impacts in compression mode, but keep increasing with the NW density in flexion mode. As shown in *Fig. 7* and *Fig. 8*, higher NW density improves the potential and electric energy generation in flexion mode. The potential increases slightly from 0.3 to 0.8 geometry ratio, which provides high tolerance to the diameter fluctuation for a steady performance. In compression mode, the potential reaches the maximum value 13 mV at R=0.4, and the electric energy density curve has a peak value at R=0.4, consisting with the potential

changing trend. At last, the analysis on the cell efficiency (*Fig. 9*) shows that the NG cell working in flexion mode has a larger efficiency for a higher NW density, and the one working in compression mode has a maximum efficiency at R=0.6 (the NW density is 1.5×10^{10} cm^{-2}).

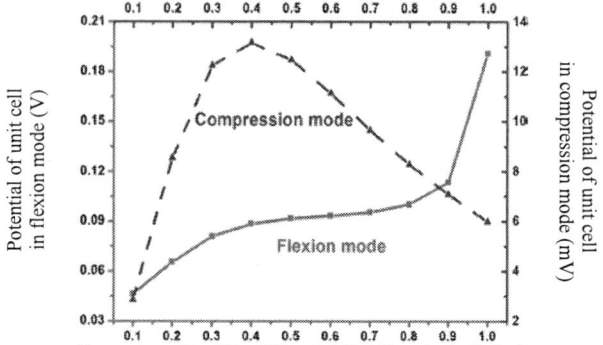

Fig. 7 The potential of an individual cell changing with the geometry ratio in flexion and compression modes.

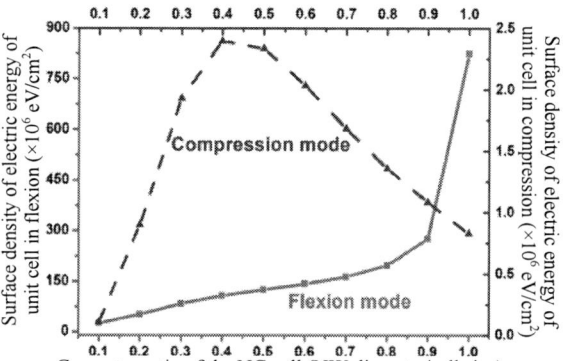

Fig. 8 Surface density of the electric energy of an individual cell changing with the geometry ratio in flexion and compression modes.

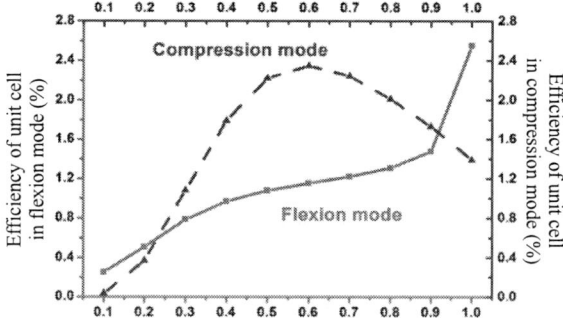

Fig. 9 Cell efficiency of an individual cell changing with the geometry ratio in flexion and compression modes.

IV. GLOBAL EVALUATION OF THE POTENTIAL GENERATION OF THE DEVICE

A. Electrodes Distribution on the Device in Flexion Mode

For a quick study in flexion mode and to avoid large aspect ratio geometry, the 2D plate model used in the simulation has a structure with 2 µm-thick ZnO layer and 1 µm-thick PMMA layer on a 25 µm-thick aluminum substrate, as shown in *Fig. 10a*. When the plate is bent by a pressure of 100 Pa, the transversal strain (strain XX component) is positive at two lateral ends and turns into negative at the central zone (*Fig. 10b*), which means that the NG cells at two end are extended while the central cells are compressed. Thus the potential generated by cells with different positions has different signs, resulting in a neutralized potential around 5 µV if the surface is covered by an electrode. To solve this problem, multiple electrodes can be integrated to improve the output potential. *Fig. 11a* is the schematic of the new design where a central electrode and two side electrodes replace the single electrode. The potential difference between the central electrode and the side electrodes generated by this design is 1.35 V (*Fig. 11b*).

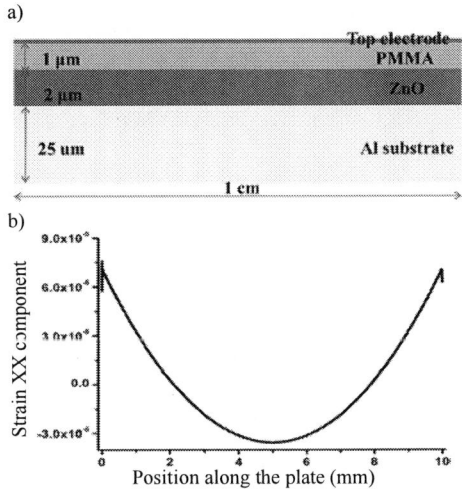

Fig. 10 a) schematic of 2D plate model for electrodes distribution study; b) transversal strain along the plate.

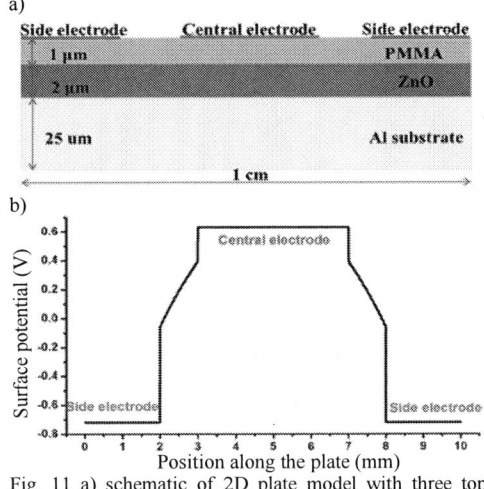

Fig. 11 a) schematic of 2D plate model with three top electrodes; b) potential distribution along the plate surface.

B. Global Evaluaton of the Device in Compression Mode

NG cell matrix under compressive forces has been studied by Hinchet et al. to investigate the performance of NG devices [8]. As the matrix size increases from 1 cell × 1 cell to 15 cells

×15 cells, the potential different between the top electrode and the bottom electrode increases from 20 mV to 70 mV at 1 MPa input pressure. Compared with individual NG cell, the matrix reduces edge effects and 3D dielectric losses and the output potential converges to a maximum value of 80 mV.

V. CONCLUSION

In this paper, FEM simulation was used to evaluate the performance of NGs based on integrated piezoelectric nanowires in terms of potential, electrical energy generation and energy conversion efficiency. The NWs were assumed to be grown on a flexible aluminum substrate. The analysis of simulation results in the two operation modes showed that the flexion leads to higher potential generated, higher sensitivity to input pressure and higher tolerance to fluctuations of NW density. This can be particularly interesting for pressure sensors. On the other hand, the compression mode provides higher energy conversion efficiency and would be better suited to energy harvesting devices.

ACKNOWLEDGMENT

The work is supported in part by French Ministry, the ANR COSCOF, and the CNRS INSIS Institute within the framework of Exploratory Projects on energy and European FEDER through FluMin3 project.

REFERENCES

[1] A Cook-Chennault, N Thambi and A M Sastry, "Powering MEMS portable devices—a review of non-regenerative and regenerative power supply systems with special emphasis on piezoelectric energy harvesting systems", Smart Mater. Struct., vol. 17 043001, 2008.

[2] I. Dakua, and N. Afzulpurkar, "Piezoelectric energy generation and harvesting at the nano-scale: materials and devices" Nanomater. Nanotechnol. August 27 2013.

[3] Y. Gao, and Z. L. Wang, "Electrostatic potential in a bent piezoelectric nanowire. The fundamental theory of nanogenerator and nanopiezotronics" Nano lett., vol. 7, pp. 2499-2505, June, 2007

[4] R. Tao, R. Hinchet, G. Ardila and M. Mouis, "Evaluation of Vertical Integrated Nanogenerator Performances in Flexion" J. Phys.: Conf. Ser. 476 012006 2013.

[5] B. A. Niemeier, Formability topics – metallic materials, 1rd ed. American Society for Testing and Materials, 1978, pp. 73.

[6] L. Lidgren, H. Drar, and J. Moller, "Strength of polymethylmethacrylate increased by vacuum mixing", Acta. Orthop. Scand., 1984.

[7] W. J. Lee, J. G. Chang, S. P. Ju, M. H. Weng, and C.H. Lee, "Structure-dependent mechanical properties of ultrathin zinc oxide nanowires", Nanoscale Research Letters , April 20 2011.

[8] R. Hinchet, S. Lee, G. Ardila, L. Montes, M. Mouis, and Z. L. Wang, "Performance optimization of vertical nanowire-based piezoelectric nanogenerators", Adv. Funct. Mater., October 8 2013.

Design of a low power wireless sensor network node for distributed active vibration control system.

M. Zieliński, F. Mieyeville, D. Navarro
Université de Lyon,
Institut des Nanotechnologies de Lyon,
Ecole Centrale de Lyon
Ecully, France
Corresponding author: mateusz.zielinski@ec-lyon.fr

O. Bareille
Université de Lyon,
Laboratoire de Tribologie et Dynamique des Systèmes,
Ecole Centrale de Lyon
Ecully France

Abstract— **This paper describes design of the wireless sensor network (WSN) node for distributed active vibration control system for the automotive application. The node using one piezoelectric element provides several features (sensing, shunting and energy harvesting). The implementation of the WSN node with designed SSHI circuit is presented. Vibration measurements and shunting the piezoelectric element capabilities are verified.**

Keywords— **Piezoelectric, Wireless Sensor Network (WSN), Active Vibration Control (AVC), Energy Harvesting.**

I. INTRODUCTION

Wireless sensor networks (WSNs) are commonly used in many applications for industrial and civilian solutions. Energy aware nodes which compose a network are distributed in big areas and provide a wireless network which is able to collect measurements, process data, and react under pre-programmed circumstances. WSNs are used to monitor physical or environmental conditions like temperature, sound, vibration, pressure, motion or pollutants [1]. WSNs are present in rapidly developing engineering fields, like structural health monitoring systems (SHMs), and can provide new functionalities compared to existing wired systems. For example they provide facility managers the possibility of adopting predictive maintenance strategies, instead of the planned maintenance, what is important following a seismic event [2].

Active vibration control (AVC) becomes a very important issue in many engineering fields (SHM, automotive, industrial). Several solutions for AVC has been proposed and tested with promising results [3]. However implementation of these systems in a WSN is a challenge for designers. The entire vibration control system must be designed in order to pass very stringent requirements of the WSN and control law [4].

II. RESEARCH ASSUMPTIONS

The usage of the active vibration control can reduce the weight of conventional passive methods, helping the push towards lighter, more fuel efficient vehicles [5]. Active methods include vibration sources which are driven by a control strategy algorithm to provide destructive interference of real vibrations. There are two possibilities to implement the control strategy: feedforward (open-loop) and feedback (closed-loop). A variety of algorithms have been used to adapt the controller; most are based on high computational adaptive filtering methods: least mean square (LMS) and filtered reference LMS (FxLMS) [5]. Conventional AVC system based on wired network (composed of sensors and actuators) with high-power controller and fast data processing needs a lot of the energy. WSNs have rather low transmission rates. Due to delays, implementation of the real-time system which is necessary to provide data processing for centralized AVC systems is not possible [4]. Replacing a big centralized (wired) system with little low power nodes can improve the AVC in the scope of costs and energy consumption.

Our distributed approach used in place of the well known centralized can be a solution. In this approach, intelligent nodes provide local action, which decrease the amount of the information to transfer comparing to the centralized method. The feasibility of distributed autonomous nodes (with sensing feature) coupled to local semi-active vibration dissipation is the aim of the work.

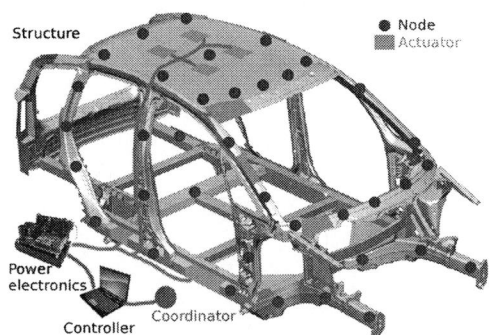

Fig. 1. System structure.

Figure 1 presents the proposed WSN system for the distributed AVC for automotive application. Wireless nodes are implemented in a body car. The nodes are in a star topology towards coordinator. Distributed system means multi-layer system. The first level contains the nodes. Each node can measure vibrations, shunt the piezoelectric element and harvest energy. The second level is the global approach. The nodes communicate with the coordinator which controls the wired high power actuators.

III. WSN NODE DESIGN

The WSN node provides several hardware features: sensing, energy harvesting and mechanical damping. Following paragraphs describe these features and provide description of chosen solutions.

A. Sensing vibrations

Various vibration noises have been already investigated in literature [6]. Body car vibrations are low frequency signals. Noises inside a car under operating conditions have low frequencies (less than 300Hz) and are mostly determined by acoustic resonances and body vibrations modes [7] [8].

Mechanical vibration velocity and amplitude are described in function of electrical and material parameters of a piezoelectric element [9]. The piezoelectric element in open-circuit provides the output voltage which corresponds to the mechanical acceleration. It makes the piezoelectric element proper for sensing vibrations.

B. Mechanical damping

Shunting the piezoelectric element can be used for damping the vibrations. The efficiency of dissipation energy in the shunt connected to the piezoelectric element is currently an issue of research. We distinguish passive and active methods. Passive methods use passive electrical elements to provide mechanical damping [10]. Active methods use non-linear circuits like negative-capacitance [11] or switching methods [12].

C. Energy harvesting

The piezoelectric element can be described as a current source with an inbuilt capacitance. The existence of this capacitance suggests using the shunt inductor to achieve maximum power flow. However, this solution can not be used in real design due to very big values of the needed induction. Several solutions for harvesting energy from the piezoelectric element have been already proposed. So-called "synchronized switch damping" (SSD) is a semi-passive method, which was developed to address the problem of vibration damping. Techniques based on SSD increase the energy flow between the piezoelectric elements and load [13] which provides efficient energy harvesting (The SSHI - switching over the inductance technique is one of the SSD methods).

D. Chosen solutions

Design of an autonomous and intelligent WSN node must consider requirements mentioned in previous paragraphs. Low frequency vibrations in a body car suggest use of energy efficient but slow and low-computing, 8bit low-power microcontroller with an internal analog-to-digital converter (ADC). It provides low power consumption. Sensing vibrations need the piezoelectric element in the open-circuit – it is a sensing requirement. Comparison of methods proposed for vibration damping and energy harvesting, and bearing in mind the condition for "sensing", allowed to choose for design the serial switching over the inductance (Serial SSHI) method. It is semi-passive technique, based on non-linear processing on the piezoelectric voltage.

IV. IMPLEMENTATION

A. Implementation of sensing vibrations

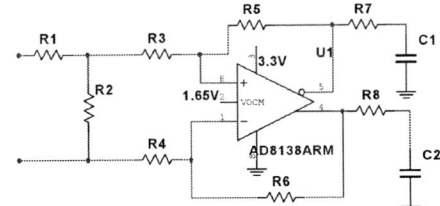

Fig. 2. Sensing circuit

Designed circuit for sensing vibrations and signal conditioning is presented on figure 2. It is composed of the voltage divider (R1 and R2) and AD8138ARM low distortion differential analog-to-digital (ADC) driver from Analog Devices. Low pass filters (R7 with C1 and R8 with C2) are used to cut-off high frequencies on the outputs of the ADC driver. The differential ADC driver also provides offset voltage (Pin 2 connected to the 1.65V). Hence, the negative and positive voltage values can be measured by microcontroller. In designed system an internal ADC of the microcontroller is used. This solution provides low energy consumption since there is no additional device to supply. The R3, R4, R5 and R6 are used to set up the gain of the differential amplifier.

B. Implementation of switching circuit

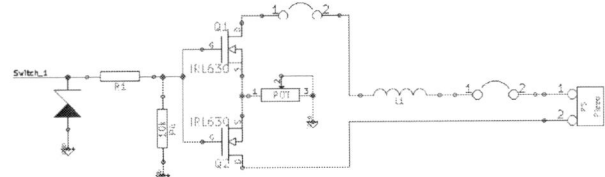

Fig. 3. SSHI circuit

Serial SSHI circuit is presented on figure 3. The zener diode D1 is used to protect the microcontroller output; the R1 resistor sets the maximum value of the current (correlated with time needed to open the switch). The R4 is used to turn off the transistor switch. Two IRL630 low-logic-level NMOS transistors with 3V threshold voltage are used. It simplifies the circuit because there is no need to use the external gate driver. High drain-source voltage (200V) allows using high power piezoelectric transducers.

C. Prototype node and measurement station

Fig. 4. Photo of the prototype WSN node

978-1-4799-4993-9/14 $31.00 © 2014 IEEE

Figure 4 presents the photo of the real prototype device. It is composed of two PCB boards. The first board contains microcontroller and supply circuit, the second provides the SSHI circuit, sensing circuit and wireless communication. Device is compatible with the Arduino shield [14]. Hence, it can be easily used with various microcontrollers, or more complex signal processing units. It provides compact and expandable platform.

For wireless communication we used the MRF24J40 radiofrequency transceiver from Microchip. It provides hardware support for IEEE 802.15.4 physical layer. The node is equipped with low-power microcontroller PIC16LF88 with internal 10bit ADC. The measurement station contains a metal plate with attached piezoelectric elements (PI-876.A15 from PI Ceramic).

V. MEASUREMENTS

Designed WSN node is connected to the piezoelectric element. Vibrations are generated by using the function generator and the electrical vibrator (The vibrations are generated at a frequency 170Hz). External accelerometer, attached on the metal plate, next to the piezoelectric element is used to provide reference measurements of the acceleration (Accelerometer sensitivity: 10,43 mV/g).

A. Verification of the sensing circuit

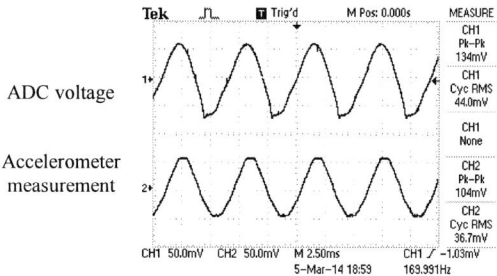

Fig. 5. Validation of the sensing circuit

Figure 5 presents validation of the sensing circuit. Transistor-switch is open and provides the piezoelectric element in open-circuit. First channel (CH1) presents the voltage over the ADC (because of the 1.65 V voltage offset, CH1 is presented in the AC mode). Second channel (CH2) presents the signal measured by the external accelerometer attached to the plate.

The resulting wave forms confirm design. Signal provided to the ADC is in phase with the acceleration. The amplitude value is well adapted to the microcontroller requirements; additionally the offset voltage provides measurements of the positive and negative values.

B. Verification of the switching circuit.

Figure 6 presents waveforms of the piezoelectric voltage (CH1) and the current in the circuit (CH2) while the transistor-switch is open. The value of the current is measured over the small serial resistance added to the switching circuit.

Figure 7 presents waveforms while the transistor-switch is closed. The CH1 presents the accelerometer measurement (in phase with the piezoelectric voltage); CH2 presents the current in the circuit. Figure 7 also shows the phase shift between voltage and current waveform (the capacitance character).

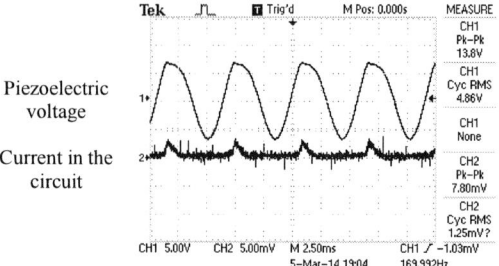

Fig. 6. Transistor-switch open

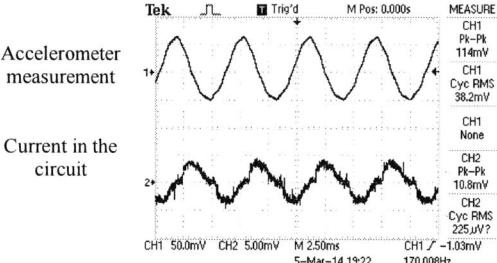

Fig. 7. Transistor-switch closed

The resulting wave forms confirm design of the bidirectional transistor switch. Measurements also confirm the use of the microcontroller to control the switching circuit.

C. Verification of the SSHI circuit

In the serial SSHI method, the inbuilt piezoelectric capacitance and external inductance creates the serial resonant circuit. This circuit is kept in the open-circuit thanks to the switch. While the extrema of the mechanical displacement is detected, the switch is closed for a half of the electrical resonant period. It causes inversion of the piezoelectric element voltage. The period of the mechanical displacement is much longer than the period of the electrical circuit.

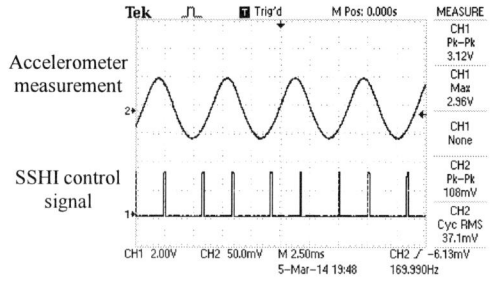

Fig. 8. Peak detector

The algorithm implemented in the microcontroller provides the peak-detection. Figure 8 presents the peak-detector output signal (CH1) and accelerometer waveform (CH2). The SSHI circuit is kept open during these measurements; the voltage over the piezoelectric element corresponds to the acceleration.

The inversion process is presented on figure 9 (The switch is closed for a half of the resonant period). CH1 presents the piezoelectric element voltage; CH2 is SSHI control signal.

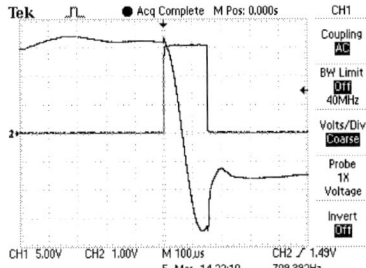

Fig. 9. Voltage inversion

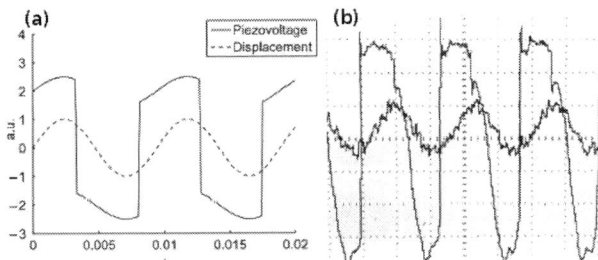

Fig. 10. (a) Theoretical SSHI waveforms[15] (b) Measurements

Figure 10a presents the theoretical waveforms for the serial SSHI method. Figure 10b contains measured waveforms obtained for the prototype node. The sinusoidal signal is acceleration; the second waveform is a piezoelectric element voltage. As it can be noticed on figure 10b the piezoelectric voltage during SSHI switching does not correspond to the displacement.

Figure 11 verifies the peak detection while the SSHI circuit is working. CH1 is the ADC voltage. CH2 presents the peak detection signal. It proves disadvantage of applied continuous peak-detection algorithm for series SSHI method for the self-powered solutions.

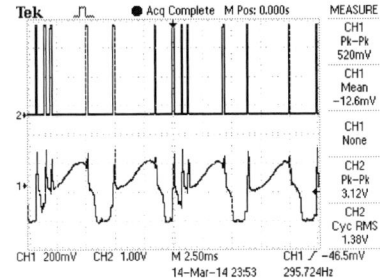

Fig. 11. Sensing voltage (CH2) and peak detection signal (CH1) measured during the SSHI switching.

VI. CONCLUSION

Big centralized and wired systems for active control are costly and use a big amount of energy. Replacing them by the energy aware WSN has been proposed. The autonomous and intelligent node for AVC system is proposed. The designed node provides sensing vibrations using the internal ADC of the microcontroller and serial SSHI circuit for energy harvesting and mechanical damping. Design of the sensing vibration and switching circuits has been presented. Usage of the 8bit low-power microprocessor for sensing vibrations and control the SSHI circuit has been validated. The disadvantage of continuous peak-detection for self-powered systems has been indicated. The measurements prove the theoretical research and confirm realization of the designed circuits.

The next step of the work is the validation of the Serial SSHI circuit (energy harvesting and mechanical damping capabilities) and possibility of the self-supply feature.

REFERENCES

[1] J. Yick, B. Mukherjee, D. Ghosal "Wireless sensor network survey" Computer Networks vol. 52 p. 2292-2330, August 2008.

[2] J. P. Lynch "An overview of wireless structural health monitoring for civil structures" Philosophical Transactions, Royal Society A, p. 345-372, 2007.

[3] F. Svaricek, T. Fueger, H. Karkosh, P. Marienfeld, C. Bohn "Automotive Applications of Active Control" Vibration Control pp. 380, September 2010, Sciyo, Croatia.

[4] F. Mieyeville, M. Ichchou, G. Scorletti, D. Navarro, W. Du "Wireless Sensor networks for active vibration control in automobile structires". Smart Mater. Struct 21, 2012.

[5] S.J. Elliott "A review of Active Noise and Vibration Control in road vehicles", ISVR Tehcnical Memorandum No 981, December 2008.

[6] C.R. Fuller, A.H. Von Flotow "Active Control of Sound and Vibration" IEEE Control Systems

[7] L. Hermans and H. Van Der Auweraer "Modal testing and analysis of structures under operational conditions: industrial applications"

[8] S.H. Kim, J.M. Lee, M..H. Sung "Structural-Acustic Modal Coupling Analysis and Application to Noise Reduction in a Vehicle Passenger Compartment". Journal of Sound and Vibration 255(5), p. 989-999, 1999.

[9] E. Lefeuvre, A. Badel, C. Richard, L. Petit, D. Guyomar "A comparison between several vibration-powered piezoelectric generators for standalone systems" Elsevier, Sensors and Actuators A 126 (2006) p. 405-416.

[10] Jin-Young Jeon, "Passive vibratin damping enhancement of piezoelectric shunt damping system using optimization approach", Journal of Mechanical Science and Technology, 23, 2009.

[11] B de Marneffe, A Preumont, "Vibration damping with negative capacitance shunts: theory and experiment", Smart Materials and Structures 17, 2008.

[12] Saber Mohammadi, Akram Khodayari, "Damping analyses of structural vibratins and shunted piezoelectric transducers", Smart Materials Research, 2012.

[13] E. Lefeuvre, A. Badel, C. Richad, L. Petit, D. Guyomar, "A comparsion between several vibration-powered piezoelectric generators for standalone systems", Sensors and Actuators 126, 2006.

[14] http://arduino.cc/en/uploads/Main/Arduino_Uno_Rev3-schematic.pdf

[15] M. Lallart, D. Guyomar "An optimized self-powered switching circuit for non-linear energy harvesting with low voltage" Smart Master, Struc. 17, 2008.

Co-design of Dual-band GSM Filtenna based on Printed-IFA for Energy Harvesting

Manh Ha Hoang[1,2], Van Hieu Nguyen[2]
[1]Faculty of Electrical & Electronics Engineering
Ho Chi Minh City University of Technology
Ho Chi Minh, Vietnam
{manh-ha.hoang, van-hieu.nguyen}@minatec.inpg.fr

Thi Quynh Van Hoang[2], Tan-Phu Vuong[2]
[2]IMEP – LAHC
Grenoble Institute of Technology
Grenoble, France
{thi-quynh-van.hoang, tan-phu.vuong}@minatec.inpg.fr

Abstract— Interest in radiofrequency energy harvesting (EH) systems for green and renewable energy schemes has increased remarkably in recent years thanks to their unlimited applications. The key element on the receiving side of an EH system is the rectenna (rectifier + antenna) which receives electromagnetic power and converts it into electric power. In order to reduce the size of the rectenna, it has been suggested to combine antenna and filter together. This kind of antenna is sometimes called "filtenna". We present in this paper a dual-band filtenna for application of harvesting RF energy from cellular network frequency bands (900 MHz and 1800 MHz). By using the principle of Printed-Invert-F-Antennas and a slot on the ground plane of the antenna, the high-order harmonics are rejected. Thus, the antenna acts as a dual-band-pass filter for the 900 MHz and 1800 MHz bands and as a stop-band filter for the higher frequencies. The measured antenna return loss over the frequency band from 2 GHz to 8 GHz is less than 1.5 dB so all the high-order harmonics are suppressed. The realized gains of the antenna are 1.9 dBi and 2.55 dBi at 900 MHz and 1800 MHz, respectively.

Index Terms--Dual-band; GSM; harmonic rejection; rectenna; wireless power transmission.

I. INTRODUCTION

Wireless power transmission has received important attention in recent decades thanks to their unlimited applications such as receiving power where physical connections are difficult or unreachable, replacing/recharging battery in RFIDs [1], creating an activation signal for wireless sensor networks (WSN) [2], and so on. When a dedicated source is introduced in the system, the power collection and conversion is defined as "wireless power transfer". When ambient energy is used as the source, e.g. TV or GSM broadcasting, we define it as "energy harvesting".

Energy harvesting uses the abundant radiofrequency (RF) energy from surrounding sources, such as nearby mobile phones, Wi-Fi, broadcast television signals, etc. and can provide unlimited energy for the lifespan of electronic devices. RF energy is collected by an antenna and converted into a usable DC power. Ambient RF energy is pervasive, especially

that from cellular networks and Wi-Fi networks. The study presented in [3] shows that ambient electromagnetic power spreads over wide frequency band (680 MHz – 3.5 GHz) and is quite constant over time. The highest density of this power is around –14.5dBm/m² (corresponding to 35.5µW/m²) at the 1800 MHz – 1900 MHz band where cellular networks operate. This level is not sufficient to power devices directly but it could be stored in a capacitor or micro-batteries. Moreover, recent technology trends in ultra-low power microcontrollers and sensors have opened the way to the development of applications that can be activated by low power ranging from a few µW to a few hundred µW [4]. Therefore, energy harvesting would enable continuous charging of low power devices, making a battery unnecessary. As a result, these devices would be connector, cable and battery free, and give the user significant mobile freedom while charging and in use.

The key component of an energy harvesting system is the rectenna which receives RF power and converts it into DC power. Since the first rectenna was presented in 1963 by Brown, various types of rectenna have been developed [5]. A conventional rectenna consists of an antenna, a band-pass filter (BPF), a Schottky diode(s), a low-pass filter (LPF) for the DC component, and a resistive load. The rectifier diode is a nonlinear element, which generates the harmonics of the fundamental frequency. These unwanted harmonics will re-radiate through the antenna and reduce RF-to-DC conversion efficiency of the rectifier. Therefore, a BPF must be added between the antenna and the diode to suppress this harmonic interference and improve system performance. In order to reduce the size of the rectenna, several rectennas with harmonic rejection antenna have been suggested that integrate the BPF and the antenna [6]-[10]. Hence, an antenna with harmonic rejection properties offers the advantages of low cost, simple design, and conversion efficiency enhancement. With respect to operating frequency, most harmonic rejection rectennas are designed for a single frequency. However, with the development of multiple frequency bands in today's wireless communication systems, the multi-band rectennas would be useful. Some conventional multi-band rectennas are presented in [11]-[13].

In this paper, we propose a filtenna which operates at dual-band GSM and simultaneously offers high-order harmonic rejection functionality for the application of energy harvesting. The outline of this paper is as follows. In section II, the co-design of dual-band filtenna is presented. In section III, the performance of the proposed filtenna is shown by simulation and measurement. A discussion is presented in section IV and section V concludes the paper.

II. MODIFIED PRINTED-IFA DESIGN

The proposed antenna was inspired by the traditional Printed-Inverted-F-Antenna (Fig. 1) which is widely used in today's wireless communication handsets [14]. The feature of this kind of antenna is its compact $\lambda/4$ resonant length. In addition, it is easily fabricated; it offers a quasi-omnidirectional radiation pattern and a good efficiency.

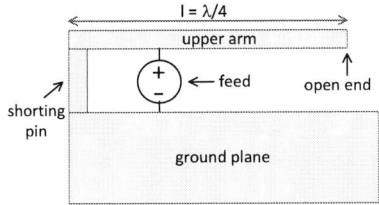

Figure 1. Traditional Printed-IFA structure.

In this paper, in order to miniature the antenna and integrate the band-stop filter behavior into the antenna, several important techniques were employed.

Firstly, with respect to the shorting pin, the proposed Printed-IFA is modified from the traditional structure. This is because for electromagnetic energy propagation on a microstrip, the conduction loss (on the copper) is many times higher than the dielectric loss (on the substrate). Therefore, a traditional Printed-IFA with the upper arm shorted directly to the ground plane possesses an important conduction loss at the shorting plane. In order to improve the radiation efficiency of the Printed-IFA, a capacitive coupling is used instead of direct connection to the ground plane. The ground plane is then placed on the other side of the IFA as shown in Fig. 3. The conductive areas C_{12} and C_{22} (Fig. 2) create the capacitive coupling with the ground plane via dielectric layer to maintain the closed current path. This coupling does not cause the important conduction loss while preserving the resonant frequency of the Printed-IFA.

Secondly, by adding a capacitor (C_{11}) to the open end of the IFA, the resonant length can be reduced. Instead of a resonant length of 22 mm for the classical 1800 MHz IFA, this value is decreased to 19 mm for the modified IFA [15].

Thirdly, to achieve a dual-band operation at 900 MHz and 1800 MHz, a double-Printed-IFA is designed. The 1800 MHz IFA is integrated inside the 900 MHz IFA. While the 1800 MHz IFA is directly fed by the 50 Ohm microstrip line, the 900 MHz IFA is fed via the coupling between itself and the 1800 MHz one. In fact, the combination of the 1800 MHz IFA and the feeding line creates the "source" for the 900 MHz IFA. This indirect "source" for the 900 MHz IFA together with the indirect shorting pin between the radiating element

and the ground plane mentioned in the first point reduce appreciably all the resonant frequencies different than the fundamental ones of the two IFAs. This property offers efficient band-stop filter behavior for the proposed antenna.

Finally, by adding a slot on the ground plane at the suitable position, the stop-band filter behavior is considerably more efficient. The dimensions of the slot and the number of slot on the ground plane were optimized in order to obtain the best behavior of the stop-band filter from 2 GHz to 8 GHz. The comparison between the antenna with and without the slot is presented in Fig. 4. We can note that for the structure without the slot, a dip point is at 7.52 GHz with an impedance of (30+j22) Ohm at the position corresponding to the slot (2.2 mm from the feed point). By adding a slot of 0.8 mm width and 12 mm length, we added an imaginary part of 176 Ohm to the impedance viewed at the slot position and then moved the antenna impedance at 7.52 GHz to a constant SWR circle much greater than the original one on the Smith chart. Thus, the reflection coefficient became much more important, and the stop-band behavior improved.

The proposed antenna was printed in the Rogers 4003 substrate with a thickness of 0.813 mm, a dielectric constant of 3.55, and a loss tangent of 0.0027. The geometry of the proposed antenna is shown in Fig. 2 with its dimensions in millimeters. The yellow part is the conductor on one side and the red part is the ground plane with its slot on the other side.

The investigated bandwidth is from 0.5 GHz to 8 GHz. For simulation in this case, a time domain solver is more efficient than a frequency domain solver, so the transient solver within the CST Microwave Studio was used. The total size of the filtenna is 60mm x 32mm x 0.813mm.

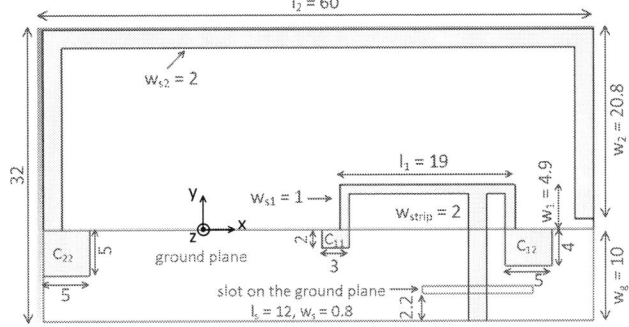

Figure 2. Geometry of the dual-band filtenna with dimensions in millimeters.

III. SIMULATION AND EXPERIMENTAL RESULTS

The Printed Circuit Board (PCB) layout of the proposed filtenna is shown in Fig. 3. The conductor thickness is 17µm.

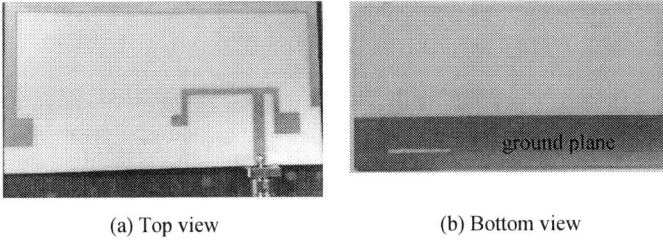

(a) Top view   (b) Bottom view

Figure 3. Photos of the fabricated proposed filtenna.

A. Reflection coefficient

Measurement was performed by using the 8510C Agilent Vector Network Analyzer. Fig. 4 presents the reflection coefficient in dB of the proposed filtenna for three cases: simulation of filtenna with the slot, simulation of filtenna without the slot and measurement of filtenna with the slot. These results show the following:

- The filtenna can effectively reject the harmonic signals generated by the diode up to 8 GHz and prevent them from re-radiating through the antenna. Indeed, a return loss less than 1.5 dB is observed over the frequency band from 2 GHz to 8 GHz.

- The slot on the ground plane offers a remarkable efficiency in "flattening" the desired stop-frequency-band. Moreover, it does not affect the original resonant frequencies of the Printed-IFAs.

- In a comparison between simulated and measured results, a slight shift in frequency is observed for low resonant frequency (887 MHz for simulation vs. 866 MHz for measurement, so 2.4% shifted) and also for high resonant frequency (1736 MHz for simulation vs. 1820 MHz for measurement, so 4.8% shifted). This can be explained either by the tolerance of measurement or by the imprecise value of dielectric constant of the Rogers 4003 substrate which is provided between 3.38 and 3.62 by the constructor. However, this slight shift in frequency is acceptable.

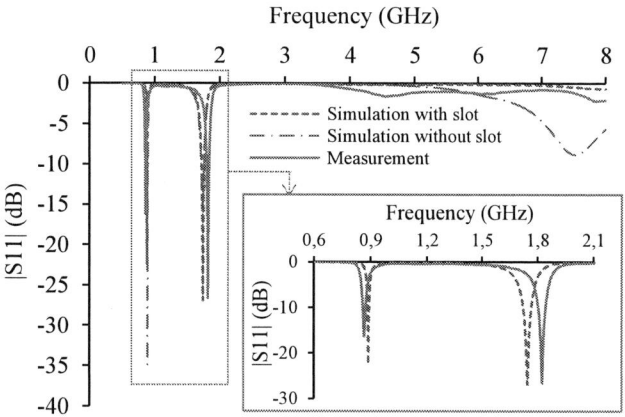

Figure 4. Simulated (CST Time domain) and measured reflection coefficient of the proposed filtenna.

B. Radiation patterns

The measurement of filtenna radiation patterns is performed in an anechoic chamber. The transmitting antenna is a horn antenna with 7.0 dBi gain at 900 MHz and 9.8 dBi gain at 1800 MHz. The largest dimension of the horn antenna is 22 cm, so the far field zone of this antenna is considered from 60 cm for the GSM frequency bands. The proposed filtenna was located at a distance of 300 cm from the horn antenna so the far-filed condition is ensured. The initial position of the filtenna is illustrated in Fig. 5 with its coordinate system.

Since our Printed-IFA does not possess a full ground plane, in order to limit the reflection effect of the metallic part (motor and support), we placed the Printed-IFA higher than this metallic part. However, in this configuration, a slight nonalignment occurred between the horn antenna and the antenna under test (AUT).

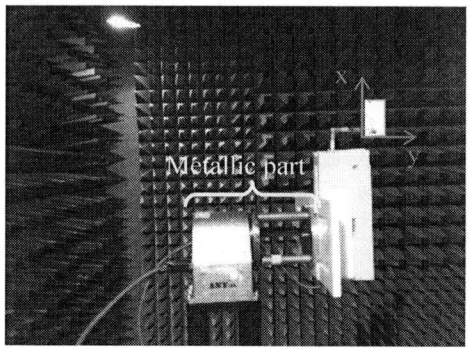

Figure 5. Measurement setup for filtenna radiation patterns.

Fig. 6 illustrates the simulated and measured radiation patterns at low and high resonant frequencies of the GSM bands for YZ-plane ($\varphi = 90°$) and XY-plane ($\theta = 90°$). It can be observed that there is a significant difference between the simulated radiation patterns and the measured ones. Although the AUT was placed higher than the metallic part to avoid its reflection effects, these effects seem to be non-negligible. The radiation on the back side seems to have disappeared. Nevertheless, the maximum gain orientation is coherent between the simulation and measurement.

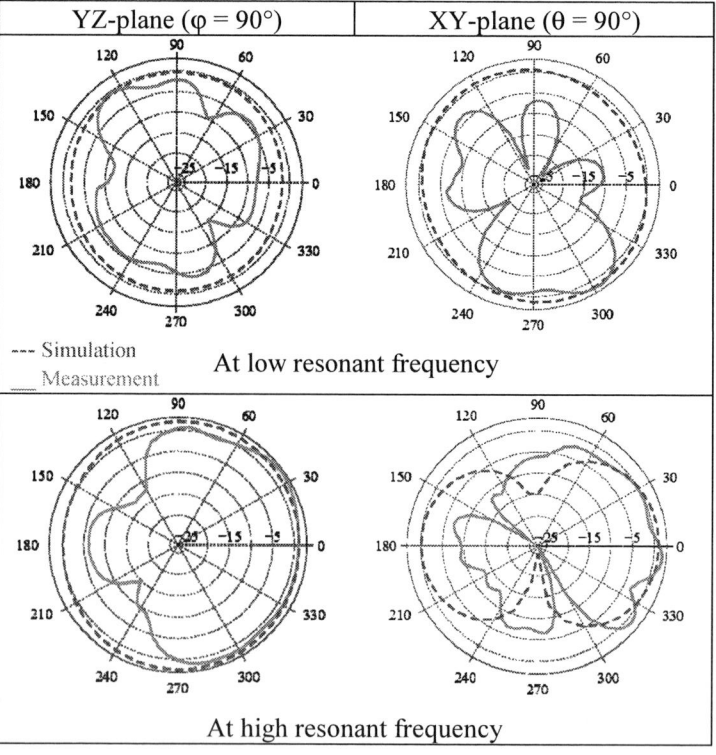

Figure 6. Radiation patterns of filtenna at two GSM frequencies.

Table I summarized the realized gain of the proposed filtenna obtained by simulation and by measurement. A good agreement was observed. However, a slightly greater measurement value compared to simulation value can be explained by the fact that: in the measurement, due to the reflecting plane of the metallic part the radiation in the back side goes towards the front side and more energy is radiated in this orientation.

TABLE I.   REALIZED GAIN OF FILTENNA IN DBI

Simulation	887 MHz	1736 MHz
	1.65	2.42
Measurement	866 MHz	1820 MHz
	1.9	2.55

IV.   DISCUSSION

This dual-band GSM filtenna will be used to harvest RF energy from cellular network frequency bands. To implement a complete rectenna, a matching circuit, a Schottky diode(s), a low-pass filter and a resistive load have to be added. The principal function of a rectenna is to convert RF energy to DC power. Then, the main design objective is to obtain high RF-to-DC conversion efficient. By collecting the maximum RF energy and deliver it to the rectifier, the RF-to-DC conversion efficiency will be improved and the DC power delivered to the load will be maximized. The antenna part has already been designed in this paper and has shown its good performance. The challenge lies now in the design of dual-band matching circuit. We prefer a microstrip structure to a discrete component structure. This is because with a microstrip structure all the parasite elements can be taken into account in the simulation process, so the simulated results would be reliable and would be close to the measured ones.

V.   CONCLUSION

In this paper, a simple dual-band filtenna design has been implemented for the GSM energy harvesting application. The traditional Printed-IFA structure has been modified in order to obtain the harmonic rejection property. Therefore, the co-design of antenna and filter were performed simultaneously. The reflection coefficient of the filtenna is greater than −1.5dB over the 2 GHz – 8 GHz band so no high-order harmonics are re-radiated through the antenna. Thus, these high-order harmonics are rejected back to the diode and remixed to generate more DC power and the RF-to-DC conversion efficiency is improved. The dual-band operation was obtained by using two Printed-IFAs. Their realized gains are 1.9 dBi and 2.55 dBi at 900 MHz and 1800 MHz bands, respectively. In future work, a complete rectenna will be designed to harvest GSM energy from cellular networks.

ACKNOWLEDGMENT

The authors would like to thank Mr. Antoine Gachon for his technical assistance, Mrs. Hong Phuong Phan for her technical advice and the CMIRA2013 project for its sponsorship.

REFERENCES

[1] Raymond E. Barnett, Jin Liu and Steve Lazar, "A RF to DC Voltage Conversion Model for Multi-Stage Rectifiers in UHF RFID Transponders," *IEEE Journal of Solid-State circuits*, vol. 44, no. 2, pp. 354–370, Feb. 2009.

[2] T. Le, K. Mayaram and T. Fiez, "Efficient far-field radio frequency energy harvesting for passively powered sensor networks," *IEEE Journal of Solid-State Circuits*, vol. 43, no. 5, pp. 1287–1302, May 2008.

[3] D. Bouchouicha, F. Dupont, M. Latrach, L.Ventura, "Ambient RF energy harvesting," *International Conference on Renewable Energies and Power Quality (ICREPQ'10)*, Granada (Spain), 23-25 Mar., 2010.

[4] Mile K. Stojčev, Mirko R. Kosanović, Ljubiša R. Golubović, "Power Management and Energy Harvesting Techniques for Wireless Sensor Nodes," TELSIKS 2009, Serbia, 7-9 Oct., 2009, pp.65–72.

[5] W. C. Brown, "The history of power transmission by radio waves," *IEEE Trans. Microw. Theory Tech.*, vol. MTT-32, no. 9, pp. 1230–1242, Sep. 1984.

[6] Y. J. Ren, M. F. Farooqui, and K. Chang, "A compact dual-frequency rectifying antenna with high-orders harmonic-rejection," *IEEE Trans. Antennas Propag.*, vol. 55, no. 7, pp. 2110–2113, Jul. 2007.

[7] J.Y. Park, S. M.Han, and T. Itoh, "A rectenna design with harmonic rejecting circular-sector antenna," *IEEE AntennasWireless Propag. Lett.*, vol. 3, pp. 52–54, 2004.

[8] Zhongkun Ma and Guy A. E. Vandenbosch, "Wideband Harmonic Rejection Filtenna for Wireless Power Transfer," *IEEE Trans. Antennas Propag.*, vol. 62, no. 1, pp. 371–377, Jan. 2014.

[9] F. Huang, T. Yo, C. Lee, and C. Luo, "Design of circular polarization antenna with harmonic suppression for rectenna application," *IEEE Antennas Wireless Propag. Lett.*, vol. 11, pp. 592–595, 2012.

[10] Z. Harouni, L. Osman, and A. Gharsallah, "Efficient 2.45 GHz CPW patch antenna including harmonic rejecting device for wireless power transmission," in *Proc. Int. Multi-Conf. Syst. Signals and Devices*, 2011, pp. 1–3.

[11] A. Costanzo, F. Donzelli, D. Masotti, and V. Rizzoli, "Rigorous design of RF multi-resonator power harvesters," in *2010 Proc. 4th EUCAP*, Barcelona, Spain, 12-16 Apr. 2010.

[12] K. Niotaki, Sangkil Kim, S. Jeong, A. Collado, and al., "A Compact Dual-Band Rectenna Using Slot-Loaded Dual Band Folded Dipole Antenna," *IEEE Antennas Wireless Propag. Lett.*, vol. 12, pp. 1634–1637, 2013.

[13] F.-J. Huang, C.-M. Lee, C.-L. Chang, L.-K. Chen, T.-C. Yo, and C.-H. Luo, "Rectenna application ofminiaturized implantable antenna design for triple-band biotelemetry communication," *IEEE Trans. Antennas Propag.*, vol. 59, no. 7, pp. 2646–2653, Jul. 2011.

[14] M. C. Huynh and W. Stutzman, "Ground plane effects on planar inverted-F antenna (PIFA) performance," in *IEE Proc. Microwaves Antennas and Propagations*, vol. 150, issue 4, pp. 209–213, Aug. 2003.

[15] D. M. Elsheakh, and E.A. Abdallah, "Compact Multiband Multifolded-Slot Antenna Loaded With Printed-IFA," *IEEE Antennas Wireless Propag. Lett.*, vol. 11, pp. 1478–1481, 2012.

978-1-4799-4993-9/14 $31.00 © 2014 IEEE

Sub-Threshold Based Power Detector for Low-Cost Millimeter-Wave Applications

Ayssar Serhan[1,2], Estelle Lauga-Larroze[1,2], Jean-Michel Fournier[1,2]

[1]Univ. Grenoble Alpes, IMEP-LAHC, F-38000 Grenoble, France
[2]CNRS, IMEP-LAHC, F-38000 Grenoble, France

Abstract—**This article presents a power detector proposed for low-cost millimeter-wave applications. Thus, the detector uses MOSFET transistor in sub-threshold region in order to realize the signal rectification benefiting from the natural exponential characteristics of MOS transistor in this regime. The detector was designed using the BiCMOS 55 nm technology from ST-Microelectronics. Thanks to this technology, the bipolar version of the detector has also been designed allowing comparison of these two cells. Each of them consume only 30 μW with an 80µm x 80µm area. Theoretical analysis and simulation results demonstrate the advantage of MOS detector in terms of input impedance which reduces the impact of the detector on the device under test (DUT). However, the bipolar detector provides faster responses due to its higher current capability. The MOS based detector has a response time of 1 ns. Its detection range and sensitivity are 20 dB and -18 dBm respectively, in the 50 GHz to 90 GHz frequency band, and are limited by the interconnections parasitic. The design and optimization steps are described in order to provide a general methodology to design high performance power detectors.**

Keywords—Power detector, Sub-threshold, CMOS, mm-Wave.

I. INTRODUCTION

To ensure robust performance of radio frequency (RF) system, each component must receive the appropriate signal level from the former component and pass the correct signal level to the succeeding component. If the signal level is too low the signal can be obscured in noise. Alternatively, if the signal level is too high, nonlinear behavior and distortion will appear. More accurately, in the transmission stage, power is managed to maintain high efficiency, and conformity to spectral regulation. Identically, in the reception stage, power is managed to maintain optimal sensitivity, selectivity, and dynamic range [1]. The key component in such systems is the power detector (PD).

The importance of power detector holds in its low frequency output signal. In a transmitting system, many applications use PDs to adjust the transmit power more accurately. These are known as automatic level control (ALC) solutions in which the detector's output is fed back to the control circuit [2]. Moreover, built-in self-test (BIST) solutions uses power detector to provide low-frequency low-cost signal of some critical point in the circuit under test (CUT). These solutions reduce the overall test cost since the low frequency measurement can be realized using low-cost instrumentation (voltmeter, oscilloscope) instead of high cost automatic test equipment (ATE) [3][4].

Diode based detector are widely used in power detection instrumentation due to its simplicity and its good performance [5][6]. Unfortunately, high-frequency diodes such as Shottky

diodes are unavailable in many advanced low-cost process [7]. The alternate solution of diode is the bipolar transistor. Power detector using bipolar transistor have been implemented in [7-9] benefiting from the natural exponential I-V characteristic of the bipolar transistor. Some of these detectors was extended to mm-Wave frequencies range. However, these detectors are not compatible with standard CMOS process. Many efforts are underway to develop mm-Wave MOSFET based detector operating with the square law nature of I_d-V_{gs} [10][11]; however they are either limited with a small detection range, or a long response time (e.g. limited bandwidth).

In this work, we have implemented a MOSFET-based power detector, operating in sub-threshold region, using BiCMOS 55 nm technology from STMicroelectronics. Simulation results show that the proposed detector has a high input impedance and reasonable response time, and is therefore suitable for power sensing and leveling in millimeter-wave power amplifier. It also offers a simple and low-power application potential as an envelope detector for direct modulation use like OKK/ASK multi-Gbps communication systems.

II. MILLIMETER WAVE PD DESIGN CHALLENGES

As mentioned before, power detector are usually used for ALC, and BIST solutions. Dealing with such applications at millimeter wave frequencies presents several challenges. For ALC solutions, the main limitation is the envelope bandwidth (e.g. envelope speed) of the mm-Wave signal which is around 1 GHz for 60 GHz WLAN systems [12].This wide bandwidth requires the detector response time to be lower than 1 ns, in order to track the envelope variation of the amplitude modulated signal like QAM or OFDM signals. In addition, the detection dynamic range (DDR) of the detector is fixed by the peak to average ratio (PAPR) of the modulated signal. This PAPR becomes more important when dealing with multicarrier modulation scheme like OFDM. Hence, the detector should have a DDR of about 10 dB, at least, for 60 GHz applications.

Finally, low power consumption, small layout footprint, high input impedance, detection sensitivity, low sensitivity to process and temperature variations, and compatibility with standard low-cost semiconductor process are common features required for mm-Wave power detector as explained in [13].

III. CIRCUIT DESCRIPTION AND THEORY

The architecture of the proposed detector is inspired from [9]. The circuit proposed by Meyer uses bipolar transistor to realize the non-linear rectification function taking advantage of the exponential I-V characteristics of bipolar transistor [9]. To realize the same function using low-cost CMOS process, we employed the sub-threshold MOSFET operation.

This work has been performed in the RF2THZ SiSoC project of the EUREKA program CATRENE in which the G-INP partner is funded by the DGCIS, France.

A. Sub-threshold operation

In weak inversion, and for $V_{ds} > 0.1\ V$, the drain current of an NMOS transistor is given by [14][15]:

$$I_D = I_{s1} \cdot e^{\frac{(V_{gs} - V_{th})}{U_T}} \qquad (1)$$

where V_{gs} is the gate-to-source voltage, V_{th} is the threshold voltage, I_{s1} is the reverse bias saturation current and is given by:

$$I_{s1} = \frac{2\mu_n C_{ox} U_T^2}{\gamma} \cdot \frac{W}{L} \qquad (2)$$

where μ_n is the electron mobility, C_{ox} is gate oxide capacitance per unit-length, γ is the temperature coefficient of V_{th}, W and L are the transistor width and length respectively, and U_T is the thermal voltage expressed by:

$$U_T = \frac{n\, K_B T}{q} \approx 25\ mV\ at\ 27\ °C \qquad (3)$$

with K_B the Boltzmann constant, T the absolute temperature in Kelvin, n the non-ideality factor and q is the elementary charge.

B. Transfer function derivation

The simplified schematic of the detector is shown in Fig. 1. The transistors M_1 and M_2 are biased in sub- threshold region. Assuming a RF input signal V_i, with amplitude V_{rf} and frequency W_0, applied at the gate of $M1$:

$$V_{GM1} = V_{dd} + V_{rf} \cos(W_0 t) \qquad (4)$$

The current equation of M1 is given by:

$$I_{D1} = I_{s1} \cdot e^{\frac{(V_{gs_{M1}} - V_{th})}{U_T}} = I_{s1} \cdot e^{\frac{(V_{dd} + V_{rf}\cos(w_0 t) - V_{o^+} - V_{TH})}{U_T}} \qquad (5)$$

By using the modified Bessel function, the drain current of M_1 can be written by

$$I_{D1} = I_{s1} \cdot e^{\frac{(V_{dd} - V_{TH})}{U_T}} e^{\frac{-V_{o^+}}{U_T}} \left[I_0(b) + \sum_{K=1}^{n} I_k(b) \cos(W_0 t) \right] \qquad (6)$$

Where $I_k(b)$ terms are the modified Bessel, and $I_0(b)$ is:

$$I_0(b) = \frac{e^b}{\sqrt{2\pi b}}\ ,\ and\ \ b = \frac{V_{rf}}{U_T}$$

I_0 represents the average value of the drain current results when applying V_{rf}. The average value of I_{D1} denoted $I_{D1_{av}}$ can be expressed by:

$$I_{D1_{av}} = I_{s1} \cdot e^{\frac{(V_{dd} - V_{TH})}{U_T}} \frac{e^{\frac{V_{rf}}{U_T}}}{\sqrt{2\pi \frac{V_{rf}}{U_T}} \cdot e^{\frac{V_{o^+}}{U_T}}} \qquad (7)$$

When the amplitude of the RF input signal V_{rf} changes, the transistor M_1 tends to change its average current according to (7). However, the current source I_1 forces the average current in M_1 so that $I_1 = I_{D1_{av}}$. Hence, the voltage V_{o^+} will change, instead of $I_{D1_{av}}$ so as to compensate V_{rf} variations and maintain a constant mean value of the gate-to-source voltage of M_1. This explains how the output V_{o^+} tracks the RF input signal V_{rf}.

To simplify the input-output relationship of the detector, the transistor M_1 and M_2, the current source I_1 and I_2 are assumed to be identical. Furthermore, since there is no RF signal at the gate of M_2 the current equation of M_2 is expressed by:

$$I_{D2} = I_2 = I_{s2} \cdot e^{\frac{(V_{dd} - V_{TH})}{U_T}} e^{\frac{-V_o-}{U_T}} \qquad (8)$$

By comparing (8) and (7), we can show that:

$$I_{s2} \cdot e^{\frac{(V_{dd} - V_{TH})}{U_T}} e^{\frac{-V_o-}{U_T}} = I_{s1} \cdot e^{\frac{(V_{dd} - V_{TH})}{U_T}} \frac{e^{\frac{V_{rf}}{U_T}}}{\sqrt{2\pi \frac{V_{rf}}{U_T}} \cdot e^{\frac{V_{o^+}}{U_T}}} \qquad (9)$$

And

$$e^{\frac{V_{o^+} - V_o-}{U_T}} = \frac{e^{\frac{V_{rf}}{U_T}}}{\sqrt{2\pi \frac{V_{rf}}{U_T}}} \qquad (10)$$

Finally, the input output relationship is obtained by applying the log function on (10), thus the transfer function is:

$$V_o = V_{o^+} - V_o- = V_{rf} + \delta_{Vo} \qquad (11)$$

Where δ_{VT} is a voltage and temperature dependent error term and is given by:

$$\delta_{Vo} = U_T \cdot \log\left(\sqrt{2\pi \frac{V_{rf}}{U_T}} \right) \qquad (12)$$

In fact, the transistor M_2 sets up an offsetting dc voltage Vo^- that cancels out the dc offset of V_{o^+}, thus the voltage difference, $V_o = V_{o^+} - V_o-$ is zero when the input V_{rf} is zero. C_2 is a decoupling capacitor that prevents V_{cc} noise from corrupting V_0^-. Capacitor C_1 is the integration capacitor whose value is set by compromising the speed and the output voltage ripple. The error term (12) tends to infinity for small V_{rf} values. Thus, the low boundary of the detection range is limited by the following condition:

$$\delta_{Vo} \ll V_{rf}$$

The choice of C_1 ,C_2, I_1, I_2, and the size of M_1 and M_2 will be explained in the next section.

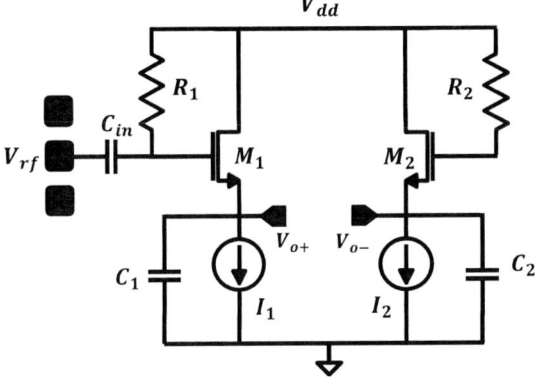

Fig. 1. Simplified schematic of the power detector

IV. DETECTOR DESIGN METHODOLOGY

The choice of the width of transistor M_1 is done as to minimize the input parallel capacitor (C_{ip}) and to maximize the parallel input resistance (R_{ip}) seen at the gate of $M1$. To determine this value, the transistor's width is swept, and the evolution of C_{ip} and R_{ip} are plotted in function of W_1. As shown in Fig. 2, the best choice of C_{ip} and R_{ip} can be obtained using small transistor size. In our case, 1 μm transistor's width was retained in order to facilitate the layout connection of the transistor. The length of the transistor is kept to its minimum value (60 nm) for the same reason.

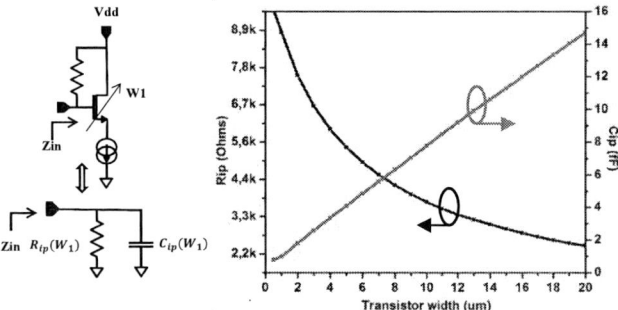

Fig. 2. Evolution of C_{ip} and R_{ip} in function of W_1

After choosing the transistor size, the bias current should be set as to force M_1 to operate in sub-threshold regime. To determine the appropriate value of I_1, the drain current of M_1 and its corresponding operation region are plotted in function of the gate-to-source voltage $V_{gs_{M1}}$, Fig. 3. A 15 μA current (e.g. $V_{gs_{M1}} = 550\ mV$) was retained as to be in the middle of the sub-threshold region of the transistor. Furthermore, the current mirror uses 3 L_{min} channel length in order to reduce the impact of the channel modulation effect and ensure robust current copy from the current source to M_1.

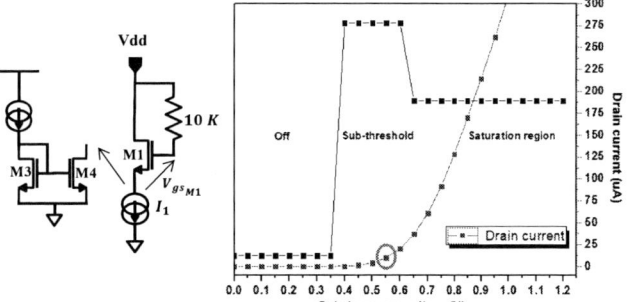

Fig. 3. Bench used to determine M_1 bias current (left), M_1 drain current and operation region in function of V_{gs} (right).

Once the bias current is selected, the minimum value of the capacitor C_1 is determined for a desired maximum voltage drop (e.g. voltage ripple) Δ_{Vo} in V_o. This voltage drop can be expressed by:

$$\Delta_{Vo} = \frac{I_1}{C_1} \Delta t$$

Where Δt approximately half of the period of the RF input signal and is given by:

$$\Delta t = \frac{T_0}{2}\ , where\ T = \frac{1}{f_0}$$

For $I_1 = 15\ \mu A$, $f_0 = 60\ GHz$ and $\Delta_{Vo} = 1\ mV$, the capacitor C_1 is:

$$C_1 = \frac{\Delta t}{\Delta_{Vo}} I_1 = 125 fF$$

Finally, the response time of the power detector is determined by the following expression:

$$\tau \approx (gm_{M1}^{-1} // R_{ds_{M4}}) . C_1 \approx \frac{C_1}{gm_{M1}} = \frac{125 fF}{134\ \mu\Omega^{-1}} = 0.92\ ns$$

Where gm_{M1} the transconductance of the transistor M_1 biased at I_1, and $R_{ds_{M4}}$ is the drain resistance of the current mirror transistor M_4. The value of $R_{ds_{M4}}$ can be neglected since it's much higher than the inverse of gm_{M1}.

The design has been implemented in the Bi-CMOS 55nm technology from STMicroelectronics. The layout is shown in Fig. 4. The power detector layout has been carefully designed to reduce the capacitive parasitic at the input stage and output stage, in order to preserve good sensitivity and short response time. Metal-Oxide-Metal (MOM) capacitors have been used for AC coupling and DC bypass. Dummy capacitor was added to make the layout fully symmetric. This results in a more robust design against process variation.

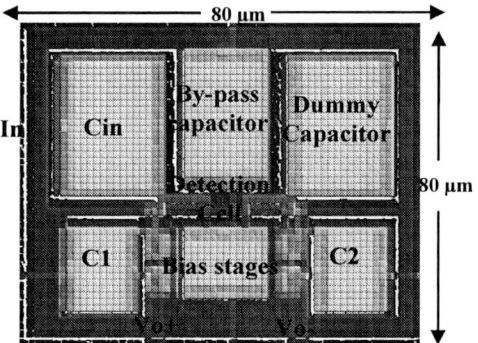

Fig. 4. Layout of the proposed detector

V. SIMULATION RESULTS

Fig. 5 shows the transfer function of the PD showing the output voltage V_{out} as function of the RF input signal voltage. It shows a linear relationship from 40 mV to 420 mV which correspond to -18 dBm and 2.4 dBm of power respectively; for a 50 Ω power source.

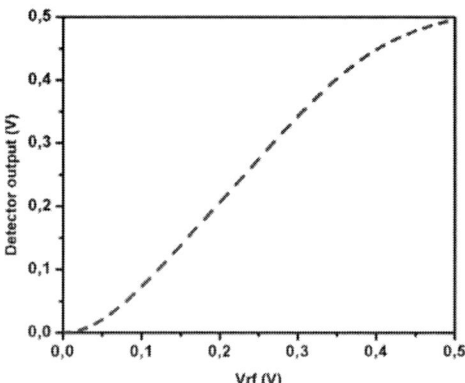

Fig. 5. Detector transfer function simulated with a 60 GHz input signal.

The effective input impedance Z_{ieff}, defined as the ratio of the applied input voltage to the resulting input current, was simulated with 60 a GHz input signal. As shown in Fig. 6, Z_{ieff} rises with increasing input power level. This high input impedance ensures that the detector does not load the circuit

978-1-4799-4993-9/14 $31.00 © 2014 IEEE

under test. Results were verified using both transient and large signal S-parameter (LPSS) simulations.

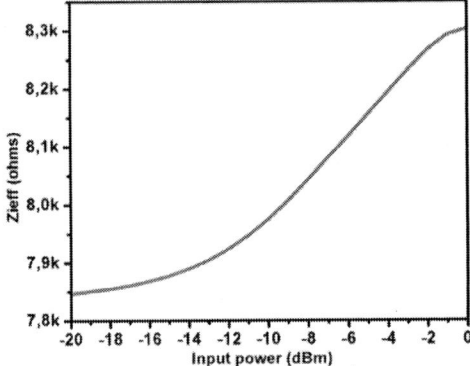

Fig. 6. Evolution of Z_{ieff} in function of input power at 60 GHz

For comparison issue, the bipolar version of the detector was designed according to the methodology presented in [9]. Table I resume the principle performances, of both bipolar and MOS version, simulated at 60 GHz. As we can notice, MOS detector shows higher input impedance while the bipolar detector has shorter response time. The higher input impedance of the MOS detector reflects its higher input quality factor, while the high speed of the bipolar detector reflects its higher trans-conductance (gm). Noting that the speed of both detectors can be enhanced at the cost of residual output voltage drop as explained in section IV. Finally, the dynamic range and detection sensitivity of both detectors is approximately similar.

TABLE I COMPARISON BETWEEN MEYER DETECTOR AND THE PROPOSED DETECTOR

Performances	MOS based detector	Bipolar based detector
DC Power (mW)	30	
Operating frequency (GHz)	60	
Active layout area	80μm x 80μm	
Response time (ns)	0.92	0.18
Input impedance (Ω)	8 K	3.8K

VI. CONCLUSION

A power detector suitable for millimeter wave applications is presented. The detector is fully compatible with standard low-cost CMOS process and presents the advantages of high input impedance and ultra-low power consumption. However, the small value of the MOS transistor's transconductance, especially when operating in sub-threshold regime, limits the detector response time to 1 ns. The detector dynamic range and

sensitivity are 20 dB and -18 dBm respectively. The detectors was sent for fabrication using the BiCMOS 55 nm technology from STMicroelectronics in order to validate the simulation and the design methodology by experimental results.

REFERENCES

[1] Hittite Microwave corporation, "Latest Advance in RF Power Detection: RMS Detector with iPAR", *Microwave Journal*, Vol. 51, no.12, p104, 2008.

[2] U.R. Pfeiffer, D. Goren, "A 20 dBm Fully-Integrated 60 GHz SiGe Power Amplifier With Automatic Level Control," *IEEE Journal of Solid-State Circuits*, vol.42, no.7, pp.1455-1463, 2007.

[3] L. Abdallah, H.-G. Stratigopoulos, C. Kelma, and S. Mir, "Sensors for built-in alternate RF test," in *Proc. IEEE European Test Symposium*, 2010, pp. 49-54.

[4] P. N. Variyam, S. Cherubal, and A. Chatterjee, "Prediction of analog performance parameters using fast transient testing," *IEEE Transactions on Computer-Aided Design of Integrated Circuits and Systems*, vol. 21, no. 3, pp. 349–361, 2002.

[5] W. Jeon, J. Rodgers, J. Melngailis, "Design and fabrication of Schottky diode, on-chip RF power detector," in *Proc. Semiconductor Device Research Symposium*, pp.294-295, 2003

[6] I.J. Bahl, "Application notes - Broadband Power Detectors," *IEEE Microwave Magazine*, vol.8, no.3, pp.82-86, 2007

[7] J. Zhang, V. Fusco, Y. Zhang, "A compact V-band active SiGe power detector," in *Proc. Microwave Integrated Circuits Conference (EuMIC)*, pp.528-531, 2012.

[8] T. Zhang, W.R. Eisenstadt, R.M. Fox, Q. Yin, "Bipolar Microwave RMS Power Detectors," *IEEE Jornal of Solid-State Circuits*, vol.41, no.9, pp.2188-2192, 2006

[9] R.G. Meyer, "Low-power monolithic RF peak detector analysis," *IEEE Journal of Solid-State Circuits*, vol.30, no.1, pp.65-67, 1995.

[10] J. Gorisse, A. Cathelin, A. Kaiser, E. Kerherve "A 60GHz 65nm CMOS RMS power detector for antenna impedance mismatch detection," in *Proc. IEEE European Solid State Circuit Conference (ESSCIRC)*, pp.172-175, 2009

[11] A. Siligaris, F. Chaix, M. Pelissier, V. Puyal, J. Zevallos, L. Dussopt, P. Vincent, "A low power 60-GHz 2.2-Gbps UWB transceiver with integrated antennas for short range communications," in *Proc. IEEE Radio Frequency Integrated Circuits Symposium (RFIC)*, pp.297-300, 2013

[12] T.S. Rappaport, J.N. Murdock, F. Gutierrez, "State of the Art in 60-GHz Integrated Circuits and Systems for Wireless Communications," In *Proc. IEEE*, vol.99, no.8, pp.1390-1436, 2011.

[13] A. Serhan, E. Lauga-Larroze,J.-M. Fournier, "A V-Band BiCMOS power detector for millimeter-wave applications," in *Proc. IEEE International Conference on Microelectronics* , pp.1-4, 2013.

[14] R.R. Harrison, "A wide-linear-range subthreshold CMOS transconductor employing the back-gate effect," in *Proc IEEE International Symposium on Circuits and Systems(ISCAS)*, vol.3, pp.727-730, 2002.

[15] S.M. Sharroush, Y.S. Abdalla, A.A. Dessouki, E.-S.A. El-Badawy, "Subthreshold MOSFET transistor amplifier operation," in *Proc. International Design and Test Workshop (IDT)*, pp.1,6, 15-17,2009

978-1-4799-4993-9/14 $31.00 © 2014 IEEE

Structured Design to Optimize the Output Power of Stacked Power Amplifiers

Elena Sobotta, Robert Wolf, David Fritsche and Frank Ellinger
Chair for Circuit Design and Network Theory
Technische Universität Dresden, 01062 Dresden, Germany

Abstract—An enhancement of an analytical algorithm to simplify and structure the design of stacked power amplifiers is presented in this work. This enhancement includes the calculation of passive networks, which compensate the parasitic capacitances of the transistors and thereby increase the distortion-free output power and the power added efficiency (PAE). As an example the algorithm is applied to a power amplifier (PA), using the IBM 180 nm CMOS process. The PA operates at 2 GHz for the long term evolution (LTE) standard. The post-layout-simulation exhibits an output power in the 1 dB compression point of 28.2 dBm, leading to a PAE of 30 %. The relative 3 dB bandwidth of the output power reaches a high value of 33 %. The PA fulfills the specifications of LTE and therewith the high requirements on linearity.

Index Terms—stacked power amplifiers, CMOS, load lines, LTE

I. INTRODUCTION

The main target of the design of a PA is to maximize the output power, the power gain and the PAE. A major constraint is the drain source or collector emitter voltage, which limits the output swing and therefore the output power.

One commonly used basic topology for a PA is a cascode. To maximize the output power, more transistors can be added, where each has to contribute equally to the output swing. The obvious idea is to stack transistors in series or parallel. Series stacking increases the resistance and therewith the optimum load whereas parallel stacking decreases the optimum load. Approaches of series stacked transistors has been presented in the past [1], [2]. Often, the design is not optimum in terms of output swing. One approach to overcome these difficulties is to design a passive network at the gates of the stacked transistors. An algorithm for the design of these networks is described in [3].

This algorithm is enhanced to determine compensation networks at the drains of the transistors. Without a compensation the maximum distortion-free output power cannot be reached. Prior works determined the components of the compensation networks only empirically [3]. This takes a lot of time in the design process and the values are only a rough approximation. Using the method presented in this article, a complete dimensioning of these elements can be calculated with one algorithm. Thus it is even possible to recalculate the values for the post-layout-simulation.

The following section describes the topology of the PA and the algorithm is introduced. In the conclusion the algorithm is verified.

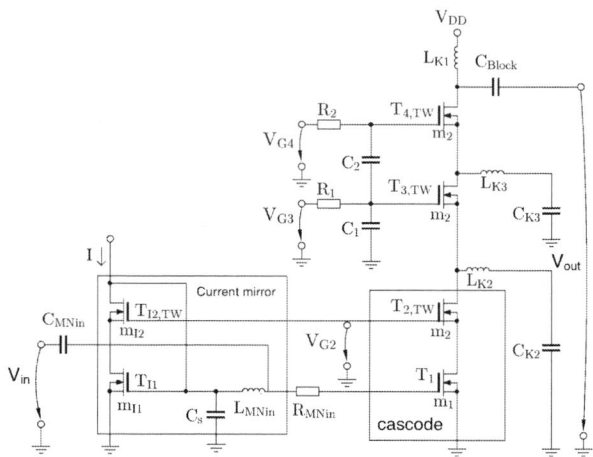

Fig. 1. Schematic of the presented stacked PA

II. DESIGN

A. Circuit description

To demonstrate the advantage of the optimization algorithm, a PA, consisting of a cascode with two stacked transistors is chosen. In Fig. 1 the single-ended version of the described pseudo-differential circuit is shown. The transistors are triple-well n-channel FETs.

Further, the input of the PA consists of a matching network, which enables conjugate complex matching. A cascode current mirror works as biasing network.

The maximum drain source voltage is limited by the hot carrier effect. A further increase of this voltage would degrade the transistor over time and decrease the life cycle of the circuit.

At the output the maximum voltage and current swing also define the optimum load. The optimum load for this circuit is selected to the system impedance of $50\,\Omega$. This impedance allows the circuit to operate without any additional matching network, thus keeping the bandwidth high. For a detailed relationship between the different parameters and the number of stacked transistors, refer to [3]. The inductors L_{K1} and $L_{K2/3}$ form the compensation network. In the following section the values of these elements will be be calculated. Also the values of the capacitors C_1 and C_2 will be determined in that section. These capacitors adjust the magnitude and phase of the drain source voltages. The resistors R_1 and R_2 are high

978-1-4799-4993-9/14 $31.00 © 2014 IEEE

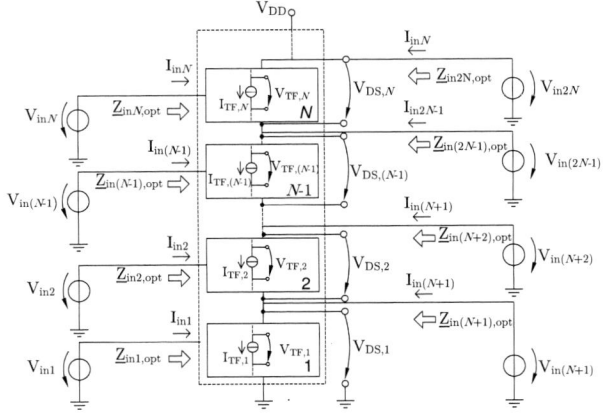

Fig. 2. AC simulation concept of a stacked PA with N stages without biasing

ohmic and feed the voltages V_{G3} and V_{G4} to the gates.

B. Calculation of the optimization networks

In this section an algorithm with two core tasks will be introduced. First, it determines a network, which ensures that all drain source voltages are equal in amplitude and phase. This guarantees that the voltage swing at the output is multiplied by the number of stacked transistors. Ideally all transistors clip at the same input voltage level. Otherwise the linearity of the amplifier would decrease. This part of the algorithm is already known and documented in [3].

The second, novel part of the algorithm describes a method to calculate the compensation network for the parasitic capacitance of the transistors. Up to this point the load lines would have to be found in an empirical way. The parasitic capacitances change the purely resistive load to a complex one. This prevents the inner transistors to work in a linear way. A clipping effect of the load lines would occur earlier and decrease the maximum possible voltage or current swing of the transistors. As a result, the maximum output power and the PAE are lowered. Load line theory is detailed in [4], [5].

For the calculation the small-signal behavior of the PA is considered. In Fig. 2 a simulation circuit of a stacked amplifier with N stages without biasing is shown. For this approach $2N$ AC voltages sources are needed to determine the amplitude and phase of the drain source voltage swing of the transistors and the optimal load. The AC voltages are connected from gate and additionally from drain of each transistor to ground. According to the description in [3], only one AC voltage source will be activated at a time. After $2N$ simulations a complex matrix can be formed.

For a better understanding, the dimensions of the matrices are noted in the upper indices. Matrices with only one number in the upper index can be considered as vectors.

To describe the optimum load lines, the drain current $\underline{I}_{TF}$ and the drain source voltage $\underline{V}_{TF}$ of the inner transistor are needed. The system can be described in dependence on the input voltage $\underline{V}_{in}$. The drain source voltage $\underline{V}_{DS}$ and the inner drain source voltage of the transistor $\underline{V}_{TF}$ can be expressed by the voltage gain $\underline{A}_{DS}$ and $\underline{A}_{TF}$, respectively.

$$\underline{V}_{DS}^N = \underline{A}_{DS}^{N \times 2N} \cdot \underline{V}_{in}^{2N} \quad (1)$$

$$\text{with} \quad \underline{A}_{DS,s,t} = \frac{V_{DS,s}}{V_{in,t}}\bigg|_{V_{in,n}=0 \ (\forall \, n \neq t)}$$

$$\underline{V}_{TF}^N = \underline{A}_{TF}^{N \times 2N} \cdot \underline{V}_{in}^{2N} \quad (2)$$

$$\text{with} \quad \underline{A}_{TF,u,v} = \frac{V_{TF,u}}{V_{in,v}}\bigg|_{V_{in,n}=0 \ (\forall \, n \neq v)}$$

$$s, u \in [1, N]; \quad t, v \in [1, 2N]$$

Similarly the currents $\underline{I}_{TF}$ and $\underline{I}_{in}$ can be derived from the admittance matrices $\underline{Y}_{TF}$ and $\underline{Y}_{in}$ and the voltage $\underline{V}_{in}$.

$$\underline{I}_{in}^{2N} = \underline{Y}_{in}^{2N \times 2N} \cdot \underline{V}_{in}^{2N} \quad (3)$$

$$\text{with} \quad \underline{Y}_{in,w,x} = \frac{I_{in,w}}{V_{in,x}}\bigg|_{V_{in,n}=0 \ (\forall \, n \neq x)}$$

$$\underline{I}_{TF}^N = \underline{Y}_{TF}^{N \times 2N} \cdot \underline{V}_{in}^{2N} \quad (4)$$

$$\text{with} \quad \underline{Y}_{TF,y,z} = \frac{I_{TF,y}}{V_{in,z}}\bigg|_{V_{in,n}=0 \ (\forall \, n \neq z)}$$

$$y \in [1, N]; \quad w, x, z \in [1, 2N]$$

With these equations the system is determined. Now the behavior of the system can be defined. Two conditions are necessary.

First, the drain source voltages of all transistors should have be equal in amplitude and phase. Here, the phase of $\underline{V}_{DS,n}$ is set to zero.

$$\underline{V}_{DS}^N \overset{!}{=} \frac{1}{\sqrt{2}} \cdot [\hat{V}_{DS,2,max} \cdots \hat{V}_{DS,2,max}]^T \quad (5)$$

Second the quotient of $\underline{V}_{TF}$ and $\underline{I}_{TF}$ should result in the optimum load R_{TF}. The optimum load is specified for each transistor and not for the whole amplifier.

$$\underline{V}_{TF}^N \overset{!}{=} -R_{TF} \cdot \underline{I}_{TF}^N \quad (6)$$

With (2) and (6) the following equation can be formed.

$$0^N = \left(\underline{A}_{TF}^{N \times 2N} + R_{TF} \cdot \underline{Y}_{TF}^{N \times 2N}\right) \cdot \underline{V}_{in}^{2N} \quad (7)$$

All these conditions and equations can be combined in one matrix $\underline{M}$:

$$\underline{V}_{out}^{2N} = \underline{M}^{2N \times 2N} \cdot \underline{V}_{in,opt}^{2N} \quad (8)$$

978-1-4799-4993-9/14 $31.00 © 2014 IEEE

$$
\begin{pmatrix}
\frac{\widehat{V}_{\mathrm{DS,2,max}}}{\sqrt{2}} \\
\vdots \\
\frac{\widehat{V}_{\mathrm{DS,2,max}}}{\sqrt{2}} \\
0 \\
\vdots \\
0
\end{pmatrix}
=
\underbrace{
\begin{pmatrix}
\underline{A}_{\mathrm{DS},1,1} & \cdots & \underline{A}_{\mathrm{DS},1,2N} \\
\vdots & & \vdots \\
\underline{A}_{\mathrm{DS},N,1} & \cdots & \underline{A}_{\mathrm{DS},N,2N} \\
\underline{V}_{\mathrm{TF},1,1} + R_{\mathrm{TF}} \cdot \underline{Y}_{\mathrm{TF},1,1} & \cdots & \underline{V}_{\mathrm{TF},1,2N} + R_{\mathrm{TF}} \cdot \underline{Y}_{\mathrm{TF},1,2N} \\
\vdots & & \vdots \\
\underline{V}_{\mathrm{TF},N,1} + R_{\mathrm{TF}} \cdot \underline{Y}_{\mathrm{TF},N,1} & \cdots & \underline{V}_{\mathrm{TF},N,2N} + R_{\mathrm{TF}} \cdot \underline{Y}_{\mathrm{TF},N,2N}
\end{pmatrix}
}_{\underline{M}}
\cdot
\begin{pmatrix}
\underline{V}_{\mathrm{in},1} \\
\vdots \\
\underline{V}_{\mathrm{in},N} \\
\underline{V}_{\mathrm{in},N+1} \\
\vdots \\
\underline{V}_{\mathrm{in},2N}
\end{pmatrix}
\tag{15}
$$

The first N rows are derived by (1). The rows $N+1$ to $2N$ consist of (7). The result is written down in (15).

The matrix $\underline{M}$ is a square matrix and can be inverted. Now it is possible to calculate the optimum input voltage $\underline{V}_{\mathrm{in,opt}}$.

$$
\underline{V}_{\mathrm{in,opt}}^{2N} = (\underline{M}^{2N \times 2N})^{-1} \cdot \underline{V}_{\mathrm{out}}^{2N} \tag{9}
$$

Using $\underline{I}_{\mathrm{in,opt}}$ the optimum impedance $\underline{Z}_{\mathrm{in,\,opt}}$ can be calculated. This impedance characterizes the voltage and current, which can be seen at the gates and the drains of the transistors, respectively. If these impedances are connected to the nodes of the AC sources, the amplifier exhibit the specified characteristics.

$$
\underline{I}_{\mathrm{in,opt}}^{2N} = \underline{Y}_{\mathrm{in}}^{2N \times 2N} \cdot \underline{V}_{\mathrm{in,opt}}^{2N} \tag{10}
$$

$$
\underline{Z}_{\mathrm{in},n,\mathrm{opt}} = \frac{\underline{V}_{\mathrm{in},n,\mathrm{opt}}}{\underline{I}_{\mathrm{in},n,\mathrm{opt}}} = R_{\mathrm{in},n,\mathrm{opt}} + j \cdot X_{\mathrm{in},n,\mathrm{opt}} \tag{11}
$$

Each impedance can be converted into an equivalent circuit, which consists of a resistive and a reactive element. In this case, the impedance will be transformed to a parallel circuit of both elements.

$$
R_n = -\frac{R_{\mathrm{in},n,\mathrm{opt}}^2 + X_{\mathrm{in},n,\mathrm{opt}}^2}{R_{\mathrm{in},n,\mathrm{opt}}} \qquad n \in [1,2N] \tag{12}
$$

$$
C_n = \frac{1}{\omega} \cdot \frac{X_{\mathrm{in},n,\mathrm{opt}}}{R_{\mathrm{in},n,\mathrm{opt}}^2 + X_{\mathrm{in},n,\mathrm{opt}}^2} \qquad n \in [1,N-1] \tag{13}
$$

$$
L_{\mathrm{K}n} = -\frac{1}{\omega} \cdot \frac{R_{\mathrm{in},n,\mathrm{opt}}^2 + X_{\mathrm{in},n,\mathrm{opt}}^2}{X_{\mathrm{in},n,\mathrm{opt}}} \qquad n \in [1,N] \tag{14}
$$

First the the equivalent networks of the AC sources 2 to 5 are considered. The calculated magnitudes of the resistances are high that the parallel reactive elements dominates the behavior of the circuit. Thus the resistors can be neglected without changing the behavior of the circuit significantly. For the equivalent networks 2 and 3 the reactive elements can be realized by capacitances. In Fig. 3 the corresponding capacitances C_1 and C_2 are depicted over frequency. The reactive elements of the equivalent networks 4 and 5 are the inductances $L_{\mathrm{K}2}$ and $L_{\mathrm{K}3}$ shown in Fig. 4. It can be seen, that the values of the capacitances and the inductances do not vary

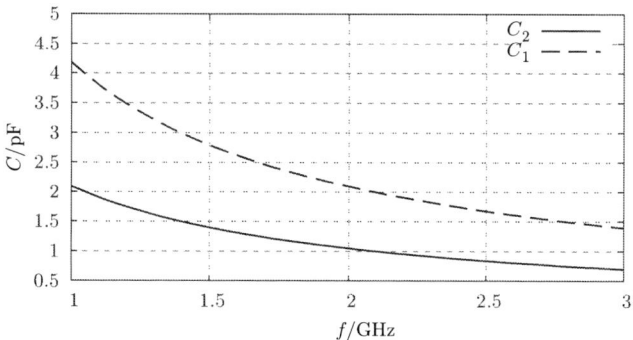

Fig. 3. Calculated capacitance values of the networks at the gates

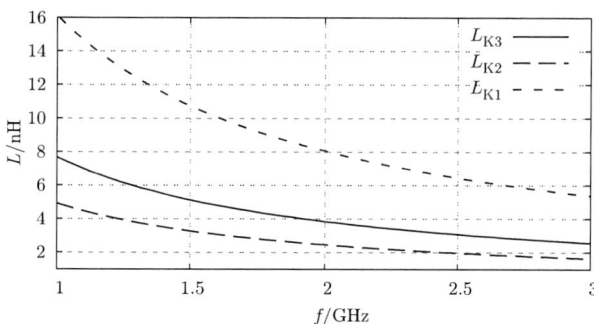

Fig. 4. Calculated inductance values of the networks at the drains

significantly over frequency. Compared to the narrow band input matching, the implementation of these compensation elements does not affect the PA's bandwidth.

The calculated equivalent network at the output consists of a resistance, which is equal to the optimum load, and an inductance. The inductor $L_{\mathrm{K}1}$ can be reused as bias inductor.

For the input equivalent network a negative resistance parallel to a negative reactance is simulated. The negative resistance is low ohmic, which ensures that real power at the input can be consumed by the PA. The reactance can be compensated by the input matching.

III. RESULTS

The circuit was implemented using an IBM 180 nm Bulk CMOS process. Below the results of the post-layout-simulation with RC extraction and real inductor models are

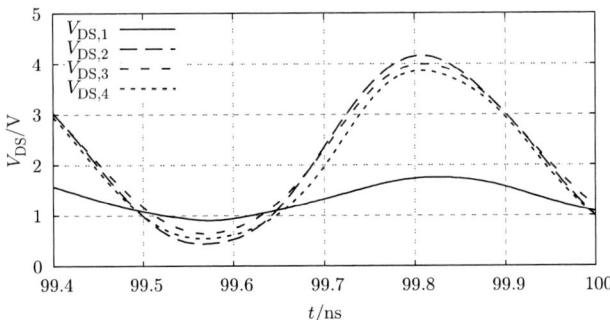

Fig. 5. Transient voltage swing of the stacked transistors

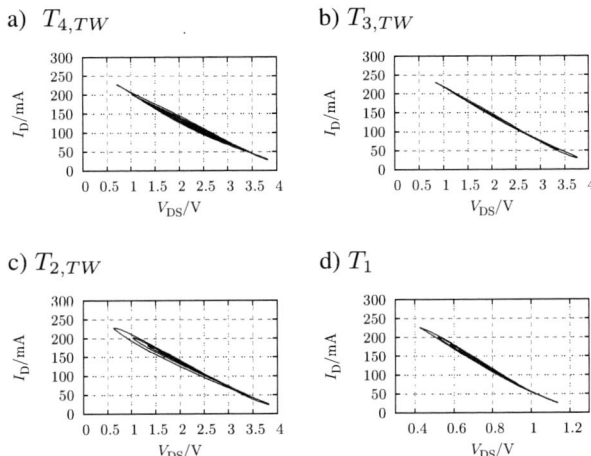

Fig. 6. Load lines after compensation

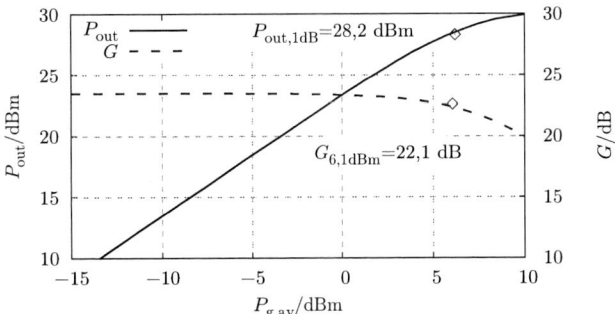

Fig. 7. Output power and power gain versus available input power

Table I compares the presented circuit with state-of-the-art stacked power amplifiers. It has to be noted, that the matching networks of the presented design are fully integrated.

TABLE I
STATE-OF-THE-ART STACKED POWER AMPLIFIERS

Ref	[2]	[1]	[6]	[3]	This work [1]
Tech.	500nm GaAs	65nm CMOS	130nm CMOS	250nm SiGe	180nm CMOS
Freq./GHz	5,2	1,8	1,9	2	2
V_{DD}/V	12	3,4	6,5	7,8	7,8
P_{out}/dBm	22,2	25,2	30,8	26,2	28,2
PAE/%	17,7	38	32,4	34	30
Fully integrated	yes	no	no	yes	yes

1 post-layout-simulation

IV. CONCLUSION

In this article an enhancement of the algorithm for stacked PAs presented in [3] is introduced. The enhancement includes the calculation of a parasitic compensation network. The algorithm could be verified by the post-layout-simulation of a stacked PA. With this algorithm it is possible to increase the PAE and the distortion-free output power to accommodate demanding standards like LTE.

REFERENCES

[1] S. Leuschner, J.-E. Mueller, H. Klar, "A 1.8GHz wideband stacked-cascode CMOS power amplifier for WCDMA applications in 65nm standard CMOS," *Proc. IEEE Radio Frequency Integrated Circuits Symposium*, pp. 1 – 4, 2011.

[2] C.-C. Shen, H.-Y. Chang, G. D. Vendelin, "Comparison of enhancement- and depletion-mode triple stacked power amplifiers in 0.5 μm AlGaAs/GaAa PHEMT technology," *Proc. European Microwave Integrated Circuits Conference*, pp. 222 – 225, 2009.

[3] D. Fritsche, R. Wolf, F. Ellinger, "Analysis and Design of a Stacked Power Amplifier with very high Bandwidth," *IEEE Transactions on Microwave Theory and Techniques*, vol. 60, pp. 3223 – 3231, 2012.

[4] S. Cripps, "A Theory for the Prediction of GaAs FET Load-Pull Power Contours," *IEEE MTT-S Int. Microw. Symp. Dig.*, pp. 221 – 223, 1983.

[5] S. Hauptmann, F. Ellinger, "Optimized Transistor Output Power - Extending Cripps Loadline Method to Cascode Stages," *IEEE Transactions on Microwave Theory and Techniques*, vol. 59, pp. 2017 – 2023, 2011.

[6] S. Pornpromlikit; J. Jeong; C. Presti; A. Scuderi, P. Asbeck, "A Watt-Level Stacked-FET Linear Power Amplifier in Silicon-on-Insulator CMOS," *IEEE Transactions on Microwave Theory and Techniques*, vol. 58, pp. 57 – 64, 2010.

described. The circuit needs a supply voltage of 7.6 V and has a die size of 1.38 mm².

First the calculated values of the inductors and the capacitors are verified. The simulated transient voltages at the 1 dB compression point of the transistors are shown in Fig. 5. According to the conditions in section II-B the voltages have almost the same amplitude and phase. Thus a constructive addition is possible and the highest overall voltage swing without clipping will be reached.

The second condition for calculating the elements was a purely resistive load line. In Fig. 6 the load lines of the transistors are plotted. They show only little deviation from linear behavior and the transistor can be excited towards its limitations without clipping too early. This figure shows the simulations of the load lines with ideal inductors.

In the following Fig. 7 the output power versus the available input power is shown. It also depicts the power gain. The large relative 3 dB bandwidth of the output power is 33 %. The PAE in the 1 dB compression point reaches the high value of 30 %.

The LTE standard requires high linearity. The limits of the LTE standard for the error vector magnitude (EVM) of 12.5 % for QAM-16 can be satisfied with this circuit at average output power of 28.2 dBm with 3.3 dB in back-off.

66-87 GHz Power Amplifier with 20dBm 1-dB compression point and 35% peak PAE in a 55nm SiGe technology

David del Rio[*†], Roc Berenguer[*†], Ainhoa Rezola[*†] and Juan Francisco Sevillano[*†]

[*]Electronics and Communications Department,
Centro de Estudios e Investigaciones Técnicas (CEIT), 20018 San Sebastián, SPAIN
Email: ddelrio@ceit.es , Phone number: +34 943 212800, Fax: +34 943 213076
[†]Electrical, Electronic and Control Engineering Department,
Technological Campus of University of Navarra (TECNUN), 20018 San Sebastián, SPAIN

Abstract—This paper presents the design of a mm-Wave PA for a transmitter of an E-Band mobile backhaul network, implemented using a 55nm SiGe BiCMOS technology. It has a 4-stage balanced CE configuration and the outputs are converted to single-ended using an integrated balun, which provides insertion losses smaller than 1dB at E-Band frequencies. The PA presents a maximum S21 of 15.5 dB at 74GHz and its 3-dB bandwidth covers from 66 to 87 GHz. Its output 1-dB compression point has a maximum of 20dBm and is above 18.75dBm across the whole E-Band, with a peak PAE of 35%. The different stages are externally biased and the Vcc voltage is 1.5V. The total DC power consumption is 275 mW.

Index Terms—Power amplifier, BiCMOS integrated circuits, SiGe, 55nm, mm-wave, E-Band.

I. INTRODUCTION

The emergence of high speed wireless devices has led to a demand of higher bandwidth, which is stressing current network infrastructures to achieve higher spectral efficiencies and capacities up to 10Gbps in the backhaul networks [1]. Microwave links are widely used as cheap and flexible mobile backhaul solutions, and these Gbps speeds can be allocated in the E-Band, which offers enough bandwidth (71-76GHz, 81-86GHz).

For this paper an E-Band transceiver using a bandwidth of 2GHz has been considered, which enables network operators to fit easily two full-duplex links in the same site. That means that a spectrally-efficient modulation as 64-QAM must be used, which in turn requires challenging specifications for the RF components, such as the output power back-off, which has to be on the order of 10dB. This makes the design of components such as Power Amplifiers challenging, as high 1dB compression points must be achieved. For the considered transceiver, the output P_{1dB} specification for the PA is of at least 18dBm. These kinds of circuits can be developed using SiGe BiCMOS technologies, which make it possible to design low-cost integrated circuits for many mm-wave applications such as E-Band backhauling or automotive radar systems [2], [3], [4]. However, they also introduce some technological limitations such as lower breakdown voltages, which decreases the maximum available output power.

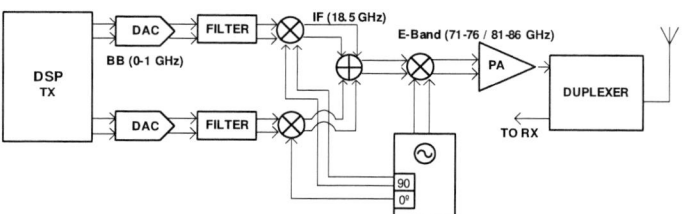

Fig. 1. Architecture of the E-Band transmitter.

This paper presents the design of a PA for an E-Band transmitter like the one shown in Figure 1, which represents a good balance between DAC sampling rate requirements and complexity of the RF circuitry. It consists of an I/Q up-converting modulator that converts the baseband phase and quadrature components to an IF of 18.5 GHz. The I and Q channels are then combined and up-converted to E-band frequencies. Both the 71-76 or 81-86 GHz sub-bands can be used depending on which one is being used for transmission on the other transceiver, in order to make the link full-duplex. Finally, a wideband PA amplifies the signal before transmission. All the blocks have a differential configuration, except the PA, which has a single ended output to facilitate the transition to the output waveguide.

Although the signal is transmitted using a 2GHz channel, this channel can be accommodated anywhere in the E-band, so that two full-duplex links can be placed on the same location. This implies that the mm-wave mixer and the PA bandwidths must cover the whole E-Band.

Section II describes the design of the PA and Section III presents post-layout simulation results. Finally, the main conclusions of the paper are summarized in Section IV.

II. CIRCUIT IMPLEMENTATION OF THE PA

A. Architecture

The PA is implemented using a 55nm SiGE BiCMOS process of STMicroelectronics, with high speed bipolar transistors of f_T/f_{MAX} 320/370 GHz and which provides 8 metal layers,

978-1-4799-4993-9/14 $31.00 © 2014 IEEE

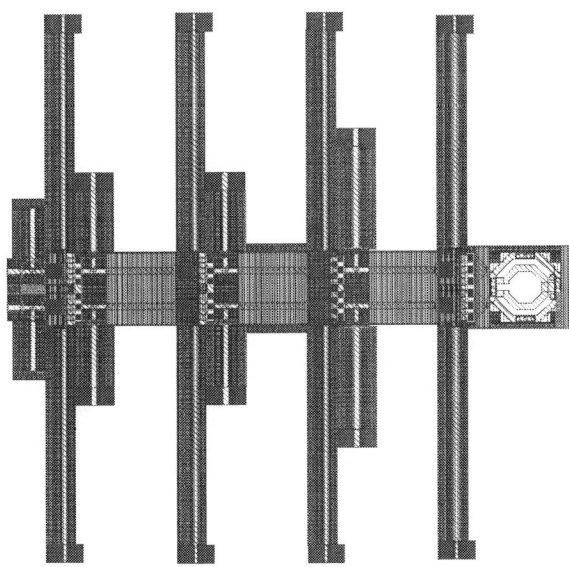

Fig. 3. Layout of the PA, without PADs.

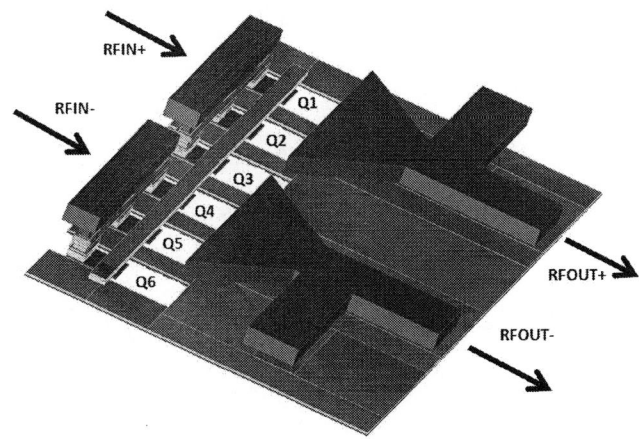

Fig. 4. 3D view view of the EM simulated transistor connections.

being 2 of them thick metals and the last one an ultra-thick metal layer. The circuit consists of two identical unit cells, combined at the output using an integrated balun. Figure 2 shows the schematic of one of these unit cells, with all the power cells and the matching networks. The core layout (without PADs) is presented in Figure 3, and occupies an area of $800 \mu m$ x $800 \mu m$.

It consists of 4 CE stages, each independently biased externally to optimally bias each stage as well as compensate for process or temperatures variations. Bias is optimized for compression at the output stages and for gain at the previous stages. Additionally, biasing is also chosen so that the gain of the first stages starts to increase when the others approach compression, which helps pushing P_{1dB} towards P_{SAT} as outlined in [5]. The biasing points are isolated from the mm-wave path through quarter-wave transmission lines, and shorted to ground through arrays of big MOM capacitors. All the transmission lines and the T-like connections between them have been EM modelled using Agilent Momentum.

B. Transistors

The high speed bipolar transistors provided in the design kit are used. They are externally biased to a value below the Vbe that achieves the maximum f_T, resulting in a current density of $0.4 - 0.5 mA/\mu m$, which provides a good balance between gain, compression and efficiency values. A Vcc of 1.5V is used to avoid exceeding the BV_{CE0} limit.

Each transistor cell is made using multiple parallel emitters to reduce the length to width ratio, and the size is increased at the output stages to increase the output power compression. The output stage is built of 3 cells with 4 x $8 \mu m$ x $0.25 \mu m$ parallel emitters each, which result in an equivalent size of $96 \mu m$ x $0.25 \mu m$. The size of the 3rd stage is reduced by 25%, placing 3 cells with 3 x $8 \mu m$ x $0.25 \mu m$ parallel emitters,

while the size of the first two stages is further reduced by another 30% using 3 cells with 3 x $7 \mu m$ x $0.2 \mu m$ parallel emitters. The cells are placed at a distance which addresses the trade off between interconnection losses, reliability and electromigration constraints for connections and good contacts between the emitters and GND, which can be critical in CE amplifiers.

The parasitic resistances and capacitances of the transistor cell layout have been extracted and taken into account for the design, while the connections between the transistors and the rest of the circuit have been electromagnetically simulated, following a procedure similar to the one that proved to give very accurate results in [6]. Once the layout is done, the transistor cells are removed and ports are added for the EM simulation. Figure 4 shows a 3D view of one of these connections.

C. Matching networks

The input has been matched to a differential 100Ω impedance at the center of the E-Band, while the inter-stage and output matching networks have been optimized to achieve a flat gain over the whole E-Band bandwidth, maintaining an output 1dB compression point bigger than 18dBm. They are implemented using transmission lines and MOM capacitors. Side-shielded Microstrip lines have been used for the shorted stubs at the input, whereas the lines used in the inter-stage matching networks are differential back- and shide-shielded coplanar waveguides. The 8th top ultra-thick metal has been used for all the conductors to achieve high quality and bigger power handling capacity, whereas the 1st and 2nd metals are stacked to provide the GND to the circuit. The MOM capacitors have been custom-made using the top metals as well, so as to achieve high quality factors ($Q \sim 15 - 20$) at mm-wave frequencies.

D. Output Balun

An integrated octogonal transformer balun is used to convert the differential output to single ended. Figure 5 shows its 3D view as introduced in the EM simulator. The primary is

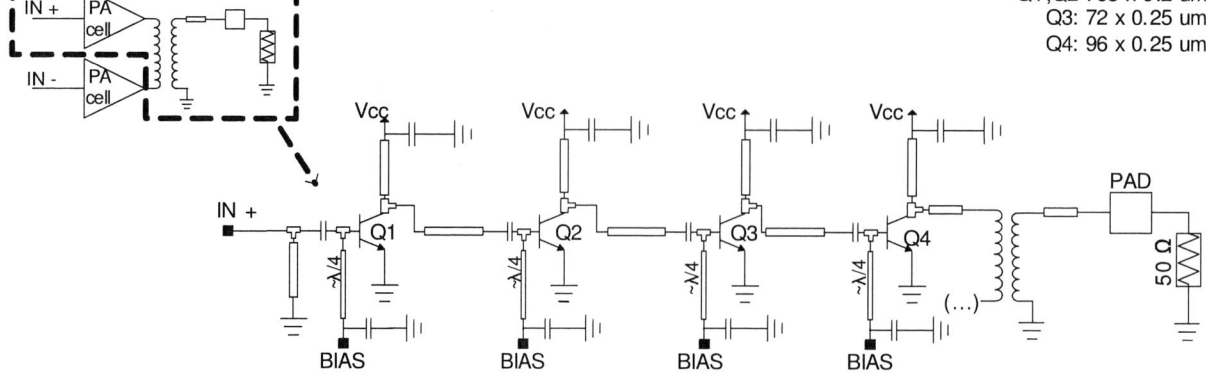

Fig. 2. Schematic view of one of the 2 parallel PA cells.

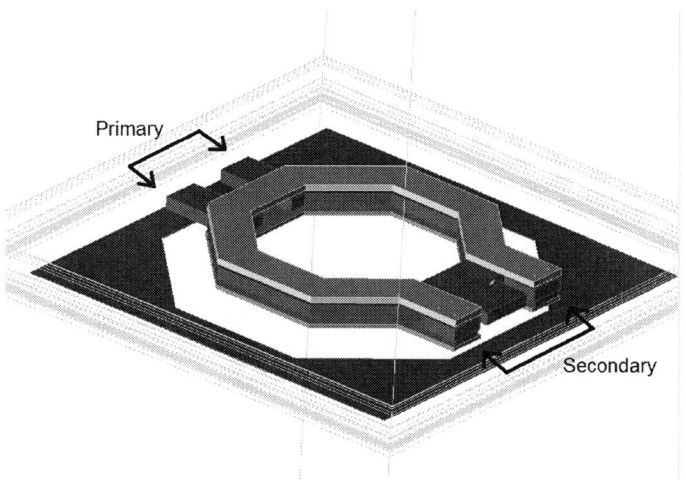

Fig. 5. 3D view of the output balun.

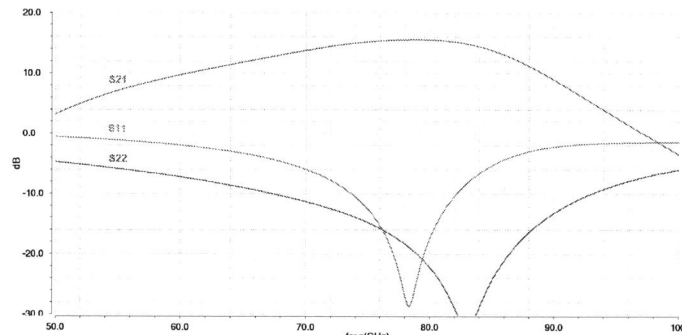

Fig. 6. Simulated S-parameters of the PA.

implemented using a single turn in M8, whereas the secondary is made using single M7 and AP turns connected in parallel. Thus, the primary is surrounded at both upper and lower sides by the secondary and the coupling is maximized. It is designed to provide minimum loss at E-Band frequencies, as well as to present a low impedance to the PA output stage, which maximizes P_{1dB}.

The impedance transformation already accounts for the PAD capacitance and the connections between the balun and the rest of the circuit, so that no additional matching network is needed.

III. SIMULATION RESULTS

All the results are based on simulations of the extracted layouts and electromagnetic simulations, as described above. Figure 6 shows the simulated S11, S21 and S22 parameters of the PA (S12 is smaller than -60dB and falls out of the scale). It can be observed that S21 peaks at 80GHz with a value of 15.5dB, with a broadband 3-dB bandwidth of 66-87GHz. The gain flatness at the E-Band is 1.23dB for the 71-76GHz sub-band and 2.34dB for the 81-86GHz sub-band.

Figure 7 presents the output power as a function of the input power at 74GHz. P_{1dB} and P_{SAT} are 20dBm and 21 dBm respectively, which results in a difference of only 1dB between both points.

Figure 8 shows the output P_{1dB} and Power Added Efficiency at compression as a function of frequency. The maximum P1dB is 20dBm at 74GHz, and it is above 18.75dBm across the whole E-Band. As for the PAE, it has a maximum value of 35% at 74GHz and it is higher than 25% across the considered band.

Additionally, the loss of the balun has been EM simulated stand-alone, resulting in a minimum loss of 0.7dB and a value better than 1dB at E-Band frequencies, as shown in Figure 9.

The main results are summarized and compared to other mm-wave PAs in Table I. It can be observed that the use of a 55nm technology makes it possible to deliver higher power at mm-wave frequencies while mantaining a wideband behaviour and high efficiency.

IV. CONCLUSIONS

The design of a PA for a E-Band backhauling transmitter has been presented. Simulation results show that its maximum S21 is 15.5dB and its 3-dB bandwidth covers the whole E-Band, with a maximum output P_{1dB} of 20dBm and a peak PAE of 35%. To the best of the author's knowledge, these are

978-1-4799-4993-9/14 $31.00 © 2014 IEEE

TABLE I
COMPARISON OF STATE OF THE ART MM-WAVE PAS

Ref.	Technology	Freq(GHz)	Peak PAE(%)	Gain(dB)	P_{1dB}(dBm)	Topology	DC Power(mW)
This work*	55nm BiCMOS	66-87	35	15.5	20	4xCE, balanced	275
[7]	120nm BiCMOS	65.15-81.85	11	21	17.6	5xCE, single-ended	560
[8]	130nm BiCMOS	79-97	15.4	14.5	18.8	3xCE, balanced	415
[9]	130nm BiCMOS	60	8.6	17.3	13	4xCE, balanced	528
[5]	250nm BiCMOS	61	19.7	18.8	14.5	2xCascode, single-ended	132
[10]	130nm BiCMOS	78.9-87.5	9	27	16	3xCB, balanced	353
[11]	40nm CMOS	70.3-85.5	22.3	18.1	17.8	2xCS, 4-way diff.	N/A

*Post-layout simulation results.

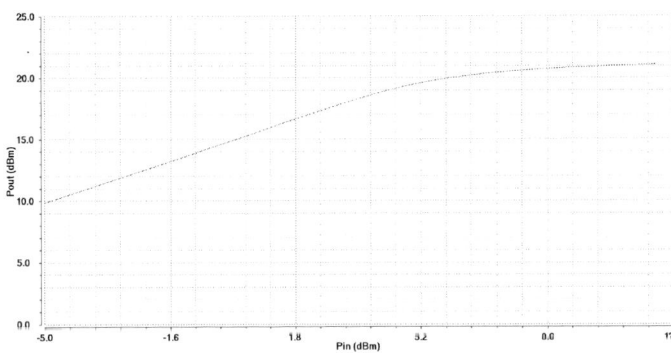

Fig. 7. Pout vs. Pin at 74Ghz.

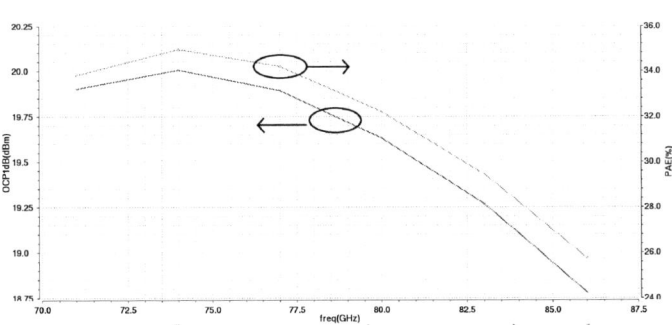

Fig. 8. Simulated output 1dB compression point and PAE at compression.

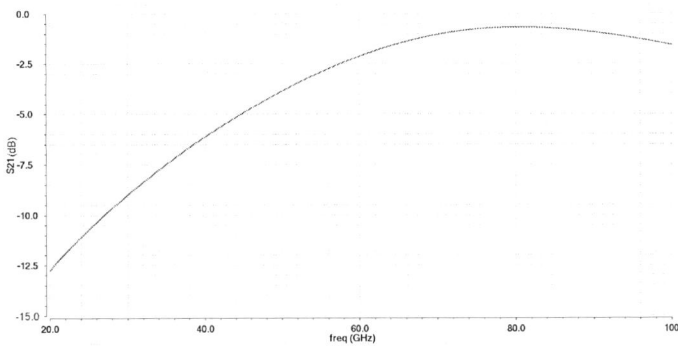

Fig. 9. EM simulated Insertion loss of the output balun.

the highes P_{1dB} and PAE reported to the date for a E-Band SiGE power amplifier.

ACKNOWLEDGMENT

The research leading to these results has received funding from the European Community's Framework Programme FP7/2007-2013 under grant agreement no. 317957. Consortium: Ceit, FhG, ALU-I, CEA, INCIDE, SiR, ST-I, Sivers IMA, OTE

REFERENCES

[1] M. G. Frecassetti, "Mobile backhaul network evolution," presented at the XXVIII Conference on Design of Circuits and Integrated Systems (DCIS), 2013.

[2] O. Katz, R. Ben-Yishay, R. Carmon, B. Sheinman, F. Szenher, D. Papae, and D. Elad, "A fully integrated SiGe E-BAND transceiver chipset for broadband point-to-point communication," in *Radio and Wireless Symposium (RWS), 2012 IEEE*, Jan 2012, pp. 431–434.

[3] R. Ben-Yishay, R. Carmon, O. Katz, B. Sheinman, D. Papae, F. Szenher, and D. Elad, "A millimeter-wave SiGe power amplifier with highly selective image reject filter," in *Microwaves, Communications, Antennas and Electronics Systems (COMCAS), 2011 IEEE International Conference on*, Nov 2011, pp. 1–5.

[4] S. Nicolson, K. H. K. Yau, S. Pruvost, V. Danelon, P. Chevalier, P. Garcia, A. Chantre, B. Sautreuil, and S. Voinigescu, "A Low-Voltage SiGe BiCMOS 77-GHz Automotive Radar Chipset," *Microwave Theory and Techniques, IEEE Transactions on*, vol. 56, no. 5, pp. 1092–1104, May 2008.

[5] V.-H. Do, V. Subramanian, W. Keusgen, and G. Boeck, "Design and Optimization of a High Efficiency 60 GHz SiGe-HBT Power Amplifier," in *Radio-Frequency Integration Technology, 2007. RFIT 007. IEEE International Workshop on*, Dec 2007, pp. 150–153.

[6] V. Subramanian, A. Hamidian, W. Keusgen, V.-H. Do, and G. Boeck, "Layout design considerations for 60 GHZ SiGe power amplifiers," in *Microwaves, Radar and Wireless Communications, 2008. MIKON 2008. 17th International Conference on*, May 2008, pp. 1–4.

[7] R. Yishay, R. Carmon, O. Katz, B. Sheinman, and D. Elad, "A 20dBm E-band power amplifier in SiGe BiCMOS technology," in *Microwave Conference (EuMC), 2012 42nd European*, Oct 2012, pp. 1079–1082.

[8] M. Chang and G. Rebeiz, "A wideband high-efficiency 79-97 GHz SiGe linear power amplifier with >90 mW output," in *Bipolar/BiCMOS Circuits and Technology Meeting, 2008. BCTM 2008. IEEE*, Oct 2008, pp. 69–72.

[9] N. Demirel, E. Kerherve, D. Pache, and R. Plana, "Design techniques and considerations for mmwave SiGe power amplifiers," in *Microwave and Optoelectronics Conference (IMOC), 2009 SBMO/IEEE MTT-S International*, Nov 2009, pp. 37–41.

[10] Y. Zhao and J. Long, "A Wideband, Dual-Path, Millimeter-Wave Power Amplifier With 20 dBm Output Power and PAE Above 15% in 130 nm SiGe-BiCMOS," *Solid-State Circuits, IEEE Journal of*, vol. 47, no. 9, pp. 1981–1997, Sept 2012.

[11] D. Zhao and P. Reynaert, "14.1 a 0.9v 20.9dbm 22.3with broadband parallel-series power combiner in 40nm cmos," in *Solid-State Circuits Conference Digest of Technical Papers (ISSCC), 2014 IEEE International*, Feb 2014, pp. 248–249.

A Linear Model of Efficiency for Switched-Capacitor RF Power-Amplifiers

Antonio Passamani[1,2], Davide Ponton[1], Gerhard Knoblinger[1], and Andrea Bevilacqua *Member, IEEE*[2]

[1]Intel Mobile Communications, Villach, Austria

[2]Department of Information Engineering, University of Padova, Italy

Abstract—In this paper, a linear model for intrinsic power and efficiency in Radio-Frequency Switched-Capacitor Power Amplifiers (SCPA) is presented. Given a target output power and frequency of operation this model enables sizing the output stage inverter for maximum efficiency by means of back-of-the-envelope equations. The model is validated by *SpectreRF* simulations for a low-power CMOS 28nm technology for different frequencies of interest.

Index Terms—RF-DAC, DPA, PA, C-DAC, power, efficiency

I. INTRODUCTION

DESIGN solutions for personal mobile communications enable small battery operated devices with robust, powerful, efficient and cost-effective transmitters. Traditional transmitter (TX) lineups (see Fig. 1a) are analog intensive [1], and do not benefit from CMOS technology scaling. The introduction of high dynamic range digital-to-analog converter (DAC) running at RF frequency, a.k.a. RF-DAC, enables a digital intensive solution [1]. The entire TX chain is implemented by digital-like circuit blocks and strongly benefits from CMOS technology scaling, except for the off-chip analog Power Amplifier (PA) as shown in Fig.1b [2]. The next step towards the software defined radios forcast by Mitola [3] is to incorporate the PA in the mixed-signal output stage of the DAC [1]. A digital TX with no external PA has already been demostrated in [4] and in [5] as sketched in Fig.1c, where a SCPA has been adopted.

This paper presents a linear model for the power and efficiency of the SCPA at full scale. This model can be used to accurately size the output stage of this new class of circuits by means of simple equations.

II. FULL SCALE MODEL

A SCPA is a Capacitive-DAC (C-DAC) capable of delivering high output power [5]. The key building blocks of the SCPA are the scalable switched capacitor networks connected to the load via the output matching network (Fig. 2), driven by the local oscillator (LO) signal at RF frequency. The PA can be arranged as an array of identical cells. Depending on the digital input code c, the array is dynamically split between n *on* cells, which switch their output stage from 0 to V_{dd} (Fig. 4), and $N-n$ *off* cells which tie their output to ground. Being

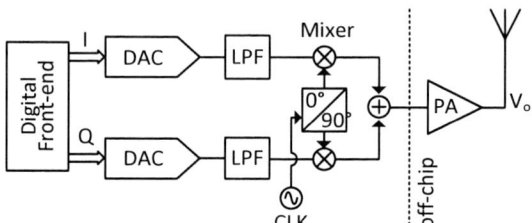

(a) Analog intensive TX

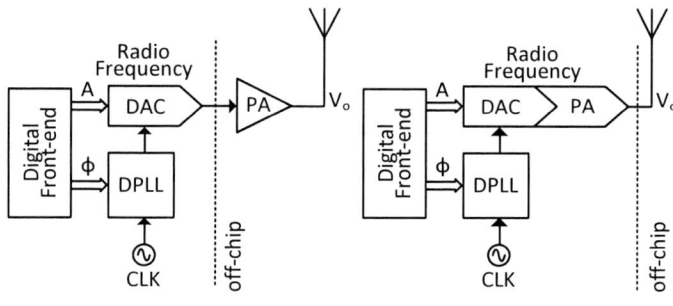

(b) Digital intensive polar TX   (c) Digital polar TX

Figure 1. TX signal chain evolution [1] to [3].

N the total number of cells in the array, the output voltage at the fundamental frequency of the LO signal is proportional to the number of cells which are in the switching *on* state [5],

$$V_o \propto \frac{n}{N} V_{dd}. \tag{1}$$

As a consequence, the power on the load resistance increases with the C-DAC output voltage

$$P_o \propto \frac{V_o^2}{R_l} \propto n^2. \tag{2}$$

Since all the cells in the array are equal, when all the cells are turned *on*, the entire array resembles a single unitary cell whose multiplicity is N, as sketched in Fig. 3.

The output stage of the equivalent full scale cell connects for half the period the output to GND through the $nMOS$ and for half the period the output to V_{dd} through the $pMOS$. The $pMOS$ devices feature approximately twice the W/L ratio of the $nMOS$ devices in order to achieve the same ON resistance $R_{on\,n,p}$ for both high and low input signals

978-1-4799-4993-9/14 $31.00 © 2014 IEEE   146

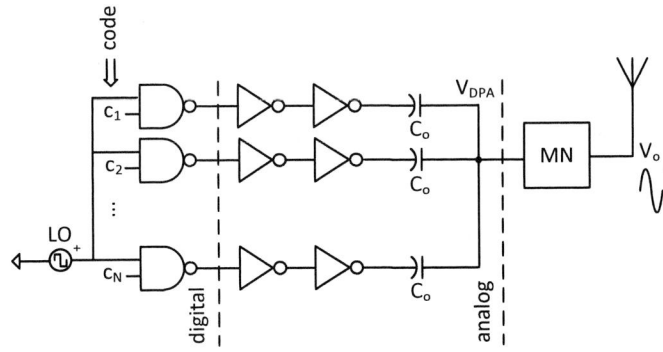

Figure 2. SCPA array circuit

$$R_{on} = R_{on\,p} = R_{on\,n}. \qquad (3)$$

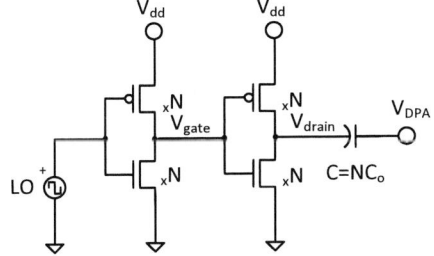

Figure 3. SCPA equivalent full scale circuit

During the transition phase, the ON resistance peaks because during that time $MOSFETs$ exit linear operation region. Assuming that the rise and fall time of the input signal is small compared to the input signal period, as in Fig. 4, then the impact of the increased output resistance during the transitions is negligible and R_{on} can be approximated as a constant.

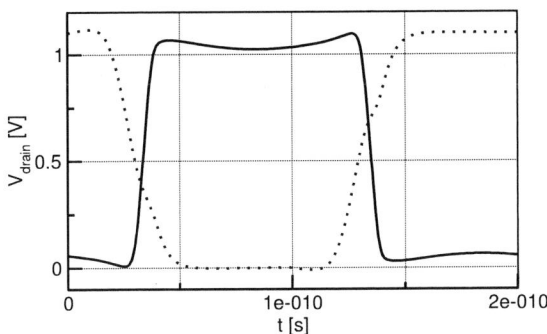

Figure 4. Example of output $MOSFETs$ drain voltage at LO operating frequency of $5GHz$. $MOSFETs$ always work in linear region except while switching. Gate voltage in dotted line.

The switching output stage is thus modeled as an input voltage commutating at the LO frequency driving the capacitance C through a constant resistance R_{on} (see Fig. 5).

C_p accounts for all the dynamic losses caused by the capacitive parasitics of the $CMOS$ devices as well as the

parasitics of the load capacitance C. $I_{sc\,avg}$ is the average short circuit current during the LO period.

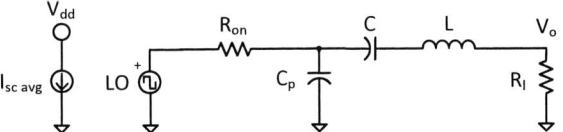

Figure 5. Linear model of the SCPA at full scale.

Since the scope of this work is to model and evaluate the efficiency of the active part of the SCPA, an ideal lossless matching network is assumed. For the sake of simplicity, the matching network is assumed to be made of a single inductor L, while R_l in Fig. 5 is the load due to the antenna. Since the input signal of the linear model is a square wave, the circuit has to be analyzed at all the harmonics. The ideal lossless inductance L series resonates the SCPA capacitance C at the fundamental frequency of the LO. Therefore, at this frequency the LC can be modeled with a short, as in Fig. 6a. The fundamental tone of the input voltage is directly transferred to the load through R_{on}.

On the other hand, the LC series appears as a high impedance at all the other harmonics. As a consequence, only the fundamental flows to the load (Fig. 6b) while $I_{sc\,avg}$ and the current through the capacitor C_p make up for the additional losses.

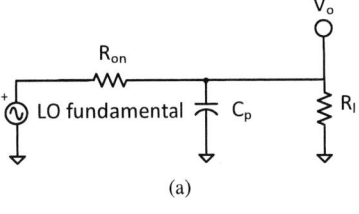

(a)

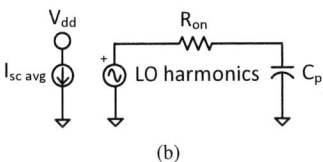

(b)

Figure 6. Single tone equivalent circuit model for the fundamental tone (Fig.6a) and the harmonics (Fig.6b) of the LO signal

By means of this simplified linear model, the various power contributions can be analyzed. The parasitic capacitor C_p is charged and discharged at the LO frequency. The power loss on C_p can be coarsely estimated as

$$P_{C_P} = f C_p V_{dd}^2 \qquad (4)$$

while the power loss due to $I_{sc\,avg}$ is

$$P_{I_{sc}} = I_{sc\,avg} V_{dd}. \qquad (5)$$

Assuming $1/(R_l C_p)$ is much larger than the fundamental frequency, the output power is given by:

$$P_{R_l} = \frac{1}{R_l} \left(\frac{V_f}{\sqrt{2}} \frac{R_l}{R_{on} + R_l} \right)^2 \qquad (6)$$

where V_f is the fundamental tone of the input square wave swinging from 0 to V_{dd}

$$V_f = \frac{2}{\pi} V_{dd}. \qquad (7)$$

The output signal current also dissipates power on R_{on}, generating an additional power loss:

$$P_{R_{on}} = \frac{1}{R_{on}} \left(\frac{V_f}{\sqrt{2}} \frac{R_{on}}{R_{on} + R_l} \right)^2. \qquad (8)$$

III. Efficiency Analysis

Load-pull analysis is used to find the optimum load impendance to achieve the highest efficiency. During the load-pull analysis L and R_l are swept until the best operating point is found (see Fig. 7).

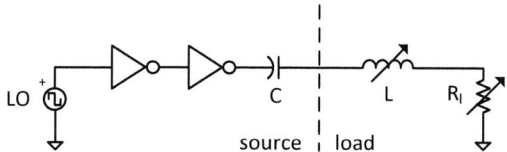

Figure 7. Load pull analysis finds the optimal load for a fixed source

The efficiency is defined as

$$\eta = \frac{P_{R_l}}{P_t} \qquad (9)$$

where P_t is total power drawn from the supply.

To make the following analysis independent from the absolute value of R_{on} we define

$$\alpha \triangleq \frac{R_l}{R_{on}}. \qquad (10)$$

Equation (9) can be rewritten as:

$$\eta = \frac{P_{R_l}}{P_{switch} + P_{R_l} + P_{R_{on}}}$$
$$= \frac{\frac{1}{\alpha R_{on}} \frac{V_f^2}{2} \left(\frac{\alpha}{1+\alpha} \right)^2}{P_{switch} + \frac{1}{R_{on}} \frac{V_f^2}{2} (1+\alpha) \left(\frac{1}{1+\alpha} \right)^2} \qquad (11)$$

where

$$P_{switch} \triangleq P_{C_p} + P_{I_{sc}} \qquad (12)$$

is the total power dissipated by the output stage at open load condition (i.e. $R_l \to \infty$).

Defining

$$\beta \triangleq \frac{2 R_{on} P_{switch}}{V_f^2} \qquad (13)$$

(11) can be rewritten in a compact fashion as:

$$\eta = \frac{\alpha}{\beta(1+\alpha)^2 + (1+\alpha)} \qquad (14)$$

where β, according to (13), is defined once P_{switch}, R_{on} and V_f are known. P_{switch} is measured as the power taken from the supply for switching the output inverter at open load condition.

It is worth to mention that β is a technology parameter independent from the CMOS inverter size. We show this by rewriting the average short circuit current as

$$I_{sc\,avg} = 2 N f T_{sc} \frac{V_{dd}}{R_{sc\,0}}. \qquad (15)$$

$R_{sc\,0}$ is defined as the average short circuit resistance offered during commutation times by one inverter. Equation (15) makes the dependance of I_{sc} from the operation frequency f explicit. For longer switching time T_{sc}, and for an increased frequency of operation f, $I_{sc\,avg}$ increases linearly.

Moreover, the capacitance C_p is proportional to the number of inverters in the output stage

$$C_p = N C_{p\,0} \qquad (16)$$

where $C_{p\,0}$ is the parasitic capacitance of one output stage and thus

$$P_{switch} = f C_p V_{dd}^2 + V_{dd} \left(N f 2 T_{sc} \frac{V_{dd}}{R_{sc\,0}} \right)$$
$$= V_{dd}^2 \left(N f C_{p\,o} + N f 2 T_{sc} \frac{1}{R_{sc\,0}} \right). \qquad (17)$$

Using (7) and (17), (13) can be rewritten as follows

$$\beta = \frac{2 R_{on\,0} f \left(C_{p\,o} + 2 T_{sc} \frac{1}{R_{sc\,0}} \right)}{\left(\frac{2}{\pi} \right)^2}, \qquad (18)$$

where $R_{on\,0} = N R_{on}$ is the on resistance of one output inverter. Equation (18) depends only on technology parameters $R_{on\,0}$, $C_{p\,0}$ and $R_{sc\,0}$, while T_{sc} and the operation frequency f are design parameters. It is worth to note that when the short circuit current can be neglected,

$$\tau_{on} = R_{on\,0} C_{p\,0} \qquad (19)$$

determines β, and thus maximum efficiency, for a given frequency of operation.

The point of maximum efficiency can be found directly from (14) by differentiation:

$$\left. \frac{\delta \eta}{\delta \alpha} \right|_{\alpha_{max}} = 0 \;\Rightarrow\; \alpha_{max} = \sqrt{\frac{\beta + 1}{\beta}}. \qquad (20)$$

Substituting (20) in (14) and approximating $\beta + 1 \approx 1$, the maximum efficiency, η_{max}, can be approximated as:

$$\eta_{max} \cong \frac{1}{1 + 2\sqrt{\beta}}. \qquad (21)$$

IV. Validation

The linear model presented in Sec. II is validated against *SpectreRF* simulations of a 28nm CMOS technology. Results refer at nominal corner and temperature $T = 27°$, $f = 2.4GHz$ and $f = 5GHz$. These are frequencies of interest for many wireless applications, such as WiFi and Bluetooth. The simulated efficiency (solid lines) is compared to the linear model results (dotted lines) in Fig. 8, showing good agreement.

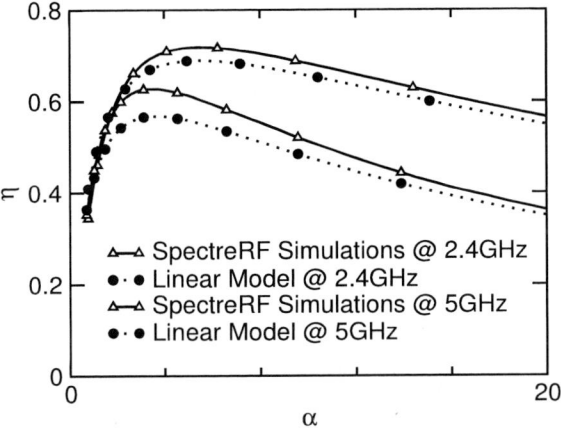

Figure 8. Comparison between linear model and *SpectreRF* simulations . Linear model results are plotted with coefficient $\beta = 0.035$ as calculated from $P_{switch} = 67.73mW$, $R_p = R_n = 0.13\Omega$, $f = 2.4GHz$ and $\beta = 0.082$ as calculated from $P_{switch} = 133mW$, $R_p = R_n = 0.155\Omega$, $f = 5GHz$. $V_{dd} = 1.1V$.

Table I compares α_{max} and η_{max} calculated with (20), (14) and (21) to the results of *SpectreRF* simulations. The linear model, i.e. (14) and (20), gives a slightly underestimated η_{max} (in this example: 3% less at $2.4GHz$), due to overestimation of P_{C_P} in (4). α_{max} is predicted accurately. The approximated model, i.e. (21), is also in good agreement with the simulation results.

Table I

COMPARISON OF MAXIMUM EFFICIENCY IN SIMPLIFIED MODEL, LINEAR MODEL AND *SpectreRF* SIMULATIONS

	Frequency	Equation (21)	Equation (14)	Simulations
α_{max}	2.4GHz	5.43	5.43	5.43
η_{max}		73%	69%	72%
α_{max}	5GHz	3.6	3.6	3.6
η_{max}		64%	57%	63%

V. Discussion

The presented model can be used for the optimimal sizing of the output stage of the SCPA.

For a given R_l and technology parameters, the value of α_{max} sets the multiplicity N of the output stage of the SCPA. If a higher level of output power is required, both R_l and the multiplicity of the output stage must be scaled by the same factor M (the multiplicity becoming $M \cdot N$), resulting in

$$P_{R_l}* = \frac{1}{\frac{R_l}{M}} \left(V_f \frac{\frac{R_l}{M}}{\frac{R_{on}}{M} + \frac{R_l}{M}} \right)^2 = M P_{R_l} \quad (22)$$

where, in the maximum efficiency conditions, we have

$$P_{R_l} \approx \frac{1}{R_l} \left(\frac{V_f}{\sqrt{2}} \right)^2 \frac{1}{(1 + \sqrt{\beta})^2} \approx \frac{1}{R_l} \left(\frac{V_f}{\sqrt{2}} \right)^2 \eta_{max}. \quad (23)$$

The scaling of R_l can be achieved by means of an impedance transformation network. Typically, the higher the transformation ratio, the larger the losses introduced by the matching network [6]. This can be a bottleneck on high power circuit implementations with low voltage power supply.

Equation (21) points out the important fact that the smaller the β, the higher the efficiency. Combining (21) with β as defined in (18) it turns out that better efficiency is obtained when the short-circuit current, i.e. (15), is small and τ_{on} as in (19) is small. As a consequence, scaled technologies featuring lower τ_{on} are expected to yield higher efficiencies. There is not any direct dependance of the intrinsic efficiency on the supply voltage. The SCPA efficiency decreases when increasing the operating frequency due to the direct proportionality between P_{switch} and f as shown in (17).

VI. Conclusions

A linear model for the SCPA at full scale power has been derived and validated by means of *SpectreRF* simulations. The model accurately predicts the point of peak efficiency and enables the sizing of the output stage of the SCPA to achieve the maximum efficiency at any target output power by means of back-of-the-envelope equations.

The work focused on the efficiency at full scale power. In a future work, the proposed linear model will be used to describe the behaviour of the efficiency at any input digital code of the SCPA as well.

References

[1] J. Walling, S.-M. Yoo, and D. Allstot, "Digital power amplifier: A new way to exploit the switched-capacitor circuit," *Communications Magazine, IEEE*, vol. 50, no. 4, pp. 145–151, April 2012.

[2] Z. Boos, A. Menkhoff, F. Kuttner, M. Schimper, J. Moreira, H. Geltinger, T. Gossmann, P. Pfann, A. Belitzer, and T. Bauernfeind, "A fully digital multimode polar transmitter employing 17b rf dac in 3g mode," in *Solid-State Circuits Conference Digest of Technical Papers (ISSCC), 2011 IEEE International*, Feb 2011, pp. 376–378.

[3] J. Mitola, "The software radio architecture," *Communications Magazine, IEEE*, vol. 33, no. 5, pp. 26–38, May 1995.

[4] H. Wang, C.-H. Peng, C. Lu, Y. Chang, R. Huang, A. Chang, G. Shih, R. Hsu, P. Liang, S. Son, A. Niknejad, G. Chien, C. Tsai, and H. Hwang, "A highly-efficient multi-band multi-mode digital quadrature transmitter with 2d pre-distortion," in *Circuits and Systems (ISCAS), 2013 IEEE International Symposium on*, May 2013, pp. 501–504.

[5] S.-M. Yoo, J. Walling, E.-C. Woo, B. Jann, and D. Allstot, "A switched-capacitor rf power amplifier," *Solid-State Circuits, IEEE Journal of*, vol. 46, no. 12, pp. 2977–2987, Dec 2011.

[6] T. H. Lee, *The Design of CMOS Radio-Frequency Integrated Circuits*. Cambridge University Press, 2004.

Analysis and Design of a High Power, High Gain SiGe BiCMOS Output Stage for Use in a Millimeter-Wave Power Amplifier

Ramses Pierco
Ghent University,
INTEC/IMEC
Ghent, Belgium
Email: ramses.pierco@intec.ugent.be

Timothy De Keulenaer
Ghent University,
INTEC/IMEC
Ghent, Belgium

Guy Torfs
Ghent University,
INTEC/IMEC
Ghent, Belgium

Johan Bauwelinck
Ghent University,
INTEC/IMEC
Ghent, Belgium

Abstract—**In this paper a high gain, high power output stage designed in a 250nm SiGe BiCMOS technology is presented. The used topology together with a discussion on the stability of the output stage is explained in detail. In order to increase the gain of the output stage and thus increases the attainable power added efficiency (PAE), positive feedback is used. Furthermore a formula predicting the input impedance of a common base transistor at high frequencies is deducted which explains and predicts the magnitude of the feedback mechanism. The output stage achieves a peak gain of 14.4dB at 31GHz with a maximum output power of 22dBm.**

I. INTRODUCTION

The continuous increase in cutoff frequencies (f_T, f_{max}) of SiGe BiCMOS technologies makes the latter a sensible option for the realization of millimeter-wave applications. However, with the increase of speed of these technologies the avalanche breakdown voltages (BV_{CEO} and BV_{CBO}) are reduced to values which make it hard to create a millimeter-wave power amplifier (PA) with high output power. The relative small output power of a single PA cell, in comparison to GaAs/GaN/InP PAs, can be countered by power combining several smaller cells both on-chip as off-chip.

Next to creating a high output power, it is also required to have a high PAE since this will lead to lower DC power consumption and thus lower chip temperature. Since the input power of an amplifier is factored into the PAE, this can only be achieved by having a high gain output stage. Moreover, higher gain also allows to go to a more efficient amplifier class like Class-B. This type of amplifier requires an increase in drive level of 6dB in comparison to a Class-A, which is only acceptable with a sufficiently high gain. Yielding the main reason why millimeter-wave power amplifiers are typically biased in Class-AB [1], [2].

The topology of the presented output stage is discussed in Section II of this paper together with how the BV_{CEO}-limit can be overcome. Section III, in turn, explains how the gain of the output stage is increased by applying positive feedback and also provides a formule which can give an estimate of the magnitude of the feedback response and the frequencies at which it occurs. When going to higher gain, special care has

to be taken to assure stability. This is discussed in Section IV. The performance of the output stage is summarized in Section V together with a conclusion.

II. CASCODE TOPOLOGY AND TRANSISTOR LIMITATIONS

The output stage topology, chosen in this work, is a differential cascode topology as shown in Fig.1. Although this type of topology needs more headroom, and thus larger DC voltage, than a simple common emitter stage, it can achieve a higher output power and higher PAE. The reason for this is the biasing of the common base cascode transistor which is biased with a more or less constant emitter current. In [3] and [4] it was shown that a common base output transistor can safely be biased at a collector emitter voltage which is significantly higher than BV_{CEO}. Also, since the base of the cascode transistor has a low impedance to ground, the peak value of the time-varying collector emitter voltage can approach the avalanche breakdown limit. The reason for this is that the holes flowing back to the base created by impact-ionization are shunted to ground. As a result the maximum usable peak collector voltage approaches BV_{CBO} [1].

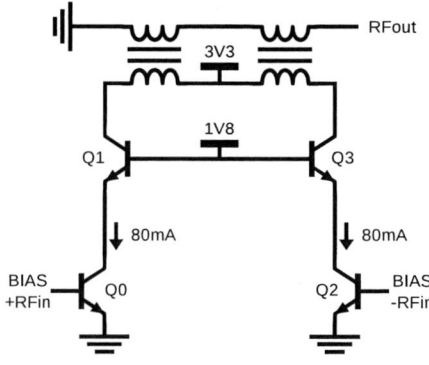

Fig. 1. Differential cascode topology with output balun used as output stage.

The 250nm SiGe BiCMOS technology used in the design of this output stage has a minimum BV_{CEO} of 2.1V and the minimum BV_{CBO} amounts to 6V. The common base output transistor is biased at a collector emitter voltage (V_{CE}) of 2.4V which is above the minimum BV_{CEO} but still safe due to the constant emitter current. Using a V_{CE} of 2.4V leads

978-1-4799-4993-9/14 $31.00 © 2014 IEEE

to a linear voltage swing of 3.8V$_{pp}$ (with a saturation voltage of 0.5V and V_{BE} = 0.9V), while using a common emitter stage biased at a V_{CE} of 2V (considering a margin of 0.1V to be safely below BV_{CEO}) would lead to a linear swing of 3V$_{pp}$. In other words, the use of a cascode topology leads to a 2.05dB increase in linear output power while the DC power consumption goes up with a factor of 1.57 (1.96dB). Although the efficiency η (P$_{out}$/P$_{DC}$) of a cascode is only marginally larger than the efficiency of a common emitter stage, the PAE ((P$_{out}$ - P$_{in}$)/P$_{DC}$) is significantly better due to the larger gain of a cascode and thus smaller required input power. This higher gain stems from the suppression of the miller capacitance at the input due to the cascode [5], however, in Section III it will be shown that this isn't entirely true. In conclusion, using a cascode helps to overcome the BV_{CEO} limitation and thus achieve a higher output power. Additionally it leads to a higher gain and PAE than a simple common emitter stage.

III. GAIN IMPROVEMENT USING POSITIVE FEEDBACK

A. Cascode and Miller capacitance

The reason why a cascode is used is the suppression of the Miller capacitance at the input of the output stage and the resulting higher gain. This suppression is caused by the unity voltage gain of the common emitter transistors in a cascode configuration which leads to a zero AC voltage between the collector and base of transistors Q0 and Q2 in Fig.1 which removes feedback over the base collector capacitance and hence kills the Miller effect. Important to note here is that the unity voltage gain comes from the input impedance of a common base transistor which is approximately $1/g_m$, multiplied with the transconductance g_m of the common emitter transistor. However, it is well known that at high frequencies the input impedance of a common base transistor becomes inductive and thus will lead to a larger than one voltage gain and a return of the Miller capacitance. Hereafter it will be shown that the input impedance not only becomes inductive but also increases with the voltage gain of the common base transistors.

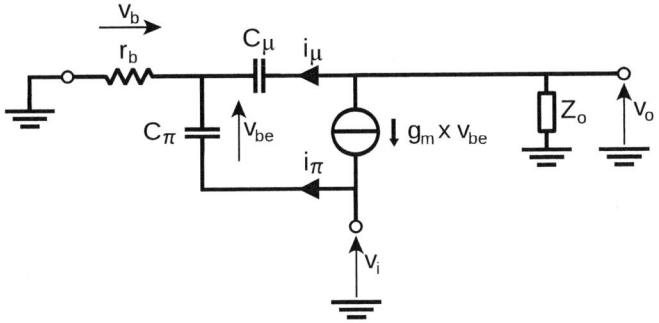

Fig. 2. Small signal representation of the common base transistor in a cascode configuration with v$_i$ the input voltage, v$_o$ the output voltage and Z$_o$ the load impedance of the cascode.

In order to calculate the input impedance of common base transistors Q1 and Q3 in Fig.1, the small signal model shown in Fig.2 is used. This is a simplified version of the complete bipolar small signal equivalent circuit [6]. The impedance Z_o is the resonant load impedance of the cascode output stage with the collector to substrate capacitance C_{CS} included.

The parasitic resistances r$_\mu$ and r$_o$ can be neglected at high frequencies in comparison to C$_\mu$ and Z$_o$. Resistances r$_{pi}$, r$_{ex}$ and r$_c$ are left out for initial analysis but are reconsidered in the final formula describing the input impedance. The input impedance of the small signal equivalent of Fig.2 equals:

$$Z_{in} = \frac{v_i}{-g_m \cdot v_{be} + i_\pi}$$

With $v_{be} = -i_\pi/sC_\pi$ this leads to:

$$Z_{in} = \frac{v_i}{\frac{g_m \cdot i_\pi}{sC_\pi} + i_\pi} = \frac{v_i}{i_\pi \cdot \left(1 + \frac{g_m}{sC_\pi}\right)} \tag{1}$$

All that is left to do, is to solve i_π in function of v_i:

$$i_\pi = (v_i - v_b) \cdot sC_\pi \tag{2}$$

The term v_b in equation (2) can be written as:

$$
\begin{aligned}
v_b &= r_b \cdot (i_\pi + i_\mu) \\
&= r_b \cdot \left(i_\pi + \frac{v_o - v_b}{1/sC_\mu}\right) \\
&\cdots \\
v_b &= \frac{v_o \cdot sC_\mu \cdot r_b + i_\pi \cdot r_b}{1 + sC_\mu \cdot r_b}
\end{aligned} \tag{3}
$$

When equation (2) and (3) are combined the following expression is found for i_π:

$$
\begin{aligned}
i_\pi &= \left(v_i - \frac{v_o \cdot sC_\mu \cdot r_b}{1 + sC_\mu \cdot r_b} - \frac{i_\pi \cdot r_b}{1 + sC_\mu \cdot r_b}\right) \cdot sC_\pi \\
&\cdots \\
i_\pi &= \frac{v_i \cdot sC_\pi \cdot \left(1 - \frac{v_o}{v_i} \cdot \frac{sC_\mu \cdot r_b}{1 + sC_\mu \cdot r_b}\right)}{\left(1 + \frac{sC_\pi \cdot r_b}{1 + sC_\mu \cdot r_b}\right)} \\
i_\pi &= \frac{v_i \cdot sC_\pi \cdot \left(1 - A_V \cdot \frac{sC_\mu \cdot r_b}{1 + sC_\mu \cdot r_b}\right)}{\left(1 + \frac{sC_\pi \cdot r_b}{1 + sC_\mu \cdot r_b}\right)}
\end{aligned} \tag{4}
$$

In equation (4) the factor A_V is the voltage gain of the common base transistor. To attain a formula for the input impedance of the common base transistor equation (4) is put into equation (1) leading to the following result:

$$Z_{in} = \frac{1 + s \cdot (C_\pi + C_\mu) \cdot r_b}{(g_m + sC_\pi) \cdot [1 + sC_\mu \cdot r_b \cdot (1 - A_V)]} \tag{5}$$

The formula for the input impedance shown in equation (5) is quite remarkable since it shows that the input impedance depends on the voltage gain A_V of the common base transistor and thus also on the impedance Z_o at the output of the cascode. Since this output impedance is a resonant load made up out of the balun inductance and the C_{CS} capacitance of the common base transistors, the input impedance of the common base will increase at exactly the frequency of interest for the output stage. This means that the Miller effect, and subsequentially the Miller capacitance, will be largest at the wanted frequency thus dampening the gain. Furthermore equation (5) shows that any parasitic resistance or inductance in series with r_b will result in a large increase in input impedance and Miller capacitance.

To get a more accurate description of the input impedance the resistances r_{ex} and r_c need to be included. With some lengthy calculus it can then be shown that the input impedance is given by:

$$Z_{in} = \frac{1 + s\left(C_\pi + C_\mu\right) \cdot r_b + sC_\mu \cdot r_c \left[1 + r_b \cdot \left(g_m + sC_\pi\right)\right]}{\left(g_m + sC_\pi\right) \cdot \left[1 + sC_\mu \cdot r_c + sC_\mu \cdot r_b \cdot \left(1 - A_V\right)\right]}$$
$$+ r_{ex} \cdot \left[1 + sC_\mu \cdot \left(r_b + r_c\right)\right] \qquad (6)$$

In DC, equation (6) results in the familiar result $Z_{in,DC} = 1/g_m + r_{ex}$. To prove that the previous analysis is accurate, the input impedance calculated by means of formula (6) is compared to the simulated input impedance of a common base in Fig.3 and Fig.4 showing respectively the real and imaginary part of the input impedance.

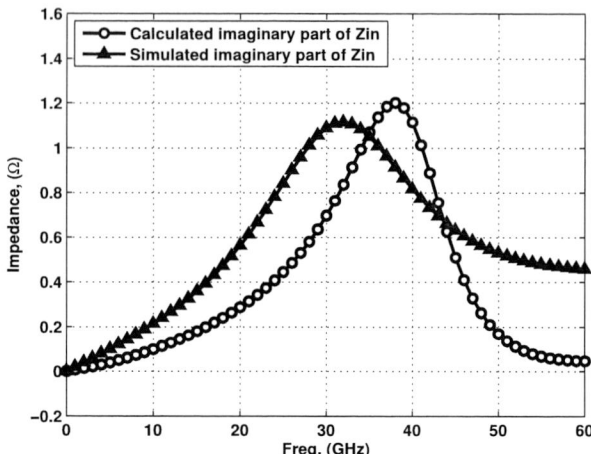

Fig. 4. Calculated imaginary part of the input impedance of a common base transistor versus the simulated values. The resonant frequency of Z_o is 31GHz in simulation.

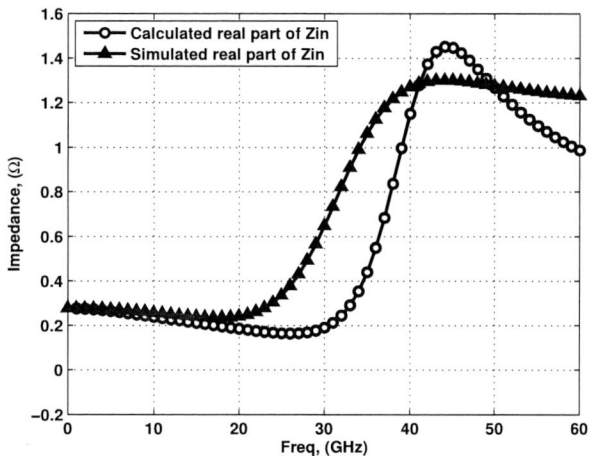

Fig. 3. Calculated real part of the input impedance of a common base transistor versus the simulated values. The resonant frequency of Z_o is 31GHz in simulation.

Figures 3 and 4 show that formula (6) is able to predict the behavior of the input impedance of a common base transistor at high frequencies. It also shows that the input impedance of a common base transistor increases at the resonant frequency of the output load and thus will have a larger input capacitance for the cascode at resonance due to the Miller effect. To counter the resulting drop in gain at resonance, positive feedback can be used as will be shown hereafter.

B. Feedback mechanism and gain improvement

The output stage described in this paper uses cross-coupled feedback capacitors C_F across the collector base junction of the common emitter transistors as shown in Fig.5.

The feedback in Fig.5 works by compensating the input capacitance of the cascode with the negative capacitance provided by the cross-coupled capacitors which exploit the Miller effect. The total capacitance at the input $C_{in,tot}$ is then given by the following formula [5]:

$$C_{in,tot} = C_{in} + \left(1 - |A_{V,CE}|\right) \cdot C_F \qquad (7)$$

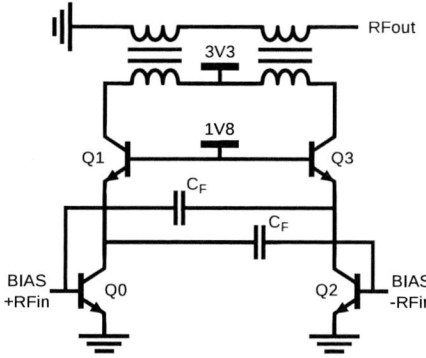

Fig. 5. Differential cascode topology with cross-coupled feedback capacitors across the common emitter transistors.

With C_{in} the input capacitance without feedback and $A_{V,CE}$ the voltage gain of the common emitter transistors. Equation (7) shows that the total input capacitance of the cascode can be decreased by adding the cross-coupled feedback capacitors.

By adding positive feedback an increase of 2dB in gain can be noticed in Fig.6. By choosing a larger feedback capacitor an even bigger increase in gain can be achieved, however care has to be taken not to cause instability (negative $C_{in,tot}$ in equation (7)). Although an increase with a mere 2dB might not seem much, it quickly becomes significant when the output stage is biased more towards Class-B operation and the gain starts to drop. Also to reduce the linearity and gain demands of a driving stage, a difference of 2dB certainly is noteworthy.

IV. STABILITY CONSIDERATIONS

An obvious source of instability comes from the cross-coupled feedback capacitors which should be limited in size to ensure stability. A much less evident source of instability stems from the parasitic capacitor between the collector and emitter of transistors Q1 and Q3 in Fig.5. This capacitor isn't present in the transistor itself but comes from layout capacitance between the collector and emitter interconnects. Especially when multiple transistors are combined in parallel

978-1-4799-4993-9/14 $31.00 © 2014 IEEE

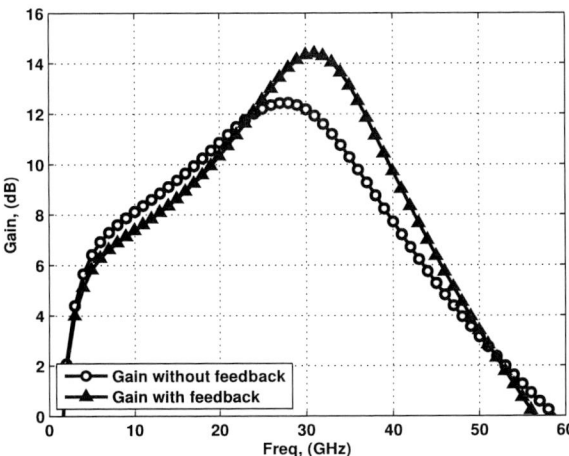

Fig. 6. Gain of the differential cascode configuration with and without the cross-coupled feedback capacitors.

for high power with small spacing to ensure bias and thermal stability, this capacitor becomes large and can cause instability. The basic mechanism behind this instability is the positive feedback between collector and emitter due to this capacitor as discussed in detail in [7].

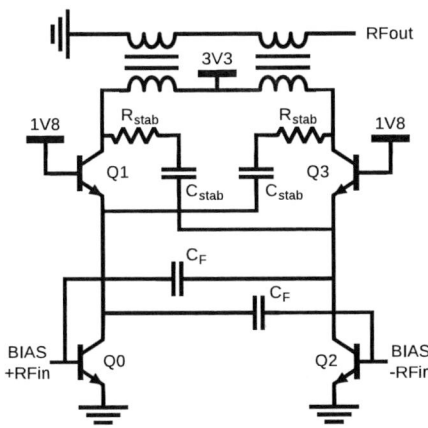

Fig. 7. Differential cascode topology with cross-coupled series resistor R_{stab} and capacitor C_{stab} across the common base transistors.

This is solved by placing a cross-coupled capacitor between the collector and emitter of the common base transistors [7]. However, this leads to a large shift in resonant frequency of the output impedance Z_o. By using a series combination of a resistor R_{stab} and capacitor C_{stab}, as shown in Fig.7, stable operation is obtained with a smaller shift in resonant frequency. The addition of the cross-coupled series combination of resistor and capacitor at the common base transistors will lead to a decrease in common-mode stability. This however is countered by the addition of the cross-coupled capacitors at the common emitter stage which ensures stable common-mode behavior.

V. CONCLUSION AND FURTHER WORK

In this paper, an analysis of an output stage design for a millimeter-wave power amplifier has been provided, high-

lighting the use of a cascode to achieve high gain and high output power. Furthermore a formula is derived for the input impedance of a common base stage at high frequencies which explains the drop in gain at the resonant frequency of the output load. It is also shown that this effect can be countered by applying positive feedback around the common emitter transistors by means of cross-coupled capacitors. The resulting gain for the output stage is more than 14dB at 31GHz in simulation with a maximum output power of 22dBm leading to a PAE of 26%.

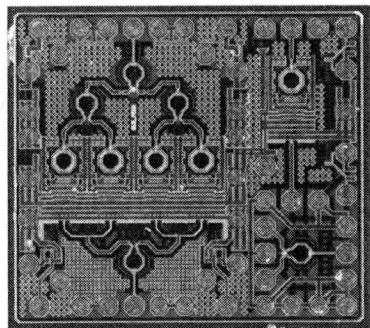

Fig. 8. Die photomicrograph of the complete amplifier together with teststructures (rightmost part of the die). The complete die measures 1670μm by 1490μm.

The output stage has been designed in a 250nm SiGe BiC-MOS technology and is used in a four-way power combining power amplifier at a frequency of about 30GHz which has been taped-out and produced. This amplifier will be mounted directly onto a heat sink and will be probed to verify the performance.

ACKNOWLEDGMENTS

The authors would like to thank Jeroen Missine and Bjorn Vandecasteele, both at the Centre for Microsystems Technology (TFCG Microsystems Lab), for their useful advice and help with the current and future assembly of the chips.

REFERENCES

[1] T. S. D. Cheung and J. R. Long, "A 21-26-GHz SiGe bipolar power amplifier MMIC," *IEEE J. Solid-State Circuits*, vol. 40, pp. 2583–2597, Dec. 2005.

[2] S. C. Cripps, *RF Power Amplifiers for Wireless Communications*. Norwood, MA: Artech House, 1999.

[3] H. Li, H.-M. Rein, T. Suttorp, and J. Böck, "Fully integrated SiGe VCOs with powerful output buffer for 77-GHz automotive radar systems and applications around 100 GHz," *IEEE J. Solid-State Circuits*, vol. 39, pp. 1650–1658, Oct. 2004.

[4] M. Rickelt and H.-M. Rein, "Influence of impact-ionization induced instabilities on the maximum usable output voltage of Si bipolar transistors," *IEEE Trans. Electron Devices*, vol. 48, pp. 774–783, Apr. 2001.

[5] E. Säckinger, *Broadband Circuits for Optical Fiber Communications*. New York, NY: John Wiley & Sons, 2005.

[6] P. R. Gray, P. J. Hurst, S. H. Lewis, and R. G. Meyer, *Analysis and Design of Analog Integrated Circuits*. New York, NY: John Wiley & Sons, 2001.

[7] Y. Zhao and J. R. Long, "A wideband, dual-path, millimeter-wave power amplifier with 20 dBm output power and PAE above 15% in 130 nm SiGe-BiCMOS," *IEEE J. Solid-State Circuits*, vol. 47, pp. 1981–1997, Sep. 2012.

Base-Station Design for Passive UHF RFID Tags with Pulse-Width Modulated Backscattering

Kerem Kapucu and Catherine Dehollain
Institute of Electrical Engineering, RFIC Research Group
Ecole Polytechnique Fédérale de Lausanne (EPFL)
CH-1015, Lausanne, Switzerland
Email: kerem.kapucu@epfl.ch

Abstract—**In this work, a base-station design is proposed for passive UHF RFID tags that use pulse-width modulation (PWM) in backscattering communication. The theoretical analysis of the PWM backscattering is presented along with the challenges it creates in the base-station design. The self-jamming issue in the base station is analyzed with a numerical case study. The theoretical analysis of the reverse communication channel is presented. A method is proposed for wired backscattering measurements, which models the path loss and reflection from the tag eliminating the need for wireless measurements in an anechoic chamber. It is shown by measurements that the base-station is capable of measuring the pulse-width of the backscattered signal with an error less than 1% at a distance of 2 meters.**

Index Terms—**Radio frequency identification (RFID), passive RFID, remote powering, UHF, backscattering, base-station**

I. INTRODUCTION

The integration of sensors with passive UHF RFID tags extends the application of RFID technology to many new fields such as environmental monitoring, healthcare, and food quality monitoring [1]. It has been shown that phase-locked loop (PLL) based sensor interfaces can be used in RFID tags with capacitive sensors to reduce the power consumption of the tags [1].

In [1], a novel method is proposed for tag-to-reader communication, where the backscattering signal is pulse width modulated. The duty-cycle of the pulse-width modulated signal carries the information about the capacitance value of the sensing element on the tag. This type of RFID tag requires a dedicated base-station design.

The biggest issue in RFID base-station design for passive tags is the carrier leakage from transmit to receive path. In order to have a long operating range, the transmitter emits the maximum power that is allowed by the regulations. This full-power carrier creates a self-jamming signal at the center of the receiver bandwidth, which is much more powerful compared to the received back-scatter signal from the RFID tag.

In order to overcome the self-jamming problem, one approach is to separate the transmit path from the receive path by using two antennas [2]. The transmit-receive isolation depends on the coupling between the two antennas. It is obvious that in order to have a better isolation, the antennas should be placed

far away from each other. The separation should be more than two wavelengths to eliminate near-field coupling.

Another approach to overcome the self-jamming problem is to use an isolating element between the transmit and receive paths, while using a single antenna. The isolating element can be a circulator or a directional coupler. In this work, the aim is to design a base-station for passive UHF RFID tags with sensors, which use pulse-width modulation (PWM) for tag-to-reader communication. In order to have a more compact base-station, the single antenna approach is chosen. A circulator is used to isolate the transmit path from the receive path. Section II gives an overview and theoretical analysis of the PWM based backscattering communication. The self-jamming issue and the reverse communication link is explained along with the design of the base station in Section III. The implementation is explained and the measurement results are presented in Section IV and the results are discussed in Section V.

II. PWM BASED BACKSCATTERING

In [1], a PLL-based capacitive sensor interface is used. The output of the sensor interface is a pulse train whose duty cycle is proportional to the variation in the sensor capacitance value. The pulse train signal controls the backscattering switches that short the antenna terminals. This way the sensor value information is pulse-width modulated and sent to the base-station (reader) by backscattering. In contrast to the standard RFID systems where the information is digitized for tag-to-reader communication, PWM approach is analog in nature.

Figure 1 shows a pulse train, in other words, a square wave of amplitude A with arbitrary duty cycle, $d = k/T$, with the pulse width k, and period $T = 1/f$. This pulse train can be stated as a sum of sine and cosine waves using the well-known Fourier synthesis equation:

$$x(t) = a_0 + \sum_{n=1}^{\infty} a_n \cos\left(2\pi f t n\right) + \sum_{n=1}^{\infty} b_n \sin\left(2\pi f t n\right) \quad (1)$$

The dc component a_0, and the amplitudes of the cosine and

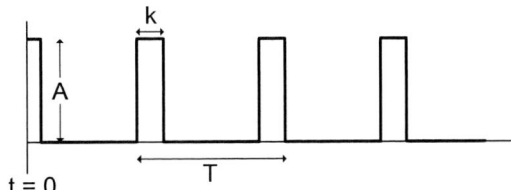

Fig. 1. The pulse-train, i.e., a square wave with an arbitrary duty-cycle $d = k/T$, period T, and amplitude A.

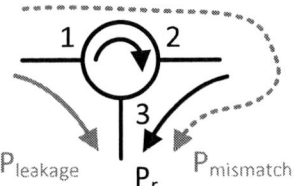

Fig. 2. The components of self-jamming signal shown around the circulator. $P_{leakage}$ is the signal power leaked through the isolation path of the circulator, $P_{mismatch}$ is the signal power that is reflected back from the base-station antenna due to mismatch, and P_r is the power of the received modulated wave backscattered from the tag.

sine waves, a_n and b_n, can be calculated as:

$$a_0 = A\,d \tag{2}$$

$$a_n = \frac{2A}{n\pi} \sin(n\pi d) \tag{3}$$

$$b_n = 0 \tag{4}$$

Consider two pulse trains with duty cycles d and $1-d$ (e.g. 20% and 80%), the amplitudes of the cosine waves for the two signals will be:

$$a_n(d) = \frac{2A}{n\pi} \sin(n\pi d) \tag{5}$$

$$a_n(1-d) = \frac{2A}{n\pi} \sin[n\pi(1-d)] \tag{6}$$

Rearranging (6) using the trigonometric identity yields:

$$a_n(1-d) = \begin{cases} \frac{2A}{n\pi} \sin(n\pi d) & \text{if } n \text{ is odd,} \\ -\frac{2A}{n\pi} \sin(n\pi d) & \text{if } n \text{ is even.} \end{cases} \tag{7}$$

$$\tag{8}$$

$$= \begin{cases} a_n(d) & \text{if } n \text{ is odd,} \\ -a_n(d) & \text{if } n \text{ is even.} \end{cases} \tag{9}$$

This shows that, in order to discriminate between square waves with complementing duty cycles, e.g. 20% and 80%, we need to have the second and even the fourth harmonics in the receiver. That condition puts forward a larger bandwidth requirement for the receiver. Note that the dc component cannot be used due to the transmit leakage.

III. BASE-STATION DESIGN

The base-station for passive RFID systems has three main functions: 1) remote powering the passive RFID tags, 2) sending the commands, i.e., reader-to-tag communication, 3) receiving the backscattered information from the RFID tags, i.e., tag-to-reader communication. In this work, the tags are "always ON," i.e., they start operating and sending back the sensor readout as long as there is enough power at the tag. Hence, the reader-to-tag communication is out of the scope of this work.

In order to have a compact base-station, the single antenna approach is chosen in this work. The transmit and the receive paths are isolated by using a circulator. The circulator is a critical component in the base-station because the amplitude of the self-jamming signal depends on the isolation performance of the circulator.

A. Self-jamming

Figure 2 shows the two components of the self-jamming. First one is the power leaked through the isolation path of the circulator, denoted as $P_{leakage}$. The second one is due to the mismatch between the circulator and the antenna. The carrier wave that is reflected back from the antenna input port passes through the circulator, whose power is denoted as $P_{mismatch}$. These two self-jamming components are added to the received signal (received modulated signal power denoted by P_r) and fed into the receiver.

Commercial antennas have reflection coefficients between -10 dB and -20 dB, which results in a relatively large $P_{mismatch}$. The reflection coefficient of the base-station antenna used in this work is -24 dB at 866 MHz [3].

The signal power at the input of the receiver can be written as:

$$P_{in} = P_r + P_{leakage} + P_{mismatch} \tag{10}$$

If we denote the output power of the power amplifier as P_{PA} and the reflection coefficient of the antenna in dB as $S_{11,ant}$, the leakage and mismatch components can be expressed in dB as:

$$P_{leakage} = P_{PA} - |S_{31,circ}| \tag{11}$$

$$P_{mismatch} = P_{PA} - |S_{11,ant}| - |S_{32,circ}| \tag{12}$$

where $S_{31,circ}$ is the isolation of the circulator from port 1 to port 3 and $S_{32,circ}$ represents the loss in the circulator from port 2 to port 3, i.e., the insertion loss. The insertion loss of the circulator can be neglected since it is generally very small.

B. The reverse link

The power received at the base-station scattered from an RFID tag in the far-field can be written using the classical radar equation as [4]:

$$P_{received} = P_{eirp} G_b \frac{\lambda^2}{(4\pi)^3 r^4} \sigma \tag{13}$$

where P_{eirp} is the radiated power from the base-station antenna, G_b is the base-station antenna gain, λ is the wavelength, and r is the distance between the base-station and the RFID tag, and σ is the radar cross-section of the RFID tag. In this formulation, the polarization losses and the multi-path effects are ignored.

978-1-4799-4993-9/14 $31.00 © 2014 IEEE

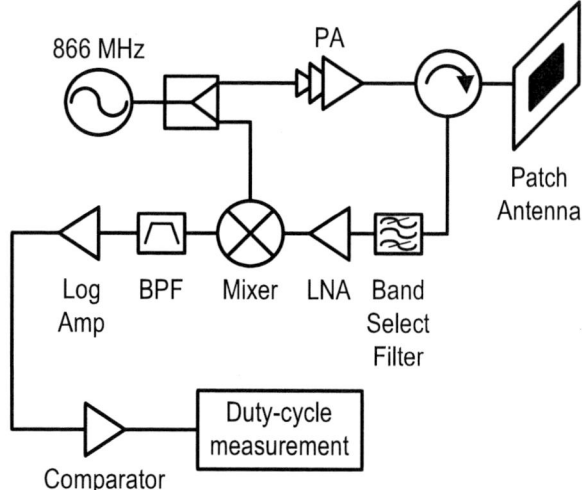

Fig. 3. The block diagram of the base-station.

If the RFID tag is backscattering by modulating the impedance at the antenna inputs, the modulated power received at the base-station can be expressed as:

$$P_r = P_{eirp} \, G_b \, \frac{\lambda^2}{(4\pi)^3 \, r^4} \, \Delta\sigma \qquad (14)$$

where $\Delta\sigma$ is the differential radar cross-section of the tag.

The radar cross-section of the tag can be expressed by using the Thevenin equivalent circuit of the RFID tag as [5]:

$$\sigma = \frac{\lambda^2 \, G_t^2 \, R_a^2}{\pi |Z_a + Z_c|^2} \qquad (15)$$

where G_t is the gain of the tag antenna, $Z_a = R_a + jX_a$ is the complex antenna impedance, and $Z_c = R_c + jX_c$ is the complex input impedance of the tag chip.

The power radiated from the base-station, P_{eirp}, can be calculated as:

$$P_{eirp} = P_{PA} \left(1 - |S_{11,ant}|^2\right) G_b \approx P_{PA} \, G_b \qquad (16)$$

neglecting the insertion loss of the circulator. For a well matched antenna ($|S_{11,ant}| < -20\,dB$), the $1 - |S_{11,ant}|^2$ term might be omitted for simplification. Thus, inserting (16) in (14) yields:

$$P_r = P_{PA} \, G_b^2 \, \frac{\lambda^2}{(4\pi)^3 \, r^4} \, \Delta\sigma \qquad (17)$$

C. Case study

The maximum allowed power in the 865.6-867.6 MHz European ISM band is 3.3 W e.i.r.p (35.2 dBm). and the base-station antenna is a 50 Ω planar patch antenna with 3.6 dBi gain at 866 MHz [3]. Hence, the output of the power amplifier is 31.6 dBm when the maximum allowed power is transmitted from the base-station. Commercially available circulators can achieve 30 dB isolation. Putting in the values for the circulator

isolation and the reflection coefficient of the antenna yields:

$$P_{leakage} = 31.6\,dBm - 30\,dB = 1.6\,dBm \qquad (18)$$
$$P_{mismatch} = 31.6\,dBm - 24\,dB = 7.6\,dBm \qquad (19)$$

The larger of these two components will be dominant in the self-jamming signal: in this case $P_{mismatch}$ is dominant. In other words, the circulator isolation itself doesn't guarantee low self-jamming signal if the antenna input reflection coefficient (S_{11}) is not low enough.

D. Transmit

The block diagram of the base-station is given in Fig. 3. The transmit section of the base-station consists of the signal source, the power splitter, the power amplifier, the circulator, and the antenna. The signal source generates the carrier signal at 866 MHz. The signal is then split into two by the power splitter, one half being fed to the power amplifier while the other is fed to mixer in the receiver. The base-station antenna emits the electromagnetic wave for the RFID tags. The passive RFID tags are powered up by the electromagnetic wave power emitted by the base-station. The regulations put a limit on the maximum RF power that can be emitted from the base-station. The phase noise of the signal source is also a critical parameter that affects the quality of reverse communication link.

E. Receive

The receive section of the base-station consists of the antenna, the circulator, the band-select filter, the low-noise amplifier (LNA), the down-converting mixer, the bandpass filter, and the logarithmic amplifier. The modulated backscattering signal from the tag is collected by the base-station antenna and fed into the receiver through the circulator. In order to filter out the unwanted parts of the electromagnetic spectrum, a band-select filter is used, which helps preventing the LNA from saturation. The filtered signal is amplified by the LNA and down-converted to DC by the mixer, in a direct conversion fashion. The reason to use this architecture is the large self-jamming signal present in the receiver. The log-amp saturates due to the self-jamming signal if it is used directly after the LNA. It is easier to filter out the self-jamming signal at the output of the mixer, since it basically creates a DC offset. A band-pass filter (BPF) filters out the DC offset, passing the wanted part of the signal to the log-amp, whose output is amplified by a comparator. Finally, the duty-cycle of the signal is measured.

The large self-jamming signal creates problems in this architecture. The LNA and the mixer can saturate due to the self-jamming signal. One method to decrease the amplitude of the self-jamming signal is to use a high-gain antenna and decrease the output power of the PA.

IV. IMPLEMENTATION AND MEASUREMENTS

The base-station is implemented according to Fig. 3 using commercially available components. Figure 4 shows the circuit that is used in the measurements to model the path loss and the

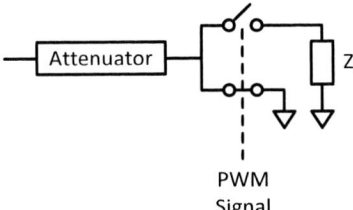

Fig. 4. The circuit that is used in measurements to model the path loss and the backscattering from the tag.

backscattering from the tag. An attenuator is used to mimic the path loss.

The value of attenuation is selected to model the path loss at 2 meters, taking into account the gain of the base-station antenna. As seen in (17), the received power is a function of the square of the base-station antenna gain. Hence, the value of attenuation can be calculated in dB as:

$$P_{att} = P_{Loss} - 2 \cdot G_b \tag{20}$$

where P_{att} is the value of the attenuation, and P_{Loss} is the path loss in free space. The required attenuation is calculated to be 30 dB.

The backscattering from the tag is modeled using a waveform generator and an RF switch. The PWM signal from the waveform generator controls the RF switch, which changes the load seen by the cable between short circuit and the impedance Z_L. The difference in the electromagnetic power reflection from the tag for the high and low states of the PWM signal is calculated and the value of Z_L is adjusted accordingly so that the change in the power reflection from the circuit between the two states is the same as the difference in the electromagnetic power reflected from the tag.

The duty-cycle of the PWM signal controlling the RF switch is swept from 20% to 80% with 5% steps and the duty-cycle is measured five times at each step at the base-station. Figure 5 shows the error in each measurement and the average measured duty-cycle at each step plotted with respect to the actual duty-cycle of the control signal. Measurements show that the base-station can measure the duty-cycle with an error less than 1%.

Figure 6 shows example voltage waveforms recorded during the measurements for a 100 kHz PWM signal with 35% duty cycle. The output of the logarithmic amplifier is shown along with the output of the comparator.

V. CONCLUSION

A base-station design is proposed for passive UHF RFID tags that use pulse-width modulation (PWM) in backscattering communication. The self-jamming problem is critical in the base-station design. It is shown theoretically that amplitude of the self-leakage signal depends on the circulator isolation and the reflection coefficient of the base-station antenna. The architecture of the proposed base-station which is designed for a single-antenna is presented. A circuit based on an attenuator

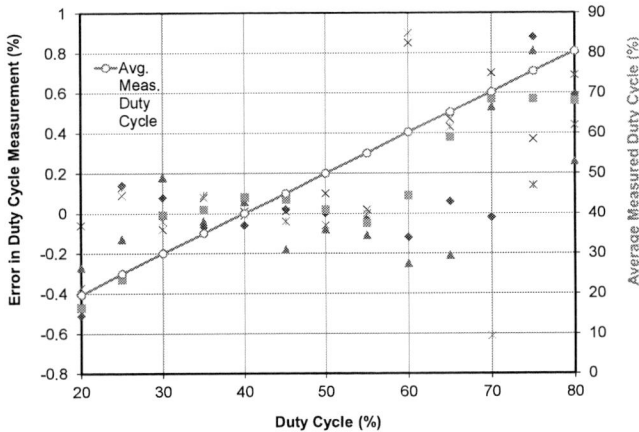

Fig. 5. The results of the duty-cycle measurements. The error in each measurement and the average measured duty-cycle are plotted with respect to the actual duty-cycle of the control signal which is swept from 20% to 80% with 5% steps.

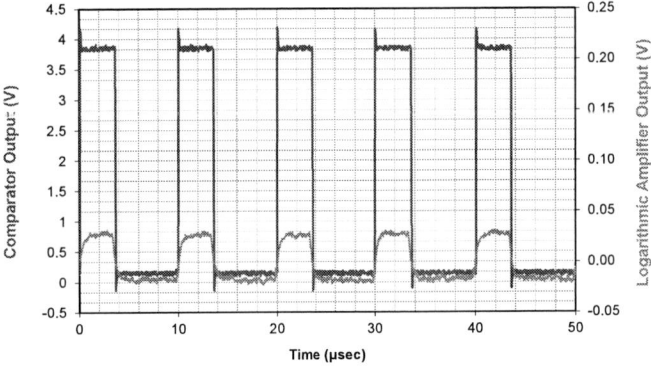

Fig. 6. The measured voltage waveforms at the output of the log-amp (red) and at the output of the comparator (blue) for a 100 kHz PWM signal with 35% duty-cycle. The $50\,mV_{p-p}$ signal at the output of the logarithmic amplifier is amplified to a rail-to-rail signal by the comparator.

and an RF switch is used in measurements to model the path loss and reflection from the tag. Measurement results show that the base-station is capable of measuring the pulse-width of the backscattered signal with an error less than 1% from a distance of 2 meters.

REFERENCES

[1] K. Kapucu, J. Panades, and C. Dehollain, "Design of a passive UHF RFID tag for capacitive sensor applications," in *9th Conference on Ph.D Research in Microelectronics and Electronics (PRIME 2013)*, 2013, pp. 213–216.

[2] L. Mayer, R. Langwieser, and A. Scholtz, "Evaluation of passive carrier-suppression techniques for UHF RFID systems," in *IEEE MTT-S International Microwave Workshop on Wireless Sensing, Local Positioning, and RFID, 2009. IMWS 2009*, Sep. 2009, pp. 1–4.

[3] K. Kapucu and C. Dehollain, "Remote powering link for a passive UHF RFID tag for capacitive sensor applications," in *2013 IEEE 20th International Conference on Electronics, Circuits, and Systems (ICECS)*, Dec. 2013, pp. 823–826.

[4] C. A. Balanis, *Antenna Theory: Analysis and Design, 3rd ed.* Wiley Interscience, May 2005.

[5] P. Nikitin and K. Rao, "Theory and measurement of backscattering from RFID tags," *IEEE Antennas and Propagation Magazine*, vol. 48, no. 6, pp. 212–218, 2006.

Backside illuminated wafer-to-wafer bonding Single Photon Avalanche Diode array

Design, fabrication and preliminary tests

Yu Zou, Danilo Bronzi, Federica Villa
Dipartimento di Elettronica, Informazione e Bioingegneria
Politecnico di Milano
Milan, Italy
yu.zou@polimi.it

Sascha Weyers
Fraunhofer IMS
Duisburg, Germany

Abstract—**We present an innovative sensor chip, exploiting backside illumination of a silicon-on-insulator (SOI) wafer integrating custom single photon avalanche diodes (SPADs), flipped and wafer-bonded on a standard CMOS wafer integrating the analog front-end circuit, in-pixel digital processing and readout electronics. Two major improvements are achieved: higher pixel density and fill-factor, since these detectors are placed on the top of the corresponding smart-pixel electronics, instead of being placed side-by-side (as in planar structures); enhanced spectral sensitivity in the near-infrared, up to 1 μm wavelength, thanks to thicker active volume within the SOI detector wafer and to the backside illumination of the active area.**

Keywords— Single-photon avalanche diode (SPAD); backside illumination; near-infrared (NIR); indirect time-of-flight; wafer-to-wafer bonding; silicon-on-insulator (SOI).

I. Introduction

A Single Photon Avalanche Diode (SPAD) is a p-n junction reverse biased well beyond its breakdown voltage. At this bias an electron hole pair generated by a single photon can trigger a macroscopic avalanche current. Compared to photomultiplier tubes (PMTs) and micro-channel plates (MCPs), SPADs are solid-state devices, very compact, insensitive to magnetic fields, require lower bias voltages, and can be easily integrated with complex electronics. The implementation of SPADs in standard CMOS technologies [1] made it possible to fabricate SPAD imaging arrays with monolithic integration of the detector itself, front-end circuitry and digital pre-processing electronics directly in-pixel [2],[3]. These arrays are suited for either 3D ranging or 2D imaging applications, achieving high sensitivity together with high frame rates. Nonetheless, the main drawback of such an approach is the poor fill-factor and the low quantum efficiency related to the narrow width of the depleted zone, resulting from typical CMOS doping concentration.

In this paper, we present a novel approach with a backside illuminated SPAD (BackSPAD) array using wafer-to-wafer (W2W) bonding 3D structure. A silicon-on-insulator (SOI) wafer, integrating just an array of 32×32 SPAD devices, is flipped and wafer-bonded, through a Solid Liquid

InterDiffusion (SLID) technique [4], to a standard 0.35 μm CMOS read-out circuitry (ROIC). Subsequently, the SOI wafer is back-thinned to the oxide, thus enabling backside illumination.

II. BackSPAD Sensor Fabrication Process

The SPAD manufacturing in the standard CMOS process on one side has the big advantage of using the mainstream technologies (availability, yield and cost); on the other side, it is associated with two major disadvantages: SPAD performance depend on p-n junctions (doping concentration, depth of p-n junctions, etc.) and are limited by the design rules of the CMOS process, which are not necessary the ideal ones for maximizing SPAD performance. Drawbacks can be overcome by employing a dedicated process for SPAD fabrication.

The BackSPADs will be fabricated in 0.35 μm SOI technology with deep trench isolation. The SOI substrate consists of the base wafer, 1 μm buried oxide and 3 μm Si film. This technology will provide a stable basis for the SPAD fabrication. Since the SOI wafer will contain SPADs only, some process parameters will be adjusted to achieve high SPAD performance.

The structure of SOI-SPAD consists of p+ shallow implantation, a deeper n-well, which defines the active area, and a surrounding p guard-ring to prevent peripheral edge

Fig. 1. Layout of a BackSPAD pixel, with surrounding deep trench isolation (yellow line) and four contacts to the cathode. The fill-factor is 75.4%.

978-1-4799-4993-9/14 $31.00 © 2014 IEEE

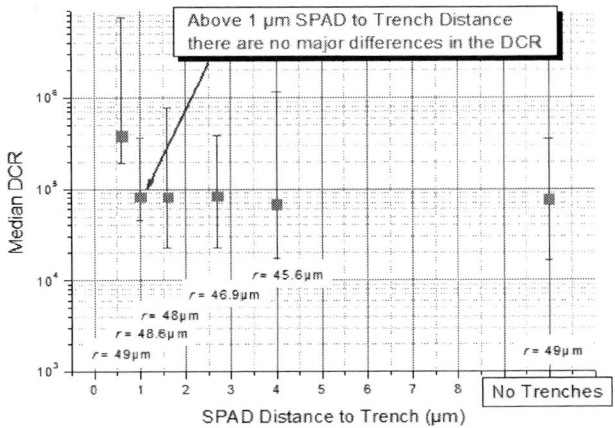

Fig. 2. Dependence of the Dark Count Rate on the distance from the SPAD to the trench, at different radius r of the SPADs. As can be seen, the deep trench does not increase the DCR for distances down to 1 μm from the SPAD edge.

breakdown (PEB). To allow a more compact design, each SPAD is surrounded by a deep trench which prevents crosstalk and punch through effects. A first BackSPAD fabrication recipe exploited the same processing already employed in a previous CMOS SPADs batch [5], so the general layouts for the active part of both SPADs are identical. Fig. 1 shows the layout of a BackSPAD pixel, with the surrounding deep trench isolation (yellow lines).

By using deep trench-isolation, adjacent n-wells are separated by dielectric isolation, which significantly reduces pixel pitch compared to the p-n junction isolation. For a given pitch, the SPAD size is mainly determined by the area that is necessary for electrical contacts to the n-well and minimum distance between active SPAD area and the deep trench. The latter is necessary to relax the impact of crystal defects generated by the deep trench process steps into the SPAD active area. In order to determine this minimum distance, in the test wafer we laid out SPADs with different diameters inside constant pitch trenches. Fig. 2 shows that if SPAD-to-trench distance is above 1 μm, there is no major impact on dark count rate (DCR), meaning that no crystal defect propagates into the SPAD active area. In Fig. 1, the BackSPAD pixel dimensions are 100 μm × 100 μm. Note that metallization does not spoil active area and fill-factor is still very high, 75.4% for a SPAD with 49 μm radius. This fill-factor is the highest value ever reached without microlenses among different two dimensional CMOS SPAD arrays.

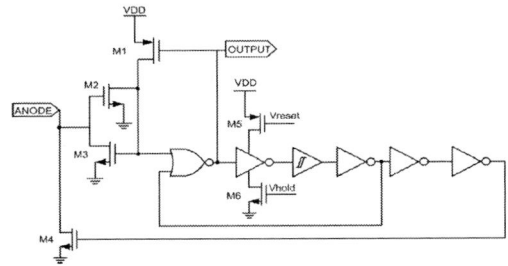

Fig. 4. Variable load queching circuit with adjustable hold-off time and reset time is designed for BackSPAD application.

III. CMOS WAFER WITH BACKSPAD ELECTRONICS

We designed the BackSPAD array with the main aim of providing fast 2D imaging and 3D ranging. Considering these specific applications, the pixel-level signal processing and the global shutter are the best choices for fast elaboration, and to avoid artifact of moving objects. The artifact may give error to the measurement of distance that is difficult to remove.

Fig. 3 shows the signal processing circuit for single pixel in order to provide 2D imaging and 3D ranging via indirect time-of-flight techniques [6]. The pitch of the pixel matches with the dimension of a single SPAD (100 μm x 100 μm). In this limited area, we have put a sensing circuit, shaping electronics, one 6-bit counter, two 5-bit bidirectional counters, internal memory and necessary output buffers to drive the column data bus.

After the SPAD is triggered, a suitable front-end electronics (known as quenching circuit) must quench the avalanche current, by lowering the bias below breakdown, and then restore the operating voltage after a defined dead time. This quenching circuit is developed based on a variable load quenching circuit in [7]. Fig. 4 shows the block diagram of quenching circuit. The hold-off time and reset time are adjustable by controlling the voltage on the two transistors M5, M6. These adjustments make the designed front-end circuit suitable for quenching and sensing of different SPADs structures, even with heavy anode capacitance. The shaping electronics modify the waveform on the VLQC output signal to correctly drive the counters. The 6-bit counter is used for acquiring 2D image and background information. Two 5-bit bidirectional counters are the key elements by using time-of-flight techniques to obtain 3D ranging [7]. The internal

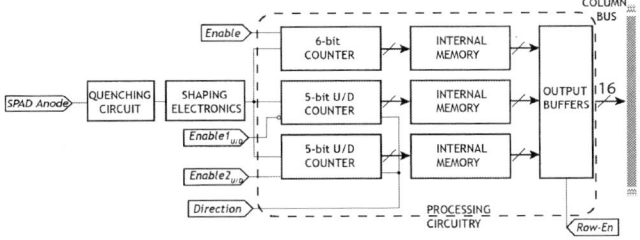

Fig. 3. Block diagram of electronics for BackSPAD pixel.

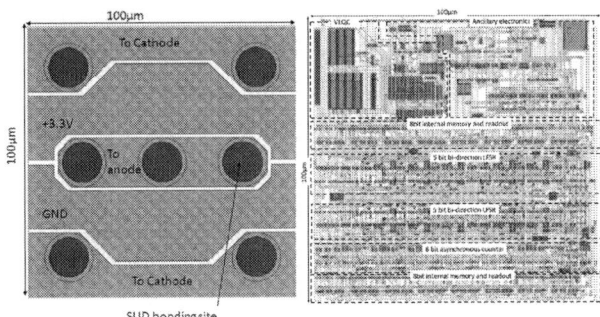

Fig. 5. Top metal of BackSPAD pixel, that will allow proper bonding of the corresponding detector (left) and layout of BackSPAD pixel (right).

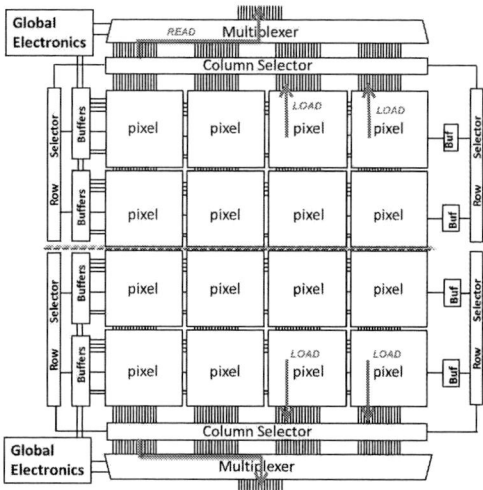

Fig. 6. Block Diagram of a 4x4-pixel array with peripheral electronics. The actural 32x32 BackSPAD array is based on the same architecture

memories are necessary to store the counts in order to allow parallel integration and readout from all pixels of a frame when global shutter is applied. Once a pixel is selected, all 16-bit data will be readout at the same time.

For the layout of BackSPAD pixel, routing is made through three metal layers, while the top fourth metal is used for proper bonding with the corresponding SOI SPAD (Fig. 5 left). The layout of the BackSPAD pixel with its main building blocks and relative dimensions is shown in Fig. 5 right.

We designed a 32x32-pixel array with horizontal access circuit (row selector), vertical access circuit (column selector) and readout electronics. To improve the readout speed, the array has been subdivided into two rectangular sub-arrays of the same size (16x32 pixels). Fig. 6 shows a 4×4 mini-array, with the same functionality as the real 32×32 array. The horizontal access is performed on both sides along two half-rows. We read out half-row while load the next half-row to the column bus, so each pixel has 16 clock periods to write data to column bus ensuring accurate data readout.

IV. WAFER TO WAFER BONDING PROCESS

A cross section of BackSPAD is sketched in Fig. 7. The SOI SPADs are fabricated on a thick film SOI wafer substrate as described before. The readout electronics are fabricated separately on a second wafer using the standard 0.35μm CMOS technology. Once both wafers are processed, the hybridization is performed by flipping down the SOI wafer and bonding it on the CMOS wafer. The intended bonding process of the two wafer employs the Solid Liquid Interdiffusion (SLID) bonding technique using copper (Cu) and tin (Sn) as the bumping materials [4].

The back thinning process of the bonded BackSPAD wafer will be done in two steps. Firstly the bulk substrate of the SOI wafer (initial thickness 725 μm) is removed partly (with a remaining thickness of about 50 -100 μm) after the bonding step by mechanical grinding. The etching of the remaining silicon layer with etch stop on the buried oxide (BOX) is performed by a highly selective isotropic etch process. This intended sequence enables the backside illumination of the SPAD array through the buried oxide. The advantage of this two-step approach (grinding + etching) is that a huge amount of silicon can be removed in a short time in comparison to a complete isotropic etch process.

In order to create bonding metal pads for the packaging, in the last step, vias are etched from the backside through the BOX and a metallization layer is deposited by sputtering. The BackSPAD is now ready for dicing and the packaging.

V. PRELIMINARY TEST OF SPADs BEFORE INTEGRATION

As we mentioned before, the SPAD devices for the BackSPAD approach were fabricated and designed starting from the CMOS processes and structures used in [4]. These devices have been called "low-voltage" LV-BackSPAD, referring to the expected breakdown voltage that should be similar to the one achieved in [5]. We conceived a variant for exploiting a fully depleted silicon film when the SPAD device is under working conditions (i.e. biased above breakdown), in order to increase the photo detection efficiency of the device at longer wavelengths. These devices have been called "high-

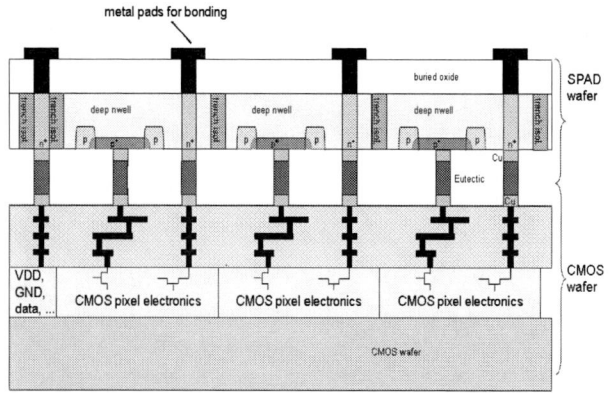

Fig. 7. Bonded SOI wafer and CMOS wafer after realization of metal bond pads on top.

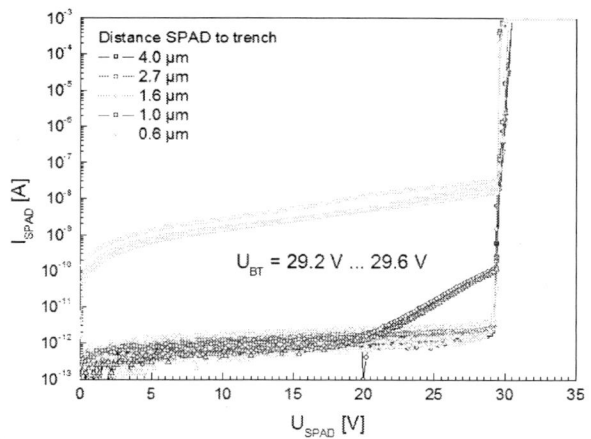

Fig. 8. Measured I-V characteristics of LV-BackSPAD devices before wafer bonding.

978-1-4799-4993-9/14 $31.00 © 2014 IEEE

voltage" HV-BackSPADs.

After fabrication of the SPAD devices on SOI substrates, preliminary electrical tests were performed on wafer level to determine the breakdown voltage and leakage current of the devices. Results of measurements for the LV-BackSPADs are shown in Fig. 8. We did three times measurements for each diameter SPADs. Looking at the leakage current below and the SPADs' breakdown voltage, devices with the smallest distance (0.6 μm) between the SPAD structure and the isolation trench show a significantly higher current level than the other devices. For the devices with 1.0 μm distance, the leakage current starts to increase at 19 V. This indicates the influence of the defects, which were generated by the trench fabrication, on the performance of the SPAD. By separating the SPAD structure 1.6 μm and more from the isolation trench, no influence of the distance on the leakage current level can be observed. For the breakdown voltage itself, no dependency on the distance between SPAD structure and trench could be observed. Therefore, even for the 0.6 μm SPAD to trench distance, the guard-ring prevents any influence of the trench on the active p-n junction of the device.

The results for the HV-BackSPADs are displayed in Fig. 9. For these devices, no influence of the SPAD to trench distance could be observed, but the leakage current level is at 0.6 nA to 0.8 nA in between the levels for the best and the worst LV-BackSPAD devices we measured before. Initially a breakdown voltage of 49 V - 50 V was detected. By performing a second measurement at the already measured structures, the breakdown voltage could increase to above 52 V and the slope at the breakdown voltage was increased. Further measurements do not change the characteristics. This difference is due to the increment of SPAD depletion width (fully depleted) when we

applied first time the high bias voltage.

VI. CONCLUSION

In this work, we presented SPAD on SOI wafer fabrication process, signal processing electronics fabricated in standard 0.35 μm CMOS wafer containing 32x32-pixel array for providing fast 2D imaging and 3D ranging application, SOI wafer to CMOS wafer bonding process and preliminary tests on SOI wafer. This BackSPAD array has the highest fill-factor with respect to the normal planar structure used in CMOS SPAD arrays (without microlenses). By backside illumination, we could also increase the photon detection efficiency at near infrared range. Due to the separated wafer structure, High-performing SPADs could be manufactured, without hindering the fabrication of the readout electronics, as it happens in dedicated CMOS-based SPAD structures. The test of bonding and whole chip will be complete once the full assembly is produced.

ACKNOWLEDGMENT

The research leading to these results was partially funded as the "MiSPiA" project within the ICT theme of the EU Seventh Framework Programme(FP7, 2007-2013) under grant agreement n° 257646.

REFERENCES

[1] A. Rochas, M. Gani, B. Furrer, P.A. Besse, R.S. Popovic, G. Riborby, G. and N. Gisin, "Single photon detector fabricated in a complementary metal-oxide-semiconductor high-voltage technology," Rev. Sci. Instrum., vol. 74, no. 7, pp. 3263-3270, Jul. 2003.

[2] D. Stoppa, L. Pancheri, M. Scandiuzzo, L. Gonzo, G.-F. Della Betta, and A. Simoni, "A CMOS 3-D imager based on single photon avalanche diode," IEEE Trans. Circuits Syst. I, vol. 54, no. 1, pp. 4–12, Jan. 2007.

[3] C. Niclass, M, Sergio, and E. Charbon, "A single photon avalanche diode array fabricated in deep-submicron CMOS Technology," in Proceedings Design Automation and Test in Europe, vol. 1, pp. 1-6, 2006.

[4] A. Munding, H. Hübner, A. Kaiser, S. Penka, P. Benkart, and E. Kohn, "Cu/Sn Solid–Liquid Interdiffusion Bonding," Wafer Level 3-D ICs Process Technology Integrated Circuits and Systems, Springer Science+Business Media, pp. 1-39, 2008.

[5] D. Bronzi, F. Villa, S. Bellisai, B. Markovic, S. Tisa, A. Tosi, F. Zappa, S. Weyers, D. Durini, W. Brockherde, U. Paschen, "Low-noise and large-area CMOS SPADs with timing response free from slow tails," in ESSDERC, 2012 Proceedings of the European, pp.230,233, 17-21 Sept. 2012.

[6] D. Bronzi, F. Villa, S. Bellisai, B. Markovic, G. Boso, C. Scarcella, A. Della Frera, A. Tosi, "CMOS SPAD pixels for indirect time-of-flight ranging," Photonics Conference (IPC), 2012 IEEE, pp.22,23, 23-27 Sept. 2012.

[7] D. Bronzi, S. Tisa, F. Villa, S. Bellisai, A. Tosi, F. Zappa, "Fast Sensing and Quenching of CMOS SPADs for Minimal Afterpulsing Effects," Photonics Technology Letters, IEEE, vol.25, no.8, pp.776-779, April15, 2013.

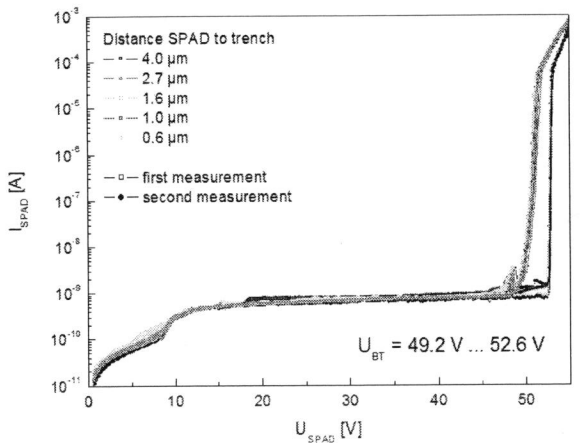

Fig. 9. Measured I-V characteristics of HV-BackSPAD devices before wafer bonding.

5x5 SPAD Matrices for the Study of the Trade-offs between Fill Factor, Dark Count Rate and Crosstalk in the Design of CMOS Image Sensors

Manuel Moreno-García, Rocío del Río, Óscar Guerra, and Ángel Rodríguez-Vázquez
Instituto de Microelectrónica de Sevilla, IMSE-CNM (CSIC/Universidad de Sevilla)
C/Americo Vespucio, s/n, 41092 Seville (Spain)
Email: moreno@imse-cnm.csic.es

Abstract—**CMOS Single Photon Avalanche Diodes (SPADs) are a dedicated type of photodetectors that are attracting increasing interest. Crosstalk and fill factor are magnitudes that become important when dealing with arrays of SPADs. There are trade-offs that involve these two magnitudes and dark count rate (DCR) which are of great interest for the implementation of image sensors. A set of 5x5 matrices of SPADs with different sizes and shapes is designed to study the relationships between FF, crosstalk and DCR, and conceive an accurate behavioural model of SPAD arrays. The testchip is fully operative and preliminary experimental results are presented.**

I. INTRODUCTION

Single Photon Avalanche Diodes (SPADs) are p-n junctions reverse biased at a voltage larger than its breakdown voltage (V_{BR}) [1]. The resulting high electric field in the depletion region causes the amount of photogenerated carriers in this zone to increase exponentially in time and a macroscopic current can be measured in practice from a single incoming photon. To use these devices as photodetectors it is necessary to extinguish the avalanche and reset the original bias conditions to enable the detection of new incoming photons [2]. The additional circuitry used with such purpose is the quenching circuit [3].

Although the trend in the design of CMOS SPADs is to go to smaller technological nodes [4] –which a priori favors the fill factor (FF)–, the increasing performance requirements imply that additional electronics must be integrated, reducing the FF. Due to their simplicity, a way to reduce the area occupied by the integrated electronics is the use of passive quenching circuits (PQC) instead of active (AQC) or mixed (MQC) ones [3]. However, SPADs with active recharge (MQC) return abruptly to their initial biasing conditions while the recharge is slower in PQCs (see Fig. 1). As a result, any incident photon –or carriers that are thermally generated or released by internal traps– could retrigger the device before the recharge phase ends. Some adverse effects of this retriggering phenomenon are the variation of the dead time (t_{dt}, time in which the SPAD is not sensitive to the arrival of new photons) or the reduction of the photon detection efficiency (PDE) [5].

Another way to improve the FF is to increase the area of the SPAD regarding the electronics within the array cell, but this results in an unwanted increase of DCR (frequency of the avalanche events caused by non photogenerated carriers) [6].

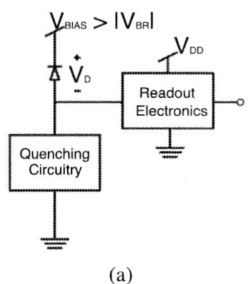

(a)

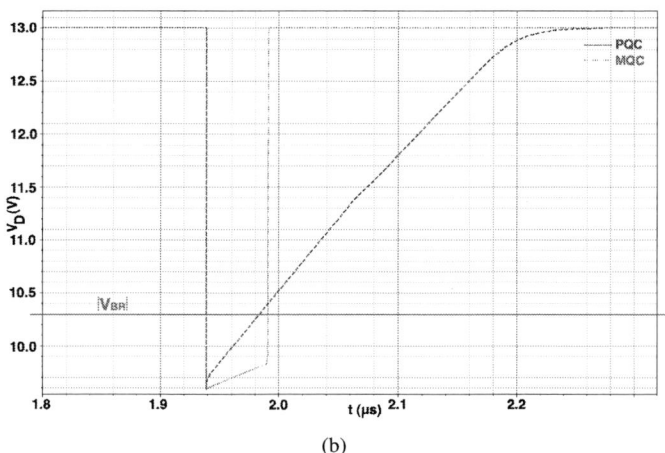

(b)

Fig. 1. (a) Conceptual diagram of the SPAD and its accompanying circuitry. (b) Transient response of the SPAD voltage during both the quenching and the recharge phases for passive and mixed quenching circuits.

Additionally, when implementing CMOS SPAD arrays, the reduction of the pixel pitch and the growth of the SPAD area would be negative in terms of crosstalk, that represents the probability of having a spurious avalanche in a pixel caused by a photon detection or dark count in a neighboring cell [6].

In this work we present a testchip for studying the existing trade-offs between FF, DCR and crosstalk, which are key when designing CMOS SPAD-based image sensors. To this end, a precise extrapolated model of the behaviour of SPAD arrays in terms of these magnitudes will be developed.

II. TESTCHIP DESCRIPTION

A standard CMOS 180nm process has been chosen for the design of the SPADs and their quenching circuits, together

978-1-4799-4993-9/14 $31.00 © 2014 IEEE

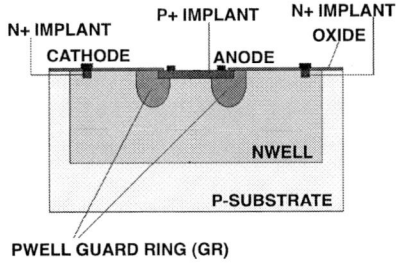

(a)

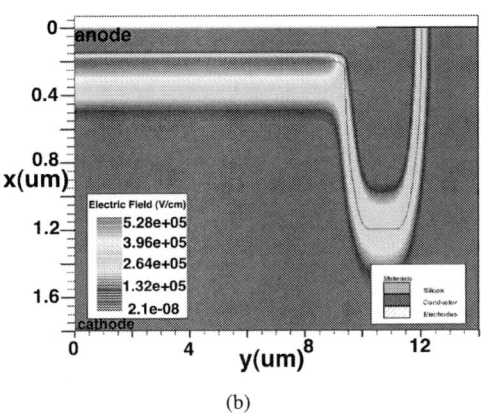

(b)

Fig. 2. (a) Cross section of the SPAD, with the p-well guard ring to avoid PEB. (b) Simulation in ATLAS of the SPAD structure.

with the accompanying test circuitry.

A. SPADs Physical Structure

A variety of SPAD structures has been published [6] to avoid premature edge breakdown (PEB) –a spurious avalanche that preferably takes place at the edges of the p-n interface of the device and reduces its sensitivity [7]. Among the possibilities, due to the limited layer availability and the design rules of the selected standard CMOS process, a SPAD structure has been chosen that includes a guard ring using a lower doped p-well around the p$^+$ region (see Fig. 2a).

The physical structure of the SPADs has been designed and validated through 2D simulations in ATLAS [8]. For this purpose, the dimensions of the implanted areas have been extracted from the technology files and the diode model parameters of the process, whereas the doping concentrations have been estimated from previous experimental results in the same technological process [9], [10]. Fig. 2b shows the efficiency of the guard ring for lowering the electric field at the edges of the active region.

B. SPADs Geometry

Fig. 3 shows the top view of the three different SPAD shapes that have been included in the testchip to obtain information about their relative performance in terms of DCR; namely, circular, cigar-shaped and ellipsoidal. In [11], a cigar-shaped structure is proposed as an interesting alternative to implement large-area SPADs without having restrictive increases of the DCR, due to the larger gettering efficiency of the geometry

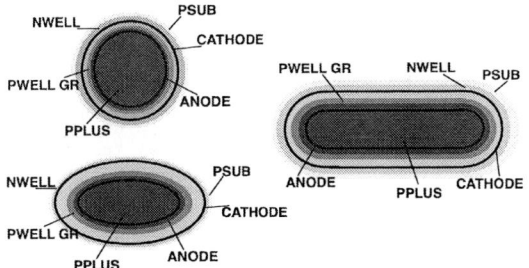

Fig. 3. Top view of the three SPAD geometries included in the testchip.

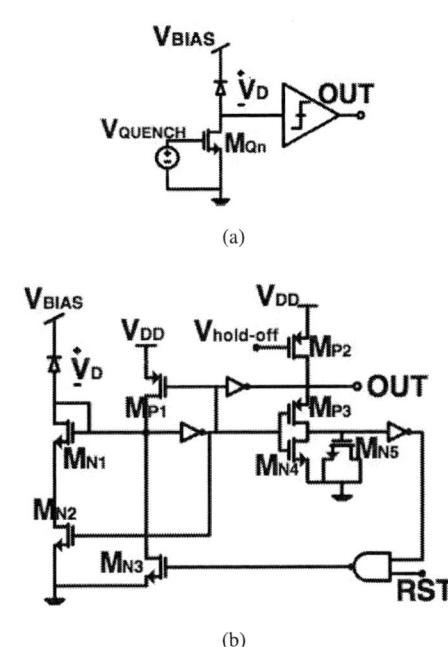

Fig. 4. (a) Passive quenching circuit. (b) Mixed quenching circuit.

to decrease the defect density in the active region [12]. As an extension of this concept, we propose the ellipse due to its resemblance to the cigar-shape and its smoother profile.

C. Crosstalk Characterization

Besides dark counts, crosstalk is another source of noise to be considered when working with arrays of SPADs. It can be caused by two different mechanisms:

- Optical crosstalk: During an avalanche process some secondary photons can be emitted from the SPAD due to electroluminiscence [13]. These photons can propagate through the chip and be detected by neighboring pixels in the array, resulting in a non-desired avalanche pulse.
- Electrical crosstalk: It is due to photons absorbed beyond the active area of the device. These photons generate free carriers that can diffuse laterally to different depletion regions where they could cause an avalanche [7].

The first one depends on the distance between SPADs and also on their size, since the parasitic capacitance influences the number of carriers flowing during the avalanche [7]. Its effect can thus be reduced by decreasing the SPAD size or increasing their separation in the array, but at the cost of a loss of FF.

The 5x5 matrices of SPADs included in the testchip allow us to characterize the dependence of crosstalk on distance and on SPAD size for modeling purposes. Due to constraints imposed by the experimental setup, a 60-μm pixel pitch has been chosen to enable the individual illumination of each cell.

D. Quenching Circuitry

The aim of quenching circuits is to extinguish the SPADs avalanche current and return them to their bias conditions, allowing the detection of new incident photons. Two different alternatives have been included in the testchip (see Fig. 4):

- PQC with an nMOS transistor operating in the triode region and a comparator with programmable threshold for readout purposes.
- MQC with passive quenching and active recharge, in which t_{dt} can be widely adjusted by varying $V_{hold-off}$.

The electrical design of both quenching circuits in Fig. 4 has been performed using the behavioural model of SPADs proposed in [14] and extended in [10], which includes the DCR as a Poisson distributed phenomenon. Behavioural model parameters have been fine tuned, taking advantage of previous experimental results in the same process [10].

E. Testchip Composition

The testchip has been designed in a 1.8-V 180-nm standard CMOS process, occupies an area of 1.5×1.5mm^2 and comprises:

- nine 5×5 SPAD arrays for the characterization of crosstalk for different sensor shapes, sizes and quenchings.
- four individual SPAD structures for studying their transient operation and measuring their IV-characteristic.
- a 1×8 array to examine alternative test configurations.

Table I summarizes the different test structures included in the testchip. Note that not only the shape and size of the SPADs vary, but also the minor to major axis ratio of cigar and ellipse shapes to check the role of this parameter. Fig. 5 shows the layout of the whole chip and highlights its main parts.

III. EXPERIMENTAL RESULTS

The testing of the chip is in progress at the time of writing this paper. Some results of the measurements performed up to this moment are presented in this section for single devices.

In order to get the breakdown voltage and the IV-characteristic, the HP4155A semiconductor parameter analyzer has been utilized. Fig. 6 (left) illustrates the variation of the current through the SPAD versus the bias voltage for two devices with different shape (circular and ellipsoidal) and size (14 and 7μm, respectively). It can be seen that for currents around 200μA (*latching* current), the avalanche is no more self-sustained and the current drops sharply to zero.

On the right side of Fig. 6, the variation of the breakdown voltage with temperature is represented. Theory [14] tells us that the breakdown voltage of SPADs vary with temperature:

TABLE I
SUMMARY OF THE INCLUDED TEST STRUCTURES.

	Test structure	Shape	Active area (μm^2)	Minor to major axis ratio	Quenching circuit
5×5 Arrays	#1	Circle	155	1	PQC
	#2	Circle	350	1	PQC
	#3	Ellipse	350	0.62	PQC
	#4	Circle	350	1	MQC
	#5	Ellipse	155	0.62	PQC
	#6	Ellipse	350	0.28	MQC
	#7	Cigar	350	0.5	MQC
	#8	Cigar	155	0.39	MQC
	#9	Ellipse	350	0.62	MQC
1×8 Array	#10	Circle	350	1	MQC
	#11	Cigar	113	0.39	PQC
	#12	Circle	40	1	MQC
	#13	Ellipse	40	0.37	MQC
	#14	Cigar	40	0.41	MQC
	#15	Ellipse	155	0.34	PQC
	#16	Cigar	155	0.39	PQC
	#17	Cigar	40	0.22	PQC
Individual Structures	#18	Circle	155	1	–
	#19	Cigar	155	0.42	–
	#20	Ellipse	155	0.52	–
	#21	Ellipse	155	0.6	–

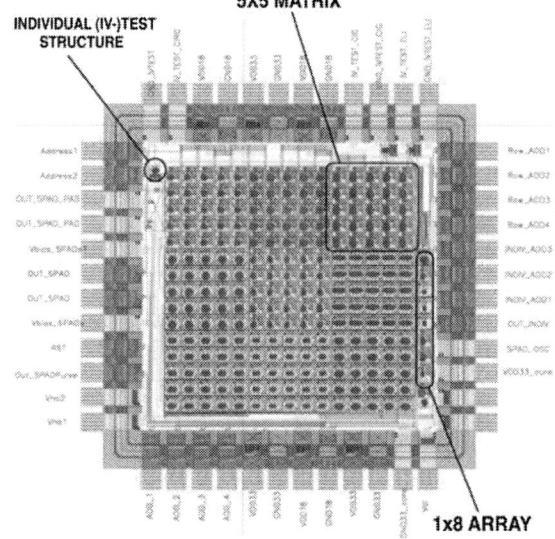

Fig. 5. Layout of the testchip.

$$V_{BR} = V_{B0} \cdot [1 + \beta \cdot (T - T_0)] \qquad (1)$$

where V_{B0} is the breakdown voltage at room temperature, and β is the temperature coefficient. In this case, the values of this coefficient are β_{CIR}=6.82×10^{-4}K^{-1} and β_{ELL}=6.79×10^{-4}K^{-1}.

Fig. 7 shows the dependence of the SPAD resistance with temperature. This magnitude is comprised of the space-charge resistance, the resistance of the neutral regions, and the ohmic resistance from the contact to the neutral region of the junction [5]. The SPAD resistance will be studied in detail because it is very important to know the current flowing through the device

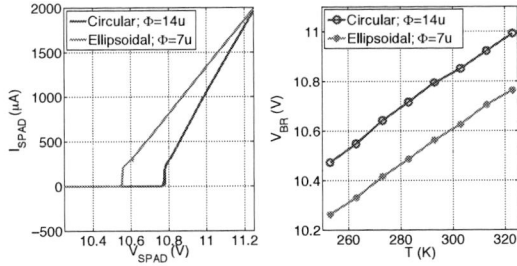

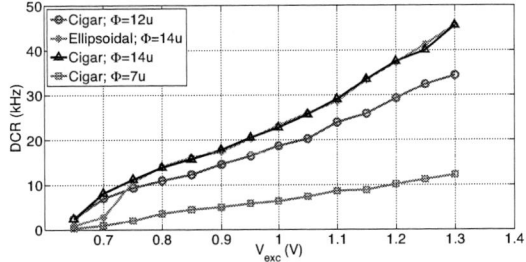

Fig. 6. IV-characteristic (left) and breakdown voltage variation with temperature (right) of the circular ($\Phi=14\mu$m) and ellipsoidal ($\Phi=7\mu$m) SPADs.

Fig. 8. Dark count rate variation with the excess bias voltage at room temperature. The test structures are those with PQC into the 1x8 array.

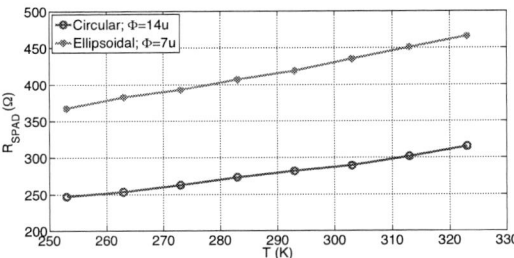

Fig. 7. SPAD resistance dependence for the circular ($\Phi=14\mu$m) and ellipsoidal ($\Phi=7\mu$m) devices.

when it comes to design. These results show a similar behavior for both SPADs and also that the smaller the area, the larger resistance of the SPAD, as expected [5].

Dark count rate versus the excess bias voltage is represented in Fig. 8 for the test structures with PQC included in the 1x8 array. In the range of excess bias voltages swept, the behavior in all the cases is quite linear. Moreover, the cigar and ellipsoidal SPADs with the same size ($\Phi=14\mu$m) present the same levels of dark counts. It is still under experimental validation if for larger active area sizes and comparing with the circular one, these two shapes represent an improvement in terms of the device noise.

Finally, the measurement of the PDE has been performed in the single-photon regime (incident optical power in the nW/mm^2 range). As an orientation, its maximum value for the circular SPAD ($\Phi=14\mu$m) is 4.2% for $V_{exc}=1$V at 540nm.

IV. CONCLUSIONS

Dark count rate, crosstalk and fill factor are key magnitudes when designing CMOS SPAD arrays. A set of distinct test structures is presented for the study of the existing trade-offs between them. This comparative study will provide important information about the design considerations to be taken into account by CMOS image sensors designers and towards the realization of a complete model for SPAD simulation. Moreover, the investigation of different shapes for the SPADs that could allow an increase in their sizes without prohibitively large increases in the DCR is of great importance since this would mean larger FF and photon sensitivity. The fabricated testchip is fully operative and its complete experimental characterization will be available at the time of the conference.

ACKNOWLEDGMENT

The authors would like to thank the Spanish National Research Council (Consejo Superior de Investigaciones Científicas, CSIC) for the JAE-PREDOC research grant that supports the work here presented. This work has been partially funded by ONR under Project N000141410355, the MINECO (Spain) under Project IPT-2011-1625-430000, the Junta de Andalucía under Project P12-TIC-2338, and the MINECO (Spain) through projects TEC2012-38921-C02-01, co-funded by the European Regional Development Fund.

REFERENCES

[1] F. Zappa et al., "Principles and Features of Single Photon Avalanche Diode Arrays," Sensors and Actuators A: Physical, vol. 140, pp. 103–112, Oct. 2007.

[2] A. Dalla Mora et al., "Single-Photon Avalanche Diode Model for Circuit Simulations," IEEE Photonics Technology Letters, vol. 19, pp. 1922–1924, Dec. 2007.

[3] A. Gallivanoni, I. Rech, and M. Ghioni, "Progress in Quenching Circuits for Single Photon Avalanche Diodes," IEEE Trans. on Nuclear Science, vol. 57, pp. 3815–3826, Dec. 2010.

[4] M. A. Karami, Deep-submicron CMOS Single Photon Detectors and Quantum Effects. PhD thesis, Technische Universiteit Delft, 2011.

[5] H. Finkelstein, Shallow-Trench Isolation Bounded Single-Photon Avalanche Diodes in Commercial Deep Submicron CMOS Technologies. PhD thesis, University of California, San Diego, 2007.

[6] G. F. Dalla Betta et al., Avalanche Photodiodes in Submicron CMOS Technologies for High-Sensitivity Imaging, ch. 11 in Advances in Photodiodes, pp. 225–248. InTech, 2011.

[7] P. Seitz and A. J. P. Theuwissen, Single-Photon Imaging. Springer Series in Optical Sciences, 2011.

[8] http://www.silvaco.com/.

[9] B. Blanco-Filgueira, P. López, and J. B. Roldán, "Experimental characterization of peripheral photocurrent in cmos photodiodes down to 65nm technology," Semiconductor Science and Technology, vol. 28, Apr. 2013.

[10] M. Moreno et al., "CMOS SPADs Selection, Modeling, and Characterization Towards Image Sensors Implementation," in Proc. of the 19th Intl. Conf. on Electronics, Circuits and Systems (ICECS), pp. 332–335, Dec. 2012.

[11] B. Nouri, M. Dandin, and P. Abshire, "Large-area low-noise single-photon avalanche diodes in standard cmos," in IEEE Sensors, Oct. 2012.

[12] A. Zanchi et al., "On-chip probes for silicon defectivity ranking and mapping," in Proc. of the 38th IEEE International Reliability Physics Symposium, pp. 370–376, 2000.

[13] R. Newman, "Visible light from a silicon p-n junction," Physical Review, vol. 100, no. 2, pp. 700–703, 1955.

[14] G. Giustolisi, R. Mita, and G. Palumbo, "Behavioral Modeling of Statistical Phenomena of Single-Photon Avalanche Diodes," International J. of Circuit Theory and Applications, vol. 40, pp. 661–679, July 2012.

Structural, magnetic and dielectric properties of non conducting nanocomposites for RF applications

H. Takacs[1,2], B. Viala[1], J.-H. Tortai[2], J. Alarcon Ramos[1], M. Bousquet[1], F. Duclairoir[3], C. Gourgon[2]

[1]CEA, LETI, MINATEC Campus, Grenoble, France
[2]LTM-CNRS-UJF, CEA, LETI, MINATEC Campus, Grenoble, France
[3]CEA, INAC, Grenoble, France
helene.takacs@cea.fr

Abstract— **This paper describes solution-cast, spin-coating and hot-press planarization process for preparing metal/polymer magnetic nanocomposites films on silicon. Non-equivalent properties combining non-conducting, high permeability and enhanced permittivity with high percolation threshold are discussed.**

Keywords—nanoparticles, nanocomposites, magnetic, dielectric

I. INTRODUCTION

Miniaturization of RF circuits has been a crucial objective for years resulting in the integration of capacitors and inductors, excepted antennas. Antennas did not follow the trend because size reduction leads to certain poor radiating performance which can be hardly compensated with historical materials as well. Indeed neither high permittivity dielectrics nor high permeability ferromagnetics have succeeded yet with antennas. They all result in detrimental reduction of bandwidth and gain. Recently ideal magnetic materials (no conductivity, no losses) have been shown advantageous for antenna miniaturization as large permeability helps to keep radiating efficiency and bandwidth [1]. These theoretical results have motivated non-conducting magnetic materials development for antennas. But conductivity with high permeability ferromagnetic materials is the technological obstacle. Nanocomposites (NCs) consisting of metallic magnetic nanoparticles (NPs) embedded in a polymer matrix become an attractive candidate.

NPs of oxides and films have been synthesized and thoroughly studied so far [2], [3], [4]. Indeed NPs undergo complete oxidation during synthesis or functionalization due to their smal size (< 50 nm). NCs of magnetic oxide have been investigated but they display ultra-low permeability which hardly matches with applications. To prevent metallic NPs from oxidation, the use of graphene shell has been reported first by [5]. This has led to recent development of metal/graphene (M/C) nanopowders. Early commercials products appeared in 2006 mainly for bio applications. Besides bio applications, a very few works have been reported on films of M/C NPs. In [6] authors show Co/C NCs bulky thermoplastics for actuator applications. In [7] first dielectric investigation of films of conductor-dielectric NCs is shown. The emerging scheme of metal-dielectric NCs in which conducting NPs (magnetic) are used embedded in a dielectric polymer matrix opens a new field for magnetic materials for antennas. High permeability (μ) and permittivity (ε) combined with no conductivity (σ) is expected close to the percolation threshold. Adjusting volume fraction and spacing between M/C NPs would allow controlling percolation phenomena and achieving optimal μ/ε ratio for high performing miniature antennas.

The aim of this work is to investigate a process, which combines solution-cast, spin-coating and hot-press planarization process for preparing uniform non-conducting magnetic metal-polymer NCs films on silicon. Two films of polystyrene (PS) NCs were prepared using commercial NPs of Co/C and Ni/C, respectively, with a similar volume fraction of ~ 20 vol. %. The originality of this work is to use a double core-shell scheme consisting of graphene (as a first shell) to prevent oxidation and a functionalizing polymer (as the second shell) to avoid percolation.

II. EXPERIMENTAL DETAILS

A. Starting materials

This work is based on metal/graphene NPs, as early proposed by [5], because graphene efficiently prevent NPs from oxidation. Additionally we use the graphene shell as a specific platform for chemical grafting of polymers. This makes chemistries independent of the nature of NPs material and useful for others applications. Commercial Co/C NPs ($\leq$ 50 nm) and Ni/C NPs ($\leq$ 20nm) are purchased from Sigma-Aldrich and Skyspring Nanomaterials, respectively. The properties of the raw nanopowders will be investigated first to allow comparison after solution-cast and after film deposition. Then two polymers are used, one for the non-conducting second shell and one for the final matrix. The functionalizing polymer for the second shell consists of pyrene-terminated PS with 5.6 kg.mol^{-1} (Py-R-PS$^{5.6k}$) provided by Polymer Source. The molecular weight has been chosen to obtain a shell thickness of ~2-4 nm to ensure electrical insulation and allow dipolar interaction between magnetic NPs so that NCs can perform close to percolation threshold and show unusual properties. The final encapsulating polymer is a commercial 35 kg.mol^{-1} PS (PS^{35k}) from Sigma-Aldrich. The solvent used for solution-cast is chloroform (CHCl$_3$). This work only used non-covalent chemistries (π-π interactions) which are compatible with clean room environment.

978-1-4799-4993-9/14 $31.00 © 2014 IEEE

B. NPs solution-cast

First 500 mg of M/C (M=Co or Ni) is dispersed in 15 mL of $CHCl_3$. The dispersion is sonicated during 30 min at 100 W by means of a VCX500 ultrasonic processor (Sonics). During sonication 4-5 mg of Py-R-PS$^{5.6k}$ is added to the solution. Pyrene groups of Py-R-PS$^{5.6k}$ interact with graphene coating through π-stacking thanks to delocalized electrons sharing. The obtained solution is called M/C//Py-R-PS$^{5.6k}$. Then 2 wt.% of PS^{35k} is added to M/C//Py-R-PS$^{5.6k}$ and the final solution is sonicated at the same power for another 30 min. PS^{35k} simply interacts with Py-R-PS$^{5.6k}$ by self-chemical compatibility. The obtained final solution is labeled M/C//Py-R-PS$^{5.6k}$/PS^{35k}.

The stability of the solutions was investigated first. Sedimentation experiments have been performed. For this purpose, the absorbance of the solutions was measured at different wavelengths just after sonication. Absorbance curves (integrated area) vs. time are shown in Figure 1 for the different steps. The first step consists in functionalization of NPs with Py-R-PS$^{5.6k}$. It results in increasing the NPs stability by a factor 5 (red curve) comparing with a raw solution (green curve). Second, when adding final PS^{35k} for encapsulation the sedimentation time increases by a factor 13 (blue curve). Thus the proposed solution-cast leads to homogeneous stable NPs solution over 75 min which is suitable for film fabrication.

C. NCs film deposition

Films of NCs are deposited on 4-inches oxidized silicon wafers (later 8-inches wafers) for microelectronics purposes. To form a film, 750 μL of NC solution is spin-coated at 1000 rpm, with an acceleration of 5000 rpm.s^{-1}. Then as-spin-coated films are annealed at 110 °C for 10 min on a hot plate for solvent evaporation. These conditions lead to uniform NCs films of 2 μm thick with typically less than 5% of uniformity. As the surface of NCs films may become rough when the volume fraction of NPs exceeds 10% a hot-press planarization process has been developed. Films of NCs are pressed at 120 °C and 40 bars for 5 min with EITRE NIL Obducat tool. Finally, planarized films of NCs are flat with negligible surface roughness (< 50 nm) and compatible with further metallization process. Therefore, printed antennas can be further realized directly on the top surface of the NCs films.

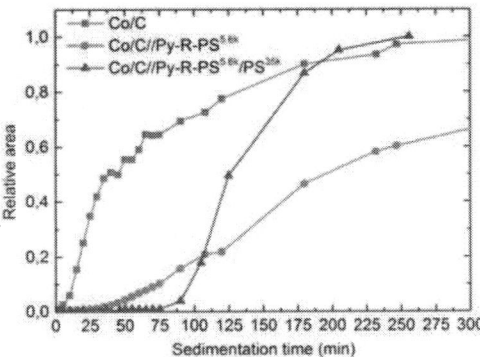

Figure 1 Relative area of integrated absorbance curves vs. time

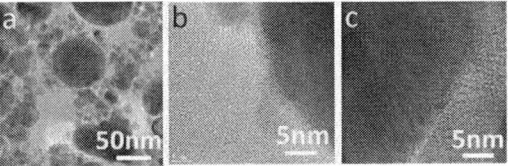

Figure 2 TEM images of commercial Co/C NPs (a&b) and a Co/C//Py-R-PS$^{5.6k}$ super-NPs inside a composite (c)

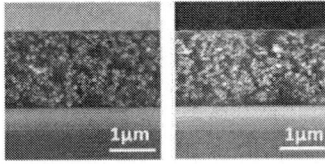

Figure 3 SEM cross section images of planarized NCs films of cobalt (left) and nickel (right)

III. RESULTS AND DISCUSSION

A. Microstructure and volume fraction of NCs films

NPs and NCs films have been observed by SEM (Zeiss Ultra Plus microscope) and TEM (FEI Osiris microscope).

First, as received commercial NPs have been observed to evaluate the average dispersity of NPs size and the quality of the graphene shell (number of layers, continuity etc.). Indeed the integrity of graphene layers is a key point as they prevent from oxidation and are used as a specific chemical grafting platform. As shown on Figure 2 (a&b) Co/C commercial NPs are quite polydispersed with diameters ranging from 10 to 50 nm. Ni/C powders (not shown here) are similar but the mean particle diameter is smaller (~ 15 nm). In both cases, graphene layers surrounding NPs are well visible, indicating a shell thickness of a few nanometers, corresponding to 5-10 graphene sheets. Figure 2 (c) is a detailed image of a super-particle of Co/C//Py-R-PS$^{5.6k}$ inside a film. So far TEM cannot contrast polymers and disable viewing both Py-R-PS$^{5.6k}$ shell and PS^{35k} matrix. However it is possible to see the graphene shell very detailed and even count the number of layers. This allows concluding that graphene has not been damaged during solution-cast and deposition and that NPs are still protected against oxidation. Therefore, aging of NCs films will be limited and long term stability will be expected. Preliminary results have shown that magnetic properties of both NPs and NCs do not change (less than few %) over 6 months.

Figure 3 shows cross section SEM images of planarized films of M/C//Py-R-PS$^{5.6k}$/PS^{35k}. Films are flat and exhibit quite homogeneous NPs volume distribution from top-to-bottom, regardless of the nature of NPs (Co or Ni). Because of magnetic dipolar interactions, super-particles of M/C//Py-R-PS$^{5.6k}$ tends to aggregate naturally. But aggregates are rather small (< 0.5 μm) and well dispersed. One notes that the size distribution of NPs is less dispersed in films than in raw powders (small NPs are less present) which can be attributed to the removal of the supernatant during solution-cast. The absence of very big aggregates (larger than the thickness) results of the sedimentation time (75 min) exceeding the duration of NCs preparation. Therefore, the proposed process for NCs films on silicon wafer which combines solution-cast, spin-coating and hot-press planarization process successfully leads to flat, uniform, and non-percolated films of metal-polymer.

In order to determine the NPs weight and volume fractions of the films, Thermo-gravimetric-analysis (TGA) have been performed from room temperature to 500 °C at 10°C.min^{-1} under N$_2$ gaz flow (Setaram TGA 92). NCs are peeled off from the substrate and the residual product is used for TGA. The results for Co/C and Ni/C films are shown on Figure 4.

TGA profile for Co/C//Py-R-PS$^{5.6k}$/PS^{35k} indicates that PS^{35k} matrix degradation starts at 250 °C and finishes at 430 °C. For Ni/C//Py-R-PS$^{5.6k}$/PS^{35k} the PS^{35k} matrix starts to degrade earlier at 225 °C and ends later at 445 °C. These numbers are consistent with PS. The faster degradation for Co/C NCs may be due to larger mean particle diameter (50 nm for Co/C in contrast to 20 nm for Ni/C). TGA profiles do not give indication about the degradation of the Py-R-PS$^{5.6k}$ shell as it can hardly be separated from the PS^{35k} matrix. Finally TGA indicates metal/polymer weight fraction for each film of NCs: 65.4 wt. % for Co and 62.1 wt. % for Ni NCs. For comparison purposes, it is more adequate to describe NCs films in terms of volume fraction (vol. %). TGA weight fraction is converted into volume fraction by means of equation (1):

$$\varphi_v = \frac{1}{1 + \frac{\rho_M}{\rho_{PS}} \cdot \frac{1}{\varphi_w - 1}} \quad (1)$$

where φ_v and φ_w are the volume and weight fractions, ρ_M and ρ_{PS} are the densities of metal and PS. The results are reported in Table 1 with 18.4 vol. % for Co and 16.3 vol. % for Ni NCs. These values are close to expected values (20 vol. %).

B. Magnetic and dielectric properties of NCs films

Quasi-static magnetic measurements have been performed at room temperature on a Microsense Vibrating Sample Magnetometer (VSM). Hysteresis loop for 2 µm thick planarized Co/C//Py-R-PS$^{5.6k}$/PS^{35k} film is shown in Figure 5 and compare to those of raw Co/C NPs. This film exhibits a remarkably high saturation magnetization (M$_s$) of 0.55 T which is consistent with a volume fraction of ~20 vol. %. Hysteresis loop also shows a coercive field (H$_c$) of 250 Oe and a magnetocristalline anisotropy field (H$_K$) of 1020 Oe which significantly differ from raw Co/C NPs.

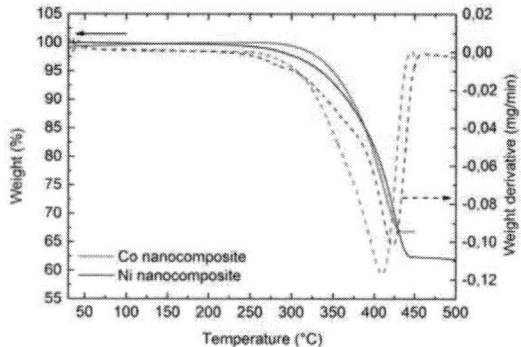

Figure 4 TGA profiles of Co/C and Ni/C films of nanocomposites

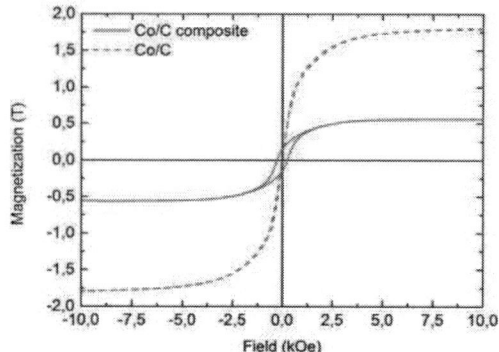

Figure 5 Hysteresis loops of Co/C film of nanocomposites and Co/C raw particles

The increase of H$_c$ and of H$_K$ indicates that Co/C NPs inside the film are magnetically coupled. In contrast raw powders of Co/C are a mixture of paramagnetic and super paramagnetic NPs due to the presence of very small NPs (< 10 nm). Thus the proposed process successfully leads to appropriate diluted ferromagnetic behavior with Co/C films.

In contrast, Ni/C//Py-R-PS$^{5.6k}$/PS^{35k} film with comparable volume fraction (20 vol. %) has shown mitigated results with M$_s$ of 0.06 T which is approximately one quarter of raw Ni/C NPs. Changes in H$_c$ (50 Oe against 20 Oe) and H$_K$ (360 Oe against 250 Oe) are less significant too. This may indicate that super-paramagnetic NPs (which do not contribute to magnetization and dipolar coupling at room temperature) remain present in the film for half of the volume fraction. Therefore the same process with Ni/C leads to more diluted magnetization and weaker ferromagnetism.

Permeability (µ) measurements have been performed up to 1 GHz by wideband impedance spectrometry using the single coil technique and Agilent PNA N5222A. Films of Co/C//Py-R-PS$^{5.6k}$/PS^{35k} show µ ~ 4 to 5 which remains flat up to 1 GHz. This indicates that eddy current do not propagate into the film (2 µm thick). This confirms the double core shell scheme successfully leads to "no" conducting film at high frequency. No magnetic resonance is observed up to 1 GHz. The magnetic resonance frequency may depends on the size and shape of aggregates inside the film (the larger the aggregates the lower the frequency). The absence of magnetic resonance below 1 GHz means that films of Co/C//Py-R-PS$^{5.6k}$/PS^{35k} are quite uniform with small aggregates (< 1 µm). Comparing with similar film of Ni/C, lower permeability is observed (µ ~ 2) as these films undergo weaker ferromagnetism.

Dielectric constant (ε) and conductivity (σ, S.m^{-1}) measurements have been performed at 1 kHz with MIM capacitors of NCs. Films have been deposited on metallized 4-inches silicon wafers (metallization layer serves as bottom electrode). Then gold has been evaporated on top of the film through a hard mask to make the top electrode. A comparison was made with MIM capacitors with pure PS^{35k}. First, the results point out those films of M/C//Py-R-PS$^{5.6k}$/PS^{35k} show very low conductivity (~10^{-6} S.m^{-1}). It is higher than pure PS^{35k} (10^{-10} S.m^{-1}) but it confirms no percolation at 20 vol. % of NPs. Second, films exhibit enhanced permittivity compared to pure PS^{35k}. Permittivity enhancement (by a factor 7) is consistent with [7] where authors observed similar increase (actually slightly less significant) for films of copolymers of PVDF filled with 20 vol% Ni NPs. As explained in [7] percolation phenomenon with metal/polymer is responsible of unusual dielectric properties. Films become conductive when the content of conducting filler is higher than the percolation threshold. This happens when the conducting filler forms conducting channels across the sample. With the original double core $M/C//Py\text{-}R\text{-}PS^{5.6k}$ super-particles percolation threshold can be significantly increased compared to usual metal-polymer composites (up to 50 vol %) which corresponds to the ultimate formation of large aggregates across the film. Therefore, high percolation threshold with double core $M/C//Py\text{-}R\text{-}PS^{5.6k}$ super-particles embedded in a matrix of PS^{35k} lead to non-equivalent properties with ''no'' conductivity, high permeability and enhanced permittivity. Therefore, on concludes these films are suitable for antenna applications.

Magnetic and dielectric results are summarized in Table 1.

Low-profile antennas (sub-wavelength) suffer from narrow bandwidth and weak radiating efficiency. According to [1], if the material surrounding the antenna is non-conducting and has high permeability and permittivity, the wavelength (λ) in such medium will be shorter by a factor equal to the square root of the product $\mu.\epsilon$. Additionally, $\mu > 1$ would help keeping appropriate bandwidth. When considering an antenna perfectly surrounding by films of $Co/C//Py\text{-}R\text{-}PS^{5.6k}/PS^{35k}$ maximum size reduction would be 1 : 7. Furthermore as films of $M/C//Py\text{-}R\text{-}PS^{5.6k}/PS^{35k}$ have been successfully deposited on flexible substrates (PEN), simple printed antennas on top of $M/C//Py\text{-}R\text{-}PS^{5.6k}/PS^{35k}$ film would benefit of intermediate miniaturization factor (up to 1 : 3) which can be a breakthrough for miniaturizing RFID tags.

IV. CONCLUSION

We report a process including solution-cast, spin coating and hot-press planarization for preparing non conducting uniform metal-polymer magnetic NCs films on silicon. Two films of $M/C//Py\text{-}R\text{-}PS^{5.6k}/PS^{35k}$ were prepared using this process with a similar volume fraction of ~ 20 vol. % using commercial NPs of Co/C and Ni/C. Structural, magnetic and dielectric characterizations have been detailed. Non-equivalent properties combining non-conducting, high permeability and enhanced permittivity with high percolation threshold are shown. We conclude that these films are suitable for antenna applications indicating theoretical size reduction of 1 : 7 (1 : 3) while keeping efficient radiating performance similar to air.

ACKNOWLEDGMENTS

This work was performed in the frame of TOURS 2015, project supported by the French ''Programme de l'Economie Numérique des Investissements d'Avenir''.

Table 1 Summary of properties of films of $M/C//Py\text{-}R\text{-}PS^{5.6k}/PS^{35k}$ (20 vol.%) and comparison with film of pure PS^{35k}

	PS^{35k}	Ni NC	Co NC
φ_w (%)	0	62.1	65.4
φ_v (%)	0	16.3	18.4
M, (T)	-	0.06	0.55
H_k (Oe)	-	360	1020
μ	1	2	5
ϵ	1.5	11.2	12.0
σ (S.m^{-1})	~10^{-10}	~10^{-6}	~10^{-6}

REFERENCES

[1] R. C. Hansen and M. Burke, "Antennas with magneto-dielectrics," *Microwave and Optical Technology Letters,* vol. 26, pp. 75-78, 2000.

[2] N. R. Jana, Y. Chen, and X. Peng, "Size- and Shape-Controlled Magnetic (Cr, Mn, Fe, Co, Ni) Oxide Nanocrystals via a Simple and General Approach," *Chemistry of Materials,* vol. 16, pp. 3931-3935, 2004/10/01 2004.

[3] Y. Wang, X. Teng, J.-S. Wang, and H. Yang, "Solvent-Free Atom Transfer Radical Polymerization in the Synthesis of Fe2O3@Polystyrene Core−Shell Nanoparticles," *Nano Letters,* vol. 3, pp. 789-793, 2003/06/01 2003.

[4] B. Frka-Petesic, J. Fresnais, J.-F. Berret, V. Dupuis, R. Perzynski, and O. Sandre, "Stabilization and controlled association of superparamagnetic nanoparticles using block copolymers," *Journal of Magnetism and Magnetic Materials,* vol. 321, pp. 667-670, 2009.

[5] T. Hayashi, S. Hirono, M. Tomita, S. Umemura, and J.-J. Delaunay, "Magnetic Thin Films of Cobalt Nanocrystals Encapsulated in Graphite-Like Carbon," *MRS Online Proceedings Library,* vol. 475, pp. null-null, 1997.

[6] N. A. Luechinger, N. Booth, G. Heness, S. Bandyopadhyay, R. N. Grass, and W. J. Stark, "Surfactant-Free, Melt-Processable Metal–Polymer Hybrid Materials: Use of Graphene as a Dispersing Agent," *Advanced Materials,* vol. 20, pp. 3044-3049, 2008.

[7] L. Zhang, W. Wang, X. Wang, P. Bass, and Z.-Y. Cheng, "Metal-polymer nanocomposites with high percolation threshold and high dielectric constant," *Applied Physics Letters,* vol. 103, pp. -, 2013.

Fabrication and Characterization of ECM Memories Based on a Ge₂Sb₂Te₅ Solid Electrolyte

C. Rebora, M. Bocquet, T. Ouled-Khachroum, M. Putero and D.Deleruyelle

Laboratoire IM2NP, Institut Matériaux Microélectronique Nanosciences de Provence
UMR CNRS 7334, Aix-Marseille Université
Avenue Escadrille Normandie Niemen, 13397 Marseille Cedex 20, France
Charles.rebora@im2np.fr

Abstract— **This work deals with the study of electrochemical metallization memory cells (ECM) also called CBRAM (Conductive Bridge RAM). Memory stacks were fabricated by sputtering onto SiO₂/Si substrates and were characterized by atomic force microscopy and a mercury drop probe. These stacks employ a Ge₂Sb₂Te₅ (GST) layer as a solid electrolyte which has been barely employed in CBRAM devices. Electrical measurements demonstrate resistance switching of the stacks due to the formation/dissolution of metallic filaments within the GST layer. However, the memory elements featuring a silver top electrode do not exhibit such switching behavior but show instead an ohmic behavior. This result is interpreted through physical analysis revealing the presence of silver in each layer of the memory devices. Finally, a physical model is presented. This model was used to interpret adequately the bipolar resistance switching phenomenon observed in the memory stacks.**

Keywords—ECM; CBRAM; characterization; model

I. INTRODUCTION

Due to their interesting properties such as high programming speed and high endurance, Electrochemical Metallization memories Cells (ECM) are increasingly studied by the scientific community. These devices are based on the resistance change of a Metal-Insulator-Metal (MIM) structure due to the formation/rupture of a conductive filament in a solid electrolyte [1-2]. The first electrode is composed of silver (Ag) or copper (Cu) and is called the active electrode (AE). The second one is consists of an electrochemically inert metal such as platinum (Pt) or tungsten (W): it is called the inert electrode (IE). The solid electrolyte sandwiched between the two electrodes may consist of different materials such as an insulator (SiO₂, HfO₂, …) or amorphous chalcogenides layers (Ge$_x$Se$_y$, Ge$_x$Sb$_x$Te$_y$, …).

The memory effect of CBRAM devices can be described as follows. In its pristine state, the cell features a High Resistance State (HRS) [fig. 1(a)] [1]. Upon the application of a positive voltage to the active electrode, an anodic dissolution process occurs and copper or silver cations penetrate into the electrolyte. Due to the electric field between the electrodes, the metallic ions migrate through the electrolyte. When they reach the inert electrode, they are reduced [fig. 1(b)]. Thus, a metallic cluster is formed at the electrolyte/inert electrode interface which initiates the growth of a metallic dendrite towards the active electrode [fig. 1(c)]. Resistance switching toward a Low Resistane State (LRS) – ON state – is achieved when the metallic filament bridges the two electrodes [fig. 1(d)]. The above mentioned HRS→LRS transition corresponds to the SET phenomenon.

By reversing the voltage polarity applied to the AE (*i.e.* negative voltage), the filament is disrupted by dissolution and the metallic cations migrate toward the active electrode. Therefore, the cell returns into a HRS or OFF state [fig. 1(e)]. The LRS→HRS switching is called the RESET phenomenon. Previous studies showed that the filament cannot be completely dissolved, leaving residual conduction paths [2]. In this case, the cell cannot returns in its pristine resistance state. Therefore the HRS is intermediate between the LRS and pristine state. The ECM cells have a bipolar behavior, characterized by two distinct resistance states that can be read unambiguously by sensing the current at low voltage.

GST is mostly used as the prototype material for phase change memory. It has the advantage of being deposited in its amorphous state which is insulating. Recently, GST has been successfully used as the solid electrolyte layer of ECM: the devices featured printed Ag electrodes on plastic substrates [3]. One advantage of GST is that it has a lower Young modulus than oxides, close to the value of plastic [4], which is beneficial for flexible electronics applications.

In this paper, we present memory cells fabricated with a Ag active electrode deposited by sputtering, as the whole memory stack. We show that the memory effects can be obtained only when the AE is beneath the GST layer. The ECM behavior is characterized by Conductive Atomic Force Microscopy (C-AFM) and by using a mercury drop probe. Finally, the growth/dissolution of the filament and the electrical characteristics are also modeled by physical compact model. A satisfactory agreement is achieved with experimental data.

II. SAMPLES FABRICATION

Memory cells were manufactured by radio frequency sputtering. The stacks were deposited on 200 mm SiO₂/Si substrates. This type of substrate allows an easy transfer toward physical or electrical characterization tools. Two types of samples were fabricated. The first one consists of memory stacks without inert electrode: they consist of GST(45 nm)/ Ag(140 nm)/ SiO₂/Si [fig. 2(a)]. In this case, the metallic tip of an AFM will be used as an inert electrode. Beyond the study of resistance switching at a nanoscale, such a model system emulates an extreme downscaling of the memory device, the cell area being defined by the contact radius between the AFM tip and the solid electrolyte. The second type of samples consists of memory devices featuring an inert Pt electrode. The cells are based on the following stack: Ag(140nm)/ GST(45nm)/ Pt-Ti(45nm)/ SiO$_2$/Si [fig. 2(b)]. In this case, the Ag is deposited through a shadow mask to define square pads with size ranging from 30 to 250 μm [fig. 2 (c)].

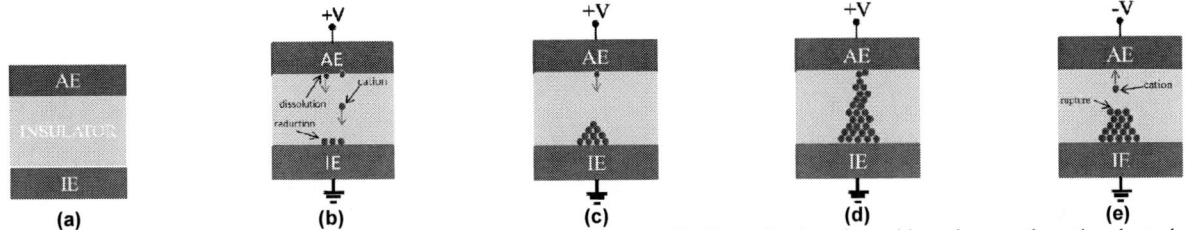

Figure 1: (a) In its pristine state, the ECM cell is in a very high resistance state (HRS). (b) The application of a positive voltage on the active electrode (AE), promotes the dissolution of the electrode and the migration of metallic cations toward the inert electrode (IE). (c) Formation of an metallic cluster at the interface insulator/inert electrode initiate the growth of a metallic filament in the direction of the active electrode. (d) When the filament bridges the AE and the IE (SET phenomenon). The cell is in a low resistance state (LRS) called ON state. (e) The application of a negative voltage to the AE dissolves the filament (RESET phenomenon). The cations migrate to the AE. The cell returns in a HRS or OFF state.

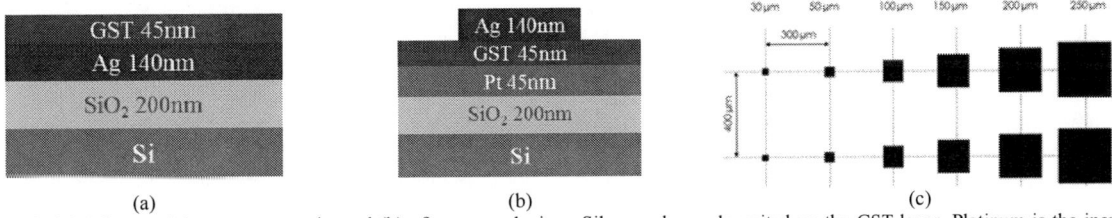

Figure 2: (a) Scheme of the memory stacks and (b) of memory devices. Silver pads are deposited on the GST layer. Platinum is the inert electrode. (c) Scheme of the mask used to fabricate the pads.

III. RESULTS

A. Characterization of memory stacks by C-AFM

C-AFM measurements were performed on a JSPM 5410 (JEOL society) Atomic Force Microscope (AFM). A tip covered with Ti-Pt is used to act as the IE. C-AFM characterization was performed following three steps.

At first [Fig. 3 (a) and 3 (b)], topographic images and current maps were measured (using a tip bias voltage of 0.25V) on the GST surface (Fig.3). As shown on the current maps, no preferential conduction pathway is evidenced at this stage.

In a second step (Fig. 4), a double voltage ramp (-1.5 V to 2 V) is applied positioning the tip randomly on the pre-characterized surface. In order not to damage the tip, the current compliance is 10 nA. I-V characteristics show that the application of a positive voltage greater than 1 V allows for switching the stack from HRS to LRS, corresponding to the formation of a silver filament (Fig. 4). However, for a voltage of 50 mV the resistance is still high with a value of 62 MΩ. It could be attributed to the formation of a conductive filament which is stop by the current compliance. Thus, there is a gap of GST between the tip and the top of the filament. During the return scan, a current drop and a return to a high resistance state is observed for a voltage of 0.8V, it corresponds to the RESET step (rupture of the filament). The use of opposite bias to the creation and dissolution of the conductive filament confirms the bipolar nature of the memory effect.

Subsequently to this full write/erase cycle, a new mapping of topography and current is performed with a zoom on the part of the surface where the IV-characteristic has been obtained. As shown in Fig. 5, a clear change in the topography and the presence of a small (i.e. diameter below 10nm) residual conductive path can be observed at the location where the spectroscopic I-V measurement was performed. This hillock can be attributed to formation/dissolution of a metallic filament.

B. Characterization of memory stack using a mercury drop probe

The memory stack was also characterized by a mercury drop probe connected to a semiconductor parameter analyzer Agilent B4156. A mercury drop (750μm of diameter) is deposited at the surface of the memory stack and serves as IE. The bias is applied to the mercury drop (Fig. 6). In this case the conductive filament is formed for negative bias voltages.

As shown in Fig. 7(a), the bipolar behavior of the observed by C-AFM was confirmed. Surprisingly, we also demonstrated that the memory stack can operate in a singular unipolar behavior [Fig. 7 (b)] in which the creation and the rupture of the filament are obtained for negative biases only. Some authors has reported the role of the electrodes in the behavior switching [5] and some of them has observed unipolar behavior for Cu/ZrO$_2$:Cu/Pt CBRAM cells [6].

It could be suppose that the electrode of mercury plays a particular role in the behavior switching of the cell.

978-1-4799-4993-9/14 $31.00 © 2014 IEEE

Thus, the mercury drop can diffuse in the GST layer and participate to a local joule heating which destructs a part of the filament. The very low voltages for filament formation (-0.8 V) suggests that the filament is never completely destructed.

C. Electrical characterization of memory devices

Memory devices were characterized using a probe station connected to a parameter analyzer Agilent B4156. The size of the devices is defined by the silver pad (Fig. 3). Since the voltage bias is applied to the top AE, formation of conductive filaments is expected to occur for positive voltages. To compare the electrical characteristics of devices of various area, the current density is plotted a function of the applied voltage in Fig. 8. It has to be noted that all the J-V curves exhibit an ohmic behavior for positive voltage biases applied to the AE. Moreover all measured devices quickly reach the current limit set to the parameter analyzer.

Whatever the surface of the pad, the same value of current density is obtained for positive bias. This means that the current is proportional to the pad surface and that conduction is uniform overall the pads. To understand the origin of the failure of these devices, Transmission Electron Microscopy (TEM) analyses were performed. Figure 9 (a) shows a cross section image performed on a pristine memory. Each layer of the memory device can be seen. We can notice that the GST layer has an inhomogeneous contrast. Energy dispersive X-ray analyses (EDX) were performed in the various layers of the stack [Fig. 9 (b)]: they reveal the presence of silver throughout the memory stack, including in the Ti-Pt layer.

This explains why the devices exhibiting an ohmic behavior, were not able to operate a resistance switching effect. The origin of the proliferation of Ag within the memory stack might be attributed to the sputtering of Ag on top of the GST surface. It seems lead to an unexpected incorporation of silver within the GST layer avoiding any return to a High Resistance State.

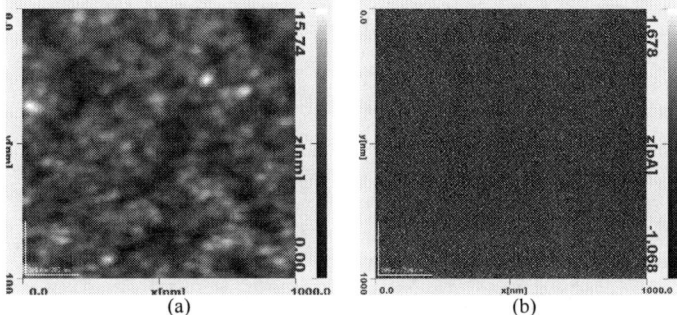

(a)   (b)

Figure 3: AFM images of the memory stack in the pristine state. (a) Topographic image of the GST surface. The average roughness is 7.8 nm. (b) Current image performed on the same area: the stack memory is initially in a HRS (order of pA) state.

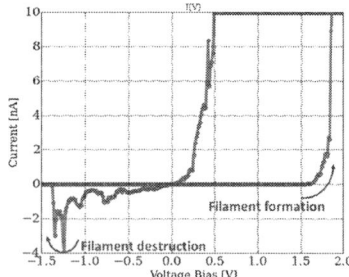

Figure 4: I-V curve of the memory stack, a double ramp voltage: -1.5V→2V→-1.5V is applied on the surface of Figure 3.

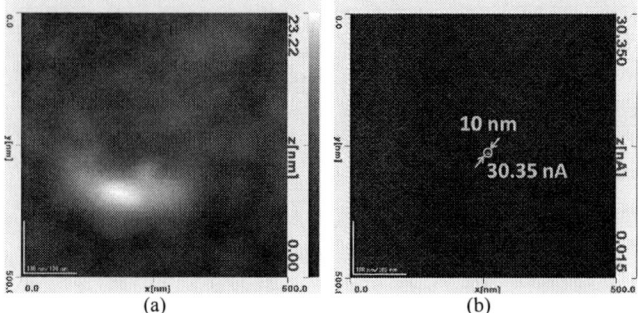

(a)   (b)

Figure 5: (a) Surface topography revealing the formation of a protrusion where the I-V cycle of Fig. 4 was achieved and (b) current map revealing a small (i.e. <10nm width) residual conductive region at the location where the conductive filament was formed.

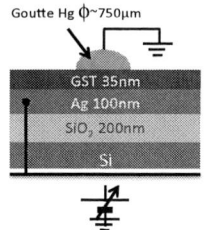

Figure 6 : Schematic view of the experimental setup used for the Hg drop measurements.

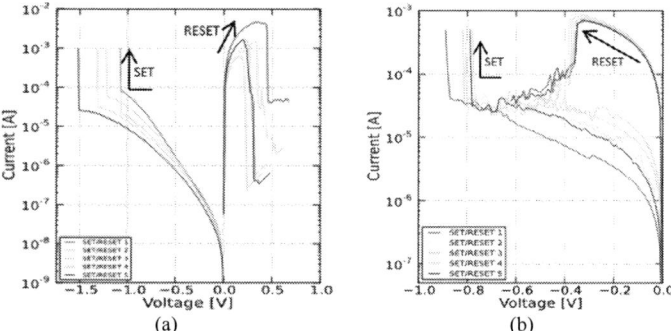

(a)   (b)

Figure 7: I-V characteristics obtained by mercury drop. The scales of the curves are in semilog. Electrical measurements revealed either a (a) bipolar behavior or a (b) unipolar behavior.

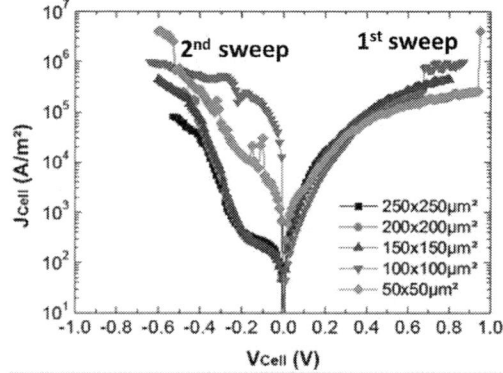

Figure 8: J-V curves of the memory devices featuring different pad size.

978-1-4799-4993-9/14 $31.00 © 2014 IEEE

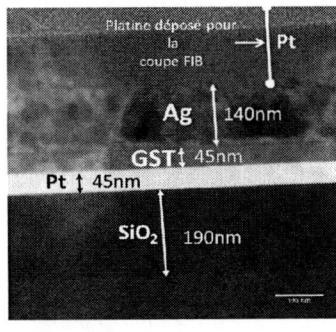

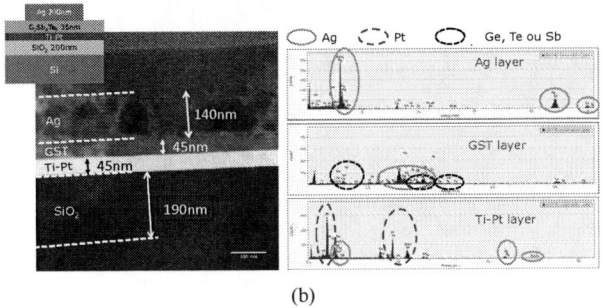

(a)

(b)

Figure 9: (a) TEM image of a pristine memory device (b) EDX analyses performed through the the Ag, GST and Ti-Pt layers.

IV. PHYSICAL MODELING

To gain more insight on the physical phenomena governing the formation/rupture of conductive filament in the GST, a physical model has been developed. This model is based on a compact model reported in Ref. 7. The model assumes that the flux of cations through the solid electrolyte is due to a Mott-Gurney hopping process. This self-accelerated process ends when the filament bridges both electrodes, achieving the SET process. For the sake of simplicity, we consider here that the filament has a cylindrical rather than conical shape [7].

Concerning the RESET process, we assume that applying a negative voltage dissolves the portion of the filament at the anode side. The corresponding dissolution process is described by the Butler-Volmer equation [7]. As soon as a gap is opened in the filament its dissolution rate is through a Mott-Gurney current describing the migration of cations toward the AE.

As shown in Fig. 10, the model showed a satisfactory agreement with electrical measurements performed on the memory stacks using an Hg drop electrode.

V. CONCLUSION

ECM memory stacks (without an inert electrode) were made by sputtering. Electrical characterizations by AFM probe and dropping mercury have confirmed the existence of a bipolar memory effect due to the formation/rupture of a metal filament in the solid electrolyte based GST.

However, the memory devices are, damaged due to diffusion of silver throughout the MIM structure and do not allow to observe memory effect of ECM types. To solve this problem, we plan to use an interstitial layer between the GST layer and the silver electrode to limit the diffusion of silver during the deposition of the electrode. Complementary manner, we also plan to reverse the order of electrodes deposition. In first, deposit the active electrode then the inert electrode.

Finally, a physical model has been developed. It is based on the ionic transport of metal cations in the solid electrolyte and the redox rate calculation. This model allowed us to interpret adequately the results on memory stacks during this study. These first results are encouraging achievement to the manufacture of non-volatile memories with low operating voltages cells.

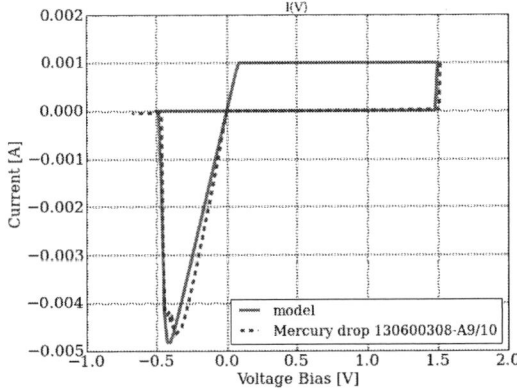

Figure 10: I-V characteristics obtained by Hg probe on a GST/Ti-Pt layer and confrontation to the physical model. A satisfactory agreement was obtained.

REFERENCES

[1] I. Valov, R. Waser and J.R. Jameson and M.N. Kozicki, "Electrochemical metallization memories—fundamentals, applications, prospects", Nanotechnology 22 (2011) 289502 (22pp).

[2] T. Tsuruoka, K Terabe and T Hasegawa and M. Aono "Forming and switching mechanisms of a cation-migration-based oxide resistive memory", Nanotechnology 21 (2010) 425205 (8pp).

[3] D. Deleruyelle, M. Putero, T. Ouled-Khachroum, M. Bocquet, M.-V. Coulet, X. Boddaert, C. Calmes and C. Muller, "Ge$_2$Sb$_2$Te$_5$ layer used as solid electrolyte in conductive-bridge memory devices fabricated on flexible substrate", Solid-State Electronics 79 (2013), p.159–165.

[4] Y. Choi and Y.-K. Lee, "Elastic Modulus of Amorphous Ge$_2$Sb$_2$Te$_5$ Thin Film Measured by Uniaxial Microtensile Test," Electron. Mater. Lett., vol. 6, no. 1, pp. 23–26, Mar. 2010.

[5] Y. Hua, D. Perello, M. Yun, D.-H. Kwon, M. Kim, "Variation of switching mechanism in TiO$_2$ thin film resistive random access memory with Ag and graphene electrodes", Microelectronic Engineering, 104 (2013), 42–47.

[6] W. Guan, M. Liu, S. Long, Q. Liu and W. Wang, "On the resistive switching mechanisms of Cu/ZrO$_2$:Cu/Pt", Appl. Phys. Lett. 93, 223506 (2008).

[7] S. YU and H.-S. Philip Wong, "Compact Modeling of Conducting-Bridge Random-Access Memory (CBRAM)", Electron Devices, IEEE Transactions on, vol. 58, no. 5, pp. 1352–1360, 2011.

978-1-4799-4993-9/14 $31.00 © 2014 IEEE

Role of Nanowire Length in Morphological and Electrical Properties of Silicon Nanonets

Pauline Serre, Céline Ternon
LTM, CNRS/UJF-Grenoble1/CEA
Grenoble, France
Celine.ternon@grenoble-inp.fr

Pierre Chapron
IMEP-LAHC, Grenoble INP - MINATEC
Université Grenoble Alpes,
Grenoble, France

Pierre Chapron, Quentin Durlin, Anaïs Francheteau, Arthur Lantreibecq, Céline Ternon
LMGP, Grenoble INP - MINATEC
Université Grenoble Alpes, Grenoble, France

Abstract— While electronic devices based on a unique nanowire are intensively studied, we present here the potential of random assemblies of silicon nanowires (SiNWs) as interesting material for electronic applications. This structure, called a nanonet (for NANOstructured NETwork), exploits the unique properties of the nanowires such as high surface area and large aspect ratio while avoiding the long, expensive and complex processing required for individual nanowire devices. In this work, the vacuum filtration method is used to elaborate homogeneous, reproducible and conducting silicon nanonets. A study of the morphological and electrical behavior of such nanonets is provided as a function of the SiNW length and density. We demonstrate that despite the complexity of the nanonet geometry, it is possible to control their morphology and their electrical properties. Besides, we show that these characteristics strongly depend on the SiNW length and density. In view of the obtained properties, we demonstrate that Si nanonets are electrically active materials with interesting morphological properties and a potential for a wide range of applications and particularly in the sensing field.

Keywords— Silicon nanowires, nanonets, percolation threshold

I. INTRODUCTION

A "nanonet" (NANOstructured NETwork) is defined as a random network of high aspect ratio nanostructures (nanowires (NWs) or nanotubes (NTs))[1, 2]. When its thickness is significantly thinner than the length scale of its individual components, the nanonet is considered as two-dimensional (2D) because such networks show 2D percolation properties. 2D nanonets have several unique properties arising either from the individual components, the NWs or NTs, or from the network structural properties such as high surface area, high transparency and high flexibility, electrical conductance, good reproducibility and fault tolerance [3]. In the literature, 2D nanonets were mainly fabricated by solution-based assembly of the NWs or NTs (for e.g. spray coating [4, 5], Langmuir-Blodgett [6–8] or vacuum filtration [9, 10]), which enables the

The research leading to these results has received funding from the European Community's Seventh Framework Programme (FP7/2007-2013) under grant agreement NANOFUNCTION n°257375.

fabrication of nanonets having a wide range of thicknesses, from sub-monolayer coverage to over 1 μm thick. Among the solution-based assembly methods for the nanonet fabrication, the vacuum filtration method is simple, versatile [10, 12, 14], low cost, and scalable to large areas and guarantees the film homogeneity. The nanonets assembled by this method can be easily deposited at room temperature onto various types of substrate: transparent or opaque, conductor or insulating, flexible or rigid. In the recent years, such nanonets based on carbon NTs, also called carbon nanonets, have been studied and integrated into chemical and biological sensors that exhibit high sensitivity to gas or biological molecules [11–13]. However, studies dealing with nanonets are mainly related to transparent conducting material using carbon NTs and silver NWs [14–20]. By contrast, little literature has been reported for Si nanonets and their possible range of applications [21–23]. Indeed, Si nanonets are more likely to be insulating materials as SiNWs are encapsulated by a SiO_2 shell that prevents the electron transfer at the NW/NW junctions. Besides, carbon nanonets are highly interesting structures for chemical and biological sensing or for flexible electronic as demonstrated with carbon nanonets. But the latter are not fully satisfying as they are composed of both metallic (1/3) and semiconducting CNTs (2/3), limiting the available densities below the metallic percolation threshold. Thus, semiconducting Si nanonets could usefully replace the carbon nanonets and therefore, it is relevant to study the Si nanonet electrical properties and the conduction mechanisms involved. A preliminary step consists in studying the role of the NW/NW junctions by working with degenerated SiNWs in order to discard the other limiting mechanisms and to investigate the percolation properties of such materials.

In this work, we show that degenerated Si nanonets behave like standard percolating media despite the numerous NW/NW junctions and the oxide shell surrounding the SiNWs. This study focuses on the determination of the percolation threshold defined as the critical NW density, d_c, under which no conduction through the nanonet is possible. As described in the literature, such a threshold is directly linked to the NW length composing the nanonet and decreases when the NW length increases [26]. In a first part, the morphological properties of the Si nanonets are studied as a function of the experimental conditions and versus the SiNW length. Then, in a second part, the network electrical properties are studied as a function of the SiNW length and density in the nanonet leading to the determination of the percolation threshold. Finally, this work demonstrates that the complex nanonets can be satisfactorily described using a simplified model structure.

II. MATERIALS AND METHODS

A. Fabrication of the silicon nanonet

Degenerated n-type silicon nanowires ($\rho \approx 10^{-3}\Omega.cm$, $N_d \approx 5.10^{19}$ at.cm^{-3}), were synthesized on <111> silicon substrate by the Vapor Liquid Solid Chemical Vapor Deposition (VLS-CVD) [27] catalyzed by gold droplets which were formed by the dewetting at 800 °C of a gold thin film (2 nm). Silane (SiH$_4$) was used as the gaseous silicon precursor and phosphine (PH$_3$) as the dopant. Hydrogen chloride (HCl) was also added during the growth in order to inhibit the lateral growth and the gold diffusion [28, 29]. This VLS-CVD allows epitaxial growth of SiNWs with a diameter monitored by the catalyst size [30]. In this study, the mean diameter of the as-grown NWs is 85 nm ± 20% while their length is varied (6 µm, 10 µm and 20 µm) and well controlled by the deposition time.

The so-fabricated SiNWs were assembled into nanonets by the vacuum filtration method. First, the SiNWs were dispersed in deionised water by ultrasonication for 5 min. Then, the SiNW solution was characterized by absorption spectroscopy in order to determine the Si amount in solution. After adjusting the Si concentration by diluting the solution until its absorbance at 400 nm equal 0.06, the latter was then filtered through a 0.1 µm porous nitrocellulose membrane [21]. As the solvent goes through the pores, the NWs were trapped on the membrane surface subsequently forming the nanonet. Small volumes of the SiNW solution (5-20 mL) were filtered in order to prepare nanonets with controllable density in the range of the percolation threshold. Finally, by dissolving the membrane in acetone, the nanonets were transferred onto a substrate composed of a 200 nm-thick Si$_3$N$_4$ layer deposited on Si.

B. Electrical test structure

In order to perform electrical measurements, an electrical test structure was elaborated as illustrated in Fig. 1. Metallic electrodes with a diameter of 200 µm and separated by 50 µm were deposited through a shadow mask. To promote a direct contact between the metallic electrodes and the SiNWs, the nanonets were treated just before the metal evaporation with hydrofluoric acid vapor during 30 s in order to remove the native oxide. Then, a trilayer stack composed of 120 nm of nickel, 180 nm of aluminum and 50 nm of gold was deposited. A SEM image of a typical nanonet device is shown in Fig. 1b.

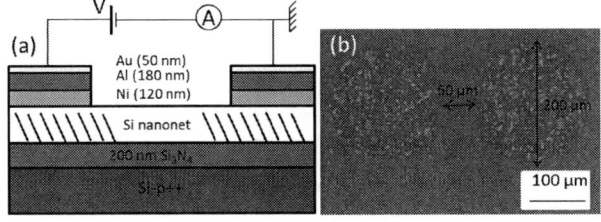

Fig. 1. Silicon nanowire networks based device. Schematic (a) and SEM top view image (b) of the processed electrical test structure based on Si nanonet.

C. Characterization

Morphological studies were performed by scanning electron microscopy (SEM) observations using a Zeiss Ultra+ Microscope. SiNW densities were deduced by analyzing SEM images with the help of Image J software. The electrical characterization of the Si nanonets were performed on a Karl Süss probe station controlled by a HP 4155 analyzer at room temperature, in the dark and in ambient air. As depicted in Fig. 1a, a voltage was applied between two electrodes of the nanonet and the resulting current between them was measured.

In this work, degenerated SiNWs were chosen to get a high conductivity from the nanowires so that the electrical behavior of the network is mainly due to the internanowire junctions. The resistor configuration was used to determine the current value as a function of the voltage for nanonets with different SiNW lengths and densities.

III. RESULTS AND DISCUSSION

A. Si nanonet morphology

The filtration method allows elaborating, at room temperature, coherent and homogeneous nanonets with a good adhesion on the substrate and a good control of the nanowire density. The SiNW density in the nanonet is directly linked to the amount of silicon in solution controlled by the solution absorbance at 400 nm and the filtered volume of SiNW solution [21]. Fig. 2a and 2b show two homogeneous nanonets elaborated from 5 mL and 20 mL of 6 µm-long SiNW solution respectively. No continuous path over long distance (>20 µm) is observed on Fig. 2a, implying that such low density nanonet is below the percolation threshold. By contrast, the numerous continuous paths present in Fig. 2b suggest that the denser nanonet is above the percolation threshold. Besides, on these images, one can notice a bimodal distribution of the NW lengths: the nanonets are composed of long NWs (mean size 5 µm) and short NWs (mean size 1 µm). The sonication step, necessary for the NW dispersion in solution, is responsible for the NW shortening [31] and the size splitting into two populations. The same observation was done for the 10 µm long NWs which split into 8 µm and 2 µm NWs, and for the 20 µm long NWs splitting into 15 µm and 5 µm NWs. As a consequence, the definition of the SiNW density is not obvious and two different methods were compared to determine the NW density. First, densities of the longer NWs (d_L) were directly deduced from SEM images by counting only the long NWs. For this purpose, for each NW length and for each filtrated volume, at least eight images were analyzed at different locations on the nanonet leading to a mean density with error bars matching with the SiNW density variabilities for a given nanonet as reported in Fig. 2c. This method is a correct estimation of the long NW population even if such a density underestimates the real density by neglecting all the small NWs. Second, an equivalent density (d_{eq}) was determined using the coverage area which is defined as the ratio between the surface covered by NWs and the total surface of the SEM images. It is assumed that the surface is covered by a unique NW population characterized by the as-grown length (L) and diameter (D). In such a way, the equivalent density is defined as the ratio between the coverage area and the surface covered by each NW ($L \times D$). Such an equivalent density overestimates the contribution of long NWs to percolation but takes into account the short NWs, while still underestimating the real density.

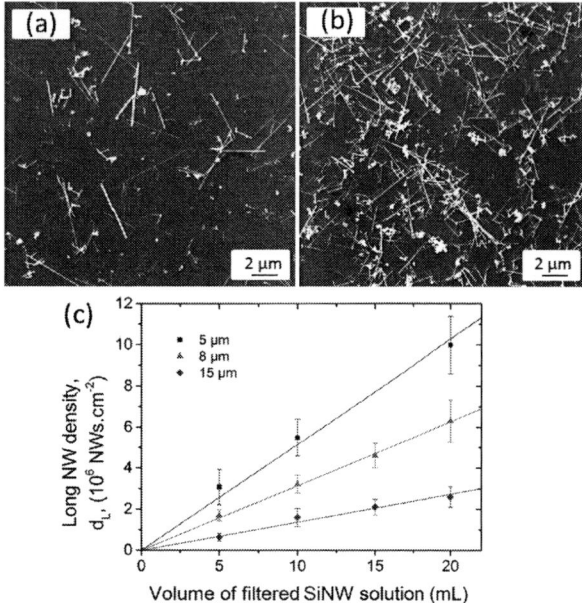

Fig. 2. SEM images of Si nanonets composed of 6 μm SiNWs obtained from (a) 5 mL and (b) 20 mL. (c) Long NW density, d_L, deduced from SEM images by counting only the long NWs as a function of the filtered volume of SiNW solution.

From SEM images, ImageJ software was used to determine the coverage area for a given nanonet. For each nanonet corresponding to a given filtered volume, the same image analysis was performed on more than eight pictures arising from different regions of the samples. The mean coverage area is plotted as a function of the filtered volume in Fig. 3a for nanonets based on different NW lengths. The extreme values of the coverage area are evidenced on Fig3a thanks to the error bars. As the absorbance spectroscopy allows monitoring the silicon amount in solution, as soon as the solution absorbance is fixed (0.06 at 400 nm), the coverage area for a given volume of filtered solution is independent from the SiNW length as shown in Fig. 3a. Then, the equivalent NW density in the nanonet was directly deduced from coverage area, by assuming that there is no dispersion in the NW length. Thus, the equivalent density, d_{eq}, is plotted as a function of the filtered volume of SiNW solution in Fig. 3b for nanonets based on different NW lengths. Thanks to the filtration process, the linearity between density and volume of filtered solution is straightforward. Then, for a given amount of silicon, it is obvious that the number of NWs required to cover a given surface is higher for shorter NWs than for longer ones. Therefore, the NW density increases quicker in the former case.

Thus, we first demonstrate that the NW density is well controlled by the filtered volume of NW solution. The statistical study of numerous images arising from different nanonets also evidences the reproducibility from one nanonet to another and finally the homogeneity inside the nanonets is illustrated by the small error bars in Figs 2 and 3.

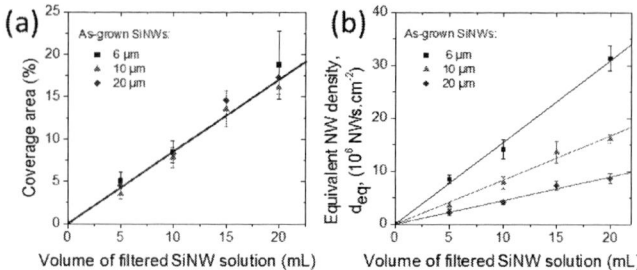

Fig. 3. (a) Coverage area as a function of the filtered volume of solution for the three NW lengths studied (6 μm, 10 μm and 20 μm as-grown). (b) Equivalent SiNW density, d_{eq}, calculated from the coverage area as a function of the filtered volume of solution.

B. Electrical properties

In order to determine the influence of the NW length and density into the nanonets on the electrical percolation threshold, the electrical behavior of the nanonets was studied in details. Fig. 4a shows bi-directional I-V characteristics for nanonets assembled from the 20 μm as-grown SiNW solution when the filtered volume varies between 5 and 20 mL. Except for the nanonet elaborated from the smallest volume (5 mL) which is non-conducting, all other nanonets are conducting. Moreover, the current intensity flowing through the nanonet increases with the filtered volume. As the experimental percolation threshold is defined as the smallest volume of solution required to obtain a measurable current (> 10pA at 5V) in every tested devices (>5), this volume is found to be 10 mL for this NW length. From this percolation volume, the equivalent density, d_{eq}, and the long NW density, d_L, at the percolation threshold were determined. In the same way, by studying the electrical behavior of the nanonets composed of smaller NWs, the equivalent density, d_{eq}, and the long NW density, d_L, at the percolation threshold are determined and reported as a function of the NW length in Fig. 4b. As develop in the literature [26], in case of nanonet composed of an homogeneous population of NWs with the fixed length L, the theoretical value of the critical density is given by the relation $d_c L^2 \approx 4.236^2/\pi$ as plotted in Fig. 4b. The theoretical evolution of critical density versus NW length is found to lie between the experimental evolutions determined using the long NW density (lower bound) and the equivalent density (upper bound).

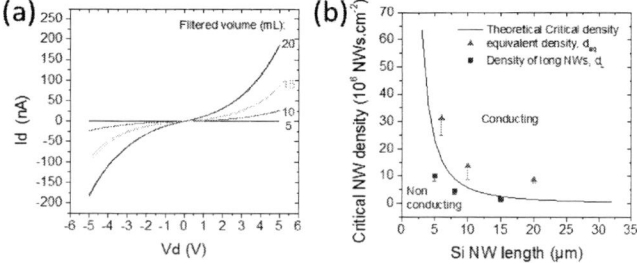

Fig. 4. (a) Bi-directional I-V characteristics for nanonets assembled from the 20 μm as-grown SiNW solution when the filtered volume varies between 5 and 20 mL. (b) Theoretical critical NW density, equivalent NW density, d_{eq}, and long NW density, d_L, at the percolation threshold as a function of the SiNW length. Note that as-grown length is used for the equivalent density, whereas the long NW length is considered for the long NW density.

This result is coherent, as the fabricated nanonets are not composed of uniform NW population with constant length. As far as the long NWs approach is considered, the NW density (d_L) is underestimated by neglecting the short NWs, which explained why the percolation appears at smaller NW density than predicted by the model. In the case of the equivalent density (d_{eq}) approach, the NWs are considered longer than reality (as-grown length) in order to take into account the shorter NWs. But, in such a way, the contribution of each NW to the conduction paths is overestimated. This explains why, in the equivalent density approach, the percolation tends to arise later than the theoretical one. Therefore, although the fabricated nanonets are not composed of NW homogeneous in lengths, their percolation behaviors are found well reproduced by theory by defining two simple approaches to evaluate the critical NW density.

IV. CONCLUSION

In summary, homogeneous, reproducible and conducting silicon nanonets were prepared by using the vacuum filtration method. The SiNW density was precisely monitored so that it was possible to study the percolation threshold as a function of the NW length. Due to the sonication step, the NWs composing the nanonets were splitted in a bimodal population composed of short and long NWs. To deal with this complex configuration, two densities were defined and compared: the equivalent density, d_{eq}, based on the as-grown NW length (before sonication) and the long NW density, d_L, based only on the longest NWs that are 15 to 25% shorter than the as-grown ones. Using these two approaches, the critical NW density required for electrical percolation was experimentally determined as a function of the NW length. It was highlighted that both densities were underestimating the actual NW density. By neglecting the small NWs contribution, the percolation threshold obtained with d_L arises earlier than the one got by d_{eq} which takes them into account. Good agreement with the theoretical value was reported. Therefore, Si nanonets constitute an electrically active macroscopic material composed of nanoscale components that behave like standard percolating media.

REFERENCES

[1] G. Gruner, "Carbon nanonets spark new electronics," *SCIAM*, vol. 17, pp. 48-55, 2007.

[2] Y. Zhao and G. Grüner, "Nanonet as a scaffold with targeted functionalities," *J. Mater. Chem.*, vol. 22, pp. 24983-24991, 2012.

[3] G. Gruner, "Carbon nanotube films for transparent and plastic electronics," *J. Mater. Chemistry*, vol. 16, p. 3533, 2006.

[4] N. Ferrer-Anglada, M. Kaempgen, V. Skakalova, U. Dettlaf-Weglikowska, and S. Roth, "Synthesis and characterization of carbon nanotube-conducting polymer thin films," *Diam. and Relat. Mater.*, vol. 13, pp. 256-260, 2004.

[5] M. Kaempgen, G. S. Duesberg, and S. Roth, "Transparent carbon nanotube coatings," *Appl. Surf. Sci.*, vol. 252, pp. 425-429, 2005.

[6] S. Acharya et al., "A Semiconductor-Nanowire Assembly of Ultrahigh Junction Density by the Langmuir–Blodgett Technique," *Adv. Mater.*, vol. 18, pp. 210-213, 2006.

[7] J. Park, G. Shin, and J. S. Ha, "Controlling orientation of V_2O_5 nanowires within micropatterns via microcontact printing combined with the gluing Langmuir–Blodgett technique," *Nanotechnology*, vol. 19, p. 395303, 2008.

[8] A. Tao et al., "Langmuir−Blodgett Silver Nanowire Monolayers for Molecular Sensing Using Surface-Enhanced Raman Spectroscopy," *Nano letters*, vol. 3, pp. 1229-1233, 2003.

[9] J. Li et al., "Organic Light-Emitting Diodes Having Carbon Nanotube Anodes," *Nano letters*, vol. 6, pp. 2472-2477, 2006.

[10] Z. Wu et al., "Transparent, conductive carbon nanotube films," *Science*, vol. 305, no. 5688, pp. 1273-1276, Aug. 2004.

[11] H. R. Byon and H. C. Choi, "Network single-walled CNT-field effect transistors with increased Schottky contact area for highly sensitive biosensor applications," *J. Am. Chem. Soc.*, vol. 128, pp. 2188-9, 2006.

[12] C. Woo et al., "Fabrication of flexible and transparent single-wall carbon nanotube gas sensors by vacuum filtration and poly(dimethyl siloxane) mold transfer," *Microelectron. Eng.*, vol. 84, pp. 1610-1613, 2007.

[13] A. Star et al., "Label-free detection of DNA hybridization using CNT network field-effect transistors," *PNAS*, vol. 103, pp. 921-926, 2006.

[14] S. De et al., "Silver Nanowire Networks as Flexible, Transparent, Conducting Films: Extremely High DC to Optical Conductivity Ratios," *ACS nano*, vol. 3, no. 7, pp. 1767-74, Jul. 2009.

[15] L. Hu, H. S. Kim, J.-Y. Lee, P. Peumans, and Y. Cui, "Scalable coating and properties of transparent, flexible, silver nanowire electrodes.," *ACS nano*, vol. 4, no. 5. pp. 2955-63, 25-May-2010.

[16] A. R. Madaria, A. Kumar, F. N. Ishikawa, and C. Zhou, "Uniform, highly conductive, and patterned transparent films of a percolating silver nanowire network on rigid and flexible substrates using a dry transfer technique," *Nano Research*, vol. 3, no. 8, pp. 564-573, Jul. 2010.

[17] D. Langley et al, "Flexible transparent conductive materials based on AgNW networks: a review," *Nanotechnology*, vol. 24, p. 452001, 2013.

[18] Y. Wang, T. Feng, K. Wang, M. Qian, Y. Chen, and Z. Sun, "A Facile Method for Preparing Transparent, Conductive, and Paper-Like Silver Nanowire Films," *Journal of Nanomaterials*, 2011.

[19] M. a Topinka, M. W. Rowell, D. Goldhaber Gordon, M. D. McGehee, D. S. Hecht, and G. Gruner, "Charge transport in interpenetrating networks of semiconducting and metallic carbon nanotubes," *Nano letters*, vol. 9, no. 5, pp. 1866-71, May 2009.

[20] L. Hu, D. S. Hecht, and G. Gruner, "Percolation in Transparent and Conducting Carbon Nanotube Networks," *Nano letters*, vol. 4, no. 12, pp. 2513-2517, 2004.

[21] P. Serre, C. Ternon, V. Stambouli-Séné, P. Periwal, and T. Baron, "Fabrication of silicon nanowire networks for biological sensing," *Sensors and Actuators B: Chemical*, vol. 182, pp. 390-395, 2013.

[22] K. Heo et al., "Large scale assembly of silicon nanowire network-based devices using conventional microfabrication facilities," *Nano Letters*, vol. 8, no. 12, pp. 4523-4527, 2008.

[23] E. Mulazimoglu et al., "Silicon nanowire network metal-semiconductor-metal photodetectors," *Appl. Phys. Lett.*, vol. 103, p. 083114, 2013.

[24] S. H. Dalal et al., "Synthesis of ZnO nanowires for Thin Film Network Transistors," *Proc. of SPIE*, vol. 7037, 2008.

[25] F. Li, P. D. Nellist, and D. J. H. Cockayne, "Doping-dependent nanofaceting on silicon nanowire surfaces," *Appl. Phys. Lett.*, vol. 94, no. 26, p. 263111, 2009.

[26] G. . Pike and C. . Seager, "Percolation and conductivity, a computer study. I," *Phys. Rev. B*, vol. 10, no. 4, pp. 1421-1434, 1974.

[27] R. S. Wagner and W. C. Ellis, "Vapor-Liquid-Solid Mechanism of Single Crystal Growth," *Appl. Phys. Lett.*, vol. 4, no. 5, p. 89, 1964.

[28] P. Gentile et al., "Effect of HCl on the doping and shape control of silicon nanowires," *Nanotechnology*, vol. 23, p. 215702, Jun. 2012.

[29] F. Oehler, P. Gentile, T. Baron, and P. Ferret, "The effects of HCl on SiNW growth: surface chlorination and existence of a 'diffusion-limited minimum diameter.'" *Nanotechnology*, vol. 20, no. 47, p. 475307, 2009.

[30] Y. Cui, L. J. Lauhon, M. S. Gudiksen, J. Wang, and C. M. Lieber, "Diameter-controlled synthesis of single-crystal silicon nanowires," *Appl. Phys. Lett.*, vol. 78, p. 2214, 2001.

[31] S. Sorel,et al., "The dependence of the optoelectrical properties of silver nanowire networks on nanowire length and diameter," *Nanotechnology*, vol. 23, p. 185201, 2012

Wavy channel thin film transistor for area efficient, high performance and low power applications

Amir N.Hanna, Galo A. Torres Sevilla, Mohamed T. Ghoneim, Muhammad M. Hussain

Integrated Nanotechnology Lab, King Abdullah University of Science and Technology, Thuwal 23955-6900, Saudi Arabia.
Email: amir.hanna@kaust.edu.sa or MuhammadMustafa.Hussain@kaust.edu.sa

Abstract— We report a new Thin Film Transistor (TFT) architecture that allows expansion of the device width using wavy (continuous without separation) fin features – termed as wavy channel (WC) architecture. This architecture allows expansion of transistor width in a direction perpendicular to the substrate, thus not consuming extra chip area, achieving area efficiency. The devices have shown for a 13% increase in the device width resulting in a maximum 2.4× increase in 'ON' current value of the WCTFT, when compared to planar devices consuming the same chip area, while using atomic layer deposition based zinc oxide (ZnO) as the channel material. The WCTFT devices also maintain similar 'OFF' current value, ~100 pA, when compared to planar devices, thus not compromising on power consumption for performance which usually happens with larger width devices. This work offers a pragmatic opportunity to use WCTFTs as backplane circuitry for large-area high-resolution display applications without any limitation any TFT materials.

Index Terms--thin film transistor, ZnO, High Performance, Low Power .

I. INTRODUCTION

Increasing output current from Thin Film Transistors (TFTs) with low total power consumption is essential for large-area high-resolution displays. This is especially important when considering TFT for backplane circuitry of Organic Light Emitting Diode (OLED) displays, where larger drive currents are required while maintaining low stand-by, or 'OFF' state, power consumption [1]. Also, large panel displays are in high demand from the consumers; therefore, achieving higher performance while maintaining low-cost integration of TFTs is critical. In accordance with the scaling trend of logic transistors, scaling the TFTs is a pursued method to increase the output current [2, 3]. However, lithographic scaling is expensive as well as it limits the transistor width resulting in reduced current. Also, as channel length is reduced Short Channel Effects (SCE) are reported such as threshold voltage shifts and higher OFF state leakage, leading to degraded overall performance [4]. Therefore, fin type feature vertically integrated wavy channel (WC) architecture can play critical role to increase the width, in the direction perpendicular to the substrate, without increasing the transistor area. We have previously shown the usefulness of this new architecture for both logic transistors, and poly-silicon thin film transistors [5,

6]. This novel architecture is shown in Figure 1(a), where it is shown vis-à-vis the planar counterpart in Figure 1(b). Both devices consume the same chip area. The extra transistor width, W_{extra}, is equal to 2× the number of fins × fin height. The TFT architecture is back-gated as shown in the schematic for both the wavy and planar devices. We chose Zinc oxide (ZnO) as the channel material since Amorphous Oxide Semiconductors (AOS) are desirable for their high mobility, low leakage, transparency and their low temperature deposition that allows integration on flexible substrates [7]. We also chose high-κ dielectric aluminum oxide (Al_2O_3) as it forms a better interface with ZnO compared to conventional SiO_2 [8].We used Atomic Layer Deposition (ALD) for both the films to achieve both uniform and conformal depositions on the fin features. The WC devices have shown higher drain currents, higher field effect mobility, similar I_{ON}/I_{OFF} ratio, and similar OFF current value when compared to planar devices that consume the same chip area. This shows the potential of achieving higher drive currents, without compromising overall power consumption. This is particularly critical for the next generation of flexible displays using low cost polymer substrates or even paper, where power consumption is an important metric [9], especially when considering Complementary Metal Oxide Semiconductor (CMOS) based logic and analog circuits, where both n and p-type semiconductors are integrated on the same substrate with thermal budget limitations [10].

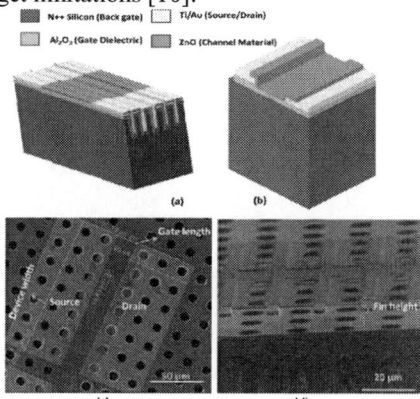

Figure 1 (a) Schematic of the fabricated wavy device , (b) planar device both consuming same chip area,(c) SEM image of wavy device that has 7 fins, 20 μm channel length and 140 μm planar device width, with 18 μm W_{extra}.(d) Tilted view showing fin height of ~1.3 μm

---

This work was supported in part by the KAUST Office of Competitive Research Funds (OCRF) under Grant CRG-1-2012-HUS-008.

978-1-4799-4993-9/14 $31.00 © 2014 IEEE

II. `DEVICE FABRICATION

Device fabrication starts by first patterning 1.3 µm tall fin features using Deep Reactive Ion Etching (DRIE) in a heavily doped n-type Si wafer that acts as a back gate followed by the deposition of 50 nm of ALD Al_2O_3, gate dielectric, at 300 °C. Source and drain of titanium based adhesion layer followed by gold as metal contact (Ti-Au) were then deposited at room temperature by sputtering process and patterned by a lift-off process. Next we performed a low-temperature (100 °C) deposition of ALD ZnO. The thickness of ZnO layer was ~40 nm. The film resistivity was confirmed by a four-point probe measurement to be ~ 10 Ω.cm. Next, devices were isolated by Reactive Ion Etching (RIE), which patterned both the active layer, ZnO, and the gate dielectric, Al_2O_3. SEM images of the fabricated devices are shown in Figure 1(c and d), where (c) shows a wavy device that has 7 fins, and (d) shows a tilted view of the wavy device showing a fin height of 1.3 µm. Holes showing in both figures are intended for releasing a relatively thin, 25-50 µm, sheet of the Silicon substrate to make flexible devices, as shown previously in our work [11]. Figure 2 shows the nano-crystalline nature of the ZnO film as illustrated in the surface SEM image.

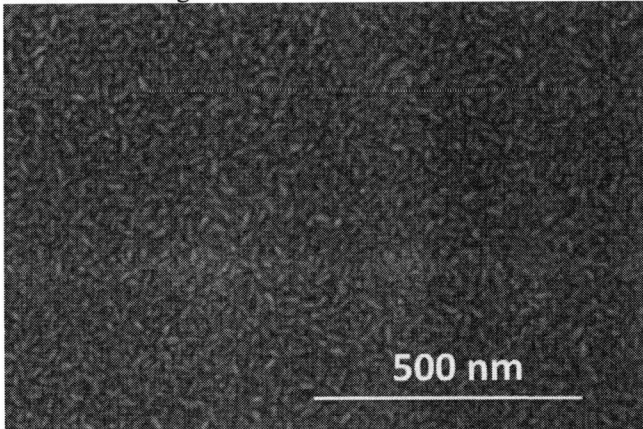

Figure 2 Surface SEM image of ZnO film showing nano-crystalline nature of the film

III. ELECTRICAL CHARACTERIZATION

The electrical characteristics of both the planar and WC devices were compared for channel lengths of 20, 15, 10 and 5 µm. Keithley 4200 semiconductor parameter analyser was used for measuring the electrical characteristics. All the reported data are averages of at least 8 different devices. Since both planar and wavy devices are on the same die, electrical characteristics comparison was prone to wafer-to-wafer or die-to-die process variability. The planar devices have channel width of 140 µm, while WC devices have an extra device width, W_{extra}, of ~18 µm making the wavy to planar device width ratio 1.13. Transfer and output characteristics of 20 µm, gate length, wavy and planar devices are shown in Figure 3 (a) and (b), respectively. Transfer curves show that planar and wavy devices have threshold voltages, V_t of 0.5 V and 0.87 V, respectively. V_t values were extracted by extrapolation from

the $\sqrt{I}_{DS}$ –V_{GS} curve at the point of its' highest first derivative [12]. Field effect electron motilities of 0.22 and 0.33cm2/V.s were extracted for planar and WC devices, respectively. These values are saturation mobility values and were calculating using equation (1):

$$\mu_{sat} = \left(\frac{2L}{W}\right)\left(\frac{1}{C_{ox}}\right)\left(\frac{d\sqrt{I_D}}{dV_{GS}}\right)^2 \qquad (1)$$

The linear region of the $\sqrt{I}_{DS}$ –V_{GS} curve was used to extract slope value, $\frac{d\sqrt{I_D}}{dV_{GS}}$, that were used for this calculation.

Figures 3(c) and (d) show the transfer and output characteristics of 5 µm gate length devices. . Transfer curves show that planar and wavy devices have threshold voltages, V_t of 0.05 V and 0.16 V, respectively. Also, Field effect electron motilities of 0.55 and 0.74 cm²/V.s were extracted for planar and WC devices, respectively.

20 µm devices show I_{on}/I_{off} ratio of 10^5, and 'OFF' current value of ~ 100 pA for both planar and WC devices as shown in Figure 3(a). Also, Figure 3 (b) shows that the average 'ON' current value of the planar device is 8.9×10^{-6} A, while WC devices have an average 'ON' current of 1.35×10^{-5} A. On the other hand, 5 µm planar devices show I_{on}/I_{off} ratio of $>10^5$ while WC devices show I_{on}/I_{off} ~ 2×10^4 devices as shown in Figure 3(c). Also, Figure 3 (d) shows that the average 'ON' current value of the planar device is 7.7×10^{-6} A, while WC devices have an average 'ON' current of 1.7×10^{-5} A.

The ratio of the wavy-to-planar drain 'ON' currents is plotted in Figure 4(a) for 20 µm devices. The mask design was such that a row of planar devices lie within 200 µm distance from a row of WC devices to ensure fair comparison when comparing 'ON' current ratio values. So, only "neighbouring" devices were considered in calculating the ratio. Figure 4(a) shows ~ 1.5× improvement in 'ON' current for WC devices over the planar counterparts, while both devices have similar OFF current value as was shown in Figure 3(a). The same ratio was also plotted for devices of 15, 10 and 5 µm channel lengths, as shown in Figure 4(b), which shows higher ratios of ~1.65× and ~2.4× increase over planar counterparts for 10 and 5 µm devices, respectively, while 15 µm devices has shown a similar 1.5× increase, all compared at the same biasing condition of V_{DS}= 5 V and V_{GS}= 10V.

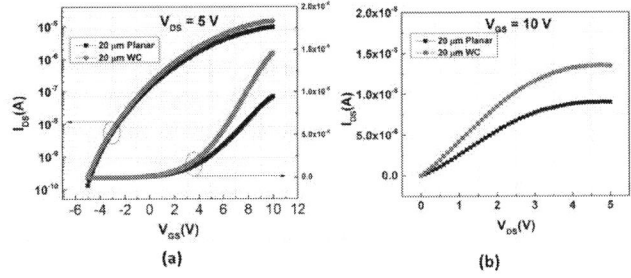

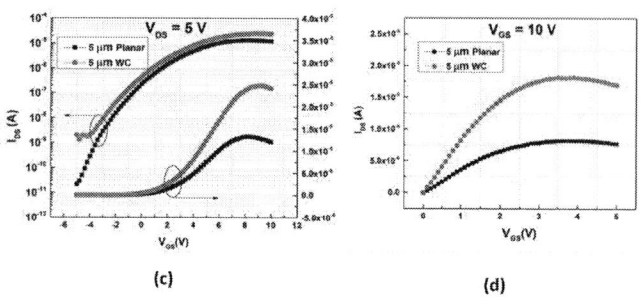

(c)   (d)

Figure 3 (a) Transfer and (b) Output characteristics of 20 μm planar and wavy channel (WC) transistors, and (c) , (d) of 5 μm devices.

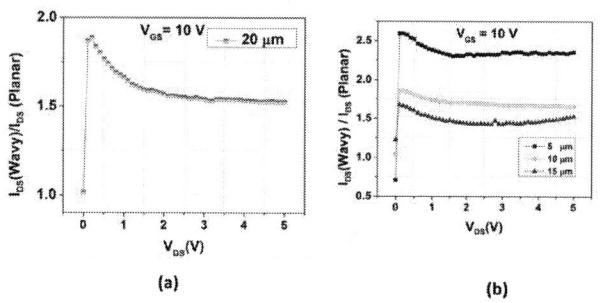

(a)   (b)

Figure 4 (a) Wavy to planar ON current ratio for 20 μm gate length devices and for showing 1.5X increase for wavy ON current over planar counterpart(b) and ~1.5X, 1.65X and 2.4X for 15,10 and 5 μm gate length devices, respectively

IV. RESULTS AND DISCUSSION

Wavy Channel (WC) have shown higher 'ON' current, and field effect mobility values compared to planar devices. This has been shown, for example, in 20 μm gate length devices as WC demonstrated 1.5× increase in 'ON' current, as well as field effect mobility when compared to planar counterparts. However, this cannot be solely attributed to the extra device width, W_{extra}, as WC devices only have 13% extra width when compared to planar counterparts. One possible explanation for both the higher 'ON' current and field effect mobility values is an enhanced electric field value in the channel due to device architecture. The WC architecture causes higher electric field around fin bottom corners since electric field have two components, one due to the planar part of the transistor and the other due to fin sidewall contribution.

To confirm this assumption, a COMSOL simulation was made to confirm this assumption. Figure 5 shows a schematic of the back gate dielectric interface under bias, as well as, a COMSOL simulation measuring the Electric Displacement Field, D, map inside Al_2O_3 dielectric for the dashed part. The simulation uses Cu as a metal back gate, Al_2O_3 (50 nm) gate dielectric, and a voltage bias, V_{GS}, of 10V. The map shows that the electric field is higher around corners, giving a D value of 0.014 C/m^2, while the D value is 0.01 C/m^2 in the planar parts. This is due to contribution from both the sidewalls as well as the planar parts leading to more than 50% higher D values around corners when compared to planar parts.

The simulation results could explain why apparently 50% higher mobility were noticed for the 20 μm WC devices compared to planar devices, namely 0.33 $cm^2/V.s$ for WC vs. 0.22 $cm^2/V.s$ for planar devices.

As for the shorter channel TFTs, L = 5 μm, we believe that a combination of electric field enhancement, and reported short-channel-effects, especially lowering of V_t, have caused the higher ratio, 2.4×, of the wavy-to-planar drain 'ON' current[4].

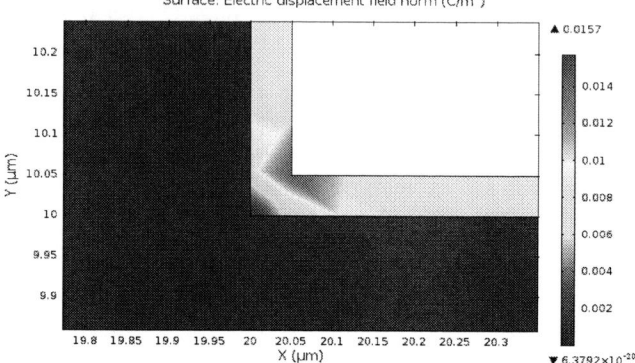

Figure 5 COMSOL simulation of Displace Electric Field, D in (C/m2), showing high D values at corners due to contribution from both the side walls and the planar part. Electric field values are 50% higher at corners due to device architecture .

V. CONCLUSION

In summary, we have shown the advantage of using the wavy architecture in terms of area efficiency, higher output current, higher field effect mobility and similar OFF current levels and Ion/off ratios to the planar counterparts. WC devices have shown 1.5×, 1.65× and 2.4× extra 'ON' current value when compared to planar counterparts for 20, 10 and 5 μm devices. The low 'OFF' current levels for WC devices,~100 pA, and high Ion/off ratios, ~10^5, insure that standby power consumption remain similar to planar counterpart, while improving 'ON' current values. This proves the significance of this new architecture for large area high resolution display applications

VI. REFERENCES:

[1] Y. Kuo, "Thin Film Transistor Technology—Past, Present, and Future," Electrochem. Soc. Interface 55 (2013).

[2] X. Guo, R. Sporea, J. Shannon, and S. R. Silva, "Down-scaling of Thin-Film Transistors: Opportunities and Design Challenges," ECS Trans. 22, 227-238 (2009).

[3] O. Bonnaud, P. Zhang, E. Jacques, and R. Rogel, "Vertical Channel Thin Film Transistor: Improvement Approach Similar to Multigate Monolithic CMOS Technology," ECS Trans. 37, 29-37 (2011).

[4] H.-H. Hsieh and C.-C. Wu, "Scaling behavior of ZnO transparent thin-film transistors," Appl. Phys. Lett. 89, 041109-041109-3 (2006).

[5] H. M. Fahad, A. M. Hussain, G. T. Sevilla, and M. M. Hussain, "Wavy channel transistor for area efficient high performance operation," in Appl. Phys. Lett. vol. 102, ed, 2013, p. 134109.

[6] A. M. H. Galo A. Torres Sevilla, Amir Hanna and Muhammad M. Hussain, "Wavy-Channel Poly-Si Thin Film Transistor," in IEEE NMDC Y. Tzeng, Ed., ed. Tainan, Taiwan, 2013.

[7] E. Fortunato, P. Barquinha, and R. Martins, "Oxide Semiconductor Thin-Film Transistors: A Review of Recent Advances," Adv. Mater. 24, 2945-2986 (2012).

[8] P. F. Carcia, R. S. McLean, and M. H. Reilly, "High-performance ZnO thin-film transistors on gate dielectrics grown by atomic layer deposition," Appl. Phys. Lett. 88, 123509-3 (2006).

[9] R. F. P. Martins, A. Ahnood, N. Correia, L. M. N. P. Pereira, R. Barros, P. M. C. B. Barquinha, et al., "Recyclable, Flexible, Low-Power Oxide Electronics," Adv. Funct. Mater. 23, 2153-2161 (2013).

[10] R. Martins, A. Nathan, R. Barros, L. Pereira, P. Barquinha, N. Correia, et al., "Complementary Metal Oxide Semiconductor Technology With and On Paper," Adv. Mater. 23, 4491-4496 (2011).

[11] J. P. Rojas, G. T. Sevilla, and M. M. Hussain, "Structural and electrical characteristics of high-k/metal gate metal oxide semiconductor capacitors fabricated on flexible, semi-transparent silicon (100) fabric," Appl. Phys. Lett. 102, 064102-4 (2013).

[12] A. Ortiz-Conde, F. García Sánchez, J. J. Liou, A. Cerdeira, M. Estrada, and Y. Yue, "A review of recent MOSFET threshold voltage extraction methods," Microelectron. Reliab. 42, 583-596 (2002).

Investigation of the optics system carbonaceous contamination induced by chemically amplified resist outgassing under e-beam radiation

A.-P. Mebiene-Engohang
*STMicroelectronics, 850 rue Jean Monnet, F-38920,
Crolles, France
e-mail address: armel-petit.mebiene@cea.fr*

M.-L. Pourteau, J.-C. Marusic, L. Pain
*CEA, LETI, Silicon Technology Division, F-38054,
Grenoble, France*

S. David, S. Labau, J. Boussey
*LTM-CNRS-UJF/CEA-Leti-Minatec, 17 rue des Martyrs, F-38054,
Grenoble, France*

Abstract — **The developing multi e-beam lithography tools face important challenges in controlling the contamination of the optics system due to the deposition of hydrocarbon layer induced by the resist outgassing under e-beam radiation and high-vacuum. In this work, we present an experimental methodology allowing the investigation of the specific silicon micromachined membranes (called mimics) carbonaceous contamination induced by resist outgassing under 5keV e-beam radiation by using a dedicated experimental setup designed in CEA-Leti. The Focus Ion Beam combined to Scanning Electron Microscopy (FIB-SEM) and X-ray Photoelectron Spectroscopy (XPS) characterization techniques were used to determine the contamination layer thickness and elementary composition, respectively. A first process-oriented conclusion from this work shows that the contamination layer growth depends on e-beam current density and induced precursor pressure in the vicinity of the mimics.**

Keywords — *Chemically amplified resist; e-beam radiation; outgassing; carbonaceous contamination; XPS; FIB-SEM*

I. INTRODUCTION

It is well known [1] that the resist outgassing under electron or photon exposures leads to the release of hydrocarbonaceous molecules in the vicinity of the exposed wafers. Currently, inside single electron beam lithography tool, the contamination of the optics projection system due to this resist outgassing is not critical. However, in emerging multi e-beams exposure tools, based on the use of several thousand of parallel electron beams, the optics projection system contamination issue can no longer be neglected. Consequently, it becomes mandatory to carefully analyze the resist outgassing mechanisms and their contribution to the contamination layer growth inside the tools. It is worth noticing that since the last decade, several research groups acting on the development of EUV and e-beam lithography techniques have been addressing similar resist outgassing issues [2–6] by using various experimental approaches. However, little was done to investigate the accurate mechanisms governing the growth of the corresponding contamination layer at the optics' level. In 1983, K. Boller's team was the first one to show that the contamination layer growth on the optics projection depends on the e-beam exposure dose, the hydrocarbon pressure inside the exposure chamber and the optics projection system's temperature [7]. This team concluded also that only the photo-emitted electrons with low energy can initiate the hydrocarbonaceous layer growth reaction. Later, R. Kurt's team [8] confirmed that the contamination layer growth depends mainly on the hydrocarbonaceous species capability to physisorb on the optics system surface. It was also shown that the contamination layer is mainly composed of carbon atoms (more than 90%), regardless of the hydrocarbonaceous precursor type initially physisorbed. The works published in 2013 [9] have confirmed this last observation by measuring 92.2% carbon atoms content in the contamination layer induced by the chemically amplified resists outgassing. In collaboration with Mapper Lithography, the main purpose of our current experimental investigations is to determine under which conditions, the contamination layer induced by resist outgassing may obstruct the mimic holes and modify, thereby, the e-beams paths characteristics. In this paper, we will investigate the contamination layer growth induced on the mimic by using a reference resist that outgassing amount was found, as reported in our previous work [10], to be equal to 2.32 molecules.electron^{-1}. Experiments will be conducted under four e-beam current densities and induced precursor pressure in the vicinity of the mimic.

II. EXPERIMENTAL DETAILS

A. Material and sample preparation conditions

For contamination growth tests, we have selected one e-beam resist (Resist C) that is a Positive Chemically Amplified Resist (PCAR) previously assessed in term of outgassing [10]. The Resist C is PCAR with PHS and acrylate hybrid polymer matrix type. It was spin-coated on 100mm silicon wafers. Spin-coating parameters were set to get 37nm resist film thickness. The Resist C was exposed to its dose to size ($2.5*D_0$, where D_0 is the dose to clear determined by resist contrast curve) using the experimental set-up shown in next section.

B. Contamination growth tests tool

Reference [11] gives a complete description of the home built Leti's outgassing test tool (cf Figure 1) in both setup configurations used: outgassing and contamination tests.

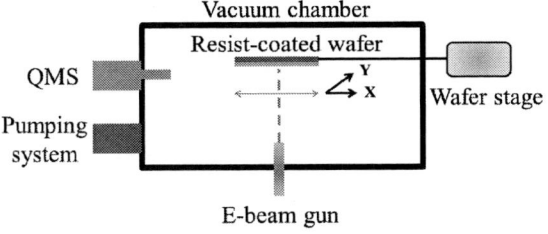

a) Resist outgassing test configuration of the tool

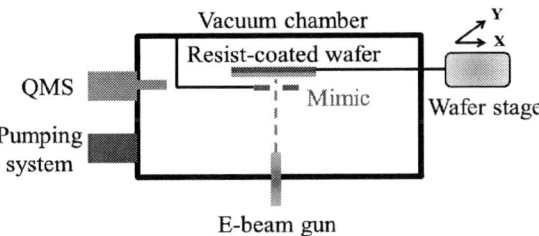

b) Contamination test configuration of the tool

Fig. 1. Schematic of Leti's 5keV outgassing test tool allowing two types of investigation – a) outgassing and b) contamination test configurations

This setup is mainly composed of a vacuum chamber connected to an "ATH1603M" turbomolecular pump that allows reaching ultra-high vacuum conditions. The steady state nominal pressure obtained was better than 3.10^{-7} Pa. An "EGG-3101" electron gun working at 5kV bias was used for resists exposure. The outgassing measurement is performed using a "Microvision" QMS positioned very close to the resist-coated wafer (cf Figure 1-a). In the tool configuration schematized in Figure 1-b, a wafer stage was implemented in order to allow 100mm silicon wafers moving along two axis. The electron beam, which is aligned with the mimic holes, is motionless in this configuration.

C. Contamination process flow

Preliminary experiments were conducted to calibrate the wafer scan speed, the e-beam gun and its raster scanning capabilities. Similarly, the effective pumping speed was properly measured according to a well-defined experimental protocol [12] adapted to the current set-up.

Before starting the contamination measurements, the stability of the chamber background was evaluated [11]. Indeed, low background noise is obviously very important to guarantee that the measured contamination layer will mainly be due to the irradiated resists.

In contamination test configuration, the wafer scan speed required to reach the desired exposure dose depends on the working e-beam current density and on the mimic design. The area exposed during contamination growth test was set equal to 25cm² by wafer because of the limited lateral extent of the wafer stage axis. Besides, the possible minimum gap between the mimic and resist coated wafer was set at 200µm and, finally, an orifice (2mm diameter and 8mm depth) were micromachined in a copper plate and placed underneath the mimic in order to have accurate control of the hydrocarbon pressure in the vicinity of the mimic as illustrated on Figure 2. The 10*10 holes (16µm diameter and 100µm depth) mimic design was used in this work as shown in Figure 2.

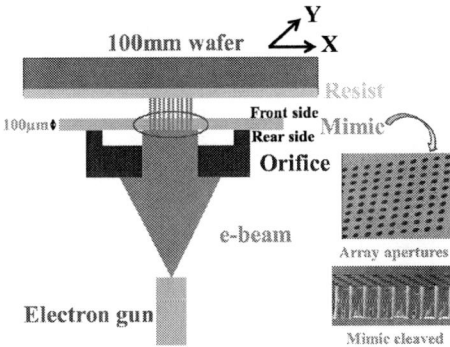

Fig. 2. Schematic of wafer stage implemented inside our experimental set-up showing the resist coated wafer mounted in front of the mimic (a silicon membrane with micro holes) and orifice part

For each e-beam current density set, 15 identical wafers were systematically exposed in order to cumulate, at the mimic level, the contamination grown layer. The duration of such exposures depends on the e-beam current density and was found to be on few hours range in this work (from 6 to 10hours).

D. Contamination layer characterization by XPS and FIB-SEM

The contamination layer is expected to grow mainly on the rear face of the mimics as well as inside its holes. At this stage, one needs to carry out experimental characterizations that can give its elementary composition and its thickness profile at each point of the mimics. As indicated previously, the contamination layer induced by the chemically amplified resists outgassing, which are typically composed of multiple hydrocarbon species, could contain more than 92% of carbon atoms. To check this commonly established fact, X-ray photoelectron spectroscopy analysis was conducted on the rear face of the contaminated silicon mimics. The tool used is Theta 300, Thermo Fisher Scientific™, UK and the X-rays source is Al Kα monochromatic source. The spot size is on the order of 100µm which allows exploring many holes at a once. In addition to this chemical analysis, morphological characterization was done using FIB-STEM tool (Helios450s NanoLab™ DualBeam™, FEI, USA) where the focused gallium beam is used to properly cleave the mimics (cf Figure 2) along a series of holes. In such a way, the SEM cross section can show the profile of the contamination layer and give access to the accurate thickness of the contamination layer.

III. RESULTS AND DISCUSSIONS

A. Elements composition identification of the contamination layer by XPS

Before being subjected to contamination experiment, virgin mimic surface was analyzed by XPS and its composition was adopted as reference. After 9.35 hours exposure at 37.5 µC•cm^{-2} and with a 200µm gap between mimic and wafer, the same mimic was taken out of the chamber and analyzed again by XPS. ; The obtained ratios were plotted on Figure 3. One can see that the contaminated mimic spot contains about 91% Carbon (C), 7% Oxygen (O), 2% Silicon (Si) while the virgin sample was found to be composed of 56% O, 33% Si and only 11% C. One has to notice that residual carbon on the surface of the bare mimic can be explained by ambient residual contamination and, on the other hand, native silicon oxide explains the oxygen amount detected.

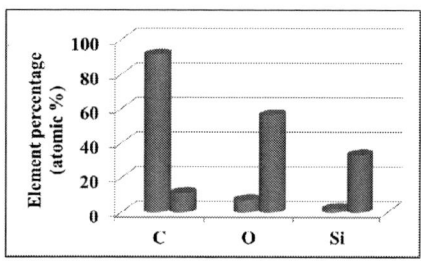

Fig. 3. Elements composition of the contaminated mimic spot and reference sample in blue and red respectively

B. Contamination layer shape observation and measurement by FIB-SEM

The contamination growth tests were done by irradiating Resist C material (characterized by a 2.32 molecules.electron^{-1} outgassing amount) with four e-beam current densities ranging from 10 up to 25A/m² through 10*10 holes mimic located at 200µm underneath the resist surface. After exposure of 15 wafers, the mimic was taken out of the chamber and cleaved by focused gallium ions beam to allow SEM observations across the individual holes.

SEM cross-section observation (Figure 4-d) shows that the contamination growth inside mimic holes (radial contamination layer growth) is higher than the contamination growth in surface of mimic: 27 and 10nm/hour growth rates with 20A/m² e-beam current density, respectively. Such a thick radial contamination layer may lead to the obstruction of the mimic holes and modify, thereby, the e-beams paths characteristics. Figure 4-a shows the contamination spot under the incident e-beam coming through the orifice (spot area around the array apertures).

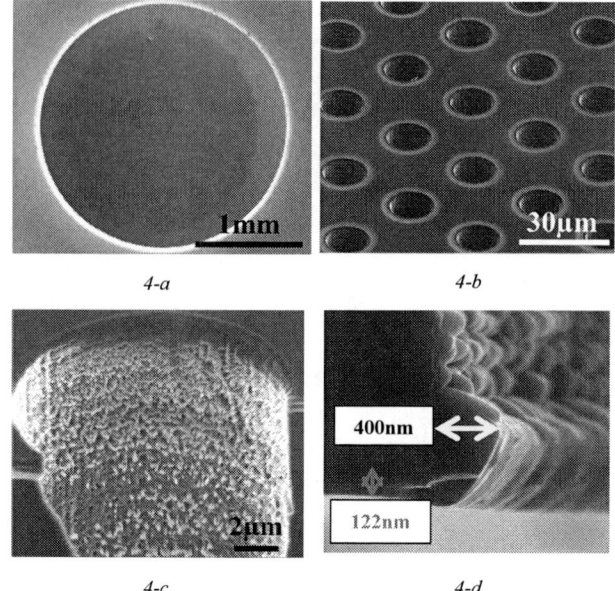

Fig. 4. SEM observations of the mimic contaminated taken on the rear side of the mimic - a) Mimic SEM top view just after contamination run, b) SEM top view of the holes array, c) SEM tilted top view and d) SEM cross-section view of a mimic hole

The growth of the contamination is also found to be more important around the mimic holes as it can be shown by the rings around the holes (Figure 4-b). Finally, it was confirmed that the contamination layer inside the hole is thicker in the rear side of the mimic facing the incident e-beam. This phenomenon can be attributed to the availability, at this location, of a high number of low energy secondary electrons (~10eV) that are remitted from the silicon wafer substrate being bombarded by the incident electrons and high hydrocarbon molecules. They are two main parameters that contribute to contamination grow. Indeed, according to the configuration of contamination test tool (Fig. 1-b & 2), the precursors conductance between mimic and wafer is low. Therefore, the hydrocarbonaceous species pressure inside the holes is higher than the precursors pressure available on the mimic surface facing the electron beam.

These observations were systematically conducted for four e-beam current densities. The contamination layer growth rate Γ (nm/hour), measured at the thicker place, is plotted as a function of e-beam current density J (A/m²) on Figure 5.

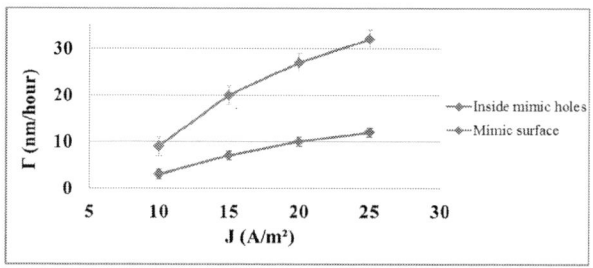

Fig. 5. Contamination growth rate as a function of the e-beam current density as obtained from Resist C irradiated at 37.5 µC•cm^{-2}

It was found that the contamination layer growth rate increases with the e-beam current density on the density range explored. Such a behavior suggests that the contamination growth occurs in the electrons limited regime as proposed by [13].

According to the experimental set-up configuration, the variation of the e-beam current density induces a variation of the local pressure value in the vicinity of the mimic. This is quiet similar to what should occur in the real multi e-beams tools because the conductance between the mimic and the wafer is set by the tool configuration and remain the same. Therefore, the results obtained on the contamination layer growth in our tool can be considered as a faithful estimation of the effect of the e-beam current density and the hydrocarbon pressure of the vicinity of the mimic. In order to make vary the pressure value in the vicinity of the mimic (with an e-beam current density given), ongoing experiments are conducted with variable gap values (from 200 μm, as in this study, up to 800μm).

IV. CONCLUSION AND PERSPECTIVES

We have investigated the contamination layer growth of one e-beam resist formulation (Resist C) dedicated to e-beam lithography. The FIB-SEM and XPS techniques were used to characterize the contamination layer thickness and its elementary composition, respectively. It has been observed that the contamination growth rate increases almost linearly with the e-beam current density on the density range explored. The XPS analysis has revealed that the contamination layer contains more than 91% of carbon atoms but the silicon and oxygen atoms were detected in tiny percentage. Currently, the tests of the hydrocarbon pressure variation in the vicinity of the mimic (by mimic and wafer gap changing) are doing to assess its impact on contamination growth rate. The next steps of this work, at short term, are to investigate the contamination growth behavior of the Resists A, B and D yet assessed in outgassing and the TC dependency on the contamination growth rate for each resist tested. Thereby, the resists outgassing and contamination growth behavior could be correlated.

ACKNOWLEDGMENT

The research leading to these results has been performed in the frame of the industrial collaborative consortium, IMAGINE. The authors acknowledge all the partners for their fruitful collaboration.

REFERENCES

[1] G. Denbeaux, Y. Kandel, G. Kane, D. Alvardo, M. Upadhyaya, Y. Khopkar, A. Friz, K. Petrillo, J. Sohn, C. Sarma, et D. Ashworth, « Resist outgassing contamination growth results using both photon and electron exposures », in *Proc. of SPIE*, 2013, p. 86790L‑8.

[2] I. Pollentier, A.-M. Goethals, R. Gronheid, J. Steinhoff, et J. Van Dijk, « Characterization of EUV optics contamination due to photoresist related outgassing », in *Proc. of SPIE*, 2010, p. 76361W‑10.

[3] Y. Nishiyama, T. Anazawa, H. Oizumi, I. Nishiyama, O. Suga, K. Abe, S. Kagata, et A. Izumi, « Carbon contamination of EUV mask: film characterization, impact on lithographic performance, and cleaning », in *Proc. of SPIE*, 2008, p. 692116‑10.

[4] S. Kobayashi, J. J. Santillan, H. Oizumi, et T. Itani, « EUV resist outgassing release characterization and analysis », *Microelectron. Eng.*, vol. 86, n° 4‑6, p. 479‑482, avr. 2009.

[5] S. Kobayashi, J. J. Santillan, et T. Itani, « EUV Resist Outgassing: Quantification and Release Mechanisms », *J. Photopolym. Sci. Technol.*, vol. 21, n° 4, p. 469‑474, 2008.

[6] N. Sugie, T. Takahashi, K. Katayama, I. Takagi, Y. Kikuchi, E. Shiobara, H. Tanaka, S. Inoue, T. Watanabe, T. Harada, et H. Kinoshita, « Comparison of Resist Outgassing Characterization between High Power EUV and EB », *J. Photopolym. Sci. Technol.*, vol. 25, n° 5, p. 617‑624, 2012.

[7] K. Boller, R.-P. Haelbich, H. Hogrefe, W. Jark, et C. Kunz, « Investigation of carbon contamination of mirror surfaces exposed to synchrotron radiation », *Nucl. Instrum. Methods Phys. Res.*, vol. 208, n° 1, p. 273–279, 1983.

[8] R. Kurt, M. van Beek, C. Crombeen, P. Zalm, et Y. Tamminga, « Radiation-induced carbon contamination of optics », in *Proc. of SPIE*, 2002, vol. 4688, p. 702‑709.

[9] S.-H. Chang, S.-F. Chen, Y.-Y. Chen, M.-C. Chien, S.-C. Chien, T.-L. Lee, J. J. H. Chen, et A. Yen, « Balancing lithographic performance and resist outgassing in EUV resists », in *Proc. of SPIE*, 2013, p. 86790O‑9.

[10] A.-P. Mebiene-Engohang, P. Michallon, R. Tiron, H. Fontaine, B. Icard, M.-L. Pourteau, J.-C. Marusic, D. Bensahel, L. Pain, et J. Boussey, « Resist outgassing assessment for multi electron beams lithography », *Microelectron. Eng.*, nov. 2013.

[11] A.-P. Mebiene-Engohang, M.-L. Pourteau, J.-C. Marusic, L. Pain, et T. Nakayama, « Investigation of the resist outgassing and hydrocarbonaceous contamination induced in multi electron beams lithography tools », in *Proc. of SPIE*, 2014, vol. 9049‑100, in Press.

[12] J. Leclerc, « Production du vide », *Tech. Ing. Génie Mécanique*, n° BM4270, p. BM4270–1, 1999.

[13] C. Tarrio, R. F. Berg, S. B. Hill, S. Grantham, N. S. Faradzhev, et T. B. Lucatorto, « Effects of varying the parameters in witness-sample-based photoresist outgas testing: dependence of the carbon growth on pumping speed and the dose, time, and area of resist exposure », in *Proc. of SPIE*, 2013, p. 867920‑11.

A Suitable Inductor Modeling for DC-DC Converters

Andrea Mocci, Alessandro Serpi, Ignazio Marongiu, Gianluca Gatto

Department of Electrical and Electronic Engineering
University of Cagliari
Cagliari, Italy
gatto@diee.unica.it

Abstract — An inductor modeling suitable for DC-DC converters is proposed in this paper. It consists of introducing an appropriate parallel resistance in order to account for inductor current ripple when averaging technique is employed. This allows the development of an improved averaged model for DC-DC converters, which takes into account inductor power losses due to current ripple phenomenon. The worth and effectiveness of the proposed modeling approach has been validated through a simulation study, which is performed in the Matlab-Simulink environment and refers to the case of a boost DC-DC converter.

Keywords — *Averaging technique; Current Ripple; DC-DC power converters; Modeling*

I. INTRODUCTION

The increasing use of electric and electronic devices supplied by power electronic converters has led to significant improvements of the latter. These are achieved by means of the development of novel converter topologies, but also due to the employment of recently-developed and high-performance semiconductor devices (MOSFETs, IGBTs, diodes, etc.), as well as of suitable control and modulation strategies.

In this context, mathematical models of power electronic DC-DC converters with a high level of detail represent a key aspect, especially in supporting the design stage. In fact, these models allow accurate estimations of converter performances and efficiency, on the basis of which the design of power electronic converters can be successfully optimized. For this purpose, a number of modeling approaches have been proposed in the literature, among which the averaging technique seems to be one of the most promising. In fact, it enables a ripple-free representation of power electronic converters, appropriately removing the non-linearity related to semiconductor devices [1]-[2]. Thus, averaging technique is widely employed for estimating power electronic DC-DC converter performances, as well as for designing appropriate voltage and current control loops.

Several averaged models of power electronic DC-DC converter have been proposed in the literature, many of which, however, neglect parasitic elements of converter components, as well as the effects of switching phenomena. Such approximations significantly ease the achievement of averaged models, but prevent accurate estimations of current and voltage evolutions, as well as the accomplishment of sound power losses and efficiency analysis [3]-[8]. In this context, inductor current ripple is almost always neglected, leading to mismatches in averaged inductor current and power losses estimations, especially when relatively low switching frequency is employed [9]-[10].

Thus, an inductor modeling is proposed in this paper. It consists of introducing an additional parallel resistance with the aim of appropriately accounting for inductor losses due to current ripple phenomenon. Thus, an improved averaged model is achieved compared to that proposed in [8]. It enables very good estimations of both inductor current ripple and averaged power losses. This is proved by means of a simulation study, which is performed in the Matlab-Simulink environment and refers to the case of a boost DC-DC converter.

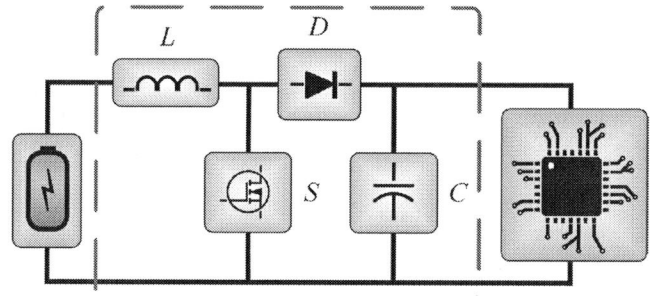

Fig. 1. Schematic representation of a boost DC-DC converter.

II. BOOST MATHEMATICAL MODELS

The schematic representation of a boost DC-DC converter is depicted in Fig. 1. It can be seen that it is made up mainly of an inductor L and a capacitor C, which are appropriately coupled by means of a switch S and a diode D. The boost DC-DC converter is supplied by a DC-source, whereas its load can be represented by an electronic device.

A. Continuous-Time Model

In order to achieve the continuous-time model of the boost converter, reference is made to Fig. 2, which summarizes its different operating states. In particular, assuming that the boost converter is in the ON state at first, the input source charges the inductor L, while the load is supplied by the capacitor C.

As soon as the switch command signal is set low, the switch TURN-OFF occurs. In this operating condition, the voltage across S (v_S) gradually increases, as well as that across the diode (v_D), until D is forward-biased.

Andrea Mocci gratefully acknowledges Sardinia Regional Government for the financial support of her PhD scholarship (P.O.R. Sardegna F.S.E. Operational Programme of the Autonomous Region of Sardinia, European Social Fund 2007-2013 - Axis IV Human Resources, Objective 1.3, Line of Activity 1.3.1.).

978-1-4799-4993-9/14 $31.00 © 2014 IEEE

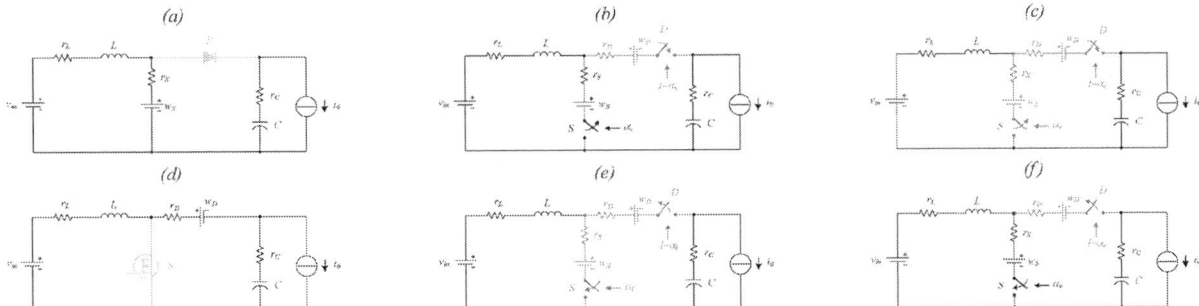

Fig. 2. Operating states of boost DC-DC converters: ON (a), TURN OFF (b,c), OFF (d) and TURN ON (e,f).

Subsequently, the current through the diode (i_D) gradually increases while switch current (i_S) decreases down to zero. Such voltage and current evolutions due to switching phenomena can be modelled by introducing the two signals α_v and α_i depicted in Fig. 3, where $(t_{r,i}, t_{r,v})$ denote the rise times of switch current and voltage respectively, $(t_{f,i}, t_{f,v})$ being the corresponding fall times.

Once TURN-OFF is accomplished, the OFF state occurs; over such an operating condition, both the input source and the inductor supply the load and recharge the capacitor at the same time. It is worth noting that if inductor current becomes zero during the OFF state, the diode turns off, leading to Discontinuous Conduction Mode (DCM). However, such an operating state is not considered in this paper, thus it is assumed that the boost converter always operates in Continuous Conduction Mode (CCM), i.e. the inductor current is always greater than zero.

When the switch command signal is set to high, TURN-ON occurs. Similarly to TURN-OFF, i_D gradually decreases, while i_S increases. As soon as i_D becomes zero, the diode is reverse biased, thus both v_D and v_S decreases. Also in this case, voltage and current evolutions due to switching phenomena can be modelled by α_v and α_i shown in Fig. 3.

On the basis of the previous considerations, the continuous-time model of the boost converter can be expressed as:

$$\dot{x} = Ax + Bu + Cw \qquad (1)$$

where x denotes the state vector, whereas u and w are the input vectors, all defined as

$$x = \begin{bmatrix} v_C \\ i_L \end{bmatrix} \quad , \quad u = \begin{bmatrix} v_{in} \\ i_0 \end{bmatrix} \quad , \quad w = \begin{bmatrix} w_S \\ w_D \end{bmatrix}. \qquad (2)$$

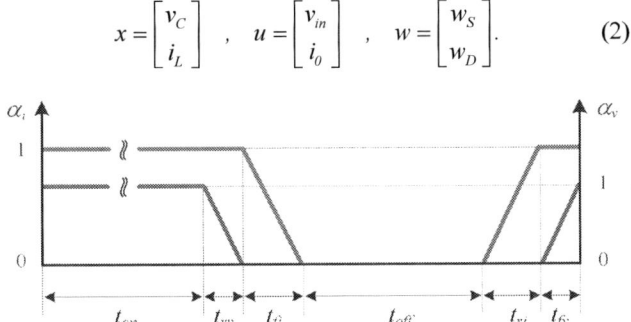

Fig. 3. The α_i (red) and α_v (blue) signals.

In particular, referring to (2), v_C and i_L denote the capacitor voltage and inductor current respectively, v_{in} and i_0 being the input voltage and load current as well. Furthermore, w_S and w_D are the voltage drops over switch and diode respectively, whereas A, B and C are defined by the following equation:

$$A = \begin{bmatrix} 0 & \dfrac{(1-\alpha_i)}{C} \\ -\dfrac{(1-\alpha_v)}{L} & -\left(\dfrac{r_L}{L} + \alpha_v \dfrac{r_S}{L} + (1-\alpha_i)\dfrac{r_D + r_C}{L} \right) \end{bmatrix} \qquad (3)$$

$$B = \begin{bmatrix} 0 & -\dfrac{1}{C} \\ \dfrac{1}{L} & (1-\alpha_v)\dfrac{r_C}{L} \end{bmatrix} \quad , \quad C = \begin{bmatrix} 0 & 0 \\ -\dfrac{\alpha_v}{L} & -\dfrac{(1-\alpha_v)}{L} \end{bmatrix}.$$

in which r_L and r_S represent the parasitic resistance of inductor and switch respectively, whereas r_D and r_C are those of diode and capacitor.

B. Averaged Model

On the basis of (1) and due to the linear time evolution of both α_v and α_i, the averaged model of the boost converter can be easily achieved as:

$$\dot{\bar{x}} = \overline{A}\,\bar{x} + \overline{B}\,u + \overline{C}\,w \qquad (4)$$

in which $\bar{x}$ denotes the averaged state vector, whereas $\bar{A}$, $\bar{B}$ and $\bar{C}$ can be derived directly from A, B and C by replacing both α_v and α_i with their corresponding averaged values, $\bar{\alpha}_v$ and $\bar{\alpha}_i$, which are defined as follows:

$$\bar{\alpha}_i = \bar{\alpha} + \frac{t_{ri} + t_{fi}}{2} f_S \quad , \quad \bar{\alpha}_v = \bar{\alpha} - \frac{t_{rv} + t_{fv}}{2} f_S \qquad (5)$$

$\bar{\alpha}$ and f_S being the ideal duty cycle and the switching frequency respectively. Referring to (4), it is possible to compute the power balance of the boost converter as

$$\bar{x}^T Q\, \dot{\bar{x}} = \bar{x}^T Q \left(\overline{A}\,\bar{x} + \overline{B}\,u + \overline{C}\,w \right) \qquad (6)$$

in which

$$Q = \begin{bmatrix} C & 0 \\ 0 & L \end{bmatrix} . \qquad (7)$$

In particular, the left-side term of (6) represents the power drawn or delivered by the inductor and the capacitor, whereas right-side terms account for all converter losses, as well as for input and output powers.

III. PROPOSED INDUCTOR MODELLING

Referring to (6), it is possible to demonstrate that Joule losses of the inductor is computed as:

$$P_{J,L} = r_L \, \bar{i}_L^2 \qquad (8)$$

Therefore, (6) does not take into account inductor losses due to current ripple occurring on i_L because this last is inherently neglected by (4). Although such an approximation does not impair the effectiveness of both (4) and (6) significantly, it may introduce detectable power balance mismatches, especially when relatively low switching frequency is employed.

Therefore, a suitable inductor model is proposed in this paper, which is shown in Fig. 4. In particular, an additional parallel resistance r_P is appropriately introduced in order to account for inductor current ripple, even at steady state operation. As a result, the inductor equation becomes

$$v = r_L \, i + L \frac{di_L}{dt} \qquad (9)$$

where v and i denote the overall inductor voltage and current respectively. This last can be expressed as

$$i = i_L + i_P \quad , \quad i_P = \frac{L}{r_P} \frac{di_L}{dt} \qquad (10)$$

where r_P has to be chosen in accordance with:

$$r_P = -r_L + \frac{48}{r_L} \left(L f_s \right)^2 . \qquad (11)$$

It is worth noting that when large inductance and/or switching frequency is employed, inductor current ripple becomes negligible, as well as its contribution to inductor power losses. This is appropriately accounted by r_P, whose value increases with the square of both L and f_s, making i_P negligible compared to i_L, as stated by (10). On the basis of

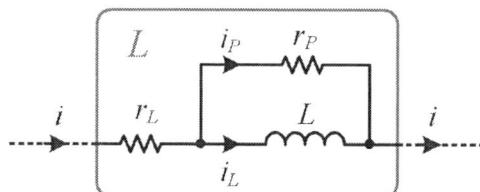

Fig. 4. The inductor model proposed in this paper.

both (9) and (10), it is thus possible to determine a more accurate mathematical model of the boost converter, which can be still expressed by (1), but defining A, B and C as

$$A = \begin{bmatrix} -\dfrac{(1-\alpha_i)r_P}{\rho_C\, C} & \dfrac{(1-\alpha_i)r_P}{\rho_C\, C} \\[2ex] -\dfrac{(1-\alpha_v)r_P}{\rho_L\, L} & -\dfrac{r_P\,(\rho_L - r_P)}{\rho_L\, L} \end{bmatrix}$$

$$B = \begin{bmatrix} \dfrac{(1-\alpha_i)}{\rho_C\, C} & -\dfrac{(\rho_C - (1-\alpha_i)r_C)}{\rho_C\, C} \\[2ex] \dfrac{r_P}{\rho_L\, L} & \dfrac{(1-\alpha_v)r_C\, r_P}{\rho_L\, L} \end{bmatrix} \qquad (12)$$

$$C = \begin{bmatrix} 0 & -\dfrac{(1-\alpha_i)}{\rho_C\, C} \\[2ex] -\dfrac{\alpha_v r_P}{\rho_L\, L} & -\dfrac{(1-\alpha_v)r_P}{\rho_L\, L} \end{bmatrix}$$

where:

$$\begin{aligned} \rho_C &= r_P + r_L + (1-\alpha_i)(r_D + r_C) \\ \rho_L &= r_P + r_L + (1-\alpha_i)(r_D + r_C) + \alpha_v r_S \end{aligned} \qquad (13)$$

In conclusion, on the basis of (1) and (12), an improved averaged model of the boost converter can be introduced. This can be still expressed by (4), but due to the non-linearity introduced by (12), $\bar{A}$, $\bar{B}$ and $\bar{C}$ have to be computed as

$$\bar{M} = \frac{1}{T_s} \int_0^{T_s} M \, dt \quad , \quad M \in \{A, B, C\} \qquad (14)$$

IV. SIMULATIONS

A simulation study has been performed in Matlab-Simulink with the aim of verifying the effectiveness of the proposed modelling approach. Thus, reference is made to the boost DC-DC converter depicted in Fig. 5, whose main details are

Fig. 5. Prototype of the boost DC-DC converter considered for simulations.

TABLE I.   BOOST PARAMETERS

	Value	Unit		Value	Unit
L	24.6	μH	r_L	20	mΩ
C	30	μF	r_C	2	mΩ
w_S	0	V	r_S	36.8	mΩ
w_D	0.6	V	r_D	10	mΩ
v_{in}	24	V	i_0	8.88	A

summed up in Table I, together with rated input voltage and load current. In addition, the switching frequency is set to 20 kHz at first, the duty cycle being constant at 0.5 for all simulations.

Simulation results are depicted in Fig. 6 through Fig. 8. In particular, firstly referring to Fig. 6, it can be seen that significant inductor current ripple occurs at 20 kHz (about 23.4 A), which is even greater than averaged inductor current (about 17.6 A). Referring to Fig. 7, averaged powers achieved in simulations are compared to those computed by means of the proposed averaged model. It can be seen that slight mismatches occur, except for $\bar{P}_{J,C}$, whose contribution to the power balance is, however, quite negligible.

Subsequently, several simulations have been carried out by increasing, from time to time, the switching frequency up to 120 kHz. As a result, the inductor losses due to current ripple only have been appropriately computed and compared with those achieved by the proposed averaged model, leading to the evolutions shown in Fig. 8. It can be seen that an almost perfect agreement between simulations and proposed averaged model is achieved, thus corroborating the effectiveness of the proposed modelling approach.

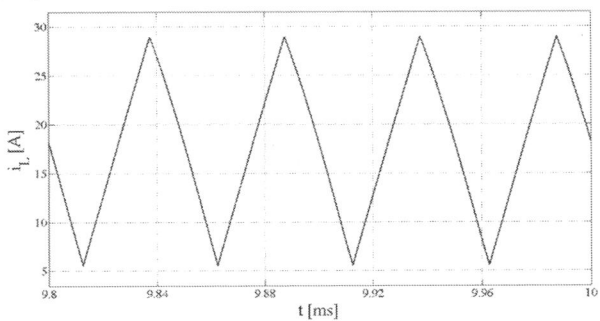

Fig. 6. Steady state evolutions of inductor current at 20 kHz over four sampling time intervals.

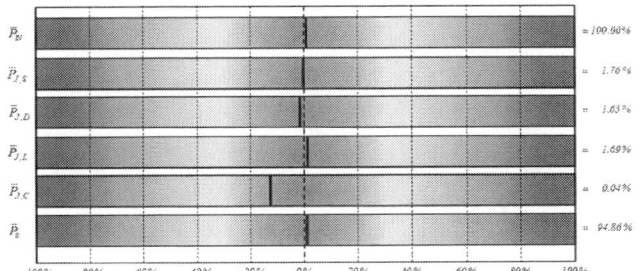

Fig. 7. Averaged power errors between simulations and proposed averaged model (the share of each power contribution is reported on the right).

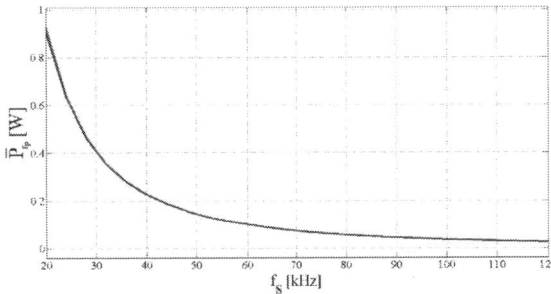

Fig. 8. Inductor losses due to current ripple achieved at different switching frequencies: simulations (red) and proposed averaged model (blue).

V. CONCLUSION

An inductor modelling suitable for DC-DC converters is proposed in this paper. It allows the achievement of an improved averaged model of boost DC-DC converters, which appropriately takes into account inductor losses due to current ripple. Simulation results highlight that the proposed averaged model enables a very good estimation of all DC-DC converter power losses, making it particularly suitable as a supporting tool for the design stage. Its effectiveness in terms of dynamic analysis of converter performances will be investigated in future works, as well as its suitability for designing appropriate control loops.

REFERENCES

[1] D. Maksimovic, A.M. Stanković, V.J. Thottuvelil, G.C. Verghese, "Modeling and simulation of power electronic converters", *Proc. IEEE*, vol. 89, no. 6, pp. 898-912, Jun. 2001.

[2] S. Chiniforoosh, J. Jatskevich, A. Yazdani, V. Sood, V. Dinavahi, J.A. Martinez, A. Ramirez, "Definitions and Applications of Dynamic Average Models for Analysis of Power Systems", *IEEE Trans. Power Del.*, vol. 25, no 4, pp. 2655-2669, Oct. 2010

[3] A. Davoudi, J. Jatskevich, "Realization of parasitics in state-space average-value modeling of PWM DC-DC converters", *IEEE Trans. Power Electron.*, vol. 21, no. 4, pp. 1142-1147, Jul. 2006.

[4] G. Gatto, I. Marongiu, A. Mocci, A. Serpi, I.L. Spano, "A novel continuous-time equivalent circuit for boost DC-DC converters", in *Proc. 39th Annual Conference of the IEEE Industrial Electronics Society (IECON 2013)*, Vienna (Austria), Nov. 10-13, 2013, pp. 262-267.

[5] K. Gorecki, J. Zarebski, R. Zarebski, "Investigations of usefulness of average models for calculations characteristics of the boost converter at the steady state", in *Proc. International Conference on Modern Problems of Radio Engineering, Telecommunications and Computer Science*, Lviv-Slavsko (Ukraine), Feb. 19-23, 2008, pp. 163-166.

[6] C.T. Rim, G.B. Joung, G.H. Cho, "Practical switch based state-space modeling of DC-DC converters with all parasitics", *IEEE Trans. Power Electron.*, vol. 6, no. 4, pp. 611-617, Oct. 1991.

[7] A. Ammous, K. Ammous, M. Ayedi, Y. Ounajjar, F. Sellami, "An advanced PWM-switch model including semiconductor device nonlinearities", *IEEE Trans. Power Electron.*, vol. 18, no. 5, pp. 1230-1237, Sept. 2003.

[8] G. Gatto, I. Marongiu, A. Mocci, A. Serpi, I.L. Spano, "An improved averaged model for boost DC-DC converters", in *Proc. 39th Annual Conference of the IEEE Industrial Electronics Society (IECON 2013)*, Vienna (Austria), Nov. 10-13, 2013, pp. 412-417.

[9] D. Czarkowski, M.K. Kazimierczuk, "Energy-conservation approach to modeling PWM DC-DC converters", *IEEE Trans. Aerosp. Electron. Syst.*, vol. 29, no 3, pp. 1059-1063, Jul. 1993.

[10] V. Vorperian, "A ripple theorem for PWM DC-to-DC converters operating in continuous conduction mode", in *Proc. 35th IEEE Annual Power Electronics Specialists Conference (PESC 04)*, Aachen (Germany), June 20-26, 2004, vol. 1, pp. 28-35.

Monolithically Integrated Voltage Level Shifter for Wide Bandgap Devices-Based Converters

Romain Grezaud[1,2], François Ayel[1]
[1] CEA, LETI
MINATEC Campus
38054 Grenoble Cedex 9, France
romain.grezaud@cea.fr

Nicolas Rouger[2], Jean-Christophe Crebier[2]
[2] Univ. Grenoble Alpes
G2Elab, CNRS
38400 Grenoble, France
nicolas.rouger@g2elab.grenoble-inp.fr

Abstract—**Power converters based on Wide Bandgap Devices (WBD) are able to operate at high-frequency and high-voltage. In such synchronous converters a very short dead-time is advised because a lot of WBD do no have a parasitic body diode to conduct current in reverse. Into a high-voltage inverter, isolators generate isolated floating gate signals from logic level input signals. Mismatch between high-side and low-side signal propagation delays involves setting a long secure dead-time. To overcome this matching issue in order to guarantee the dead-time duration and time location, we propose other input signal propagation paths. Both input signals pass through a single low-side 2-channel digital isolator and one isolated signal is translated to the high-side gate driver by a high-speed monolithically integrated level-shifter. The proposed voltage level translator structure has been implemented with a high-voltage diode-less SiC vertical JFET and low voltage MOSFETs. At 240V power supply voltage and 100 kHz switching frequency, the level-shifter propagation delay is only 10 ns with a 1.85 mA supply current. Moreover an effective built-in current regulator prevents the circuit from any current spikes, overvoltage and overconsumption. Such topology allows to safely set a very short 23 ns dead-time and improves WBD-based converter operations while reducing size and price.**

Keywords—dead-time reduction, diode-less synchronous converter, level-shifter, monolithically integrated, propagation delay matching, voltage level translator, wide bandgap devices.

I. Introduction

Wide Bandgap Devices (WBD) as GaN HEMTs or SiC JFETs overcome some theoretical limitations of conventional silicon devices. They are expected to address higher temperature [1]-[2], higher power density [3]-[4] and higher frequency operations [4]-[6]. Although WBD can be used into converters in place of silicon transistors, they are structurally different. Indeed GaN HEMTs and some SiC JFETs do not have a parasitic body diode [3]-[4], [7] generating extra losses during dead-time into an inverter leg structure. To take full advantage of their outstanding performances the gate driver and the dead-time have to be carefully adapted [4], [8]-[9].

Each power transistors into a high-voltage inverter leg structure are driven between gate and source by a floating gate driver. This gate drive circuit receives an isolated input signal from an external microcontroller unit (MCU). Several signal isolation techniques have been already developed such as

photo-coupler [10], wireless pulse transformer [11] or electromagnetic resonant coupler [12]. Because of high-voltage power supply, the high-side and low-side floating input signals have to be generated by two separate isolators. Mismatch between two separate isolation circuits being important [10]-[11] a high-voltage diode-less WBD-based inverter leg cannot safely operate with an appropriate very short dead-time [4], [8].

To overcome mismatching issue while getting the lowest cost and size, we propose a topology based on a single digital 2-channel isolator and one high-speed high-voltage monolithically integrated level-shifter with a built-in current limitation. Previous works on high-speed level-shifter report low power circuits and built-in short circuit protections but only low voltage operations [13]-[14]. Other works [15]-[16] report high-speed high-voltage level-shifters but additional high-voltage devices are required and no built-in current protection are implemented.

The present paper is organized as follow. Section II introduces the high-voltage WBD-based inverter leg. The proposed level-shifter structure is detailed in section III. Section IV presents the experimental results.

II. Wide Bandgap Device Half-Bridge

A. Diode-Less Wide Bandgap Devices

WBD are made of semiconductor with higher bandgap energy and critical field than silicon ones. These material characteristics confer outstanding performances on WBD. For the same breakdown voltage WBD have a better specific ON state resistance Ron·S. Optimization of Ron·S and loss densities result in smaller WBD with better switching characteristics. WBD being smaller, turn-on energies are smaller too. Because of fast transitions and low turn-on energy, very low impedance package and gate driver are recommended to prevent WBD from faulty turn-on and destructive overvoltage [9].

Some WBD as GaN HEMT [4], [8] and vertical SiC trench JFET [7] do not have a parasitic body diode between drain and source. The current can even flow between drain and source in any directions through their channel. When the gate-source voltage V_{gs} is over the threshold voltage V_{th} operations of diode-less WBD in the first and third quadrant are symmetrical as well as MOSFET. When V_{gs} is below V_{th}, characteristics in

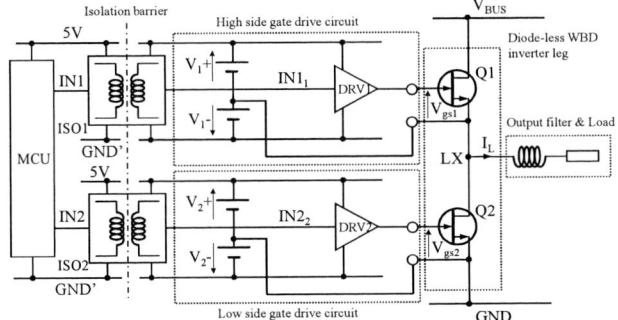

Fig. 1. Schematic of a conventional high-voltage synchronous converter based on diode-less WBD

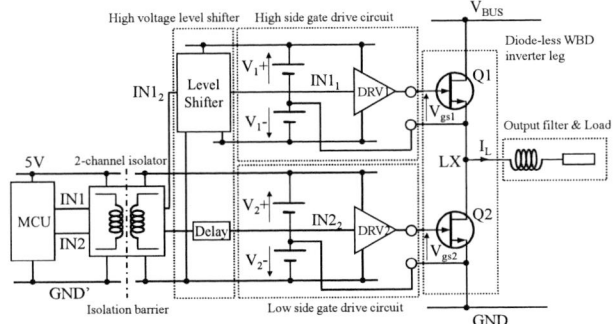

Fig. 2. Schematic of an half-bridge topology with a single 2-channel isolator and a high-voltage level shifter

the third quadrant are depending on the gate voltage [4], [7]-[8]. Further V_{gs} is from the threshold voltage, higher is the source-drain voltage drop V_{ds} and so higher the reverse conduction losses are.

B. Diode-Less Inverter Leg Operations

The elementary inverter leg structure is present in many types of synchronous power converters as AC/DC, DC/DC or DC/AC converters. This elementary structure shown on Fig. 1 is made of two power transistors Q1 and Q2 connected in series between power supply terminals. Both power transistors are driven between gate and source by a floating gate driver. This gate driver is supplied with a positive V+ and a negative V- isolated power supply. V+ and V- levels are set in agreement with the threshold voltage V_{th} in order to correctly turn-on and turn-off the power transistor. Each gate drivers receive an isolated input signal from an external microcontroller unit (MCU) to alternately switch Q1 and Q2.

Synchronous converters based on diode-less WBD can properly operate without any additional antiparallel diodes [3]-[4], [7]-[8]. In that way cost and size are reduced but also switching loss. Indeed there are no more reverse recovery charge and parasitic capacitance of diode. But during dead-time, both power transistors are off and the inductance current can no more flow through a freewheeling diode. The current flows through Q1 or Q2 channel generating extra dead-time loss because of the particular reverse conduction mode [4], [8]. Therefore in many case a very short dead-time improves operation of pure diode-less synchronous converters.

C. Mismatch between High-Side and Low-Side

The MCU generates two opposite signals IN1 and IN2 with a secure non-recovery time. These input signals are isolated and translated to signals $IN1_1$ and $IN2_2$ by isolators ISO1 and ISO2. Into conventional high voltage synchronous buck, the high-side and low-side floating input signals have to be generated by two separate isolators.

Many isolation techniques have been developed such as photo-coupler [10], wireless pulse transformer [11] or electromagnetic resonant coupler [12]. Some of them [10]-[11] are commercially available. Datasheets of these industrial isolation circuits point out an important dispersion on propagation delay. Especially in the photo-coupler case [10]

the propagation delay dispersion is over 150 ns. While in the digital pulse transformer case [11] the typical propagation delay is 27 ns and mismatch between two units can be reduced to 10 ns with a special top grade.

Even with expensive special matching units, 10 ns isolator propagation delay skew does not allow to safely set a dead-time as short as 10 ns because of an inherent cross-conduction risk. Considering an additional 10 ns mismatch between high-side and low-side gate drive circuits, a dead-time larger than 30 ns is more appropriate.

III. A MONOLITHICALLY INTEGRATED LEVEL-SHIFTER

A. An Half-Bridge Topology with a Single 2-Channel Isolator

In order to ensure proper operations of high-voltage synchronous converters with a very short dead-time we propose the topology shown on Fig. 2. Input signals IN1 and IN2 are isolated by a single 2-channel digital isolator. One input signal is shifted to the high side gate driver input by a high-voltage level translator while the other signal is delayed by a time corresponding to the estimated level-shifter propagation delay.

If propagation delays are difficult to match between two different parts because of process dispersion, it is easier to obtain when signals pass through the same chip. Into ADuM240xCRWZ parts [11], a 2 ns maximum channel-to-channel matching is reached. In the same way if the monolithically integrated level-shifter keeps a good side-to-side matching, a very short dead-time can be safely set while reducing volume, price and potentially common mode signal propagation by removing an isolator.

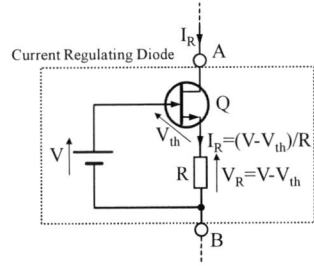

Fig. 3. A circuit schematic of a current regulator diode example

B. The Current Regulator Diode

A Current Regulator Diode (CRD) [17] operates as a current limiter diode. It consists of a transistor Q, a power supply V and a resistance R. The current I_R shown on Fig. 3 can only flow through the CRD from terminal A to terminal B. When I_R rises, the resistance voltage V_R rises until gate-source voltage V_{gs} of Q is equal to the threshold voltage V_{th}. When the threshold voltage is reach the current can no more rise. It is so regulated to a value given by (1) depending on V_{th}, R and V.

$$I_R = \frac{V - V_{th}}{R} \qquad (1)$$

One can notice that a normally-on or normally-off device Q can be used. With a negative threshold voltage V_{th} and so a normally-on transistor, the power supply V can be remove. The current is simply limited to a value equal to the ratio of V_{th} by R. If transistor Q is a high-voltage device, such CRD constitutes an easy way to limit current between two high-voltage terminals.

C. An Integrated Level-Shifter with a Current Regulator

The topology of the proposed high-speed high-voltage level-shifter is described on Fig. 4. It is based on the CRD principle to translate signal from low-side to high-side though the use of a controllable and limited current I_R. The voltage level translator is made of three parts monolithically integrated to high-side/low-side gate drivers and high-voltage WBD inverter leg. The implemented controllable CRD consists of low-voltage power supplies V_2+, V_2-, resistance R2, switch N1 and a high-voltage transistor Q3.

When the floating input signal $IN1_2$ referenced to the low side is high, N1 is switched on and a regulated current I_{R2} given by (2) flows from the high-side to low-side. In the general case high-side and low-side being symmetrical, voltage levels V_1+ and V_1- are respectively equals to V_2+ and V_2-. By properly setting the R1/R2 ratio as a function of (V_2- - V_2- - V_{th}), the input translated signal $\overline{IN1_1}$ referenced to the high-side and given by (3) is null when $IN1_2$ is high. The low-voltage high-side gate driver is theoretically protects against overvoltage and overcurrent because of the regulated current flowing through it

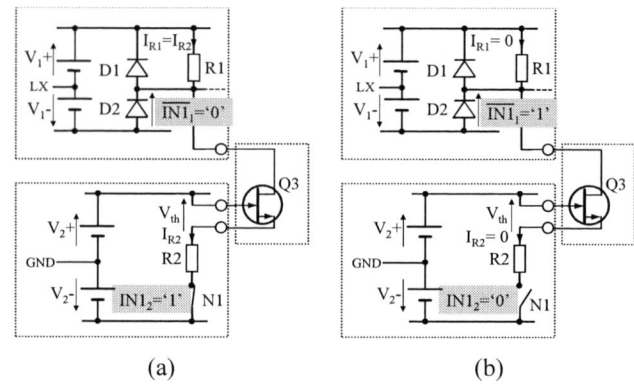

Fig. 5. Voltage shifting of the input signal $IN1_2$ referenced to the low side in both cases: (a) $IN1_2$ is high and (b) $IN1_2$ is low.

and integrated clamping diodes D1 and D2.

$$I_{R2} = I_{R1} = \frac{V_2+ - V_2- - V_{th}}{R2} \qquad (2)$$

$$\overline{IN1_1} = V_1+ - V_1- - \frac{R1}{R2} \cdot (V_2+ - V_2- - V_{th}) \approx 0\,V \qquad (3)$$

When the floating input signal $IN1_2$ referenced to the low side is low, N1 is switched off and no more current is flowing from high-side to low-side. The voltage drop at the resistance R1 is null and so the input translated signal $\overline{IN1_1}$ given by (4) is high.

$$\overline{IN1_1} = V_1+ - V_1- \qquad (4)$$

The proposed topology allows quick input signal shift from low-side to high-side and low power consumption thanks to an efficient current regulation. Performances can still be improved by sending pulses through the level-shifter rather than a continuous signal as on Fig. 6. In that case a regulated current is flowing from high-side to low-side just during short pulses. Pulses are generated from the low-side and captured on the high-side by a negative-edge-triggered D flip-flop.

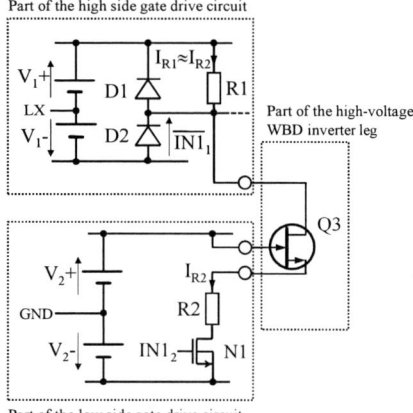

Fig. 4. The high-voltage level shifter monolithically integrated to gate drivers and WBD inverter leg with a built-in current limiter

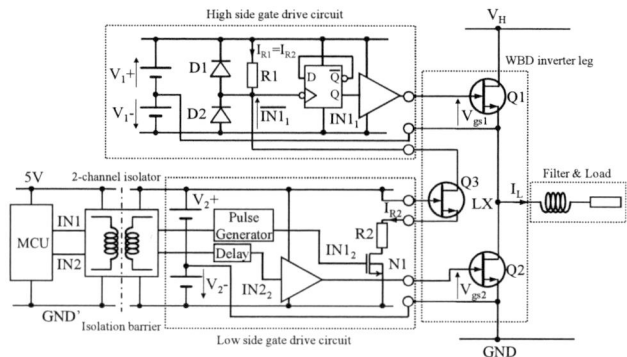

Fig. 6. Schematic of the proposed monolithically integrated high-voltage level shifter into a WBD inverter leg

978-1-4799-4993-9/14 $31.00 © 2014 IEEE

TABLE I.   MEASURED LEVEL-SHIFTER SPECIFICATIONS

Parameter	Symbol	Value	Unit
Power supply voltage	V_H	240	V
High-side driver power supply voltage	V1-	-12	V
Switching frequency	f_{SW}	100	kHz
Pulse width	t_{pulse}	250	ns
Supply current	I_{V1-}	1.85	mA
Regulated current	$I_{R2}=I_{R1}$	110	mA
Propagation delay	t_{delay}	10	ns

IV.   EXPERIMENTAL VALIDATION

The proposed high-voltage high-speed level-shifter topology that can be monolithically integrated has been implemented with discrete devices as a 1200V diode-less normally-on SiC vertical JFET Q3 and a 50V NMOS N1. Q1-Q3 being normally-on devices, negative single driver power supplies V- are used. Experimental results and associated waveforms of the voltage translator are depicted in Table I and Fig. 7. Vd_{Q3} and Vd_{N1} are the drain potentials of Q3 and N1. The built-in current regulator limits the current flowing from high-side to low-side during pulses to 110mA with R1 and R2 equal to 101Ω. Without current regulator and for the chosen JFET this current can quickly reached a critical value of 75A. At 240V power supply voltage, 100 kHz switching frequency with 250 ns pulse width, the level-shifter propagation delay is only 10 ns while it consumes 1.85 mA on the high-side driver power supply. This delay is measured from the 50% level of the $IN1_2$ rising edge to the 50% level of the $\overline{IN1_1}$ falling edge. To match propagation delays between both sides a fixed delay is added on the low side on Fig. 6. This one is equal to 10 ns plus 3ns plus 3ns corresponding to the typical propagation delay through the level-shifter, the D flip-flop and the pulse generator. With 33% maximum process variation on the level-shifter, the D-flip-flop, the pulse generator and the delay, the maximum mismatch between $IN1_1$ and $IN2_2$ signals is 11 ns. Considering additional 10 ns maximal mismatch between high-side and low-side gate drivers and 2 ns isolator channel-to-channel matching, a very short 23 ns dead-time can be safely set by guaranteeing no V_{gs1} and V_{gs2} overlapping.

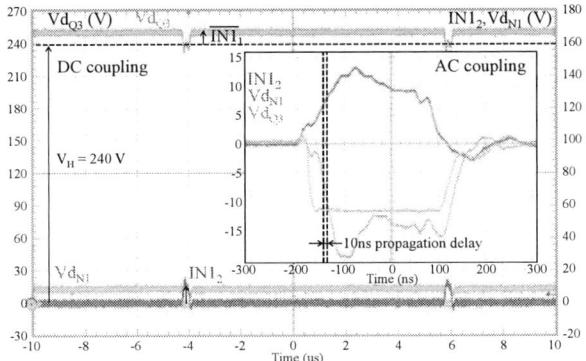

Fig. 7. Experimental waveforms of the high-speed voltage level-shifter

V.   CONCLUSION

In order to take full advantage of diode-less WBD outstanding performances we proposed a high-speed high-voltage monolithically integrated voltage level translator topology with a built-in current regulator. The structure has been implemented with a 1200V diode-less SiC JFET and a 50V NMOS. At 240V power supply voltage and 100 kHz switching frequency, the level-shifter propagation delay is only 10 ns with a 1.85 mA supply current. By using a single 2-channel isolator and the proposed level-shifter instead of two separate isolation circuits a very short 23 ns dead-time can be safely set improving WBD-based converter operations while reducing volume and price.

REFERENCES

[1] B. Wrzecionko, D. Bortis, and J. W. Kolar, "A 120 °C Ambient Temperature Forced Air-Cooled Normally-off SiC JFET Automotive Inverter System," *IEEE Trans. on Power Electron.*, vol. 29, no. 5, pp. 2345–2358, May 2014.

[2] Thierry Lebey, et al., "High temperature high voltage packaging of wideband gap semiconductors using gas insulating medium," in *Proc. IEEE Int. Power Electron. Conf.*, 2010, pp. 180-186.

[3] T. Morita, et al., "99.3% Efficiency of three-phase inverter for motor drive using GaN-based Gate Injection Transistors", in *Proc. IEEE Appl. Power Electron. Conf.*, 2011, pp. 481-484.

[4] S. Ji, D. Reusch, and F. C. Lee, "High frequency high power density 3D integrated gallium nitride-based point of load module design," *IEEE Trans. Power Electron.*, vol. 28, no. 9, pp. 4216–4226, Sep. 2013.

[5] F. C. Lee and Q Li, "High-Frequency integrated point-of-load converters: Overview," *IEEE Trans. Power Electron.*, vol. 28, no. 9, pp. 4127–4136, Sep. 2013.

[6] J. Delaine, P. Olivier, D. Frey, and K. Guepratte, "High frequency DC-DC converter using GaN device," in *Proc. IEEE Appl. Power Electron. Conf.*, 2012, pp. 1754–1761.

[7] D. C. Sheridan, K. Chatty, V. Bondarenko, and J. B. Casady, "Reverse conduction properties of vertical SiC trench JFETs," in *Proc. IEEE Int. Symp. on Power Semicond. Devices and ICs*, 2012, pp. 385-388.

[8] L. Hoffmann, C. Gautier, S. Lefebvre, and F. Costa, "Optimization of the driver of GaN power transistors through measurement of their thermal behavior," *IEEE Trans. on Power Electron.*, vol. 29, no. 5, pp. 2359-2366, May 2014.

[9] X. Youhao, M. Chen, K. Nielson, and R. Bell, "Optimization of the drive circuit for enhancement mode power GaN FETs in DC-DC converters," in *Proc. IEEE Appl. Power Electron. Conf.*, 2012, pp. 2467–2471.

[10] "HCPL-2200: Very High CMR, Wide Vcc Logic Gate Optocouplers datasheet: www.agilent.com.".

[11] "ADUM2400: Quad-Channel Isolator datasheet: www.analog.com.".

[12] S. Nagai, et al., "A DC-isolated gate drive IC with drive-by-microwave technology for power switching devices, " in *Proc. IEEE Int. Solid-State Circuits Conf.*, 2012, pp. 404–406.

[13] Y.-M. Li, C.-B. Wen, B. Yuan, L.-M. Wen, and Q. Ye, "A high speed and power-efficient level shifter for high voltage buck converter drivers, " in *Proc. IEEE Int. Conf. Solid-State Integr. Circuit Tech*, 2010, pp. 309-311.

[14] S. Ali, S. Tanner, and P.-A. Farine, "A robust, low power, high speed voltage level shifter with built-in short circuit current reduction, " in *Proc. Euro. Conf. Circuit Theory Design*, 2011, pp. 142-145.

[15] M. Berkhout, "An integrated 200-W class-D audio amplifier, " *IEEE J. Solid-State Circuits*, vol. 38, no. 7, pp. 1198-1206, Jul. 2003.

[16] D. R. H. Carter and R. A. McMahon, "An integrated level shifter for use in high frequency half-bridges, " in *Proc. IEEE New Dev. Power Semiconductor Devices*, 1996, pp. 9/1 -9/8.

[17] "CR160: Current Regulator Diodes datasheet: www.vishay.com.".

978-1-4799-4993-9/14 $31.00 © 2014 IEEE

Extensive Electro-Thermal Simulation Methodology for Automotive High Power Circuits

Adrian-Gabriel Bajenaru

Infineon Technologies Romania
Blvd. Dimitrie Pompeiu 6, 20335, Bucharest
University Politehnica of Bucharest
Blvd. Iuliu Maniu 1-3, Bucharest
adrian-gabriel.bajenaru-ee@infineon.com

Cristian Mihai Boianceanu

Infineon Technologies Romania
Blvd. Dimitrie Pompeiu 6, 20335, Bucharest
cristianmihai.boianceanu@infineon.com

Fabio Ballarin

Infineon Technologies Italia
Automotive Standard VREG
Via N. Tommaseo 65/B, 35131 Padova
fabio.ballarin@infineon.com

Prof. Gheorghe Brezeanu

University Politehnica of Bucharest
Blvd. Iuliu Maniu 1-3, Bucharest
gheorghe.brezeanu@dce.pub.ro

Abstract—**This paper presents a method of electro-thermal co-simulation for power electronic circuits made in a high voltage BiCMOS technology for automotive applications. The method allows the complete and correct modeling and optimization of high power analog circuits and guarantees robustness in fault conditions. The method was successfully applied on a low dropout linear voltage regulator, emphasizing the elements that influence the electro-thermal behavior of the circuit and its protection circuitry.**

Keywords—*electro-thermal; co-simulation; methodology; high power circuits; linear voltage regulator; protection circuits.*

I. INTRODUCTION

The purpose of this paper is to present a method for a thorough verification of the electro-thermal performance of high power circuits and their protection circuitry: current limiting, Safe Operation Area (SOA), thermal shutdown. The method is applied for a low dropout linear voltage regulator, but can be also applied for other power circuits. The accent is placed on the elements that influence the electro-thermal-behavior of the circuits [1] and must be included in the simulation setup and how to improve the performance of the circuit by correct layout placement of the protection circuitry. Section II describes a typical circuit and the technology for which the method was developed, then in section III the actual method of electro-thermal investigation is described, first identifying the prerequisites for an accurate simulation and then presenting the simulation steps and the corresponding results used to optimize the circuit behavior.

II. CIRCUIT DESCRIPTION

The simulation methodology was developed for a High Voltage Bipolar-CMOS technology (HV-BiCMOS) for automotive applications. Although the simulation methodology is universal and theoretically can be applied to any technology, the factors that have to be considered and their individual impact are technology specific. The HV-BiCMOS technology integrates low and high voltage bipolar transistors and CMOS devices, and the lateral isolation is made by trenches.

A typical circuit designed in this technology for which we are interested to perform an electro-thermal analysis is a Low

DropOut (LDO) linear voltage regulator, a circuit widely used in automotive systems to provide other circuits with a stable and reliable voltage supply. A block diagram and a floorplan of the LDO regulator presenting the main building blocks is shown in Fig. 1. It consists of a bandgap voltage reference, an error amplifier, an enable and internal supply block, the PNP pass device (T_{pass}) and the protection circuitry. The pass device occupies more than $1/3^{rd}$ of the total chip area and is positioned towards the bottom-right corner. The protection circuitry assures an output characteristic in short circuit as shown in Fig.

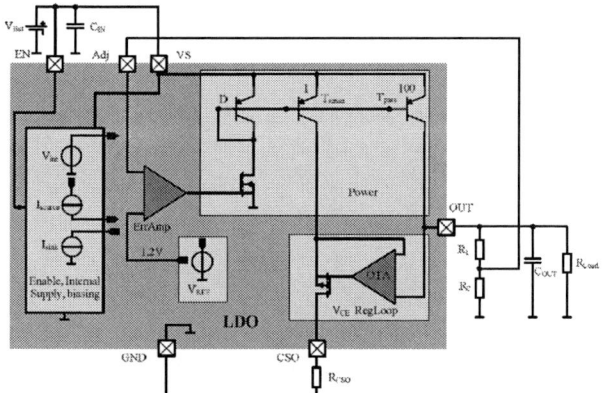

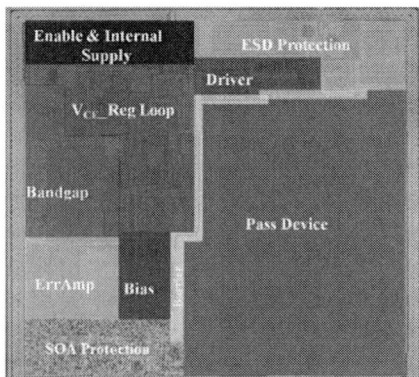

Fig. 1. Block diagram (top) and floorplan (bottom) of the low dropout linear voltage regulator.

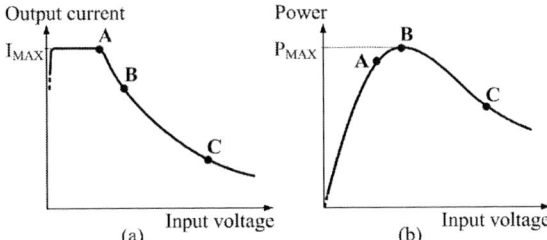

Fig. 2. Output characteristic (SOA) of the linear voltage regulator in short circuit condition: (a) output current; (b) dissipated power.

2, keeping the power device in the SOA. A detailed description of the voltage regulator can be found in [2].

III. SIMULATION METHODOLOGY

The purpose of the electro-thermal analysis is to investigate the circuit behavior when placed in a fault condition like a short circuit at the output pin, with an emphasis on the zone where both the current limiting and the SOA protection are active (points A, B, C in Fig. 2). The points were chosen using the following considerations: A − corresponds to the point with the highest voltage drop that still delivers the maximum output current in short circuit; B − corresponds to the point with maximum power dissipation; and C − corresponds to high voltage drop and lower output current, where as shown in [1] the device is most likely to be damaged due to electro-thermal causes. To investigate the electro-thermal behavior of the circuit a method of co-simulation as presented in [1, 3-5] is proposed. This implies coupling a 3D thermal simulator (as the one described in [6]), with a SPICE like circuit simulator. The flow-chart describing this method is presented in Fig.3. The method has the advantage of high accuracy if the setup is modeled correctly, but can also be computationally intensive and attention must be paid to what is taken into account for simulation.

A. Simulation Prerequisites

The junction to ambient thermal impedances for the given circuit, package and different PCB mountings, according to the JDEC-JESD51 standard, are shown in Fig. 4 (a). Using the junction-to-ambient thermal resistance value (the steady state thermal impedance) results in a maximum temperature well above 175°C for all the three power values A, B and C in Fig. 2. This implies that in all operating conditions on the SOA

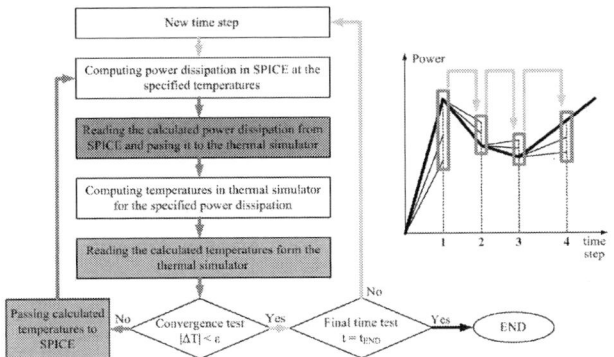

Fig. 3. Flow-chart of the electro-thermal co-simulation method based on the relaxation method, and the evolution in time of the power.

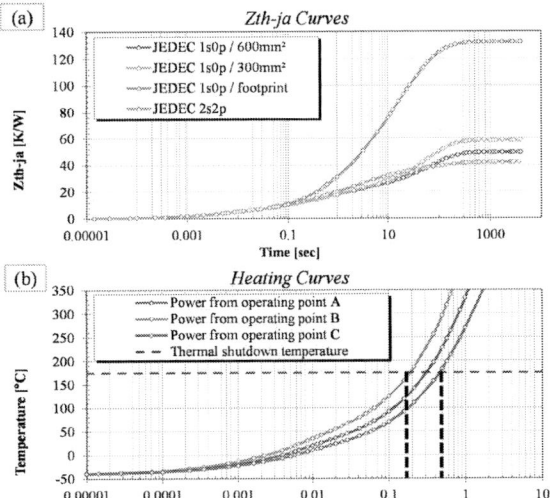

Fig. 4. (a) Thermal impedances according to JDEC-JESD51 standard, and (b) heating curves for the operating points A, B and C in Fig. 2.

curve, during the short circuit condition the thermal shutdown event will be triggered. Using the thermal impedance again, we can estimate the heating curve and the delay of the thermal shutdown event as being in the hundreds of milliseconds range, when starting from a −40°C ambient, as shown in Fig. 4 (b). This can be used only as a rough estimation of our simulation time, because the thermal impedances are calculated using a uniform power distribution on the pass device. The actual simulation time could extend in the 1 second range and by looking at the heating curve we can deduce that the simulation setup for the thermal simulator must include not only the die and package information, but also the PCB mounting, whose influence cannot be neglected at this timescale.

An important aspect that has to be considered is the temperature range of the models. The automotive temperature range is -40°C to 150°C, but the models must cover a much wider interval. For the electro-thermal simulation the temperature range of the power transistor has to be −40°C to 300°C, to correctly cover the modeling of potentially destructive fault conditions (e.g. a short circuit) and highlight the triggering moment of the thermal shutdown (typically at 175°C) and the subsequent circuit turn-off behavior.

The power PNP structure used for the pass device of the circuit presented in section II is constructed using a standard parameterized cell (pcell), which can sometimes result in an electrical model that does not take into account some aspects important for the electro-thermal analysis. An approach is to construct an equivalent schematic of the pass device by splitting the device in many elementary PNP cells and annotating the parasitic elements like series resistances and leakage diodes.

1) Elementary PNP cell model

The fitting of the electrical model for the power transistors is usually done on several test structure of different areas and different form factors. This means that with one model and a minimum set of parameters regarding the geometry, the model tries to cover all the possible cases. Thus, the model also captures some of the parasitic elements (metallization, base resistance, leakage). To proper split the power stage in elementary cells, a correct modeled elementary cell is needed, without the mentioned parasitic elements. The elementary cell can be as small as one emitter disc or can be a transistor equivalent to several emitter discs, depending on the desired granularity in the pass device splitting. Then the parasitic elements can be annotated, resulting in an equivalent schematic of the power stage much closer to the real device.

2) Parasitic resistances

An important aspect that must not be neglected in modeling a power stage is the metallization resistance, especially on the emitters of the power transistor. The series resistances of the metallization is annotated, based on the geometry that is used, after the power stage was split in the elementary cells. This way also the temperature variation coefficient of the resistance can be taken into account. The parasitic base resistance can play an important role in the distribution of the current in a large power stage, especially in current limit condition, when the power stage is working at high current densities and the forward current gain of the device is low. As in the case of the metallization resistance, we annotate the distributed parasitic base resistance after the split of the power stage in several sections, and create a network of base resistances that connects to the bases of each device. Like in the case of the metallization resistance, here also the temperature variation of the resistance is not neglected.

3) Leakage modeling

The leakage to the p-type substrate in a trench isolated technology has two main components dependent on the area and the perimeter of the power device. To model this for each elementary cell a diode connected between the base and the substrate with a leakage proportional to the cell area has to be added. Also on the perimeter of the device, where the trench is situated, additional diodes have to be added to account for the perimeter leakage.

B. Electro-Thermal Co-Simulation

After setting up the power device in the electro-thermal simulation setup, including all the parasitic elements, the next step is to determine all other devices that play a role in the electro-thermal behavior and that must also be included in the simulation setup. Although an approach in which all the devices from the electric schematic are included in the simulation is possible, it is cumbersome and unnecessary. For the circuit presented in section II the elements that have to be included, in addition to the pass device, are: the thermal shutdown sensor and thermal shutdown circuit, the current limiting and SOA protection circuits, bias block and the driver block. The bias block has to be annotated because it can influence the thermal behavior of the device and the temperature at which the thermal shutdown takes place. The driver block has to be included because the power density is large enough to increase its temperature and heat up also the devices nearby.

The three main aspects that concern us during the design and verification phase are the power distribution on the powerstage (and the corresponding temperature profiles on the chip), the overall output current behavior during a short circuit event and the thermal coupling of the thermal shutdown sensor to the hotspot. The power distribution on the powerstage depends on the dimensions, placement and splitting of the power device, but also on its parasitic elements. The optimization of these elements is constrained by the initial floorplan, bondpads placement, chip size and chip aspect ratio (constrained by the package) and other routing constraints (especially in a single metal technology as the one used here), but it is not addressed in this paper.

1) Current sensing element positioning

When we refer to the output current behavior during short circuit we are interested if the current is increasing, constant or decreasing in time, as the temperature on the chip is rising. In other words we are interested if the output current is temperature compensated or has a negative temperature variation coefficient. A positive temperature variation coefficient is not desired because it indicates the circuit is more vulnerable to short circuit conditions. To optimize the output current behavior the best placement of the current sensing element must be determined. For this, during the first electro-thermal simulations the current sense element is thermally decoupled from the powerstage in the electro-thermal simulation (the temperature increase is not annotated for the sense element). The circuit behavior is simulated for the three operating conditions A, B and C in Fig. 2. The simulations are stopped when the peak temperature on the powerstage reaches 250°C. The average, maximum and minimum power densities on the powerstage elementary cells are presented in Fig. 5. The average power density indicates also the output current behavior, because the device is operated at a constant input voltage. Because the current sensing element is thermally decoupled from the powerstage, the output current is increasing in time and so does the spread between the minimum and maximum power densities. The worst case corresponds to point C, where due to the high power dissipation but lower current

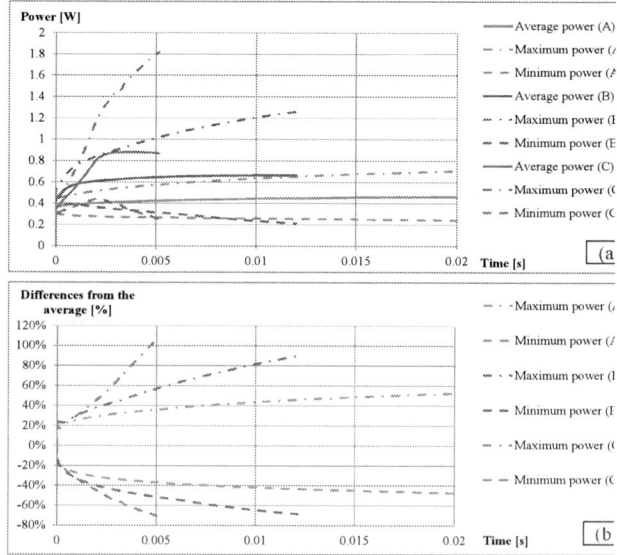

Fig. 5. Power dissipation on the powerstage elementary cells (a) and difference in percentages from the average value (b), with the current sensing element thermally decoupled.

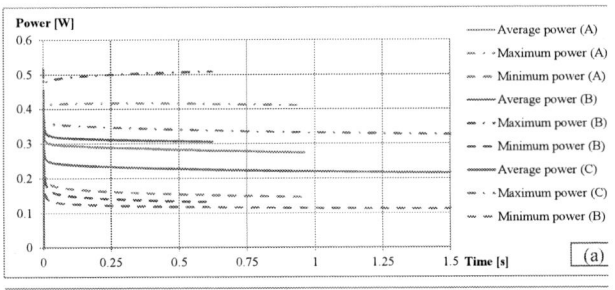

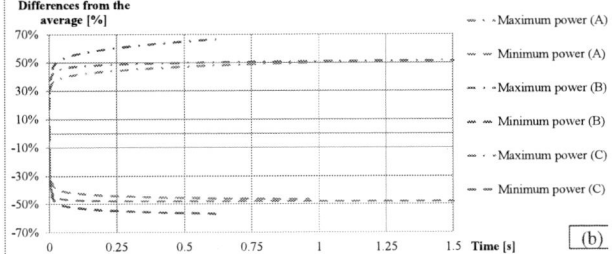

Fig. 6. Power dissipation on the powerstage elementary cells (a) and difference in percentages from the average value (b), with the current sensing element placed near the hotspot.

values the intrinsic ballasting resistors of the PNP cells are less efficient in keeping the power distribution uniform. The electro-thermal effects are more pronounced in this operating point and lead to a more confined hotspot formation, as shown in the temperature map in Fig. 7. (c1).

Using the temperature profiles at the moment when the peak temperature is reached (shown in Fig. 7. (a1), (b1), (c1)) and the power densities for each element, the current sensing element optimum position is chosen to be as close as possible to the maximum temperature and the highest power density in all three cases. The results with the current sensing placed in the optimum position are shown in Fig. 6. The average power density (thus also the output current) is decreasing in time. The spread between the maximum and minimum power densities is smaller than in the previous case and much more constant. The peak temperature of 250°C is reached in a longer interval of hundreds of milliseconds, compared to milliseconds when the sensing element is decoupled.

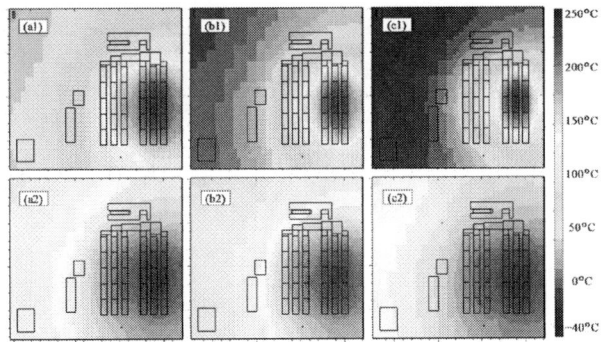

Fig. 7. Temperature profiles when the peak temperature reaches 250°C. (a1), (b1), (c1) – current sensing element decoupled, operating points A, B and C, respectively; (a2), (b2), (c2) – current sensing element optimally placed, operating points A, B and C, respectively.

2) Thermal shutdown sensor positioning

The temperature profiles at the moment when the peak temperature is reached, with the current sensing positioned in the optimal place are shown in Fig. 7. (a2), (b2), (c2). Based on these profiles an optimum position for the temperature shutdown sensor can be found. The temperature is much more uniform across the chip surface than in the case with the sensing element decoupled. Therefore, there is a larger degree of freedom in choosing the sensor position, because the sensor will be anyhow much better coupled to the peak temperature on the chip.

IV. CONCLUSIONS

The presented simulation methodology, based on electro-thermal co-simulation, allows the complete and correct modeling and optimization of high power analog circuits (including current limiting, SOA and thermal shutdown functions) to guarantee robustness under fault conditions. Using this methodology we presented the steps needed during the design phase for the optimization of a linear voltage regulator's robustness under short circuit conditions, ensuring correct operation for different points on its SOA characteristic. The results shown can only be obtained correctly using the presented electro-thermal modeling of the circuit's behavior. Other methods that rely only on the analysis based on pure electrical simulators, even though they can capture some of the thermal aspects for the proposed system, they cannot insure the correct functioning in all situations.

ACKNOWLEDGMENT

The work has been funded by Infineon Technologies Romania and by the Sectoral Operational Programme Human Resources Development 2007-2013 of the Romanian Ministry of Labour, Family and Social Protection through the Financial Agreement POSDRU/107/1.5/S/76903.

REFERENCES

[1] Bajenaru, A.-G.; Boianceanu, C.; Brezeanu, G., "Investigation of electro-thermal behaviour of a linear voltage regulator and its protection circuits by simulator coupling," Semiconductor Conference (CAS), 2013 International , vol.2, pp.237-240, 14-16 Oct. 2013

[2] Schroter, P.; Jahn, S.; Klotz, F.; Ballarin, F.; Gini, F.; Piselli, M., "EMI resisting LDO voltage regulator with integrated current monitor," Electromagnetic Compatibility of Integrated Circuits (EMC Compo), 2013 9th Intl Workshop on, pp.107-112, 15-18 Dec. 2013

[3] Petrosjanc, K.O.; Ryabov, N. I.; Kharitonov, I. A.; Kozynko, P. A., "Electro-thermal simulation: a new subsystem in Mentor Graphics IC Design flow," Thermal Investigations of ICs and Systems, 2009. THERMINIC 2009. 15th International Workshop on, pp.70-74, 7-9 Oct. 2009

[4] Attar, S.; Yagoub, M. C E; Mohammadi, F., "New Electro-Thermal Integrated Circuit Modeling using Coupling of Simulators," Electrical and Computer Engineering, 2006. CCECE '06. Canadian Conference on, pp.1218-1222, May 2006

[5] Wunsche, S.; Clauss, C.; Schwarz, P.; Winkler, F., "Electro-thermal circuit simulation using simulator coupling," Very Large Scale Integratssion (VLSI) Systems, IEEE Transactions on , vol.5, no.3, pp.277-282, Sept. 1997

[6] Pfost, M.; Boianceanu, C.; Lohmeyer, H.; Stecher, M., "Electrothermal Simulation of Self-Heating in DMOS Transistors up to Thermal Runaway," Electron Devices, IEEE Transactions on , vol.60, no.2, pp.699-707, Feb. 2013

Two-Dimensional Optical Beam Induced Current measurements in 4H-SiC bipolar diodes

H. Hamad[1], P. Bevilacqua[2], C. Raynaud[3], D. Planson[4]

Université de Lyon, INSA de Lyon, Ampère Laboratory - UMR 5005. 21, Avenue Jean Capelle
69621 VILLEURBANNE CEDEX, France
[1]hassan.hamad@insa-lyon.fr, [2]pascal.bevilacqua@insa-lyon.fr, [3]christophe.raynaud@insa-lyon.fr,
[4]dominique.planson@insalyon.fr.

Abstract–**This paper illustrates the photon's absorption phenomenon in 4H-SiC. It shows two-dimensional Optical Beam Induced Current measurements (2D-OBIC) in 4H-SiC bipolar diodes. Two different diode structures were studied: the first one is a circular MESA protected avalanche diode with an optical window, and the second structure is a PN diode protected with a junction termination extension layer. The results provided an image of the electric field distribution in the diode surface. These measurements validate the efficiency of the used protection. The effect of radius at the periphery of the diode was also studied. Structural defects are explored by a variation of photo-current at this local point.**

Keywords – 4H-SIC; bipolar diodes; photon's absorption; 2D-OBIC.

I. INTRODUCTION

In the last few decades, the use of wide bandgap (WBG) semi-conductors (4H-SiC, GaN, etc...) has become popular in the domain of power electronics. The physical properties of WBG semi-conductors exceed that of the traditional Si, mainly their critical electric field (E_c) and high thermal conductivity. These properties allow producing new electronic components with smaller size or to fabricate components with higher breakdown voltages [1]. Nevertheless, local breakdown was observed at electrical fields that are significantly lower than E_c. Thus, additional research is needed in order to overcome all the shortcomings and improve the overall performance of these WBG semi-conductors.

Methods that utilize an optical beam are used to study the properties of semi-conductors [2]. OBIC analyses are employed to determine the minority carrier lifetime [3]. In addition, the multiplication coefficient is then studied from the variation of OBIC in terms of reverse voltage [4]. Moreover, the latter is then used to extract the ionization rates of WBG semi-conductors. Furthermore, an image of the electric field distribution is obtained by scanning the surface of the power component.

II. THEORETICAL OBIC

When a semi-conductor is illuminated with an appropriate wavelength, electron-hole pairs (EHPs) are generated by absorbing the incident photons. In order to localize generated EHPs, the $P^+/N/N^+$ junction shown in Figure 1 is reverse biased to create high electric field region, then a laser beam illuminates perpendicularly the junction to generate EHPs. The EHPs generated in the space charge region (SCR), which is dominated by high electric field, are spited, the electron towards the cathode, the hole towards the anode. These carriers will acquire kinetic energy and will be accelerated. They will reach the edge of SCR and an induced current is measured. That is named Optical Beam Induced Current (OBIC). If their kinetic energy is high enough, they may enter in collision with other particles in the semi-conductor leading to generate new EHPs; when the charge carriers portion become large, avalanche takes place. The latter action happens when reverse voltage is near breakdown voltage when electric field becomes very high and generated EHPs have great kinetic energies. If photons are absorbed far outside the SPR, there is no electric field to separate the generated EHPs, so that the electrons and holes recombine before reaching the electrodes and no current is measured. If photons are absorbed near outside the SCR, i.e. at a distance i smaller than the minority carrier diffusion length, the carriers may reach the SCR and then lead to non-null OBIC, as described above.

To study the nature of absorption process, the comparison between photonic energy (E_Φ) and material bandgap E_G is mandatory. If E_Φ is greater than E_G ($E_\Phi \geq E_G$), the absorption of only one photon leads to generate an EHP [5]. In other

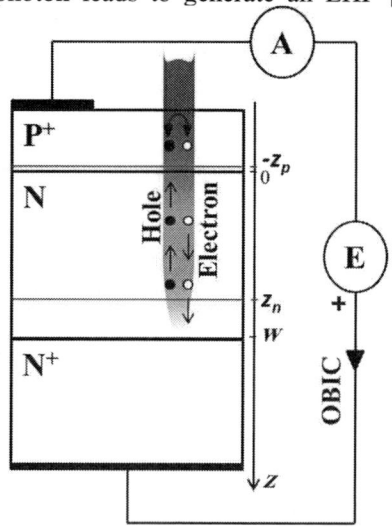

Figure 1: Schematic View of OBIC principle.

cases, when E_Φ becomes too low compared to E_G ($E_\Phi \ll E_G$), two or more photons are absorbed to generate one EHP [6][7]. In this work, since the used wavelength is 349nm, the photonic energy is then 3.55eV. The bandgap of 4H-SiC is 3.36eV, and then mono-photonic absorption dominates in the realized OBIC measurements.

III. EXPERIMENTAL SETUP

A UV pulsed laser is used to generate EHPs, the wavelength of incident light is 349nm, the repetition rate can be modified between 100 Hz and 5 kHz (Figure 2). The pulse energy can be adjusted up to 120 µJ. The pulse duration varies between 3 and 7ns according to frequency and pulse energy. An optical bench consisting of two semi-reflecting mirrors and a focusing lens is controlled with LabView to move the position (x, y and z) of the focal point. At the focal point the spot diameter is about 20µm and the beam power density is up to 100 GW.cm^{-2}. In this work, the repetition frequency is set to 1 kHz and all measurements are realized under very low pulse energy (smaller than 1 µJ/pulse). This energy is high enough to generate EHPs so OBIC will be non-null.

Two different diode structures are studied. First one is a MESA protected avalanche diode produced by French-German Institute of Saint Louis (ISL), its breakdown voltage V_{BR} is of 59V (Figure 3a). Optical window of 100×100 µm^2 is performed using SIMS technique [8] through metallization allowing the comparison of electric field at the center and the periphery of the test diode. The second structure is a P$^+$/N/N$^+$ diode protected with a 200µm length junction termination extension (JTE) layer (Figure 3b), its breakdown voltage is about 800V. Reverse I-V characteristics are realized foremost to select compatible diodes for OBIC measurements. Leakage current must be low to not mask OBIC. OBIC measurements are performed by scanning the surface of a diode with laser beam. The diode is reverse biased by using Keithley 237 Source Measurement Unit (SMU) which resolution is lower than 1pA. The SMU measurements are synchronized with the position of the spot via Lab View and the results are assembled in a table to be studied later.

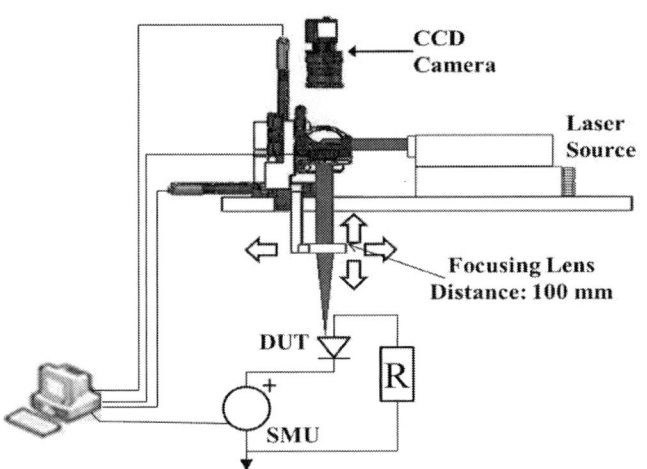

Figure 2: Schematic of OBIC bench.

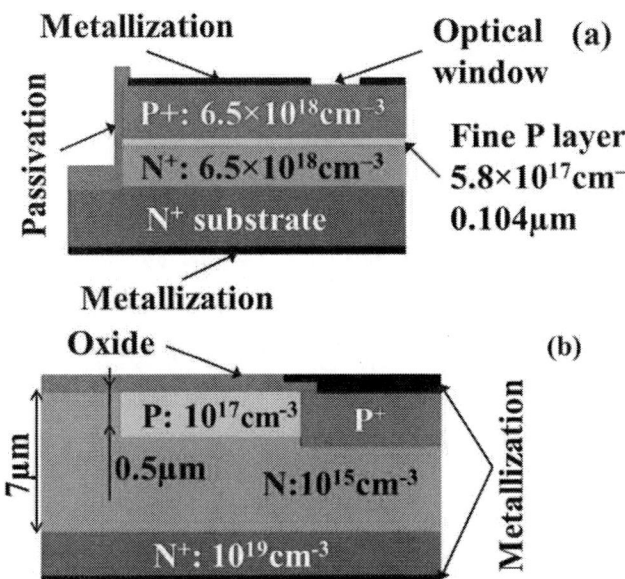

Figure 3: Cross section view of studied diode. (a) Avalanche diode, (b) P$^+$/N/N$^+$ diode.

IV. RESULTS AND DISCUSSION

OBIC scans are realized on a part of avalanche diode. This scan includes the optical window and a MESA etching of the diode as shown in figure 4a. The figure 4b displays the studied part of P$^+$/N/N$^+$ diode with OBIC measurements. OBIC scan covers a corner of the diode.

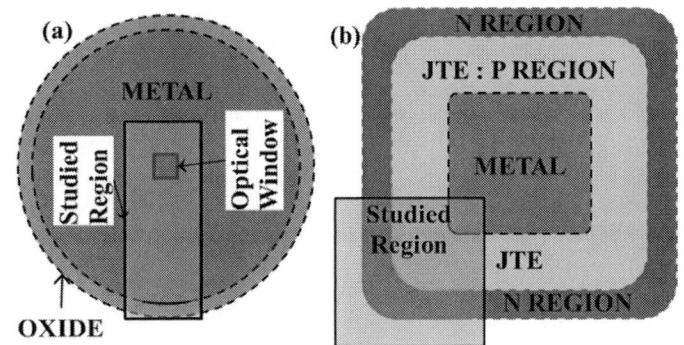

Figure 4: Over view of test diodes showing the OBIC scanned region of each diode: (a) Avalanche diode, (b) P+/N/N+ diode.

Figure 5 displays OBIC measurements on avalanche diode. The results show highest OBIC signal when the laser beam passes through the optical window. When the laser beam illuminates the metallization, that is a reflecting surface, the induced current is null. When the laser beam illuminates the edge of MESA etching, an induced current appears, but it is smaller than at the center of the diode. OBIC signal is null outside. As mentioned before, OBIC signal is an image of electric field. Since the induced current is the highest at the center of the diode, this means that also, electric field reaches its maximum at the center of the component for different voltage levels (especially when V_{BR} is approached). This result shows that the realized MESA protection is efficient and the avalanche does not take place locally in the diode.

Other OBIC measurements are realized on the $P^+/N/N^+$ JTE protected diode. These measurements are performed on about a quarter of a diode for voltage levels between 0 and 800V (Figure 6). The induced current is null when illuminating either the metallization or outside JTE. It is important when illuminating the JTE. OBIC signal has almost the same look for different reverse voltage levels, this explains that the JTE layer is well realized, and at the corner of the diode, the current is about the same as at the middle of the JTE. This shows that the radius of the curve at the corner of the diode is efficient.

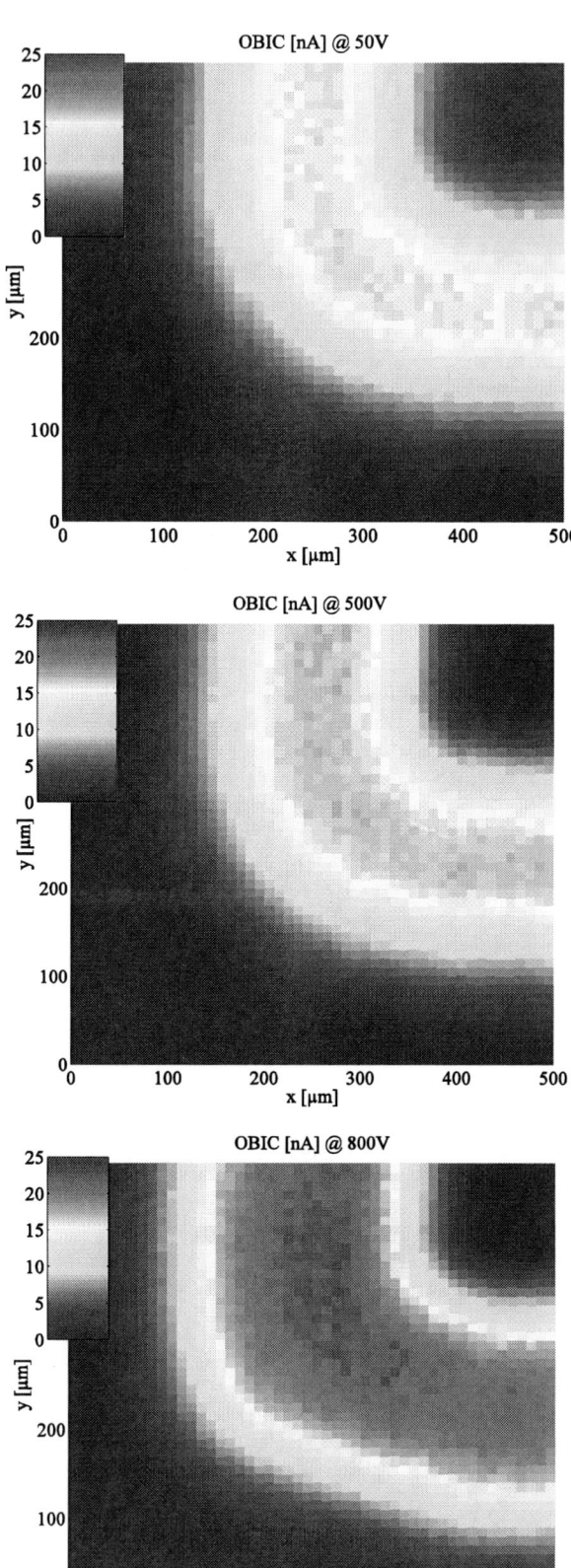

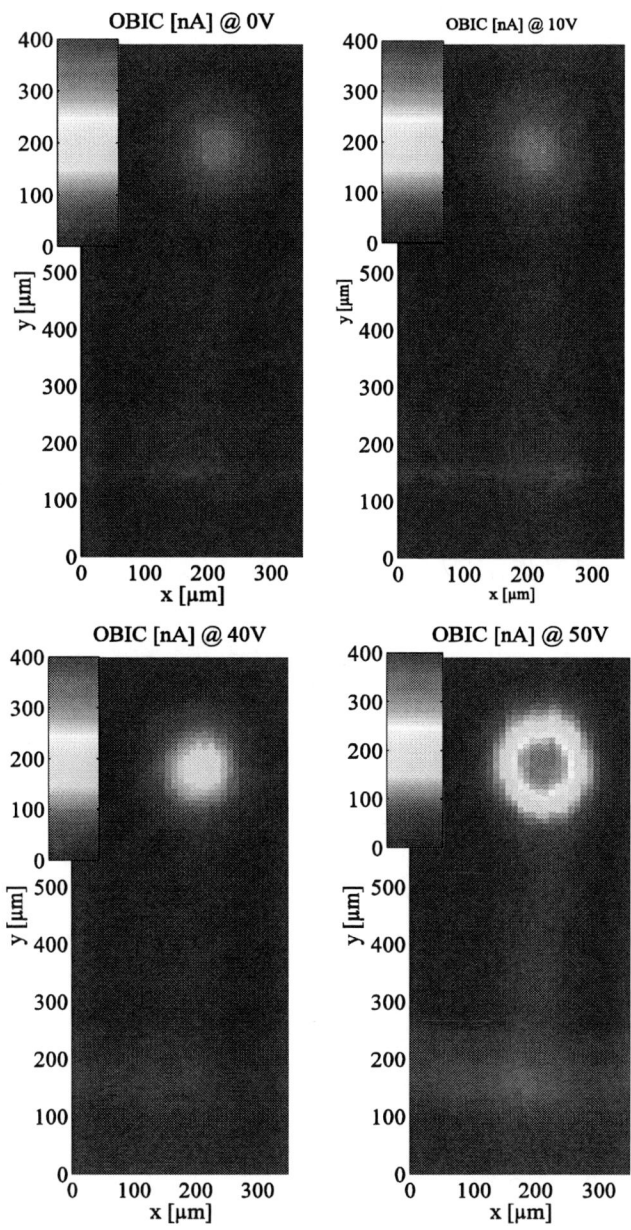

Figure 5: 2D-OBIC on a part of an avalanche diode for $V_R = 0, 10, 40, 50V$ with a resolution of 10μm (color bar unit is nA).

Figure 6: 2D-OBIC on a par of $P^+/N/N^+$ diode at 50, 500 and 800V (color bar unit is nA).

Some diodes presenting a V_{BR} smaller than 800V were studied with OBIC. Figure 7 shows an OBIC statement at 100V for a diode presenting an important leakage current (about 100nA at 300V). The result shows a non-homogeneous OBIC distribution in the JTE. A structure defect is detected using OBIC method. The induced current is higher at the corner of the JTE. This defect may be at the origin of the high leakage current of the test diode.

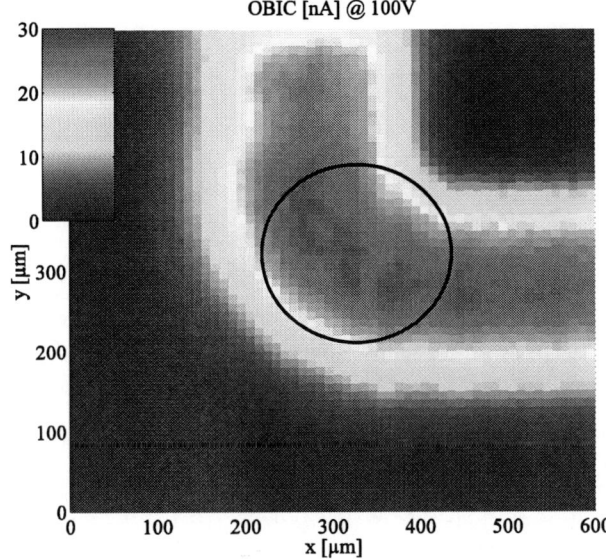

Figure 7: 2D-OBIC statement for a diode presenting a defect (color bar unit is nA).

V. CONCLUSION

In this paper, the photon absorption in 4H-SiC is used to study electronic characteristics of this material. A very low beam power is enough to generate electron-hole pairs since the photonic energy is greater than the bandgap of 4H-SiC. OBIC method (current measurement) combined with Semiconductor Device Software Sentaurus-TCAD will be used to determine the electric field distribution in the surface of a diode. It shows the efficiency of peripheral protection for the diode. It can be used to detect component defects at low voltage biasing. Two different protections were studied. Both methods work well and help avoiding local breakdown in the diodes. It is also shown that for low voltage components, JTE protection is not required and a simple MESA etching is sufficient to avoid local breakdown.

VI. ACKNOWLEDGEMENT

Authors thank Dr. Bertrand VERGNE from ISL to permit us to realize many measurements at ISL and to provide test diodes to realize this work. We thank also TRACE program for financial support.

VII. REFERENCES

[1] "Status of silicon carbide (SiC) as a wide-bandgap semiconductor for high-temperature applcations: a review", Casady J. B. and Johnson R. W., Solid State Electron, vol.39, p. 1409, 1996.

[2] "OBIC analysis for 1.3 kV 6H–SiC P⁺N planar bipolar diodes protected by Junction Termination Extension", C. Raynaud, S. R. Wang, D.

Planson, M. Lazar and J. P. Chante, Diamond & Related Materials, vol. 13, p. 1697, 2004.

[3] "Determination of minority-carrier lifetime and surface recombination velocity by optical-beam-induced-current measurements at different light wavelengths", T. Flohr and R. Helbig, Journal of Applied physics, vol. 66, no. 7, p. 3060, 1989.

[4] "Temperature dependence of hole impact ionization coefficients in 4H and 6H-SiC", R. Raghunathan and B.J. Baliga. Solid-State Electronics 43, pp.199-211, 1999.

[5] "OBIC measurements on avalanche diodes in 4H-SiC for the determination of impact ionization coefficients", D.M.Nguyen, C.Raynaud, M.Lazar, G.Pâques, S.Scharnholz, N.Dheilly, D.Tournier and D.Planson. Materials Science Forum, vol. 717-720, p.545, 2012.

[6] "Optical Beam Induced Current measurements based on two-photon absorption process in 4H-SiC bipolar diodes", H. Hamad, C. Raynaud, P. Bevilacqua, D. Tournier, B. Vergne and D. Planson, Applied Physics Letter, vol. 104, p. 082102, 2014.

[7] "Comparison of one- and two-photon optical beam-induced current imaging", C. Xu and W. Denk, Journal of Applied Physics, vol. 86, no. 4, p. 2226, 1999.

[8] "SIMS analyses applied to open an optical window in 4H-SiC devices for electro-optical measurements", M. Lazar, F. Jomard, D.M. Nguyen, C. Raynaud, G. Pâques, S. Scharnholz, D. Tournier and D. Planson. Materials Science Forum, vol. 717-720, p. 885, 2012.

Towards the Use of Functionality-Enhanced Devices : A Transversal Design Approach

Pierre-Emmanuel Gaillardon

EPFL (Swiss Federal Institute of Technology), Lausanne, Switzerland

Exploiting unconventional physical properties, several nanodevices showed an alternative to Moore's Law by the increase of their functionality rather than the pure scaling. Innovative device behaviors transduce to new circuit/architecture opportunities. Here, we focus on a novel class of computation devices that exhibit controllable-polarity property. At advanced technology nodes, Schottky contacts at channel interfaces are becoming challenging to avoid. Hence, devices face an ambipolar behavior, i.e., that the device exhibits n- and p-type characteristics simultaneously. Such a property is desirable for logic computation. Indeed, it has been recently demonstrated by EPFL that by constructing independent double-gate structures on Vertically stacked nanowires FETs (NWFETs), the device polarity can be electrostatically forced to be either n- or p-type. Controllable-polarity devices are logical bi-conditional on both gate values and enable a compact realization of XOR-based logic functions, which are not implementable in CMOS in a compact form. Hyper regular architectures and new EDA tools are then needed to leverage the intrinsic properties of controllable-polarity devices from an application perspective.

In this talk, I will cover the different aspects of the design with controllable-polarity devices ranging from device fabrication to logic synthesis tools, and I will emphasize on the work organization and importance for interdisciplinary teams in the field of emerging technologies.

Safe Operation Region Characterization for Quantifying the Reliability of CMOS Logic Affected by Process Variations

Usman Khalid, Antonio Mastrandrea, Mauro Olivieri
Department of Information, Electronics and Telecommunication Engineering
Sapienza University of Rome, Italy
{khalidu, mastrandrea, olivieri}@diet.uniroma1.it

Abstract— **Technology parameter variations combined with voltage noise can become a major cause of logic errors in digital circuits. This presentation brings in the idea of "safe operation region" to permit a robust analytical Monte Carlo evaluation of the reliability of logic circuits in a given technology, avoiding time-consuming SPICE-level or device-level Monte Carlo simulations. The application of the approach is demonstrated for the case of a 22 nm bulk CMOS process.**

Keywords— *Failure probability, input-signal variation, process variations, , nano-CMOS circuits.*

I. INTRODUCTION

Stochastic deviations in device process parameters have increased the matter of variability in digital circuits' speed and leakage power, on which extensive research have already been conducted in recent years [1-4]. The reliable logic operation issues increases with deep submicron CMOS technologies, this concludes that the random fluctuation of noisy logic errors in the digital circuit's output – generally minimal – can be tremendously raised by process variations [4-8].

The immense numbers of device or circuit level iterative Monte Carlo simulations of the required circuit should be executed, in terms of estimating the output error probability, as an input of standard deviations by using compact models. There is a huge number of iterative simulations are required in order to capture the realistic probability values, therefore such techniques are infeasible in terms of extensive time consumption. This research paper demonstrates an approach for analyzing the failure probability of CMOS logic in a given technology by using safe operation regions' Monte Carlo analysis.

As for other technology-related performance parameters, such as minimum propagation delay and energy-delay product, the reference circuit for analyzing the reliability associated to a given technology is the basic inverter logic gate [9]. The application of the proposed approach is demonstrated for the case of an inverter gate in 22 nm bulk-CMOS technology, subject to variations on effective length *Leff*, effective width *Weff*, oxide thickness *Tox* and doping concentration *Ndep*, for different values of standard deviations of the parameters.

II. BACKGROUND DEFINITIONS

It is generally agreed that a logic circuit operates fallaciously when its output voltage does not equate to the input logic values and by the logic value determined by the function it implements. Let's take the case study of the inverter cells, the conventional definition of valid output voltage is shown in Fig. 1. Generally in all CMOS technologies, the output high logic of inverter A is representing V_{GND} as corresponding nominal voltage *VoutA*, whereas V_{DD} is the case for logic low output. The voltage shifts from its nominal value if malfunctioning occurs which may induce the erroneous logic value to inverter B. The threshold voltages for correct logic value representation *VIH* and *VIL* in inverter B are specified as the points where the gain of voltage slope of the voltage transfer characteristics (VTC) of the inverter cell is equivalent to negative one. The values of *VIH* and *VIL* are widely described in data sheets of standard cells. Thus, the value of logic input of inverter B, i.e. *VoutA*, is not valid when it falls within *VIL* and *VIH* threshold values [9].

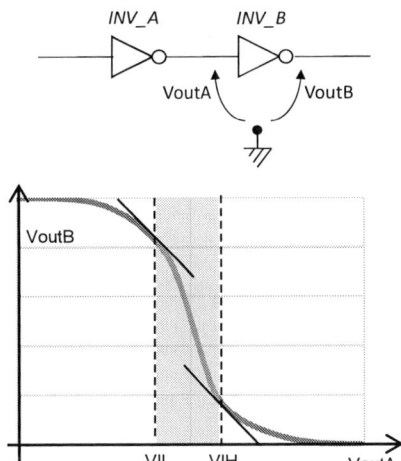

Fig. 1. Top: Two inverter cells. Bottom: Logic thresholds. Grey area denotes invalid *VoutA* logic value.

III. METHODOLOGY IMPLEMENTATION

The logic thresholds in 22 nm metal gate/high-k bulk CMOS technology [9][10] are computed by direct SPICE simulations of the VTC with titular values, getting high noise margin value *VIH* at 0.64 V and low noise margin value *VIL* at 0.33 V with 1 V supply.

The following simulation setup for right operation analysis is represented in Fig. 2. The ideal values of V_{GND} or V_{DD} could forcefully turns-on the pull-down or pull-up device by the varying input voltage when it is supposed to be turned-off. This phenomenon finally deteriorating the literal output voltage provided the large enough variation on input signal. Such faulty output is also deteriorated by process-induced fluctuations of the P-type and N-type device threshold voltages from their titular values.

The joint effect of noise-induced input voltage variation and process-induced threshold voltage fluctuations can be thoroughly characterized by SPICE simulation by driving predetermined deviations, and guiding to a plot of the output voltage as a function of the literal fluctuations. Fig. 3 represents the case of threshold voltage V_{TP} only variations, taking V_{TN} fixed at the titular value, and input voltage V_{IN} variation, in the target technology, for high and low anticipated output logic value. One can form a similar diagram between V_{TP} and V_{IN} by fixing V_{TP} value as well.

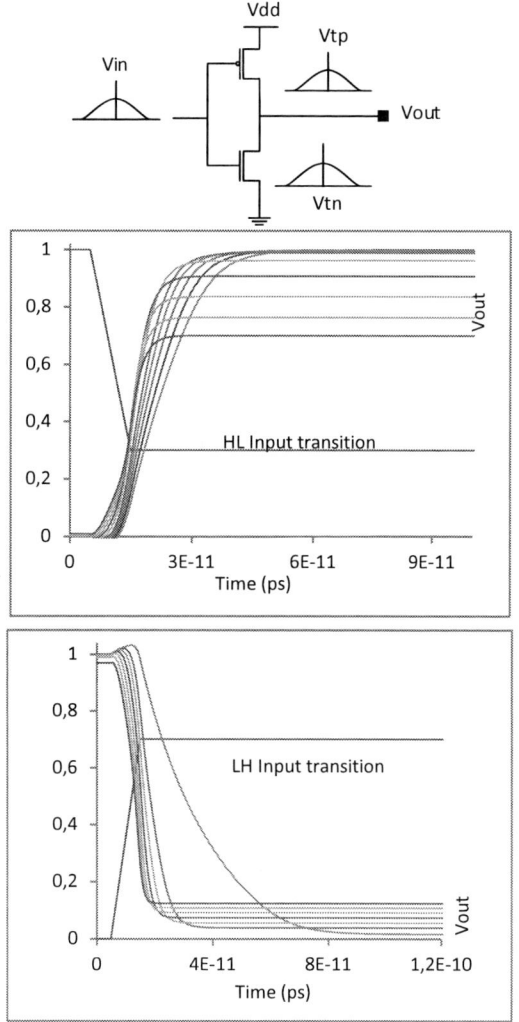

Fig. 2. Top: Simulated circuit; Bottom: Output voltages for selected V_{THP} variations for HL and LH inputs respectively.

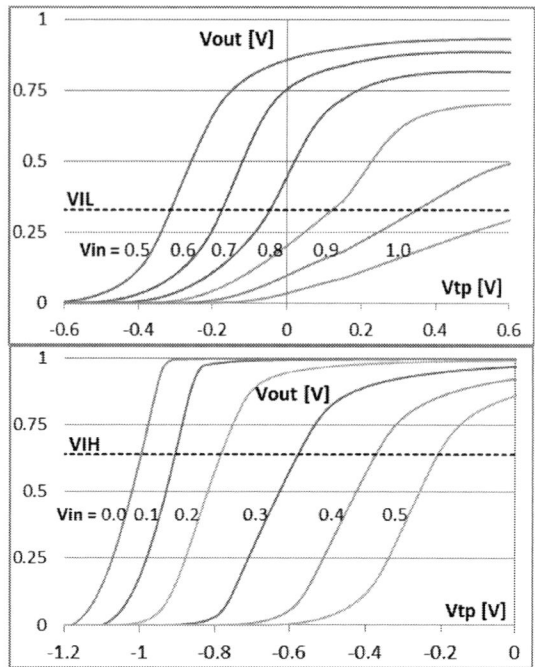

Fig. 3. Fluctuations between V_{THP} versus V_{IN}. Top: V_{IN} High; Bottom: V_{IN} Low.

We can set up a diagram by finding the intersections of the V_{OUT} versus V_{TP} and V_{IN} with boundaries *VIL* and *VIH* voltages, where the curves $V_{TP} = G(V_{IN})$ specifies the boundary for safe operation region when V_{TN} is nominal. To take into account also V_{TN} variations, various $G(V_{IN})$ curves can be produced for low and high V_{IN} according to the matching V_{TN} values, and correspondingly we can characterize the boundary for safe operation region as a function $V_{TP} = G(V_{IN}, V_{TN})$. Hence, Fig. 4 illustrates the combined variation of V_{TN}, V_{TP} and V_{IN} which leads to a 3-D safe operation region.

Now, having characterized the values of $G(V_{IN}, V_{TN})$ for the range of interest of V_{IN}, V_{TN} (we targeted a 6-sigma range of variation around the nominal values), we can generate a sufficiently large set of pseudo-random numbers of V_{TP}, V_{TN} and V_{IN} and execute a Monte Carlo computation of the circuit's probability operation in order to fall the samples up or down of the safe operation region.

The threshold voltage values in P-type and N-type MOSFET devices are evaluated with extensive detail in the BSIM4 compact model as a heuristic parameters and mathematical function of technology by using V_{TH} equation declared in [11] and also has been used in [12] for BSIM and PD-SOI models' threshold voltage comparison analysis.

Ultimately, it is likely to produce a haphazardly big vector of V_{TP} and V_{TN} numbers as an input of process parameters variations. The selected process parameters and their corresponding nominal values are briefly described in the result section.

978-1-4799-4993-9/14 $31.00 © 2014 IEEE

IV. RESULTS

Without loss of generality we limited the analysis on the variations of oxide thickness *Tox*, channel doping *Ndep*, effective length *Leff* and effective width *Weff*. In the target 22 nm technology model, the nominal (i.e. mean) values are shown in Table I. We considered a Normal distribution of the selected process parameters.

TABLE I. NOMINAL VALUE OF ANALYZED PROCESS PARAMETERS IN THE TARGET 22 NM TECHNOLOGY

	Tox	Ndep	L_{eff}	W_{eff}
PMOS	$6.7 \cdot 10^{-10}$	$4.4 \cdot 10^{18}$	$1.93 \cdot 10^{-8}$	$1.20 \cdot 10^{-8}$
NMOS	$6.5 \cdot 10^{-10}$	$1.2 \cdot 10^{19}$	$1.93 \cdot 10^{-8}$	$1.20 \cdot 10^{-8}$

A subset of the configurations was used as a verification test set, for which the proposed calculation approach was verified versus SPICE level Monte Carlo simulations assuming the same technology parameter variations. For the verification purposes, we fixed input voltage standard deviation at $\sigma V_{IN} = 0.25$ V, a rather large value, in order to have SPICE Monte Carlo simulation results significant in a relatively low number of simulation iterations. We performed 10^6 Monte Carlo iterations, enough to obtain convergence of the resulting probability of failure to a stable value.

The resulting speedup of the proposed approach over SPICE in computation run time was always above 10^4. Accuracy results are shown in Table II, along with diverse standard deviation values assumed for process parameter (*L*, *W*, *Tox*, and *Ndep*) referring to both high and low logic level for V_{IN}. The relative error between the proposed approach and SPICE Monte Carlo simulation is below 17% relative error for high V_{IN} and below 5% relative error for low V_{IN} respectively, on the calculated failure probabilities. Note that for the purpose of evaluating the general reliability of a given technology in certain noise condition, even predicting the correct order of magnitude of the failure probability would be significant.

Table III shows another set of results for the same standard deviations of process parameters, but with four different standard deviation values for voltage input voltage noise, namely $\sigma V_{IN} = 0.15, 0.20, 0.25$ and 0.30. The results show the increased probability of failure according to both process parameters and input higher standard deviation values.

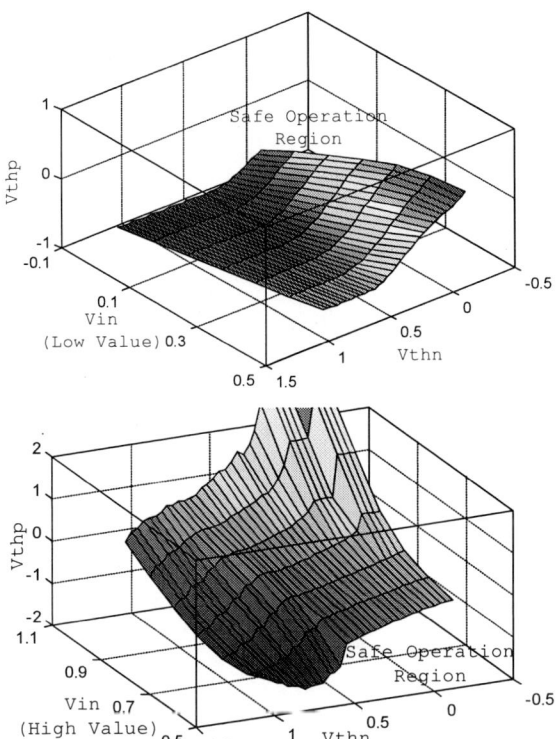

Fig. 4. 3-D Safe operation regions in 22 nm CMOS inverter cell, Top: Low V_{IN}; Bottom: High V_{IN}.

We furthermore explored a total of 256 experiments resulting from 4 process parameters, 4 standard deviation values and 4 input voltage standard deviation values. The experiments addressed the analysis of the dependency of failure probability on the different process parameters. The results identified oxide thickness *Tox* as the most critical parameter for variations, having the highest relative effect on the failure probability for lowest values of input noise standard deviation. The detail of the impact of *Tox* standard deviation on the probability of failure is reported in Fig. 5, for 0.15 V input noise standard deviation.

TABLE II. FAILURE PROBABILITY COMPARISON BETWEEN PROPOSED APPROACH AND SPICE MONTE CARLO SIMULATIONS

	σ Tox	σ Ndep	σ L_{eff}	σ W_{eff}	Safe Operation Region Monte Carlo	SPICE level Monte Carlo	Relative Error
High Vin	$15 \cdot 10^{-12}$	$20 \cdot 10^{16}$	$25 \cdot 10^{-10}$	$30 \cdot 10^{-10}$	$3.10 \cdot 10^{-2}$	$3.41 \cdot 10^{-2}$	- 09.09 %
	$20 \cdot 10^{-12}$	$25 \cdot 10^{16}$	$30 \cdot 10^{-10}$	$15 \cdot 10^{-10}$	$3.14 \cdot 10^{-2}$	$3.49 \cdot 10^{-2}$	- 10.02 %
	$25 \cdot 10^{-12}$	$30 \cdot 10^{16}$	$15 \cdot 10^{-10}$	$20 \cdot 10^{-10}$	$3.30 \cdot 10^{-2}$	$3.85 \cdot 10^{-2}$	- 14.28 %
	$30 \cdot 10^{-12}$	$15 \cdot 10^{16}$	$20 \cdot 10^{-10}$	$25 \cdot 10^{-10}$	$3.37 \cdot 10^{-2}$	$3.99 \cdot 10^{-2}$	- 15.53 %
Low Vin	$15 \cdot 10^{-12}$	$20 \cdot 10^{16}$	$25 \cdot 10^{-10}$	$30 \cdot 10^{-10}$	$3.18 \cdot 10^{-2}$	$3.24 \cdot 10^{-2}$	- 01.85 %
	$20 \cdot 10^{-12}$	$25 \cdot 10^{16}$	$30 \cdot 10^{-10}$	$15 \cdot 10^{-10}$	$3.20 \cdot 10^{-2}$	$3.31 \cdot 10^{-2}$	- 03.32 %
	$25 \cdot 10^{-12}$	$30 \cdot 10^{16}$	$15 \cdot 10^{-10}$	$20 \cdot 10^{-10}$	$3.34 \cdot 10^{-2}$	$3.37 \cdot 10^{-2}$	- 00.89 %
	$30 \cdot 10^{-12}$	$15 \cdot 10^{16}$	$20 \cdot 10^{-10}$	$25 \cdot 10^{-10}$	$3.39 \cdot 10^{-2}$	$3.54 \cdot 10^{-2}$	- 04.23 %

TABLE III. FAILURE PROBABILITY RESULTS FOR SELECTED CASES OF PROCESS PARAMETERS' STANDARD DEVIATION VALUES WITH INPUT VOLTAGE NOISE HIGH V_{IN}

σ Tox	σ Ndep	σ Leff	σ Weff	σ Vin=0.15	σ Vin=0.20	σ Vin=0.25	σ Vin=0.30
$15 \cdot 10^{-12}$	$20 \cdot 10^{16}$	$15 \cdot 10^{-10}$	$25 \cdot 10^{-10}$	$1.170 \cdot 10^{-2}$	$9.950 \cdot 10^{-2}$	$3.107 \cdot 10^{-2}$	$6.032 \cdot 10^{-2}$
$15 \cdot 10^{-12}$	$20 \cdot 10^{16}$	$25 \cdot 10^{-10}$	$30 \cdot 10^{-10}$	$1.170 \cdot 10^{-2}$	$9.960 \cdot 10^{-2}$	$3.101 \cdot 10^{-2}$	$6.033 \cdot 10^{-2}$
$15 \cdot 10^{-12}$	$25 \cdot 10^{16}$	$20 \cdot 10^{-10}$	$15 \cdot 10^{-10}$	$1.170 \cdot 10^{-2}$	$9.940 \cdot 10^{-2}$	$3.106 \cdot 10^{-2}$	$6.030 \cdot 10^{-2}$
$15 \cdot 10^{-12}$	$30 \cdot 10^{16}$	$20 \cdot 10^{-10}$	$20 \cdot 10^{-10}$	$1.170 \cdot 10^{-2}$	$9.960 \cdot 10^{-2}$	$3.107 \cdot 10^{-2}$	$6.024 \cdot 10^{-2}$
$20 \cdot 10^{-12}$	$15 \cdot 10^{16}$	$15 \cdot 10^{-10}$	$25 \cdot 10^{-10}$	$1.470 \cdot 10^{-2}$	$1.103 \cdot 10^{-2}$	$3.146 \cdot 10^{-2}$	$6.125 \cdot 10^{-2}$
$20 \cdot 10^{-12}$	$20 \cdot 10^{16}$	$25 \cdot 10^{-10}$	$30 \cdot 10^{-10}$	$1.470 \cdot 10^{-2}$	$1.101 \cdot 10^{-2}$	$3.146 \cdot 10^{-2}$	$6.126 \cdot 10^{-2}$
$20 \cdot 10^{-12}$	$25 \cdot 10^{16}$	$20 \cdot 10^{-10}$	$15 \cdot 10^{-10}$	$1.470 \cdot 10^{-2}$	$1.102 \cdot 10^{-2}$	$3.148 \cdot 10^{-2}$	$6.125 \cdot 10^{-2}$
$20 \cdot 10^{-12}$	$30 \cdot 10^{16}$	$30 \cdot 10^{-10}$	$25 \cdot 10^{-10}$	$1.460 \cdot 10^{-2}$	$1.100 \cdot 10^{-2}$	$3.146 \cdot 10^{-2}$	$6.121 \cdot 10^{-2}$
$25 \cdot 10^{-12}$	$15 \cdot 10^{16}$	$15 \cdot 10^{-10}$	$25 \cdot 10^{-10}$	$1.630 \cdot 10^{-2}$	$1.166 \cdot 10^{-2}$	$3.307 \cdot 10^{-2}$	$6.225 \cdot 10^{-2}$
$25 \cdot 10^{-12}$	$15 \cdot 10^{16}$	$25 \cdot 10^{-10}$	$30 \cdot 10^{-10}$	$1.630 \cdot 10^{-2}$	$1.165 \cdot 10^{-2}$	$3.307 \cdot 10^{-2}$	$6.224 \cdot 10^{-2}$
$25 \cdot 10^{-12}$	$25 \cdot 10^{16}$	$25 \cdot 10^{-10}$	$25 \cdot 10^{-10}$	$1.640 \cdot 10^{-2}$	$1.165 \cdot 10^{-2}$	$3.308 \cdot 10^{-2}$	$6.221 \cdot 10^{-2}$
$25 \cdot 10^{-12}$	$30 \cdot 10^{16}$	$30 \cdot 10^{-10}$	$25 \cdot 10^{-10}$	$1.640 \cdot 10^{-2}$	$1.165 \cdot 10^{-2}$	$3.306 \cdot 10^{-2}$	$6.221 \cdot 10^{-2}$
$30 \cdot 10^{-12}$	$15 \cdot 10^{16}$	$20 \cdot 10^{-10}$	$25 \cdot 10^{-10}$	$2.020 \cdot 10^{-2}$	$1.267 \cdot 10^{-2}$	$3.377 \cdot 10^{-2}$	$6.249 \cdot 10^{-2}$
$30 \cdot 10^{-12}$	$20 \cdot 10^{16}$	$25 \cdot 10^{-10}$	$30 \cdot 10^{-10}$	$2.010 \cdot 10^{-2}$	$1.268 \cdot 10^{-2}$	$3.376 \cdot 10^{-2}$	$6.244 \cdot 10^{-2}$
$30 \cdot 10^{-12}$	$25 \cdot 10^{16}$	$20 \cdot 10^{-10}$	$15 \cdot 10^{-10}$	$2.010 \cdot 10^{-2}$	$1.269 \cdot 10^{-2}$	$3.378 \cdot 10^{-2}$	$6.246 \cdot 10^{-2}$
$30 \cdot 10^{-12}$	$30 \cdot 10^{16}$	$30 \cdot 10^{-10}$	$30 \cdot 10^{-10}$	$2.000 \cdot 10^{-2}$	$1.267 \cdot 10^{-2}$	$3.375 \cdot 10^{-2}$	$6.241 \cdot 10^{-2}$

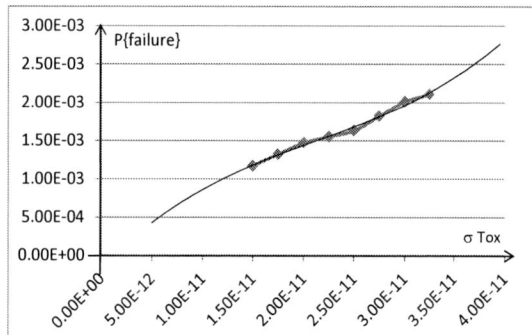

Fig. 5. Detail of the dependency of the probability of failure from *Tox* standard deviation, for noise standard deviation σ*Vin* = 0.15 V.

V. CONCLUSIONS

The characterization of safe operation regions has been introduced, to analyze the behavior of the probability of failure in logic circuits fabricated in nano-CMOS technologies, resulting from noise-induced voltage variations and process-induced device parameter fluctuations. A wide range of verification of the safe operation region approach versus SPICE Monte Carlo simulations has been conducted. An exploration of the impact of different parameter variations on logic level reliability has identified *Tox* as the most critical device parameter for variations. The approach shows $>10^4$ speedup with respect to SPICE Monte Carlo analysis and therefore can be the baseline for efficiently characterizing the expected reliability of digital nano-CMOS processes, as functions of process parameter variability and voltage noise. Further research is addressing the extension of the approach to the evaluation of failure probabilities in specific circuits composed of multiple standard cells including memory elements.

REFERENCES

[1] Abbas Z., Genua V., Olivieri M., "A novel logic level calculation model for leakage currents in digital nano-CMOS circuits," 7th IEEE Conference on Ph.D. Research in Microelectronics and Electronics (PRIME), Trento, Italy, pp. 221-224, 3-7 July 2011.

[2] Drego N., Chandrakasan A., Boning D., "Lack of Spatial Correlation in MOSFET Threshold Voltage Variation and Implications for Voltage Scaling," IEEE Trans on Semiconductor Manufacturing, vol.22, no.2, pp. 245-255, May 2009.

[3] Abbas Z., Khalid U., Olivieri M., "Sizing and optimization of low power process variation aware standard cells," IEEE International Integrated Reliability Workshop Final Report (IRW), pp.181-184, 13-17 Oct. 2013.

[4] Intl. Technology Roadmap for Semiconductors 2011 Ed. Executive Summary, at http://www.itrs.net/Links/2011ITRS/Home2011.htm

[5] Olivieri M., Mastrandrea A., "Logic Drivers: A Propagation Delay Modeling Paradigm for Statistical Simulation of Standard Cell Designs," IEEE Transactions on Very Large Scale Integration (VLSI) Systems, vol.22, no.6, pp. 1429-1440, June 2014.

[6] A. Asenov, "Statistical device variability and its impact on design" (keynote), Intl Symp. Asynch. Circ. & Syst. ASYNCH'08, Newcastle upon Tyne, UK, pp. xv–xvi, Apr. 2008.

[7] Jun Yin, Xiaokang Shi, Ru Huang, "A new method to simulate random dopant induced threshold voltage fluctuations in sub-50 nm MOSFET's with non-uniform channel doping," Solid-State Electronics, Vol. 50, Issues 9–10, pp. 1551-1556, Sep-Oct. 2006.

[8] U. Khalid, A. Mastrandrea, M. Olivieri, "Novel Approaches to Quantify Failure Probability due to Process Variations in Nano-scale CMOS Logic", 29th IEEE International Conference on Microelectronics, Belgrade, Serbia, pp. 371-374, 12-14 May 2014.

[9] S.M. Kang, Y. Leblebici, CMOS Digital Integrated Circuits: Analysis and Design, McGraw Hill, New York, USA, 2003.

[10] Predictive Technology Models, Latest Models [Online]. Available: http://ptm.asu.edu/latest.html

[11] BSIM4 MOSFET Model User's Manual. [Online]. Available: http://www-device.eecs.berkeley.edu/bsim/?page=BSIM4_LR

[12] Jimenez-P A., De La Hidalga-W F. J., "Analysis of the Threshold Voltage BSIM-Model for a Short Channel PD-SOI DTMOS," 4th International Conference on Electrical and Electronics Engineering (ICEEE), pp. 381-384, 5-7 Sept. 2007.

978-1-4799-4993-9/14 $31.00 © 2014 IEEE

Brain Inspired High Performance Electronics on Flexible Silicon

Galo Andres Torres Sevilla, Jhonathan Prieto Rojas and Muhammad Mustafa Hussain
Computer, Electrical and Mathematical Engineering Division. King Abdullah University of Science and Technology.
Thuwal, Mekkah, Saudi Arabia
Email: MuhammadMustafa.Hussain@kaust.edu.sa

Abstract— **Brain's stunning speed, energy efficiency and massive parallelism makes it the role model for upcoming high performance computation systems. Although human brain components are a million times slower than state of the art silicon industry components [1], they can perform 1016 operations per second while consuming less power than an electrical light bulb. In order to perform the same amount of computation with today's most advanced computers, the output of an entire power station would be needed. In that sense, to obtain brain like computation, ultra-fast devices with ultra-low power consumption will have to be integrated in extremely reduced areas, achievable only if brain folded structure is mimicked. Therefore, to allow brain-inspired computation, flexible and transparent platform will be needed to achieve foldable structures and their integration on asymmetric surfaces. In this work, we show a new method to fabricate 3D and planar FET architectures in flexible and semitransparent silicon fabric without comprising performance and maintaining cost/yield advantage offered by silicon-based electronics.**

Keywords—flexible electronics; mono-crystallline silicon; 3D flexible electronics; field effect tranisistor.

I. INTRODUCTION

Today's flexible electronics research is mainly based on the fabrication of solid-state devices on organic substrates, which show extreme flexibility and transparency [2-5]. However, the low temperature requirements to processes organic substrates and their inherited low electron mobility hinder their potential for truly high performance flexible electronics. Some other approaches have been investigated recently involving transfer of monocrystalline silicon on polymers [6-11]. Although these techniques exhibit high yield and feature definition, they are incompatible with very critical state of the art processes such as mask alignment. Also thinning techniques have been used to produce mechanically flexible silicon chips [12,13]. However, their major drawback consists in the use of extremely abrasive processes that may damage the on-hip devices. To mitigate all these weaknesses, in the recent past we have demonstrated transformational electronics where we transform the traditional silicon electronics into flexible and transparent one with already fabricated devices retaining their status-quo performance, efficiency, ultra-large-scale-integration density, thermal budget and finally cost.

Using this low-cost generic batch process, we make trenches in the unused areas of a silicon circuitry, followed by vertical sidewall protection layer (spacer) formation and finally carrying out an isotropic etching at the bottom of the deep trenches to form 20 mm caves in the silicon substrate to release an ultra-thin (>5 mm), fully flexible (5 mm bending radius) and semi-transparent (12% transmittance) silicon fabric with pre-fabricated devices. After the removal of the top layer of whole wafers in this way, we perform chemical mechanical polishing to planarize the remaining bottom substrates to recycle it and then fabricate next set of devices on it followed by the repetition of the same process to form up to 6 layers of silicon fabric with devices from a standard 0.5 mm thick silicon wafer. Following a similar process, here we demonstrate, semiconductor industry's most advanced architecture FinFET and planar MOSFETs with the most advanced material set: high-k/metal gate stacks.

FinFET is a new generation device architecture which has been adopted by semiconductor giant Intel Corporation from 2011 in their microprocessors. FinFET a member of multi-gate FET family offers non-planar 3D topology where the channels are vertically aligned in arrays of ultra-thin silicon fins bordered by multiple gates (in our case it is two gates) to ensure higher electrostatic control so the short channel effects can be mitigated as well as performance can be enhanced. In addition to the advanced topology, we have also integrated semiconductor industries' most advanced high-k/metal gate stacks to make our device fully state-of-the-art.

II. FABRICATION PROCESS

A. Flexible FinFETs [14]

Figure 1 shows the basic steps to release FinFETs from the carrier substrate and transfer them to a polymer substrate. Each of the fabricated wafers was diced into 2.5 cm x 3 cm pieces in order to process each die separately. It is to be noted that although complete wafer release can be easily performed, die processing was chosen in order to perform different etching conditions on each die. At this point, dies were processed with research level lithography to create the etch hole patterns in the inactive areas of the fabricated devices (Fig. 1b). The distance between the holes depends on the selectivity of the isotropic etchant between silicon and buried oxide (BOX) (>1000:1) and the thickness of the bottom oxide layer. The next step consists in removing the interlayer dielectric (ILD) from the holes to allow access for xenon di-fluoride (XeF_2) isotropic etchant. Next, while keeping the photoresist, the dies were placed in XeF_2 chamber to isotropically remove the silicon from the bottom of the BOX and create caves (Fig. 1c), once these caves meet with each other, the SOI and the BOX, is completely released from the bulk substrate allowing us to peel-off the devices (Fig. 1d) and transfer them to a flexible polyimide carrier substrate (Kapton) with a thin uncured PDMS spin coated film to enhance adhesion between the peeled devices and the carrier substrate (Fig. 1e). Finally the PDMS layer is cured. It is to be noted that the holes did not represent any constraint in the design of the fabricated devices due to high selectivity between thermal oxide and silicon under XeF_2 etching conditions.

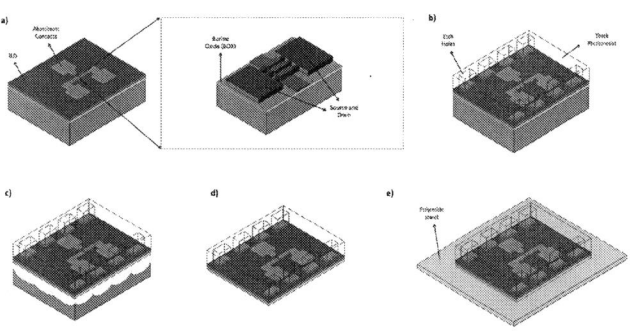

Fig. 1.   Flexible FinFET fabrication flow.

B. Planar MOSFETs [15]

Figure 2 shows the fabrication process for planar MOSFETs on flexible silicon. The fabrication begins with standard 4" silicon wafers (100). All the devices are built using research grade lithography and industry compatible etching and deposition processes. First the active definition is done using one lithographic step and one etch step to remove the silicon dioxide (SiO_2) from the area where the transistor is built. Next, the gate is deposited (TaN + Al_2O_3) and patterned using regular lithography and reactive ion etching. Finally the devices are implanted in the source and drain regions and contacted using aluminum contacts. At this point, the fabrication of the transistors is completed and we can proceed to the release of each individual die. The dies are taken for a final lithography and patterned with the etch holes which are done with deep RIE. Finally, spacers are deposited in the walls of the deep trenches and the dies are taken into XeF_2 chamber in order to isotropically etch the bottom silicon and release the devices from the bulk substrate.

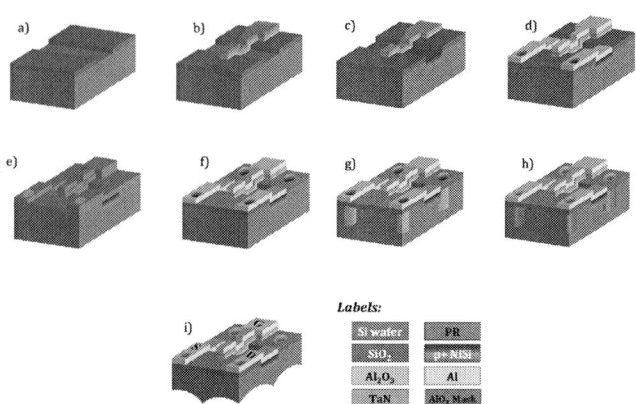

Fig. 2.   Flexible planar MOSFETs fabrication flow.

III. RESULTS

In order to have a complete understanding of the electrical behavior of released and unreleased devices, the same transistor were characterized before and after release process. The obtained results are described below and show very low changes in the main parameters of the devices.

A. Flexible FinFETs [14]

Electrical results are summarized in figure 3 while the fabrication results can be found in figure 3. To study the behavior of the FinFETs, we first started measuring the I-V characteristics. Figure 3 (a – f) shows a comparison for the same PMOS and NMOS devices before and after peel-off processing. The measured NMOS and PMOS devices have gate length (L) 250 nm and channel width (W) 3.6 mm. The current at saturation with VGS = 1.5V for NMOS and VGS = -1.5V for PMOS at VDS = 1V and -1V respectively are 549 mA/mm for NMOS and 110 mA/mm for PMOS, and 58.48 mA/mm and 7.73 mA/mm at VDS = 50mV and -50mV for NMOS and PMOS respectively. In the case of NMOS, Ion/Ioff ratio is 4.6 decades and for PMOS is 4.78 decades. Ion/Ioff ratio was calculated using Intel Corporation's method, which states that Ioff should be calculated at V_{th} minus one third of VDS for non-optimized gate stacks. Gate leakage was also found to be 3.6 A/cm^2 for NMOS and 1.18 A/cm^2 for PMOS at VGS = 1.5V and VGS = -1.5V respectively. It is to be noted, there is a small current reduction of 7% for NMOS and 12% for PMOS in the saturation region of the transistors.

V_{th}^{sat} for released and unreleased NMOS was 0.325V and 0.345V respectively and V_{th}^{sat} for released and unreleased PMOS was 0.66V and 0.713V respectively. The difference in V_{th}^{sat} does not represent a significant change (only 6% for NMOS and 8% for PMOS), therefore confirming the consistency of the process. The comparison between sub-threshold swing for released and unreleased samples also indicates an insignificant change of only 3.4% for NMOS and 2.8% for PMOS. In summary, apart from the 7% for NMOS and 12% for PMOS reduction in the on state current, the devices behave very similarly when comparing released NMOS and PMOS with unreleased NMOS and PMOS.

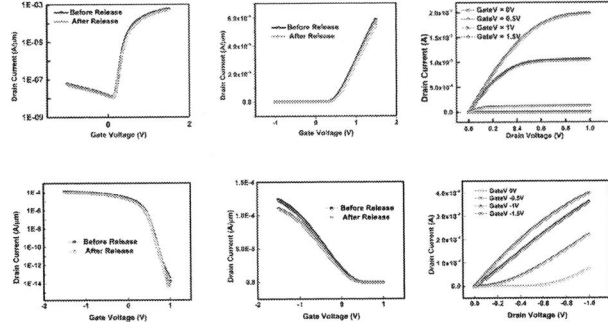

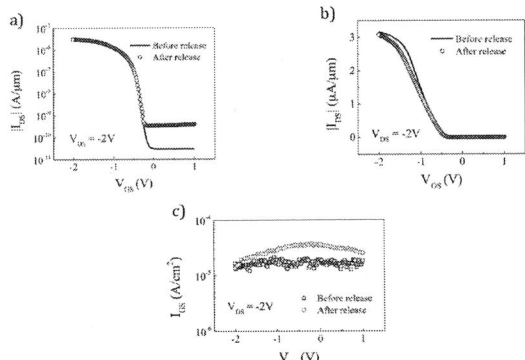

Fig. 3. Flexible FinFET electrical characteristics.

It is to be noted that PMOS devices were not optimized for V_{th} during the fabrication process. This causes a large shift of V_{th} towards the positive side of the gate voltage; however, the difference between released and unreleased PMOS devices differ in only 8% of the original threshold voltage value, showing the consistency of the release process.

B. Planar MOSFETs [15]

In order to study the main characteristics of our MOSFETs we measured the I-V characteristics in both the sub-threshold and linear regions. The results are shown in Figure 6 for a representative device. The current at saturation with V_{GS} = -2 V is -4.4 μA/μm for V_{DS} = -2 V and -0.3 μA/μm for V_{DS} = -100 mV. I_{on}/I_{off} ratio is 3.7 decades and the gate leakage is as small as 70 μA/cm2 at V_{GS} = -1 V. The measured devices have a gate length (L) of 8 μm and a channel width (W) of 5 μm.

In order to have a better understanding of the implications associated with the release method and the related sub-processes, we have compared the electrical performance of a sample before and after release processing. Figure 4 compares the I_D-V_G transfer characteristics in the saturation and sub-threshold regime of the same transistor before and after release. As can be observed there is a small current reduction, around 2.6%, in the saturation region of the transistor (Figure 4a). In fact, due to the presence of strained oxide on top the silicon for insulation of devices, there is a small residual strain which can contribute to the reduction in current and other parameters, as will be discussed in the upcoming section. The increase in I_{off} current, up to one decade, from ~30 pA/mm to ~300 pA/mm, can be related to the reduced substrate thickness, which makes the sample more susceptible to noise or thermal excitation allowing more off-state leakage. Gate leakage current for unreleased and released samples is shown in Figure 4c. We observed a small increase in leakage although it does not represent a significant difference. Additionally we have extracted the threshold voltage (V_{th}) and the sub-threshold swing (S) from both released and unreleased samples. The threshold voltage in the released sample showed a reduction around 7 %, most likely related to the current reduction, from -0.48 V in the unreleased sample to -0.45 V in the released one. The sub-threshold swing, on the other hand, showed an insignificant change, increasing from 75.24 mV dec^{-1} in the unreleased sample to 80.3 mV dec^{-1} in the released one, representing a 6.7 % increase.

Fig. 4. Planar MOSFETs elctrical characteristics.

In summary, besides the off-current, less than 10 % variation in main parameters results from the additional processing related to the release method. This can be attributed to the residual strain as a result of the oxide and other deposited layers on the fabric. On the other hand, even though the I_{off} current, and therefore I_{on}/ I_{off} ratio, are significantly affected, the values remain at an acceptable and competitive level.

IV. CONCLUSION

State of the art FinFET and planar MOSFET devices have been demonstrated on a flexible monocrystalline silicon platform starting from standard 8" silicon-on-insulator wafer and 4" bulk silicon wafer respectively. Using only industry standard processes, the flexibility shown is extremely high achieving a minimum-bending radius of 5mm. The results set a major step toward the integration of high performance devices on flexible platforms while keeping the performance and more important the cost yield advantage obtained with silicon based fabrication.

REFERENCES

[1] Brain Computation: http://www.cs.utexas.edu/~dana/Book1.pdf (last accessed 24th February 2014)

[2] Rogers, J. A., Bao, Z., Baldwin, K., Dodabalapur, A., Crone, B., Raju, V. R., Kuck, V., Katz, H., Amundson, K., Ewing, J. & Drzaic, P. Paper-like electronic displays: Large-area rubber- stamped plastic sheets of electronics and microencapsulated electrophoretic inks. Proc. Natl. Acad. Sci. USA. 98, 4835-4840 (2001).

[3] Gelinck, G. H., Huitema, H. E. A., Van Veenendaal, E., Cantatore, E., Schrijnemakers, L., Van Der Putten, J., Geuns, T. C. T., Beenhakkers, M., Giesbers, J. B., Huisman, B. H., Meijer, E. J. Benito, E. M., Touwslager, F. J., Marsman, A. W., Van Rens, B. J. E. & De Leeuw, D. M. Flexible active-matrix displays and shift registers based on solution-processed organic transistors. Nat. Mater. 3, 106-110 (2004).

[4] Lin, P. & Yan, F. Organic Thin-Film Transistors for Chemical and Biological Sensing. Adv. Mater. 24, 34–51 (2012).

[5] Baca, A., Ahn, J. H., Sun, Y., Meitl, M., Menard, E., Kim, H. S., Choi, W., Kim, D. H., Huang, Y. & Rogers, J. Semiconductor Wires and Ribbons for High- Performance Flexible Electronics. Angew. Chem. Int. Edit. 47, 5524-5542 (2008).

[6] Ahn, J. –H. et al. High-Speed Mechanically Flexible Single-Crystal Silicon Thin-Film Transistors on Plastic Substrates. IEEE Elect. Dev. Lett. 27 (6), 460 (2006).

[7] Sun, L. et al. 12-GHz Thin-Film Transistors on Transferrable Silicon Nanomembranes for High-Performance Flexible Electronics. Small 6 (22), 2553 (2010).

[8] Ahn, J. –H. et al. Bendable integrated circuits on plastic substrates by use of printed ribbons of single-crystalline silicon. Appl. Phys. Lett. 90, 213501 (2007).

[9] Kim, H. –S. et al. Self-assembled nanodielectrics and silicon nanomembranes for low voltage, flexible transistors, and logic gates on plastic substrates. Appl. Phys. Lett. 95, 183504 (2009).

[10] Tae-il Kim. et al. Deterministic assembly of releasable single crystal silicon-metal oxide field-effect devices formed from bulk wafers. App. Phys. Lett. 102, 182104 (2013).

[11] Lee, K. J. et al. Fabrication of microstructured silicon (μs-Si) from a bulk Si wafer and its use in the printing of high-performance thin-film transistors on plastic substrates. J. Micromech. Microeng. 20, 075018 (2010).

[12] Loher, T., Seckel, M., Pahl, B., Bottcher, L., Ostmann, A., Reichl, H. Highly integrated flexible electronic circuits and modules, 3rd Intl. Microsy. Packag., Assembly. Circ. Technol. Conf., 86 (2008).

[13] Zhai, Y., Mathew, L., Rao, R., Xu, D., Banerjee. S. K. High-Performance Flexible Thin-Film Transistors Exfoliated from Bulk Wafer. Nano Lett. 12(11), 5609 (2012).

[14] Torres Sevilla, G., Rojas, J. P., Hossain, M. F., Hussain, A. M., Ghanem, R., Smith, C. E. and Hussain, M. M. Flexible and Transparent Silicon Based Sub-100 nm Non-planar 3D FinFET CMOS for Brain-inspired Computation. Adv. Mat. (2014)

[15] Rojas, J. P. , Torres Sevilla, G. A. and Hussain, M. M. Can We Build a Truly High Performance Computer Which is Flexible and Transparent?. Nature Sci. Rep. (3), 2609

Towards Formal Verification of Reset Sequence in Fully Asynchronous Digital Circuits

Oleksandr Melnychenko
Institute of Computer Engineering
Vienna University of Technology
Vienna, Austria
Email: oleksandr.melnychenko@tuwien.ac.at

Hans-Peter Kreuter
Infineon Technologies Austria AG
Villach, Austria
Email: hanspeter.kreuter@infineon.com

Abstract—**We propose a method for the formal reset sequence verification for digital asynchronous circuits. First the traditional approach for the reset verification is discussed and the need for a novel solution is shown. The proposed method is based on the extension of the standard logic types with a multi-value logic type and a source code instrumentation method. The method is finally applied to an exemplary circuit fragment showing promising results.**

I. INTRODUCTION

The advent of recent power semiconductor technologies are leading to increase the complexity of analogue mixed-signal circuits [1], [2]. A classical design approach is to apply discretization techniques at the border of digital and analogue domains. Consequently, the digital subsystem becomes a fully synchronous finite state machine and can be implemented using the state-of-the-art synchronous circuits design methodology [3]. The advantage of this approach is obvious—the synchronous circuits design methodology is well-established, there are many efficient synthesis tools available, and, finally, the verification of synchronous circuits is straightforward [4].

Unfortunately, standard synchronous circuits design methods are not well-suited for all kinds of mixed-signal circuits due to several reasons. Registers may be shared within the analogue and digital domains. More complex power modes for both the digital and the analogue domains are required combined with low power consumption [1].

As synchronous circuits do not meet all the requirements, an asynchronous design method can be applied [5]. More precisely, these techniques assume that asynchronous inputs of flip-flops and latches, namely set and reset signals, can be driven not only by primary inputs of the digital subsystem, but also by some internal signals. The same is true for clock pins of the flip-flops.

Removal of synchronism restrictions allows more freedom. However, an asynchronous design requires much more verification and testing efforts compared to a synchronous one [4]. A lot of questions arise in asynchronous circuits that would be implicitly solved by design in synchronous circuits.

One of such questions is soundness of the reset procedure. When a circuit starts its operation, certain signals should be applied to certain pins in a particular order. We call this routine a reset sequence. The purpose of the reset sequence is to ensure that all internal memory elements (flip-flops and latches) in the

circuit get particular well-defined values (0 or 1) before the operations starts. Having at least one memory element with metastable or even well-defined random value might cause unexpected results during further operation [5].

This paper addresses the problem of formal reset sequence verification for digital asynchronous circuits. More precisely, a formal reset verification at register transfer level (RTL) is considered. Authors assume that VHDL description of the circuit is available [6]. With the proposed method it is possible to prove the reset sequence correctness for a particular VHDL design and to guarantee the reset sequence to function correctly. The method does not consider delays existing in the final implementation, i.e. timing is out of scope in this research.

II. RESET PROBLEM IN DETAIL

There is no need to discuss the importance of a reset sequence verification—reset pins are used in vast majority of digital circuits [3]. Moreover, the task of reset sequence verification is not a trivial one. In synchronous circuits the functional properties of a reset sequence are defined by design. The remaining verification task is to meet the timing requirements, i.e. we have to ensure that synthesis and place-and-route phase do not violate the desired timing constraints [4].

However, if we consider asynchronous circuits as discussed above, even functional behaviour regardless timing is not that simple. Consider a fragment of a digital circuit in the Fig. 1. It contains three combinatorial blocks: and-gate, or-gate and 2-to-1 multiplexer, and two memory blocks: RS-latch and D flip-flop. They are labeled correspondingly. The output of each block is on the right, the rest of the pins are inputs. The reset pins of RS-latch and D flip-flop are on the bottom side, and the set pin of RS-latch is on the top side. The clock pin of the D flip-flop is marked with a triangle. The address pin of the multiplexer is on the bottom side. A circle in front of an input denotes a negation. It is assumed that signals A, C, D, L_1, L_2, L_3 and S_1 are driven by some other internal components or are primary input pins. Absence of global master clock and reset signals makes this circuit fully asynchronous.

The condition for the D flip-flop in Fig. 1 to be reset is $(Q_1 \wedge L_1) \vee L_2 = 0$. However, we do not know the values of Q_1, L_1 and L_2. Moreover, they can change during reset sequence several times. Since these signals could be driven by any internal signals, e.g. outputs of some other circuit components, it is not obvious to prove the reset condition.

978-1-4799-4993-9/14 $31.00 © 2014 IEEE

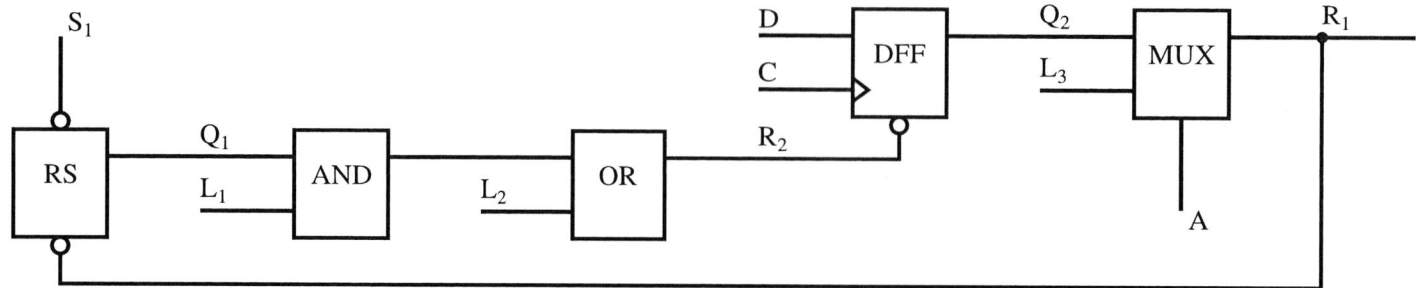

Fig. 1.   Example of fully asynchronous digital circuit

The same is true for RS-latch and all other memory elements in general.

To demonstrate a problem with reset sequence, suppose that $S_1 = L_1 = 1$ and $L_2 = 0$ during the whole reset sequence. This means that $R_2 = Q_1$. Suppose also that signal Q_2 is always transferred to the output of multiplexer, i.e. $R_1 = Q_2$. In this case the reset sequence does not perform well. Immediately after power-on each of the values Q_1 and Q_2 is random or even metastable [5]. For Q_1 to get a well-defined value it is required to provide $S_1 = 0$ or $R_1 = 0$. However, S_1 is always 1 due to our assumptions and $R_1 = Q_2$, where Q_2 is undefined. Thus, we cannot guarantee that Q_1 gets a well-defined value. A similar reasoning works for the value of Q_2. Due to our assumptions these arguments work during the whole reset phase. Finally, we conclude that the shown reset sequence does not work correctly, and we have to find another reset sequence or fix the circuit.

We stick to this circuit example (Fig. 1) in the paper for the sake of simplicity and clearness. Later on we present our method for detection of such problems or proving their absence. Discussions of the method application in a real industrial project as well as strengths and weaknesses of the method follow.

III. TRADITIONAL SOLUTION

In a traditional design flow environment simulations are performed to find various bugs in VHDL descriptions of digital circuits including reset problems [4]. For further discussions we refer to the VHDL source (Fig. 2) of the proposed example circuit fragment.

In most cases VHDL designers use *std_logic*, *std_ulogic* or similar standard types instead of *bit* type [7], [6]. This makes a lot of sense since *std_logic* and similar types are supported for synthesis as well as *bit* type, but they have more features for simulations. One of the important features is that they have more than two logical values which allows to simulate unknown or undefined values. Synthesizeable VHDL code should not contain default signal value assignments [8]. The simulation environment sets all these signals to uninitialized values before each simulation run. Consequently, the verification engineer should observe during the reset sequence the transition from uninitialized values to the desired logical states.

This traditional approach leads to "perfect" results for synchronous circuits. However, the same approach cannot be applied for asynchronous circuits. Moreover, following the

---

signal A, C, D, L1, L2, ...,   S1: *std_logic* ;

RS: *process* (R1, S1)
begin
  if   R1 = '0'   *then* Q1 <= '0';
  elsif S1 = '0'   *then* Q1 <= '1';   *end if* ;
end process;

DFF: *process* (R2, C)
begin
  if   R2 = '0'   *then* Q2 <= '0';
  elsif *rising_edge* (C) *then* Q2 <= D;   *end if* ;
end process;

CL: R2 <= (Q1 *and* L1) *or* L2;

MUX: *process* (A, Q2, L3)
begin
  case A *is*
    when '1'   => R1 <= Q2;
    when others => R1 <= L3;   *end case*;
end process MUX;

Fig. 2.   VHDL source code corresponding to the circuit example

traditional approach can be dangerous since it provides too optimistic results.

Consider our example in Fig. 1. When the simulation starts all signals get the value 'U' (uninitialized). Suppose that the signal A does not get a defined value for some time. Suppose also that L_1 gets the value '1' and L_2 and L_3 get the value '0' at some moment during the reset process, while A still stays 'U'. The expected multiplexer behaviour implies that R_1 has the value 'U' since the address bit A and one of the inputs Q_2 are undefined.

Consider now a VHDL source code in Fig. 2 that corresponds to the circuit example. If we look at the multiplexer process we find out that R_1 gets value '0' instead of the expected 'U'. This happens because of the *case* statement. Here it does not describe adequately the multiplexer behaviour regarding undefined values of inputs. As a result, the RS-latch gets reset. Finally, the output value of the RS-latch propagates to the D flip-flop reset pin and it gets reset as well. A short summary is that we have two memory elements that erroneously get reset during traditional simulation, although the VHDL code fully fulfills synthesis standards [8].

978-1-4799-4993-9/14 $31.00 © 2014 IEEE   212

The next step in the traditional verification methodology are simulations at the gate level that are performed after synthesis [4]. Correctness of reset sequence is checked again. At this step the presented reset problem will be found. However, there are disadvantages of finding problems at this level of abstraction. First of all, synthesis requires time and human effort and cannot be done after any small change in the design. Second, it could be problematic to map bugs at the gate level back to the bugs at the RTL level, i.e. require a lot of effort as well. It is reasonable to develop a method for fixing reset problems at the RTL level before synthesis in order to further reduce the development effort.

IV. PROPOSED SOLUTION

A. Tri-State Logic for Reset Verification

Consider again our VHDL example (Fig. 2) and multivalued *std_logic* type for verification. Although, the problem with the multiplexer was demonstrated before, we observe that this logic represents adequately the behaviour of the circuit parts that are coded explicitly with logical operators like *and*, *or*, *not*, etc. An example of such a combinatorial logic block is labeled "CL" in Fig. 2. When *std_logic* type is used at each logical operator it can output 'U' depending on its arguments, e.g. '1' $\wedge$ 'U' = 'U'.

This multi-valued logic represented by *std_logic* type is mostly what we require for the reset sequence verification. In fact, we need only the subset that contains the '0', '1' and 'U' values. However, there are some problems that cannot be solved with a multi-valued logic types defined in the VHDL standard [7]. In real industrial projects it is not usual to apply explicit logical operator coding style. There are VHDL coding constructs that have larger expressive power and help to produce better structured and clearer code in shorter time. One of them is the *if-then-else* operator. These operators use the *boolean* data type for condition definition [6]. Comparing two *std_logic* type values produces the *boolean* type results and is source for inconsistency between simulation outcome and the expected behaviour (Table I).

TABLE I.   INCONSISTENCY IN COMPARISON SIMULATION

Expression	Expected Result	Simulation Result
'1' = 'U'	UNDEF	FALSE
'U' = 'U'	UNDEF	TRUE

The *boolean* data type has only two values. However, a third (undefined) value is required to represent the result of comparison. The same is true for other functions and operators with the *boolean* return type, e.g. *rising_edge*. The authors propose to define a new data type *tri_bit* with three logical values '0', '1' and 'U' and use this type instead of all types representing digital signals and *boolean* type. In the situations where *boolean* or digital signal type value is required by VHDL syntax, special conversion functions should be applied.

The advantage of *tri_bit* type over *std_logic* is the possibility to define all standard operators like *and*, *or*, *not*, =, etc. in arbitrary way. The authors implemented the proposed type in a package (Fig. 3) as a wrapper around an enumerated type *t_bit*. This trick allows to use the implicit equality operator together with the user-defined one. The user-defined equality operator

```
package tri_logic  is
    type t_bit      is  ( 'U', '0', '1') ;
    type tri_bit    is  record
        value : t_bit ;
    end record  tri_bit ;
    constant tri_U :    tri_bit   := ( value => 'U');
    ...
end package  tri_logic ;

package body tri_logic   is
    function  is_U (a : tri_bit ) return  boolean is  begin
        if  a. value = 'U' then return   true ;
        else  return  false ;  end if ;
    end function  is_U;
    ...
    function  to_bit  (a : tri_bit ) return  bit  is  begin
        if  is_U(a)  then
            assert  false  report  "tri conv." severity  failure ;
        elsif  is_1 (a) then return  '1';
        else  return  '0';  end if ;
    end function  to_bit ;
    ...
    function  "or" (a,  b : tri_bit ) return  tri_bit    is begin
        if      is_1 (a)  or  is_1 (b) then return  tri_1 ;
        elsif  is_U(a)  or  is_U(b) then return  tri_U;
        else    return  tri_0 ;  end if ;
    end function  "or";
    ...
    function  "=" (a,  b : tri_bit ) return  tri_bit    is begin
        if      is_U(a)  or  is_U(b)  then return  tri_U;
        elsif  is_1 (a)  xor  is_1 (b) then return  tri_0 ;
        else    return  tri_1 ;  end if ;
    end function  "=";
    ...
    function  rising_edge  (signal  s :  tri_bit )
    return  tri_bit   is  begin
        if  s' event  then
            return  (s' last_value  = tri_0 )  and (s = tri_1 );
        elsif  is_U(s)  then return  tri_U;
        else  return  tri_0 ;  end if ;
    end function  rising_edge ;
    ...
end package body tri_logic  ;
```

Fig. 3.   VHDL package for formal reset verification

returns *tri_bit* value and is supposed to be called in the source code. The implicit equality operator of the type *t_bit* returns *boolean* value and is used in the package implementation.

The package contains convenient constants *tri_U*, *tri_1* and *tri_0*, functions *is_U*, *is_1* and *is_0* that test *tri_bit* for a particular value, functions *to_bit* and *to_boolean* that convert *tri_bit* to standard types, logical operators *and*, *or*, etc., comparison operators such as equality and other special functions like *rising_edge*. Similar functions are omitted in Fig. 3. Their implementation is obvious.

B. Source Code Instrumentation

The proposed package introduces a principally new logical framework for reset sequence verification. However, it cannot

```
signal  A, C, D, L1, L2, L3, Q1, Q2, R1, R2, S1: tri_bit ;
...
DFF: process (R2, C)
begin
    if      is_U(R2 = tri_0 )        then Q2 <= tri_U;
    elsif   is_1 (R2 = tri_0 )       then Q2 <= tri_0;
    elsif   is_U(rising_edge (C))    then Q2 <= tri_U;
    elsif   is_1 (rising_edge (C))   then Q2 <= D;      end if ;
end process ;
...
MUX: process (A, Q2, L3)
    variable  V: bit ;
begin
    if  is_U(A) then  R1 <= tri_U; else  V := to_bit (A);
      case  V is
          when '1'    => R1 <= Q2;
          when others => R1 <= L3;   end case;  end if ;
end process  MUX;
```

Fig. 4.   Instrumented VHDL source code

be directly applied to the original VHDL source code. Recall, we consider only synthesizable code.

Let us analyze the example with erroneous reset in the section III. When the signal A is 'U' then the multiplexer output results in a defined value. The problem is that the original source code (Fig. 2) does not describe the behaviour of the multiplexer when the address input is undefined. This happens because the synthesis tools cannot process the information about undefined signals and states. Consequently, VHDL synthesis standard requires to avoid undefined values.

Similarly, our source code does not define what happens if the reset or clock input of the D flip-flop is undefined. The same is true for the RS-latch. The problem is existing for sequential as well as combinatorial logic. The reason is always the same: we do not provide some details about the circuit behaviour due to the synthesis requirements.

On the other hand, we require the missing behavioural details that consider undefined values of signals. The authors propose to add this details explicitly in the source code in order to perform an accurate verification of fully asynchronous circuits. Modification of the original code is well-known technique in software verification and is called source code instrumentation [9]. Fig. 4 presents the required source code instrumentation for the multiplexer and D flip-flop in the discussed example (Fig. 1, 2). The instrumentation of the RS-latch description is done in a similar fashion. The combinatorial logic block "CL" that uses only logical operators does not require instrumentation.

V.   Experimental Results

The authors applied the proposed method for the reset sequence verification during the development of test structures for asynchronous interface of a mixed-signal power circuit [1]. The device has a digital module with fully asynchronous interconnections. The available VHDL source code has been instrumented to be in line with the proposed package for reset verification. The package contains operators and functions working with scalar and array types. The code supporting

one-dimensional arrays of tri_bit is not presented in Fig. 3 for the sake of simplicity and compactness. Though, the implementation is quite obvious.

Simulations with the instrumented code and developed package discovered a problem in the reset sequence. In contrast to this result, simulations with the original uninstrumented source code and std_logic type could not find the bug. The problem was similar to the discussed example in the section III: several memory elements were in the loop of undefined reset chains, while traditional simulations erroneously assigned some logical values to them. With the help of simulator tools the author were able even to trace the origin of the problem and suggest a fix for the circuit.

VI.   Summary and Outlook

The work presents a method for formal reset sequence verification. The method mostly concentrates on fully asynchronous circuits, because they do not enjoy a particular design structure and methodology that would make reset properties correct by design. The method was successfully applied in an industrial product development.

The proposed logical framework together with the source code instrumentation allow to perform the desired functional check of reset properties at the register transfer level. The verification is performed by running a single simulation test and checking the signals values afterwards. Traditional simulation approaches will most likely produce too optimistic results, which are unsafe.

This work is a step towards the formal verification of fully asynchronous circuits. There are much more challenging problems in this area, like reset verification with respect to timing properties or checking timing in basic operational mode. They are still open for the future.

Acknowledgment

This work was jointly funded by the Federal Ministry of Economics and Labor of the Republic of Austria and the Carinthian Economic Promotion Fund (KWF).

References

[1] B. Murari, F. Bertotti, and G. A. Vignola, *Smart Power ICs—Technologies and Applications*.   Springer Verlag, 1995.

[2] R. Ploss, A. Mueller, and P. Leteinturier, "Solving automotive challenges with electronics," in *VLSI Technology, Systems and Applications, 2008. VLSI-TSA 2008. International Symposium on*.   IEEE, 2008, pp. 1–2.

[3] M. Arora, *The Art of Hardware Architecture. Design Methods and Techniques for Digital Circuits*.   Springer Verlag, 2012.

[4] N. Weste and D. Harris, *CMOS VLSI Design: A Circuits and Systems Perspective*, 4th ed.   Addison-Wesley, 2010.

[5] J. M. Rabaey, A. Chandrakasan, and B. Nikolic, *Digital integrated circuits. A design perspective*, 2nd ed.   Prentice Hall, 2004.

[6] *IEEE Standard VHDL Language Reference Manual*, IEEE Std. 1076-1987, 1988.

[7] *IEEE Standard Multivalue Logic System for VHDL Model Interoperability (Stdlogic1164)*, IEEE Std. 1164-1993.

[8] *IEEE Standard for VHDL Register Transfer Level (RTL) Synthesis*, IEEE Std. 1076.6-1999.

[9] J. Seyster, K. Dixit, X. Huang, R. Grosu, K. Havelund, S. A. Smolka, S. D. Stoller, and E. Zadok, "Interaspect: Aspect-oriented instrumentation with gcc," *Formal Methods in System Design*, vol. 41, no. 3, pp. 295–320, 2012.

978-1-4799-4993-9/14 $31.00 © 2014 IEEE

A New Circuit Topology for Floating High Voltage Level Shifters

Dawei Liu
Department of Computer Science

University of Bristol
Bristol, UK
dawei.liu@bristol.ac.uk

Simon J. Hollis
Department of Computer Science

University of Bristol
Bristol, UK
simon@cs.bris.ac.uk

Bernard H. Stark
Department of Electrical and
Electronic Engineering
University of Bristol
Bristol, UK
bernard.stark@bristol.ac.uk

Abstract—**A novel and simple circuit topology is presented for high-speed, floating, high voltage level shifters. It uses a current mirror plus latch circuit composed of two inverters. Simulations based on AMS 0.18 μm High Voltage (HV) CMOS Technology show this circuit to combine high speed, low power dissipation, and small layout area. The simulation results show the propagation delay to be below 150 ps for a transition from 1.8 V to 13.8 V.**

Keywords—high speed, low power, floating, high voltage, level shifter, HV CMOS, BCD

I. INTRODUCTION

Driven by the market, the development of integrated circuit design techniques and the triple-well HV-CMOS/BCD (Bipolar CMOS and DMOS) technologies, more and more multi-power-rail systems are integrated on a single chip to realise complicated functions, improve system reliability, and decrease the cost of the system. As the communication bridge between different power rails, level shifters are widely used in DC/DC BUCK Converters, display drivers, EEPROMs, MEMS and so on. Floating high voltage level shifters are used to shift the potential of control signals from circuits powered by low voltage power rails to the potential of circuits with floating power and ground rails. The literature [1]-[6] presents several floating high voltage level shifters' design techniques based on High Voltage (HV) CMOS Technologies. In this paper, a floating high voltage level shifter is proposed with a view to combining beneficial features of the floating level shifters presented in [1-2] and [4-6]. This novel level shifter is shown, via simulation, to contain the positive features of the reviewed level shifters, and to provide a better trade-off between speed, layout area and power dissipation.

This paper is organised as follows: Section II reviews three different kinds of floating level shifters. In Section III the design procedure of the proposed floating high voltage level shifter is presented. Section IV shows the simulation results and discusses these. Finally, Section V draws conclusions.

II. REVIEW OF FLOATING HIGH-VOLTAGE LEVEL SHIFTERS

HV-CMOS level shifters can be broadly categorised into three types, as illustrated in Fig. 1, where the red dashed boxes show isolation areas provides by deep N-wells; all transistors are placed in deep N-wells, including the floating LV transistors. Fig. 1(a) shows the conventional low-to-high voltage (LV to HV) level shifter [1,2]. These level shifters use the cascaded HV NMOS to protect and clamp the LV input related transistors, and HV PMOS to protect and clamp the output related floating LV transistors. As graphically analysed in [3], this kind of floating high voltage level shifter cannot operate at high speed, and consumes a large layout area. The level shifter presented in [3] makes significant improvements in speed and layout area, but at the expense of additional complexity and a control signal to set the initial state, which may not suitable in some applications. Fig 1(b) shows a second type of HV level shifter [4,5]. This topology uses the diode connected floating LV PMOS transistors to clamp the voltage potential at nodes N1 and N2 to one gate-to-source (V_{GS}) voltage drop from the floating high voltage rail VDDH. This voltage clamping technique allows the level shifters of [4,5] to operate at high speed, but the drawback is continuous power dissipation due to the alternately turn on of HNM1 and HNM2.

The third kind of floating high level shifter [6] is illustrated in Fig. 1(c). It uses narrow pulse triggers as input signals to decide the output state. This trigger method gives low power dissipation, operates at sub GHz frequencies, and saves layout area. However, this circuit uses the diodes with their anodes connected to the floating low voltage rail VSSH to clamp the voltage potential at nodes N1 and N2. This voltage clamp technique complicates the provision of supply voltage to the floating LV circuits that follow the level shifter, due to the clamping diodes forwarding biasing voltage drop, this will be illustrated in the simulation results section. Another disadvantage is the parasitic capacitor and the reverse recovery charge effect of the clamped diodes, which limits the operational speed to over 1 GHz in current technologies. Overall, this pulse triggered method appears to provide a better trade-off between operational frequency, layout area, and power dissipation. In the next section, we propose a novel floating high voltage level shifter that is based on the pulse triggered method, however which avoids aforementioned weaknesses.

This work was supported by EPSRC grant # EP/K021273/1

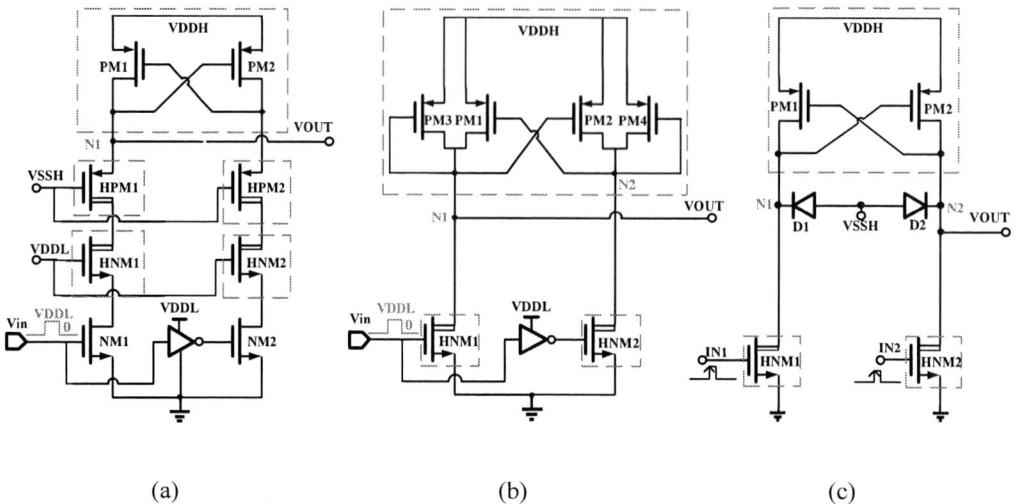

Fig. 1. Three HV floating voltage level shifters with different floating low voltage clamp techniques: (a) biased HV PMOS clamping (b) diode connected PMOS clamping (c) diode clamping. Red dashed boxes are the independent deep N-wells. VDDH is the floating power supply rail, and VSSH the floating ground rail, VDDL is the low voltage supply rail.

III. THE PROPOSED FLOATING HIGH-VOLTAGE LEVEL SHIFTER

A. Design Approach

The diode connected PMOS in the level shifter of Fig. 1(b) clamps output nodes N1 and N2 preferably, and the pulse triggered method operates at high speed and consumes low power. It is therefore desirable to merge these two aspects into one design.

In order to explore this idea, first consider the gate voltage clamping circuit, as redrawn in Fig. 2. The gate voltage clamping circuit clamps the gate voltage so that VG = VDDH - V_{GS}. When VIN goes high, a current I_{in} will flow through PM1 and HNM1 to ground. Therefore, the current I_{in} indicates that the input VIN is high. If this current signal can be used to trigger and latch the output signal, then a level shifter could be realized in a simple and attractive way.

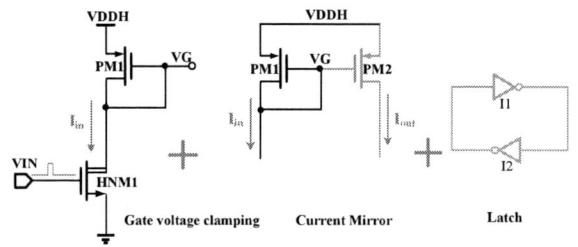

Fig. 2. Gate voltage clamping, current mirror and latch circuit

The current I_{in} can be transferred to I_{out} using a current mirror circuit, as indicated in Fig. 2. I_{out} is then used to trigger a latch, as illustrated in Fig. 2. In this method the diode connected PMOS PM1 has two functions: clamping its gate voltage, and detecting the input high voltage pulse. The current

mirror circuit copies the input current information, and the latch circuit latches the output state accurately.

B. Realization

The proposed floating high voltage level shifter circuit is shown in Fig. 3. VDDH is the floating power supply rail, and VSSH is the floating ground rail. The red dashed box is the deep N-well that is biased by VDDH. The inverters in Fig. 3 are also biased by VDDH and VSSH.

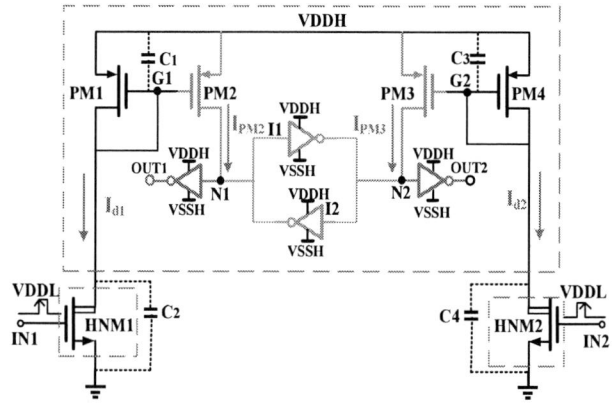

Fig. 3. The proposed floating high-voltage level shifter

When the input IN1 goes high, then HNM1 switches on, and PM2 mirrors the current flow through PM1, meaning that node N1 is pulled up by PM2. As the voltage at node N1 exceeds the trigger voltage of the latch composed of I1 and I2, N2 will be set to VSSH. The positive feedback of the latch will accelerate node N1 to rise to VDDH at the same time, and lock the output state at nodes N1 and N2. Then outputs OUT1 and OUT2 will be held at VSSH and VDDH respectively, even when IN1 goes low again. Thus the input pulse signal at node IN1 can trigger the latch to lock N1 to VDDH and N2 to VSSH. To change the state of N1 and N2, a pulse signal can be applied to IN2. The pulse signal at node IN2 will then trigger N2 to VDDH and N1 to VSSH.

The level shifter propagation delay is dependent on the drain currents of HNM1 and HNM2, the transconductance (g_m) of PM1 and PM3, and the parasitic capacitors at nodes G1 and G2. As the HV NMOS transistors HNM1 and HNM2 operate in the switched mode, and they can be driven by the LV transistors, the time from the input pulses arriving to the drain currents flowing is short. The minimum length for HNM1 and HNM2 can be chosen to guarantee the large drain current in a small layout area, and to minimize parasitic capacitance.

The majority of the propagation delay is the response time of the current mirror. Taking node G1 for example, the propagation delay is dominated by the time from the onset of current flow I_{d1} in HNM1, to the voltage V_{G1} at G1 dropping enough to trigger the latch circuit. Fig. 4 provides a simple model to calculate this response time. C_{G1} is the parasitic capacitance at node G1, as indicated in Fig. 3 by the parallel connected C1 and C2. Gm_{PM1} is the g_m of PM1, and V_{G1} is the gate voltage of PM1.

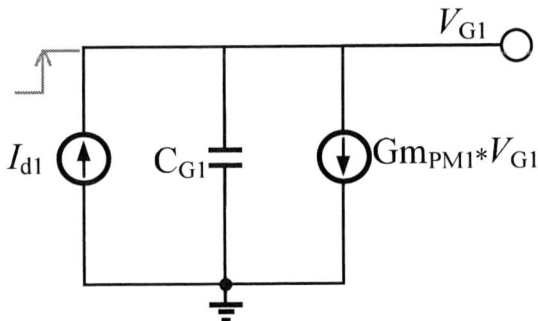

Fig. 4.   The model of the current mirror with a step input current

Using the Laplace transform, the step response transfer function from I_{d1} to V_{G1} is given by:

$$V_{G1} = \frac{I_{d1}}{Gm_{PM1}} \left(\frac{1}{1 + \dfrac{S}{\dfrac{Gm_{PM1}}{C_{G1}}}} \right) \quad (1)$$

As PM1 and PM2 have the same transconductance ($Gm_{PM1} = Gm_{PM2}$) the current through PM2 is given by:

$$I_{PM2} = Gm_{PM2} * V_{G1} = I_{d1} \left(\frac{1}{1 + \dfrac{S}{\dfrac{Gm_{PM1}}{C_{G1}}}} \right) \quad (2)$$

Equation (2) shows an interesting result. The current I_{PM2} to trigger the latch is determined by I_{d1}, Gm_{PM1} and C_{G1}. I_{d1} can be set to the maximum available value with V_{GS} of PM1 not exceeding 1.8 V. As C_{G1} is determined by the gate-to-source capacitance of PM1 and PM2, Gm_{PM1}/C_{G1} approximates to half of the unity current gain frequency (f_T). This is a natural feature of a large bandwidth current mirror, and the minimum value of length for all the transistors can be chosen to guarantee maximum f_T; Based on AMS 0.18μm parameters, a operation frequency of 3 GHz should be achievable by our design. This is validated by simulation in Section IV.

It can be concluded that the proposed level shifter has the following advantages by design: First, the circuit can operate at ultra high frequency. Second, it does not use HV PMOS transistors, saving much layout area. Third, it avoids continuous current flow. We now investigate a simulated implementation to establish if this circuit combines the benefits of diode-connected and pulse-triggered level shifters without introducing any undesirable trade-offs.

IV.   SIMULATION RESULTS

The proposed level shifter and the level shifters in [1-6] can be realized in a standard HV CMOS or BCD Technology. In this section the level shifter in Fig. 1(c) [6] is compared with our proposed design by simulation.

The AMS 0.18 μm HV CMOS Technology is chosen to be the demonstration technology. In this technology, the HV NMOS parameters are: $V_{GS} = 1.8$ V, $V_{DS} = 20$ V; the LV transistors and floating LV transistors parameters are: $V_{GS} = 1.8$ V, $V_{DS} = 1.8$ V. Two schottky diodes ($V_f = 0.45$ V) are used as the clamp diodes in the circuit of Fig. 1(c). VDDH and VSSH are set to 13.8 V and 12 V respectively, and the input high voltage VDDL is 1.8 V. The input HV NMOS of these two level shifters are chosen to be same size, so that their power dissipation is similar. The input signals at the IN1 and IN2 nodes in both level shifters are exactly the same, and input pulse width is 1 ns; the frequency is 200 MHz. The simulation results are shown in Fig. 5.

In Fig. 5, it can be observed that the VOUT in Fig. 1(c) [6] is 0.6 V below VSSH, when IN2 is high voltage. This is due to the diode forward voltage drop clamping VOUT. This effect means the maximum voltage swing of the VOUT connecting to the following buffer will be 2.4 V (1.8 V + 0.6 V), which may cause reliability problems or decrease the life-time, since PM1, PM2 and the following buffer transistor operational voltages are 1.8 V. To avoid this problem, the differential voltage between VDDH and VSSH should be set less than 1.2 V to guarantee VOUT voltage swing is within 1.8 V. However this will limit the following buffer output swing range to 1.2 V. In [6], the level shifter is used to drive a buck converter high-side switch FET. The limited differential voltage between VDDH and VSSH means the high-side switch FET cannot fully turn on, which will lower the buck converter efficiency and induce heat dissipation problems. After IN2 turns low, due to the half-latch feature, VOUT increases slowly to approach 12 V, since the VOUT becomes a high resistance point after IN2 turns low. The high resistance state at node VOUT is more sensitive to the noise or other transient signals' disturbance.

978-1-4799-4993-9/14 $31.00 © 2014 IEEE

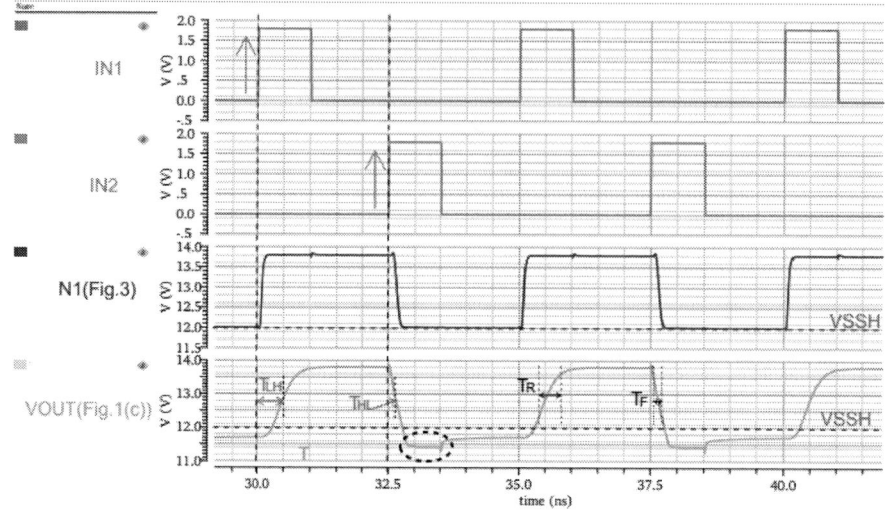

Fig. 5.   Transient simulation results of the proposed level shifter and the level shifter in Fig. 1(c) [6]

On the contrary, the proposed level shifter output node N1 avoids these problems. Since N1 is triggered by PM1 and latched by a full-latch circuit (shown in Fig. 2), the voltage range at node N1 is just between VDDH and VSSH. Whether the output is VDDH or VSSH, N1 is still a low resistance node due to the full latch circuit.

Table 1 summarizes the performance of these two kinds of level shifters.

TABLE I.   THE COMPARISON OF THESE TWO LEVEL SHIFTERS

Parameters	Circuit in Fig 1(c) [6]	Proposed Level shifter
VDDH (V)	13.8	13.8
VSSH (V)	12	12
Output swing (V)	2.4	1.8
Rise time (ps): T_R	400	76
Fall time (ps): T_F	250	150
Rise delay (ps): T_{LH}	480	73
Fall delay (ps): T_{HL}	180	120

As explained in Section III, the proposed level shifter can operate at ultra high frequency. The propagation delay to be below 150 ps for a transition from 1.8 V to 13.8 V. Table 1 indicates clearly that rise time of the level shifter in [6] is more than 5 times longer than that of the proposed level shifter, and the rise delay is more than 6 times longer.

V.   CONCLUSION

This paper proposes a novel floating high voltage level shifter which uses pulses as input signals, and which uses a current mirror circuit to sense the input current and trigger the latch circuit to decide the output state. Implemented in standard HV CMOS/BCD technology, it can operate at high speed with propagation delay of less than 150 ps. The simulated implementation achieves an improved trade-off in speed, layout area and power dissipation, when compared to the

known floating high voltage level shifters. Further work will be carried out to optimize the proposed level shifter, and the level shifter will be taped out to verify the analysis and simulation results.

ACKNOWLEDGEMENT

This project is funded through the UK Engineering and Physical Sciences Research Council (EPSRC) grant number EP/K021273/1, "Pulse quietening at source for higher-frequency power and signal switching".

REFERENCES

[1]   B. Choi, "Enhancement of current driving capability in data driver ICs for plasma display panels," IEEE Trans. Consumer Electron., vol. 55, pp. 992–997, Aug. 2009.

[2]   M. Khorasani, M. Benham, L. van den Berg, C. J. Backhouse, and D. G. Elliott, "High-voltage CMOS controller for microfluidics," IEEE Trans. Biomed. Circuits Syst. , vol. 3, no. 2, pp. 89–96, April 2009.

[3]   Y. Moghe, T. Lehmann, and T. Piessens, "Nanosecond delay floating high voltage level shifters in a 0.35 µ m HV-CMOS technology," IEEE J. Solid-State Circuits , vol. 46, no. 2, pp. 485–496, Feb. 2011.

[4]   M. J. Declerq, M. Schubert, F. Clement, "5 V-to- 75 V CMOS Output Interface Circuits, " ISSCC, pp.162-163, Feb. 1993.

[5]   J. F. Richard, B. Lessard and R. Meingan, S. Martel, and Y. Savaria, "High Voltage Interfaces for CMOS/DMOS Technologies," Proceedings of the IEEE Northeast Workshop on Circuits and Systems, June 2003.

[6]   Yan-Ming Li, Chang-Bao Wen, Bing Yuan, Li-Min Wen, Qiang Ye "A High Speed and Power-Efficient Level Shifter for High Voltage Buck Converter Drivers," Solid-State and Integrated Circuit Technology (ICSICT), 2010 10th IEEE International Conference on pp 309-311, Nov. 2010.

Probabilistic Saboteur-based Simulated Fault Injection Techniques for Low Supply Voltage Interconnects

Sergiu Nimara   Alexandru Amaricai   Oana Boncalo   Mircea Popa

"Politehnica" University of Timisoara
Timisoara, Romania
sergiu.nimara@gmail.com, alexandru.amaricai@cs.upt.ro, oana.boncalo@cs.upt.ro, mircea.popa@rectorat.upt.ro

Abstract— Probabilistic behavior of logic gates represents one of the main reliability problems associated to CMOS circuits supplied at very low supply voltages. This paper aims to analyze the impact of probabilistic faults in interconnects, by means of HDL saboteur-based simulated fault injection (SFI). We propose four types of saboteurs: the simplistic probabilistic type, a switching type - aware and two data dependent types. We have analyzed the behavior of the Wishbone bus in the presence of probabilistic errors. Several sets of simulations have been performed, by injecting probabilistic faults on address, control signals and data components of the bus. The performed simulations indicate that the simulation time for a SFI campaign is 1.7x higher with respect to the gold circuit.

Keywords—simulated fault injection, saboteurs, interconnects, probabilistic circuits;

I. INTRODUCTION

The quest for lower power consumption has led to dramatic down scaling of the supply voltage to sub and near threshold regimes. Coupled with the scaling of transistor sizes to nanometer levels and with process and temperature variations, this has led to important reliability issues in logic devices. These logic circuits exhibit a probabilistic behavior. The probabilistic behavior may become more acute in interconnects; this is due to systematic and random process variations, including metal, dielectric barriers and low-*k* dielectrics, combined with crosstalk noise [1]. Therefore, in addition to the effects caused by transient faults at gate-level, the probabilistic error occurrence in interconnects must be taken into account.

The reliability attributes of digital systems can be efficiently determined using fault injection techniques, which are classified in three main categories: hardware-based, software-based and simulated fault injection [2],[3],[4]. The simulation-based fault injection technique is preferred over the others because it offers the possibility of an early diagnosis of the circuit under test (CUT), during the design phase [4]. Simulated fault injection can be performed using techniques which do not require source code intervention (based on

simulator commands and scripts) and techniques which alter the CUT's source code (saboteurs and mutants). Regarding the former, these rely heavily on the employed HDL simulator's capability and limited fault modeling capability. Regarding the latter, the overhead given by the source code modification is compensated by the increased fault modeling capability.

This paper proposes probabilistic saboteur based techniques for reliability analysis of interconnects. We focus on reliability issues of signals transmitted on interconnects and not on logic gates and memory elements. We propose several types of saboteurs. The most simple relies on performing a probabilistic bit-flip on a signal of the interconnect. The most accurate takes into consideration data dependency: the probabilities for each signal depend on the data values transmitted on the interconnect. This type of saboteur captures in the most accurate way due to the importance of the crosstalk in interconnects' behavior (crosstalk is data dependent). We have performed our analysis on the open-source Wishbone bus. We have analyzed the impact of probabilistic faults on different types of signals of the bus: data signals, address lines, control and handshaking signals.

This paper is organized as follows: Section II is dedicated to reliability issues of interconnects; the probabilistic saboteur-based simulated fault injection technique is described in Section III, while Section IV presents the simulation campaigns. Some concluding observations are stated in the last section of this paper.

II. RELIABILITY ISSUES IN INTERCONNECTS

The main factors that lead to reliability issues in interconnects are process variation and crosstalk induced faults. Regarding the process variations, the most frequent forms of it are represented by: device geometry variations, device material and electrical parameter variations, interconnect geometry and material parameter variations [5]. These variations will have an effect on the metal thickness or length, dielectric thickness, contact and via size, metal resistivity or dielectric constant. Thus, the resistance,

978-1-4799-4993-9/14 $31.00 © 2014 IEEE

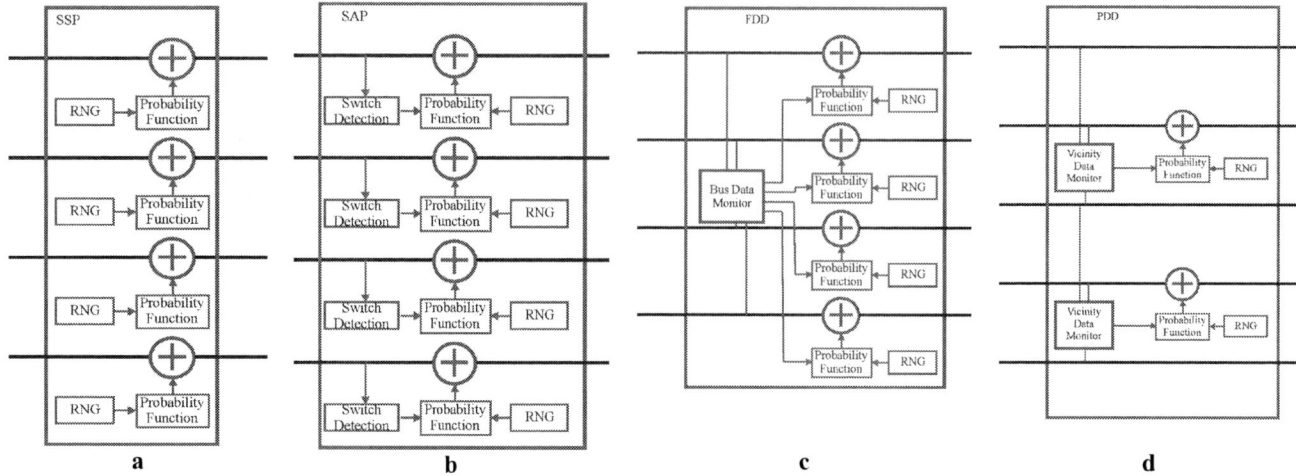

Fig. 1 - The saboteurs' architectures according to four fault models (a – SSP, b – SAP, c – FDD, d – PDD)

capacitance or inductance parameters of a wire are affected. Process variation in interconnects may alter the timing characteristics of the signals. Thus, an erroneous result at the moment when a certain signal is sampled may appear due to increased resistance or ground capacitance of the wire.

Crosstalk faults are most probably the result of an inappropriate interconnect routing scheme, rather than manufacturing defects [6], and they are strongly data-dependent. For the interconnection lines, cross-talk induced faults result from an undesired inductive or capacitive coupling between two or more signal lines, producing both timing alterations and / or noise (like glitches) on those signals [7]. These parasitic couplings determine an energy transfer from one wire to another, depending on the driver strength and they result in crosstalk faults [6]. The authors in [6] realize a classification of crosstalk faults into crosstalk induced glitches and delays. Crosstalk induced glitches appear on a static victim (affected) line when one or more aggressor lines switch their logic value, while crosstalk induced delays occur when aggressor and victim signals change their logic state simultaneously [6]. The most dominant effect is represented by the capacitive crosstalk: this affect only the neighboring line [8]. The inductive crosstalk has a smaller influence with respect to the capacitive one; however, the inductive effects may span across multiple lines [6],[11].

In this paper, we address the probabilistic occurrence of faults caused by either process variation or crosstalk effect, affecting the interconnects of a digital system. The circuit under test is described in the next section, along with the fault injection method employed.

III.   PROBABILISTIC SABOTEURS FOR INTERCONNECTS

Simulated fault injection can be achieved without source code intervention (such as the ones based on simulator commands) or by utilization of techniques which alter the CUT's HDL description. The techniques which rely on source code modification present two advantages: (i) they are

independent of the HDL simulator (ii) they present high fault modeling capability. Two techniques are widely used for simulated fault injection: mutants and saboteurs. The mutant represents a component description which replaces the correct one. In VHDL, mutants are implemented using multiple architectures for the same entity. In Verilog, the mutant represents another module description. The saboteur is a special component (entity or module) which alters the value or timing characteristics of a signal [4]. The goal of our analysis is to perform reliability evaluation (by modifying the values) for different groups of signals contained by the interconnect. This makes the saboteur the natural candidate for our analysis. Our methodology has been implemented in Verilog; however, it can be easily adapted to VHDL. Several types of saboteurs have been proposed, such as [4], [9]: serial simple unidirectional saboteur, serial simple bidirectional, serial complex saboteur, serial complex bidirectional saboteur, *n*-bit unidirectional serial saboteur, *n*-bit bidirectional serial saboteur, parallel saboteur. According to this classification, the saboteurs employed in this paper can be considered *n*-bit unidirectional serial saboteurs.

For probabilistic interconnects, we have developed four types of saboteurs:

1. *Standard Signal Probabilistic* (SSP) saboteur. It represents the simplest one because it only flips the logic value of a certain signal with a given bit-independent probability. It doesn't take into account the last type of transition that took place on that line, nor the data pattern. The SSP model-based saboteur contains a fault insertion module, which is triggered according to the desired probability of failure and to the output provided by a random number generator.

2. *Switching-Aware Probabilistic* (SAP) saboteur. This model considers probabilistic behavior for a signal only when switching is taking place. It models accurately timing faults: the switching for a line does not respect a given timing constraint. The most simplistic type of SAP considers the same probability for both types of

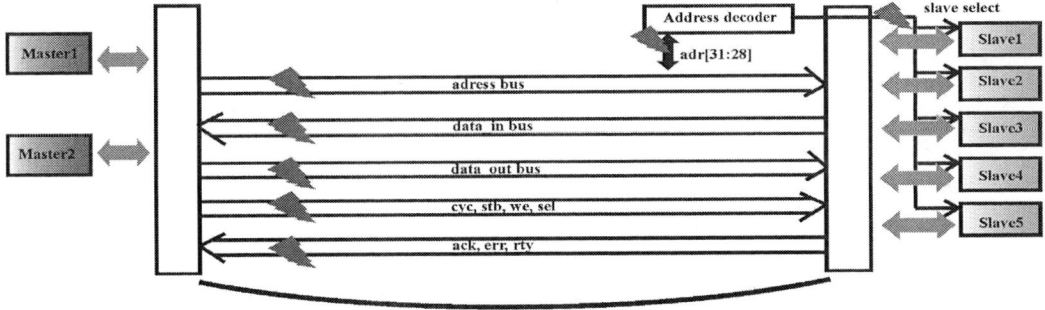

Fig. 2 - The wishbone's signal groups that are the subject of fault injection campaigns

switching; a more accurate considers different probabilities for charging and discharging processes. The architecture for this saboteur contains a switching detector (or switching type).

3. *Full Data Dependent* (FDD) saboteur. For this model, the probabilities for a line are dependent on the data configuration on the entire bus. This represents the most accurate model, as the timing and value characteristics for a wire are affected by crosstalk (which is data dependent). Although this model is the most accurate, it has very poor scalability: for an *n*-bit bus, 2^{2n} probabilities for a single line are derived (the crosstalk noise manifests when the bus switches, therefore).

4. *Partial Data Dependent* (PDD) saboteur. This model represents a simplification of the previous one. The probabilities for a line are dependent on the data configuration on vicinity (1-wire vicinity or 2-wire vicinity). The 1-wire vicinity model is based on the fact that the capacitance effect (which is dominant) manifests only on the neighbor line.

Fig. 1 presents the architectures for four types of saboteurs. All saboteurs consist of a random number generator (which is used to compute the probability of an error). The SAP incorporates a switch detection module, while the PDD and FDD monitor the data on the lines.

IV.   THE FAULT INJECTION CAMPAIGNS

We performed several simulation campaigns, each of them consisting of 1000 runs and data transmitted was chosen randomly for each run. The simulations have been carried out using Modelsim 10.3 commercial HDL simulator on desktop computer with Intel Core 2 Duo at 2.4 GHz and 2 GB of main memory, with Windows XP OS.

The circuit under test has been the open-source Wishbone bus, designed in Verilog HDL and available on the OpenCores site [10]. The system was simulated in the particular case of 2 master units and 5 slave units, with 32-bit data and address buses. We have simulated conventional read and write cycles. The sabotaged signals have been grouped into the following:

- Data write signals (the 32-bit unidirectional data bus from master to slave)

- Data read signals (the 32-bit unidirectional data bus from slave to master)

- Address signals – a distinction between the first 4 address bits (the ones used to select the slave) and the rest of the address bits (which are used to address within the slave)

- Master control and handshaking signals (*we, cyc, stb* and *sel*)

- Slave handshaking signals (*ack, rty, err*)

The analyzed system is depicted in Fig. 2.

The simulation campaigns and simulation times are presented in Table I. Regarding the simulation times, a simulation set consisting of 1000 executions requires less than 2 s. The gold circuit simulation requires about 1 s.

Regarding the reliability analysis of the Wishbone bus, the following conclusions can be drawn:

1. Faults affecting the most significant signals of the address line have a dramatic effect on the overall signal reliability, as these signals are used for slave selection. Therefore, an error on these signals will result in selecting a wrong slave.

2. Faults affecting master to slave control and handshaking signals (*cyc, stb* and *we*) have the following effects: wrong type of transaction (read instead of write or vice-versa), no transaction is performed (because the bus arbiter cannot grant the bus to the master which had asserted the *cyc* signal or the slave to take into consideration the request from a master), prematurely terminated transactions (due to errors on an ongoing transaction on *cyc* and *stb* signals – these signals are activated throughout an entire transaction);

3. Faults affecting the slave to master handshaking have the following effects: the bus may enter into a stand-still, as the master does not de-asserts the *cyc* signals because he has not received any *ack, rty* or *err*; a transaction may be terminated before, as the master receives a wrong ack, err, or *rty* – in case an error affects *ack*, the master may read the wrong data; longer transaction when errors appear on the *rty* signal (usually a master restarts the transaction for a *rty*).

4. Faults that affect *sel* lines and data signals affect only the data transmitted on the bus. They do not affect the transaction timing or flow.

978-1-4799-4993-9/14 $31.00 © 2014 IEEE   221

Thus, regarding the reliability of the bus, the most critical signals are the most significant bits in the address line and the control and handshaking signals.

V. CONCLUSIONS

This paper proposes SFI based reliability methods for interconnects affected by probabilistic faults. The SFI techniques used are based on saboteurs. We have developed four types of saboteurs: the simplistic probabilistic type, the switching-aware probabilistic saboteur, the full data-dependent saboteur and the partial data-dependent saboteur. The full data dependent and the partial data-dependent saboteurs are the most accurate ones, as the crosstalk noise which affects the interconnects is data dependent. We have performed several simulation campaigns analyzing the effects of probabilistic faults affecting interconnects on a Wishbone bus system, consisting of 2 masters and 5 slaves. The simulations have indicated the most critical signals in the overall reliability.

ACKNOWLEDGMENT

This work has been supported by the project "*i-RISC: Innovative Reliable Chip Designs from Low-Powered Unreliable Components*" – FP7 Grant Number 309129.

REFERENCES

[1] N. S. Nagaraj, "Interconnect process variations: theory and practice", Proceedings of the 19th International Conference on VLSI Design (VLSID), 2006

[2] H. R. Zarandi, S. G. Miremadi, A. Ejlali, "Dependability Analysis using a Fault Injection Tool Based on Synthesizability of HDL Models", Proceedings of the 18th IEEE International Symposium on Defect and Fault Tolerance in VLSI Systems (DFT), 2003

[3] J. Gracia, J. C. Baraza, D. Gil, P.J. Gil, "Comparison and Application of Different VHDL-Based Fault Injection Techniques", Proceedings of the IEEE International Symposium on Defect and Fault Tolerance in VLSI Systems (DFT), 2001

[4] J. C. Baraza, J. Gracia, D. Gil, P.J. Gil, "Improvement of Fault Injection Techniques Based on VHDL Code Modification", Tenth IEEE International High-Level Design Validation and Test Workshop, 2005

[5] D. Boning, S. Nassif, "Models of process variations in device and interconnect", Design of High-Performance Microprocessor Circuits, chapter 06, pp. 98-116

[6] S. Hasan, A. K. Palit, W. Anheier, "Fault Diagnosis of Crosstalk Induced Glitches and Delay Faults", 13th IEEE International Symposium on Design and Diagnostics of Electronic Circuits and Systems (DDECS), 2010

[7] M. Favalli, C. Metra, "TMR Voting in the Presence of Crosstalk Faults at the Voter Inputs", IEEE Transactions on Reliability, vol. 53, no. 3, september 2004

[8] A. Sanyal, A. Pan, S. Kundu, "A study on impact of loading effect on capacitive crosstalk noise" Proc. Int. Symp. On Quality Electronic Design (ISQED), 2009

[9] E. Jenn, J. Arlat, M. Rimen, J. Ohlsson, J. Karlsson, "Fault Injection into VHDL Models: The MEFISTO Tool", Proceedings 24th Annual International Symposium on Fault Tolerant Computing Systems (FTCS-24), 1994, pp 66-75

[10] OpenCores website: http://www.opencores.org

[11] K. Agarwal, D. Sylvester, D. Blaauw, "Modeling and analysis of crosstalk noise in coupled RLC interconnects", IEEE Trans. on Computer-Aided Design of Integrated Circuits and Systems, vol. 25, no. 5, 2006

TABLE I.   SIMULATION RESULTS FOR ALL CAMPAIGNS

Fault model type	Victim signal	Probability of failure	Runtime [ms]
SSP during WRITE cycle	sel	3%	1828
	sel and data	3%	1765
	adr[31:28]	3%	1812
	adr[31:28]	5%	1750
	adr[31:28]	10%	1750
	cyc, stb, we, sel	3%	1750
SSP during READ cycle	ack, err, rty	3%	1703
SAP during WRITE cycle	adr[31:28]	5% for 0->1 3% for 1->0	1766
	adr[31:28]	10% for 0->1 5% for 1->0	1766
	cyc, stb, we, sel	5% for 0->1 3% for 1->0	1750
	cyc, stb, we, sel	10% for 0->1 5% for 1->0	1781
	data	5% for 0->1 3% for 1->0	1766
	data	10% for 0->1 5% for 1->0	1782
SAP during READ cycle	ack, err, rty	5% for 0->1 3% for 1->0	1703
	ack, err, rty	10% for 0->1 5% for 1->0	1703
PDD during WRITE cycle (1-wire vicinity)	adr[31:28]	[3% ÷ 20%], depending on the transition pattern	1797
	adr[27:0]		1797
	slave select signals		1766
	cyc, stb, we, sel		1766
	ack, err, rty		1765
	data		1782
PDD during READ cycle (1-wire vicinity)	ack, err, rty	[3% ÷ 20%], depending on the transition pattern	1782
	data		1906
Gold circuit – WRITE cycle	NO fault injection	0%	1078
Gold circuit – READ cycle	NO fault injection	0%	1046

Design of a secure architecture for scalar multiplication on elliptic curves

Simon PONTIE, Paolo MAISTRI

Univ. Grenoble Alpes, TIMA, F-38031 Grenoble
CNRS, TIMA, F-38031 Grenoble
{simon.pontie, paolo.maistri}@imag.fr

Abstract—**Embedded systems support more and more features. Authentication and confidentiality are part of them. These systems have limitations that put the public-key RSA algorithm at a disadvantage: Elliptic curve cryptography (ECC) becomes more attractive because it requires less energy and less area. A lot of attacks exploit physical access on cryptographic hardware device: power analysis attacks (SPA, DPA), or timing analysis attacks. The coprocessor presented here supports all critical operations of an ECC cryptosystem and has been secured against side channel attacks.**

Keywords—Elliptic curve cryptography, side channel analysis, scalar multiplication.

I. INTRODUCTION

When a large number of entities want to securely exchange messages, asymmetric cryptography simplifies the infrastructure. This solution is usually only for complex systems. The increasing number of embedded systems encourages implementation of asymmetric cryptography in this environment. Elliptic curves in asymmetric cryptography [1] are particularly compatible with a constrained environment because their implementations need fewer resources than the RSA algorithm [2]. An ECC (Elliptic Curve Cryptography) key size of 163 bits has a comparable level of security than an RSA key size of 1024 bits [3].

We designed a secured crypto-processor that supports ECC. To explain its architecture, the paper is structured as follows. The use of Elliptic curves in cryptography is explained in the next section. In Section III, we detail three passive attacks that exploit side channel leakage. Section IV presents our countermeasure based on randomized windows in the scalar

multiplication. The architecture of the crypto-processor is presented in Section V. Our processor is compared with four others designs in Section VI. Finally, Section VII concludes the paper.

II. ELLIPTIC CURVES IN CRYPTOGRAPHY

All cryptographic protocols using elliptic curves are based on the discrete logarithm problem: reversing the scalar multiplication. Elliptic curves are based on a hierarchy of four levels (figure 1). The coprocessor design is built on this hierarchy.

In cryptography, a curve is defined by: its equation, the underlying field, and a generator point. There are two usable fields: the prime field and the binary field. Binary field support is optional in the majority of protocols, but the prime field has to be supported. In this paper, only GF(p) (prime field) is described but our processor supports both. In GF(p), the laws of addition and multiplication are the usual laws but modulo the prime number p. *0* and *1* are respectively the null and the unit elements, neutral to addition and multiplication, respectively. These field operations build the first level of the elliptic curves hierarchy. Point coordinates x and y are elements of GF(p). All the Weierstraß curves in this field are isomorphic with a reduced equation (1).

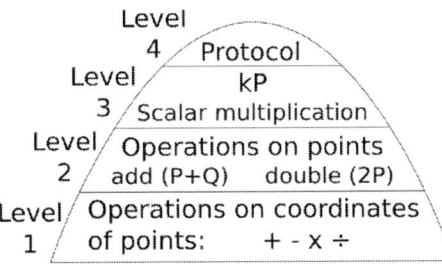

Fig. 1. Hierarchy of the ECC protocol.

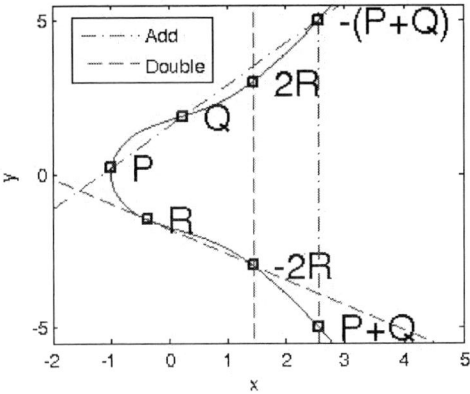

Fig. 2. Addition and doubling of point on $y^2 = x^3 + 2x + 3$.

$$y^2 = x^3 + a_4 x + a_6 \qquad (1)$$

The addition of two points on the curve can be geometrically defined. A curve on the real field (figure 2) illustrates this geometric construction. To add two points (P and Q) of the curve, the intermediate point have to be identified: this point is at the third intersection between the curve and the straight line passing through P and Q. The result is the symmetrical point of the intermediate point with respect to the abscissa axis. If P and Q are the same point, the straight line passing through P and Q is replaced by the tangent in the same point: this operation is a doubling of point, shown in Figure 2 with the doubling of the point R. These two operations build the second level of the elliptic curve hierarchy (see again figure 1). Multiple additions or doubling on the generator point may generate a certain number of points, which constitutes the *order* of the curve: it is a group. The neutral element is named infinite point.

The scalar multiplication fills the third level. This operation is the product between a scalar k and a point P. There are lots of algorithms to perform the scalar multiplication by using the two basic operations: addition and doubling of points. We present the simplest.

The algorithm "Double and Add" is used to calculate kP. It begins by the initialisation of an accumulator with the infinite point. The scalar k is scanned from the most significant bit to the least significant. For each bit, the accumulator is doubled; if the bit is a 1 then P is added to the accumulator. This algorithm consumes $n-1$ doubling and $n/2$ additions. n is the bit size of the scalar and depends on the curve *order*.

The fourth and last level is the protocol level. There are lots of cryptography protocols that use elliptic curves. All exploit the discrete logarithm problem because the group generated by the elliptic curve points is one of the few groups with no known algorithm to solve this problem in polynomial time. The secrete key is a scalar used in the scalar multiplication during the deciphering. This operation has to be secure, because it is directly linked to the secret key value. In the next section, we briefly present side channel attacks that may compromise simple designs of the scalar multiplication.

III. SIDE CHANNEL ATTACKS

In this section we present three side channel attacks. We use the weakness of two scalar multiplication algorithms to explain these attacks.

The computation time information is the easiest side channel to exploit in order to speed up the discrete logarithm problem [4]. In the algorithm "Double and Add" presented before, the number of additions is the Hamming weight of the scalar. It is easy to recover this weight from the computation time because the time needed to perform point doubling is an invariant and the number of additions is the only origin of the time variation.

There is another attack using the distinction between an addition and a doubling. This difference causes two distinct patterns of power consumption. An attacker can recover the addition and doubling order with pattern identification on a single power trace during a deciphering (figure 3). If the scalar multiplication algorithm is the "Double and Add", the attacker can recover the key. This attack is named SPA (Single Power Analysis) [5]. The figure 3 shows the power consumption of an ASIC crypto-processor using the "Double and Add" algorithm on a technology 0.35 µm. It is easy to identify the two patterns and to recover the key (179).

There are lots of scalar multiplications algorithms secure against these two attacks. For example, the Montgomery ladder, unlike "Double and Add", uses a scalar scanning right-to-left. To compute kP, it uses two intermediate points, where one is the addition between the other and P. This algorithm is secure against SPA and timing analysis attacks because it uses the same number of additions and doubling in the same order. However, this algorithm is not secure against another attack exploiting the power consumption: the DPA (Differential Power Analysis) attacks [5]. It is a statistical attack using a lot of power consumption traces of scalar multiplications between the secret key and different points P.

Others attacks exist [6], the next section presents a secure algorithm against the three attacks presented before.

IV. A SECURE SCALAR MULTIPLICATION

Our algorithm scans left-to-right the scalar with a window method. The secrete key is read digit by digit from the MSB to the LSB with a size larger than one bit. These larger digits allow saving additions. However, filling a precompute table is required before starting the scalar multiplication. This table includes all dP points, for each possible digit d.

The window method can improve the safety against statistical attacks (DPA) by randomly selecting the size of each window. However, window sizes can be recovered by counting the doublings between the additions with a SPA attack. Thus, dummy additions are inserted during kP computing to hide window sizes. Dummy additions and random window sizes break the synchronization of DPA traces. To be secure against timing analysis, each scalar multiplication always computes the same number of additions and doublings by using dummy operations.

Each window size is randomly selected between two thresholds: *WMIN* and *WMAX*. *WMAX* is the largest bit size of a digit. The precompute table length is linked to this size because the table includes dP points for each digit d between 1 and $2^{WMAX}-1$. *WMIN* is the smallest bit size of a window. The largest number of additions occurs when the scalar is cut in

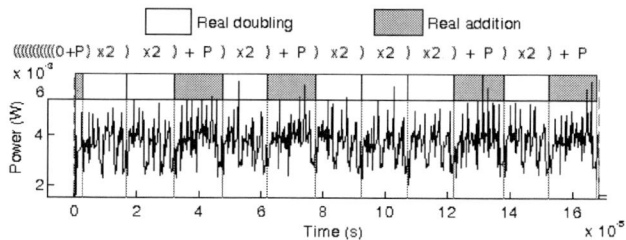

Fig. 3. A power trace of the "double and add" algorithm to calculate 179P.

windows with sizes of *WMIN* bits. This cutting is the worst-case. To always compute the same number of addition, each scalar multiplication is compared with the worst-case to count in real time the number of saved additions. All these additions, saved by the window method, will be spent by dummy additions. It is required to compute dummy additions inside the scalar multiplication computation because they are required to hide the size of the windows. This is achieved by using a counter named *dummy_cnt* which contains the number of dummy additions to be inserted during the real *kP* computation. This counter is incremented when an addition is saved, and decremented when a dummy addition is inserted. After each operation, we try to insert dummy additions: the probability to succeed increases when *dummy_cnt* is high. This probability distribution is chosen to have a small value in *dummy_cnt* at the scalar multiplication end.

V. ARCHITECTURE

Our processor architecture follows the elliptic curve hierarchy (figure 1). The first level implements the GF(*p*) arithmetics. It uses one adder-subtractor that support operands with the same sizes than the bigger prime *p* supported. A Montgomery multiplier [7] uses the adder-subtractor to compute C:

$$C = A \times B \times 2^{-r} \text{ modulo } p \qquad (2)$$

r is the bit size of *p*. It is an efficient multiplier but we need to calculate in the Montgomery domain to get rid of the 2^{-r} term. The image of *a* in this domain is *A* and is computed as:

$$A = a \times 2^{r} \text{ modulo } p \qquad (3)$$

In this domain, the Montgomery multiplier computes a classic modular multiplication because with operands *A* and *B*; the result is the image of *a* x *b* named C:

$$A \times B \times 2^{-r} = a \times 2^{r} \times b \times 2^{r} \times 2^{-r} = (a \times b) \times 2^{r} = c \times 2^{r} = C \quad (4)$$

To get into the Montgomery domain, we use the Montgomery multiplier with the two operands *a* and 2^{2r}. To get back into the classical domain, we use the operands *A* and *1*. The modular adder and the modular subtractor use the same adder-subtractor than the multiplier. They use the carry out to compare the result and the prime *p*. Adder-subtractor and multiplier on GF(*p*) are implemented in the same component to share internal registers.

The component *curve register* contains four *n* bit-size registers. The first stores *p*. The second is contains the value 2^{2r} modulo *p* to get into the Montgomery domain. The two last registers contain the curve coefficients a_4 and a_6 in the Montgomery domain. At the loading, the 2^{2r} value is checked with two multiplications in the Montgomery multiplier.

The level 2 is in Figure 4. This component computes the addition or the doubling of points. The operating part is made of the two components presented before and few registers. The control part implements addition and doubling by using formulas on points in a Jacobian representation (*X,Y,Z*). To recover the conventional affine coordinate, an inversion and few multiplications are required:

$$(x, y) = (X / Z^2, Y / Z^3) \qquad (5)$$

A third coordinate is necessary with Jacobian representation [8], but only one inversion is required (to recover the final point affine coordinates) because addition and doubling formulas are inversion free. The affine system would require one modular inversion (a very time consuming operation) for each point operation.

The components of level 2 (figure 4) include 3 registers to store the coordinates of the point *Q*, which is the accumulator point. This component also includes 5 registers to implement the two operations $Q = Q + P_i$ and $Q = 2 \times Q$. To compute these operations only one instance of the adder-subtractor-multiplier is used. Three others operations are possible: initialisation of *Q*, getting *Q* into the Montgomery domain and getting back *Q* from this domain. It is possible to compute addition and doubling with *Q* write back disabled in order to compute dummy operations. After a doubling or an addition of points, an error signal can be returned if one of the operands was the infinite point or if the two operands of the addition were equal (a doubling should be computed).

The level 3 contains four components: a pseudo-random generator, a precompute table, the operative part *kP_op*, and the control part *kP_cp* of the scalar multiplication, respectively.

The secret key is stored in a shift register within *kP_op*. At each doubling, this register is shifted to the left. At the window end, overflow bits constitute the digit *d*. This digit is used to select the *dP* point from the precompute table. Before computing the doublings of the next window, *dP* is added to the accumulator point *Q*. All these operations are implemented in *kP_cp*, which also implements the counter *dummy_cnt*. On the other hand, it is the control part that chooses the window sizes and the insertion of dummy operation insertions.

The control part *kP_cp* fills the precompute table just after loading the point *P* by using *kP_op*. All precompute points are stored in Jacobian representation and in the Montgomery

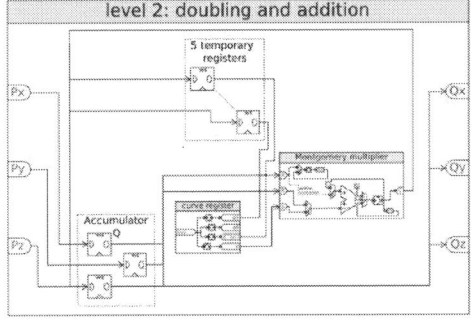

Fig. 4. Level 2 and level 1 components.

domain. At the scalar k loading, the multiplication starts. After each window, kP_cp choose the next window size by using the pseudo-random generator and launches the computation of the next digit in kP_op. After each real addition, real doubling, dummy addition or dummy doubling, the control part tries to insert a dummy addition if *dummy_cnt* is not empty. The probability to succeed increases if *dummy_cnt* value is high. Dummy additions are implemented as additions between the accumulator point Q and a point randomly chosen in the precompute table. To hide the window size of the first window and always compute the same number of doublings, dummy doublings are also inserted at the start of the scalar multiplication. Before launching the computation, the accumulator Q is initialized to a point randomly chosen in the precompute table for dummy doublings.

VI. DISCUSSION

In this part we compare our coprocessor with four others implementation on FPGA target (figure 5). Time of computation, frequency and area are normalized in reference to our coprocessor on the same FPGA with parameters as close as possible to the other.

It is possible to implement SPA protection with unified addition and doubling formulas [10], but this countermeasure is not safe against timing analysis attacks. Many implementations of ECC processors choose the Montgomery ladder algorithm to compute the scalar multiplication because it is safe against SPA and timing analysis attacks [11][12]. On the other hand, DPA attacks [5] succeed against this algorithm. The DPA attacks can be made more difficult by field operations with random computation to obtain representations of the points in a random Montgomery domain [12]. Our countermeasure is at the algorithm level and can be mixed with random representations of points.

Our coprocessor use more slices than the most of the others implementations: 14886 slices to support all curves on $GF(p)$ with a p size of 256 bits or less. The big area cause is mainly the precompute table that occupies 53% of the total are. This area can be reduced by decreasing the *WMAX* value (here 4 bits) at the cost of robustness against DPA attacks. With the same consequences, the computation time of the scalar multiplication can be reduced by increasing the *WMIN* value (here 2 bits). Dummy additions increase the number of additions but this is limited by *WMIN*.

We do not use specific components existing only in FPGA targets (DSP, BRAM), in order to keep compatibility and easy portability to the ASIC technology. To minimize the area, we use only one multiplier. The computation time can be decreased by using several multipliers but the area cost is high. Another way [9] to increase performances is to use DSP components or further optimization by supporting only few curves.

VII. CONCLUSIONS

This new implementation uses a window algorithm for the scalar multiplication to secure an ECC coprocessor against side channels attacks (SPA, DPA and timing analysis). This is a

modular solution because we can choose the secure level against DPA attacks by forcing area (via *WMAX*) or forcing time of computation (via *WMIN*). We will implement real attacks on FPGA target to measure the number of required traces to success with a DPA attacks. We will modify the $GF(p)$ arithmetic with pipelined additions to reduce the critical path and the time of computation.

ACKNOWLEDGMENT

This work has been partially supported by Labex PERSYVAL-LAB (ANR—11-LABX-0025).

REFERENCES

[1] Koblitz, Neal. "Elliptic curve cryptosystems." Mathematics of computation 48, no. 177 (1987): 203-209.

[2] Rivest, Ronald L., Adi Shamir, and Len Adleman. "A method for obtaining digital signatures and public-key cryptosystems." Communications of the ACM 21, no. 2 (1978): 120-126

[3] Lenstra, Arjen K., and Eric R. Verheul. "Selecting cryptographic key sizes." Journal of cryptology 14, no. 4 (2001): 255-293.

[4] Kocher, P. C. (1996, January). Timing attacks on implementations of Diffie-Hellman, RSA, DSS, and other systems. In Advances in Cryptology—CRYPTO'96 (pp. 104-113). Springer Berlin Heidelberg.

[5] Kocher, P., Jaffe, J., & Jun, B. (1999, January). Differential power analysis. In Advances in Cryptology—CRYPTO'99 (pp. 388-397). Springer Berlin Heidelberg.

[6] Fan, Junfeng, et al. State-of-the-art of secure ECC implementations: a survey on known side-channel attacks and countermeasures. Hardware-Oriented Security and Trust (HOST), 2010 IEEE International Symposium on. IEEE, 2010.

[7] Montgomery, P. L. (1985). Modular multiplication without trial division. Mathematics of computation, 44(170), 519-521.

[8] Cohen, H., Frey, G., Avanzi, R., Doche, C., Lange, T., Nguyen, K., & Vercauteren, F. (Eds.). (2005). Handbook of elliptic and hyperelliptic curve cryptography. CRC press.

[9] Güneysu, T., & Paar, C. (2008). Ultra high performance ECC over NIST primes on commercial FPGAs. In Cryptographic Hardware and Embedded Systems–CHES 2008 (pp. 62-78). Springer Berlin Heidelberg.

[10] Morales-Sandoval, M., et al. "A reconfigurable GF (2^M) elliptic curve cryptographic coprocessor." Programmable Logic (SPL), 2011 VII.

[11] Mahdizadeh, H., & Masoumi, M. (2013). Novel Architecture for Efficient FPGA Implementation of Elliptic Curve Cryptographic Processor Over GF (2^{163}).

[12] Lee, Jen-Wei, et al. "An efficient countermeasure against correlation power-analysis attacks with randomized montgomery operations for DF-ECC processor." Cryptographic Hardware and Embedded Systems–CHES 2012. Springer Berlin Heidelberg, 2012. 548-564.

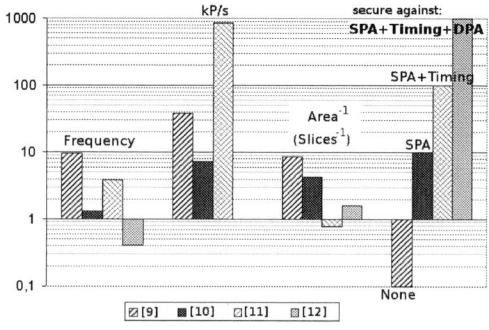

Fig. 5. Comparison between our coprocessor and four others.

ASIC design of a Phoneme Recogniser based on Discrete Wavelet Transforms and Support Vector Machines

Cutajar M., Gatt E., Grech I., Casha O.
Department of Microelectronics & Nanoelectronics
Faculty of ICT, University of Malta, Msida, Malta
michelle.cutajar.05@um.edu.mt

Abstract—This paper presents the design of an ASIC for the task of multi-speaker phoneme recognition in continuous speech environments. The phoneme recogniser is based on DWTs for feature extraction and the One-against-one SVM method, along a priorities scheme, for classification. The ASIC design was fabricated on an AMS 0.35μ CMOS C35B4C3 chip. The final ASIC design resulted into a chip size equal to 43.35mm², with the requirement of an external memory storage of size 18.25Mb. Moreover, the ASIC design of the phoneme recogniser is approximately 4 times faster than the equivalent software-based approach and consumes 12.5mW, making it appealing to mobile devices. The performance results obtained from the ASIC design confirmed that this system is a promising basis for future hardware ASR systems.

Keywords—ASIC design; phoneme recognition; speaker-independent; discrete wavelet transforms; support vector machines

I. INTRODUCTION

Automatic Speech Recognition (ASR) is becoming increasingly more popularly used in most applications in today's technologies. Various applications where ASR is, or can be employed, vary from simple tasks to more complex ones. Some of these include speech-to-text input, ticket reservations, air traffic control, security and biometric identification, gaming, home automation and automobile sectors [1][2]. Additionally, disabled and elderly persons can highly benefit from advances in the field of ASR.

During the past years, a lot of research has been carried out, with the current trend being speech recognition in speaker-independent continuous speech environments with large vocabulary. However, most of the research carried out so far focused on software applications, and not much work has been carried out on the design of hardware recognisers, which is also necessary in order to attain further improvements in the field of ASR. Over the past years, ASR was not widely used in potential applications, due to a number of limitations, such as processing power and limited hardware resources [3]. However, with today's advances in customised hardware, the design of ASR systems on-chip has become more feasible.

In this research, a phoneme recognition system based on Discrete Wavelet Transforms (DWTs) and Support Vector Machines (SVMs) was designed on a dedicated chip, in order to evaluate its potential into becoming a portable and efficient system which can be employed in portable battery-powered devices. This paper is organised as follows: Section II presents the phoneme recognition system, and Section III deals with the implementation and details of the corresponding Application-Specific Integrated Circuit (ASIC) design. The results obtained are presented and discussed in Section IV. Finally, Section V presents any concluding remarks, and future directions.

II. PHONEME RECOGNITION SYSTEM

The phoneme recognition system was designed for the task of multi-speaker phoneme recognition in continuous speech environments. The proposed system consists of three stages:- pre-processing stage, feature extraction stage, and classification stage. First, the pre-processing stage divides each phoneme instance into frames of 256 samples, and the corresponding outer frames of each phoneme instance are removed. Afterwards, the feature extraction stage extracts an eight-element feature vector of frequency band power percentages, through the utilisation of a DWT, which employs the Daubechies wavelet of order 8, and carries out seven decomposition levels on each phoneme frame. Finally, the classification stage is based on the One-against-one SVM method, along with the priorities scheme, which outputs the three most probably phoneme representations for the presented input phoneme frame, as shown in Fig. 1 [4] - [7].

The phoneme recognition system was evaluated with the TIMIT corpus [8]. The phoneme classes available in the TIMIT corpus were grouped, through the use of a confusion matrix during the recognition process, in order to identify those phonemes which are being highly mistaken with each other. As a result, the number of possible classifications was reduced to 42, some representing phonemes and others representing groups of phonemes. The accuracy achieved from the software implementation of the proposed phoneme recognition design was equal to 75.41%. For further details on the proposed phoneme recognition system, interested readers are referred to [4] - [7].

III. ASIC DESIGN

The proposed phoneme recogniser was designed as a digital ASIC design, and fabricated on an AMS 0.35μ CMOS C35B4C3, 4M/2P/HR/5V IO chip. The hardware architecture includes training data, obtained from the software design, and hence only the testing phase is executed by the hardware

---

The research work disclosed in this publication is partially funded by the Strategic Educational Pathways Scholarship Scheme (Malta). The scholarship is part-financed by the European Union – European Social Fund.

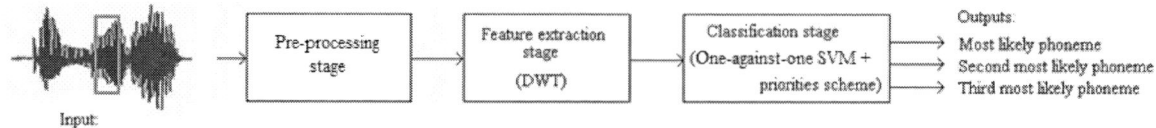

Fig. 1.   Phoneme recognition architecture

design. Apart from this, the pre-processing stage was implemented as software [9].

For the design of the hardware architecture, a fixed-point numeric representation was used. Starting with the classification stage, since the hardware-based phoneme recognition system has to recognise between 42 phoneme classes, this results in the design of 861 SVM classifiers for the One-against-one SVM method. However, if all the 861 SVM classifiers were implemented in parallel, this would have resulted in large chip size. Hence, the implementation of all the classifiers was not feasible, and instead the option taken up was the implementation of only two classifiers in parallel, and these classifiers would be used to process all of the 861 classifiers [10].

Although only two classifiers were included in the SVM architecture, all of the support vectors, their corresponding weight values, and the bias values, of all the 861 classifiers had to be stored [10]. After finding the optimal data representation for the classification stage [10], those weight values which resulted into a zero value were removed. As a result, the total amount of weights, and corresponding support vectors, were both reduced to an amount equal to 104,643, which results in a decrease of 88.9% with respect to the total amount. Nonetheless, this still resulted in a huge memory space, which could not be successfully synthesised on the ASIC design. Hence, an external memory storage of size equal to 16Mb and 2.25Mb was necessary, to store all the support vectors and the corresponding weight values, respectively. However, the ROM storing the bias values could be synthesised on-chip along with the SVM architecture, due to its small size.

Once the hardware design of the classification stage was ready, the feature extraction stage was designed based on the lifting-scheme approach [9]. The Daubechies wavelet of order 5 was considered for the design of the lifting-scheme architecture, since better accuracies were obtained when considering the latter wavelet order. Apart from this, the consideration of order 5, resulted also in a reduction in the required hardware resources, as well as in computational time, since the lifting-scheme architecture of order 5 is 1.8 times smaller than that of order 8. The processing of all the seven decomposition levels was then carried out with only one lifting-scheme block, computing all seven decomposition levels output serially. Subsequently, once the DWT function was ready, and the wavelet coefficients were obtained, the corresponding percentage energy of each DWT band was calculated resulting in an eight-element frequency band power percentage [9].

Afterwards, both hardware architectures of the feature extraction stage and the classification stage were merged together to form the finalised design of the phoneme recognition system. However, the requirement of an external memory storage led to the phoneme recognition ASIC design being pad-limited. In view of this fact, the classification stage was changed such that the two SVM classifiers had to access the external memory storage one at a time. In this manner, the amount of input pins to read from the external memory storage was reduced by half. Therefore, the resultant area consumption was now significantly reduced, as the number of pads was decreased. However, this came at the expense of an increase in computational time by approximately two times as much, since most of the parallelism of the One-against-one SVM architecture was now lost.

The finalised ASIC design had a total number of 233 I/O pins. However, from these pins, 163 pins were allocated to access the external memory storages, and 20 of these I/O pins were auxiliary testing I/O connections.

The ASIC design for the phoneme recognition system is shown in Fig. 2. The dimensions of the ASIC design are 6.54mm by 6.63mm, which result in a total area consumption equal to 43.35mm^2. A 257 pin Ceramic Pin Grid Array (CPGA257) was then chosen for packaging, due to the amount of pins the ASIC design has.

IV.   RESULTS AND DISCUSSIONS

For the analysis of the fabricated ASIC chip, the HTG-V6-PCIE Xilinx Virtex 6 SX475T FPGA (HTG-V6-PCIE-SX475T) evaluation board was utilised for the storage of the support vectors and weight values of the One-against-one SVM architecture, as well as for the input of signals to the ASIC design.

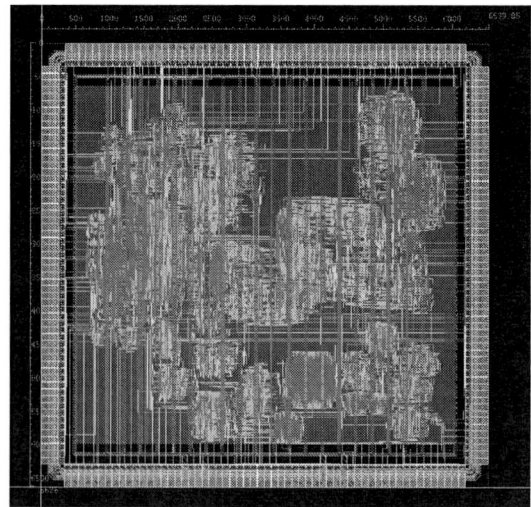

Fig. 2.   ASIC design of phoneme recognition system

The phoneme recognition system implemented on the ASIC was verified for correct performance by choosing two randomly selected phoneme test frames, and inputting these for recognition. The selected phonemes were test sample numbers 35 and 4223, representing phonemes /ih/ and /ao/ respectively. Test sample number 35, was not correctly recognised by both the software- and hardware-based approaches, whereas the other phoneme sample was correctly recognised by both implementations. During the software implementation, each of the 42 grouped phoneme classes was assigned a number. Starting with test sample number 35, which corresponded to a phoneme frame sample of phoneme /ih/, the assigned phoneme class number was 25. However, the three phoneme representations obtained from the hardware design were phoneme class numbers 34, 32 and 27, representing phonemes /ix/, /er/, and /ey/, respectively. All of these phonemes, along with phoneme /ih/, are from the category of vowels. Hence, since these phonemes are from the same category, these have similar frequency spectra, and therefore can be easily misrecognised with each other [5]. The results obtained from the ASIC chip, for test sample number 35 are shown in Fig. 3. Considering now the second example, which was test sample number 4223, this test sample represented a phoneme frame of phoneme /ao/, and the assigned phoneme class number was 28. This test sample was correctly recognised by the hardware design of the phoneme recognition system, as shown in Fig. 4.

The total area consumption of the finalised ASIC design was equal to $43.35mm^2$, with the requirement of an external memory storage of size 18.25Mb, for the storing of the support vectors, and the corresponding weight values, of the One-against-one SVM architecture. As regards to the power consumption, both static and dynamic power consumptions were evaluated directly from the ASIC chip. From the results obtained, the static power of the ASIC chip was found to be negligible, whereas the dynamic power consumption was found to be equal to 12.5mW. These readings were taken when the ASIC design was operating at maximum frequency, which was found to be equal to 6.38MHz. Hence, these results show that the ASIC design is adequate for use in battery-powered devices due to the low power consumption it requires. Moreover, comparing the ASIC design to the software approach, the laptop's power specifications are DC 19V 4.22A, with an Intel Core i7 2.67GHz processor, 6GB RAM, and running a Windows OS 64-bit platform. Hence, the ASIC design was also able to achieve lower power consumption than the software implementation.

Additionally, a number of compromises had to be taken, due to the requirement of an external memory storage. This was done mostly at the expense of an increase in the final computational time. However, the ASIC design was still able to obtain a lower computational time when compared to the software design, which was based on the use of MATLAB R2009b (64-bit), on a Windows OS 64-bit platform, Intel Core i7 2.67GHz processor, 6GB RAM. The computational time achieved by the phoneme recognition system designed on chip was equal to 16.6ms, approximately half of the computational time the software-based approach requires to classify a phoneme frame. However, the hardware design can actually be approximately 4 times faster than the software-based approach, since most of the parallelism in the One-against-one SVM architecture was lost, in order to reduce the amount of I/O pads to access the external memory storage, resulting in a lower area consumption and chip cost.

However, the accuracy obtained from the hardware architecture of the phoneme recognition system resulted in a decrease of 19.6%, with respect to the software design. This difference in recognition rate can be attributed to the One-against-one SVM architecture, and the considerable amount of support vectors a SVM architecture needs, for this to achieve adequate performance [10]. Hence, a method which reduces the amount of support vectors during the training process, but which still obtains adequate recognition rate from the multiclass SVM architecture, needs to be found. This would then allow the accuracy attained from the respective hardware-based approaches to be more approximate to those achieved from the corresponding software implementations. Furthermore, with the reduction of support vectors, and the advance of ASIC memory model generators, the inclusion of support vectors as internal memory on-chip, as well as of their corresponding weight values, will be more feasible. As a result, lower computational times and power consumption will be achieved, due to the elimination of any latency related to the utilisation of external memory storage.

A fair comparison between the designed hardware recogniser and other hardware designs of ASR systems is very difficult, due to the consideration of different hardware technologies, implementation styles, and lack of complete information. The phoneme recognition ASIC design presented here, is compared to two hardware ASR systems, one based on ANN, and another one based on HMM, as shown in Table I. Comparing these two research designs to the proposed phoneme recogniser, the power dissipation of the proposed phoneme ASIC design is lower. Nonetheless, the hardware designs proposed in [11] and [12], were evaluated as word recognisers, and thus an equitable comparison with respect to the power consumption cannot be concluded. However, it is also important to point out, that the research work presented in [11], considered only the design of the output probability calculation on hardware, and not the whole HMM architecture. Additionally, even the semi-custom design presented in [12], considered only the implementation of Discriminative Bayesian Neural Network (DBNN). However, in this research paper, not only the classification stage of the phoneme recognition system, which was based on a multiclass SVM method, was implemented on hardware, but also the DWT algorithm for feature extraction. As regards to the speed performance, both hardware designs presented in [11] and [12], achieved a higher operating frequency. However, the computational time of the ASIC design presented in this paper, had to be reduced approximately by half, in order to reduce the total area consumption, and hence cost, which resulted from the significant amount of I/O pads which were initially required in order to access the external memory storage. Hence, the actual maximum frequency at which the hardware-based design of the proposed phoneme recognition system can operate, is approximately twice the 6.38MHz.

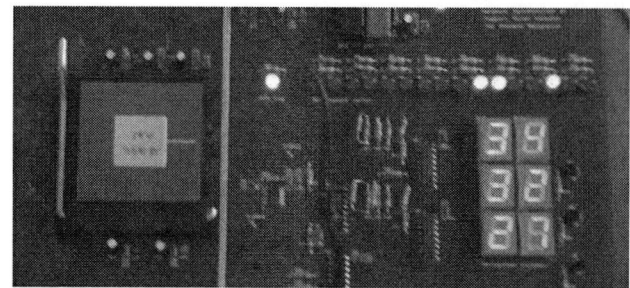

Fig. 3. ASIC chip result for test sample no. 35, phoneme /ih/

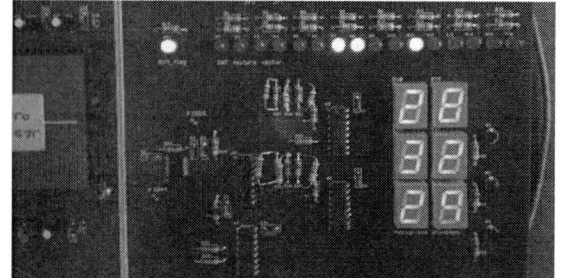

Fig. 4. ASIC chip result for test sample no. 4223, phoneme /ao/

TABLE I.   COMPARISON TO OTHER ASR HADRWAREW DESIGNS

Ref no.	Research work	ASIC technology	ASIC design	Area consumption	Real-time operation frequency	Power consumption
Proposed design	phoneme recogniser	AMS 0.35µ CMOS C35B4C3, 4M/2P/HR/5V IO chip	DWT, along with One-against-one SVM and a priorities scheme	43.35mm^2, external memory storage 18.25Mb	6.38MHz	12.5mW
[11]	word recogniser	coprocessor: Xilinx Spartan-3A DSP XC3SD3400A or IBM 0.13µm and a microcontroller: SAMSUNG S3C44b0X, ARM7 processor	Output probability calculation of the CHMM	coprocessor: IBM - 3.24mm^2	coprocessor: FPGA - 10MHz microcontroller: 40MHz	coprocessor: FPGA - 15.2mW microcontroller: 10mW
[12]	word recogniser	semi-custom design; 0.35µm 4 metal layer CMOS	DBNN architecture	20.29 mm^2, external memory storage 2.5Mb	10MHz	38mW

Hence, from the above comparison, it can be concluded, that the proposed phoneme hardware recogniser compares favourably well to current hardware ASR systems.

V. CONCLUSIONS

In this paper, the design on an ASIC of the phoneme recognition system, which was proposed in previous research papers [4] - [7], is presented. The ASIC design was fabricated on an AMS 0.35µ CMOS C35B4C3 chip. The finalised ASIC design resulted into a total chip size equal to 43.35mm^2, with an external memory storage of size 18.25Mb. Furthermore, the ASIC design of the phoneme recognition system can provide a speed approximately 4 times faster than the corresponding software-based approach, and consumes only 12.5mW, making it appealing to mobile devices.

All of the results obtained lead to the conclusion, that the ASIC design of the proposed phoneme recognition system already showed great potential for future hardware designs of ASR systems. However, first and foremost, a method which reduces the amount of support vectors during the training process of SVM algorithms, and still attains satisfactory recognition rates, needs to be found. This will result in lower computational time, due to the elimination of any propagation delays related to the utilisation of external memory storage, as well as in the power consumption. Additionally, the accuracy obtained from the hardware-based approach will approximate more to that achieved from the software implementation, since up to this stage a decrease in accuracy of 19.6% was obtained, when compared to the software implementation.

REFERENCES

[1] S. Ranjan, "A discrete wavelet transform based approach to Hindi speech recognition," in *Signal Acquisition and Processing, 2010. ICSAP '10. International Conference on* , Bangalore, 2010, pp. 345-348.

[2] S.B. Junior, R.C. Guido, S. Chen, L.S. Vieira, and F.L. Sanchez, "Improved Dynamic Time Warping based on the Discrete Wavelet Transform," in *Multimedia Workshops, 2007. ISMW '07. Ninth IEEE International Symposium on*, Taichung, Taiwan, 2007, pp. 256-263.

[3] L.D. Paulson, "Speech recognition moves from software to hardware," *Computer*, vol. 39, no. 11, pp. 15-18, November 2006.

[4] M. Cutajar, E. Gatt, I. Grech, O. Casha, and J. Micallef, "Comparison of different multiclass SVM methods for speaker independent phoneme recognition," in *Communications, Control and Signal Processing (ISCCSP), 2012 5th International Symposium on*, Rome, 2012.

[5] M. Cutajar, E. Gatt, I. Grech, O. Casha, and J. Micallef, "Support Vector Machines with the Priorities Method for Speaker Independent Phoneme Recognition," in *Signal Processing and Information Technology (ISSPIT), 2011 IEEE International Symposium on*, Bilbao, 2011, pp. 409-414.

[6] M. Cutajar, E. Gatt, I. Grech, O. Casha, and J. Micallef, "Discrete wavelet transforms with multiclass SVM for phoneme recognition," in *EUROCON, 2013 IEEE*, Zagreb, Croatia, 2013, pp. 1695-1700.

[7] M. Cutajar, E. Gatt, I. Grech, O. Casha, and J. Micallef, "A study on pitch variation on the use of DWT with SVM for speaker independent phoneme recognition," in *Communications, Control and Signal Processing (ISCCSP), 2012 5th International Symposium on*, Rome, 2012.

[8] (1990, October) The DARPA TIMIT Acoustic-Phonetic Continuous Speech Corpus. NIST Speech Disc CD1-1.1.

[9] M. Cutajar, E. Gatt, I. Grech, O. Casha, and J. Micallef, "Design of a Hardware-based Discrete Wavelet Transform Architecture for Phoneme Recognition," in *ISCCSP'14 Proceedings*, Athens, Greece, 2014.

[10] M. Cutajar, E. Gatt, I. Grech, O. Casha, and J. Micallef, "Hardware-based support vector machine for phoneme classification," in *EUROCON, 2013 IEEE*, Zagreb, Croatia, 2013, pp. 1701-1708.

[11] Peng Li and Hua Tang, "Design of a Low-Power Coprocessor for Mid-Size Vocabulary Speech Recognition Systems," *Circuits and Systems I: Regular Papers, IEEE Transactions on*, vol. 58, no. 5, pp. 961-970, May 2011.

[12] Jhing-Fa Wang, Jia-Ching Wang, An-Nan Seun, Chung-Hsien Wu, and Fan-Min Li, "VLSI Architecture and Implementation for Speech Recogniser Based on Discriminative Bayesian Neural Network," *IEICE Transactions on Fundamentals of Electronics, Communications and Computer Sciences*, vol. E85-A, no. 8, pp. 1861-1869, August 2002.

ESD Co-Design methodologies for RF and mmW circuits

Roc Berenguer, PhD

Senior RFIC designer
CEIT, Spain

One of the bottlenecks of introducing CMOS RF and millimeter wave (mmW) circuits to the market is their susceptibility to electrostatic discharge (ESD). It is due to both gate oxide breakdown and junction degradation related problems, caused by the thinner gate oxide thickness and increased doping levels of state-of-the-art CMOS processes. On the other hand, the increasing operation frequency range makes the ESD protection design a great challenge. The parasitic effects of the on-chip ESD devices on the RF path often degrade the power gain, noise figure and linearity of the RF and mmW front-ends, especially the low noise amplifier (LNA) input stage. Therefore, there is a strong motivation to simultaneously optimize RF performance and provide ESD protection for RF and mmW circuits. The talk will introduce to RFIC designers the state-of-the-art techniques, such as inductive cancellation or impedance isolation techniques, to simultaneously fulfill the requirements of both the RF performance and ESD protection.

Two design cases will be presented. First, the design of a decreasing-sized □-model electrostatic discharge (ESD) protection structure applied to protect against ESD stresses at the RF input pad of an ultra-low power CMOS front-end operating in the 2.4-GHz industrial–scientific– medical band. This structure can sustain a human body-model ESD level higher than 16 kV and a machine-model ESD level higher than 1 kV without degrading the RF performance of the front-end. Second, an area efficient Electrostatic Discharge (ESD) protection structure to protect the RF input PAD of a 77 GHz low noise amplifier (LNA) in a 65nm CMOS process. The proposed RF-ESD protection co-design using an inductive cancellation method can handle Transmission Line Pulse (TLP) ESD currents up to more than 2.7A without RF performance degradation, which corresponds to an equivalent 4.05 kV voltage level of the human body model (HBM).

978-1-4799-4993-9/14 $31.00 © 2014 IEEE

Filterless millimetre-wave optical generation using optical phase modulators without DC bias

Rabiaa Guemri, Frédéric Lucarz, Daniel Bourreau,
Camilla Kärnfelt, Jean-Louis de Bougrenet de la
Tocnaye

Optics Department, Télécom Bretagne
Technopôle Brest-Iroise CS 83818
29238 Brest Cedex 3, France

E-mail : Rabiaa.Guemri@telecom-bretagne.eu

Trevor Hall

Centre for Research in Photonics, PTLab,
University of Ottawa, 800 King Edward Avenue,
Ottawa, Ontario K1N 6N5, Canada

Abstract—A tunable millimetre-wave generator using optical phase modulators with no DC bias and no filters (neither RF nor optical filters) is proposed in this paper. A 60 GHz RF signal is optically generated by multiplying the frequency of an input RF signal at 7.5 GHz by a factor of 8. The electrical suppression ratio is around 70 dB as shown by simulations.

Keywords—millimetre-wave, optical generation, phase modulators, radio-over-fibre

I. INTRODUCTION

The evolution of millimetre-wave communications requires efficient methods to generate and distribute high-frequency signals in wireless access networks. Electrical generation of millimetre-wave (mm-wave) signals suffers from limitations due to high propagation losses in the air, as well as in waveguides. Besides, electrical techniques are not particularly well adapted to fibre-supported wireless networks. Over the last decade, many research activities have been conducted to optically generate and transport mm-wave signals [1] to take advantage of huge bandwidths and low attenuations of optical fibres in the context of Radio-over-Fibre (RoF) [2]-[5]. The concept of optical generation is to heterodyne two optical signals spectrally separated from each other by the required mm-wave frequency, so as to generate an RF signal at the desired frequency upon photo-detection. A plurality of photonic techniques has been proposed so far, such as: *dual-mode lasers* [6], *optical phase locking* [7], *optical injection locking* [8], *optical injection phase locking* [9], *four-wave mixing* [10]-[11] *and stimulated Brillouin scattering* [12]. The most attractive solution to generate two optical signals to be heterodyned is based on the external modulation of a single laser source, mainly for its simplicity and tunability in frequency. Moreover, as the heterodyned signals are issued from the same laser source, they are advantageously coherent, thereby resulting in reduced phase noise. Many variants of this technique were studied achieving different multiplication factors i.e the ratio between the frequencies of the RF output signal and that of the RF input signal. A frequency doubler based on Mach-Zehnder modulators (MZM) was first demonstrated in 1992 [13]. Then, higher multiplication factors (4 and 6) were achieved [14]-[20]. In order to reduce the costs and use low-frequency components, some techniques wherein RF input frequency is multiplied by a factor of 8 were studied [21]-[23].

These techniques usually use optical Notch filters to suppress optical carriers or RF filters to suppress unwanted harmonics that arise from heterodyning, otherwise the suppression ratio would be limited. Moreover, such techniques rely on MZMs that need DC bias and whose half-wave voltage (V_π) depends on the frequency, thus hindering the frequency tunability of the system. In addition, these systems need control circuits to readjust DC bias voltage, which shifts or drifts during operation time of modulators.

In this paper, we propose, by simulation, a filterless method without any DC bias to optically generate millimetre-wave signals, using an input RF signal at a low frequency that will be naturally multiplied by a factor of 8 at the output of the system with a suppression ratio of 70 dB.

II. SIMULATION SETUP AND RESULTS

The proposed system setup was simulated under VPI TransmissionMaker^{TM} as illustrated in Fig. 1. This system is based on the use of four optical phase modulators (PM) arranged in parallel. Each PM has one RF input and one optical input but with no input for DC bias. In this setup, a continuous wave optical signal provided by a laser source is split in four parts, so that each part is injected at the optical input of each PM.

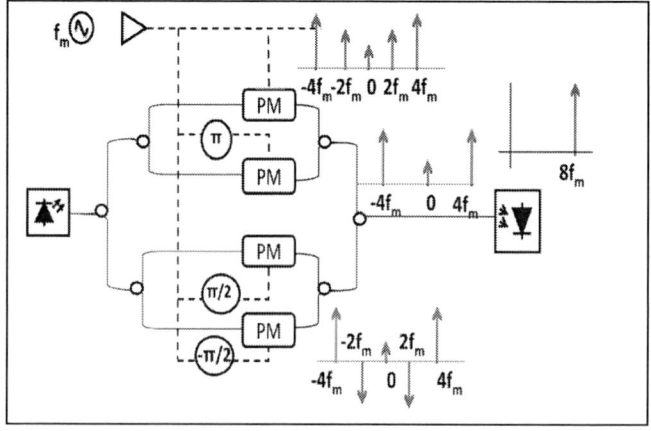

Fig. 1. Setup of an optical millimetre-wave generator by multiplication by a factor of 8, as simulated under VPI Transmission Maker^{TM}

An input 7.5 GHz RF signal (f_m) is split into four signals with equal powers. As depicted in Fig. 1, these four signals are then phase-shifted with respect to each other, as follows: the second signal is shifted by π with respect to the input signal, the third by $\pi/2$ and the third by $3\pi/2$, all of which are injected to the RF input of the four phase modulators respectively. The outputs of the four optical modulators are then combined together (Fig. 1) to produce an optical field at the input of the photodiode, whose amplitude E is given by the following formula:

$$E = \frac{1}{2}(1-\alpha)E_0 \sum_{n=-\infty}^{+\infty} J_{4n}(\varphi V_{RF}) \cdot \{\cos(\omega_0 t + 4n\,\omega_M t) + \cos(\omega_0 t - 4n\,\omega_M t)\}$$

With,
E_0 Amplitude of the optical signal

α Insertion loss

ω_M Angular frequency of electrical signal

ω_0 Angular frequency of optical signal.

With the above configuration, harmonics of orders ± 1, ± 2, ± 3 are suppressed at the input of the photodiode. Thus, the optical signal is composed out of $\pm 4^{th}$ order harmonics situated at 30 GHz symmetrically around the optical carrier, as shown in the optical spectrum represented in Fig. 2. This signal is then injected to the photodiode.

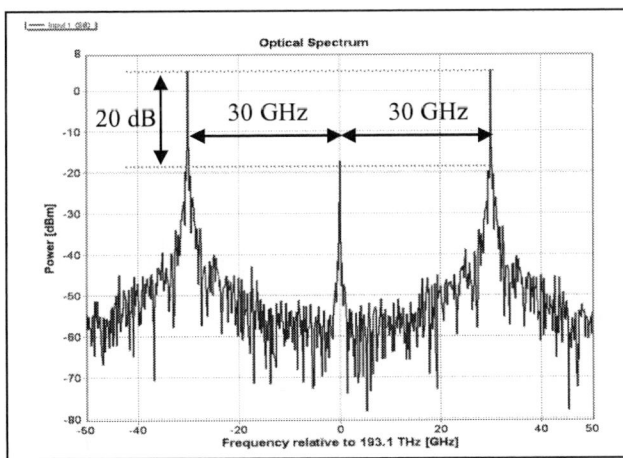

Fig. 2. Simulated optical spectrum at the input of the photodiode

According to simulations and mathematical calculations, the optical carrier should be reduced, in order to maximize the power of the 60 GHz RF signal at the output of the photodiode and minimize the amplitude of unwanted harmonics. This is done by minimizing the value of the zero-order Bessel function (J_0) while maximizing the value of the 4th order of the same function (J_4). The value of the Bessel function order depends on the value of the RF input power as explained in the previous equation. For a given value of input power, J_4 could be maximum while J_0 is minimum as shown in Fig. 3.

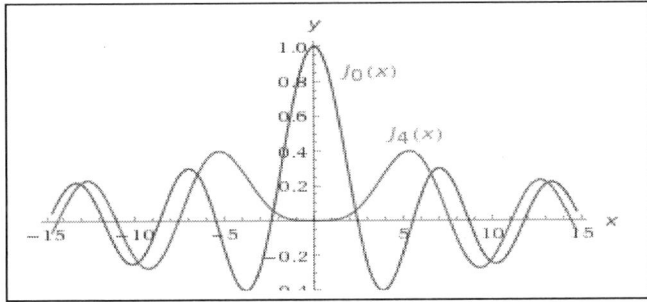

Fig. 3. Bessel function for different orders

With the proposed model, the optical carrier is rejected by 20 dB (Fig. 2). After photo-detection, the electrical spectrum is composed of an RF signal at 60 GHz with a harmonics suppression ratio of 70 dB with respect to 30 GHz harmonic (Fig. 4). Even though the optical carrier is not completely suppressed as we did not use any optical filter, an electrical suppression ratio of 70 dB is obtained, without any RF filtering either.

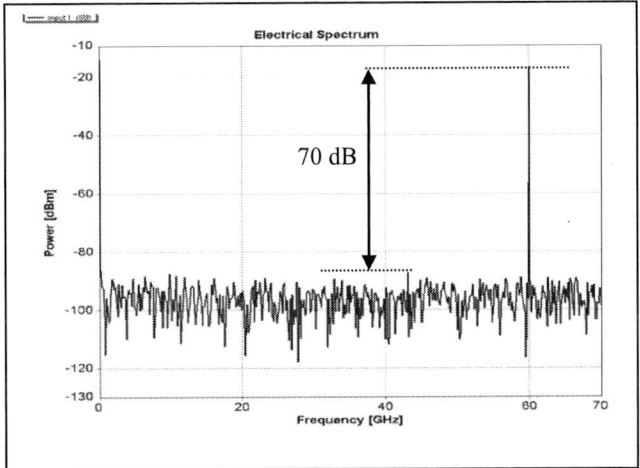

Fig. 4. Simulated electrical spectrum at the output of the photodiode

These results are obtained in ideal conditions. In practice, as RF couplers are used [24]-[25], power imbalance (due to the errors in the splitting ratio) and phase errors in hybrid couplers can occur. To study the impact of power imbalance, the amplitude of the four harmonics (15 GHz, 30 GHz, 45 GHz, 60 GHz) is presented for a splitting ratio comprised between 50%-50% to 60%-40% as shown in Fig. 5. If there is no imbalance between the branches (50%-50%), a harmonic suppression of 70 dB is achieved. With a splitting ratio of 52%-48%, the suppression ratio reduces to 20 dB, due to the 45 GHz harmonic that becomes important. In commercial hybrid couplers, values of splitting ratio are of the order of 48%-52% [26].

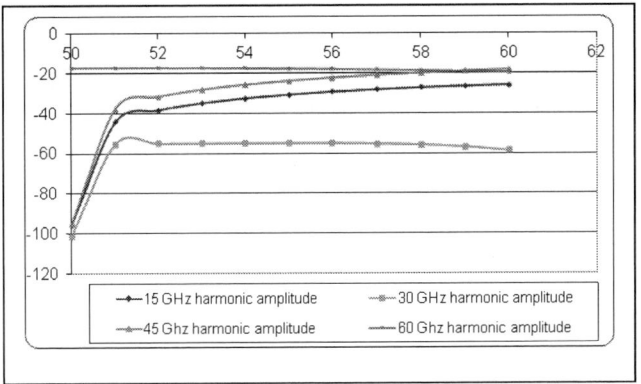

Fig. 5. Impact of input power imbalance on the generated harmonics

To study the impact of phase errors on this system, a phase error of ± 5 degree is introduced in the simulations. In Fig. 6, the impact of a phase error from 0 to 5 degree is presented. For a negative phase error (-5 to 0 degree), the behaviour is symmetric. With a phase error of 0.5 degree, a suppression ratio of more than 30 dB is obtained. In commercial hybrid couplers, typical values of phase errors are about ~2 degrees

[26]. With a phase error of more than 2 degrees, the suppression ratio starts to become less than 20 dB.

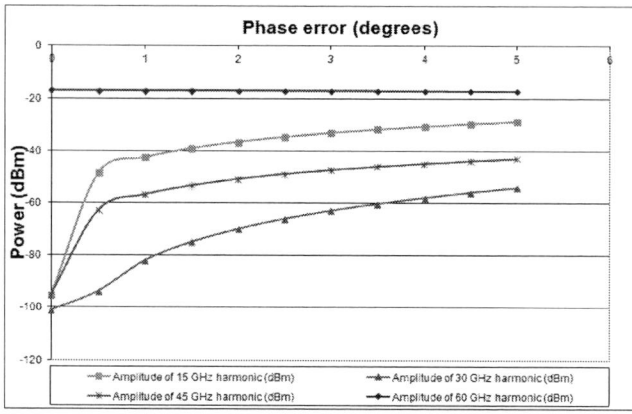

Fig. 6. Impact of the phase error on the generated harmonics

CONCLUSIONS

The proposed millimetre-wave generator can be easily tuned by adjusting its input RF frequency that is fed to the optical phase modulators. Furthermore, it is filterless in that it does not require any form of filtering (neither optical nor RF) and does not need any DC bias. This system enables to generate a 60 GHz signal by multiplying the frequency of an input RF signal (7.5 GHz) by a factor of 8, with a harmonic suppression ratio of 70 dB if no phase error or power imbalance is considered.

Sensitivity to phase errors, and input power imbalance were simulated showing that with 2% of power imbalance or with 2-degree phase errors, a suppression ratio of 20 dB can still be obtained. Experimental work including spectral analysis and phase noise measurements will be carried out shortly in order to confirm these results.

REFERENCES

[1] J. Yao, "Microwave Photonics", Journal of Lightwave Technology, Feb. 2009, vol. 27, no. 3, pp. 314-335.

[2] T. Hall, R. Maldonado-Basilio, S. Abdul-Majid, J. Seregely, R. Li, I. Antoine-Pérez, H. Nikkhah, F. Lucarz, J.L De Bougrenet De La Tocnaye, B. Fracasso, P. Pajusco, C. Karnfelt, D. Bourreau, M. Ney, R. Guemri, Y. Josse, H. Liu, "Radio-over-fibre access for sustainable digital cities", Annales des Télécommunications, Jan. 2013

[3] R. Guemri, H. Liu, I. Jäger, D. Bourreau, C. Kärnfelt, F. Lucarz, "Radio-over-Fibre transmission of multiple wireless standards for digital cities: exploiting the new tramway infrastructure", 3rd International Conference on Access Networks, June 2012.

[4] C. Kärnfelt, M. Ney, D. Bourreau, A. Bikiny, G. Guevel, Y. Paugam, F. Gallée, S. Meyer, J. Guillory, A. Pizzinat, B. Charbonnier, O. Bouffant, G. Delas, H.W. Li, E. Tanguy, M. Brunet, G. Lirzin, A. Chousseaud, C. Algani, A.L Billabert, J.L Polleux, C. Capena, G. Gougeon and V. Gouin, "A 60 GHz radio-over-fiber home area network project", International Symposium on Green Radio-over-Fibre and All-optical Technologies for Wireless Access Networks GROWAN, 15-17 June 2011.

[5] H. Liu, C. Kärnfelt, F. Lucarz and D. Bourreau, "Pico-cellular system using radio-over-fibre technology for wireless high bit rate communications", International Symposium on Green Radio-over-Fibre and All-optical Technologies for Wireless Access Networks, 15-17 June 2011.

[6] X. Chen, Z. Deng, and J. P. Yao, "Photonic generation of microwave signal using dual-wavelength single longitudinal mode fiber ring laser," IEEE, Microwave Theory and Techniques, Feb. 2006, vol.54, no. 2, pp. 804–809.

[7] U. Gliese, T. Nielsen, M. Bruun, E. Christensen, K. Stubkjaer, S. Lindgren, and B. Broberg, "A wideband heterodyne optical phase locked loop for generation of 3–18 GHz microwave carriers," IEEE Photonics Technology Letters, Aug. 1992, vol. 4, no. 8, pp. 936–938.

[8] L. Goldberg, H. Taylor, J. Weller, and D. Bloom, "Microwave signal generation with injection locked laser diodes," Electronics Letters, Jun. 1983, vol. 19, no. 13, pp. 491–493.

[9] A. Bordonaalli, C. Walton, and A. Seeds, "High-performance phase locking of wide linewidth semiconductor lasers by combined use of optical injection locking and optical phase-lock loop," Journal of Lightwave Technology, Feb. 1999, vol. 17, no. 2, pp. 328–342.

[10] Q. Wang, F. Zeng, H. Rideout, and J. Yao "Millimeter-wave generation based on four-wave mixing in an SOA", Microwave Photonics, 2006, pp. 1-4.

[11] L. Xu, C. Li, S. Lo and H. Tsang, "Millimeter wave generation using four wave mixing in silicon waveguide", OptoElectronics and Communications Conference (OECC), 2010, pp. 860-861.

[12] M. Junker, T. Schneider, K. Lauterbach, R. Henker, M. Ammann and A. Schwarzbacher, "High Quality Millimeter Wave Generation via Stimulated Brillouin Scattering", Lasers and Electro-Optics, 2007, pp. 1-2.

[13] J. O'Reilly, P. Lane, R. Heidemann, and R. Hofstetter, "Optical generation of very narrow linewidth millimetre wave signals," Electronics Letters,1992, vol. 28, no. 25, pp. 2309–2311.

[14] G. Qi, J. P. Yao, J. Seregelyi, S. Paquet, and C. Belisle, "Generation and distribution of a wide-band continuously tunable millimeter-wave signal with an optical external modulation technique," IEEE Microwave Theory and Techniques, Oct 2005, vol. 53, no. 10, pp. 3090-3097.

[15] J. Zhang, H. W. Chen, M. H. Chen, T. L. Wang, and S. H. Xie, "A photonic microwave frequency quadrupler using two cascaded intensity modulators with repetitious optical carrier suppression," IEEE Photonics Technology Letters, Aug 2007, vol. 19, no. 13-16, pp. 1057-1059.

[16] C. T. Lin, P. T. Shih, J. Chen, W. Q. Xue, P. C. Peng, and S. Chi, "Optical millimetre-wave signal generation using frequency quadrupling technique and no optical filtering," IEEE Photonics Technology Letters, June 2008, vol. 20, no. 9-12, pp. 1027-1029.

[17] J. Yu, Z. Jia, T. Wang, and G. Chang, "Centralized lightwave radio over-fiber system with photonic frequency quadrupling for high-frequency millimeter-wave generation", IEEE Photonics Technology Letters, Oct. 2007, vol. 19, no. 19, pp. 1499–1501.

[18] J. P. Yao and H. Chi, "Frequency quadrupling and upconversion in a radio over fiber link," Journal of Lightwave Technologies, Aug. 2008, vol. 26, no. 15.

[19] M. Mohamed, X. Zhang, B. Hraimel and K. Wu, "Frequency sextupler for millimeterwave over Fiber systems", Optics Express, June 2008, vol. 16, no. 14, pp. 10141–10151.

[20] S. L. Pan and J. P. Yao, "Tunable sub-terahertz wave generation based on photonic frequency sextupling using a polarization modulator and a wavelength-fixed notch filter", IEEE Microwave Theory and Techniques, Jul. 2010, vol. 58, no. 7, pp. 1967-1975.

[21] Y. Zhang and S. Pan, "Experimental Demonstration of Frequency-Octupled Millimeter-wave Signal Generation Based on a Dual-Parallel Mach-Zehnder Modulator", Microwave Workshop Series on Millimeter Wave Wireless Technology and Applications (IMWS), Sept. 2012, pp.1-4.

[22] J. Ma, X. Xin, J. Yu, C. Yu, K. Wang, H. Huang, and L Rao "Optical millimeter wave generated by octupling the frequency of the local oscillator", Journal of Optical Networking, 2008, vol. 7, no. 10, pp. 837-845.

[23] W. Li, and J. Yao, "Microwave Generation Based on Optical Domain Microwave Frequency Octupling", IEEE Photonics Technology Letters, Jan. 2010, vol. 22, no. 1, pp. 24 – 26.

[24] B. Della, E. Daniel, C. Person, D. Bourreau, S. Toutain, "High Performance Lange Coupler", Electronics Letters, Oct. 1992, pp. 1997-1998.

[25] V.K. Velidi, G.Shankar, K. Divyabramham, S. Sanyal, "Compact coupled line quadrature hybrid coupler with enhanced balance bandwidth", Applied Electromagnetics Conference (AEMC), Dec. 2011, pp. 1-4.

[26] Microwave Power Dividers and Couplers Tutorial Overview and Definition of Terms
http://www.markimicrowave.com/Assets/appnotes/microwave_power_dividers_and_couplers_primer.pdf

A Digitally Controlled Threshold Adjustment Circuit in a 0.13μm SiGe BiCMOS Technology for Receiving Multilevel Signals up to 80Gbps

Timothy De Keulenaer
Ghent University,
INTEC/IMEC
Ghent, Belgium
Email: timothy.dekeulenaer@intec.ugent.be

Guy Torfs
Ghent University,
INTEC/IMEC
Ghent, Belgium

Ramses Pierco,
Johan Bauwelinck
Ghent University,
INTEC/IMEC
Ghent, Belgium

Abstract—**In this paper, a high bandwidth digitally controlled threshold adjustment circuit is proposed which can be used for demodulating high-speed multi-level signals. Simulations of the bandwidth are presented together with measurements of the control currents to indicate the threshold adjustment capability. A bandwidth above 80GHz in a 0.13μm SiGe BiCMOS technology and a threshold tunable between ±160mV in steps of 0.6mV is achieved, allowing very precise control of the threshold level. This allows the circuit to accurately position the threshold on the eye-crossing of a high speed multi-level signals. By applying this circuit to demodulate a duobinary signal over a 40GHz channel, a data rate of up to 80Gbps can be achieved.**

I. Introduction

Advances in the data rate of modern communication systems has lead to the need for an increase in speed of inter-chip communication. Typically, a non-return-to-zero (NRZ) signal is transmitted over a PCB transmission line (microstrip, grounded coplanar waveguide), but, with rising speed, the limited bandwidth of the channel (the transmission line) and the maximum bandwidth that can be achieved by the chip technology (indicated by the maximum transition frequency f_t) impose a limitation on the maximum data rate. To counter this limitation, multi-level signalling such as PAM-4 or duobinary have been used in recent papers [1]–[3] to demonstrate data-rates up to 25Gbps across electrical backplanes.

However, although more advanced modulation schemes require less bandwidth, the processing needed on both the transmitting and the receiving chip significantly increases. Among others, a threshold adjustment circuit (TAC) operating at high speed is needed at the receiving end to seperate different symbol levels. This paper presents a 80GHz TAC designed in a 0.13μm SiGe BiCMOS technology using a current DAC for threshold adjustment. The large bandwidth and the fine threshold control of the presented circuit allows duobinary transmission up to 80Gbps when signalling over a 40GHz channel.

The structure of this paper is the following: in Section II the topology of a standard non-clocked duobinary receiver is discussed, Section III presents the circuit that is used to achieve a high speed threshold adjustment, the DACs used in the control of the threshold level are shown in Section IV,

Section V discusses the fabrication of the chip and further work that has to be done and Section VI summarizes the results and concludes this paper.

II. Non-clocked Duobinary Receiver

One of the more promising modulation schemes to reduce the required channel bandwidth without adding too much chip complexity is the duobinary scheme discussed in [4]. A receiver structure as well as the typical 3-level eyediagram of this modulation scheme is shown in Fig.1. Due to the high speeds, a fully differential channel is needed, although, for simplicity, only the single-ended signaling is drawn. The wideband input amplifier needs a bandwidth comparable to the channel bandwidth, for a 80Gbps duobinary transmission this corresponds to approximately 40GHz. An input buffer achieving this performance is shown in [5]. After the input buffer, the signal is split and compared with both a lower and an upper threshold voltages. To do this comparison, the differential signal is first shifted using a TAC. This will position the zero level of the signal in the middle of the lower and upper eye respectively. This signal is then applied to a limitting amplifier stage which regenerates a signal with digital levels, i.e. logic high and logic low above and below the threshold respectively, necessary for the XNOR gate, which will demodulate the duobinary signal.

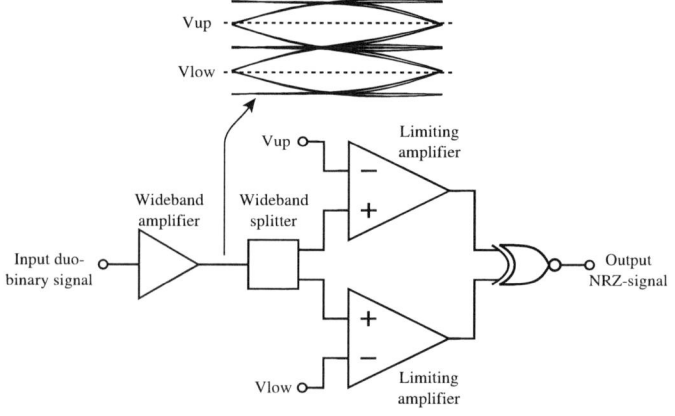

Fig. 1. Standard non-clocked duobinary receiver as described in [4].

III. EMITTER FOLLOWER THRESHOLD ADJUSTMENT CIRCUIT

The TAC, as shown in Fig.2 consists of two distinct parts: an emitter follower with threshold adjustment, and a cascoded output driver. Both circuits will be discussed in greater detail in the following sections.

A. Emitter follower threshold adjustment

The input of the TAC uses emitter followers (EF), shown as transistors Q_0 and Q_1 in Fig.2. The transistors are biased with a constant current of 3.2mA, which is optimized for maximal bandwith (max f_t biasing). The use of EFs has two main benefits: Firstly, they provide a low output impedance, and as a result allow for a higher bandwidth when driving the capacitive input of the cascoded output stage. Secondly, the voltage relationship between base (fixed at 2.4V DC by the wideband input amplifier) and emitter (given the constant emitter current) is fixed. This results in an equal DC voltage at the emitters of transistors Q_0 and Q_1. To introduce a shift in DC voltage, and hence in threshold, a series resistor (R_0 and R_1) is added between the output of the EFs and the input of the cascoded output stage. The biasing current of the EF is split into two parts, one directly connected to the emitter, one connected through this series resistor. By changing the ratio of these two current sources (varying α between 0mA - 1.3mA), the amount of current flowing through the resistor and hence the DC level at the input of the next stage can be controlled. By varying the DC voltage of the positive and negative input in the opposite direction, the threshold can be adjusted.

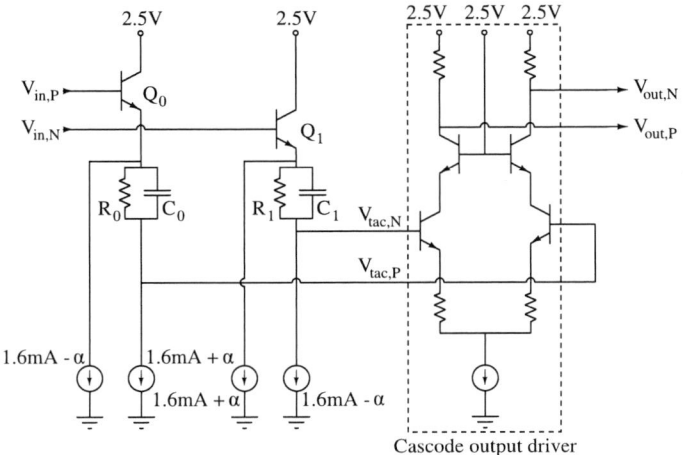

Fig. 2. Threshold adjustment circuit consisting of emitter followers biased with a constant current and the cascade output driver. The threshold level is adjusted by means of α ranging from 0mA - 1.3mA.

Furthermore, capacitors C_0, C_1 are added in parallel with resistors R_0,R_1 in Fig.2. This provides a low impedance path at higher frequencies between the output of the emitter followers and the input of the cascode, and hence, increase the bandwith. The capacitors used in this circuit are metal-insulator-metal (MIM) capacitors with their bottom plate connected to the emitter of the EFs and their top plate connected to the input of the cascode output driver. This reduces the parasitic capacitance at the input of the cascoded amplifier and increases the bandwidth. The use of this topology leads to shifting of signals with a bandwidth of more than 80GHz as shown in

the post layout simulation results of Fig.3. The variation on the bandwidth is less than 5% across the whole shifting range. The shifting resistors R_0 and R_1 have a value of 80Ω in this design. The following trade-off exists: larger values give a larger maximum threshold adjustment but a smaller resolution and the need for a higher bypass capacitance while smaller values mainly deteriorate the maximum threshold varation. Parallel with these 80Ω resistors a 1pF MIM capacitor was implemented.

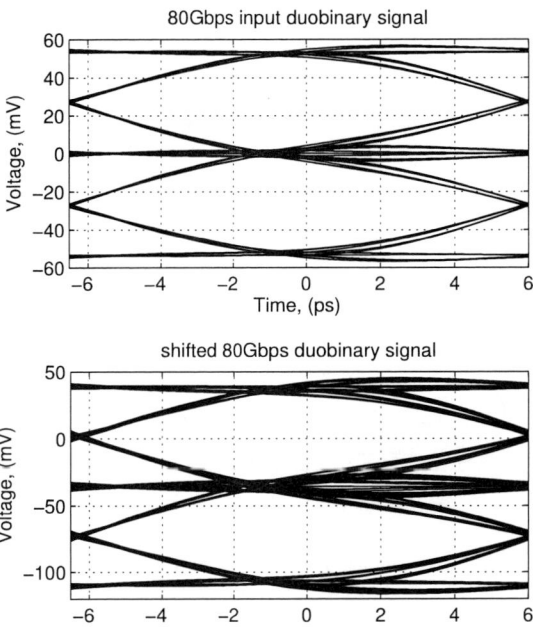

Fig. 3. Input eye-diagram of a 80Gbps duobinary input signal together with the eye-diagram of the shifted version at the output of the threshold adjustment circuit.

The current mirrors providing the current to the EFs are implemented as cascoded current mirrors to reduce the effect of the varying collector voltages of the mirror transistor. This allows us to have a current resolution of less than 5μA resulting in a voltage shifting resolution of less than 1mV.

B. Cascoded output driver

The switching output driver of which the threshold is adjusted, consists of a degenerated differential cascode as shown in Fig.2. By using a cascode, the miller capacitance at the input can be diminished [6] and the maximally allowed voltage swing at the output is higher since the peak voltage at the collector of a common base stage can go up to BV_{CBO} [7], [8]. Feedback by means of resistive degeneration is applied to increase the bandwidth of the circuit by trading in gain. Furthermore, it makes the cascode less sensitive to differences in DC-voltage at the input which limits the accuracy that is needed for the TAC. The biasing of the cascode is such that the transistors achieve their maximum f_t. The load resistor is a 80Ω poly resistor chosen to allow enough bandwidth when loaded with the input capacitance of the next stage. The tail current is 5mA.

In Fig.3 the output eye-diagram shows a zero-volt level that corresponds to the eye-crossing of the upper duobinary eye.

By adding a limiting amplifier to this output the wanted NRZ-signal for the upper half of the duobinary signal is obtained.

C. Bandwidth and linearity

In Fig.4 the gain and the input referred 1dB compression point are simulated in function of the frequency. The -3dB bandwidth of the circuit is above 80GHz, to ensure that the level shifting stage will not limit the bandwidth in a 80Gbps data system. The simulated input referred 1dB compression point is around -4dBm.

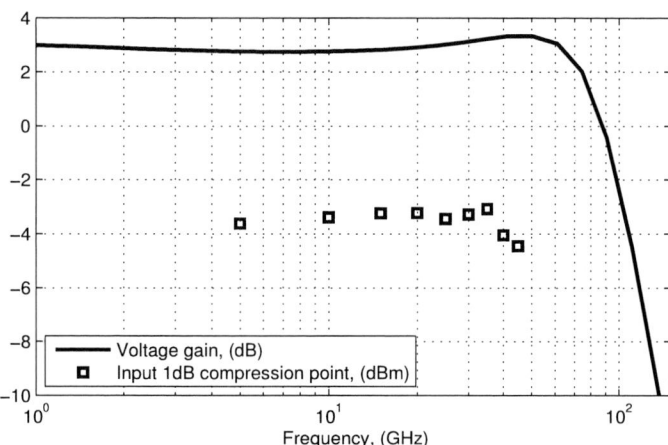

Fig. 4. Post layout simulation results for the voltage gain and input referred 1dB compression point of the TAC in function of the frequency.

The 1dB compression point was simulated by driving the power in a 50Ω input buffer connected in front of the TAC and measuring the input power of the TAC and the output power of the TAC. Since the linearity of the input buffer is greater than the linarity of the TAC this is a valid measure for the linearity. At higher frequencies it is no longer possible to include a sufficient number of harmonics in the simulation, which leads to a lower simulated 1dB compression point, as can be noticed from the two rightmost 1dB compression points in Fig.4. Verification of the linearity at high frequencies was subsequentially done by checking the in- and output diagram at different input power levels.

IV. DIGITALLY CONTROLLED CURRENT DACS

The circuit used to generate the currents for the EFs is shown in Fig.5, both currents are derived from one variable current source I_{var}. This source is implemented as a traditional binary weighted current mirror with switchable gates to control the total current. The variable current is mirrored and amplified five times. This amplification allows to reduce the current in the DAC as well as the area consumed by the DAC. However, care has to be taken to meet the matching requirements for the 7 bit DAC resolution used in this design.

In a second stage the amplified variable current is added and substracted from a constant current source I_0. Using this technique the current through the EFs will always be two times I_0. The last stage of the current generator is the implementation of a switch which allows to change the shifting direction.

Measurements on the fabricated die showed that the current

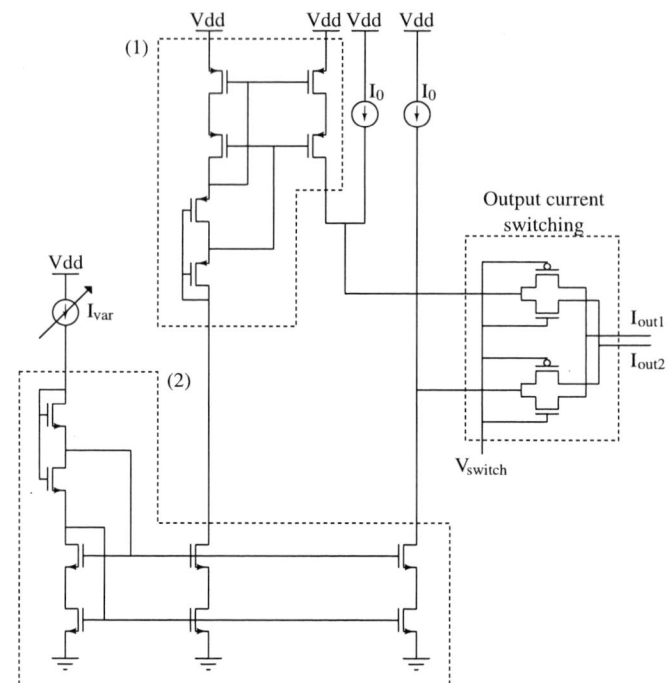

Fig. 5. Circuit used to generate complementary output currents I_{out1} and I_{out2} from the DAC current I_{var}. Sooch cascode current mirrors, subcircuits (1) and (2), are used to get accurate copies of the DAC current. The digital signal V_{switch} interchanges the output currents.

switching circuit has a minimum of $263\mu A$ and a maximum of 1.26mA, variable in steps of $4\mu A$. Taking into account the 80Ω resistor that converts this current into the level shifting voltage and the times two multiplication in the EF current mirrors, we can shift the level 160mV up or down in steps of 0.64mV. The current running through the EFs is constant around 3mA.

V. FABRICATION AND FURTHER WORK

A chip implementing the discussed threshold adjustment circuit has been fabricated in a $0.13\mu m$ SiGe BiCMOS technology with an f_t beyond 200GHz. The part of the chip with the discussed circuit is shown in the micrograph of Fig.6. Although the proposed circuit can not be tested on its own, preliminary measurements show that an eye at the input can be shifted over the complete input range.

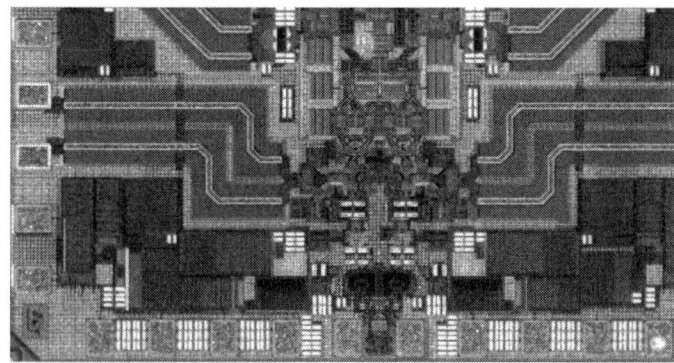

Fig. 6. Die micrograph of the TAC circuit.

Further testing of the chip implementing the discussed circuit, will include the reception and demodulation of a 80Gbps signal over a 40GHz channel. With improvement of the channel bandwidth even higher data rates should be attainable since the bandwidth of the TAC is as high as 80GHz. This means that with a sufficient bandwidth for the input buffer and the channel, a data rate up to 160Gbps becomes possible.

VI. CONCLUSION AND RESULTS

In this paper a broadband threshold adjustment circuit capable of level shifting a duobinary data stream with a bandwidth up to 80GHz is presented. The threshold adjustment circuit can shift a differential input signal up or down by 160mV with a resolution of 0.64mV. It operates from a single 2.5V power supply, consumes only 35mW in a 130μm SiGe BiCMOS technology and still has an input referred 1dB compression point of approximately -4dBm. Using this circuit it is possible to demodulate a multitude of modulation schemes (e.g. duobinary, PAM-4) at high bandwidths allowing to achieve higher data rates over channels with limited bandwidth while keeping the added circuit complexity low.

ACKNOWLEDGMENTS

The authors would like to thank the Agency for Innovation by Science and Technology in Flanders (IWT) for supporting this work.

REFERENCES

[1] D. Correia, V. Shah, C. Chua, and P. Amleshi, "Performance comparison of different encoding schemes in backplane channel at 25gbps+," in *Electromagnetic Compatibility (EMC), 2013 IEEE International Symposium on*. IEEE, 2013, pp. 306–311.

[2] J. Lee, M.-S. Chen, and H.-D. Wang, "Design and comparison of three 20-gb/s backplane transceivers for duobinary, pam4, and nrz data," *Solid-State Circuits, IEEE Journal of*, vol. 43, no. 9, pp. 2120–2133, 2008.

[3] A. Adamiecki, M. Duelk, and J. Sinsky, "25 gbit/s electrical duobinary transmission over fr-4 backplanes," *Electronics Letters*, vol. 41, no. 14, pp. 826–827, 2005.

[4] J. H. Sinsky, M. Duelk, and A. Adamiecki, "High-speed electrical backplane transmission using duobinary signaling," *Microwave Theory and Techniques, IEEE Transactions on*, vol. 53, no. 1, pp. 152–160, 2005.

[5] T. De Keulenaer, Y. Ban, Z. Li, and J. Bauwelinck, "Design of a 80 gbit/s sige bicmos fully differential input buffer for serial electrical communication," in *International Conference on Electronics, Circuits and Systems (ICECS-2012)*, 2012, pp. 237–239.

[6] E. Säckinger, *Broadband Circuits for Optical Fiber Communications*. New York, NY: John Wiley & Sons, 2005.

[7] H. Li, H.-M. Rein, T. Suttorp, and J. Böck, "Fully integrated SiGe VCOs with powerful output buffer for 77-GHz automotive radar systems and applications around 100 GHz," *IEEE J. Solid-State Circuits*, vol. 39, pp. 1650–1658, Oct. 2004.

[8] M. Rickelt and H.-M. Rein, "Influence of impact-ionization induced instabilities on the maximum usable output voltage of Si bipolar transistors," *IEEE Trans. Electron Devices*, vol. 48, pp. 774–783, Apr. 2001.

A 60 GHz down-conversion mixer using a novel topology in 65 nm CMOS

Wang Chong, Li Zhiqun, Li Qin, Liu Yang, Cao Jia, and Wang Zhigong

Institute of RF- & OE-ICs, Southeast University, Nanjing 210096, China
Engineering Research Center of RF-ICs & RF-Systems,
Ministry of Education, Southeast University, Nanjing 210096, China
zhiqunli@seu.edu.cn

Abstract—A 60 GHz down-conversion mixer used in the unlicensed 60 GHz band system in 65-nm CMOS technology is presented in this paper. Based on the double-balanced Gilbert cell, the mixer comprises a cross-coupled pair to rise conversion gain and two series *LCR* network resonating at IF frequency to enhance bandwidth. As a result, both high gain and broad bandwidth are achieved. From the simulation results, the conversion gain exceeds 10 dB and 3-dB IF bandwidth is from 8 GHz to 16 GHz, the OP_{1dB} is –6 dBm and noise figure is below 12 dB in band of interest. The mixer consumes 5 mA from a 1.2 V supply without buffer, and the chip area is 1×0.75 mm^2 with pads.

Keywords—60 GHz; MMW; mixer; Gilbert cell; conversion gain; broadband.

I. INTRODUCTION

The desire for high data-rate wireless communication has drawn the millimeter-wave (MMW) technology to enter the civil field. In the past few years, many countries have released the band around 60 GHz for applications such as Wireless Personal Area Networks (WPAN), the frequency range is usually from 57 GHz to 66 GHz, and the bandwidth of each channel exceeds 2 GHz, which guarantees milti-Gb/s data-rate. In the past, III-V semiconductor technologies were adopted in most of the MMW applications. Fortunately, as CMOS technology is scaled into the nanometer range, it is possible to realize 60 GHz transceivers by CMOS, which is more attractive for its low cost and high integration [1].

The mixer is the second active block in the receiver. It should be designed with high conversion gain and sufficient bandwidth to guarantee a flat gain in the interesting band. Moreover, the linearity, noise figure and isolation between ports should also be considered. To satisfy so many performances, the Gilbert cell is a proper choice, which has become the most popular structure in mixer design [2]. For broadband and MMW systems such as 60 GHz application, the Gilbert cell is also a candidate, but there are some differences with low-GHz design, new problems emerge and corresponding considerations must be performed. The mixer in this paper makes some improvement base on Gilbert cell, letting it suitable to broad band and MMW applications.

This paper is organized as follows. Section II presents the 60 GHz receiver scheme, points out the challenge in the mixer design for this scheme, and then a novel mixer topology is proposed. Section III analyzes the novel topology in this design. Section IV shows the simulation results while section V draws conclusion.

II. TOPOLOGY OF THE PROPOSED MIXER

A. The 60 GHz Receiver Scheme

In our 60 GHz receiver, a twice conversion slide-IF architecture is used. Fig.1 illustrates the receiver Block diagram. A 48 GHz LO is needed for the first conversion and 12 GHz quadrature LO for the second conversion. A 24 GHz PLL is designed, which provides 48 GHz LO by a doubler and 12 GHz quadrature LO by divider. The advantage of this architecture is only one oscillator is needed. The block we discussed in this paper is the first mixer in the figure.

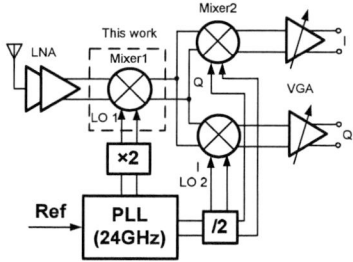

Fig. 1. Block diagram of the 60 GHz receiver

B. The Main Design Challenge

The main challenges that realizing MMW Gilbert mixers by CMOS technology are the low transconductance of the transistors and Q factor of the inductors, due to low electron mobility and high substrate loss of CMOS technology, respectively, which greatly constraint the conversion gain [3]. Moreover, the system architecture proposes other strict requirement to the first mixer according to the 802.15.3c standard: in 60 GHz communication system, a channel bandwidth is 2.16 GHz, it is broadband relative to 12 GHz IF frequency, so the 60 GHz mixer should exhibit broad IF bandwidth. For a Gilbert cell topology, achieving both high conversion gain and broad IF bandwidth demand the real part of the load admittance, the conductance, maintain constant with a small value in a wide band near the IF frequency, while the

---

Project supported by the National High Technology Research and Development Program in China (No. 2011AA010202).

978-1-4799-4993-9/14 $31.00 © 2014 IEEE

imaginary part of the load admittance, the susceptance, equal or close to zero in the same band. Unfortunately, the types of loads usually used in Gilbert cell do not exhibit these characters. A resistor or a transistor used as active load can provide a constant conductance in a wide band, but the susceptance component increasing sharply at high IF frequency such as 12 GHz due to parasitic capacitance, which result in sharply dropped gain. An inductor must be used to resonate out the capacitance, but usually the *LC* load resonating at 12 GHz can't provide a bandwidth of more than 2.16 GHz, except deliberately decrease the Q factor of the load, at the expense of lower gain, but this is not our intention. So, new topology to load stage of Gilbert cell must be proposed for broadband operation.

Now we face a situation that both conversion gain and IF bandwidth are worse than we expect. We can't solve this problem by simple tradeoff between them. For example, in order to increase the gain, a negative conductance parallel with the *LC* load can be used to reduce the conductance, in turn the gain is enhanced, and the negative conductance can be easily realized by a cross-coupled pair. However, it will sacrifice the bandwidth in that the Q factor of the load is raised actually.

C. Topology of the Proposed Mixer

We can notice for a parallel *LCG* network, the input admittance can be written as:

$$Y(\omega) = G + j\omega C - j\frac{1}{\omega L} \tag{1}$$

From (1), we can see the conductance is a constant, while the bandwidth is depended on the susceptance component. Susceptance varies with frequency due to the *LC* frequency features. At the resonant frequency, susceptance reaches to zero, when deviated from the resonant point, susceptance increases proportion to the frequency offset, one side at the zero point is plus and other side is minus. At frequencies the susceptance value is great enough, it will bypass more current injected to the *LCG* network, while the current flow through conductance is less, so the voltage across the cell is lower, and by this means the bandwidth is limited.

From this view, we can suppose another *LC* network also resonate at the same frequency, and exhibit contrary susceptance frequency character with the original network, so the susceptance can be offset in a wide band if parallel them together. Fortunately, a simple series *LC* network possess this property, the whole susceptance will be near zero in a wide band after the series *LC* network is paralleled with the original one. Then, a negative conductance can be paralleled with the two *LC* network to reduce the whole conductance. Finally, both high gain and broad bandwidth can be achieved through this method. Fig. 2 shows the whole circuits we proposed.

At the RF and LO port, two on-chip transformer baluns are designed to convert the input single-ended signal to differential output. The two baluns are same and they work well from 45 GHz to 70 GHz. In order to enhance the

performance of the transconductance stage, two important improvements are performed based on the traditional Gilbert topology. First, a PMOS transistors are stacked on the NMOS one to rise the overall transconductance of the input stage, furthermore, the current flow though the switch stage accordingly reduce due to the paralleled PMOS transistor, so the switch can commutate more sharply, it is beneficial to the mixer's conversion gain and noise figure [4]. Second, the parasite capacitance at the drain of the transconductance stage or the common-source of the switch pairs, as the C_d in fig. 2, will severely degrade the performance. Hence, two inductors, L_1 and L_2 are paralleled at the nodes to resonate out the capacitance at the desired band and two bypass capacitors connect with them to the ground.

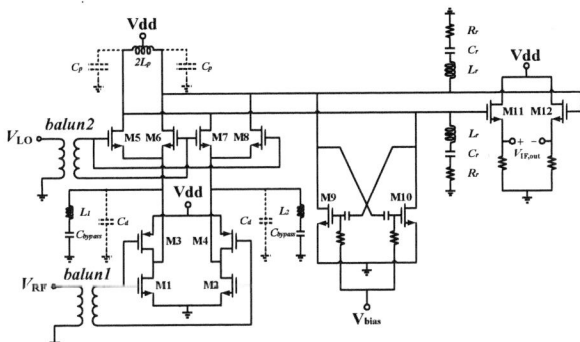

Fig. 2. Schematic diagram of the proposed mixer

The switch pairs convert the input RF current to IF frequency by LO action. The IF output current see a parallel *LCG* load from the drain of the switch transistors, and the conversion gain and IF bandwidth of the mixer are decided directly by the *LCG* load. As mentioned before, a series *LC* network parallel with the load can extend bandwidth significantly, while a negative conductance also parallel with the load, which is introduced by a simple cross-coupled pair, to enhance the conversion gain. We should notice a resistor must be added to the series *LC* network, otherwise, it will be shorted at the resonating frequency. The negative conductance can be adjusted by changing the bias voltage of the cross-coupled pairs, so by this means the conversion gain can be adjusted. Finally, a source follower used as a buffer is added at the output of the core mixer for measurement.

After a brief description of the mixer topology, the theoretical analysis of the proposed novel topology will be performed in next section.

III. ANALYSIS OF THE NOVEL TOPOLOGY

In this section, let's analyze the novel topology for bandwidth extension with another *LCR* network. We denote L_p, C_p and G_p as the component values of original parallel *LCG* load. Where L_p is the load inductance, C_p represents the total capacitance to ground at the switch pair output node, including parasitic capacitance and input capacitance of the buffer, G_p express the conductance introduced by loss of the load and output conductance of the switch transistors. L_p and C_p resonant at IF center frequency ω_0. Then, we parallel a series LCR network with the original load, denoting L_r, C_r and R_r as the corresponding component values, L_r and C_r also resonant at ω_0, form a new load, its equivalent model is shown in Fig. 3.

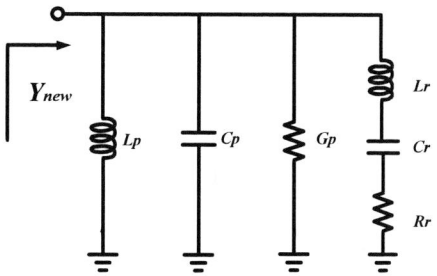

Fig. 3. Equivalent model of the new load

Now, the new load's input admittance can be written as:

$$Y_{new}(\omega) = G_p + j\omega C_p - j\frac{1}{\omega L_p} + \frac{1}{j\omega L_r - j\dfrac{1}{\omega C_r} + R_r} \quad (2)$$

Substituting ω with $\omega_0 + \Delta\omega$, where $\Delta\omega$ denotes the offset from ω_0, after some derivation, we can obtain the input admittance at $\Delta\omega$ is:

$$Y_{new}(\Delta\omega) = G_p + \frac{R_r}{R_r^2 + (2\Delta\omega L_r)^2} + j2\Delta\omega C_p - j\frac{2\Delta\omega L_r}{R_r^2 + (2\Delta\omega L_r)^2} \quad (3)$$

It can be seen that the susceptance of Y_{new} may be equal to or near zero and its conductance remain constant in a wide band if the value of L_r and R_r are properly selected. If they satisfy:

$$R_r^2 \gg (2\Delta\omega L_r)^2 \quad (4)$$

$$C_p = L_r / R_r^2 \quad (5)$$

The load admittance will not vary with frequency. Accordingly, a flat conversion gain is achieved in a wide band. Then, the parallel cross-coupled pair introduce a minus conductance, $-G_n$, in the right of formula (3):

$$Y_{new} = G_p + \frac{1}{R_r} - G_n \quad (6)$$

The conversion gain of this topology can be written as:

$$Gain_{conversion} = \frac{2}{\pi}(g_{mn} + g_{mp}) / (G_p + \frac{1}{R_r} - G_n) \quad (7)$$

Where g_{mn} and g_{mp} is the transconductance of input NMOS and PMOS transistor, respectively, the coefficient $2/\pi$ is introduced in frequency translation process. If it satisfies: $G_p + (1/R_r) > G_n > 1/R_r$, the circuit will not self-oscillate while the conversion gain can be higher than the topology without both cross-coupled pair and series LCR network.

After we are clear about the principle of this topology, the layout and simulation results will be shown in next section.

IV. LAYOUT AND SIMULATION RESULTS

The mixer is designed in a 65nm 1P9M CMOS technology, which top metal, M9, is most thick one, so all the passive components are performed by M9. Its layout is shown in Fig. 4. In the design, in order to consider all the high frequency effects, we use electromagnetic (EM) tool, HFSS, to model and simulate all the transmission lines, inductors and baluns. Microstrip line is selected for connection in the layout, M9 is used as the signal line and M1 is the ground (because of size scaling, M1 is invisible in the layout).

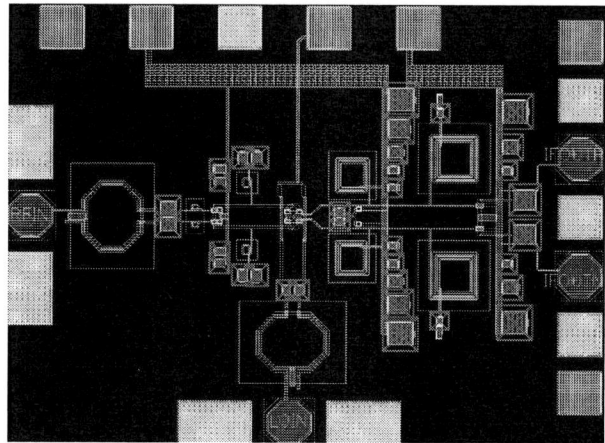

Fig. 4. Layout design of the proposed mixer

The die occupies an area of 1×0.75 mm^2, including pads. The single-ended RF and LO input from the left and bottom GSG pads, respectively. The PGSGSGP pads at the right output differential IF signals, as well as the top five DC pads provide the power supply and bias needed in the circuit.

For a fixed LO frequency of 48 GHz, the IF range from 8 GHz to 16 GHz, correspondingly RF range from 56 GHz to 64 GHz, its conversion gain is depicted in Fig. 5. Besides the conversion gain curve of the proposed topology, other two curves are depicted in the same figure for comparison, one is the conversion gain of topology without both cross-coupled pair and series LCR network, another is the conversion gain of topology with cross-coupled pair but not series LCR network. We can see the topology without both cross-coupled pair and series LCR network has both low gain and narrow bandwidth, when cross-coupled pair are added, the gain can be enhanced greatly, but the bandwidth will be narrower the same ratio correspondingly. For the proposed topology, high gain and broad bandwidth are achieved simultaneously. In this design, the conversion gain is higher than 10 dB and IF 3-dB bandwidth range covers 8 GHz to 16 GHz, while consuming 5 mA current from a 1.2 V supply without buffer. Both conversion gain and bandwidth are excellent for 60 GHz down-conversion mixers.

TABLE. I COMPARISON WITH STATE OF ART MIXERS

REFERENCE	[6]	[7]	[8]	this work
technology	90 nm CMOS	65 nm CMOS	65 nm CMOS	65 nm CMOS
conversion gain (dB)	15.5	9.1	2	10.4
IF bandwidth (GHz)	5.4	4	1.3	8
noise figure (dB)	12.8	12	12	11.2
OP$_{1dB}$ (dBm)	-	-5	-16	-6
power dissipation (mW)	17	2.8	6	6

V. CONCLUSION

A 60 GHz down-converting mixer with a novel topology in a 65 nm CMOS is proposed in this paper. Comparing with the traditional Gilbert cell, a cross-coupled pair and two series *LCR* networks are added on the load stage. Finally, high conversion gain and broad bandwidth are achieved simultaneously. Consuming 5 mA current from 1.2 V supply, the conversion gain exceeds 10dB and IF 3dB-bandwidth covers 8 GHz to 16 GHz. The noise figure is below 12 dB at the interesting band and the output 1dB compression point is -6dBm, are also quite good.

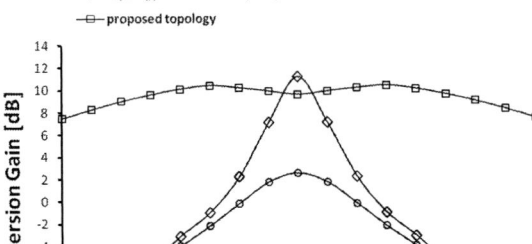

Fig. 5. Conversion gain and bandwidth comparison for three topologies

The noise figure and 1 dB compression point are shown in Fig. 6 and Fig. 7, respectively. The noise figure is below 12 dB in the IF band. The input 1 dB compression point achieves -16 dBm, and the corresponding output 1 dB compression point is -6 dBm.

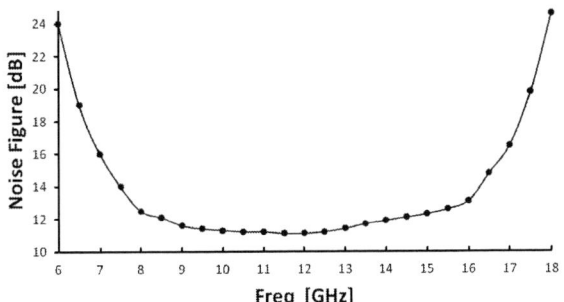

Fig. 6 Noise figure of the designed mixer

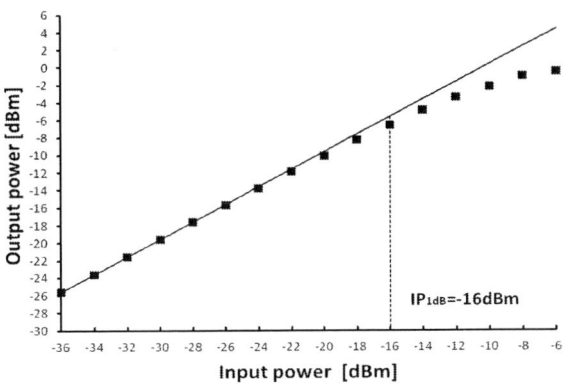

Fig. 7 Output power vs. input power

In order to evaluate our mixer design, a comparison with other state of art circuits is performed in TABLE. I. The conversion gain, IF bandwidth, noise figure, OP$_{1dB}$ and power dissipation are considered. From the comparison, we can see a good performance is achieved for our mixer.

REFERENCES

[1] Doan. C.H, Emami. S, Sobel. D.A, Niknejad. A.M and Brodersen. R.W, "Design Considerations for 60 GHz CMOS Radios," IEEE Communications Magazine, vol.42, pp. 132-140, Dec 2004.

[2] P. J. Sullivan, B. A. Xavier and W. H. Ku, "Low voltage performance of a microwave CMOS Gilbert cell mixer," IEEE J. Solid-state Circuits, vol.32, pp. 1151-1155, July 1997.

[3] Nouri. N, Nezhad-Ahamdi. M.R, Mirabbasi. S, and Safavi-Naeini. S, "A double-balanced CMOS mixer with on-chip balun for 60-GHz receivers," IEEE Conference on NEWCAS, 2010 8th International.

[4] Razavi. B, "A 60-GHz CMOS receiver front-end," IEEE J. Solid-state Circuits, vol.41, pp. 17-22, Jan 2006.

[5] Doan. C.H, Emami. S, Sobel. D.A, Niknejad. A.M and Brodersen. R.W, "Millimeter-wave CMOS design". IEEE J. Solid-state Circuits, vol.40, pp. 144-155, Jan 2005.

[6] Lee. J.H, and Lin. Y.S, "60 GHz CMOS downconversion mixer with 15.46 dB gain and 64.7 dB LO-RF isolation". IEEE Electronics letters, vol.49, pp. 264-266, Feb 2013.

[7] Kraemer. M, Ercoli. M, Dragomirescu.D and Plana. R, "A wideband single-balanced down-mixer for the 60 GHz band in 65 nm CMOS". Proceedings of Asia-Pacific Microwave Conference 2010.

[8] Sakian. P, Mahmoudi. R, van Zeijl. P, Lont. M and van Roermund. A, "A 60-GHz double-balanced homodyne down-converter in 65-nm CMOS process". Microwave Integrated Circuits Conference, 2009. EuMIC 2009. European.

A High Conversion Gain Millimeter-Wave Frequency Doubler in 65nm CMOS

Liu Yang[1,2], Li Zhiqun[1,2+], Li Qin[1,2], Wang Chong[1,2], Wang Zhigong[1,2]

[1]Institute of RF- & OE-ICs, Southeast University, Nanjing, China
[2] Engineering Research Center of RF-ICs & RF-Systems, Ministry of Education, Southeast University, Nanjing, China
[+]Corresponding author: zhiqunli@seu.edu.cn

Abstract— **This paper presents a high conversion gain doubler-balanced active frequency doubler for millimeter-wave application. The frequency doubler contains an improved push-push structure, two quarter-wavelength transmission lines, and output power enhancement using negative resistor. The 3-dB band of the frequency doubler is 19~28 GHz of input frequency, the maximum conversion gain reaches −5.3 dB, the fundamental rejection is above 55 dB, and the power consumption is 17 mW under 1.2V VDD. The frequency doubler is designed in 65nm CMOS process.**

Keywords—**frequency doubler, 65nm CMOS, milllimeter-wave(MMW), conversion gain.**

I. INTRODUCTION

Frequency Doubler is a widely-used and important component in microwave and millimeter-wave frequency synthesizer systems.

A common MMW phase-lock-loop (PLL) frequency synthesizer system is shown in Fig. 1 (a); a voltage-controlled-oscillator and a frequency divider are used in this system. Designing the VCO and frequency divider become more difficult as the output frequency goes high. In IEEE 802.15 standard, the carrier frequency ranges from 59~64GHz, which means the LO frequency of a transmitter might reach 40~50GHz or higher, which is quite a challenge in VCO and frequency divider designing in CMOS technology.

To reduce the risk of designing high frequency VCO and frequency divider, a frequency synthesizer based on PLL and frequency doubler could be used, shown in Fig.1 (b). A frequency doubler and a low frequency VCO are used instead of the high frequency VCO and divider [1][2].

Frequency doubler could be implemented with these major structures: mono-transistor [3][4][5], diodes [6], push-push [7][8][9][10], Gilbert cell [11][12], and injection locking [13]. Push-push structure is widely used because of the simple structure and good performance in fundamental rejection. This paper reports a doubler-balanced active frequency doubler based on improved push-push structure, and output power enhancement using negative resistor.

The paper is organized as follows. Section II describes the circuit design. Section III shows the post-layout simulation results and comparison with other frequency doublers. Section IV draws conclusions.

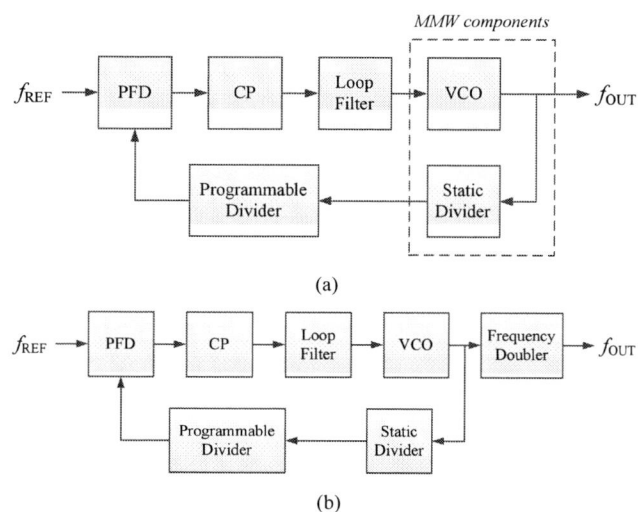

(a)

(b)

Fig.1. Frequency synthesizers for MMW

(a) Common MMW PLL frequency synthesizer

(b)Frequency synthesizer using frequency doubler

II. CIRCUIT DESIGN

The schematic of proposed frequency doubler is shown in Fig.2. There are three main parts in this design, part 1 is improved push-push structure frequency doubler, and part 2 is quarter-wavelength transmission line, and part 3 is output enhancement with negative resistor.

A. Improved push-push structure frequency doubler

Push-push structure is a widely-used structure is frequency doubler design, Fig.3 (a) shows its schematic. It can be described as two parallel common-source amplifiers sharing the same load. [7][8][9][10]

Let the differential input signals $v_{in}(t)+ = A\cos(\omega_{in}t)$ and $v_{in}(t)- = -A\cos(\omega_{in}t)$. Since the nonlinearity of the amplifier, the output signals of the two amplifiers are:

National High Technology Research and Development Program (No. 2011AA010202)

978-1-4799-4993-9/14 $31.00 © 2014 IEEE

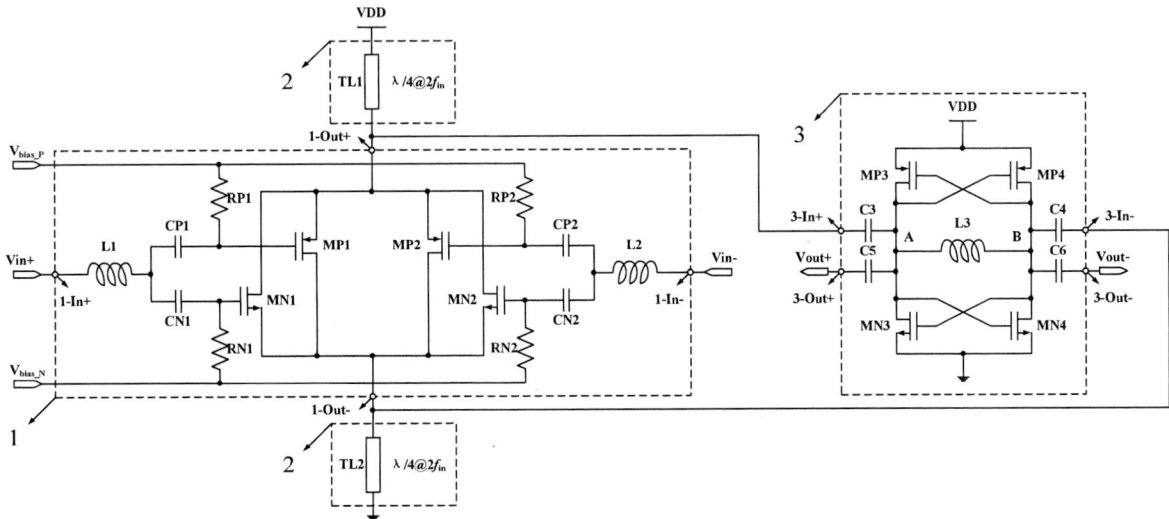

Fig.2. Frequency doubler based on improved push-push structuure and output enhancement with negative resistor

$$v_{out}(t)+ = \frac{1}{2}\alpha_2 A^2 + \left(\alpha_1 A + \frac{3}{4}\alpha_3 A^3\right)\cos(\omega_{in}t)$$
$$+\frac{1}{2}\alpha_2 A^2 \cos(2\omega_{in}t) + \frac{1}{4}\alpha_3 A^3 \cos(3\omega_{in}t) + \cdots \quad (1)$$

$$v_{out}(t)- = \frac{1}{2}\alpha_2 A^2 - \left(\alpha_1 A + \frac{3}{4}\alpha_3 A^3\right)\cos(\omega_{in}t)$$
$$+\frac{1}{2}\alpha_2 A^2 \cos(2\omega_{in}t) - \frac{1}{4}\alpha_3 A^3 \cos(3\omega_{in}t) + \cdots \quad (2)$$

Where α_n is the gain of the n_{th}-order harmonic of the amplifiers. The final output of the frequency doubler is the sum of (1) and (2):

$$v_{out}(t) = v_{out}(t) + + v_{out}(t) - = \alpha_2 A^2 + \alpha_2 A^2 \cos(2\omega_{in}t) + \cdots (3)$$

As shown in (3), the fundamental and odd harmonics are cancelled in the output signal, only DC component and even harmonics left. The DC component could be blocked by a capacitor, and the 4th or higher harmonics could be easily filtered out because of their low amplitude and high frequency.

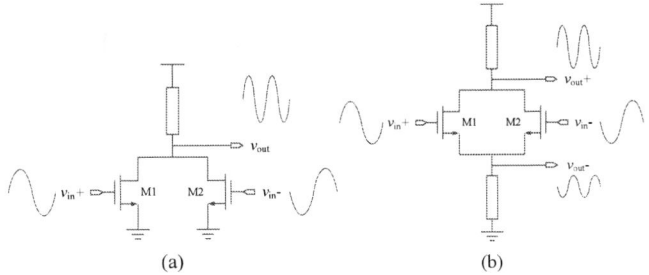

Fig.3. Schematic of push-push structure frequency doubler

(a)single-ended output (b)differential output

The common push-push structure in Fig.3 (a) uses a pair of differential signals to generate a single-ended output signal. Differential outputs could be obtained by connecting both the two drains and the two sources of the two transistors together, as shown in Fig.3 (b). As the gain differs in the drain and the source of a transistor, the differential outputs are not balanced because the amplitudes are not the same. The proposed frequency doubler employs an improved push-push structure to obtain balanced differential outputs, as shown in Fig.2. MP1 and MP2 form one common push-push structure, MN1 and MN2 form another. The two parts operate in parallel, which can generate balanced differential output signals by setting the bias voltage properly. Inductors L1 and L2 are used for input impedance matching.

B. Quarter-wavelength transmission line

The quarter-wavelength transmission line is a CPW transmission line. It is modeled in HFSS, as shown in Fig.4. Fig.5 shows the S22 of the transmission line. When one end is connected to the VDD or GND, this line presents high impedance at output frequency $2f_{in}$ which prevents the output signal of the frequency doubler from entering the VDD or GND, and low impedance at $4f_{in}$ which filters the 4th-order harmonic out.

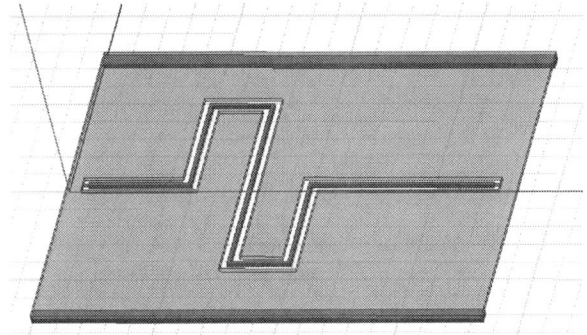

Fig.4. Quarter-wavelength transmission line built in HFSS

978-1-4799-4993-9/14 $31.00 © 2014 IEEE

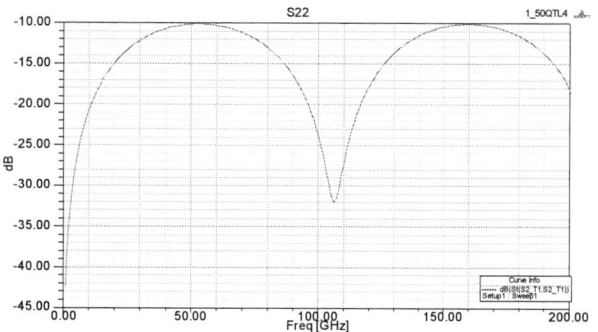

Fig.5 S22 curve of the quarter-wavelength transmission line

C. Output power enhancement using negative resistor

The common push-push structure frequency doubler has low conversion gain because the gain of 2nd-order harmonic is low, and the output signal might not have enough power to drive the load such as mixers. The proposed frequency doubler employs an output power enhancement technology based on negative resistor, which can effectively boost the output power.

The principle of negative resistor enhancement is shown in Fig.6 (a). There is a pair of differential output current i_1 and i_2, R_{L1} and R_{L2} are the load resistors. A negative resistor –R is inserted between output node A and B. –R generates a current i_{neg} from node B to A, which would enhance the current in the loads. The final output currents are:

$$i_{L1} = i_1 + i_{neg} \qquad (4)$$

$$i_{L2} = i_2 - i_{neg} \qquad (5)$$

The negative resistor consists of four complementary cross-coupled MOSFETs and an inductor, as shown in Fig.6 (b).

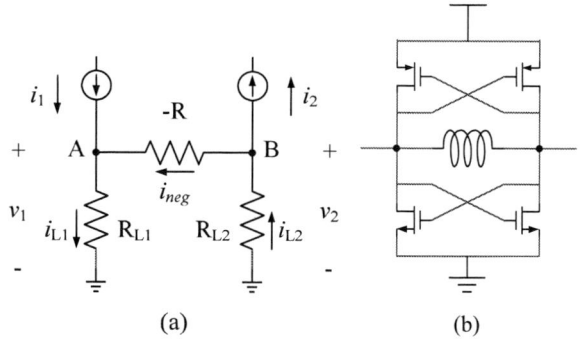

(a)　　　　　　　　(b)

Fig.6. (a) Principle of output power enhancement using negative resisto

(b) Negative resistor schematic

III.　SIMULATION RESULTS

The layout of the proposed frequency doubler designed in 65nm CMOS process is shown in Fig.7. The scale of the whole chip is 865μm by 510μm. The post-layout simulation results are shown below.

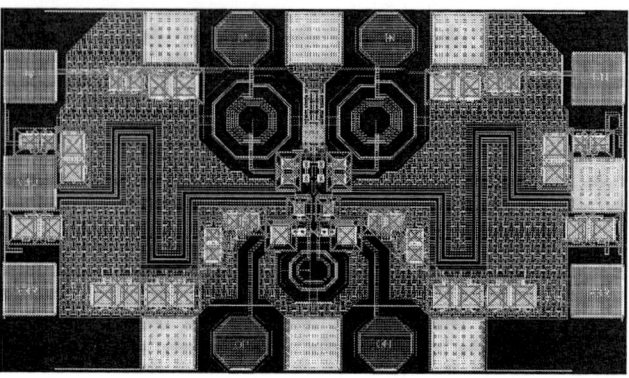

Fig.7. Layout of proposed frequency doubler

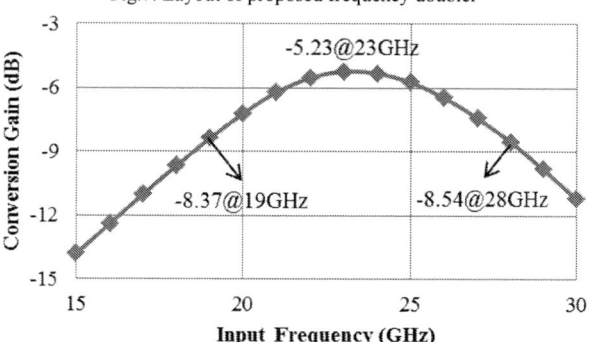

Fig.8. Conversion gain curve

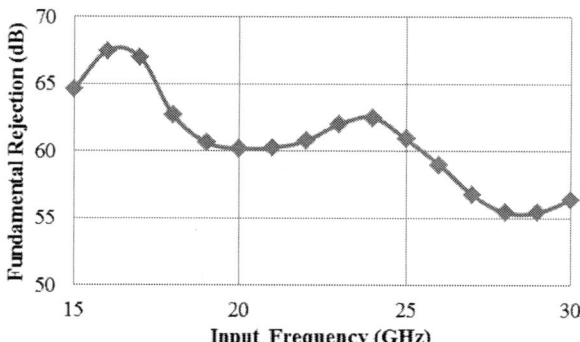

Fig.9. Fundamental rejection curve

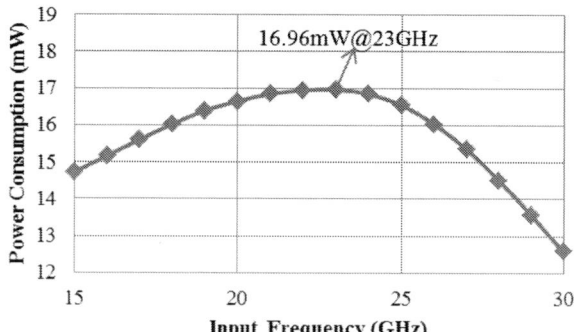

Fig.10. Power consumption curve

978-1-4799-4993-9/14 $31.00 © 2014 IEEE

TABLE I. PERFORMACE SUMMARY AND COMPARISON

Refs.	Output Frequency (GHz)	Conversion Gain (dB)	Fundamental Rejection (dB)	Power Consumption (mW)	V$_{DD}$ (V)	Tech.
[2]	5.2 ~ 5.8	0 ~ 2.7	> 20	18	1.8	0.18μm CMOS
[4]	50 ~ 70	−10	10 ~ 25	4.5	1.5	90nm CMOS
[6]	25 ~ 75	−15.5 ~ −11	32 ~ 59	0	passive	0.18μm CMOS
[7]	75 ~ 110	−16	20 ~ 30	13.8	1.2	65nm CMOS
[8]	95 ~ 150	−8	30 ~ 42.4	19.2 ~ 22.8	1.2	65nm CMOS
[9]	4	−13.05	> 25	26.9	1.8	0.18μm CMOS
This work	38 ~ 56	−8.54 ~ −5.23	> 55	17	1.2	65nm CMOS

Fig. 8 shows the conversion gain of the proposed frequency doubler. The maximum conversion gain reaches −5.23dB at 23GHz input frequency. The 3-dB band is 19~28GHz of input frequency, in which the conversion gain ranges from −8.54dB to −5.23dB. The fundamental rejection curve is shown in Fig.9, and the minimum value is 55dB. The power consumption is shown in Fig.10, the supply voltage is 1.2V, and the maximum value is less than 17mW.

Table 1 shows the comparison of the proposed frequency doubler and others of the references in CMOS process. This work presents the highest conversion gain and competitive fundamental rejection and power consumption in MMW frequency band.

IV. CONCLUSION

A double-balanced active frequency doubler in 65nm CMOS for millimeter-wave application has been presented in this work. The frequency doubler employs improved push-push structure and output power enhancement using negative resistor. The frequency doubler provides a conversion gain from −8.45 to −5.23dB and a fundamental rejection above 55 dB in 19~28GHz band of input frequency. The power consumption is less than 17mW under 1.2V voltage supply.

ACKNOWLEDGMENT

The work is supported by the National High Technology Research and Development Program (No. 2011AA010202).

REFERENCES

[1] Ja-Yol Lee, Haecheon Kim, Hyun-Kyu Yu, "A 52GHz Millimeter-Wave PLL Synthesizer for 60GHz WPAN Radio", Proceedings of the 3rd European Microwave Integrated Circuits Conference, 2008, pp.155-158.

[2] Kazuya Yamamoto, "A 1.8-V Operation 5-GHz-Band CMOS Frequency Doubler Using Current-Reuse Circuit Design Technique", IEEE Journal of Solid-State Circuits, Vol. 40, No. 6, June 2005, pp.1288-1295.

[3] Yong-Chae Jeong, Jong-Sik Lim, "A Novel Frequency Doubler Using Feedforward Technique and Defected Ground Structure", IEEE

Microwave and Wireless Components Letters, Vol. 14, No. 12, December 2004, pp.557-559.

[4] Jixin Chen; Pinpin Yan; Wei Hong, "A 50–70GHz frequency doubler in 90nm CMOS," Microwave Workshop Series on Millimeter Wave Wireless Technology and Applications (IMWS), 2012 IEEE MTT-S International ,18-20 Sept. 2012, pp.1-3.

[5] Pekka Kangaslahti, Petteri Alinikula, Veikko Porra, "Miniaturized Artificial-Transmission-Line Monolithic Millimeter-Wave Frequency Doubler", IEEE Transactions On Microwave Theory And Techniques, Vol. 48, No. 4, April 2000, pp.510-518.

[6] Tsung-Yu Yang, Hwann-Kaeo Chiou, "A 25–75 GHz Miniature Double Balanced Frequency Doubler in 0.18-μm CMOS Technology", IEEE Microwave and Wireless Components Letters, Vol. 18, No. 4, April 2008, pp.275-277.

[7] Mikko Varonen, Mikko Kärkkäinen, Dan Sandström, Kari A. I. Halonen, "A 100-GHz Balanced FET Frequency Doubler in 65-nm CMOS", Proceedings of the 6th European Microwave Integrated Circuits Conference, pp.105-107.

[8] Ping-Han Tsai, Yu-Hsuan Lin, Jing-Lin Kuo, Zuo-Min Tsai, Huei Wang, "Broadband Balanced Frequency Doublers With Fundamental Rejection Enhancement Using a Novel Compensated Marchand Balun", IEEE Transactions on Microwave Theory and Techniques, Vol. 61, No. 5, May 2013, pp.1913-1923.

[9] Stanley S. K. Ho, Carlos E. Saavedra, "Frequency Doubler Employing Active Fundamental Cancellation in CMOS", Sarnoff Symposium, 2009. March 30 -April 1, 2009, pp.1-4.

[10] Jiangtao Sun, Qing Liu, Yong-Ju Suh, Takayuki Shibata, Toshihiko Yoshimasu, "A Low DC Power High Conversion Gain Frequency Doubler IC for 22-29GHz UWB Applications", Proceedings of Asia-Pacific Microwave Conference 2010, pp.944-947.

[11] Jiankang Li, Yong-Zhong Xiong, Wang Ling Goh, Wen Wu, "A 27–41 GHz Frequency Doubler With Conversion Gain of 12 dB and PAE of 16.9%", IEEE Microwave and Wireless Components Letters, Vol. 22, No. 8, August 2012, pp.427-429.

[12] Austin Ying-Kuang Chen, Yves Baeyens, Young-Kai Chen, Jenshan Lin, "A 36–80 GHz High Gain Millimeter-Wave Double-Balanced Active Frequency Doubler in SiGe BiCMOS", IEEE Microwave and Wireless Components Letters, Vol. 19, No. 9, September 2009, pp.572-574.

[13] Enrico Monaco, Massimo Pozzoni, Francesco Svelto, Andrea Mazzanti, "Injection-Locked CMOS Frequency Doublers for μ-Wave and mm-Wave Applications", IEEE Journal of Solid-State Circuits, Vol. 45, No. 8, August 2010, pp.1565-1574.

978-1-4799-4993-9/14 $31.00 © 2014 IEEE

Integrated Multi-band Fractional-N PLL for FMCW Radar Systems at 2.4 and 5.8 GHz

Niko Joram, Bastian Lindner, Jens Wagner and Frank Ellinger
Chair for Circuit Design and Network Theory
Technische Universität Dresden, 01062 Dresden, Germany

Abstract—This article presents a fractional-N phase-locked loop (PLL) for the use in frequency-modulated continuous wave (FMCW) radar systems. The presented design supports division ratios from 59 to 4092 with a maximum input frequency of 7 GHz, covering the 2.39 GHz to 3.28 GHz and 4.79 GHz to 6.55 GHz bands using a dual-band voltage-controlled oscillator (VCO) with a frequency resolution of 0.6 Hz. This corresponds to a large relative bandwidth of more than 31 %. Reference spur levels are lower than −65 dBc while phase noise is at −103 dBc/Hz at 1 MHz offset frequency. A key feature for radar applications is the automatic chirp waveform generation. The complete circuit including VCO consumes less than 122 mW and is implemented using an IBM 180 nm SiGe BiCMOS process.

Index Terms—Phase locked loops, chirp radar, SiGe BiCMOS

I. INTRODUCTION

Local positioning systems (LPS) using the FMCW approach enjoy great popularity because of the low complexity hardware and the signal processing in frequency domain. Recent research has been done to develop systems to work in multiple industrial science medical (ISM) bands to enhance resolution in complex environments [1]. Such systems benefit from radio frequency front-end components, which can concurrently handle all the necessary bands. As a result, the positioning tags will get more competitive, because they are smaller in size and have reduced power consumption and cost.

The task of the PLL in a FMCW radar system is to generate the chirp. It translates a stable, but low reference frequency, e.g. from a crystal, to an unstable high frequency oscillator, typically a VCO. To achieve good radar performance, the PLL has to be optimized for large possible chirp bandwidth, gradient and linearity. Range resolution of a FMCW radar system increases with increasing chirp bandwidth. Furthermore, it should be able to handle a large frequency range at the RF input to allow operation in several bands. In-band phase noise and spurious emissions must be low. Discrete frequency steps for chirp generation have to be small enough, such that the resulting chirp after filtering in the loop is smooth and linearity errors are reduced. Consequentially, the PLL has to use a fractional-N frequency divider. The divider has to be programmable to select different channels and include automatic frequency stepping to generate the chirp. For the use in LPS, the coverage range of the system should be high, which calls for using frequency bands below 10 GHz, namely the ISM bands at 2.4 GHz to 2.5 GHz and 5.725 GHz to 5.875 GHz.

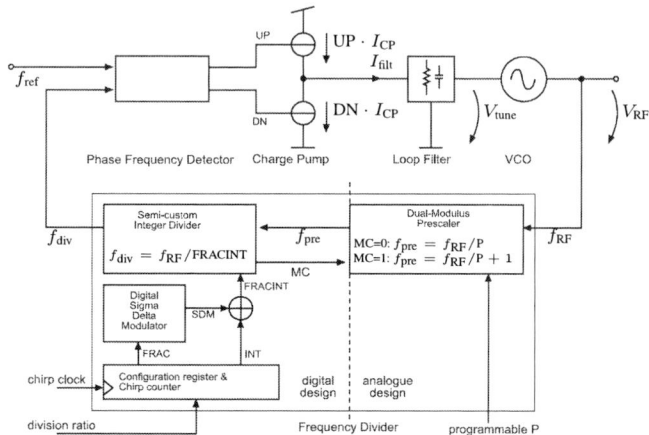

Fig. 1. PLL block diagram

There are FMCW PLLs available which are designed for the use in radar distance sensors and concentrate on operation in one band, mostly around 77 GHz [2], [3] with relative bandwidths below 10 %. Some sub-10 GHz ISM wide-band PLLs have also been presented primarily for the use in communications systems [4], [5], which show good phase noise performance, but are not optimized for bandwidth, small frequency steps or modulation capabilities like chirp generation. This article therefore presents a design combining a wideband architecture with chirp generation capability suitable for FMCW LPS. Section II describes major components of the PLL with special attention to isolation between digital and analogue blocks, section III presents measurement results of the integrated circuit.

II. SYSTEM DESCRIPTION

A. Overview

A block diagram of the PLL is shown in Fig. 1. It is a basic charge pump PLL architecture, which is commonly used in communications today. The reference frequency f_{ref} and the frequency at the fractional divider output f_{div} are compared by a phase frequency detector (PFD), which generates pulses proportional to the phase difference between the two input signals. If f_{div} lags f_{ref}, UP pulses are generated and current is injected into the passive loop filter by the charge pump, increasing the tuning voltage V_{tune} and thus frequency f_{RF}. If f_{div} leads f_{ref}, DN pulses are generated and consequently f_{RF}

978-1-4799-4993-9/14 $31.00 © 2014 IEEE

is decreased. For the VCO, the dual-band circuit described in [6] is employed.

B. Frequency Divider

The frequency divider consists of a full-custom dual-modulus prescaler running at f_{RF}, semi-custom integer divider running at the divided RF clock f_{pre} and sigma delta modulator (SDM). The programmable prescaler divides the RF frequency before the integer divider. The integer divider is implemented with digital standard cells. With the specified semiconductor technology, the maximum frequency of a synthesized divider was determined to approximately 400 MHz.

Using a programmable modulus P in the prescaler enables the PLL to process RF frequencies of both ISM bands. The modulus control signal MC controlled by the integer divider is used to switch the division ratio of the prescaler between P and $P + 1$. The prescaler output then results in

$$f_{pre} = \begin{cases} f_{RF}/P & \text{MC=0} \\ f_{RF}/(P+1) & \text{MC=1} \end{cases}. \quad (1)$$

Any integer division ratio can be reached by dividing a certain number of cycles of f_{pre} by P and another number by $P + 1$, which is done by the integer divider through switching of the MC signal. The integer divider is implemented using the well-known pulse-swallow counter [7]. To extend the divider to fractional operation, a SDM is used to map the fractional frequency control word FRAC to a number stream, whose average over time lies between 0 and 1. The SDM value is then added to the integer part of the frequency control word INT, resulting in FRACINT, which represents the total division ratio. Consequentially, the division ratio is changing with every cycle, providing the fractional value through averaging in the loop. The output of the frequency divider block which is fed to the PFD then calculates as

$$f_{div} = f_{RF}/\text{FRACINT}. \quad (2)$$

Two basic chirp waveforms (triangular and sawtooth-shaped) can be programmed using an externally clocked up/down counter, which increments or decrements FRACINT by a previously programmed value, allowing different chirp gradients. Furthermore, by using an external clock for this counter, synchronization issues with the rest of the radar system can be avoided, since the start of the chirp becomes deterministic.

C. Dual-Modulus Prescaler

A schematic of the programmable dual-modulus prescaler is depicted in Fig. 2. Since this block has to process high frequencies directly from the VCO, standard CMOS gates cannot be used with the specified technology. Thus, it is implemented with custom-designed emitter-coupled logic (ECL) gates using bipolar transistors. Furthermore, all signals are laid out differentially.

The prescaler is an extension of the design from [7] and it is composed of two parts: a synchronous 4/5 divider based on a Johnson counter consisting of the flip-flops DF1, DF2 and

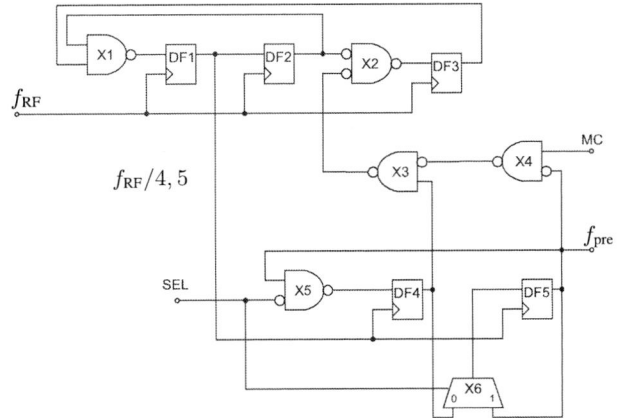

Fig. 2. Programmable dual modulus prescaler

DF3 and a switchable divide-by-2 or divide-by-4 using DF4 and DF5.

The Johnson counter includes a pulse stretcher circuit consisting of X2 and DF3, which is controlled by the modulus control signal MC. If enabled, the chain divides f_{RF} by 5, otherwise by 4. The divided frequency is then passed to the second divider circuit represented by DF4 and DF5. The SEL signal determines the modulus, which can be $P = \{8, 16\}$. When the SEL signal is set to 0, the multiplexer X6 connects DF4 and DF5, forming a divide-by-4 circuit. The lower divider is also connected to the pulse stretcher via X3 and X4, such that the total division factors can be 8, 9, 16 or 17, depending on SEL and MC.

The output signal f_{pre} is fed via an ECL-to-CMOS level shifter to the CMOS semi-custom part of the frequency divider.

D. PFD and Charge Pump

As PFD, the basic two flip-flop design with reset delay is employed [4]. The charge pump used in this design has a very large output voltage range of 83.3 %, corresponding to 0.2 V to 2.7 V, which makes excellent use of the frequency range of the VCO [8]. It then amounts to 4.79 GHz to 6.55 GHz or 2.39 GHz to 3.28 GHz with enabled divide-by-2. Due to its low mismatch between source and sink currents of below 2.1 %, static phase offsets and reference spurs are highly suppressed.

E. Sigma Delta Modulator

The used SDM has a MASH 1-1-1 architecture with added dithering by means of a linear feedback shift register in the third stage [9]. The frequency resolution of the PLL is determined by the modulus of the SDM. It calculates as

$$\Delta f = \frac{f_{ref}}{2^m}, \quad (3)$$

where m is the number of bits of the SDM accumulators. To reach a high frequency resolution, a 25 bit modulator and 20 MHz reference frequency were selected, yielding a resolution of 0.6 Hz. Because of the large modulus, the SDM had to be pipelined to allow synthesizing the design with a constrained clock of more than 20 MHz.

978-1-4799-4993-9/14 $31.00 © 2014 IEEE

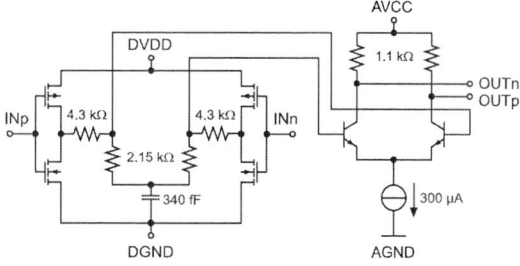

Fig. 3. CMOS to ECL converter

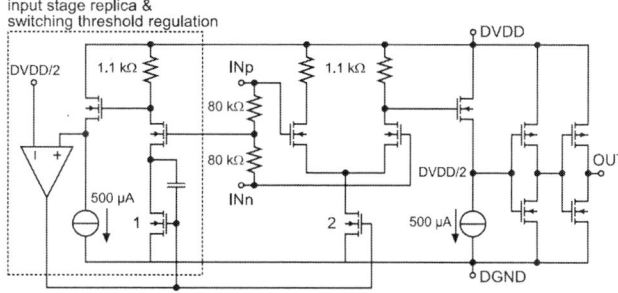

Fig. 4. ECL to CMOS converter

Fig. 5. PLL and VCO micrograph, PLL area amounts to $0.65\,\text{mm}^2$.

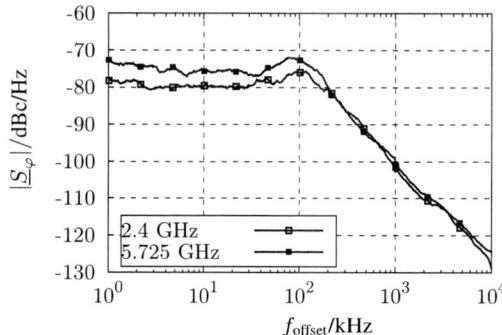

Fig. 6. Measured phase noise of the PLL in the two bands

F. Isolation Concept

Since this PLL is a mixed-signal design containing sensitive analogue blocks like charge pump and VCO together with a digital frequency divider on a single chip, an isolation concept has to be devised. It was decided to place the charge pump and digital dividers into separate isolated p-wells, which allows isolating their grounds from each other and from the VCO. The different grounds are then connected off-chip to a low-impedance ground plane. This prevents switching noise from entering the analogue blocks, which would eventually lead to increased spur and noise level. This procedure however brings up the problem of crossing with signals between circuits with different grounds.

The key to solving this problem is to sense the signals always differentially on the receiving side and rejecting the common-mode part. The proposed circuit to translate from CMOS to ECL signals is shown in Fig. 3. The pseudo-differential CMOS input signal is buffered by inverters and reduced in swing by the resistor network to accommodate to the following differential pair stage. High-frequency common mode disturbances are shunted to ground by the capacitor. The differential amplifier buffers the signal again and translates the levels to ECL for the prescaler. This circuit is used to translate the SEL and MC signals.

The circuit translating from ECL to CMOS is presented in Fig. 4. It consists of a differential pair input stage, followed by a common-drain stage as level shifter and two CMOS inverters used for signal shaping. The voltage at the input of the CMOS inverters has to be set such that they are always biased in the vicinity of their switching threshold. Hence, they will amplify the still small signal swing from the differential amplifier. To set the switching threshold to the desired value, e.g. DVDD/2, one half of the input stage is replicated and a regulation loop is added, which sets the tail current of the replica and the input stage to achieve DVDD/2 also at the inverter inputs. The circuit works at frequencies of more than $400\,\text{MHz}$ and is therefore used to translate the prescaled clock f_{pre}.

III. MEASUREMENT

The PLL was integrated into a FMCW transceiver chip, implemented in an IBM 180 nm SiGe BiCMOS process. It was measured in-system bonded onto a circuit board. A micrograph of all involved blocks is presented in Fig. 5. The passive third-order loop filter was implemented off-chip. The area of the PLL without VCO and buffers amounts to $0.65\,\text{mm}^2$.

Fig. 6 shows the phase noise in the two ISM bands. The reference source was a signal generator with $-120\,\text{dBc/Hz}$. In-band phase noise is less than $-75\,\text{dBc/Hz}$ and is dominated by the loop filter, PFD and charge pump. Out-of-band phase noise is at $-103\,\text{dBc/Hz}$ at 1 MHz offset, which is about 5 dB worse than that of the VCO alone, indicating a small contribution from the SDM.

The spectral performance is shown in Fig. 7 and was recorded with a real-time spectrum analyzer. Due to the large modulus of the SDM and dithering, fractional spurs fade into broadband noise. There are some spots with slightly increased signal power at 10 MHz and 20 MHz offsets, indicating a fractional spur at $-59\,\text{dBc}$ and a reference spur at $-69\,\text{dBc}$.

Table I compares the presented PLL to recent publications

978-1-4799-4993-9/14 $31.00 © 2014 IEEE

TABLE I
COMPARISON WITH SUB-10 GHz STATE-OF-THE-ART INTEGRATED FRACTIONAL-N PLLs

Ref.	Tech.	Freq. range (GHz)	Rel. band-width (%)	Freq. step (Hz)	Reference spur (dBc)	In-band PN @10 kHz (dBc/Hz)	Out-of-band PN @1 MHz (dBc/Hz)	P_{DC} (mW)
[10]	130 nm CMOS	2.3-2.8	19.6	-	-	-83	-113	4.2
[11]	180 nm CMOS	2.2-2.6	16.6	4.88 k	-39	-70	-90	22
[12]	90 nm CMOS	1.7-2.5	38	-	-	-75	-115	1.13
[4]	130 nm CMOS	2.17-2.95, 4.35-5.9	30.4	<1 k	-	<-80	<-108	51
[5]	500 nm SiGe	2.4, 4.9-5.8	16.8	781 k, 468 k	-	-98	<-120	99
This	180 nm SiGe	2.29-3.34, 4.57-6.69	31	0.6	<-65	<-75	-103	<122 (12 VCO)

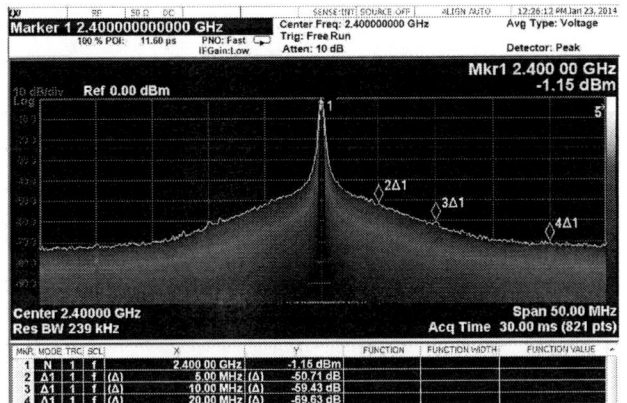

Fig. 7. Measured spectral performance of the PLL

of sub-10 GHz PLLs. The current work, although having worse phase noise performance, covers a large bandwidth of more than 31 % in two bands with a very small frequency step size.

IV. CONCLUSION

Presented in this paper was a wide-band PLL covering the license-free 2.4 GHz and 5.8 GHz ISM bands with a large relative bandwidth of more than 31 %. Having built-in chirp generation capabilities and small possible modulation frequency steps, it is suitable for the use in high-resolution FMCW-based local positioning systems.

ACKNOWLEDGMENT

The research leading to these results has received funding from the European Community's Seventh Framework Programme (FP7/2007-2013) under grant agreement n°242411 (E-SPONDER).

REFERENCES

[1] N. Joram, B. Al-Qudsi, J. Wagner, A. Strobel, and F. Ellinger, "Design of a multi-band FMCW radar module," in *10th Workshop on Positioning Navigation and Communication (WPNC)*, 2013, pp. 1–6.

[2] H. Matsumura, M. Sato, A. Mineyama, T. Suzuki, and N. Hara, "Ultra-low phase noise 76-81 GHz PLL synthesizer for FMCW radar in 65 nm CMOS," in *Asia-Pacific Microwave Conference Proceedings (APMC)*, Dec 2012, pp. 649–651.

[3] H. Ng, A. Fischer, R. Feger, R. Stuhlberger, L. Maurer, and A. Stelzer, "A DLL-Supported, Low Phase Noise Fractional-N PLL With a Wide-band VCO and a Highly Linear Frequency Ramp Generator for FMCW Radars," *IEEE Transactions on Circuits and Systems I: Regular Papers*, vol. 60, no. 12, pp. 3289–3302, Dec 2013.

[4] A. Bonfanti, C. Samori, and A. Lacaita, "A multistandard Σ-Δ fractional-N frequency synthesizer for 802.11a/b/g WLAN," in *33rd European Solid State Circuits Conference (ESSCIRC)*, Sept 2007, pp. 480–483.

[5] J. Rogers, F. Dai, M. Cavin, and D. Rahn, "A multiband ΔΣ fractional-N frequency synthesizer for a MIMO WLAN transceiver RFIC," *IEEE Journal of Solid-State Circuits*, vol. 40, no. 3, pp. 678–689, March 2005.

[6] N. Joram, R. Wolf, and F. Ellinger, "A SiGe Wideband VCO and Divider MMIC with Low Gain Variation for Multi-band Systems at 2.4 and 5.8 GHz," in *9th Conference on Ph.D. Research in Microelectronics and Electronics (PRIME)*, 2013, p. on CD.

[7] Y. Kado, T. Ohno, M. Harada, K. Deguchi, and T. Tsuchiya, "An ultralow power CMOS/SIMOX programmable counter LSI," *IEEE Journal of Solid-State Circuits*, vol. 32, no. 10, pp. 1582–1587, Oct 1997.

[8] N. Joram, R. Wolf, and F. Ellinger, "High swing pll charge pump with current mismatch reduction," *Electronics Letters*, vol. 50, no. 9, pp. 661–663, April 2014.

[9] V. Gonzalez-Diaz, M. Garcia-Andrade, G. Flores-Verdad, and F. Maloberti, "Efficient Dithering in MASH Sigma-Delta Modulators for Fractional Frequency Synthesizers," *IEEE Transactions on Circuits and Systems I: Regular Papers*, vol. 57, no. 9, pp. 2394–2403, Sept 2010.

[10] W. Lee and S. Cho, "A 2.4-GHz reference doubled fractional-N PLL with dual phase detector in 0.13-um CMOS," in *Proceedings of 2010 IEEE International Symposium on Circuits and Systems (ISCAS)*, May 2010, pp. 1328–1331.

[11] C.-L. Ti, Y.-H. Liu, and T.-H. Lin, "A 2.4-GHz fractional-N PLL with a PFD/CP linearization and an improved CP circuit," in *IEEE International Symposium on Circuits and Systems (ISCAS)*, May 2008, pp. 1728–1731.

[12] M. Vidojkovic, Y.-H. Liu, X. Huang, K. Imamura, G. Dolmans, and H. de Groot, "A fully integrated 1.7-2.5GHz 1mW fractional-N PLL for WBAN and WSN applications," in *IEEE Radio Frequency Integrated Circuits Symposium (RFIC)*, June 2012, pp. 185–188.

A Low Power, Small Area, Fully Integrated 5.5GHz CMOS LC-VCO

Shaahin Haddadinejad
Institut fuer Elektronische Bauelemente
und Schaltungstechnik
Technische Universitaet Braunschweig
Email: shaahin@nst.ing.tu-bs.de

Achim Noculak
Institut fuer Theoretische Elektrotechnik
RWTH Aachen University

Michael Hinz
and Bernd Meinerzhagen
Institut fuer Elektronische Bauelemente
und Schaltungstechnik
Technische Universitaet Braunschweig

Abstract—**An integrated 5.5 GHz low power voltage-controlled oscillator (VCO) has been designed and implemented in a 250 nm CMOS process. This cross-coupled LC VCO achieves measured phase noise of about 90 dBc/Hz at an offset frequency of 1MHz and a power consumption of less than 2.1 mW. The output frequency of the VCO can be tuned from 5.19 GHz to 6.12 GHz, which correspond to a 16.9% tuning range, obtained by tuning the control voltage of the varactor. The VCO output power remains nearly constant over the entire tuning range.**

I. INTRODUCTION

Due to the increasing demand and the fast development of data communication, low cost and high performance RF front-end integrated circuits consuming low DC power are required. Fully integrated CMOS voltage-controlled oscillators (VCOs) are very important and challenging building blocks in RF trancievers.

The typical performance parameters of a VCO are phase noise, tuning range, output power and DC power consumption. Ring oscillators are known to provide wider frequency tuning ranges compared to VCOs with LC tanks but the poor phase noise performance of ring oszillators is usually a serious concern for designers [1]-[2]. The large area required by the LC-tanks is normally considered a drawback for LC-VCOs but with the current advancements in IC technology, moderate and acceptable sized inductors can be used to design the LC-tanks without compromising on the quality factors (Q).

Moreover it has been demonstrated that the Cross-Coupled Double-Switch architecture for LC-VCOs can achieve 6dB better phase noise compared to the Single-Switch architecture, when both oscillators are operated at the same bias current [3]-[4].

In this paper, a 5.5 GHz Double-Switch CMOS LC-VCO with fairly low phase noise, low DC power consumption and large tuning range is presented. Section II describes the design of the LC-VCO which includes the topology and circuit structure used in the design. In section III measurement results of the LC-VCO are shown and compared to the respective simulation results. Moreover the VCO presented in this paper is compared to similar VCOs presented in recent publications. Finally a conclusion is drawn in Section IV.

II. CIRCUIT DESIGN

The circuit schematic of the cross-coupled differential LC-VCO including the differential buffer at the output is shown in

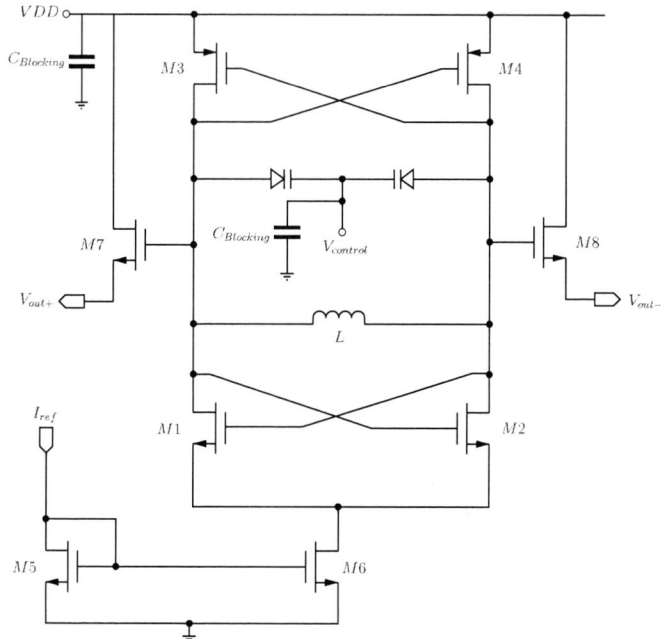

Fig. 1.   Cross-Coupled Double-Switch LC-VCO

Fig. 1. Cross-coupled NMOS transistors (M1, M2) and Cross-coupled PMOS transistors (M3, M4) are forming the VCO core. This Double-Switch structure is a popular choice among designers since it allows a larger differential voltage swing compared to the Single-Switch structure which results in a better phase noise performance. A proper choice of transistors dimensions yields a common mode level at the output of the VCO core around $\frac{VDD}{2}$ which maximizes the tuning range [5]. On the other hand a disadvantage of the Double-Switch structure is that the PMOS transistors become fairly wide and contribute large parasitic capacitances which restrict the tuning range. Table I shows the widths of the transistors and all transistors have the minimum channel length.

The circuit was fabricated using the 250nm RF-CMOS technology of IHP-Microelectronics in Frankfurt/Oder, Germany. The predesigned special accumulation MOS varactors and the predesigned spiral inductor with inductance $L = 2.8nH$ available in this CMOS technology are used in the VCO design presented here. The VCO core inductor L has a rather small area and a good quality factor.

978-1-4799-4993-9/14 $31.00 © 2014 IEEE

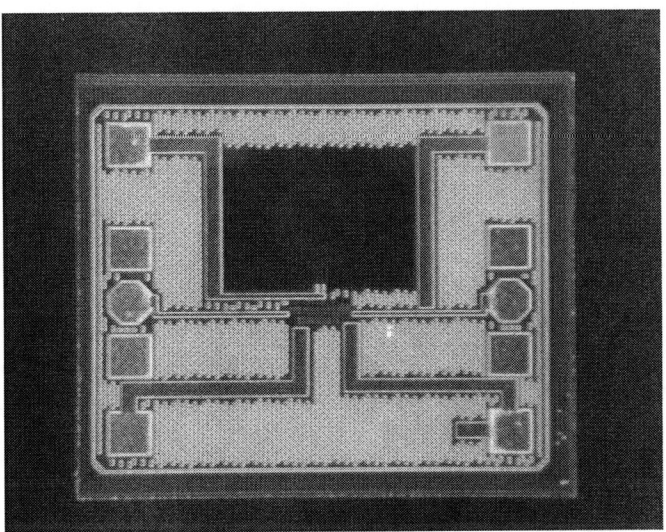

Fig. 2.   Chip Photograph of LC-VCO

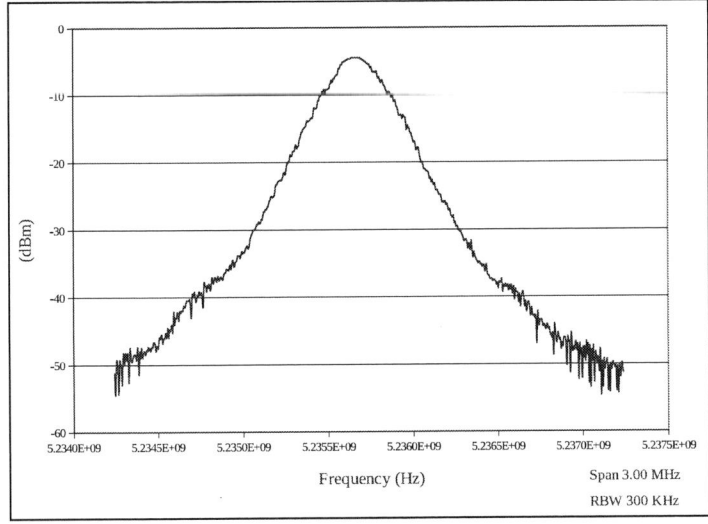

Fig. 3.   Output spectrum at the center frequency of the VCO for $V_{control} = 0.75$V and $P_{DC} = 2.1$mW.

The Inductor L and the varactors supplemented by the parasitic capacitances of the transistors and the inductor from the high Q tank circuit, which resonates at frequency f_o. Realistic tank circuits have losses that can be well represented at least close to the resonance frequency by a parasitic resistance $2 \cdot R_{loss}$ parallel to the inductor L. The cross coupled transistor pairs M1, M2, and M3, M4 form a negative resistance compensating the losses of the resonator. Given the transconductance of the PMOS transistors g_{mp} and the transconductance of the NMOS transistors g_{mn} and small signal conditions the negative resistance R_{neg} can be calculated as:

$$R_{neg} = -\frac{1}{g_{mp} + g_{mn}} \qquad (2)$$

The VCO starts to oscillate due to noise if the following condition holds:

$$\left| \frac{1}{R_{neg}} \right| > \left| \frac{1}{R_{loss}} \right|$$

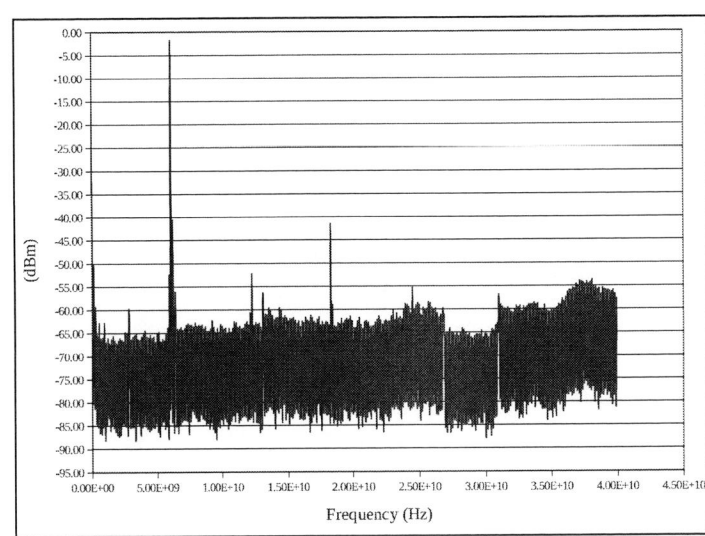

Fig. 4.   Harmonics spectrum of the VCO for $V_{control} = 0.75$V and $P_{DC} = 2.1$mW.

If $g_{mp} = g_{mn} = g_m$ holds, this is equivalent to

$$g_m > \frac{1}{2R_{loss}}. \qquad (3)$$

Transistors M5 and M6 form the tail current source for the VCO core. A constant tail current $I_{tail,M6}$ is forced through the VCO core by the current mirror circuit. I_{ref} is injected off chip. Transistors M7 and M8 are applied as common drain output buffers, because of their large input impedances and their DC current is provided through the bias-T circuits connected to the source contacts of the buffers devices during measurement.

III.   Measurement Results

Circuit and layout design and optimization were performed using the Cadence software and the design kits provided by IHP-Microelectronics. Fig. 2 shows the photograph of the fabricated VCO. The chip size is 850 μm ×685 μm including the pads. The layout was made as symmetrical and compact as possible to ensure differential operation and reduce parasitic inductance and capacitance. This is important since any asymmetry in the tank may cause a significant shift of the center frequency. Besides that, the width of the metal lines connecting the LC tank have been laid out properly to withstand the large current due to resonance.

For all measurements shown here the conditions $V_{DD} = 2.5$V and $I_{tail,M6} = 0.83$mA hold leading to a DC power consumption of less than $P_{DC} = 2.1$mW.

Fig. 3 shows the output spectrum close to the center frequency of the VCO. The output power at oscillation frequency is about -5 dBm.

The harmonic characteristic of the VCO is given in Fig. 4. As can been seen the second and third harmonics are kept below -55 dBm.

The measured and simulated oscillation frequency vs. tuning voltage characteristics are shown in Fig. 5. The relative

978-1-4799-4993-9/14 $31.00 © 2014 IEEE   253

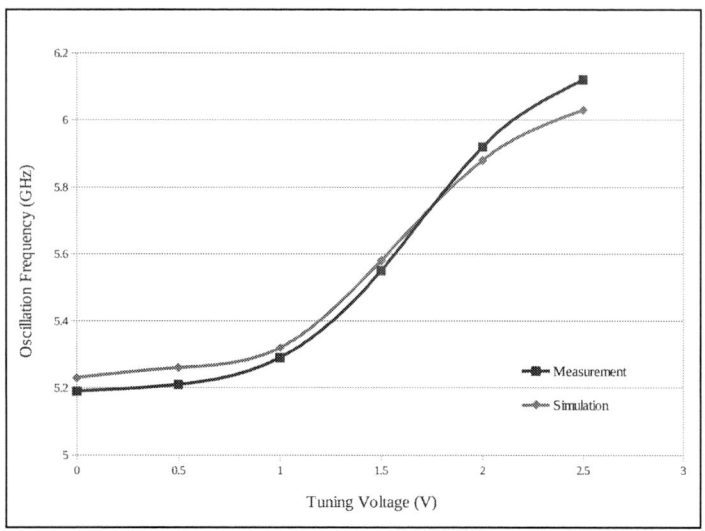

Fig. 5. Oscillation Frequency vs. Tuning Voltage ($P_{DC} = 2.1$mW)

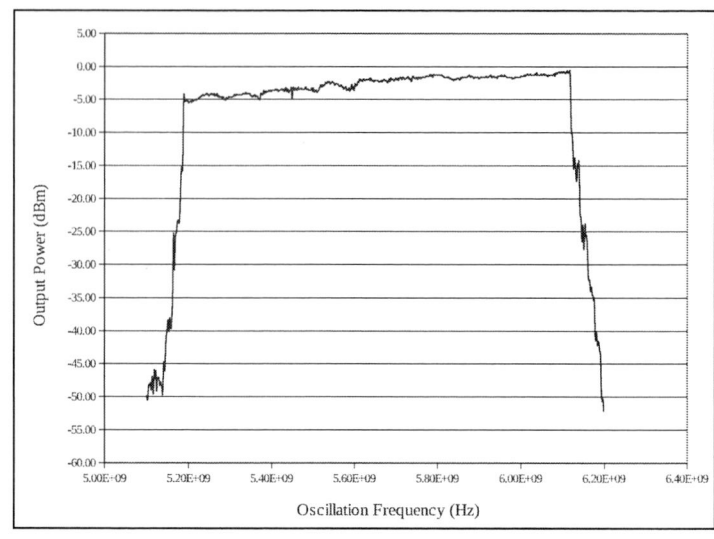

Fig. 6. Output power vs. Oscillation frequency range ($P_{DC} = 2.1$mW)

difference between the measured and simulated oscillation frequency is less than 2% which demonstrates an impressive simulation accuracy. Varying $V_{control}$ between 0 and 2.5 Volt the oscillation frequency can be tuned between 5.19 GHz and 6.12 GHz which results in a tuning range of 930 MHz.

Fig. 6 shows the output power vs. oscillation frequency characteristic. Over the tuning range the output power increases from -5dBm to -1 dBm which is a variation of just 4 dB. It is know that larger variations of the output amplitude lead to limitations of tuning range [5]. Therefore this small variation of 4 dB indicates that the VCO is properly designed.

Figure 7 shows the measured phase noise power characteristic of the oscillator in dBm for a measurement bandwidth of 300 KHz. This yields a measured phase noise of about 90 dBc/Hz at 1 MHz offset. This differs significantly from the simulated phase noise of about 111 dBc/Hz at 1 MHz offset. The reason for this large difference has not yet been clarified.

For characterizing the VCO performance, the widely used figures of merit FOM and FOM_T, which includes the tuning range as well, are adopted in this work. FOM and FOM_T are defined as follows [6]:

$$FOM = L\{\Delta f\} - 20 \log(\frac{f_o}{\Delta f}) + 10 \log(\frac{P_{DC}}{1mW}) \quad (4)$$

$$FOM_T = L\{\Delta f\} - 20 \log(\frac{f_o}{\Delta f} \cdot \frac{TR\%}{10}) + 10 \log(\frac{P_{DC}}{1mW}) \quad (5)$$

Here $L\{\Delta f\}$ is the VCO phase noise, Δf is the offset frequency, f_o is the oscillation frequency, $TR\%$ is tuning range and P_{diss} is the DC power consumption. Measured FOM and FOM_T for the VCO presented in this paper are about -162 dBc/Hz and -166 dBc/Hz, respectively.

Table II shows the performance summary and the comparison with previously published papers.

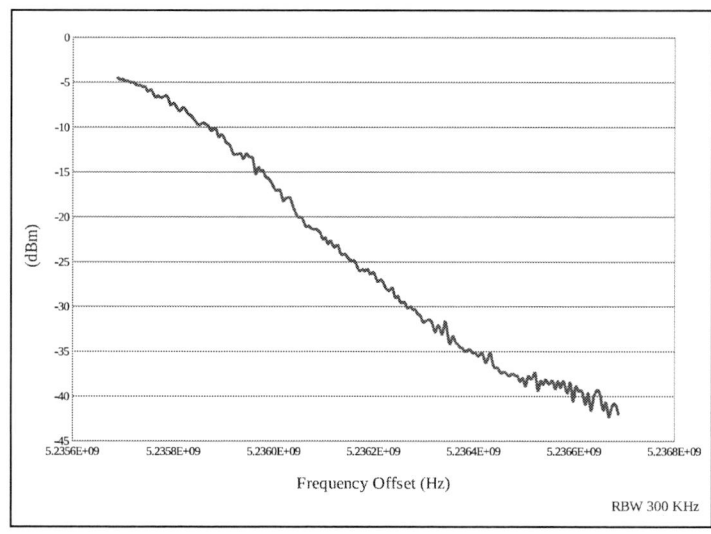

Fig. 7. Measured phase noise characteristic @ 1 MHz Offset for $P_{DC} = 2.1$mW and $V_{control} = 0.75$V

IV. CONCLUSION

An integrated CMOS LC VCO using double-switch cross-coupled transistors, a spiral inductor and an accumulation MOS varactors is presented. The VCO is integrated using a 250 nm RF-CMOS process. The LC VCO has a tuning range of 16.9% between 5.19 GHz and 6.12 GHz. It consumes only 2.1 mW for a 2.5V DC supply voltage. The measured phase noise of the VCO is about -90 dBc/Hz at 1 MHz offset. The measured FOM and FOM_T figures for the VCO are about -162 dBc/Hz and -166 dBc/Hz, respectively.

ACKNOWLEDGMENT

This research was partially supported by the Deutsche Forschungsgemeinschaft (DFG-GZ INST 188/277-1 FUGG). Moreover the authors like to thank IHP-Microelectronics, Frankfurt/Oder, Germany, for the chip fabrication.

978-1-4799-4993-9/14 $31.00 © 2014 IEEE

TABLE I.   WIDTHS OF THE TRANSISTORS

	M1	M2	M3	M4	M5	M6	M7	M8
Width (um)	15	15	22	22	50	500	180	180

TABLE II.   PERFORMANCE SUMMARY AND COMPARISON OF PARAMETERS

	This Work	[7]	[8]	[9]	[10]
Process (nm)	250	180	180	180	180
Oscillation freq. (GHz)	5.5	5.44	4.25	4.2	5.1
Core current (mA)	0.83	9	4	6	15.38
Supply voltage (V)	2.5	1.1	2	1	1.8
Power Consumption (mW)	2.075	9.9	8	6	27.7
Tuning range (%)	16.9	3.4	30	42	17
Measured Phase noise (dBz/Hz)	-89.77 @1MHz	-124.4 @1MHz	-114 @1MHz	-116.3 @1MHz	-134 @1MHz
FOM	-162	-189.1	-176	-181	-192
Output Power (dBm)	-3	-0.8	-2	-7.8	–
Chip Area (mm^2)	0.58	0.61	0.55	1.4	3.42

REFERENCES

[1] J.Long and R.J Weber, A 2.4GHz Low-Power Low-Phase-Noise CMOS VCO Using Spiral Inductors and Junction Varactors, International Symposium on Circuits and Systems, vol. 4, pp. 545-548, May 2004.

[2] T.I. Ahrens and T. H. Lee, A 1.4-GHz 3-mW CMOS LC Low Phase Noise VCO using Tapped Bond Wire Inductances, International Symposium on Low Power Electronics and Design, pp. 1619, 1999

[3] P. Andreani and A. Fard,A 2.3 GHz LC-tank CMOS VCO with optimal phase noise performance, in IEEE Int. Solid- State Circuits Conf.(ISSCC) Dig. Tech. Papers, 2006, pp. 194-195.

[4] P. Andreani and A. Fard, More on the 1/f 2 phase noise performance of CMOS differential-pair LC-tank oscillators, IEEE J. Solid-State Circuits, vol. 41, no. 12, pp. 2703-2712, Dec. 2006.

[5] B. Razavi, RF Microelectronics, Second Edition, 2011: Pearson Education, Inc.

[6] J. C. Chien and L. H. Lu,Design of wide-tuning-range millimeter-wave CMOS VCO with a standing wave architecture, IEEE J. Solid-State Circuits, vol. 42, no. 9, pp. 1942-1952, Sep. 2007.

[7] S. L. Jang, S. H. Huang, C. C. Liu, and M. H. Juang, CMOS Colpitts quadrature VCO using the body injection-locked coupling technique, IEEE Microw. Wireless Compon. Lett. , vol. 19, no. 4, pp. 230232, Apr. 2009.

[8] S. Y. Lee and C. Y. Chen, Analysis and design of a wide-tuning-range VCO with quadrature outputs, IEEE Trans. Circuits Syst. II, Exp. Briefs, vol. 55, no. 12, pp. 12091213, Dec. 2008

[9] S. Rong and H. C. Luong, A 1 V 4 GHZ-and-10 GHz trasformer-based dual-band quadrature VCO in 0.18 m CMOS, in IEEE CICC 2007, pp. 817820.

[10] C.-W. Yyao and A. N. Willson, Jr., A phase-noise reduction technique for quadrature LC-VCO with phase-to-amplitude noise conversion, in IEEE ISSCC Dig. Tech. Papers, 2006, pp. 196197

978-1-4799-4993-9/14 $31.00 © 2014 IEEE

Comparative Analyses of Phase Noise in Differential Oscillator Topologies in 28 nm CMOS Technology

Ilias Chlis[1,2], Domenico Pepe[1], and Domenico Zito[1,2]

[1]Tyndall National Institute, "Lee Maltings", Dyke Parade, Cork, Ireland
[2]Department of Electrical and Electronic Engineering, University College Cork, Cork, Ireland
Email: {ilias.chlis, domenico.pepe, domenico.zito}@tyndall.ie

Abstract— This paper reports comparative analyses of phase noise in common-source cross-coupled differential pair, differential Colpitts, Hartley and Armstrong LC oscillator topologies designed in 28 nm CMOS technology for 10 GHz operations. The Impulse Sensitivity Function is used to carry out qualitative and quantitative analyses of the phase noise exhibited by each circuit component in each topology. The analyses show that the lowest phase noise is exhibited by the differential Armstrong topology. Additionally, the results show the impact of flicker noise on phase noise performances.

Keywords—impulse sensitivity function, oscillators, phase noise.

I. INTRODUCTION

One of the most critical components of modern communication transceivers is the local oscillator. Its phase noise (PN) performance directly affects the bit-error rate (BER) of the overall system [1-3]. Despite decades of impressive development in high-frequency IC design, there are no complete studies on how to choose the best oscillator topology for the operating frequencies of interest. Usually, owing to its reliable start-up, common-source cross-coupled differential pair topology is chosen, with no or little further considerations.

A possible way of getting an insight about comparative analysis of phase noise in oscillator topologies is through the use of the impulse sensitivity function (ISF), represented as Γ(x) [4]. ISF can provide the white, flicker and total PN for each device in the oscillator [5]. Thereby, valuable intuition can be gained from the identification of dominant noise sources in the circuit.

In our previous work [5, 6], common-source cross-coupled differential pair topology was compared in terms of its PN performance with single-ended Colpitts and Hartley topologies, at 10 GHz. Under the adopted design conditions, the cross-coupled topology was found to give lower PN than Colpitts, with the highest PN given by Hartley.

Despite the results in [6] provide interesting terms of comparison, it would be more interesting and fair extending the comparison of the common-source cross-coupled differential pair topology to Colpitts and Hartley differential topologies.

This work is supported by Science Foundation Ireland (SFI) under grants 11/RFP/ECE3325 and 07/SK/I1258.

Colpitts and Hartley have long been used in their single-ended configuration in discrete circuit design. However, with the emergence of silicon technology, differential versions have been developed, since the number of transistors is no longer a key cost determining factor. Assuming perfect symmetry, differential topologies are ideally immune to common-mode noise. Examples are the noise coming from the supply rails or from the bias-network. Therefore, it is worth expanding the PN comparison carried out in [5, 6] also to the differential versions of Colpitts and Hartley topologies.

Moreover, the comparative analyses could be interestingly extended to another oscillator topology which may deserve our attention such as Armstrong. Recent implementation of this circuit shows potential for very low power operation, while achieving high output signal spectral purity [7]. Moreover, the exploitation of a transformer coupling allows for smaller inductors, leading to reduced area and higher quality factor (Q) for the inductors [8, 9].

This paper reports comparative analyses of PN in the differential common-source cross-coupled pair, Colpitts, Hartley and Armstrong oscillator topologies. The oscillator circuit topologies have been analysed under the same common design conditions, such as power consumption, supply voltage, transistor current density and sizing (area, aspect ratio, finger width), inductance and quality factor of the integrated spiral inductors, coupling factor of the integrated transformers, and considering the full models of the transistors available within the process design kit, including all their parasitic components related to their actual size, but excluding the layout interconnections, since the additional parasitic components introduced by the layout implementation could mask the results of the topological investigations. The results could drive the designer through the best choice among the examined topologies.

This paper is organized as follows. Section II describes the design of the circuit topologies. Section III reports the comparison of their PN performance. In Section IV the contributions from each noise source to the PN spectrum are presented. Finally, in Section V the conclusions are drawn.

II. OSCILLATOR CIRCUIT TOPOLOGIES

Fig. 1 shows the oscillator topologies designed in 28 nm bulk CMOS technology with 1 V supply. The same figure shows the current impulsive sources acting in parallel to the

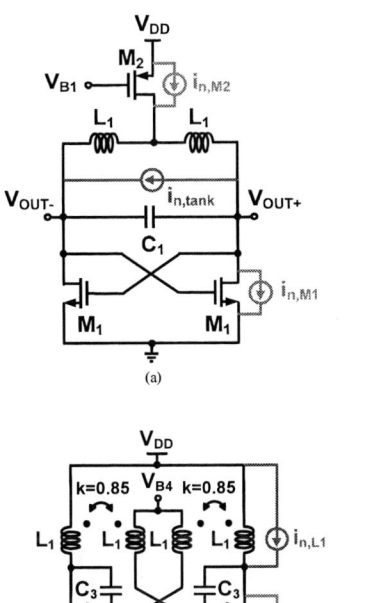
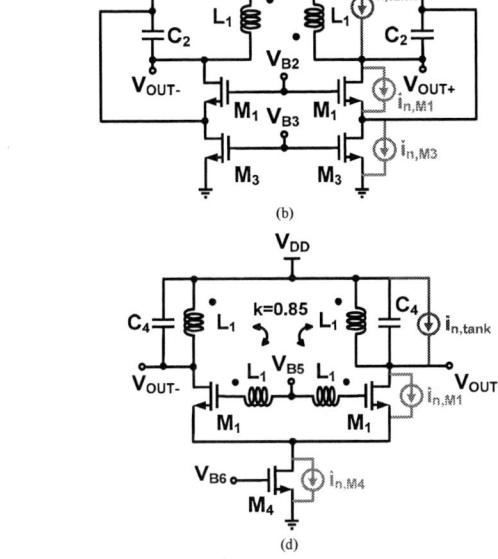

Fig. 1. Schematic of the oscillator circuit topologies: a) common-source cross-coupled differential pair; b) differential Colpitts; c) differential Hartley; d) differential Armstrong with transformer coupling. V_{B1}, V_{B2}, V_{B3}, V_{B4}, V_{B5}, and V_{B6} are the DC bias voltages.

inherent current noise sources and used for calculating the ISF. Based on the findings in [5], transient simulations were performed in Cadence-SpectreRF for an injected current amplitude of 1 µA. Simulation time was significantly reduced by using OCEAN scripts [10].

The sizes of the active and passive devices used are reported in Table I. Capacitors are considered ideal, whereas the quality factor (Q) of all the spiral inductors is assumed to be 10, i.e. a feasible value for 10 GHz [11, 12]. In addition, a coupling factor of 0.85 is assumed for the transformers. Still for a fair comparison, bias currents are chosen such that the total power consumption is 6.3 mW in all cases. Furthermore, all the topologies have been implemented with peer-size transistors for gain (M_1), as well as for the overall bias current.

The cross-coupled pair in Fig. 1 (a) provides the negative resistance needed for oscillation start-up. A p-MOS transistor is chosen as a current source since it exhibits lower flicker noise [13]. The transformer coupling in the differential Colpitts in Fig. 1 (b) helps suppressing common-mode oscillations [14]. Moreover, for lower PN, two separate current sources are used for biasing, as in [15]. Regarding the differential Hartley in Fig. 1 (c), transformer-coupling is used in order to reduce the area on chip occupied by the inductors [16]. Finally, in the differential Armstrong of Fig. 1 (d), the transformer coupling between gate and drain is considered for the same reasons.

TABLE I.   DEVICE SIZING

Transistor Width [µm]				Capacitor value [fF]				Inductor value [pH]
M_1	M_2	M_3	M_4	C_1	C_2	C_3	C_4	L_1
15	30	15	30	229	527	132.5	469	500

III.   PHASE NOISE COMPARISON

Fig. 2 reports the PN obtained through the ISF and the PN obtained by direct plots from periodic steady state (PSS) and periodic noise (Pnoise) simulations. ISF allows us to determine the flicker and thermal noise contributions to PN, also reported in Fig. 2. Table II provides the results for a 1 MHz frequency offset from the carrier. As seen, the PN predicted by ISF matches well (within 1.6 dB) with the values obtained by means of SpectreRF simulations.

From Table II it can be observed that under the adopted design conditions, common to all topologies, for an oscillation frequency of 10 GHz, differential Armstrong topology shows the lowest PN. From SpectreRF results, cross-coupled and differential Hartley are characterized by similar PN performance while differential Colpitts shows the worst spectral purity compared to the others. Thereby, this topological investigation shows that when PN performance is critical, differential Armstrong reported here should be the topology of choice, despite its larger area with respect to the cross-coupled and differential Colpitts topologies.

978-1-4799-4993-9/14 $31.00 © 2014 IEEE   257

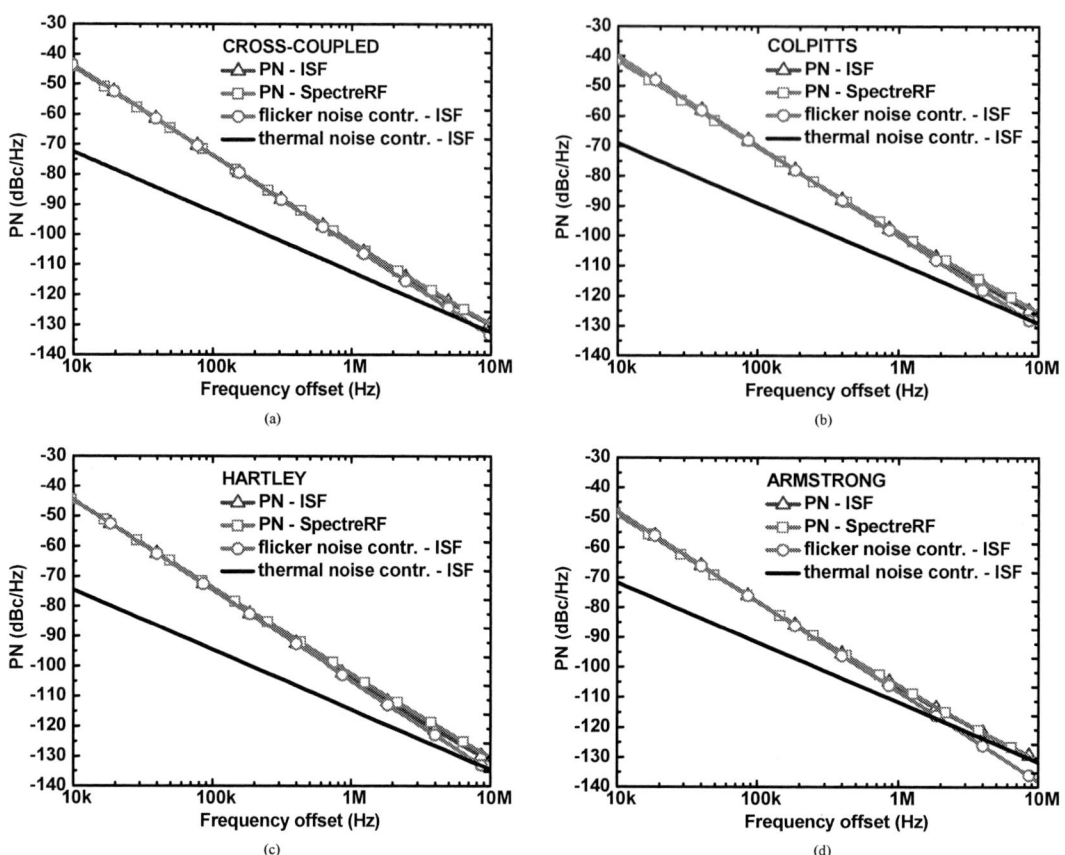

Fig. 2. PN vs. frequency offset obtained through the ISF for a 1 μA current impulse and direct plot from PSS and Pnoise SpectreRF simulations, for the oscillation frequency of 10 GHz. a) common-source cross-coupled differential pair. The $1/f^3$ PN corner is at the frequency offset of 7.5 MHz; b) differential Colpitts. The $1/f^3$ PN corner is at the frequency offset of 8 MHz; c) differential Hartley. The $1/f^3$ PN corner is at the frequency offset of 10 MHz; d) differential Armstrong. The $1/f^3$ PN corner is at the frequency offset of 2.1 MHz.

TABLE II. SUMMARY OF PN PERFORMANCE

Topology	PN [dBc/Hz]	
	SpectreRF	ISF
Cross-coupled	-102.66	-102.69
Colpitts	-99.56	-100.26
Hartley	-102.59	-104.15
Armstrong	-106.12	-106.57

TABLE III. SUMMARY OF NOISE SOURCE CONTRIBUTIONS TO PN

Topology	Devices	Flicker contr. (%)	Thermal contr. (%)	Total contr. at 1 MHz offset (%)
Cross-coupled	M_1-pair	98.9	42.7	85.5
	M_2	1.1	52.9	13.4
	LC tank	-	4.4	1.1
Colpitts	M_1-pair	10.6	19.6	11.6
	M_3-pair	89.4	60.2	86.1
	LC tank	-	20.2	2.3
Hartley	M_1-pair	100	85	98.6
	LC tank	-	15	1.4
Armstrong	M_1-pair	16.4	41	23.9
	M_4	83.6	49.2	73.1
	LC tank	-	9.8	3.0

IV. NOISE CONTRIBUTIONS

Table III reports the percentage contributions of the active and passive devices to the flicker and thermal noise components of the PN spectrum. Table III shows also the total device contributions to PN, at an offset of 1 MHz from the carrier frequency. Several important observations can now be derived from these results.

In the common-source cross-coupled differential pair topology, the cross-coupled pair is the principal contributor to the flicker noise PN component. As a result, due to the fact that at 1 MHz offset the oscillator output spectrum is still at the $1/f^3$ region according to Fig. 2 (a), the cross-coupled pair is the dominant source of PN. Regarding the differential Colpitts topology, the flicker contribution of the tail current transistor prevails. Thereby, for the same reason as above, the noise source of the tail current transistor is responsible for the largest portion of PN at 1 MHz offset. The pair of transistors in differential Hartley is almost the sole source of noise. This is because the noise generated by the parasitic resistance of the inductors, is at least one order of magnitude less than the flicker and thermal noise of the active devices. Finally, the noise source of the tail current transistor in the differential Armstrong is mostly responsible for the total flicker noise of the oscillator output spectrum. Therefore, it presents the main contribution of PN at 1 MHz offset since the $1/f^3$ PN corner is at 2.1 MHz as depicted in Fig. 2 (d).

From the previous discussion we can conclude that in all four topologies, at 1 MHz offset from the carrier frequency, we are still at the $1/f^3$ region of the PN spectrum. This confirms the rising role of flicker noise in oscillators in nanoscale CMOS technology. This means that the devices with the largest contribution to the flicker noise component of PN will also dominate PN. Thereby, the designer should try to minimize if possible both flicker noise generation as well as flicker noise up-conversion due to these sources. For example, increasing the width of the cross-coupled devices of Fig. 1 (a) would reduce the flicker noise produced by the transistor pair. On the other hand, flicker noise up-conversion gain would be increased. This is due to the increased small-signal loop gain which would in turn cause a higher distortion of the voltage output [17, 18]. Thereby, the cross-coupled devices should be sized to assure enough gain for the start-up conditions without exceeding.

Finally, on the basis of the common conditions adopted for the comparison, we can note that the lower flicker noise appearing at the output spectrum of differential Armstrong is due to the lower flicker noise up-conversion effect.

V. CONCLUSIONS

Comparative analyses of phase noise have been carried out for common-source cross-coupled differential pair, as well as for differential Colpitts, Hartley and Armstrong topologies. The circuits were designed in 28 nm CMOS technology, operating at an oscillation frequency of 10 GHz. For a fair comparison, the design of the circuit topologies was carried under the same conditions in terms of power consumption, quality factor of all the spiral inductors, and in particular those of the LC tank. Furthermore, the phase noise results from PSS and Pnoise simulations were compared with phase noise predictions obtained by the impulse sensitivity function. Finally, the contributions from each device to the flicker and thermal noise components of phase noise and to the total phase noise at 1 MHz offset from the carrier frequency were reported and discussed.

In particular, the results show that, under the adopted common design conditions, the differential Armstrong topology considered in this study is superior in terms of phase noise performance to the other topologies under evaluation. On the other hand, the worst performance is exhibited by the differential Colpitts topology. These results would suggest the opportunity to reconsider the Armstrong topology for further investigations and implementations in nano-scale CMOS technology.

Finally, the analysis of the results obtained by the impulse sensitivity function allowed us to identify the dominant noise contributions for each oscillator topology. It can be observed that in all cases, the flicker noise from the active devices is the component with the most significant effect in terms of phase noise on the oscillator output spectrum at 1 MHz offset from the carrier frequency.

REFERENCES

[1] D.Pepe, and D. Zito, "System-level simulations investigating the system-on-chip implementation of 60 GHz transceivers for wireless uncompressed HD video communications," in Applications of MATLAB in Science and Engineering, chapter 9, pp. 181-196, InTech Open Access, Vienna, Austria, 2011.

[2] D. Zito, D. Pepe, and A. Fonte, "13 GHz CMOS LC active inductor VCO," IEEE Microwave and Wireless Components Letters, vol. 22, issue 3, pp. 138-140, 2012.

[3] M. Voicu, D. Pepe, and D. Zito, "Performance and trends in millimetre-wave CMOS oscillators for emerging wireless applications," International Journal of Microwave Science and Technology, vol. 2013, article ID 312618, 6 pages, 2013.

[4] A. Hajimiri, and T.H. Lee, "A general theory of phase noise in electrical oscillators," IEEE Journal of Solid-State Circuits, vol. 33, No. 2, pp. 179-194, Feb. 1998.

[5] I. Chlis, D. Pepe, and D. Zito, "Phase noise comparative analysis of LC oscillators in 28 nm CMOS through the impulse sensitivity function," IEEE Proc. of the 9th Conference on Ph.D. Research in Microelectronics and Electronics (PRIME '13), pp. 85-88, Villach, Austria, 2013.

[6] I. Chlis, D. Pepe, and D. Zito, "Comparative analyses of phase noise in 28 nm CMOS LC oscillator circuit topologies: Hartley, Colpitts, and common-source cross-coupled differential pair," The Scientific World Journal, vol. 2014, article ID 421321, 13 pages, 2014.

[7] T. Nguyen, and J. Lee, "Ultralow-power Ku-band dual-feedback Armstrong VCO with a wide tuning range," IEEE Transactions on Circuits and Systems II, vol. 59, no. 7, pp. 394-398, 2012.

[8] A. Niknejad, "Electromagnetics for high-speed analog and digital communication circuits," Cambridge University Press, p. 147, 2007.

[9] L. Aluigi, F. Alimenti, D. Pepe, L. Roselli, and D. Zito, "MIDAS: Automated approach to design microwave integrated inductors and transformers on silicon," Radioengineering, vol. 22, issue 3, pp. 714-723, 2013.

[10] S. Levantino, P. Maffezzoni, F. Pepe, A. Bonfanti, C. Samori, A. Lacaita, "Efficient calculation of the impulse sensitivity function in oscillators," IEEE Transactions on Circuits and Systems II, vol. 59, no. 10, October 2012, pp. 628-632.

[11] D. Zito, A. Fonte, and D. Pepe, "Microwave active inductors," IEEE Microwave and Wireless Components Letters, vol. 19, issue 7, pp. 461-463, 2009.

[12] D. Pepe, and D. Zito, "50 GHz mm-wave CMOS active inductor," IEEE Microwave and Wireless Components Letters, vol. 24, issue 4, pp. 254-256, 2014.

[13] A. Antonopoulos, M. Bucher, K. Papathanasiou, N. Mavredakis, N. Makris, R. K. Sharma, P. Sakalas, and M. Schroter, "CMOS small-signal and thermal noise modeling at high frequencies," IEEE Transactions on Electron Devices, vol. 60, no. 11, pp. 3726-3733, 2013.

[14] P. Andreani, X. Wang, L. Vandi, and A. Fard, "A study of phase noise in Colpitts and LC-tank CMOS oscillators," IEEE Journal of Solid State Circuits , vol. 40, no. 5, pp. 1107-1118, 2005.

[15] L. Dauphinee, M. Copeland, P. Schvan, "A balanced 1.5 GHz voltage controlled oscillator with an integrated LC resonator", IEEE International Solid State Circuits Conference, Dig. of Technical Papers, pp. 390-491, 1997.

[16] M. Bao, and Y. Li., "A compact 23 GHz Hartly VCO in 0.13μm CMOS technology," 3rd European Microwave Integrated Circuits Conference, pp. 75-78, 2008.

[17] A. Bonfanti, F. Pepe, C. Samori, and A. Lacaita, "Flicker noise up-conversion due to harmonic distortion in Van der Pol CMOS oscillators," IEEE Transactions on Circuits and Systems I, vol. 59, no. 7, pp. 1418-1430, 2012.

[18] F. Pepe, A. Bonfanti, and A. Lacaita, "A fast and accurate simulation method of impulse sensitivity function in oscillators," IEEE Proc. of 35th International Convention MIPRO, pp. 66-71, 2012

Pulsed oscillations generator based on initial conditioned and switched cross-coupled MOS oscillator. Application to the synchronization of the pulsed oscillations.

Clément Jany, Alexandre Siligaris, Pierre Vincent
CEA-TECH, LETI,
17, rue des Martyrs
38054 Grenoble, France
Email: clement.jany@cea.fr

Philippe Ferrari
Université de Grenoble-Alpes,
CNRS, IMEP-LAHC,Minatec,
38016 Grenoble, France

Abstract—**This work focuses on the analysis of a periodically switched cross-coupled nMOS oscillator when periodic initial conditions are applied. Based on a classical Van der Pol model of the oscillator, an analytical solution is used that takes into account the time dependency of the instantaneous amplitude and frequency. This leads to a model for pulsed oscillations generated by the initial conditioned and the periodically switched oscillator. An analysis is performed on the synchronization of the pulsed oscillations depending on the initial condition. It is shown that when periodic initial conditions are applied at oscillation start, the pulsed oscillations are synchronized on the input signal. The observed synchronization phenomenon is called pseudo-locking.**[1]

I. INTRODUCTION

Ultra-Wide-Band (UWB) circuits are widely used in todays applications such as Wireless Sensor Networks, Gigabit WLAN, automotive radar [1-2]. UWB relies on the use of pulsed oscillations that are composed of a succession of oscillations wagons separated by blank intervals and can be obtained switching on-and-off the supply of a cross-coupled nMOS oscillator [2]. This is equivalent to periodically switching the nonlinear function in the Van der Pol oscillator. This particular oscillator has been intensively studied in the last century, but no exact analytical solution has been found yet. In this paper, a novel analytical solution of the Van der Pol oscillator is proposed that takes into account the time-dependent amplitude and frequency behavior for the pulsed oscillations and leads to a novel model. It is shown that the resulting pulsed oscillations can be synchronized to the switching signal when applying specific initial conditions at each oscillating start. Thus, when synchronization occurs, the output signal copies the phase and frequency properties of the input switching signal. The circuit architecture and the related differential equation are presented in section II. Section III gives the analytical equations of the pulsed oscillations and the model is compared to numerical simulations showing a very good accuracy. In section IV it is shown that the output signal can be synchronized to the switching signal and thus copy the phase and frequency properties.

II. PERIODICALLY REPEATED OSCILLATION TRAIN ARCHITECTURE AND CHARACTERISTIC EQUATION

In this work we consider the oscillator depicted in Fig. 1.b. It is composed of a nMOS cross-coupled transistor pair and an LC tank. The cross-coupled pair is switched on and off by the T_{PRP}-periodic signal v_{in} via transistor T_3. Transistor T_4 is used to inject an initial condition (IC) in the oscillator at every oscillation start as described in figure 1.a.

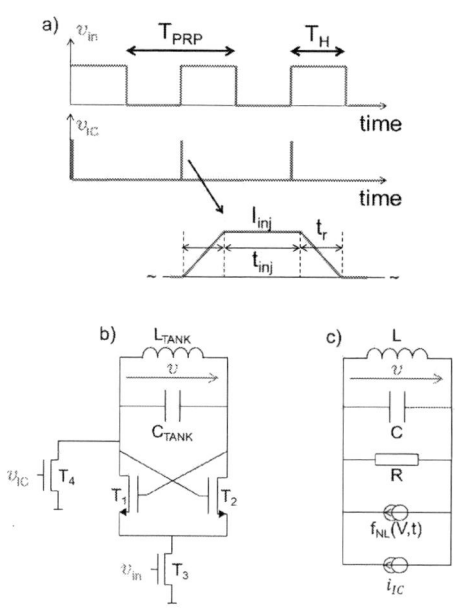

Fig. 1.   **a)** Waveforms of the input signal v_{in} and the initial condition. **a)** Oscillator composed of a cross-coupled nMOS transistor pair, an LC tank and a current source T_3 driven by the input signal v_{in}. **c)** The simplified schematic where $f_{NL}(v,t)$ stands for the equivalent nonlinear I-V polynom of the cross-coupled nMOS pair.

According to [3], the cross-coupled pair can be modeled as a third order non-linear I-V characteristic:

$$i(v) = f_{NL}(v) = -\alpha v + \gamma v^3. \qquad (1)$$

This is the Van der Pol function [4]. Switching periodically the cross-coupled pair with the v_{in} signal (see Fig. 1.a) introduces a periodic dependency into the nonlinear parameters α and γ. These parameters are constant during T_H duration and zero otherwise:

$$f_{NL}(V,t) = \begin{cases} -\alpha V + \gamma V^3 & \text{, if } v_{in} = \text{"1"} \\ 0 & \text{, if } v_{in} = \text{"0".} \end{cases} \quad (2)$$

The intrinsic elements of the MOS transistors (g_D, C_{GS}, C_{DS}, C_{GD}) are taken into account in the total resistance R and the total capacitance C. Hence, the oscillator depicted in Fig. 1.b is simplified into the schematic in Fig. 1.c. Applying Kirchhoff's laws on the simplified schematic shown in Fig. 1.c, one obtains the following non-linear ordinary differential equation:

$$\ddot{v} + \omega_0^2 v = \frac{1}{C}\left(\frac{df_{NL}(v,t)}{dv} - \frac{1}{R}\right)\dot{v} + \frac{1}{C}\frac{di_{IC}}{dt}, \quad (3)$$

with $\omega_0^2 = \frac{1}{LC}$.

Equation (??) describes the operation of the system depicted in Fig 1.c. Equation (??) has been numerically simulated with realistic circuit parameters, as depicted in Fig. 2.

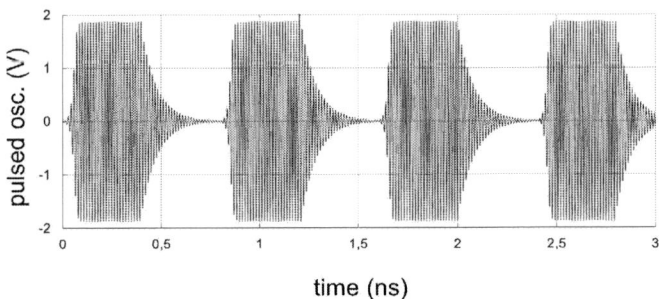

Fig. 2. Numerical simulation of equation (??).

III. ANALYTICAL STUDY OF OSCILLATIONS

Pulsed oscillations are composed of periodically repeated starting ($v_{in} = 1$) and vanishing ($v_{in} = 0$) of oscillations. Equation (2) is the simplification of (??) for the starting of oscillations.

$$\begin{cases} v_S(0) = v_V(end); \dot{v}_S(0) = \dot{v}_V(end) \\ \ddot{v}_S + \omega_0^2 v_S = \frac{1}{C}\left(\alpha - \frac{1}{R} - \gamma v_S^2\right)\dot{v}_S + \frac{1}{C}\frac{di_{IC}}{dt}. \end{cases} \quad (4)$$

The first line of this equation deals with the initial condition of this equation which are the final condition for the previous vanishing. This is the free Van der Pol equation [4], which finds no exact analytical solution, as of yet.

Equation (3) is the simplification of (??) for the vanishing of oscillations.

$$\begin{cases} v_V(0) = v_S(end); \dot{v}_V(0) = \dot{v}_S(end) \\ \ddot{v}_V + \frac{1}{RC}\dot{v}_V + \omega_0^2 v_V = 0. \end{cases} \quad (5)$$

This is a linear second order ordinary differential equation.

Approximated solutions for the starting of oscillation and the exact analytical solution for the vanishing of oscillations are presented in the next subsections.

A. Initial condition of each oscillation wagon

Let's consider that the initial condition signal i_{IC} length ($t_{inj} + 2t_r$) is short compared to the period of v_{in} (T_{PRP}). Then the influence of i_{IC} is modeled as a new initial condition of the differential equation (2). Thus, (2) becomes :

$$\begin{cases} v_S(0) = IC; \dot{v}_S(0) = \dot{IC} \\ \ddot{v}_S + \omega_0^2 v_S = \frac{1}{C}\left(\alpha - \frac{1}{R} - \gamma v_S^2\right)\dot{v}_S. \end{cases} \quad (6)$$

With IC and $\dot{IC}$ the influence of the input signal v_{IC}. Those values are calculated as the solution of equation (2) at $t = t_{inj} + 2t_r$. This corresponds to the birth of oscillation where the oscillations are small, so that γ is neglected. Hence, equation (2) is simplified into:

$$\begin{cases} v_B(0) = v_V(end); \dot{v}_B(0) = \dot{v}_V(end) \\ \ddot{v}_B + \omega_0^2 v_B = \frac{1}{C}\left(\alpha - \frac{1}{R}\right)\dot{v}_B + \frac{1}{C}\frac{di_{IC}}{dt}. \end{cases} \quad (7)$$

From the waveform of i_{IC} in Fig. 1, the waveform of its first derivative is deduced :

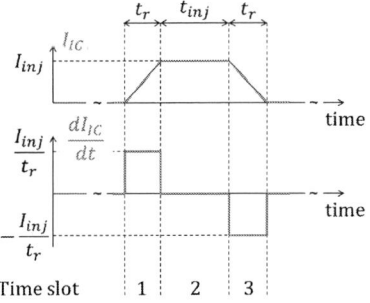

Fig. 3. Waveform of the initial condition signal and of its first derivative.

It's obvious that this function $\frac{di_{IC}}{dt}$ is constant in each of the three time slots pictured in Fig. 2. The general solution of the second order linear ordinary differential equation 5 in the time slot i is :

$$v_{B_i}(t) = A_i e^{\frac{t}{\tau_B}} \sin(\omega_B t + \phi_i), \quad (8)$$

with

$$\begin{cases} \frac{1}{\tau_B} = \frac{1}{2C}\left(\alpha - \frac{1}{R}\right), \\ \omega_B = \frac{1}{2}\sqrt{\frac{4}{LC} - \frac{1}{C^2}\left(\alpha - \frac{1}{R}\right)^2}. \end{cases} \quad (9)$$

and A_i and ϕ_i the initial conditions of the ith time slot, and are calculated from the boundary conditions :

$$\begin{cases} v_{B_1}(0) = v_V(end); \dot{v}_{B_1}(0) = \dot{v}_V(end) \\ v_{B_2}(0) = v_{B_1}(t_r); \dot{v}_{B_2}(0) = \dot{v}_{B_1}(t_r) \\ v_{B_3}(0) = v_{B_2}(t_{inj}); \dot{v}_{B_3}(0) = \dot{v}_{B_2}(t_{inj}) \end{cases} \quad (10)$$

Finally, one obtains :

$$\begin{cases} CI = v_{B_3}(t_r) \\ \dot{CI} = \dot{v}_{B_3}(t_r) \end{cases} \quad (11)$$

and the oscillation start is fully described by equations (4) and (7).

B. Modeling the starting and the vanishing of oscillations

The approximated solution for equations (4) and (3) can be found in [7]. It is written in the form :

$$\begin{cases} v_S(t) = A_S(t) \sin \left(\omega_S(t) \, t + \phi_{S_0} \right), \\ v_V(t) = A_{V_0} e^{-\frac{t}{\tau_V}} \sin \left(\omega_V + \phi_{V_0} \right), \end{cases} \quad (12)$$

with

$$A_S(t) = \frac{A_{S_0} e^{\frac{\varepsilon \omega_0 t}{2}}}{\sqrt{1 + \left(\dfrac{A_{S_0}}{A_{SS}} \right)^2 \left(e^{\varepsilon \omega_0 t} - 1 \right)}}, \quad (13)$$

$$\omega_S(t) = \omega_B + K A_S(t)^2 \frac{d A_S(t)}{dt} + \frac{\omega_{SS} - \omega_B}{A_{SS}} A_S(t). \quad (14)$$

$$\begin{cases} \tau_V = 2RC \\ \omega_V = \dfrac{1}{2} \sqrt{\dfrac{4}{LC} - \dfrac{1}{(RC)^2}}. \end{cases} \quad (15)$$

$$\begin{aligned} A_{SS} &\approx 2 \sqrt{\dfrac{\alpha - 1/R}{\gamma}}, \\ \omega_{SS} &\approx \dfrac{\omega_0}{1 + \frac{1}{16}\varepsilon^2}, \end{aligned} \quad (16)$$

$$\varepsilon = \frac{\alpha - 1/R}{C \omega_0} \quad (17)$$

C. Pulsed oscillations model

As explained in [7], the pulsed oscillation model consists in a succession of starting and vanishing of oscillations. Let us rewrite equations for the starting and the vanishing (8) of oscillations as a function of both time and initial conditions:

$$v(t) = \begin{cases} v_S\left(t_k, A_{S0}^k, \phi_{S0}^k \right) & , t_k \in [0; T_H) \\ v_V\left(t_k - T_H, A_{V0}^k, \phi_{V0}^k \right) & , t_k \in [T_{PRP} - T_H; T_{PRP}), \end{cases} \quad (18)$$

with $t_k = t \, modulo \, T_{PRP}$. $\left(A_{S0}^k, \phi_{S0}^k \right)$ is the initial condition of the kth oscillations starting and is calculated from IC and $\dot{IC}$. The initial condition of the kth oscillations vanishing $\left(A_{V0}^k, \phi_{V0}^k \right)$ are calculated from the final condition of the previous oscillations starting. Comparison between this model (14) and numerical simulation of (**??**) is shown in Fig. **??**.

IV. SYNCHRONIZATION ON THE INPUT SIGNAL

According to [7], the synchronization on the input signal depends on the convergence of a recursive sequence that is calculated form the boundary conditions between the vanishing and starting regimes. Indeed, the final state at the end of the kth vanishing depends on its initial state, that corresponds the final state of the kth starting of oscillation which depends on its initial state, etc etc. Hence, it was proved that the convergence of this recursive sequence depends on the system parameters, on the input signal parameters and on the very first initial condition. The convergence of this sequence, that is equivalent to the T_{PRP}-periodicity of the pulsed oscillations, is called pseudo-locking.

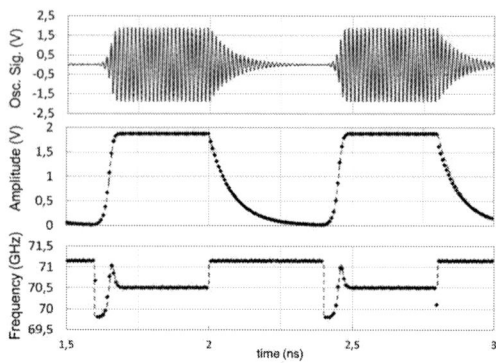

Fig. 4. Comparison between numerical simulation of (**??**) and the proposed model (14). For the numerical simulation of (**??**), discrete amplitude values are considered at the maximum of each oscillation, and the frequency is obtained using a classical zero-crossing method.

In this work, an initial condition signal i_{IC} is used at the beginning of every starting of oscillation. Assuming that the oscillation residue at the beginning of the starting of oscillations is negligible compared to the provided initial condition, then IC and $\dot{IC}$ no longer depend on the oscillation residue but depend only on the external initial condition i_{IC}. It follows that the initial state of any starting of oscillation does no longer depend on the final state of the previous vanishing, and the pseudo-locking is guaranteed.

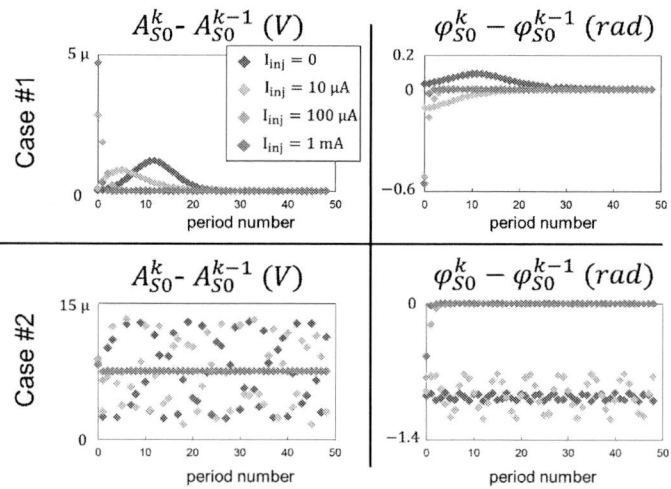

Fig. 5. Numerical simulations of the model presented in 14. At each oscillations start, the differential amplitude $A_{S0}^k - A_{S0}^{k-1}$ and phase $\phi_{S0}^k - \phi_{S0}^{k-1}$ are plotted for various values of I_{inj}. The first case treats a set of the oscillator parameters for which the pulsed oscillations converge for any initial condition. In the second case, the convergence is achieved only when I_{inj} is high enough.

Fig. 4 shows at each starting of oscillations the differential amplitude $A_{S0}^k - A_{S0}^{k-1}$ and phase $\phi_{S0}^k - \phi_{S0}^{k-1}$ for various values of I_{inj}. Two cases are treated with two different sets of the system parameters $(R, L, C, \alpha, \gamma)$.

In the first case, synchronization is observed for all the values of the initial condition signal amplitude (I_{inj}). Increasing I_{inj} results in a faster synchronization. Indeed, when $I_{inj} = 1mA$, the pulsed oscillations are synchronized on the input signal in less than 10 periods, which is equivalent to $5nsec$ time for an input period of $500psec$. In the second case, the system parameters are set so that no pseudo-locking is achieved without the initial condition signal (I_{inj}). Hence, the red symbols do not converge. Increasing I_{inj} results into the synchronization of the pulsed oscillations, as it is shown by the green symbols (I_{inj}=100 μA and I_{inj}=1 mA). Moreover, the bigger the initial condition amplitude (I_{inj}) is, the faster the oscillations start. Hence, the initial condition parameters (I_{inj},t_r and t_{inj}) can be set by the designer to adjust the system performances.

V. CONCLUSION

In this work, a theoretical analysis based on the Van der Pol equation is performed, allowing to model pulsed oscillations produced by an initial conditioned and periodically switched nMOS cross-coupled oscillator. A compact analytical model is proposed that takes into account the time-dependence of the amplitude and frequency of oscillations. It is shown that the use of this external periodic initial condition signal guaranties the synchronization of the pulsed oscillations on the input switching signal. The analysis presented in this paper highlights the potentialities of such a circuit for UWB applications.

REFERENCES

[1] A. Siligaris, et al. "A low power 60-GHz 2.2-Gbps UWB transceiver with integrated antennas for short range communications," *IEEE Radio Frequency Integrated Circuits Symposium*, pp. 297-300, June 2013.

[2] M. Pelissier, et al. "A 112Mb/s full duplex remotely-powered impulse-UWB RFID transceiver for wireless NV-memory applications," *IEEE Symposium Very Large Scale Integration, Circuit*, pp. 25-26, June 2010.

[3] N. Deparis, et al.,"Ultra low consumption UWB pulsed-ILO RF front-end transmitter at 60 GHz in 65-nm CMOS-SOI,"*IEEE International Symposium on Personal, Indoor and Mobile Radio Communications*, pp.1652-1656, Sept. 2009.

[4] B. Van der Pol, "The Nonlinear Theory of Electric Oscillations," *Proceedings of the Institute of Radio Engineers*, pp.1051-1086, Sept. 1934.

[5] N. Bogoliubov and J. Mitropolsky, 'Les méthodes asymptotiques en théorie des oscillations non linéaires' (in French), *Monographies internationales de mathématiques modernes, 2*, Gauthier-Villars, 1962.

[6] C. Jany et al., "A novel harmonic selection technique based on the injection of a periodically repeated oscillations train into an oscillator", *IEEE MTT International Microwave Symposium*, june 2014.

[7] C. Jany et al.,"A novel approximated solution for the Van der Pol oscillator. Application to pulsed oscillations modeling in switched cross-coupled MOS oscillators," *IEEE Midwest Symposium on Circuit And System 2014*, august 2014.

High Precision Bidirectional Chopper Instrumentation Amplifier With Negative and Positive Input Common Mode Range

Matei Nicolae Stan[i,ii], Laurentiu Creosteanu[ii], Gheorghe Brezeanu[i]

[i] Electronics, Telecommunications and Information Technology
University Politehnica of Bucharest
Bucharest, Romania
Matei.Stan@onsemi.com

[ii] ON Semiconductor Romania
Bucharest, Romania

Abstract — **A current feedback instrumentation amplifier (CFIA) with extended common mode input voltage range is presented. Referred to input offset of 70μV is obtained firstly by using chopping and gain enhancement techniques in order to increase the DC gain and secondly by adding an offset reduction loop (ORL). The proposed circuit is implemented in a high voltage 0.5μm SOI technology. A common mode rejection ratio (CMRR) of 260dB has been obtained by using single ended three stage architecture with a common mode input voltage from -18V to 80V. The stability of the CFIA is ensured for a close loop gain of 20V/V.**

Keywords—chopping, offset, high common mode input voltage

I. INTRODUCTION

In many current sensing applications, amplifying very small differential signals over a wide common mode input voltage is needed. The prevailing error sources in these applications are offset, 1/f noise and drift. There are three offset compensation techniques: trimming, auto-zeroing and chopping and combinations of these.

Trimming is generally used within an operational amplifier with resistive feedback in order to compensate the mismatch between resistors.

The second approach is auto-zeroing which has two different phases: measuring the offset and then subtracting it.

Unlike auto-zeroing, chopping is a continuous time modulation technique in which the offset is moved at higher frequencies and then filtered by a low pass filter. Although the auto-zeroing technique is also used in noise and dc offset cancelation, the chopping technique has the advantage that using continuous time modulation instead of discrete-time modulation does not cause noise folding back with the trade off of larger output ripple.

II. INSTRUMENTATION AMPLIFIERS WITH CHOPPING TECHNIQUE

The basic architecture for the current feedback instrumentation amplifier is presented in Fig. 1.

The first transconductance stage (gm_{in}) converts the input differential voltage (V_{id}) into a differential current (I_1). The voltage feedback (V_{fb}) is collected from the R_2 resistor. Gm_{fb} transconductance transforms V_{fb} into the I_2 differential current. The output stage A regulates V_{out} so as to set the sum of I_1 and I_2 to zero.

$$V_{out} = \frac{gm_{in}}{gm_{fb}} \frac{R_1 + R_2}{R_2} V_{id} + V_{ref} \qquad (1)$$

The output voltage as a function of V_{id} is shown in Eq. 1. The gain accuracy is dependent on the matching of gm_{in} and gm_{fb}.

One of the main advantages of this structure is that high CMRR is obtained by using the gm_{in} input transconductor in order to isolate the input common mode voltage by converting the input differential voltage in differential currents[1]-[5].

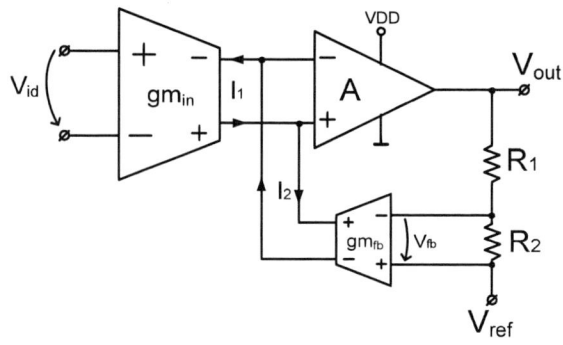

Fig. 1.   Current feedback instrumentation amplifier.

978-1-4799-4993-9/14 $31.00 © 2014 IEEE

III. CFIA BLOCK DIAGRAM AND CHOPPING STRATEGY

The block diagram of CFIA is shown in Fig. 2. The first stage is composed by gm_{in} transconductor and A1 and the second and third stage is symbolized by the A$_2$ block.

The gm_n and gm_p transconductor are used to sense positive and respectively negative input common mode voltages. A$_2$ represents a coscode structure that sums the currents from gm_{in} and gm_{fb}. The second stage is a folded cascode amplifier followed by an AB class amplifier in order to drive a 50kΩ resistive load.

Other papers have proved that using two different frequencies for chopping the input and intermediate stage shows a 200 times smaller 1/f noise than if chopping had been used only on the first stage [6].

Moreover, the input stages have a DC gain of 170dB to supply enough loop gain and eliminate the noise from the intermediate stage. In order to assure the stability of the three-stage amplifier, for a closed loop gain of 20V/V, a nested Miller frequency-compensation was used. Since two chopping frequencies are used, the amplitude of the resulting chopper ripple can have quite high values, in the range of hundreds of mV [1].

Fig. 2 highlights the offset reduction loop (ORL), which synchronously demodulates the output ripple from AC to DC to obtain an average value of the offset. This value is then converted to I_{out} currents in order to attenuate the offset. The ORL is composed of a sense capacitor C_{orl}, an integrator A_{orl} with C_{int}, a chopper $CH4$ and a transconductor gm_I. The sense capacitor C_{orl} will, in the start-up condition, convert the amplifier output ripple $V_{out, ripple}$ into an AC current I_{orl}. The AC current is afterwards demodulated by chopper CH4. The DC current I_{dc} resulting from this process will further be integrated by an integrator and will generate a DC compensation voltage V_{orl}. The latter is proportional with the ripple amplitude. It will be fed back via transconductance gm$_I$ to the outputs of gm_{in} and gm_{fb} and will inject a current that compensates for the offset between these two.

For high negative and positive common mode input voltage two level shifters are used to drive input chopper CH1 in order to limit the Vgs voltage of the switches.

IV. INPUT STAGE

Eq. 1 shows that the closed loop gain is also given by the ratio between gm_{in} and gm_{fb}. The transconductor gm_{fg} must sense the V_{fb} voltage which is around the ground potential, hence gm_{fb} level must have the input pair implemented with PMOS transistors. For positive and negative input CM, transconductors gm_p and gm_n must have the same transconductance and it must be equal to gm_{fb} in order to maintain the ratio $gm_{in}/gm_{fb}=1$ [7].

Fig. 3 shows, at transistor level, the way in which the same transconductance is obtained for two complementary stages with the purpose of maintaining a better gain accuracy. Low voltage transistors M_{21} and M_{22} shift the input voltage V_{sense} with V_{gs} and force these to drop across the resistors R_{n1} and R_{n2}. The amplifying transistors M_{23} and M_{24} are driven by the high voltage transistors M_{19} and M_{20}. This way the transconductance of gm_{in} is precisely set by the value of R_{n1} and R_{n2}. Since the maximum input CM voltage is 80V and the V_{ds} of M_{19}, M_{20}, M_{23} and M_{24} is close to V_{cm}, a high voltage process is used.

Gm$_p$ transconductor is designed in the same fashion as gm_n with complementary transistors in order to sense V_{sense} for negative common mode range. Instead of injecting, PMOS transistors M_5 and M_6 sink I_1 and I_2 currents in folded cascode stage. A folded cascode composed by the tranzistors M_9 to M_{12} is used in order to source the differential currents into the signal path. For a voltage supply of 5V the slight transition between these two complementary stages gm_p and gm_n is made at 2.5V. When the V_{cm} voltage goes above 2.5V, gm_n transconductor starts working whilst gm_p transconductor works properly for V_{cm} smaller than 2.5V. The sum of the output currents I_{p1}, I_{p2}, I_{n1} and I_{n2} is constant. For a V_{cm} higher than 2.5 V, I_{n1} and I_{n2} currents are injected in a fully differential folded cascode. However, when V_{cm} is lower than 2.5V the I_{p1} and I_{p2} currents source in the same way. Since the folded cascode is a fully differential structure, a common-mode voltage feedback is used in order to set the output voltages V_{out+} and V_{out-} at the halfway of the supply rails [3].

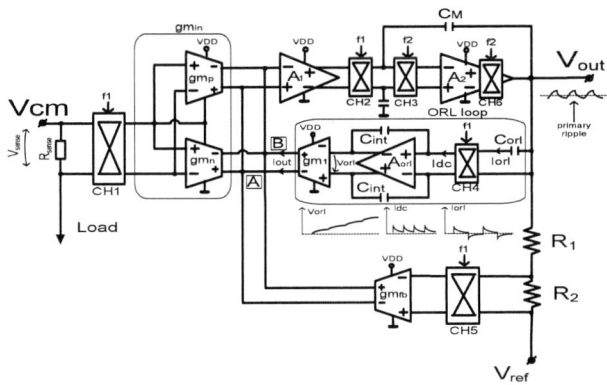

Fig. 2. CFIA Block diagram and chopping strategy

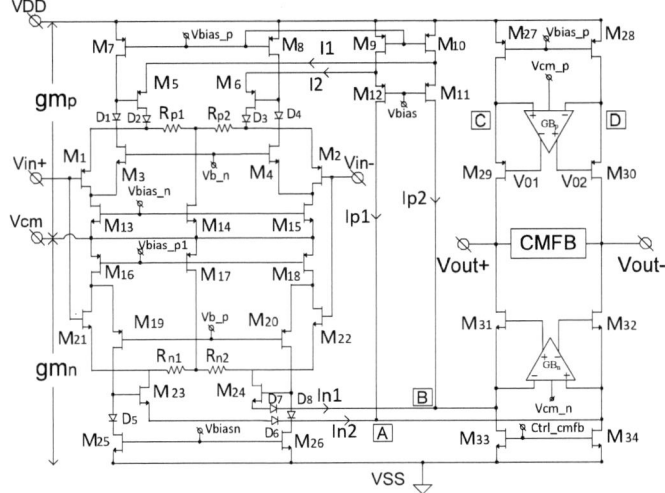

Fig. 3. Schematic of the input stage

978-1-4799-4993-9/14 $31.00 © 2014 IEEE

V. SECOND AND THIRD STAGE

Since the output of the first stage is at 2.5V, the second stage is composed of a folded cascode with PMOS input transistor followed by a class AB output stage [4]. The simplified transistor level implementation for intermediate and output stage is shown in Fig. 4. Class-AB output stage was implemented in order to obtain rail to rail and output current capability. A translinear loop composed of the M_{46} to M_{49} transistors, which set the output quiescent current, is highlighted in Fig. 4. To achieve better AC performance M_{49} and M_{50} transistors are bootstrapped. The CH3 and CH5 choppers, which are used for demodulating the f_2 frequency, are situated at a low impedance node in order to obtain a better settling time.

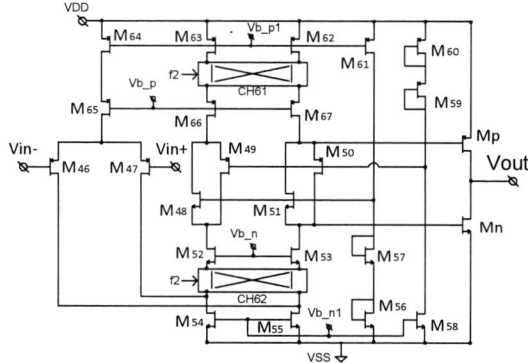

Fig. 4. Schematic of the the second and third stage

VI. AUXILIARY BLOCKS

A. Feedback transconductor

Fig. 5 describes the implementation of gm_{fb} which is similar to that of transconductors gm_p and gm_n. M_{35} and M_{36}, which are NMOS input transistors, behave as voltage followers and their purpose is to force the input voltage V_{id} along resistors R_{fb1} and R_{fb2}. The inverting amplifiers realized with M_{15} and M_{16} are driven by NMOS folded cascode transistors M_{13} and M_{14} that are connected to the drain of the transistors M_{11} and M_{12}.

The linearity of these stages is assured by I*R product. Considering that the circuit has to work between -25° and 125°C, the bias generator must have complementary variation characteristic to the resistors, over the temperature range [2].

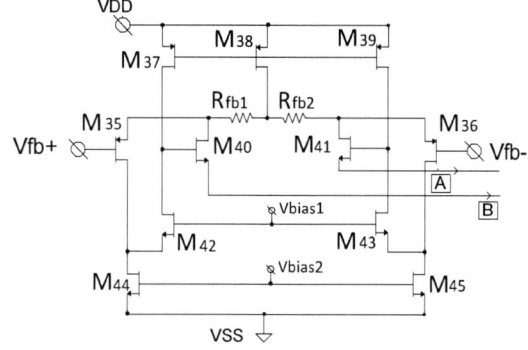

Fig. 5. Transistor level gmfb transconductor

B. Gain boosted amplifiers

The noise and gain accuracy of the CFIA is determined by the input and feedback stage. A fully differential gain boosted (GB) topology is used for the input stage, providing a high gain in order to suppress noise and nonlinearity from the next stages. Fig. 6 shows the implementation of GB_n which regulates the gate of M_{28} and M_{29} in order to set the C and D node voltage at V_{cm_p}. In this way a very high output resistance and therefore high gain is obtained. GB_p amplifier is designed similarly, the only difference being the use of complementary transistors to maintain M_{33} and M_{34} in the saturation region.

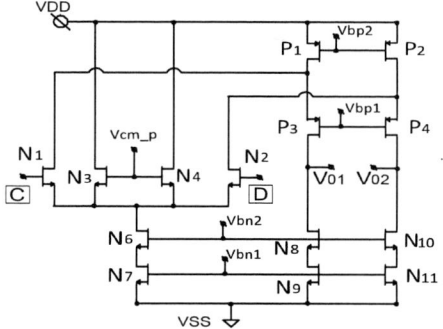

Fig. 6. Boost amplifier GB_n

VII. SIMULATION RESULTS

The current feedback instrumentation amplifier was simulated in a high voltage 0.5 μm SOI technology with low threshold voltage transistors and high-resistivity poly resistors.

A comparison with and without the offset reduction loop is shown in Fig. 7. As it can be seen, the ORL acts like a notch filter having the corner frequency close to 30kHz and therefore the output waveform is smoothed. The waveform of V_{out} with ORL enabled shows a much lower output ripple.

All the simulations performed have V_{REF}=2V, V_{DD}=5V.

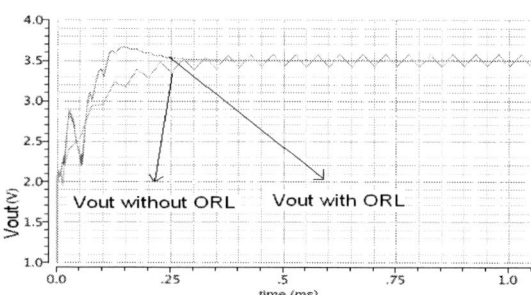

Fig. 7. Simulation results - output voltage with and without ORL
for V_{sense} = 75mV

Given the fact that the amplifier is bidirectional, Fig. 8 illustrates the V_{out} waveform for positive and negative values of V_{sense} while the common mode voltage is -18V. For this simulation purpose we added an additional offset source V_{OS} of 10mV to show the CFIA ability to reduce its initial offset. We present for each V_{sense} voltage two simulation results: the upper one with V_{OS}=10mV added and the bottom one with V_{OS}=0mV.

978-1-4799-4993-9/14 $31.00 © 2014 IEEE   266

The same simulations are performed at common mode voltage of 50V. The results are presented in Fig. 9.

We have observed that at high common mode voltage the gain error is larger because there is a worse matching between the gm_{in} and gm_{fb}.

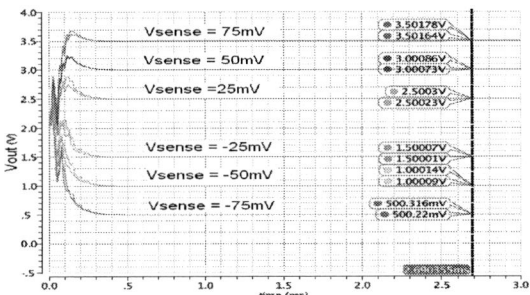

Fig. 8.   Simulation results - output voltage for
Vcm = - 18V an Vref = 2V

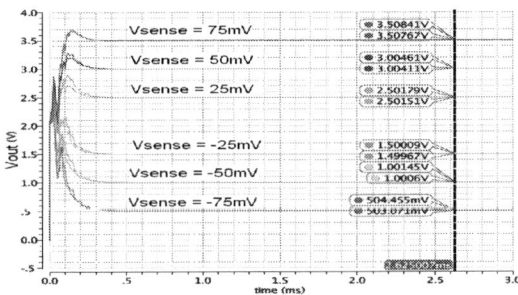

Fig. 9.   Simulation results - output voltage for
Vcm = 50V and Vref = 2V

The gain accuracy of the topology is defined as the ratio between output resistor R_1 and R_2 and the mismatch between the input and feedback transconductance. Monte Carlo simulations were performed using Spectre®. The variations of these parameters were compensated using proprietary trimming procedure in order to maintain the gain error in proposed specifications.

In Fig. 10.   output voltage variation is presented due to the process variation and mismatch.

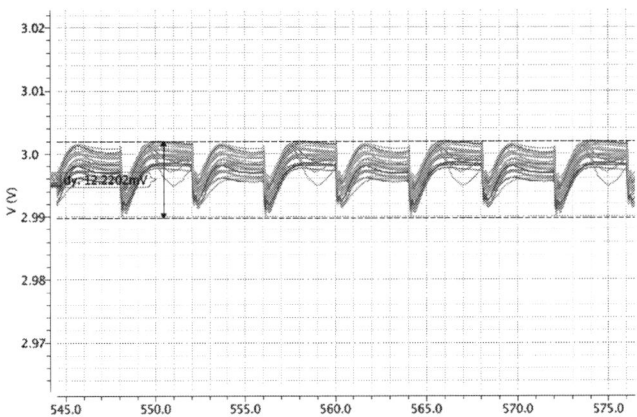

Fig. 10. Output voltage for V_{ref}=2V and V_{sense}=50mV

Table 1 illustrates the proposed specifications of the circuit along with the simulation results.

Table 1. Proposed performance vs. simulation results

Parameter	Unit	Proposed Performance	Simulation Results		
			Min	Typ	Max
RTI offset voltage	μV	<200	60	160	200
Input CM range	V	-18 to 80	-18 to 80		
Input DF rage (Vsense)	mV	+ 150	+ 150	+ 150	+ 150
Gain error	%	+ 0.4	+ 0.15	+ 0.2	+ 0.35
CMRR	dB	>210	220	240	-
Quiescent current	μA	<500	390	410	420

VIII.   CONCLUSION

The paper presents a new strategy for implementing a bidirectional high accuracy current sense amplifier where the common mode voltage V_{cm} can independently range from -18 to 80 V while the supply voltage is at 5V.

Compared to a classical OA with resistive feedback, which denotes four large resistors in order to minimize current consumption from the input terminals, this topology implies only two small resistors and thus a better matching. Also, better CMRR figures are obtained with less demanding from the technology. Given that the structure performs bidirectional current measurements across a shunt resistor the output voltage V_{out} is referred to an external voltage reference V_{ref}.

The amplifier achieves a $70\mu V$ referred to input offset while drawing a $120\mu A$ from V_{cm} and $420\mu A$ from the power supply. These good performances are obtained over a common mode range that includes the -18V whilst the state of the art is ranging starting from 0V.

Acknowledgment

This work has been funded by the Sectoral Operational Program Human Resources Development 2007-2013 of the Ministry of European Funds through the Financial Agreement POSDRU/159/1.5/S/134398.

REFERENCES

[1]   J. H. Huijsing, R. Wu, K. A. A. Makinwa, "Current feedback instrumentation amplifiers", Patetnt no.: US 8,179,195, B1, May 2012.

[2]   J. H. Huijsing, B. Shahi, "Accurate voltage to current converters for rail sensing current-feedback instrumantation amplifiers", Patent no.: US 7,202,738 B1, April 2007.

[3]   J. F. Witte, K. A. A. Makinwa, J. H. Huijsing, Dynamic Offset CMOS Compensated Amplifiers, Springer, 2009.

[4]   J. H. Huijsing, Operational Amplifiers- Theory and Design, 2nd ed., Springer, 2011.

[5]   J. F. Witte , J. H. Huijsing, K. A. A. Makinwa, "A current feedback instrumentation amplifier with 5μV offset for bidirectional high-side current-sensing", IEEE Journal of Solid-State Circuits, Vol 43, no. 12, December 2008, pp. 2769-2775.

[6]   R. Wu, K. A. A. Makinwa, J. H. Huijsing, "A chopper current-feedback instrumentation amplifier with a 1 mHz 1/f, noise corner and an AC-coupled ripple reduction loop", IEEE Journal of Solid-State Circuits, Vol 44, no. 12, 2009.

[7]   R. Wu, Precision Instrumentation Amplifiers and a Read-Out IC for Sensor Interfacing,Springer, August 2012.

Comparative Study of a Fully Differential Op Amp in FinFET and Planar Technologies

Sébastien Morrison[1,2], Bertrand Parvais[1], Gerd Vandersteen[2], Kenichi Miyaguchi[1], Abdelkarim Mercha[1], Piet Wambacq[1,2]

[1]imec, Leuven, Belgium
[2]Vrije Universiteit Brussel, Dept. of Electronics and Informatics (ETRO), Brussels, Belgium

Abstract — **In the race to deliver ever smaller and faster devices, bulk FinFETs are seen as a viable alternative to planar bulk technologies. With that in mind, a new benchmarking scheme is implemented in order to effectively and fairly compare, in simulation, a 10nm FinFET technology with a 28nm planar CMOS one on a 100 MHz gain-bandwidth operational amplifier. For identical phase margins, the 10nm design consumes 99 μA compared to over 123 μA in 28nm, yielding a substantial decrease in power consumption in favor of the FinFET-based design.**

Keywords — **Op amp; benchmarking; 10nm FinFET; stacked transistors; 28nm planar; CMOS**

I. INTRODUCTION

Downscaling trends will require important changes in device architecture, bringing new opportunities and limitations to circuit designers [1]. Because new technologies are optimized for digital applications, it is important to assess at an early stage of their development how they also affect the performance of certain critical analog circuits. To this end, the case of an operational amplifier (op amp), designed in a 10nm bulk FinFET technology, is examined in this paper.

FinFET transistors are nonplanar devices in which the gate wraps around the channel on 3 sides, improving gate control of the channel and reducing short channel effects. This not only enables further downscaling, but also higher intrinsic gains, achievable in more power efficient ways due to better electrostatics. However, studies have shown FinFETs suffer from higher parasitics (source resistance, extrinsic capacitors) compared to planar devices [2].

Therefore, the goal of this paper is to see how, in simulation, an op amp, designed using bulk FinFET devices, performs compared to one designed using a current, commercially available, 28nm planar technology. The same design strategy was used for both technologies in which choices were made to take advantage of device specificities in order to see whether one technology outperforms the other in certain fields of application.

The rest of this paper is structured as follows. Single transistors in each technology are first studied (Section II), then the op amp topology used for the design is described (Section III) before developing a design strategy (Section IV) for the circuits. The simulation results are presented and discussed in Section V, yielding the remarks in Section VI that conclude the paper.

II. SINGLE TRANSISTOR STUDY

In order to take full advantage of the specificities of both technologies (N10 FinFET and N28 planar), single NMOS and PMOS transistors were examined first; their main characteristics are reported below for both minimum length (20nm in N10 and 30nm in N28) and long gate (200nm) devices. It is worth mentioning that in the FinFET-based technology, long-channel transistors are obtained by stacking multiple minimum-length devices with their gates tied together, rather than using individual transistors with physically longer gates as in the planar 28nm technology. Stacking transistors is likely to become the norm in the next generation of technologies, as the realization of longer gates brings about complications for the lithography and the processing. Furthermore, gate stacking has the advantage it avoids issues linked to pitch irregularities between gate fingers or nonplanar surfaces long gate devices can sometimes suffer from. However, these advantages come at the price of increased parasitics compared to conventional long gate devices.

Information on the technologies used in this study is given in Table 1.

Table 1: Technology characteristics

	Minimum gate length	Gate pitch	Supply voltage	Fin height / width / pitch
N10 (low V_t)	20nm	64nm	0.7V	30nm/6.7nm/36nm
N28 (low V_t)	30nm	130nm	0.9V	-

FinFET devices are able to meet a spec on g_m at much lower inversion levels than their planar equivalents, especially for short gate devices (Fig. 1). This characteristic should be exploited when designing with FinFETs as it is at low inversion levels that the highest power efficiencies and intrinsic gain values can be achieved (Fig. 2 and 3). However, the advantage N10 (FinFET) has over N28 (planar) for short gate devices becomes less significant as gate length increases.

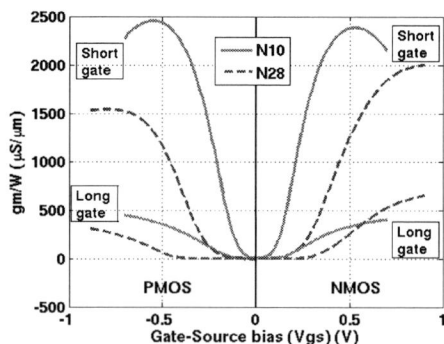

Fig. 1 Normalized transconductance (g_m) for $|V_{DS}| = V_{DD}$. Left side of plot represents g_m of PMOS devices ($V_{GS} < 0$); right side concerns NMOS devices ($V_{GS} > 0$). Higher g_m values at lower overdrives can be obtained in N10, especially in the case of PMOS devices.

Fig. 3 shows that FinFETs offer much higher intrinsic gains (g_m/g_{ds}) than transistors in N28, so long as low inversion biases are used. This is due to the better electrostatic control of the channel FinFETs have over planar devices, especially for short gate devices. This is not the case in N28 where intrinsic gain values rapidly drop when gate lengths scale down. From the figures discussed above, it is apparent that transistors have (at low frequencies) more desirable characteristics at low inversion levels. However, for a given current or g_m value, the width of a transistor increases as the overdrive applied is reduced, which increases the surface it occupies and its parasitic capacitance. There is therefore a trade-off between a transistor's biasing level and its dimensions.

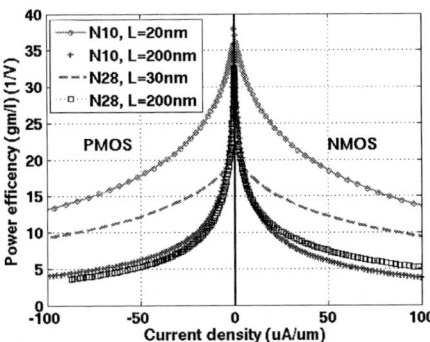

Fig. 2 Power efficiency of NMOS and PMOS devices as a function of current density, $|V_{DS}| = V_{DD}$. Curves show N10 FinFET has higher g_m/I ratios than N28 planar, but this efficiency gap reduces as gate length increases.

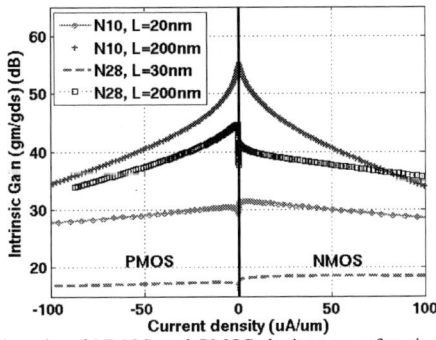

Fig. 3 Intrinsic gain of NMOS and PMOS devices as a function of current density, $|V_{DS}| = V_{DD}$. At low current densities, higher intrinsic gains are achieved in N10. Even short gate devices can attain over 30dB of gain, which is not the case in N28.

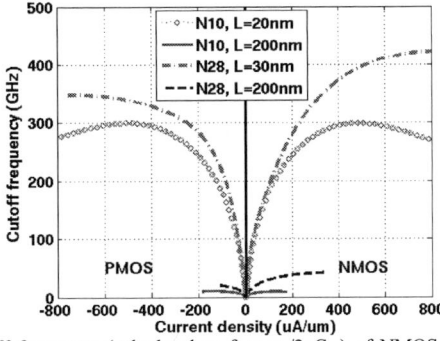

Fig. 4 Cutoff frequency (calculated as $f_T \approx g_m/2\pi C_{gg}$) of NMOS and PMOS devices as a function of current density, $|V_{DS}| = V_{DD}$. N28 planar devices are able to reach higher cutoff frequencies because they suffer from less parasitics, both resistive and capacitive.

One of the effects of high parasitic capacitance is illustrated in Fig. 4 which plots the cutoff frequency $f_T \approx g_m/2\pi C_{gg}$ as a function of current density. In this respect, FinFETs perform less well than the planar devices we compare them to. Although their g_m is higher for a given current, the ratio of that g_m to the gate capacitance (C_{gg}) of the device is lower in N10 because of extrinsic resistive and capacitive parasitics. Fig. 4 is an important figure of merit in high frequency applications, but the highlights it brings concerning device parasitics have impacts on the stability behavior of the op amps examined here, as will be pointed out in the relevant sections.

These characteristics will be used to design the op amp described in the next section.

III. CIRCUIT AND SPECIFICATIONS

The operational amplifier that will be used for benchmark purposes in this study is a fully differential transconductance amplifier (OTA) with common mode feedback (Fig. 5). It is made up of an amplification stage, which is a perfectly symmetric 2-stage Miller compensated OTA and a second amplifier used to stabilize variations in the common mode bias. From hereon, this work will focus on the amplification stage of the design.

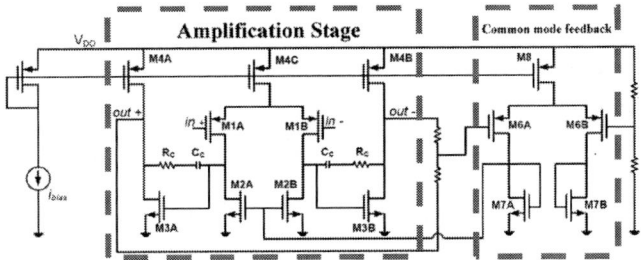

Fig. 5 The 2-stage fully differential op amp circuit used for this study. This paper focuses on the amplifying stage, but care was also taken to ensure the common mode feedback stage stabilizes common mode variations, at least up to the gain-bandwidth (GBW).

The first stage of the amplifier is a differential pair with PMOS input transistors (M1A and M1B) and NMOS transistors used as active loads (M2A and M2B). The low-frequency voltage gain of this stage is given by the ratio of the input transistors' transconductance (g_m) and the total conductance (g_{ds}) seen at the output of the stage, or

$$A_{V1} = \frac{g_{m1}}{g_{ds1} + g_{ds2}} \quad (1)$$

where g_{m1} denotes the transconductance of M1A or M1B and likewise for what regards g_{ds1} and g_{ds2}.

The second stage of the design is a common source one comprising an NMOS transistor (M3A/B) loaded with a PMOS transistor (M4A/B) used as an active load. Its DC gain is

$$A_{V2} = \frac{g_{m3}}{g_{ds3} + g_{ds4}} \quad (2)$$

and the overall low frequency gain of the op amp is given by the product of the gains of each individual stage. The 2 stages are connected to each other by a Miller capacitor (C_C) and a resistor (R_C) used to guarantee the closed loop stability of the design and ensure sufficient phase margin (PM). The first 2 poles of the design lie at the outputs of the first and second stages and their values are given by [3]

$$p_{n1} = -\frac{g_{ds1} + g_{ds2}}{C_n + (C_c + C_{gd3})\left(1 + \frac{g_{m3}}{g_{ds3} + g_{ds4}}\right)} \quad (3)$$

$$p_{n2} = -\frac{g_{m3}(C_c + C_{gd3})}{(C_c + C_{gd3})C_L + (C_c + C_{gd3})C_n + C_L C_n} \quad (4)$$

respectively, with C_L, the capacitance that loads the circuit in Fig. 5, $C_n = C_{gg3} + C_{gd1,2} + C_{db1,2}$, with C_{gg3}, the total gate capacitance at the gate of M3A/B and $C_{xy1,2} = C_{xy1} + C_{xy2}$ where C_{xy} represents the capacitance between nodes x and y. A rough approximation of these poles is given by $p_{n1} = -(g_{ds1} + g_{ds2})/C_C A_{V2}$ and $p_{n2} = -g_{m3}/C_L$. Because $g_{ds1} + g_{ds2} < g_{m3}$ and $C_C A_{V2} > C_L$, p_{n1} occurs before p_{n2}, so the pole at the output of the first stage is the dominant one.

The Miller capacitance strengthens a shunt feedback loop between the gate and drain of M3A/B, which increases the stability and also shifts the design's 3rd pole to higher frequencies as will be discussed later. The resistor R_C is used to compensate the second pole (p_{n2}) by a positive zero, creating a pole-zero doublet. However, it also leads to an additional pole given by

$$p_{n3} = -\frac{G_C}{C_n} \quad (5)$$

where $G_C=1/R_C$ is chosen such that a pole-zero doublet occurs at the frequency of the second pole, leading to

$$G_C = \frac{C_c^2 g_{m3}}{C_c^2 + C_c C_L + C_c C_n + C_L C_n} \qquad (6)$$

From (3), (4) and (6) we notice it is advantageous to choose a large Miller capacitance for stability reasons as the poles in the design will be further apart from one another.

The frequency behavior of the design is captured by its gain-bandwidth product (GBW) and is given by A_V*p_{n1}, or

$$GBW \approx \frac{g_{m1}}{C_c + C_{gd3}} \qquad (7)$$

From (8) we notice on the contrary that a low value of C_C should be chosen in order to maximize GBW.

Based on the topology described above, op amps were designed in N10 and N28 to meet a specification on the GBW of 100 MHz with load (C_L) and Miller (C_C) capacitors of 1pF and 500fF respectively. In both cases, a DC gain above 60 dB and a PM of roughly 80° should be achieved, while current consumption should be kept to a minimum.

IV. DESIGN METHODOLOGY

In this section, we describe the design strategy that was used to develop the circuits in this study. Trade offs are illustrated in the case of a 100 MHz GBW design, but the same principles apply for any set of specifications.

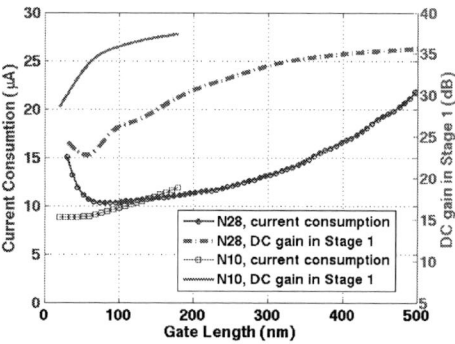

Fig. 6 Gain-current trade-off in Stage 1 for a fixed g_m value imposed by the constraint on GBW. Both in N10 and in N28, increasing the gate length leads to more gain, but at the cost of a higher current consumption.

The first stage of the amplifier needs to be designed such that the specification on GBW is met. Indeed, (7) requires that $g_{m1}=2\pi f_{GBW}C_C$ so the g_m of the input PMOS (M1A/B) is fixed. Furthermore, most of the overall gain should be achieved in this stage, because, due to a constraint on the circuit's settling time which will be explained later, the second stage cannot produce as much gain as the first one. Bearing in mind these considerations and the device characteristics plotted in Section II, transistors should be made as wide and as long as possible so as to bias them at low inversion levels, enabling the highest power efficiency levels and intrinsic gains to be achieved. However, there is a limit to the dimensions the differential pair input transistors can have because they determine the frequency of the pole (p_{n4}) at the op amp's input node when it is used in a closed loop configuration. Indeed, the frequency of this pole is given by $f_{p4} = G_{input}/2\pi C_{gg1}$ and is therefore inversely proportional to M1A/B's total gate capacitance (C_{gg1}) which increases with the transistor's dimensions ($C_{gg1} \propto W*f(L,V_{in})$). In order for this pole to lie above the GBW, there is a certain value C_{gg1} should not exceed. For this constraint on the input pole to be satisfied, it is not possible to use transistors that are both long and wide; the larger the transistor is made, the narrower it will have to be and vice-versa. This condition leads to a trade-off between the gain the stage is able to achieve and its current consumption. Indeed, to maximize the gain, long transistors should be used, but this reduces the maximum width they may have so as not to exceed the limit imposed on their total gate

capacitance. Consequently, the transistors need to be biased at higher inversion levels increasing the current consumption that is necessary in order to satisfy the specification on g_{m1} required for GBW purposes. Inversely, reducing the gate length will therefore reduce the current consumption, but at the price of less gain in the stage. The plot in Fig. 6 shows the DC gain and current consumption depending on gate length for transistors that are as wide as they can be to respect the constraint on C_{gg}. The trends described above clearly appear on the graphs[1], and have guided the design choices made for the first amplifying stage. In N10, a gain of over 35 dB can be achieved with a 100nm gate length device, yielding a current consumption under 9µA, while in N28, a 300nm gate length device, consuming just over 13µA is necessary to achieve just under 35 dB of gain.

The other transistors in this stage (M2A/B and M5) are not constrained by any specs and were designed such that a current equal to M1 (in the case of M2) and twice that (in the case M5) flows through them. The width of M2 was chosen to be 1.5 times that of M1 so that both these transistors have similar capacitive parasitics.

The second stage of the op amp is described next. Its purpose is twofold: it increases the total gain of the design (which should be at least 60 dB), and increases its output swing (which is becoming an increasingly important concern as supply voltages scale down as technologies advance).

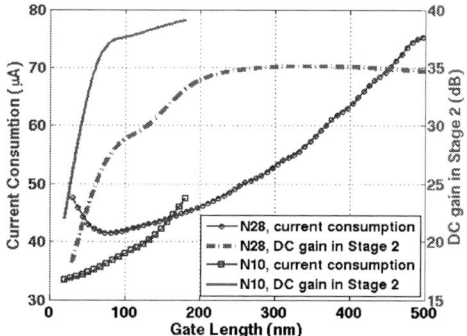

Fig. 7 Gain-current trade-off in Stage 2 for a fixed g_m value imposed by the constraint on settling time. There is a tougher constraint on gain in this stage for the N28 technology because less gain could be obtained in the 1st stage, resulting in a higher current consumption.

Two important constraints are associated with the design of this stage. The first one concerns the settling time. It can be shown [4] that in order not to degrade its value, any pole-zero doublet should lie beyond the GBW. In this design, the pole at the output of the second stage is compensated by a zero, so its frequency should exceed the GBW by a certain factor to respect the above rule on settling time. Because this pole is given by (4), this condition fixes a constraint on g_{m3}. The second design requirement has to do with the PM, for which a spec is given. The PM is linked to how far apart 2 uncompensated poles lie from each other. In this design, it is therefore linked to the frequency of the 3rd pole created by R_C and given by (5). For a given value of g_{m3} (see above), the frequency of this pole depends only on C_n which is dominated by C_{gg3}. (It is to guarantee this last point that M2 was not made more than 1.5 times larger than M1 in the design of the first stage). Therefore, the lower C_{gg3}, the higher the frequency of the 3rd pole and the larger the PM. However, this constrains the dimensions of M3 requiring higher overdrives to be used to meet the spec on g_{m3}. Much like in the design of the first stage, a trade off lies between the PM and the current consumption as the higher overdrives required to increase the PM reduce the transistor's power efficiency.

---

[1] In N28, for gate lengths below 90nm, the plot shows a different behavior to that expected concerning current consumption. This is due to the fact that for the maximum width they may have, the PMOS' biasing level is below the one yielding the maximum power efficiency: in deep weak inversion, g_m/I drops and so current consumption increases. Consumption could therefore be lowered by reducing the width (i.e. increasing the overdrive) at the cost of less gain. Consequently they are bad design choices because longer gate devices enable more gain to be achieved for the same current consumption.

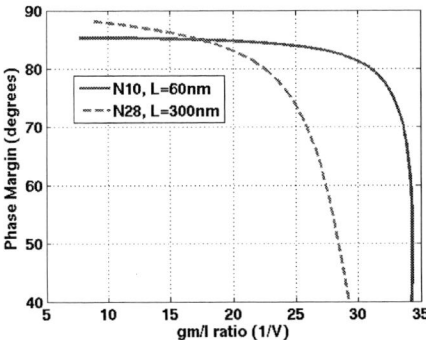

Fig. 8 Power efficiency (of M3) versus (differential) phase margin (of the op amp design). The figure illustrates the trade-offs there are between stability margins and the transistor's g_m/I ratio.

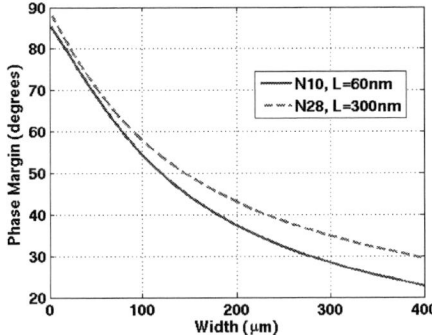

Fig. 9 Width (of M3) versus (differential) phase margin (of the op amp design). The wider M3 is made, the deeper in weak inversion it can be biased, increasing its gain and g_m/I ratio, but at the cost of less phase margin.

As well as these constraints, the overall design must meet the spec on the gain of at least 60 dB. This is more of a concern in the N28 design where less gain is achieved in the first amplifying stage. From Fig. 7, we notice that to meet the spec on the designs' gains, gate lengths of at least 40 and 250nm need to be chosen in N10 and N28 respectively; a margin was taken, and gate lengths of 60 and 300nm were selected. Based on these choices of gate lengths, the compromises between gate width, PM and power efficiency were examined by means of the plots presented in Fig. 8 and 9 which show how the PM changes as a function of g_m/I and width respectively. For a given spec on PM, the transistors' widths could be chosen in this way, for both the technologies under study.

V. RESULTS AND DISCUSSION

The strategy described above was used to size the transistors in the design before simulating it using Spectre; the results in table 2 show discrepancies of only a few dB with the flow's predictions.

For almost identical gains, stability margins and GBW, the N10 design consumes only 99 µA when 123.4 µA are necessary in N28. Bearing in mind that the supply voltage in N10 is 0.7V and 0.9V in N28, this leads to a 37.6% reduction in power consumption in favor of N10. This outcome is entirely due to short channel FinFETs' better electrostatics at low inversion levels. Indeed, it is because substantial (> 30 dB) gains can be achieved already with 60nm gate length devices that the 2nd stage of the FinFET-based op amp is able to meet the spec on g_{m3} in a very power-efficient way (g_m/I = 30.9 in N10 versus 25.2 in N28), considering the constraint on PM which limits the width of the device. Conversely, it is because little gain is achieved below 200nm using the N28 technology that longer, less efficient, devices are necessary in both stages. The above results would be even better in N10 if it wasn't for the high parasitic capacitances FinFETs suffer from. For given W and L values, C_{gg}, C_{gd}, and C_{db} are all higher in N10 which limits how deep in weak inversion a transistor may be biased. These parasitics also explain

why g_{m1} needs to be higher in N10 than in N28 for the same GBW, and why transistor widths in N10 must be smaller than those of their planar equivalents to meet PM requirements, even though their gate lengths are smaller.

The noise and mismatch behavior of both designs was also examined. Concerning mismatches, A_{vt} constants of 1mV/um and 1.8mV/um were used for N10 and N28 respectively. Using Pelgrom's law [5] ($\sigma_{Vt}=A_{vt}/\sqrt{(WL)}$) and the data from Table 2, this leads to threshold voltage shifts ($\Delta V_{th} = 3\sigma_{Vt}$) of 2.49mV (in N10) and 2.47mV (in N28). For what regards the low-frequency noise analysis, the ITRS [6] predicts lower K_f/C_{ox} constants as technologies scale; however, because WL products are smaller in N10, higher Flicker noise levels (12.2uV/$\sqrt{Hz}$ versus 4.8uV/$\sqrt{Hz}$ of equivalent input noise at 1Hz) are observed compared to N28. This last point also explains why the N10 design suffers from a slightly higher ΔV_{th} than the N28 one.

Table 2: Simulation results

	N10 (FinFET)	N28 (Planar)
GBW	108 MHz	109.8 MHz
PM (differential)	82°	81°
DC gain	63 dB	62 dB
Consumption	99 µA	123.4 µA
Mp1: W/L	14.5µm/100nm	16µm/300nm
Mp1: $g_m/I/g_{ds}$	370.2µS/11.3µA/3.1µS	353µS/14.7µA/2.9µS
Mn3: W/L	17.2µm/60nm	51µm/300nm
Mn3: $g_m/I/g_{ds}$	1mS/32.4µA/13µS	1mS/39.7µA/14.4µS

Consequently, it appears the FinFET technology is superior to the planar one based on the results from this low GBW op amp. However, these conclusions do not hold for higher specs on GBW, only achievable using higher overdrive biases. A few other designs with higher GBW were also developed; they will not be presented here but some conclusions concerning them will be listed. The effects of the (resistive) parasitics mentioned above worsen as current densities increase, reducing the efficiency gap between the 2 devices. Furthermore, the lower supply voltage in N10 makes it more challenging to keep all transistors saturated, particularly in the differential pair, requiring other compromises to be made. Therefore, for a certain GBW onwards, the planar technology outperforms the FinFETs; finding out the point when such a change occurs could be the object of a future study.

VI. CONCLUSIONS

In this paper, 2 op amps designed using 2 different bulk CMOS technologies – a 10nm FinFET-based and a 28nm planar one – are compared using the same design methodology and the same set of specs. Results show the FinFET technology is able to deliver more gain using less current, but suffers from higher parasitics compared to planar devices. These parasitics constrain the phase margins and increase with current densities which might limit the effectiveness of FinFETs as specifications on GBW become higher.

ACKNOWLEDGMENTS

The authors would like to thank the members of imec's INSITE program and the VUB (Vrije Universiteit Brussel) for their support.

REFERENCES

[1] C. Hu, "New Sub-20nm Transistors – Why and How," in *Proc. Design Automation Conf.*, June 2011.

[2] V. Subramanian et al. "Planar Bulk MOSFETS Versus FinFETs: An Analog/RF Perspective," in *IEEE Trans. Electron Devices*, Dec. 2006.

[3] P. Gray et al. "MOS operational amplifier design – a tutorial overview" in *IEEE Journal of Solid-State Circuits*, 1982.

[4] W. Sansen, *Analog Design Essentials*. Springer, 2007.

[5] M. Pelgrom et al. "Matching properties of MOS transistors" in *IEEE Journal of Solid-State Circuits*, 1989.

[6] International Roadmap for Semiconductors. http://www.itrs.net/, 2014.

A novel architecture for current-feedback instrumentation amplifiers with rail-to-rail input range

F. Del Cesta, A. N. Longhitano, P. Bruschi
Dipartimento di Ingegneria dell'Informazione
University of Pisa
Pisa - Italy
francesco.delcesta@for.unipi.it

M. Piotto
IEIIT - Pisa
CNR
Pisa, Italy

Abstract— **This paper presents a fully-differential current-feedback instrumentation amplifier with rail-to-rail input common-mode (CM) voltage range. Gm mismatches due to input CM variations have been reduced exploiting an original common-modes equalization loop. Chopping modulation is adopted to improve the performances of the amplifier in terms of offset and low frequency noise, while the intrinsic low-pass transfer function of the amplifier is exploited for the reduction of the offset-ripple. A prototype has been designed using the UMC 0.18um MM/RF CMOS process. Simulations, performed with a supply voltage of 1.5 V, showed that a maximum relative gain error of nearly ±1.5 %, against rail-to-rail input CM voltage variations, can be achieved.**

Keywords—instrumentation amplifier, current-feedback, rail-to-rail common-mode voltage range, sensor interface

I. INTRODUCTION

In the world of sensor interfaces instrumentation amplifiers (in-amps) are suitable for interfacing a wide variety of sensors [1]. Among the possible amplifier topologies, current feedback architectures have significant advantages in sensor applications, due to high CMRR, power efficiency, and the possibility of obtaining rail-to-rail input common-mode voltage range (CMVR) [2].

A major limitation to in-amps accuracy and resolution derives from noise and offset. This problem is particularly severe in CMOS implementation. Chopper modulation is a very efficient technique for reducing offset and low frequency noise contributions, since, differently from correlated double sampling and auto-zero techniques, it does not suffer from noise fold-over [3]. Numerous examples of applications of chopper modulation to current-feedback instrumentation amplifiers can be found in the literature [2,4]. The typical configuration is shown in Fig 1.

In order to interface the widest set of sensors, the amplifier should have an input stage with rail-to-rail input CMVR. On the other hand, optimum conditions for interfacing with following stages (e.g. ADC) and maximization of the output swing require the output CM voltage to be constant. These constraints can lead to a mismatch between transconductances G_{m1} and G_{m2}, since the first receives the input CM voltage (V_{ICM}), and the second receives the output CM voltage (V_{OCM}).

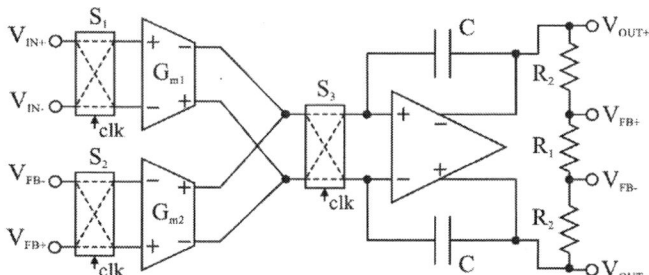

Fig. 1. Block diagram of a two-stage fully-differential current-feedback instrumentation amplifier exploiting chopper modulation. S1, S2 and S3 are chopper modulators.

This mismatch translates into a gain error, proportional to the ratio G_{m1}/G_{m2} [4]. The error can be particularly significant with complementary *p-n* input stages, since G_m variations up to 50% are possible, when the input CM voltage approaches supply rails. This effect can be mitigated using constant-gm topologies [5], but the residual gm variation is still too large for precision applications. An effective solution proposed in the literature employs a capacitively-coupled chopper topology, however it is based on a complex architecture and requires a careful layout for element matching [6]. In this paper a simple approach, in which G_m mismatches are reduced exploiting equalization of the input and the feedback CM voltages, is proposed.

A further problem to deal with is the offset ripple, resulting from the chopper modulation. A classic approach consists in cascading the amplifier with an analog low-pass filter. This solution, however, is prone from the offset and noise introduced by the filter itself. Other approaches have been proposed in the literature, such as an auto-correction feedback loop [7], a continuous-time ripple reduction loop [8] and a multi-path architecture [2]. In this work, we adopted a solution exploiting the intrinsic first-order low-pass transfer function of the classic two-stage configuration of Fig 1, whose cut-off frequency was properly lowered. A second order implementation of this idea is shown in [4].

The effectiveness of the proposed architecture is proven by means of simulations performed on a prototype cell, designed using CMOS devices from the UMC 0.18um CMOS process.

978-1-4799-4993-9/14 $31.00 © 2014 IEEE

II. AMPLIFIER TOPOLOGY

A. System Architecture

Fig. 2 shows the block diagram of the proposed architecture. Note that, compared to the classic architecture shown in Fig.1, the input stage is formed by the cascade of a differential difference amplifier (G_{m1}, G_{m2}, R) and the transconductor G_{m3} (see paragraph C for details).

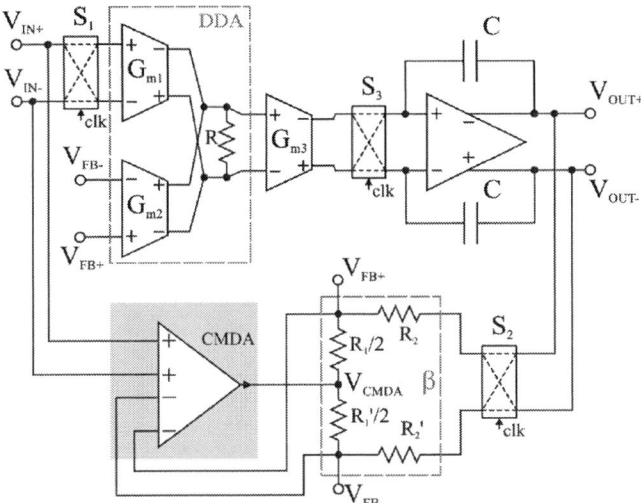

Fig. 2. Block diagram of the proposed architecture. The block CMDA equalizes the common-mode voltages of transconductors Gm1 and Gm2.

In order to achieve a rail-to-rail input CMVR, transconductors G_{m1} and G_{m2} have been designed with complementary *p-n* input stages. Unfortunately, by adopting this configuration, mismatches between G_{m1} and G_{m2} (due to different input CM voltages for transconductors) could lead to a large error on the differential gain of the amplifier. The occurrence of this circumstance has been avoided by means of a common-mode equalization loop, based on the common-mode differential amplifier CMDA, that amplifies the difference between the input CM voltage (V_{ICM}) and the feedback CM voltage (V_{FBCM}). In the nominal case, the operation of the loop is described by following equations:

$$V_{FBCM} = \beta_{Nom} \cdot V_{OCM} + (1 - \beta_{Nom}) \cdot V_{CMDA} \quad (1)$$

$$V_{CMDA} = A_{CMDA} \cdot (V_{ICM} - V_{FBCM}) \quad (2)$$

where A_{CDMA} is the gain of the CMDA, V_{OCM} is the CM output voltage, V_{CMDA} is the output voltage of the CMDA, and β is the transfer function of the feedback resistive net, whose nominal value (for $R_1=R_1'$, $R_2=R_2'$) is:

$$\beta_{Nom} = \frac{R_1}{2R_2 + R_1} \quad (3)$$

From (1), if $\beta_{Nom} << 1$:

$$V_{FBCM} \cong V_{CMDA} \quad (4)$$

Eq. (2) and (4) show that, if $A_{CDMA}>>1$ and $\beta_{Nom}<<1$, the loop forces V_{FBCM} to the same value of V_{ICM}, with an approximate relative error $1/A_{CMDA}$, strongly reducing mismatches between G_{m1} and G_{m2}. However, input common-mode equalization has an important drawback. In fact, it would be desirable that the CMDA does not affect the differential feedback voltage (V_{FBD}), which should only depend upon the differential output voltage V_{OD}:

$$V_{FBD} = \beta \cdot V_{OD} \quad (5)$$

leading to a nominal amplifier gain of:

$$A_{D_{Nom}} = \frac{1}{\beta_{Nom}} \quad (6)$$

Unfortunately, it can be demonstrated that, in presence of a difference between V_{OCM} and V_{CMDA}, the unavoidable mismatch between R_1/R_2 and R_1'/R_2' ratios would result in adding a parasitic differential term (ΔV_{FBD}) to V_{FBD}:

$$\Delta V_{FBD} = \gamma \cdot (V_{OCM} - V_{CMDA}) \quad (7)$$

where γ depends on resistor mismatch. This term contributes to the overall input offset voltage of the in-amp, degrading its performances. In order to avoid this drawback, the modulator S_2 has been placed between the output voltage and the feedback loop, instead of between the feedback loop and transconductor G_{m2}, as in Fig 1. As a result, the offset from (7) is not modulated at the input of G_{m2} and is then rejected through modulator S_3, such as the offset from G_{m1} and G_{m2}.

B. Topology of the CMDA

In order to obtain a high value for A_{CMDA}, a two-stage architecture has been adopted. Besides, the input stage has been designed with complementary *p-n* input pairs, in order to allow the amplifier to properly work over a wide V_{ICM} range (see Fig. 3).

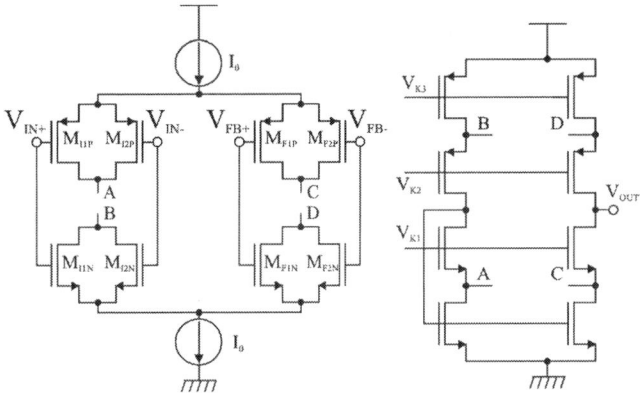

Fig. 3. Topology of the input stage of the CMDA.

C. Input Stage

The overall cut-off frequency f_0 of the proposed in-amp is given by:

$$f_0 = \frac{\beta}{2\pi} \cdot \frac{G_{m3} \cdot R \cdot G_{m1,2}}{C} \qquad (8)$$

where $G_{m3} \cdot R \cdot G_{m1,2}$ is the compound transconductance deriving from the cascade of the DDA and G_{m3} (Fig. 2). If $R \cdot G_{m1,2} >> 1$, it can be demonstrated that the input-referred thermal noise PSD (S_{Vth}) is practically dominated by $G_{m1,2}$, according to:

$$S_{Vth} = m \cdot 4KT \cdot \frac{1}{G_{m1,2}} \qquad m > 1 \qquad (9)$$

As stated in the introduction we required that $f_0 << f_{chop}$, where f_{chop} is the chopper frequency, in order to provide effective offset-ripple rejection. With f_{chop} of the order of tens of kHz, f_0 should be equal or less than 1 kHz. According to (8) this result is compatible with on-chip capacitors only using very low compound transconductance values. Since $G_{m1,2}$ is determined by the specification on the input thermal noise PSD through (9), the requirement of low compound transconductance is satisfied by choosing a value of G_{m3} low enough. This degree of freedom is not present in the classical in-amp of Fig. 1, since it uses a single-stage architecture for input transconductors. Further details about the optimization of two-stage low-G_m transconductors can be found in [9].

1) Topology of the differential difference amplifier

The differential difference amplifier (DDA), composed by the transconductor G_{m1}, the transconductor G_{m2}, and the resistor R, has been designed as a folded-cascode amplifier with two complementary p-n input stages, as shown in Fig. 4.

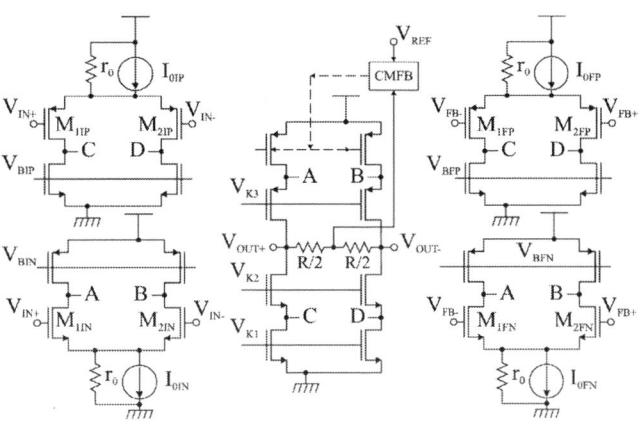

Fig. 4. Simplified circuit diagram of the DDA formed by the transconductors Gm1 and Gm2 and the resistor R.

Since the aim of this work was to realize an in-amp for applications where a true rail-to-rail input CMVR is required, the DDA was carefully designed to be extremely robust with respect to changes in the input CM voltage.

Complementary p-n input stages are typically prone to wide operating point changes, caused by turning off of either the p or the n pair. In classic cascode architectures, these changes would alter the current balance at the output nodes by up to 50 % of the quiescent current value. This would result in a shift of the output CM voltage that cannot be easily handled by only the CMFB circuit.

In order to avoid this phenomenon, each of the four input differential pairs of the DDA has been self-biased by a dummy pair, whose current tracks the tail current variations of the corresponding input pair. In this way, for CM signals, input stages are completely separated from the common-gate stage, greatly facilitating the operation of the CMFB. The bias scheme for a single complementary input port is shown in Fig. 5.

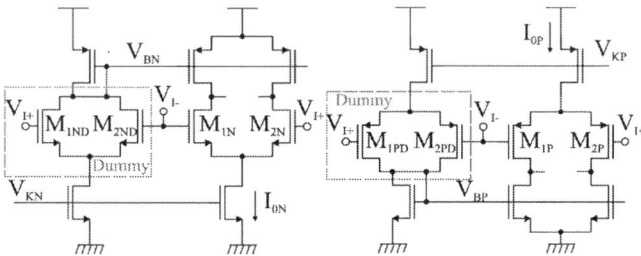

Fig. 5. Bias scheme for a single complementary input port.

D. Output Stage

Transconductor G_{m3} and the Miller Integrator have been merged into a single two-stage Gm-OpAmp architecture. Pseudo-differential pairs have been adopted in order to increase linearity, as shown in Fig. 6.

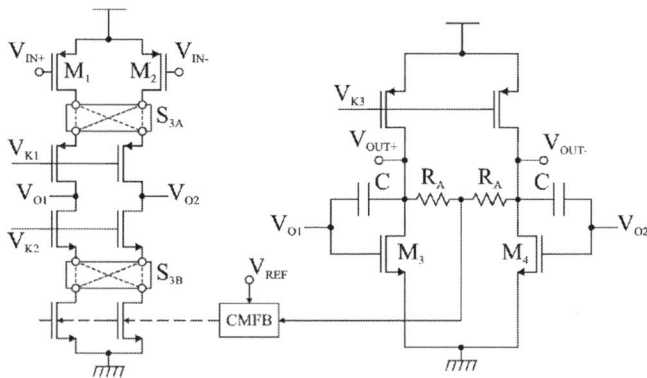

Fig. 6. Simplified circuit diagram of the Gm-OpAmp block.

III. SIMULATIONS

The effectiveness of the proposed approach has been proven by means of a preliminary prototype, designed using the 0.18μm CMOS MM/RF process of UMC, with a power supply voltage of 1.5 V. In this preliminary design, current consumption has not been optimized and the nominal in-amp gain has been set to 200.

In order to reduce flicker noise components, a chopping frequency f_{chop} of 50kHz has been chosen, while the cut-off frequency f_0 of the amplifier, given by (8), has been set to 1kHz, so that an attenuation of the offset-ripple of about 36dB is achieved. Note that a value of 100 has been chosen for $R \cdot G_{m1,2}$, following the optimization criteria described in [8].

Simulations have been performed using spectreRF, which allows extending AC and NOISE analysis to circuits in periodic steady-state conditions. The effectiveness of the common-mode equalization loop has been verified by performing a set of 15 PAC (periodic steady-state AC) analysis with input CM voltages varying across both supply rails. As shown in Fig. 7, the error on the nominal value of the amplifier gain stays below ±1.5% for values of the input CM voltage from 10 mV to Vdd (1.5 V). The large $G_{m1,2}$ variation is reflected only into the cut-off frequency variation, according to (8).

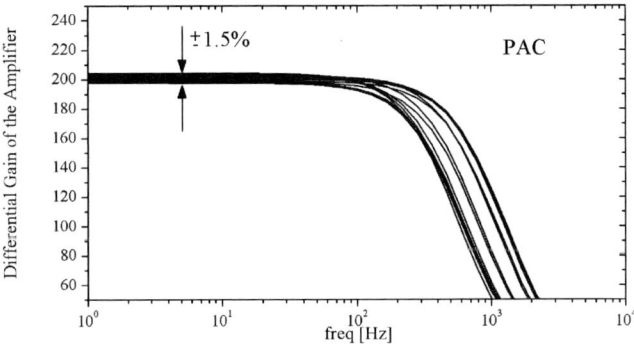

Fig. 7. PAC analysis with input CM voltage varying from 10 mV to Vdd in 15 steps.

The PSD of the input-referred thermal noise was found to be about 25 nV/√Hz, according to PNOISE (periodic steady-state ac noise) simulations, as shown in Fig.8.

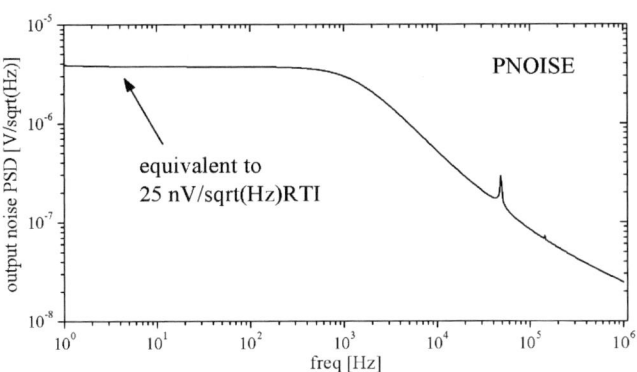

Fig. 8. Result of the PNOISE simulation with V_{ICM}=Vdd/2. Note the attenuation of the offset-ripple at f_{chop}=50kHz.

The main performances of the amplifier are summarized in Table 1 and compared to previous designs. Because of a lack of Monte-Carlo models for the low-threshold devices used in this design, CMRR and PSRR have not been evaluated.

TABLE I.   CHARACTERISTICS OF THE PROPOSED CIRCUIT COMPARED WITH PREVIOUS WORKS.

Ref	*This work (simulations)*	*[5]*	*[2]*	*[8]*
Year	2014	2011	2012	2010
Technology	0.18 μm	65 nm	0.7 um	0.32 um
Rail-to-Rail	Yes	Yes	No	No
Input Noise	25 nV/√Hz	60 nV/√Hz	21 nV/√Hz	24 nV/√Hz
Gain	200	100	100	200
Gain Accuracy	±1.5%	0.16%	0.53%	<0.5%
Supply Current	500 uA	1.8 uA	143 uA	1200 uA

IV.   CONCLUSIONS

The performed simulations have shown the effectiveness of the proposed approach. A true rail-to-rail input CMVR has been obtained, with a relative gain error of only ±1.5% for a rail-to-rail input CM voltage excursion. Besides, the amplifier effectively exploits its low-pass transfer function to reject its modulated offset, and the input noise PSD makes the in-amp adequate for low-noise applications. The present prototype was mainly aimed to demonstrate the validity of the proposed principle. Many parts of the amplifier have not been optimized for power consumption, so that significant improvements of this figure can be envisioned.

REFERENCES

[1] R. Wu, J.H. Huijsing and K. A. A. Makinwa, "Precision instrumetation amplifiers and read-out integrated circuits" Analog Circuits and Signal Processing Springer Science+Business Media New York 2013

[2] Q. Fan, J. H. Huijsing and K. A. A. Makinwa, "21 nV/√Hz chopper-stabilized multi-path current-feedback instrumentation amplifier with 2μV Offset" IEEE Journal of Solid-State Circuits, vol. 47, no. 2, February 2012

[3] C. C. Enz and G. C. Temes, "Circuit techniques for reducing the effects of Op-Amp imperfections: Autozeroing, correlated double sampling, and chopper stabilization," Proc. IEEE, vol. 84, no. 11, pp. 1584–1614, Nov. 1996.

[4] F. Butti, P. Bruschi, M. Piotto, "A chopper modulated low noise instrumentation amplifier for MEMS thermal sensors interfacing", Ph.D. Research in Microelectronics and Electronics (PRIME), 2011 7th Conference on

[5] Ferri G. and Sansen W., "A rail-to-rail Constant gm low-voltage CMOS operational transconductance amplifier", IEEE JSSC, Vol. 32, No. 10, pp. 1563-1567, October 1997

[6] Q. Fan, F. Sebastiano, J. H. Huijsing and K, A. A. Makinwa, "A 1.8μW 60 nV/√Hz capacitively-coupled chopper instrumentation amplifier in 65 nm CMOS for Wireless Sensor Nodes" IEEE Journal of Solid-State Circuits, Vol. 46, no. 7, July 2011

[7] Y. Kusuda, "A 5.9 nV/√Hz chopper operational amplifier with 0.78μV maximum offset and 28.3 nV/°C offset drift," IEEE ISSCC Dig. Tech. Papers, pp. 242–243, Feb. 2011.

[8] R. Wu, K. A. A. Makinwa, and J. H. Huijsing, "A chopper current feedback instrumentation amplifier with a 1 mHz noise corner and an AC-coupled ripple-reduction loop," IEEE J. Solid-State Circuits, vol. 44, no. 12, pp. 3232–3243, Dec. 2009.

[9] F. Butti, M. Dei, P. Bruschi, M. Piotto, "A compact instrumentation amplifier for MEMS thermal sensor interfacing", Analog Integrated Circuits and Signal Processing September 2012, Volume 72, Issue 3, pp 585-59

A Bootstrap Transimpedance Amplifier for High Speed Optical Transcutaneous Wireless Links

Tianyi Liu, Zhicheng Cai, Jens Anders and Maurits Ortmanns

Institute of Microelectronics, University of Ulm, Ulm D-89081

Email: {tianyi.liu, zhicheng.cai, jens.anders, maurits.ortmanns}@uni-ulm.de

Abstract—We report on a prototype of an experimental bootstrap transimpedance preamplifier (BTA) for high speed transcutaneous data transfer. Due to the strong scattering effect of the human skin and the corresponding widely spread optical power distribution, a large size photodiode is required at the external receiver in order to increase the transmission efficiency as well as the misalignment tolerance in the transcutaneous optical telemetric link (TOTL). However, the increased junction capacitance associated with a larger size photodiodes makes the design of high speed transimpedance amplifiers (TIA) challenging. To mitigate this tradeoff, in this paper, we use bootstrapping technique to achieve a large TIA bandwidth even in the presence of large-size photodiodes. Thereby, an emitter follower is employed as the bootstrapping buffer to minimize the junction capacitance seen by the transimpedance stage. The proposed circuit is tested on circuit board level in combination with a 10 mm² photodiode with a parasitic capacitance as large as 350 pF and achieves a 61.5 dBΩ transimpedance gain over a bandwidth of 61 MHz with an input referred current noise density of 158 pA/√Hz. Therewith, a transcutaneous optical link with 75 MBps is achieved.

I. INTRODUCTION

Motor paralysis, which is among the most serious neural diseases, results in reduced mobility or sensation of the patients. In the past few decades, the Brain-Machine-Interface (BMI) has been proposed and later proved to restore purposeful movement in people affected by paralysis. As an example, in 2014 [1], a neural prosthesis was presented, which uses the neural signals from a monkey to control limb movements of a second functionally paralyzed primate avatar.

Despite these great advances, in order to put BMIs into clinical application, many challenges remain to be solved. For example, wireless telemetry should be incorporated into cortical array implants to avoid the potential infection caused by transcutaneous wires (Fig. 1(a)). The transcutaneous optical telemetric link (TOTL) system presents a promising option for the high speed transcutaneous biomedical data transfer, due to its large achievable bandwidth, its low power consumption as well as its robustness against interference. With a large communication bandwidth, recorded data from a larger number of neurons can be transferred to the external data processing unit simultaneously, which enables the control of prosthetic limbs with more degrees of freedom [2]. Here, the significant power reduction by more than two decades reported for TOTL system over the last few decades [3][4] renders them operable from a wireless power source as it has to be used in

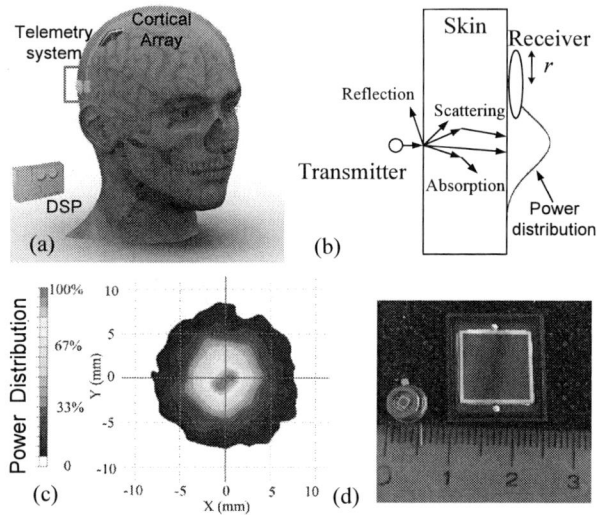

Fig. 1. (a) Brain-Machine-Interface with a optical transcutaneous telemetric link; (b) light/tissue interaction in the TOTL system; (c) The power distribution at the receiver's side with a tissue thickness of 5 mm; (d)The comparison of the small size photodiode and big size photodiode.

implanted systems. Additionally, compared to electrical RF communication links, optical links are not affected by radio spectrum regulation and they are immune to any RF noise source, including ambient RF noise and inductive power links. The latter is widely used in the BMIs, transferring power to the implants.

Despite all these advantages, the achievable performance of the TOTL system is in practice limited by the strong scattering of light with power in the near infrared (NIR) spectrum inside human tissues. To illustrate this diffusion effect, Fig. 1(c) shows the simulated power distribution of light with a wavelength of 850 nm having traversed a distance of 5 mm through tissue. The simulation was performed using the Monte Carlo Method proposed in [5].

This diffused power distribution prevents the use of small-size photodiodes because they would result in a low transmission efficiency. Besides, it it hard to apply a lens or mirror to focus the transmitted photons due to the anisotropic scattering effect of the light/tissue interaction [5]. Therefore, choosing a photodiode with a larger active area increases the transmission efficiency which in turn decreases the power by the implanted transmitter. Unfortunately, the junction capacitance of the photodiode increases with its size. As an example, the left

978-1-4799-4993-9/14 $31.00 © 2014 IEEE

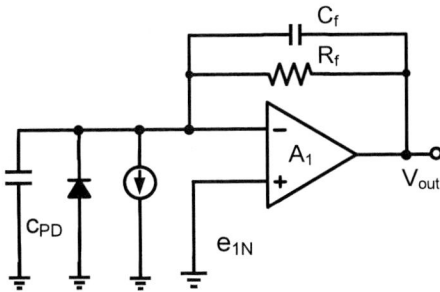

Fig. 2. Schematic of the conventional shunt-shunt feedback TIA

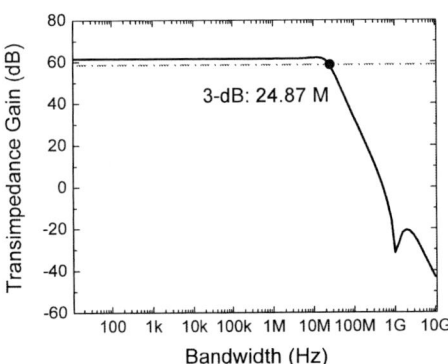

Fig. 3. Simulation of the conventional TIA frequency response

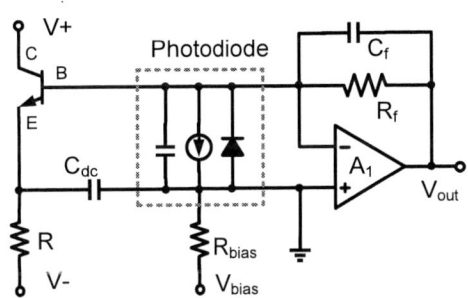

Fig. 4. Schematic of the bootstrap transimpedance amplifier.

photodiode in Fig. 1(d) displays a small parasitic capacitance of only 3 pF, while that large diode on the right is 350 pF. If a 50 Ω TIA resistor is used in combination with the large photodiode, the resulting bandwidth of the optical front-end becomes 9 MHz only, which is far less than required [6]. From this example it is evident that such a large photodiode capacitance imposes a difficult challenge for the design of a transimpedance amplifier (TIA) having to drive such a large capacitance at the same time achieving the required large bandwidth.

To circumvent this problem, in this paper, we propose the use of a bootstrap transimpedance amplifier (BTA) [7] in the TOTL system. Applying the bootstrapping technique, greatly increases the achievable TIA bandwidth, compared to a conventional shunt-shunt feedback TIA. In section II, the principle of the BTA is introduced and the achievable bandwidth is compared to that of a conventional TIA. The measurement setup and results are then discussed in section III, before the paper is concluded with a summary and a brief outlook on future work in section IV.

II. THEORETICAL ANALYSIS

The circuit diagram of a well-known conventional shunt-shunt feedback TIA is shown in Fig. 2. The photodiode with its potentially large parasitic capacitance is connected to the inverting input of the core amplifier LHM6624. The signal current from the photodiode flows into a 1.2 kΩ feedback resistor R_f. Since the input of the amplifier remains at a virtual ground, the output of the TIA responds as: $V_{out} = i \times R_f$. The 1 pF feedback capacitor is used to ensure stability.

According to [8], the bandwidth of this TIA can be approximated as:

$$f_{3dB} \approx \sqrt{\frac{GBW}{2\pi R_f C_T}} \qquad (1)$$

where GBW is the gain bandwidth product of the core amplifier, C_T is the total input capacitance which is dominated by C_{PD}. Compared with the optical fiber link where the photodiodes have a smaller junction capacitance (< 1 pF), the photodiode used in this TOTL system can be 350 pF to maximize transmission efficiency. Therefore, the achievable bandwidth of the TIA greatly decreases.

The AC simulation in Fig. 3 shows that the 3-dB of such a TIA is only 25 MHz, which is not sufficient for a BMI

incorporating up to hundreds of microelectrodes [2]. One possible way to increase the bandwidth is to reduce the value of the feedback resistor. However, the thermal noise of the feedback resistor will be increased. Moreover, with the decrease of the transimpedance gain, the noise from the following circuit, typically the limiting amplifier, will become important and cannot be ignored anymore [9]. Both effects eventually decrease the achievable receiver sensitivity.

Another technique to extend the TIA bandwidth, known as bootstrapping, is shown in Fig. 4. An emitter follower is placed across the photodiode, monitoring the signal voltage at the diode's cathode and driving the anode with the same signal. In the ideal case, the AC voltage across the photodiode is reduced to zero which in turn also reduces the effective capacitance to zero. In reality, the gain of the emitter follower is always slightly less than one, which leaves a residual signal voltage across the photodiode. Nonetheless, the junction capacitance seen by the transimpedance stage is significantly decreased and the bandwidth is greatly extended. The bootstrapping circuitry introduces two new capacitors into the system. The base emitter capacitance C_{EB} of the transistor appears in parallel with C_{PD}, and it is therefore also been bootstrapped. The collector-base capacitance C_{CB} forms a voltage divider with C_{PD} [10]. It decreases the bandwidth of the BTA. Therefore, a transistor with a low collector-base capacitor is preferred to maximize the achievable bandwidth and the transistor BFG25A/X from NXP has been chosen in this work. The simulated frequency response of the BTA is shown in Fig. 5. As expected, the closed-loop bandwidth has been boosted compared to the

978-1-4799-4993-9/14 $31.00 © 2014 IEEE

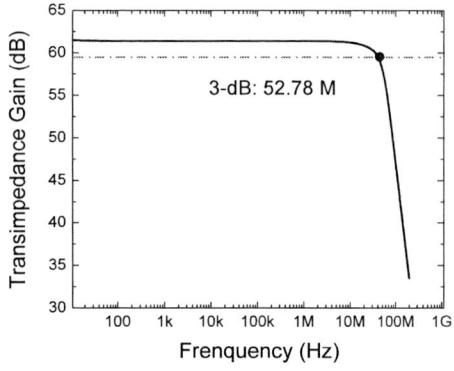

Fig. 5. Simulation of the BTA frequency response.

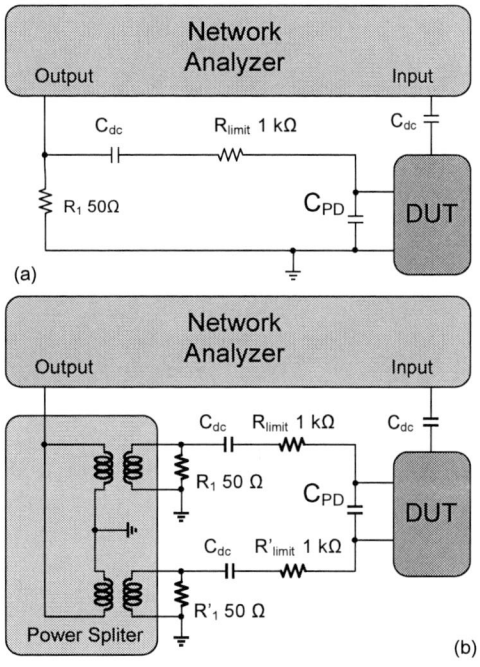

Fig. 6. Measurement setup for the bandwidth test of the conventional TIA and proposed BTA.

conventional TIA from 25 MHz to a value of 53 MHz.

III. EXPERIMENT RESULTS

A. Bandwidth

The bandwidth test setup for the conventional TIA is shown in Fig. 6(a). Since the built-in signal generator of the network analyzer is a voltage source and the input of the TIA is a current source, an equivalent circuit of the photodiode has to be used. This passive circuit delivers a low-level current signal from a capacitive source impedance, thus emulating the photocurrent and the input capacitance to the TIA. R_1 is set to be 50 Ω to match the impedance. C_{PD} is served as the junction capacitance of the photodiode and C_{dc} block the DC signal. The network analyzer used in the experiment is ROHDE&SCHWARZ ZVL (9 KHz -13.6 GHz). This test circuit eliminates the effect of the photodiode and only tests the bandwidth of the TIA itself.

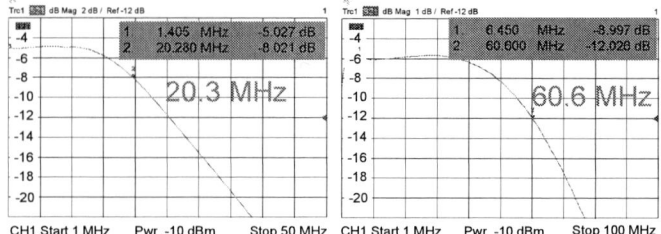

Fig. 7. Measured S parameters for the conventional TIA and proposed BTA

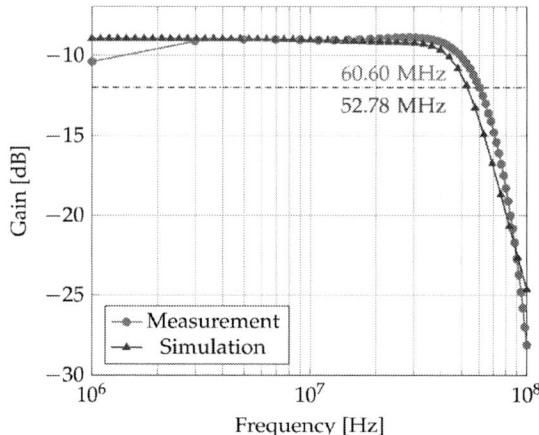

Fig. 8. Comparison of the measured and simulated frequency response of the BTA

The test circuit in Fig. 6(a), however, cannot be directly applied to the proposed BTA circuit. This is because the anode of the photodiode is an AC ground to the conventional TIA, while in the proposed BTA, both the anode and the cathode are floating. Therefore, a floating current source is required to perform the frequency response analysis. In this work, a two way 180° power splitter (ZFSCJ-2-4+) is used to decouple the grounds of the network analyzer and the DUT (device under test). The modified test interface is shown in Fig. 6(b).

Fig. 7 compares of the frequency response of the conventional TIA with that of the proposed BTA. Both circuit incorporate the same feedback resistor (R_F=1.2 kΩ) and input capacitance (C_{PD}=350 pF). Thanks to the bootstrapping, the 3-dB bandwidth of the proposed BTA (61 MHz) is almost three times larger than that of the conventional TIA (20 MHz).

It should be noted that the measured S21 parameter only indicates the frequency response of the TIA and according to [11], the real transimpedance gain can be calculated as:

$$R_T \approx \frac{V_{out}}{Vin} \cdot R_{limit} = 2 \cdot S21 \cdot R_{limit} \qquad (2)$$

Fig. 8 compares the simulated BTA frequency response with the measured one. The measured 3-dB bandwidth is slightly larger than the simulated one. That is due to a small overshoot near the 3-dB bandwidth, which indicates a slightly underdamped design but at the same time enhances the effective bandwidth.

978-1-4799-4993-9/14 $31.00 © 2014 IEEE   278

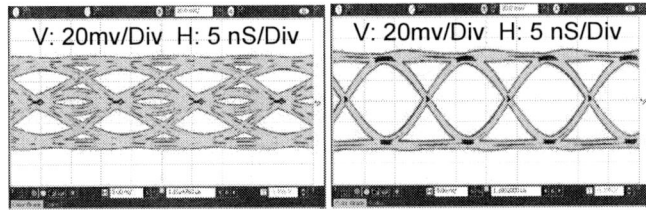

Fig. 9. Measured 75 Mbps eye diagram for the conventional TIA and proposed BTA

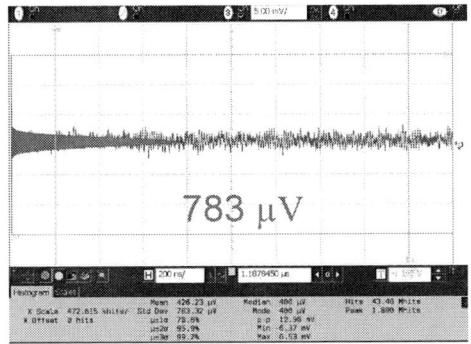

Fig. 10. Measured integrated output voltage noise of the BTA

B. Eye diagram

Fig. 9 shows a $2^7 - 1$ PRBS (Pseudorandom binary sequence) eye diagram obtained at 75 Mbps with an estimated input optical power of 130 μW. In the left eye diagram corresponding to the conventional TIA, the eyes are almost closed due to the limited available bandwidth. In contrast to this, in the right eye diagram corresponding to the BTA, the increased bandwidth ensures widely open eyes. Clearly, the improved eye diagram is beneficial for the CDR (clock and data recovery) circuit to recovery the digital data correctly.

C. Noise performance

The integrated output voltage noise from the BTA, measured using the histogram function of an oscilloscope (Agilent MSO8104A) in the absence of an input signal, is shown in Fig. 10. The measured noise floor is around 272 μV_{rms}. After deconvolving the noise floor, the integrated output voltage noise for the conventional TIA and the proposed BTA are estimated at 292 μV and 739 μV, respectively. The integrated input referred current noise density can be calculated as:

$$i_{\text{n\_TIA}} = \frac{2\sqrt{(i_{\text{n\_measured}})^2 - (i_{\text{n\_floor}})^2}}{R_{\text{f}}} = 1.23 \ \mu A \qquad (3)$$

and the average input-referred noise current density is

$$i_{\text{n\_TIA\_avg}} = \frac{i_{\text{n\_TIA}}}{\sqrt{\text{BW}}} = 158 \ \text{pA}/\sqrt{\text{Hz}} \qquad (4)$$

Table I compares the conventional TIA with the proposed BTA. Both circuits have the same transimpedance gain and input capacitance. The use of the bootstrapping greatly increases bandwidth by a factor of three, while the input noise spectral density only increases by about 40%.

TABLE I
COMPARISON OF THE SPECIFICATIONS OF THE CONVENTIONAL TIA AND PROPOSED BTA

TIA specification	TIA	BTA
Transimpedance gain (Ω)	1.2 k	1.2 k
Input Capacitance (pF)	350	350
3-dB Bandwidth (Hz)	20 M	61 M
Input noise current (pA/$\sqrt{\text{Hz}}$)	115	158
Input RMS noise current (μA)	0.48	1.23

IV. CONCLUSION

In this paper, we presented a prototype of a bootstrap transimpedance amplifier for high speed optical transcutaneous wireless communication. Utilizing the bootstrapping technique, the achievable receiver bandwidth has been significantly increased. The conventional TIA and proposed BTA were fabricated and tested. The BTA achieves a transimpedance gain of 1.2 kΩ (61.5 dBΩ) over a 3-dB bandwidth of 61 MHz in the presence of an input an input capacitance of 350 pF. Open eyes were measured at 75 Mbps PRBS data. Noise measurements indicate an input referred noise current density of 158 pA/$\sqrt{\text{Hz}}$, which presents a 40 % increase compared to the conventional benchmark TIA. The proposed BTA enables designers to use large-size the photodiodes, increasing the transmission efficiency of the TOTL system, which in turn allows for a reduction of the transmit power in the implantable transmitter.

REFERENCES

[1] M. M. Shanech, R. C. Hu, and Z. M. Williams, "A corticalspinal prosthesis for targeted limb movement in paralysed primate avatars," *Nature Communications*, vol. 5, 2014.

[2] M. a. Lebedev, A. J. Tate, T. L. Hanson, Z. Li, J. E. O'Doherty, J. a. Winans, P. J. Ifft, K. Z. Zhuang, N. a. Fitzsimmons, D. a. Schwarz, A. M. Fuller, J. H. An, and M. a. L. Nicolelis, "Future developments in brain-machine interface research.," *Clinics (São Paulo, Brazil)*, vol. 66 Suppl 1, pp. 25–32, Jan. 2011.

[3] K. Guillory, A. Misener, and A. Pungor, "Hybrid rf/ir transcutaneous telemetry for power and high-bandwidth data," in *Engineering in Medicine and Biology Society, 2004. IEMBS '04. 26th Annual International Conference of the IEEE*, vol. 2, pp. 4338 –4340, Sept. 2004.

[4] T. Liu, U. Bihr, J. Anders, and M. Ortmanns, "Performance evaluation of a low power optical wireless link for biomedical data transfer," in *Circuits and Systems, 2014. ISCAS 2014. IEEE International Symposium on*, Accpeted 2014.

[5] L. Wang and D. Ph, "Monte Carlo Modeling of Light Transport in Multi-layered Tissues in Standard C," 1992.

[6] T. Liu, U. Bihr, S. Anis, and M. Ortmanns, "Optical transcutaneous link for low power, high data rate telemetry," in *Engineering in Medicine and Biology Society (EMBC), 2012 Annual International Conference of the IEEE*, pp. 3535–3538, 2012.

[7] A. Street, P. Stavrinou, D. Edwards, and G. Parry, "Optical preamplifier designs for ir-lan applications," in *Optical Free Space Communication Links, IEE Colloquium on*, pp. 8/1–8/6, Feb 1996.

[8] E. Sackinger, *Broadband circuits for optical fiber communication*. John Wiley & Sons, 2005.

[9] J. Quarfoot, "Noise , slew rate , and unity gain stability tradeoffs in high speed amplifiers," 1999.

[10] P. C. D. Hobbs, "Photodiode Front Ends - The Real Story," 2001.

[11] "Sa5211, transimpedance amplifier product specification," 1998.

High Accuracy Current Sense Amplifier With Extended Input Common Mode Range

Răzvan Puşcaşu[1,2], Pavel Brînzoi[2], Laurenţiu Creoşteanu[2], Gheorghe Brezeanu[1]

[1]Electronics, Telecommunications and Information Technology
University Politehnica of Bucharest
Bucharest, Romania
razvan.puscasu@ymail.com

[2]ON Semiconductor Romania
Bucharest, Romania

Abstract—**A high accuracy current sense amplifier with input common mode range extending beyond the positive supply rail is presented. The offset voltage is minimized using an auto-zero topology with a fully differential internal path. Characteristics of the presented amplifier include an input voltage range of 30V, independent of the supply rail, a fixed gain with a typical gain error of 0.05% and a referred-to-input offset voltage of less than 10μV in the −40°C to 125°C temperature range. The proposed architecture was implemented in a 0.5μm CMOS process and its performances were confirmed by post-layout simulation results.**

Keywords— *current sense amplifier, auto-zero, extended input common mode range*

I. INTRODUCTION

Power management in battery-powered portable applications, automotive or telecom equipment requires precision current sensing circuits in order to achieve the desired accuracy. Load current is measured through a low-value resistor placed in series with the power supply and the proportional voltage is then amplified and processed [1]-[4]. Lowering the resistor value (and thus power dissipation) increases the overall efficiency therefore a low offset and high common mode rejection ratio (CMRR) architecture is required. Furthermore, the ability to sense common mode voltages above the supply rail makes the system suitable for low voltage applications. Various sense amplifier architectures with extended input common range have been proposed in literature [2]-[4]. Desired precision and wide input common mode range are however obtained using complex circuit designs which lead to high current consumption, large die areas and therefore higher fabrication costs [3], [4]. This paper presents a high accuracy sense amplifier for bidirectional high and low side current measurement, using a continuous time auto-zero topology. The proposed circuit uses a simple method to cover a wide input common mode range between -0.2V and 30V, independent of the supply rail. The supply voltage can vary from 2.5V to 30V, with a low quiescent current consumption. The circuit was implemented in a standard CMOS BCD process, with a die area of 0.9 mm².

II. CIRCUIT DESCRIPTION

A block diagram of the proposed architecture is shown in Fig. 1. The amplifier's fixed gain of 100V/V is given by R and R_f integrated resistors. The output voltage has the expression

$$V_{OUT} = V_{REF} + \frac{R_f}{R}(V_{IN+} - V_{IN-}) \qquad (1)$$

V_{OUT} is referred to an external user defined reference voltage which eases the processing of the output signal with a data acquisition system, enabling bidirectional current sensing. There are two signal paths connected in parallel. The main path, through A_M dictates the frequency response of the system and has an open-loop voltage gain greater than 120dB. The second signal path, from A_N through B_M, establishes the required precision, with a DC gain greater than 180dB. There are two main aspects of concern when setting the DC gain of the internal amplifiers: firstly, the offset contributions of the amplifiers are minimized by properly selecting the gain ratios between the signal paths and by choosing the gain value of B_N stage [5]. Secondly, if the mismatch between the gain resistors is ignored then the closed-loop gain error (and the overall accuracy) is mainly influenced by the open-loop gain variation. The minimum open-loop gain of the internal amplifier that needs to be obtained in order to meet the required gain error specifications can be expressed as

$$A_{OL} \approx \frac{R_f}{R} \frac{100}{GAIN\_ERROR} \qquad (2)$$

where GAIN_ERROR has a value of 0.05. Consequently, the minimum DC open-loop gain is approximately 106 dB, value which needs to be achieved by both signal paths, over temperature. A_N, B_N, A_M and B_M gains are calculated from the input pins to the output pins of the nulling and main amplifiers respectively, including internal intermediate stages.

Referred-to-input offset voltage is reduced by using an auto-zero architecture that allows a continuous signal operation. The chosen topology consists of a main amplifier (MAIN AMP), which is always connected in the signal path and a nulling amplifier (NULL AMP) for offset correction.

This work was funded by POSDRU/159/1.5/S/132395

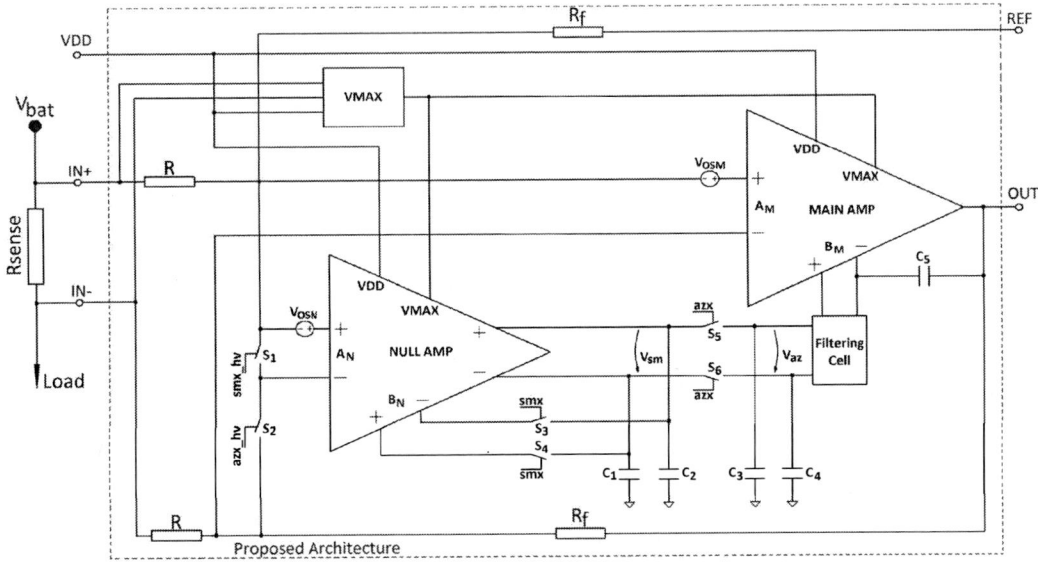

Fig. 1. Block diagram of the proposed architecture

The offset reduction process consists of two phases: in the *smx* phase (sampling phase), S_1, S_3 and S_4 switches are on and S_2, S_5 and S_6 switches are off (see Fig.1). The nulling amplifier is disconnected from the signal path. Its offset voltage appears at its output (V_{sm}) and is stored as a correction voltage for the next phase, on C_1 and C_2 capacitors [5]. In the auto-zero time interval (*azx* phase), S_1, S_3 and S_4 switches are off and S_2, S_5 and S_6 switches are on (see Fig.1). The nulling amplifier is connected in parallel with the main amplifier's inputs. The output voltage of the nulling amplifier (V_{az}) is stored on C_3 and C_4 capacitors and acts as a correction voltage for the main amplifier. Taking into account the correction voltages obtained throughout the auto-zero process the output voltage can be written as

$$V_{OUT} = V_{SENSE}(A_M + A_N B_N) + A_M V_{OSM} + \frac{A_M B_M}{1 + B_N} V_{OSN} \tag{3}$$

where $V_{SENSE} = V_{IN+} - V_{IN-}$. V_{OSM} and V_{OSN} are the inherent offset voltages of the main and nulling amplifiers. By considering the particular case in which $A_M \equiv A_N$ and $B_M \equiv B_N \gg 1$, the expression of the output voltage becomes

$$V_{OUT} = A_N B_N (V_{SENSE} + \frac{V_{OSM} + V_{OSN}}{B_N}) \tag{4}$$

Equation (4) shows that the offset voltage of the main and null amplifiers is divided by the gain of B_N stage. Optimizing the input stages by means of gain and gain ratios can reduce the input offset from several millivolts to the microvolt level [5], [6].

S_1-S_6 switches are designed as small CMOS transfer gates for reducing charge injection and clock switching errors. Proper auto-zero operation requires that S_1 and S_2 switches driving voltages must be referred to the input common mode value (*smx_hv* and *azx_hv*, see Fig.1). However, for input voltages exceeding the supply rail, only the PMOS transistors will function, since the NMOS transistors would need a gate voltage higher than the input common mode. The *azx* and *smx* control phases are designed taking into account the delays introduced by level shifting, ensuring a sufficient non-overlapping time for good circuit performance. S_1 and S_2 switches need to be isolated from the die substrate in order to function suitably. Minimizing the charge injection and clock feedthrough errors require small devices with low input capacitance, therefore high voltage transistors could not be used. The transfer gates were designed using isolated NMOS devices built in an N-buried layer tub.

Extension of the common mode range is done using the VMAX block (see Fig.1). When the input common voltage exceeds the positive supply rail, the biasing voltage of the main and null amplifiers is switched from VDD to VMAX, which generates the maximum voltage between V_{IN+}, V_{IN-} and VDD. The comparison between V_{IN+} and V_{IN-} voltages is necessary due to the low gain of the amplifier, which allows a relatively high differential input voltage ($\pm300mV$) and because of the bidirectional current sensing architecture. However, the current drawn from the input pins is considerably smaller than the load current therefore the measurement accuracy is not affected. Furthermore, the proposed method does not introduce additional voltage errors because the VMAX cell is connected directly to the input pads, before the input resistors R.

A simplified schematic of the nulling amplifier is presented in Fig.2. The two input pairs A_N and B_N consist of a rail-to-rail configuration with *MP0 - MP3*, *MN0 - MN3* as PMOS and NMOS devices respectively, which are then summed and fed into a folded cascode structure. The PMOS input pair is supplied directly from VDD while the NMOS differential pair is biased from VMAX cell, allowing A_N driving voltage to extend over VDD. The Schotky diode *DS0* isolates the common mode voltage from the biasing circuit, protecting the VDD supplied devices. Exceeding the supply

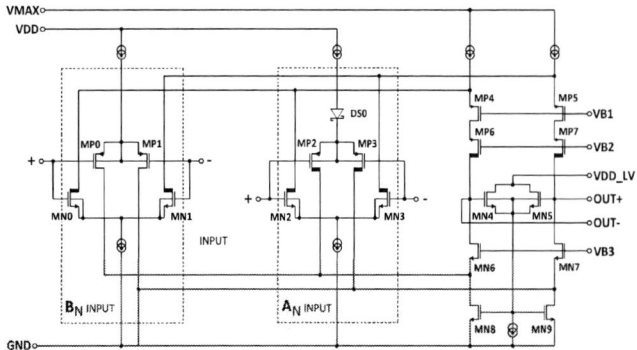

Fig. 2. Simplified schematic of the nulling amplifier

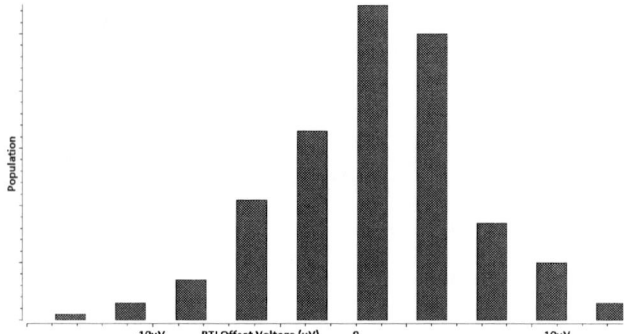

Fig. 3. RTI offset voltage distribution from MC simulations including PEX; VDD = 5V, V_{CM} = 12V, V_{REF} = 2.5V, V_{SENSE} = 0V

voltage would forward bias and damage the substrate diodes of the tail current mirror of *MP2* and *MP3*. Moreover, additional circuitry is required for ensuring that the gate-to-source voltage of the PMOS pair does not damage the devices when $V_{CM} \gg$ VDD. These caution measures are nevertheless not necessary for the B_N stage, since the value of its input common mode always remains in the low voltage domain. The bolded drains on *MN0-MN3*, *MP6* and *MP7* transistors represent the drain extension which allows for high voltage operation. All other transistors are low voltage devices. *MN4* and *MN5* transistors form the common mode feedback circuit which sets V_{OUT+} and V_{OUT-} to approximately 1.5V and is biased from an internal regulator. *MP6* and *MP7* transistors maximize the DC gain by increasing the output impedance, since the overall referred-to-input offset voltage is directly affected by the gain of the nulling amplifier. The main amplifier has a similar architecture, combined with a rail-to-rail class AB biased output stage for driving heavy loads.

III. SIMULATION RESULTS

The proposed architecture was simulated and implemented in a 0.5μm BCD CMOS process with low threshold transistors, drain extension capability and high resistivity resistors. An active die area of 0.9mm² was obtained, resulting in low fabrication costs. Simulations were performed using Spectre® and included post-layout parasitic extraction (PEX). The supply voltage range is from 2.5V to 30V (with a quiescent current consumption of 57μA) and the input common mode can vary from −0.2V to 30V, independent of the supply rail. The amplifier's closed-loop gain is set at 100V/V using R and R_f resistors (see Fig. 1) which have values of 10kΩ and 1MΩ respectively. The required accuracy and gain error are obtained using a proprietary trimming procedure combined with a continuous time auto-zero process. Simulation setup included a 10kΩ resistor in parallel with a 10pF capacitor load.

The referred-to-input (RTI) offset voltage distribution is presented in Fig.3. The figures were obtained by performing 100 Monte Carlo (MC) runs in the −40°C to 125°C temperature range. Simulations included extracted post-layout parasitic elements but the gain resistors mismatch was not taken into consideration since it will be compensated by

the trimming mechanism. The distribution is not zero-centered due to the systematic offset of the internal amplifier.

The filtering cell smoothes the output ripple by reducing charge injection and clock feedthrough errors from the high gain signal path. A comparison between the output waveforms with and without the filtering cell is presented in Fig. 4. It can be observed that the output ripple is significantly reduced using the filtering cell, from 1.54mV$_{pk-pk}$ to under 500μV$_{pk-pk}$ and thus the V_{OUT} waveform is smoothed. The 500μV$_{pk-pk}$ is translated into a 5μV$_{pk-pk}$ referred-to-input offset voltage, including the higher sampling phase spike. However, the output ripple cannot be entirely removed by internal filtering. The higher spike value ("A" point) is given by the charge injection errors stored on C_1 and C_2 capacitors in the sampling phase [5], errors that have a significant contribution to the overall offset. The lower amplitude spike ("B" point) is given by switching errors in the auto-zero phase. Special care must be taken in the layout process in order to maintain the internal differential path as symmetrical as possible [7], by ensuring that the critical nets are loaded by close-valued parasitic resistors and capacitors. The same principle applies for the clock phases routing also.

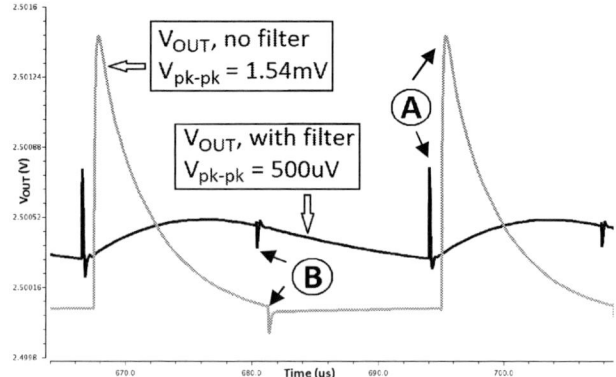

Fig. 4. Simulated output waveform comparison with and without the filtering cell; VDD = 5V, V_{CM} = 12V, V_{REF} = 2.5V, V_{SENSE} = 0V

Figure 5 illustrates the variation of the common mode rejection curve over the trimming process. The internal amplifier has a common mode rejection of over 120dB.

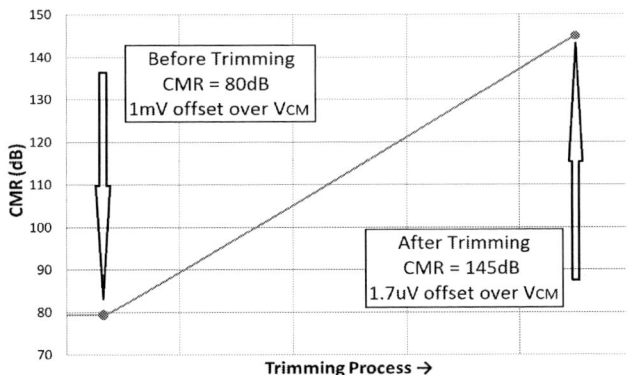

Fig. 5. Simulated input bias current;
VDD = 5V, V_{CM} = 0V to 30V, V_{REF} = 2.5V, V_{SENSE} = 0V

However, the main cause of CMR degradation in this type of circuits is the gain resistors mismatch. Since CMR is inversely proportional with the imbalance between R and R_f resistors, a proprietary trimming procedure is used in order to meet the required accuracy and precision. As an example, the offset of the amplifier (excluding the gain resistors) is initially $17\mu V$ (random Monte Carlo run). Due to resistor mismatch, the CMR drops to 80dB and the offset variation exceeds 1mV over the input range. The trimming process compensates the gain resistor mismatch so that after the trimming procedure the referred to input offset error drops under $5\mu V$ ($3\mu V$ to 4.7 μV, over the entire common mode range) and the CMR increases to 145dB.

A summary of the performances achieved by the proposed architecture is presented in Table I, along with a comparison with state of the art.

TABLE I.   PERFORMANCE SUMMARY AND COMPARISON
WITH STATE OF THE ART

Parameter	This Work			[2]	[3]	[4]	Unit
	Min	Typ	Max				
Supply Voltage Range (VDD)	2.5		30	2.8 to 5.5	±18	2.7 to 26	V
Input CM range (V_{CM}) [a]	−0.2		30	1.9 to 30	10 to 36	0 to 26	V
Input DF range (V_{SENSE})			±300	±150	±20mV to ±10V	±260	mV
RTI Offset Voltage		3	±10	5	20	±1	µV
PSR	120	133		121	123	140	dB
Gain		100		20	1/8 to 128	100	V/V
Gain Error		±0.05	±1	0.1	0.058	0.02	%
CMR	125	140		143	120	140	dB
Quiscent Current	50	57	63	3250	3000	65	µA
Die Area		0.9		2.5	8.6	-	mm²

[a] Independent of the supply voltage

Bolded figures represent values obtained over the entire temperature range, from -40°C to 125°C with post-layout Monte Carlo simulations, in order to better predict silicon results. The referred-to-input offset voltage is measured at the amplifier's inputs, without taking into account the resistor mismatch. All presented simulation results were obtained with VDD = 5V, V_{REF} = 2.5V, V_{SENSE} = 0V and V_{CM} set to 12V, except CMR and PSR figures, which were obtained for the entire input common mode and supply ranges. In addition the gain error was simulated for differential input signals between −5mV and 5mV.

IV.   CONCLUSION

In this paper, an accurate current sense amplifier with extended input common mode range was presented. The proposed architecture has a fixed gain of 100V/V and the reduction of the input offset error is done using a continuous time auto-zero technique. Offset voltages less than $10\mu V$ over the entire common mode range yield CMR figures of over 120dB, which corroborated with low gain errors allows the use of small resistors in bidirectional, high or low side current sense configurations. Using a simple method, the input common mode voltage can vary from −0.2V to 30V, independent of the supply rail and is only limited by the high voltage capabilities of the technology. A chip area of 0.9mm² was obtained and the quiescent current consumed from VDD is 57 µA. Rail-to-rail output capability together with the user defined reference voltage ease the integration of the proposed current sensing architecture with data acquisition systems, using a minimal number of external components.

ACKNOWLEDGMENT

This work was funded by the Sectoral Operational Program Human Resources Development 2007-2013 of the Ministry of European Funds through the Financial Agreement POSDRU/159/1.5/S/132395.

REFERENCES

[1] Shalmany, S.H.; Draxelmayr, D.; Makinwa, K.A.A., "A micropower battery current sensor with ±0.03% (3σ) inaccuracy from −40 to +85°C," Solid-State Circuits Conference Digest of Technical Papers (ISSCC), 2013 IEEE International , vol., no., pp.386,387, 17-21 Feb. 2013

[2] Witte, J.F.; Huijsing, J.H.; Makinwa, K. A A, "A Current-Feedback Instrumentation Amplifier with 5µV Offset for Bidirectional High-Side Current-Sensing," Solid-State Circuits Conference, 2008. ISSCC 2008. Digest of Technical Papers. IEEE International , vol., no., pp.74,596, 3-7 Feb. 2008

[3] Schaffer, V.; Snoeij, M.F.; Ivanov, M.V.; Trifonov, D.T., "A 36 V Programmable Instrumentation Amplifier With Sub-20 uV Offset and a CMRR in Excess of 120 dB at All Gain Settings," Solid-State Circuits, IEEE Journal of , vol.44, no.7, pp.2036,2046, July 2009

[4] Texas Instruments, INA214 Data Sheet, Revision E, 2013, Accessed on March 10, 2014 <http://www.ti.com/lit/ds/symlink/ina210.pdf>

[5] Danchiv, A.; Bodea, M., "Residual offset optimization for a continuous time auto-zero amplifier," Signals and Electronic Systems, 2008. ICSES '08. International Conference on , vol., no., pp.281,284, 14-17 Sept. 2008

[6] Finvers, I.G.; Haslett, J.W.; Trofimenkoff, F.N., "A high temperature precision amplifier," Solid-State Circuits, IEEE Journal of , vol.30, no.2, pp.120,128, Feb 1995

[7] Burt, R.; Zhang, J., "A Micropower Chopper-Stabilized Operational Amplifier Using a SC Notch Filter With Synchronous Integration Inside the Continuous-Time Signal Path," Solid-State Circuits, IEEE Journal of , vol.41, no.12, pp.2729,2736, Dec. 2000

Test and Diagnosis of FPGA Cluster Using Partial Reconfiguration

Saif Ur Rehman, Mounir Benabdenbi, and Lorena Anghel
Grenoble Alpes University, TIMA Laboratory, Grenoble, France

Abstract—FPGA undergoes a large number of test configurations for faults detection and diagnosis which requires a significant amount of test time and off-chip memory to store test configuration bits. Some FPGAs support partial reconfiguration in which a portion of FPGA can be reconfigured without reconfiguring the remaining portions. This partial reconfiguration approach can be utilized for FPGA testing, to reduce the amount of test time and configuration bit storage. In this paper, we propose a Built-In Self-test (BIST) scheme utilizing partial reconfigurability of a mesh FPGA. To implement the BIST scheme, automated tools are developed to produce the required configuration bitstream using standard FPGA CAD flow. A comparative analysis of partial and full reconfiguration is presented which shows that test time and storage size for configuration bitstream reduce to half using partial reconfiguration approach.

I. INTRODUCTION

FPGAs are gaining a significant share in IC industry due to their reconfigurability and shorter time-to-market. FPGAs can be reprogrammed by the user to have an arbitrary design. Due to their high performance and low power features, FPGAs are used to implement complex digital systems. To ensure the reliability of a device, exhaustive testing for manufacturing defects is of critical importance. As the device should be tested in all modes of operation, exhaustive FPGA testing requires a large number of configurations.

FPGA testing is a time consuming process in which configuration time dominates the overall test time. To program an FPGA for an application, dedicated configuration bits are loaded into the FPGA memory array (SRAM cells). These configuration bits are stored outside the chip and grouped into the frames of a certain bit-width. For loading an application, these bits are accessed one frame at a time and written into the SRAM cells. Therefore, exhaustive testing becomes very costly for those FPGAs which require full reconfiguration to be loaded each time to implement an application. Usually test cost is evaluated in term of test time (i.e. configuration time) and the memory to store the required configuration bitstream.

Some FPGAs provide the facility to configure some of their modules, keeping the configuration of others unchanged. In this case of partial reconfiguration, only the configuration bits needed to be modified are loaded into the FPGA. Thus, the number of frames required to have a new FPGA configuration is reduced. This aspect of partial reconfiguration can be efficiently utilized in the FPGA testing where test time reduction can be achieved by using less number of frames per configuration. In this paper, we propose a test scheme to detect and locate/diagnose stuck-at faults in the logic and intra-cluster

interconnect in a mesh FPGA. For this purpose, a sequence of test configurations is developed using partial reconfiguration. The impact of partial reconfiguration on test time reduction is investigated and a comparative analysis is presented by considering the test time required in the case of partial as well as full reconfiguration approaches. The remainder of this paper is organized as follows. Next section gives a brief overview of the related work done in the past. Section III highlights some key features of the FPGA considered in this work. Section IV describes the proposed test scheme. Experimental results are presented in section V. Finally, section IV concludes the paper.

II. RELATED WORK

In most of the previous studies, FPGA testing is carried out using Built-In Self-Test approach [1], [2]. BIST is considered as the most efficient technique for FPGA testing as it exploits very well the FPGA reconfigurability and its regular structure. In BIST, a part of FGPA is configured as tester which tests the other part that is configured to be tested [3]. In [4], an FPGA BIST approach is presented in which test time reduction is achieved at the cost of an area overhead by adding partially reconfigurable structures. An online testing is presented in [5] where partial reconfiguration is used for testing spare resources without affecting the application running on the other part of FPGA. BIST for Virtex and Spartan FPGAs using partial reconfiguration is given in [6]. Although BIST is a generic technique, test configurations are architecture specific. A sequence of dedicated configurations are required to perform testing of different module of the FPGA (i.e. logic, interconnect, SRAM, etc). Most of the prior work done in this context, target FPGAs having fully populated interconnect. These interconnect provide full routability among logic resources but does not exploit FPGA logic equivalence, thus, incur more area and long propagation delay [7]. Therefore, test schemes are needed to be worked out for the FPGAs having depopulated interconnect. In our previous work [8], test scheme for an FPGA having depopulated interconnect is presented. However, this scheme utilizes full reconfiguration approach for fault detection and diagnosis. This paper focuses on the development of a test scheme using partial reconfiguration for logic and depopulated interconnect resources in a mesh-based FPGA.

III. OVERVIEW OF THE CLUSTER ARCHITECTURE

In modern FPGAs, logic blocks are grouped together in the form of a cluster. Clustering is aimed to reduce the interconnect area and signal propagation delay. Clusters are typically arranged in a grid and are surrounded by horizontal and vertical routing channels. The routing resources are

978-1-4799-4993-9/14 $31.00 © 2014 IEEE

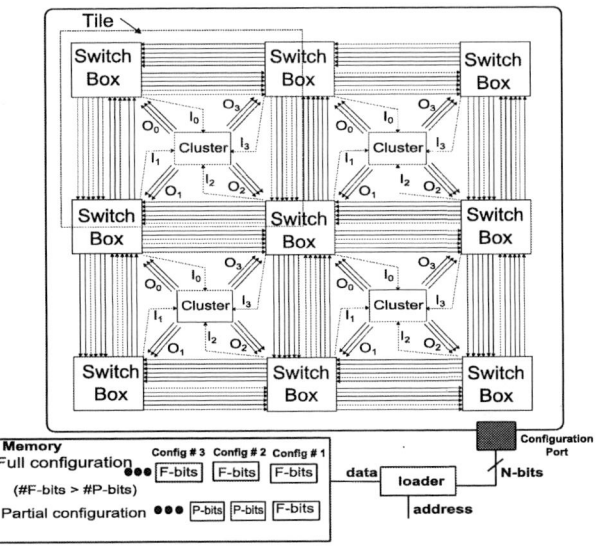

Fig. 1. Mesh of clusters FPGA

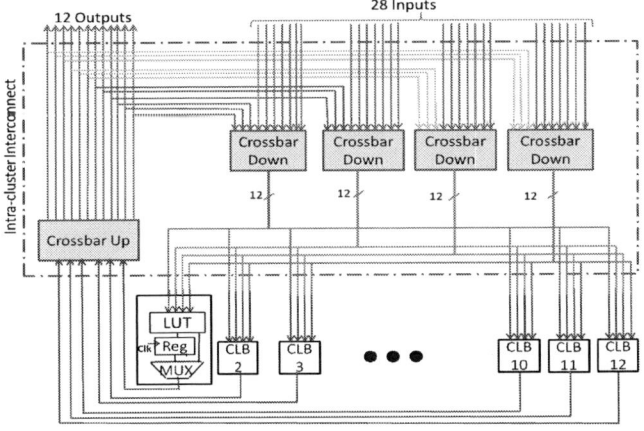

Fig. 2. Structure of a cluster

the number of frames required in partial reconfigurations are reduced.

In this FPGA architecture, each cluster is surrounded by four switch boxes. The cluster has 28 inputs and 12 outputs with 12 Configurable Logic Blocks (CLBs) as shown in Fig. 2. All the cluster outputs are also uniformly distributed as feedback to the crossbar down in order to re-utilize logic computation results and hence allow locality of the application being mapped on the FPGA. Each crossbar down receives three among the twelve cluster outputs. Each CLB contains one 4-input LUT and a register. A multiplexer is used to provide either registered or unregistered output of LUT at the CLB output (cf. Fig. 2).

The intra-cluster crossbar is a multiplexer based structure. The down linking block consists of 4 crossbars, each of them is made up of 12 multiplexers 10:1. They link the cluster inputs to the logic blocks. An up link block (crossbar up) links the CLB's outputs to the CLBs of the same cluster through crossbar down. The structure of crossbar up is similar to that of a crossbar down and is composed of 12 multiplexers 12:1.

IV. TEST METHODOLOGY

In this section, we describe the proposed BIST scheme for logic and intra-cluster interconnect using partial reconfiguration.

A. BIST Architecture

Usually FPGA BIST is performed in sessions. In a session, some clusters/CLBs are programmed to have the functionality of Test Pattern Generators (TPGs), some are configured as Output Response Analyzers (ORAs) and some as Block Under Test (BUTs). TPG produces test patterns which are applied to BUTs. The BUTs output response to these test patterns is analyzed in ORAs. To perform an exhaustive testing of a BUT, its all modules (e.g. RAMs, logic, interconnect) need to be tested in all modes of operation. For that purpose, a sequence of configurations is required to test each module. Usually each type of BUT module is tested separately using a dedicated sequence of configurations. Similarly, the type of test patterns produced by TPG depends on the BUT module being tested. Therefore, TPG is also required to be reconfigured accordingly while testing a BUT. When BUT test is complete, next test session starts in which BUTs are configured to function as TPG/ORA which then test the TPG and ORA of the previous session.

A simplified BIST structure is shown in Fig. 3 where clusters are configured as TPG/ORA or BUT. To produce pseudo-exhaustive test patterns, n-bit Linear Feedback Shift Register (LFSR) is implemented in TPG, where 'n' is defined by the number of inputs of the module under test. Here, we employ a comparison-based analyzer, where any mismatch between the outputs of BUTs is stored and recovered at the end of test sequence in each test configuration. Each CLB of the ORA cluster acts as an independent comparator which compares the output of two CLBs each from different BUTs. To recover the results from ORA, memory read-back approach is utilized which allows to read the contents of ORA flip-flop at the end of each test sequence.

composed of wiring segments and programmable switches. A connection box is used to provide connection between logic blocks and adjacent routing channels. A set of switches used to connect horizontal and vertical routing channels is called switch box.

In this paper, we consider the SRAM-based mesh FPGA architecture proposed in [9] where depopulated interconnect structure is used both in the cluster and switch box. In this topology, connection box is avoided by connecting clusters directly to switch box. This architecture is composed of arrays of tiles. Each tile comprises a cluster and external interconnect as shown in Fig. 1. FPGA configuration bits stored in the external memory are loaded into the FPGA through dedicated port. An exemplary scenario is depicted in Fig. 1 in which the FPGA is fully configured using 'F-bits'. In case of full reconfiguration, 'F-bits' are loaded each time for a new configuration. In partial reconfiguration, FPGA is fully programmed only for the first configuration. In the following configurations, only the configuration bits ('P-bits') required by the module (e.g. cluster) to be configured are loaded. Thus,

978-1-4799-4993-9/14 $31.00 © 2014 IEEE   285

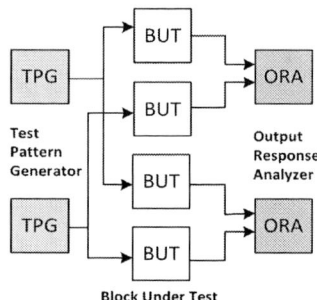

Fig. 3.   BIST Structure

In BIST structure, we use two TPGs producing identical test patterns. A single TPG is used to feed multiple BUTs. Each BUT is compared with two different BUTs having identical test patterns but from different TPGs. This approach helps in avoiding any fault masking that may occur due to multiple even faults in a BUT or faulty TPG/ORA. The only assumption made in this procedure is that the SRAM cells and I/O blocks of the FPGA are fault free to avoid any fault masking, as the target is to detect and diagnose faults in the cluster. Usually SRAM cells and I/O blocks are tested in earlier stages using dedicated BIST schemes.

B. Logic Block Test Configurations

In the first configuration of BIST implementation, whole FPGA is fully configured to have all BIST components i.e. TPG, BUT and ORA. This initial full configuration is then followed by partial configurations in which different modules of BUT are tested and diagnosed for stuck-at faults. The LUT in a CLB is basically a series of multiplexers with SRAM cells at the input of these multiplexers. To test a CLB in LUT mode of operation, LUT is configured to have the functionality of exclusive-OR and then exclusive-NOR. During these configurations, unregistered LUT output is selected. Similarly, to test the path through the flip-flop, another set of configurations is required.

To apply the test patterns at CLB inputs (select signals of the LUT multiplexers), each crossbar down is configured to receive a unique input of the cluster (see Fig. 2). In this way, all CLBs in a cluster are tested simultaneously using identical test patterns. To observe each CLB at the cluster output, crossbar up is configured accordingly. Throughout the CLB testing, crossbar up and crossbar down are not required to be reconfigured.

C. Intra-cluster Interconnect Test Configurations

For fault diagnosis at the resolution of a multiplexer in the crossbar structures, we divide the interconnect testing in two phases. In the first phase, crossbar up is tested for all possible configurations. Test patterns are applied at the cluster inputs and the response is observed at the cluster output. Therefore, crossbar down and CLBs are required to be configured to pass these test patterns to crossbar up. Figure 4 shows a schematic of the first test configuration of crossbar up. The active paths in this configuration are shown as solid lines. In this configuration, each crossbar down receives 4 bits of test pattern. CLBs are configured accordingly to select the correct input (cf. Fig. 4). During crossbar up testing, only crossbar up

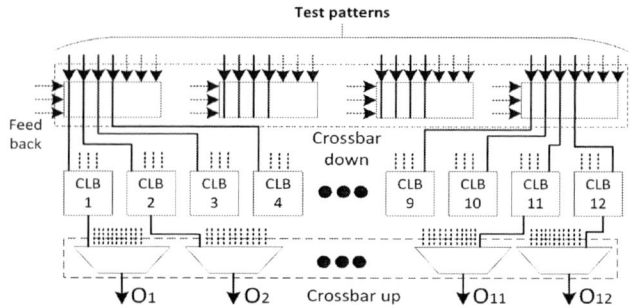

Fig. 4.   A test configuration of a crossbar up

is partially reconfigured whereas CLB and crossbar down are configured once and kept unchanged.

In the second phase, crossbar down is tested thoroughly while crossbar up is fixed at its last fault free configuration. To attain a high diagnostic resolution, one crossbar down among four is tested at a time. During crossbar down testing, logic block and crossbar up are configured such that the targeted paths in the crossbar down can be observed at the cluster output. CLBs are required to be reconfigured only at the beginning of a crossbar down testing, otherwise, CLBs configurations are kept constant throughout the testing of that crossbar down.

V.   EXPERIMENTAL EVALUATION

To implement the test scheme mentioned above, FPGA configuration bitstream is required. For that purpose, standard CAD flow for FPGA (shown in Fig. 5) is used to produce the required bitstream.

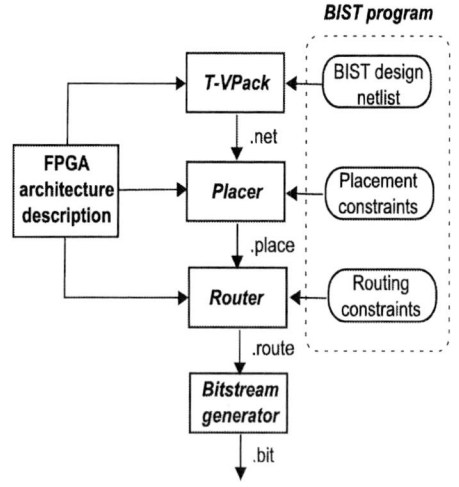

Fig. 5.   FPGA CAD flow

Clustering of CLBs is performed using T-VPack tool [10] according to the FPGA architecture parameters. Design netlists for TPG, ORA and BUT, defining the functionality of these modules are given to T-VPack. To place the BIST module at the desired position in the FPGA, placement constraints are given to the *Placer* in the form of files. Similarly, routing constraints for each configuration are given to the *Router* which define the targeted path in that configuration. Each

978-1-4799-4993-9/14 $31.00 © 2014 IEEE

TABLE I.   RESULTS FOR 100% FAULT COVERAGE OF
CLB AND CLUSTER INTERCONNECT

FPGA block	Using full reconfiguration		Using partial reconfiguration		Gain (%)
	Config #	Test cycles	Config #	Test cycles	
Crossbar Down	60	3360	60	1440	57%
Crossbar Up	24	1344	24	720	46%
CLB	4	224	4	128	42.85%

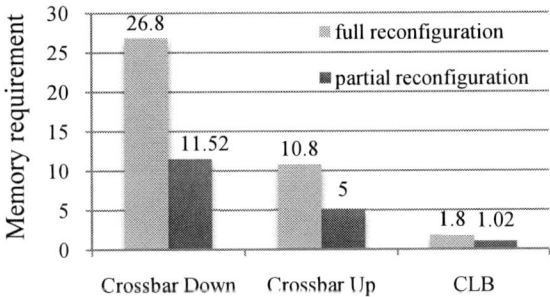

Fig. 6.   Memory requirement for full and partial test
reconfigurations of a cluster

configuration corresponds to a unique routing/placement constraint. Since the presented BIST scheme is scalable to any FPGA array size, a set of automated tools is developed to generate these files required to produce the sequence of test configuration bitstreams for a user given FPGA array size.

To analyze the efficiency of the proposed test scheme, faults are injected at the gate level cluster architecture. For that purpose, most commonly used single stuck-at fault model is employed using *Synopsys Tetramax* to define the list of all potential faults in the cluster interconnect. Efficiency of the test scheme is calculated in terms of fault coverage which is defined as the ratio between detected faults and total number of potential faults in a given circuit. Table I shows the number of test configurations required to achieve 100% stuck-at fault coverage in CLB and intra-cluster interconnect. A comparative analysis is presented using the results obtained with full configuration test scheme given in [8]. The number of test cycles required to configure a cluster for a complete test procedure is also presented. For the same number of test configurations in both full and partial reconfiguration schemes, test time reduction is achieved by reducing the number of configuration bit frames. Hence, it reduces the number of required test cycles to load the configuration bitstream.

Another important advantage of partial reconfiguration technique is the reduction in memory size to store the test configurations. It is depicted in Fig. 6 where a comparison is given in terms of memory requirement for CLB and cluster interconnect testing using full and partial reconfiguration schemes. The normalized results are obtained using actual memory size requirement for crossbar up, down and CLB. As

can be seen from the results, memory requirement is reduced to half by using partial reconfiguration for test schemes.

VI.   CONCLUSION

We developed a BIST scheme using partial reconfiguration for the configurable logic block and intra-cluster interconnect in a mesh-based FPGA. In the beginning of a BIST sequence, a full configuration is loaded. Later on, FPGA is partially reconfigured throughout the testing sessions. We classify the interconnect testing in two phases, based on the functionality of the modules. In one phase, targeted interconnect are partially reconfigured throughout their testing, keeping the other's configurations unchanged. We showed that 100% stuck-at fault coverage can be obtained using partial reconfiguration with a gain of 43-57% in test time. Similarly, partial reconfiguration during FPGA testing reduces the amount of memory storage for configuration bits to half as compared to full reconfiguration approach. To implement the proposed BIST scheme, automated tools were developed and integrated into the standard design flow for the required bitstream generation.

In the future, we aim to develop test and diagnosis schemes for switch box interconnect using partial reconfiguration. In these schemes, bridging and delay faults will also be considered along with stuck-at.

ACKNOWLEDGMENT

This work was performed in the framework of Robust FPGA project ANR 11 INS-02 funded by The National Research Agency of France, The Pole Systematic and The Pole Minalogic.

REFERENCES

[1] B. Dutton, C. Stroud, "Built-in Self-Test of configurable logic blocks in Virtex-5 FPGAs", Proc. IEEE Southeastern Symp, on System Theory, pp.230-234, 2009.

[2] J. Yao, B. Dixon, C. Stroud and V. Nelson," System-level Built-In Self-Test of global routing resources in Virtex-4 FPGAs", Proc. IEEE Southeastern Symp, on System Theory, pp.29-32, 2009.

[3] C. Stroud, A Designer's Guide to Built-in Self-Test, Kluwer Academic Publishers, Boston, 2002.

[4] Z. Jianfeng, H. Hu, W. Dong and P. Liyang, "A cost-efficient self configurable BIST technique for testing multiplexer-based FPGA interconnect," Journal of Electronic Testing, pp. 647-655, Oct. 2011.

[5] Suthar," Built-In Self-Test of FPGAs with provable diagnosabilities and high diagnostic coverage with application to on-line testing, Proc. VLSI Test Symposium, 2006.

[6] S. Dhingra, S. Garimella, A. Newalker, and C. Stroud, "Built-In Self-Test of Virtex and Spartan-II FPGAs using partial reconfiguration," Proc. North Atlantic Workshop, pp. 7-14, 2005.

[7] Z. Marrakchi, H. Mrabet, C. Masson, and H. Mehrez, "Mesh of tree: unifying mesh and MFPGA for better device performances," Proc. First International Symposium on Networks-on-Chip, pp. 243-252, 2007.

[8] S. Rehman, M. Benabdenbi and L. Anghel, "BIST for logic and local interconnect resources in a novel mesh of cluster FPGA," Int. symposium on defect and fault tolerance in VLSI and nanotechnology systems, pp. 296-301 Oct. 2013.

[9] Z. Marrakchi, H. Mrabet, and H. Mehrez, "Programmable gate array, switch box and logic unit for such an array," Patent US 7 795 911 B2, 2010.

[10] A. S. Marquardt, V. Betz, and J. Rose, " Using cluster-based logic blocks and timing-driven packing to improve FPGA speed and density," in Proc. Int. Symposium on Field Programmable Gate Arrays, New York, USA. ACM, pp. 37-46, 1999.

A New Hardware Implementation of The Advanced Encryption Standard Algorithm for Automotive Applications

Riccardo Cassettari
PhD student
University of Pisa
Pisa, 56122, Italy
Email: riccardo.cassettari@for.unipi.it

Luca Fanucci
Professor
University of Pisa
Pisa, 56122, Italy
Email: l.fanucci@iet.unipi.it

Giorgio Boccini
Electronic Engineer
University of Pisa
Pisa, 56122, Italy
Email: giorgio.boccini@ing.unipi.it

Abstract—**Modern cars are no longer mere mechanical devices and they are dominated by a large number of IT systems that guide a wide number of embedded systems called Electronic Control Unit (ECU). While this transformation has driven major advancements in efficiency and safety, it has also introduced a range of new potential risks. After a brief introduction of the security in automotive environment we investigate how the automotive community approached this problem.**

In order to ensure some security aspects in automotive environment, it is needed a hardware implementation of the Advanced Encryption Standard (AES) algorithm with higher speed throughput than existing solutions. For this purpose, a new hardware implementation of this cryptographic algorithm is presented.

The implementation results are compared with previous works.

I. INTRODUCTION

Information Technology (IT) is the application of computers and telecommunications equipment to store, retrieve, transmit and manipulate data. IT security has gained central importance for many new automotive applications and services. On the production side, we observe that the cost for electronics and IT is approaching the 50% threshold of all manufacturing costs. Perhaps more importantly, there are estimates that already today more than 90% of all vehicle innovations are centred around software and hardware (admittedly not only digital hardware, though) [1]. IT systems in cars can roughly be classified into three main areas:

- Basic car functions, e.g., engine control, steering, and braking.
- Secondary car functions, e.g., window control and immobilizers.
- Infotainment applications, e.g., navigation systems, music and video entertainment, and location-based services.

Future cars are becoming even more dependent on IT security due to the following developments:

- IT is predicted that an increasing number of ECU will be reprogrammable and this process must be protected.
- Many cars will communicate with the environment in a wireless fashion, which makes strong security a necessity.

- New business models ,e.g. time-limited flash images or pay-per-use infotainment content, will become possible for the car industry, but will only succeed if abuse can be prevented.
- There will be an increasing number of legislative demands, which can only be solved by means of modern IT security functions, such as tamper resistant tachographs, secure emergency call functions, secure road billing etc.
- Increasing networking of cars will allow the collection of data for each driver (e.g., driving behaviour, locations visited), which will put high demands on privacy technology.
- Future cars will often be personalized, which requires a secure identification of the driver.
- Electronic anti-theft measures will go beyond current immobilizers, e.g., by protecting individual components.

The paper is organized as follows. In section II is analysed the State of Art of the security features in current automotive microcontrollers. In section III first, the AES algorithm is described and then a high speed hardware implementation is proposed. Implementation results and comparison with previous solutions are presented in section IV. Finally the conclusions are drawn in Section V.

II. STATE OF ART

Today most vehicle manufacturer incorporate security features as a design requirement. However, realizing dependable IT security solutions in a vehicular environment considerably differs from realizing IT security for typical desktop or in general for others environments. In a typical vehicular attack scenario a malicious user, for instance, has extended attack possibilities (i.e., insider attacks, physical attacks, offline attacks) and could have many different attack incentives and attack points (e.g., tachometer manipulations by the vehicle owner vs. theft of the vehicle components vs. industrial espionage). Thus, just porting standard security solutions to the, moreover, very heterogeneous IT environment usually will not work. However, there already exist some first automotive-capable (software) security solutions. Besides, especially with regard to potential internal and physical attackers, these software solutions have to be protected against manipulations as well. In order to reliably enforce the security of software security mechanisms, the application of hardware security modules is one effective countermeasure as these modules:

- protect software security measures by acting as trusted security anchor,
- securely generate, store, and process security-critical material shielded from any potentially malicious software,
- restrict the possibilities of hardware tampering attacks by applying effective tamper-protection measures,
- accelerate security measures by applying specialized cryptographic hardware,
- reduce security costs on high volumes by applying highly optimized special circuitry instead of costly general purpose hardware.

Many workgroup, projects and studies are born in the recent years to realize a standard solution for the security problem. How we can see in table I there is a general trend to implement a common standard solution. Naturally, the benefits of using a single, consolidated solution are manifold.

Company	Family	Security module	Reference Standard	
Freescale	32-bit Qorivva microcontrollers	CSE Option (crypto service engine)	SHE functional specification Version 1.1	■
	32-bit Qorivva microcontrollers	Hardware Security Module (HSM)	EVITA project	❑
Fujitsu	Fujitsu Cortex R4 Family	Secure hardware Extension (SHE)	SHE functional specification Version 1.1	■
Infineon	AUDO MAX	Secure hardware Extension (SHE)	SHE functional specification Version 1.1	■
	AURIX	Hardware Security Module (HSM)	EVITA project	❑
ST	SPC56 32-bit MCUs	Cryptographic Service engine (CSE) module	SHE functional specification Version 1.1	■
Analog devices	Blackfin	Lockbox Secure Technology	Proprietary solution	■
TI	OMAP Applications Processors	TI's M-Shield	Proprietary solution	■
Renesas	RH850/F1x Series	ICU-S/M (intelligent cryptographic unit)	M= EVITA project, S=SHE	❑

TABLE I

COMPARISON MANUFACTURE SOLUTIONS. ■=*in production,* ❑=*Pre-Production*

In the following, main solutions are presented, among those adopted by industrial manufactures in automotive field.

1) SHE: HIS work-group [2], composed by some vehicle manufacturers like Audi, BMW, Daimler, Porsche, and Volkswagen, was the first that addressed the security problem. They performed a system analysis focusing on:

- Possible attacks
- Benefit for the attacker
- Damage for the user and so on

After this, HIS group developed a hardware module called Secure Hardware Extension (SHE), that together with a software application ensure different security aspects. The SHE was the first attempt to create a standard solution to solve the security problem in the automotive environment, however the SHE gives an optimized solution limited to Car to Car (C2C) communication. Moreover, the SHE does not provide all the security features. For those and others reasons, many solutions available today on the market do not make use of this module.

2) EVITA Project: The objectives of E-safety Vehicle Intrusion proTected Application (EVITA), a project co-funded by the European Commission, are to design, to verify, and to prototype an architecture for automotive on-board networks where security-relevant components are protected against tampering and sensitive data are protected against compromise. Thus, EVITA provides a basis for the secure deployment of electronic safety aids based on vehicle-to-vehicle and vehicle-to-infrastructure communication. Focussing on on-board network protection, EVITA complements other e-safety related projects that focus on protecting the communication of vehicles with the outside. Based on the security requirements and the automotive constraints, they designed a secure on-board architecture, which includes a Hardware Secure Module (HSM), and secure on-board communications protocols. The security functions are partitioned between software and hardware. The root of trust is placed in the HSM that should be realised as extension of automotive microcontrollers. As a result of the EVITA analysis, in table II, is shown which algorithms have to be implemented through the HSM for different levels of security, compared with the SHE features.

	HSM *Full*	*Medium*	*Light*	SHE
Cryptographic algorithms				
ECC/RSA	■/■	■/■	❑/❑	❑/❑
AES/DES	■/❖	■/❖	■/❑	■/❑
WHIRPOOL/SHA	■/■	■/■	❑/❑	❑/❑
Hardware acceleration				
ECC/RSA	■/❑	❑/❑	❑/❑	❑/❑
AES/DES	■/❑	■/❑	■/❑	■/❑
WHIRPOOL/SHA	■/❑	❑/❑	❑/❑	❑/❑
Security features				
Secure/authenticated boot	■/■	■/■	❖/❖	■/❑
Key AC per use/bootstrap	■/■	■/■	■/❖	❑/■
PRNG with TRNG seed	■	■	■	■
Monotonic counters 32/64 bit	■/■	■/■	❑/❑	❑/❑
Tick/UTC-synced clock	■/■	■/■	■/■	❑/❑
Internal processig				
Programmable/preset CPU	■/❖	■/❖	❑/❖	❑/■
Internal V/NV (key) memory	■/■	■/■	❖/❖	■/■
Asynchronous/parallel IF	■/❖	■/❑	■/❑	■/❑

TABLE II

EVITA HSM COMPARISON: ■=AVAILABLE, ❑=NOT AVAILABLE, ❖=PARTLY OR OPTIONALLY

From the previous analysis, it is clear that a cryptographic algorithm implementation is always required in order to cover several security aspects such as privacy, integrity and authenticity. SHE and EVITA solutions, suggest the AES algorithm [3] for this purpose. [4] has proposed a hardware implementation, based on the results of EVITA project, of the AES algorithm for automotive environment which reaches a theoretical throughput of 242 Mbps at 100 MHz of clock frequency. However the infotainment applications require more and more computational resources even also with regard of the security aspect and thus, for the specific AES implementation, higher throughput is needed.

III. CRYPTOGRAPHIC ALGORITHM DESIGN

The AES algorithm implementation presented, has as main goal, the achievement of high speed data processing. In the following, after a brief description of the algorithm, the hardware implementation is described.

A. AES Algorithm

The AES cipher is almost identical to the block cipher Rijndael [5]. The Rijndael block and key size vary between 128, 192 and 256 bits. However, the AES standard only calls for a block size of 128 bits. Hence, only Rijndael with a block length of 128 bits is known as the AES algorithm. AES consists of so-called layers. Each layer manipulates all 128 bits of the data path. The data path is also referred to as the state of the algorithm. There are only three different types of layers. Each round, with the exception of the first, consists of all three layers as shown in figure 1: the plaintext is denoted as x, the cipher text as y and the number of rounds as n_r, which are 10 for AES-128. Moreover, the last round n_r does not make use of the *MixColumn* transformation, which makes the encryption and decryption scheme symmetric. Each transformation is briefly described below.

- *Key Addition layer*: A 128-bit round key, or sub key, which has been derived from the main key in the key schedule, is XORed to the state.
- *Byte Substitution layer*: Each element of the state is non-linearly transformed using lookup tables with special mathematical properties. This introduces confusion to the data, i.e., it assures that changes in individual state bits propagate quickly across the data path.
- *ShiftRows layer*: it permutes the data on a byte level.
- *MixColumn layer*: it is a matrix operation, which combines (mixes) blocks of four bytes.

ShiftRows and *MixColumn* transformations perform diffusion principle over all state bits.

For each round, *Key Addition layer* exploits a different sub-key. All sub-key are generated by a *Key Scheduler* starting from the *Master Key*. The number of sub-keys is equal to the number of rounds, which is 10, for a block-cipher size of 128 bit.

For the decryption process, inverse transformations of the previous one are needed. Moreover, the order of using the sub-keys is reversed.

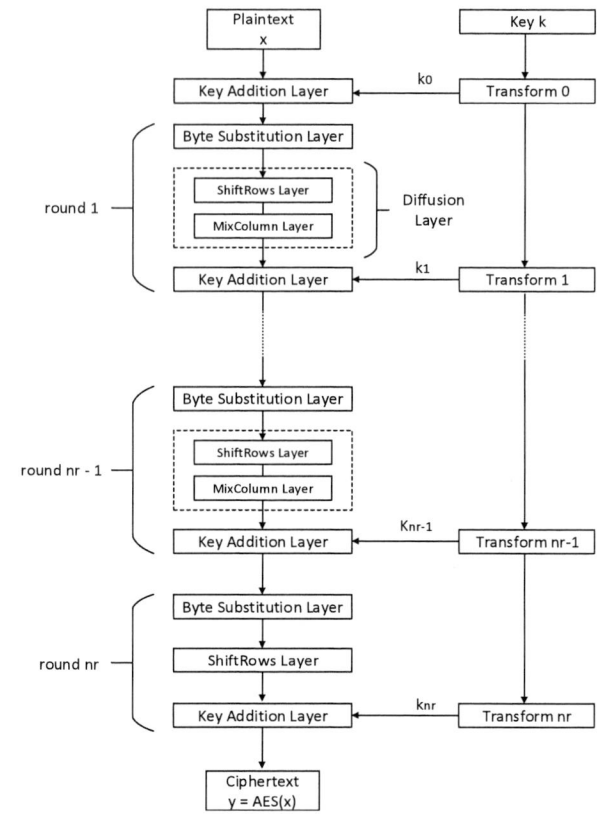

Fig. 1. AES flow chart

B. Implementation

The architecture implemented is depicted in figure 2. It implements the the Rijndael algorithm with 128 bit block size. Longer encryption keys such as 192 and 256 are not necessary according to EVITA results. The key schedule block computes

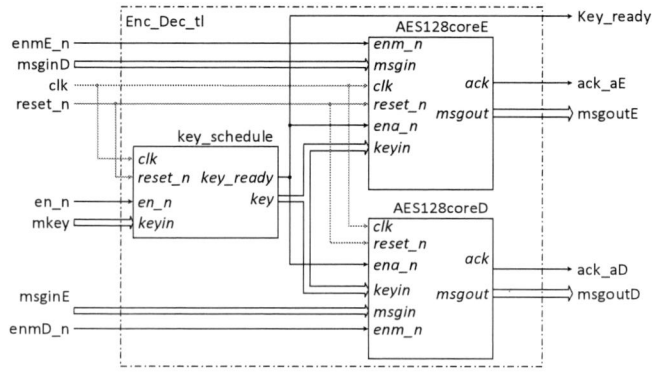

Fig. 2. Architecture implemented

the sub-keys starting from the master key provided as input on 128 bit and store them into a RAM. This step takes 80 clock cycles and must be executed before performing encryption/decryption processes. Every times a new master key is provided, the sub-keys must be re-generated. Encryption and decryption blocks can work simultaneously and both process

978-1-4799-4993-9/14 $31.00 © 2014 IEEE   290

input blocks of 128 bit. The state is implemented through a matrix of registers, for both encryption and decryption. Each transformation acts on these registers, using the current values as input and updating the content with the output of the transformation process. Once master key and sub-keys are stored in RAM, the processing of a single block requires 40 clock cycles. The S-box sub-block is implemented by a ROM solution that is shared between the key scheduler block and the encryption block since this two modules never work at the same time. The module was described in Verilog language and synthesized onto a FPGA Xilinx Virtex 5 *XC5VFX70T-1FF1136C*.

IV. RESULTS

The maximum frequency achieved was 212 MHz, that means a theoretical throughput of 678 Mbps, considering 40 clock cycles for a single cryptographic operation. However, in the worst case, assuming that the Master Key changes every times a new encryption/decryption operation is performed, other 81 clock cycles have to be considered in the throughput computation, that means a performance degradation of 66%. Area occupation is reported in table III. In the same table,

	Registers	LUTs	Block RAM	Throughput (Mbps)
[4] Solution	279	1137	0	242
Our Solution	396	1187	4	320

TABLE III
IMPLEMENTATION RESULTS

we have compared our solution with [4] implementation. The throughput reported for this implementation, is computed for a working frequency of 100 MHz, which is used in [4].

As shown, it was achieved an improvement in term of maximum throughput respect to the solution [4] of 32%. This result is paid in term of area occupation. In order to insert this cryptographic block as a hardware peripheral of an automotive MicroController Unit (MCU), a preliminary synthesis was performed on a 65 nm technology, resulting in an area of 36 Kgate.

V. CONCLUSION

In the first part of this paper was presented the results of an analysis on the security features available in current automotive microcontrollers. An important result of this analysis is the need of a hardware implementation of the AES algorithm in order to ensure a minimum level of security in automotive environment. Although few solutions already exist,they do not reach high speed data processing, which is required for infotainment application. In order to cover this requirement, a high throughput hardware implementation was presented. The architecture was described in Verilog language and was synthesised onto a Xilinx Virtex 5 XC5VFX70T-1FF1136C FPGA. The maximum theoretical throughput achieved was 678 Mbps.

REFERENCES

[1] K. Lemke, C. Paar, and M. Wolf, *Embedded Security in Cars*. Springer Berlin Heidelberg, 2006.

[2] A. AG, D. A. BMW AG, P. AG, and V. AG, *HIS AK Security*, HerstellerInitiative Software Std., June 2010.

[3] *ADVANCED ENCRYPTION STANDARD (AES)*, Federal Information Processing Standards Publications Std., November 2001.

[4] M. Wolf and T. Gendrullis, "Design, implementation, and evaluation of a vehicular hardware security module," *Lecture Notes in Computer Science*, 2012.

[5] J. Daemen and V. Rijmen, "The block cipher rijndael," in *Smart Card Research and Applications*, ser. Lecture Notes in Computer Science, J.-J. Quisquater and B. Schneier, Eds. Springer Berlin Heidelberg, 2000, vol. 1820, pp. 277–284.

FPGA Design of the Decoding Functions in the Physical Layer Adaptation Subsystem of the XG-PON Optical Network Unit/Terminal

Georgios Georgis
Charalambos Tzeranis
and Dionysios Reisis
Electronics Lab, Faculty of Physics
National and Kapodistrian University of Athens
Athens, Greece
Email: dreisis@phys.uoa.gr

George Synnefakis
R&D Center,
InAccess Networks S.A.,
Athens, Greece

Abstract—**The XG-PON standard for Passive Optical Networks (PONs) has imposed requirements for high performance processing in the architectures of network equipment. Especially, the designs of the 10Gbps receiver terminals and the network units (ONTs and ONUs) can become quite demanding. The current paper focuses on the XG-PON ONT/ONU receiver and presents an FPGA design realizing the decoding functions of the XG-PON physical adaptation layer: The scrambling, the RS(248,216) decoding and the Hybrid Error Correction (HEC) architectures, which are designed to communicate through a 64-bit bus. This work describes the components' features and validates the results by showing the design's performance on a Xilinx Kintex 7 FPGA.**

I. INTRODUCTION

The evolving PON technologies and the design of Fiber-To-The-Curb (FTTC) network architectures target the ability of universal communication with increased capacity and a large number of connected users. Moreover, they offer enhanced security and services. To this respect, the latest XG-PON standard [1] has required the network down/upstream traffic to reach 10Gbps and 2.5Gbps respectively. Hence, the corresponding XG-PON systems, which realize the functions of the Optical Line Terminal (OLT), the Network Unit (ONU) and the Terminal (ONT), have to accomplish calculations at high speed on data having a width of 32 bits. Solutions for the ONT and ONU architectures have been proposed by researchers [6], [7] and engineers [8], [9] based on various approaches.

Aiming at providing an effective architecture for realizing the functions of the XG-PON physical adaptation layer, this work presents an approach for designing the components which implement the major decoding functions of the layer. The paper focuses mainly on the receiver, which includes more demanding circuits due to the 10Gbps downstream traffic. We describe the design and the architectural features of the Reed-Solomon RS(248,216) decoder approach, which is the most

This work has been funded by the NSRF (2007-2013) Synergasia- II/EPAN-II Program "Asymmetric PON for xDSL and FTTH Access," GSRT, Ministry of Education, Religious Affairs, Culture and Sports. (09SYN-71-839).

demanding circuit with respect to the computations involved, required resources and the delay that it induces. Furthermore, we present the design of the parallel scrambler used at the receiver's input and the HEC decoder used in the GPON Encapsulation Method (GEM) headers and Allocation structures. The proposed design targets at minimizing the latency induced by the units realizing the above functions. Improving the latency of the downstream decoding functions is important because it increases the number of cycles, which are available to the ONU/ONT architecture for preparing the upstream frame. Our approach is based on using components with 64 bit parallel input and pipeline structure, which achieve low latency and minimize the memory size required for accommodating the layer.

The paper is organized as follows: Section II highlights the features of the frames given as input to the ONU/ONT and the input (de)scrambler circuit. Sections III and IV show the design and details of the proposed modules. Section V presents details of the FPGA implementation and finally, section VI concludes the paper.

II. INPUT AND SCRAMBLER

The receiver of the XG-PON physical adaptation layer accepts input frames of constant size and duration (125 μsec) which, at the rate of 9.95328 Gbit/s, translates to 38880 words of 32 bits width. The bit stream uses the non-return to zero (NRZ) scrambled line coding. Excluding the first 6 words of each frame all the other data (words) are encoded and scrambled. The first 6 uncoded words of each frame are unscrambled to synchronize the receiver. The proposed architecture uses a Serializer/Deserializer (SerDes) to shape the input bit stream into a 2-word (64 bit parallel) format and communicates these 2-word structures to the other components through a 64 bit parallel bus which operates at 155.52 MHz.

The scrambling process is used to induce pseudo-randomization on the transmitted data, without adding any redundancy. Scrambling in the G.987.3 standard is based on

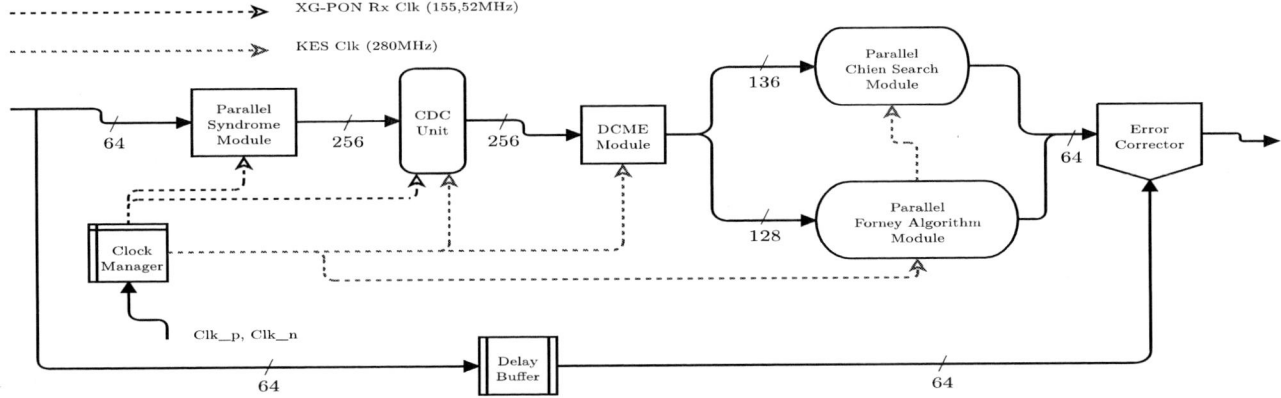

Fig. 1. Overview of the 8-parallel RS(248,216) Decoder.

the $x^{58}+x^{39}+1$ polynomial, represented by a 58-bit linear-feedback shift register (LFSR). The XG-PON side-stream scrambler is synchronized at the start of each frame, by parallel loading of the LFSR with a 58-bit seed. The latter consists of the 51-bit superframe counter (*SFC* field loaded into the 51 MSBs) and a 7-bit sequence of ones, (remaining LSBs).

According to the serial implementation of the XG-PON scrambler, bits 39 and 58 are modulo-added and fed back to the input of the LFSR. Additionally, each data input bit is modulo-added to the 58th bit of the LFSR [1]. Practical implementation of the scrambling process at the 9.95328Gb/s line rate, requires modification of the standard scrambler to operate in parallel [2]. Therefore, in order to design an n-bit parallel scrambler we need to pre-determine the LFSR state as well as the scrambler output after n cycles. In our case, $n = 64$ as the operating frequency is 155,52MHz. Initially and at the end of the first cycle, the LFSR states are:

$$x_0^0 = 0, \ x_0^1 = 1, \ \ldots, \ x_0^{57} = x^{57} \tag{1}$$
$$x_1^0 = x^{57} + x^{38}, x_1^1 = 0, \ldots, x_1^{57} = x^{56} \tag{2}$$

After 64 clock cycles the LFSR state and the output become :

$$x_{63}^0 = x^{52} + x^{14}, \ldots, x_{63}^{57} = x^{51} + x^{32} \tag{3}$$
$$out_{63}^0 = in^0 + x^{57}, \ \ldots, \ out_{63}^{63} = in^{63} + x^{33} + x^{52} \tag{4}$$

To verify the scrambler operation, we set the input vector to zero, initialize the seed with 0x7F (corresponding to $SFC = 0$) and compare the output to the vectors provided in [1]. The implemented XG-PON scrambler processes data continuously, as long as the enable signal is set to 1.

III. ERROR DECODING

The decoding process presented by the G.987.3 standard is based on the RS(248,216) code, which is a truncated version of the (255,223) Reed-Solomon error correcting code (ECC) over the $GF(2^8)$ field. Each symbol r_k, $(0 \leq k \leq 255)$ of the code is comprised of 8-bits. Therefore, performing the calculations for the ECC at the required rate imposes the processing of the received codeword bytes in parallel. Taking into account the

G.987.3 PHY frame size and boundaries, an XGPON word consisting of 32 bits or a 2-word structure of 64 bits would be the ideal candidates. A technique processing 4 symbols on every cycle will require an operating frequency of 311.04MHz, a constraint which is considered difficult to meet first, because this is a relatively high frequency for the majority of commercially available FPGAs and second because the amount of the FPGA resources required to realize the decoder and the remaining downstream logic is large, a fact leading to long routing paths and limiting the effective operating frequency. The remaining choice is the 2-word structure consisting of 64 bits, which corresponds to a processing rate of 8 codeword symbols in every clock cycle at a frequency of 155.52MHz ($XG-PON_{clk}^{Rx}$). The blocks of the decoder architecture are depicted in Fig. 1. The following subsections describe details of the decoder's blocks.

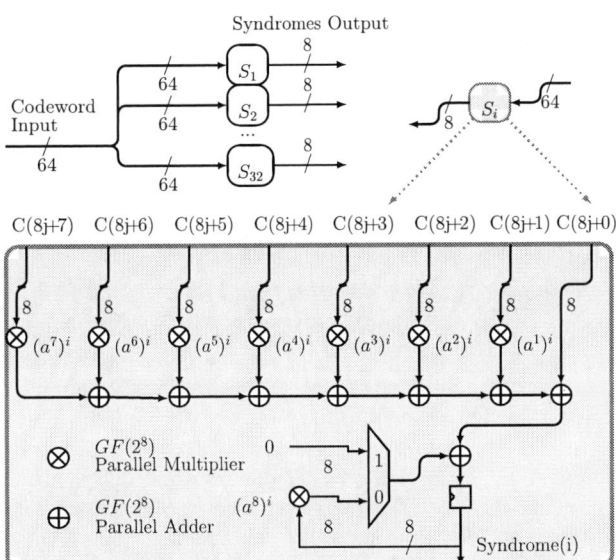

Fig. 2. RS(248,216) decoder syndromes computation module: overview (top) and cell details (bottom) - $\mathcal{C}(8j+m)$ is the codeword fraction where $0 \leq j \leq S_{calc}^{clocks}-1$, $0 \leq m \leq 7$.

978-1-4799-4993-9/14 $31.00 © 2014 IEEE

A. Syndrome Calculation Unit

The parallel syndrome calculation unit (Fig. 2), is based on the architecture described in [4], which is adapted to process the RS(248,216) truncated code. Compared to RS(255,239), the XG-PON ECC is of higher complexity, because it has an error correction capability t of 16 errors. Therefore, the degree $2t$ of the syndrome polynomial is equal to 32 and the unit contains 32 syndrome cells. A 64-bit codeword fraction is forwarded simultaneously as the input to all of the syndrome cells. Each cell S_i computes a part of the received polynomial solution, for a specific subset of the generator polynomial roots α^i, $i \leq 1 \leq 32$. The syndrome calculation for the truncated codeword of 248 bytes requires 31 clock-cycles (S_{calc}^{clocks}) to be completed. At the 31st cycle, the multiplexer in Fig. 2 is switched to 1, to allow the storage of the initial computation for the leading 64-bit fraction of the next codeword.

B. Clock Domain Crossing (CDC) Unit

The majority of decoder implementations forward the syndromes serially to the key equation solver (KES) in order to minimize the area requirements. In our case, a problem arises due to the value of t, and the XG-PON downstream requirements: $2t > S_{calc}^{clocks}$. We addressed this issue by increasing the frequency of the KES and the Chien/Forney units to 280MHz (KES_{clk}^{Rx}) and hence, output the syndromes in 18 $XG-PON_{clk}^{Rx}$ cycles. The CDC unit is utilized as the interface between the clock domains. It uses a simple handshake pulse protocol to register the 256-bit parallel syndrome output to the KES domain. The corresponding start pulse acts as the handshake pulse, which registers the parallel output after safely reaching the second clock domain. Therefore, the logic which serializes the syndromes and parallelizes the locator/evaluator polynomial output is implemented in the DCME block.

C. Key Equation Solver (KES) Unit

The KES module is based on the Degree-Computationless Modified Euclidean (DCME) algorithm in [3] and its pipelined implementation in [5]. The reasons supporting this choice are: a) the target high clock frequency and b) the low complexity of the DCME algorithm. Compared to [5], inclusion of 32 processing elements was directed by both the XG-PON continuous downstream flow and the code specification.

D. Chien Search/Forney Algorithm Unit

At the end of the iterative DCME process, the KES module computes the error locator ($\sigma_0...\sigma_{16}$) and the error evaluator ($\omega_0...\omega_{15}$) polynomial coefficients. These values are simultaneously provided as input to the blocks performing the Chien Search and the Forney's Algorithm. These blocks operate in parallel and receive a 64-bit input from the delay buffer. Approximately 10 $XG-PON_{clk}^{Rx}$ cycles are required to convert the serial DCME PE output to the parallel input format. The implementation of these blocks, follows the architecture presented in [4], which is modified to involve rearranged coefficients/cells for the higher polynomial degrees. A series of 8 LUTs compute the inverse of the error locator polynomial

TABLE I
FPGA UTILIZATION: XG-PON Rx MODULES ON THE
XC7K325T-2FFG900C BOARD.

Implementation Results	Slice LUTs	Slices	Slice Registers	Frequency (MHz)	RAM Blocks
HEC	1922	612	329	333.443	0
Scrambler	122	31	122	479.616	0
Syndromes	3338	1018	637	295.247	0
KES	8983	3138	6548	316.156	0
Chien/Forney	4397	1470	2799	325.203	4

derivative $(\sigma'(\alpha^k))^{-1}$ required in the error evaluation process. Moreover, a toggle register controls a swap unit which exchanges the most with the least significant bytes of the 64-bit corrector word at each cycle. Finally, the mask bits produced are modulo added to each 64-bit codeword section read from the delay buffer. The reading process of the 27, 8-symbol words (64-bits each) requires 15 $XG-PON_{clk}^{Rx}$ cycles. Fig. 3 depicts the circuit realizing the Chien search (left), the circuit for the Forney algorithm (right) and the error correction circuit (top). Let us also note that all RS decoder components were implemented using fully parallel GF(2^8) Galois multipliers, in an effort to minimize latency.

IV. HYBRID ERROR CORRECTION (HEC)

The XG-PON HEC unit performs the realization of a two phase process: a) error decoding based on the BCH(63,12,2) code (generator polynomial $x^{12}+x^{10}+x^8+x^5+x^4+x^3+1$) and b) even parity detection. Note here that the BCH code does not include the parity bit when calculated, though the parity bit requires the decoded results. The same code is used to protect 51-bit structures in the XG-PON allocations (the BW-map structure) and the XGEM header [1] as well as 19-bit structures in the entire XGTC frame header (after zero pre-pending). The design of this unit allows single and double errors to have a unique 12-bit syndrome and hence, these errors can be detected and corrected. Triple errors are simply detected and excluded. The BCH decoder architecture, depicted in Fig. 4, follows a fully parallel 64-bit approach, in order to reduce latency and comply with the other circuits' data width. The HEC module consists of a syndrome calculation unit and a LUT to decode the computed syndrome. It features a latency of 3 cycles and it outputs: a) either a 63-bit corrected codeword plus a parity bit or b) the input 64-bit vector along with a pulse denoting the occurrence of an unrecoverable error.

V. IMPLEMENTATION AND PERFORMANCE

All the implemented XG-PON modules were verified separately on the Xilinx KC705 board using a 280MHz clock frequency produced by the board's 200MHz differential clock. We note that apart from the synchronizer's GTX transceivers and the clock manager, all blocks were designed using synthesizable Verilog code; no other Xilinx cores were utilized.

Table I displays the logic utilization of distinct modules on the Kintex 7 FPGA board. We note that the synthesizer is configured for speed optimization and normal effort. The reported frequency corresponds to the minimum achieved period following the post place and route static timing analysis.

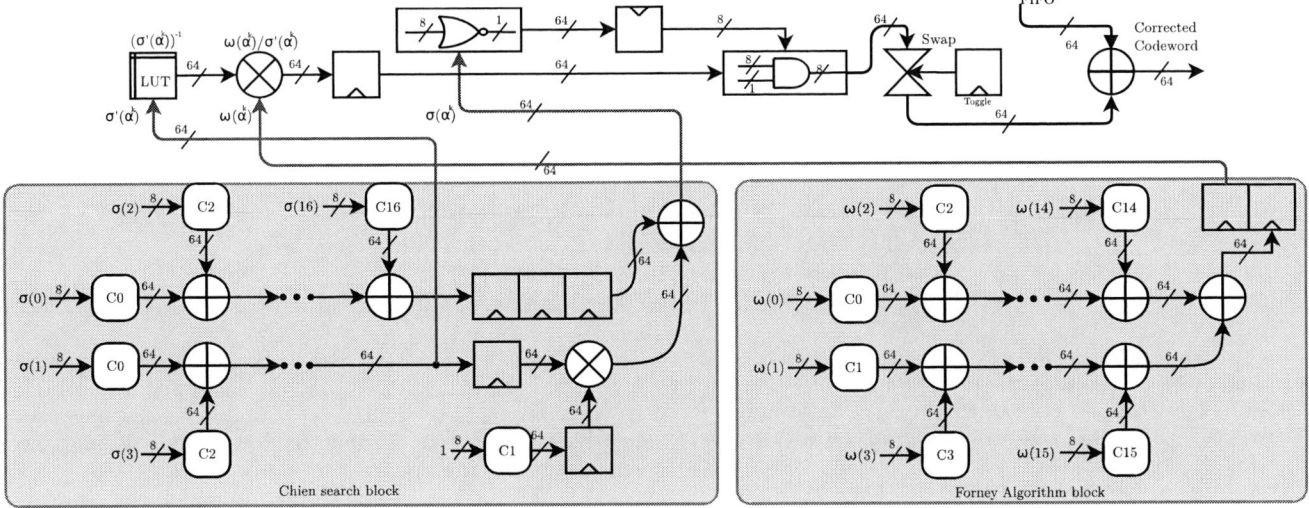

Fig. 3. RS Decoder module sub-circuits: Chien search (bottom-left), Forney's algorithm (bottom-right) and Error Correction (top) units.

VI. CONCLUSIONS AND FUTURE WORK

The current paper presented an approach for the architecture of the XG-PON physical adaptation layer decoding components focusing mainly on the 10 Gbps receiver circuits. In the future, our research will be based on the techniques presented in this paper and it will focus on producing a methodology for parallelizing the KES module in order to reduce latency, implementation cost and eliminate the need for a second clock domain. These improvements will merit high-throughput passive optical networks, such as the NG-PON2 (40 Gbps).

REFERENCES

[1] *10-Gigabit-capable passive optical networks (XG-PON): Transmission convergence (TC) layer specification*, ITU-T Telecom. Std. Sector Recommendation G.987.3, Rev. 10/10, 2010.

[2] D. R. Stauffer et al, *High Speed Serdes Devices and Applications.* 233 Spring Street, NY: Springer, 2008.

[3] J. H. Baek and M. H. Sunwoo, "New Degree-Computationless Modified Euclid Algorithm and Architecture for Reed-Solomon Decoder," *IEEE Trans. Commun.*, vol. 14, no. 8, pp. 915–919, Mar. 2006.

[4] H. J. Ahn, C. S. Choi, and H. Lee, "High-Throughput Variable-Length Reed-Solomon Decoder for High-Rate WPAN Applications," in *Proc. IEEE 54th International Midwest Symposium on Circuits and Systems (MWSCAS)*, Seoul, Aug. 2011, pp. 1–4.

[5] S. Lee and H. Lee, "A High-Speed Pipelined Degree-Computationless Modified Euclidean Algorithm Architecture for Reed-Solomon Decoders," *IEICE Transactions on Fundamentals of Electronics, Communications, and Computer Sciences*, vol. E91-A, no. 3, pp. 830–835, Mar. 2008.

[6] B. Schrenk, C. Stamatiadis, I. Lazarou, A. Maziotis, G. de Valicourt, J. Lázaro, J. Prat, and H. Avramopoulos, "On an ONU for Full-Duplex 10.5 Gbps/λ with Shared Delay Interferometer for Format Conversion and Chirp Filtering," in *Optical Fiber Communication Conference*, Los Angeles, CA, Mar. 2011, pp. 1–3.

[7] B. Schrenk, G. de Valicourt, J. Lázaro, and J. Prat, "FSK + ASK/ASK Operation for Optical 20/10 Gbps Access Networks with Simple Reflective User Terminals," in *National Fiber Optic Engineers Conference*, Los Angeles, CA, Mar. 2011, pp. 1–3.

[8] H. Wu and M. Zhao, "From GPON to 10G GPON," *Huawei Communicate*, vol. 57, pp. 49–51, Sep. 2010. [Online]. Available: http://www.huawei.com/en/about-huawei/publications/communicate/hw-081018.htm

[9] R. L. J. Smith and B. Rao, "The migration to 10G GPON," *FTTH Prism*, vol. 7, pp. 19–25, Sep. 2010.

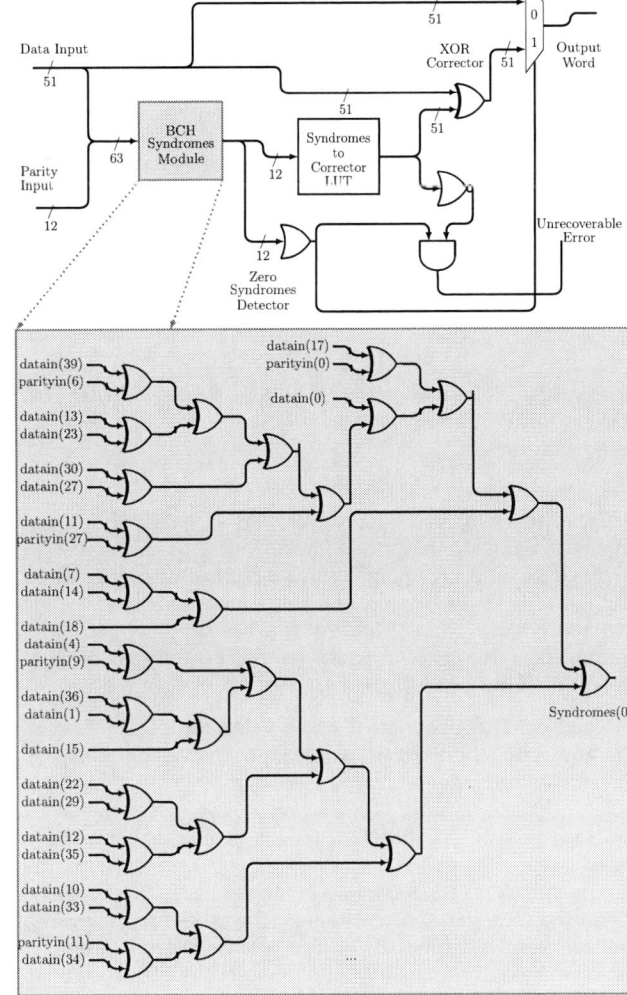

Fig. 4. The XG-PON HEC decoder module.

Fast Register Criticality Evaluation in a SPARC Microprocessor

Kais Chibani, Michele Portolan, Régis Leveugle

Univ. Grenoble Alpes, TIMA Laboratory, F-38031, Grenoble, France
CNRS, TIMA Laboratory, F-38031, Grenoble, France
{Kais.Chibani, Michele.Portolan, Regis.Leveugle}@imag.fr

Abstract—**Many applications impose safety and/or security constraints which require protections against the effects of transient faults. The most critical elements must be identified to achieve good efficiency and cost trade-offs when selective hardening is necessary. In embedded microprocessor-based systems (i.e. most of SoCs) the system dependability is strongly correlated with internal register criticality since external memories are protected by error correcting codes. The robustness analysis of these systems consists in precisely assessing the criticality of internal registers used by the application program. The evaluation often aims at selecting a minimum number of registers to be protected. At the same time, the accurate assessment is complicated in the case of recent processors because of the evolution of architectures and the implementation of new mechanisms to improve performance (e.g., pipeline, forwarding mechanisms,...). Classical fault-injection approaches require long experimental times to determine the most critical set of registers. This paper presents an approach based on modeling the effect of transient faults taking into account the micro-architectural features and proposes a new methodology to refine and accelerate evaluations of register criticality. This new approach is compared with fault injections. The results show the effectiveness of the prediction algorithm.**

Keywords—Robustness, Dependability, SEEs, SPARC V8, LEON3, Criticality, Fault injection.

I. INTRODUCTION

A growing number of applications are based on the use of complex integrated circuits and, more generally, on the use of embedded systems including hardware and software (operating system and software application) implemented in a single integrated circuit. The confidence that can be placed in these systems depends on their ability to react safely in case of transient disturbances.

Because of the continued device size scaling in CMOS technology, reliability has become an increasing concern. Reduced feature size, increased chip density, reduced supply voltage, shrinking nodal capacitances have resulted in an increased susceptibility of these designs to single event effects (SEEs) due to ionizing radiations [1, 2, 3]. These effects can result in soft errors (i.e., spurious bit flips in registers) that can further lead to either malfunctioning of a part of a design or a complete system failure. Therefore, efficient techniques are necessary to analyze, quantify and mitigate the effects of such events, especially, if the design is to be used in safety-critical applications. However, a circuit can contain hundreds of

thousands of flip-flops. In consequence, a complete protection is very costly and is not affordable for many applications. That is why different robustness analysis approaches have been proposed, aiming at limiting the protections to the most sensitive blocks in a circuit.

In systems using microprocessors with performance-oriented design, the risk of dysfunction in case of disruption heavily depends on the use of internal micro-architecture registers by the application program. The evaluation of the robustness therefore requires an accurate assessment of the criticality of these registers. In this context, our work aims at proposing a new approach to achieve a fast but accurate criticality analysis taking into account the micro-architecture characteristics. The case study focuses on a pipelined execution unit of the LEON3 microprocessor, available as a synthesizable VHDL model.

After a brief presentation of the LEON3 and its basic characteristics in section II, we will present in section III our approach to robustness analysis including an algorithm for predicting critical registers. The experimental results will be presented in section IV and compared with those obtained using a classical method of fault injection.

II. THE LEON3 MICROPROCESSOR

A. Presentation

The LEON3 is a Register Transfer Level (RTL) VHDL model of a 32-bit processor compliant with the SPARC V8 architecture [4].The VHDL model is fully synthesizable and can be implemented on both FPGAs and ASICs. LEON3 model is designed for embedded applications, combining high performance with low complexity and low power consumption. Its basic features are the seven pipeline stages of the integer unit, and the separate units of multiplication and division, floating point, memory management unit and MAC function unit. Also, it includes implemented interfaces for AMBA 2.0 AHB bus, coprocessor and on-chip debugging. The integer unit, the general purpose register file, the caches and their controllers, along with the floating point unit and the coprocessor are all regarded as the processor's core.

We will focus in the following on the integer unit, which pipeline is implemented in the following seven stages: Instruction Fetch, Decode, Register Access, Execute, Memory, Exception and Write Back. The cache subsystem is

978-1-4799-4993-9/14 $31.00 © 2014 IEEE

implemented on the basis of Harvard architecture, with two separate instruction and data units. The architectural block diagram of LEON3 is shown in Figure 1.

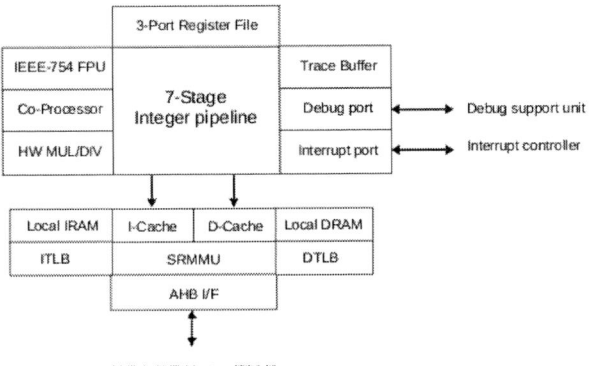

Fig. 1.   LEON3 processor core block diagram [5].

B. Configuration

The LEON3 processor is highly configurable, very suitable for SoC designs through the use of VHDL generics, and does not rely on any global configuration package. It is thus possible to instantiate several processor cores in the same design with different configurations. The LEON3 template designs can be configured using a graphical tool. This allows new users to quickly define a suitable custom configuration. The configuration tool does not only configure the processor, but also other on-chip peripherals such as memory controllers and network interfaces.

In this work, we have used a standard configuration without floating point unit or coprocessor. In addition, the multiplier/divider unit was disabled and we focused on the integer unit with only the arithmetic-logic unit. Another important characteristic in the SPARC V8 architecture is the number of "Register windows" in the register file, since the SPARC architecture is based on a sliding register window to manage how programs access the available registers. The number of LEON3 register windows is configurable within the limits of the SPARC standard (2 - 32). We used the default setting of 8 windows which makes, according to the SPARC architecture, a total of 136 registers that form the register file.

III.   APPROACH TO ROBUSTNESS ANALYSIS

Robustness evaluation often aims at identifying the most critical elements in the circuit, i.e., the error locations having the highest probability to lead to failures for a given application program. The goal of such an evaluation is to ensure effective protection and at the same time to reduce the costs by limiting the redundancy to the most vulnerable blocks. We will first give a definition for the criticality of a register. Then we will present a new method for computing this criticality in a pipelined microprocessor, including a prediction algorithm.

A. Criticality of a Register

The criticality of a register will be defined here as the probability that a fault in this register will result in a visible error in the final output of a given program. For example, if the register "R" has a criticality of 50% for a given application program, half of the transient faults occuring in "R" will on an average lead to a final outcome of the program that will be wrong.

B. Criticality Analysis Methodology

The assessment of the criticality for the different registers contained in the integer unit requires a dynamic analysis of the registers behavior during the program execution. The purpose of this analysis is to describe the information transfer across internal registers while taking into account the characteristics of the microarchitecture. The method used consists in determining, for each execution cycle, the state of the integer unit pipeline. Then, knowing the internal architecture of this unit and the different mechanisms implemented, we can estimate the criticality of registers.

The state of the pipeline is obtained through a single execution trace. This trace is obtained without additional cost for the designer, since it corresponds to the simulation to be made for functional validation. For our experiments, it was obtained thanks to a modification of the RTL source file of the LEON3 integer unit, adding a process producing a text file describing the evolution of the LEON3 state during the simulation. The same information could be obtained without any source code modification by a simple configuration of the VHDL simulator. In both cases, the generated file logs the activity of each pipeline stage during the program execution. For a given stage, the data shown are:

- The instruction in hexadecimal format,
- The address of the instruction in hexadecimal format,
- The window number which is specified by the current window pointer (CWP) field in the processor state register (PSR). According to the chosen configuration, this number varies between 0 and 7,
- The validity of each instruction. A value of 1 indicates that the instruction is valid in the corresponding stage whereas a value of 0 indicates that the instruction has been canceled (thus invalid) at this level. This lets us know if we should take into account the information of the pipeline stage.

Figure 2 is an extract from the file that we named "capture". The simulation was performed using Modelsim from Mentor Graphics. Note that in some particular cases of complex Sparc instructions, a single instruction can in some cases be processed by two pipeline stages at the same cycle.

978-1-4799-4993-9/14 $31.00 © 2014 IEEE   297

INSTRUCTION	ADDRESS	CWP	VALIDITY	STAGE
8610E370	400033E0	5	1	DECODE
0710001A	400033DC	5	1	REGISTER-ACCESS
820860FF	400033D8	5	1	EXECUTE
83306010	400033D4	5	1	MEMORY
83306010	400033D4	5	1	EXCEPTION
C207BFEC	400033D0	5	1	WRITE-BACK

CYCLE N

Fig. 2. Extract from the generated file after simulation of the LEON3 processor running the target program.

C. Prediction Algorithm

The prediction algorithm analyzes the "capture" file obtained after the simulation. Figure 3 summarizes the overall analysis process.

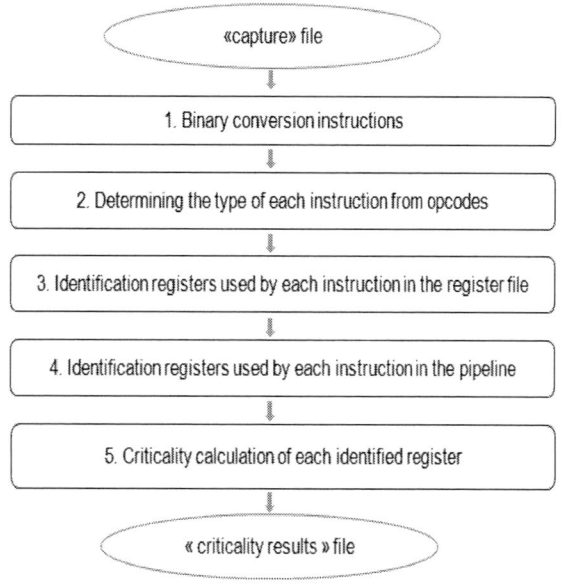

Fig. 3. Process for calculating the criticality.

The algorithm first performs the conversion of instructions encoded in hexadecimal format to binary code. A treatment is made on this binary code to identify source registers and the destination register of the instruction. It involves the identification of the logical registers which are used. Then, from the window number, the algorithm can identify physical registers located in the register file. The next step consists in identifying the internal registers used by each instruction. To achieve this, the algorithm first interprets the "opcode" and determines the type of the instruction. Once the instruction is known, it is associated with a predefined model that indicates which internal registers are critical for this type of instruction. These models were developed using simulation in order to get, for a given instruction of the SPARC architecture, the registers

actually used at each pipeline stage. The calculation of the criticality is based on the different registers used in each cycle as identified in the previous steps. The lifetime for each register is computed by summing the critical cycles. The criticality is the percentage of critical cycles during the application execution (computation details are not reported here due to the available space). The output file presents the percentages of criticality for all internal registers.

IV. VALIDATION OF THE DEVELOPED ALGORITHM

The effectiveness of the prediction algorithm was analyzed by comparing its results with those acquired through different fault injection campaigns that were performed on the LEON3 running several application programs.

A. Fault Injection

Fault injection is a classical technique to analyze the dependability of systems. It consists in observing the system's behavior in presence of faults (or errors) that are voluntarily induced in the system. The fault injection techniques have been recognized for a long time as necessary to validate the dependability of a system by analyzing the behavior of the devices when a fault occurs. Several types of techniques have been developed in the last twenty years for injecting faults in behavioral descriptions, including simulation-based or emulation-based techniques. Simulation-based techniques can be applied at several levels, from pure behavioral levels (e.g., in Transaction-Level Models) down to gate-level descriptions or even lower. Simulations at the higher levels lack accuracy since the actual registers are not included in the model. Simulations at the lowest levels are very slow. In consequence, most early dependability evaluations are performed at Register Transfer Level (RTL), when the actual registers are known and the evaluation can be fast enough.

In order to speed-up the evaluation with respect to simulation, emulation can be used on FPGA-based pltaforms. For our experiments, we have chosen to use the emulation technique based on partial reconfiguration of a circuit prototype implemented on a SRAM-based FPGA [6, 7]. Thus, the LEON3 microprocessor was implemented on a Virtex V device. The partial reconfiguration is performed by an embedded processor included in the same FPGA. This has the advantage of considerably reducing the number of data to be exchanged between the FPGA component and the host PC, thus accelerating the fault injection process.

B. Experimental Set-up and Analysis of Results

Experiments have been carried out mainly on the Mibench benchmark suite [8]. In the following, we will only present the results for two applications but the same conclusions hold for the others. The first example is a classical application in the field of data security known as Advanced Encryption Standard (AES). The second is an algorithm computing the discrete Fourier transform, that is especially used in the field of digital signal processing and is known as Fast Fourier Transform (FFT).

Using the LEON cross compiler with the standard optimization option, we compiled each application in order to simulate it and to run the same version on the LEON3

978-1-4799-4993-9/14 $31.00 © 2014 IEEE

processor that was synthesized in a Virtex V FPGA (reference LX110TXUPV5 [9]). On the one hand, we performed several campaigns of emulation-based fault injections in the registers used to implement the LEON3 integer pipeline. On the other hand, we launched the prediction program on the execution trace obtained after simulation. Figures 4 and 5 illustrate for the two applications the criticality of some internal registers.

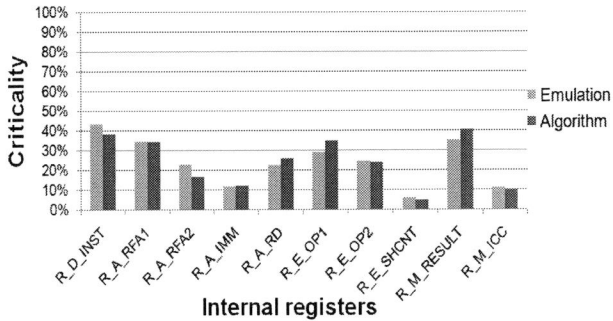

Fig. 4. Comparison between prediction algorithm and emulation-based fault injection for AES application.

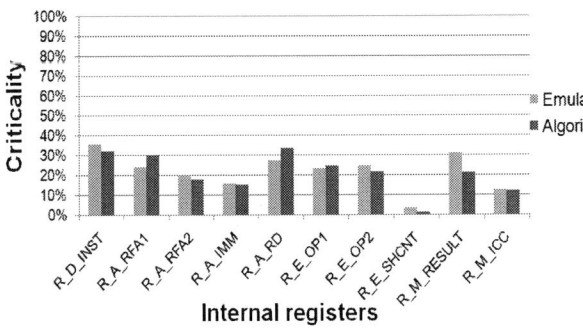

Fig. 5. Comparison between prediction algorithm and emulation-based fault injection for FFT application.

The emulation-based fault-injection campaign allowed us to perform about 6,000 injections for a given pipeline register. According to studies carried out in [10], the margin of error is 5% with 95% confidence. We can notice that the results of the algorithm are in agreement with those of injections and respect the margin of error that was fixed.

TABLE I.   TIME NEEDED FOR CRITICALITY EVALUATION

Application	I-Emulation	II-Algorithm	Speed-up factor
AES	670 min	15 min	45
FFT	820 min	17 min	48

Table 1 summarizes for both techniques, and for each studied application, the average time required to compute the regsiter criticality. The machine on which we performed our experiments has an Intel dual-core processor clocked at 3.16 GHz with 4 GB of RAM. Compared with fault injection,

we can see that the developed algorithm reduces the analysis time by a factor of 40 or more. This acceleration factor makes possible iterative assessments of robustness for different versions of the same application, even much more complex than the examples cited here. A similar improvement of the robustness of complex applications using only fault injection is often not possible, even by exploiting the speed of emulation.

V.   CONCLUSION

We have proposed a new methodology to evaluate the actual robustness of a microprocessor-based system running a given software. The approach is based on a prediction algorithm which allows the designer to compute the criticality of all internal registers. Emulation-based fault injections have shown that the proposed approach can achieve a very good evaluation level and at the same time reduce the experimental time by a factor of about 40. This allows a designer to compare several versions of the same application program, while even emulation-based fault-injections cannot in many cases offer such an opportunity. In addition, using our approach does not imply the implementation of the prototyping set-up, that can be tricky and time-consuming. With this approach, a developer can more easily decide the most critical registers to protect and/or the best version of the program to use e.g., by changing some compilation options. Further work includes a generalization to other processors.

REFERENCES

[1] R. Baumann, "Soft Errors in Advanced Computer Systems," IEEE Design & Test of Computers, vol. 22, no. 3, May 2005, pp. 258–266.

[2] P. Shivakumar, M. Kistler, S.W. Keckler, D. Burger, and L. Alvisi, "Modeling the Effect of Technology Trends on the Soft Error Rate of Combinational Logic," Proc. Int'l Conf. Dependable Systems and Networks, 2002, pp. 389-398.

[3] T. Karnik, B. Bloechel, K. Soumyanath, V. De, and S. Borkar, "Scaling trends of cosmic ray induced soft errors in static latches beyond 0.18u," VLSI Symposium, 2001, pp. 61-62.

[4] SPARC international inc. The SPARC Architecture Manual. Version 8. 1991, 1992.

[5] Aeroflex GAISLER, "GRLIB IP Core User's Manual," 2013, chapter 68, http://gaisler.com/products/grlib/grip.pdf.

[6] L. Sterpone, M. Violante, "A new partial reconfiguration-based fault-injection system to evaluate SEU effects in SRAM-based FPGAs," IEEE Transactions on Nuclear Science, vol. 54, issue 4, part 2, August 2007, pp. 965-970.

[7] M. Ben Jrad, R. Leveugle, "Comparison of FPGA platforms for emulation-based fault injections using run-time reconfiguration," 27th Conference on Design of Circuits and Integrated Systems, November 2012, pp. 184-188.

[8] M. Guthaus, J.S. Ringenberg, D. Ernst, T.M. Austin, T. Mudge, and R.B. Brown, "Mibench: A free, commercially representative embedded benchmark suite," presented at the IWWC, 2001, pp. 3-14.

[9] http://www.xilinx.com/univ/xupv5-lx110t-bsb.htm.

[10] R. Leveugle, A. Calvez, P. Maistri, P. Vanhauwaert, "Statistical Fault Injection: Quantified Error and Confidence," Design, Automation and Test in Europe, 2009, pp 502-505.

Challenges and benefits of microelectronics for power electronics: from integrated optical driving to optimized power semiconductor switches

Nicolas Rouger[1,2]

[1]Univ. Grenoble Alpes, G2Elab, Grenoble (France)
[2]CNRS, G2Elab, Grenoble (France)

In the view of driving power semiconductor devices such as MOSFET, IGBT or HEMT high voltage transistors, several auxiliary functions are required (floating supply, gate charge/discharge control, isolation, sensors and protections). These auxiliary functions are based on discrete devices, optimized separately from the power transistor, with different technologies and packaging techniques. While novel power semiconductors can be more efficient, faster and operated at higher temperatures, the association of power transistors and their required drivers are still limiting the overall performances.

In this context, an innovative approach is introduced where both the power switch and their dedicated electronics are designed at the same time, sharing design constraints and requiring a whole system-level approach. This integration technique necessitates however to control the fabrication and design of both the power transistors and their associated electronics functions.

We will first present the limits of monolithically integrated circuits, with different attempts to integrate both a high voltage power transistor and part of its driving circuits. Then, the benefits and challenges of custom IC design on CMOS platforms are presented. Among others, the question of integrating the galvanic insulation unit for the transfer of gate signal orders is discussed. Several designs and characterizations are demonstrated for an integrated optical driving or electromagnetic driving through coreless transformers. Key perspectives towards the integration in the context of novel wide bandgap power devices and multi transistor based power converters will be discussed.

978-1-4799-4993-9/14 $31.00 © 2014 IEEE

An Improved DC-Link Voltage Equalization for Three-Level Neutral-Point Clamped Converters

Mario Porru, Alessandro Serpi, Ignazio Marongiu, Alfonso Damiano
Department of Electrical and Electronic Engineering
University of Cagliari
Cagliari, Italy
alfio@diee.unica.it

Abstract—A DC-link voltage equalization algorithm (DCL-E) for Three-Level Neutral-Point Clamped converters (NPCs) is proposed in this paper. It consists of appropriately regulating DC-link currents with the aim of equalizing DC-link voltages as fast as possible, minimizing capacitor voltage and current ripple and prioritizing NPC load requirements at the same time. This goal is achieved by means of suitable PWM patterns, which also guarantee an appropriate DC-link capacitor exploitation over both transient and steady state operation. The effectiveness of the proposed DCL-E is verified through a simulation study, which refers to the case of a three-phase NPC that feeds a Surface-Mounted Permanent Magnet Synchronous Machine.

Keywords—*Current control, Neutral-point clamped converter, Pulse width modulation, Three-level converter.*

I. INTRODUCTION

The increasing development of power electronic devices, together with the great availability of powerful and low-cost processing units, has made multi-level converters feasible and suitable solutions for several applications, such as electric vehicles, wind and photovoltaic power plants, static VAR compensators and distribution systems. This is due to a number of advantages compared to traditional two-level converters, such as reduced power losses and voltage rating of semiconductor devices, as well as improved power quality and reduced filter requirements [1].

Several multi-level converter topologies have been proposed in the literature, among which the Three-Level Neutral-Point-Clamped Converter (NPC) is surely the most frequently used in industrial applications, as either rectifier or inverter [2,3]. One of the most critical issues of NPC is DC-link voltage equalization, which should be guaranteed in order to avoid dangerous voltage stresses on capacitors and switching devices [4]. Although self-balancing may be achieved, non-ideal components and some operating conditions may result in large capacitor voltage drifts [5]. Hence, a number of DC-link voltage equalization algorithms for NPC have been proposed in the literature, which aim to equalize capacitor voltages for any load, possibly without affecting NPC output voltage and current waveforms [6-11]. In this context, much less attention is given to capacitor stress in terms of current ripple due to the equalization process.

Thus, an improved DC-link voltage equalization algorithm (DCL-E) is proposed in this paper. It is developed on the basis of [12] with the aim of improving the DC-link voltage

Mario Porru gratefully acknowledges Sardinia Regional Government for the financial support of her PhD scholarship (P.O.R. Sardegna F.S.E. Operational Programme of the Autonomous Region of Sardinia, European Social Fund 2007-2013 - Axis IV Human Resources, Objective l.3, Line of Activity l.3.1.)

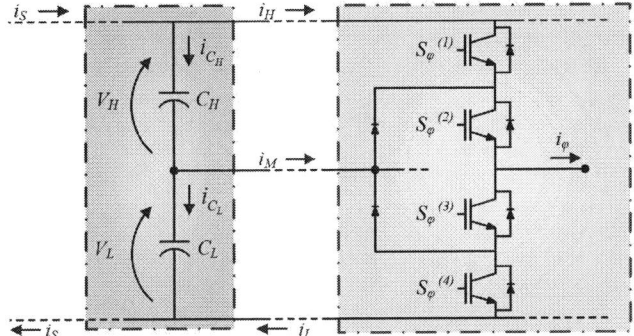

Fig. 1. Schematic representation of DC-link (red box) and NPC (orange box).

equalization process, especially in terms of voltage and current ripple. This is achieved by an appropriate exploitation of NPC flexibility in terms of PWM patterns. The proposed DCL-E consists of synthesizing reference DC-link currents on the basis of DC-link voltage unbalance and NPC load requirements. Such reference profiles are then tracked as well as possible by selecting the most suitable PWM pattern in each sampling time interval. The performance improvement achievable by means of the proposed DCL-E is highlighted through a simulation study, which refers to the case of a Surface-Mounted Permanent Magnet Synchronous Machine fed by a three-phase NPC.

II. NPC MATHEMATICAL MODEL

Referring to the schematic representation of an NPC shown in Fig. 1, it can be seen that each of its leg consists of four switches and two clamping diodes mainly. Furthermore, the DC-link consists of high-side and low-side capacitors, whose voltages are denoted by V_H and V_L respectively. Therefore, the leg states {H,M,L} can be introduced in accordance with the switching states S, as pointed out in Table I, V_φ being the voltage of the generic phase φ.

TABLE I SWITCHING AND LEG STATE

Leg State	Switching States				V_φ
	$S_\varphi^{(1)}$	$S_\varphi^{(2)}$	$S_\varphi^{(3)}$	$S_\varphi^{(4)}$	
H	1	1	0	0	V_H
M	0	1	1	0	0
L	0	0	1	1	$-V_L$

TABLE II   VOLTAGE AND CURRENT SECTORS

σ_v, σ_i	I	II	III	IV	V	VI
s_{ab}, s_a	0	0	1	1	1	0
s_{bc}, s_b	1	0	0	0	1	1
s_{ca}, s_c	1	1	1	0	0	0

TABLE III   VOLTAGE AND CURRENT INDEXES

σ_v	I	II	III	IV	V	VI
u	a	c	b	a	c	b
v	b	a	c	b	a	c
w	c	b	a	c	b	a

Referring to a three-phase NPC, it is possible to define voltage sectors (σ_v) and current sectors (σ_i), on the basis of the signs of reference chain voltages $\{s_{ab}, s_{bc}, s_{ca}\}$ and actual phase currents $\{s_a, s_b, s_c\}$ respectively, as summed up in Table II. Consequently, a new set of indexes $\{u, v, w\}$ can be introduced, which alternatively denotes the phase terminals $\{a, b, c\}$ over each voltage sector, as pointed out in Table III. As a result, p.u. chain voltages and DC-link currents can be expressed by (1) and (2) respectively:

$$v_{uv} = (x_H - y_H)v_H - (x_L - y_L)v_L$$
$$v_{vw} = (y_H - z_H)v_H - (y_L - z_L)v_L \qquad (1)$$
$$v_{wu} = (z_H - x_H)v_H - (z_L - x_L)v_L$$

$$i_H = x_H \cdot i_u + y_H \cdot i_v + z_H \cdot i_w$$
$$i_M = x_M \cdot i_u + y_M \cdot i_v + z_M \cdot i_w \qquad (2)$$
$$i_L = -(x_L \cdot i_u + y_L \cdot i_v + z_L \cdot i_w)$$

where $\{x, y, z\}$ are the equivalent switching states, whereas v_H and v_L denotes the p.u. capacitor voltages as

$$v_H = \frac{V_H}{V_H + V_L} \quad , \quad v_L = \frac{V_L}{V_H + V_L}. \qquad (3)$$

Thus, it is possible to state that v_{uv} always represents the voltage of maximum magnitude, v_{vw} and v_{wu} being always opposite to v_{uv}, as pointed out in [12].

III.   DC-LINK EQUALIZATION ALGORITHM

Referring to Fig. 1, the time derivative of both high-side and low-side capacitor voltages can be expressed as:

$$\frac{dV_H}{dt} = \frac{i_S - i_H}{C} \quad , \quad \frac{dV_L}{dt} = \frac{i_S - i_L}{C} \qquad (4)$$

in which i_S denotes the current supplied by the DC source and C is the capacity of each DC-link capacitor. As a consequence, subtracting (4) from each other yields

$$\frac{dV_M}{dt} = \frac{i_M}{C} \qquad (5)$$

where V_M and i_M denote the DC-link voltage and current unbalance respectively, which are defined as follows:

$$V_M = V_H - V_L$$
$$i_M = -(i_H - i_L). \qquad (6)$$

Therefore, DC-link voltage unbalance can be suppressed by means of i_M, as evidenced by (5). This can be achieved resorting to a hysteresis regulator [12], which, however, implies i_M maximization or minimization, even when DC-link voltage equalization is accomplished. This produces unsuitable current and voltage ripple on DC-link capacitor, leading to their overexploitations. On the basis of this, an improved DC-link voltage equalization algorithm (DCL-E) is developed, whose equivalent control scheme is depicted in Fig. 2. In particular, $i_M{}^*$ is synthesized on the basis of V_M through a simple proportional regulator, whose gain can be set in order to guarantee adequate dynamic performances. In particular, only a proportional regulator is needed because $i_M{}^*$ has to be held constant at zero once DC-Link voltage equalization is achieved.

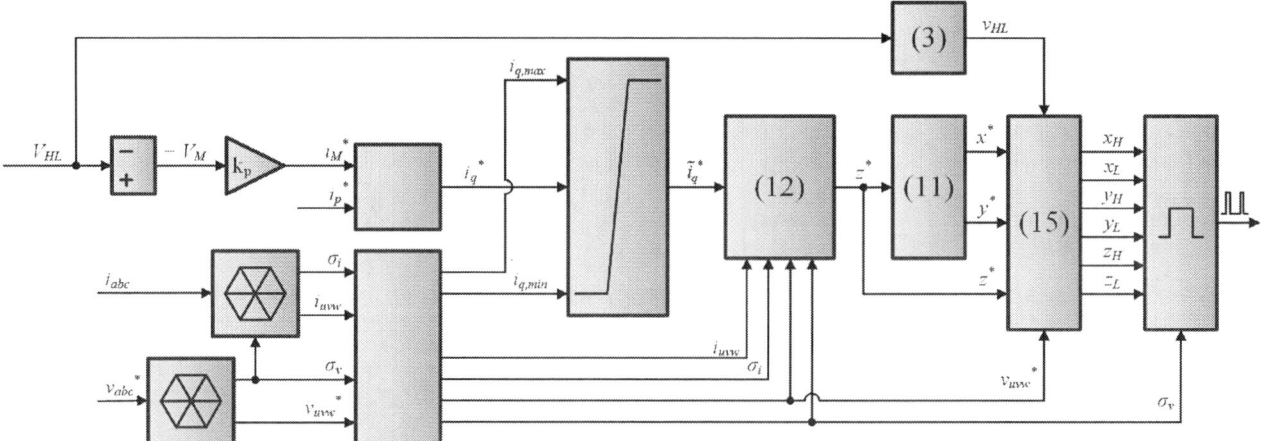

Fig. 2.   Equivalent control scheme of the proposed DCL-E.

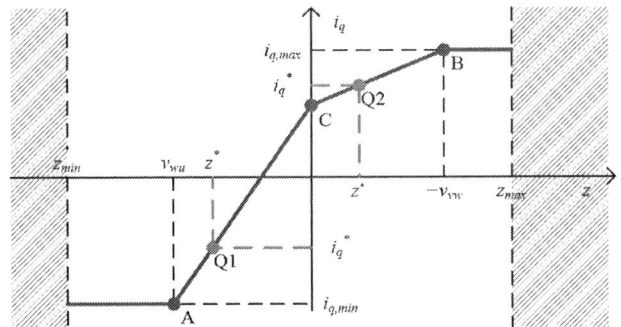

Fig. 3. The i_q locus in correspondence of a generic odd voltage sector and of the first current sector.

Subsequently, both i_H and i_L have to be driven in order to satisfy NPC load requirements and DC-link voltage equalization needs at the same time. This dual purpose requires the introduction of two equivalent input currents, i_p and i_q, which are defined as

$$
\begin{aligned}
i_p &= v_H i_H + v_L i_L \\
i_q &= v_H i_H - v_L i_L .
\end{aligned}
\qquad (7)
$$

In particular, i_p accounts for NPC output power, thus its reference value is imposed by load requirements. Whereas, on the basis of (6) and (7), the reference i_q value (i_q^*) can be chosen in accordance with both i_p and i_M, this last accounting for DC-link voltage equalization needs. Thus, by appropriately combining (2) with (7), i_q can be better expressed as:

$$
i_q = |x| \cdot i_u + |y| \cdot i_v + |z| \cdot i_w
\qquad (8)
$$

where:

$$
\psi = \psi_H v_H - \psi_L v_L \quad , \quad \psi \in \{x,y,z\} .
\qquad (9)
$$

In particular, (8) states that i_q depends on NPC phase currents through the overall switching states $\{x,y,z\}$, which are defined by (9). These last can vary in accordance with

$$
-v_L \leq \psi \leq v_H \quad , \quad \psi \in \{x,y,z\} .
\qquad (10)
$$

As a consequence, by substituting (9) in (1), the following expression is achieved:

$$
\begin{aligned}
x &= -v_{wu} + z \\
y &= v_{vw} + z \quad .
\end{aligned}
\qquad (11)
$$

Thus, the substitution of (11) in (8) and (10) yields respectively

$$
i_q = |-v_{wu} + z| \cdot i_u + |v_{vw} + z| \cdot i_v + |z| \cdot i_w
\qquad (12)
$$

$$
z_{min} \leq z \leq z_{max}
\qquad (13)
$$

where

$$
\begin{aligned}
z_{min} &= max\{v_{wu}, -v_{vw}\} - v_L \\
z_{max} &= min\{v_{wu}, -v_{vw}\} + v_H
\end{aligned}
\qquad (14)
$$

On the basis of (12) and (13) it is possible to select the i_q value nearest to i_q^* and, thus, to calculate the corresponding z^* value, as shown in Fig. 3. Once z^* is determined, it is possible to compute both x^* and y^* by means of (11), as well as all the equivalent switching states as

$$
\psi_H = \frac{|\psi| + \psi}{2v_H} \quad , \quad \psi_L = \frac{|\psi| - \psi}{2v_L} \quad , \quad \psi \in \{x,y,z\} .
\qquad (15)
$$

IV. SIMULATIONS

The proposed DC-Link voltage equalization algorithm is verified through a simulation study, which is performed in the Matlab-Simulink environment. Reference is made to the case of a Surface-Mounted Permanent Magnet Synchronous Machine (SPM, 8.25 kW, 2500 rpm) fed by an NPC, whose DC-link is supplied by a DC-source, as shown in Fig. 4. In addition, the NPC switching frequency is set to about 20 kHz, the sampling time interval being 51.2 µs. In order to highlight the performance improvement achievable by means of the proposed DCL-E (case A), it is compared with the DC-link equalization algorithm proposed in [12] (case B).

Simulation results are shown in Fig. 5 through Fig. 8. In particular, a step reference torque is imposed at the SPM start-up, thus the rotor speed increases, steady state operation being reached in about 0.1 s. Referring to Fig. 5, it can be seen that a significant initial voltage unbalance is imposed in both cases. The corresponding capacitor current evolutions are shown in Fig. 6, whereas V_M and i_M evolutions are depicted in Fig. 7, equivalent switching states evolutions being depicted in Fig. 8. It is worthy of note that all the evolutions achieved in cases A and B are almost superimposed during the equalization process. This is due to the fact that i_M is appropriately minimized in both cases, in order to equalize DC-link voltages as fast as possible, in accordance with SPM requirements.

As soon as DC-link voltage equalization is achieved, significant differences between case A and case B occur. In fact, referring to case B at first, unsuitable capacitor voltage and current ripple arise, which do not significantly occur in case A. This does not affect DC-link voltage equalization

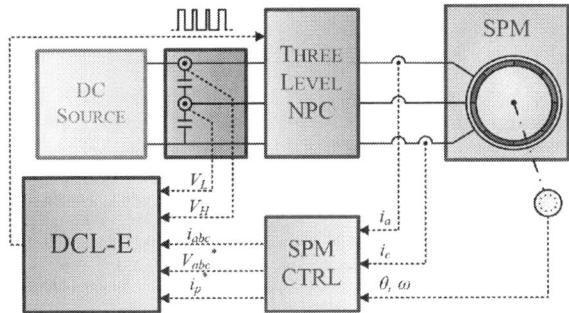

Fig. 4. Simulation set-up.

978-1-4799-4993-9/14 $31.00 © 2014 IEEE

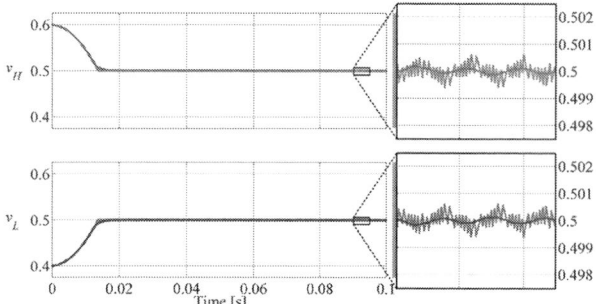

Fig. 5. The v_H (red) and v_L (blue) evolutions in cases A and B (lighted).

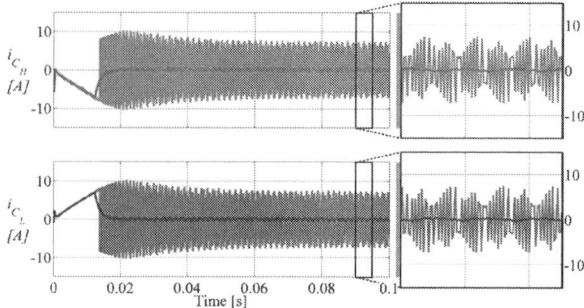

Fig. 6. The i_{C_H} (red) and i_{C_L} (blue) evolutions in cases A and B (lighted).

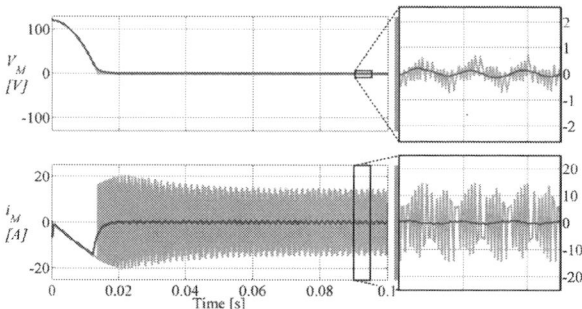

Fig. 7. The V_M and i_M evolutions in cases A and B (lighted).

because DC-link voltage ripple is quite negligible in both cases, as highlighted in Fig. 5. However, large current ripple on both capacitors occurs in case B, as well highlighted by Fig. 6. This is due to the lack of an appropriate control on i_M evolution in case B, as shown in Fig. 7. Such a current ripple leads to improper exploitation of DC-link capacitors, which is appropriately avoided in case A, in correspondence of which current ripple is about 15 times less than that occurring in case B. Such an improvement is also highlighted by Fig. 8, which shows the appropriate selection of PWM patterns occurring in case A, by means of which i_M is successfully held constant at zero once DC-link equalization is achieved.

V. CONCLUSION

An improved DC-link voltage equalization algorithm (DCL-E) for Three-Level Neutral-Point-Clamped converters has been proposed in this paper. It allows the equalization of DC-link voltages, successfully preserving NPC load performances at the same time. In addition, simulation results have highlighted that the proposed DCL-E does not lead to unsuitable DC-Link capacitor overexploitations, by appropriately minimizing their voltage and current ripple at steady state operation. This novel approach enables the development of novel control strategies involving the NPC.

REFERENCES

[1] H. Abu-Rub, J. Holtz, J. Rodriguez, G. Baoming, "Medium-Voltage Multilevel Converters: State of the Art, Challenges, and Requirements in Industrial Applications", *IEEE Trans. Ind. Electron.*, vol. 57, no. 8, pp. 2581-2596, Aug. 2010.

[2] A. Nabae, I. Takahashi, H. Akagi, "A New Neutral-Point-Clamped PWM Inverter," *IEEE Trans. Ind. Appl.*, vol. IA-17, no. 5, pp. 518–523, Sep. 1981.

[3] J. Rodriguez, S. Bernet, P. K. Steimer, I. E. Lizama, "A Survey on Neutral-Point-Clamped Inverters," *IEEE Trans. Ind. Electron.*, vol. 57, no. 7, pp. 2219-2230, Jul. 2010.

[4] N. Celanovic, D. Boroyevich, "A comprehensive study of neutral-point voltage balancing problem in three-level neutral-point-clamped voltage source PWM inverters", *IEEE Trans. Power Electron.*, vol. 15, no. 2, pp. 242-249, Mar. 2000.

[5] J. Shen, S. Schröder, R. Rösner, S. El-Barbari, "A Comprehensive Study of Neutral-Point Self-Balancing Effect in Neutral-Point-Clamped Three-Level Inverters", *IEEE Trans. Power Electron.*, vol. 26, no. 11, pp. 3084-3095, Nov. 2011.

[6] J. Shen, S. Schroder, B. Duro, R. Roesner, "A Neutral-Point Balancing Controller for a Three-Level Inverter With Full Power-Factor Range and Low Distortion", *IEEE Trans. Ind. Appl.*, vol. 49, no. 1, pp. 138-148, Jan. 2013.

[7] Y. Zhang, Z. Zhao, J. Zhu, "A Hybrid PWM Applied to High-Power Three-Level Inverter-Fed Induction-Motor Drives", *IEEE Trans. Ind.*

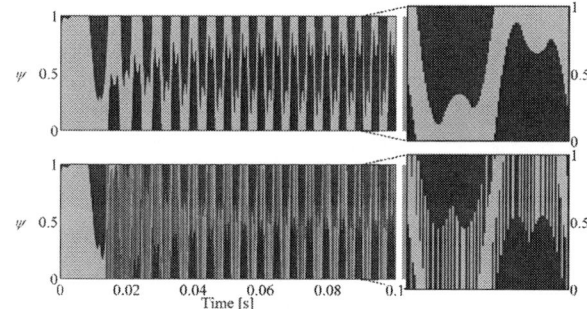

Fig. 8. Equivalent switching state evolutions on an NPC leg in case A (on the top) and case B (on the bottom): ψ_H (red), ψ_M (green) and ψ_L (blue).

Electron., vol. 58, no. 8, pp. 3409-3420, Aug. 2011.

[8] G. I. Orfanoudakis, M. A. Yuratich, S. M. Sharkh, "Nearest-Vector Modulation Strategies With Minimum Amplitude of Low-Frequency Neutral-Point Voltage Oscillations for the Neutral-Point-Clamped Converter", *IEEE Trans. Power Electron.*, vol. 28, no. 10, pp. 4485-4499, Oct. 2013.

[9] S. Busquets-Monge, J. Bordonau, D. Boroyevich, S. Somavilla, "The nearest three virtual space vector PWM - a modulation for the comprehensive neutral-point balancing in the three-level NPC inverter", *IEEE Power Electron. Lett.*, vol. 2, no. 1, pp. 11-15, Mar. 2004.

[10] J.-H. Cho, N.-J. Ku, J.-T. Han, R.-Y. Kim, D.-S. Hyun, "A simple control method for neutral-point voltage oscillation reduction of three-level Neutral-Point-Clamped inverter", *in Proc. 39th Annual Conference of Industrial Electronics Society (IECON 2013)*, Vienna (Austria), Nov. 10-13, 2013, pp. 304-309.

[11] A. H. Bhat N. Langer, "Capacitor Voltage Balancing of Three-Phase Neutral-Point-Clamped Rectifier Using Modified Reference Vector", *IEEE Trans. Power Electron.*, vol. 29, no. 2, pp. 561-568, Feb. 2014.

[12] A. Damiano, I. Marongiu, M. Porru, A. Serpi, "A suitable PWM for DC-link voltage equalization of Three-Level Neutral-Point Clamped Converters", *in Proc. 39th Annual Conference of Industrial Electronics Society (IECON 2013)*, Vienna (Austria), Nov. 10-13, 2013, pp. 328-333.

Simplified Review of DCDC Switching Noise and Spectrum Contents

Adnan FARES, IEEE Student Member

Sami AJRAM, IEEE Member
PMIC Design Department
SL3J SYSTEMS
Gardanne, France
adnan.fares@sl3j.com, sami.ajram@sl3j.com

Guy CATHÉBRAS, IEEE Member

Microelectronic Department
LIRMM Laboratory
Montpellier, France
guy.cathebras@lirmm.fr

Abstract —**This paper presents an overview of the switching noise analysis in a DCDC converter by considering the first order parasitic elements involved in the noise generation. Simplified theory is provided in order to help circuit designers selecting the optimum components for low noise operation. Theory versus simulation and measurement are carried out.**

I. INTRODUCTION

THE trends observed over the last 5 years in the power management IC development area, especially in handheld and ultra-portable devices such as cell phones and PDAs, were clearly oriented toward increasing the switching frequency up to the 10MHz range in order to achieve higher density of integration and smaller size for the passive components [1], [2],[5]. 2 to 6 MHz DCDC converters are now commonly employed in Dynamic Voltage Biasing circuits for WCDMA RFPAs [3], [4] where designers are concerned by the output voltage ripple or the switching noise and how it translates into sideband spectrum spurs impacting the Adjacent Channel Leakage Ratio (ACLR) performance of the transmitted RF Signal as well as the noise power shifted up to the RX bands.

Reducing the voltage ripple of DCDC converters can be achieved by increasing the inductance value, increasing the filtering capacitor value or stacking various technology capacitors in order to reduce the Equivalent Series Inductance (ESL) and resistance (ESR). However, this systematically results in an increase of the converter size and limits the overall system performance in point of view load transient, step response, and input noise discrimination or line regulation.

A very low ripple is required for DCDCs supplying RF Front End circuits or RF Transmitters in order to minimize the impact on the RX or TX Bit Error Rates of the "polluted" channels. The best approach - when possible - is to operate the DCDC at switching frequency that falls out of the RF bands as proposed in [1] knowing that this would not apply to low load conditions where a DCDC operates in a variable Pulse-Frequency Modulation mode.

Further needs are sought for the new generation of DCDCs supplying 4G transceivers due to the significant increase of the RF signal bandwidth (up to 20MHz in LTE [6]) [7], [8], which pulls more noise spurs in the main transmitted channels, due to the increased DCDC output noise caused by the high dynamic in the RF signal envelop

(QAM 64 [6]). Such applications require fast transient large loop bandwidth DCDCs, assisted with large bandwidth linear regulators.

Various papers have attempted to provide the compromise between size and performance, by focusing only on the impact of the passive components on the ripple amplitude and ignoring the other performance parameters like the overshoot, the higher order noise harmonics, and the loop bandwidth. Papers such as [9] and [11] provide an analytic approach for optimum inductor calculation allowing low ripple conditions, and other application oriented literatures such as [12] provide simplified analyses for selecting the best passive components depending on their rated performance and electrical equivalent models.

Regarding the solutions proposed for cancelling or reducing the effect of the DCDC noise, one can cascade the DCDC with a high PSRR LDO [13], or use interleaved multiphase architecture followed by a high bandwidth high PSRR LDO, as used in Envelop Elimination Restoration RFPAs [14], knowing that such techniques do not eliminate the direct effect of the DCDC switching ripple but only the load transient related ripple.

The purpose of this paper is to provide a review of the existing theory that engineers constantly need to consider for achieving the best compromise between the DCDC noise performance and the selection of the passive components. Section 2, is focused on the selection of the filtering capacitance illustrated in the case of a buck converter, while sections 3 and 4 provides the detailed theory allowing to estimate the output voltage ripple and the corresponding spectrum. Finally, simulation result and spectrum measurements are compared to illustrate the theory.

II. POSITION OF THE PROBLEM

A. Sizing the filtering capacitor

In a DCDC converter such as a step down or buck converter operating in PWM mode (Figure 1), the sizing of the output capacitor depends on three parameters:

1) The voltage overshoot ΔV_{OS} occurring at the end of the start-up phase especially when the inductor current hits the current limit before settling to permanent regime. (Figure 2)

2) The loop bandwidth which is dominated by the double

978-1-4799-4993-9/14 $31.00 © 2014 IEEE 305

poles formed by the LC filter

3) The voltage ripple induced by $\frac{\Delta I}{2}$, the amplitude of the inductor current ripple

Most often the size of the capacitor is determined by the voltage overshoot involving the current limit as it constitutes the worst case condition for energy exchange between the inductor and the capacitor and worst case perturbation for the regulation loop. This actually allows designers selecting the technology of the capacitor (multilayer ceramic MLCC, tantalum, or electrolytic) which indirectly sets the magnitudes of the main parasitic elements ESL and ESR to be considered in the estimation of the output voltage ripple.

Designers can use filter stacking or cascading techniques to leverage some high frequency spur discrimination but it is fundamental for a given filter technology to accurately determine the ripple waveform and its corresponding spectrum contents.

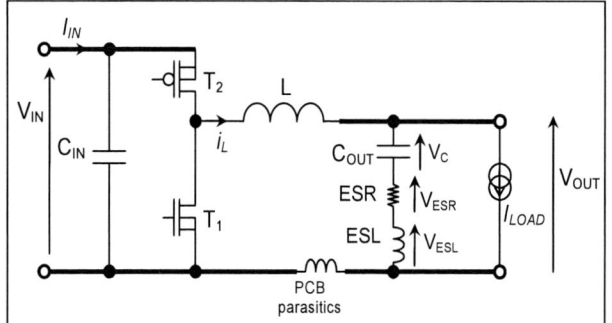

Figure 1.   Output stage of a buck converter showing the parasitic elements of the output filter

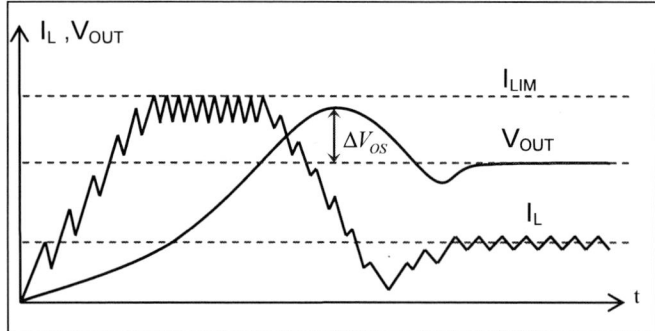

Figure 2.   Voltage overshoot occurring at the end of the start-up phase in a DCDC converter.

We can give a fairly good estimation of the voltage overshoot magnitude occurring at the end of the start-up phase and thus determine the value of the capacitor required at the output. We assume that, during the start-up, the current in the inductor is limited to a maximum value I_{LIM} in order to prevent magnetic field saturation or prevent overheating. When the output voltage reaches a regulated value, the excessive energy stored in the inductor is entirely transferred to the output capacitor which results in a voltage overshoot as shown on Figure 2.

Equation (1) represents the energy exchange law between L and C_{OUT} in a worst case condition where <u>no current is absorbed by the load</u>.

$$\frac{1}{2}C_{OUT}\left(V_{OUT}+\Delta V_{OS}\right)^2=\frac{1}{2}C_{OUT}V_{OUT}^2+\frac{1}{2}L\,I_{LIM}^2 \qquad (1)$$

Where: C_{OUT} is the output capacitance, L is the inductance, V_{OUT} is the output voltage, I_{LIM} is the current limit and ΔV_{OS} is the overshoot in the output voltage of the converter.

In other terms, and assuming ΔV_{OS} is small compared to V_{OUT}, C_{OUT} can be written as follows

$$\frac{\Delta V_{OS}}{V_{OUT}}=\frac{1}{2}\frac{L\,I_{LIM}^2}{C_{OUT}V_{OUT}^2} \qquad (2)$$

$$C_{OUT}=\frac{1}{2}\frac{L\,I_{LIM}^2}{V_{OUT}^2\left(\dfrac{\Delta V_{OS}}{V_{OUT}}\right)} \qquad (3)$$

The circuit designer can use equation (3) to select the value of C_{OUT} depending on the desired ΔV_{OS} and I_{LIM}.

For instance, a 3.6V to 1.2V; 0.5A buck operating at 6MHz, with a current limit of 1A worst case, using a 470nH inductor and tolerating an overshoot of 10%, would require a filtering capacitor of 1.6µF.

Regarding the loop bandwidth, C_{OUT} impacts the base position of the poles of the LC filter as follows:

$$F_{LC}=\frac{1}{2\pi\sqrt{L\times C_{OUT}}} \qquad (4)$$

Which will require a particular attention if larger bandwidth is needed.

At last, the DCDC output ripple represented by the peak to peak value ΔV_{OUT_PP} can be written as follow:

$$\Delta V_{OUT\_PP}=\frac{\Delta I.T_0}{8\times C_{OUT}}=V_{OUT}\frac{\pi^2(1-D)}{2}\left(\frac{F_{LC}}{F_0}\right)^2 \qquad (5)$$

Where: ΔI is the inductor's current ripple, D is the duty cycle, F_0 is the switching frequency and T_0 is the switching period.

In our example, the bandwidth of the converter would be around 180 KHz and the ripple voltage would be about 1.2mV.

The output ripple might appear to be acceptable by the designers while the bandwidth would not, but require further improvements in order to achieve better loop response. In such case, designers have to implement additional features such as current injection and high frequency zero compensation in order to enlarge the bandwidth.

Specific applications like analog front end or RF Power Amplifiers might tolerate some increase of the overshoot during transient phases but are deemed very sensitive to the spectrum contents of the DCDC especially because they fall into the Intermediate Frequency (IF) bandwidth and systematically impact the Signal to Noise Ratio.

Furthermore, we found relevant to deeply analyze the output voltage waveform, estimate the magnitude of the voltage ripple and provide a theoretical basis for determining the harmonic contents.

III.   DETAILED ANALYSIS OF THE OUTPUT VOLTAGE RIPPLE

In stationary regime, the DC component of the inductor current is assumed to be equal to the average load current

978-1-4799-4993-9/14 $31.00 © 2014 IEEE   306

while the AC component is absorbed by the filtering capacitor creating the output voltage ripple.

The output voltage ripple can be estimated as the sum of three components: V_{ESR} the voltage induced in the ESR which has the same waveforms of the inductor current, V_{ESL} the voltage induced in the ESL which is square wave as the inductor current is assumed to be triangular, and finally V_C which is the result from the cyclic charging and discharging of the output capacitor (Figure 3).

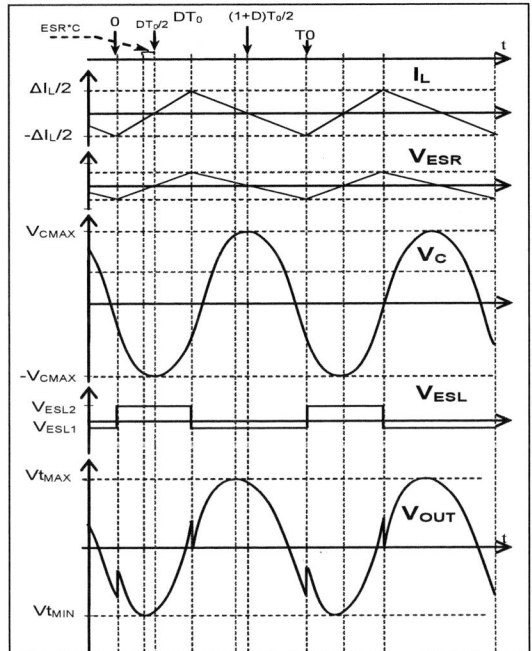

Figure 3. Detailed contributions of the parasitic elements to the output voltage ripple in a DCDC converter.

The corresponding waveforms shown in Figure 3 highlight the effect of each component on the overall ripple. The assumption in this illustration is that the ESR effect has a smaller order of magnitude than the voltage produced by the capacitor itself, while the ESL induced voltage has much more effect on the capacitor voltage; this fairly applies to MLCC capacitors used in high frequency DCDCs today. More interestingly, it is shown that the ESL induced ripple results in changing the magnitude of the output ripple while it introduces sharp voltage variations and increases the high frequency and undesirable spectrum contents.

The total amplitude of the output ripple is given by:

$$\Delta V_{OUT\_PP} = \frac{\Delta I \times T_0}{8 \times C} + \frac{ESR^2 \times C \times \Delta I}{2T_0 D(1-D)} - \frac{ESL \times \Delta I}{D(1-D)T_0} \quad (6)$$

Which highlights the role of ESL in reducing the ripple magnitude.

IV. SPECTRUM OF THE OUTPUT RIPPLE

In the case of RF transmitters the ripple voltage of the DCDC converter interferes with the carrier frequency due to the nonlinear behavior of the RFPA. The interference is mostly linked to the 3rd intermodulation effect IM3 such as presented in [15]. However one has to consider the direct mixing effect ($F_{Carrier} \pm F_0$), ($F_{Carrier} \pm 2F_0$), ($2F_{Carrier} \pm F_0$) which could not be neglected due to the large "distance" between the carrier frequency (F_{TX}) and the switching frequency of the DCDC (F_0).

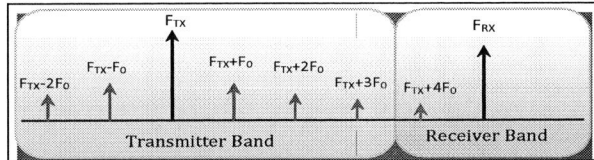

Figure 4. Transfer of the DCDC Noise to the RFPA band.

As a constraint, the designer has to prevent any interference with the adjacent channels where an RFPA has to show a rejection of 33dBc (first channel) [6] such as for 3G and 4G standards.

The designer would need to employ our theory to model the DCDC behavior and provide a better estimation of the ACLR.

The spectrum of the ripple voltage can be written using Fourier transform as follows:

$$V_{OUT}(nF_0) = \frac{\Delta I \sin c(\pi n D)}{1 - D} \times$$
$$\sqrt{\left(\frac{ESR}{\pi n}\right)^2 + \left(\frac{T_0}{2\pi^2 n^2 C} - \frac{2 \times ESL}{T_0}\right)^2} \quad (7)$$

Where: $V_{ESR}(nF_0)$, $V_{ESL}(nF_0)$, $V_C(nF_0)$ are the discrete Fourier transforms of the ESR, the ESL, and the Capacitor voltages respectively.

One can notice again from the spectrum study that the ESL reduces amplitude of the ripple voltage.

V. EXPERIMENT:

An experiment set-up using a 2MHz buck converter was built and measurements have been compared to simulations carried out using Cadence®Spectre.

TABLE I.   PARAMETERS OF THE DCDC CONVERTER

V_{IN}	V_{OUT}	F_0	C_{OUT}	L	ESL	ESR
3.6	1.85	2MHz	1.6 µF	2.6 µH	1.3 nH	5.2 mΩ

Where the values of C_{OUT} and L are measured, the value of ESR is extracted from the datasheet and the value of the ESL is estimated from the sharp edge in the output voltage.

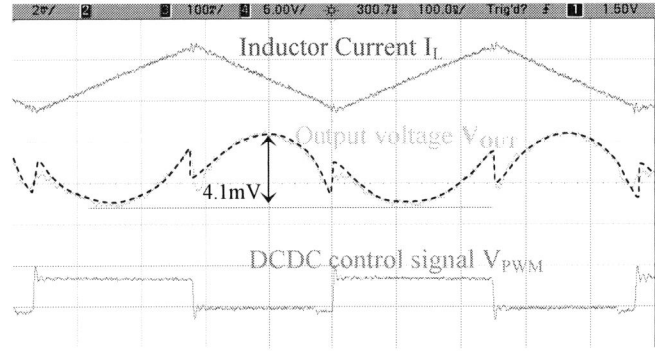

Figure 5. Experimental results for the buck converter. Dashed line is the expected waveform of the output voltage.

978-1-4799-4993-9/14 $31.00 © 2014 IEEE   307

The detailed waveforms, shown in Figure 6, are extracted from the simulation.

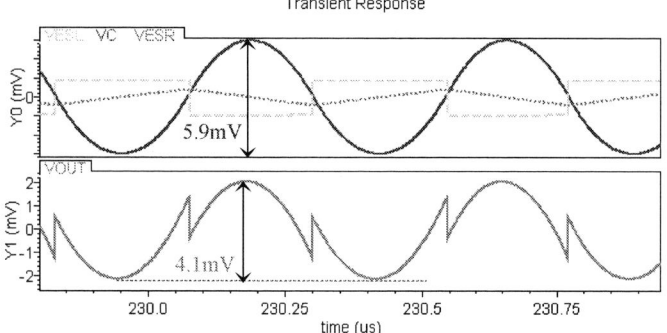

Figure 6. Simulated waveforms (AC mode) of the output voltage (V_{OUT}) and detailed contributions of each element of the capacitor model (V_C, V_{ESL}, V_{ESR}).

The ripple voltage across C_{OUT}, without considering the ESL and the ESR would be 5.9mV while in practice we observed 4.1mV.

Finally the spectrum of the ripple voltage (Figure 7) is extracted from the same simulation test bench:

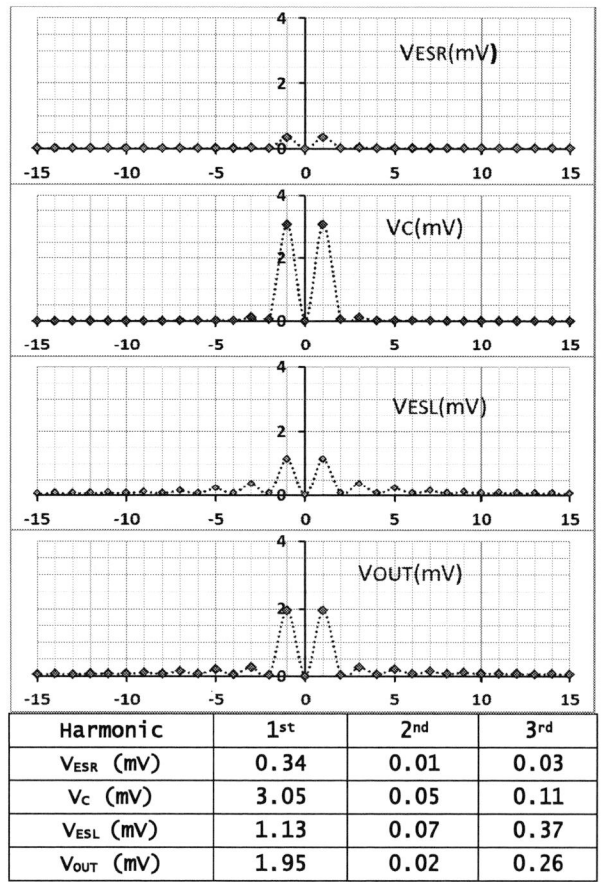

Harmonic	1st	2nd	3rd
V_{ESR} (mV)	0.34	0.01	0.03
V_C (mV)	3.05	0.05	0.11
V_{ESL} (mV)	1.13	0.07	0.37
V_{OUT} (mV)	1.95	0.02	0.26

Figure 7. Detailed spectrum contents of the output voltage (V_{OUT}) with the contributions of each element of the capacitor model (V_C, V_{ESL}, V_{ESR}).

Once more, the spectrum shows a lower magnitude of the first harmonic compared to the ideal C_{OUT}.

Nevertheless, the contribution of the 2nd harmonic deemed very low compared to the 3rd harmonic which is dominated by the ESL.

VI. CONCLUSION

In this paper we have described the influence of the output capacitor on the overshoot, the bandwidth and the output ripple voltage of the DCDC converter. We then extended the study of the output ripple and described the impact of the output capacitor including its parasitic elements on that. The ESL of the capacitor was shown to have the dominated effect on the ripple voltage where it reduces the amplitude of the ripple and adds high frequency components in the spectrum. Finally we have validated the analysis by an experiment and a simulation.

REFERENCES

[1] S. Ajram and S. Kuran, "Dynamic selection of oscillation signal frequency for power converter," US Patent 7,826,815, Nov. 2, 2010.

[2] J. Hannon, R. Foley, J. Griffiths, D. O'Sullivan, K. G. McCarthy, and M. G. Egan, "A 20 MHz 200-500 mA Monolithic Buck Converter for RF Applications," *IEEE Appl. Power Electron. Conf. Expo.*, pp. 503–508, Feb. 2009.

[3] Fairchild, "800mA Buck converter for 3G RFPAs," FAN5902 datasheet, May 2009.

[4] National Semiconductor, "650 mA Miniature adjustable, Step-Down DC-DC Converter for RF Power Amplifiers," LM3218 datasheet, April 2006.

[5] B. Sahu, G.A. Rincon-Mora, "System-level requirements of DC-DC converters for dynamic power supplies of power amplifiers," *ASIC, 2002. Proceedings. 2002 IEEE Asia-Pacific Conference on,* vol., no., pp.149,152, 2002.

[6] 3rd Generation Partnership Project, "Evolved Universal Terrestrial Radio Access (E-UTRA), user Equipment (UE) radio transmission and reception," *3rd Generation Partnership Project*, TS 36.101 V11.1.0, 2012, [online]. Available: http://www.3gpp.org/ftp/Specs/archive/36_series/36.101/36101-b10.zip [accessed: 11,2013]

[7] Nujira, "Envelope Tracking: unlocking the potential of CMOS PAs in 4G smartphones," Nujira, white paper, Feb. 2013. [Online]. Available: https://www.nujira.com/pages/files/Papers_and_Presentations/Nujira_Envelope_Tracking_with_CMOS_PAs_v1.0.pdf

[8] M. Vasic, O. Garcia, J. A. Oliver, P. Alou and J. A. Cobos, "Survey of architectures and optimizations for wide bandwidth envelope amplifier," *Power Electronics and Motion Control Conference (EPE/PEMC)*, 2012.

[9] L. Shulin, L. Yan and L. Li, "Analysis of Output Voltage Ripple of Buck DC-DC converter and its design," *Proc. 2nd Int. Conf. on Power Electronics and Intelligent Transportation System (PEITS)* , vol.2, no., pp.112,115, 19-20 Dec. 2009.

[10] S. Liu, J. Liu, and J. Zhang, "Research on output voltage ripple of boost DC/DC converters," *Proc. Int. MultiConf. of Engineers and Computer Scientists.* Vol. 2. 2008.

[11] M. Laflin, "High Frequency Implications for Switch-mode DC-DC Converter Design," Enpirion, 2007 [online]. Available: EETimes, http://www.eetimes.com/document.asp?doc_id=1273157

[12] D. Schelle and J. Castorena "Buck-Converter Design Demystified," Maxim, white paper, June, 2006. [Online]. Available: http://powerelectronics.com/site-files/powerelectronics.com/files/archive/powerelectronics.com/mag/606PET25.pdf

[13] P. M. Cheng, O. García, M. Vasić, P. Alou, J. A. Oliver, G. Montoro, and J. A. Cobos, "Envelope Amplifier Based on a Hybrid Series Converter With The Slow-envelope Technique*, " Energy Conversion Congress and Exposition (ECCE),2012 IEEE,* pp. 1-5.

[14] P. M. Cheng, M. Vasic, O. Garcia, J. A. Oliver, and P. A. J. A. Cobos, "Multiphase buck converter with minimum time control strategy for RF envelope modulation", *Applied Power Electronics Conference* APEC, *IEEE* 2011

[15] J. Vuolevi and T. Rahkonen, *Distortion in RF Power Amplifiers*, Artech House, 2003

978-1-4799-4993-9/14 $31.00 © 2014 IEEE   308

Quantifying the Figures of Merit of Graphene-Based Adiabatic Pass-XNOR Logic (PXL) Circuits

Valerio Tenace, Andrea Calimera, Enrico Macii, Massimo Poncino

Dipartimento di Automatica e Informatica, Politecnico di Torino, Torino, Italy

Email: {valerio.tenace, andrea.calimera, enrico.macii, massimo.poncino}@polito.it

Abstract—In this work we quantify the figures of merit of p-n junction based adiabatic graphene circuits implemented through a new logic design style, the *Adiabatic Pass-XNOR Logic* style (Adiabatic-PXL). First we show how graphene p-n junctions naturally implement transmission gates with embedded XNOR functionality (the *Pass-XNOR* gate); second, we present a dedicated logic synthesis flow for integrating those gates into adiabatic logic circuits with ultra low-power features. Simulation results have shown that Adiabatic-PXL circuits are 4.2X to 5.5X more energy efficient than non-adiabatic counterparts, still with significant amount of area savings (67% less devices on average).

I. INTRODUCTION

Graphene was isolated for the first time in 2004 [1]. Since its discovery, a consistent share of research activities focused on the study of the electrical properties of this material. High carrier mobility and high saturation velocity have been proven to be key characteristics of graphene, which is pointed as a potential candidate to replace silicon in the next generation of electronic devices. Unfortunately, when used in the digital domain, graphene shows an indisputable limit, that is, the lack of an energy bandgap between conduction and valence bands. This characteristic prevents the material to show an high ON/OFF current ratio and, hence, to implement digital devices.

Different solutions have been proposed to overcome this drawback. Among them, one of the first appeared relies on patterning graphene sheets into narrow stripes of width W. The resulting ribbons, called Graphene Nanoribbons (GNRs), show an energy band-gap $\propto 1/W$. Narrow GNRs exhibit an energy gap similar to that of todays' semiconductors, thereby allowing the implementation of graphene transistors (GNR-FET[2]). However, the patterning process may sensible alter the level of disorder of the material, thereby imposing significant degradation of the electrical properties of the devices [3]. As an alternative to GNRs, *electrostatic doping* has been proposed as a less aggressive, yet efficient strategy to implement equivalent p-n junctions, the key element behind any electronic device. For instance, in [4], the authors make use of two back-faced p-n junctions to implement a Reconfigurable-Multiplexer (the RG-MUX) that serves as basic logic primitive for complex logic circuits.

In this paper we propose an alternative way to use graphene p-n junctions for the implementation of ultra low-power circuits. As will discussed later in the next sections, a graphene p-n junction behaves as a voltage controlled resistor. More specifically, its electrical behavior resembles that of MOS transmission gates with enhanced logic functionality, the XNOR logic function; when the control signals are concordant (i.e., same logic values), the gate is ON (i.e., low in-to-out resistance), when discordant (i.e., opposite logic values), the gate is OFF (i.e., high in-to-out resistance). We refer the graphene p-n junction as a *Pass-XNOR gate*. A circuit implemented by

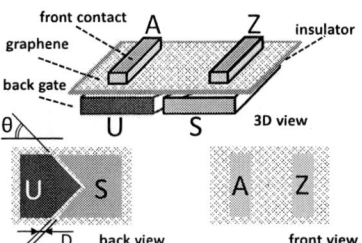

Figure 1. Graphene p-n junction.

means of series/parallel connections of pass-XNOR gates is what we refer as Pass-XNOR Logic (PXL) circuit. It can be seen as middle way between Pass-Transistor Logic (PTL) and Complementary-MOS logic (CMOS). Like PTL the information is carried out by means of signal propagation through root-to-sink paths rather than charges stored in parasitic capacitance; like in CMOS, connections of series and parallel branches of logic devices implement AND and OR logic conjunctions. When a PXL circuit is powered following the adiabatic-charging principle [5], its power budget drastically reduces, reaching almost zero in an ideal case.

As main contribution, we quantify the amount of power savings achieved with *Adiabatic PXL* circuits obtained using a dedicated One-Step Synthesis strategy based on *Biconditional Binary Decision Diagrams* [6]. Simulation results show that PXL is 4.2X to 5.5X more energy efficient w.r.t. non-adiabatic graphene circuits, still guaranteeing significant area reduction (67% less devices on average).

II. BACKGROUND

A. Graphene p-n junction

Figure 1 depicts the structure of a graphene p-n junction. It consists of a graphene sheet on top of which two metal-to-graphene contacts, A and Z, serve as signal input and output respectively, and a thick layer of oxide that isolates the two back-gates, S and U, from the graphene itself.

Exploiting the principle of electrostatic doping, voltage potentials on terminals S and U work as a control knob to tune the Fermi Energy (E_F) of the overlapping graphene regions [7]; a negative voltage shifts down E_F in the valence band leading to p-type doping, whereas a positive voltage shifts E_F up in the conductance band leading to n-type doping. When symmetric voltages are concurrently applied on S and U terminals, i.e., $V(S) = +V$ and $V(U) = -V$, the device implements the p-n junction. As described in [7], under such configuration, carriers transmission from the p-region towards the n-region is subject to the transmission probability $T(\theta)$:

$$T(\theta) = \cos^2(\theta) e^{-\pi k D \sin^2(\theta)} \quad (1)$$

where θ is the angle between the electron's wave vector $\vec{k}$ and the normal of the junction (45° as imposed by the triangular

978-1-4799-4993-9/14 $31.00 © 2014 IEEE

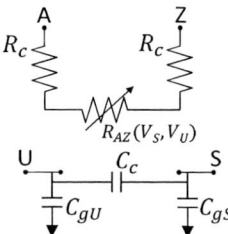

Figure 2. p-n junction electrical model

shape of the back-gates, see Figure 1), and D is the width of the metal gap between the back-gates, assumed to be $18nm$. It is worth noticing that $T(\theta) = 1$, i.e., 100% of carrier transmitted, when voltages applied at the back-gates are concordant (n-n or p-p doping configuration).

B. p-n junction electrical model

Figure 2 introduces the electrical model of the p-n junction. There are two R_C resistors connected to pins A and Z representing the parasitic resistance of metal-to-graphene contacts. The resistor R_{AZ} models the resistive path across graphene between the input A and the output Z; its analytical expression is given as $R_{AZ} = \frac{R_0}{N_{ch}T(\theta)}$, where $R_0 = \frac{h}{4q^2}$ is the quantum resistance per propagation mode, N_{ch} is the number of excited propagation modes, and $T(\theta)$ is the transmission probability described by (1). As reported in [4], values of R_{AZ} ranges from $R_{ON} = 300\Omega$, (under n-n or p-p configuration), to $R_{OFF} = 10^7\Omega$ (p-n or n-p configuration). The model also integrates the coupling capacitance C_C between the two metal split gates, and two lumped capacitances connected to the back-gates S and U, namely, C_{gS} and C_{gU} respectively, which consist of the series of the oxide capacitance and the quantum capacitance of the graphene sheet[1].

C. Reconfigurable Graphene MUX

The Reconfigurable Graphene MUX (RG-MUX) represents the first example of logic gate implemented with graphene p-n junctions [4], [8]. As depicted in Figure 3, it consists of two back-faced p-n junctions: J_1, controlled though the back-gates S and $\bar{U}$, and J_2, controlled through back-gates S and U. Notice that U and $\bar{U}$ are always driven by symmetric voltages. Let us consider the case $U = $ '1' and $\bar{U} = $ '0'. When $S = $ '0' a p-p junction is formed on J_1, i.e., R_{ON} between A and Z, and a p-n junction is formed on J_2, i.e., R_{OFF} between B and Z; hence the output Z follows the signal on A. On the contrary, when $S = $ '1', J_1 is configured as p-n, i.e., R_{OFF} between A and Z, whereas J_2 as n-n, i.e., R_{ON} between B and Z; hence the output Z follows the signal on B. This exactly implements the function of a 2-to-1 Multiplexer. For more details on the electrical model of an RG-MUX please refer to [9].

III. PASS-XNOR LOGIC

A. Pass-XNOR Gate

The Pass-XNOR gate simply consists of a p-n junction where the back-gates U and S are fed with digital input signals, while the front contacts A and Z work as source and drain of a stimulus ramp pulse used to evaluate the logic function.
As shown in Figure 4, when U and S have same logic value, the equivalent in-to-out front resistance R_{AZ} is set to R_{ON} leaving the input pulse on A passing through the junction and

---

[1]For an exhaustive discussion on the p-n junction and its electrical model, interested readers can refer to [4].

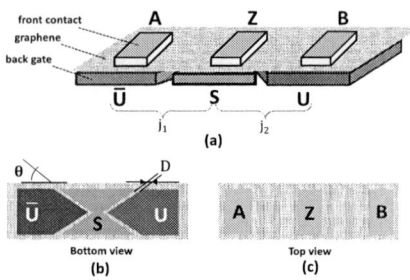

Figure 3. Reconfigurable Graphene MUX structure.

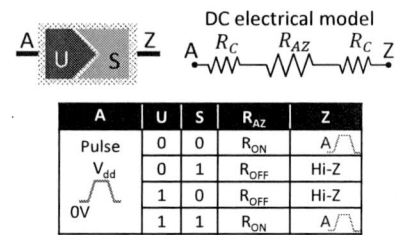

Figure 4. Functional behavior of the Pass-XNOR logic gate.

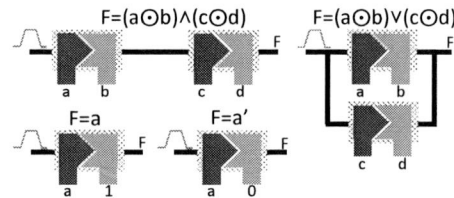

Figure 5. Basic logic functions using the Pass-XNOR gate.

reaching Z; this represents the '1'-logic at the output. Opposite logic values, on the contrary, set the in-to-out front resistance R_{AZ} to R_{OFF} ($\gg R_{ON}$), forcing Z in a high-impedance state; this represents the '0'-logic at the output.

B. Building Complex Functions

The Pass-XNOR gate shows a higher expressive power compared to the CMOS counterparts as it requires a smaller number of devices to implement XNOR/XOR-dominated logic functions [10]. Unfortunately, the XNOR operator per se is not *functional complete* and other logic connectives are needed.
Figure 5 shows possible network topologies used to implement the *AND/OR* conjunctions, the logical *Identity* and the *Complement*. Series connections of the front contacts perform the logic AND ($\wedge$); parallel connections of the front contacts perform the logic OR ($\vee$); connecting one of the back-gates to '1' or '0' allows to implement logical *identity* or *complement* ($\neg$) respectively.

C. Logical Computation through a Pass-XNOR Logic Network

The logic computation of a PXL Logic Network consists of two distinct phases: the *configuration phase* and the *evaluation phase*. In the configuration phase the primary logic inputs, i.e., the literals composing the logic function, are fed to the back-gates of the Pass-XNOR gates. At the end of this phase the doping profile of each and every device is fixed and the resistive paths of the network set up. In the evaluation phase, the input ramp pulse is injected in the network through the front input, i.e., the root of the network, and eventually

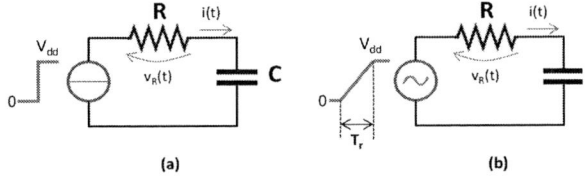

Figure 6. Non-adiabatic (a) and adiabatic (b) RC circuit.

propagated to the front output. A pulse detected at the front output evaluates the implemented function as TRUE.

Notice that the front input is unique, therefore, only one single front signal is shared among all the parallel branches of the network. As for the CMOS technology, multiple-output functions need distinct logic networks.

IV. ADIABATIC CHARGING PRINCIPLE

Let us consider the RC circuit in Figure 6(a). Suppose we apply a step-signal voltage source, e.g., from $0V$ to V_{DD}; the energy supplied by the voltage source is $E_S = CV_{DD}^2$. We also know that, when the supply signal reaches its maximum value, the energy accumulated in the capacitor C is given by $E_C = \frac{1}{2}CV_{DD}^2$, while the other half E_R is dissipated across the resistor R in the form of heat.

Now let us now consider a supply ramp-signal with a slow-rising slope, as depicted in Figure 6(b). We refer the time the input ramp goes from the 20% to the 80% of the V_{DD} as the Transition Time (T_r) of the signal. If T_r is big enough, then the capacitor charges virtually instantaneously with zero-voltage drop across R.

Given $v_R(t)$ the voltage drop across R, the energy E_R dissipated on R over time T is given by:

$$E_R = \int_0^T v_R(t)i(t)dt = \int_0^{T_r} R\frac{C^2V_{DD}^2}{T_r^2}dt = \frac{RC}{T_r}CV_{DD}^2 \tag{2}$$

For $T_r = 2RC$, E_R is equal to E_C, while for $T_r > 2RC$, E_R will shrink accordingly, thus leading to more energy efficient switching, i.e., the *adiabatic switching*.

As illustrated in Section III-A, a Pass-XNOR gate is naturally implemented through a single p-n junction which, in turn, behaves as a resistor between input A and output Z. When a slow input ramp is fed as input at the front contact A, the energy consumed to charge the load capacitance at Z follows the same adiabatic law described in Equation 2. Hence, during the evaluation phase of Pass-XNOR logic networks, the adiabatic charging principle does apply.

V. ONE-STEP LOGIC SYNTHESIS FOR PXL

In One-Step Synthesis, logic optimizations and technology mapping are carried out at the same time. That's possible if we rely on a data structure that allows to appropriately represent the logical structure of the network. For this purpose we adopted the recently introduced Biconditional Binary Decision Diagram [6], which improves the concept of Binary Decision Diagram to achive more compact representations of a logical structure.

A. Biconditional Binary Decision Diagrams

A Biconditional Binary Decision Diagram (BBDD) is a BDD [11] where the Shannon expansion is replaced by the *biconditional expansion*. This concept is an extension of the (x_i, p)-decomposition and was recently introduced in [6]. According

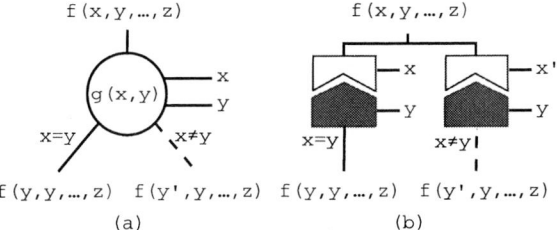

Figure 7. Non-terminal node in BBDD (a) and Pass-XNOR logic (b).

to [12], the (x_i, p)-decomposition of a generic Boolean function $f(\chi)$ is given by (3), where $x_i, p \in \chi$ and $x_i \neq p$.

$$f(\chi) = (\bar{x}_i \oplus p)f|_{x_i=p} + (x_i \oplus p)f|_{x_i \neq p} \tag{3}$$

Hence, it is straightforward to derive the biconditional expansion (4) from (3), as:

$$f(\chi) = (x_i \oplus p)f|_{x_i \neq p} + (x_i \odot p)f|_{x_i=p} \tag{4}$$

In this paper, we implemented a slightly different version of (4) as a building block of the BBDD structure, called the *XNOR expansion* (5).

$$f(\chi) = (\bar{x}_i \odot p)f|_{x_i \neq p} + (x_i \odot p)f|_{x_i=p} \tag{5}$$

Formally, a BBDD of a function F is defined as a rooted, directed acyclic graph defined as $G = (V \cup \Phi \cup \Theta, T, E)$. In such structure, we denote three different sets of nodes: *(i)* non-terminal nodes $v \in V$, depicted in Figure 7(a) each of them labeled with a function $g(x, y)$, where $x, y \in \chi$ and function g being the $XNOR$ operation between the two variables; *(ii)* terminal nodes $\theta \in \Theta$, with outdegree 0, which represent the value 1 or 0 assumed by F; *(iii)* function nodes $\phi \in \Phi$, with indegree 0 and outdegree 1, which represent the main function nodes. Each non-terminal node v has two out-edges denoted with T and E, as in the standard BDD structure, and they are connected to child nodes representing the positive cofactor ($f|_{x=y}$) and the negative cofactor ($f|_{x \neq y}$) respectively.

It is worth noticing that the BBDD structure can be mapped directly in Pass-XNOR logic by replacing each non-terminal node with two appropriately configured PXL gates, as depicted in 7(b).

Reduction rules that apply to a classical BDD structure also holds for the BBDD structure. In particular, since we want to produce devices with the smallest area, we can apply the following rules to reduce the size of a BBDD: *(i)* remove isomorphic subgraphs; *(ii)* remove empty levels [11].

VI. EXPERIMENTAL RESULTS

In this section we show the benefits achieved adopting Adiabatic Pass-XNOR logic over traditional, non-adiabatic ones. Results are collected using a few benchmarks coming from the ACM/SIGDA suite, plus two hand-made xnor-rich benchmarks, i.e., $xnor9$ and $xnor11$, that compute the $XNOR$ operator among 9 and 11 primary inputs respectively. The experimental setup is composed of three different synthesis flows, described as follows.

* **PXL**: the BBDD structure is obtained with an in-house C program. Once the benchmark's BBDD is optimized each node is replaced with the corresponding device and interconnections are established.

* **RG-MUX**: A TCL script manages the BDD structure (obtained with a C program that relies on the CUDD library) and replaces each node with an RG-MUX [8] device.

Table I.   BENCHMARKS CHARACTERISTICS

	Inputs	Outputs	PXL	RG-MUX	CMOS
xnor9	9	1	9	19	35
xnor11	11	1	11	23	51
9symml	9	1	26	35	78
parity	16	1	18	33	91
misex1	8	7	38	49	55
avg.			**20.4**	**31.8**	**62**

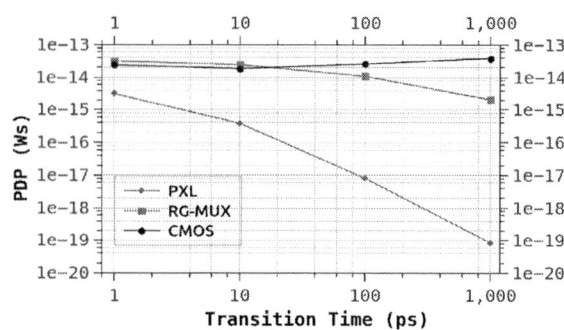

Figure 9.   Average Power Delay Product vs. Transition time

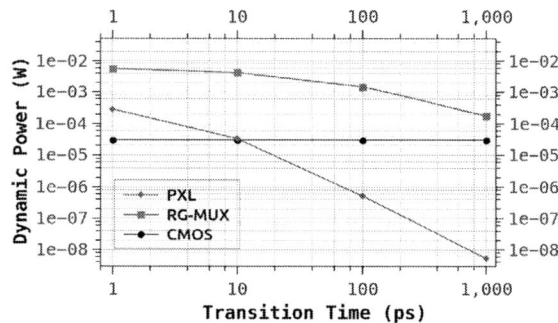

Figure 8.   Average Dynamic Power vs. Transition time.

* **CMOS**: classical standard-cell based design flow with Synopsys' Design Compiler. Synthesis is done using minimum-sized basic logic gates, i.e., AND, INV, OR, XNOR and BFX, belonging to an industrial 45nm CMOS technology library. Delay and power measurements are done in Synopsys HSPICE for PXL and RG-MUX circuits [13], [14], while we used timing and power reports of Synopsys' Design Compiler for CMOS implementation. Table I reports, for each benchmark, the total number of inputs and outputs (columns $Inputs$ and $Outputs$), and the number of devices after synthesis for the three implementation styles under analysis (columns PXL, $RG-MUX$, $CMOS$). Thanks to the higher expressive power of Pass-XNOR gates, PXL requires 36% less devices w.r.t. RG-MUX logic and 67% less devices w.r.t. the CMOS counterpart. Concerning power consumption, Figure 8 shows the average (over all the benchmarks) Dynamic Power consumption for different Transition Time (T_r) of the input signals. As can be seen from the plot, for $T_r < 10ps$, CMOS (circle marker) is the most efficient. For $T_r > 10ps$, namely, when the adiabatic charging principle comes into play, PXL (diamond marker) outperforms CMOS (breakeven at 10ps). In the best case, i.e., $T_r = 1ns$, PXL logic consumes 3.5 orders less power than CMOS, almost 4 orders less w.r.t. RG-MUX (square marker); the latter resulting the most power hungry implementation. To better quantify the benefits of PXL, we also report the average Power-Delay Product (PDP) as function of T_r, Figure 9. As one can observe, PXL circuits achieve very small PDP when compared to RG-MUX and CMOS, namely 4.2 and 5.5 orders less at $T_r = 1ns$.

VII.   CONCLUSIONS

This paper quantifies the potential of graphene p-n junctions for implementing logic circuits that operate according to the principle of adiabatic computation, i.e., the Adiabatic Pass-XNOR Logic (PXL) circuits. Results, obtained through SPICE level simulation of circuit netlists obtained with dedicated synthesis tools, show that PXL circuits can operate with about 3.5 orders of magnitude lower power than their CMOS

counterpart, overcoming a limitation of p-n junction based graphene circuits implemented in previous works (RG-MUX).

REFERENCES

[1] K. S. Novoselov, A. K. Geim, S. Morozov, D. Jiang, Y. Zhang, S. Dubonos, I. Grigorieva, and A. Firsov, "Electric field effect in atomically thin carbon films," *Science*, vol. 306, no. 5696, pp. 666–669, 2004.

[2] M. C. Lemme, T. Echtermeyer, M. Baus, and H. Kurz, "A graphene field-effect device," *Electron Device Letters, IEEE*, vol. 28, no. 4, pp. 282–284, April 2007.

[3] D. Basu, M. Gilbert, L. Register, S. K. Banerjee, and A. H. MacDonald, "Effect of edge roughness on electronic transport in graphene nanoribbon channel metal-oxide-semiconductor field-effect transistors," *Applied Physics Letters*, vol. 92, no. 4, p. 042114, 2008.

[4] S. Tanachutiwat, J. Ung Lee, W. Wang, and C. Y. Sung, "Reconfigurable multi-function logic based on graphene p-n junctions," in *Design Automation Conference (DAC), 2010 47th ACM/IEEE*.   IEEE, 2010, pp. 883–888.

[5] J. Koller and W. Athas, "Adiabatic switching, low energy computing, and the physics of storing and erasing information," in *Physics and Computation, 1992. PhysComp'92., Workshop on*.   IEEE, 1992, pp. 267–270.

[6] L. Amarú, P.-E. Gaillardon, and G. De Micheli, "Biconditional bdd: a novel canonical bdd for logic synthesis targeting xor-rich circuits," in *Proceedings of the Conference on Design, Automation and Test in Europe*.   EDA Consortium, 2013, pp. 1014–1017.

[7] V. V. Cheianov and V. I. Falko, "Selective transmission of dirac electrons and ballistic magnetoresistance of np junctions in graphene," *Physical review b*, vol. 74, no. 4, p. 041403, 2006.

[8] S. Miryala, A. Calimera, M. Poncino, and E. Macii, "Exploration of different implementation styles for graphene-based reconfigurable gates," in *IC Design & Technology (ICICDT), 2013 International Conference on*.   IEEE, 2013, pp. 21–24.

[9] S. Miryala, M. Montazeri, A. Calimera, E. Macii, and M. Poncino, "A verilog-a model for reconfigurable logic gates based on graphene pn-junctions," in *Proceedings of the Conference on Design, Automation and Test in Europe*.   EDA Consortium, 2013, pp. 877–880.

[10] V. Tenace, A. Calimera, E. Macii, and M. Poncino, "Pass-xnor logic: A new logic style for pn junction based graphene circuits," in *Design, Automation and Test in Europe Conference and Exhibition (DATE), 2014*.   IEEE, 2014, pp. 1–4.

[11] R. E. Bryant, "Symbolic boolean manipulation with ordered binary-decision diagrams," *ACM Computing Surveys (CSUR)*, vol. 24, no. 3, pp. 293–318, 1992.

[12] A. Bernasconi, V. Ciriani, G. Trucco, and T. Villa, "On decomposing boolean functions via extended cofactoring," in *Proceedings of the Conference on Design, Automation and Test in Europe*.   European Design and Automation Association, 2009, pp. 1464–1469.

[13] S. Miryala, A. Calimera, E. Macii, and M. Poncino, "Delay model for reconfigurable logic gates based on graphene pn-junctions," in *Proceedings of the 23rd ACM international conference on Great lakes symposium on VLSI*.   ACM, 2013, pp. 227–232.

[14] S. Miryala, A. Calimera, E. Macii, and M. Poncino, "Power modeling and characterization of graphene-based logic gates," in *Power and Timing Modeling, Optimization and Simulation (PATMOS), 2013 23rd International Workshop on*.   IEEE, 2013, pp. 223–226.

978-1-4799-4993-9/14 $31.00 © 2014 IEEE

3D Modeling of CNT Networks for Sensing Applications

Simone Colasanti, Vijay Deep Bhatt and Paolo Lugli
Institute of Nanoelectronics
Technical University Munich (TUM)
Munich, Germany
Email: simone.colasanti@nano.ei.tum.de

Abstract—A novel numerical 3D-model for simulations of random networks of carbon nanotubes is presented. This new algorithm takes into account the real 3D nature of these networks, allowing a better understanding of their electrical properties. CNTs are modeled as stiff cylinder with geometrical properties derived accordingly to different distributions. In this model CNT are allowed to bend in order to adapt in a realistic fashion to the surrounding environment. The electrical behavior of the network is simulated with a SPICE program. A temperature analysis has been performed, showing a good match between the simulations and the experiments. The present approach constitutes the basic building block for the simulation of a variety of sensors on rigid and flexible substrates.

Keywords—*Carbon Nanotubes, CNT Networks, 3D-Modeling, SPICE, Sensors.*

I. INTRODUCTION

In the last few years, nanoscale materials have been intensively studied due to their unique properties and potential applications. Among them, carbon nanotubes (CNTs) have received a lot of interest since they have been discovered in 1991 [1]. The realization of devices based on a single CNT has been proved to be very difficult, since it requires a fine control of the positioning as well as all the properties of the CNT. In contrast, random network of CNTs required easy and low cost processes, granting at the same time good performance in comparison to other organic devices [2][3]. Recently, solution-based film depositions have become a very interesting fabrication process. For these CNTs network the most attractive deposition methods are spin-coating or spray-coating technique [4]. In this case, CNTs are dispersed in a liquid solution where there is no a real control on the chirality of the single CNTs, leading thus to a mixture of different nanotubes with different properties. It is well known that a CNT can behave as a semiconductor or a metal, according to its chiral vector [5]. Statistically, if the nanotubes are not post-processed and sorted, a random network will contain 2/3 semiconducting and 1/3 of metallic CNT.

Lots of efforts have been dedicated to trying to model these complex networks, and so far only 2D models have been presented [6][7]. In such models the number of junctions between different CNTs is overestimated and the current flow can be far away from the real behavior.

In this paper we present a real full-3D model for randomly aligned CNTs network. All the characteristics of the CNTs, both physical and electrical, are extracted randomly according to some statistical distributions. Once the network is generated, the program converts the network in a netlist for a SPICE simulation. The network is composed of resistors, and their values are computed and stored by the program. Every quantity has a given temperature, in order to be able to simulate temperature sensors made with such films.

This paper is organized as follows. Section II introduces the implemented algorithm, giving an overview of the network generation. The simulations are presented in Section III, together with the discussion of the results. Finally, the conclusion of our study is drawn in the last section.

II. NETWORK MODELING

In our study every CNT is represented in a 3D space as a cylinder with a certain radius and a length that are randomly chosen accordingly to a particular distribution. This distribution is deduced from a CNT film fabricated by a spray deposition technique, showing a lognormal behavior with a mean of 744 nm and a variance of 95 nm [8]. The diameter is kept constant at a value of 1.2 nm. The first step is the position of the center of mass of the nanotube. With the information of the dimensions of the tube, and a randomly chosen orientation, the first nanotube is deposited on the substrate. Before the second nanotube will be deposited, the information of its nature, i.e. semiconducting or metallic, is stored in the same matrix that already contains the coordinates of the nanotube. The process is repeated for a number of CNTs that represent a certain density.

In our novel approach when a new nanotube finds another nanotube already deposited, the algorithm calculates first the point of intersection between the two tubes and then it bends the new nanotube according to some geometrical conditions. The process of bending is repeated every time the nanotube finds a new resting point on the network; this happens especially with high density films. In Fig. 1 the situation of two intersecting tubes is shown.

With this method the number of the junctions between CNTs is much closer to reality and the current distribution in the network is more precise as well. With a 2D model this effect is neglected and a new nanotube can intersect also with

978-1-4799-4993-9/14 $31.00 © 2014 IEEE

nanotubes which are not really in contact. This leads to an overestimation of the number of junctions that are actually the highest resistances in the network. The morphologies of two different films are represented in Fig. 2, showing that for higher density the film is actually thicker. The thicknesses obtained are in good agreement with those observed experimentally for these kinds of films.

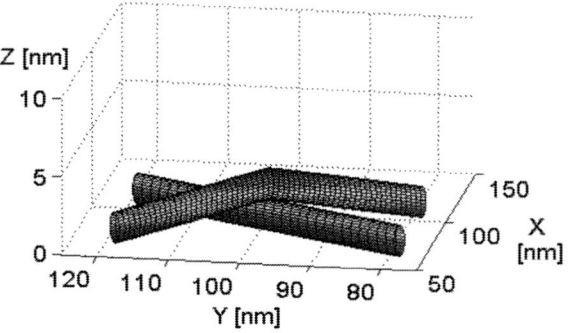

Fig. 1. Intersection between two tubes with consequently bending of the new nanotube.

Every CNT and every junction is modeled as a resistor, with electrical properties depending on its nature. The dependence on the temperature has been included in order to simulate sensors based on these films.

A. Metallic CNTs

Metallic CNTs are modeled according to

$$R_M = R_0 \left(1 + \frac{L}{\lambda}\right) [1 + \alpha (T - T_0)] \tag{1}$$

where R_0 represents the quantum resistance due to the sub-bands of the nanotube and is equal to 6.5 kΩ, L is the length of the tube segment, λ is the mean free path, α is a temperature coefficient and T_0 a temperature reference [9][10].

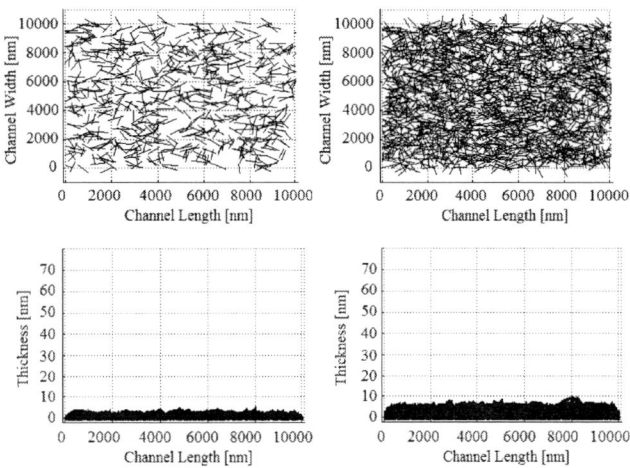

Fig. 2. Plots of two different morphologies. On the left side a low density network (10 CNT/μm²) both from top view and side view; on the right side same plots for a higher density network (25 CNT/μm²).

B. Semiconducting CNTs

For the semiconducting CNTs we used the following approach. In every semiconductor the resistance can be represented in this way

$$R_S = \frac{L}{\sigma \Sigma} \tag{2}$$

where L is the length of the tube segment, σ is the conductivity and Σ is the section of the tube. The conductivity is calculated from

$$\sigma = q \left(n \, \mu_n + p \, \mu_p\right) \tag{3}$$

The values for the carrier concentrations are obtained from

$$n = N_C \exp\left(-\frac{E_C - E_F}{kT}\right) \tag{4a}$$

$$p = N_V \exp\left(-\frac{E_F - E_V}{kT}\right) \tag{4b}$$

The expressions for the densities of states are

$$N_{C,V} = \left(\frac{2 \pi m_{n,p}^* k T}{\hbar^2}\right)^{3/2} \tag{5}$$

The values for mobilities and effective masses are taken from the literature [11]. Since these values vary with the chirality, they are randomly chosen according to a certain distribution in order to reproduce the differences in chirality between the nanotubes. The only unknown variable is the Fermi level energy. The problem arises because the CNTs films obtained experimentally using printing techniques are not intrinsic and show a very strong p-type behavior. This is due to different unintentionally doping mechanisms during the growth of the single nanotube and during the device fabrication process. Furthermore there are several interactions with the solvents in the solution, with the substrate and with the environment. The result is an uncertainty on the position of the Fermi level. To represent that, the Fermi level of a single nanotube is chosen according to a log-normal distribution centered in the lower part of the energy gap, in order to reproduce the p-type behavior. In such a way is it possible to obtain reasonable values of the semiconductors resistances. With this method will be also possible in the future to study the effect of gas molecules on the electrical properties of the nanotubes.

C. CNTs Junctions

For the CNTs junctions we choose values from the literature [12]. It is worth to notice that while for metal-metal junctions and semiconducting-semiconducting junctions the behavior can be well represented by a resistance, for the metal-semiconducting junction this is not true anymore. Some experimental studies showed that the real I-V characteristic of these junctions is basically the one of a Schottky diode [12]. In this work we are still assuming a resistive behavior, with a constant value around 10 MΩ, while for metal-metal junctions and semiconducting-semiconducting junctions we use values respectively around 250 kΩ and 500 kΩ. The dependence respect to the temperature is assumed to be significant only for the semiconducting-metal junctions, and it follows

$$R_{MS}(T) = R_{MS0} \left(\frac{T_0}{T} \right)^2$$

(6)

where R_{MS0} is the nominal value and T_0 is a temperature reference. We are currently working on the implementation of the Schottky diode model for the metal-semiconducting junctions, since there are a few limitations with our approximation. With the resistive approach we are neglecting the rectifier behavior of the Schottky barrier and also the temperature dependence of the junction might be improved with a more complex model.

III. SIMULATIONS

After generating the complete morphology a netlist has to be created. The program assigns a value of a resistor between every couple of nodes of the generated network by checking a table that contains all the information about the nodes properties and their connections. Afterwards, a SPICE program reads the netlist, simulates the network and gives back the data to the program for post processing.

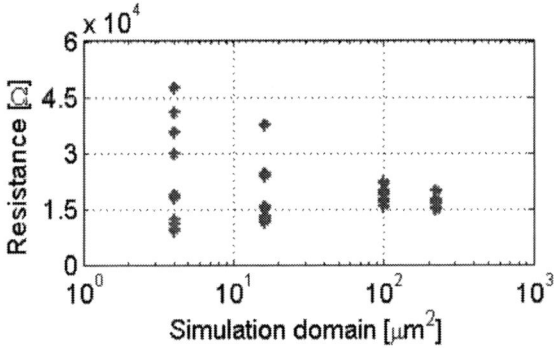

Fig. 3. Resistance versus simulation domains; all these networks have same nominal characteristics in term of density (10 CNT/μm²), ratio between semiconducting and metallic nanotubes (2:1) and so on. For simulation domains greater than 100 μm², the variance is small enough.

There are some computational limitations due to the dimension of the matrix that contains all the nodes and how they are connected. In this 3D model, two CNTs touching each other require already a 6x6 matrix; if we consider simulation cases with ten thousand or more CNTs, it is easy to imagine how that can be become computationally untreatable.

Furthermore, sufficiently large simulation domains are needed, in order to avoid "border effects" and consequently big variance of the results. This is an unavoidable problem that does not depend on the simulation itself but it lies in the nature of this type of stochastic problem. Also a very precise deposition process, if performed on a small area, will give every time a different device with different morphologies. This problem is shown in Fig. 3, where the variance of the simulation output from 10 different networks with the same nominal parameter can be clearly seen. As soon as the simulation domain reaches a sort of threshold, the variance becomes very small. It is important to notice that this

threshold depends on the density of CNTs of the film; it seems reasonable to think that this threshold of the simulation domain is strictly related to the percolation threshold of the network.

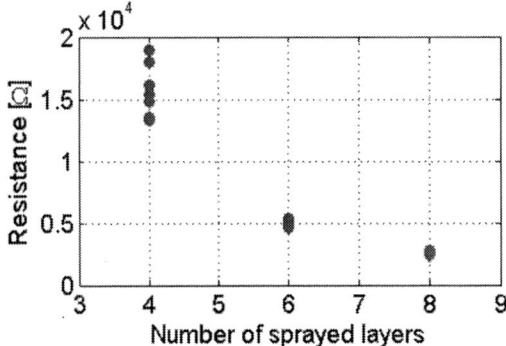

Fig. 4. Resistance versus density of CNTs, here represented as number of sprayed layers in analogy to the spray-coating setup used. By increasing the density of the film, the resistance decrease as well as the variance of the simulation outputs.

This can be seen also in Fig. 4, where we have simulated networks with different density of CNTs and with the simulation domain kept constant. The first effect is that the resistance of the network is reduced, since by increasing the density we are actually increasing the number of percolation paths. The second effect however is related to the simulation domain threshold, and in fact more dense networks show less variance of the results.

Because of the noise in the results, is important to simulate several time different morphologies and obtain sufficient statistics from the results to be able to predict the behavior of a particular network. For the simulation of the temperature sensors we followed this approach and we compare the results of our simulations with five different sensors with the same nominal properties.

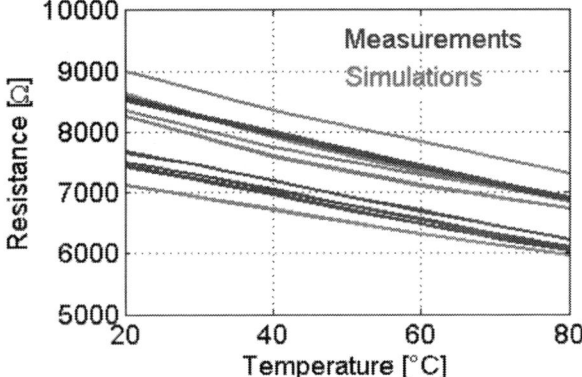

Fig. 5. Simulation of CNT based temperature sensors (red curves) compared to the experimentally measured characteristics (blue curves) of 5 different sensors with the same nominal characteristics.

The sensors consists of a thin film of CNTs with the semiconducting-to-metallic ratio 2:1. The density of the film is around 25 CNT/μm^2, deposited on a SiO_2 substrate and electrically connected via an interdigitated electrode structure. The simulation domain has been chosen equal to 10x10 μm^2, since it is a good trade-off between low variance of the results and low computational time consumption. Furthermore, every simulated curve is the result of the average between 20 simulations. The results of our simulations are presented in Fig. 5, showing a good fit with the experimentally measured values. The resistance of the network decreases with the increase of the temperature, showing thus a semiconducting behavior as expected. The resistance-temperature coefficient (RTC) of the sample can be calculated by

$$RTC = \frac{R_{change}}{\Delta T} = \frac{R - R_0}{R_0 \Delta T} \cdot 100 \qquad (7)$$

and is equal approximately to -0.3 %°C^{-1}.

Because of the spreading of the measured results it is still quite difficult to evaluate the agreement between the experiments and the simulations in order to validate the model with a better accuracy.

IV. CONCLUSION

In this work we presented a new 3D algorithm for the modeling of random CNTs networks. This new algorithm allowed us to get morphologies much closer to the reality respect to any 2D model. Our program chooses and assigns parameters and properties to the network in a stochastic way, according to some distributions found in literature. The netlist generated is then used in a SPICE program that simulates the electrical transport of the network. The model has been validated through a temperature analysis, by comparing the simulations with the results of the experiment for different devices with the same nominal characteristics. Since the process of deposition itself is a stochastic phenomenon, different devices lead to different electrical properties. This behavior has been replicate with good approximation by our simulations. The simulation framework we have presented here is the first step of a project that will provide us with a tool to design sensors on different substrates. In fact, although in the present study, the substrate is treated as rigid, a bendable one can be implemented. In addition to the temperature dependence of the film, the influence of pressure, gases and biomolecules will be calculated in the future.

ACKNOWLEDGMENT

This work was partly supported by the European Commission under grant agreement PITN-GA-2012-317488-CONTEST and the TUM Graduate School.

REFERENCES

[1] S. Iijima, "Helical Microtubules of Graphitic Carbon", Nature 354, 56 London, 1991.

[2] E. S. Snow, J. P. Novak, P. M. Campbell and D. Park, "Random networks of carbon nanotubes as an electronic material", Applied Physics Letters, 82(13), 2145, 2003.

[3] N. Rouhi, D. Jain, P. J. Burke, ACS Nano Article ASAP, "High-performance semiconducting nanotube inks: progress and prospects", 2011 American Chemical Society.

[4] A. Abdellah, A. Yaqub, C. Ferrari, B. Fabel, P. Lugli, G. Scarpa, "Spray Deposition of Highly Uniform CNT Films and Their Application in Gas Sensing", IEEE Nano 2011, pp. 1118-1123, Aug. 2010.

[5] L. Hu, D. S. Hecht and G. Grüner, "Carbon Nanotube Thin Films: Fabrication, Properties and Applications", Chem. Rev., 110, pp. 5790-5844, 2010.

[6] L. P. Simoneau, J. Villeneuve, C. M. Aguirre, R. Martel, P. Desjardins and A. Rochefort, "Influence of statistical distributions on the electrical properties od disordered and aligned carbon nanotube networks", Journal of Applied Physics, 114, 114312, 2013.

[7] G. Sassine, F. Martinez, F. Pascal and A. Hoffmann, "Numerical Simulation of Non-Homogeneous 2D-CNT Structures", 2013 22nd IEEE International Conference on Noise and Fluctuations (IEEE-ICNF), 1-4, Jun. 2013.

[8] E. Albert, A. Abdellah, G. Scarpa and P. Lugli, "Electronic transport modeling with HSPICE in random CNT networks", 2012 12th IEEE International Conference on Nanotechnology (IEEE-NANO), 1–4, 2012.

[9] D. A. Jack, C-S. Yeh, Z. Liang, S. Li, J. G. Park, J. C. Fielding, "Electrical conductivity modeling and experimental study of densely packed SWCNT networks", Nanotechnology, 2010.

[10] S. Kumar, J. Y. Murthy, M. A. Alam, "Percolating Conduction in Finite Nanotube Networks", Phys. Rev. Lett., Aug. 2005.

[11] B. Xu, J. Yin and Z. Liu, "Phonon Scattering and Electron Transport in Single Wall Carbon Nanotube", Physical and Chemical Properties of Carbon Nanotubes, Dr. Satoru Suzuki Ed., 2013.

[12] M. S. Fuhrer, J. Nygård, L. Shih, M. Forero, Young-Gui Yoon, M. S. C. Mazzoni, Hyoung Joon Choi, J. Ihm, S. G. Louie, A. Zettl, P. L. McEuen, "Crossed Nanotube Junctions", Science, vol. 288, 5465, pp. 494-497, 2000.

A Quantitative Approach to Testing in Quantum dot Cellular Automata: NanoMagnet Logic Case

Giovanna Turvani, Fabrizio Riente, Mariagrazia Graziano and Maurizio Zamboni
Electronics and Telecommunications Department, Politecnico di Torino, Italy

Abstract—**With the approaching of CMOS scaling limits the interest on emerging technologies is rapidly growing. Among emerging technologies, Quantum dot Cellular Automata (QCA) is one of the most studied. Particularly the magnetic implementation, NanoMagnet Logic (NML), offers very low power consumption and it combines logic and memory on a unique device. Despite the advantages of these technologies, QCA and NML working principle relies on the electric or magnetic interaction among neighbor cells, so it is very sensitive to process variations. The behavior of circuits is therefore largely affected by defects and fabrication variations.**

To effectively design circuits with these technologies, proper tools for testing circuits are necessary. In this work we present an innovative test environment for NML technology. The test algorithm is integrated in ToPoliNano, our design and simulation tool for emerging technologies, and it is specifically tailored to support the analysis of faults in large complexity circuits. Thanks to this tool it is possible to design and test complex NML circuits considering the effect of process variations in terms of Yield and Output Error Rate. The approach gives then feedback to the technologists, remarkably helping the future development of this technology. Moreover, notwithstanding the methodology is applied here to NML circuits only, it can also be successfully applied to QCA technology in general, greatly enhancing the value of the work we proposed here.

I. INTRODUCTION

According to the ITRS Roadmap [1], CMOS transistors scaling is reaching its unavoidable physical limits and therefore alternative technologies must be developed. Among these emerging nanotechnologies Quantum dot Cellular Automata (QCA) [2] is one of the most studied. To represent logic values in QCA, different charge configurations stored on identical cells [3] are used instead of voltage levels as it happens in CMOS. Currently, two are the types of QCA implementations that attract most the attention of researchers: Magnetic QCA or NanoMagnet Logic (NML) [4] and Molecular QCA [5] [6] (MQCA). NML technology is interesting for its very low power consumption [7] and the intrinsic memory ability, while molecular QCA provides very high clock frequencies and extremely reduced feature sizes [8].

In our work we focus mainly on NML technology, because it is the only QCA implementation which has an extensive experimental validation [10] [11] [7], though attention to faults in MQCA implementation has been recently assured as well. The digital information is represented using single domain magnets with a rectangular shape; typical dimensions of the nanomagnets were approximately 90 (l) 60 (w) 20 (th) nm^3. If magnets size is sufficiently small they can assume only two possible stable states: "0" and "1" (Fig. 1 (b)) [12] [13]. Signal propagation is obtained placing magnets one nearby the other, exploiting magnetic interaction among neighbor cells.

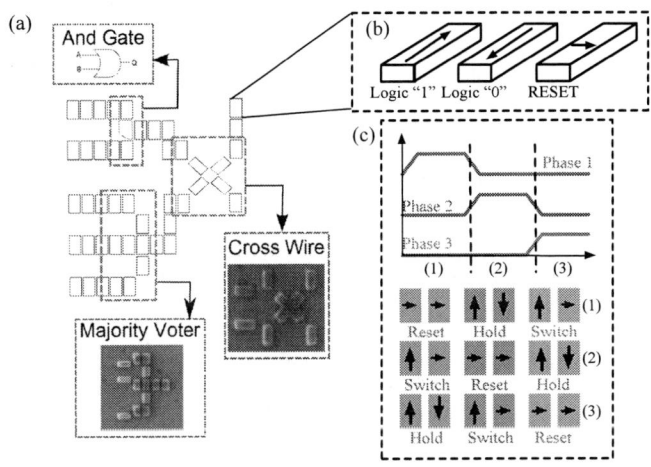

Figure 1: (a) Example of NML circuit layout. The main logic gates are highlighted, a AND, a majority voter [9] and a crosswire, a particular block that allows to cross two wires on the same plane. (b) Nanomagnets logic values. (c) Multiphase clocking scheme used with NML.

Since the magnetic field generated by a single magnet in not sufficient to alter the state of its neighbor, a clocking mechanism must be used [14]. This field forces magnets in an unstable state ("RESET" case, see Fig. 1 (b)). When the magnetic field is removed, magnets realign themselves following the input element in an anti-ferromagnetic sequence. Due to the influence of thermal noise the maximum number of magnets that can be cascaded without incurring in errors during the signals propagation is limited [15]. In order to design real magnetic circuits a multiphase clock system is therefore necessary, as, for example, we proposed in [16]. Circuits are divided in areas (clock zones) made by a limited number of magnets. At every clock zone a different clock signal is applied. The behavior is shown in Fig. 1 (c), where three clock signal are applied to the circuit. Thanks to this mechanism when magnets of a clock zone are switching (SWITCH in Fig. 1 (c)), magnets on the left clock zone are in the HOLD state and act as inputs while magnets on the right clock zone are in the RESET state and have no influence on switching magnets. The sequence of RESET, SWITCH and HOLD phases guarantees a correct signal propagation as can be understood from Fig. 1 (c) observing the temporal sequence (1), (2) and (3). The clock frequency is typically in the order of 100 MHz. Logic computation is obtained using elementary gates, such as majority voters and cross-wires, depicted in Fig. 1 (a), where an example of gates we fabricated is shown.

Since NML technology is still in the research phase nano-

978-1-4799-4993-9/14 $31.00 © 2014 IEEE

magnets are normally fabricated using electron beam lithography (EBL), but it is possible to achieve the required resolution also with ultra violet optical lithography [17]. The magnetic material is deposited on the substrate through sputtering or evaporation, the magnetic layer is patterned with lithography and then the geometry is obtained through the selective removal of material with an etching process. Regardless of the lithographic technique used, the fabrication process will introduce some variations that may affect the functionality of the final circuit. Possible causes of defects can be: 1) The presence of defects on the substrate, 2) micro-movements of the substrate during the lithography phase, 3) over-exposure or under-exposure in the etching phase or 4) electrons scattering that can cause exposure of the resist outside the desired region. All these variations can alter the relative position of magnets and therefore the magnetic interaction among them, causing possible malfunctioning.

In order to reliably study and evaluate the competitiveness of this technology, it is necessary to design and to examine circuits of a reasonable complexity keeping into account the effects of process variations. The effect of process variations can be evaluated using low level physical simulators, as shown in [9] and [18]. However only small circuits can be analyzed with such simulators due to extreme requirement in terms of computational power. We developed a new algorithm to analyze the faults derived by process variations in complex circuits and we have integrated this algorithm in our design and simulation tool for emerging nanotechnologies, ToPoliNano [19] [20]. ToPoliNano is a CAD tool, which allows working with different emerging technologies, including NML. The whole software is developed in C++ and it can be used under different platforms such as Linux, Mac OS X and Windows. It has a rich graphical interface, which helps users throughout the design process. The simulation algorithm is based on the results of physical level simulations we already executed and discussed in [18]. It allows the estimation of the effects of faults generated by displacements in the relative magnets position. Both output-error-rate and yield of circuits implemented and simulated in presence of faults are the outcome of the simulations.

Our contribution allows to understand the reliability of NML and to give feedbacks to the technologists for in terms of both qualitative and quantitative directions toward which it is preferable to improve the fabrication process. The work here proposed, then, remarkably helps the future development of this emerging technology.

II. NML FAULT ANALYSIS

Since magnetic interaction strongly depends on the distance among neighbor magnets and on their sizes, it is important to better underline how the process variations affect the logical behavior of circuits. The model of interaction between magnet has been extracted from a finite element simulation; the approach we adopted is as follows: 1) starting from an already synthesized, placed an routed NML circuits based on nanomagnets we associate to magnets position a certain variability as discussed in the following and according to information obtained by the technological processes; 2) we run simulations using the algorithm implemented in ToPoliNano for NML enriched with the capability to take into account

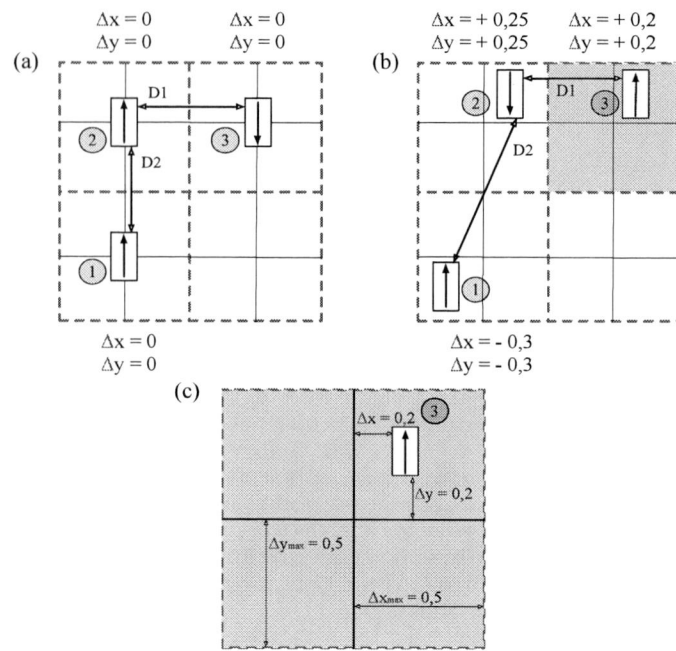

Figure 2: (a) Ideal behavior of a sample circuit, magnets are placed evenly across the plane with regular spacing. (b) Example of circuit affected by process variations; near each quadrant are specified the Δx and Δy shifts. In this non-ideal case the signal is not propagated correctly starting from the second magnet. (c) Zoom of a quadrant in which are marked the maximum allowed shifting values

the new positions and introducing proper criteria and decision mechanisms to define whether the information is correctly propagated or not; 3) we compare the simulation results to defect-free simulation outputs and detect the presence of possible errors. In the following we describe in details this method.

In our design approach circuits are seen as matrix in which each node is occupied by a single magnet, in order to obtain very regular structures. We want to evaluate the effect of non-uniform distances among magnets, so we introduced in ToPoliNano the possibility to inject irregularities in the final circuit layout. Fig. 2 (a) shows an example of an ideal NML circuit, where each magnet is evenly spaced among its neighbors. Horizontal and vertical distances are equal to 20 nm according to the technological reference we have [9]. In this ideal case the magnetization of each magnet evaluated with ToPoliNano corresponds to the exact physical behavior.

Fig. 2 (b) and (c) show part of a circuit matrix where magnets are shifted from their original position. Details of the model are represented in Fig. 2 (c) in which the Δx and Δy shifts from the central node are highlighted. The shift values Δx and Δy represent the relative displacements respect to the reference node. The maximum possible shift is the absolute value of 0.5 (both in vertical and in horizontal), this means that each magnet can moves with a displacement of $\pm$ 50% w.r.t. the reference coordinate. As an example in the case of Fig. 2 (c) magnets can move within the horizontal coordinates $1, 5 < x < 2, 5$ and the vertical coordinates $0, 5 < y < 1, 5$.

978-1-4799-4993-9/14 $31.00 © 2014 IEEE   318

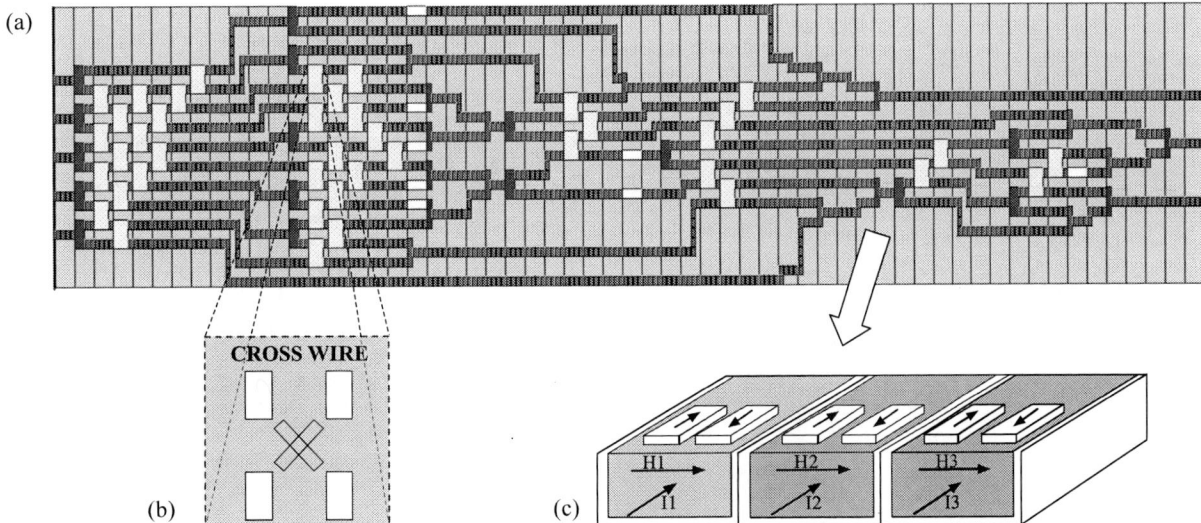

Figure 3: (a) Layout of a 2 bits Ripple Carry Adder in which are highlighted clock zones. (b) Enlargement of a sample block: the CrossWire component is largely employed in NML technology and it is used in order to realize interconnections. (c) Example of clock zone structures: magnets are placed on a wire on propagates information horizontally.

Fig. 2 (b) shows different possible combinations of Δx and Δy. For each quadrant containing magnets the relative displacements are reported. Values of $D1$ and $D2$ are different with respect to the previous ideal case (Fig. 2 (a)), in particular $D2$ proves to be higher than the maximum distance that permits correct signal propagation.By performing several micromagnetic simulations in [9] [18] we obtained the threshold distance values that guarantee a correct magnetization both horizontally and vertically. Our tool uses this parameters to establish the logic state of each device; thus, in case $D2$ is greater than the threshold value, then the ferromagnetic interaction between magnet 1 and magnet 2 is not correct and an error is generated.

At the time of writing, the two values defining a magnet displacement are loaded by an external file which contains a pair of Δx and Δy randomly generated for each node of the circuit. In an extension of this work we plan to be able to include displacements maps derived by real physical implementations, even though, from a methodological point of view nothing would be changed. Once the position is updated for all the magnets in the circuit, ToPoliNano calculates the magnetization of each magnet according to its simulation algorithm (referring to Fig. 2): I) The logic state of magnet 1 is calculated II) before evaluating the state of the second magnet, D2 must be calculated: if this value is greater than the fixed threshold the final state will be the same of the previous ideal case. Otherwise, if the distance is too high, we fixed a percentage of probability that the information will not be propagated correctly. In this case it is possible to notice from Fig. 2 (b) that vertical ferromagnetic interaction between the first two magnets is not respected. III) The same approach is repeated in order to evaluate the magnetization of the third cell, in this case, since D1 is lower than the fixed threshold the information is propagated correctly, so the antiferromagnetic horizontal interaction happens correctly.

Since displacement values are loaded from an external file,

different distributions of distances and their effects on the circuit behavior can be analyzed: thus real data from defects due to fabrication steps associated to specific zones of the circuit can be imported in the analysis.

III. Results

The layout shown in Fig. 3 (a) depicts a two bits Ripple Carry Adder (RCA 2) automatically generated by ToPoliNano, starting from a VHDL description (for a detailed explanation on how this result is achieved refer to [19] [20]). Generally, our approach can be applied to circuits of any complexity, in this case we have chosen to present a RCA since from an architectural point of view, the circuit is implemented with few basic blocks. As an example Fig. 3 (b) shows a zoom of the crosswire [11], a particular block that allows to cross two wires on the same plane without interferences. Fig. 3 (c) shows the working principle of the clock mechanism: magnets are placed over a wire where a current run through generating the magnetic field. This layout is chosen accordingly to the theoretical and technological constraints related to the fabrication process, which was demonstrated experimentally in [10]. The RCA2 circuit was therefore simulated considering firstly the ideal magnets positions and then taking into account also faults derived from process variations. The obtained waveform in the ideal case is depicted in Fig. 4 (b). For what concerning the simulation duration, ToPoliNano employs 1640 ms in order to perform the ideal simulation, and 3251 ms to complete the simulation which takes into account faults. The adder was then tested applying the fault analysis algorithm and changing the magnets positions. As an example, Fig. 4 (c) shows that, for example, with an input configuration of A = 11, B = 00, Input Carry = 0 the resulting output is incorrect (S = 10 and Output Carry = 1). The circuit has been tested varying Δx and Δy within different ranges according to Table I. For each combination 1000 iterations were performed exploiting a MonteCarlo-like approach. These results are used to obtain the

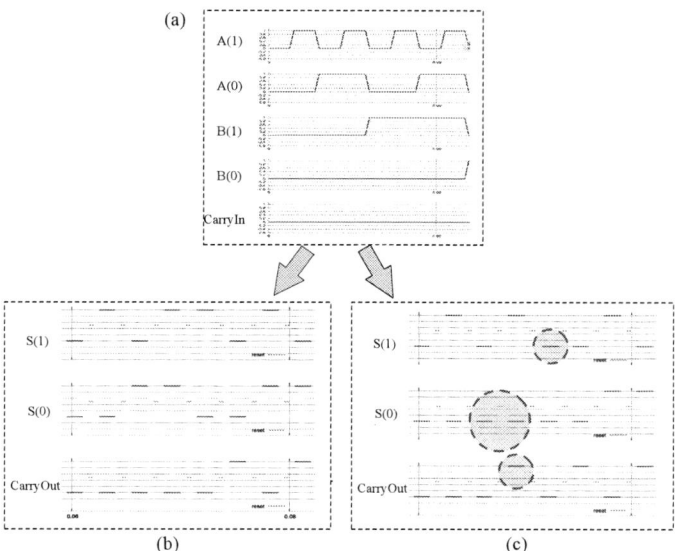

Figure 4: (a) Input signals for the RCA2 circuit. (b) Output signals obtained performing the ideal logic simulation. (c) Output signals obtained performing the logical simulation which takes into account faults derived from process variations.

Δx	Δy	$OER_{avg_{1000}}$	Yield
± 0.12	0.12	0.8728	0
± 0.1	0.1	0.1364	0.604
± 0.07	0.07	0	1
± 0.05	0.05	0	1
± 0.1	0	0	1
± 0	0.1	0	1
± 0.2	0	0.8703	0
± 0	0.2	0 .8625	0

Table I: Simulation results of Output Error Rate and Yield considering different nanomagnets shift

mean value of Output Error Rate $OER_{Ave_{1000}}$ and the number of fully working circuits (Yield). As an overall consideration, we notice that, with shift values smaller than 0.1, the mean yield is 1. This means that the lithographic process permits to tolerate variations in the magnets position around 10%.

IV. CONCLUSIONS

We studied how process variations due to non-uniform spacing between magnets may introduce faults which affect the logical behavior of NML based circuits. We also demonstrate that this analysis can be performed even on complex circuits exploiting our algorithms integrated in the ToPoliNano CAD tool. This work represents a remarkably innovative approach in the study of emerging nanotechnologies since it allows a systematic design and analysis similar to what can be obtained with CMOS technology.

As future works we are extending the analysis to more complex distribution of faults. We will also introduce in ToPoliNano a physical level simulation algorithm based on a simplified version of the LLG equation which rules the micromagnetic dynamics. Thanks to this new algorithm it will be possible to obtain more accurate results without the need

to use external computationally intensive low level simulations to evaluate the error probability.

REFERENCES

[1] "International Technology Roadmap of Semiconductors," 2012, http://public.itrs.net.

[2] C. Lent, P. Tougaw, W. Porod, and G. Bernstein, "Quantum cellular automata," *Nanotechnology*, vol. 4, pp. 49–57, 1993.

[3] A. Csurgay, W. Porod, and C. Lent, "Signal processing with near-neighborcoupled time-varying quantum-dot arrays," *IEEE Transaction On Circuits and Systems*, vol. 47, no. 8, pp. 1212–1223, 2000.

[4] W. Porod, "Magnetic Logic Devices Based on Field-Coupled Nano-magnets," *Nano & Giga*, 2007.

[5] C. Lent and B. Isaksen, "Clocked Molecular Quantum-Dot Cellular Automata," *IEEE Transactions on Electron Devices*, vol. 50, no. 9, pp. 1890–1896, Sep. 2003.

[6] A. Pulimeno, M. Graziano, D. Demarchi, and G. Piccinini, "Towards a molecular QCA wire: simulation of write-in and read-out systems," *Solid State Electronics*, vol. 77, pp. 101–107, 2012.

[7] M. Vacca, M. Graziano, A. Chiolerio, A. Lamberti, M. Laurenti, D. Balma, E. Enrico, F. Celegato, P. Tiberto, and M. Zamboni, "Electric clock for NanoMagnet Logic Circuits ," *Anderson, N.G., Bhanja, S. (eds.), Field-Coupled Nanocomputing. LNCS. Springer, Heidelberg*, vol. vol. 8280, 2014 (forthcoming).

[8] M. L. C. L. Y. Lu, "Molecular electronics - from structure to circuit dynamics," in *Sixth IEEE Conference on Nanotechnology*. Cincinnati-Ohio, USA: IEEE, 2006, pp. 62–65.

[9] M. Vacca, D. Vighetti, M. Mascarino, L. Amaru, M. Graziano, and M. Zamboni, "Magnetic QCA Majority Voter Feasibility Analysis ," *2011 7th Conference on Ph.D. Research in Microelectronics and Electronics (PRIME)*, pp. 229–232, 2011.

[10] M. Alam, M. Siddiq, G. Bernstein, M. Niemier, W. Porod, and X. Hu, "On-chip Clocking for Nanomagnet Logic Devices," *IEEE Transaction on Nanotechnology*, 2009.

[11] M. Niemier and al., "Nanomagnet logic: progress toward system-level integration," *J. Phys.: Condens. Matter*, vol. 23, p. 34, Nov. 2011.

[12] A. Imre, L. Ji, G. Csaba, A.O. Orlov, G. Bernstein, and W. Porod, "Magnetic Logic Devices Based on Field-Coupled Nanomagnets," *2005 International Semiconductor Device Research Symposium*, p. 25, December 2005.

[13] R. Cowburn and M. Welland, "Room temperature magnetic quantum cellular automata," *Science*, vol. 287, pp. 1466–1468, 2000.

[14] M. Graziano, A. Chiolerio, and M. Zamboni, "A Technology Aware Magnetic QCA NCL-HDL Architecture." Genova, Italy: IEEE, 2009, pp. 763–766.

[15] G. Csaba and W. Porod, "Behavior of Nanomagnet Logic in the Presence of Thermal Noise," in *International Workshop on Computational Electronics*. Pisa, Italy: IEEE, 2010, pp. 1–4.

[16] M. Graziano, M. Vacca, A. Chiolerio, and M. Zamboni, "A NCL-HDL Snake-Clock Based Magnetic QCA Architecture," *IEEE Transaction on Nanotechnology*, vol. 10, no. 5, pp. 1141–1149, Sep. 2011.

[17] D. Bisero, P. Cremon, M. Madami, S. Tacchi, G. Gubbiotti, G. Carlotti, and A. Adeyeye, "Nucleation and Propagation of Vortex States in Dense Chains of Regular Particles," *Magnet2011, 2nd Italian conference on magnetism*, february 2011.

[18] M. Vacca, M. Graziano, and M. Zamboni, "Majority Voter Full Characterization for Nanomagnet Logic Circuits," *IEEE T. on Nanotechnology*, vol. 11, no. 5, pp. 940–947, Sep. 2012.

[19] S. Frache, D. Chiabrando, M. Graziano, F. Riente, G. Turvani, and M. Zamboni, "ToPoliNano: Nanoarchitectures Design Made Real," *IEEE International Symposium on Nanoscale Architectures (NANOARCH)*, pp. 160–167, 2012.

[20] M. Vacca, S. Frache, M. Graziano, F. Riente, G. Turvani, M. R. Roch, and M. Zamboni, "ToPoliNano: NanoMagnet Logic Circuits Design and Simulation," *Anderson, N.G., Bhanja, S. (eds.), Field-Coupled Nanocomputing. LNCS. Springer, Heidelberg*, vol. vol. 8280, 2014 (forthcoming).

978-1-4799-4993-9/14 $31.00 © 2014 IEEE

Compact Model for Phase Change Memory Cells

Erika Covi*[†], Athanasios Kiouseloglou*[‡], Alessandro Cabrini*, Guido Torelli*

*Dipartimento di Ingegneria Industriale e dell'Informazione, University of Pavia, Via Ferrata 1, 27100 Pavia, Italy
[†]present address Laboratorio MDM, IMM - CNR, Via C. Olivetti 2, 20864 Agrate Brianza (MB), Italy
[‡]CEA-LETI, MINATEC Campus, 17 rue des Martyrs, F-38054 GRENOBLE Cedex 9, France

Abstract—In this paper, a compact model for PCM cells with a physical approach, which evaluates the state of the cell during and after a programming operation, is described. The model is able to simulate the state of the active material of the cell during the whole programming operation. It takes into account the dynamic of the amorphous cap growth during a RESET operation, as well as the crystallization process, expressed in terms of Crystal Fraction (CF) during a SET operation. The model was validated through comparison with experimental data.

I. INTRODUCTION

In the semiconductor market, memories are becoming more and more important thanks to the increasing number of portable devices that need a memory chip (cameras, music players, smartphones, laptops, tablets, etc.) and to the introduction of concepts as Cloud computing. A key challenge in this scenario is developing new memory technologies which keep on enhancing their performance while decreasing their energy consumption. In this respect, new kinds of non-volatile memories are being studied. Among them, Phase Change Memories (PCMs) turned out to be the most promising technology, which may be able to replace Flash technology [1].

The PCM storage element (Fig. 1) is a chalcogenide alloy, typically $Ge_2Sb_2Te_5$ (GST), which presents two states: amorphous, a disordered state which features high resistivity (RESET), and (poly-)crystalline, an ordered lattice featuring low resistivity (SET). The transition between the two states is fast, reversible, and thermally induced.

Having a reliable model of the behaviour of PCM cells during programming is very important to facilitate the design work, also aiming at optimizing programming algorithms, especially when multilevel programming is concerned [2]. Although other models have been presented in literature [3], [4], [5], the proposed model aims at combining the advantages of a compact model [6], and of a physical approach [7], to determine the evolution of the amorphous and crystalline phases during both a RESET (transition from a low resistance state to a high resistance state) and a SET (transition from a high resistance state to a low resistance state) operation. The model is sufficiently simple to be easily implemented in a circuit simulator and it can thus be considered a compact model. In addition, it makes use of equations which describe the physical evolution of GST during programming operations, and it is therefore based on a physical approach.

II. COMPACT MODEL IMPLEMENTATION

The basic idea of the proposed model is shown in the flow-chart in Fig. 2. When a pulse is applied to the cell, a

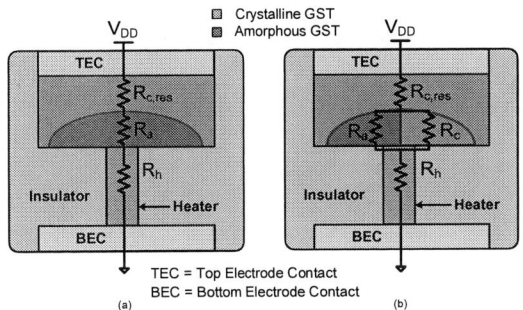

Fig. 1. Equivalent resistance of GST after a RESET (a) or a partial-SET (b) operation. The heater is substantially a resistor that acts as a heating element (exploiting Joule effect) during programming. The TEC and the BEC allow the cell to be contacted by external metal lines.

temperature increase takes place at the heater-GST interface, due to Joule effect. This temperature determines whether a RESET operation (high temperature increase), a SET operation (intermediate temperature increase) or a reading operation (negligible temperature increase) is being performed: the cell resistance is derived from the obtained temperature. To compare model results with measurements, the reading current, which is the current flowing through the cell when a low-voltage reading pulse is applied and which gives information about the state of the cell, is derived. The model continuously updates the cell resistance and the reading current and hence the state of the cell, during the whole simulation.

During the RESET operation, the GST starts to amorphize. Since not all the GST volume can amorphize, the portion which may be affected by the phase change is referred to as active region. At the end of the RESET operation, the GST resistance is modelled as the series of the amorphous cap resistance, R_a, which is dependent on the thickness of the amorphous cap, and the resistance of the residual crystalline portion, $R_{c,res}$ (Fig. 1(a)).

In contrast, during the SET operation, the amorphous cap starts crystallizing, thus increasing the GST Crystal Fraction (CF), which is the volumetric percentage of the crystallized material with respect to the total amorphous volume at the beginning of the crystallization process. In this operation, a phenomenon typical of amorphous chalcogenide materials occurs, i.e. electronic switching [8], which is an abrupt drop of the resistance of the memory cell when the applied voltage exceeds a voltage named threshold voltage and results in an

978-1-4799-4993-9/14 $31.00 © 2014 IEEE

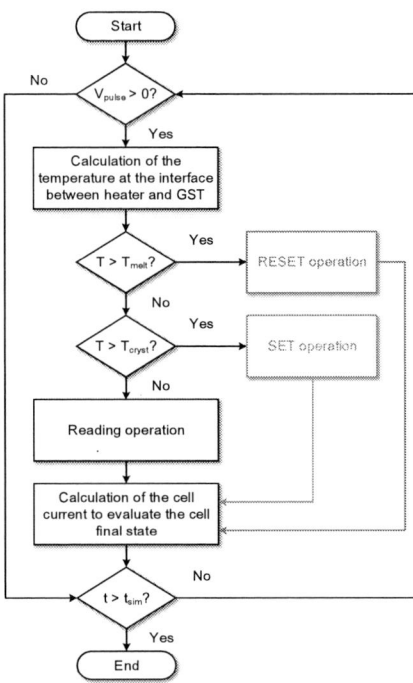

Fig. 2. Flow-chart describing the basic idea of the model.

S-shape of the I-V characteristic of the cell (the resistance of the GST becomes negligible). This resistance drop allows providing the power required to achieve the state transition. At the end of the SET operation, the GST resistance is evaluated considering that a portion of the amorphous cap can still be present. The resistance of the GST can be modelled as the series of the resistance of the residual crystalline portion and the parallel connection of the resistance of the residual amorphous cap and the resistance of the re-crystallized portion, R_c (Fig. 1(b)) [9].

The cell behaviour is described in the model by mainly using the Johnson-Mehl-Avrami-Kolmogorov (JMAK) theory [10] (SET operation) and the amorphization equation (RESET operation) [7]. The implemented model is a dynamic system conceived with the purpose of simulating the evolution of the GST state during the whole programming operation.

The model can be conceptually broken into four main parts:
1) calculation of the temperature at heater-GST interface;
2) calculation of the thickness of the amorphous cap;
3) calculation of the CF (during both crystallization and amorphization processes);
4) evaluation of the cell state.

In simulations, the above four parts cooperate to derive the state of the cell during a programming operation.

A. Calculation of the temperature at the heater-GST interface

The asymptotic increase of the temperature, ΔT_J, at the heater-GST interface is due to Joule heating and is derived as

$$\Delta T_J = R_{th} \frac{V_{pulse}}{R_h}(V_{pulse} - V_H) \tag{1}$$

where V_{pulse} is the programming voltage, R_h is the resistance of the heater, R_{th} is the equivalent thermal resistance of the memory cell (whose value depends on the GST phase), and V_H is the holding voltage [11] of the S-shaped I-V characteristic of the cell. Once ΔT_J is known, it is possible to evaluate which portion of the GST volume crystallizes or amorphizes, depending on the GST thermal profile.

Since the modelled system is a dynamic one, the temperature part was implemented according to (1), taking the temperature dynamic into account. The temperature dynamic can be modelled as a first-order Ordinary Differential Equation (ODE) so as to reproduce the transient of the cell temperature:

$$\frac{dT_{if}}{dt} = \frac{\Delta T_J - T_{if}}{\tau_T} \tag{2}$$

where τ_T is a time constant which models the temperature dynamic and T_{if} is the instantaneous increase of the temperature at the heater-GST interface. The thermal profile over time of the temperature at the heater-GST interface is obtained by solving (2), whereas the instantaneous temperature at the heater-GST interface, T_{inst}, is found by adding T_{if} to the operating temperature, T_{op} ($T_{inst} = T_{if} + T_{op}$).

B. Calculation of the thickness of the amorphous cap (RESET modelling)

During a RESET operation, the active GST volume is heated above its melting point. The chalcogenide starts melting and, after a quick quenching of the GST, the atoms are rearranged into a chaotic configuration, typical of the amorphous state. T_{inst} determines the maximum thickness of the amorphous cap. Since this thickness directly depends on temperature, it is also dependent on the programming pulse voltage amplitude. The dynamic of the amorphization process is therefore affected by two contributions: the temperature at the heater-GST interface and the amorphous cap growth. In the proposed model, both these contributions are taken into account.

Since the PCM cell has a thermal capacitance which delays the amorphization process, a dynamic model must account for both the maximum thickness of the amorphous cap ($z_{A,max}$) and the time the cap takes to reach this asymptotic value. The maximum thickness of the amorphous cap is expressed as a function of the temperature at heater-GST interface [12] as

$$z_{A,max} = h\frac{T_{inst} - T_{melt}}{T_{inst} - T_{op}} \tag{3}$$

where h is the thickness of the GST layer and T_{melt} is the GST melting temperature ($\approx 600\ °C$).

The instantaneous thickness, z_A, of the amorphous cap depends on $z_{A,max}$ and K_a, which is an amorphization constant determining the speed of the process. Also in this case, a first-order ODE can describe the evolution of z_A over time:

$$\frac{dz_A(t)}{dt} = K_a\ (z_A - z_{A,max}) \tag{4}$$

Finally, the resistance of the GST after a RESET operation ($R_{GST,rst}$) and, hence, the reading current, are derived:

$$R_{GST,rst} = \frac{\rho_A\ z_A}{S} + \frac{\rho_C\ (h - z_A)}{S} \tag{5}$$

where S is the area of the GST layer and ρ_C and ρ_A are the resistivities of the crystalline and the amorphous phases.

C. Calculation of the Crystal Fraction (SET modelling)

The crystallization process is characterized by a re-arrangement of the atoms of the material in an ordered structure. The JMAK theory is typically used to quantitatively describe the crystallization process because of its flexibility and simplicity. The JMAK model describes the evolution of the CF over time as $CF = 1 - e^{-K_c}$, where K_c is named reaction constant and it depends on temperature (T), crystallization activation energy (E_a), and crystallization rate (K_0) following the Arrhenius-like equation [13]

$$K_c = K_0 \ e^{-\frac{E_a}{k_B \ T}} \tag{6}$$

where k_B is the Boltzmann constant.

To describe the CF evolution, the JMAK law is completed by accounting for the amorphization equation, $\frac{dCF}{dt} = -K_a \ CF$. Both the amorphization and the crystallization phenomena are thus included in the same ODE:

$$\frac{dCF}{dt} = K_c \ (1 - CF) - K_a \ CF \tag{7}$$

Equation (7) was implemented to describe the evolution of the CF over time. It is worth noticing that K_a can be treated as a constant, since it has no dependence on time or temperature, whereas K_c is variable both over time and temperature. More specifically, during a SET operation, K_a is set equal to 0, since amorphization only takes place in the RESET operation decreasing the crystal fraction. Similarly, K_c is set equal to 0 during a RESET operation. Moreover, we experimentally observed that K_0 shows an exponential dependence on temperature

$$K_0 = 10^{\ a \ T_{inst}^2 + b \ T_{inst} + c} \tag{8}$$

where a, b, and c are three coefficients that can be obtained through a fitting operation on experimental data.

The crystallization of the amorphous GST in any part of its volume depends on the phase of the neighbouring material. The probability of phase change is higher in regions close to crystalline grains than in a fully amorphous portion. It can be assumed that a crystalline filament can be formed in the amorphous cap during SET programming, which explains the abrupt resistance drop during a SET operation [11]. In fact, the formation of a crystalline path in the amorphous region is electrically equivalent to connecting two resistors in parallel, one of which has a resistance much lower than the other. The final resistance of the cell thus decreases abruptly.

In this case, the GST conductance (and, hence, the reading current) is calculated rather than its resistance

$$G_{GST,set} = G_{SET} \ CF + G_{RST} \ (1 - CF) \tag{9}$$

where G_{SET} and G_{RST} are the conductances of the full-SET and the full-RESET states, respectively. These two conductances are equal to $G_{SET} = \sigma_C \frac{S}{h}$ and $G_{RST} = \frac{\sigma_C \ \sigma_A \ S}{\sigma_A \ (h - z_A) + \sigma_C \ z_A}$, where σ_C and σ_A are the conductances of the crystalline and amorphous phases.

D. Evaluation of the cell state

When a pulse V_{pulse} is applied to the cell, the temperature T_{inst} at the heater-GST interface is calculated, thus determining which operation is being performed (RESET, SET, or read). If a RESET operation is taking place, the CF decreases according to (7) and the maximum thickness of the amorphous cap and the dynamic of its growth are calculated through (3) and (4). Then, the value of the GST resistance, R_{GST}, is updated by using (5). If a SET operation is being performed, the CF grows with a dynamic given by (7), then the value of the GST conductance, G_{GST}, is updated by using (9). After calculating $R_{GST} = \frac{1}{G_{GST}}$, the cell reading current I_{read} and, hence, the cell state, are easily calculated as

$$I_{read} = \frac{V_{read}}{R_{GST} + R_h} \tag{10}$$

where V_{read} is the reading voltage applied to the PCM cell.

III. MODEL VALIDATION

In this Section, the model is validated through comparison with experimental data. All parameters in the model were found through a fitting operation with measurements. The memory device used, which includes an array of PCM μtrench cells, was fabricated in a 180 nm CMOS technology [14]. Both RESET and SET operations were validated through single pulse programming algorithms [2]. Before applying a programming pulse, the cell was always brought in the same initial condition (crystalline phase before a RESET operation and amorphous phase before a SET operation). The same algorithm was then used to simulate the modelled cell.

It is expected that the RESET operation is very well repeatable, since the major contribution to the GST resistance is given by the amorphous cap. The model can be thus validated by using a single set of measurements on a cell. Pulses with different amplitudes (from 4 V to 5 V, step $\Delta V = 250$ mV), and durations (from 50 ns to 400 ns, variable step), were applied to the memory cell, then the cell was simulated by means of the proposed model applying the same sequence of programming pulses. The comparison between model and measurements is shown in Fig. 3. From Fig. 3(a), it is apparent that, for any value of pulse time duration, the higher the programming voltage, the lower the reading current (and, hence, the higher the GST resistance), as expected from (3). From Fig. 3(b), it is observed that, for any value of pulse amplitude, the reading current reaches a stable value for long pulses, as described in (4). Both Figs. 3(a) and 3(b) show very good agreement between model and measurements.

When compared to RESET operation, SET operation is more sensitive to the microstructure of GST. Therefore, in Fig. 4, simulations results with the proposed model are compared to five sets of measurements carried out on the same cell. In order to validate the model, error bars (length $\pm\sigma$) were added in both Figures. Figs. 4(a) and 4(b) were obtained by applying SET pulses with different amplitudes (from 1.7 V to 2.9 V, step $\Delta V = 300$ mV), and different time durations (from 50 ns to 500 ns, step $\Delta t = 50$ ns). From Fig. 4(a), it

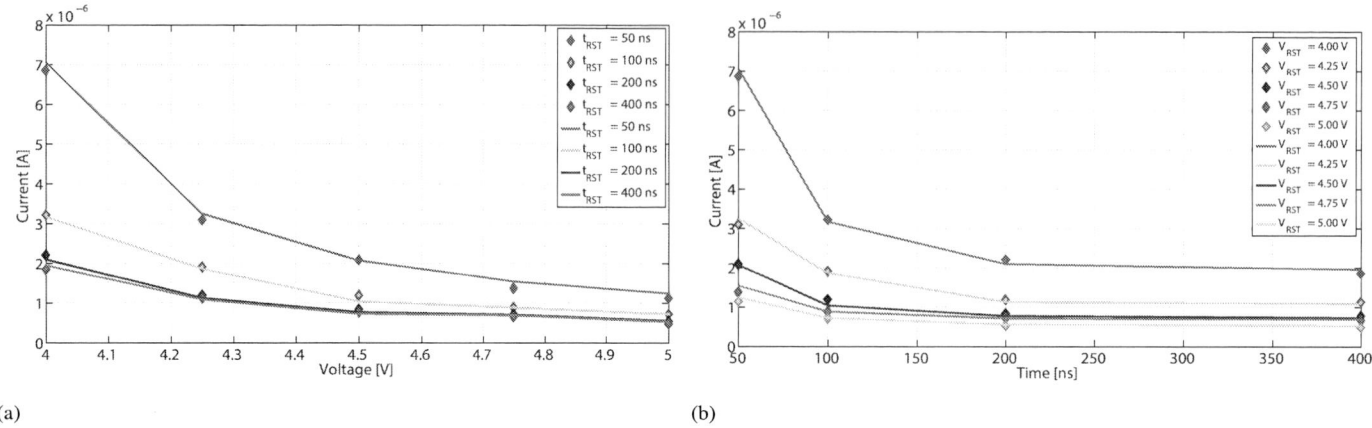

(a)   (b)

Fig. 3. RESET dynamics: comparison between simulations (solid line) and measurements (dots and error bars) for RESET pulses with different (a) time duration and (b) amplitude.

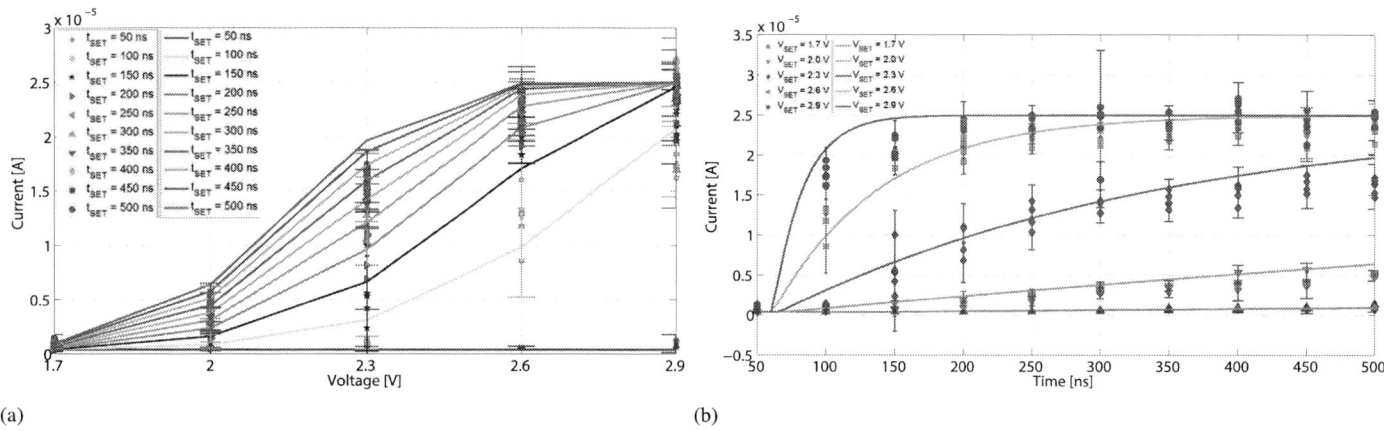

(a)   (b)

Fig. 4. SET dynamics: comparison between simulations (solid line) and measurements (dots and error bars) for SET pulses with different (a) time duration and (b) amplitude.

is clear that the crystallization process speed increases with increasing amplitude of the programming voltage and, hence, with the temperature at the heater-GST interface, as expressed in (6) and (8). The effect of the decrease of the GST resistance, which follows equation (9), is shown in Fig. 4(b). Moreover, when the crystallization process starts (t = about 60 ns), if the temperature is sufficiently high (e.g. V_{SET} = 2.6 V or V_{SET} = 2.9 V), a more significant increase of the reading current in the first 200 ns is observed. This effect is ascribed to the dependence of the probability of forming a crystalline path upon temperature. If the spread in the SET operation is considered, the model results mostly lie within the error bars in both Fig. 4(a) and 4(b), thus demonstrating the effectiveness of the proposed model also for SET programming.

IV. CONCLUSION

This paper presented a continuous-time compact model for PCMs, which makes use of a physical approach. The model was implemented so as to describe the programming operations as a dynamic system. After the model parameters were found through a fitting operation with experimental data,

it was validated by comparison with measurements on μtrench PCM cells.

ACKNOWLEDGMENT

The Authors wish to thank P. Malcovati, L. Vendrame, M. Rizzi and D. De Martini for fruitful discussions on the topic. The Authors also acknowledge M. Fanciulli and S. Spiga for the strong support and sponsorship.

REFERENCES

[1] R. Bez et al., in IEEE IMW, 2013, pp. 13–16.
[2] S. Braga et al., in IEEE MTDT, 2009, pp. 3–6.
[3] P. Fantini et al., in SISPAD, 2006, pp. 162 – 165.
[4] Y.-B. Liao et al., in EDSSC, 2007, pp. 625 – 628.
[5] K. C. Kwong et al., in ICSICT, 2008, pp. 492 – 495.
[6] D. Ventrice et al., IEEE EDL, vol. 28, no. 11, pp. 973–975, 2007.
[7] S. Braga et al., in ESSDERC, 2008, pp. 154–157.
[8] A. Pirovano et al., IEEE TED, vol. 51, no. 3, pp. 452–459, 2004.
[9] D. Ielmini et al., IEEE EDL, vol. 25, no. 7, pp. 507 – 509, 2004.
[10] M. Avrami, J. of Chemical Physics, vol. 9, no. 2, pp. 177–184, 1941.
[11] A. Lacaita et al., in SISPAD, 2005, pp. 267 – 270.
[12] S. Braga et al., Semic. Sci. and Tech., vol. 24, no. 11, 2009.
[13] N. Mehta and A. Kumar, Mat. Chem. and Phys., vol. 96, no. 1, pp. 73–78, 2006.
[14] F. Bedeschi et al., IEEE JSSC, vol. 40, no. 7, pp. 1557–1565, 2005.

High Resolution Current-Mode CCO-Based Continuous Time Delta-Sigma Modulators for Sensor-Array Applications

Anouar Laifi, M. Adib Al Abaji, and Roland Thewes

Faculty of Electrical Engineering & Computer Science, Chair of Sensors and Actuator Systems
Berlin Institute of Technology (TU Berlin)
Berlin, Germany
anouar.laifi@tu-berlin.de, m.alabaji@tu-berlin.de, roland.thewes@tu-berlin.de

Abstract — A current-mode delta-sigma modulator is presented for electrochemical sensor arrays utilizing a Current Conveyor and a Current Controlled Oscillator (CCO). Second order noise shaping is achieved although a very simple topology is used with only one integrator. The impact of oscillator non-linearity is kept low thanks to a pseudo-differential design. Simulations predict an SNDR of 102 dB at 10 kHz bandwidth, 5 MHz sampling frequency, oversampling ratio of 256, and a CCO non-linearity of 1%. Even at 5% oscillator non-linearity an SNDR of 72 dB is achieved. A feasibility check is accomplished by designing the proposed topology on transistor-level.

Keywords — *Delta-Sigma modulator; current-mode; CCO-based quantizer; non-linearity; continuous-time; ADC; CCII*

I. INTRODUCTION

Sensor arrays with in-sensor-site A/D conversion are used in imaging, biochemical sensing, and further application areas [1-5]. Many of them operate under the boundary conditions of low frequency input signals, current-domain signal representation, and limited area per sensor site. In this context, Delta-Sigma ($\Delta\Sigma$) ADCs are attractive, since at low-frequency input signals they can be operated at high oversampling ratios translating into high resolution in spite of only moderate clock frequencies. Moreover, the moderate requirements of $\Delta\Sigma$ ADCs concerning their analog circuit blocks allow the realization of the entire circuit within comparably small sensor-site areas.

In applications as described above, oscillator-based - and in particular current-controlled oscillator-based - A/D conversion principles offer advantages like small area, low power consumption, and good responsiveness to small signals. Furthermore, oscillator-based quantizers provide high immunity to external noise [6] and suitability for low supply voltages [7]. Both properties make Current-Controlled Oscillators (CCOs) suitable to be integrated within arrays with relatively small sensor site pitches, where the supply voltage may vary from site to site and decrease towards the center or the edges of the array because of IR voltage drops on the power supply net. Whereas these boundary conditions make CCOs excellent candidates to implement the $\Delta\Sigma$ conversion principle in sensor arrays, current-mode circuits moreover

exhibit a good performance regarding bandwidth and power consumption.

In this work we present a current-mode $\Delta\Sigma$ modulator topology which is highly suitable for sensor array operation. The proposed structure is designed on system-level, modeled, and investigated by means of simulation. Furthermore, the effects of oscillator non-linearity and clock jitter are studied. The modulator resolution is considered under the condition of 10 kHz bandwidth at 5 MHz sampling frequency, which corresponds to an oversampling ratio equal to 256.

II. BACKGROUND

A. Current Controlled Oscillator

A CCO converts an input current into a frequency. The CCO output frequency as well as its phase depend on its free running frequency f_c, on its gain K_c, and on its input current $i(t)$. In this work we focus on the oscillator output phase θ_{cco}, which can be written as [8]:

$$\theta_{cco}(t) = 2\pi \int_0^\infty (f_c + K_c i(t))\, d\tau \qquad (1)$$

Practical CCOs, however, suffer from non-idealities, namely non-linearity, which will be considered later.

B. Current Coveyor

Due to the advantages of current-mode circuits like higher gain-bandwidth product compared to op-amp circuits operated with signals in the voltage domain, low-voltage capability, and low-power consumption, the Current Conveyor II (CCII) approach is attractive for analog signal processing purposes [9].

There, current is conveyed between two terminals with significantly different impedance levels. The current conveyor as such is a three terminal device: A voltage applied to a first input terminal Y appears on the second input terminal X, while zero current flows through Y, which thus represents a high impedance node. The current flowing through X is conveyed to the output Z. As the current through Z can flow in both directions, two types of CCIIs can be distinguished, a positive one (CCII+) and a negative counterpart (CCII-), respectively [9].

978-1-4799-4993-9/14 $31.00 © 2014 IEEE

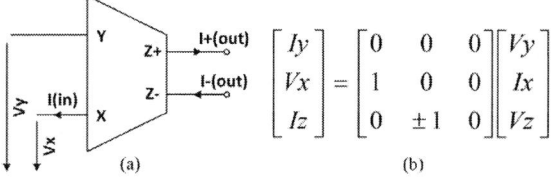

Fig. 1. (a): Dual output current conveyer II block. (b): Transfer matrix.

III. BLOCK-LEVEL DESIGN AND MODELING

To study the feasibility of the proposed ADC topology, the functionality of each circuit involved is modeled. However, an in-depth study and the exploration of the performance potential also require the consideration of the main non-idealities. For this purpose a block-based model with configurable non-ideal characteristics is elaborated.

A. Integrator Model

The Dual Output CCII (DOCCII) is able to deliver a positive and a negative output current (Fig.1), which is exploited to realize a pseudo-differential configuration. The DOCCII-based integrator model is shown in Fig. 2 [9].

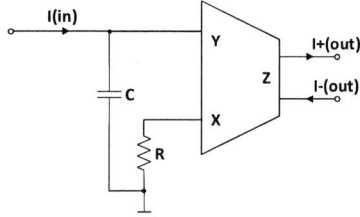

Fig. 2. CCII integrator configuration.

B. CCO-based Quantizer

1) Non-linearity

One of the main characteristics of the CCO is the conversion gain K_c, which stands for the ratio between output frequency and input current. Ideally, the CCO gain should remain constant over its whole dynamic range. Unfortunately however, the CCO gain in the real world is a non-linear function of the input signal leading to second, third, and higher order harmonics in the phase output. Consequently, the SNDR at the oscillator output and thus also at the output of the entire ADC is reduced. On the other hand, the inherent Sinc filter in the oscillator phase output attenuates higher order terms [9]. This Sinc filter function appears thanks to the integration property of the oscillator and results from the integration of a sinusoidal input signal.

The calculated attenuation for the 4th harmonic already reaches 54 dB, so that a suitable CCO model is achieved taking into account only the 2nd and 3rd harmonic. Consequently, the CCO gain is modeled as a 3rd order function:

$$K_c(i) = K_c i + \alpha i^2 + \beta i^3 \qquad (2)$$

The coefficients included in the model are calculated using the optimization toolbox in MATLAB®. To verify the amount of non-linearity, the gain function is swept over the dynamic range and the non-linearity is calculated according to [8]:

$$nonlinearity(\%) = \left(1 - \left|\frac{\left|\left(\frac{f_{max}-f_{min}}{2}\right)+f_{min}\right|-f_x}{\left(\frac{f_{max}-f_{min}}{2}\right)+f_{min}}\right|\right) \cdot 100 \quad (3)$$

There, f_x is the real output frequency where maximum deviation from the ideal value is obtained, while f_{max} and f_{min} are the boundaries of the frequency range of interest.

2) Quantization method

Quantization of the oscillator output can take place in the time- or in the frequency-domain. In our circuit the latter method is chosen, since it leads to a noise shaped output signal [11]. This choice implies the requirement:

$$f_{clk} \geq 2 f_{cco,max} , \qquad (4)$$

where f_{clk} is the sampling clock frequency and $f_{cco,max}$ is the maximum oscillation frequency, respectively. In this case the quantization principle is based on the detection of a transition of the oscillator output during the sampling time window. This task is accomplished by a simple digital circuit (cf. Fig. 3) consisting of two D-flip flops and a XOR-gate, a so-called Frequency-to-Digital Converter (FDC) [12].

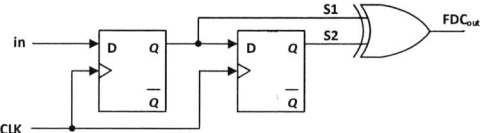

Fig. 3. Frequency to digital convertor (FDC).

IV. ADC TOPOLOGY

The basic ADC consists of an oscillator followed by a counter [11]. The analog information is converted to the time and hence to the frequency domain, so that the resulting signal is of digital nature (when considered in the voltage domain) and can be easily post-processed. Unfortunately, an open loop configuration as described above is very sensitive to oscillator non-idealities, especially the oscillator non-linearity [13]. For that reason we chose a closed loop configuration and integrate the oscillator output signal in a $\Delta\Sigma$-modulator. As a reference point for the following discussion we consider an (ideal) classic 1st order $\Delta\Sigma$-modulator using one integrator followed by a comparator. It reaches an SNR of approximately 70 dB at an oversampling ratio of 256. In this work, we operate a CCO-based quantizer in a simple $\Delta\Sigma$ modulator and optimize its topology to enhance the resolution and to ensure high stability and low sensitivity to non-linearity.

A. Single-ended CCO-based $\Delta\Sigma$-modulator

A 2nd order noise shaping is obtained if the comparator in the 1st order modulator is replaced by a CCO-based quantizer as shown in Fig. 4 [11, 13].

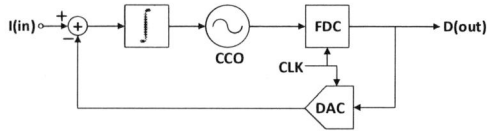

Fig. 4. Single-ended CCO-based $\Delta\Sigma$-modulator with 2nd order noise shaping despite using a single integrator only.

978-1-4799-4993-9/14 $31.00 © 2014 IEEE

B. Pseudo-differential topology

Though the feedback decreases the impact of non-linearity on the modulator resolution, non-linearity is still one of the main limiting factors. A way to mitigate the effect of non-linearity is to increase the integrator gain [13] in order to reduce the quantization noise in the signal-band. A further way to reduce non-linearity is to use a pseudo-differential structure, which cancels out the even harmonics [6]. The combination of this measure with feedback leads to the pseudo-differential ΔΣ modulator presented in Fig. 5. The integrator is realized on the basis of a CCII with differential outputs. Each output is connected to an independent oscillator whose outputs are fed into a digital subtractor.

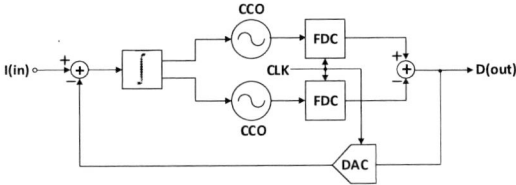

Fig. 5.   Pseudo-differential CCO-based ΔΣ-modulator.

Another advantage is given by the feedback configuration. The insertion of the subtractor adds additional 0.5 bits of resolution through the feedback loop. On the DAC-level this translates into a third output value equal to 0.

V.   TRANSISTOR-LEVEL IMPLEMENTATION

The circuit is designed on transistor level and simulated in a 180 nm standard CMOS technology.

A. DOCCII-based integrator

The DOCCII shown in Fig. 6 is based on a miller opamp topology. The complementary current signals are generated by adding a current mirror to the original CCII.

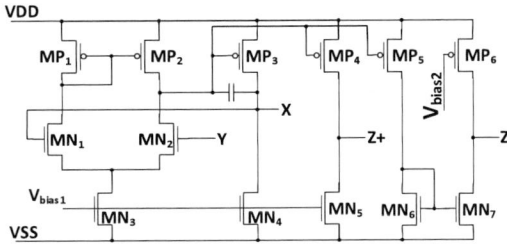

Fig. 6.   Dual Output Current Conveyor II (DOCCII).

B. CCO

The relaxation oscillator reported in [14] and illustrated in Fig. 7 is suitable for low supply voltage operation. It is designed such that its oscillation frequency does not exceed 2.5 MHz at maximum input current. Its operation principle is based on the integration of the input current on the capacitors C until V_{int1} (when V_{out2} is low) or V_{int2} (when V_{out1} is low) reaches the respective inverter's threshold voltage and the latch state changes. Since the oscillator allows only one current direction, an offset current is needed to ensure proper operation of both quantization paths.

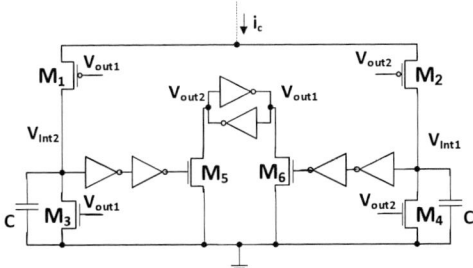

Fig. 7.   Relaxation oscillator.

C. Current Steering Feedback DAC

The DAC is realized using cascode current mirrors. Cascode Transistors as well as operating the switches by means of low swing signals reduce the switching artifacts.

VI.   SIMULATION RESULTS

The performance improvements expected from the presented ΔΣ modulator are investigated by simulation on system-level. For this purpose, different simulations are performed using Verilog-A block modeling and Spectre. As input signal, a sinusoidal current with 1 kHz frequency and a 5 MHz clock signal are chosen. The chosen band of interest is 10 kHz motivated by the typical applications mentioned before.

The ideal single-ended ΔΣ modulator achieves a second order noise shaping with an SNDR of about 93 dB. Further simulations with different levels of oscillator non-linearities (Table I) reveal severe performance degradation. For example, 2% oscillator non-linearity decreases the resolution to 72 dB. The impact of the 2nd and 3rd harmonic for this case are highlighted in Fig. 8. Also the effect of jitter is considered. E.g. at 1% clock jitter (3σ value referred to the clock period time), the performance of the pseudo-differential modulator is about 15 dB better thanks to the extra feedback level.

TABLE I.   COMPARISON OF 2ND ORDER SINGLE-ENDED AND PSEUDO-DIFFERENTIAL STRUCTURE

	SNDR Single-ended structure [dB]	SNDR Pseudo-differential structure [dB]
0% non-linearity, 1-phase oscillator, 0% clock jitter[a]	93	98
0% non-linearity, 3-phase oscillator, 0% clock jitter[a]	99	105
1% non-linearity, 1-phase oscillator, 0% clock jitter[a]	86	95
1% non-linearity, 3-phase oscillator, 0% clock jitter[a]	79	102
2% non-linearity, 1-phase oscillator, 0% clock jitter[a]	72	92
2% non-linearity, 3-phase oscillator, 0% clock jitter[a]	62	98
5% non-linearity, 1-phase oscillator, 0% clock jitter[a]	45	71
0% non-linearity, 1-phase oscillator , 1% clock jitter[a]	45	60
0% non-linearity, 3-phase oscillator, 1% clock jitter[a]	60	75

a. = 3σ/clock period. Clock jitter is assumed to be normally distributed.

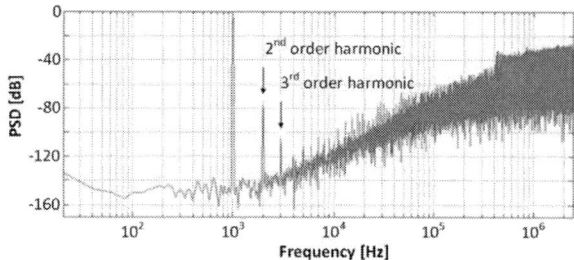

Fig. 8. FFT output spectrum for the single-ended structure with 2% non-linearity.

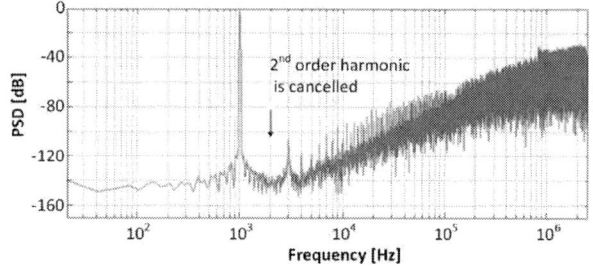

Fig. 9. FFT output spectrum for the pseudo-differential structure with 2% non-linearity.

In the output spectrum of the non-ideal pseudo-differential $\Delta\Sigma$ modulator with 1.5 bit feedback, the 2nd harmonic disappears. Fig. 9 shows that this structure cancels the even harmonics as expected. An SNDR of 92 dB is still reached at 2% of oscillator non-linearity. Even with 5% non-linearity, this structure still has a resolution of about 72 dB.

The transistor-level circuit described in section V achieves an SNDR of 94 dB and consumes 22 µW at a supply voltage of 1.8V. Table II compares the performance of the proposed modulator with another CCII-based modulator [15] and with a recently proposed electrochemical $\Delta\Sigma$-modulator.

TABLE II. MODULATOR PERFORMANCE COMPARISON

	Tech (m)	P (W)	OSR	BW (Hz)	SNDR (dB)	Ord.	FOM [15] (J/conv)
[15]	90 n	0.61m	64	78 k	60.2	2	4.67 p
[16]	2.5 µ	25 µ	256	2	61.8	1	6.25 n
This work	180 n	39.6 µ	250	10 k	94	1	49 f

Once a real circuit is manufactured, mismatch will affect the conversion gain, the center frequency, and the non-linearity of the oscillators. In case of gain mismatch, the signal power after the subtractor decreases and consequently the resolution, too. For instance, 5% gain mismatch reduces the resolution achieved under ideal conditions from 100 dB to 94 dB. Center frequency mismatch adds a dc-component to the output signal, which can be interpreted as a shift of the modulator input dynamic range. This problem can be solved by enlarging the input dynamic range and removing the dc-component digitally. In case of non-linearity mismatch, the even harmonics are not canceled completely.

VII. CONCLUSION

A current-mode CCO-based $\Delta\Sigma$-modulator topology using a CCII-based current-mode integrator and a CCO-based quantizer has been presented, modeled, and investigated . It is simple, robust, and suitable for low-voltage operation, especially in electrochemical Sensor Arrays. Although only one integrator is used, the presented modulator achieves 2nd order noise shaping. Thanks to its pseudo-differential structure and 1.5 bit feedback, it has a better performance in presence of clock jitter, and the impact of oscillator non-linearity is low: At an oversampling ratio of 256, 3 oscillator phases, and at 1% CCO non-linearity, the proposed $\Delta\Sigma$-modulator achieves an SNDR of 102 dB.

REFERENCES

[1] F. Heer et al., "CMOS electro-chemical DNA-detection array with on-chip ADC", Tech. Dig. IEEE ISSCC, pp. 168-604, 2008.

[2] M. Schienle et al., "A fully electronic DNA sensor with 128 positions and in-pixel A/D conversion", IEEE J. Solid-State Circuits, pp. 2438-2445, 2004.

[3] S. Kleinfelder et al., "A 10kframe/s 0.18µm CMOS digital pixel sensor with pixel-level memory", Tech. Dig. IEEE ISSCC, pp. 88–89, 2001.

[4] U. Ringh et al., "CMOS RC-oscillator technique for digital read out from an IR bolometer array", Tech. Dig. Int. Conf. Solid-State Sensors and Actuators, pp. 138–141, 1995.

[5] M. Nazari et al., "CMOS Neurotransmitter Microarray: 96-Channel Integrated Potentiostat with On-die Microsensors", IEEE Transactions on Biomedical Circuits and Systems, pp. 338-348, 2013.

[6] Y. Yoon et al., "A linearization technique for voltage-controlled oscillator-based ADC", Proc. IEEE Int. SoC Design Conf. (ISOCC), pp. 317-320, 2009.

[7] W. Gaber et al., "Systematic design of continuous-time $\Sigma\Delta$ modulator with VCO-based quantizer", Proc. IEEE ISCAS'10, pp. 29-32, 2010.

[8] J. Kim et al., "A time-based analog-to-digital converter using a multi-phase voltage controlled oscillator", Proc. IEEE ISCAS'06, pp. 3934-3937, 2006.

[9] G. Ferri, 'Low voltage, low power CMOS current conveyors', Springer, 2003.

[10] H. Venkatram et al., "Least mean square calibration method for VCO non-linearity", Proc. IEEE Int. Conf. on Microelectronics (ICM), pp. 1-4, 2010.

[11] A. Iwata et al., "An architecture of Delta-Sigma A-to-D converters using a voltage controlled oscillator as a multi-bit quantizer", Proc. IEEE ISCAS'98, pp. 389-392, 1998.

[12] M. Hovin et al, "A narrow-band delta-sigma frequency-to-digital converter", Proc. IEEE ISCAS'97, pp. 77-80, 1997.

[13] M.Z. Straayer et al., "A 12-bit, 10-MHz bandwidth, continuous-time $\Sigma\Delta$ ADC with a 5-bit, 950-MS/s VCO-based quantizer", IEEE J. Solid-State Circuits, pp. 805-814, 2008.

[14] J. Zhao and C. Wang, "CMOS current-controlled oscillators", Proc. IEEE ISCAS, pp. 929-932, 2007.

[15] H. Balasubramaniam and K. Hofmann, "A new 60.2 dB 2nd order delta sigma modulator using open loop buffers based on current conveyors", Proc. IEEE SCD, pp. 1-3, 2011.

[16] S. Sutula et al., "A 25-µW All-MOS Potentiostatic Delta-Sigma ADC for Smart Electrochemical Sensors", IEEE Trans. on Circuits and Systems I, Vol. 61, pp. 671-679, 2014.

A 32-Channel 12-bits 65nm Wilkinson ADC for CMS Central Tracker

Tommaso Vergine[*†], Marcello De Matteis[*‡], Andrea Baschirotto[†], A. Marchioro[§]

* University of Pavia Italy. Department of Electrical Computer and Biomedical Engineering.
†Dept. of Physics, University of Milano-Bicocca, Milano, Italy
‡ University of Salento, Lecce Italy. Department of Innovation Engineering.
§ CERN, 1211 Geneve 23, Switzerland.
tommaso.vergine, marcello.dematteis, andrea.baschirotto@unimib.it.

Abstract—**The ADC proposed in this paper is part of a larger VLSI (Very Large Scale Integration) circuit, called Detector Control Unit (DCU), whose aim is to monitor some critical quantities in the High-Energy Physical experiments inside the Large Hadron Collider (LHC). In particular it has been developed for the CMS (Compact Muon Solenoid) central tracker. The damage caused by radiation in such an environment requires that some key-parameters like leakage currents, local temperatures and supply voltages are carefully monitored in order to reduce failure events and improve overall performance. Indeed, CMOS integrated circuits, if exposed to very high radiation levels can experience large leakage currents and significant voltage/temperature variations. For this reason this monitoring system has been realized to provide real time information about the electrical/physical scenario of detectors. The ADC here shown, based on a Wilkinson single-slope architecture, has a resolution of 12 bits and is able to manage 32 input analog channels. Simulation results have shown a definitive 11 bit accuracy and a power consumption of about 500μW. The ADC has been designed in CMOS 65nm technology. It operates with a 40 MHz clock frequency. The final signal sample rate is about 5.5kHz.**

Keywords: A-to-D Converter, ADC, Low-Power, Radiation Hardness, High-Energy Experiments.

I. INTRODUCTION

CMOS integrated microelectronic circuits are more and more used in High-Energy Physic experiments. Even thought this approach is characterized by lacking reconfigurability and portability, it brings several advantages. First of all costs reduction and performance improvement. When CMOS read-out circuits are exposed to very high radiation levels performance degradation and breakdown events appear. Thus, in the CMS central tracker a monitoring system (named DCU[1]), has been implemented in order to provide real-time information about the status of detectors. An electrical/physical quantities monitoring (i.e. temperature, voltages and leakage currents) can increase by far the life-time and the performance of integrated circuits. Notice that, as it happens for detectors, also the monitoring system can be also exposed to radiation. Hence it has to be radiation-hard. Typically using devices wider than 0.8-1 μm can increase the radiation tolerance (the shift of the threshold voltage is drastically reduced) but, when narrow channels are needed, proper layout techniques,for better immunity, can be considered. In this design both approaches have been used: transistors greater than a certain (minimum) area and with an annular gate layout[4]. The ADC is based on a

single-ramp (Wilkinson) architecture[2][3], although the large conversion time typically required by these converters is not an issue because the voltage supply, leakage and temperature time variations are more slower than the ADC conversion rate. The ramp generator circuit, one of the most important blocks in a Wilkinson ADC, has been carefully designed in order to provide an accurate and linear ramp signal over all the input full-scale. Also the comparator design has been challenging. It has to manage an input full-scale of 1 V with a voltage supply of 1.2 V. This paper is organized as follow. In Section II the DCU system is described, while in Section III the most important ADC specifications are addressed. Section IV presents all analog blocks that constitute the ADC. Finally, Section V reports the simulations results, and at the end of the paper Conclusions are drawn.

II. SYSTEM OVERVIEW

The DCU is a VLSI circuit designed for the CMS central tracker. Its aim is to monitor some key-parameters (both electrical and physical quantities) for the life-time of detectors. The silicon micro-strip detectors in the CMS tracker, when exposed to high level of radiation undergo several damaging phenomena. The two most important effects are an increase of the detector leakage current and a change in the depletion voltage. Thus, in order to extend life-time and to maintain integrity of detectors, a careful monitoring of environmental conditions is needed. This is done by the DCU.

III. ADC SPECIFICATIONS

The most important ADC converter specifications are obtained by system-level considerations and are presented in Tab. I . The ADC has to manage 32 analog inputs. The quantization error has to be lower than 1.5 LSB. The analog input full-scale is from 0 V up to 1 V. The maximum allowed power consumption, including power of digital circuits, has to be lower than 2 mW (from a single 1.2 V supply voltage) and the maximum conversion time has to be lower than 1 ms.

IV. ADC ARCHITECTURE

In a Wilkinson ADC the conversion is performed comparing the input signal with a linearly increasing voltage, generated on a capacitor by a fixed current. If the charging current is accurate and the capacitor is stable the voltage measured is simply proportional to the time needed to charge

Parameter	Value
# of channels	32
Resolution	12 bit
Input Range	0V ÷ 1V
Quantization Error (max)	1.5 LSB
Power Consumption	< 2mW
Conversion Time	< 1ms
Voltage Supply	1.2V

Table I: ADC Specifications

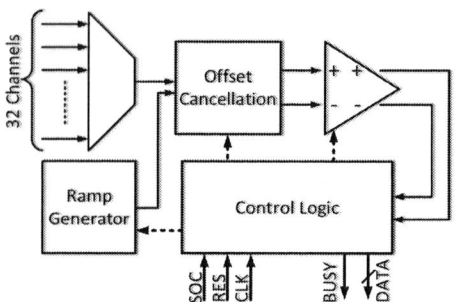

Figure 1: ADC Block Scheme

the capacitor. In other words a voltage vs time conversion is carried out. The block scheme of the proposed ADC is shown in Fig. 1. As said previously, being the conversion algorithm based on a comparison with a ramp signal, every slope variation can critically affect the overall ADC performance. For this reason a careful design of the Ramp Generator circuit is here developed. In order to allow the digital conversion of several input signals, an analog multiplexer has been placed as first block of the ADC. Considering the input full-scale a design effort is required for Ramp Generator and the following blocks design because they have to manage signals very close to the voltage supply. An important design specification, to be taken into account, is that the external biasing current, essentially used to bias the ramp generator and the comparator, can be periodically tuned. This allows to compensate variations due to the technological spread and mismatch. On the other hand, it is very important to design the internal current mirrors able to manage the exact reference current without significant degradation due to the MOS transistor process/mismatch variations. The third component in the ADC is the Offset Cancellation block. If properly controlled during a conversion it allows to evaluate the comparator offset voltage and to obtain its conversion value. This is done because a digital offset compensation technique has been implemented. Regarding the comparator, it has been implemented with two complementary input stages (in order to manage the input full-scale without any biasing problems) and a latched output (in order to improve the compatibility between analog and digital blocks and to reduce static power). The last block is the Control Logic that is in charge of the ADC overall timing and control. The CLK signal is the external clock while SOC is the asynchronous start of conversion signal. To control the Offset Cancellation block, provide a controlled (stopped) clock (to reduce power consumption) to the comparator and give the 12 bits output converted value (DATA) are the main functions of the Control Logic block. The BUSY signal is an ADC activity flag. It indicates when the digital conversion result can be read and when the ADC is ready for another conversion.

A. Analog Multiplexer

High-Energy Physics experiments are often characterized by a large number of signals to be processed. For this reason a multi-channels multiplexer, shown in Fig. 1, has been introduced. A decoder structure (controlled by 5 bits) has been used in order to minimize the unwanted effects of the unselected channels (e.g. leakage current and noise). Each switch has been designed as complementary switch. This approach allows to minimize the on-resistance, and consequently the input settling time, keeping a low circuital complexity.

B. Clocked Comparator

The comparator plays a key role in the ADC performance. First of all, it has to be able to sense a difference at least of 1 LSB ($\approx 244\mu V$) between its input signals to guarantee a resolution of 12 bit for the ADC. In addition the parasitic (non-linear) capacitance, seen from its inputs, has to be negligible if compared with the ramp generator capacitance. To guarantee low parasitic input capacitances it is enough to reduce the area of the input pair transistors but this typically leads to an increase of the comparator offset. However, this does not represent a issue because a proper conversion algorithm allows to cancel-out the offset contribution. For this reason the fixing of the parasitic capacitance issue has been preferred. The schematic of the comparator is shown in Fig. 2[6]. It has a rail-to-rail structure with two complementary input stages that share the same output circuit. In this way it is possible to manage the input signals, close to ground and voltage supply, also in the worst case scenario. On the other hand, without any equalization circuit, this comparator can easily exhibit a variable offset, as function of the input signal. Another consequence of the complementary input stages could be the cross-over distortion but it can be neglected, because this circuit is used as comparator. The output stage is latched instead and reset to ground with the positive edge of the clock. To reduce the kickback noise each input stage has been drawn in a continuous-time version (no-latched).

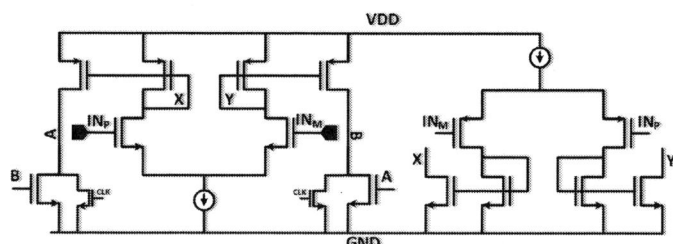

Figure 2: Rail to Rail Comparator

C. Offset Cancellation Block

The basic idea for the offset cancellation technique is to remove the contribution of the comparator offset directly into the digital domain (all other contributions to the offset voltage, like residual error in the ramp reset/sample phases or the charge injection of switches are negligible). To do this the offset voltage is initially treated as an input signal and it is converted in a digital word using a suitable ramp signal (see Ramp Generator section). Thus, the Offset Cancellation Block essentially swaps the polarity of the comparator input signals to allow, jointly with the ramp generator, the offset evaluation (absolute value and sign) from the ADC Control Logic. For example a positive offset, modelled as a DC voltage source in series with the positive input of the comparator, has been assumed. As shown in Fig. 3, placing the input voltage on the comparator positive input and using a ramp signal, starting from V_{IN} on the comparator negative input, it is possible to evaluate the digital conversion result of the offset voltage, V_{OFFSET}[LSB]. On the other hand, using the same polarity of the comparator inputs but with a ramp signal starting from 0 V the conversion result will be V_{IN}[LSB] + V_{OFFSET}[LSB]. Thus, obtaining these two information, it is possible to cancel out the offset contribution, subtracting its value from the digital conversion of the input signal. If a negative offset is assumed the algorithm continues to work properly just swapping the polarity of the comparator inputs. Since the offset is not constant each input signal conversion is preceded by an offset calibration procedure. This implies a longer conversion time but it does not represent an issue because the maximum time required for a complete conversion (offset and analog input) is about $200\mu s \ll 1ms$ (specification). Finally, considering that the ADC has to be able to digitalize $V_{IN} + V_{OFFSET}$ also with $V_{IN} = 1V$, the internal counting registers have been made using 13 bit to avoid overflows and the charging period has been extended.

D. Ramp Generator

Any deviation in the ramp signal, like wrong slope or poor linearity, can easily degrade the ADC overall performance. For this reason the ramp generator block has to be carefully designed. To realize a ramp signal from 0 V to 1 V (the input full scale) a grounded capacitor has been charged with a constant accurate current. This current source has to exhibit an output resistance in the GΩ order to minimize the charging current lost in the output resistance of the generator and to obtain a linear ramp signal. A Matlab model has provided 5GΩ as minimum needed output resistance. To obtain such a value a regulated low-voltage cascode current mirror has been designed, as shown in Fig. 4a [5]. The introduced negative loop increases the output resistance and keeps stable the operating point of the transistor M3, when M4 drain changes between 0V and 1V. This guarantees an output resistance of about 6GΩ. In order to reduce power consumption the operational amplifier has been designed with the minimum needed performance (i.e. 55 dB of dc gain and 1.3 MHz of bandwidth). The achieved phase margin is about 80°. The ramp generator has been completely validated with Monte Carlo tool, including also process and temperature variations. This design guarantees an error less than 5% on the ramp slope. Nevertheless, this deviation falls in the tunable range that the external biasing current can compensate. As said in the previous section, it is possible to use the same conversion algorithm for the input signal and the comparator offset. To convert the offset and store its value in the control logic a suitable ramp signal has to be used. As shown in Fig. 3, the offset contribution can be evaluated changing the starting point of the ramp signal

(sampling V_{IN} voltage on the ramp capacitor) and performing an analog to digital conversion. This allows a correct offset evaluation, as function of V_{IN}. To obtain a ramp with a variable starting point a set of switches have been introduced. These deviate the charging current during the sampling and reset phases (controlling the ϕ signal) and pre-charge with V_{in} or reset to ground the ramp signal. Note that the ramp charging period of the offset conversion can be made shorter than that of the input signal conversion, with the benefit of a faster conversion. Only few hundreds clock cycles (enough to convert the offset in the worst case) have been reserved for this phase. The ramp generator circuit is shown in Fig. 4b.

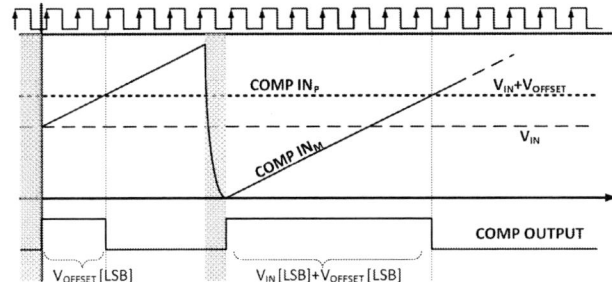

Figure 3: Offset Conversion vs Input Conversion

V. SIMULATION RESULTS

Being the simulation time quite long to characterize the ADC for all input values, two simulation sets have been performed. In particular, nominal and transient noise simulations, with three input signals: $\simeq$1.3 mV (5.5 LSB), $\simeq$490 mV (2007.5 LSB) and $\simeq$976 mV (4000.5 LSB). As illustrated in Fig. 5, 6, 7, the first set of simulations has provided results within ±0.5 LSB from the ideal results. In Fig. 5 and 7 the offset sign is positive. Since the Control Logic starts assuming a positive offset, only one conversion is needed to correctly digitalize its information. In addition, the Control Logic automatically connects the input and the ramp signal directly to the comparator with the proper polarity. On the other hand, in Fig. 6, the comparator offset has a negative sign. In this condition two conversion steps are needed for the offset conversion and the polarity of the comparator inputs is inverted. A slight degradation of the ADC performance is

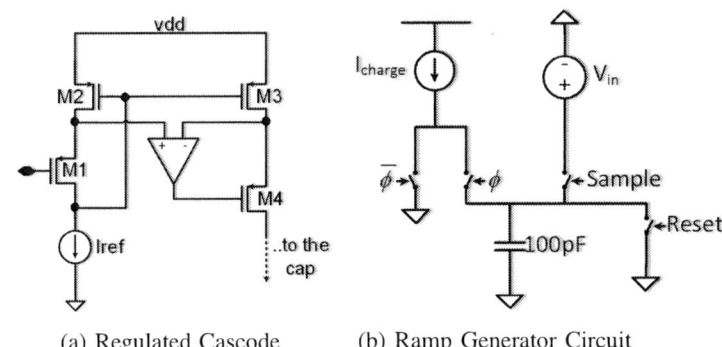

(a) Regulated Cascode   (b) Ramp Generator Circuit

Figure 4: Ramp Generator and Low Voltage Current Mirror

978-1-4799-4993-9/14 $31.00 © 2014 IEEE

observed in Fig. 8, where the converter has been simulated using Transient Noise tool. Using the same input levels ten different simulations have been run demonstrating that the worst case accuracy is 11bits. The ADC power consumption is around 600 μW, including dynamic power due to the digital circuits.

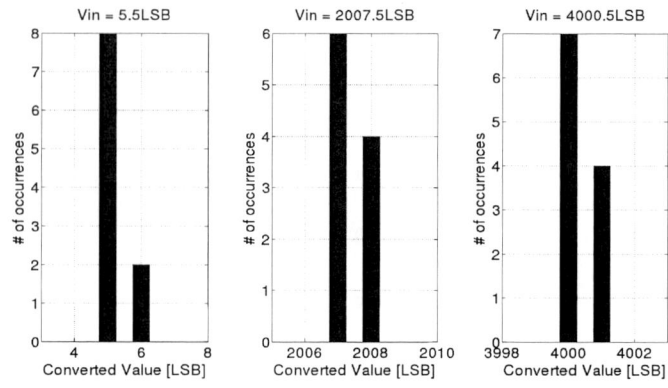

Figure 8: Conversion results in Transient Noise simulations

VI. CONCLUSION

In this paper a 32 Channel 12 bits Wilkinson ADC has been presented. It has been designed in 65 nm CMOS technology. Particular attention has been focused on the ramp generator and the comparator design. The ramp voltage has been obtained charging a grounded capacitor with a very accurate current mirror. The overall deviation of on the ramp slope, due to the process and mismatch, falls into the range that the external current can tune. The comparator has a rail-to-rail design with two complementary input stages that share the same output (latched) stage. The maximum quantization error, in transient noise simulation, is 1 LSB. The ADC performance are summarized in Tab. II.

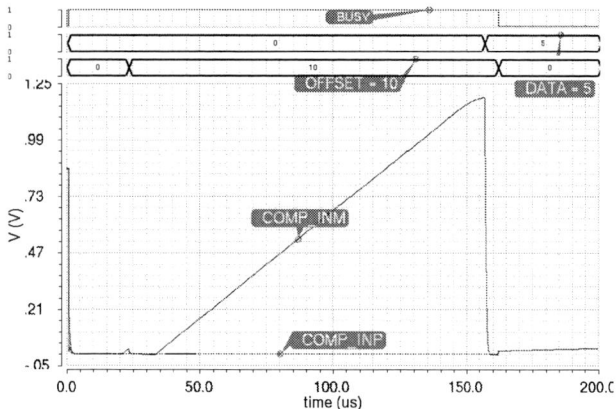

Figure 5: V_{in} = 5.5 LSB and V_{offset} = 10 LSB

Parameter	Value
# of channels	32
Resolution	12 bit
Input Range	0V ÷ 1V
Quantization Error (max)	1 LSB
Power Consumption	≃ 600 μW
Conversion Time	≃ 200 μs

Table II: ADC Performance

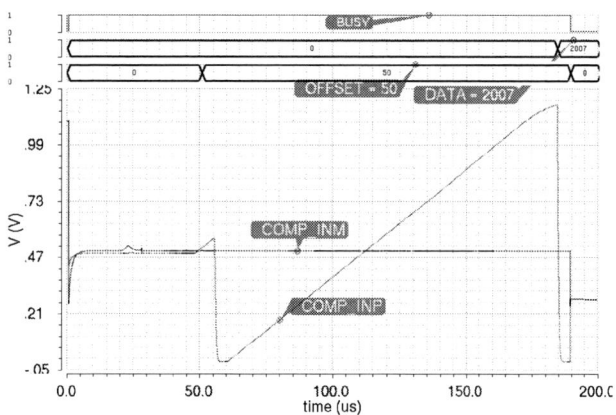

Figure 6: V_{in} = 2007.5 LSB and V_{offset} = -50 LSB

REFERENCES

[1] G. Magazzú et al., "The detector control unit: an ASIC for the monitoring of the CMS silicon tracker," Nuclear Science, IEEE Transactions on , vol.51, no.4, pp.1333,1336, Aug. 2004

[2] T. Vergine et al., "A 32-channel 12-bits single slope A-to-D converter for LHC environment," IC Design Technology (ICICDT), 2013 International Conference on , vol., no., pp.139,142, 29-31 May 2013

[3] T. Vergine et al., "An automatic calibration circuit for 12-bits single-ramp A-to-D converter in LHC environments," Ph.D. Research in Microelectronics and Electronics (PRIME), 2013 9th Conference on , vol., no., pp.45,48, 24-27 June 2013

[4] Fan Xue et al., "Gate-enclosed NMOS transistors, " 2011 J. Semicond. 32 084002

[5] J. Ramirez-Angulo et al., "Low supply voltage high performance CMOS current mirror with low input and output voltage requirements " IEEE Transactions on Circuits and Systems-II Express Briefs, Vol. 51, No. 3, March 2004

[6] Sung-Min Chin et al.,"A new rail-to-rail comparator with adaptive power control for low power SAR ADCs in biomedical application," Circuits and Systems (ISCAS), Proceedings of 2010 IEEE International Symposium on , vol., no., pp.1575,1578, May 30 2010-June 2 2010

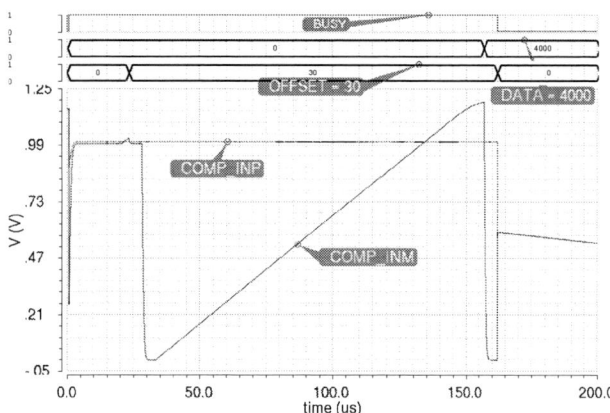

Figure 7: V_{in} = 4000.5 LSB and V_{offset} = 30 LSB

Design of a Low-Power Calibratable Charge-Redistribution SAR ADC

Soheil Aghaie, Jan Henning Mueller, Ralf Wunderlich, Stefan Heinen

Chair of Integrated Analog Circuits and RF Systems
RWTH Aachen University, Aachen, Germany
Email: ias@rwth-aachen.de

Abstract—This paper describes the implementation of a calibratable charge-redistribution (CR) based successive-approximation-register (SAR) analog-to-digital converter (ADC). This differential 13-bit CR SAR ADC is implemented with a unit capacitance of 1.54 fF which is much smaller than the capacitance used in most novel SAR ADCs. As a result, the area and the power consumption are reduced significantly. In the targeted 65-nm technology with a supply voltage of 1.2 V, an optimistic FOM of about 2.76 fJ/conversion-step is achieved through simulations. Since the larger mismatch between such small capacitors of the ADC degrades its linearity significantly, a foreground calibration technique based on trimming each capacitance with smaller switchable capacitors is developed to compensate the mismatch. Also, for the purpose of mismatch detection, an offset-calibratable double-tail latch comparator is designed to achieve an offset below 70 µV.

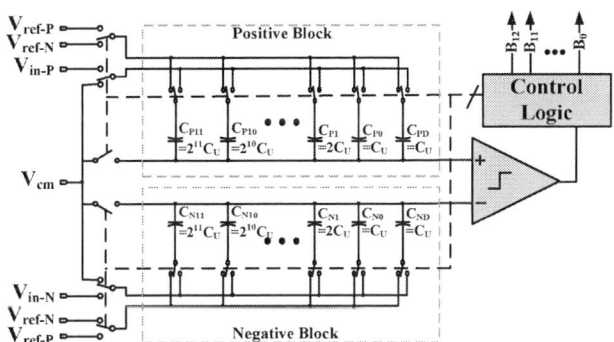

Fig. 1: Structure of a differential 13-bit CR SAR ADC.

I. INTRODUCTION

Modern applications of ADC blocks push the focus of the research towards higher resolutions and sampling rates along with lower area and power consumption [1]. CR SAR ADCs are of special interest in many applications with medium-to-high sampling rates due to their area and power efficiency. These circuits consist of a number of switchable capacitors which generate different voltage levels according to the ratio of capacitors [2]. By using a smaller unit capacitance for the CR digital-to-analog converter (DAC), the power and area of the ADC can be decreased. However, as the mismatch is increased with smaller sizes of capacitors, a small unit capacitance limits the achievable effective resolution of the ADC by introducing non-linearity [3]. In order to solve this problem, calibration techniques have been introduced to compensate the error caused by non-idealities like mismatch of capacitors. This paper presents a new calibration technique for improving the linearity of a CR SAR ADC with a very small unit capacitance. The calibration technique modifies the capacitance of each capacitor of the ADC by adding very small trimming capacitors in parallel to it in order to reduce its mismatch with other capacitors.

II. BACKGROUND

A. Differential CR SAR ADC

A differential 13-bit CR SAR ADC which is the base of this work is shown in Fig. 1 [2]. The capacitors in each of the positive and negative capacitor blocks are arranged in a binary

weighted scheme. During the sampling phase, the inputs of the comparator are connected to the common-mode voltage (V_{cm}) while bottom-plates of capacitors are connected to the corresponding positive or negative input signal (V_{in-P} or V_{in-N}). As a result, the capacitors are charged with the input signal. After the sampling phase, the inputs of the comparator are disconnected from V_{cm} and in the first conversion cycle, in order to determine the reference voltage of each capacitor block, the sign of the sampled input voltage is decided by connecting bottom-plates of all capacitors to V_{cm}. In the next conversion cycles, bottom plates of capacitors are connected to either of the reference voltage (V_{ref-P} or V_{ref-N}) or V_{cm} according to the binary word from the control logic which performs a successive-approximation binary search. As a result of charge conservation at each of the comparator's input nodes, their voltage would be the output of the DAC minus the corresponding input signal.

B. Mismatch of Capacitors

The larger mismatch of smaller capacitors degrades the linearity of the ADC. A maximum differential non-linearity (DNL) of 0.5 is considered as the criteria for linearity [4]:

$$DNL = \left| \frac{Code\_Width_m - \triangle}{\triangle} \right| < \frac{1}{2}. \qquad (1)$$

As the worst-case of DNL of a CR DAC occurs for the code 0111... , a maximum allowed mismatch can be calculated for the MSB capacitor:

$$\left| \triangle C_{mismatch\_(N-2)} \right| < \triangle C_{allowerd\_(N-2)} = \frac{C_U}{2}, \qquad (2)$$

978-1-4799-4993-9/14 $31.00 © 2014 IEEE

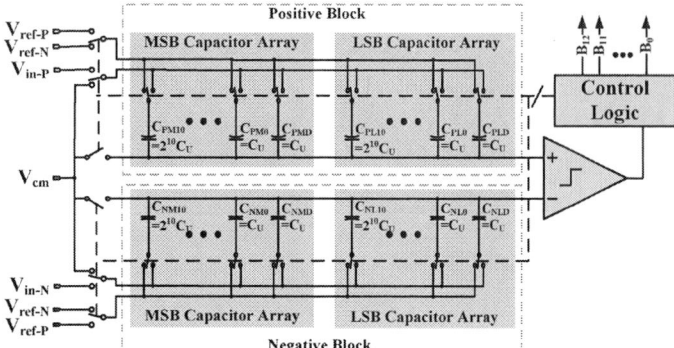

Fig. 2: Differential CR SAR ADC with split MSB array.

where C_U is the unit capacitor and $\triangle C_{mismatch\_(N-2)}$ is the mismatch between the MSB capacitor, C_{N-2}, and all its smaller capacitors including the dummy capacitor (C_D). Since each capacitor C_i consists of 2^i unit capacitors (C_U) and the variation of capacitors have normal distributions, a relative maximum allowed mismatch can be also calculated for other capacitors as well:

$$|\triangle C_{mismatch\_i}| < \triangle C_{allowerd\_i} = \frac{C_U}{2^{\frac{N-i}{2}}}. \qquad (3)$$

As a result, the unit capacitance of a CR SAR ADC must be large enough to fulfill (3). It must be mentioned that another limitation for using smaller capacitors is the thermal noise during the sampling phase. However, for a resolution of 13 bit, it can be neglected compared to the mismatch limitation.

C. Differential CR SAR ADC with MSB Capacitor Splitting

MSB Capacitor splitting introduced by [5] is a technique for reducing the power consumption of the CR SAR ADC by omitting the switching energy consumption during down-transitions. Fig. 2 shows the structure of the differential 13-bit CR SAR ADC with MSB capacitor splitting which is used in this work. Instead of a single-switched MSB capacitor in each block, it consists of an array of capacitors just like the array of other smaller capacitors. In the decision phase corresponding to the MSB bit, all the capacitors of the MSB array are connected to V_{ref}. In the next conversion phases, for each up-transition the corresponding capacitor is connected to the reference voltage just like it is done in the conventional switching scheme. On the other side, if a down-transition occurs for bit i, instead of discharging C_i and charging the next capacitor (C_{i-1}), only $C_{M(i-1)}$ is discharged. In this way, the average switching energy during a number of A/D conversions is reduced by 37% [5].

III. REALIZATION OF THE CALIBRATION TECHNIQUE

As mentioned earlier, a calibration technique is developed in this work which is used to eliminate the mismatch between capacitors by adding smaller trimming capacitors.

A. Matching Scheme and the Trimming Sequence

In a simple CR SAR ADC without capacitor splitting for the MSB bit, each capacitor C_i must be matching the sum of

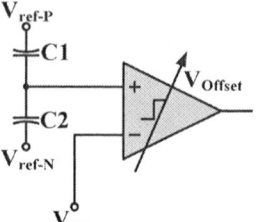

(a) Mismatch detection through capacitive voltage division.

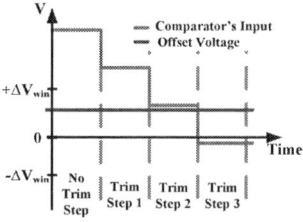

(b) Illustration of trimming with an offset calibrated comparator.

Fig. 3: Mismatch detection and trimming with the comparator.

all its smaller capacitors including C_D. With MSB capacitor splitting, the same matching conditions as in the simple CR DAC must be fulfilled for the LSB arrays. This condition for C_{Li} can be summarized as below:

$$\left| C_{Li} - \left(\sum_{k=0}^{i-1} C_{Lk} + C_{LD} \right) \right| < \triangle C_{allowed\_i}. \qquad (4)$$

In the MSB array each capacitor C_{Mi} remains connected to the reference voltage unless bit $i+1$ is reset to '0' due to a down-transition. In other words, capacitor C_{Mi} corresponds to bit $i+1$ of the binary word and it is considered to be matching smaller LSB capacitors:

$$\left| C_{Mi} - \left(\sum_{k=0}^{i-1} C_{Lk} + C_{LD} \right) \right| < \triangle C_{allowed\_i+1}. \qquad (5)$$

The calibration begins with the positive block and after all capacitors in this block are matched, it goes through the negative one. In each block, the matching begins with C_{L0} and then continues with other capacitors of the LSB array respectively. After the mismatch of the LSB array is compensated, the calibration is continued in the MSB array with a chronological order.

However, with this trimming method, each capacitance can be only increased. Therefore, in order to guarantee matching between capacitors, it is required to begin the calibration with a larger unit capacitance (C_{UL}) instead of C_U for C_{LD}.

B. Mismatch Detection and Trimming With the Comparator

A simple capacitive division configuration of capacitors along with the comparator can be used for detecting whether the undesired mismatch as illustrated in Fig. 3a.

If both capacitors are first discharged, the difference of the voltage of the middle node (V_x) to V_{cm} which is denoted by $\triangle V_{win}$, would be:

$$\triangle V_{win} = V_X - V_{cm} = (C_1 - C_2) \cdot \frac{\frac{1}{2} V_{ref\_diff}}{C_1 + C_2 + C_P}, \qquad (6)$$

where C_p is the parasitic capacitance at the input node of the comparator.

Therefore, by considering the maximum allowed mismatch a voltage window can be defined that V_x must be inside it. A comparator with an adjustable offset can be used for detecting the mismatch by checking both of the window borders as done in [3]. However, as values of $\triangle V_{win}$ are very small for

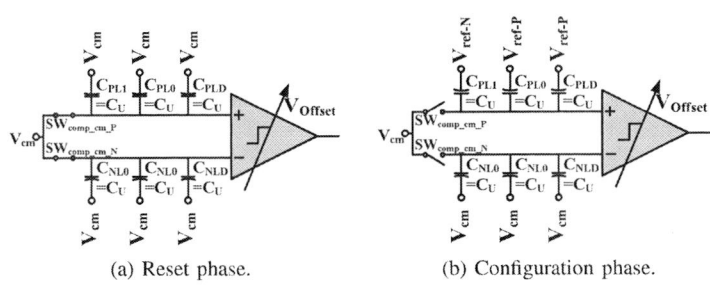

(a) Reset phase.   (b) Configuration phase.

Fig. 4: Example of a full mismatch detection phase for C_{PL1}.

different mismatch detection phases, it requires a huge effort to adjust the offset of the comparator precisely. Furthermore, C_p which affects the values of $\triangle V_{win}$ cannot be predicted before fabrication. For these reasons, the window mismatch detection method is not practical for all cases of mismatch detection. Instead, a method in combination with trimming of capacitors is used. In this alternative method which is illustrated in Fig. 3b, the comparator is calibrated to have a positive offset lower than the smallest value of $\triangle V_{win}$. In order to match two capacitors, one of them is trimmed with trimming steps smaller than the corresponding $\triangle V_{win}$ until achieving the smallest mismatch which makes the comparator's output '0'.

In the calibratable CR SAR ADC of this work, the mismatch detection configuration is formed by disconnecting the capacitors which are not taking part in the mismatch detection. As illustrated in Fig. 4 for trimming of C_{PL1}, each trimming phase consists of two parts. First, during a reset phase, all the connected capacitors in both blocks are discharged. Then, the inputs of the comparator are disconnected from V_{cm} and with a short delay the targeted capacitive divider is formed in the corresponding capacitor block. Keeping the same capacitor cells connected to the other comparator's input helps to compensate the undesired effects of the charge injection and leakage currents.

A problem that rises here is that for smaller capacitors, required trimming steps which must be less than the corresponding $\triangle V_{win}$ would be very small. Designing and using such small trimming capacitors cannot be achieved through a finger-shaped structure with a desired minimum size for its fingers. Therefore, it is decided not to trim small capacitors up to C_{L3} in the LSB array and up to C_{M2} in the MSB array. Instead, a selection method can be used in which a larger number of capacitor cells are provided and the final used ones are selected out of them in a way that they match. The unit capacitance for all of the provided selection cells must be C_{UL} to guarantee one-sided calibration. However, since smaller capacitors of the ADC have less effect on its non-linearity, their mismatch can be even left uncompensated. Thus, the option for disabling the selection process is also considered in implementation of the control logic.

IV. CIRCUIT IMPLEMENTATION

The ADC is designed in a 65-nm technology with a supply voltage of 1.2 V and a sampling rate of about 1.42 MHz. The

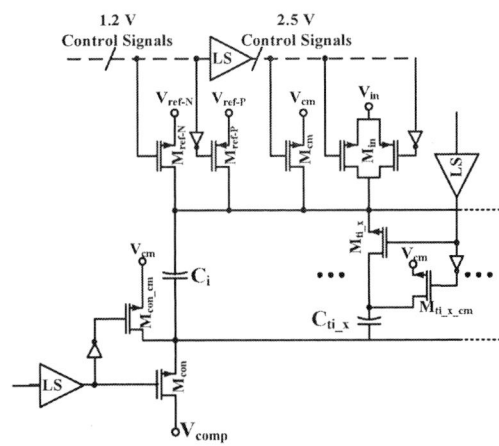

Fig. 5: Implementation of capacitor cells with MOS switches.

TABLE I: Specifications of the trimming capacitor cells. (Units are in aF and μV.)

i	$\triangle C_{allowed}$		$\triangle V_{win}$		C_{tr_U}		# bits	
	LSB	MSB	LSB	MSB	LSB	MSB	LSB	MSB
3	–	68.09	–	1657	–	69	–	4
4	68.09	96.29	828.6	1171	69	85	5	5
5	96.29	136.2	586.0	828.6	85	85	6	6
6	136.2	192.6	414.3	586.0	85	180	7	6
7	192.6	272.3	293.0	414.3	180	180	7	7
8	272.3	385.2	207.2	293.0	180	330	8	7
9	385.2	544.7	146.5	207.2	330	330	8	8
10	544.7	770.3	103.6	146.5	330	330	9	9

control logic of the circuit is implemented through a VHDL code and then it is synthesized to be used along with the other parts of the circuitry which are described in this section.

A. Implementation of Capacitors and Switches

Fig. 5 shows the structure of each capacitor cell. For each of the trimming cells, a number of binary-weighted trimming capacitors are considered. If a capacitor cell is disconnected from the comparator, both its plates would be connected to V_{cm}. Furthermore, since even a small leakage current can charge the small trimming capacitors quickly, they are also connected to V_{cm} when not used. Level-shifters are also used to generate 2.5-V control signals for conducting mid-level voltages.

A capacitance of 1.54 fF with a relative standard deviation of 0.5 % is used as C_U for this circuit. It is designed through an MOM structure with 6 fingers in each of metal layers M5 and M6 of the targeted 65-nm technology while the width of each finger and the spacing between fingers are both 100 nm.

C_{UL} which has a capacitance of 1.622 fF is also designed through the same structure as C_U while the spacing between its fingers is increased to 130 nm. Furthermore, the trimming capacitors are designed on a single metal layer—M6 in this work—with 1 or 2 fingers for each trimming unit capacitor.

Tab. I shows the values of $\triangle C_{allowed}$, $\triangle V_{win}$, trimming unit capacitance (C_{tr_U}), and the number of required trimming bits for the trimming cells.

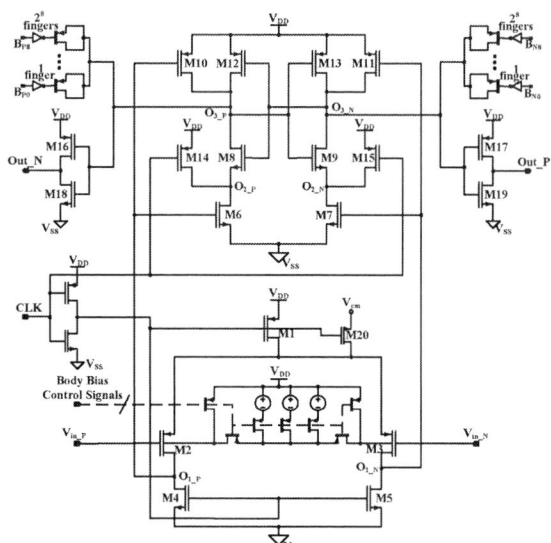

Fig. 6: Double-tail latch with 11 bits of offset compensation.

B. Offset-Calibratable Comparator

For having a low-power comparator which can achieve a very small offset of less than 100 µV, a double-tail latch is used which is shown in Fig. 6 [6]. The input stage is biased with a small current to optimize the offset while the current of the second stage is larger to have a high-speed operation. With the optimized sizing of transistors, the comparator's output settles in less than 2 ns and Monte-Carlo simulations show that the 1-σ input referred offset without any offset modification would be 10.5 mV.

Two 11-bit words (BN0-BN10 and BP0-BP10) are provided for the comparator and each modifies the offset in one direction. In each of these words, bits 10 and 9 are considered for shifting the body bias of the input transistors by 50-mV steps. Since the offset is directly dependent on the mismatch of the threshold voltage (V_{th}) of the input transistors, such shifts that changes V_{th} due to the body effect, modifies the offset. Simulations show that by applying these two offset modification bits, the offset can be limited to below 10 mV in all worst-case process and temperature corners.

The 9 remaining bits of the offset-modification words increase the capacitance of one of the output nodes of the second stage. The capacitance is increased, by applying a 0-V voltage to PMOS gate fingers with a unit size of $W/L = 120n/150n$. Also, an additional capacitance of 80 fF is connected to each of the outputs so that the offset can be modified linearly. With this sizing, each offset modification step would be between 25 µV and 70 µV in different worst-case corners. Therefore, the offset can be calibrated to below 70 µV.

For the purpose of offset calibration in each trimming configuration, a sequence like the one in Fig. 4 is used. First, all of the connected capacitors are discharged and then the inputs of the comparator are disconnected from V_{cm} while the bottom-plates of capacitors remain connected to V_{cm}. As a result, the same voltage is applied to both inputs of the comparator and the comparison result determines the sign of the offset. In the next phases, a successive-approximation binary search is performed on offset modification bits to modify the offset in one direction depending on the offset sign.

V. SIMULATION RESULTS

Simulations show that the average switching energy consumption of the ADC for each A/D conversion would be 2.4 pJ while the average energy consumed for controlling switches is 15.2 pJ. With an ideal ENOB of 13 bit and a sampling frequency of 1.42 MHz, the figure-of-merit (FOM) is expected to be about 2.15 fJ/conversion-step. On the other hand, the energy consumption of the comparator during one complete comparison which depends on the value of the differential input can be from 300 fJ up to 1 pJ in the worst-case condition. In the typical operating conditions and the room temperature, the comparator would increase the FOM of the ADC by 0.61 fJ/conversion-step. Therefore, an overall FOM of 2.76 fJ/conversion-step is expected for this CR SAR ADC in the ideal case.

VI. CONCLUSION

A foreground calibration technique to reduce the mismatch between capacitors has been developed. Thereby, the area and power consumption of the ADC can be reduced by using small capacitors while their mismatch is compensated. Such an ADC with a relatively high resolution and a very small unit capacitance is implemented in a 65-nm technology and it is expected to show an FOM of about 2.76 fJ/conversion-step. However, since the mismatch is not totally canceled by this calibration technique and the effect of noise and the mismatch between trimming capacitors cannot be estimated through simulations in a limited time, considering an ENOB of 13 bits is too optimistic. Despite this fact, it is still expected to achieve an FOM less than 5 fJ/conversion-step.

The circuit is going be taped out and then, a number of measurements will be performed so that the performance of the ADC with the developed calibration technique can be verified.

REFERENCES

[1] B. Murmann. A/D converter trends: power dissipation, scaling and digitally assisted architectures. In *Custom Integrated Circuits Conference, 2008. CICC 2008. IEEE*, pages 105–112, Sept 2008.

[2] J.L. McCreary and P.R. Gray. All-MOS charge redistribution analog-to-digital conversion techniques. i. *Solid-State Circuits, IEEE Journal of*, 10(6):371–379, Dec 1975.

[3] J.H. Mueller, S. Strache, L. Busch, R. Wunderlich, and S. Heinen. A calibratable capacitance array based approach for high resolution CR SAR ADCs. In *Mixed Design of Integrated Circuits and Systems (MIXDES), 2012 Proceedings of the 19th International Conference*, pages 183–188, May 2012.

[4] R.J. Baker. *CMOS: Circuit Design, Layout, and Simulation*. IEEE Press Series on Microelectronic Systems. Wiley, 2nd revised edition, 2008.

[5] B.P. Ginsburg and A.P. Chandrakasan. An energy-efficient charge recycling approach for a SAR converter with capacitive DAC. In *Circuits and Systems, 2005. ISCAS 2005. IEEE International Symposium on*, pages 184–187 Vol. 1, May 2005.

[6] M. Miyahara, Y. Asada, D. Paik, and A. Matsuzawa. A low-noise self-calibrating dynamic comparator for high-speed ADCs. In *Solid-State Circuits Conference, 2008. A-SSCC '08. IEEE Asian*, pages 269–272, Nov 2008.

A 10 bit 12.8 MS/s SAR Analog-to-Digital Converter in a 250 nm SiGe BiCMOS Technology

Johannes Digel, Markus Grözing, Manfred Berroth
Institute of Electrical and Optical Communications Engineering
University of Stuttgart, Germany

Abstract—This paper presents a 10 bit 12.8 MS/s successive approximation register (SAR) analog-to-digital converter (ADC) implemented in a 250 nm SiGe BiCMOS technology. An energy-efficient switching algorithm with top-plate sampling is applied which reduces the total input capacitance by 50%. High-impedance inputs with emitter followers and internal reference voltage generation make it suitable for applications that require precise on-chip voltage monitoring.

For a low-frequency input signal, measured SNDR and SFDR of the presented SAR ADC are 48.7 dB and 57.8 dB. The effective resolution bandwidth (ERBW) is 19 MHz. The ADC draws 17.4 mA from a 2.6 V supply including reference voltage generation, clock drivers and emitter follower buffers for input and reference voltages. The die area is $2.1 \times 0.7\,\text{mm}^2$ with the ADC core occupying $1 \times 0.5\,\text{mm}^2$.

A formula for relating static nonlinearity (INL) measurements with dynamic SNDR/ENOB measurements is derived. From output codes recorded with constant input voltages, the distortion power caused by nonlinearity and the noise of the reference voltage source and the comparator are determined. After adapting them to sinusoidal inputs, the expected impact on SNDR and ENOB is derived.

I. Introduction

Medium-resolution medium-speed ADCs are suitable for a variety of applications. In wireless communication systems an ADC digitizes the received signal before it is passed on to a digital signal processor. Especially narrowband sub-1 GHz wireless communication systems can be implemented with the presented kind of converter. Other applications are sensor systems where analog sensor measurements are to be converted into digital domain to be sent via a bus system. Furthermore such ADCs can be integrated into more sophisticated circuits in an assistive manner, e.g. for calibration. In a high-speed, low-resolution time-interleaved ADC a medium-resolution, medium-speed sub-ADC can be applied to calibrate each single channel. Another application is to monitor analog voltages inside an integrated circuit.

High input impedance and internal reference voltage generation qualify the ADC to be used as an independent assistive block inside complex systems. Its fully differential design enables the integration together with CMOS logic without being prone to supply or substrate noise. The extra amount of supply power can be tolerated because assistive circuits are usually not required permanently and can be switched off when not needed.

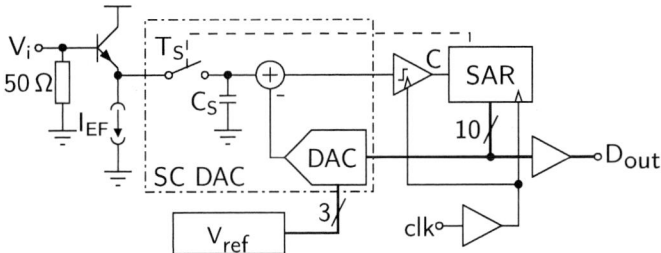

Fig. 1. Block diagram of the ADC (analog signals drawn single-ended for simplicity)

Section II introduces the applied SC-switching algorithm and explains circuit implementation details. Measurement results with constant and sinusoidal input signals are given in Section III, Section IV relates these results by investigating the impact of noise and nonlinearity on the dynamic effective resolution. Section V concludes the paper.

II. Circuit Design and Layout

The proposed ADC works with successive approximation based on a charge redistribution principle similar to the split capacitor algorithm [1]. The block diagram in Fig. 1 shows the components required for the successive approximation. The bit values that are sequentially derived by the comparator are stored in the SAR. They are fed back to a digital-to-analog converter (DAC) and the generated analog corresponding voltage is subtracted from the input signal. Track-and-Hold circuit, subtractor and DAC are included in a switched-capacitor (SC) DAC composed of binary weighted capacitors and MOSFET switches. The signal input is matched to $50\,\Omega$ (not shown) and drives an emitter follower. Three reference voltages V_{CM}, $V_{\text{ref}+}$ and $V_{\text{ref}-}$ are generated on-chip and fed to the SC DAC. The entire circuit is designed in a differential manner to reduce distortion due to substrate or supply noise.

The differential input pins are connected to emitter followers to drive the capacitive input impedance of the SC DAC. The reference voltages are generated by an operational amplifier (op-amp) with common-mode control [2] and resistive feedback. Two resistor voltage dividers generate the desired common-mode voltage and another voltage controlling the reference voltage difference. The output impedances of the common-mode and reference voltage generators are decreased using emitter followers as output buffers.

978-1-4799-4993-9/14 $31.00 © 2014 IEEE

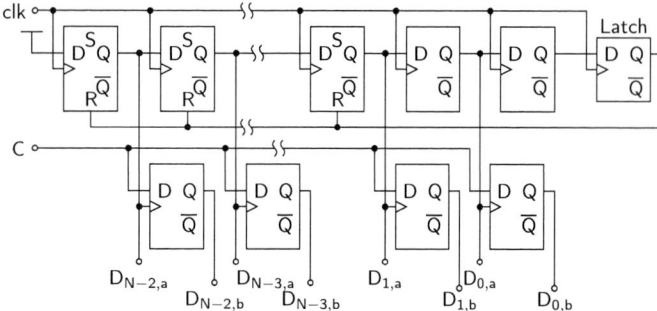

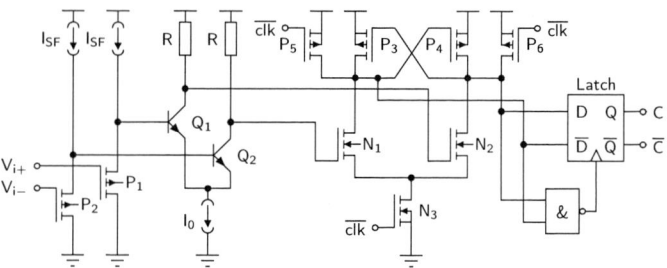

Fig. 4. Three-stage comparator

Fig. 2. Successive approximation register

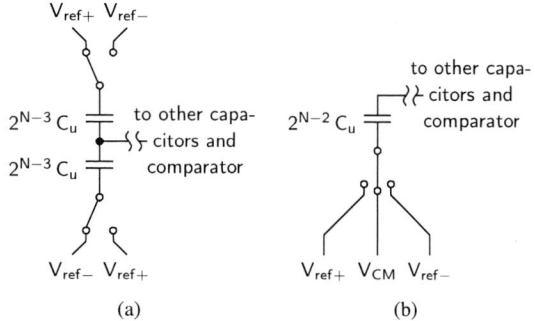

Fig. 3. Capacitor topology of an N-bit SC DAC for the MSB with (a) split capacitors forming a capacitive voltage divider [1] and (b) the implemented structure with the common-mode voltage V_{CM} as additional reference voltage

A. Successive approximation register

The SAR in Fig. 2 controls the temporal sequence of the successive approximation process and switches the SC DAC according to the determined bit values. The upper line of delay-flip-flops (DFFs) is connected as a shift register which is reset after each conversion cycle by an internally generated signal. Each conversion cycle has a length of twelve clock periods. The data register, i.e. the lower line of DFFs, stores the bit values received from the comparator via the input C. The ADC operates synchronously with the clock signal clk having a frequency of twelve times the sampling rate. The outputs $D_{i,\mathrm{a}}$ and $D_{i,\mathrm{b}}$ control the SC DAC. All components are designed pseudo-differentially, i.e. for every logical signal its inverse is generated, too.

B. Switched-capacitor digital-to-analog converter

The SC DAC is implemented with a differential array of binary weighted capacitors. The capacitor weight is set by a parallel connection of switched capacitor unit cells containing a unit capacitor of $C_{\mathrm{u}} = 5.3\,\mathrm{fF}$ and a three-input switch connecting the bottom plate to one of three voltages. A metal-insulator-metal (MIM) capacitor with a thin insulator layer is used as unit capacitor, the three-input switch is composed of two cascaded transfer gates.

The SC DAC uses top-plate sampling [3] which reduces the input capacitance by 50% and operates similar to the algorithm with split capacitors [1] where each capacitor can be connected to a positive or a negative reference voltage

$V_{\mathrm{ref}+}$ or $V_{\mathrm{ref}-}$ as shown in Fig. 3(a). However the binary weighted capacitors are not split but can be connected to the common-mode voltage $V_{\mathrm{CM}} = \frac{1}{2}(V_{\mathrm{ref}+} + V_{\mathrm{ref}-})$ instead, Fig. 3(b). The voltage transition caused by the switching operation after the MSB decision can be explained by Fig. 3. It shows the unit cell array controlled by the MSB, $N - 2$ binary downscaled unit cell arrays and a reference capacitor which are connected to the central node in parallel are not shown. With split capacitors, either the upper switch toggles to $V_{\mathrm{ref}-}$ or the lower one to $V_{\mathrm{ref}+}$, depending on the MSB. This causes the voltage across the serial capacitors to change by $\pm(V_{\mathrm{ref}+} - V_{\mathrm{ref}-})$. Considering the voltage division by two and the weight of the MSB capacitor array of 0.5 with respect to the total capacitance, the central node voltage changes by $\pm\frac{1}{4}(V_{\mathrm{ref}+} - V_{\mathrm{ref}-})$. In the proposed topology, the bottom plate of the capacitor is switched between V_{CM} and one of the reference voltages, i.e. by $\pm\frac{1}{2}(V_{\mathrm{ref}+} - V_{\mathrm{ref}-})$. Taking the capacitor weight of 0.5 into account, the central voltage changes by $\pm\frac{1}{4}(V_{\mathrm{ref}+} - V_{\mathrm{ref}-})$ as well, generating the $\frac{1}{4}$ and $\frac{3}{4}$ threshold levels for the second decision.

After reset, the bottom plates of all 512 unit capacitors are connected to the common-mode voltage V_{CM}. According to the comparator decisions, the connections are changed to $V_{\mathrm{ref}+}$ to increase the DAC output voltage or to $V_{\mathrm{ref}-}$ to decrease it. The SAR outputs $D_{i,\mathrm{a}}$ select either V_{CM} or one of the reference voltages, the outputs $D_{i,\mathrm{b}}$ decide which of them is connected to the bottom plate.

In the layout, the switched capacitor unit cells are placed in periodic structures surrounded by unconnected dummy cells for both of the differential arrays. This arrangement provides the same environment for each cell which reduces mismatch and distributes parasitics uniformly.

C. Comparator

The differential comparator shown in Fig. 4 determines one bit per step. It has three-stages, the first of which is a pair of PMOS source followers preventing a static current flow from the capacitive DAC and shifting up the common-mode level. The second stage is a current-mode-logic (CML) amplifier with NPN HBT input transistors Q_1 and Q_2. Its voltage gain of $23.7\,\mathrm{dB}$ improves the comparator's sensitivity and isolates the inputs from the regenerative nodes of the cross-coupled p-channel MOSFETs P_3 and P_4 in the third stage. These transistors drive the outputs of the third state regeneratively to

978-1-4799-4993-9/14 $31.00 © 2014 IEEE   338

Fig. 5. Chip photograph and layout

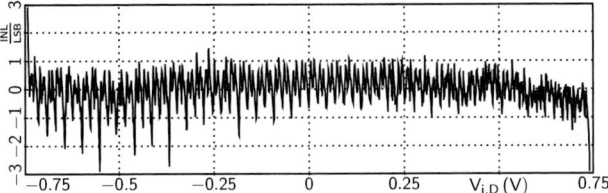

Fig. 6. INL at a sampling rate of $f_S = 12.8\,\mathrm{MS/s}$

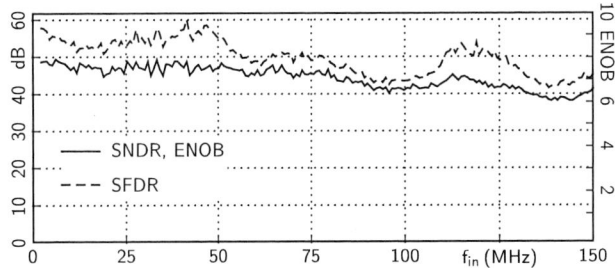

Fig. 7. SFDR, SNDR and ENOB at $12.8\,\mathrm{MS/s}$ over the input signal frequency f_{in}

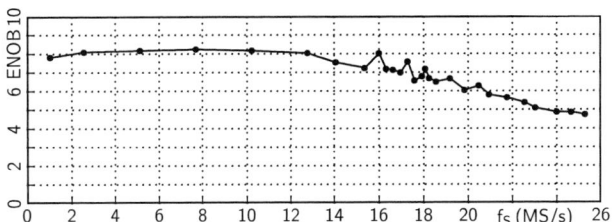

Fig. 8. ENOB over the sampling rate f_S for $f_{\mathrm{in}} \approx 2\,\mathrm{MHz}$

rail-to-rail levels after the inverted clock signal $\overline{clk}$ has become low [4]. A NAND gate is connected to the latch outputs and acknowledges the completed decision. Its output triggers a clocked latch holding the comparator decision during the reset phase.

D. Layout

A die photograph is shown in the upper half and the chip layout in the lower half of Fig. 5. The die size is $2.1 \times 0.7\,\mathrm{mm}^2$. The ADC core occupies approximately $0.5\,\mathrm{mm}^2$ in the left half of the chip. Two differential arrays of unit capacitors and switches in the upper and lower half form the SC DAC. SAR and comparator are placed in the gap between these arrays, the output register follows to the right. The reference voltage generator and the emitter followers are placed in the left part of the ADC core.

III. MEASUREMENT RESULTS

This section presents the measured DC and AC characteristics of the ADC. Measurements are performed with a DC voltage source and a sinusoidal signal generator connected to the ADC inputs and a clock source providing the twelvefold sampling rate. The digital outputs are recorded by a logic analyzer. At a conversion rate of $12.8\,\mathrm{MS/s}$, the ADC draws $17.4\,\mathrm{mA}$ from a $2.6\,\mathrm{V}$ supply.

Fig. 6 shows the integral nonlinearity (INL) at a sampling rate of $f_S = 12.8\,\mathrm{MS/s}$ [5]. It is worse for negative differential input voltages and approximately remains below $\pm 1\,\mathrm{LSB}$ in the positive range. For smaller sampling rates, e.g. $1\,\mathrm{MS/s}$, the tracking period is 12.8 times longer and the INL is smaller than $0.6\,\mathrm{LSB}$. This indicates that the sampling switches dominate nonlinearity compared to capacitor matching at $12.8\,\mathrm{MS/s}$. Assuming equal probability for all codes, the rms value of the the INL is $\sigma_{\mathrm{INL,eq}} = 0.54\,\mathrm{LSB}$.

The spurious-free dynamic range (SFDR), signal-to-noise-and-distortion ratio (SNDR) and the effective number of bits (ENOB) at $f_S = 12.8\,\mathrm{MS/s}$ are given in Fig. 7. For low-frequency inputs, the SFDR and the SNDR are $57.8\,\mathrm{dB}$ and $48.7\,\mathrm{dB}$. Near Nyquist frequency, the SFDR is $55.4\,\mathrm{dB}$ and the SNDR is $49.5\,\mathrm{dB}$. ENOB reaches 7.9 at DC and remains over 7 up to an input frequency of $f_{\mathrm{in}} = 80\,\mathrm{MHz}$. The effective resolution bandwidth (ERBW) is $19\,\mathrm{MHz}$ which exceeds the third Nyquist band.

Fig. 8 shows that the ADC can operate at higher sampling rates than $12.8\,\mathrm{MS/s}$ but its performance decreases. For lower sampling rates and an input signal frequency of $f_{\mathrm{in}} \approx 2\,\mathrm{MHz}$, the ENOB reaches a maximum of 8.2 from $4\,\mathrm{MS/s}$ to $10\,\mathrm{MS/s}$. For sampling rates exceeding $13\,\mathrm{MS/s}$, the performance is limited by incomplete sampling due to the bandwidth of the sampling switch resistance and input capacitance.

IV. RELATIONSHIP OF DC AND AC MEASUREMENTS

Nonlinearity error and noise deteriorate the dynamic performance of the ADC. In this section, the INL measurement is analyzed and its expected value and variance, hence nonlinearity and noise, are related to the maximum achievable SNDR and ENOB.

The INL derivation is based on records of 200 samples for a set of equidistant, constant input voltages $V_{\mathrm{i,D}}$ [5]. Subtracting the mean values of these records from an ideal straight line representing an ideal transfer characteristic of an infinite-resolution ADC yields the INL given in Fig. 6. The standard deviations $\sigma_{\mathrm{Code}}(V_{\mathrm{i,D}})$ given in Fig. 9 show the amount of random variations of the output codes for a constant input voltage $V_{\mathrm{i,D}}$. For input voltages near zero, the mean of $\sigma_{\mathrm{Code}}(V_{\mathrm{i,D}})$ has a minimum and it increases towards the edges of the voltage range. These random variations are mainly caused by two sources: comparator noise and reference voltage noise.

Comparator noise has the same impact on all comparator decisions, hence its power σ_{cmp}^2 effects the conversion range

978-1-4799-4993-9/14 $31.00 © 2014 IEEE

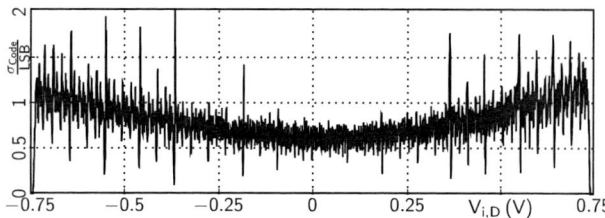

Fig. 9. Standard deviation $\sigma_{\text{Code}}(V_{i,D})$ of codes in INL measurement

uniformly. Noisy reference voltages directly disturb the outer edges of the input voltage range but have no influence on the middle level. Thus the contribution of the differential reference voltage noise to the standard deviation of measured codes depends linearly on the differential input voltage magnitude. The noise power $\sigma_{\text{Code}}^2(V_{i,D})$ that appears for a constant input voltage difference $V_{i,D}$ is given by

$$\sigma_{\text{Code}}^2(V_{i,D}) = \left(\sigma_{\text{ref}} \frac{V_{i,D}}{V_{i,D,\text{max}}}\right)^2 + \sigma_{\text{cmp}}^2 \qquad (1)$$

where the maximum allowed input voltage magnitude $V_{i,D,\text{max}}$ is approximately $0.74\,$V. In Fig. 9, the mean value of $\sigma_{\text{Code}}(0)$ gives the comparator's noise contribution $\sigma_{\text{cmp}} \approx 0.6\,$LSB. From $\sigma_{\text{Code}}(\pm V_{i,D,\text{max}})$ the noise contribution from the reference voltage generator $\sigma_{\text{ref}} \approx 0.9\,$LSB can be determined.

For measuring the SFDR, SNDR or effective resolution of an ADC, a sine signal is applied to the input. In this case, the output codes of the ADC are not equally distributed but follow the probability density function (pdf) of a sine signal. The pdf of an unscaled sine wave is

$$f_X(x) = \frac{1}{\pi \sqrt{1 - x^2}} \qquad (2)$$

for $x \in (-1, 1)$ and $f_X(x) = 0$ elsewhere. This pdf can be scaled to fit the input voltage range by replacing x by $\frac{V_{i,D}}{V_{i,D,\text{max}}}$.

The distortion power contained in the digitized sine signal due to nonlinearity can be estimated by the INL values of Fig. 6 weighted by a scaled version of Equ. (2). Summing up the weighted nonlinearity distortion power yields

$$\sigma_{\text{INL}}^2 = (0.55\,\text{LSB})^2. \qquad (3)$$

The amount of random excess noise can be estimated in a similar way by weighting σ_{Code} with Equ. (2) and determining its mean square. For the measurement shown in Fig. 9 this results in

$$\sigma_N^2 = (0.91\,\text{LSB})^2. \qquad (4)$$

The SNDR of an ADC with the signal power P, quantization noise power $\sigma_Q^2 = \frac{\text{LSB}^2}{12}$ and uncorrelated excess noise and distortion power $\sigma_E^2 = \sigma_{\text{INL}}^2 + \sigma_N^2$ is given by

$$SNDR = 10 \lg \frac{P}{\sigma_Q^2 + \sigma_E^2} = 10 \lg \frac{P}{\sigma_Q^2} - 10 \lg \frac{\sigma_Q^2 + \sigma_E^2}{\sigma_Q^2}. \qquad (5)$$

From the INL measurement, the distortion power and excess noise $\sigma_E^2 = \sigma_{\text{INL}}^2 + \sigma_N^2$ have been derived in Equ. (3) and

(4). Thus nonlinearity and random noise degrade the signal-to-noise-and-distortion ratio by $10 \lg(\sigma_Q^2 + \sigma_{\text{INL}}^2 + \sigma_N^2)/\sigma_Q^2$ dB. With these results, the expected effective resolution from DC measurements is

$$ENOB_{\text{ex}} = 10 - \frac{10 \lg \frac{\sigma_Q^2 + \sigma_{\text{INL}}^2 + \sigma_N^2}{\sigma_Q^2}}{6.02} = 8.07 \qquad (6)$$

which fits well to the ENOB measurements. The slightly worse measured ENOB is mainly caused by additional nonlinearity of the sampling switch for changing input voltages.

The ENOB could be improved by optimizing the noise bandwidth of the reference voltage source and finding a better compromise between comparator speed and noise.

V. Conclusion

An ADC suitable for the application as an assistive circuit is presented. Having high-impedance inputs and generating its own reference voltages, it can be used to monitor an internal voltage or for calibration of an analog or mixed-signal circuit. The relatively large power consumption of $45.24\,$mW is dominated by emitter follower buffers and the op-amp that generates the reference voltages. It can be tolerated because the ADC used as an assistive circuit can be turned off during normal operation of the surrounding circuit. The clock frequency determines the flexible conversion rate which only has minor impact on the performance up to $12.8\,$MS/s. Thanks to the fully differential design, the ADC can be integrated together with logical circuits without being prone to substrate or supply noise.

At $12.8\,$MS/s and low frequency input signals, the effective resolution is $7.9\,$bit and remains over $7\,$bit up to $80\,$MHz. For lower sampling rates, the effective resolution near DC input reaches $8.4\,$bit. This agrees well with INL measurements at lower sampling rates showing less nonlinearity distortion error.

A method to relate the results of DC-input measurements to dynamic measurements of SNDR and ENOB is presented. This method reliably predicts the ADC's dynamic performance from nonlinearity measurements.

VI. Acknowledgement

The authors thank J.-C. Scheytt and IHP for providing chip area in IHP's $250\,$nm SiGe BiCMOS technology and supporting the circuit design.

References

[1] B. Ginsburg and A. Chandrakasan, "An energy-efficient charge recycling approach for a SAR converter with capacitive DAC," ISCAS 2005, vol. 1, May 2005, pp. 184–187.

[2] J. Digel, M. Masini, M. Grözing, M. Berroth, G. Fischer, O. Sonom, H. Gustat, and J.-C. Scheytt, "An Integrating Digitizer for an IR-UWB Receiver," ANALOG'11, Nov. 2011.

[3] C.-C. Liu, S.-J. Chang, G.-Y. Huang, and Y.-Z. Lin, "A 0.92 mW 10-bit 50-MS/s SAR ADC in 0.13 μm CMOS process," Symposium on VLSI Circuits 2009, Jun. 2009, pp. 236–237.

[4] J. Digel, M. Grözing, M. Berroth, H. Gustat, and J.-C. Scheytt, "High-speed comparators for SAR ADCs in 130 nm BiCMOS," PRIME 2010, Jul. 2010.

[5] "IEEE Standard for Terminology and Test Methods for Analog-to-Digital Converters," *IEEE Std 1241-2010 (Revision of IEEE Std 1241-2000)*, pp. 1–139, Jan 2011.

Continuous Time Analog Filters Design in Ultra-Deep-sub-1µm CMOS Technologies

M. De Matteis[1], A. Baschirotto[1]

[1]Physics Department. University of Milano Bicocca. Italy.
marcello.dematteis@unimib.it, andrea.baschirotto@unimib.it

Abstract— Analog filters are widely used in several kinds of integrated mixed-signal systems, since they are intrinsically needed for in-band signal selection, out-of-band noise rejection and anti-aliasing for the A-to-D (D-to-A) conversion.

CMOS technological scaling down has already entered in nm-range scenario, leading to severe degradation of some of the most important MOS device performance for analog design. Among them, lower V_{DD}/V_{TH} ratio (supply voltage and threshold voltage, respectively), higher power consumption, and transistor intrinsic gain decreasing. On the other side the increasing MOS transistor transconductance (g_m) and transition frequency (f_T) enables tens of GHz applications, which take advantage of cheaper and smart CMOS processes.

Specific continuous-time analog filters will be presented in this paper, with the aim to introduce novel circuital topologies or optimizations and techniques suitable to mitigate the severe issues present in sub-90nm CMOS technological nodes, and at the same time to exploit the opportunity of higher technological scaling-down.

Keywords—*Analog Filters; Continuous-Time; nm CMOS Technology; sub-1V; Low Power.*

I. INTRODUCTION.

Nanometer range CMOS technologies are a key opportunity for smart and efficient mixed-signal systems implementation. Integration of analog and digital circuits in the same die area is then sustained by the technological scaling-down, since lower power consumption can be achieved in mostly-digital systems due to the lower supply voltage[1].

Even though digital signal processing is replacing several analog functions, the first interface of the mixed-signal systems with the external world is intrinsically analog. In addition mixed-signal systems require a conversion step from analog (digital) to digital (analog) domain. For this reason analog filters play a key role for signal conditioning, in terms of bandwidth selection, out-of-band noise/interferes rejection, or anti-aliasing. As a consequence, in order to achieve efficient and reliable circuits in nm-range CMOS technologies, analog designers must be able to manage the poor analog performance of the MOS transistors and at the same time to take advantage of the opportunities given by the increasing technological scaling-down

The scaling-down of the physical (length, oxide thickness, etc) and electrical (supply voltage) parameters leads to a significant improvement for digital circuits in terms of speed and power. This trend is not so automatic for analog circuits, since they experience severe design drawbacks, like V_{DD}/V_{TH} decreasing (supply voltage and threshold voltage, respectively), higher sensitivity to Process-Voltage-Temperature (PVT) variations, transistor intrinsic gain decreasing and in general higher current consumption.

Nevertheless, increasing transition frequency in CMOS nm-range technologies is an evident opportunity for high-data rate wireless transceivers implementation[4], and as a consequence analog designers must be able to operate in such contrasting scenario, where wide bandwidth analog signals have to be processed with acceptable frequency mask accuracy, and maintaining the Signal-to-Noise ratio.

The paper is organized as follows. Section 2 introduces a WLAN filter operating at 0.55V supply voltage. A proper common mode circuit and specific design choices have been used to achieve a stable operating point for the Opamps, overcoming the V_{DD}/V_{TH} issues so low as 2. The filter has been designed in CMOS 0.13µm technology, with V_{TH}=0.35V[2].

In Section 3 and Section 4, two analog filters circuits implemented in CMOS 45nm and 28nm processes, currently under development, will be presented for operating in next generation 60GHz transceivers. In particular the first one is a complete design of a closed-loop 4th order Active-RC low-pass filter, at 1GHz -3dB frequency[3]. The second filter is based on a super-buffer architecture[4] adequately improved for synthesizing biquadratic cell and amplifiers.

II. A 0.55V 4TH-ORDER ANALOG BASEBAND FILTER.

A 0.55V supply voltage 4th-order low-pass continuous-time filter is presented. The low-voltage operating point is achieved by an improved bias circuit that uses different opamp input and output common-mode voltages. The 4th–order filter architecture is composed by two Active-Gm-RC biquadratic cells, which use a single opamp per-cell with a unity-gain-bandwidth comparable to the filter cut-off frequency. The -3dB filter frequency is 12MHz and it can be adjusted by means of a digitally-controlled capacitance array. In a standard 0.13µm CMOS technology with V_{THN}≈0.3V and V_{THP}≈0.35V, the filter operates with a supply voltage as low as 0.55V. The filter consumes 3.4mW. A 8dBm-in-band IIP3 and a 13.3dBm-out-of-band IIP3 demonstrate the validity of the proposal.

A. 0.55V Filter Design.

The complete schematic of a 4th order filter meeting the WLAN receivers specifications is illustrated in Fig. 1. The filter is composed by the cascade of two biquadratic cells. Each biquadratic cell is implemented using the Active-Gm-RC topology[5], where the Opamp unity gain bandwidth is used as filter transfer function parameter. The low-pass transfer function of the biquadratic cell is given by:

$$\text{eq. 1} \qquad T(s) = G \cdot \frac{1}{s^2 \cdot C_1 \frac{R_2}{\omega_{ugbw}} + s \cdot \frac{1}{\omega_{ugbw}}\left(1 + \frac{R_2}{R_1}\right) + 1}$$

where G is the biquadratic cell dc-gain, R_1-R_2-C_1 are the passive components in Fig. 1, and ω_{ugbw} is the Opamp unity gain bandwidth.

The current source pair composed by M_{B1}-M_{B2} (M_{B3}-M_{B4}) transistors allows to lower the Opamp input common mode voltage to 0.1V. As a result, the current (I_{B12}) flowing by the M_{B1}-M_{B2} fixes the Opamp input common mode voltage ($V_{inOP,DC}$) as indicated in eq. 2:

$$\text{eq. 2} \qquad V_{inOP,DC} = \frac{V_{DD}}{2} - I_{B12} \cdot \left(\frac{R_1 \cdot R_2}{R_1 + R_2}\right)$$

The novelty of this filter is the presence of the Input-Common-Mode-Feedback-circuit. The main task of the I-CMFB is to reduce the sensitivity of the biquadratic cell operating point with respect to process-voltage-temperature (PVT) variations. Since in CMOS process the physical and electrical characteristics of the devices change due to the process and temperature, the Opamp input common mode voltage (in theory fixed by M_{B1}-M_{B2} current) could divert from nominal 0.1V biasing value. The simple resistor variation modifies the $V_{inOP,DC}$ value, as reported in eq. 2.

The operating point margin for Opamp input differential pair at 0.55V is very limited, so that every unwanted variation can be detrimental for the circuit operating point. The main task of the I-CMFB is to sense eventual common mode variations and to reregulate the M_{B1}-M_{B2} current in order to maintain the Opamp dc

978-1-4799-4993-9/14 $31.00 © 2014 IEEE

input voltage approximately equal to 0.1V. The Input-CMFB circuit is a closed-loop circuit, whose main tasks are:
- To sense the variation of the Opamp input common mode voltage (fixed at 0.1V in nominal conditions).
- To increase/decrease the MB1-MB2 biasing current as a function of the sensed Opamp input CM variation.

The sensing task is performed connecting the source of the fully-differential Opamp input MOS with the plus node of the single-ended Opamp, as illustrated in Fig. 1(see V_{CM_SENSE} node). In this way no additional components connected at the Opamp input nodes are needed for implementing the sensing.

B. 0.55V Filter Experimental Results.

A prototype of the proposed filter has been realized in a 0.13μm CMOS technology. The chip photo is then shown in Fig. 3. The core area is 0.45 mm^2 and is widely dominated by the capacitances. As illustrated in Fig. 1, capacitor arrays are used in order to compensate the CMOS process deviation. The generic schematic of the Opamps used in both biquadratic cell is shown in Fig. 2, where the input stage bias circuit is also shown (left side). It correlates the transconductance variations of the opamp input pair transistors (due to PVT spread) to the variation of the resistances of each cell and forces the opamp input stage g_m to be proportional to R_M.

Fig. 4 shows the filter gain frequency response for V_{DD}=0.55V, V_{DD}=0.525V, and V_{DD}=0.5V. A 11.3MHz cut-off frequency is obtained, performing only 5% of deviation with respect to the nominal -3dB frequency, as reported in Table I.

The dc gain is -1.4dB. In fact each biquadratic cell introduces some drop at the dc, dependent on the finite dc gain of low voltage opamps. In addition the frequency responses are plotted also for values below V_{DDmin} (525mV and 500mV). The selectivity of the filter is approximately maintained also at 0.5V supply voltage. Unfortunately, a dc-gain drop appears for 0.5V supply voltage, due to the reduced Opamps gain at lower V_{DD}. A 0.5dBm 1dB-Compression Point is achieved, the in band IIP3 has been also evaluated and is equal to 10dBm.

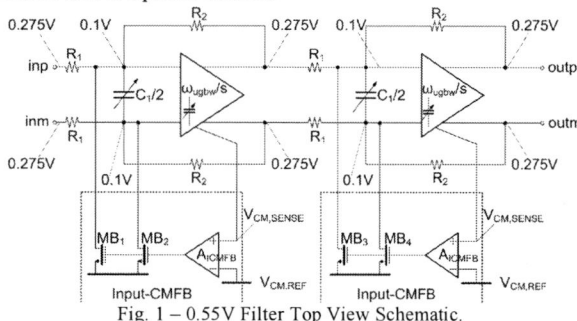

Fig. 1 – 0.55V Filter Top View Schematic.

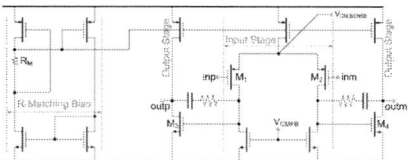

Fig. 2 – 0.55V Opamp Top View Schematic.

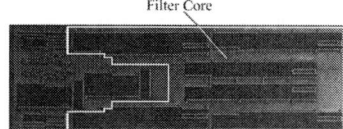

Fig. 3 – Prototype Filter Photo.

Parameter	Value
Filter Order	4th
G	0dB
Cell1/Cell2 Poles Frequency – $f_{@-3dB}$	11.3MHz
V_{DD}	0.55V
CMOS	0.13μm (V_{THNMOS}=0.3V)
Power Cons.	3.5mW
In-band IIP3	10dBm
1dBcP	0.5dBm
Output Noise	110 μV$_{rms}$
SNR – THD@40dBc	60DB

Table 1 – 0.55V Filter Performance.

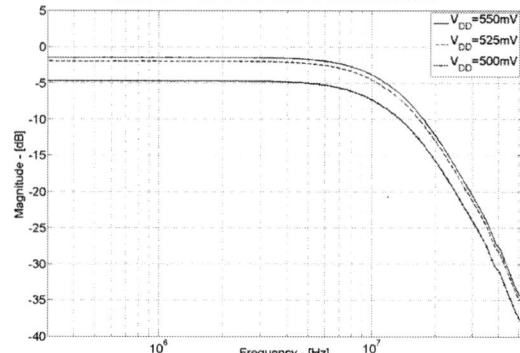

Fig. 4 – Filter transfer function vs. V_{DD}.

III. A 54DB-DR 1-GHz-BANDWIDTH CONTINUOUS-TIME LOW-PASS FILTER WITH IN-BAND NOISE REDUCTION.

The Active-RC topologies are often selected for continuous-time filter design, due to the capability to guarantee large linearity and high frequency response accuracy, comparing with g_m-C filters where power consumption and noise are typically lower, while linearity and frequency response can be critical. Actually, a typical design issue in broad band Active-RC filters is thermal noise power spectral density, since when it is integrated over the entire bandwidth it can affect critically the Signal-To-Noise-Ratio. This issue is further stressed in 57-66GHz transceivers, where base-band chain bandwidth requirements are in the GHz order. The bandwidth request imposes deep integration scale, so that typically CMOS 65nm and down nodes are required, since they exhibit larger transition frequency. In this scenario the analog filter here presented, proposes a circuital solution, suitable to reduce the in-band integrated noise power, and to perform 55dB SNR. The key concept of this filter is to synthesize a 5th-order 1GHz-bandwidth low-pass filter using a single compact Active-RC closed-loop cell. Most of the thermal noise sources are high-pass filtered, while maintaining the low-pass filtering of the in-band signal. A prototype of the filter has been designed in CMOS 45nm technology, with 1GHz –3dB-bandwidth.

A. 5th-Order 1GHz-Bandwidth Filter Scheme.

Fig. 5 shows the main scheme of the filter. There are three main blocks: an integrator (OP_1-R_{in}-C_1) and two biquadratic cells (by OP_2 and OP_3). The integrator is fully defined by its unity gain frequency.

The two biquadratic cells must be adequately sized in order to comply with the transfer function requirements. Two feedbacks are then needed to synthesize a 5th order low-pass transfer function.

B. Filter Transfer Function.

The most common implementation approach for a 5th order filter is the cascade of three Active-RC cells (one for the real pole and one cell for each complex poles pair). At 0dB dc-gain every cell contributes to the overall thermal noise power.

978-1-4799-4993-9/14 $31.00 © 2014 IEEE

In these conditions, large power is expected for (low) resistive load driving, and Opamp input stage g_m (for noise power reduction).

The presented design approach uses a single compact cell for an overall 5th order transfer function, where in-band noise reduction is implemented, and power consumption is 9mW.

The integrator is designed by using a single-Opamp topology, and the two biquadratic cells are implemented in Active-G_m-RC configuration. That allows to synthesize a complex poles pair using only one additional Opamp, comparing with the most popular 2nd-order Tow-Thomas biquad. The filter ideal transfer function is:

$$\text{eq. 3} \qquad T(s) = \frac{G}{1+\frac{s}{\omega_0}} \cdot \frac{1}{1+\frac{s}{\omega_0 \cdot Q_1}+\left(\frac{s}{\omega_0}\right)^2} \cdot \frac{1}{1+\frac{s}{\omega_0 \cdot Q_2}+\left(\frac{s}{\omega_0}\right)^2}$$

The transfer function is composed by two complex poles pairs (with the same frequency ω_0, and quality factor Q_1-Q_2) and one negative real pole (having the same frequency ω_0). The filter transfer function can be expressed as a function of the design parameters: R-C values, and OP_2-OP_3 unity gain bandwidth (ω_{u2} and ω_{u3}, since in Active-G_m-RC cell the Opamp is used as integrator). The resulting transfer function based on the circuit in Fig. 5 is:

$$\text{eq. 4} \qquad T(s) = \frac{R_6}{R_{in}} \cdot \frac{1}{a_{rc5} \cdot s^5 + a_{rc4} \cdot s^4 + a_{rc3} \cdot s^3 + a_{rc2} \cdot s^2 + a_{rc1} \cdot s + 1}$$

where a_i parameter are function of R-C values, and OP_2-OP_3 unity gain bandwidth (ω_{u2} and ω_{u3})[3].

C. Noise.

The only thermal noise sources affecting the overall filter noise power spectral density are dependent on R_i, R_6 and OP1. In fact, due to the presence of C_1 capacitance, no feedback is present between the OP1 output and its inverting input (at dc). So that if the signal is not coming from v_{in}, i_6 current is always zero. As a result, every internal loop noise contribution is totally high-pass filtered out by the same circuital topology. The final expression of the in-band IRN for the filter (IRN$_{FILTER}$) in Fig. 5 is given by eq. 5:

$$\text{eq. 5} \qquad IRN_{FILTER}^2 = 8 \cdot k \cdot T \cdot \left(R_{in} + R_6 \cdot \left(\frac{1}{G}\right)^2\right) + IRN_{OP1}^2 \cdot \left(1+\frac{1}{G}\right)^2$$

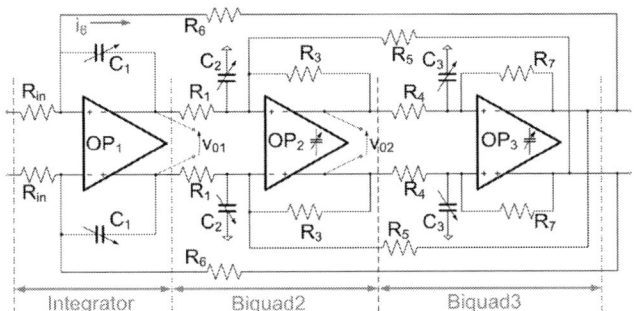

Fig. 5 – 5th Order Filter Top-View Schematic.

Obviously, increasing frequency, the C_1 impedance decreases and the inside-loop noise contributions become critical. However, the in-band noise is strongly attenuated.

IV. SUPER-BUFFER-BASED BASEBAND FOR 60GHz TRANSCEIVERS.

The object of this section is a compact reconfigurable base band section for 60GHz transceivers. The base band is composed by the cascade of two macro-blocks. The first one is a reconfigurable gain filter (5dB and 25dB), and the second one is a 4th-order Butterworth low-pass filter. Both blocks features variable bandwidth, 880MHz and 1.76GHz as in 60Ghz telecommunication standard.

The base band section is designed in 28nm CMOS technology and it is currently under development, so that the main performance will be summarized as simulation results.

One of the main ideas of this base band section is to exploit a single compact cell as key building block for filtering and amplification. Such cell is the super-buffer circuital topology, whose basic schematic is shown in Fig. 6 [4].

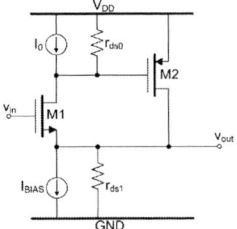

Fig. 6 – Super-Buffer.

The M2 transistor implements a local loop, where at dc the loop gain is given by eq. 6 (the M1 drain-source resistance has been neglected, r_{ds0} and r_{ds1} are the overall resistance connected at M1 drain and source respectively).

$$\text{eq. 6} \qquad G_{loop} \cong -\frac{g_{m1} \cdot r_{ds1} \cdot g_{m2} \cdot r_{ds0}}{1+g_{m1} \cdot r_{ds1}} \cong -g_{m2} \cdot r_{ds0}$$

The dc-gain of the super-buffer topology results improved with respect to the simple follower, as indicated in eq. 7:

$$\text{eq. 7} \qquad G_{dc} \cong \frac{g_{m1} \cdot r_{ds1} \cdot g_{m2} \cdot r_{ds0}}{1+g_{m1} \cdot r_{ds1} \cdot g_{m2} \cdot r_{ds0}} \cong 1$$

In addition, the output resistance ($1/g_m$ in simple follower implementation) is here a factor $g_m \cdot r_{ds}$ lower, as in eq. 8:

$$\text{eq. 8} \qquad R_{OUT} \cong \frac{r_{ds1}}{1+g_{m1} \cdot r_{ds1}+g_{m1} \cdot r_{ds1} \cdot g_{m2} \cdot r_{ds0}} \cong \frac{1}{g_{m1}} \cdot \frac{1}{g_{m2} \cdot r_{ds0}}$$

A. Filter and Amplifier.

Starting from the circuital topology in

Fig. 6, a 2nd-order biquadratic cell can be synthesized as shown in Fig. 7. C_2 is connected between drain and source of M1, and C_1 is connected between M1 source and ground. The biquadratic cell in Fig. 7 is in pseudo-differential topology. The poles frequency and quality factor of the single complex poles pair are given by:

$$\text{eq. 9} \qquad \omega_0^2 = \frac{g_{m2} \cdot g_{m1}}{C_1 \cdot C_2}$$

$$\text{eq. 10} \qquad Q = \sqrt{\frac{C_2}{C_1} \cdot \frac{g_{m1}}{g_{m2}}}$$

Resuming the v_{out}/v_{in} transfer function of the filter is:

$$\text{eq. 11} \qquad T(s) = \frac{1}{1+s \cdot \frac{C_1}{g_{m1}}+s^2 \cdot \frac{C_1 \cdot C_2}{g_{m2} \cdot g_{m1}}}$$

The same super-follower cell can be used for amplifier implementation, as shown in eq. 12, where the transfer function at low frequency is:

$$\text{eq. 12} \qquad T(s) = 1 + \frac{R_2}{R_1}$$

These two cells have been used to design a 7th-order low pass filter with programmable gain and bandwidth in CMOS 28nm technology at 1.1V supply, whose main schematic (in single ended version) is shown in Fig. 7. The base band has been definitively implemented in fully differential topology, and it is reconfigurable in terms of gain (5dB and 25dB) and bandwidth (0.88GHz and 1.76GHz). There are two main macro blocks. The first one is composed by the cascade of three amplifiers (1st-order filters), with 5dB, 10dB, and 10dB gain for each one. PGA1-PGA2-PGA3 have about 3GHz -3dB-bandiwdth, and their main purpose is to perform a low-noise amplifications of the weak input signal at the base band chain input nodes. The second macro is a 4th-order low-pass filter with programmable cut-off frequency (0.88MHz and 1.76GHz).

Frequency programmability is implemented at constant quality factor by acting on C_1-C_2 capacitors. An overall resume of the base band performance is reported in Table 2. Proper switches configurations are set to optimize power as a function of the required gain. For 5dB gain PGA2 and PGA3 are switched off, reducing power and noise. The SNR varies accordingly to dc-gain from 36dB, at 25 dB dc-gain case, up to 57 dB at 0dB dc-gain. The input referred noise remains stable at 5nV/√Hz.

V. CONCLUSIONS.

The main parameters for analog filters comparison are power, linearity, noise and bandwidth. Telecommunication standards typically impose severe requirements to the base band analog filters, and for this reason the intermodulation product performance vs. noise are taken into account for linearity. One of the most usual approach, in order to correctly compare different filters for different bandwidth and standards, is to use the following Figure-of-Merit[1]:

$$eq.\ 13 \quad FoM = 10 \cdot log_{10} \frac{IMFDR_3 \cdot f_{-3dB} \cdot N}{P_W}$$

where PW is the total power consumption, f_{-3dB} is the cut-off frequency, N is the number of poles, and $IMFD_3$ is the spurious-free IM3 and it is calculated as follows:

$$eq.\ 14 \quad IMFDR_3 = \left(\frac{IIP3}{V_{N,in}}\right)^{4/3}$$

IIP_3 is the 3rd order input intercept point, while $V_{N,in}$ is the in-band integrated input referred noise. Both IIP_3 and $V_{N,in}$ are expressed in V_{rms}. The Figure-of-Merit in eq. 13 is plotted vs. Power-per-Poles in Fig. 8.

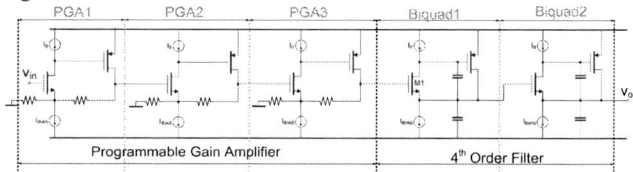

Fig. 7 – Super-Buffer-Based 7th order filter.

Bandwidth	1.76GHz		870MHz	
dc-gain	5dB	25dB	5dB	25dB
-3dB Bandwidth	1.87GHz	1.81GHz	889MHz	868MHz
IRN	5.01nV/√Hz	5.07nV/√Hz	5nV/√Hz	4.7nV/√Hz
In-Band Noise (100kHz÷2GHz)	285μVrms	2.99mVrms	283μVrms	2.3mVrms
$V_{OUT,SWING}$ (THD≥30dBc)	0.3V_{0-peak}	0.28V_{0-peak}	0.3V_{0-peak}	0.2V_{0-peak}
SNR	57.13dB	36.5dB	57.4dB	35.77dB
Power	13mW	20mW	5mW	7mW
Output IP3	7dBm	2dBm	8dBm	3dBm

Table 2 – Base band Section Performance.

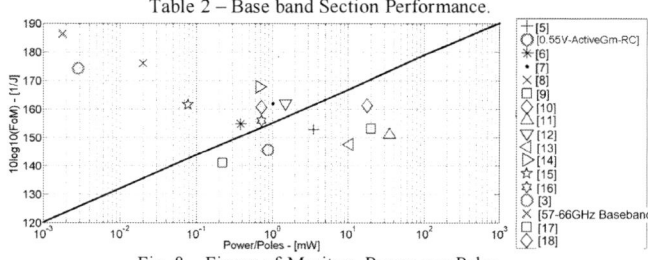

Fig. 8 – Figure-of-Merit vs. Power-per-Poles.

VI. REFERENCES

[1]   Sansen, W. "Analog design challenges in nanometer CMOS technologies". Solid-State Circuits Conference, 2007. ASSCC '07. IEEE Asian. Publication Year: 2007 , Page(s): 5-9.

[2]   De Matteis, M. ; D'Amico, S. ; Baschirotto, A. "A 0.55 V 60 dB-DR Fourth-Order Analog Baseband Filter". Solid-State Circuits, IEEE Journal of. Volume 44, Issue: 9. Publication Year: 2009. Page(s): 2525-2534.

[3]   De Matteis, M. ; D'Amico, S. ; Cocciolo, G. ; De Blasi, M. ; Baschirotto, A. "A 54dB-DR 1-GHz-bandwidth continuous-time low-pass filter with in-band noise reduction". Circuits and Systems (ISCAS), 2013 IEEE International Symposium on . Publication Year: 2013. Page(s): 1280-1283.

[4]   Wambacq, P. ; Giannini, V. ; Scheir, K. ; Van Thillo, W. ; Rolain, Y. "A fifth-order 880MHz/1.76GHz active low pass filter for 60GHz communications in 40nm digital CMOS". ESSCIRC, 2010 Proceedings of the. Publication Year: 2010. Page(s): 350-353.

[5]   S. D'Amico, V. Giannini, A. Baschirotto, "A 4th-order active Gm-RC reconfigurable (UMTS/WLAN) filter". Solid-State Circuits, IEEE Journal of Volume 41, Issue 7, July 2006 Page(s):1630 – 1637.

[6]   S. D'Amico, M. De Blasi, M. De Matteis, A. Baschirotto "A 255 MHz Programmable Gain Amplifier and Low-Pass Filter for Ultra Low Power Impulse-Radio UWB Receivers" Transactions on Circuits and Systems –I: Regular Paper, vol. 59, no.2, February 2012, pages 337-345.

[7]   S. D'Amico, M.Conta, A. Baschirotto "A 4.1-mW 10-MHz Fourth-Order Source-Follower-Based Continuous-Time Filter With 79-dB DR" IEEE Journal of Solid-State Circuits. Volume 41, no. 12, December 2006, pages 2713-2719.

[8]   S. D'Amico, M. De Matteis, A. Baschirotto "A 6th-Order 100μA 280MHz Source-Follower-Based Single-loop Continuous-Time Filter" International Solid-State Circuits Conference 2008. Digest of Technical Papers. Page(s): 72-73.

[9]   S. Pavan, T. Laxminidhi "A 70-500MHz Programmable CMOS Filter Compensated for MOS Nonquasistatic Effects" Proc. of ESSCIRC 2006.

[10]J. Harrison, N. Weste "A 500MHz CMOS Anti-Alias Filter using Feed-Forward Op-amps with Local Common-Mode Feedback" Proc. of ISSCC 2003.

[11]Tien-Yu Lo, Chung-Chih Hung "A 1 GHz OTA-Based Low-Pass Filter with A High-Speed Automatic Tuning Scheme"Proc. of ASSCC 2007.

[12]H. Amir-Aslanzadeh, E. J. Pankratz, E. Sánchez-Sinencio "A 1-V +31 dBm IIP3, Reconfigurable, Continuously Tunable, Power-Adjustable Active-RC LPF" IEEE Journal of Solid-State Circuits vol. 44, no. 2, February 2009, pages 495-508.

[13]M. Mobarak, M. Onabajo, J. Silva-Martinez, E. Sánchez-Sinencio "Attenuation-Predistortion Linearization of CMOS OTAs With Digital Correction of Process Variations in OTA-C Filter Applications" IEEE Journal of Solid-State Circuits vol. 45, no. 2, February 2010, pages 351-367.

[14]Le Ye, Huailing Liao, Congyin Shi, Junhua Liu, and Ru Huang "A 2.3mA 240-to-500MHz 6th-order Active-RC Low- Pass Filter for Ultra-Wideband Transceiver", Proc. of ASSCC 2010.

[15]L. Ye, C. Shi, H. Liao, R. Huang, Y. Wang "Highly Power-Efficient Active-RC Filters With Wide Bandwidth-Range Using Low-Gain Push-Pull Opamps" IEEE Transactions on Circuits and Systems—I: regular papers, vol. 60, no. 1, January 2013 pages 95-107.

[16]M. S. Oskooei, N. Masoumi, M. Kamarei, H. Sjöland "A CMOS 4.35-mW +22-dBm IIP3 Continuously Tunable Channel Select Filter for WLAN/WiMAX Receivers" IEEE Journal of Solid-State Circuits, vol. 46, no. 6, June 2011, pages 1382-1391.

[17]S. Chatterjee, Y. Tsividis and P. Kinget. "A 0.5V filter with PLL-based tuning in 0.18μm CMOS," ISSCC Dig. Tech. Papers, pp. 506-507, Feb. 2005.

[18]B. Drost, M. Talegaonkar, P. K. Hanumolu. "A 0.55V 61dB-SNR 67dB-SFDR 7MHz 4th-Order Butterworth Filter Using Ring-Oscillator-Based Integrators in 90nm CMOS" ISSCC Dig. Tech. Papers, pp. 360-361, Feb. 2012.

978-1-4799-4993-9/14 $31.00 © 2014 IEEE

A compact low-noise fully differential bandgap voltage reference with intrinsic noise filtering

A. N. Longhitano, F. del Cesta, P. Bruschi
Dipartimento di Ingegneria dell'Informazione
University of Pisa
Pisa - Italy
aurelio.longhitano@for.unipi.it

R. Simmarano
Sensichips srl
Latina - Italy

roberto.simmarano@sensichips.com

Abstract—**A new architecture for differential bandgap voltage references is presented. The system is based on a switched capacitor amplifier that performs correlated double sampling to cancel offset and reduce flicker noise while maintaining a valid output voltage throughout the clock cycle. The circuit noise is filtered by an intrinsic discrete time low-pass function with tunable cut-off frequency. A prototype designed with 0.18 um CMOS process is described. Preliminary performances are estimated by means of periodic noise analysis carried out with the SpectreRF simulator.**

Keywords— differential bandgap; switched capacitor; correlated double sampling; offset cancellation

I. INTRODUCTION

Voltage references are circuits that are practically mandatory in all mixed signal systems for data acquisition. A voltage reference should exhibit low dependence from both temperature and supply voltage. In a CMOS circuit, voltage references, based on the bandgap approach, can be implemented using either the substrate BJTs, or MOSFETs in weak inversion [1]. The diffusion of fully-differential architectures has been urging the development of voltage references with balanced output, i.e. with differential output and constant common mode voltage. A possible solution [2] that can be implemented in a standard CMOS process (n-well type) is shown in Fig.1. Briefly, the negative feedback operated by amplifier A sets the voltage V_e to zero. In these conditions [2] the output voltage is equal to the usual bandgap expression:

$$V_{BG} = |V_{BE1}| + V_T \frac{R}{R_1} \ln(m) \qquad (1)$$

where *m* is the ratio of Q2 to Q1 areas and V_T is equal to $k_B T/q$ with *T* the absolute temperature, k_B the Boltzmann constant and *q* the electron charge. With a proper choice of the ratio R/R_1 a voltage, which is nearly temperature independent in the temperature interval of interest, can be obtained. A serious problem that afflicts such an architecture is noise. The input voltage noise (and offset) of the amplifier is transferred to the output with a gain equal to $1/\gamma$, where γ is the small signal transfer ratio between the V_{BG} and V_e. Since $\gamma \ll 1$, the input noise voltage is multiplied by a large factor. In CMOS circuits,

characterized by large flicker noise densities, acceptable noise performances can be obtained only at the cost of very large area occupation, which limits the integration of the reference block into complex systems-on-a-chip (SoCs). Bandgap circuits, that use well known dynamic techniques [3] to reduce the flicker noise density, have been proposed.

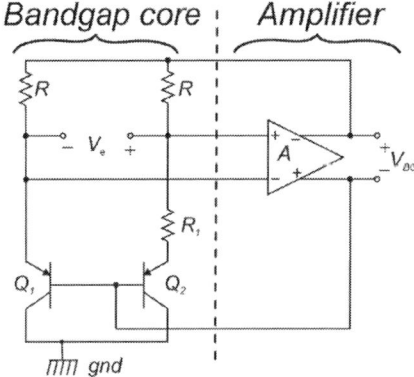

Fig. 1. Floating bandgap reference [2]

Chopper modulation can be applied to the amplifier A in order to suppress the offset and low frequency noise components. Nevertheless, chopper amplifiers require complex strategies to filter the typically large offset ripple present at the output. In [4] a switched capacitor notch filter is used to perform this operation. Solutions based on correlated double sampling (CDS) or auto-zero (AZ) techniques are generally simpler, but their noise performances are worse due to aliasing. Furthermore, typical CDS and AZ switched capacitor implementations require that the amplifier is periodically involved in an input offset sampling phase, where the output voltage gets unavailable [5] or imprecise [6]. This characteristic makes the reference block not suitable for complex systems, where several sub-circuits access the reference voltage with different paces and phases.

In this work, we propose a switched capacitor architecture that implements the CDS approach and produces an output voltage that is available during the whole clock cycle. In addition, noise is filtered by the intrinsic low pass transfer function of the system, with significant advantages in terms of total output rms noise.

This work has been partially financed by Sensichips srl.

978-1-4799-4993-9/14 $31.00 © 2014 IEEE

II. SYSTEM ARCHIECTURE

A. Principle od operation

The proposed architecture is based on the floating BG voltage reference, shown in Fig. 1 [2], combined with an original amplifier architecture aimed to reduce noise and offset related errors. The proposed architecture, illustrated in Fig. 2, consists of a two stages-two phase switched capacitors amplifier. Blocks A1 and A2 are fully differential operational amplifiers, designed to have both a large differential-to-differential gain $A_{dd}>0$ and a large common mode-to-common mode gain $A_{cc}<0$. In particular, the input-output characteristics for A1 and A2 can be written as:

$$V_{od} = A_{dd}\left(V_{id} - V_{io}\right); \; V_{oc} = A_{cc}\left(V_{ic} - V_{ref}\right) \qquad (2)$$

where, V_{od}, V_{oc} are the differential and common mode output voltages, respectively, V_{id}, V_{ic} are the differential and common mode input voltages, respectively, while V_{io} is the input referred offset voltage. Voltage V_{ref} indicates the common mode reference, set to $V_{dd}/2$.

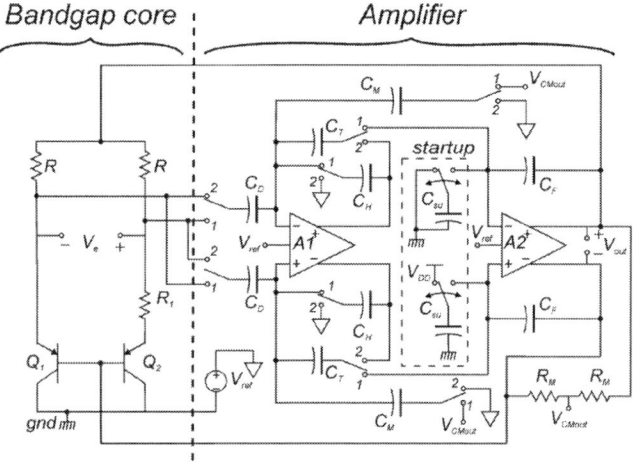

Fig. 2. Schematic of the proposed architecture.

The first stage, based on op-amp A1, samples the error voltage V_e through capacitors C_D and the difference voltage $V_{ref} - V_{CMout}$ through capacitors C_M. During phase 2 a differential charge proportional to V_e plus a common mode charge proportional to $V_{ref} - V_{CMout}$ is stored into capacitors C_T. Capacitors C_H are used to maintain feedback of A1 during phase 2, according to the approach described in [6] that allows reduction of the error due to A1 finite gain. In phase 1 of the following clock cycle, the error charge is integrated by A2, that, together with capacitors C_F, forms a Miller integrator. During normal operation, there is no need of resetting the integrator, so that the output voltage is always available. Capacitors C_{su} are active only at start-up: their operation is described later.

The amplifier configurations in the two phases are detailed in Fig. 3. Due to (2), with $|Add|$, $|Acc|>>1$, we will suppose that

generalized virtual ground is present at A1 and A2 inputs, i.e. $V_{ic}=V_{ref}$, $V_{id}=V_{io}$.

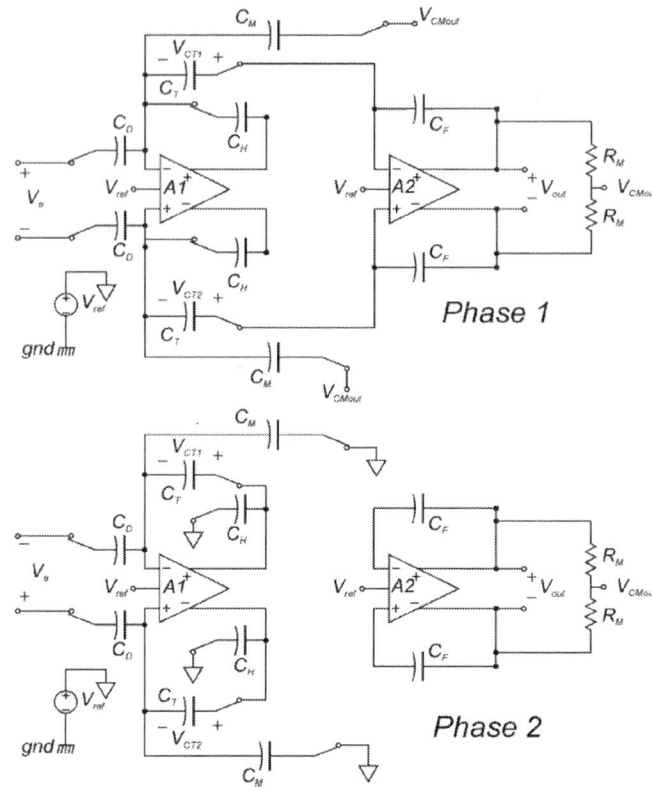

Fig. 3. Connections in phase 1 and phase 2 for the two stage amplifier. Start-up capacitors are omitted.

Starting from phase 1, it is important to observe that capacitors C_T sample a differential voltage ($V_{CT1}-V_{CT2}$) equal to:

$$V_{CTd} = V_{io2} - V_{io1} \qquad (3)$$

while their common mode voltage is zero. In the transition from phase 1 and 2, capacitors C_D experience a differential voltage variation equal to $2V_e$, with zero common mode voltage variation. Conversely, capacitors C_M undergo a common mode voltage variation equal to $V_{CMout}-V_{ref}$, while their differential voltage is constantly zero. The differential mode ($Q_{CT-diff}$) and the common mode (Q_{CT-cm}) charge injected into capacitors C_T, in the phase 1 to phase 2 transition, is then:

$$\begin{cases} \Delta Q_{CT-diff} = 2V_e C_D \\ \Delta Q_{CT-cm} = C_M\left(V_{CMout} - V_{ref}\right) \end{cases} \qquad (4)$$

It can be easily noted that the offset of amplifier A1 does not contribute to these charge transfers. In phase 1 of the next clock cycle, capacitors C_T are connected back between A1 and A2 input ports. Therefore, they are discharged again to the voltage indicated in (3). Therefore, only the charge accumulated in the phase 1/phase 2 transition, given in (4), is transferred into capacitors C_F of the second stage. Thus, also this charge transfer is offset free.

Indexing with k the clock cycles, it is possible to write the difference equations that characterize the common mode and differential mode behavior of the amplifier as

$$V_{outD}(k)=V_{outD}(k-1)-2\frac{C_D}{C_F}V_e(k-1) \qquad (5)$$

$$V_{outCM}(k)=V_{outCM}(k-1)+\frac{C_M}{C_F}\left[V_{ref}-V_{CMout}(k-1)\right] \qquad (6)$$

Equations (5) and (6) show that the amplifier operates as a discrete time integrator, so that the DC gain would theoretically tend to infinity and is practically limited only by the finite gain of the amplifiers. Therefore, if the system is stable, the dc value of V_e tends to zero as required for correct operation of the bandgap core. As a result, the output differential voltage of the amplifier assumes the correct value V_{BG}, given by (1). In addition, (6) guarantees that the output common mode voltage is correctly set to V_{ref} at steady state. Note that the bandgap core produces $V_e=0$ also for $V_{outD}=0$. In order to prevent the system from being trapped into this unwanted operating point, the start-up circuit, represented by capacitors C_{su} in Fig. 2, injects a differential charge into the integrator, making the output voltage increase until it reaches approximately $V_{BG}/2$.

B. Stability

Stability can be studied by analyzing discrete time relationships (5) and (6). Using Z-transform it is possible to show that (6), that rules the common mode behavior, is stable for $C_M/C_F<2$. In order to study the differential mode it is necessary to linearize the V_e vs V_{outD} transfer function around the operating point $V_{outD}=V_{BG}$, obtaining, after tedious calculations:

$$v_e=\gamma v_{outD} \quad \text{with} \quad \gamma=\frac{r\ln^2(m)}{(r\ln(m)+1)[(r+1)\ln(m)+1]} \qquad (7)$$

where $r=R/R_1$. Substituting this expression of v_e into (5), a small signal discrete time equation is obtained, which is stable for $\gamma C_D/C_F<1$.

C. Noise

The simplified analysis performed in previous paragraph was aimed to illustrate the operating principle and demonstrate that the behavior is actually that of an offset free integrator. Two important aspects for the estimation of the output noise have been neglected:

1) Besides the offset voltage, noise is present at the amplifier inputs. Only correlated components, that can be considered practically constant across the clock period are effectively reduced. Foldover of high frequency components occurs due to sampling.

2) Continuous-time noise components are also present on the output. Indicating with v_{n1} and v_{n2} the input referred noise

sources of A1 and A2, the output noise is v_{n2} in phase 2 and $v_{n2}+(v_{n2}-v_{n1})C_T/C_F$ in phase 1. These noise components, once referred to the input (v_e voltage) of the two stage amplifier are reduced by the amplifier gain. Since this gain is high at low frequencies (due to the integrator behavior), low frequency noise components (i.e. flicker noise) are strongly rejected. A rigorous noise analysis is rather complex and is beyond the aim of this paper. A simplified model can be obtained by adding an input referred noise to the linearized version of (5), obtaining:

$$v_{outD}(k)=v_{outD}(k-1)-2\frac{C_D}{C_F}\gamma v_{outD}(k-1)+v_n(k-1) \qquad (8)$$

where $v_n(k)$ is dominated mainly by sampling of A1 input referred noise and is therefore affected by foldover. Applying the Z-transform to (8) it is possible to show that the output noise is filtered by a first order discrete time low pass function with cut-off frequency nearly equal to $f_{ck}\gamma C_D/\pi C_F$. This effect contributes to reduce to total output noise amplitude.

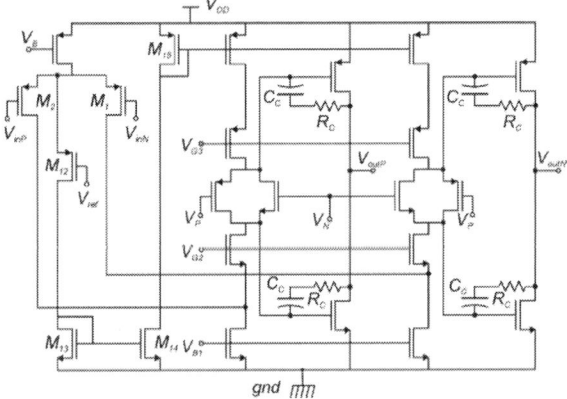

Fig. 4. Schematic of operational amplifier A2.

D. Topology of the operational amplifiers

Fig. 4 shows a simplified schematic view of A2, consisting in a two stage fully-differential amplifier with folded cascode input stage and class AB common source output stage. Standard Miller compensation (C_C, R_C groups) is used to obtain closed loop stability. The two stage structure of A2, together with the class AB output stage is introduced to allow driving of low resistive loads. The output common mode voltage of both A1 and A2 is stabilized by the local AC capacitive feedback network and by the discrete time feedback represented by (6). For this to occur a large common mode-to-common gain is required. This is obtained adding transistor M12 (Fig. 4) with the gate connected to V_{ref}. M12 is sized twice M1 and M2, so that M12 and M1-M2 act as a differential pair with respect to the common mode signal. If the drain of M12 was connected to ground, the common mode gain would be positive and feedback connection would not possible. For this reason the drain current of M12 is mirrored to the common gate stage with a gain sufficient to reverse the sign of the common mode gain. Amplifier A1 has a single stage structure similar to the first stage of A2. In this case an inverting common mode gain is obtained by simply connecting M12 drain to ground. V_P and V_N are biasing voltages.

III. SIMULATION

Simulations were carried out on a prototype designed with the UMC 0.18 μm CMOS process. Supply voltage was set to 3 V unless differently specified. The following values were chosen for capacitors and resistors: C_D=2 pF, C_T=1 pF, C_M=500 fF, C_H=1 pF, C_F=4 pF, C_{su}=500 fF R=523.7 kΩ, R_1=53.76 kΩ, R_M=200 kΩ. Areas of bipolar transistors were sized to have a ratio m=8. The corresponding value of γ is 0.0855. According to preliminary layout, area occupation is 0.072 mm². The bandgap core was sized to achieve a zero temperature coefficient TC at 26°C, resulting in a floating output voltage of 1.2225 V. Fig. 5 shows the differential output voltage as a function of temperature: a total variation of 3mV in the range from -40°C to 100°C can be estimated. The quiescent supply current is 60μA, while a maximum output current of 10 mA can be delivered with less than 0.1 % voltage drop. The circuit is stimulated by a 100 kHz clock and is able to work correctly with power supply voltage from 1.8 V to 3.3 V with a line regulation of 10 ppm/V. In order to limit charge injection of the switches, small area switches and dummy switch structures [3] were used. Both amplifiers A1 and A2 can be switched off by enable inputs (not shown in figures); in addition a reset signal that activates switches that short circuit capacitors C_F, is introduced.

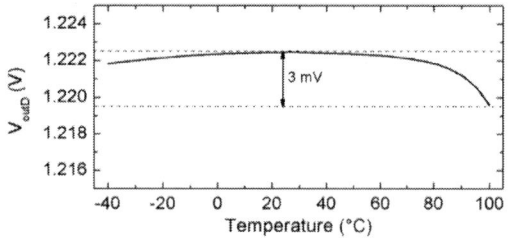

Fig. 5. Output bandgap voltage versus temperature.

Since Monte Carlo models were not available for the devices used, offset voltage of amplifiers was simulated using two differential voltage sources at the input of both amplifiers. Fig. 6 shows transient simulation results, where the instants at which the circuit is enabled and the reset signal is released are highlighted. Curves with and without offset sources (10 mV on both A1 and A2) are shown. Curves differ by only 80 μV at steady state.

Fig. 7 (a) shows the output power spectral density simulated by periodic noise analysis. It should be observed that power spectral density is actually low pass filtered as discussed in the previous section. The power supply rejection (PSR), obtained by periodic AC simulations is shown in Fig. 7 (b).

IV. CONCLUSIONS

An alternative architecture for bandgap voltage reference, capable to provide, for the whole clock cycle, an offset free and low noise output has been proposed. This characteristic, together with the intrinsic tunable noise low pass filter, makes the circuit suitable for integration in fully differential systems.

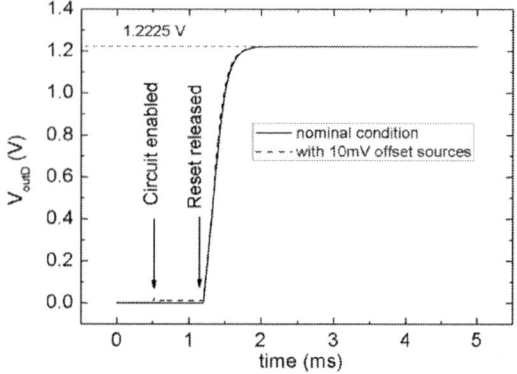

Fig. 6. Nominal transient (soldi line) and transient with 10mV offset sources (dashed line).

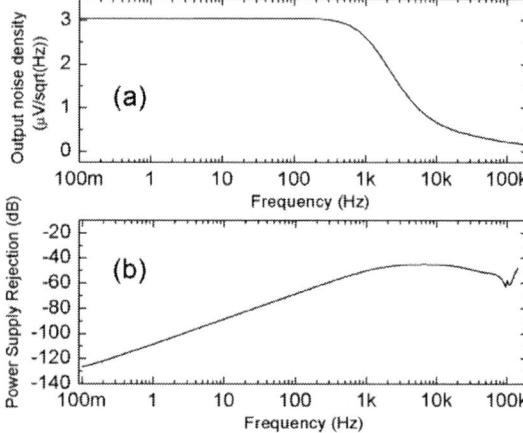

Fig. 7. Output noise density (a) and power supply rejection (b) obtained from steady state simulations (PNOISE and PAC, respectively).

REFERENCES

[1] C. J. B. Fayomi, G. I. Wirth, H. F. Achigui, A. Matsuzawa, "Sub 1 V CMOS bandgap reference design techniques: a survey," Analog Integr. Circ. Sig. Process, Springer, vol. 62, pp. 141-157, 2010.

[2] M. Ferro, F. Salerno, and R. Castello, "A floating cmos bandgap voltage reference for differential applications," IEEE J. Solid-State Circuits, vol. 24, pp. 690-697, no. 3, June 1989.

[3] C.C. Enz and G.C. Temes, "Circuit techniques for reducing the effects of op-amp imperfections: autozeroing, correlated double sampling, and chopper stabilization", Proc. of the IEEE, vol. 84, pp. 1584-1614, November 1996.

[4] G. Ge, C. Zhang, G. Hoogzaad, and K. A. A. Makinwa, "A single-trim CMOS bandgap reference with a 3σ inaccuracy of ±0.15% from -40°c to 125°c," IEEE J. Solid-State Circuits, vol. 46, pp. 2693-2701, no. 11, November 2011.

[5] G. Nicollini and D. Senderowicz, "A CMOS bandgap reference for differential signal processing," IEEE J. Solid-State Circuits, vol. 26, pp. 41-50, no. 1, January 1991.

[6] K. Martin, L. Ozcolak, Y. S. Lee, and G. C. Temes, "A differential switched-capacitor amplifier," IEEE J. Solid-State Circuits, vol. SC-22, pp. 104-106, no. 1, February 1987.

A 65nm CMOS Technology Radiation-Hard Bandgap Reference Circuit

Tommaso Vergine[†*], Stefano Michelis[§], Marcello De Matteis[*‡], Andrea Baschirotto[*]

[†] University of Pavia Italy. Department of Electrical Computer and Biomedical Engineering.
[*]Dept. of Physics, University of Milano-Bicocca, Milano, Italy
[‡] University of Salento, Lecce Italy. Department of Innovation Engineering.
[§] CERN, 1211 Geneve 23, Switzerland.
tommaso.vergine, marcello.dematteis, andrea.baschirotto@unimib.it.

Abstract—This paper presents a BandGap reference circuit with low sensitivity to temperature and to the voltage supply variations. It has been designed to be Radiation-Hard up to 1 GRad. This voltage reference has been developed in a commercial 65nm CMOS technology with 1.2 V of nominal voltage supply. A current-mode architecture has been chosen to allow the low-voltage operation. Particular attention has been dedicated to circuit radiation hardness, in order to provide a stable voltage signal also with high radiation levels, like that of high-energy physics experiments. One of the advantages of CMOS scaling-down process is that the effects, due to radiation exposure, steadily decrease making circuits more and more robust. It follows that, in a conventional BandGap circuit, the most critical aspect could regard the diodes, or in general, the sensing elements. This design has been preceded by a series of measurements of two different sensing device in order to use that with the better radiation hardness .The BandGap reference circuit has been simulated with temperature range from -10 °C to 50 °C. The output value is around 330 mV with a curvature error of 0.05% in nominal conditions. The maximum output deviation in the absolute value is about ±1.1% and ±1.6% under process and mismatch respectively. The integrated noise from 0.01 Hz to 100 MHz is about 180 μV and the power consumption is 240μW. The radiation effects have been simulated modifying the models of devices according to measurements. In this case, thanks to a proper sizing, the output voltage shift is of a few millivolts.

Keywords: BandGap voltage reference, Low-voltage, CMOS, Radiation Hardness, Enclosed Layout, High-Energy Experiments.

I. INTRODUCTION

Reference voltage generating circuits are widely used in analog and mixed-signal systems. Their aim is to provide an accurate voltage, independent on temperature and supply variations. Traditionally, the BandGap reference circuits generated the output voltage from the sum of the built-in voltage of a diode and the thermal voltage of kT/q, multiplied by a constant term but this precluded the low-voltage operation since the supply voltage would have to be larger than the 1.25 output voltage. The technological scaling-down process has forced designers to find alternative solutions to overcome the low-voltage supply issue. One of the most common approach, in sub-micron technologies, is the so-called current-mode [1]. The output voltage is obtained from the sum of two currents: the first is proportional to the built-in voltage of a diode and the second is proportional to the thermal voltage. This BandGap

design has been developed starting from the latter architecture. A particular attention has been also focused on the BandGap radiation-hardness. This is the additional specification required to work in high-energy physics experiments [2][3] or in the aerospace activities. Deep sub-micron technology are inherently highly tolerant to radiation, mainly due to the reduced oxide thickness[4]. Recombination of the radiation-induced charges is favoured by quantum tunnelling [5]. In a BandGap reference circuit it is possible to neglect the contribution of the radiation effects of transistors on the output voltage by a proper sizing. Nevertheless, just the effects of radiation on diodes can shift the output voltage of a BandGap of few hundreds millivolts. This can easily compromise, for example, the correct behaviour of all other analog blocks. To reduce the radiation damage effects on diodes a proper layout technique can be adopted. Indeed, surrounding the diode's p+ diffusion with thin oxide (instead of shallow trench isolation field oxide - STI), it is possible to avoid the TID (Total Ionization Dose) radiation effects. This technique is called enclosed layout [5][7][8]. A resistive voltage divider has been inserted in the main circuit of the BandGap to allows the use of an operational amplifier (opamp) with the input stage of type PMOS. The voltage across diodes is typically about 700-800 mV and, if directly applied to the opamp input, could generate biasing problems (especially in the worst case scenario - VDD 10% lower than nominal value and higher threshold voltage). A PMOS type opamp provides better performance especially in terms of unity gain bandwidth and flicker noise. This paper is organized as follow. In Section II the enclosed layout diode and its TID robustness are shown while in Section III the BandGap architecture is presented. Section IV shows the simulation results and at the end of this paper the Conclusions are drawn.

II. ENCLOSED LAYOUT VS CONVENTIONAL DIODE

A flux of high energetic charged particles and gammas leads to total ionization dose (TID) effects in MOS transistors [5]. Radiation creates, inside the oxide, free electron-hole pairs and the holes can be trapped at the interface between SiO_2 and Si. The electric field in the channel is changed by these holes presence with the consequence of a threshold voltage shift and leakage current. When the oxide thickness is lower than 12nm [6] the tunnelling of free electrons from channel into the gate oxide reduces drastically the number of holes, before they can convert to interface trapped states. This is why, in sub-micron technology, the main source of TID effects is the thick isolation oxide all around MOS devices. An enclosed

978-1-4799-4993-9/14 $31.00 © 2014 IEEE

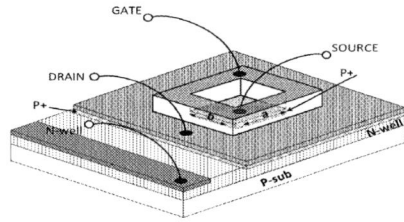

(a) EL Diode 3-D view.

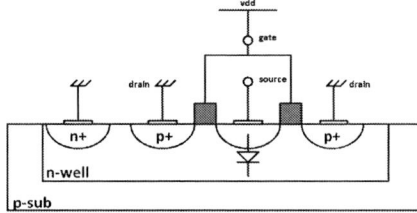

(b) EL Diode vertical section.

Figure 1: Enclosed Layout (EL) Diode

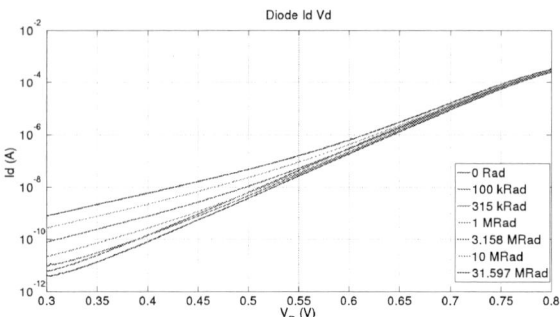

(a) Conventional Diode Current.

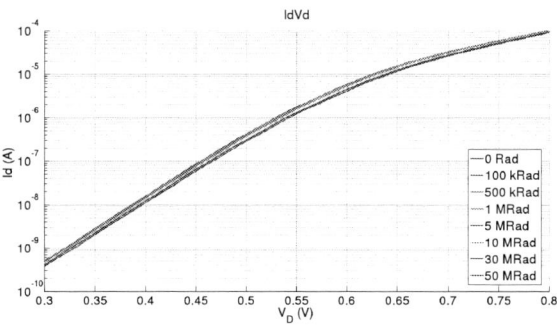

(b) EL Diode Current.

Figure 2: Current vs TID

layout can drastically reduce this problem because, as shown in Fig. 1a , the gate thin oxide of a transistor with an annular layout completely surrounds the source diffusion, distancing the STI oxide [7][8]. Fig. 1b shows how it is possible to obtain a diode from a PMOS with a enclosed layout. The gate contact, biased to VDD, avoids to the PMOS to turn on. The source diffusion becomes the anode and the n-well contact the cathode. In Fig. 2 a comparison between the I_D vs V_D of a conventional diode and an enclosed layout diode is shown at different TID values. The diffusion area is the same for both devices and is $0.5 \mu m^2$. The test of sensing devices has been done at the SEIFERT RP149 Xray machine installed at CERN. As shown in 2a, the characteristic of the conventional diode is strongly changed by radiation. The damage is larger for smaller values of bias current. In the past the BandGap circuits were designed using high biasing current to reduce the effects of radiation (introduction of leakage current), but this approach led to higher power consumption. In Fig. 2b the current of an enclosed layout diode is reported, for different TID values. The radiation degradation effects are much lower than the conventional diode, especially considering that it has been exposed up to 50 MRad (i.e. 20 MRad more than the conventional diode). This diode has been therefore used in the BandGap circuit design.

III. BANDGAP VOLTAGE REFERENCE CIRCUIT

The basic idea in a current-mode BandGap reference circuit is to combine two currents, with different temperature dependences, in order to obtain a stable output current/voltage. The area of the diodes plays a key role in the temperature dependence. In general, the temperature dependence of a semiconductor diode can be approximated as reported in Eq. 1. In other words, the voltage across a diode is IPTAT (Inversely Proportional To the Absolute Temperature with a proportional factor α) and its maximum value, occurring at 0 °K, is the BandGap voltage of silicon.

$$V_D = V_{BG} - \alpha\,T \qquad (1)$$

where α is given by $\alpha = \mathrm{k}\,\ln\left(\mathrm{const}/J\right)$. Thus, assuming the same current level through the two diodes and being designed with different areas, they work with different current densities. As consequence, they exhibit a different temperature trend. In addition, the temperature sensitivity can be increased using different biasing currents per each diode. The BandGap circuit is shown in Fig. 3. Eight replicas of D_1 have been used to realize D_2 and the transistor M_1 has been sized eight time larger than the transistor M_2 (better temperature sensitivity). Being the voltage across D_1 (for the chosen biasing current level) around 800 mV, two resistive voltage dividers have been necessary in order to use, without biasing problems, a PMOS type operational amplifier. The opamp keeps the two nodes V'_a, V'_b at equal voltages and consequently V_a and V_b (all voltage divider resistors have the same value). The current in the transistor M_2 can be expressed as reported in Eq. 2.

$$I_{M2} = I_{Rs} + I_{R2} + I_{D2} = \frac{2\,R_s + R_2}{2\,R_s\,R_2}\,V_{D1} + \frac{\Delta V}{R_3} \qquad (2)$$

with $\Delta\mathrm{V} = \mathrm{V_{D1}} - \mathrm{V_{D2}}$ and where the first term is IPTAT while the second is Proportional To the Absolute Temperature (PTAT). Changing the value of resistances it is possible to obtain an output current independent from the temperature variations. To work properly it is necessary that the ratio of the current densities of the two diodes (J_{D2}/J_{D1}) is independent from temperature. Assuming the current in M_1

978-1-4799-4993-9/14 $31.00 © 2014 IEEE

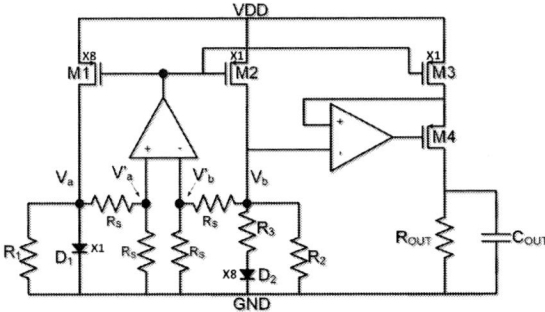

Figure 3: BandGap Voltage Reference Circuit.

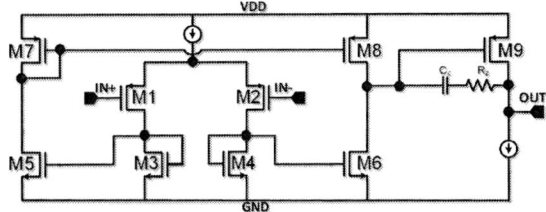

Figure 4: Operational Amplifier.

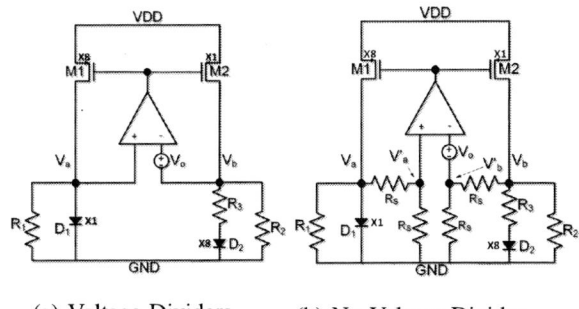

(a) Voltage Dividers. (b) No Voltage Divider.

Figure 5: Opamp Offset Contribution.

and M_2 independent from temperature. As said before, the current that flows through D_2 (I_{D2}) is PTAT. The current of R_2 is thus IPTAT. The resistor R_1, placed in parallel with D_1, deviates the current that passes through R_2 from D_1 making the current of D_1 PTAT too, as that of D_2. Being M_1 eight times greater than M_2, R_1 will be eight times lower than R_2. Assuming a 1:1 mirroring factor between M_2 and M_3 the output voltage is given by Eq. 3. A regulated cascode current mirror has been used in order to increase the accuracy of the output current [9]. Contrarily to the opamp in the main BandGap sub-circuit, this has to have an NMOS type input differential pair because the input voltage is next to 800 mV. The opamp in the output stage has been designed with the minimum DC gain and bandwidth, in order to reduce power consumption. In particular, it exhibits about 60 dB of DC gain and 150 kHz of unity gain bandwidth. The output capacitance C_{OUT} allows to filter-out some noise and increase power supply rejection.

$$V_{OUT} = \left[\frac{2\,R_s + R_2}{2\,R_s\,R_2} V_{D1} + \frac{\Delta V}{R_3} \right] R_{OUT} \qquad (3)$$

A. Operational Amplifier and Offset Contribution

Particular attention has to be focused on the operational amplifier and its offset voltage contribution on the output of the BandGap circuit. A symmetrical operational amplifier, compensated by Miller technique, has been chosen. Its schematic is shown in Fig. 4. This architecture allows to improve the BandGap overall power supply rejection since the opamp output node follows the voltage supply variation. In this way the V_{GS} of transistors M_1, M_2 and M_3 of Fig. 3 remains constant. Also in this case it has been sized in order to provide the minimum required DC gain and bandwidth ($\simeq$65 dB and 1.5 MHz). Considering the BandGap circuit without resistive divider networks, as shown in 5b. The nodes V_a and V_b represent the effective operational amplifier input. Considering also the opamp offset as a DC voltage in series with the negative input. The output current becomes as reported by Eq. 4. To make negligible the offset contribution the first two terms have to be greater than the last one. To do this the ratio A_2/A_1 (areas of diodes) or the mirroring factor between M_1 and M_2 can be increased. Typically the latter is preferred, especially in radiation environments, because increasing the area of devices increases also the susceptibility to radiation effects.

$$I_{OUT} = \frac{V_{D1}}{R_2} + \frac{\Delta V}{R_3} - V_o \left(\frac{1}{R_3} + \frac{1}{R_2} \right) \qquad (4)$$

Considering now the circuit of Fig. 5a. The output current is given by Eq. 5. This means that the opamp offset is more important in the output current value. Thus, it has to be carefully evaluated and sized, if a certain precision in the output current is needed. The input pair has been sized taking into account the Eq. 6 to minimize the offset voltage due to the MOS threshold mismatch [10]. In addition, the current mirrors of Fig. 4 have been sized in order to work in strong inversion, minimizing in this way the offset quantity linked to current mismatch. All transistors inside the BandGap circuit have been sized in order to minimize the effects due to the threshold voltage shift up to 1 GRad. PMOS transistors (with W or L > $5\mu m$) exhibit a threshold shift of 60 mV while NMOS transistors (with W or L > $3\mu m$) exhibit a shift of 20 mV.

$$I_{OUT} = \frac{2\,R_s + R_2}{2\,R_s\,R_2} V_{D1} + \frac{\Delta V}{R_3} - V_o \left(\frac{1}{R_S} + \frac{2}{R_3} + \frac{2}{R_2} \right) \qquad (5)$$

$$\sigma_{Vth} \sim \frac{1}{\sqrt{W\,L}} \qquad (6)$$

IV. SIMULATION RESULTS

Fig. 6 shows the nominal simulation of the BandGap output voltage, for a temperature swept from -10 ^oC to 50 ^oC. The mean value is about 331.6 mV and the curvature error is only

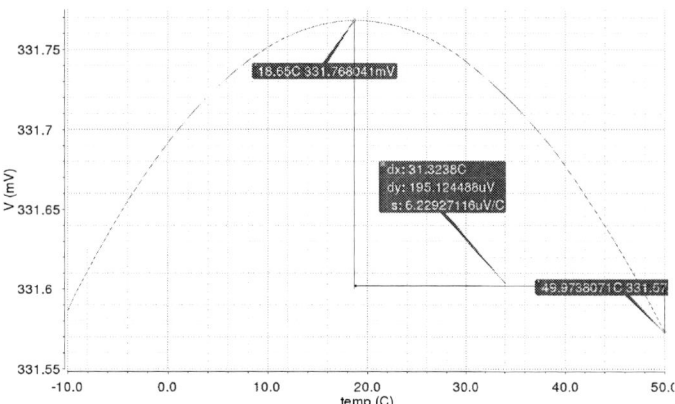

Figure 6: BandGap Output Voltage in Nominal Conditions.

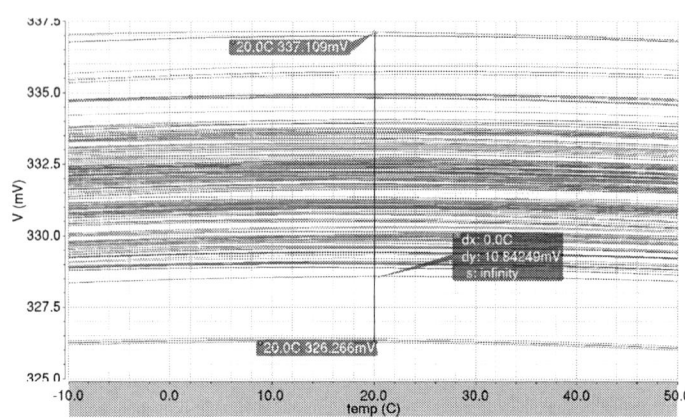

Figure 8: BandGap Output Voltage under Mismatch Simulations.

about 0.06%. The BandGap has been simulated also under Process-Voltage variations, according to the Tab. I. The output voltage is shown in Fig. 7. Also in this case the maximum curvature error is less than 0.1% and the maximum spread among all output values is about ± 1.1%. Fig. 8 shows the BandGap output voltage under Mismatch simulations. The maximum deviation is about ± 1.6%, mainly due to the offset voltage of the operational amplifier. It has been possible to simulate the radiation effects modifying the MOS threshold voltage in the model files of technology increasing the absolute value variation of the ouptu of about 2%.

Parameter	Value
Process	ss, ff, fs, sf, tt
Voltage Supply	1.2, 1.08, 1.32

Table I: PV Simulation Corners

oxide all around diodes with thin oxide (Gate Oxide), much more efficient under radiation. The BandGap design has been made in a commercial 65nm CMOS Technology and the main performance is resumed in Tab. II.

Parameter	Nominal	Process-Voltage	Mismatch
$<V>$	331.6 mV	331.8 mV	331.9 mV
σ	-	1.3 mV	1.64 mV
Curvature Error	0.06 %	0.15 %	0.4 %
Power Consumption	$\simeq 200\mu W$	$\simeq 240\mu W$	$\simeq 230\mu W$
Noise(0.01Hz÷100MHz)	170 μV	188 μV	190 μV

Table II: Summarized Performance

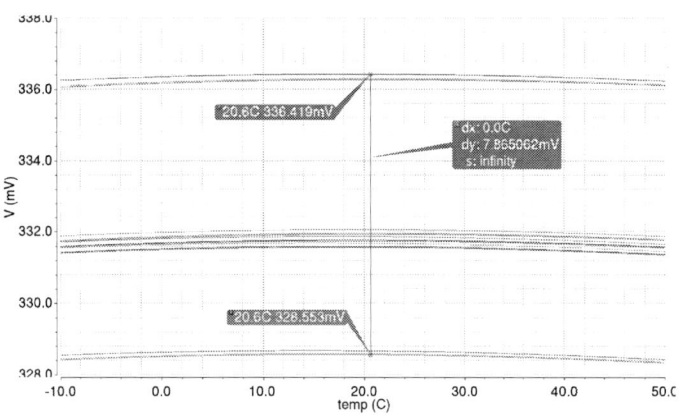

Figure 7: BandGap Output Voltage under P-V Variations.

V. CONCLUSIONS

A Low-Voltage Rad-Hard BandGap voltage reference circuit has been proposed in this paper. The current-mode architecture has been chosen to allow low-voltage operation. Particular attention has been focused on the design of the sensing elements, i.e. the diodes to reduce drastically the TID effects (mainly leakage current and threshold shift). A proper layout technique has been adopted in order to replace the STI

REFERENCES

[1] Banba, H., et al. "A CMOS bandgap reference circuit with sub-1-V operation," Solid-State Circuits, IEEE Journal of , vol.34, no.5, pp.670,674, May 1999 doi: 10.1109/4.760378

[2] Magazzú, G., et al. "The detector control unit: an ASIC for the monitoring of the CMS silicon tracker," Nuclear Science, IEEE Transactions on , vol.51, no.4, pp.1333,1336, Aug. 2004

[3] Vergine, T., et al. "A 32-channel 12-bits single slope A-to-D converter for LHC environment," IC Design Technology (ICICDT), 2013 International Conference on , vol., no., pp.139,142, 29-31 May 2013

[4] C. Claeys and E. Simoen, "Radiation Effects in Advanced Semiconductor Materials and Devices," New York: Springer-Verlag, 2002, pp. 2024.

[5] Gromov, V., et al. "A Radiation Hard Bandgap Reference Circuit in a Standard 0.13 m CMOS Technology," Nuclear Science, IEEE Transactions on , vol.54, no.6, pp.2727,2733, Dec. 2007 doi: 10.1109/TNS.2007.910170

[6] Saks, N.S., et al. "Radiation Effects in MOS Capacitors with Very Thin Oxides at 80K," Nuclear Science, IEEE Transactions on , vol.31, no.6, pp.1249,1255, Dec. 1984 doi: 10.1109/TNS.1984.4333491

[7] D. R. Alexander et al., "Design issues for radiation tolerant microcircuits in space " in Proc. 1996 IEEE NSERC Short Course, 1996, pp. V-1 - V-54

[8] Lacoe, R.C., "Improving Integrated Circuit Performance Through the Application of Hardness-by-Design Methodology," Nuclear Science, IEEE Transactions on , vol.55, no.4, pp.1903,1925, Aug. 2008 doi: 10.1109/TNS.2008.2000480

[9] J. Ramirez-Angulo, et al. "Low supply voltage high performance CMOS current mirror with low input and output voltage requirements " IEEE Transactions on Circuits and Systems-II Express Briefs, Vol. 51, No. 3, March 2004

[10] W. Sansen, "Analog Design Essentials," 2006.

978-1-4799-4993-9/14 $31.00 © 2014 IEEE   352

A modified CMOS nano-power resistorless current reference circuit

Shailesh Singh Chouhan, Kari Halonen
SMARAD-II, Department of Micro and Nano Sciences,
Aalto University School of Electrical Engineering, Finland
Email: shailesh.chouhan@aalto.fi, kari.halonen@aalto.fi

Abstract—**In this work, all MOS current reference circuit is proposed using a standard 0.18 μm technology and the simulations were performed using the Cadence Spectre simulator. The proposed current reference circuit is based on, the resistorless current reference circuit suggested by Oguey and Aebishcher. The Oguey's circuit is capable of generating the reference current in a nanoampere range, but with the high temperature coefficient (TC). The reason for high TC, that we found, is the lack of control over the gate-source voltage of an active resistor used in the design. This gate-source voltage is one of the controlling parameters responsible to obtain adequate thermal compensation for the reference current. In the proposed work, we modified the architecture to limit the variation of the gate-source voltage with the temperature and hence controls the thermal behaviour of the reference current. The working supply voltage of the proposed circuit ranges from 1.25 V to 2 V. The temperature coefficients of the reference current generated from the proposed and conventional architectures are 39.8 ppm/$^\circ$C and 545.12 ppm/$^\circ$C respectively at a supply voltage of 1.25V for the temperature ranging from -60°C to 85°C. The maximum power consumption of proposed and conventional architecture is 624.8nW and 468.59nW at a supply voltage of 2 V with the layout area of 0.0013μm^2 and 0.001μm^2 respectively.**

Index Terms—**Current reference, CMOS circuits, low power, temperature coefficient.**

I. INTRODUCTION

A current reference circuit is one of the most important building blocks for an analog, digital, and mixed-signal circuit systems.

It generates the temperature and the supply independent current, which is used for the biasing of an on-chip analog components like operational amplifiers, ADCs and DACs.

Therefore, various approaches have been proposed to design the current reference circuits.

These designs are broadly classified into two approaches.

First, the use of control voltage to generate the temperature compensated reference current [1], [2], [3] and [4].

In a second, the sum of the currents/voltages with the positive and negative temperature coefficients [5], [6] and [7] is used to generate the reference-current.

Generally, in most of the designs the temperature compensation is obtained by the use of the resistors.

The dimension of the temperature compensating resistors increases with reduction in the value of the reference current.

As a solution, various architectures based on resistor-less implementations are available in the literature [8], [9], [10] and [11] for the current reference circuit design.

The proposed work is an extension of the work presented in [11].

By using simulations, we found that the circuit of [11] has the high temperature coefficient for the reference current.

One of the reasons for it is an inadequate temperature gradient of the gate-source voltage of NMOST which is implemented as an active resistor.

In this work, we are extending the circuit of [11] with a simple circuit arrangement which controls the variation of V_{gs} with the temperature and hence the resultant reference current acquires a low temperature coefficient.

This paper is organized as follows: in Sections II the principle of the proposed current reference circuit is explained while the design implementation is explained in Section III; the simulation results are shown in Section IV and finally the conclusions are made in Section V.

II. PROPOSED CIRCUIT

A. Conventional current reference circuit

In [12] the authors proposed a well-known beta multiplier circuit with MOSTs and only one passive resistor, shown in Fig.1(a). The use of a passive-resistor in a circuit [12] restricts the reference current in the micro-ampere range since with reduction in reference current value, the size of passive-resistor will increase. The work presented in [11] is extension of work presented in [12] to generate current in a nano ampere regime.

The authors of [11] replaced a passive resistor with an active resistor by using NMOST (Mn3) operating in the deep triode region as shown in Fig.1(b). As suggested in [11] we used the cascode arrangement to get a high PSRR value shown in Fig.1(c). In the circuit (Fig.1(c)), the gate-source voltage of diode-connected NMOST(Mn4) keep NMOST (Mn3) in the deep triode region. By applying Kirchhoff's voltage law (KVL) in a loop formed by NMOSTs Mn3 and Mn4,

$$V_{gs3} = V_{gs4} \tag{1}$$

The NMOST Mn4 is in saturation region hence,

$$V_{gs3} = V_{gs4} = V_{thn} + \sqrt{2I_{ref}/K_n S4} \tag{2}$$

where V_{thn} is the threshold voltage , I_{ref} is the reference current, K_n is the transconductance parameter and S_4 is the aspect ratio of NMOST Mn4. By differentiating (2) with

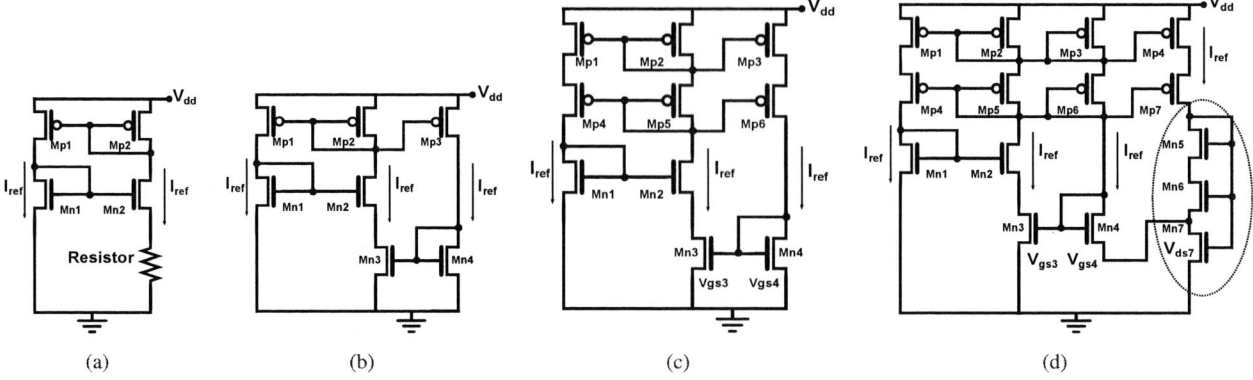

Figure 1. (a) Beta multiplier circuit [12] (b) Resistorless beta multiplier circuit [11] (c) Resistorless beta multiplier using cascode biasing to obtain high PSRR (d) Proposed current reference circuit.

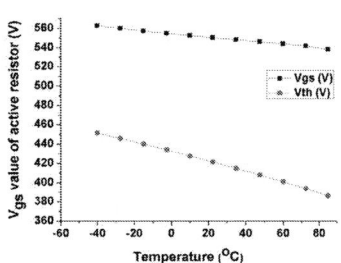

Figure 2. The variation of the gate-source voltage (V_{gs3}) and the threshold voltage (V_{thn}) with the temperature for the supply voltage of 1.25V. The variation of V_{gs3} is not according to the variation of V_{thn} with the temperature.

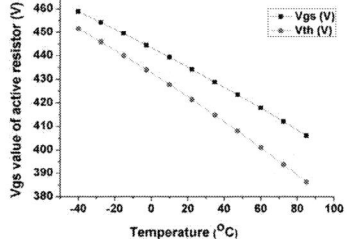

Figure 3. The variation of the gate-source voltage (V_{gs3}) and the threshold voltage (V_{thn}) with the temperature for the supply voltage of 1.25V in the proposed circuit. The variation in V_{gs3} with respect to the temperature is almost equal to the V_{thn} variation.

respect to the temperature (T) we get,

$$(\partial V_{gs3}/\partial T)_{\mathbf{a}} = (\partial V_{thn}/\partial T)_{\mathbf{b}} + \underbrace{\sqrt{I_{ref}/2K_n S4} \left(\frac{1}{I_{ref}} \partial I_{ref}/\partial T \right)}_{\mathbf{c}}$$
(3)

It is clear from (3), that the change of gate-source voltage with the temperature (a) is a function of the temperature variation of the threshold voltage (b) and the temperature coefficient (TC) of the reference current (c). Thus the condition required to get zero TC of I_{ref} is,

$$\partial V_{gs3}/\partial T = \partial V_{thn}/\partial T \tag{4}$$

In practice, the threshold voltage (V_{thn}) decreases with the temperature [13], hence the gate-source voltage (V_{gs3}) should also decrease with a same slope to get temperature-independent reference current (I_{ref}). In order to verify (4) we performed a DC simulation on the circuit (Fig.1(c)) for the temperature range of $-40°C$ to $+85°C$ at a supply voltage of 1.25V. The simulation result (Fig.2) shows that the both V_{gs3} and V_{thn} are decreasing with temperature but with different slopes. The slope of V_{gs3} is ≈ 0.196mV/°C, while the slope of $V_{thn} \approx 0.522$mV/°C, thus this difference in slopes results high TC of reference current (I_{ref}).

B. Proposed current reference circuit

It is clear from Fig.2 that the thermal gradient of V_{gs3} is less than the thermal gradient of V_{thn}. Thus, to equalize them we need an extra voltage term, with opposite temperature behavior

due to which slope of V_{gs3} will become almost equal to the slope of V_{thn} with respect to the temperature. The proposed circuit arrangement to generate this voltage is designed by using three NMOSTs Mn5, Mn6 and Mn7 respectively as highlighted in Fig.1(d). In this composite NMOSTs array Mn5 is operating in the saturation, while Mn6 and Mn7 are in the linear region of operation. By applying KVL in the loop formed by NMOSTs Mn3, Mn4 and Mn7,

$$V_{gs3} = V_{gs4} + V_{ds7} \tag{5}$$

We performed a DC simulation on the proposed circuit (Fig.1(d)) for the temperature ranges from $-40°C$ to $+85°C$ at a supply voltage of 1.25V. The simulation result (Fig.3) shows that the slope of V_{gs3} improves to ≈ 0.424mV/°C, due to PTAT behavior of V_{ds7}.

III. DESIGN IMPLEMENTATION

In the proposed circuit the NMOST Mn1 and Mn2 are in subthreshold saturation, hence the I_{ref} will be given by,

$$I_{ref} = I_0 S_1 e^{(V_{gs1} - V_{thn})/\eta V_T} \tag{6}$$

where I_0 is a technology parameter, S_1 and V_{gs1} represents the aspect ratio and the gate-source voltage of NMOST Mn1 respectively, V_{thn} is the threshold voltage, η is the subthreshold slope and V_T is the thermal voltage. By using (6), expression of the gate-source voltages of Mn1 and Mn2 will be,

$$V_{gs1} = V_{thn} + \eta V_T ln\left[I_{ref}/S_1 I_0\right] \tag{7}$$

978-1-4799-4993-9/14 $31.00 © 2014 IEEE

$$V_{gs2} = V_{thn} + \eta V_T ln\left[I_{ref}/S_2 I_0\right] \tag{8}$$

By applying KVL in a loop formed by NMOSTs Mn1, Mn2 and Mn3, and using (7) and (8), we can derive,

$$V_{ds3} = \eta V_T ln\left[S_2/S_1\right] \tag{9}$$

The NMOST Mn4 is in saturation region hence,

$$V_{gs4} = V_{thn} + \sqrt{2I_{ref}/S4} \tag{10}$$

In NMOSTs stack, Mn5 is in the saturation region, Mn6 and Mn7 are in a linear region. Hence by applying KVL in a loop formed by Mn5, Mn6 and Mn7,

$$V_{gs7} = V_{ds5} + V_{ds6} + V_{ds7} \tag{11}$$

Since Mn5 is in saturation region hence,

$$V_{gs5} = V_{ds5} = V_{thn} + \sqrt{2I_{ref}/K_n S_5} \tag{12}$$

The NMOST Mn6 is linear region hence,

$$I_{ref} = S_6 K_n \left(V_{gs6} - V_{thn}\right) V_{ds6} \tag{13}$$

The gate source voltage V_{gs6} can be written as,

$$V_{gs6} = V_{ds5} + V_{ds6} \tag{14}$$

By substituting (14) and (12) in (13) a quadratic equation for V_{ds6} will form as,

$$V_{ds6}^2 + V_{ds6}\sqrt{2I_{ref}/K_n S_5} - (I_{ref}/K_n S_6) = 0 \tag{15}$$

Solving (15) using quadratic formula will result,

$$V_{ds6} = 0.5\left[-\sqrt{2I_{ref}/K_n S_5} + \sqrt{(2I_{ref}/K_n)(S_6 + 4S_5/S_6 S_5)}\right] \tag{16}$$

Similarly, Mn7 is in linear region,

$$2I_{ref} = S_7 K_n(V_{gs7} - V_{thn})V_{ds7} \tag{17}$$

Using (11), (12) and (16) in (17) will result a quadratic equation for V_{ds7} as shown,

$$V_{ds7}^2 + 0.5\left(V_{ds7}\sqrt{2I_{ref}/K_n S_5}\right)$$
$$+ 0.5\left(\sqrt{(2I_{ref}/K_n)(S_6 + 4S_5/S_6 S_5)}\right) - 2(I_{ref}/K_n S_7) = 0 \tag{18}$$

On solving (18) using quadratic formula,

$$V_{ds7} = -0.25\left[\sqrt{(I_{ref}/K_n S_5)} + \sqrt{(I_{ref}/K_n S_x)}\right]$$
$$+ 0.5\sqrt{(2I_{ref}/K_n)\left[(1/S_5 + 2/\sqrt{S_5 S_x} + 1/S_x + 4/S_7\right]} \tag{19}$$

where

$$S_x = (S_6 S_5)/(S_6 + 4S_5) \tag{20}$$

The NMOST Mn3 is in deep triode linear region hence I_{ref} will be given by,

$$I_{ref} = S_3 K_n(V_{gs3} - V_{thn})V_{ds3} \tag{21}$$

By using (5), (9), (10) and (19) in (21), equation of the reference current will be given by,

$$I_{ref} = 2S_3^2 K_n \eta^2 V_T^2 ln^2(S_2/S_1)\left[(1/\sqrt{1/S_4}) + S_x/2\right]_b^2 \tag{22}$$

A. Temperature Dependence of I_{ref}

The NMOST(Mn3) is in deep triode region and the drain current equation is given in (21). To find the temperature behavior of the reference current, differentiating (21) with respect to the temperature (T),

$$\partial I_{ref}/\partial T = K_n S_3 V_{ds3}(\partial V_{gs3}/\partial T - \partial V_{thn}/\partial T)_a$$
$$+ K_n S_3(V_{gs3} - V_{thn})_b(\partial V_{ds3}/\partial T)_c \tag{23}$$

Referring to simulation result shown in Fig.3, the slope of the V_{gs3} and V_{thn} are almost equal. Thus, we can say that the terms (a) and (b) in (23) will result small values. This will cause less deviation in the reference current (I_{ref}) with the temperature. Hence we can say that, I_{ref} will be almost independent with the temperature. Alternatively, from (22) it is clear that I_{ref} is mainly function of aspect ratios of composite array NMOSTs. Hence, by proper selection of aspect ratios a second order thermal compensation can achieved, and this will make generated reference current (I_{ref}) almost independent of the temperature.

IV. SIMULATION RESULTS

The current reference circuit based on [11] and the proposed current reference circuit shown in Fig.1(c) and Fig.1(d) respectively, designed using a standard 0.18 μm CMOS technology, and simulated using spectre simulator. The targeted reference current for both of the architectures were 80 nA. The simulation environment for both of the architectures were kept same. Fig.4, shows the variation of reference current with

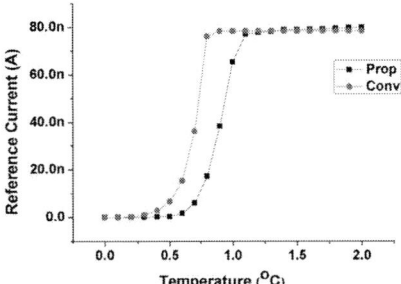

Figure 4. The variation of the reference current with the supply voltage at the room temperature 27°C. The proposed circuit started working from ≈1.2V. The change in reference current value for the given supply voltage range in the proposed and the conventional circuits are 0.2 nA and 0.1 nA respectively.

the supply voltage ranging from 0 to 2V at the room temperature 27°C. The proposed and conventional current reference circuits achieved the current value of 78.1 nA and 78.4 nA for the selected supply voltage range at room temperature. The minimum allowed supply voltage for the proposed circuit is 1.25 V. The simulation result of the reference currents against temperature is in Fig.5 for both circuits. The simulation performed from −60°C to 85°C at supply voltage of 1.25 V for both architectures. The simulated temperature coefficient (TC) of the proposed circuit is 39.8 ppm/°C while for the conventional circuit the value of TC is 545.1 ppm/°C. Thus the simulated value of TC for the proposed circuit is ≈92%,

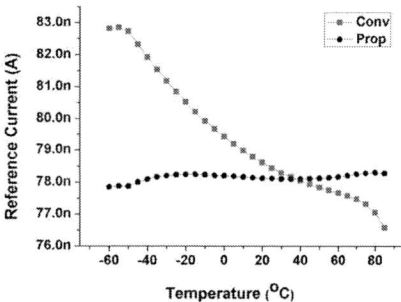

Figure 5. The variation of the reference current with the temperature ranges from −60°C to 85°C at supply voltage of 1.25 V. The proposed and the conventional circuit shows a change of 0.45 nA and 6 nA in reference current for the given temperature range. This shows that the second order thermal compensation generated from composite NMOSTs stack is working effectively.

less than the simulated value of TC achieved by conventional architecture., which confirms the theoretical analysis. The simulated results for different process corners for the proposed and the conventional current reference circuits are in Fig.6(a) and Fig.6(b) respectively.

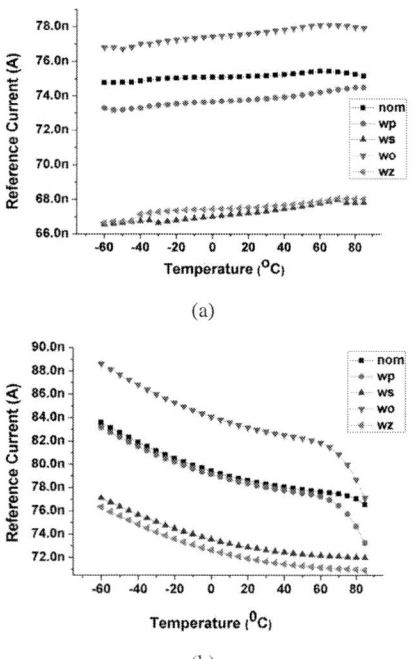

(a)

(b)

Figure 6. The simulated results for corner analysis of reference current generated from (a) the proposed and (b) the conventional current reference circuits.

V. Conclusion

In this work, we propose an extension work for the classical resistorless current reference circuit. The reference current circuits have implemented using a standard 0.18 μm CMOS technology. The simulations were carried out for the proposed and the conventional current reference circuit for the temperature ranges from −60 °C to 85 °C for a supply voltage range of 1.25 V to 2 V. The simulated temperature coefficients of the

proposed and the conventional circuits are 39.8 ppm/°C and 545.1 ppm/°C respectively. The maximum power consumption of the proposed and the conventional current reference circuits at room temperature are 624.8nW and 468.59nW respectively at a supply voltage of 2 V. Thus the proposed current reference circuit offers 92% less temperature coefficient by using 25% more power consumption than the conventional resistorless current reference circuit. Hence, the use of proposed current reference is suitable for low power applications.

VI. Acknowledgment

This publication has been partially funded by the project AUTOVOLT of Academy of Finland EFFINANO of Aalto University.

References

[1] A. Bendali and Y. Audet, "A 1-V CMOS current reference with temperature and process compensation," *Circuits and Systems I: Regular Papers, IEEE Transactions on*, vol. 54, no. 7, pp. 1424–1429, July 2007.

[2] K. Ueno, T. Hirose, T. Asai, and Y. Amemiya, "A 1-μ W 600-ppm/° C current reference circuit consisting of subthreshold CMOS circuits," *Circuits and Systems II: Express Briefs, IEEE Transactions on*, vol. 57, no. 9, pp. 681–685, Sept 2010.

[3] Q. Khan, S. Wadhwa, and K. Misri, "A low voltage switched-capacitor current reference circuit with low dependence on process, voltage and temperature," in *VLSI Design, 2003. Proceedings. 16th International Conference on*, Jan 2003, pp. 504–506.

[4] M. Paavola, M. Saukoski, M. Laiho, and K. Halonen, "A micropower voltage, current, and temperature reference for a low-power capacitive sensor interface," in *Circuits and Systems, 2007. ISCAS 2007. IEEE International Symposium on*, May 2007, pp. 3067–3070.

[5] C.-H. Wang, C.-F. Lin, W.-B. Yang, and Y.-L. Lo, "Supply voltage and temperature insensitive current reference for the 4 MHz oscillator," in *Integrated Circuits (ISIC), 2011 13th International Symposium on*, Dec 2011, pp. 35–38.

[6] T. Hirose, Y. Osaki, N. Kuroki, and M. Numa, "A nano-ampere current reference circuit and its temperature dependence control by using temperature characteristics of carrier mobilities," in *ESSCIRC, 2010 Proceedings of the*, Sept 2010, pp. 114–117.

[7] A. Amaravati, M. Dave, M. Baghini, and D. Sharma, "800 nA process-and-voltage-invariant 106-dB psrr ptat current reference," *Circuits and Systems II: Express Briefs, IEEE Transactions on*, vol. 60, no. 9, pp. 577–581, Sept 2013.

[8] N. Talebbeydokhti, P. Hanumolu, P. Kurahashi, and U.-K. Moon, "Constant transconductance bias circuit with an on-chip resistor," in *Circuits and Systems, 2006. ISCAS 2006. Proceedings. 2006 IEEE International Symposium on*, May 2006, pp. 4 pp.–2860.

[9] W. Liu, W. Khalil, M. Ismail, and E. Kussener, "A resistor-free temperature-compensated cmos current reference," in *Circuits and Systems (ISCAS), Proceedings of 2010 IEEE International Symposium on*, May 2010, pp. 845–848.

[10] M. Lukaszewicz, T. Borejko, and W. Pleskacz, "A resistorless current reference source for 65 nm cmos technology with low sensitivity to process, supply voltage and temperature variations," in *Design and Diagnostics of Electronic Circuits Systems (DDECS), 2011 IEEE 14th International Symposium on*, April 2011, pp. 75–79.

[11] H. Oguey and D. Aebischer, "CMOS current reference without resistance," *Solid-State Circuits, IEEE Journal of*, vol. 32, no. 7, pp. 1132–1135, Jul 1997.

[12] E. Vittoz and J. Fellrath, "CMOS analog integrated circuits based on weak inversion operations," *Solid-State Circuits, IEEE Journal of*, vol. 12, no. 3, pp. 224–231, Jun 1977.

[13] I. Filanovsky and A. Allam, "Mutual compensation of mobility and threshold voltage temperature effects with applications in cmos circuits," *Circuits and Systems I: Fundamental Theory and Applications, IEEE Transactions on*, vol. 48, no. 7, pp. 876–884, Jul 2001.

Towards the determination of GaN HEMT large signal model parameters by Time Domain Reflectometry method

M. Bernát, A. Šatka, A. Chvála, J. Kováč, Ľ. Sládek, D. Donoval

Institute of Electronics and Photonics
Slovak University of technology
Ilkovičova3, 812 19 Bratislava, Slovakia
e-mail: marian.bernat@stuba.sk

Abstract— **In this paper we report a development of the large signal model parameters extraction technique for gallium nitride (GaN) high electron mobility transistors (HEMT) from transient characteristics. An approach to nonlinear large-signal model parameter extraction of intrinsic model parameters namely capacitors C_{GS}, C_{DS} and C_{GD} and extrinsic resistance R_{GS} and R_{DS} has been investigated. The extraction procedure is based on electronic circuit parametric simulation of the transient characteristics of the investigated equivalent circuit and evaluation of corresponding RC time constants. This technique is improved using appropriate polarity of the input pulses in dual-head Time Domain Reflectometry (TDR) system, configured also for Time Domain Transmission (TDT) measurements.**

Index Terms— *GaN HEMT; model parameters; time domain reflectometry and transmission*

I. INTRODUCTION

GaN and related compounds represent an attractive choice among other semiconductors. With tremendous progresses made during the last decade in material quality and device processing, wide bandgap GaN HEMTs have been improved significantly in both dc and RF performances and are emerging as promising candidates for next-generation microwave and millimeter wave device's applications as a microwave power amplifiers [1]-[4]. There are also requirements to reveal behavior of measured transistor at switching mode. In that case the stepped pulses have been applied on the device due to the dynamic parameters and mainly transient characteristics and determination. The time constant of rise and falling edge of the pulse is one of the important factors of devices used in fast switching application. Determination of this dynamic parameters leads to predict behavior of transistor at complex scheme where the speed of switching operation is a key role.

Therefore an accurate GaN HEMT model generation is becoming a progressively more important part of the computer-aided-design process. Effective electronic circuit design pay attention of large-signal models capable of accurately predicting the performance of nonlinear circuits [5], [6]. Therefore it is necessary to fit intrinsic and extrinsic parameters for accurate application of the device model.

Traditionally, large signal transistor models parameters are determined from a combination of DC and S-parameter measurements under different bias conditions using small signal measurement systems [7]. All frequency independent, linear parasitic elements of the transistor are therefore determined from small-signal S-parameter measurements where relatively low input powers are applied. The bias-dependency of the time constants doesn't act.

This technique shows experimental results of the direct extractions method of the non-linear model's parameters from large-signal measurements in additions. In this paper, influence of the parameters of the HEMT large-signal equivalent circuit such as gate-source and drain-source capacitances (C_{GS}, C_{GD}), leakage resistances (R_{GS} and R_{GD}) and contact resistances on transient characteristics measured by TDR/TDT method is investigated by electronic circuit parametric simulation and analysis.

II. MODEL OF GAN HEMT TRANSISTOR

The equivalent electrical circuit for simulation of transient characteristics of GaN HEMT as measured by TDR and/or TDT method is shown in Figure 1.

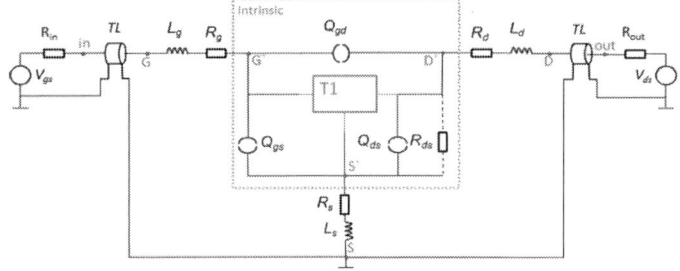

Figure 1. Large-signal equivalent circuit model of GaN HEMT measured by TDR/TDT method, with sub-circuit *T1* representing DC parameters of the HEMT transistor.

Identify applicable sponsor/s here. *(sponsors)*

978-1-4799-4993-9/14 $31.00 © 2014 IEEE

Impulse sources are represented by voltage sources V_{GS} and V_{DS} with impedance $R_{in} = R_{out} = 50\Omega$ equal to characteristic impedance of transmission line (*TL*). Both impulses sources are pulsing synchronously, the pulse amplitude of both sources is 250mV, simulating real experiments using dual head TDR sampling units of Tektronix DSA8200. The HEMT model includes both extrinsic and intrinsic HEMT elements. In this equivalent circuit, the DC parameters of investigated GaN HEMT transistor are substituted by an adapted Angelov DC equivalent model as a sub-circuit *T1* in Figure 2. The model parameters were extracted from DC characteristics of measured HEMT.

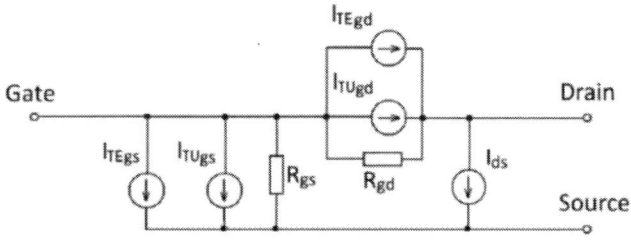

Figure 2. Equivalent static model of HEMT implemented as sub-circuit *T1* [8].

The bias-independent extrinsic elements, R_G, R_D, and R_S represent contact and semiconductor bulk resistances while L_G, L_D, and L_S model combined effect of metallization, bond wire, and package inductances. The intrinsic elements include bias-dependent C_{GS}, C_{DS} and C_{GD}. The original Angelov model includes also additional thermal sub-circuit to simulate trap and self-heating induced dispersion [9],[10] but these are not included in the simulations. Also, the bias-independent parasitic capacitances of the gate and drain pad contacts represented by C_{GP} and C_{DP}, respectively are neglected since on-chip measurements.

The R_{GS}, R_{GD} and R_{DS} resistances of the intrinsic transistor *T1* are dependent on voltages applied on the intrinsic transistor. They are implemented as current sources controlled by terminal voltages but they are assumed as voltage independent for a given operation point at a given TDR measurement. This allows a parametrical change of their value and investigation of their influence on measured transient characteristics. Charge sources Q_{GS}, Q_{GD} and Q_{DS} controlled by voltages applied on intrinsic transistor represent charging of the gate-source, gate-drain and drain-source parasitic capacitors C_{GS}, C_{GD} and C_{DS} parametrically changed during the simulations due to analysis requirements.

III. EXPERIMENTS

The TDR probe station comprises a digital serial analyzer DSA-8200 and two TDR sampling head modules 80E02 manufactured by Tektronix. Microwave air coplanar probes GSG125 (Cascade Microtech) are used for on-chip contacting of measured device, with typical insertion losses below 0.7dB in the range from DC to 40GHz. Measured device is supplied via PE1607 (Pasternack) bias tees (Figure 3). Relative

position of microwave probes and sample is carefully set by micro-positioners (Figure 4).

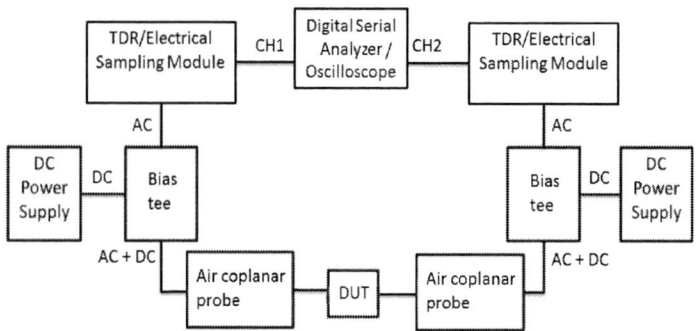

Figure 3. Block diagram of the TDR/TDT measurement setup.

TDR measurement system offers true-differential TDR measurements up to 30 GHz bandwidth with 12 ps incident rise time and 15 ps reflected rise time. In presented experiments incident pulses of the same polarity generated by these sampling heads were time aligned and synchronously fed to the device, which allowed measurement of particular waves, and evaluate also the transmission of the incident wave from the input to output of the device [11].

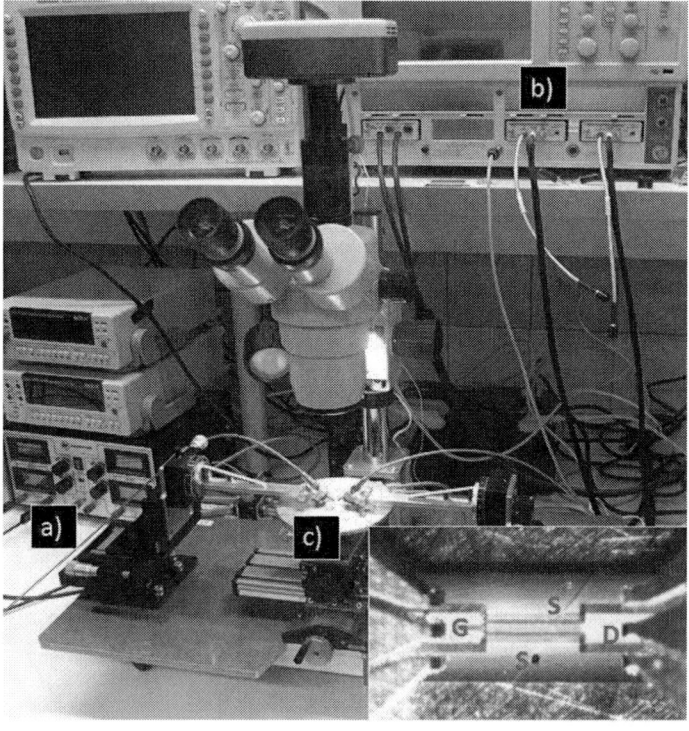

Figure 4. TDR measurement setup. The operation point of the HEMT is set by DC power supply (a); signals are measured by digital serial analyzer DSA8200 (b). DUT is contacted by air coplanar microwave probes (c).

IV. RESULTS

Measured data at $V_{GS} = 0.25$V and $V_{DS} = 0.25$V in common-source (*S*) configuration in Figure 5b clearly reveals resistive character of the output impedance with

978-1-4799-4993-9/14 $31.00 © 2014 IEEE

$(R_D+R_{DS}+R_S) = 12\ \Omega$ set by reflected voltage waveforms calculated in impedance profile. Also visible is the effect of capacitance C_{GD} resulting in charge transmission of the input pulse to the output, marked in Figure 5a by arrow [12].

Simulated TDR response waveforms reflected from the gate (G) and drain (D) of InAlN/GaN HEMT in common-source (S) configuration are shown in Figures 6, 7. In presented simulation experiments the circuit elements of the GaN HEMT large signal model were parametrically changed and amplitudes of reflected waves at *in* and *out* nodes were analyzed. The influence of these elements on transient characteristics was consequently investigated.

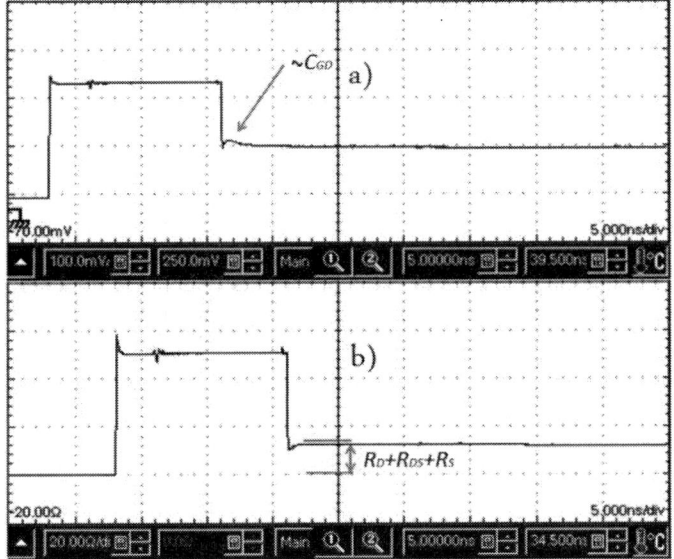

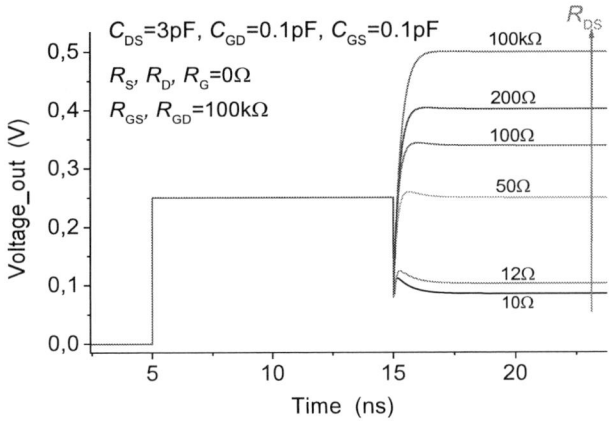

Figure 5. Measured rectangular pulse reflected from the drain of GaN HEMT (a) and calculated output impedance profile (b).

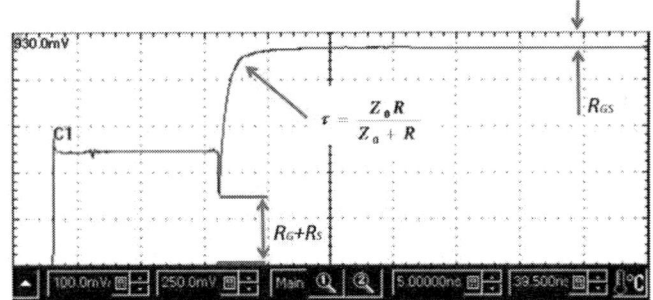

Figure 6. The dependence of the waves reflected at the *out* node on R_{DS}.

Reflected pulses at the *out* nodes of the measurement setup in dependence on parametric changed values of resistance R_{DS} are shown in Figure 6. It is evident that R_{DS} value could be extracted from the settled value of transients measured for time large enough from the reflected wave at the *out* node. This R_{DS} value can be used in next step at

particular conditions for C_{DS} determination from the transient characteristic measured from the initial part of the wave reflected at the *out* node.

Transient characteristics at the *out* node parameterized by C_{GD} are presented in Figure 8. It is clear that charging/discharging of capacitor C_{GD} influence a reflected wave after 15 ns.

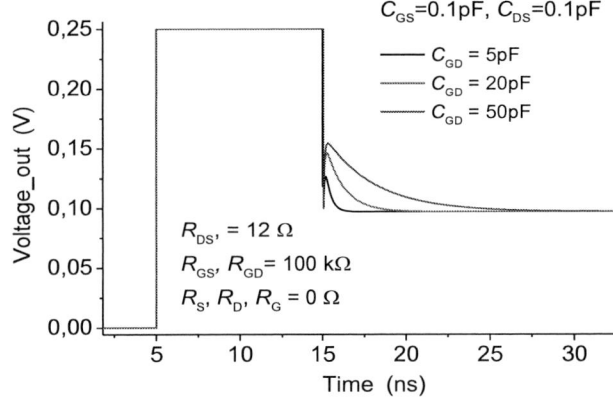

Figure 7. Simulated waves reflected at the *out* node in dependence on C_{GD}.

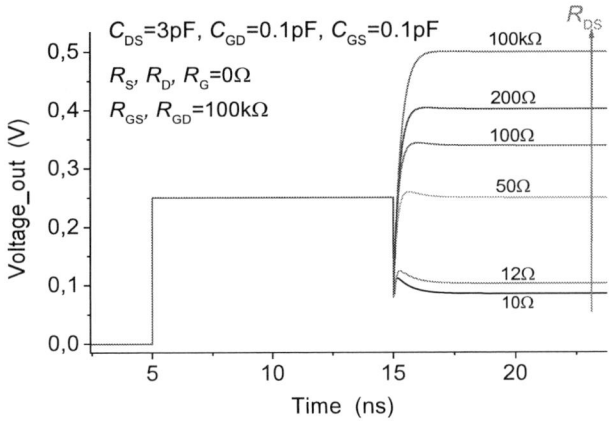

Figure 8. Measured TDR response waveforms reflected from the GaN HEMT gate in common-source configuration.

Measured TDR response waveform reflected from the gate of GaN HEMT transistor in common-source configuration is shown in Figure 8. The waveform corresponds to charging of the input capacitance C_{GS} of the transistor connected in parallel with gate leakage resistance R_{GS}. The $R_{GS} = 1850\ \Omega$ is determined from settled value of transient, whereas input capacitance $C_{GS} \cong 18$ pF is determined from transient part of the response waveform using formula (1)

$$\tau = \frac{Z_0 R}{Z_0 + R} C, \tag{1}$$

where characteristic impedance $Z_0 = 50\ \Omega$. Simulated amplitudes of the wave reflected at the *in* node on stepped capacitance C_{GS} are shown in Figure 9a. Simulated transient characteristics reveal expected capacitive character of the input impedance of the transistor, dependent on the parametric value of C_{GS}. In the initial part of the response at $t = 15$ns, when the leading edge of the pulse reaches the capacitor C_{GS}, the capacitor behaves as an electrical 'short' resulting in reflected wave of opposite polarity and zero voltage at the *in* node. As capacitor is charged up, its impedance increases up, resulting in increasing voltage at the *in* node for $t > 15\text{ns}$. Similarly, the example of amplitudes time evolutions of the reflected waves at the *out* node simulated for stepped C_{GS} are shown in Figure 9b, but the influence of C_{GS} is negligible due to the low value of this parameter. Influence of C_{DS} on input and output reflected waves is practically opposite to C_{GS}. Analyzing Figure 9 reveals that it is possible to clearly determine C_{GS} and C_{DS} values respectively.

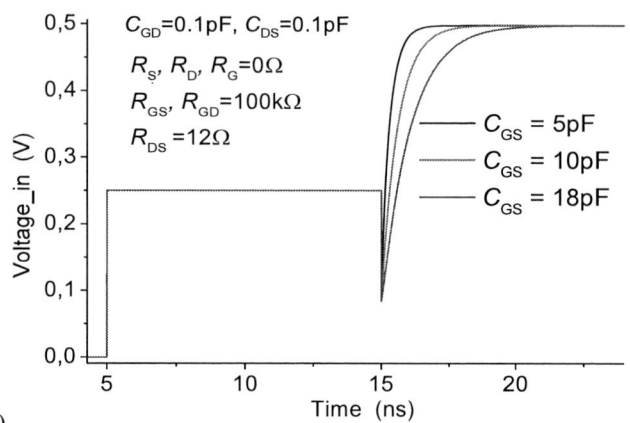

a)

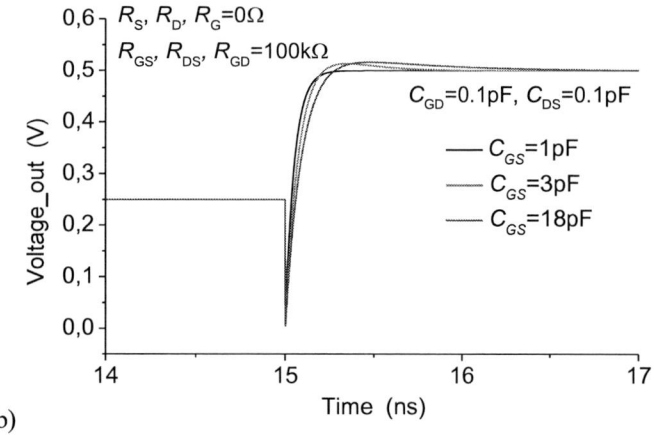

b)

Figure 9. Simulated amplitudes of reflected wave at the *in* node (a) and detail of wave reflected to the *out* node (b), in dependence on C_{GS}.

For the purpose of frequency analysis the S-parameters were pursued for this model (Figure 10).

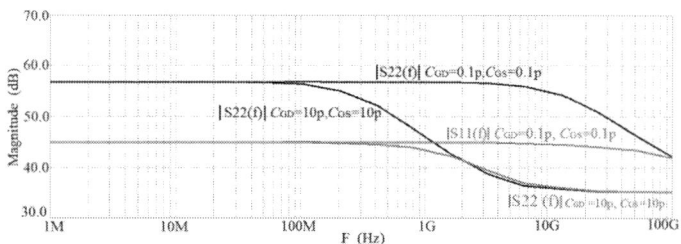

Figure 10. Frequency-domain characteristics of the circuit in Figure 1, analyzed at the same conditions as for transient analysis.

V. CONCLUSION

The simulation of the real TDR/TDT measurements reveal that influence of C_{GD} and R_{GD} on transient characteristics measured by TDR will be supressed when both voltage sources in TDR measurements have the same polarity and value of the pulses. At these circumstances, capacitances C_{GS} and C_{DS} can be determined from transient characteristics using R_{GS} and R_{DS} values estimated from settled part of transients. The simulations indicate the possibility to determine parameters of the large signal equivalent circuit from selected transient characteristics measured using appropriate polarity of the input pulses by dual-head TDR system, configured also for TDT measurements.

ACKNOWLEDGMENT

This work was supported by Slovak Research and Development Agency under the contract No. APVV-0367-11 and the Slovak Grant Agency project VEGA 1/0921/13.

REFERENCES

[1] Shen L, Heikman S, Moran B, R. Coffie, N.-Q. Zhang, D. Buttari, I. P. Smorchkova, S. Keller, S. P. DenBaars, U. K. Mishra "AlGaN/AlN/GaN high power microwave HEMT", IEEE Electron Device Lett, 2001, 22(10): 457

[2] O. Jardel, G. Callet," Performances of AlInN/GaN HEMTs for Power Applications at Microwave Frequencies, " in: Proceedings of the 5th European Microwave Integrated Circuits Conference, 2010, pp.49-52

[3] T.Kachi, "Current status of GaN power devices" in IEICE Electronic Express, Vol. 10, No.21, 1-12, 2013

[4] Kohn E, et al. "T. Switching behavior of GaN-based HFETs: thermal and electronics transients," Electron Lett 38, 2002, pp. 603–5

[5] F. van Ray, G. Kompa, "A new on-wafer large-signal waveform measurement system with 40 GHz harmonic bandwidth," IEEE MTT-S International Microwave Symposium Dig., 1992, pp. 1435-1438

[6] A.Jarndal, P.Aflaki, L. Degachi, A. Birafanec, A. Kouki, R. Negra, F. M. Ghannouchi " Large-signal model for AlGaN/GaN HEMTs suitable for RF switching-mode power amplifiers design", in: Solid-State Electronics 54, 2010, pp. 696–700

[7] Jarndal A, Kompa G. "A new small-signal modeling approach applied to GaN devices", IEEE Trans Microwave Theory Tech 53, 2005; pp. 3440–8

[8] A. Chvala, D. Donoval, A. Šatka, J. Kováč, J. Marek, M. Molnár, P. Príbytný "A new equivalent circuit model of InAlN/GaN E-HEMT," in: WOCSDICE, 2013, pp. 133-134

[9] I. Angelov, L. Bengtsson, M. Garcia, "Transactions on microwave theory and techniques," 1996, vol. 44, 1664-1674.

[10] I. Angelov, V. Desmaris, K. Dynefors, PA. Nilsson, N. Rorsman, H. Zirath, "On the large-signal modelling of AlGaN/GaN HEMTs and SiC MESFETs," in: Gallium arsenide and other semiconductor application symposium, 2005, pp. 309–12.

[11] Y.H Md Thayoob, S. Sulaiman, A. Mohd Ariffin, "Analysis of Wave Propagation in Time Domain Reflectometry Circuit Simulation Model", 2010, in: International Conference on Power and Energy 2010, pp. 276-281

[12] L. Sládek, A. Šatka, M. Bernát, D. Donoval, J. Kováč, "Experimental set-up for on-chip characterization of electronic devices by Time Domain Reflectometry", in: ASDAM, 2012, pp. 219-222

A fast and functional technique for the noise figure measurement of differential amplifiers

Yogadissen Andee, Jérôme Prouvée
CEA, LETI
17 rue des Martyrs
38054 Grenoble, France
Email: yogadissen.andee@cea.fr

François Graux
Rohde&Schwarz
9-11, rue Jeanne Braconnier
92366 Meudon-la-Forêt, France

François Danneville
IEMN, UMR CNRS 8520
Av Poincaré, BP 60069
59652 Villeneuve d'Ascq cedex, France

Abstract—**This paper presents an original technique to measure the noise figure of differential amplifiers with a four-port network analyzer. The approach is fast and simple as the S-parameters and the output noise powers are measured directly with the analyzer. There is no need of hybrid couplers or baluns. The measurement procedure and the test set-up are detailed in the paper. And to illustrate the usefulness of this new approach, measurement results of a radio-frequency differential amplifier are presented and compared to the results obtained with state-of-the-art techniques.**

Keywords—Differential amplifier, noise figure measurement, four-port Network Analyzer

I. Introduction

Differential circuits are increasingly designed for radio-frequency and other high-frequency applications, taking advantage of their immunity against common-mode noise and interference. While the small-signal behavior of these circuits can be easily characterized using 4-port Network Analyzers [1], it is more difficult to determine their noise performances. The noise figure measurement of differential circuits is a challenging subject, particularly due to the correlation of output noises. In literature, authors of [2] have dealt with the subject, but their approach uses extra equipment like hybrid couplers or baluns that complicate the measurements. An interesting technique that does not require any coupler is presented in [3]. The correlation between the output noises emerging is removed by referring the noise sources to the amplifier's inputs. This technique is very interesting but it is quite time-consuming as it requires a lot of single-ended measurements. In this paper, an original technique for the noise figure measurement of differential amplifiers is proposed. It uses a 4-port network analyzer for direct measurements of the output noises of a differential amplifier. This approach has the advantages of being fast and functional as it requires a very simple measurement setup.

II. Differential Noise Figure Measurement Techniques

A. Definition of the differential noise figure

The noise figure measurement of single-ended circuits is easily performed using the Y-factor method [4]. It is performed usually by using a noise source that delivers two different levels of noise powers to the single-ended circuit. The noise figure

of the circuit is determined by the measurement of the noise powers at the output port with a traditional signal analyzer. The noise figure is calculated using an IEEE definition [5]: For a two-port, the noise figure at a given frequency is the ratio of total output noise power per unit bandwidth to the portion of the output noise power that is due to the input noise, for the case where the input noise power is kT_0.

There is however no-well established definition of the differential noise figure. A definition analogue to the single-ended case is used. The differential noise figure is defined as the ratio of total output noise power of the differential-mode to the portion of the output noise power of the differential-mode that is due to the input noise sources. The issue about differential noise figure is the measurement of the differential-mode noise power at the outputs of a differential circuit. The difficulty that arises is due to the fact that the noises at the outputs of a differential circuit are correlated. And the correlation of these output noises cannot be directly determined with available instruments. It is therefore important to find techniques for the measurement of the differential noise figure without having to measure the correlation.

B. Measurement techniques

The classical approach for the measurement of differential amplifiers is given in [2]. The principle is to assess the differential noise figure from a single-ended configuration that can be easily characterized. The solution is to connect baluns or 180° hybrid couplers at the input and output ports of the differential amplifier, as shown in Fig. 1.

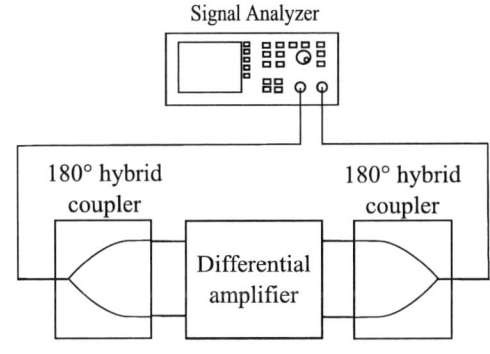

Fig. 1. Test setup of the noise figure measurement technique [2].

The noise figure of this single-ended system is then determined using the Y-factor method with a signal analyzer. A procedure for de-embedding the differential noise figure is demonstrated. It is based on a modified version of Friis equation [6] for the noise figure of the cascade of baluns which surround the differential amplifier. This technique does not require the determination of the correlation of the output noise. It uses couplers that allow the direct assessment of the differential mode noise power. For this reason, it is the most used approach. It has however the disadvantage of requiring extra equipment like baluns or couplers that need extra characterization. Moreover, these extra equipment limit the bandwidth of measurements.

An original approach is presented in [3] for the noise figure measurement of differential amplifiers without the need of couplers or baluns. This technique also deals with single-ended measurements with a signal analyzer. The correlation of output noises is addressed in different manner than in [2]. A study of the architecture of a differential amplifier is performed. A differential amplifier can be considered as 2 single-ended amplifiers connected by a common network, as shown in Fig. 2.

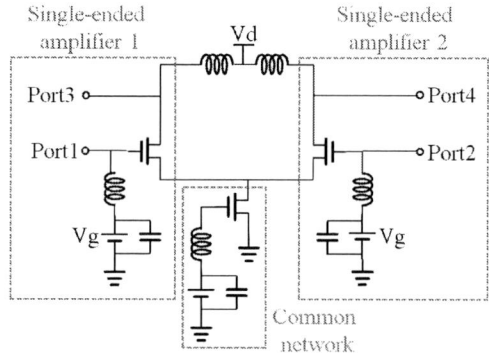

Fig. 2.   Simplified architecture of a differential amplifier.

The noises generated in each single-ended amplifier are not correlated; they can be represented as 2 input-referred noise sources which are not correlated. The common network might bring some correlation but it generates noises that appear only in the common-mode. These noises have little impact at the outputs as they are eliminated by a differential amplifier having a good common-mode rejection ratio (CMRR). Consequently, the differential amplifier can be represented as a noise-free 4-port connected to 2 input-referred noise sources as shown in Fig. 3.

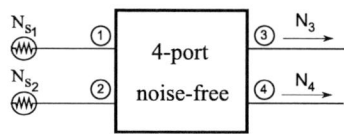

Fig. 3.   Noisy differential circuit represented by a noise-free 4-port circuit connected to 2 input-referred noise sources N_{s1} and N_{s2}. N_{si} is the sum of the noise power from the source at port i and the input-referred noise power

Using the definition of the differential noise figure cited in II.A, a new expression of the differential noise figure is given in terms of the input-referred noise powers.

$$F_{diff} = \frac{N_{s1} + N_{s2}}{2kT_0\Delta f} \qquad (1)$$

where N_{s1} and N_{s2} are calculated from the single-ended gains and the output noise powers N_3 and N_4:

$$N_{s1} = \frac{G_{42}N_3 - G_{32}N_4}{G_{31}G_{42} - G_{41}G_{32}} \qquad (2)$$

$$N_{s2} = \frac{G_{31}N_4 - G_{41}N_3}{G_{31}G_{42} - G_{41}G_{32}} \qquad (3)$$

III.   VALIDATION BY SIMULATION

The expressions of II.B are verified by simulations in Agilent's Advanced Design System (ADS). A differential amplifier in the X band is designed and simulated on ADS. It has a maximum differential gain of 11.5dB at 9.4GHz, as shown in Fig. 4. A simplified schematic diagram of the amplifier is depicted in Fig. 2.

An S-parameter simulation is performed with the input ports 1&2 and the output ports 3&4 connected to 50Ω terminations. Each termination has a noise temperature of $T_0 = 290K$. The S-parameters of the amplifier and the output noise powers are measured. The input-referred noise powers are calculated using simulation data and equations (2) and (3).The input-referred noise powers are then injected in (1) for the determination of the differential noise figure. This noise figure is shown in Fig. 4 and is compared to the differential NF obtained with technique [2] using 2 ideal 180° hybrid couplers. Fig. 4 shows that the 2 noise figures are in close agreement and therefore validates the equations given in the previous section.

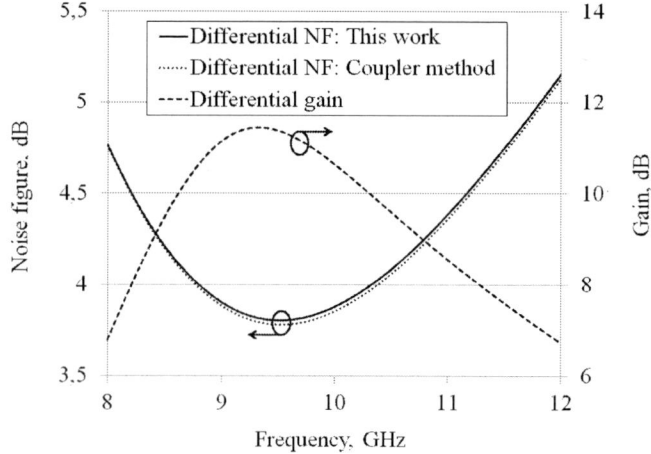

Fig. 4.   The differential gain of the simulated amplifier and the differential noise figures determined by our method and by technique [2] using ideal 180° hybrid couplers

978-1-4799-4993-9/14 $31.00 © 2014 IEEE

IV. DEVELOPMENT OF A NEW MEASUREMENT TECHNIQUE

In [3], the output noise powers used for the determination of the differential noise figure are derived from a set of 4 single-ended noise figure and 4 single-ended gain measurements. This proves to be quite time-consuming. And this long procedure yields an accumulation of measurement uncertainties, as explained in section IV. This paper gives a description of a fast and precise measurement technique with a 4-port Vector Network Analyzer (VNA). The approach relies on the direct measurement of the noise powers at the output ports of a differential amplifier. It makes use of a technique [7] that was developed for the noise figure measurement of single-ended amplifiers with a classical Network Analyzer. In [7], a test setup is described where a single-ended amplifier is connected to a VNA. The input port of the amplifier is connected a 50Ω source of the VNA. This source delivers a continuous wave (CW) signal and a noise power of kT_0. The output port is connected to a 50Ω receiver which measures the noisy signal $(CW + noise)$ coming out of the amplifier. The noise power is extracted from the noisy signal by the use of RMS and AVG detectors. The noise power is given by:

$$N_{mes} = \frac{P_{RMS} - P_{AVG}}{2} \qquad (4)$$

where P_{RMS} and P_{AVG} are respectively the root mean square power and the average power. The factor 2 is due to the single conversion concept without image conversion of the network analyzer.

This paper deals with the extension of this technique to the differential noise figure measurement. The test setup, Fig. 5, consists of a differential amplifier, a 4-port Network Analyzer and two pre-amplifiers.

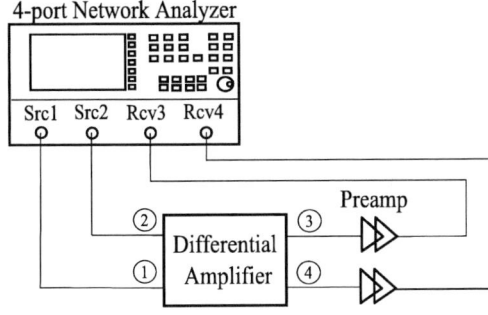

Fig. 5.   Test setup for the noise measurement of the differential amplifier

Measurements are made with a Rohde&Schwarz (R&S) ZVA24 Network Analyzer [8] on R&S VHF differential amplifier. The input ports of the amplifier are connected to two independent internal 50Ω sources of the analyzer. The low noise pre-amplifiers are placed between the amplifier and the receivers of the VNA so as to improve the precision of the measurements. As the receivers have a high noise figure, they tend to mask the noise generated by the amplifier. The pre-amplifiers are consequently used to reduce the effective noise figure. Thus, the sensitivity of the VNA increases by approximately the value of the gain of the pre-amplifiers.

After a noise calibration has been performed, the output noise powers, Fig. 6, are determined using the internal AVG and RMS detectors of the receivers. The input-referred noise powers are then calculated with (2) and (3) using the measured noise powers and the single-ended gains. These gains are easily measured on the 4-port VNA.

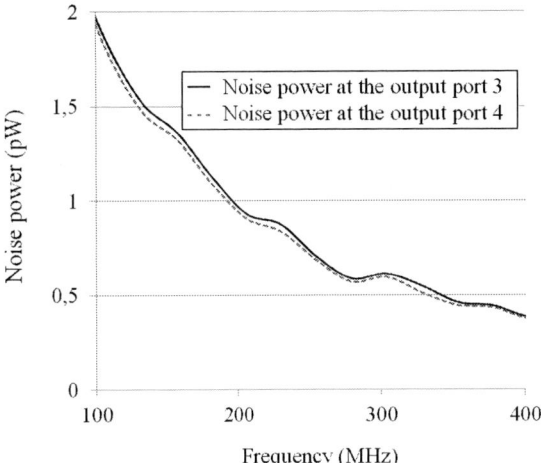

Fig. 6.   The noise powers measured at the output ports of the differential amplifier. These powers were measured with a Rohde&Schwarz (R&S) ZVA24 4-port Network Analyzer

The differential noise figure is determined using the measurement data with (1). For completeness, it is compared to the noise figures obtained with the techniques described in [2] and [3]. Fig. 7 shows the noise figure of the R&S differential amplifier determined with these three different methods.

V. EVALUATION OF MEASUREMENT RESULTS

As expected, the noise figures obtained with the three methods are of the same order of magnitude. There are some differences which are due both to the assumptions made during the development of the methods and to the measurement uncertainties. These assumptions and uncertainties must be understood in order to validate our new technique. Consequently, a brief analysis of the approaches described in [2] and [3] needs to be done.

Firstly, approach [3] deals with a series of single-ended gain and noise figure measurements based on the Y-factor method [4]. The measurements were made using an R&S FSV13 Signal Analyzer and a Noisecom's NC346D Noise Source. The difference between the noise figure given by our work and the one obtained with [3] comes mainly from the measurement uncertainties. Indeed, each single-ended gain and noise figure measured using approach [3] is given with uncertainties which depend on multiple factors such as the noise figure of the receiver, the characteristics of the amplifier, the Excess Noise Ratio (ENR) of the noise source, etc. The measurement procedure described in our work also yields its own uncertainties. The noise figures of Fig. 7 do not agree perfectly owing to the accumulation of these errors during the measurements.

978-1-4799-4993-9/14 $31.00 © 2014 IEEE   363

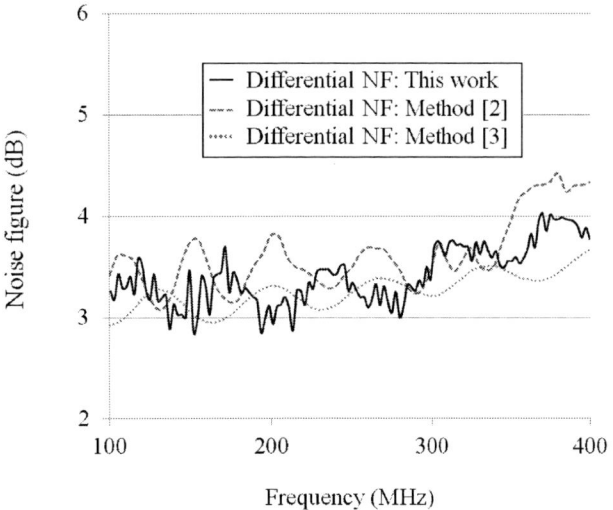

Fig. 7. Comparison between the noise figures determined with the method developed in this paper and those obtained with the techniques described in [2] and [3]

Secondly, the technique described in [2] de-embeds the differential noise figure from single-ended gain and noise figure measurements. These measurements are made using the same Signal Analyzer and Noise Source cited previously. The characterization of lossy components like the couplers yields important uncertainties. The difference between the noise figure given by our work and the one given by [2] is due not only to these uncertainties but also to the assumptions made in [2]. Indeed, the calculations in [2] are done by assuming symmetries of the losses of the couplers and of the gains of the differential amplifier. In practice, the amplifier and the couplers have some dissymmetries that cause some errors in the determination of the differential noise figure. These dissymmetries have a more important impact at higher frequencies and cause larger deviations in the measured differential noise figure.

To sum up, the analysis of the different methods explains the differences between the 3 noise figures of Fig. 7. According to the measurement results, the validity of our new technique is verified. The procedure was also validated by simulation of a differential amplifier on Agilent's Advanced Design System (ADS). Our new approach has some advantages compared to the existing methods. It offers for instance a simpler test setup than method [2] as it does not require any hybrid coupler. It is also more functional than method [3] as the noise powers and S-parameters are measured directly with a 4-port VNA.

VI. CONCLUSION

This paper has presented a fast, simple and precise approach for the noise figure measurement of differential amplifiers. The method is based on the measurement of the output noise powers with RMS and AVG detectors of a 4-port network analyzer. Its validity has been proved with the noise figure measurement of a differential RF amplifier. The measurement results have been verified by comparing them with the results obtained with the state-of-art methods.

REFERENCES

[1] Bockelman, D., and Eisenstadt, W.: 'Combined differential and common-mode scattering parameters: theory and simulation', *IEEE Trans. Microw. Theory Techn.*, 1995, **43**, (7) pp. 1530- 1539

[2] Abidi, A., and Leete, J.C.: 'De-embedding the noise figure of differential amplifiers', *IEEE J. Solid-State Circuits*, 1999, **34**, (6), pp. 882-885

[3] Belostotski, J., and Haslett, J.: 'A technique for differential noise figure measurement of differential LNAs', *IEEE Trans. Instrum. Meas.*, 2008, **37**, (7), pp. 1298-1303

[4] Agilent Technologies: 'Noise Figure Measurement Accuracy: The Y-Factor Method', *Application Note 57-2*, 2004

[5] Haus, H.A et al, 'IRE standards on methods of measuring noise in linear two ports', *Proc. IRE.*, 1960, **48**, pp. 60- 68

[6] Friis, H.T, 'Noise figures of radio receivers', *Proc. IRE.*, 1944, **32**, (7), pp. 419- 422

[7] Rohde&Schwarz:'Noise figure measurement without a noise source on a Vector Network Analyzer', *Application Note*, 2010

[8] Rohde&Schwarz:'R&S ZVA Vector Network Analyzer Specifications', *Datasheet*, 2012

Half-Thru De-embedding Method for Millimeter-Wave and Sub-Millimeter-Wave Integrated Circuits

Vipin VELAYUDHAN, Emmanuel PISTONO and Jean-Daniel ARNOULD

Univ.Grenoble Alpes, IMEP-LAHC, F-38000 Grenoble, France
CNRS, IMEP-LAHC, F-38000 Grenoble, France
Vipin.Velayudhan@minatec.grenoble-inp.fr

Abstract— **An accurate de-embedding method for millimeter-wave and sub-millimeter-wave integrated circuits is presented. In this "Half-Thru" de-embedding method, the pad-interconnects parasitics effects are modeled as a Half-Thru structure from both parts of the device under test. Several de-embedding methods over millimeter and sub-millimeter wave frequencies are compared in integrated technology by considering S-CPW transmission lines as device under test. From these comparisons we propose an effective way to de-embed transmission lines. The S-CPW transmission line model and results are obtained from full-wave electromagnetic simulations in BiCMOS 55-nm technology.**

Keywords— *De-embedding methods, characterization, millimeter wave, sub-millimeter wave , integrated circuits, S-CPW transmission lines.*

I. INTRODUCTION

Nowadays, measurement and characterization of devices in millimeter and sub-millimeter wave frequency range remain a challenge. The applications of millimeter and sub-millimeter wave frequency circuits (Video-streaming 57-66 GHz, 76-81 GHz automotive radar, medical imaging 140 GHz ...) are among the main research areas in communications domain. By considering the high frequency and thus the small size of the devices, efficient de-embedding methods must be considered to obtain accurate measurement results.

For the measurement of microwave devices with vector network analyzer, an indispensable step, called calibration, is needed. The general measurement set-up for the measurement of microwave on-wafer device under test (DUT) is shown in Fig. 1. Typically, the LRRM method [1] is considered for the millimeter-wave frequencies to eliminate, in particular, the effects of the external interconnects from the (coaxial cables or waveguides) and of the microwave probes.

Then, the Pads parasitic effects and on-wafer interconnects must be taken into account. Thus, a de-embedding step must be performed to obtain the intrinsic parameters of the DUT by removing effects of pads and on-wafer interconnects. These de-embedding methods [2]-[9] induce modifications of the design of passive and active circuits in the millimeter and sub-millimeter frequencies. Today, no solution provides a reliable and reproducible measurement of circuits in the silicon integrated technology for sub-millimeter wave frequencies beyond the 100 GHz. Indeed above this frequency, environmental measurement setup around the DUT is very critical and the effects of the pads and the substrate are no longer simple localized parasitic elements.

The de-embedding methods can be classified [2] into three types according to the size of the DUT and range of the frequency:

1. Lumped equivalent circuit model
2. Cascaded matrix based model
3. Cascaded matrix+ lumped circuit mixed model

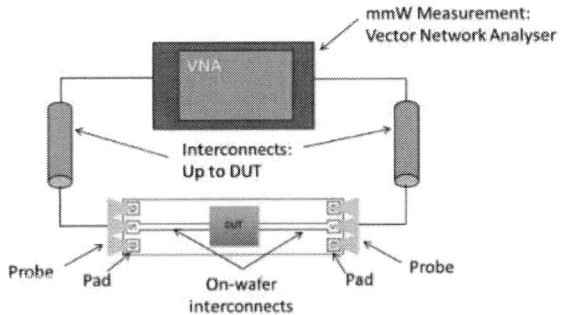

Fig. 1. Measurement setup of a device under test

Lumped equivalent circuit methods [2],[3] are used to de-embed small feeding lines lengths (as compared to considered wavelengths) and remain effective to de-embed the devices at low frequency since transmission lines are considered as lumped models. Thus, these methods are no more valid at high frequencies since feeding lines can be longer than λ/10. In the lumped circuit equivalent method, the parasitic effects of the pads and interconnects are modeled as lumped elements as exhibited for example in [2]-[4].

Cascaded equivalent models are more accurate since feeding lines are considered as transmission lines. In the cascaded matrix based methods, for example in [4], the whole test structure is considered as a cascaded network, which is more suitable for higher frequencies.

Mixed cascaded matrix+lumped circuit methods [5]-[7] considered a combination of both cascaded matrices and lumped equivalent model circuits. These methods are used to accurately de-embed both the feeding transmission lines and couplings between input/output DUT devices.

All these methods have been investigated for frequencies up to 65 GHz or 170 GHz but not more. In our study, we are comparing the accuracy of the Half-Thru de-embedding method with other de-embedding methods up to 250 GHz. After briefly explaining the Half-Thru de-embedding method, the comparison results are obtained from full-wave electromagnetic simulations with Ansoft HFSS.

978-1-4799-4993-9/14 $31.00 © 2014 IEEE

II. HALF-THRU DE-EMBEDDING METHOD

Half-Thru de-embedding is a method based on matrix calculation without any electrical model. The model of a measured DUT taking the probe pads + interconnects lines into account is shown in Fig. 2. In this method the pad-interconnects parasitics are modeled as half-thru sections. The half-thru parasitics are calculating directly from the S–parameters.

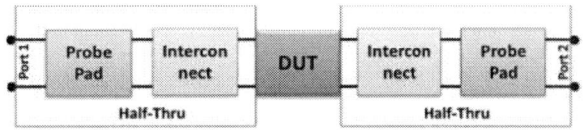

Fig. 2. Model of a measured DUT considering the probe pads and interconnects into account.

The advantage of the Half-Thru de-embedding is that there are no assumptions of symmetry, and other equivalent models for pad-interconnects parasitics. The aim of this method is to well take the parasitic effects of the half-thru from both parts of the DUT into account in order to obtain the measured circuit electrical parameters. Three test fixtures must be considered to perform our de-embedding as shown in: TL_1, TL_2 and a loaded half-thru structure. Hence, it is possible to derive the equivalent model of the half-thru (pad+access line) of the DUT as detailed below.

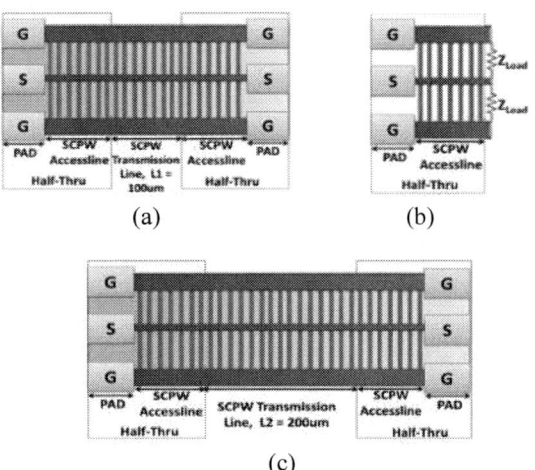

(a)   (b)

(c)

Fig. 3. Test fixtures circuits: (a) TL_1, (b) Loaded Half-thru, (c) TL_2. Access lines and central transmission lines are considered in this study as S-CPW.

The test fixtures TL_1 and TL_2 are designed to obtain the thru S-parameters. These test fixtures consist of a central transmission line with half-thru structures on both parts. The central transmission line has the same characteristics than the interconnect lines, the length of the central transmission line of TL_2 being equal to twice that of TL_1 ($L_2=2.L_1$). In this study, the access lines and the DUT are S-CPW lines [10].

The equivalent model of a thru can be derived from the [ABCD] Matrix of TL_1 and TL_2 by converting the [S] matrices of TL_1 and TL_2 into [ABCD] matrices [12]. This procedure is illustrated in the following equation

$$\begin{bmatrix} A & B \\ C & D \end{bmatrix}_{Thru} = \begin{bmatrix} A & B \\ C & D \end{bmatrix}_{L1} \begin{bmatrix} A & B \\ C & D \end{bmatrix}_{L2}^{-1} \begin{bmatrix} A & B \\ C & D \end{bmatrix}_{L1} \quad (1)$$

Then, to derive the equivalent model of the half-thru (pad + access line) of the DUT, it is necessary to measure the reflection coefficient Γ of a half-thru loaded by a well-known load Z_L different from 50 Ω. From the Signal flow graph theory (Masons Rule) [12], we can extract the effects of the pad and the access line from the thru and load.

$$S_{22} = \frac{S_{11L} - S_{21T} \cdot \Gamma - S_{11T}}{S_{11L} \cdot \Gamma - S_{11T} \cdot \Gamma - S_{21T}} \quad (2)$$

$$S_{21} = S_{12} = \sqrt{S_{21T} \cdot (1 - S_{22}^2)} \quad (3)$$

$$S_{11} = S_{11T} - S_{21T} \cdot S_{22} \quad (4)$$

S_{11L} is the reflection coefficient of the Load through the access line (Fig. 3.b). $S_{11T}=S_{22T}$ and $S_{21T}=S_{12T}$ is the S-parameters of the Thru derived from (1).

III. EXTRACTION OF LOAD VALUE

The most challenging part of the Half-Thru de-embedding is the extraction of the load value in the load de-embedding test structure. This load value must be different from 50 Ω but can vary as a function of the frequency. In practice, a fixed value for a resistance is difficult to obtain in millimeter wave frequencies. For extracting the load values, we must de-embed the load test fixture. For de-embedding the load values, we can use different de-embedding methods such as Open, Open-Short, thru de-embedding [2] which all promises good de-embedding, since the parasitics effects are considered as a small pad. The de-embedding structures for open-short de-embedding for extracting the load value is shown in Fig. 4.

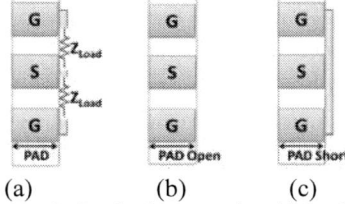

(a)   (b)   (c)

Fig. 4.Test fixtures for Load value extraction: (a) Load as DUT, (b) Open (c) Short.

The S–parameters of the de-embedding structures are converted into Y parameters in order to obtain the Z_{LOAD} as

$$Z_{LOAD} = (Y_{DUT} - Y_{OPEN})^{-1} - (Y_{SHORT} - Y_{OPEN})^{-1} \quad (5)$$

IV. SIMULATION RESULTS AND DISCUSSION

The Half-Thru de-embedding and the comparisons of other de-embedding methods are performed by using Ansys HFSS for a BiCMOS 55 nm integrated technology. The DUT consists in a S-CPW-transmission line having a length of 400 μm. Slow-wave coplanar waveguides [10],[11] are based on conventional coplanar CPW transmission lines with a patterned floating shield, including floating metallic strips under the line, as shown in Fig. 5.

978-1-4799-4993-9/14 $31.00 © 2014 IEEE

The dimensions of the coplanar strips are given by a signal width of the S-CPW $W = 4\ \mu m$, a ground width $W_g = 12\ \mu m$ and a gap between the signal and ground $G = 40\ \mu m$. The fingers have strip width of $SL = 0.16\ \mu m$ and are separated by a distance of $SS = 0.2\ \mu m$. The characteristic impedance of the line is about 70 Ω.

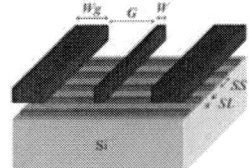

Fig. 5.S-CPW Transmission line.

The measurement setup of the device is shown in Fig. 6. The S-CPW DUT is connected with pad and access line interconnects of $100\ \mu m$ length. A ground metal plane situated underneath of the Ground-Signal-Ground PAD to connect both the ground pads.

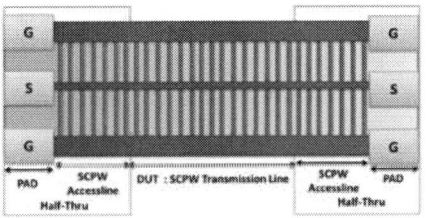

Fig. 6. Measurement setup of Half-Thru De-embedding

The de-embedded results for S-CPW transmission line are plotted in Fig. 7 to Fig. 9 by comparing the Half-thru de-embedding method with other existing methods (Vandamme [3], Mangan [5], L2L Methods [4],[7], Thru only de-embedding method [6] and TRL [8]). The red curve corresponds to electromagnetic simulations of the DUT without any access lines and PADs in order to know the true parameters of this DUT. The S-parameters and the characteristic impedance of the S-CPW transmission line are calculated.

The Half-Thru de-embedding is more accurate than other ones, since the pads and interconnects are not approximated with lumped circuit elements and no assumptions of symmetry. Moreover, our method takes the transition between the pad and the access lines into account whereas this is not the case for the cascade matrices methods.

From the S-parameter de-embedded results (Fig. 7 to Fig. 9), the Half-Thru de-embedding method allows to obtain better results than other de-embedding methods especially concerning the characteristic impedance which is dropping off when the frequency increases. Concerning the three step method of Vandamme [3], the limitations appear at very low frequencies, due to lumped circuit modelling. In Mangan [5] method the pad is assumed as a parallel admittance inducing too bad results above 100 GHz. In the cascaded matrix-based method [4], no approximation of pads and interconnects is done, but the method is good for only symmetrical pad

structures and even if symmetrical pads are considered herein, poor results are obtained above 100 GHz especially concerning the characteristic impedance because of the discontinuity between pads and access lines. TRL calibration technique [8] promises good results over the 200 GHz. But the limitation of TRL calibration is that multiple lines are required to cover wide band of frequencies band. Finally, the Half-Thru de-embedding is more accurate than other ones, since the pads and interconnects are not approximated with lumped circuit elements and no assumptions of symmetry. Moreover, this method takes the transition between the pad and the access lines into account whereas this is not the case for most of the cascade matrices methods except for TRL one. Besides, our method allows obtaining S-parameters referenced to 50 Ω whereas TRL one gives results referenced to the characteristic impedance of the line which is unknown *a priori*.

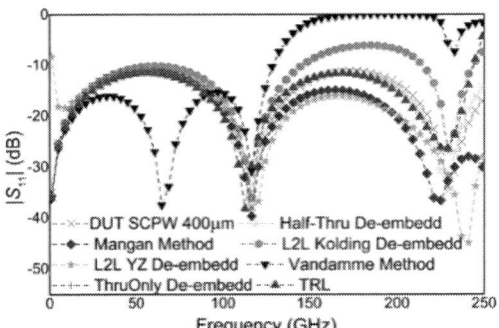

Fig. 7.Reflection coefficient of S-CPW line

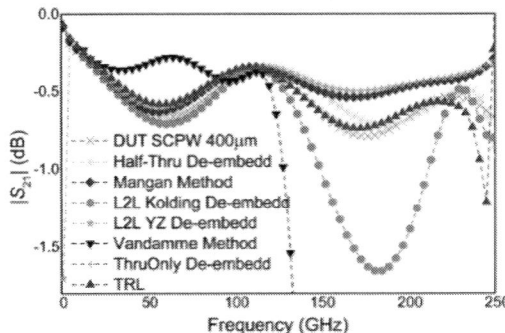

Fig. 8.Transmission coefficient of S-CPW line

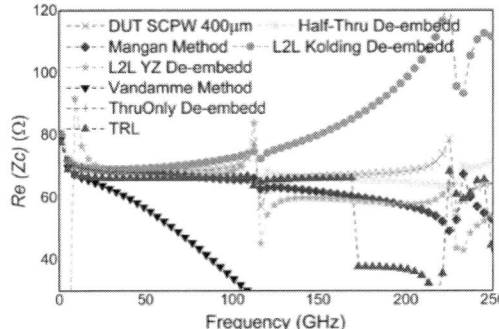

Fig. 9.Characteristic impedance of the S-CPW transmission line

V. DE-EMBEDDING WITH AND WITHOUT S-CPW ACCESSLINE

There are two kinds of de-embedding devices in the literature; ever, when the DUT is directly connected to the PAD [5],[7] or, when the DUT is connected to the PAD with interconnecting lines [4],[6]. In this part we are comparing the de-embedding with and without the access lines. From the plots of $|S_{21}|$, $|S_{11}|$ and the characteristic impedance Z_c vs frequency (Fig. 10 to Fig. 12), it is clear that a better de-embedding is obtained from the DUT when accesslines are considered. Indeed, if the DUT is directly connected to the pad without interconnects, the discontinuity between the PAD and the DUT will not be well taken into account. This will affect the accuracy of the de-embedding, especially for very high frequencies: a good de-embedding requires a continuity of propagated waves in between the half-thru and the DUT.

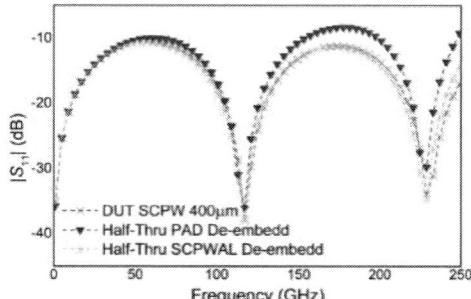

Fig. 10.Comparison of reflection coefficient

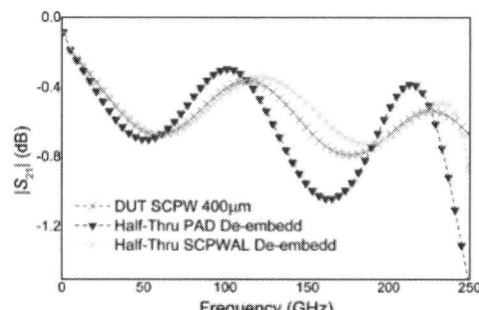

Fig. 11.Comparison of transmission coefficient

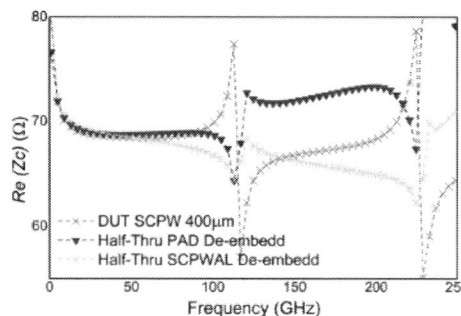

Fig. 12.Characteristic impedance of the S-CPW Transmission line vs Frequency

VI. CONCLUSIONS AND PERSPECTIVE

An accurate de-embedding method (Half-Thru De-embedding) was proposed at millimeter and sub-millimeter wave frequencies in the integrated technology. The S-CPW transmission line was considered as DUT. The simulated results and comparisons with other de-embedding methods were presented in order to point out the performance of our method. The Half-Thru de-embedding is more accurate than most of the other methods ever if TRL one gives too good results. The main advantages of our method are (i) that there is no approximation for the parasitics of the pads and interconnects, (ii) no need to have symmetrical half-thru (iii) and the results are referenced to 50 Ω. Finally, from the results we conclude that the better de-embedding is obtained by considering access lines having the same topology as the near- and far- ends of the DUT in order to have good continuities of the propagated wave through DUT.

Now, we need to realize and measure these devices to consolidate our electromagnetic simulation results.

VII. ACKNOWLEDGEMENT

The work presented here has been performed in the RF2THZ SiSoC project of the EUREKA program CATRENE in which the French partners are funded by the DGCIS.

REFERENCES

[1] F. Purroy, L. Pradell, "New theoretical analysis of the LRRM calibration technique for vector network analyzers," IEEE Trans. Instrum. Meas., vol. 50, no. 5, pp.1307-1314, Oct. 2001.

[2] B. Zhang, Y. Z. Xiong et al., "On the De-Embedding Issue of Millimeter-Wave and Sub-Millimeter-Wave Measurement and Circuit Design," IEEE Trans. Components, Manufact. Tech., vol. 2, no. 2, pp.1361-1369, Aug. 2012.

[3] E. P. Vandamme , D. Schreurs and C. Van Dinther "Improved three-step deembedding method to accurately account for the influence of pad parasitics in Si on-wafer RF test-structures", IEEE Trans. Electron Devices, vol. 48, no. 4, pp.737 -742, Apr. 2001

[4] T. E. Kolding "On-wafer calibration techniques for giga-hertz CMOS measurements", Proc. IEEE Int. Microelectronic Test Structures Conf., vol. 12, pp.105 -110, Mar. 1999

[5] A. M. Mangan , S. P. Voinigescu , Y. Ming-Ta and M. Tazlauanu "De-embedding transmission line measurements for accurate modeling of IC designs", IEEE Trans. Electron Devices, vol. 53, no. 2, pp.235 -241, Feb. 2006

[6] G. Yosuke, N. Youhei, and F. Minoru. "New On-Chip De-Embedding for Accurate Evaluation of Symmetric Devices." Japanese Journal of Applied Physics, vol. 47, no. 4, pp.2812–2816, Apr. 2008.

[7] N. Li, K. Matsushita. "Evaluation of a Multi-Line De-Embedding Technique up to 110 GHz for Millimeter-Wave CMOS Circuit Design." IEICE Trans. Fundamentals , vol. E93-A, no. 2, pp.431–439, Feb. 2010.

[8] G. F. Engen and C. A. Hoer "Thru-reflect-line: An improved technique for calibrating the dual six-port automatic network analyzer", IEEE Trans. Microwave Theory Tech., vol. 27, no.12, pp.987 -993, Dec. 1979.

[9] Z. Deng "The 'Load-Thru'(LT) de-embedding technique for the measurements of mm-wave balanced 4-port devices", IEEE Radio Frequency Integr. Circuits Symp. Dig., pp.207 -210, May 2010

[10] A.-L. Franc , E. Pistono , D. Gloria and P. Ferrari "High-performance shielded coplanarwaveguides for the design of CMOS 60-GHz band-pass filters", IEEE Trans. Electron Devices, vol. 59, no. 5, pp.1219 - 1226, May 2012

[11] A-L. Franc, E.Pistono, N. Corrao, D. Gloria, P. Ferrari "Compact high-Q, low-loss mmW transmission lines and power splitters in RF CMOS technology," Microwave Symposium Digest (MTT), pp. I-4, Jun. 2011

[12] Microwave Engineering, 4th Edition David M. Pozar, Dec. 2011

Design of passive filters using dual-mode embedded dielectric resonator

Ursula Martinez-Iranzo, Bahareh Moradi, Joan Garcia-Garcia
Department of Electronic Engineering
Universidad Autónoma de Barcelona
Cerdanyola del Vallès, Spain
ursula.martinez@uab.cat

Abstract—A novel design combining embedded high dielectric constant dielectric resonator with planar technologies is presented in this paper. The utilization of dual-mode embedded resonators with high dielectric constant allows interesting miniaturization possibilities while maintaining a competitive filter performance. The proposed design is compatible with LTCC and, in general, any planar technology such as microstrip or coplanar waveguide.

Index terms— dielectric resonator; microstrip; passive microwave filters; resonant modes.

I. INTRODUCTION

Miniaturization of passive microwave filters is one of the ongoing requirements in microwave communications applications. A number of different approaches have been reported to achieve different miniaturization level such as the utilization of sub-wavelength resonators [1], multilayer structures [2], slow-wave effects [3], meandering [4], lumped elements [5] and dual-mode resonators [6]. Designs using multiple resonator modes, are aimed to multiband filters due to the fact that the resonant frequencies of metallic layers based resonators are usually separated by wavelength multiples. The dielectric resonator (DR) resonant frequency depends on the volume and geometry as well as of the dielectric constant value. Consequently the resonant spectrum exhibits a non-linear pattern. A proper modification of the design parameters can eventually produce the degeneration of modes allowing the creation of wideband filters with a single resonator.

Examples of embedded DRs (EDRs) in PCB and planar structures can been found in the literature [7] [8]. Nevertheless in this paper it is proposed by first time the generation of the EDR using high dielectric constant paste, providing a great flexibility in the structure design and a fractional bandwidth (FWB) bandpass around 20%. This advantage can be used, for instance, to inhibit certain resonant modes fixing geometric boundary conditions [9].

Two different commercial high dielectric constant pastes have been tested to fabricate the EDR: CREATIVE 122-06, and Koartan 5483 high-K capacitor dielectric paste. Both exhibits similar permittivity values at the operating frequencies, however it would be desirable to have pastes with dielectric constants higher than 100 to obtain a better definition of the resonances and to improve the miniaturization level.

The design of dielectric embedded structures requires the utilization of a full 3D electromagnetic solver to optimize the design. 2.5D simulators (such as Momentum in Agilent ADS software) are not able to describe layers with mixed dielectric substrates. Normally, the basic design of the structures has been done in CST software due to the excellent description of the field distribution. ADS and EmPro has been used to fit the equivalent circuit model response with the experimental data.

In section II, the particularities of the proposed resonator are described, proposing a possible equivalent circuit model based on the physical characteristics. In section III it is presented a filtering structure prototype based on the EDR. Comparison between experimental and EM simulations are compared in this point to analyse the filtering structure performance. Finally, the main conclusions of the paper are highlighted in Section IV.

II. EMBEDDED RESONATOR

The EDR is produced by depositing high dielectric constant paste in a cavity milled in the substrate. A drying process is needed to obtain the EDR however the procedure is compatible with LTCC technology. The design miniaturization level increases with the value of the dielectric constant of the EDR, unfortunately commercial dielectric pastes offer epsilons below 100. CREATIVE 122-06 pad-printable high dielectric constant paste, with a value of ε_r=45 has been used to fabricate the resonator prototype. The host line is a microstrip line fabricated in a RO3010 25 mil Rogers substrate characterized by a loss tangent δ=0.0022 and dielectric constant ε_r=10.2. To facilitate the resonator excitation a microstrip ring around the EDR has been introduced to the design. Two small gaps have been introduced symmetrically in the ring to avoid low frequency transmission. The thickness of the excitation ring and the width of the symmetric gaps in the ring introduce additional freedom in the design and have been used to control the resonator response. The symmetric gaps are introduced in the equivalent circuit model (Fig. 2) through two pi-sections in the ports and the effect of the ring in the resonance will be considered as a modification of the central parallel LC tank. The microstrip surface layout of the proposed structure can be observed in Fig. 1.

The EDR cavity has been milled in the diameter space in the centre of the Fig. 1 layout. A laser milling machine LPKF

978-1-4799-4993-9/14 $31.00 © 2014 IEEE

ProtoLaser 200 has been used to generate the surface profile. Finally, creative 122-06 dielectric paste has been deposited into the cavity and the structure has been dried in a conventional oven at 100ºC for two hours.

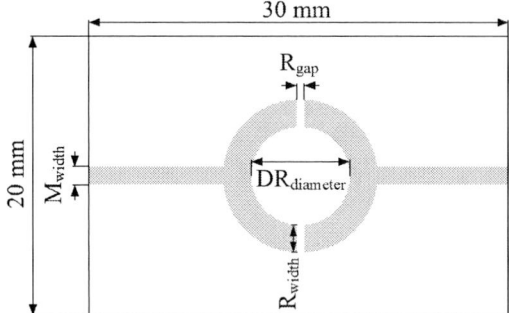

Fig. 1. Layout of the fabricated resonator prototype corresponding with a DR embedded in a substrate of a microstrip transmission line. Substrate corresponds to a RO3010 with 25 mil width.

Fig. 2 shows the equivalent circuit model of the proposed EDR. The microstrip transmission line of the input and output ports have been included since they produce phase shift important for the measure fitting. The two capacitor pi-sections models the ring gaps [5] and the LC parallel tank the EDR resonance, considering the boundary conditions fixed by the ring. Final values for the equivalent circuit parameters and for the substrate are reported in Table 1.

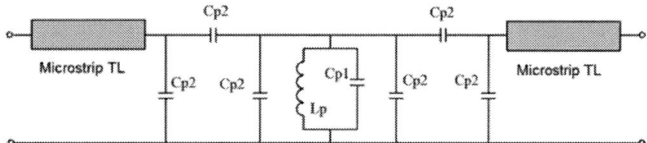

Fig. 2. Equivalent circuit model to the fabricated prototype. Gap is modelled by capacitances (C_{p2}) pi model and EDR is modelled by parallel L-C branch (L_p and C_{p1}). P1 corresponds to port 1 and P2 with port 2.

The fitting of the equivalent circuit model parameters is a complex problem with a non-unique solution. However the utilization of the tuning and optimize tools allows to achieve the excellent agreement between both measured and equivalent circuit model response shown in Fig. 3. Geometric values of the fabricated structure corresponds to M_{width}=1.316 mm, R_{gap}=200 µm, $DR_{diameter}$=7 mm and R_{width}=2 mm. As can be observed, the fitted parameters values adjust the equivalent circuit response in a wide range of frequency. The substrate parameters (loss tangent and relative permittivity) have been modified inside the known tolerance limits to fit the electromagnetic simulation with the measured results.

TABLE I.   EQUIVALENT CIRCUIT MODEL PARAMETERS

Parameter	Value
L_p	0.15 nH
C_{p1}	0.01 pF
C_{p2}	1.33 pF
Loss tangent (δ)	0.0022
$\varepsilon_{r(ROGERS3010)}$	9.9
$\varepsilon_{r(Epoxy)}$	43

Fig. 3 shows comparison between the measured, EM simulated and equivalent circuit model response using the parameters included in Table 1. In order to maximize the fitting, 3D simulation has been done with dielectric constants values of $\varepsilon_{r(Epoxy)}$=43 and $\varepsilon_{r(ROGERS3010)}$=9.9. In the range between 0 to 10 GHz it can be detected the presence of three different resonant frequencies: a non-excited resonance around f_1=3.5 GHz, a matched resonance at f_2=6.4 GHz and an unmatched resonance at f_3=8 GHz. Each of these frequencies are characterized by a specific electric and magnetic field distribution pattern. The measured maximum insertion loss is 1.8 dB at 6.6 GHz and the simulated maximum insertion loss is 0.4 dB at 6.4 GHz.

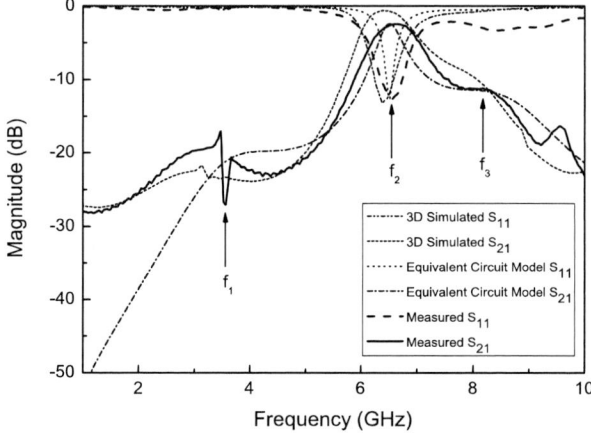

Fig. 3. Measured, 3D simulated S-parameters and equivalent circuit model results of the proposed structure. Values f_1, f_2 and f_3 correspond to 3.5, 6.4 and 8 GHz. Simulation has been done using Empro software.

The relative position of the resonant frequencies is affected by the value of physical parameters such as the value of EDR and microstrip host line permittivity (1) [11], but also of geometrical parameters such as the R_{width} or the R_{gaps}.

$$f_{nml} = \frac{c}{2\pi\sqrt{\mu_r\varepsilon_r}} \cdot a \qquad (1)$$

where f_{nml} is the EDR resonant frequency of TE_{nml} or TM_{nml} mode, μ_r is EDR relative permeability, ε_r is EDR relative permittivity and a is a value that depends on EDR geometrical values and resonant mode.

According to scalability principles, the size of a microwave filter is proportional to the guided wavelength at which it operates. Fig. 4 shows the effects of the EDR permittivity variation. It can be observed that as much higher is the permittivity value the structure spectrum is compressed allowing high miniaturization levels. An increment of 35 in the EDR dielectric constant, result in 1 GHz reduction of resonant frequency, this implies a miniaturization factor of 15 %. The available commercial materials are limited at the moment to permittivity values below 100. However a considerable effort is being dedicated to develop inks and pastes with considerably higher, as far as this goal is achieved, the miniaturization of the proposed structures will arise considerably [11]. Fig. 4 shows an example of EDR with dielectric constant of 70; dielectric materials with this permittivity value would produce 1.3 GHz decrease resonant frequency value and a miniaturization factor of 20 %.

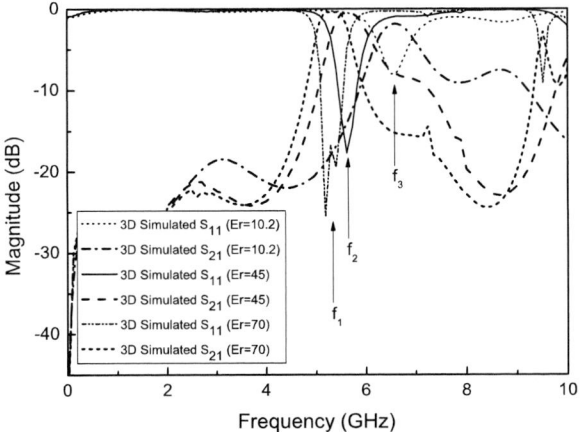

Fig. 4. 3D simulated S-parameters results of the proposed structure (ε_r=45), the same structure without EDR (dielectric constant substrate ε_r=10.2) and an example with a high dielectric constant EDR (ε_r=70). Values f_1, f_2 and f_3 correspond to 5.3, 5.6 and 6.6 GHz Simulation has been done using Empro software.

One characteristic of the EDR spectrum is a non-linear distribution of the resonant frequencies. By the modification of the geometry and the boundary conditions of the resonator it is possible to selectively modify the position of certain modes, being possible even degenerate some of them. Fig. 5 shows for instance the effect of R_{gap} that allows modifying the position of the 6.4 GHz previous resonance between 5 GHz and 6.6 GHz.

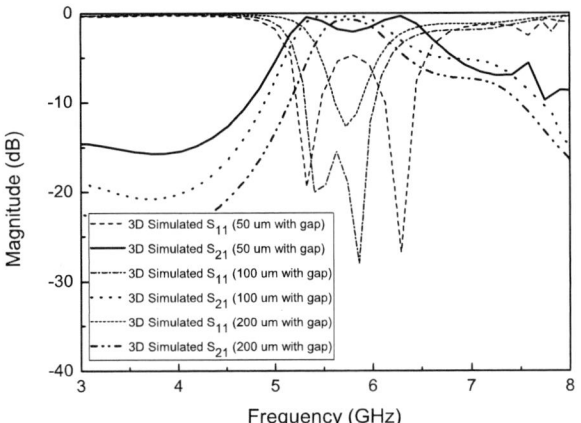

Fig. 5. S-Parameters results of the proposed structure with different 3D simulated R_{gap}. Simulation has been done using Empro software.

As far as the different modes can be shifted in the resonator spectrum it becomes possible to degenerate some of them by modifying geometrical parameters of the layout. This implies that high bandwidth can be obtained which could be a major interest to the filter design process. This feature has been used for the prototype filter design presented in this paper and explains in Section III.

III. FILTERING STRUCTURE

Taking advantage of the properties described below, a prototype bandpass filter has been designed and fabricated, by the degeneration of the f2 and f3 modes (Fig. 3) as shown in Fig. 6. Koartan 5483 high-K capacitor dielectric paste has been used to fabricate the filtering structure. The variation of R_{width} and the R_{gap} dimensions has been used as design parameters. ADS tuning and optimization tools have been used to determine the optimum values (M_{width}=1.316 mm, R_{gap}=50 µm, $DR_{diameter}$=7 mm and R_{width}=2 mm), to obtain a 20% FWB filter prototype centred at 6.3 GHz. The center frequency depends of the diameter of the DR keeping the height of the EDR fixed to the substrate thickness 0.635 mm.

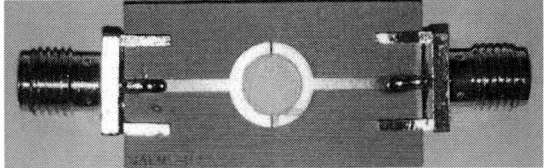

Fig. 6. Picture of the fabricated filter prototype.

Results of equivalent 3D model and measure S-parameters of the fabricated prototype are shown in Fig. 7. 3D simulation has been performed with dielectric constants values of $\varepsilon_{r(Epoxy)}$=45 and $\varepsilon_{r(ROGERS3010)}$=9.9.

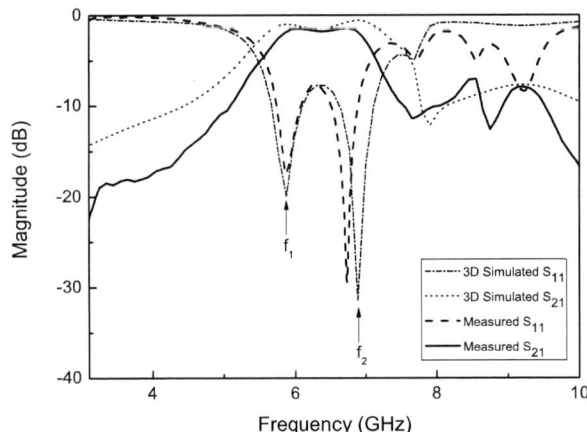

Fig. 7. Measured and 3D simulated S-parameters results of the proposed filtering structure. Values f_1 and f_2 corresponds to 5.7 and 6.9 GHz Simulation has been done using Empro software.

Discrepancies between measured and equivalent 3D model are due to parameters tolerances as dielectric constant, quality factor or loss tangent factor values. The losses in the bandpass oscillate between 0.9 and 1.4 dB, been compatible with the typical values of the ohmic lose in microstrip filters. The spurious at 8.3 GHz and 8.7 GHz represent notches around 8 dB. These spurious bands could be easily removed applying techniques based on sub-wavelength resonators acting in the microstrip input ports [13].

Different resonant modes are recognizable by observing the distribution of electric field of f_1 and f_2 (Fig. 6). Fig. 7 shows the amplitude pattern distribution observed for the two interesting excited modes in the electric field of Fig.6, whose frequencies are f_1=5.7 GHz and f_2=6.9 GHz.

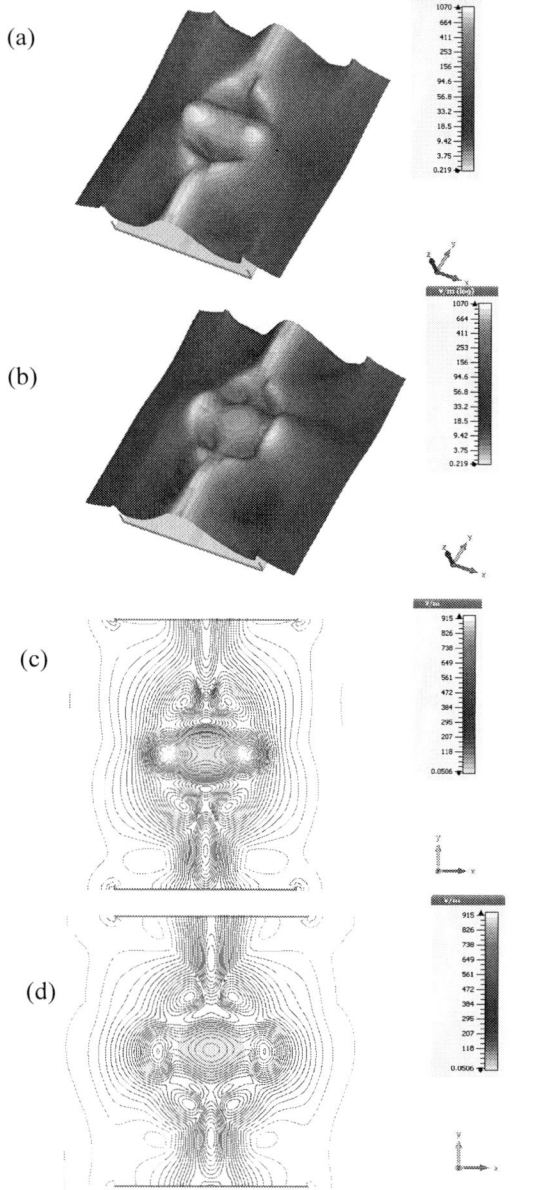

Fig. 8. Electric field total amplitude of each node. (a) carpet 3D at 5.7 GHz, (b) carpet 3D at 6.9 GHz, (c) isoline 2D at 5.7 GHz and (d) isoline 2D at 6.9 GHz. Simulation has been done using CST software.

The number of nodes of electric field indicates that any of these frequencies are the fundamental mode and therefore, that in this design; fundamental mode has not been exited. Different geometric distribution of electric and magnetic field nodes between f_1 and f_2 of Fig. 6 have been observed. According to this, two excited modes (f_1 and f_2) correspond to different modes. Spatial distribution of nodes by simulation of electric and magnetic field obtained in Fig. 8 provides information about modes excitation. This information could be utilized to inhibit or excite resonant modes to filter design.

I. CONCLUSION

The results presented in this work point out that the utilization of EDR provides miniaturization possibilities as well as an increment of the design flexibility. It has been designed a multi-mode filter prototype based on EDR with 20% FWB at 6.1 GHz. Design parameters have been identified, that allow degenerate modes in a EDR microstrip waveguide structure. It has been note that the use of conductive pastes with epsilon around 50 allow miniaturization factor of planar structures on the order of 16%.

ACKNOWLEDGMENT

This work has been supported by Ministerio de Ciencia y Educación of the Spanish government under the project TEC2010-16060.

REFERENCES

[1] J. García-García, J. Bonache, I. Gil, F. Martín, M.C. Velazquez-Ahumada, and J. Martel, "Miniaturized microstrip and CPW filters using coupled metamaterial resonators," IEEE Trans. Microwave Theory Tech., vol. 54, pp. 2628-2635, June 2006.

[2] M. Tamura, T. Yang, and T. Itoh, "Very compact and low-profile LTCC unbalanced-to-balanced filters with hybrid resonators," IEEE Trans. Microwave Theory Tech., vol. 59, pp. 1925-1936, August 2011.

[3] J.-S. Hong and M.J. Lancaster, "Theory and experiment of novel microstrip slow-wave open-loop resonator filters," IEEE Trans. Microwave Theory Tech., vol. 45, pp. 2358-2365, December 1997.

[4] S.M. Wang, C.H. Chi, M.Y Hsieh, and C.Y Chang, "Miniaturized spurious passband suppression microstrip filter using meander parallel coupled lines," IEEE Trans. Microwave Theory Tech., vol. 53, pp. 747-753, February 2005.

[5] J.-S. Hong and M.L. Lancaster, Microstrip Filters for RF/Microwave Applications. John Willey & Sons, Inc., New York 2001.

[6] A. Torabi, K. Forooraghi, O. Manoochehri, and S. Abbasiniazre, "Compact microstrip bandpass filtering using dual-mode stub-loaded resonators and capacitive/inductive sourece-load coupling," Microw. Opt. Technol. Lett., vol. 53, pp. 2111-2115, September 2011.

[7] W. Han, E. Hoppenjans, and J. Chappell, "Embedded dielectric resonator filter in layered polymer packaging," in IEEE MTT-S Int. Dig., pp. 261-264, June 2005.

[8] R. Chair, A.A. Kishk, and K.F. Lee, "Low profile wideband embedded dielectric resonator," IEEE Microw. Antennas Propag. vol. 1, pp. 294-298, April 2007.

[9] J. Garcia-Garcia, J. Ocampo, C. Martinez, J. Alonso, "Thick film high dielectric constant resonators," IEEE International Conference on Tel Aviv, pp. 1-3, November 2011.

[10] D. M. Pozar, "Microwave engineering," John Willey & Sons, Inc., New York 1998.

[11] K. A. O'Connor and R. D. Curry, "High dielectric constant composites for high power antennas," IEEE Pulsed power conference (PPC), pp. 212-217, June 2011.

[12] K. Chang and L.-H. Hsieh, "Circuits and Related Structures," John Willey & Sons, Inc., New Jersey 2004.

[13] J. García-Garcia, F Martín, F. Falcone, J. Bonache, I. Gil, T. Lopetegi, M.A.G. Laso, M. Sorolla and R. Marqués, "Spurious passband suppression in microstrip coupled line band pass filters by means of split ring resonator," IEEE Microw. Wireless Comp. Lett., vol. 14, pp. 416-418, September 2004.

The Impact of the Q-Factor of the Parasitic Capacitances of RF Transistors on their Load Modulation Capabilities

David Seebacher, Wolfgang Bösch
Graz University of Technology
Email: {david.seebacher, wolfgang.boesch}@tugraz.at

Peter Singerl, Christian Schuberth
Infineon Technologies Austria AG
Email: {peter.singerl, christian.schuberth}@infineon.com

Abstract—**With the high number of users the wireless spectrum has become a valuable resource. In order to use it efficiently sophisticated coding schemes with high crest factors are used. As a result the power amplifier is operated far below maximum output power for most of the time, which results in rather low efficiency of traditional designs. Therefore efficiency enhancement methods such as the Doherty or Chireix combiner are experiencing a revival. Their performance heavily depends on the load modulation properties of the used RF transistors and unfortunately often falls behind the predicted theory, especially for high frequency operation. This paper analyses the impact of the different loss mechanisms of RF power transistors on their load modulation capabilities. Special attention was paid to the frequency dependent losses due to the limited Q-factor of the parasitic capacitance. Based on the example of a Doherty amplifier the importance of the load modulation properties for the efficiency in back off is highlighted.**

I. Introduction

With the ever rising number of wireless users and services the available spectrum has become a valuable resource. In order to use the spectrum efficiently modern communication standards, such as LTE, employ sophisticated coding schemes, leading to signals with high crest factors. This leads to the fact that the power amplifier (PA) is operated far below maximum power for most of the time, which results in rather low efficiency of conventional designs. Therefore efficiency enhancement techniques such as the Doherty [1] or the Chireix PA [2], which rely on load modulation, are experiencing a revival. For high frequency applications their performance often falls behind the predicted theory, which is often due to the load modulation capabilities of the used RF transistors.

Thus this paper is dedicated to the analysis of the different loss mechanisms of RF power transistors and their impact on the load modulation capabilities. Based on a generalized transistor model the influence of the ON resistance, the biasing current, the parasitics Q-factor, as well as the gain are investigated.

When performing a load pull, amplifier parameters like efficiency output power and gain are evaluated over a range of different load impedances. Therefore the load impedance is modified as depicted in Fig. 1(a). The resulting contours are usually plotted in a smith chart to easily identify the spots

of maximum power and efficiency. Fig. 1(b) depicts the basic transistor model considered. It consists of the gate input with its parasitic capacitance C_{GS}' and series resistance R_G. The gate voltage V_G controls the drain current source I_D which forms together with its R_{ON} the intrinsic transistor. Additionally the drain source capacitance C_{DS}' and the drain resistor R_D are considered. The higher order harmonics are assumed to be short circuited such that class B operation occurs [3].

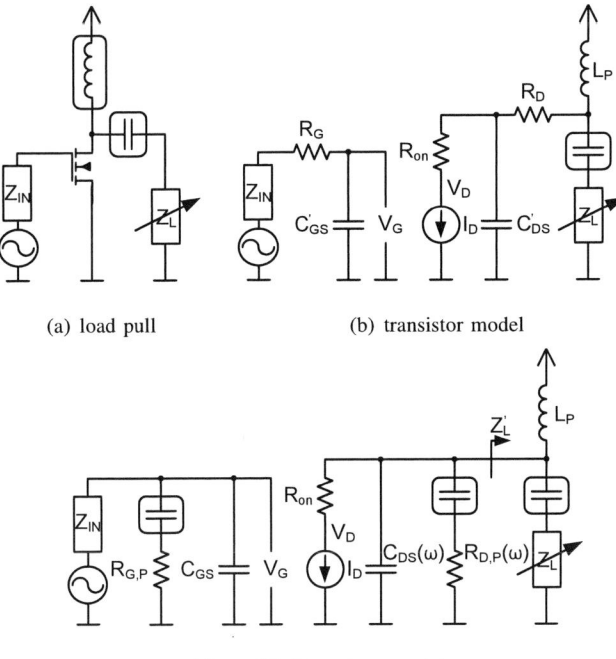

(a) load pull

(b) transistor model

(c) modified transistor model

Fig. 1. Load pull setup and transistor model.

In order to simplify the calculations the parasitic capacitance C_{DS}' and its series drain resistor R_D, which defines the Q-factor of the capacitor, can be transformed to a parallel structure [4]. The resulting schematic is provided in Fig. 1(c), where the series drain resistor R_D is converted into a (DC decoupled) parallel resistor $R_{D,P}$. Substrate losses [5] are also modeled as an equivalent parallel resistor, but are neglected in this

978-1-4799-4993-9/14 $31.00 © 2014 IEEE

analysis, as they cause the same effects as the parallel resistor $R_{D,P}$ due to the limited Q-factor of the parasitic capacitance.

In section II the efficiency calculation for the circuit in Fig. 1(c) is derived. The impact of the different loss mechanisms on the load pull contours is discussed in section III and section IV provides an analysis of the efficiency under different operational conditions. Finally the frequency behavior of a Doherty amplifier based on the derived model properties is presented and the importance of the load modulation capability of the RF power transistor is pointed out.

II. Transistor Model

In this section the efficiency of the general linear amplifier (class A to class C) as depicted in Fig. 1(c) is derived.

Fig. 2 provides the general definition of the drain current and voltage time domain waveforms

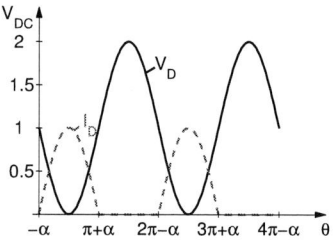

Fig. 2.   General drain current and voltage time domain waveforms.

Its drain current is defined by

$$I_D(\theta) = \begin{cases} I_{D,M}\left((1-k)\,n\,\sin\left(\theta\right) + k\right) & -\alpha \leq \theta \leq \pi+\alpha \\ 0 & \text{else} \end{cases},$$
(1)

where $I_{D,M}$ denotes the maximum drain current of the transistor, k is the normalized bias current and n is the normalized current amplitude, respectively normalized gate voltage. In order to generate only a positive drain current the (integration) border α is given by

$$\alpha = \min\left(\sin^{-1}\left(\frac{k}{(1-k)\,n}\right), \frac{\pi}{2}\right).$$
(2)

The restriction to a maximum of $\pi/2$ is required to also cover the case that the load drain current is never zero (eg. class A operation).

The drain voltage is given by

$$\begin{aligned} V_D(\theta) = V_{DC}\,(1 &- N\,\sin\left(\theta\right)\cos\left(\phi\right) \\ &- N\,\cos\left(\theta\right)\sin\left(\phi\right)), \end{aligned}$$
(3)

where V_{DC} denotes the drain bias voltage, N the normalized drain voltage swing and ϕ the phase angle of the load. As all higher order harmonics of the drain voltage are short circuited, only the fundamental of the drain current is of importance. It can be calculated by performing a Fourier Analysis of the time domain signal (1) which yields

$$I_L = \frac{1}{\pi}\int\limits_{-\alpha}^{\pi+\alpha} I_D(\theta)\sin(\theta)d\theta =$$
$$\frac{I_{D,M}}{\pi}\left((1-k)\frac{n}{2}\left(\pi + 2\alpha - \sin(2\alpha)\right) + 2k\cos\left(\alpha\right)\right),$$
(4)

which in general results in a nonlinear relation between the load current and the normalized drain current amplitude. The required load current for maximum voltage swing is calculated by taking the supply voltage and the magnitude of the load impedance into account ($I_L = V_{DC}/|Z_L|$). By resolving (4), using numerical methods considering $\alpha(n)$, the corresponding normalized drain current amplitude n can be calculated.

The effect of R_{ON}, respectively the knee voltage can be considered by

$$\eta_{knee} = \frac{R_L}{R_L + R_{ON}f(k,n)}.$$
(5)

The factor $f(k,n)$ for the effective R_{ON} depends on the biasing (k) and the normalized drive level (n), and only shows for the class B case ($f(k,n) = 2$) linear behavior. Therefore it is easier to reduce the normalized drain voltage swing N and current n accordingly to account for the drain efficiency degradation.

The resulting power dissipation in the transistor depends on the drain voltage and current as

$$P_T = \frac{1}{2\pi}\int\limits_{-\alpha}^{\pi+\alpha} I_D(\theta)V_D(\theta)d\theta,$$
(6)

which yields

$$\begin{aligned} P_T = \frac{I_{D,M}V_{DC}}{2\pi}&\left[\left((1-k)\,n - k\,N\cos(\phi)\right)2\cos(\alpha)+ \right.\\ &\left. k\left(\pi + 2\alpha\right) - N\left(1-k\right)\frac{n}{2}\cos(\phi)\left(\pi + 2\alpha - \sin(2\alpha)\right)\right]. \end{aligned}$$
(7)

The output power delivered to the load can be calculated by

$$P_L = \frac{I_L V_L \cos(\phi)}{2},$$
(8)

where the magnitude of the load voltage V_L is given by

$$V_L = |I_L Z_L| = V_{DC}N.$$
(9)

The drain efficiency is given by

$$\eta_D = \frac{P_L}{P_L + P_T}$$
(10)

and depends on the output power and the power dissipated in the transistor.

The equivalent parallel load resistance $R_{D,P}$ is frequency dependent and can be calculated by

$$R_{D,P} = R_D + \frac{1}{\omega^2 C_{DS}^2 R_D} \approx \frac{1}{\omega^2 C_{DS}^2 R_D}$$
(11)

Its contribution to the efficiency degradation is given by

$$\eta_P = \frac{R_{D,P}}{R_{D,P} + R_{L,P}},$$
(12)

where $R_{L,P}$ denotes the effective parallel load resistance of Z_L.

The influence of the power gain G_P on the Power Added Efficiency (PAE) can be accounted for by

$$\eta_G = 1 - \frac{1}{G_P}. \tag{13}$$

III. Influence of Different loss Mechanisms

In this section the impact of the different loss mechanisms on the load pull contours are evaluated considering a carrier frequency of 2.65 GHz. For the results presented a normalized capacitance C_{DS} per maximum current of 1.65 pF/A and a Q-factor of 40 for C_{DS} were considered. The knee voltage was assumed to have 10 % of the nominal supply voltage (28 V). The biasing inductance L_P in Fig.1(c) is set to resonate C_{DS} out such that the resulting load pull contours are symmetrical around the real axis.

The influence of the different effects on the load pull contours is depicted in Fig. 3. Ideally any load resistance large enough to get maximum voltage swing would results in the theoretical class B efficiency of 78.5 % (Fig. 3(a)). But the ON resistance of the transistor reduces the efficiency, especially for smaller resistor values (higher power). Thus the ON resistance of the transistor moves the points of maximum power and efficiency apart. This effect can be seen in Fig. 3(b), which depicts the impact of the ON resistance, respectively knee voltage, on the drain efficiency for class B biasing. It can be seen that for very large resistance values towards the open circuit maximum efficiency is achieved.

Considering a realistic biasing with 1 % quiescent current (or leakage) reduces the efficieny for very high resistance values. Fig. 3(c) illustrates this effect.

Beside the biasing the influence of the parallel resistor $R_{L,P}$, which accounts for the losses due the limited Q-factor of C_{DS}, significantly reduces the range of impedances for high efficiency operation. In Fig. 3(d) the influence of this effect is shown. It can be seen that the region of highest efficiency has shrunk and its maximum is reduced. The Q-factor of C_{DS} might in practice be further reduced by any series resistance (eg. bondwires). Also the biasing inductor will have a limited Q in practice. It shall be noted that the Q-factor of 40 was chosen as an example and may vary according to transistor technology and operational frequency.

When considering the gain, which has an influence on the PAE, the efficiency contours are shifted again towards higher values, as can be seen in Fig. 3(e).

The drain biasing with L_P is in practice included in the matching network and usually considered as a part of the load impedance. Fig. 3(f) shows its influence on the position of the load pull contours.

It can be summarized that the ON resistance of the transistor and the gain basically move the point of maximum efficiency towards higher resistance values. This behavior is beneficial for systems that rely on load modulation such as the Doherty amplifier. These systems require high efficiency over a wide range of load impedances to work properly. Unfortunately the biasing current and especially the Q-factor of parasitic

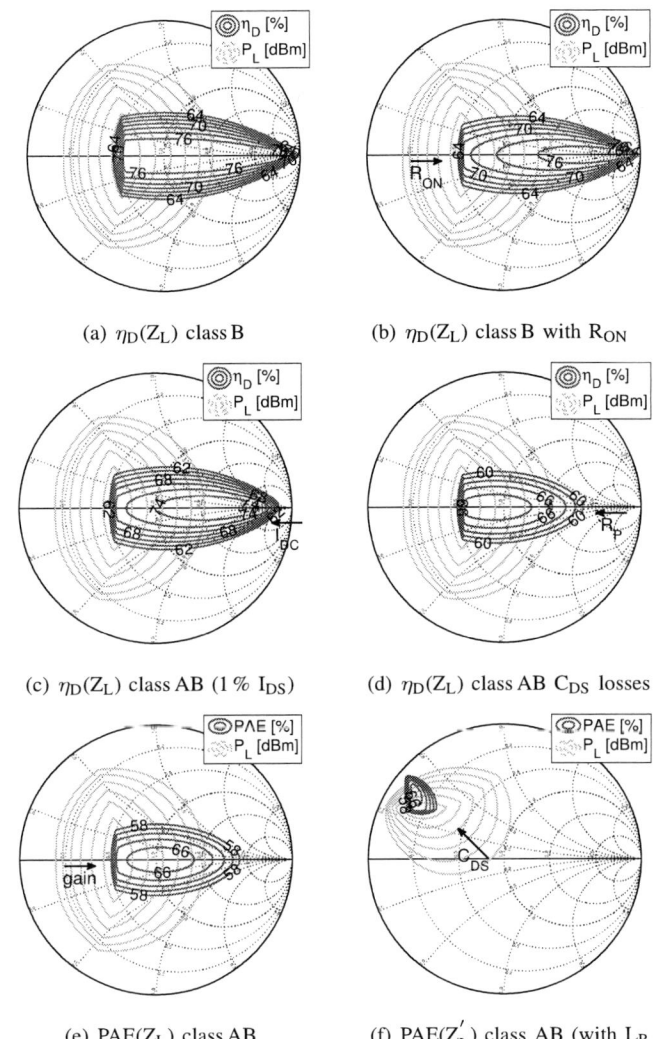

(a) $\eta_D(Z_L)$ class B

(b) $\eta_D(Z_L)$ class B with R_{ON}

(c) $\eta_D(Z_L)$ class AB (1 % I_{DS})

(d) $\eta_D(Z_L)$ class AB C_{DS} losses

(e) PAE(Z_L) class AB

(f) PAE(Z_L') class AB (with L_P

Fig. 3. Theoretical load pull contours and impact of loss mechanisms for a frequency of 2.65 GHz ($Z_{REF} = 10\Omega$).

drain capacitance C_{DS} limit this range significantly. Thus it can be said that the Q-factor of the parasitic capacitance plays an important role for the load modulation capabilities of a transistor and it imposes a limit for the high efficiency region.

IV. Dependence on Operation Conditions

During the design of a PA, parameters like current biasing condition and supply voltage can be selected, while others such as the operational frequency are fixed. Therefore the impact of these parameters on the efficiency will be discussed in this section.

A well known method to increase the efficiency is to reduce the conduction angle from class B to class C operation by modifying the biasing current. Fig. 4 shows its influence on the PAE. It depicts the efficiency along the real axis of the smith chart for a carrier frequency of 2.65 GHz versus different biasing conditions. It can be seen that the efficiency maximum occurs for slight class C biasing. For class B operation (k=0) the efficiency is only slightly smaller, while it is significantly reduced for class A (k=0.5) operation. From Fig. 4 it can be concluded that deep class C biasing is not useful, as the C_{DS} related losses ($R_{D,P}$) and the reduced gain lower the PAE.

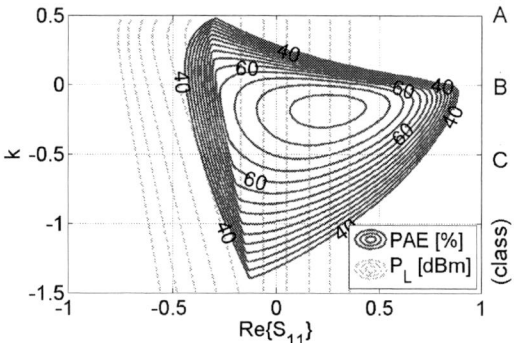

Fig. 4. PAE at 2.65 GHz versus biasing conditions ($Z_{REF} = 10\Omega$).

Beside the behavior at a single frequency the evolution of the PAE versus frequency is interesting to observe. Therefore Fig. 5 depicts the PAE along the real axis of the smith chart versus frequency. It can be seen that for very low frequencies the transistor operates efficiently over a wide range of impedances. In this region the transistor may be oversized to reduce the influence of the knee voltage and to achieve better efficiency. But when moving to higher frequencies the maximum efficiency region is significantly reduced, moving it close to the maximum output power. This trend makes the use of larger transistors inefficient and for a certain output power the optimum transistor size for maximum efficiency and power moves closely together. The fact that the region of highest efficiency is reduced for higher frequencies reduces the useful load modulation range and thus limits the achievable efficiency enhancement.

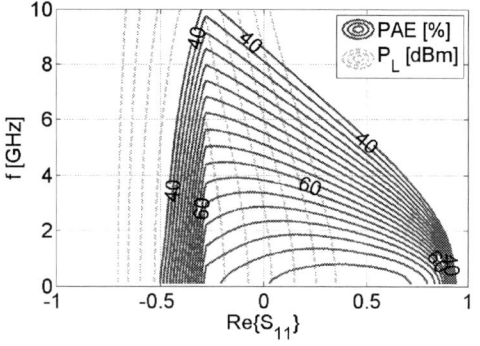

Fig. 5. PAE versus frequency ($Z_{REF} = 10\Omega$).

V. DOHERTY PA PERFORMANCE

In order to illustrate the importance of high efficiency over a wide impedance range its impact on the performance of a Doherty PA designed for different frequencies using the same devices is studied. For the main PA the efficiency behavior presented in Fig. 5 was considered and the load impedance was

designed such that maximum output power and best efficiency in back off can be achieved. The peak PA was considered to be biased in class B, such that it does not consume any power, when being inactive. The corresponding results for three different design frequencies are plotted in Fig. 6

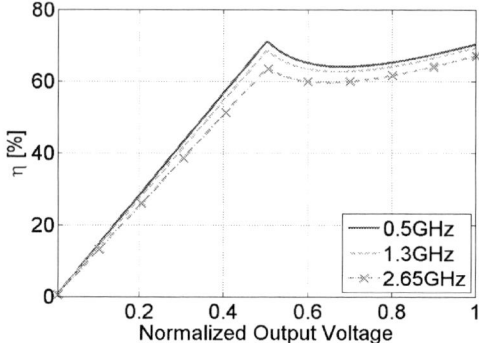

Fig. 6. Extracted Doherty PAE for different frequencies.

It can be seen that for the lowest design frequency (0.5 GHz) the best performance can be achieved. The efficiency in back off is even slightly higher than for maximum output power. This is due to the fact, that the impedance seen at the main PA in back off is increased, and the transistor still offers high efficiency over a wide impedance range for this frequency. When increasing the frequency to 2.65 GHz not only the efficiency at maximum output power is decreased, but also the efficiency in back off is heavily affected. The efficiency in back off is now smaller than the efficiency at maximum power. This is mainly caused by the (frequency) dependent losses due to the limited Q-factor of the parasitic capacitance, which significantly reduce the available efficient load modulation range for higher frequencies.

VI. CONCLUSION

The impact of the different loss mechanisms of RF power transistor has been investigated. Their individual impact on the load pull contours and furthermore on the load modulation capabilities has been discussed. In addition the efficiency behavior with respect to biasing condition, supply voltage and operational frequency is presented. Based on the example of a Doherty PA the impact of the losses due to the limited Q-factor of the parasitic capacitance and their importance for efficiency has been pointed out.

REFERENCES

[1] W. H. Doherty, "A New High Efficiency Power Amplifier for Modulated Waves," *Proceedings of the Institute of Radio Engineers*, vol. 24, no. 9, pp. 1163–1182, 1936.
[2] H. Chireix, "High Power Outphasing Modulation," *Proceedings of the Institute of Radio Engineers*, vol. 23, no. 11, pp. 1370–1392, 1935.
[3] S. Cripps, *RF Power Amplifiers for Wireless Communications*. Artech House, 2006.
[4] F. van Rijs and S. Theeuwen, "Efficiency improvement of LDMOS transistors for base stations: towards the theoretical limit," in *International Electron Devices Meeting*, 2006, pp. 1–4.
[5] J. Ankarcrona, K.-H. Eklund, L. Vestling, and J. Olsson, "Simulation and modeling of the substrate influence on the high frequency performance for RF LDMOS," in *GHz2003 Symposium*, Nov. 2003.

978-1-4799-4993-9/14 $31.00 © 2014 IEEE

Comprehensive Analysis of traps in InGaP/GaAs HBT by GR noise.

Ahmad AL HAJJAR, Jean-Christophe NALLATAMBY, Michel PRIGENT

XLIM - C²S² - Université de Limoges-16, Rue Jules Valles, Brive la Gaillarde 19100, France.

Email: ahmad.al-hajjar@xlim.fr

Abstract-In this paper, we will present a comprehensive analysis of low frequency noise in InGaP/GaAs HBT transistor, which is developed in two main axes: first, we report a study of the low frequency noise characteristics of InGaP/GaAs HBT. Our measurements were performed over the frequency range from 100 Hz to 10 MHz, under different biasing conditions and over the temperature range from 300°K to 375°K at low as well as high injection levels. Low frequency (LF) generation recombination (GR) noise measurements revealed an electron trap with activation energy of 0.536 eV. Secondly, from a rigorous physics-based noise simulation using the Langevin approach within the framework of Green's function, traps detected by temperature-dependent experimental observation are located at the heterointerface δ-InGaP/GaAs, responsible for the GR noise sources. Comparisons between LF noise measurement and numerical physics-based device noise simulation of base TLM, base-collector junction and base-emitter heterojunction allow us to locate precisely the origin of LF noise of InGaP/GaAs HBT.

Keywords-LF Noise Measurement; Numerical simulation; Noise corner frequencies; InGaP/GaAs heterojunction; Traps.

I. INTRODUCTION

In recent devices, the most important phenomenon is the trap assisted generation recombination (GR) noise, which depends on the lattice temperature of the device. In order to minimize this noise contribution in the circuit, an accurate low frequency (LF) noise model for semiconductor devices is needed. So, coherent tools for physics-based simulation and characterization for noise in semiconductor devices are crucial to improve the behavior of these components. Different setups for LF noise measurement of semi-conductor devices have been proposed in the literature, based on the use of low-noise amplifiers [1] [2] [3]. An improved version of the LF noise measurement setup is proposed in this paper allowing noise experiments at various temperatures. In order to investigate semiconductor noise mechanism, a numerical simulation of trapping noise in InGaP/GaAs heterojunction devices will be used. This powerful post-processing [4] tool has been implemented on the mathematical SCILAB software package coupled with a specified commercially available physics-based device simulator (eg: Sentaurus from Synopsys), which presents considerable advantages in performing accurate deterministic DC, AC and transient simulations of semiconductor devices, including many fundamental and parasitic effects.

This paper discusses the experimental characterization of LF equivalent short-circuit current noise sources at different temperatures and high current densities, and the use of experimental data to simulate 2D physics-based numerical

noise simulation of InGaP/GaAs HBT structure. To the best of our knowledge, this is the first paper, which shows a full characterization and helpful comparison of experimental and simulated noise in these devices at different temperatures and at operational injection levels.

To illustrate the analysis, we will present LF noise data issued from mono-finger InGaP/GaAs junctions: 1x2x30 μm², GaAs collector-base homojunction and GaAs resistor base. These devices are supplied by UMS foundry.

II. BENCH MEASUREMENT SETUP OF LF NOISE

The simplified LF noise measurement setup is shown in Figure 1.

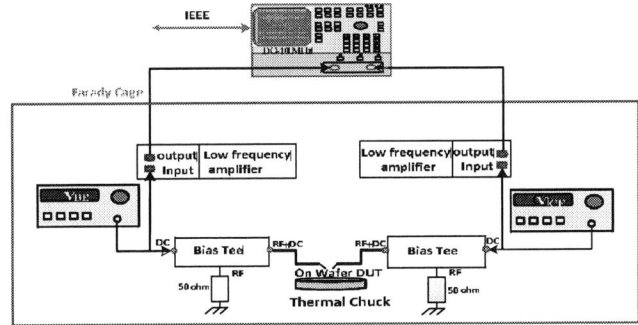

Figure 1: LF noise bench measurement.

The LF noise measurements were performed on wafer. In order to take the thermal effects into account, a thermal chuck was added to the bench and a calibration procedure has been done for the temperature. To avoid oscillations, the transistor is biased with a bias-tee loaded by 50 ohms on the RF access. We have used EG&G 5184 low noise voltage amplifiers (VAs) with a frequency range from 0.5 Hz to 1 MHz and 60 dB gain, then, a SA-230F5 amplifier from NF Corporation with 1 kHz-100 MHz bandwidth and 46 dB gain. The noise measurement setup was used in a shielded room and the lead cell batteries were used as power sources in order to minimize the disturbance at 50 Hz. A discrete fast Fourier transformation HP89410A vector signal analyzer limited to 10 MHz was used in our experiments [1].

III. CHARACTERIZATION AND EXTRACTION OF LOW FREQUENCY NOISE

The equivalent short-circuit noise current and the impedance attributed to the collector current have been characterized for different current densities in the frequency and temperature ranges already mentioned. In this case, the setup used is shown in Figure 2, along with the equivalent circuit in terms of noise and impedance analysis.

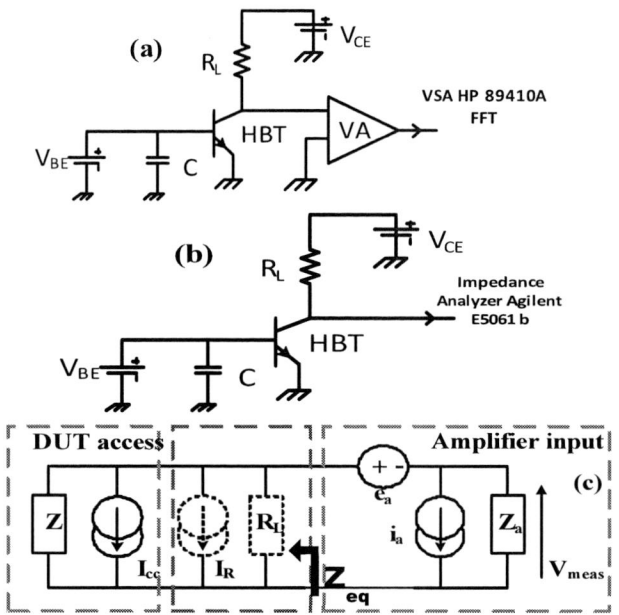

The low frequency noise characterization leads to a global evaluation of the noise but does not allow an investigation of the internal noise generators. The spatial location of the dominant noise sources is revealed by combining the rigorous trap-assisted GR noise simulation method allowing to simulate real-world devices working at high level injection [4] and trap signatures obtained from the temperature dependence of GR noise peak frequencies. Note that the temperature will be changed manually for each simulation.

V. 2-D SIMULATION AND MEASUREMENT OF LF NOISE IN InGaP/GaAs HETEROJUNCTION

2-D simulations were performed on InGaP/GaAs junctions and base layer TLM structure, in order to determine the LF noise origins. The results are compared to the corresponding measurements performed with the LF noise measurement setup.

A. Device description, and deterministic simulation

The analyzed structure of HBT for a 2-D deterministic simulation is shown on Figure 3. Half of the structure is shown and the y-axis of symmetry is located at the abscissa x=0. The two main semiconductor layers are the base layer (thickness=100 nm, $Na=4.10^{19}$ cm^{-3}) and the emitter layer (thickness=40 nm, $Nd=3.10^{17}$ cm^{-3}).

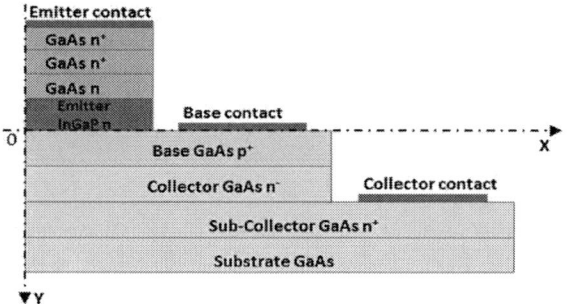

Figure 2: (a) Transistor configuration for collector (output) noise measurements, (b) Transistor configuration for collector (output) impedance measurements, (c) LF noise equivalent circuit.

To measure the LF noise current spectral density, two steps must be achieved: the direct measurements of the noise voltage by the use of VA's and the impedance of the DUT within the frequency range of interest (Figure 2(a) (b)). The DUT is represented by a current noise source (I_{cc}) in parallel with an impedance (Z). The voltage amplifier is represented by an input impedance (Z_a), a current noise source (i_a) and a voltage noise source (e_a), which have been already characterized (Figure 2(c)).

Then, by using simple Thévenin-Norton equivalent circuits and assuming that the resistor R_L generates only thermal noise, the equation that relates the voltage noise spectral density measurement $<V^2_{meas}>$ to the current noise spectral density component $<I^2_{cc}>$ is given by:

$$<I^2_{cc}>=<V^2_{meas}>\times\left|\frac{Z_{eq}+Z_a}{Z_a\times Z_{eq}}\right|^2-\frac{e^2_a}{\left|Z_{eq}\right|^2}-\frac{4KT}{R_L} \quad (1)$$

Where, Z_{eq} represents the equivalent impedance of a resistor R_L in parallel with Z.

IV. Numerical TCAD Simulation

The simulation was performed in 2D, Drift-Diffusion (D.D.) transport model, including diffusion and trap-assisted GR noise sources. In the frame of a D.D. transport model, the most rigorous method of noise analysis in semi-conductor devices associates the Langevin noise sources and Green's functions. In the frame of trap assisted GR noise, the local noise sources are generated by spontaneous fluctuations of the GR rates of the carriers. In the frame of diffusion noise, the local noise sources are generated by spontaneous fluctuations of the carrier velocities [5].

In the frequency domain, the scalar Green functions can be defined as transfer functions relating the internal noise sources to the noise response at the output terminals of the device [6].

Figure 3: Schematic of epilayers stacking of the one finger InGaP/GaAs HBT.

For each temperature, the collector current noise spectral density was simulated at a given DC value of V_{CE} for different applied V_{BE} voltages. The value of the activation energy $E_a=0.536$ eV and capture cross sections ($\sigma_p=1\times10^{-13}$ cm^2 and $\sigma_n=1\times10^{-19}$ cm^2) of traps found by GR noise measurement in InGaP/GaAs HBT have been taken for the noise simulation. We used thermionic emission current at the heterojunction interface and D.D. transport model in the bulk [7], in the deterministic simulation with Sentaurus simulator from Synopsys.

For the transport modeling, Fermi-Dirac carrier statistics, Shockley-Read-Hall, optical carrier recombination and band gap narrowing for the p-type region were used. In the simulation, the Scharfetter concentration dependent on lifetime model and the nonlinear negative differential mobility model are also introduced for electrons and holes [8].

B. Noise simulation and measurement results

By subtracting 1/f and shot noises from measurement, the product of the noise curve by the frequency $S_{fnoise}=S_{Icc}.f$ is plotted in Figure 4 for a current density equal to 10 kA/cm² and for the temperature range from 300°K to 375°K.

978-1-4799-4993-9/14 $31.00 © 2014 IEEE

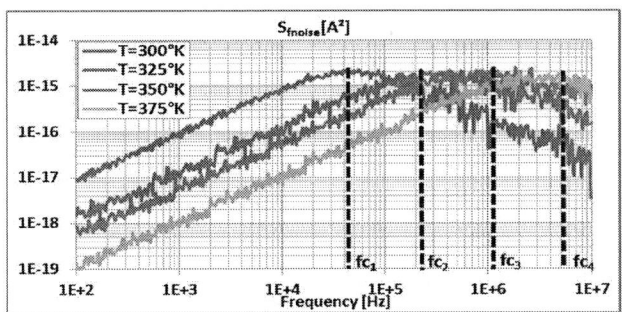

Figure 4: S_{fnoise} versus frequency at different temperature for current density equal to 10 kA/cm² (corner frequencies: f_{c1}=3.54E4 Hz, f_{c2}=2.1E5 Hz, f_{c3}=1.05E6 Hz, f_{c4}=4.12E6 Hz).

The cut-off frequency τ^{-1} of trapping and detrapping of GR noise is obtained where the GR component shows a maximum at its corner frequency. Due to the thermal activated nature of the GR process, the peak positions in the Lorentz-shaped are shifted towards the high frequency on the log (S_{fnoise}) when the temperature is increased. After the extraction of the cut-off frequencies f_{c1}, f_{c2}, f_{c3}, f_{c4} for each temperature, an Arrhenius plot was made by using the following equation [9]:

$$\frac{f_c}{T^2} = \frac{1}{\tau T^2} = \frac{\sigma V_{th} N_c}{g T^2} e^{-E_a/kT} \quad (2)$$

where f_c is the noise corner frequency, τ is the time constant, k is the Boltzmann constant, T is the measurement temperature of the device, V_{th} is the thermal velocity, N_c is the effective density of states in the conduction band, g is the level degeneracy factor (here, assumed as g=1) and E_a the activation energy. Figure 5 shows an Arrhenius plot of $\ln(\tau T^2)$ versus $1/kT$.

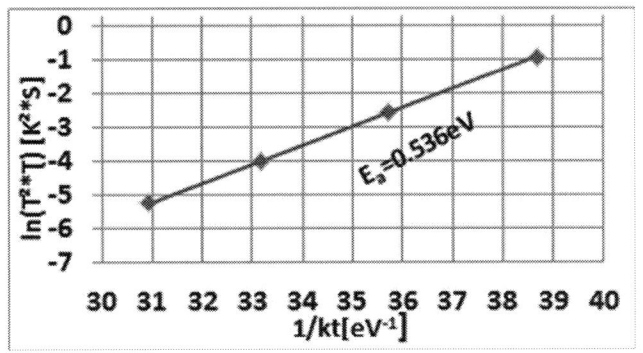

Figure 5: Arrhenius of GR noise, for temperatures varying from 300°K to 375°K for 1x2x30 µm² InGaP/ GaAs HBT.

From this plot, the apparent activation E_a and the apparent cross-section σ of the trap signatures are respectively 0.536 ± 0.05 ev and 10^{-13} cm^2. In [10] [11], the authors have shown that a potential major defect in InGaP/GaAs heterojunction is an electron trap with DLTS results.

In order to analyze the possible contribution of the R_b base resistor on the LF noise, we simulate a base TLM and measure the LF noise at room temperature. Simulations of GR noise have been carried out in presence of the EL2 defect [11]. Simulated short circuit GR current noise spectral density is masked by diffusion noise contribution. Measured and simulated results in Figure 6 correspond to the thermal noise contribution given by: S_{I-in}=4kT/R.

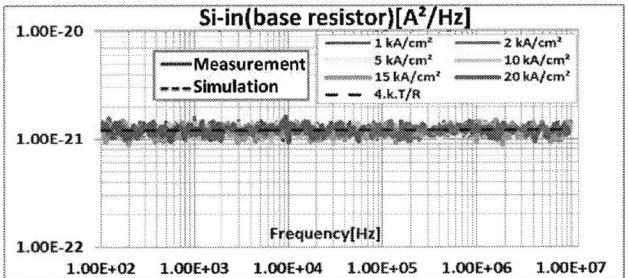

Figure 6 : Comparison of measured and simulated short circuit noise current source spectral density for 1x2x30 µm² base resistor.

We also simulate and measure the noise in base-collector junction. We present in Figure 7 the results obtained in the collector-base region normally biased with negative voltages.

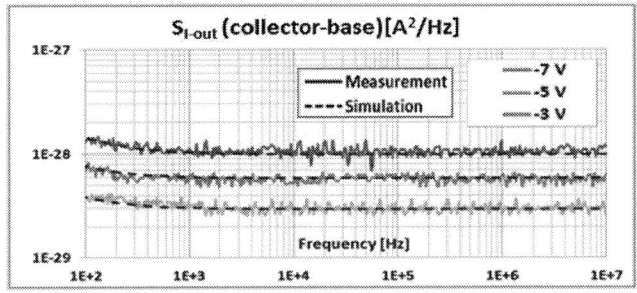

Figure 7: Comparison of measured and simulated short circuit noise current source spectral density for 1x2x30 µm² GaAs collector-base junction.

Both in simulation and measurements, no GR noise appears in base-collector junction. Finally, Figure 8 shows the simulated results of the short circuit noise current source spectral density of trap assisted GR noise in emitter base junction, superimposed on measurements for a 1x2x30 µm² size of the heterojunction at room temperature for applied current densities in the range 1 kA/cm²-20 kA/cm².

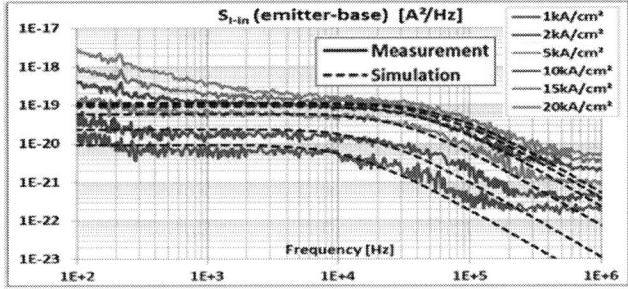

Figure 8: Comparison of measured and simulated short circuit noise current source spectral density of GR noise for 1x2x30 µm² InGaP/GaAs junction.

The simulation results of the short circuit noise current source spectral density of trap assisted GR noise are superimposed on the simulated diffusion noise and the empirical model of 1/f noise [12] for a 1x2x30 µm² size of the heterojunction for applied current densities from 1 kA/cm² to 20 kA/cm² and for two different temperatures 325°K and 350°K, as shown in Figure 9(a and b).

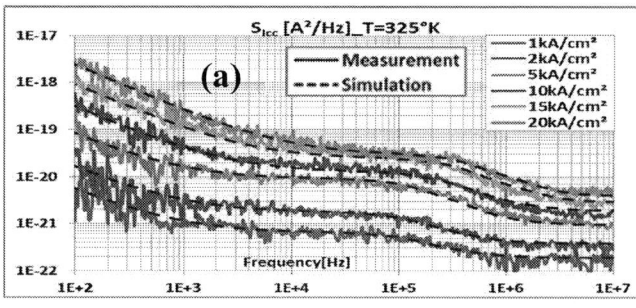

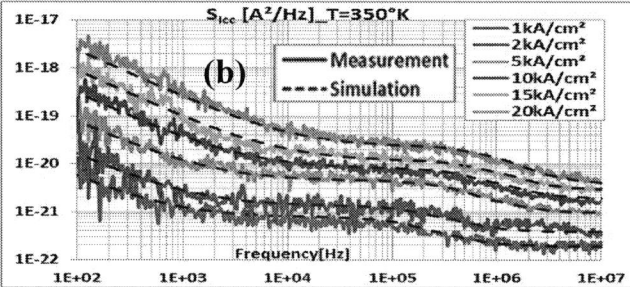

Figure 9: Comparison of measured and simulated short circuit noise current source spectral density for 1x2x30 μm² InGaP/ GaAs HBT at different currents (1 kA/cm², 2 kA/cm², 5 kA/cm², 10 kA/cm², 15 kA/cm² and 20 kA/cm²) at temperatures 325°K and 350°K.

Simulated results show clearly that this defect was found to be located in the emitter InGaP layer of the junction. The only variable parameter will be the density of traps in our simulation NT=2.10^{16} cm^{-3}. Very good agreement is obtained between measured and simulated results in the range of temperature and bias current density mentioned above which validate first our physical simulation and give us a depth understanding on LF noise sources. We distinguish between three types of noise: flicker noise ($1/f^{\alpha}$), GR noise and shot noise. GR noise is sensitive to trap and defect in the device whereas 1/f noise source is temperature-independent. When temperatures increase, the GR noise level decreases to attain the shot noise level at certain temperature while, the corner frequency increases. Let us note that the noise level increases with increasing current, while the noise level decreases with increasing temperature.

VI. CONCLUSION

The measured and simulated short circuit GR current noise spectral density shows that the bulk base emitter heterojunction is a prime contribution to the LF noise on InGaP/GaAs HBT. We can note also that the R_b base resistor and the base collector junction do not generate GR noise in linear regime operation. These results confirm and validate the realistic compact modeling of the given InGaP/GaAs technology process proposed in [13].

The low frequency measurements and simulation of InGaP/GaAs HBT performed for 300°K to 375°K temperature range for the frequencies between 100 Hz and 10 MHz with current density from 1 kA/cm² to 20 kA/cm² indicate that the deep trap (acceptor trap) with an apparent activation energy E_a=0.536 eV is the dominant source of the Lorentzian components low frequency noise behavior.

The characterization that we have presented in this paper, takes advantage of the accuracy of the measurement setup which allows us to analyze LF noises in devices up to high injections levels coupled with a TCAD simulation tools. To our knowledge, it is the first time that a complete characterization in the presence of thermal effect was performed and then, measured results have been compared successfully with 2-D simulated results of trapping noise in InGaP/GaAs heterojunction at high injections levels. The most interesting applications of the proposed method are that it can be linked with DLTS setup allowing a complete characterization of semiconductor devices in term of LF noise.

ACKNOWLEDGMENT

Dr. Didier FLORIOT and Dr. Laurent FAVÈDE from UMS are gratefully acknowledged for providing the wafers, valuable information and for helpful discussions. The authors would like to thank Dr. Khaled ABDEL HADI for the useful discussions.

REFERENCES

[1] A. AL HAJJAR, K. ABDEL HADI, J.-C. NALLATAMBY and M. PRIGENT, "Low Frequency Noise of InGaP/GaAs HBT at High Injection:A Comprehensive Analysis of Base-Collector,Base-Emitter Junctions and Base Layer TLM," in *Microwave Mediterranean Symposium*, Lebanon, 2013.

[2] S. JARRIX, C. DELSENY, F. PASCAL and G. LECOY, "Noise correlation measurements in bipolar transistors. I. Theretical expressions and extracted current spectral densities," *Journal of Applied Physics,,* pp. Vol.81, Issue 6, pp. 2651-2657, March 1997.

[3] M. BORGARINO, "Full direct low frequency noise characterization of GaAs heterojunction bipolar transistors," *Solid-State Electronics,* pp. Vol 49, Issue 8, pp. 1361-1369, August 2005.

[4] J.-C. NALLATAMBY, K. ABDELHADI, J.-C. JACQUET, M. PRIGENT, D. FLORIOT, S. DELAGE and J. OBREGON, "Numerical simulation and characterization of trapping noise in InGaP–GaAs heterojunctions devices at high injection," *Solid-State Electronics,* vol. 81, p. 35–44, 2013.

[5] K. M. VAN VLIET, "Noise and admittance of the generation—Recombination current involving SRH centers in the space-charge region of junction devices," *IEEE Transactions on Electron Devices,* vol. 23, no. 11, pp. 1236 - 1246, 1976.

[6] F. BONANI and G. GHIONE, Noise in Semiconductor Devices: Modeling and Simulation, Berlin Heidelberg: Springer-Verlag, 2001.

[7] K. HORIO and H. YANAI, "Numerical modeling of heterojunctions including the thermionic emission mechanism at the heterojunction interface," *IEEE Transactions on Electron Devices,,* vol. 37, no. 4, pp. 1093 - 1098 , 1990.

[8] J. BARNES, R. LOMAX and G. HADDAD, "Finite-element simulation of GaAs MESFET's with lateral doping profiles and submicron gates," *IEEE Transactions on Electron Devices,,* vol. 23, no. 9, pp. 1042 - 1048 , September 1976.

[9] G. MENEGHESSO, A. PACCAGNELLA, Y. HADDAB, C. CANALIi and E. ZANONI, "Evidence of interface trap creation by hot electrons in AlGaAs/GaAs high electron mobility transistors," *Appl.phys.lett.,* vol. 69, no. 10, pp. 1411-1413, 2 Sept 1996.

[10] K. CHERKAOUI, M. MURTAGH, P. KELLY, G. CREAN, S. CASETTE, S. DELAGE and S. BLAND, "Defect study of GaInP/GaAs based heterojunction bipolar transistor emitter layer," *Journal of Applied Physics,* vol. 92, no. 5, pp. 2803 - 2806, september 2002.

[11] A. MITONNEAU, A. MIRCEA, G. M. MARTIN and P. D., "Electron and hole capture cross-sections at deep centers in gallium arsenide," *Revue de Physique Appliquée,* vol. 14, pp. 853-861, 1979.

[12] F. HOOGE, "1/f noise is no suface effect," *Physics Letters,* vol. 29, no. 3, pp. 139-140, April 1969.

[13] J.-C. NALLATAMBY, M. PRIGENT, M. CAMIADE, A. SION, C. GOURDON and J. OBREGON, "An advanced Low-Frequency Noise Model of GaInP-GaAs HBT, for Accurate Prediction of Phase Noise in Oscillators," *IEEE Transactions on Microwave Theory and Techniques,* vol. 53, no. 5, pp. 1601-1612, May 2005.

Optimization of Low-Resistance State Performance in Ge-rich GST Phase Change Memory

A. Kiouseloglou*†, G. Navarro*, A. Cabrini†, G. Torelli† and L. Perniola*

*CEA-LETI, MINATEC Campus, 17 rue des Martyrs, F-38054 GRENOBLE Cedex 9, France.
†Dipartimento di Ingegneria Industriale e dell'Informazione, University of Pavia, via Ferrata 1, 27100 Pavia, Italy.
Email: athanasios.kiouseloglou@cea.fr

Abstract—In this paper, we propose a novel programming technique, named R-SET pulse, in order to optimize the Low-Resistance State (LRS) performance of Ge-rich phase change materials by overcoming the decrease of crystallization speed caused by Ge enrichment of $Ge_2Sb_2Te_5$. The R-SET pulse is capable of bringing the cell to its LRS at a lower switching threshold voltage than in the case of conventional programming pulses, thus protecting the cell from potential current overshoots during switching. The functionality of the circuit conceived to generate the R-SET pulse, which operates on a time reference scheme, is discussed. Simulations highlight the tunability of the produced R-SET pulse characteristics.

I. INTRODUCTION

Phase Change Memory (PCM) is considered among the most promising next generation non-volatile memory technologies. Thanks to its unique set of features, such as short read and write time, multi-level storage capability, and good data retention and endurance performance, PCM demonstrates the capability of entering the broad memory market and becoming a mainstream memory technology for the next decade [1].

In PCM applications, a chalcogenide alloy, such as $Ge_2Sb_2Te_5$ (GST), is reversibly switched between a high-resistivity amorphous state (RESET state) and a low-resistivity crystalline state (SET state). The transition between the two states is induced by an adequate current flow through a layer of chalcogenide alloy placed between two conductive electrodes, which causes a structural change of the active portion of the material due to Joule heating [2].

During the RESET operation, a voltage pulse of large amplitude is applied to the cell in order to melt its active region. This pulse is then quenched abruptly, which brings the active material to the amorphous state. In order to SET the cell to its Low-Resistance State, a voltage pulse of intermediate amplitude is applied to heat the material above its crystallization temperature. Threshold switching of the cell, which consists in an abrupt resistance decrease when the voltage applied across the cell overcomes a given value, referred to as threshold voltage, V_{th}, is a key phenomenon to allow a sufficient current flow through the cell ("ON regime") with reasonable programming voltages [2]. Increasing the fall time of the SET pulse ensures a better crystallization of the active material. The required duration of the SET pulse depends on the crystallization speed of the alloy, which determines the write speed performance of PCM technology [3].

In order for PCM to become a viable technology for high-volume manufacturing, its reliability has to be brought to a level similar to that of existing non-volatile memory technologies. The low crystallization temperature of GST results

in poor thermal stability and, thus, in limited data retention performance. Another downside of GST is the high current required to bring the material to the amorphous phase from the well ordered, crystalline one. Moreover, resistance drift remains one of the most important reliability issues of this technology. It has been observed that the resistivity of the amorphous phase of chalcogenide materials increases over time [4], adversely affecting the multilevel storage potential of PCM, since stochastic shifts of closely spaced programmed resistance levels may cause them to overlap and therefore lead to decoding errors [5].

Reliability of PCM can be optimized by innovative materials, such as Ge-rich GST. While Ge enrichment has a positive impact on PCM performance [6], it also leads to an increase in the resistivity of the crystalline state and a consequent SET state resistivity drift. Suppression of this phenomenon can be achieved if the cell is brought to a minimum resistance level by means of a pulse sequence of decreasing amplitude, referred to as Staircase Down (SCD) [7]. Even though the SCD approach leads the cell to an optimum crystallization of the phase change material, it is generally a long procedure, incapable of crystallizing the cell in industrially compatible times. Therefore, a new pulse sequence, named R-SET pulse, has been developed in order to successfully deal with the drift problem in a short amount of programming time.

In this paper, the features of the R-SET pulse are presented and the circuit tailored to generate this pulse is analysed. The circuit operates based on a time reference scheme and allows all pulse parameters to be externally tuned. Circuit simulation results in Cadence environment for 180 nm CMOS technology are provided to show circuit functionality.

II. R-SET PULSE

PCM takes advantage of the large programming window between the amorphous and the crystalline resistivity in phase change materials. The low-field resistance increase over time of the amorphous fraction of the memory cell affects the stability of the electrical behavior of the device and, thus, the reliability of multilevel storage. The drift phenomenon has been studied in the literature and has been found to follow the empirical power-law equation $R(t) = R_0(t/t_0)^\nu$, where t_0 is a normalizing time value, R_0 is the resistance at time t_0, and ν is the drift coefficient.

It has been recently demonstrated that Ge enrichment of GST as well as the introduction of dopants such as C and N display exceptional data retention properties. However, in order to deal with the lower crystallization speed of these materials and, at the same time, be able to well crystallize the PCM

978-1-4799-4993-9/14 $31.00 © 2014 IEEE

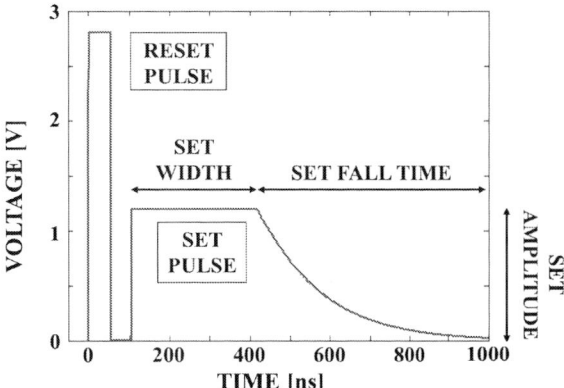

Fig. 1. The R-SET pulse is a fast sequence of a RESET pulse and a SET pulse with a long fall time.

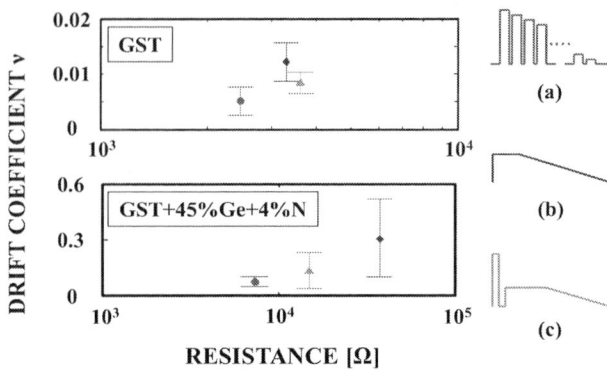

Fig. 2. Measured drift coefficient vs. programmed resistance value for GST and N doped Ge-rich GST. The cells are brought to their Low-Resistance State by means of (a) an SCD sequence, (b) a single SET pulse with a long fall time and (c) an R-SET pulse.

cells and reduce the undesired SET state drift phenomenon, an appropriate current vs. time profile has to be provided to the cell [8].

Since the drift dynamics strongly depend on the programmed resistance level [9], programming a memory cell to a minimum resistance value results in a low drift coefficient. An SCD programming sequence is able to provide a minimum SET state resistance and a correspondingly low drift coefficient. However, for a sequence of 50 pulses of width $t_w = 300$ ns, the total duration of the procedure exceeds 15 µs, thus rendering the application of the SCD sequence inappropriate for industrial applications.

Single programming pulses do not always guarantee a minimum resistance SET state, especially for Ge-rich materials. In this case, a standard SET pulse ($t_w = 300$ ns) is not able to switch the memory cell by applying voltages lower than the switching threshold voltage V_{th}, and, hence, the programming current achieved in this case is not sufficient to provide crystallization of the active material of the cell. Applying a voltage V $> V_{th}$ with a large load resistance R_{load} in series with the studied 1R memory cell leads to low current in the ON regime but, at the same time, increases parasitic contributions, thus limiting the achievable pulse width and not favoring the study of the material speed. Nevertheless, it has been demonstrated that the threshold voltage required to switch the cell can be decreased if we take advantage of the threshold voltage lowering which occurs immediately after the application of a RESET pulse [10]. During switching, the current overshoot is proportional to V_{th}/R_{load}. Keeping the threshold voltage as low as possible is important in order to minimize the current overshoot and thus cause less stress on the device. To this end, a new programming technique, named R-SET pulse, is proposed.

The R-SET pulse (Fig. 1) is a fast sequence of a RESET pulse and a SET pulse with a long fall time. The decrease in the threshold voltage of the cell that follows the application of the RESET pulse enables the possibility to achieve a SET state by applying a successive SET pulse of lower amplitude than in conventional SET programming. Increasing the fall time t_f of the SET pulse of the sequence can bring the memory cell to a lower resistance value, whereas an increase of the SET pulse width t_w leads to a higher probability to switch the cell at a lower voltage [8].

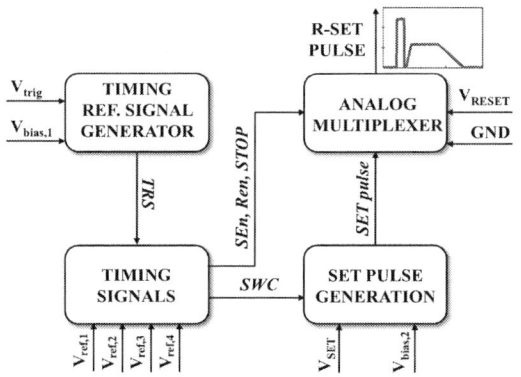

Fig. 3. Block diagram of the proposed R-SET pulse generator.

The R-SET pulse is capable of providing an appropriate current vs. time profile to the memory cell, enabling an optimum crystallization by accurately controlling the amplitude and the fall time of the SET programming voltage pulse. It is thus possible to bring the cell to a low-resistance value in industrially compatible times (as short as 1 µs), as opposed to the case of time-consuming SCD pulse sequences. R-SET pulse programming is specifically oriented for innovative PCM materials with low crystallization speed [8].

In order to evaluate the results achieved with this programming technique, we integrated GST and N-doped Ge-rich GST (GST+45%Ge+4%N) in state-of-the-art "Wall" structure PCM cells [11]. The results of the application of the R-SET pulse on GST and N-doped Ge-rich GST can be seen in Fig. 2. The devices were brought to their SET state by means of (a) an SCD sequence, (b) a single SET pulse with a long fall time ($V_{ampl} = 1.5$ V, $t_w = 300$ ns, $t_f = 1$ µs) and (c) an R-SET pulse ($t_{w,RESET} = 50$ ns, $V_{ampl,SET} = 1.2$ V, $t_{w,SET} = 300$ ns, $t_{f,SET} = 500$ ns), and their resistance value evolution was then measured at room temperature.

In the case of GST, the three procedures bring the devices to a similar average resistance value, and the dispersion of the drift coefficient is also on the same order, thus confirming the equivalence of the SET states achieved in the different cases as well as the non-detrimental effect of the R-SET procedure on the cell.

978-1-4799-4993-9/14 $31.00 © 2014 IEEE

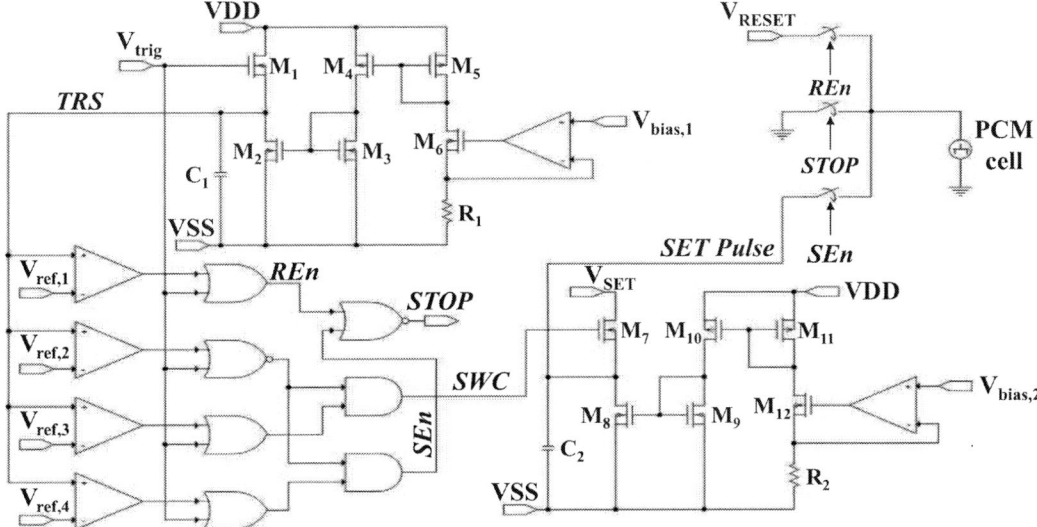

Fig. 4. Proposed circuit schematic of the R-SET generator.

In the case of N-doped Ge-rich GST, a standard SET pulse with long fall time results in the largest programmed resistance as well as in the largest drift coefficient dispersion. The improvement brought by the application of the R-SET pulse to the drift coefficient dispersion is markedly important, especially if one considers the similarly low SET resistance achieved by means of a time-consuming SCD sequence.

The R-SET pulse is essentially meant to be used for characterization of PCM devices. In order to be able to provide the pulse sequence and study the results of its application to a wide variety of innovative phase change materials, a dedicated circuit was designed, as described in the next Section. The R-SET pulse generator and the memory devices to be investigated will be assembled on a printed circuit board with chip-on-board techniques enabling the behavior study of memory cells.

III. R-SET PULSE GENERATOR

The block diagram of the circuit proposed to generate the R-SET pulse is depicted in Fig. 3. The circuit works based on a time reference scheme and is capable of fully controlling the pulse sequence characteristics. A rampdown Time Reference Signal, TRS, is generated under the control of programmable voltage $V_{bias,1}$ and gives rise to timing signals SWC, SEn, REn, and $STOP$ through comparison with predetermined reference voltages ($V_{ref,1}$ to $V_{ref,4}$). The obtained timing signals control the generation of the SET pulse (whose amplitude and fall time are tunable through voltages V_{SET} and $V_{bias,2}$, respectively) as well as the analog multiplexer, which feeds the output terminal of the circuit with the RESET voltage, the SET pulse, or ground in order to obtain the desired R-SET pulse sequence. The pulse sequence is triggered by signal V_{trig}.

The circuit diagram of the R-SET pulse generator is shown in Fig. 4. An external voltage pulse (V_{trig}), active high, initializes the circuit. Transistor M_1, which acts as a switch, rapidly charges capacitor C_1. After V_{trig} is brought back to zero, C_1 is discharged at a constant rate by transistor M_2, which replicates the current through R_1, $V_{bias,1}/R_1$, by means of current mirrors M_5-M_4 and M_3-M_2, thus generating the rampdown time reference signal, TRS.

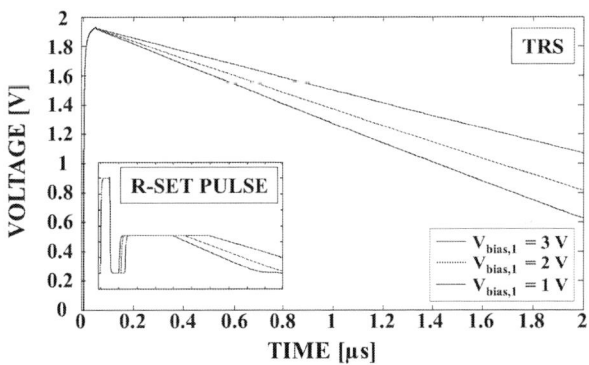

Fig. 5. Ramp down Time Reference Signal TRS. The slope of this signal is controlled by voltage $V_{bias,1}$. Different values of $V_{bias,1}$ (for the same values of $V_{ref,i}$) lead to differently timed R-SET pulses, as can be seen in the inset of the figure.

The amplifier which biases the gate of transistor M_6 consists of a PMOS cascode differential pair and an NMOS active load that performs differential to single-ended conversion. The value of bias voltage $V_{bias,1}$ controls the slope of signal TRS and hence, the timing of the obtained R-SET pulse, as can be seen in Fig. 5. The comparison of rampdown voltage TRS with four reference voltages $V_{ref,1}>V_{ref,2}>V_{ref,3}>V_{ref,4}$ produces four digital signals which are then processed by a simple combinational logic in order to generate four Timing Control signals (REn, $STOP$, SWC, and SEn) for pulse generation (Fig. 6), as detailed below.

A circuit similar to the one implemented to obtain TRS is used for the SET pulse. The generation of this pulse is triggered by signal SET Width Control (SWC), and its fall slope is controlled by means of bias voltage $V_{bias,2}$.

Signals RESET Enable (REn) and SET Enable (SEn) enable the RESET pulse and the SET pulse, respectively, to be fed to the output terminal of the circuit, whereas signal $STOP$ determines the time intervals during which the output voltage is kept at 0 V. The output of the circuit can be forced to zero in the time interval between the RESET and the SET

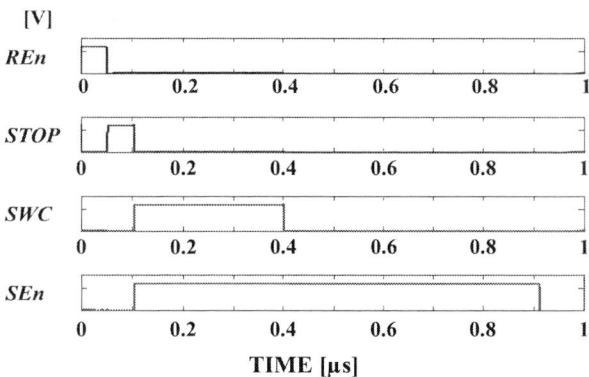

Fig. 6. Timing signals that control the analog multiplexer for the generation of the R-SET pulse.

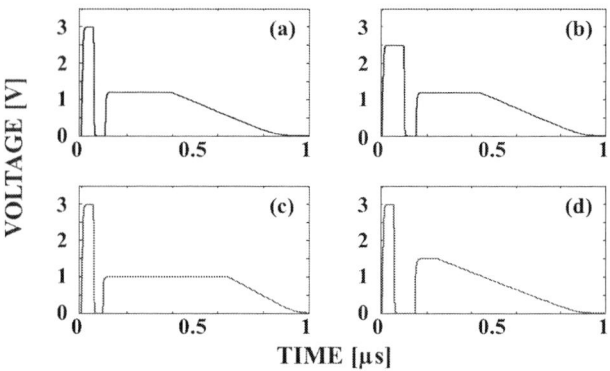

Fig. 7. Cadence simulation of the output waveform of the proposed circuit. The pulse characteristics can be externally tuned to produce (a) an R-SET pulse similar to the one applied to program the devices in Fig. 2, (b) an R-SET pulse with a different RESET pulse (c) an R-SET pulse with a different SET pulse and (d) an R-SET pulse with a SET pulse of higher amplitude and longer fall time and a longer wait time between the pulses.

pulse of the R-SET sequence and in the general case when no programming pulse has to be provided to the memory cell.

The analog voltage signals (RESET pulse, SET pulse, and ground) are fed to an analog multiplexer, which selects the signal to be delivered to the output terminal and, hence, to the PCM cell under the control of timing signals REn, SEn, and $STOP$. A CMOS analog switch is used to pass V_{RESET} and the SET pulse, whereas an NMOS switch is sufficient to pass the ground level. It is worth noting that, for a final application, a buffer should be included between the output of the R-SET generator and the PCM cell for adequate current driving capability.

IV. CIRCUIT SIMULATION

The circuit schematic in Fig. 4 was simulated in Cadence environment in 180 nm CMOS technology, which is compatible with typical fabrication technology for peripheral circuitry of non-volatile memory chips due to the voltage to be fed to the cells. The supply voltage, V_{DD}, of the circuit was set to 6 V. High-voltage devices were used. Simulated waveforms at the output of the pulse generator are shown in Fig. 7, where the tunability of the obtained pulse characteristics is apparent.

Fig. 7(a) shows an R-SET pulse consisting of a RESET pulse ($V_{RESET} = 3$ V) of width $t_{w,RESET} = 50$ ns and a subsequent SET pulse ($V_{SET} = 1.2$ V) with a width $t_{w,SET} = 300$ ns and a fall time $t_f = 500$ ns. The wait time, t_{wait}, between the two pulses was set to 50 ns. The linearity of the voltage pulse fall guarantees an optimum crystallization of the PCM cell. The amplitudes of the RESET and the SET pulse of the R-SET sequence are determined by voltages V_{RESET} and V_{SET}, respectively. Fig. 7(b) shows a RESET pulse with a lower voltage amplitude and a larger width ($V_{RESET} = 2.5$ V, $t_{w,RESET} = 100$ ns).

Fall time t_f can be changed by properly adjusting bias voltage $V_{bias,2}$, as can be seen in Fig. 7(c) ($V_{SET} = 1$ V, $t_{w,SET} = 700$ ns, $t_f = 250$ ns) and Fig. 7(d) ($t_{wait} = 100$ ns, $V_{SET} = 1.5$ V, $t_{w,SET} = 100$ ns, $t_f = 700$ ns). More specifically, decreasing $V_{bias,2}$ results in an increase of t_f, while the linearity of the voltage pulse fall slope is still maintained. It should be pointed out that when a longer fall time is desired, reference voltage $V_{ref,4}$ has to be selected accordingly in order to ensure an adequate length of signal SEn.

The wait time, t_{wait}, between the two pulses is determined by the values of $V_{ref,1}$ and $V_{ref,2}$, which control the time

instant where the fall of the RESET pulse and the rise of the SET pulse begin, respectively. Finally, voltages $V_{ref,3}$ and $V_{ref,4}$ determine the beginning and the end of the fall slope of the SET pulse. It is thus apparent that the proposed generator allows full control of the R-SET pulse characteristics.

V. CONCLUSION

In this paper, we discussed the benefits of the R-SET technique on innovative Ge-rich GST, which exhibits a crystallization speed lower than conventional GST and is also affected by an increase in the SET state resistivity over time. The R-SET pulse, which consists of a fast sequence of a RESET pulse and a SET pulse with a long fall time, takes advantage of the switching voltage decrease that occurs immediately after the application of a RESET pulse and is capable of bringing the memory cell to the SET state with a reduced programming voltage, thus protecting the cell from possible current overshoots during switching. The circuit that was designed to generate the R-SET pulse was presented and simulated in 180 nm CMOS technology. The circuit operates based on a time reference scheme and enables the control of the parameters of the generated R-SET sequence in order to fit the crystallization demands of all materials under study. The proposed technique achieves optimal crystallization of phase change materials in industrially compatible times.

ACKNOWLEDGMENT

The authors would like to thank Erika Covi for fruitful discussions on the topic.

REFERENCES

[1] R. Bez, et al., *IEEE Int. Memory Workshop*, pp.13-16 (2013).

[2] A. Redaelli, et al., *IEEE Electr. Dev. Let.*, vol.25, pp.684-686, Oct. 2004.

[3] G.W. Burr, et al., *J. Vac. Sci. Technol.*, vol.28, pp.223-262, Mar. 2010.

[4] J. Li, et al., *IEEE IRPS*, pp.6C.1.1-6C.1.6 (2012).

[5] N. Papandreou, et al., *IEEE Int. Memory Workshop*, pp.1-4 (2011).

[6] H.Y. Cheng, et al., *IEEE IEDM*, pp.3.4.1-3.4.4 (2011).

[7] F. Bedeschi, et al., *Microelectron. J.*, vol.38, no.10-11, pp.1064-1069, Oct./Nov. 2007.

[8] G. Navarro, et al., *IEEE IEDM*, pp.21.5.1-21.5.4 (2013).

[9] S. Kostylev, et al., *EPCOS*, pp.1-8 (2008).

[10] I.V. Karpov, et al., *J. Appl. Phys.*, 102, 124503 (2007).

[11] G. Servalli, *IEEE IEDM*, pp.1-4 (2009).

Design Considerations for Monolithically Integrated Fully-Depleted CMOS Image Sensors

J.B. Lincelles, O. Marcelot, P. Magnan
Integrated Image Sensor Laboratory
ISAE
Toulouse, France
Email: jean-baptiste.lincelles@isae.fr

O. Saint-Pé
Airbus Defense and Space
Toulouse, France

Abstract—**We present a design study for the fabrication of a fully depleted CMOS image sensor integrated on high-resistivity epitaxial layer. Both models and simulations are used and show that the maximal depleted thickness and the punch-trough current are dependent on the photo-diode cathode length. From these considerations, achievable performances are estimated.**

Keywords—*CMOS image sensor; fully-depleted; high resistivity silicon*

I. INTRODUCTION

Light absorption in silicon follows a Beer-Lambert law and is wavelength dependent as shown Fig. 1. To efficiently collect photo-generated charges from NIR, a detector must have a thick photo-active volume filled by an electrical field as for fully depleted Charge Coupled Devices (CCD) [1] [2].
CMOS Image Sensors (CIS) are now competing with traditional CCDs and are used for scientific applications [3] [4]. Quantum Efficiency (QE) and cross-talk are critical in such applications and could be improved by the use of fully-depleted CIS.
This paper presents design concepts for the development of monolithically integrated fully depleted CIS. The first part introduces the technological choices to fulfill this study. The second part describes phenomena induced by the depletion volume extension and their dependence on design. Predicted performances are presented in the final part.

II. CONCEPTS AND TECHNOLOGY FOR FULLY DEPLETED CMOS IMAGE SENSORS

A. Conventional CIS limitation for near-infrared detection

One of the simplest CIS pixel described Fig. 2 is composed of a PN junction and three transistors integrated in an adjacent P-well. The charge to voltage conversion is mainly performed on the junction capacitance. Pixel arrays are usually integrated on several micrometers thick moderately doped ($\approx 10^{15}/cm^3$) epitaxial layer grown on a highly doped substrate. Collected charges mainly come from the undepleted epitaxial layer due to their long lifetime in this region. The highly doped substrate is mainly considered as optically dead. QE in NIR is then very poor and field free charges diffusion induces cross-talk. To increase the maximal depleted thickness, the depletion width W from (1) leaves two action possibilities: reverse bias increase and lower doping concentration.

This work is supported by Airbus Defense and Space

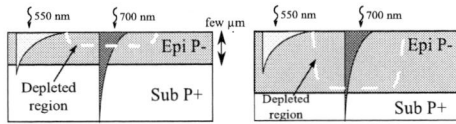

Fig. 1. Light absorption in epitaxial silicon wafer for a thin non-depleted layer, and a thick fully depleted layer.

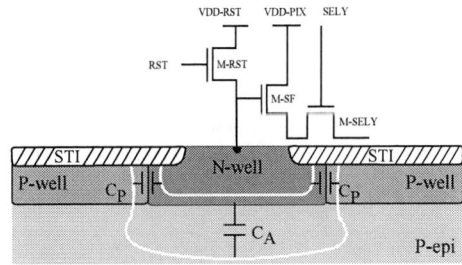

Fig. 2. 3T pixel schematic showing the three transistors and the PN photo-diode with depletion edge as well as the different capacitance contributions.

$$W = \sqrt{\frac{2\epsilon_{si}}{q}\frac{N_a + N_d}{N_a N_d}(V_r + \phi_{bi})} \quad (1)$$

With q the electron charge, N_a and N_d the doping in the P and N type region respectively, ϵ_{si} the silicon permittivity and V_r and ϕ_{bi} the reverse bias and built-in voltage respectively. The impact of the depletion width on QE was investigated by photo-current measurements.

B. Quantum Efficiency dependence on the depletion depth

PN photo-diodes were used to monitor photo-current variations with increased reverse bias. $30 \times 30 \mu m^2$, $50 \times 50 \mu m^2$ and $100 \times 100 \mu m^2$ photo-diodes were tested on a $1.0 \times 10^{15} B/cm^3$ P-type, $5 \mu m$ thick epitaxial layer grown on highly doped substrate. A test bench was used to illuminate the devices with wavelengths ranging from 450nm to 940nm, both with flat-field illumination using an Ulbricht sphere or a 15 μm wide spot centered on the photo-diodes by means of a pinhole and lens. A picoammeter was used to measure the photo-current with increasing reverse biases at room temperature. Triaxial wires and a guard potential system were used to limit the voltage dependent leakage current of the set-up.

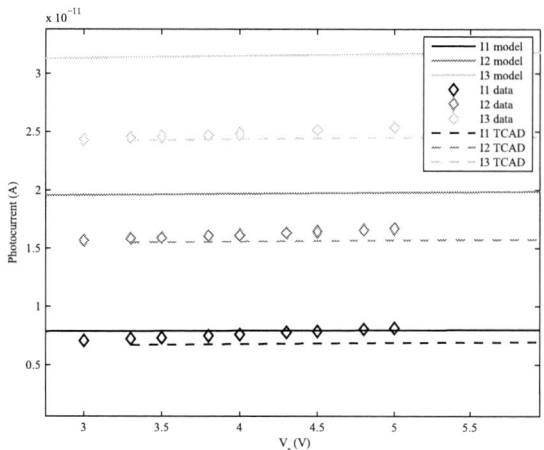

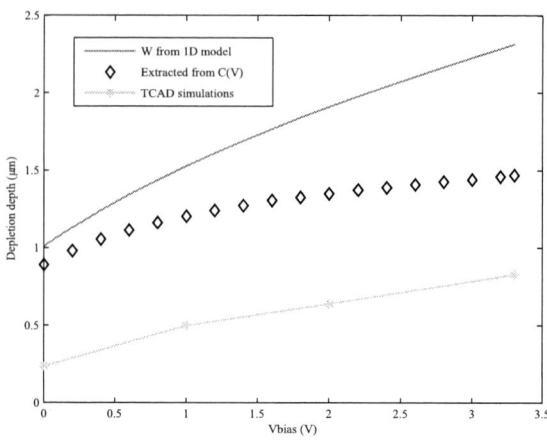

Fig. 3. Photo-current vs. reverse bias from measurements, TCAD simulations and a model using (2) fitted to the data. Presented measures were obtained with a 50x50 μm^2 photo-diode with $N_a = 1.0 \times 10^{15} B/cm^3$ illuminated with various intensity I1<I2<I3 at λ= 650 nm.

Fig. 4. Comparison of the depletion depth between TCAD simulation, 1D model from (1) and capacitance extraction using multivariate linear regression from C(V) measurements of 7 photo-diodes with different geometries.

Fig. 3, shows no improvement in the collection efficiency with increasing voltages. The same observations occurred with the other experimental configurations. This result is explained in [5, p. 180] by the small depleted depth compared to the carrier diffusion length in the photo-current expression (2) with G_L the generation rate, L_n the diffusion length of electrons.

$$I_n \propto q G_L (L_n + W) \qquad (2)$$

The measured QE is therefore slightly affected by the depleted depth but strongly dependent on the thickness of the photo-active layer. Full depletion is then mainly necessary to limit cross-talk [1]. Besides, note that in conventional CMOS processes, available voltages are limited by MOSFET voltage supply. The use of high resistivity silicon is then mandatory for depletion width improvement.

C. Use of high resistivity silicon and consequences

High resistivity silicon (1 kΩ - 10 kΩ) allows extending the collecting electric field several micrometers deep in the photo-active volume, even for conventional CMOS voltages. However, the large depletion extension produces unusual phenomena for a CIS.
First, punch-through current can occur if the depletion regions from adjacent pixels merge. This current and the conditions for its establishment have been studied on low [6] and moderately doped silicon [7] as well as isolation techniques.
Second, the large depletion depth reduces the capacitance from the cathode/substrate junction, leaving the peripheral cathode/P-well junction as the main contributor for the total photo-diode capacitance. The charge to voltage conversion factor (CVF) being inversely proportional to the collecting node capacitance, sensitivity will be mainly dependent on the photo-diode perimeter. Besides, the CVF can further be tailored by a P-Well recession from the cathode as presented Fig. 9. To understand such effects and to estimate the maximal achievable depletion depth, the shape of the depleted volume must be studied.

III. DEPLETED VOLUME EXTENSION ON HIGH-RESISTIVITY SILICON

A. One dimensional depletion model and its limitation

The depletion depth W given by (1) is related to the junction capacitance $C_a = \epsilon_{si}/W$. The node capacitance is mainly the sum of a perimeter and surface related capacitance density terms C_a and C_p as shown Fig. 2. The total capacitance is $C_{PN} = C_a A + C_p P$ where A and P are the surface and the perimeter respectively. The surface capacitance density C_a can be extracted from capacitance measurements on photo-diodes with different aspect ratios and by using multivariate linear regression to finally estimate the depleted depth as in Fig. 4. Despite the offset, TCAD simulation and extraction from measurements follow the same trend unlike the 1D model. Such measurements could be used to estimate the maximal depletion depth. Yet, observations from TCAD simulations schematically presented Fig. 5 show strong three dimensional effects on the depletion volume of a PN junction using low doped P side. As a consequence, the 1D model gives $W(N_a = 1.0 \times 10^{12}) \approx 60 \mu m$ and $W(N_a = 1.0 \times 10^{13}) \approx 22 \mu m$ which, compared to results from Fig. 6, is a wrong estimation.

B. Development of a three dimensional depletion model

TCAD tools require technological data, which are usually not available, and are not as fast as analytical models. A model was developed to quickly provide a depletion size estimation on high resistivity and to use it in a custom program dedicated to QE and cross-talk calculations [8]. To take into account the 3D effect, we approximated the depleted volume by a spherical cap. This spheroid shown Fig. 5 is defined by the depleted volume V_{dep} and the depletion edge continuity at the P-Well/Substrate interface. To calculate V_{dep}, we suppose that the depleted volume V_d in the cathode is still well approximated by the 1D model due to its relatively high impurity concentration. The total charge neutrality principle gives : $V_{dep} = N_d \times V_d / N_a$. Using the sphere radius

978-1-4799-4993-9/14 $31.00 © 2014 IEEE

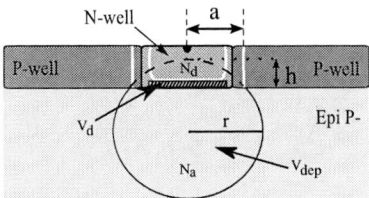

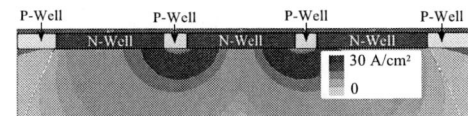

Fig. 7. Punch-through current density between photo-diodes on $1.0 \times 10^{12} B/cm^3$ silicon. The central cathode is biased at 3V, the two other cathodes are grounded.

Fig. 5. Schematic describing the analytical model for the 3D depletion volume estimation. h is the height of the spherical cap removed from the sphere of radius r and a is the distance between the cathode center and the depletion length in the P-Well. V_d is the depleted volume in the cathode used to calculate V_{dep}.

TABLE I.   TCAD RESULTS OF PUNCH-THROUGH CURRENT FROM A 3V POTENTIAL DIFFERENCE FOR VARIOUS PIXEL PITCH AND CATHODE SPACING.

Cath. length (μm)	Spacing between cathodes (μm)		
	1.5	3.5	5
1.5	5.1×10^{-9} A	3.3×10^{-12} A	4.5×10^{-13} A
5	4.0×10^{-7} A	9.0×10^{-9} A	6.1×10^{-10} A
7.5	5.5×10^{-7} A	2.8×10^{-8} A	3.7×10^{-9} A
10	6.2×10^{-7} A	4.5×10^{-8} A	8.1×10^{-9} A

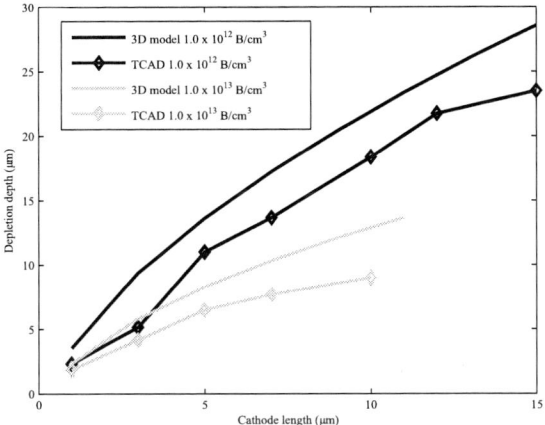

Fig. 6. Depletion depth comparison between the proposed 3D analytical model and TCAD results for an abrupt PN photo-diode on $1.0 \times 10^{12} B/cm^3$ and $1.0 \times 10^{13} B/cm^3$ doped silicon.

$r = (a^2 + h^2)/2h$, h is found by numerically solving (3).

$$V_{dep} = \frac{4}{3}\pi \left(\frac{a^2 + h^2}{2h}\right)^3 - \frac{1}{6}\pi h(3a^2 + h^2) \qquad (3)$$

It appears Fig. 6 that for cathode length smaller than the theoretical 1D depletion length for high resistivity and for conventional CMOS voltages, this model gives a better estimation than (1). Moreover, Fig. 6. shows that due to the spherical shape, the maximal depletion depth is proportional to the cathode length. These results are essential for the design of fully depleted CIS on high resistivity silicon.

C. Punch-through impact on CIS

Fig. 7 shows an example of punch-through current density for two neighboring photo-diodes built on high resistivity substrate. Consequences are studied through CIS reset and integration phases.

During the reset, a potential is applied via the reset transistor to deplete the collecting node. If punch-through occurs between pixels, the current flow will reset the entire array. However, this should not be an issue as the power source can sustain such current (few mA). During integration, each node is floating and photo-charges collection leads to the decrease of the node potential. If a potential difference between two nodes is above

the punch-through voltage, a current appears decreasing one node potential and increasing the other until the two potentials balance out leading to contrast loss in the captured images.

D. Isolation between pixels

A typical solution to limit punch-through is to increase the distance between cathodes. Yet, the photo-diode size influences the lateral extension and the depth of the depleted region. Thus, for a given spacing between two adjacent cathodes, the punch-through current is expected to depend on the cathode length. Pixel pitch is therefore a key parameter to prevent punch-through. As can be seen in Table I, TCAD simulation shows that the punch-through can be dramatically enhance by the cathode length for a given spacing. A less conventional solution is the use of deep-P-well to increase the potential barrier deeper in the substrate than with conventional P-well but its impact on detector performances must be studied as it may also reduce the depletion depth.

IV.   PERFORMANCES ESTIMATION

A. Quantum Efficiency and cross-talk

In order to estimate the quantum efficiency and electrical cross-talk at pixel level on thick high resistivity epitaxial layer, an analytical tool originally developed in [8]. IT solves the diffusion equations, and is combined with the 3D depletion model presented in this paper. Fig. 8 shows the QE and cross-talk estimation as a function of the epitaxial layer thickness for a pixel array with 10 μm pitch and 8 μm photo-diode length. The depleted depth is estimated to be around 20 μm for $N_a = 1.0 \times 10^{12} B/cm^3$. While the epitaxial layer is fully depleted, its thickness increase is beneficial as the electrical field collects charges deeper and reduces the amount of charges diffusing in the substrate field free region below the depleted volume, improving therefore cross-talk. When the epitaxial layer is no more fully depleted, diffusion in field free region occurs in the low doped silicon with high recombination lifetime, increasing the cross-talk. QE variations for flat field illumination seem to be only dependent on the epitaxial thickness, not on the depletion depth, confirming the results from Fig. 3.

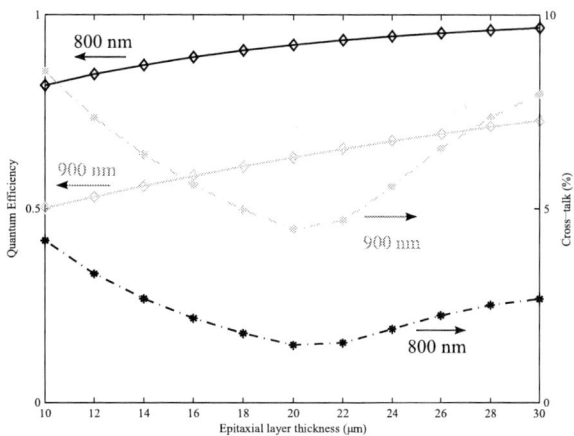

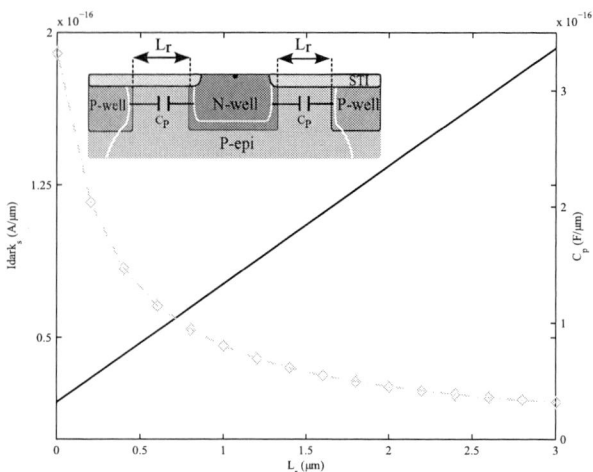

Fig. 8. Quantum Efficiency and Cross-talk vs. epitaxial thickness for a 10 μm pitch and 8 μm cathode on $1.0 \times 10^{12} B/cm^3$ for λ= 800 nm and 900 nm. Depletion depth is estimated to be 20 μm.

Fig. 9. Dark-current and perimeter capacitance density as a function of P-Well recession. Dark-current density is obtained with (4) and $s_0 = 5cm/sec$ and capacitance density with (1).

B. Dark current

Dark current represents the reverse current of the detectors photo-diodes. Impurity concentration in these detectors is low enough to prevent tunneling or impact ionization. Reverse current is then proportional to the SRH generation in the depleted volume and STI interface:

$$I_{gen}(V_r) \propto \frac{1}{2}qn_i s_0 A_{STI} + \frac{1}{2}q\frac{n_i}{\tau_0}V_{dep} \qquad (4)$$

with A_{STI} the STI area in the depleted region, s_0 the surface recombination velocity and τ_0 the effective lifetime, [5, p. 301]. Because of the surface recombination velocity, a P-Well recession increases the dark current by increasing the amount of STI interface generation centers in the depleted region. Fig. 9 shows the dark current increase and capacitance reduction with the P-Well recession. Yet, in radiative environments, dark current from STI increases with total ionizing dose [9]. Thus the recess of P-Well for scientific missions in radiative environment should be experimentally studied.

C. Test chip development

A test chip was designed and is currently in foundry for fabrication. It consists of a float zone silicon wafer and includes a 384x384, 3T pixels matrix with $10\mu m$ and $20\mu m$ pixel pitch, multiple cathode sizes and P-Well recession variations. Several test structures for capacitance extraction of photo-diodes, interface state density estimation and punch-through isolation performances were included.

V. CONCLUSION

Through measurements, the development of an analytical model and TCAD simulations, this work described critical design parameters for the development of fully depleted CIS. On high resistivity silicon, pixel pitch and cathode length determine the maximal depleted thickness and punch-through triggering conditions. A mismatch between pixel pitch, silicon

doping and epitaxial thickness leads to non-optimal QE and cross-talk which are major performances parameters for scientific applications. On such device, tailoring of the collecting node CVF is achievable through P-well recess at the expense of dark signal. These general guidelines can be used for the development of fully-depleted CIS.

ACKNOWLEDGMENT

The authors would like to thank M. Breart de Boisanger and F. Larnaudie from Airbus Defense and Space and M. Estribeau from ISAE for their support.

REFERENCES

[1] S. E. Holland, D. E. Groom, N. P. Palaio, R. J. Stover, and M. Wei, "Fully depleted, back-illuminated charge-coupled devices fabricated on high-resistivity silicon," *IEEE Trans. Electron Devices*, vol. 50, no. 1, pp. 225–238, Jan. 2003.

[2] H. Suzuki *et al*, "Development of the fully-depleted thick back-illuminated CCD by Hamamatsu," in *IEEE NUCL SCI CONF R*, vol. 6. IEEE, 2007, pp. 4581–4585.

[3] M. Bréart de Boisanger, F. Larnaudie, and O. Saint-Pé, "CMOS image sensors optimised for GEO observation," in *Proc. of SPIE*, vol. 8889, 2013.

[4] W. Dulinski *et al*, "Thin, fully depleted monolithic active pixel sensor based on 3D integration of heterogeneous CMOS layers," in *IEEE NUCL SCI CONF R*. IEEE, 2009, pp. 1165–1168.

[5] A. S. Grove, *Physics and technology of semiconductor devices*. New York: Wiley, 1967.

[6] J. Ellison *et al*, "Punch-through currents and floating strip potentials in silicon detectors," *IEEE Trans. Nucl. Sci.*, vol. 36, no. 1, pp. 267–271, 1989.

[7] G. Langfelder, "Isolation of highly doped implants on low-doped active layers for CMOS radiation drift detectors," *IEEE Trans. Electron Devices*, vol. 56, no. 8, pp. 1767–1773, Aug. 2009.

[8] I. Djite *et al*, "Theoretical models of modulation transfer function, quantum efficiency, and crosstalk for CCD and CMOS image sensors," *IEEE Trans. Electron Devices*, vol. 59, no. 3, pp. 729–737, Mar. 2012.

[9] V. Goiffon, P. Magnan, O. Saint-Pé, F. Bernard, and G. Rolland, "Total dose evaluation of deep submicron CMOS imaging technology through elementary device and pixel array behavior analysis," *IEEE Trans. Nucl. Sci.*, vol. 55, no. 6, pp. 3494–3501, Dec. 2008.

978-1-4799-4993-9/14 $31.00 © 2014 IEEE

TIA optimization for optical network receivers for multi-core systems-in-package

Robert Polster, Jose-Luis Gonzalez Jimenez
CEA, LETI, MINATEC Campus
Universite Grenoble Alpes
Grenoble, France
Email: robert.polster@cea.fr

Eric Cassan
Institut d'Electronique Fondamentale
Universite Paris Sud CNRS UMR 8622
Orsay, France

Abstract—**Transimpedance amplifiers (TIAs) are crucial elements in optical links. A simulation supported optimization study for different single ended TIAs was performed. Based on ST65nm CMOS technology, utilization scenario dependent solutions are presented, as well as an optimized solution for on-package multi-core optical network receivers. A key result of our study is the superiority of the push-pull topology over the common source topology and we argue for optimizing TIAs for designated DC input currents.**

I. INTRODUCTION

Future high performance computer (HPC) systems will have two major bottlenecks: interconnection bandwidth density and power consumption. Silicon photonic technology has been proposed recently as a cost-effective solution to tackle these issues [1,2]. Currently, copper connections are replaced by optical links at rack and board level in HPCs and data centers. The next step is the interconnection of multi-core processors which are placed in package on silicon interposers and define the basic building blocks of these computers. Several works have demonstrated the possibility of integrating all elements needed for the realization of short optical links on a silicon substrate. These elements include Ge photodiodes (PD) with capacities in the range of a few tens of fF, waveguides, other passive optical structures, and different kinds of modulators. Nevertheless, the laser source remains external or needs to be molecularly bonded onto the substrate. It has been proven that electronics can be integrated into the same substrate as optics (e.g. Luxtera). However, the most cost effective solution for multi-core on-package systems, required for HPC systems, is the realization of the electronics on a dedicated die. This die is usually produced by CMOS technology and then wire-bonded or flip-chipped onto the optical substrate [2].

The figure of merit for short range optical interconnects is the power consumption in terms of energy per bit. To be competitive with copper connections on the mm scale, the energy per bit has to drop below 1 pJ/b [7]. In order to meet the constraints given above, many different link architectures have been explored in literature, resulting in a wide range of specifications for the transimpedance amplifier (TIA), the first element in the receiver chain. The four most important parameters for TIAs are the sensitivity, the output swing, the bandwidth and the input capacitances they were optimized for. Recent works show optical sensitivities ranging from -25 dBm [3] to -4.7 dBm [4], output swings from 10 mV [4] to 100 mV [5], bandwidths from 3 GHz [3] up to 25

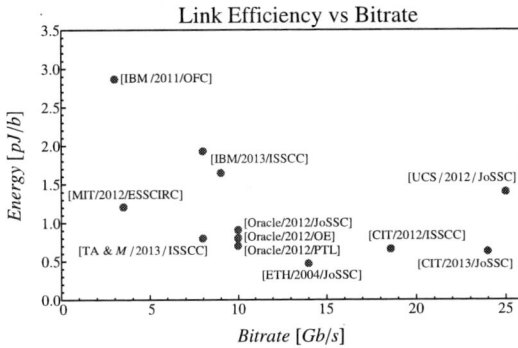

Fig. 1. Study of published receivers and their efficiencies. The energy consumption accounts for the laser, the calibration of ring modulators and the receiver.

GHz [6] and input capacitances from 5 fF [7] to 200 fF [4]. These works use a variety of TIA topologies. In fact, the shunt-shunt topology followed by a decision stage [2] is still one of the most efficient implementations. However, receivers based on integrators or even receiverless approaches followed by clever post processing stages [3,4,7] are just as common in recent literature. In this context, this paper presents a comprehensive optimization study for TIA topology selection and sizing. The TIAs are optimized in terms of power consumption for low-capacity Ge photodiodes based on silicon photonic technology for optical network receivers for multi-core systems-in-package.

II. OPTIMAL BIT RATE AND TIA TOPOLOGIES

It has been shown that for the energy efficiency of photonic links the bandwidth of a single channel in a multi-channel high data rate link is a very important parameter. Therefore studies

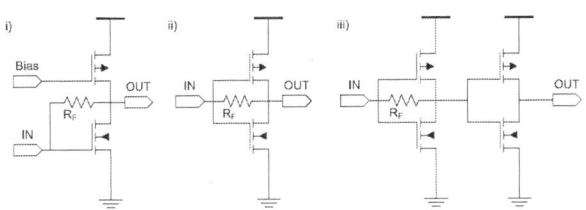

Fig. 2. Receiver topologies under study. i) Common source inverter ii) Push pull inverter iii) Double stage push pull inverter

978-1-4799-4993-9/14 $31.00 © 2014 IEEE

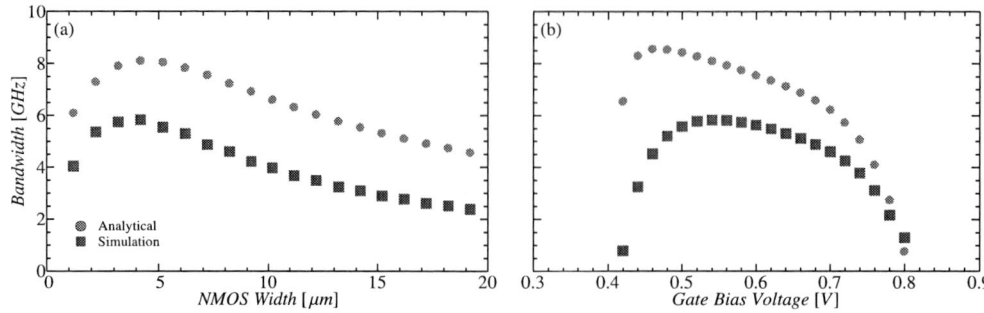

Fig. 3. The comparison between the analytical and simulated dependance of the bandwidth on different design parameters proofs, that analytical approaches based on standard small signal models of transistors lead to too high errors and do not allow to judge over the superiority of one topology over the next.

have been carried out to understand analytically which data rate might be best for different scenarios [1]. They suggest that a data rate of 4-8 GBs per channel is optimal for total data rates from 64 GB/s to 1024 GB/s. Our analysis of recently published optical link receivers (2004, 2011-2013) and their associated transmitters is shown in Fig.1. The energy consumption is calculated by the sum of the consumption of the receiver, the minimal laser energy computed from the receiver's sensitivity and a constant tuning power for each channel. For the laser, an efficiency of 20% is assumed. The tuning power stems from the need to tune ring modulators and is assumed to be 4mW. Our results indicate that an optimum bit rate is found around 10 Gb/s, in agreement with [1].

Following these results, our work studies in detail the possibilities of TIAs in a range from 1 Gb/s to 15 Gb/s. Here, different single ended TIA topologies are explored and reveal in which scenario which topology is superior to the others. The topologies under study are a common source inverter topology (CS, Fig.2.i), a push-pull inverter topology (PP, Fig.2.ii) and a push-pull inverter followed by another push-pull inverter topology (PP-PP, Fig.2.iii). All of them use resistive feedback.

III. METHODOLOGY

The first step in designing electrical circuits is always an analytical understanding and optimization of the problem. In [5] and [1] analytical optimizations of TIAs and/or entire optical links are presented. In order to show the problems with analytical optimization, we used a common source topology and computed its bandwidth based on a complete small signal model of the circuit (a π model was assumed for the transistors). The results are compared to transistor level simulations. The bandwidth over different circuit parameters is shown in Fig.3. The absolute error of roughly 10% itself is already too big to determine in most cases which topology is superior. Another problem is that although the trends look similar in both cases the maximal bandwidth is achieved for different parameters in each case. These differences are probably caused by the fact that the analytical model does not account for the frequency dependence of the transistor's capacitances. Therefore, our study uses a simulated database, based on Agilents ADS simulator (AC, noise, and DC analysis) and ST65nm CMOS technology device models. To build up the database extensive sweeps were run over the topologies under study. The load was in every case a 10 fF ideal capacitor, representing the input impedance of the stage after the TIA. It is a reasonable value if it is considered that the TIA will be

integrated with the rest of the digital receiver in the same die. Bandwidth, transimpedance, power consumption and noise are obtained from the simulation results. For every topology the input capacitance was swept from 10 fF to 200 fF, representing usual capacities of silicon photonic Ge PDs. In each case, the transistors widths were kept proportional and a single size parameter was swept: the width of the NMOS (W_N) transistor (0.135 μm 20 μm) along with the biasing point V_{bias} (0.4 V 0.8 V) for the CS topology. For the PP topology the PMOS transistor was sized so that the trip point is set to $V_{dd}/2$. The feedback resistance (R_F) was swept from 0.5 kΩ to 40 kΩ. The first step in the database analysis is to eliminate all realizations that have an integrated input noise level higher than 0.9 μArms, which corresponds to a Bit Error Rate (BER) of 10^{-12} for a -18 dBm optical input signal on a PD with a responsivity of 0.8 A/W. The database was then used to find for each triplet of bandwidth, transimpedance and input capacitance the least energy consuming implementation (W_N, V_{bias}, and R_F) for all mentioned topologies.

IV. RESULTS

The first result obtained is that the CS topology is inferior to the PP topologies in each of the studied points. Therefore the rest of our analysis will focus on the single stage PP and double stage PP-PP topologies. To visualize the results Fig.4-6 show the power consumption depending on the bit rate, transimpedance or input capacitance for different situations. The solid lines represent the PP topology while the dashed lines represent the PP-PP topology. This paper can only discusses some interesting cases of the results, but Fig.4-6 are presented for the reader to choose the optimal TIA for their configuration. Fig.4 gives us several pieces of information. Firstly, for a very small input capacitance of 10 fF and a small transimpedance of 1 kΩ the highest directly achievable bit rate is 18 Gb/s (Fig.4.a), for a BER of 10^{-12} with the ST65nm technology. The PP topology reaching this bit rate uses $R_F =$ 1.5 kΩ, $W_N = 1.635$ μm and consumes 0.24 pJ/bit. 18 Gb/s is derived from the system bandwidth of 14 GHz and a bandwidth to bit rate factor of around 1.25 [5] which accounts for the fact that some bandwidth can be sacrificed before inter symbol interference (ISI) will take effect. Nevertheless, higher bit rates can be achieved when Decision Feedback Equilibration (DFE) is used. In [3] the input bandwidth into the DFE system is 1.5 GHz and a bit rate of up to 9 Gb/s is achieved. If higher transimpedances are required, the bandwidth goes down naturally. If the input capacitance stays low the PP

978-1-4799-4993-9/14 $31.00 © 2014 IEEE

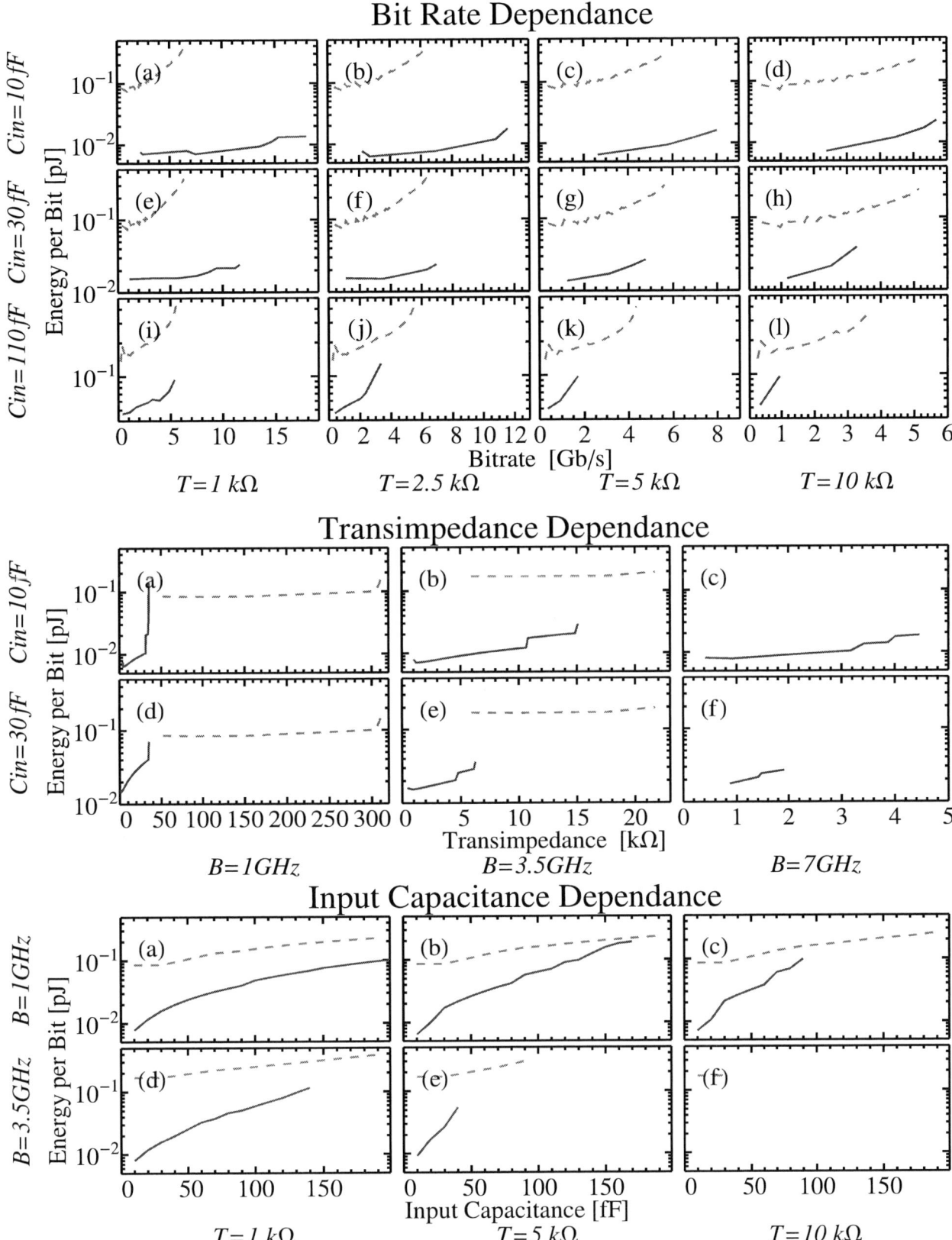

Fig. 4. Based on a extensive sweeps a database was build up to compute for the topologies under study (PP topology (solid line) and PP-PP topology (dashed line)) the minimal energy consumption per bit in dependence of bandwidth (B),

Fig. 5. transimpedance (T),

Fig. 6. and input capacitance (Cin), allowing us to find the right topology and parameters for all utilization scenarios.

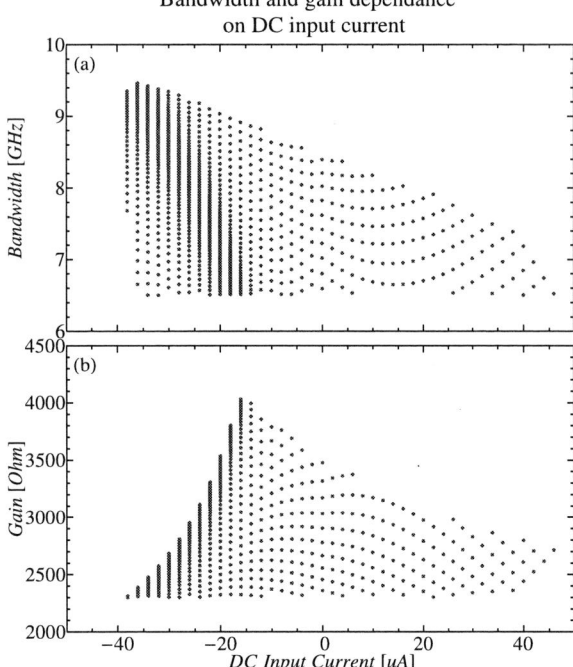

Bandwidth and gain dependance on DC input current

(a)

(b)

Fig. 7. Using an NMOS transistor as a feedback resistor introduces the possibility to adjust the TIA to different situations by changing the NMOS biasing. Not only the gain bandwidth tradeoff can be addressed but also the acceptance of different DC input currents. Our TIA shows a wide acceptance of input currents from -40μA to 45 μA

topology remains the faster topology. Nevertheless in a system containing a bonded PD the input capacitance rapidly exceeds 100 fF and the PP-PP topology becomes the faster one. The fastest configuration of PP-PP with an input capacitance of 110 fF and a transimpedance of 10 kΩ reaches a bit rate of 3.6 Gb/s using W_{N1} = 3.885 μm, R_F = 1.5 kΩ and W_{N2} = 4.885 μm. Intuitively, a single stage solution is in most cases more power efficient than a two stage solution. Nevertheless, in the particular case of low bandwidth and high transimpedance, in order to avoid a decision stage, the PP-PP topology is the more power efficient solution. As depicted in Fig.5.a the PP topology already consumes 0.20 pJ/b at a transimpedance of 35 kΩ while the PP-PP topology consumes the same power at 310 kΩ for C_{IN} = 10 fF and a bandwidth of 1 GHz.

V. OPTIMIZATION FOR OPTICAL NETWORK RECEIVERS FOR MULTI-CORE SYSTEM-IN-PACKAGE

Following our analysis in chapter II we have chosen a data rate of 8Gb/s and a minimal output swing of at least 30 mV for the receiver. This output swing corresponds to a gain of 2300 Ω, if we stick to the small optical input power of -18 dBm. Using our database the optimal TIA would be a push-pull inverter where W_N = 0.635 μm, W_P = 1.46 μm and R_F = 5000 Ω with an energy consumption of 90 fJ/b. As the receiver is supposed to be part of a network it has to be capable of receiving signals with different strength and extinction ratios. Hence, it can not be guaranteed that the DC current flowing through the TIA is zero and the TIA stays perfectly biased. Recent receiver topologies address this problem by adding a feedback loop around the TIA that steers a current source in

parallel with the TIA in order to keep the DC current flowing through the TIA equal to zero.

Modulators based on silicon photonic links can reach very high extinction ratios if high enough controlling voltages are applied. The highest possible voltage in the ST65nm technology is 2.5 V. Therefore, the expected extinction ratios of the ring modulators are between 2 and 10. This appears to be a big scale but at low optical input powers this corresponds to DC currents of 7 μA to 19 μA, if the receiver reaches a sensitivity of -18 dBm. This means that if we adapt our receiver to be optimally biased at an DC input current of 13 μA it may still work in the two extreme cases. This can be achieved by changing the factor between the PMOS and NMOS width. In order to widen the working zone even further we introduced a new configurable parameter by changing the feedback resistor into a NMOS transistor with controlled biasing. If we use the biasing voltage of the feedback transistor to adapt the TIA to the DC current our TIA can handle input currents from -40 μA up to 45 μA. Fig. 7 shows for our TIA the gain and bandwidth depending on the DC offset current. All points correspond to configurations in which the the gain is superior to 2300 Ω, the bandwidth higher than 6.5 GHz, the noise lower than 0.9 Arms and the required biasing voltage between 0.8 and 1.2 V. It is clearly visible that the optimal operation point of the TIA is in terms of gain at the desired -13 μA, while if maximal bandwidth is desired it is even further shifted to -37 μA. Our TIA consumes 38 pJ/b uses W_N = 1.805 μm, W_P = 1.8 μm and a feedback NMOS width of 0.8 μm and is currently in fabrication.

VI. CONCLUSION

It was shown that push-pull inverter based topologies in case of single ended TIAs are superior to common source inverter based topologies. Furthermore, it has been demonstrated that if the system imposes high input capacities double stage receivers are not only superior in terms of capabilities but also in terms of power consumption. Nevertheless, nowadays, as technology scales down and interchip communication becomes an active research topic, input capacities are in the range of 10 fF. This leads to single stage TIAs with very low power consumptions even for simple receiver topologies like the feedbacked push-pull inverter. It has been further shown that introducing a transistor for the signal feedback not only allows us to control the gain but also to tolerate a wide range of DC currents running through the TIA.

REFERENCES

[1] Georgas, Adressing Link-Level Design Tradeoffs for Integrated Photonic Interconnects, in CICC, pp.1-8, 2011.

[2] Liu, 10-Gbps, 5.3-mW Optical Transmitter and Receiver Circuits in 40-nm CMOS, in JSSC 47, 2049-2063(2012).

[3] Proesel, Optical Receivers using DFE-IIR Equalization, in ISSCC, pp.130-132,2013.

[4] Nazari, A 24-Gb/s Double-Sampling Receiver for Ultra-Low-Power Optical Communication, in JSSC 48, 344-357(2013).

[5] Van Blerkom,Transimpedance receiver design optimization for smart pixel arrays, JLT 16, 119-126, 1998.

[6] Takemoto, A 4x 25-to-28Gb/s 4.9mW/Gb/s -9.7dBm High-Sensitivity Optical receiver Based on 65nm CMOS for Board-to-Board Interconnects, in ISSCC, pp. 118-120, 2013.

[7] Georgas,A Monolithically-Integrated Optical Receiver in Standard 45-nm SOI, in JSSC 47, 1693-1702(2012).

Characterization and modeling of low frequency noise in 0.13 μm BiCMOS SiGe:C heterojunction bipolar trasnsistors

M. Seif, F. Pascal, B. Sagnes
IES, Université Montpellier 2, Place E. Bataillon
34095 Montpellier cedex, France
Email: marcelino.seif@ies.univ-montp2.fr

S. Haendler
STMicroelectronics, 850 rue Jean Monnet
38926 Crolles Cedex, France

Abstract—Low frequency noise (LFN) in 0.13 μm BiCMOS SiGe:C was characterized both as a function of base current bias I_B and emitter area A_e. The LFN exhibits typical behavior of 1/f noise for frequencies up to 1 KHz after which the shot noise 2qI is visible, in transistors with large emitter area ($A_e > 1\mu m^2$). The 1/f noise is modeled following the SPICE compact model, and the LFN parameters K_F and A_F were calculated. The extracted figure of merit $K_B = K_F * A_e$, has an excellent value of $1.10^{-10}\mu m^2$. The transistors with small emitter area ($A_e < 1\mu m^2$) can be affected by the presence of generation-recombination (G-R) components. In some cases, where generation-recombination presents large amplitude and high cut-off frequency, it is related to random telegraph signal (RTS).

Index Terms—BiCMOS, Heterojunction Bipolar Transistor (HBT), SiGe:C, Low frequency noise, 1/f noise, RTS.

I. INTRODUCTION

At present, the development of mobile or wireless systems requires high-performance RF circuits. Therefore, the components need to be increasingly faster, and possess high frequency criteria. However, the components architecture is not the only important feature for improving the frequency performance. It is also necessary to reduce the components dimensions. In this case, a major problem associated with this reduction is the increase of noise produced by these smaller components geometries, especially the low frequency noise. The SiGe Heterojunction Bipolar Transistors (HBTs), present high current gain, high unity-gain frequency and low noise properties and these characteristics make the SiGe HBTs a perfect candidate in RF and THz applications. Moreover, when it is associated with standard MOS technology (i.e. BiCMOS), low cost systems can be achieved.
The LFN properties in SiGe HBTs have been investigated in many publications [1]–[5]. Usually, the LFN in SiGe HBTs exhibits a 1/f noise for frequencies up to 1 kHz followed by shot noise 2qI for high frequencies f > 1 kHz. Few publications mentioned the presence of generation recombination components in transistors with small emitter area $A_e < 1\mu m^2$ [6]–[8]. In this work, we present the study of LFN in SiGe Heterojuncion Bipolar Transistor HBTs with different emitter

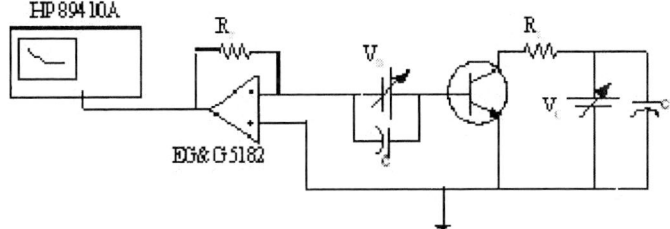

Fig. 1. LFN experimental set-up used for the measurement of the input base current spectral density: high impedance configuration

areas.
First of all in section II, we will present a brief description of the devices and of the measurement setup. In section III, we will present and discuss some I-V characteristics. In the same section, the main part of this paper will deal with the measurement and the modeling of the 1/f noise component followed by the characterization of the G-R/RTS noise component.

II. DEVICE STRUCTURE AND MEASUREMENT SETUP

A. Device structure

The studied HBTs are supplied by STMicroelectronics Crolles and are issued from a complete 0.13 μm SiGe BiCMOS technology dedicated to millimeter applications. This technology is characterized by a Double-Polysilicon fully self-aligned base-emitter structure provided by the selective epitaxial growth of the SiGe:C base. More information can be found in [9].

B. Measurement setup

The static characteristics are measured using a Keithley 4200 semiconductor characterization system. For low frequency measurements, the HBTs are biased in a common emitter configuration. A low noise current-voltage amplifier (EG&G 5182) and a spectrum analyzer (HP 89410A) are used to measure directly the current noise spectral density at the base, S_{IB}, of the HBTs. The on-wafer contacts are realized using coplanar probes. The circuit is placed into a

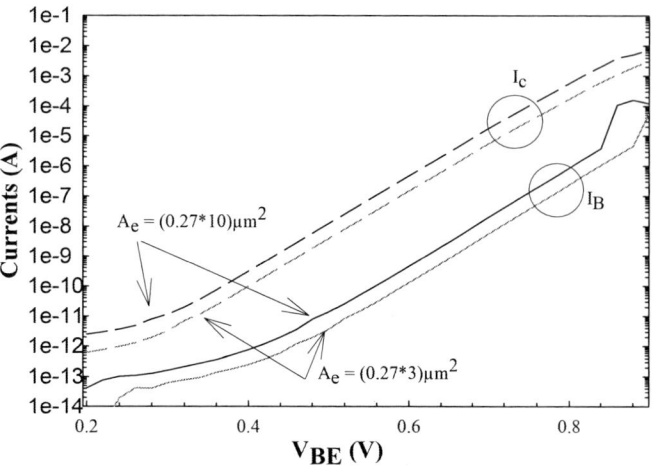

Fig. 2. Gummel plot for HBTs with $A_e = 0.27 * 3\mu m^2$ and $A_e = 0.27 * 10\mu m^2$.

Faraday cage and biased by batteries to eliminate any external disturbance. The low-frequency noise experimental set-up is presented in fig.1.

III. RESULTS AND DISCUSSION

A. First order characteristics

Fig. 2 shows the Gummel plot for transistors with emitter areas $A_e = 0.27 * 3\mu m^2$ and $A_e = 0.27 * 10\mu m^2$. For $V_{BE} < 0.6V$ the components of generation recombination and tunnel currents dominate on the base current. While for $0.6V < V_{BE} < 0.8V$ the components of diffusion current dominate. The low frequency noise will be measured for base current biases from 100 nA up to 5 μA, this current range corresponds to operating points of the HBTs where the static gain is closed to its maximum ($\beta \approx 1300$).

B. 1/f noise characteristics

The low frequency noise was studied on HBTs with different emitter areas A_e. The spectra are mainly composed of a 1/f component and of the shot noise, $2qI_B$. Nevertheless, for the smallest area, some Lorentzian shapes associated to G-R/RTS noise components can be observed (see next paragraph for the presentation and the analysis of these components).

In HBTs with large A_e the base current spectral density S_{IB} exhibits mainly a 1/f noise for frequencies up to 1 kHz followed by shot noise for frequencies > 1 kHz. An example of such noise spectra is reported in fig. 3(a).

For all transistors, the 1/f noise component level showed a quadratic evolution with the base current I_B. Fig. 3(b) shows the evolution of S_{IB} at 1 Hz versus I_B.

The 1/f noise was then modeled following the SPICE model and the base current noise spectra are the sum of 1/f and shot noise.

$$S_{I_B} = K_F \frac{I_B^{A_F}}{f^{\gamma}} + 2qI_B \quad (1)$$

K_F and A_F represent the SPICE parameters. From fig. 3(b),

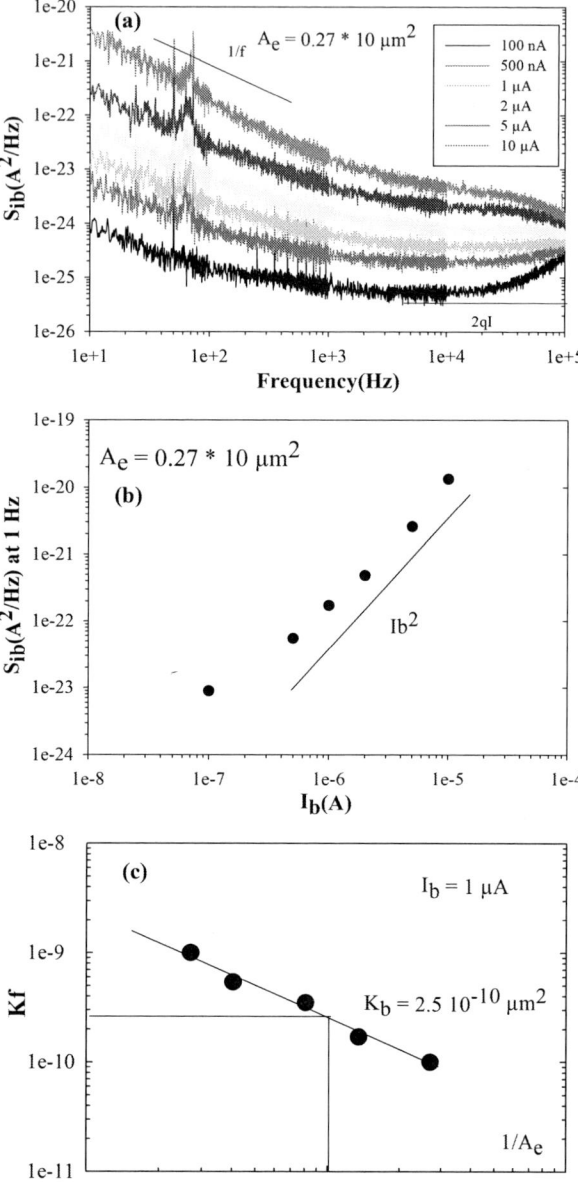

Fig. 3. (a) Base current spectral density S_{IB} versus frequency at different base current biases I_B. (b) S_{IB} at 1 Hz versus I_B, $A_e = 0.27*10\ \mu m^2$, (c) K_F versus emitter area A_e.

A_F was found equal to 2 due to the quadratic evolution of S_{IB} with I_B. Then K_F was calculated and found to be inversely proportional to the emitter area A_e, as can be seen in fig. 3(c). The values of the figure of merit $K_B = K_F * A_e$ were within the range of $1.10^{-10}\mu m^2$ to $5.10^{-10}\mu m^2$ (different dies have been studied), which is among the best published values so far in SiGe HBTs and in improvement of our past published results [4]. The quadratic evolution of the 1/f noise base current spectral density, $S_{IB}^{1/f}$, with base current I_B is typical for SiGe HBTs at low and high current biases. In this case, the dependence $S_{IB}^{1/f} \sim I_B^2$ observed in the studied HBTs as well as its A_e^{-1} evolution indicate that the main 1/f noise sources are located in the intrinsic base emitter region [6]. Concerning the exact location and the origin of the noise sources associated to this 1/f noise component, different assumptions and associated models can be found in [10]. For instance the role of the poly-silicon to silicon interface is often bring to the fore.

C. RTS and generation recombination noise characteristics

To complete the characterization of the low frequency noise in SiGe HBTs, we have studied the noise in very small submicron sized transistors, which is a possible tool to study the nature of noise.

Usually, the generation recombination noise is due to impurities deposited during the fabrication process, these impurities cause trapping-detrapping mechanisms for electron and holes that affect the number of carriers.

Fig. 4(a) and 4(b) show the noise spectra at different base currents for 2 HBTs with different emitter areas: $A_e = 0.27 * 1 \mu m^2$ and $A_e = 0.27 * 3 \mu m^2$.

In fig. 4(a), the spectral densities exhibit a 1/f dependence coupled with a G-R component bump around f = 50 Hz. In some submicron transistors where each spectrum shows a clear and unique component of G-R, with large lorentzian amplitude and high cut-off frequencies, as reported in fig. 4(b), random telegraph signal RTS is observed. Fig. 4(c) presents the RTS signature in the time domain associated to the G-R noise observed in fig. 4(b). In this case, the RTS amplitude, ΔI_B, is around 25 nA, 2.5 % of I_B.

When a lorentzian bump is present, spectra can be fitted by:

$$S_{I_B} = \frac{b}{f} + \frac{P}{1 + (\frac{f}{f_c})^2} + 2qI_B \qquad (2)$$

where P is the plateau of the lorentzian and f_C the cut-off frequency.

As can be seen in fig. 4(b), f_C increases with the base current I_B. f_C can be extracted in the frequency domain from fig. 4(b) and in the time domain according to the following forms:

$$f_C = \frac{1}{2\pi\tau} \qquad (3)$$

$$\frac{1}{\tau} = \frac{1}{\tau_{low}} + \frac{1}{\tau_{high}} \qquad (4)$$

where τ_{low} and τ_{high} are the low and high level duration of the RTS signal respectively. For instance, for $I_B = 1\mu A$, $f_C \approx 800$ Hz is obtained from fig. 4(b) and from time domain (fig. 4(c)) $\tau_{high} \approx 250\mu s$ and $\tau_{low} \approx 1.3$ ms (average values on several measurements) lead to $f_C \approx 760$ Hz. A good agreement between the two f_C extraction methods is found.

τ_{low} and τ_{high} are linked to carrier capture and emission phenomena by traps. Many analyses can be found in the literature, depending the nature of the carrier, electron or hole, and the bias and temperature evolutions of these two characteristics [8]. The increase of f_C with I_B, as can been seen in fig. 4(b), leads to a decrease of the high and low times. In our case, we have observed $\tau_{low} > \tau_{high}$ for each bias.

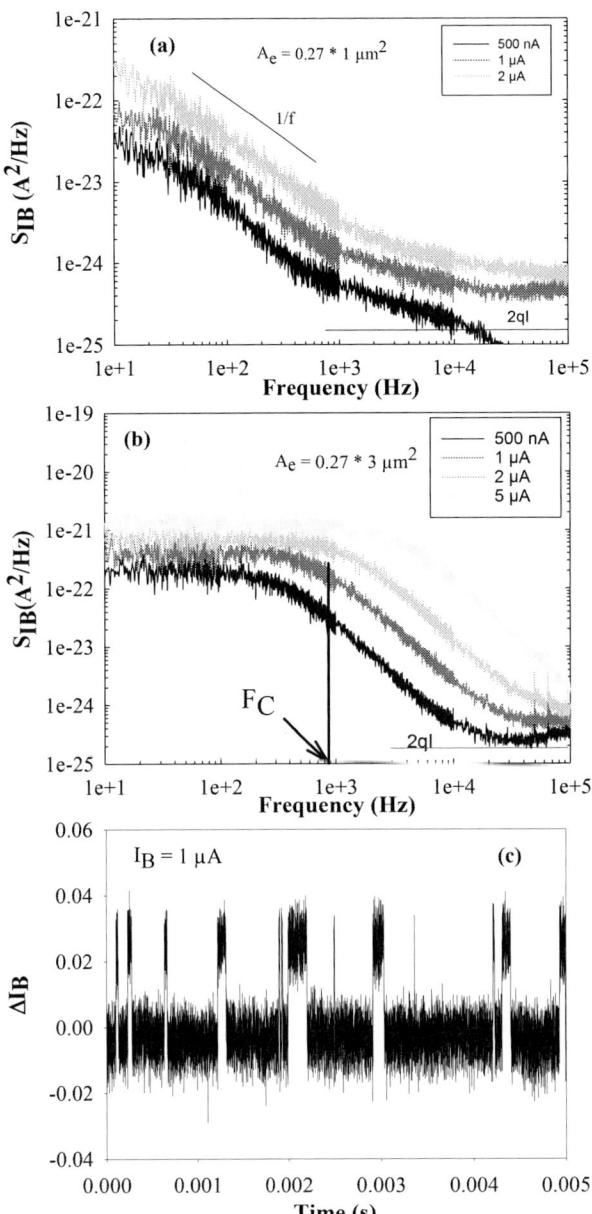

Fig. 4. (a) Base current spectral density versus frequency for HBT with $A_e = 0.27\mu m^2$. (b) Base current spectral density versus frequency for HBT with $A_e = 0.81\mu m^2$. (c) The RTS noise (time domain) in HBT with $A_e = 0.81\mu m^2$.

In order to try to locate the source of RTS noise, we studied the evolution of ΔI_B with the base current I_B (Fig. 5). From [8], a linear evolution of ΔI_B with I_B indicates that the main source producing RTS is located in the emitter base region or in the extrinsic base. Moreover, as can be seen in fig. 5, the RTS have a weaker current dependence than ΔI_B with I_B. This evolution could indicate that the main sources producing RTS are located in the spacer oxide at the emitter periphery [8], which is uncorrelated with the location of the main sources producing 1/f noise.

M. von Haartman [8] has also observed such RTS charac-

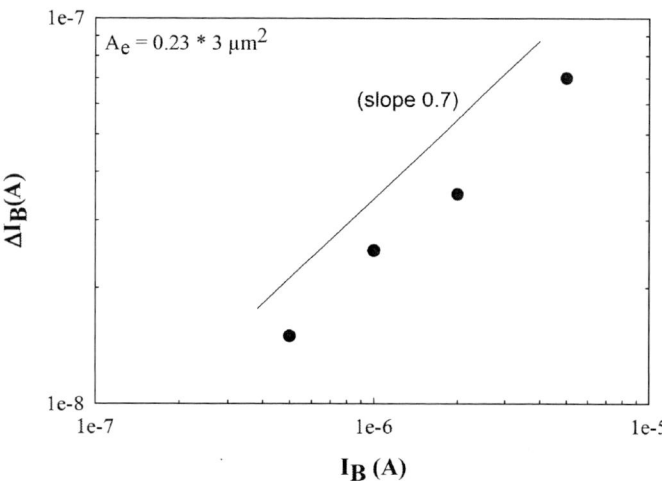

Fig. 5. ΔI_B versus I_B for HBT with $A_e = 0.81 \mu m^2$.

teristics and has proposed that τ_{high} corresponds to electron capture by traps in the spacer oxide whereas τ_{low} corresponds to hole capture or electron emission.

Nevertheless, several other hypothesis concerning the localization and the origin of the RTS noise source have been proposed: dislocations along the emitter periphery [11], defects associated to the implantation of the P+ extrinsic base [12], or traps located in the base collector space charge region due to the constraint SiGe layer [13].

IV. CONCLUSION

The low frequency noise in SiGe Heterojunction Bipolar Transistors issued from a 0.13 μm BiCMOS technology was studied.

For HBTs with large emitter area $A_e > 1 \mu m^2$, low frequency noise exhibits typical behavior of 1/f noise followed by shot noise $2qI_B$. The base current spectral density S_{IB} at 1 Hz showed a quadratic evolution with base current I_B and was inversely proportional to the emitter area A_e. The 1/f noise figure of merit K_B has an excellent value of $110^{-10} \mu m^2$. The main 1/f noise sources were found to be located at the poly-silicon to silicon interface.

Generation – recombination noise was found in submicron sized HBTs. G-R noise with both high amplitude and cut-off frequency were related to RTS noise. The RTS amplitude ΔI_B showed a 0.7 evolution with I_B. The results obtained show that trapping-detrapping processes by traps present in the oxide spacer at the emitter periphery could be the main sources producing RTS noise.

The measurement and characterization of the RTS noise in SiGe HBTs can be used as a diagnostic tool for technology process (i.e. the quality and reliability of the spacer oxide at the emitter periphery). But we have to keep in mind that this peripheral effect is mainly sensitive for the smallest transistors and not for all of them. A statistical study of the dispersion of this noise component as well as the 1/f one is in progress.

ACKNOWLEDGMENT

The HBTs are supplied by ST-Microelectronics Crolles and the research was done in the frame of the RF2TH SiSoC project of the EUREKA program CATRENE and funding of the French Ministry of Economy, Finance and Industry.

REFERENCES

[1] L. Escotte, J.-P. Roux, R. Plana, J. Graffeuil, and A. Gruhle, "Noise modeling of microwave heterojunction bipolar transistors," Electron Devices, IEEE Transactions on, vol. 42, pp. 883-889, 1995.

[2] L. S. Vempati, J. D. Cressler, J. A. Babcock, R. C. Jaeger, and D. L. Harame, "Low-frequency noise in UHV/CVD epitaxial Si and SiGe bipolar transistors," Solid-State Circuits, IEEE Journal of, vol. 31, pp. 1458-1467, 1996.

[3] R. Gabl, K. Aufinger, K. Beock, and T. F. Meister, "Low-Frequency Noise Characteristics of Advanced Si and SiGe Bipolar Transistors," in Solid-State Device Research Conference, 1997. Proceeding of the 27th European, 1997, pp. 536-539.

[4] F. Pascal, J. Raoult, B. Sagnes, A. Hoffmann, S. Haendler, and G. Morin, "Improvement of 1/f noise in advanced 0.13 μm BiCMOS SiGe: C Heterojunction Bipolar Transistors," in Noise and Fluctuations (ICNF), 2011 21st International Conference on, 2011, pp. 279-282.

[5] M. Seif, F. Pascal, B. Sagnes, A. Hoffmann, S. Haendler, P. Chevalier, and D. Gloria, "Low frequency noise measurements of advanced BiCMOS SiGeC Heterojunction Bipolar Transistors used for mm-Wave to terahertz applications," in Noise and Fluctuations (ICNF), 2013 22nd International Conference on, 2013, pp. 1-4.

[6] M. J. Deen, S. L. Rumyantsev, and M. Schroter, "On the origin of 1/f noise in polysilicon emitter bipolar transistors," Journal of Applied Physics, vol. 85, pp. 1192-1195, 1999.

[7] L. Militaru, A. Souifi, M. Mouis, and G. Brémond, "RTS noise in submicron SiGe epitaxial base bipolar transistors," Microelectronics Reliability, vol. 40, pp. 1585-1590, 2000.

[8] M. von Haartman, M. Sandén, M. Östling, and G. Bosman, "Random telegraph signal noise in SiGe heterojunction bipolar transistors," Journal of Applied Physics, vol. 92, pp. 4414-4421, 2002.

[9] G. Avenier, P. Chevalier, G. Troillard, B. Vandelle, F. Brossard, L. Depoyan, M. Buczko, S. Boret, S. Montusclat, A. Margain, S. Pruvost, S. T. Nicolson, K. H. K. Yau, D. Gloria, D. Dutartre, S. P. Voinigescu, and A. Chantre, "0.13 μm SiGe BiCMOS technology for mm-wave applications," in Bipolar/BiCMOS Circuits and Technology Meeting, 2008. BCTM 2008. IEEE, 2008, pp. 89-92.

[10] M.J. Deen and F. Pascal, "Review of low-frequency noise behaviour of polysilicon emitter bipolar junction transistors" IEE Proceedings-Circuits, Devices and Systems, 151, 125-137, 2004.

[11] X. L. Wu, A. Van Der Ziel, A. N. Birbas, and A. D. V. Rheenen, "Burst-type noise mechanisms in bipolar transistors", Solid-State Electronics, vol. 32, n 11, pp. 1039-1042, 1989

[12] S. Jouan, H. Baudry, D. Dutartre, C. Fellous, M. Laurens, D. Lenoble, M. Marty, A. Monroy, A. Perrotin, P. Ribot, G. Vincent, and A. Chantre, "Investigation of the effectiveness of buried carbon later to suppress transient diffusion effects in SiGe HBTs", 30th European Solid-State Device Research Conference, Cork, Ireland, 11-13 Septembre 2000.

[13] J.-Q. Lü and F. Koch, "Random telegraph noise in advanced self-aligned bipolar transistors", Japanese Journal of Applied Physics, vol. 35 part I, n 2B, pp. 826-832, 1996.

Towards Flexible and Conformable Electronics

Ravinder S. DAHIYA

Electronics and Nanoscale Engineering, School of Engineering, University of Glasgow, G12 8QQ, UK

The birth of microelectronics and subsequent miniaturization have revolutionized computing and communications. Yet, as revolutionary as the microelectronics technology has been, in its current form, it cannot address issues like realizing sensitive electronic systems on unconventional substrates such as plastics or paper that can be wrapped around curved surfaces such as the body of robots or artificial limbs. Early attempts to achieve conformable electronic systems primarily followed the flexible printed circuit boards (PCB) route, offering a limited degree of mechanical flexibility. Recent efforts to address these challenges include fabricating sensing and electronic components directly on the flexible substrates or on thin silicon wafers. A variety of solutions, ranging from TFTs to printed electronics have appeared using a wide variety of materials, including organic and inorganic semiconductors. Ultra-thin bendable chip is another interesting route complementing the recently explored micro-/nano-wires approach. The advent of fully flexible electronic systems will be a great leap in technology, as it will open the door to the next-generation electronic environment based on bendable and wearable devices. This lecture will present these developments with a focus on the high-performance bendable and conformable electronics.

ACKNOWLEDGMENT

This work was supported in part by the European Commission under grant agreements PITN–GA–2012–317488–CONTEST, PCOFUND–GA–2008–226070–Trentino and EPSRC Fellowship for Growth – Printable Tactile Skin (PRINTSKIN).

REFERENCES

[1] R. S. Dahiya, M. Valle, Robotic Tactile Sensing – Technologies and System. Dordrecht: Springer, 2013.

[2] R. S. Dahiya, P. Mittendorfer, M. Valle, G. Cheng, V. Lumelsky, "Directions Towards Effective Utilization of Tactile Skin -- A Review," IEEE Sensors J., vol. 13, pp. 4121 - 4138, 2013.

[3] R. S. Dahiya, S. Gennaro, "Bendable Ultra-Thin Chips on Flexible Foils," IEEE Sensors J., vol. 13, pp. 4030-4037, 2013.

[4] R. S. Dahiya, A. Adami, C. Collini, and L. Lorenzelli, "POSFET Tactile Sensing Arrays using CMOS Technology," Sensors and Actuators - A, vol. 202, pp. 226-232, 2013.

[5] R. S. Dahiya, A. Adami, C. Collini, and L. Lorenzelli, "Bendable ultra-thin silicon chips on foil," in 2012 IEEE Sensors, 2012, pp. 1-4.

[6] R. S. Dahiya, A. Adami, C. Collini, and L. Lorenzelli, "Fabrication of Single Crystal Silicon Mirco-/Nanostructures and Transferring them to Flexible Substrates," Microelectronic Engineering, vol. 98, pp. 502-507, 2012.

[7] R. S. Dahiya, D. Cattin, A. Adami, C. Collini, L. Barboni, M. Valle, L. Lorenzelli, R. Oboe, G. Metta, and F. Brunetti, "Towards Tactile Sensing System on Chip for Robotic Applications," IEEE Sensors J., vol. 11, pp. 3216-3226, 2011.

[8] G. Cannata, R. S. Dahiya, M. Maggiali, F. Mastrogiovanni, G. Metta, and M. Valle, "Modular skin for humanoid robot systems," presented at the 4th International Conference on Cognitive Systems (CogSys), Zurich, 2010.

[9] www.contest-itn.eu. CONTEST - Collaborative Network for Training in Electronic Skin Technology.

[10] Z. Celik-Butler, R. S. Dahiya, M. Q.-Lopez, Y. Xu, S. Wagner "Guest Editorial Special Issue on Flexible Sensors and Sensing Systems," IEEE Sensors J., Vol. 13 (10), pp 3854 – 3856, 2013.

[11] R. S. Dahiya, "Epidermal Electronics: Flexible Electronics for Biomedical Application," Chapter 19 (Part IV) in Handbook of Bioelectronics - Directly interfacing electronics and biological systems (ISBN: 9781107040830), S. Carrara and Kris Iniewski Eds. Cambridge University Press, 2015.

[12] P. Cosseddu, L. Seminara, L. Pinna, S. Lai, R. S. Dahiya, M. Valle, M. Capurro, A. Bonfiglio, "Tactile sensors based on the integration of a piezoelectric polymer with organic thin-film transistors" IEEE SENSORS 2014, The 13th IEEE Conf. on Sensors, Valencia, Spain, Nov 2014.

[13] S. Tinku, C. Collini, R. S. Dahiya, L. Lorenzelli, "Smart Contact Lens with Passive Structures," IEEE SENSORS 2014, The 13th IEEE Conf. on Sensors, Valencia, Spain, Nov 2014.

[14] S. Khan, R. S. Dahiya, S. Tinku, L. Lorenzelli, "Conformable Tactile Sensing using Screen Printed P(VDF-TrFE) and MWCNT/PDMS Composites" IEEE SENSORS 2014, The 13th IEEE Conf. on Sensors, Valencia, Spain, Nov 2014.

[15] S. Khan, R. S. Dahiya, L. Lorenzelli, "Flexible Thermoelectric Generator Based on Transfer Printed Si Microwires," in ESSDERC 2014, the 44th European Solid-State Device Research Conference, Venice, Italy, Sept 2014.

[16] S. Khan, R. S. Dahiya, L. Lorenzelli, "Screen Printed MWNT/PDMS Composite for Piezoresistive Flexible Sensors," in PRIME 2014 - 10th Int. Conf. on PhD Research in Microelectronics & Electronics, Grenoble, France, July 2014.

[17] S. Khan, R. S. Dahiya, L. Lorenzelli, "Screen Printed Flexible Sensors Electronic Skin," in ASMC 2014 - The 25th Annual Advanced Semiconductor Manufacturing Conference, NY, USA, May 19-21, 2014.

[18] R. S. Dahiya, A. Adami, C. Collini, M. Valle, L. Lorenzelli, "POSFET Tactile Sensing Chips using CMOS Technology," in IEEE SENSORS 2013, The 12th IEEE Conf. on Sensors, Baltimore, Maryland, USA, 4-6 Nov 2013, pp 1-4.

Integrated Low-Noise Current Amplifier for Glass-Based Nanopore Sensing

P. Ciccarella, M. Carminati, G. Ferrari

Dipartimento di Elettronica, Informazione e Bioingegneria
Politecnico di Milano
Milano, Italy
giorgio.ferrari@polimi.it

R. Fraccari, A. Bahrami

Department of Chemistry
Imperial College London
London, UK

Abstract—Nanopores realized with glass nanopipettes allow the electrical characterization of DNA molecules. A low-noise, custom-designed transimpedance amplifier has been developed to detect the current pulses (hundreds of pA) due to the features of the translocating molecule, while handling large DC currents (up to 100 nA). It is based on a CMOS current preamplifier designed to achieve high resolution with a wide bandwidth, thanks to the low parasitic capacitance of glass nanopipettes. With the aim of reducing the shot noise of the integrated amplifier, an auxiliary feedback network has been implemented. The electronic characterization of the system confirms that fast current pulses modulating a large DC current are detected. Validation experiments with DNA molecules demonstrate the actual resolution improvement with respect to the state-of-the-art.

Keywords—Low-noise current preamplifier; nanopore sensors; transimpedance amplifier.

I. INTRODUCTION

Nanopores have been attracting a widespread scientific, technological and also economic interest as compact and versatile platforms enabling the electrical detection of the features of various molecules translocating through the nanometric pore, in particular as one of the most promising candidates for automatic, fast and low-cost DNA sequencing on a lab-on-chip scale [1].

Although the majority of artificial solid-state nanopores are silicon-based (for instance fabricated drilling with an ion or an electron beam a thin silicon nitride membrane), their integration with CMOS platforms still present several technological challenges. From the point of view of the readout circuitry, the presence of the conductive silicon substrate causes a large (pF to nF) input stray capacitance, both in the case of standard ionic and transversal tunneling sensing configurations [2], thus limiting the noise performance.

In this paper we present the design and the experimental characterization of an integrated low-noise amplifier, custom designed for operating with low-parasitics glass nanopipettes [3] (tailored for rejecting the large current drop when the molecule starts to enter the pore and the tiny modulation due to different features along the molecule backbone), enabling an improved resolution-bandwidth trade-off with respect to state-of-the-art bench-top instrumentation, especially at high

Fondazione CARIPLO and The Royal Society are gratefully acknowledged for partial financial support.

frequency (10 kHz - 1 MHz).

II. CIRCUIT DESIGN

A. Design Specifications

The current drop signal ΔI, due to the pore blockade by the DNA molecule translocation, depends on the ionic concentration and varies from hundreds of pAs, for biological pores, up to nA for solid-state nanopores [1]. In glass-based nanopores, the current steps are expected to be of the same order of magnitude. In order to identify different DNA molecules that move throughout the pore, the details of the entire waveform should be discriminated, and thus a better resolution is required (the smallest current feature corresponding to a single base pair change is reported to be 0.9 pA [4]).

The resolution depends on the sensitivity-bandwidth trade-off because the equivalent input current noise, integrated over the desired bandwidth, sets the smallest signal step detectable. With the aim of detecting fast DNA translocation in nanopipettes, a current resolution better than 10 pA$_{r.m.s}$ with a bandwidth up to 100 kHz represents a challenging target.

Thanks to a low parasitic capacitance at the input node (a few picofarads), the best resolution can be achieved with low-noise CMOS current preamplifiers [5]. This solution minimizes high frequency noise, which is proportional to the total input capacitance, in order to reach higher sensitivity than previous solutions [2]. A suitable circuit architecture for handling a large DC current (up to 100 nA) and for minimizing the electronic shot noise is required, while a DC current monitor is useful for tracking the translocation event and for nanopipette device characterization purposes.

B. Low-noise Current Preamplifier

The front-end electronics is based on an original integrated current amplifier designed in CMOS 0.35 µm standard technology [6]. It consists of a cascade of two stages, both designed with low-noise and wide bandwidth criteria. Linear gain is guaranteed from DC to 1MHz thanks to a well-balanced network that establishes the current transfer function. The scheme of the electronic system, with the input current amplifier, is shown in Fig. 1. In DC, pairs of matched transistors (M_{n2} - M_{n1} and M_{p2} - M_{p1}) with the same operating point, set the current gain = N. All the gate terminals are biased at the most stable potential (i.e. ground), while the source terminals are connected together for each couple. The drains are equipotential (except for the offset voltage) so that each pair has the same V_{ds} and V_{gs}. M_{n2} and M_{p2} are designed by exactly replicating M_{n1} and M_{p1} N times, thus fixing the current gain. This solution overcomes the non-linear behavior of MOS transistors operating in sub-threshold regime for input signals sweeping the entire dynamic (± 25 nA). At high frequency, the linear gain depends on capacitance ratio. C_2 is realized replicating C_1 N times ($N = C_1/C_2$) for the best matching performance, while transfer function fluctuations, from DC to the maximum operating frequency, are negligible (less than 1.5 %). Optimal sizing is achieved by means of Monte Carlo simulations to ensure linearity performance and pole/zero compensation. C_2 is added parallel with M_{n2} and M_{p2}, introducing a zero that compensates the pole due to C_1. The first stage closed loop bandwidth f_1 = GBWP|$_{OP1}$·C_1/(C_1 + $C_{OP1|in}$ + C_{tot}) depends on total input capacitance $C_{tot} = C_{ext} + C_{in}$. Considering a $C_{in} \sim 2$ pF (input parasitic capacitance), C_1 is sized to a small value (300 fF) in order to maximize gain, bandwidth and minimize noise. OP1 is a low noise OpAmp with a GBWP up to 100 MHz to guarantee $f_1 \geq 10$ MHz. The preamplifier is made up by a pMOS differential pair with resistive load for high noise performance. Capacitive matching results has been taken into consideration for OP1 input stage sizing. The input referred current noise power spectral density (PSD), considering only the CMOS current amplifier, is given by the following expression:

$$S_i = S_v|_{OP1}(2\pi \cdot f)^2 \cdot (C_{tot} + C_{OP1|in} + C_1)^2 + 2qI \quad (1)$$

where $Sv|_{OP1}$ is the voltage noise PSD of the first operational amplifier ($Sv|_{OP2}$ can be neglected), while I is the DC current

flowing through M_{n1} and M_{p1}. Even minimizing the leakage current down to 25 nA, the preamplifier shot noise strongly affects noise performance, giving a contribution of about (89 fA/√Hz)^2. Due to a high DC current coming from the pore, up to 100 nA, an auxiliary feedback network (FN) implemented with discrete components has been introduced, as described in the next section.

C. High SensitivityTransimpedance Amplifier.

The scheme of the entire circuit that implements a transimpedance amplifier with capability to reject DC current is pictured in Fig. 1. The gain of the current preamplifier (CP) is $I_{out|CP}/I_{in} = G_1 = N_1 \cdot N_2 = 990$ and the amplified current signal is read by an external output transimpedance amplifier with a feedback resistor R_5 = 470 kΩ, giving a total transfer resistance $T = R_5 \cdot G_1 = 470$ MΩ.

Current pre-amplification breaks the noise/bandwidth trade-off of conventional transimpedance amplifiers. In the absence of the input stages, a single component R_5 would have set the noise performance ($S_i|_{R5} = 4kT/R_5$), total gain and bandwidth limited by its parasitic parallel capacitance (approximately 0.2 pF for a 0805 footprint component).

The noise of the output stage can be neglected because the contribution of R_5 is reduced by a factor G_1^2 ($S_i|_{R5} = 4kT/R_5 \cdot 1/G_1^2 \sim (0.2$ fA/√Hz)^2) and the series noise of the output transimpedance amplifier is completely negligible ($S_{i|OP5} \approx (0.5$ fA/√Hz)^2 at 400 kHz). The feedback resistor is properly chosen to satisfy gain and bandwidth requirements while, taking into account R_5C_5 pole time constant, C_5 must be < 0.8 pF, to guarantee a bandwidth higher than 400 kHz.

OP3 has GBWP $\simeq 200$ MHz and its total input capacitance ($C_{out|CP}$ + $C_{cm|OP3}$ + $C_{diff|OP3} \simeq 16$ pF + 3.9 pF +3.9 pF = 23.9 pF), introduces an additional pole at f_{5cl} = GBWP·C_5/($C_{out|CP}$ + $C_{cm|OP5}$ + $C_{diff|OP5}$) $\simeq 6.5$ MHz.

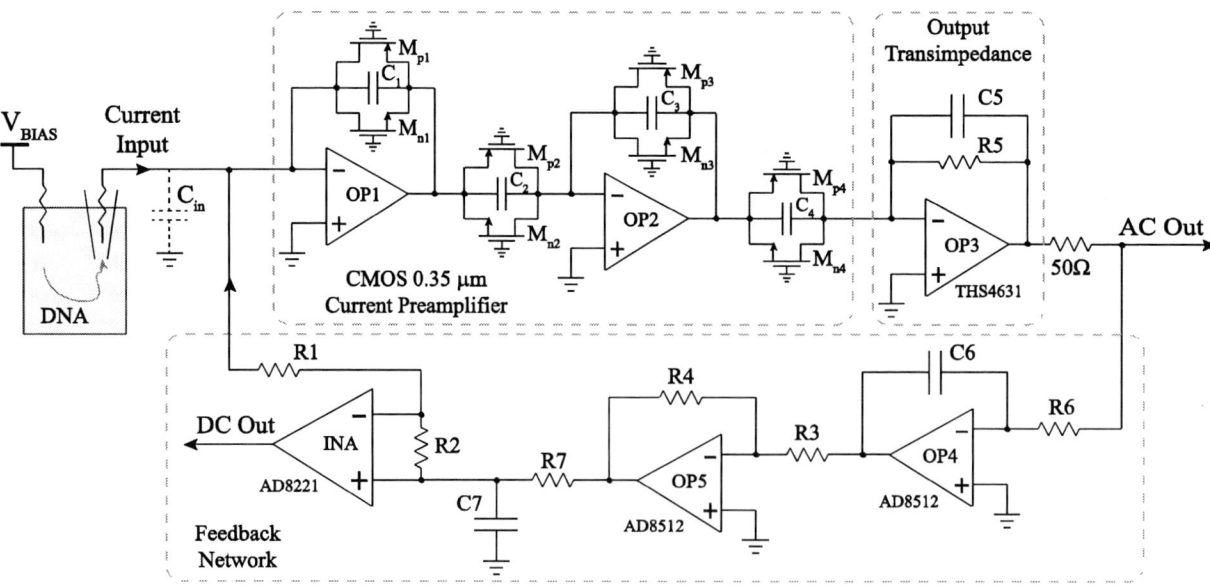

Fig. 1. Scheme of the proposed high sensitivity transimpedance amplifier: the low-noise CMOS current preamplifier allows achieving an improved resolution while an external feedback network is implemented for minimizing the amplifier shot noise and for handling a large input DC current flowing in the nanopipette.

The low frequency feedback network (FN) allows to operate with high leakage current up to 100 nA and it implements a high pass response, thus rejecting low frequency fluctuations. FN has been realized with an active integrator with the aim of maximizing DC loop gain (Gloop$_0$) and fixing a cut off frequency (f_0). An inverting buffer circuit is inserted for feeding back negatively and a further low pass filter is adopted to suppress high frequencies spurs. The cut off frequency $f_0 \approx 1/(2\pi \cdot R_6 C_6) \cdot 1/R_1 \cdot G_1 R_5$ is < 10 Hz and the loop gain is given by the following expression:

$$\text{Gloop}(s) = \text{Gloop}_0 \cdot 1/(1+s\,A_{0|OP4} \cdot R_6 C_6) \cdot 1/(1+sR_7 C_7) \quad (2)$$

where $\text{Gloop}_0 = -(1/R_1) \cdot G_1 R_5 \cdot A_0|_{OP4}$. The instrumentation amplifier (INA) reads the DC current that flows through FN via resistor R_2 and its DC output is employed to reveal nanopipettes leakage performances. The resistor R_1 is the lowest noisy component able to tap the input leakage current and its value (100 MΩ) gives an input referred noise PSD $Si|_{R1} = 4kT/R_1 = (12.9 \text{ fA}/\sqrt{Hz})^2$. The pore DC current dictates the low frequency noise performance since the higher is the leakage, the lower is R_1 with the same voltage dynamics. In conclusion, the overall input current noise approximately depends only on the input impedance and series noise of CMOS current preamplifier

$$S_i \approx S_v|_{OP1}(2\pi \cdot f)^2 \cdot (C_{tot} + C_{OP1|in} + C_1)^2 + 4kT/R_1 \quad (3)$$

while FN noise contributions can be neglected, as briefly discussed below.

In DC, the noise due to R_6 and OP_4 are directly transferred to the output node (i.e. the input current noise proportional to $(1/f)^2$ for high pass transfer function) but with appropriate sizing, their contributions can be completely neglected ($R_6 << R_5$ and $S_v|_{OP3} << 4kTR5$). For $f > f_0$ the noise due to R_6 is transferred at the output in the same way of the signal travelling forward along the loop, therefore it is low pass filtered by the active integrator and by the following passive filter. Similarly, $Sv|_{OP4}$ can be easily brought to the output like a signal applied to OP_4 plus terminal, amplified by forward gain. The latter noise source, and the same applies to all noise sources relied to the inverting buffer and INA, can also be considered applied, opportunely weight by forward transfer function, directly on R_1, thus obtaining FN contributions to input current noise that are completely negligible.

The front-end electronics board is cased in a very compact shielded housing (11.5 cm x 6.5 cm x 3 cm) and the circuit generates internally ±12 V and ±1.65 V (chip supply) from ±15 V heavy filtered power supply, by using voltage linear regulators. The AC output is used to zoom on the signal details (single base pair current variations), while the DC output provides the nanopipette leakage information. Low capacitance input connections have been carefully designed to achieve better resolution, minimizing parasitics and thus taking profit of the low-noise CMOS current preamplifier.

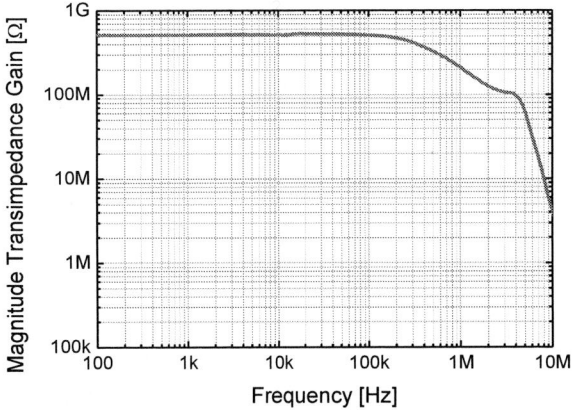

Fig. 2. Measurement of the AC transfer function: the gain is ~ 470 MΩ.

III. EXPERIMENTAL RESULTS

A. Transfer function

The magnitude of the transimpedance gain, measured by means of the Agilent E5061 network analyzer, is reported in Fig. 2. The bandwidth at -3dB is about 400 kHz, due to $R_5 C_5$ time constant, while the in-band gain is slightly higher than 470MΩ (due to the tolerance of the components). The high roll-off slope at high frequency can be explained by taking into consideration two poles of the current preamplifier at 12 and 4.5 MHz. and the closed loop bandwidth (f_{5cl} = 6.5 MHz) of the THS4631.

B. Noise Characterization

The PSD has been measured to evaluate the noise performance and the equivalent input current noise is shown in Fig. 3. As expected, at low frequency the noise is limited by the 100 MΩ resistor R1, while at high frequency the noise of the current preamplifier dominates. The series noise slightly depends on the input capacitance: for this reason at high frequency it does not scale with its value only. When the input impedance is in the same order of magnitude of the gate-

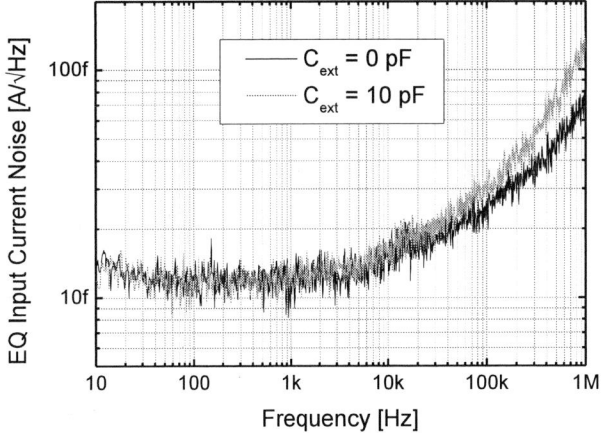

Fig. 3. Measurement of the input-referred PSD current noise for different values of the external device capacitance C_{ext}.

source capacitance of the input transistor, the current tail noise is not negligible. However, an improved noise performance, with respect to previous setup, is evident [2] and the current resolution evaluated for different filter bandwidths is reported in Tab. 1.

C. Amplifier Operation

The prototype circuit is simultaneously stimulated (Agilent 81150A) with a low frequency, high swing square wave and with fast 1 nA amplitude pulses. As depicted in Fig. 4, at the initial time, the AC output saturates, while low frequency loop reacts approximately with a time constant $1/(2\pi \cdot f_0)$. After a recovery time of about 70 ms, the current pulses are clearly visible and FN compensates the new value of the DC current input.

D. DNA Sensing and Performance Comparison

Experiments with a glass nanopipette and DNA molecules (250 pM dissolved in 1M KCl ionic solution) have been performed to validate the system developed in real condition. DNA translocations have been detected as depicted in Fig. 5, comparing the Axopatch 200B (the state-of-the-art current amplifier) with the proposed amplifier. The pore conductance is ~96 nS, extracted from I-V measurements and a potential of -400 mV has been applied to drive the negative-charged DNA through the pore. An external Bessel analog filter with a cut-off frequency of 70 kHz has been adopted and the improved resolution achieved is clearly visible.

TABLE I.   MEASURED CURRENT RESOLUTION

External Capacitance C_{ext}	Measured input-referred RMS current noise			
	2.5 kHz	25 kHz	75 kHz	100 kHz
0	0.63 pA	2.5 pA	5.34 pA	6.47 pA
10 pF	0.63 pA	2.66 pA	6.05 pA	7.57 pA

IV.   CONCLUSIONS

The design of a transimpedance amplifier based on a low-noise CMOS current preamplifier for glass-based nanopores

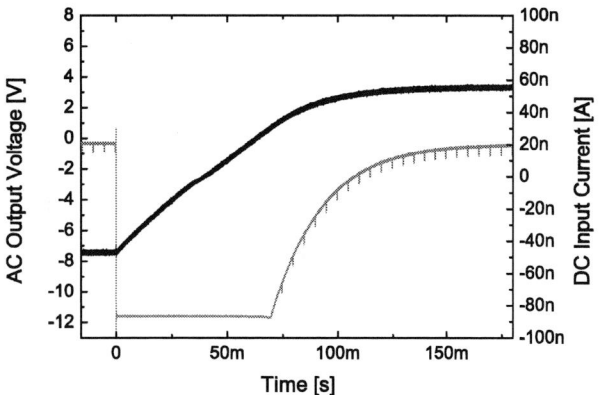

Fig. 4.   Fast current pulses (red line) superimposed to a large DC leakage current (black line) performing a large swing transition (100 nA) are measured to underline the separate detection of AC and DC signal components.

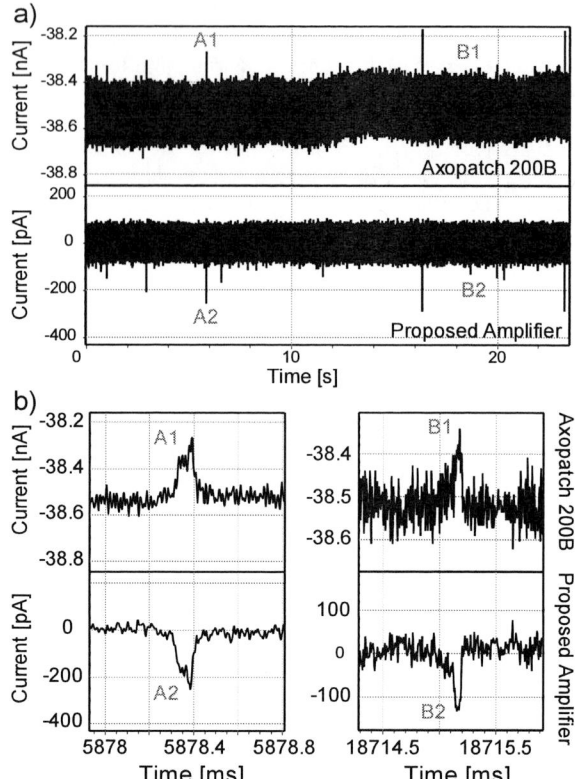

Fig. 5.   DNA translocation events have been detected by the proposed amplifier with a better resolution (with the same 70 kHz bandwidth) with respect to the state-of-the-art commercial workstation Axopatch 200B.

has been presented. Improved noise performance, thanks to the low (pF) parasitic capacitance of nanopipettes and the integration of the front-end electronics, enables achieving a higher resolution and a wider bandwidth than the state-of-the-art Axopatch 200B commercial low-noise current amplifier.

ACKNOWLEDGMENT

Dr. Tim Albrecht and the Royal Society are acknowledged.

REFERENCES

[1]   W. Timp, U. Mirsaidov, D. Wang, J. Comer, A. Aksimentiev, A, and G. Timp, "Nanopores sequencing: Electrical measurements of the code of life," IEEE Trans. Nanotechnol., vol. 9, no. 3, pp. 281-294, 2010.

[2]   M. Carminati, G. Ferrari, A. P. Ivanov, T. Albrecht, M. Sampietro, "Design and characterization of a current sensing platform for silicon-based nanopores with integrated tunneling nanoelectrodes," Analog Integr. Circ. Sig. Proc., vol. 77, pp. 333-343, 2013.

[3]   T. R. Gibb, A. P. Ivanov, J. B. Edel, and T. Albrecht, "Single Molecule Ionic Current Sensing in Segmented Flow Microfluidics," Anal. Chem, vol. 86, pp. 1865-1871, 2014.

[4]   A. Balijepalli et al., "Quantifying short-lived events in multistate ionc current measurements," ACS Nano, vol. 8, no. 2, pp. 1547-1553, 2014.

[5]   M. Crescentini, M. Bennati, M. Carminati, M. Tartagni, "Noise Limits of CMOS Current Interfaces for Biosensors: A Review," IEEE Trans. Biomed. Circuits Syst., vol 8, no. 2, pp. 278-292, 2014.

[6]   G. Ferrari, M. Farina, F. Guagliardo, M. Carminati and M. Sampietro, "Ultra-low-noise CMOS current preamplifier from DC to 1 MHz." Electronics Letters, vol. 45, no. 25, pp. 1278-1280, 2009.

Bendable Piezoresistive Sensors by Screen Printing MWCNT/PDMS Composites on Flexible Substrates

Saleem Khan[1,2], L. Lorenzelli[2]

[1]University of Trento, 38123, Italy
[2]Fondazione Bruno Kessler, Trento, 38123, Italy
skhan@fbk.eu, lorenzel@fbk.eu

R. S. Dahiya[3]

[3]Electronics and Nanoscale Engineering Research Division,
School of Engineering, University of Glasgow,G128QQ,UK
Ravinder.Dahiya@glasgow.ac.uk

Abstract—This paper presents piezoresistive sensors array, developed by screen printing of polymer nanocomposites based on multiwall carbon nanotubes (MWCNT) in poly (dimethyl-siloxane) (PDMS) matrix. The sensors are printed on 25 μm thick flexible polyethylene terephthalate (PET) substrate in the form of segmental arrays with parallel plate structures, whereby, MWCNT/PDMS nanocomposites layer is sandwiched between two printed silver plates of 1×1 mm² area each. Force concentrator structures, printed on separate PET sheet, are adhered to the bottom substrate with proper alignment to the sensory cells. Three different polymer nanocomposites (by wt. %) of carbon nanotubes (MWCNT) in PDMS matrix have been investigated here for possible application as a tactile sensor in an electronic skin for robots and general purpose applications. The change in resistance value is observed when structures are subjected to normal compressive forces as well as when they are bent.

Keywords—*MWNT, PDMS, Composites, Piezoresistance, Screen Printing, Flexible sensors, Tactile Sensing*

I. INTRODUCTION

Printed macro-electronics on polymer substrates is one of the interesting fields explored now-a-days due to advantages like low-cost and possibility of printing multifunctional electronics over areas bigger than standard wafer sizes. Macroelectronics has motivated us to explore new avenues for processing various materials and develop devices and systems which could not be realized easily by the conventional fabrication techniques on large surfaces [1-2]. Amongst macroelectronic systems, research in the advancing pressure sensors for electronic skin has gained interest. Piezoresistance is one of the modalities used for developing pressure sensors [3-4], due to advantages like simple low-cost electronics. Several materials and mechanisms explored to exploit piezoresistance for sensor include change in resistance of metallic strain gauges or change in mobility of semiconductors materials based devices [5]. These materials and structures have low gauges factors and to overcome this obstacle the conductive polymer composites have been explored by researchers to develop piezoresistive devices.

Conductive polymers have almost similar characteristics to some metals and inorganic semiconductors while maintaining typical polymer properties of flexibility, easy processing and synthesis. Incorporating multiwall carbon nanotube (MWNT) in poly (dimethyl-siloxane) (PDMS) results in a material, which possess a number of exciting properties that can successfully be harnessed in sensors and actuators [4, 6]. The actuation mechanism is based on the geometry and interconnection paths of nanotubes in the polymer matrix which vary upon application of force. This leads to a change in bulk resistivity of the composite layer. A low percolation threshold is desired to retain the static as well as dynamic mechanical, physical and electrical properties of MWCNT/PDMS composites [7-8]. Detailed description of the research and applications of nanocomposites based on MWCNT/PDMS are given in [9-10]. Different techniques for deposition and structures of MWCNT/PDMS nanocomposites like microchips, micro-heaters, strain and temperature sensors are reported [11-16]. Uniform dispersion of nanofillers is of prime importance and contributes mainly to the performance of sensors [17-18]. Although conductive polymer composites have the disadvantages of non-linearity, hysteresis and temperature drifts, they are effective for large strains and are simple and cost-effective to fabricate. These advantages make them a better choice of sensing elements for applications requiring complaint materials such as electronic skin, electronic textiles and other large deformation measurements [19]. For this reason, the work presented here employs the multiwall carbon nanotubes (MWCNT) in poly (dimethyl-siloxane) (PDMS) matrix for piezoresistive sensor arrays.

In order to deploy sensor arrays on large surfaces, it is necessary to micro-pattern the composite material in an efficient and cost-effective way. Therefore, an all screen printed piezoresistor structure is the main focus of our current study. Screen printing is used to pattern conductive plates for the effective area as well as interconnect lines and MWCNT/PDMS composites on the same conductive plates. Using the same technology for all the layers will help in minimizing processing time and improve manufacturability. Besides this the use of printing technologies will lead to reduce material waste and low cost of manufacturing [20].

This paper is organized as follows: Synthesis of polymer composite and experiments for manufacturing are presented in section II. Section III describes the physical and electrical characterization of printed structures. Finally the conclusion and future scope of this work are given in section IV.

II. MATERIAL AND PRINTING METHOD

A. MWCNT/PDMS composite synthesis

The materials used for developing piezoresistive sensors array consist of MWCNT mixed with three different weight ratios (i.e. 1%, 3% and 5%) in PDMS matrix. MWCNT sample

978-1-4799-4993-9/14 $31.00 © 2014 IEEE

purchased from Sigma Aldrich have > 95 % carbon, with nanotubes having outer diameters of 6-9 nm, length 5µm and ~2.5g/mL density at room temperature. MWCNT are first dispersed in chloroform by using mechanical stirrer for 5 minutes. The solution was kept in ultrasonic bath at 40 kHz for 30 minutes. After uniform dispersing of nanotubes in the chloroform, PDMS (Dow Corning Sylgard 184) was mixed with the solution followed by mechanical stirring for 10 minutes. Composite solution was then kept in ultrasonic bath at 40 kHz for 3 hours. Cross linking agent was added in 10:1 into the composite and degased completely in desiccator. Same steps were followed for developing all the solution samples. After degassing steps nanocomposites were immediately screen printed on already printed electrode, steps discussed in more detail in the following section.

B. Screen printing experiments

The parallel plate structure is followed to realize the piezoresistive sensors array. The MWCNT/PDMS nanocomposites layer was sandwiched between the two printed silver metal electrodes. Attractive feature of this approach is that, all the processing steps are performed by using a single printing technology i.e. screen printing. Figure 1. (a)-(b) shows the schematic and internal structure of the sensing structures.

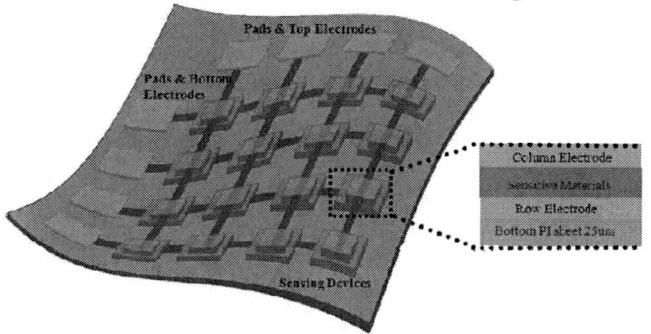

Fig. 1. (a) Schematic of the printed piezoresistor structure. (b) Side view of the structures

Silver (Ag) based paste (DuPont-5028) is used for the top and bottom electrodes. Paste viscosity is in the range of 15-30 Pa. S. Conductive tracks for the bottom electrodes are divided into 4 module, each containing 4×4 array of sensory devices. Active area of the sensing cell using printed silver plates is 1×1 mm² which are connected through printed interconnected lines 100µm wide. Distance between consecutive sensors are kept 5.6 mm in order to reduce crosstalk between neighboring sensors. For the readout signals, 2×2 mm² pads are also printed with first step of printing bottom electrodes. Printed metal patterns were sintered at 120 °C for 1 hour. A separate stencil mask with 3×3 mm² coverage area, overlapping on each side of the bottom and top electrodes was used for printing the piezoresistive material. Solutions of MWCNT/PDMS composites were printed on bottom electrode aligned by adjusting screen printing stage parameters. Deposited layers were sintered in vacuum at 80-°C for 5 hours and kept overnight. After complete polymerization, counter electrode was printed on top of composite area using the same stencil but with 90° orientation. Top and bottom electrodes are in good alignment and no short circuiting was found after checking all the devices. Force concentrator structures (3×3 mm²) are printed on a separate substrate and laminated on the bottom substrate, which also serves as an encapsulant from any environmental affects to the sensors. Force concentrators were printed using UV-curable dielectric ink supplied by DuPont. Figure. 2 (a)–(b) shows the final arrays of sensors printed on PET substrates and the final assembled devices with force concentrator structures.

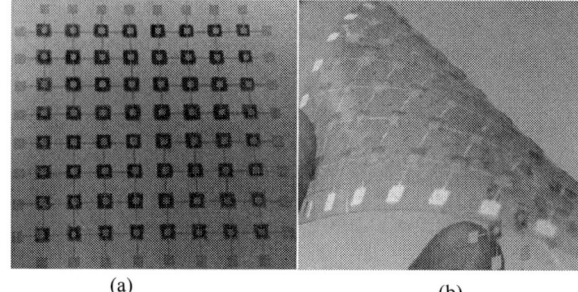

(a) (b)

Fig.2. (a) Screen Printed sensor arrays. (b). Final assembled device with force concentrator structures.

III.   RESULTS AND DISCUSSIONS

Conductive electrodes in sensitive areas and interconnect wires are first investigated to check for their use in reliable sensor applications. Sheet resistance of the printed conductive layers checked with four point probe is 14.15 Ω/sq, which is in the close range of expected sheet resistivities of silver paste after sintering. Screen printing deposits thick layers as compared to other printing tools, which is an essential requirement for printed circuits and particularly to piezoresistive bulk layers. Adhesion of conductive patterns to polymer substrates and subsequently to MWCNT/PDMS layer at different environmental conditions is of prime importance.

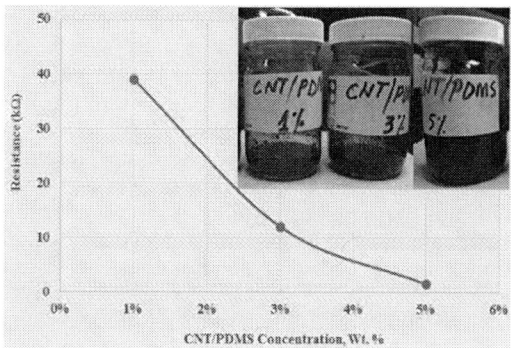

Fig. 3. Resistance change with varying filler concentration

Physical characteristics and resistive changes at different applied forces are investigated and discussed in detail in the following sub-sections.

A. Bulk resistance at different filler concentrations.

The resistance of the sensors presented here changes with filler concentration as shown Figure 3. This change in the initial resistance values for bulk piezoresistive composites at different filler concentration is mainly due to different number of interconnection paths and random distribution of MWCNT within the polymer matrix.

Filler concentration is not only the major parameter for resistance change but is also critical for printing process. Solution becomes dense with increased amount of fillers and beyond a certain limit it becomes difficult to screen print uniformly. Thus, an optimum range of filler concentration is required for controlled patterning and readable resistive response. At low concentrations close to 1%, the initial resistance is very large restricting resistance change within close limits. Also the solution is less viscous and flows out after printing which deteriorates the shape of the patterned structures. The isolation of individual sensing devices is not maintained and very irregular layer of MWCNT/PDMS is achieved. Avoiding such situation requires increase in the filler concentration. This also improves both physical and electrical properties of the device. With increased concentration, operating envelope of the device is enlarged as initial resistance decreases. The operating envelope of the sensing device becomes responsive to an increased range of forces. By observing the resistance value in Fig. 3 and also from screen printing experiments, 3 wt. % concentrations is found to be very close to the optimum desired concentration.

B. Response of printed MWCNT/PDMS composite sensors..

The response of MWCNT/PDMS composites is obtained by applying compressive forces on the sensor device. The observed change in the resistance of sensors is due to (a) micro-nanoscale changes in the carbon nanotubes as a result of mechanical deformation, and (b) the formation of conductive paths within the matrix. Prepared samples were put on a rigid surface and force was applied on top of force concentrator structures that are aligned with the sensory cells. Piezoresistive behavior was confirmed by the observed change in resistance with respect to applied force. As shown in Fig. 4, the resistance increases with increasing forces. The increase in resistance is more for sample with lesser concentration of MWCNTs. That could be due to immediate breakdown of the less number of conduction paths established in the polymer matrix. The MWCNT/PDMS samples with 1 and 3 wt. % concentrations show an approximate linear response which is often desired in pressure sensors. In case of 5 wt. % an abrupt non-uniform change in resistance occurs above forces of around 6N. This range is much higher than the force (0.001-1N) experienced by humans in daily tasks [4].

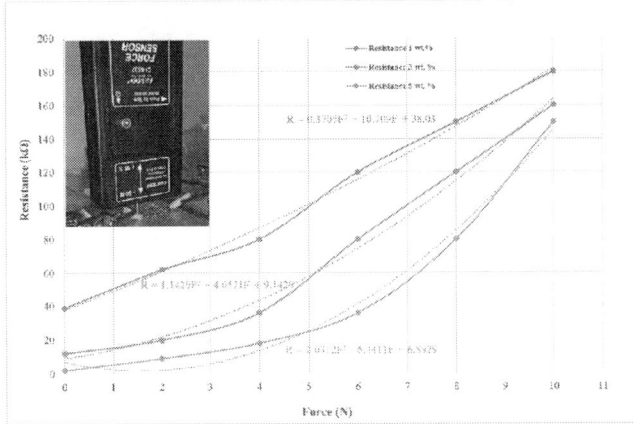

Fig. 4. Resistance change vs. applied compressive force

The sensors were also evaluated for cyclic force to check their restoration behavior to the original resistances. Restoration of the resistance values after removing the applied stress was investigated and it was observed that the sample with 3 wt. % recovered much faster than the samples with other two concentrations. For 3 wt. %, the restoration time is of the order of few seconds, while other two concentrations took few minutes to reach back to its initial value. The trend of increase in resistance values is not uniform, which might be due to nonuniform dispersion and random conductive paths made by aggregates of nanotubes within the polymer matrix. Agglomeration of MWCNTs at various locations of the layers was observed after printing as shown in the microscopic image below in Fig.5. The agglomeration of MWCNTs occurred only in 3 and 5wt. % solutions. Agglomeration of nanotubes is more evident in less concentrated solution as compared to higher concentrations. Fluctuation in resistance values in bulk sample was observed at normal conditions without applying any stress or strain, which is possible due to the fast shifting of different conductive paths generated in the bulk MWCNT/PDMS composite (especially in higher concentrations and agglomerated sites). These fluctuations in resistance were observed even under compressive force on the devices.

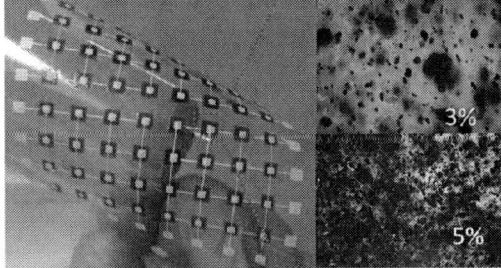

Fig. 5. Microscopic image of the printed sensors

Increase in resistance for all the three concentrations is detected when the substrates are wrapped in convex shape around a cylinder (15mm radius). This increase is caused by the bending induced strain. The conductive paths established during the normal position are enlarged which results in an increased resistance. Alternatively, decrease in resistance is recorded when the substrates are bent in concave shape. In this case the conductive paths are pushed more closer which results in increase in conductivity of the composite. This is interesting for robotic skin when mounted especially at joints, where contraction and relaxation during the movement can be monitored and controlled by using such type of strain sensors.

C. Adhesion test of MWCNT/PDMS and top electrode.

Plasma oxidation of polymer substrate was performed to promote the adhesion of MWCNT/PDMS with substrate. The plasma oxidation makes the polymer surface hydrophilic and improves the adhesion. The plasma oxidation was performed before printing the bottom electrodes. As the total coverage area of piezoresistive material (3×3 mm^2) is greater than bottom electrode (1×1 mm^2), MWCNT/PDMS makes strong bond all around the electrodes after sintering. Polymer nanocomposites are soft material owing to the intrinsic properties of base polymer matrix. When microstructures of these composites are printed onto the plasma oxidized substrate, the interface is tightly secured by strong bonding between composites and the plastic substrate. Tape test was performed for checking adhesion of Ag tracks on top of printed MWCNT/PDMS. Adhesion of Ag is also found to be dependent on filler concentration. For 1wt.% MWCNT/PDMS, about 70% of the Ag tracks got detached after peeling-off the tap. Adhesion became stronger for MWCNT/PDMS samples with higher concentrations, (i.e. with 3% and 5% filler concentrations), as about 50% and 10% of the Ag layer got detached respectively. This is because of the interaction of Ag paste on molecular level with MWCNTs in the PDMS matrix.

IV. CONCLUSION

Screen printed arrays of piezoresistive flexible pressure sensors are developed by sandwiching the MWCNT/PDMS composite layer between two patterned Ag metal layers. Fine patterning and uniform deposition of the composites has been achieved on flexible substrate. A total of 64 sensors have been fabricated in one flow by screen printing technique. Three different composite solutions based on wt. % of MWCNT in PDMS matrix have been presented. The samples have been compared on the basis of printing processability and piezoresistive response of the final devices. Based on the screen printing experiments 3wt.% solutions of MWCNT/PDMS was found to be close to the viscosity ranges required for screen printing. Also the agglomeration in 3wt% is less as compared to higher concentrations which leads to approximately linear response and less fluctuations in resistance values of the sensors. Due to low viscosity it was difficult to print the 1wt% solution with fine patterns. Same way the 5wt.% was too viscous for printing. Devices were tested under compressive force and an increase in resistance was recorded at increasing applied forces. Change in resistance is also observed in convex and concave orientations bending of flexible sensors.

Future work will involve finding threshold for optimum concentration, enhancing dispersion and reducing agglomeration of MWCNT. The characterization of whole array of devices and analysis of interference or cross talk will be investigated in future. Whole package of screen printed foldable pressure sensor is targeted for development of low cost electronic skin applications. The pressure mapping device could be enabled for local differences of pressure by utilizing the patterning capability of composite materials. Single step processing of flexible pressure sensor will be attractive for realizing large area electronic transducer system.

ACKNOWLEDGMENT

Authors are thankful to Mrs. Sajina Tinku, Severino Pedrotti and Stefano Girardi for their valuable support during manufacturing and characterization of the devices. This work was supported by the European Commission under grant agreement PITN-GA-2012-317488-CONTEST.

REFERENCES

[1] Roar R. Søndergaard, Markus HƐosel, Frederik C. Krebs, "Roll-to-Roll Fabrication of Large Area Functional Organic Materials", Journal of Polymer Science Part B: Polymer physics, 2013, 51, 16-34.

[2] A. Nathan, A. Ahnood, M. T. Cole, L. Sungsik, Y. Suzuki, P. Hiralal, F Bonaccorso, T. Hasan, L. Garcia-Gancedo, A. Dyadyusha, S. Haque, P. Andrew, S. Hofmann, J. Moultrie, C. Daping, A. J. Flewitt, A. C. Ferrari, M. J. Kelly, J. Robertson, G. A. J. Amaratunga and W. I. Milne "Flexible electronics: The next ubiquitous platform", Proc. IEEE, vol. 100, pp.1486 -1517 2012.

[3] R. S. Dahiya, P. Mittendorfer, M. Valle, G. Cheng, and V. Lumelsky, "Directions Towards Effective Utilization of Tactile Skin -- A Review," IEEE Sensors Journal, vol. 13, pp. 4121 - 4138, 2013.

[4] R. S. Dahiya, M. Valle, Robotic Tactile Sensing – Technologies and System, Springer Science, pp 1-248, 2013. (ISBN:978-94-007-0578-4)

[5] R. S. Dahiya and S. Gennaro, "Bendable Ultra-Thin Chips on Flexible Foils," IEEE Sensors Journal, vol. 13, pp. 4030-4037, 2013.

[6] Engel J, Chen J, Chen N, Pandya S, Liu C. Multi-walled carbon nanotube filled conductive elastomers: materials and application to micro transducers. In: Proceedings of the 19th IEEE international conference on MEMS; 2006. p. 246–9.

[7] Wen-Pin Shih, Li-Chi Tsao, Chian-Wen Lee, Ming-Yuan Cheng, Chienliu Chang, Yao-Joe Yang and Kuang-Chao Fan, "Flexible Temperature Sensor Array Based on a Graphite-Polydimethylsiloxane Composite," Sensors 2010, 10, pp. 3597-3610.

[8] Chung-Lin Wu, Hsueh-Chu Lin, Jiong-Shiun Hsu, Ming-Chuen Yip and Weileun Fang, "Static and dynamic mechanical properties of polydimethylsiloxane/carbon nanotube nanocomposites" Thin Solid Films 517 4895–4901, 2009.

[9] Liu, C.-X.; Choi, J.-W, "Improved dispersion of carbon nanotubes in polymers at high Concentrations," Nanomaterials 2012, 2, 329–347.

[10] Mohamed Imbaby1 and Kay Gottschalk, "Fabrication of micro pillars using multiwall carbon nanotubes/polymer nanocomposites," J. Micromech. Microeng. 23, pp. 1-6, 2013.

[11] Shruti Nambiar, John T.W. Yeow, "Conductive polymer-based sensors for biomedical applications," Biosensors and Bioelectronics 26 (2011) 1825–1832.

[12] Bianco S, Ferrario P, Quaglio M, Castagna R and Pirri C F, "Nanocomposites based on elastomeric matrix filled with carbon nanotubes for biological applications Carbon Nanotubes—From Research to Applications" ed Dr S Bianco 978-953-307-500-6 (Rijeka: InTech)

[13] Xize Niu, Suili Peng, Liyu Liu, Weijia Wen, and Ping Sheng, "Characterizing and Patterning of PDMS-Based Conducting Composites", Adv. Mater. 2007, 19, 2682–2686.

[14] M. Lu, A. Bermak and Y.-K. Lee, "Fabrication technology of piezoresistive conductive PDMS for micro fingerprint sensors," IEEE MEMS 2007, pp.251-254.

[15] N. Serraa, T. Maedera, P. Lemaireb, P. Rysera, "Formulation of Composite Resistive Pastes for Micro-Heater Manufacturing," Procedia Chemistry 1 (2009) 48–51.

[16] Chao-Xuan Liu, and Jin-Woo Choi, "Strain-Dependent Resistance of PDMS and Carbon Nanotubes Composite Microstructures", IEEE Transactions on Nanotechnology, Vol. 9, no. 5 2010.

[17] Kunmo Chu, Dongouk Kim, Yoonchul Sohn, Sangeui Lee, Changyoul Moon, and Sunghoon Park, "Electrical and thermal properties of carbon-nanotube composite for flexible electric heating-unit applications", IEEE Electron Device Letters, Vol. 34,no. 5, 2013.

[18] Hwang, J. Jang, K. Hong, N.K. Kim, J.H. Han, K. Shin, "Poly(3-hexylthiophene) wrapped carbon nanotube/poly(dimethylsiloxane) composites for use in finger-sensing piezoresistive pressure sensors", Carbon, 49 (2011), pp. 106–110

[19] Reza Rizvi, Brendan Cochrane, Elaine Biddiss and Hani Naguib, "Piezoresistance characterization of poly(dimethyl-siloxane) and poly(ethylene) carbon nanotube composites", Smart Mater. Struct. 20 (2011) 094003 (9pp).

[20] Saleem Khan, L. Lorenzelli, R.S. dahiya, "Screen Printed Flexible Pressure Sensors Skin", Manuscript accepted, ASMC 2014, May 19-21, Saratoga Springs, New York, USA (In Press).

Thickness effects of ZnO thin films on flexible ozone sensors

M. Acuautla, S. Bernardini, M. Bendahan

Aix – Marseille University
CNRS, IM2NP – UMR 7334
Marseille, France
marc.bendahan@im2np.fr, sandrine.bernardini@im2np.fr

E. Pietri

Genes'Ink
Rousset, France

Abstract— **ZnO nanoparticles thin films were deposited with different thicknesses on Kapton flexible substrates by drop coating method. The flexible platform consists in Ti/Pt interdigitated electrodes for gas detection and a heater device fabricated by Magnetron Sputtering and photolithography process. The thickness effects of the ZnO thin film on the gas sensing properties by ozone exposure at 200 °C has been studied. In order to test a deposit methodology used in large scale industrial production, an ultrasonic spray deposition was done. It was found that the sensitivity towards ozone gas is strongly dependent of the ZnO thin film thicknesses.**

Keywords—Zinc oxide; thin film; ozone; flexible substrate; gas sensor.

I. INTRODUCTION

The gas sensor's progress, to monitor toxic and hazard gases, is quickly developing due to safety industrial requirements and environmental pollution concerns. Gas sensors based on ZnO metal oxide thin film have been found as one of the most potential candidates for detecting toxic gases because of their low cost, high sensitivity, high chemical stability and ability to detect large number of gases [1-4].

Many studies have been dedicated to control the ZnO thin film properties. Indeed, it has been proved that the gas sensing properties of the metal oxide thin films are affected by different parameters such as the morphology, the sensing temperature and the film thickness [5 - 8]. In the present work, the thickness influence over the sensing properties of ZnO thin films on flexible substrate were investigated. Ozone at different concentrations is used as gas as one of the six principal pollutants considered harmful for the public health [9, 10].

II. EXPERIMENTAL PROCEDURES

A. Design and materials

The top schematic design of the flexible platform employed as micro gas sensor is illustrated in Fig. 1. Polyimide Kapton

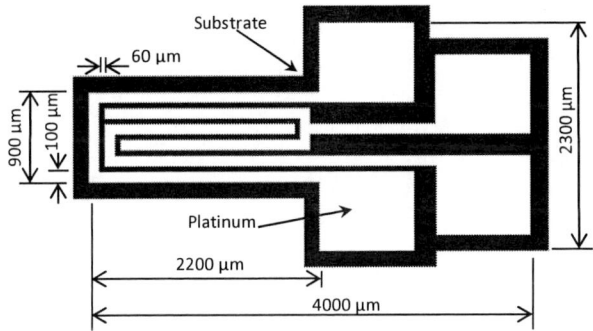

Fig. 1 Layout design of the flexible sensor.

with thickness of 75 µm is used as flexible substrate where interdigitated Ti/Pt electrodes were sputtered for gas detection with thickness of 5 nm and 100 nm respectively, and a Pt electrode around is used as heater device.

The sensor has a total area of less than 4000 µm x 2500 µm and a sensitive area of 2200 µm x 900 µm.

B. Thermal simulation and electrical calibration

Kapton as flexible substrate presents important characteristics to be used as gas sensor such as low cost, excellent thermal stability, solvent resistant, good mechanical and electronic properties in a wide range of temperature from - 269 °C to +400 °C. In order to validate the platform, a thermal simulation by finite elements and an electrical calibration have been done.

For the thermal simulation, three parameters were used and taken into account: an electrical conductivity σ=3.01E+06 S/m, an electrical resistivity ρ= 3.33E-07 Ω·m and a thermal conductivity λ= 21.94 W/m·K. These parameters were obtained by an experimental calibration at 25 °C of the platinum as thin film. Fig. 2 illustrates a homogenous temperature under the sensitive area. The linear experimental response of the temperature versus the source power for the micro heater is showed in Fig. 3.

978-1-4799-4993-9/14 $31.00 © 2014 IEEE

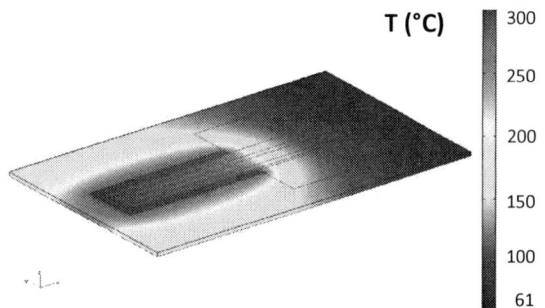

Fig. 2 Thermal simulation of the flexible platform

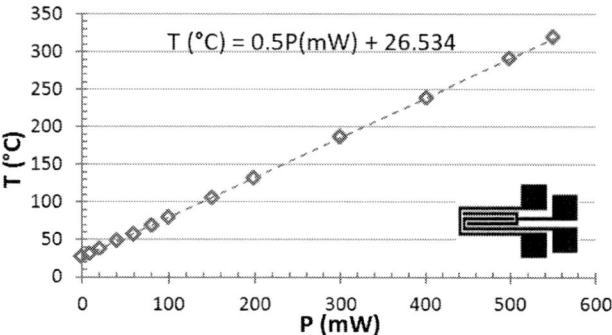

Fig. 3 Electrical calibration of the heater device on the flexible substrate

C. Fabrication of the flexible micro sensor

For the flexible sensor fabrication, the substrates were cleaned ultrasonically with acetone and ethanol, and treated by oxygen plasma to remove impurities and improve the metal film adhesion. Afterwards a Ti/Pt film with thickness of 5 nm and 100 nm was deposited by Magnetron Sputtering. Then the circuit patterns were made by standard photolithography.

To compare the thickness effects in the final gas sensing properties of the sensors, ZnO nanoparticles ink from Genes' Ink was deposited by drop coating with different amount of material to obtain four thicknesses (100 nm, 190 nm, 275 nm and 385 nm). Compared to other deposition techniques, drop coating method has the advantage of low cost, low substrate temperature deposition and is a very simple process.

Finally, the ZnO thin films deposited on flexible substrate were annealed by 3 hours at 300 °C under environmental conditions to improve the film density, the grain growth, the quality and stability of the sensitive material. The final flexible sensor fabricated is illustrated in Fig. 4.

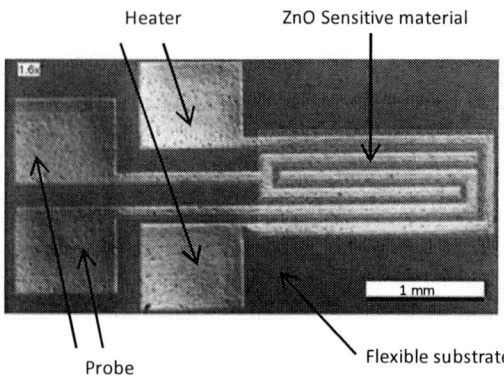

Fig. 4 Sensors fabricated on flexible substrate by Photolithography process.

The thickness measurements were carrying out by a profilometer Dektak 6M, this equipment uses a diamond tip stylus that maintains a constant stylus force over the samples surface without material damage.

The sensor sensitivities towards ozone were studied by measuring the resistance through the sensitive material and the normalized responses were calculated using the relation R_{gas}/R_0 where R_{gas} is the sample resistance measured under ozone and R_0 is the resistance measured in dry air. The measurement were carry out in a closed chamber using a programmable power supply to control the temperature of the sample, a source meter Keithley 6430 and an ozone calibration source InDevR.

III. RESULT AND DISCUSSIONS

As previously mentioned, the thermal and electrical calibrations of the flexible sensor present a linear behavior of the temperature versus the source power in the heater device and a homogenous temperature around the sensitive area; both properties help us to validate the platform.

The gas sensitivity properties were realized using different ozone concentrations from 5 ppb to 300 ppb at 200 °C with an ozone exposure during 1 minute. Fig 5 illustrates the good sensitivity of the ZnO thin film as ozone sensor over a wide range of concentrations and the dependence of the normalized response with the sample thickness.

978-1-4799-4993-9/14 $31.00 © 2014 IEEE   408

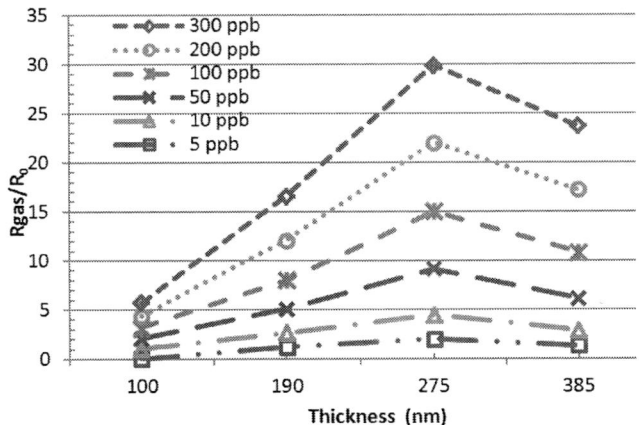

Fig. 5 Sensor responses at 200 °C with four ZnO thin film thicknesses over a wide range of ozone concentration.

The ozone responses present an increment of sensitivity with the thickness, but a maximum is found in the dependence of the sensor response. This behavior was found also by R. Ferro for NO_2 sensor based on ZnO films on alumina substrate [11], with a NO_2 concentration of 5 ppm at 275 °C it was found a maximum with a film thickness of 220 nm.

For thickness below our maximum thickness around 275 nm, the increment of the permeability associated with the increment of porosity is predominant in the sensor response, while for the thickness above 275 nm the effect of the grain size determines the behavior. According to the traditional behavior found in the literature [11 - 14], the sensor response should increases with the thickness diminution due to the increment of the active area and reduction of the grain size.

The responses and the recovery time of the sensor for an ozone concentration of 500 ppb at 200 °C are showed in Fig. 6 for various ZnO thin film thicknesses.

The recovery time increment with the thickness is an expected result ought to different studies report an increment of porosity and grain size with the increment of the thickness [12, 13]. However, the response time for the samples with thickness of 190 nm and 275 nm is faster than the ones with 100 nm and 385 nm. The obtained results should be associated with the importance modification of the thin film morphology when the amount of nanoparticles increases.

The sensor reproducibility and stability were tested using the samples with thickness of 275 nm. For an ozone concentration of 500 ppb, it was found a normalized response around $R_{gas}/R_0 = 10.7 \pm 0.3$ with fast response and recovery time (Fig. 7).

In order to test a deposit methodology usable for large scale industrial production, an ultrasonic spray deposition with thickness of 100 nm has been done performed on industrial equipment. This thickness is currently used by photovoltaic manufactures, therefore the deposit parameters such as ink viscosity, pressure, speed, etc. are well known for this thickness. This method also offers the advantage of a quick process (around 30 seconds for an A4 format) once the calibration is made.

Fig. 8 illustrates that the sensor with 100 nm film thickness follows the same behavior of the ones deposited by drop coating for the different ozone concentrations. Our drop coating studies highlight that to increase the sample sensitivity fabricated by ultrasonic spray, a thicker film around 200 nm and 300 nm should be used. These thicknesses need more studies to find the new deposit parameters.

The sensor response increment in the sample fabricated by spray deposition with thickness of 100 nm over its similar deposited by drop coating could be related by the increment of surface homogeneity in the samples fabricated by ultrasonic spray.

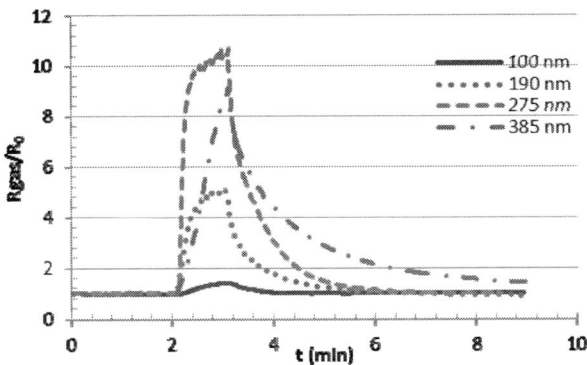

Fig. 6 Sensor responses at 200 °C with different ZnO thin film thicknesses over an ozone concentration of 500 ppb.

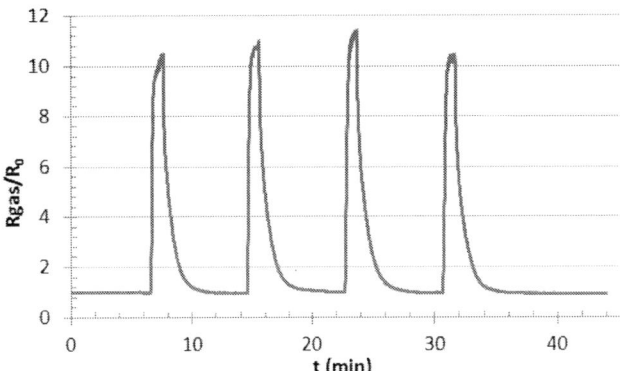

Fig. 7 Reproducibility responses of the sensor fabricated by photolithography process with thickness of 275 nm towards 500 ppb of O_3.

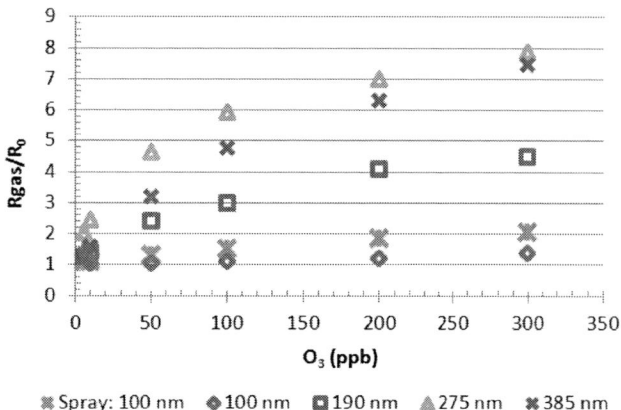

Fig. 8 Sensor responses at 200 °C with different ZnO thin film thicknesses deposited by drop coating and ultrasonic spray deposition.

IV. CONCLUSIONS

Flexible ozone sensors based on ZnO thin films were presented. The micro sensors were fabricated using Magnetron sputtering for the metal films and standard photolithography to fabricate the electrode paths. The sensitive material depositions were performed by drop coating and thicknesses of 100 nm, 190 nm, 275 nm and 385 nm were obtained, also a thickness of 100 nm fabricated by ultrasonic spray was done in order to compare the gas properties of the device using a large scale industrial deposited process.

The sensing properties were also studied towards NO_2 as oxidizing gas; however no significant response was observed at operational temperatures from 25 °C to 350 °C.

It was found that the film thickness represents a critical factor in the gas responses, and the effects of the grain size and the film porosity should be taken into account. For thicknesses below 275 nm the permeability determines the sensor behavior increasing the response with the thickness while with thickness of 385 nm there is a contrary effect ought to the increase of the grain size.

ACKNOWLEDGMENT

The authors would like to acknowledge to A. Combes for his technical support in this work, in the same manner M. Acuautla would like to acknowledge the research grant of CONACyT-MX (214895/310187).

REFERENCES

[1] R. Martins, E. Fortunato, P. Nunes, I. Ferreira, A. Marquez, M. Bender, N. Karsarakis, V. Cimalla, G. Kiriakidis, Zinc oxide as an ozone sensor, Journal of Applied Physics 96, pp. 1398-1408, 2004.

[2] S. Pati, S. B . Majumder, P. Banerji, Role of oxygen vacancy in optical and gas sensing characteristics of ZnO thin films, ournal of Alloys and Compound 541, pp.376-379, 2012.

[3] N. Le Hung, E. Ahn, H. Jung, Synthesis and gas sensing properties of ZnO nanostructures, Journal of the Korean Physical Society 57, pp. 1784-1788, 2010.

[4] A. Wei, L. Pan, W. Huang, Recent progress in the ZnO nanostructures based sensors, Material Science and Engineering B 176, pp. 1409-1421, 2011.

[5] P. S. Shewale, G. L. Agawane, S. W. Shin, A. V. Moholkar, J. Y. Lee, J. H. Kim, M.D. Uplane, Thickness dependent H_2S sensing properties of nanocrystalline ZnO thin films derived by advanced spray pyrolysis, Sensors and Actuators B 177, pp. 687-702, 2013.

[6] G. Korotcenkov, B. K. Cho, Thin film SnO_2 - based gas sensor: Film thickness influence, Sensors and Actuators B 142, pp. 321-330, 2009.

[7] J. F. Chang, H. H. Kuo, I. C. Leu, M. H. Hon, The effects of thickness and operation temperature on ZnO: Al thin films CO gas sensor, Sensors and Actuators B 84, pp. 258-264, 2002.

[8] K. Vijyalakshmi, K. Karthick, P. Deepak Raj, M. Sridharan, Influence of thickness of MgO overlayer on the properties of ZnO thin films prepared on c-pale sapphire for H_2 sensing, Ceramic International 40, pp. 827-833, 2014.

[9] J. McCarthy: Ozone Air Quality Standards. Congressional Research Service 2010.

[10] Information on http://www.epa.gov/air/criteria.html

[11] R. Ferro, The effects of the material morphology on the responses of the NO_2 sensor based on ZnO thin film, Sensors and Actuators B 143, pp. 99-102, 2009.

[12] J. F. Chang, H. H. Kuo, I. C. Leu, M. H. Hon, The effects of thickness and operation temperature on ZnO: Al thin films CO gas sensor, Sensors and Actuators B 84, pp. 258-264, 2002.

[13] K. Kakati, S. H. Jee, S. H. Kim, J. Y. Oh, Y. S. Yoon, Thickness dependency of sol – gel derived ZnO thin films on gas sensing behaviors, Thin Films 519, pp. 494-498, 2010.

[14] J. Courbat, D. Briand, L. Yue, S. Raible, N. F. de Rooij, Drop coated metal oxides gas sensor on polyimide foil with reduced power consumption for wireless applications, Sensors and Actuators B 161, pp. 862-868, 2012.

Temperature Study of High-Drive Capability Buffer for Phase Change Memories

A. Kiouseloglou*†, E. Covi†‡, G. Navarro*, A. Cabrini†, L. Perniola* and G. Torelli†

*CEA-LETI, MINATEC Campus, 17 rue des Martyrs, F-38054 GRENOBLE Cedex 9, France.
†Dipartimento di Ingegneria Industriale e dell'Informazione, University of Pavia, via Ferrata 1, 27100 Pavia, Italy.
‡present address Laboratorio MDM, IMM - CNR, Via C. Olivetti 2, 20864 Agrate Brianza (MB), Italy.
Email: athanasios.kiouseloglou@cea.fr

Abstract—**Phase Change Memory (PCM) is a non-volatile memory technology with wide programming window and continuously improving data retention performance. In order to drive the variable load PCM exhibits, the amplifier providing programming pulses to the cell must be able to accurately control pulse parameters. In this paper, we present a unity gain buffer capable of driving resistive loads varying up to three orders of magnitude. The buffer can replicate voltage pulses of amplitude up to 4.5 V with minimum rise and fall times. The study of the circuit behaviour in high-temperature environments demonstrates its accuracy over a temperature range from -50 °C up to 200 °C, enabling programming of PCM based on innovative materials in applications requiring reliable operation at high temperatures.**

I. Introduction

Phase Change Memory (PCM) is emerging as one of the most promising technologies to replace DRAM and Flash memories in large-scale memory systems. Due to the advantages it demonstrates, such as high write speed, good scalability, and compatibility with current silicon-based fabrication, PCM offers the potential to enable new applications and memory architectures in a great variety of systems [1].

PCM technology is based on the reversible transition of a chalcogenide alloy, usually $Ge_2Sb_2Te_5$ (GST), between an amorphous, high-resistivity phase and a crystalline, low-resistivity one. The difference in resistivity between the two phases, which can extend up to three orders of magnitude, is the basis for non-volatile storage. Voltage pulses can be used to perform write and read operations. A voltage pulse with a large amplitude and an abrupt fall is applied to the cell to bring it to its high-resistance (RESET) state. To bring the cell to its low-resistance (SET) state, a voltage pulse of intermediate amplitude is applied. Threshold switching of the cell, which consists in an abrupt resistance decrease when the applied voltage exceeds a given value, is a key phenomenon during SET operation, as it allows a sufficient current flow through the cell [2]. Finally, read voltage must be kept within a safe value in order not to disturb the programmed state of the cell [3].

The experimental study of the behaviour of the PCM cells under different programming conditions has led to optimization of cell architecture and materials in order to extend the data retention performance of the technology [4], [5]. In particular, the enhanced data retention performance demonstrated by PCM cells based on materials alternative to GST enables the use of this technology in high-temperature environments, such as automotive or aerospace applications, where accurate programming is of utmost importance [6].

Since the final resistance state of a PCM cell depends on the applied programming pulse [7], programming pulse parameters should be optimized in order to address final applications, and an accurate pulse control is necessary over the whole required temperature range. This is even more true during the research and development phase, when different cell architectures and materials are investigated in order to optimize PCM technology for specific applications.

When it comes to programming a PCM device, the wide change in cell resistance must be taken into account during the design phase, while it is essential that programming pulses are applied to the cell with sufficient accuracy. In this aspect, an analog voltage buffer, which is a key building block in mixed signal designs, should be able to provide adequate current to enable an optimum programming of the cell. It is thus evident that the buffer that drives the memory cell plays a major role in the final resistance state of the cell.

A highly desirable approach during the research phase conducted by research centers specializing in the development of memory cell architectures and materials is to integrate the programming pulse generator and the memory cells to be investigated in two different chips. This approach makes it possible to thoroughly analyse memory cells fabricated with specific technologies that do not include CMOS front-end manufacturing steps. In this case, the two chips including the pulse generator and the memory cells, respectively, will be assembled on a printed-circuit board with chip-on-board (COB) techniques, and the capacitive load to be driven by the buffer can be on the order of several pF.

This paper presents a buffer developed for the above purpose. In Section II, the buffer architecture is outlined. In Section III, simulation results of the buffer performance at room temperature are demonstrated. In Section IV, the behaviour of the buffer over a wide temperature range is analyzed. Finally, conclusions are drawn in Section V.

II. Buffer Architecture

The buffer must be capable of reproducing voltage pulses with particular specifications. In order to bring a memory cell to its high-resistance state (RESET programming), the buffer must be able to feed a highly variable resistive load (R_L ranging from 10 kΩ to 10 MΩ, which correspond to a typical resistance level of a cell in its SET and RESET state, respectively) with pulses having an amplitude V_{ampl} up to 4 V, a width t_w down to 50 ns, and a fall time t_f down to 15 ns. For programming the cell to its low-resistance state (SET programming), pulse specifications are

978-1-4799-4993-9/14 $31.00 © 2014 IEEE
411

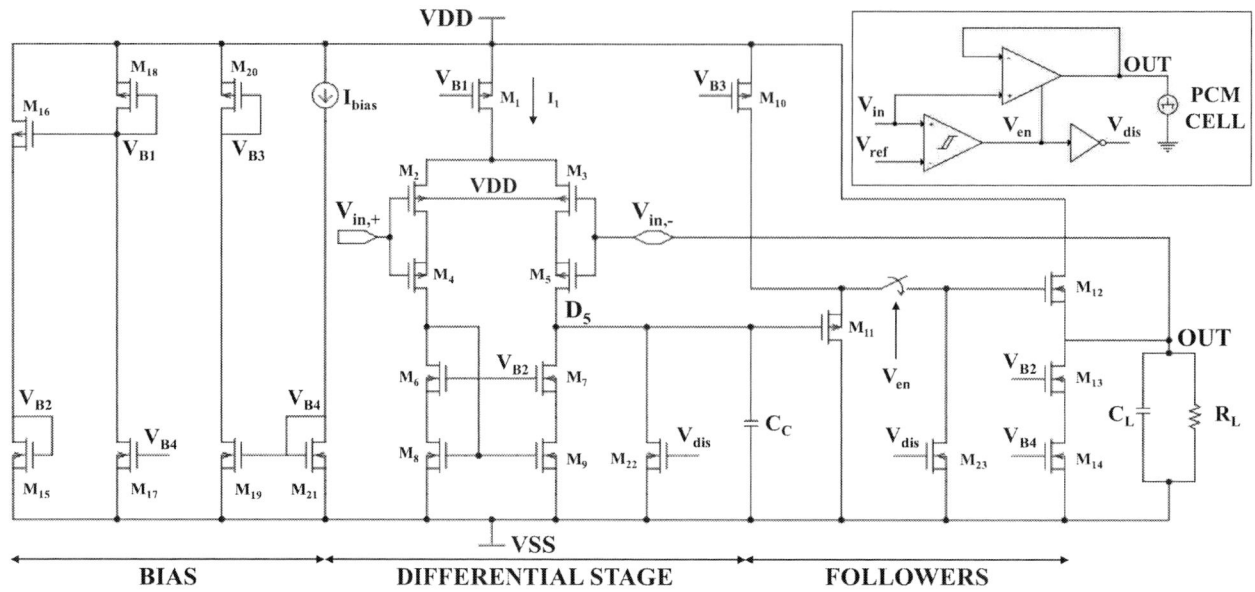

Fig. 1. Circuit schematic of the proposed unity gain buffer. The buffer is enabled when signal V_{en} is high. When input signal V_{in} is lower than reference oltage V_{ref} (set to 200 mV in our case), a comparator drives V_{en} low, thus disabling the buffer, as can be seen in the inset.

TABLE I.   COMPONENT SIZES

Transistor name	$\frac{W}{L}$ [$\frac{\mu m}{\mu m}$]	Transistor name	$\frac{W}{L}$ [$\frac{\mu m}{\mu m}$]
M_1, M_{16}, M_{20}	$\frac{40}{0.8}$	M_{11},M_{12},M_{13},M_{14}	$\frac{60}{0.8}$
M_2, M_3, M_4, M_5	$\frac{32}{0.8}$	M_{15}, M_{17}, M_{19}	$\frac{4}{0.8}$
M_6, M_7, M_8, M_9	$\frac{20}{0.8}$	M_{21}	$\frac{8}{0.8}$
M_{10}, M_{18}	$\frac{80}{0.8}$	M_{22}, M_{23}	$\frac{0.8}{0.8}$
C_L	5 pF	C_C	500 fF

less stringent. A SET pulse is a wider voltage pulse with a wer amplitude and larger rise and fall times compared to a RESET pulse. An error of ± 2.5% between the buffer input and output voltage amplitudes is considered adequate for the target application. Overshoots and ringing transients in the generated program pulse must be minimized so that the final state of the programmed cell is not affected, especially in the case of SET programming [3]. Moreover, the voltage provided to the memory cell should be controlled at least down to a safe oltage of 400 mV, which is not able to (unintentionally) alter the programmed cell resistance state [8]. The capacitive load v be driven was set to 5 pF in order to take into account the overall interconnection capacitance (including two pads) present in the case of the above mentioned two-chip COB application. A supply voltage V_{DD} of 6 V was chosen to deal with the wide input/output voltage range required of the buffer. Therefore, High-Voltage (HV) devices were used. Bias current as was set to 400 µA. In our application, this current can provided externally and is kept substantially constant over temperature, for instance by using a resistor with zero power coefficient. It is worth pointing out that power consumption of the amplifier is not a concern in our case.

The buffer schematic is based on a previously proposed topology (Fig. 1) [9]. The buffer is made up of the cascade a differential input stage, a p-type follower, and an n-type follower. The input stage consists of a PMOS cascode differential pair (devices M_2 to M_5 biased by current source

M_1) and an NMOS active load (devices M_6 to M_9), which performs differential to single-ended conversion. The p-type differential stage allows the minimum required input common mode level specifications to be met. The chosen supply voltage allows operation of the buffer with input voltages up to 4.5 V.

The cascode branch made up by devices M_3 and M_5 takes advantage of the threshold voltage difference between the two transistors due to body effect: saturated operation of both devices is ensured over the whole allowed input common-mode range even though their gates are biased with the same voltage (V_{in}^-) [10]. The same topology is used for the complementary branch M_2, M_4 for matching purposes.

The output n-type follower (transistor M_{12} biased by cascode current source M_{13}, M_{14}) allows meeting the programming current requirements (up to 0.45 mA) at the buffer output. The PMOS follower (transistor M_{11} biased by current source M_{10}) provides a level shift between the voltages at the output of the differential pair and the input of the final follower, so as to ensure saturated operation of the differential pair even when the buffer is connected in non-inverting unity-gain configuration. The branches at the left side of the differential stage provide biasing voltages of the amplifier.

In order to ensure fast settling of the output pulse and avoid large overshoots and ringing transients, an adequate Unity Gain Frequency (UGF) and a sufficiently large phase margin of the amplifier must be guaranteed. The first and the second pole of the amplifier are associated to the output nodes of the differential stage (node D_5) and the n-type follower (node OUT), respectively. Capacitor C_C provides frequency compensation. The value of current I_1 was determined from the Slew Rate (SR) requirements of the circuit, taking into account that SR = $I_1/C_{out,1}$, where $C_{out,1}$ is the total capacitance associated to node D_5. Transistor sizes of M_3 and M_{12} were chosen in order to meet settling time and stability requirements, assuming dominant-pole approximation.

Signal V_{en} (active high) enables the buffer when input voltage V_{in}^+ has to be applied to the memory cell. When V_{en} is low, the gate of M_{12} is shorted to ground through M_{23}, thus allowing the output node of the buffer to be grounded through M_{13} and M_{14}. As can be seen in the inset of Fig. 1, a comparator drives V_{en} low when the input signal becomes lower than a reference voltage ($V_{ref} = 200$ mV), thus preventing any risk of applying unintentional voltages to the memory cell. Component sizes can be found in Table I. The length of transistor gates was chosen equal for all transistors due to matching purposes, and was set bigger than the minimum allowed value for HV devices (L_{min} equal to 700 nm and 500 nm for nMOS for pMOS transistors, respectively).

III. SIMULATIONS

The buffer presented in Section II was designed in 180 nm CMOS technology featuring HV devices. In order to evaluate the performance of the circuit at room temperature, simulations with a load capacitance C_L of 5 pF and load resistances R_L of 10 kΩ and 10 MΩ were performed. The input DC voltage was set at mid-supply for the open-loop gain magnitude and phase simulations of the amplifier.

From the Bode diagrams in Fig. 2, the DC gain is 49.2 dB for $R_L = 10$ kΩ and 50.4 dB for $R_L = 10$ MΩ. UGF ranges from 48.4 MHz ($R_L = 10$ kΩ) to 49.2 MHz ($R_L = 10$ MΩ), while the phase margin ranges from 65.5° ($R_L = 10$ MΩ) to 66° ($R_L = 10$ kΩ). The gain of the amplifier in closed-loop buffer configuration is -27 mdB ($A_{CL,Buffer} = A_{OL}/(A_{OL}+1)$).

Fig. 3 demonstrates the transient simulations of a RESET pulse ($V_{ampl} = 4$ V, $t_w = 50$ ns, $t_f = 15$ ns) applied to the load. Since the rise time t_r of the voltage pulse does not affect the final resistance state, its value was set to 15 ns in all simulations. Fig. 3(a) shows the pulse characteristics and a comparison of the input and the output pulses for a load of $R_L = 10$ MΩ and $R_L = 10$ kΩ. Fig. 3(b) shows the output pulse ($R_L = 10$ kΩ) when the additional inductance due the bonding wire of the COB connection between the buffer and the memory cell to be characterized is taken into account. The overshoot observed is not significantly different than in the case when no inductance is present, as can be seen in greater detail in Fig. 3(c). For this reason, no parasitic inductance will be assumed in the simulations provided in the following. From Fig. 3, it is apparent that the buffer is able to accurately reproduce the input voltage (the error between the input and the amplitude voltage is smaller than $\pm 1\%$). The use of a capacitor C_C equal to 500 fF enables fast rise and fall times of the reproduced pulse, with a voltage overshoot below 2% (1.6% for $R_L = 10$ kΩ) of the pulse voltage amplitude.

IV. HIGH-TEMPERATURE STUDY

It has been recently demonstrated that N-doped Ge-rich GST is capable of retaining stored information beyond the soldering reflow temperature, preserving a resistance difference of at least one order of magnitude [5], opening the way for high-temperature applications, which could utilize PCM technology based on innovative materials. For this reason, we decided to study the behaviour of the proposed buffer in a wide temperature range, extending from -50 °C to 200 °C. The reliability of the simulation models is guaranteed for

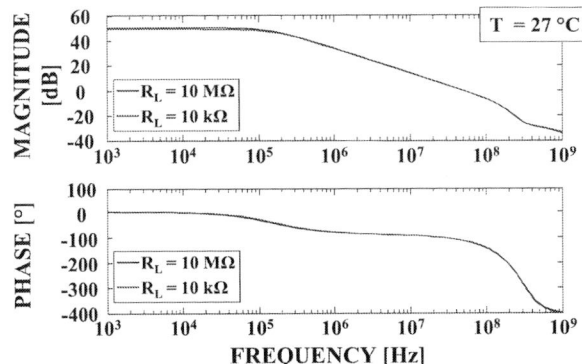

Fig. 2. Open-loop frequency response of the designed amplifier at room temperature.

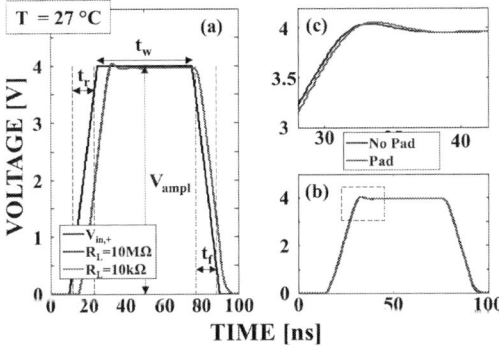

Fig. 3. Transient response of the buffer at room temperature. (a) Pulse characteristics and comparison of input and output voltage pulses for $R_L = 10$ kΩ and $R_L = 10$ MΩ. (b) Simulation of the output pulse ($R_L = 10$ kΩ) when an additional inductive load due to wire bonding is included. The response within the dashed rectangle in (b) can be seen in greater detail in (c).

temperatures ranging from -100 °C up to 200 °C by the chip manufacturer.

The frequency response of the buffer as temperature increases, exhibits a decrease in gain and UGF as a result of the decrease in the transconductance g_m of the devices (due to carrier mobility decrease), and the phase delay curve is shifted towards high frequencies due to the corresponding shift of the dominant pole (the output impedance at node D_5 is proportional to g_m). The overall effect is a better phase margin, which ensures an enhanced stability of the amplifier. As a matter of fact, it can be observed in Fig. 4 ($R_L = 10$ kΩ) and

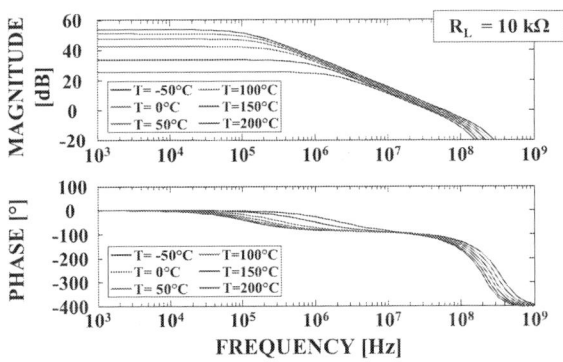

Fig. 4. Open-loop frequency response of the amplifier in a temperature range from -50 °C to 200 °C for $R_L = 10$ kΩ.

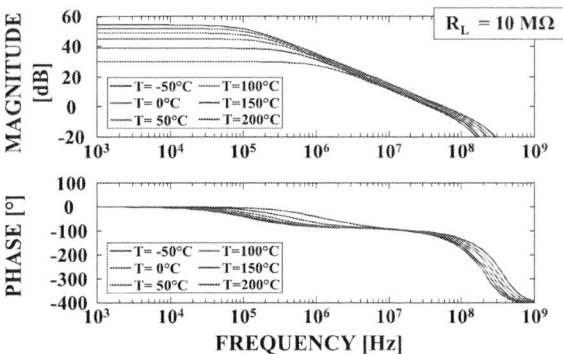

Fig. 5. Open-loop frequency response of the amplifier in a temperature range from -50 °C to 200 °C for R_L = 10 MΩ.

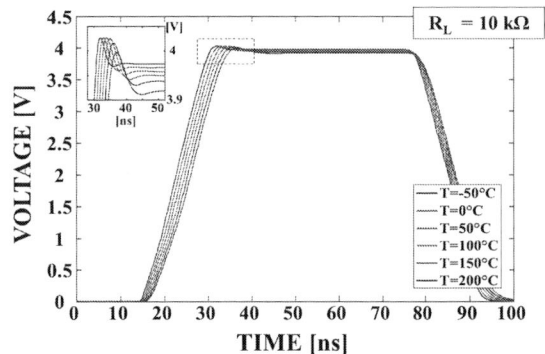

Fig. 6. Buffer transient response in a temperature range from -50 °C to 200 °C for a RESET pulse applied to a load resistor R_L = 10 kΩ. The overshoot can be seen in greater detail in the inset of the Figure.

Fig. 5 (R_L = 10 MΩ), that the open-loop DC gain of the buffer decreases (30 dB for R_L = 10 MΩ and 25 dB for R_L = 10 kΩ at 200 °C) and the decrease of the UGF combined with a shifted phase response gives rise to a better phase margin. R_L = 10 kΩ represents the heaviest load condition for the circuit, and will thus be considered in the transient analysis in the rest of the paper.

As can be seen in Fig. 6, increasing temperature slightly increases the rise time of the output voltage pulse (the input voltage pulses were considered stable in temperature and identical to the ones described in Section III) as well as the overshoot, which is kept within 2% of the final output voltage, and is therefore not a concern during cell programming [11].

The reduced pulse amplitude, due to the decrease of the DC gain, is kept within target specifications. The increase in the pulse fall time is smaller when compared to the rise time increase, and is still within the desired limit, thus allowing fast quenching of the phase change material. The consequent pulse width decrease, especially at high temperatures, is negligible and still allows inducing the melting of the active chalcogenide volume of the PCM cell. Thus, even though the DC gain decreases over temperature, the applied RESET pulse is still capable of bringing the cell to its High-Resistance State. Similar results were obtained by simulating a SET pulse.

In order to study the behaviour of the buffer at different temperatures in detail, simulations were performed in Cadence environment from -50 °C to 200 °C with a step of 10 °C. As far as the DC gain of the buffer is concerned (Fig. 7(a)), it is worth noting that while an almost linear decrease takes place

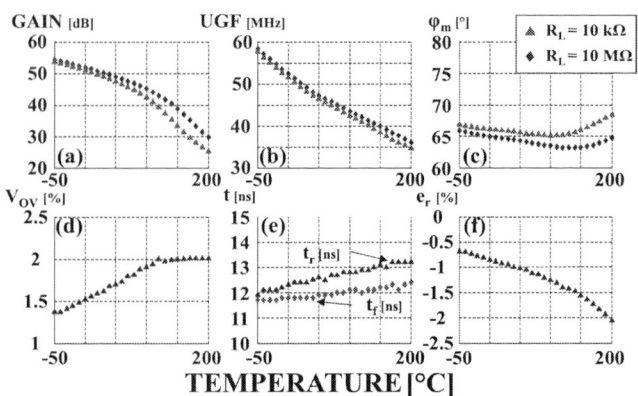

Fig. 7. Simulation results of the behaviour of (a) gain, (b) UGF, (c) phase margin, (d) overshoot, (e) pulse rise and fall time and (f) input-to-output error as temperature increases from -50 °C to 200 °C.

for both resistive loads up to 150 °C, a more abrupt decrease is observed when exceeding this temperature. The UGF of the amplifier shows an almost linear decrease for increasing temperatures (Fig. 7(b)). The phase margin (Fig. 7(c)) increases at high temperatures, whereas it shows a slight decrease at lower temperatures. Nonetheless, for both resistive loads, the phase margin never decreases below 62°, ensuring stability even at the extreme temperatures under study.

Regarding the transient characteristics of the buffer, the overshoot V_{OV} for R_L = 10 kΩ when a RESET pulse is applied (Fig. 7(d)), increases with temperature to nearly 2% whereas it remains substantially constant for temperatures above 150 °C. The rise and fall times of the pulse applied to the load (Fig. 7(e)) demonstrate a slight, linear increase with temperature. Last but not least, the input-to-output error e_r (Fig. 7(f)) ranges between -0.5% (T = -50 °C) and -2% (T = 200 °C), always remaining within the limits for successful PCM programming operation.

V. CONCLUSION

In this paper, we presented a unity gain buffer amplifier capable of providing adequate voltage pulses to resistive loads varying up to three orders of magnitude. The buffer behaviour has been investigated through Cadence simulations, first at room temperature and later on, for temperatures exceeding the usual operating temperature ranges. The accuracy and the stability of the proposed circuit demonstrate its potential to be used in automotive and aerospace applications, enabling an optimum programming of PCM cells even at high temperatures.

REFERENCES

[1] B. De Salvo, et al., *Int. Symp. on VLSI Tech. Syst. Appl.*, pp.1-2 (2012).

[2] A. Pirovano, et al., *IEEE Trans. Electron Devices*, vol.51, no.3, pp.452-459, Mar. 2004.

[3] G.W. Burr, et al., *J. Vac. Sci. Technol.*, vol.28, pp.223-262, Mar. 2010.

[4] L. Perniola, et al., *IEEE IEDM*, pp.18.7.1-18.7.4 (2012).

[5] G. Navarro, et al., *IEEE IEDM*, pp.21.5.1-21.5.4 (2013).

[6] A.E.I. Mehdi, et al., *Int. Conf. on High Temp. Electron.* (2006).

[7] S. Braga, et al., *PRIME*, pp.108-111 (2009).

[8] A.L. Lacaita, *SISPAD*, pp.267-270 (2005).

[9] E. Covi, et al., *ICICDT*, pp.127-130 (2013).

[10] D. Devecchi et al., *U.S. Patent*, 4 952 885, Aug. 28, 1990.

[11] E. Covi, et al., *IEEE Trans. on Semicond. Manuf.*, vol.27, no.2, pp.134-150, May 2014.

Large Bandwidth Tunable Analog Equalizers Based on an InP DHBT Differential Pair Amplifier Cell for 100-GBaud Communication Systems

R. Mettetal[1,2], J.-Y. Dupuy[2], A. Ouslimani[1] and J. Godin[2]

[1]ECS-Lab, ENSEA, Avenue du Ponceau, 95014 Cergy-Pontoise, France
[2]III-V Lab, a joint lab between 'Alcatel-Lucent Bell Labs France', 'Thales Research and Technology' and 'CEA Leti'
Route de Nozay, 91460 Marcoussis, France
ronan.mettetal@3-5lab.fr

Abstract—**We report the design, simulations and measurements of two tunable analog equalizers. These circuits have been realized using indium phosphide heterojunction bipolar transistors reaching f_T/f_{max} of 370/340 GHz. These analog equalizers provide a bandwidth of more than 100 GHz, depending on the settings. A varactor is used to tune the frequency response to equalize various types of 100-GBaud communication system channels.**

Keywords—equalization; optical communications; InP DHBT; differential pair amplifier cell; varactor

I. INTRODUCTION

With the continuous growth of data communications, a bandwidth bottleneck is progressively being reached [1-2]. Through last years, rapid innovation in technology has permitted to keep pace with the Internet demand, but the linear evolution of the technology does not match anymore the exponential growth of data traffic. The bandwidth limitation of current fiber optic interfaces leads to intersymbol interference (ISI) hence a restriction of data rates. One of the possible solutions is to use an equalizer to improve the bandwidth of the whole channel thanks to high frequency peaking. Many equalizers have been developed, mainly using CMOS technology to achieve low power consumption and to make them easily tunable [3-5]. However, the main drawback of standard CMOS technologies is limited f_T/f_{max}, and as a result bandwidth limitation. Moreover, to our knowledge, state-of-the-art equalizer circuits based on BiCMOS technologies have not yet addressed the 100-GHz-class bandwidth [6-9] which is considered here.

In this paper, we demonstrate two differential pair amplifier cells targeting 100-GBd-class operation. These circuits are fabricated in a 0.7-μm indium phosphide (InP) double heterojunction bipolar transistor (DHBT) technology with f_T/f_{max} reaching 370/340 GHz, achieving a maximal bandwidth of more than 110 GHz, and a maximal peaking of 5 dB at 75 GHz.

This paper is organized as follows. Section II briefly presents the InP DHBT technology used. The first part of the section III describes the design of the varactor used for tunability, the second part is dedicated to the design of the differential pair amplifier cells, and the last part shows the simulation results. On wafer measurements results are presented in section IV. Conclusions are drawn in section V.

II. InP DHBT TECHNOLOGY

The tunable analog equalizers were designed using III-V Lab's 0.7-μm InP DHBT technology [10]. This technology provides 5-, 7- and 10-μm emitter-length transistors. Figure 1 shows the f_T/f_{max} of the transistors in function of collector current at a collector-emitter voltage (V_{CE}) of 1.6 V, for the three emitter lengths.

III. TUNABLE ANALOG EQUALIZERS DESIGN

The main reasons why differential implementation has been adopted are to provide superior insensitivity to power supply parasitics and to double the available swing at the output. The two analog equalizers are composed of a differential pair amplifier cell with a matching network composed of a single pair of emitter followers.

A. Differential Pair Amplifier Cell (DPAC) Design

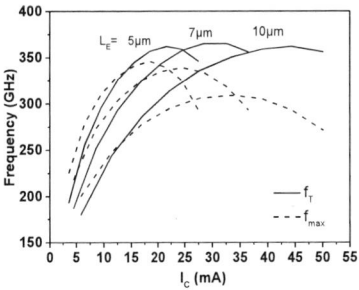

Figure 1. f_T/f_{max} versus collector current of 5-, 7- and 10-μm emitter-length transistors

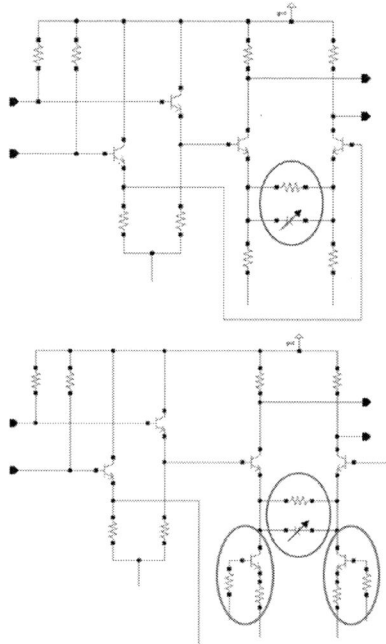

Figure 2. Two analog equalizer architectures

The difference between the two equalizers is the biasing of the differential pair. In the first one, the biasing is done through a resistor with a negative supply voltage, while in the second one, it is made through a transistor-based current source to ensure a maximal and constant current at the collector of the differential pair amplifier cell as well as to get a better common mode rejection ratio.

As shown in figure 2, the DPAC design uses emitter degeneration by a RC filter where the capacitor is a varactor to make the circuits' frequency response tunable. The varactor acts as a shortcut of the resistor at high frequency. The value of the resistor sets the gain of the circuit at low frequency; the varactor has been designed to have sufficient peaking at high frequency for all the possible settings. The design of the varactor is described in details in the next part.

Input broadband matching is done with a 50-Ω resistor in parallel with the high-impedance input emitter follower; at the output, it is done with the 50-Ω collector load resistor.

B. Varactor

Figure 3 shows the varactor used, which is composed of two 10-μm emitter-length transistors in order to have a large capacitive value; collectors and emitters of the two transistors

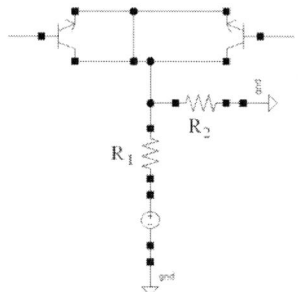

Figure 3. Schematic of the varactor

are interconnected and short-circuited to form a single node. The equivalent capacitance is seen between the two bases and the biasing is done through a DC supply with a matching network. It is more complex to get a pure variable capacitance with InP DHBT than with CMOS technologies, because of the non-negligible variation of the junction resistance [11].

The equivalent capacitance is modified by applying a voltage between collector and base. As a result, we modify the depletion and the diffusion capacitances of the two PN junctions (emitter-base and collector-base of the two transistors). The drawback of this technique is that we also modify the junctions' resistance; hence the equivalent capacitance is strongly degraded at high frequency.

We inserted a high impedance between the DC supply and the varactor as a matching network. This impedance is a voltage divider designed to have optimum results up to the maximal possible negative DC supply; in our case this negative voltage is -6 V due to the decoupling capacitor of the measurement equipment. The collector voltage is dependent on the ratio between the 2 resistors. The values 700 Ω and 3200 Ω, respectively for R_1 and R_2, have been chosen for two reasons: layout implementation compromise and parasitic impedance optimization between collector and ground.

In the final design, we inserted in the two circuits three varactors in parallel in order to have sufficient peaking at high frequency as required.

C. Simulation Results

The simulations were performed on Cadence Virtuoso® on which a model of III-V Lab's 0.7-μm InP DHBT devices has been implemented.

Electrical interconnections effects are taken into account in the simulations, but not the coupling between interconnections. Figure 4 and 5 shows the S-parameters results for five values of varactor DC supply. There are four combinations of input and output return loss because of the differential implementation, but the variation between the 2 input or output ports is negligible. Figure 6 shows transient differential results with two values of varactor DC supply. We can clearly observe the pre-emphasis effect on the simulated 100-Gb/s eye diagrams.

With the transient simulation, we notice the difference of amplitude between the two values of varactor DC supply, which is confirmed watching mixed-mode differential S-parameters simulation results. Indeed, the gain is higher at low frequency with a varactor DC supply of -6 V than 0 V. This difference of gain can be explained if we take into account the varactor behavior. When the potential between the collectors and the bases is higher, the current increases. This action reduces the dynamic resistance of the junctions. This phenomenon implies that the total value of the degeneration resistance decreases, which leads to an increasing gain at low frequency.

978-1-4799-4993-9/14 $31.00 © 2014 IEEE

IV. ANALOG EQUALIZERS MEASUREMENT

A microphotograph of the fabricated analog equalizers is shown in Figure 7. Both chips' size is 1.5 x 1.2 mm².

On wafer S-parameters measurements were performed with an Anritsu VectorStar® network analyzer, from 70 kHz to 110 GHz. We directly process the results to find the differential and the common-mode transmission parameters, respectively referenced to as S_{dd21} and S_{cc21} [12-13] :

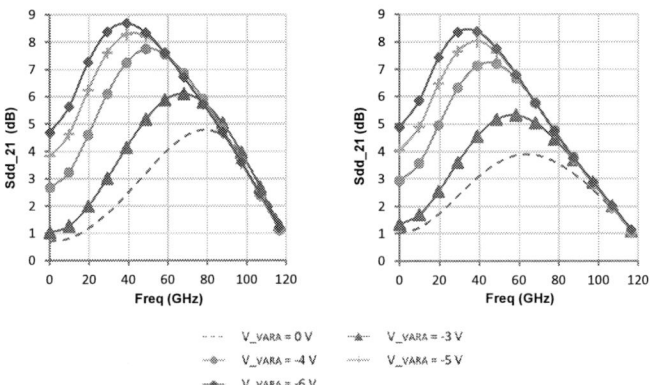

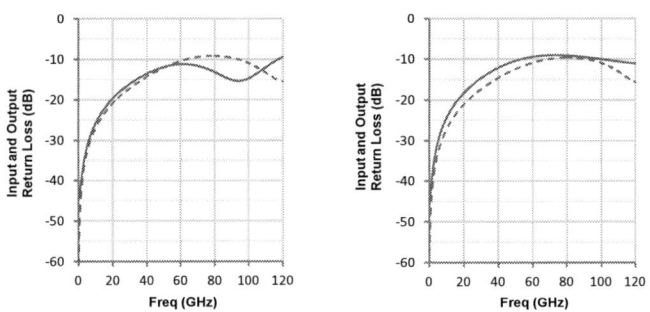

Figure 4. DPAC with a resistor biasing (left) and with a transistor biasing (right) mixed-mode differential S-parameters simulation results

Figure 5. Input (dash) and output (solid) return loss of the DPAC with resistor biasing (left) and with a transistor biasing (right)

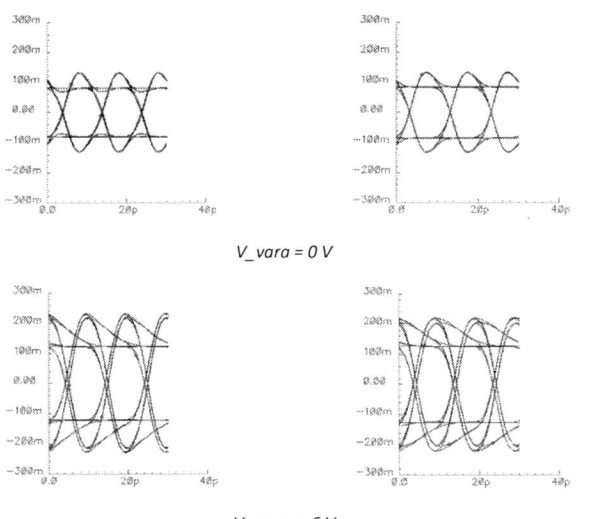

Figure 6. Output differential transient simulation results for DPAC with a resistor biasing (left) and with a transistor biasing (right) with visible pre-emphasis effect at 100-Gb/s

$$S_{dd21} = \frac{1}{2} \cdot (S_{21} - S_{23} - S_{41} + S_{43}) \tag{1}$$

$$S_{cc21} = \frac{1}{2} \cdot (S_{21} + S_{23} + S_{41} + S_{43}) \tag{2}$$

Figure 8 and 9 present the results for the two circuits. It clearly appears that the differential pair amplifier cell with the biasing made through a transistor presents a better common mode rejection ratio over the whole bandwidth.

Moreover and as explained in the previous section, with a sufficient varactor DC supply, these analog equalizers can also amplify a signal while equalizing.

We state that there is an input return loss difference between the 2 input ports for the DPAC with biasing resistor (fig. 9). This is due to a difference of layout implementation after the 2 input ports which is caused by a compromise of DC supply ports on the circuit.

We get various values of peaking because of the non ideal variable capacitance of the varactor; there is a variation from a peaking of 5 dB at 75 GHz to a peaking of 2 dB at 40 GHz for the resistor biasing and from a peaking of 3 dB at 66 GHz to a peaking of 1.5 dB at 40 GHz for the transistor biasing.

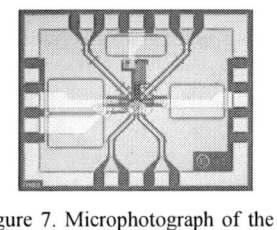
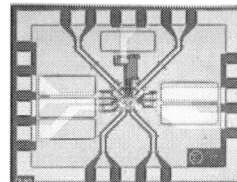

Figure 7. Microphotograph of the analog equalizers with a resistor biasing (left) and with a transistor biasing (right)

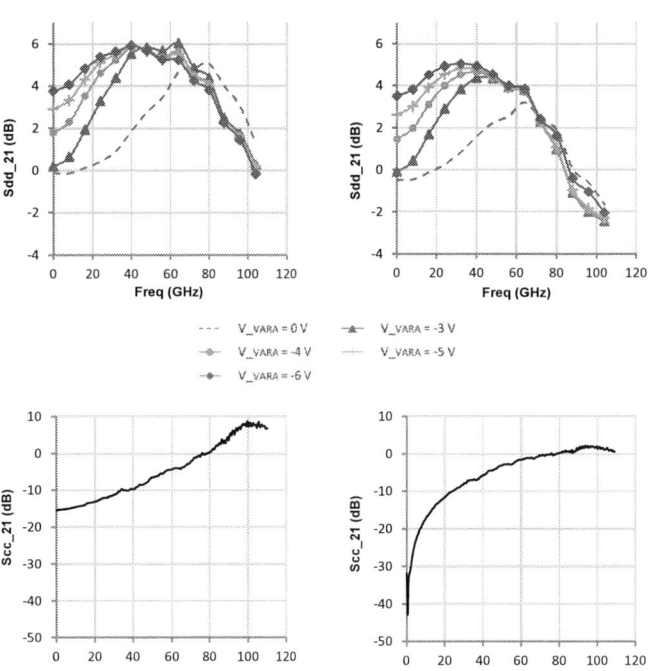

Figure 8. Resistor biasing (left) and transistor biasing (right) mixed-mode differential (up) and common (down) S_{21}

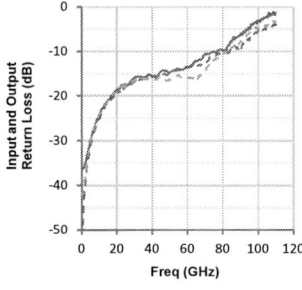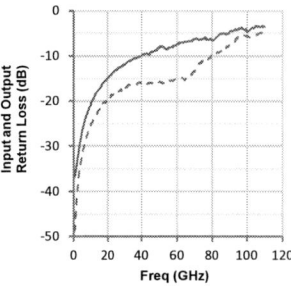

Figure 9. Input (dash) and output (solid) return loss of the DPAC with resistor biasing (left) and with transistor biasing (right)

We clearly observe differences between simulations and measurements. The coupling between interconnections which is not taken into account in the simulations could explain those differences. Better results in terms of peaking and bandwidth are obtained with DPAC with biasing resistor at the cost of worse common mode rejection ratio.

V. CONCLUSION

In this paper, we presented the design, simulations and measurements of two tunable analog equalizers with a large bandwidth. The tunability of the circuit allows equalization for various types of channels. Moreover, the simulations and the measurements demonstrated an amplification capability. Future steps include assessing the equalization effect on a filtered eye diagram with transient measurements. Another step will be to try to improve the common mode rejection ratio over a 100-GHz bandwidth.

ACKNOWLEDGMENTS

The authors want to thank V. Nodjiadjim, III-V Lab, for transistor characterization, B. Saturnin, III-V Lab, for epitaxy, O. Drisse, III-V Lab, for e-beam lithography, M. Riet, P. Berdaguer and H. Aubry, III-V Lab, for process, F. Jorge, III-V Lab, for his help with on wafer measurements, A. Konczykowska, III-V Lab, for fruitful discussions and T. Johansen from Technical University of Denmark, for transistor modeling.

REFERENCES

[1] "Cisco Visual Networking Index: Global Mobile Data Traffic Forecast Update, 2013-2018," Cisco White Paper, (http://www.cisco.com/c/en/us/solutions/collateral/service-provider/visual-networking-index-vni/white_paper_c11-520862.pdf), February 5, 2014.

[2] Weller, D. and B. Woodcock (2013), "Internet Traffic Exchange: Market Developments and Policy Challenges," OECD Digital Economy Papers, No. 207, OECD Publishing.

[3] A. Momtaz and M. M. Green, "An 80 mW 40 Gb/s 7-Tap T/2-Spaced Feed-Forward Equalizer in 65 nm CMOS," IEEE Journal of Solid-State Circuits, vol. 45, No. 3, March 2010.

[4] R. A. Aroca, P. Schvan and S. P. Voinigescu, "A 2.4-Vpp 60-Gb/s CMOS Driver With Digitally Variable Amplitude and Pre-Emphasis Control at Multiple Peaking Frequencies," IEEE Journal of Solid-State Circuits, vol. 46, No. 10, October 2010.

[5] J. F. Bulzacchelli, C. Menolfi, T. J. Beukema, D. W. Storaska, J. Hertle, D. R. Hanson, H. Ping-Hsuan, S. V. Rylov, D. Furrer, D. Gardellini, A. Prati, T. Morf, V. Sharma, R. Kelkar, H. A. Ainspan, W. R. Kelly, L. R. Chieco, G. A. Ritter, J. A. Sorice, J. D. Garlett, R.

Callan, M. Brandli, P. Buchmann, M. Kossel, T. Toifl and D. J. Friedman, "A 28-Gb/s 4-Tap FFE/15-Tap DFE Serial Link Transceiver in 32-nm SOI CMOS Technology," IEEE Journal of Solid-State Circuits, vol. 47, No. 12, December 2012.

[6] H. Wu, J. A. Tierno, P. Pepeljugoski, J. Schaub, S. Gowda, J. A. Kash and A. Hajimri, "Integrated Transversal Equalizers in High-Speed Fiber-Optic Systems," IEEE Journal of Solid-State Circuits, vol. 38, No. 12, December 2003.

[7] A. Hazneci and S. P. Voinigescu, "A 49-Gb/s, 7-tap transversal filter in 0.18-μm SiGe BiCMOS for backplane equalization," in Proc. IEEE Compound Semiconductor Integrated Circuit Symposium, Monterey, CA, Oct. 2004, pp. 101-104.

[8] R. A. Aroca, S. P. Voinigescu, "A Large Swing, 40-Gb/s SiGe BiCMOS Driver With Adjustable Pre-Emphasis for Data Transmission Over 75 Ω Coaxial Cable," IEEE Solid-State Circuits, vol. 43, No. 10, October 2008.

[9] I. Sarkas, S. P. Voinigescu, "A 1.8 V SiGe BiCMOS Cable Equalizer with 40-dB Peaking Control up to 60 GHz," in Proc. IEEE Compound Semiconductor Integrated Circuit Symposium, La Jolla, CA, Oct. 2012, pp. 1-4.

[10] J. Godin, V. Nodjiadjim, M. Riet, P. Berdaguer, O. Drisse, E. Derouin, A. Konczykowska, J. Moulu, J.-Y. Dupuy, F. Jorge, J.-L. Gentner, A. Scavennec, T. Johansen and V. Krozer, "Submicron InP DHBT Technology for High-Speed High-Swing Mixed-Signal for ICs," in Proc. IEEE Compound Semiconductor Integrated Circuits Symposium, Monterey, CA, Oct. 2008.

[11] S. Lucyszyn, I. D. Robertson and A. H. Aghvami, "Microwave Modelling of Varactor Diodes Fabricated Using Heterojunction Based Technologies," in Proc. IEEE Int. Workshop High-Performance Electron Devices Microwave Optoelectron. Applicat., London, U.K, Oct. 1993.

[12] D. E. Bockelman, W. R. Eisenstadt, "Combined Differential and Common-Mode Scattering Parameters : Theory and Simulation," IEEE Transactions on Microwave Theory and Techniques, vol. 43, No. 7, July 1995.

[13] W. Fan, A. C. W. Lu, L. L. Wai and B. K. Lok, "Mixed-mode S-parameter characterization of differential structures," Electronics Packaging Technology, 2003 5th conference, pp. 533-537

978-1-4799-4993-9/14 $31.00 © 2014 IEEE

Low Power Inductor-less CML Latch and Frequency Divider for Full-Rate 20 Gbps in 28-nm CMOS

Laszlo Szilagyi, Guido Belfiore, Ronny Henker, Frank Ellinger
Dresden University of Technology
Chair for Circuit Design and Network Theory
Dresden, Germany
Laszlo.Szilagyi@tu-dresden.de

Abstract—**The design methodology of low power current mode logic (CML) latches is described and the implementation of a D flip-flop (DFF) is presented in 28 nm CMOS technology. The DFF can work up to 22 Gbps full-rate with a bit error rate better than 10^{-12} and with a power consumption of only 880 µW. Since the circuit is inductor-less the area of the circuit is only 25 µm × 10 µm. As a further implementation of the CML latches a very low power static frequency divider with quadrature outputs in 28 nm is presented. It divides the clock signal up to 26 GHz and has only 880 µW power consumption. To our knowledge, with 0.034 mW/GHz, this static frequency divider has one of the best figure of merit reported to date.**

Keywords—*current mode logic, D flip-flop, sampling latch, static frequency divider.*

I. INTRODUCTION

Rail-to-rail logic poses difficulties in the tens of gigabits data rate (DR) domains. These impediments are given first of all by the mobility of carriers in PMOS devices, which is four times lower in highly scaled CMOS technologies than that of NMOS transistors. Thus, the output rise and fall times limit the speed of these latches. Further issues are given by the high required voltage swings that complicate the design of the voltage controlled oscillator (VCO). Instead, current mode logic (CML) uses only a small voltage swing and provides gain in the store mode [1].

The CML latch is used in several components of a broadband communication channel such as the phase detector (PD), frequency divider or sampling latch of the clock-data-recovery (CDR) circuit [2][3][4]. The D flip-flop (DFF) consists of two latches: the first one samples while the second one holds the data. Since the DFF is the bottle-neck of the CDR system, parallel architectures such as half-rate, quarter-rate or phase rotator based architectures are continuously developed [2].

The eye diagram in Fig. 1 can be sliced either on the amplitude scale, where the decision is defined by the voltage threshold V_{Sth} or on the time scale where the decision is controlled by the sampling instant t_{Samp} [5]. Since in optical communication systems the samplers are preceded by limiting amplifiers, which provide a high input swing, well above the input sensitivity, the sampling threshold, V_{Sth} is of less concern. The sampling instant is on the other hand the most important parameter when designing sampling latches. The operating speed of the latches is limited by their setup (t_s) and hold (t_h) times. The limited slew rate caused by the parasitic capacitances and resistances result in a decision uncertainty. Thus, the probability of t_{Samp} to be in the decision point where the V_{Sth} is maximum is decreased by the long t_s and t_h. These rise and fall times become dominant above the timing jitter. The clock phase margin (CPM) is defined as the bit time, named unit interval (UI) in the following, minus the jitter at a given (desired) bit error rate (BER). Since the setup and hold time dominate the jitter it can be re-written as CPM= 1 UI - (t_s + t_h). In order to achieve high DRs the two times need to be decreased.

This paper presents a sampling flip-flop designed in 28 nm CMOS which can work up to full-rate DR of 22 Gbps with a BER of less than 10^{-12}. The DFF consumes 880 µW with a supply of 1 V and has 50 Ω matched inputs and outputs. With the additional driver at the output and clock buffers the complete circuit has a power consumption 6 mW.

With modifications from [2] the DFF can work at half-rate clock, achieving a doubled DR. A similar application of the DFF is presented in section II, a static frequency divider (FD) with quadrature outputs (I/Q FD). The FD provides a division up to 26 GHz.

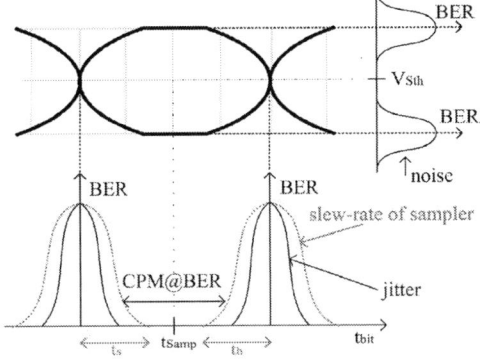

Fig. 1 Output data eye diagram with vertical and horizontal slices.

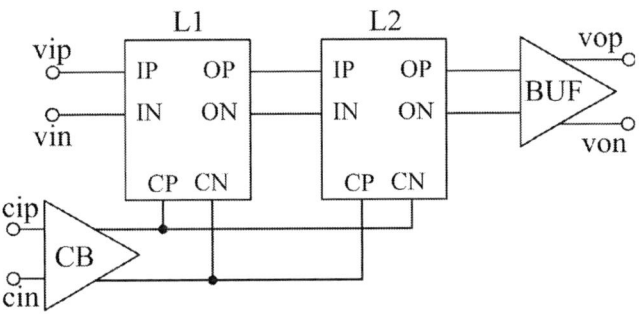

Fig. 2 Block diagram of the DFF.

978-1-4799-4993-9/14 $31.00 © 2014 IEEE

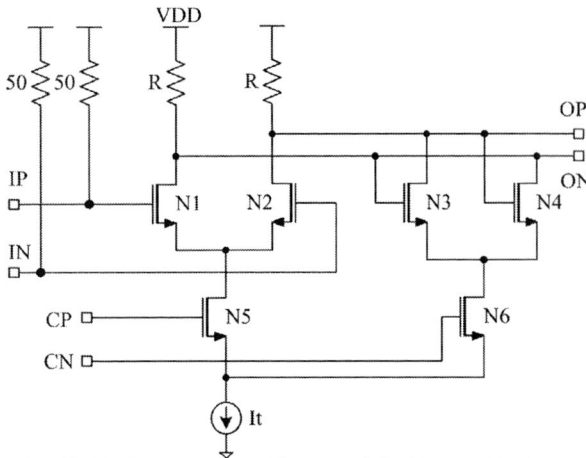

Fig. 3 Schematic of CML latch L1.

II. DESIGN OF THE CIRCUITS

A. The D-Flip-Flop

The block diagram of the DFF chip is presented in Fig. 2. It contains the CML latches L1 and L2, the clock buffer CB and output buffer BUF. These buffers are needed to match the DFF with the measurement equipment on one hand. On the other hand this logic is intended to be used in an on-chip system, therefore it needs to be isolated from the parasitic capacitances of the pads and long transmission lines while testing.

B. The CML Latches

The schematic of L1 is shown in Fig. 3. The difference between L1 and L2 is that the latch at the input needs to be matched to the signal generator. This is done with a 50 Ω resistor connected between I_P and I_N to VDD.

Since the supply voltage is 1 V, the CML voltage levels are chosen to be between VDD and VDD - 0.3 V. The output swing is also limited by the threshold voltage of the NMOS devices ($V_{th,N}$) which is around 400 mV resulting in the need of as high as possible common mode (CM) voltage. In order to meet the output voltage swing requirements and the high CM a load resistor of 700 Ω and a tail current (I_t) of 440 µA is chosen. In order to be able to operate the tracking transistors N1, N2, the hold- transistors N3, N4 in saturation and to allow a $V_{DS,ON}$ voltage drop on the clock switches, the saturation

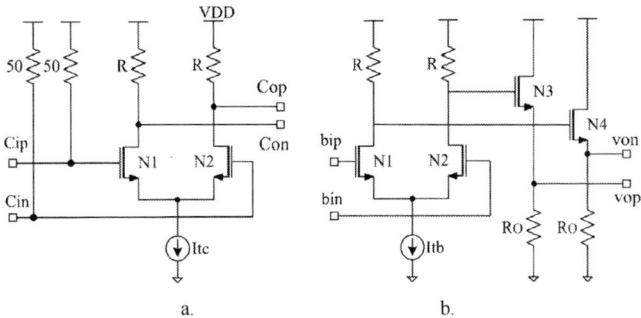

Fig. 4 Schematic of the a. clock buffer b. output buffer.

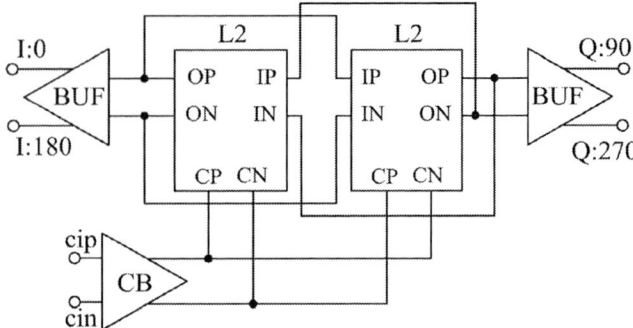

Fig. 5 Block diagram of the frequency divider.

voltage ($V_{DS,sat}$) of the transistor, which implements I_t, needs to be made small. Meanwhile, given the strong short channel effects noticed in this technology, the length of the devices with current source function should not be less than three times the minimum length, thus at least 84 nm. The resulting width/length (W/L) of the transistor with I_t current source function is rather large, 24 µm / 84 nm.

A compromise needs to be made in case of the clock switches N5 and N6 of Fig. 3. As already mentioned, the high $V_{th,N}$ of the NMOS transistors requires the saving of voltage drop on all devices. This means that these transistors need to be made as large as possible in order to decrease the on-resistance (r_{ON}). On the other hand a large W/L ratio would decrease the slew rate of the clock signals. This means that the clock buffer driving capability would need to be increased. An optimum width was found with 6 µm / 28 nm devices.

The circuit has two phases: track and hold. When the clock p is at logic 1, transistor N5 is on, while transistor N6 is off. O_P-O_N tracks the input signal with the delay defined by the RC constant in the output nodes O_N and O_P. This delay is expressed in (1), where the load capacitance C_L is $C_{g,N1}$ of L2 when calculating the τ_{track} of L1.

$$\tau_{track} = R\left(C_{g,N3} + C_{s,N3} + C_{s,N4} + C_L + C_{par}\right) = RC_{OP} \qquad (1)$$

Then, in the hold phase the clock p changes to logic 0 level, disables the differential pair of N1 and N2 and enables the latch formed by N3 and N4. The voltage difference V_{track} which was created in the tracking phase between O_P and O_N, is now regenerated to the full CML voltage V_{CML}. The latching time is

$$\tau_{latch} = \frac{C_{OP}}{g_{m,N3}} ln\left(\frac{V_{CML}}{V_{track}}\right). \qquad (2)$$

τ_{latch} depends on the voltage difference from the previous phase V_{track}, the capacitance in node O_P and the transimpedance gain, g_m of N3. It can be concluded from (2) that $g_{m,N3,N4}R > 1$ must be fulfilled in order to store the state of the cross-coupled pair indefinitely [1]. In this case, the $g_{m,N3,N4}$ needs to be larger than 1.4 mS. This is achieved by 3 µm wide N3 and N4. As it can be seen in equations (1) and (2) the gate capacitance of these transistors increases the rise and fall times of the latch. Thus they should be kept as small as possible. To further keep a low

Fig. 6 Chip micrograph of a. DFF b. frequency divider.

capacitance in the nodes O_P and O_N the width of the tracking transistors N1 and N2 should be made not more than equal to $W_{N3,N4}$, but is even better if they are sized $W_{N3,N4} / 2$ as in [2]. Transistors N1, N2, N3 and N4 were dimensioned in this case with W/L of $3\,\mu / 28\,nm$. The area of a latch is only $10\,\mu m \times 10\,\mu m$.

C. Input and Output Buffers

The DFF system was provided with a clock buffer (CB) in order to ensure the clock levels and common mode are correct at the latches and there is no significant delay between the two phases. Furthermore the clock buffer realizes a 50 Ω low impedance matching and guarantees the functionality of the circuit in case only one clock phase is connected. The schematic of the buffer is depicted by Fig. 4 a. The 50 Ω resistors in the inputs set the common mode and ensure matching with the signal source. The tail current I_{tc} is 550 µA while the output common mode voltage is 750 mV. The clock buffer provides a differential gain of 2.5 dB with a bandwidth of 41 GHz.

The output buffer depicted by Fig. 4 b. is built only for measurement purposes, driving the 50 Ω output impedances. It has a total current consumption of 4 mA. The buffer consists of the differential pair N1 and N2 with 770 µA tail current I_{tb} followed by the single-ended source followers N3 and N4 which have resistors R_O as load. The output impedance is given by $R_O\|g_{m,N3,N4}$ and is dimensioned to be 50 Ω. This solution is preferred instead an entirely differential output because in case of connecting only one output the unbalancing of the circuit will not be significant. The bandwidth of this driver stage is 29 GHz and the gain is -2 dB.

D. The Frequency Divider

Using the blocks designed in sections II B and II C a static frequency divider was designed. The block diagram shown in Fig. 5 consists of a DFF built by using twice L2 of Fig. 2 having the outputs inverted and fed back to the inputs. The outputs of these two latches realize four divided clock phases of 0°, 90°, 180° and 270°. These outputs are connected to two

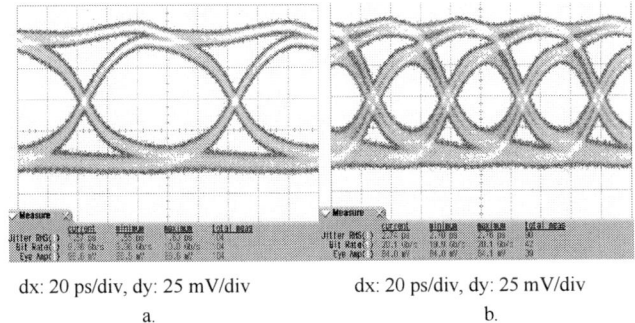

dx: 20 ps/div, dy: 25 mV/div   dx: 20 ps/div, dy: 25 mV/div

a.   b.

Fig. 7 Eye diagrams of the DFF at a. 10 Gbps b. 20 Gbps.

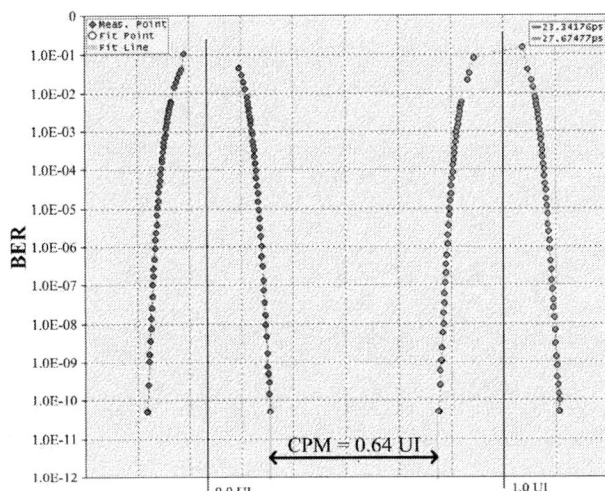

Fig. 8 Bathtub plot at 20 Gbps: BER as a function of horizontal eye opening.

differential output buffers of from section II C.

The static frequency divider has the advantage over dynamic frequency divider that it can operate in a broad band frequency range and provides more phases which can be used in different CDR topologies. On the other hand the division frequency is much lower because of the large capacitance at the outputs of the latches [6].

This I/Q FD was designed with double the size latches than that of Fig. 3, thus $W/L_{N1,N2,N3,N4}$ is 6 µm / 28 nm. The total current is 9.6 mA while the FD realizes a division up to 30 GHz. The layout was realized with great attention to symmetry in order to avoid mismatch of the signal edges and jitter. The complete clock divider circuit has an area of $500\,\mu m^2$.

III. MEASUREMENT RESULTS

The circuits were manufactured in 28 nm CMOS technology. The chip micrographs are depicted by Fig. 6. The area of each chip is defined by the pad-frame of 800 µm × 500 µm on which the active area occupies an insignificant part.

A. D-Flip-Flop

The complete DFF occupies an area of only 490 µm² and has a total power consumption of 5.6 mW. For the measurement of the DFF the supply voltage of 1 V and the reference current of 55 µA were supplied to the chip. In order to have a synchronous clock with the data, this signal came from the *clk* output of the pseudo random bit sequence (PRBS) generator and is a 0 dBm level sinusoidal signal, so the worst case of clock slew-rate is taken into account. The data stream is a $2^{31}-1$ PRBS with 400 mV$_{pp}$ voltage swing. The clock, data input as well as the data output were single-ended signals. The differential signals are calculated as twice as the measured signal on one of the channels. The eye diagrams in Fig. 7 show

TABLE I.   MEASUREMENT SUMMARY OF THE DFF

DR	Eye Ampl.	Eye width @ BER 10^-12	Jitter RMS	Q
[Gbps]	[mV]	[UI]	[ps]	
10	97	0.91	1.56	28.6
15	91	0.69	2.0	8.8
20	84	0.64	2.7	15.8

978-1-4799-4993-9/14 $31.00 © 2014 IEEE

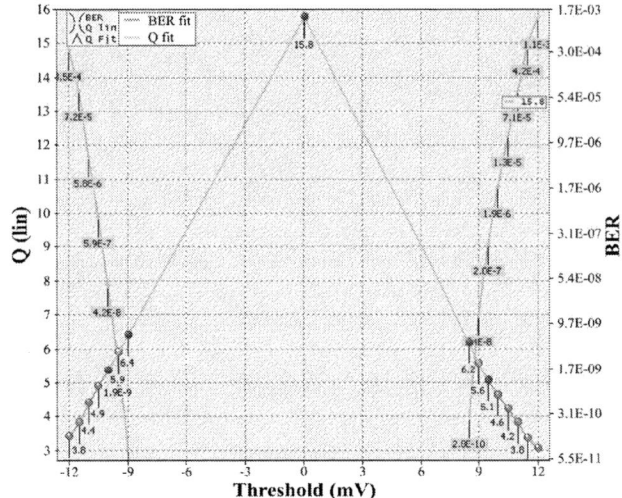

Fig. 9 20 Gbps: BER as a function of vertical eye opening.

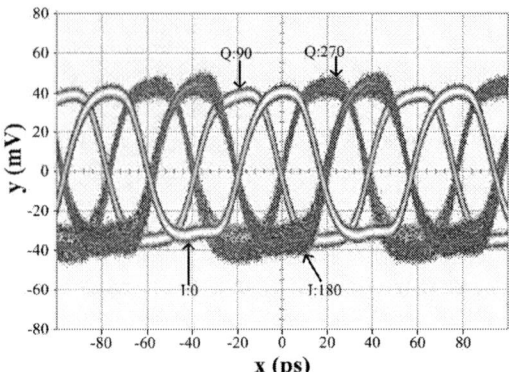

Fig. 10 The four phases of the I/Q frequency divider at 26 GHz.

the functionality of the DFF at 10 Gbps and 20 Gbps, full-rate sampling. A summary of the measurements on the DFF is presented in TABLE I.

Using a sampling oscilloscope the amplitude of the eye and the RMS jitter was measured at 10, 15 and 20 Gbps full-rate operation. Furthermore, using the bit error rate tester (BERT) the bathtub test was performed in order to measure the CPM as explained in Fig. 1. The plot in Fig. 8 shows a CPM of the DFF from Fig. 2 of 0.64 UI for a BER of 10^{-12}. Another parameter measured with the BERT is the Q parameter which quantizes the value of the sampling threshold (V_{Sth} in Fig. 1). The BER dependence on V_{Sth} for 20 Gbps can be seen in Fig. 9. A Q of 15.8 is measured for single-ended output. These measurement results with the BERT can be found as well in TABLE I.

B. Frequency Divider

The FD was supplied also with 1 V and 55 µA reference current. At the clock input a single-ended sinusoidal signal was fed, with up to 3 dBm power. The frequency divider proved to be functional up to 26 GHz. As only a two-channel sampling scope was available, the four phases were plotted in groups of two in Fig. 10. The sampling was stopped when changing the outputs. The output signal swing is as expected 80 mV$_{pp}$ single-ended. A peak-to peak jitter of 2.2 ps and an rms jitter of 0.3 ps was measured at 20 GHz.

IV. CONCLUSIONS

A sampling CML DFF was designed in 28 nm CMOS technology with a very low power consumption and small area. The top data rate of 22 Gbps full-rate was achieved. The high

DR is achieved by an optimum dimensioning of the devices as well as a symmetric and compact layout. The design is comparable with [2] and [3], but with a slightly higher power efficiency, smaller area and higher full-rate sampling speed.

Using the blocks of the DFF a static I/Q frequency divider was built. The functionality of the circuit was measured up to 26 GHz. When a figure-of-merit (FOM) of power consumption per maximum frequency division is introduced, the current design achieves the best mW power/GHz frequency in comparison with other CMOS I/Q frequency dividers as presented by TABLE II. To our knowledge, with 0.034 mW/GHz, this static frequency divider has one of the best FOM reported to date.

ACKNOWLEDGMENT

This work was partly supported by DFG in the framework of CRC 912 HAEC; by BMBF project CoolRF28 (13N11486), by ESF in the framework of the Young Investigators Group 3DCSI and by the Saxon EFRE program under the framework of project CoolOptics.

REFERENCES

[1] B. Razavi, *Design of integrated circuits for optical communications*, McGraw-Hill, 2003, pp. 344-346.

[2] G. von Büren, L. Rodoni, C. Kromer, H. Jäckel, „Low power sampling latch for up to 25 Gb/s 2x oversampling CDR in 90-nm CMOS", *ESSCIRC*, pp. 106-109, 2006.

[3] L. Rodoni, G von Buren, A. Huber, M. Schmatz, H. Jackel, „A 5.75 to 44 Gb/s quarter rate CDR with data rate selection in 90 nm bulk CMOS", *JSSC*, vol. 44, no. 7, pp. 1927-1941, 2009.

[4] S. Yan, Y. Chen, T. Wang, H. Wang, „A 40-Gb/s quarter rate CDR with 1:4 demultiplexer in 90-nm CMOS technology", *ICCT*, pp. 673-676, 2010.

[5] E. Säckinger, *Broadband circuits for optical fiber communication*, Wiley, 2005, pp. 47-95

[6] C. Kromer, G. Von Buren, G. Sialm, T. Morf, F. Ellinger, H. Jackel, „A 40-GHz static frequency divider with quadrature outputs in 80-nm CMOS", *Microwave Components and Letters*, vol. 16, no. 10, pp. 564-566, 2006.

[7] H. Knapp, H.-D. Wohlmuth, M. Wurzer, M. Rest, „25GHz static frequency divider and 25Gb/s multiplexer in 0.12/spl mu/ CMOS", *ISSCC*, vol. 1, pp. 302-468, 2002.

[8] C. Cao, K.K. O, „A power efficient 26-GHz 32:1 static frequency divider in 130-nm bulk CMOS", Microwave and Wireless Components Letters, vol. 15, no. 11, pp. 721-723, 2005.

[9] M. K. Ali,A. Hamidian, R. Shu, A. Malignaggi, G. Boeck, "45 GHz low power frequency divider in 90 nm CMOS", IEEE RFIT, pp. 65-67, 2012.

TABLE II. COMPARISON WITH OTHER I/Q FREQUENCY DIVIDERS

	Technology CMOS	Max freqency	Power	FOM
	[nm]	[GHz]	[mW]	[mW/GHz]
[7]	120	25	22.5	0.9
[8]	130	26	3.9	0.15
[6]	80	44	3.2	0.73
[9]	90	45.5	6.9	0.152
This work	28	26	0.88	0.034

A 2.4 GHz Fast Settling Wake-Up Receiver Frontend

Christoph Tzschoppe, Robert Kostack, Frank Ellinger, *Senior Member, IEEE*
Chair for Circuit Design and Network Theory
Technische Universität Dresden, EUI, IEE
01069 Dresden, Germany

Abstract—**This paper proposes the design of an integrated fast settling analog frontend for application in wireless wake-up receivers. The chip includes a low noise amplifier with matching transformer and integrated balun, a multi-hyperbolic tangent gilbert cell mixer with Sallen-Key filter, a limiting amplifier and a LC-cross coupled digital-controlled oscillator. The chain exhibits a measured conversion gain of 29 dB while only consuming an overall current of 3.3 mA in its on-mode from a 2.5 V supply. The measured input return loss is -19 dB at 2.4 GHz with a corresponding double sideband noise figure of 7.6 dB at an intermediate frequency of 51 MHz. The measured input referred 1 dB-compression point is at -44 dBm. All circuits enable on/off switching which allows operation within on-times of 200 ns. The component chain is optimized to settle in less than 60 ns until the output obtains -1 dB of its steady state amplitude. The proposed chip is fabricated in IHP 130 nm-SiGe-BiCMOS process.**

Keywords—*Low-noise amplifiers, Mixers, Digital-controlled oscillators, Wireless sensor networks, Transformers*

I. INTRODUCTION

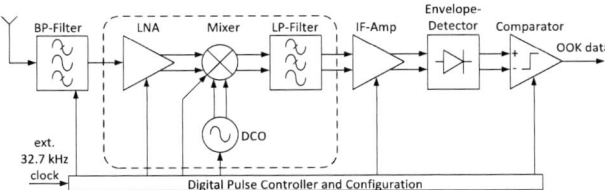

Fig. 1: Simplified block diagram wake-up receiver

In times with a rising amount of wireless sensor networks the continuous monitoring of a wireless channel is mandatory. The commonly used technique is a permanently active receiver observing transmitted data which is not efficient. To enable long battery lifetimes, an ultra-low power wake-up receiver concept based on a novel fast-sampling method was presented in [1]. The average current is reduced due to a periodic on/off switching with a duty-cycle < 0.1 %. This cycle mode requires each circuit to be turned on/off and be able to settle during small on-times. For application in such wake-up receivers an analog BiCMOS frontend with fast turn-on ability to operate within on-times of 200 ns is designed. The circuits are optimized for low current consumption because the RF-circuits are the most current consuming components in the OOK-receiver shown in Fig. 1. The RC-time constants of the bias networks are minimized and the cutoff-frequencies of the signal chain are properly chosen to enable the pulsed operation. In section II the circuit implementation is described more in detail, followed by section III that covers all simulation and measurement results.

II. CIRCUIT IMPLEMENTATION

In this section the building blocks circuit implementation is investigated. As already mentioned the analog frontend consists mainly of five circuit blocks as depicted in Fig.2.

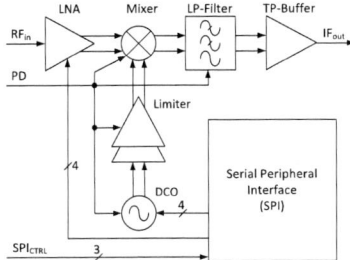

Fig. 2: Block diagram of proposed frontend

The serial peripheral interface (SPI) is used to configure digital control bits of the analog circuits. The LNA is narrowband ($\mathrm{BW}_{-3\,\mathrm{dB}}$ = 230 MHz) due to the LC-load. A tuning of the gain peak to the correct center frequency is mandatory due to process variations. This is also mandatory to adjust the center frequency f_{LO} of the local oscillator. All control signals to the SPI are provided via an Atmega 324A micro-controller. The power down (PD) enable is applied to every building block with the exception of the buffer to perform on/off switching (Fig. 2).

A. Low Noise Amplifier With Balun

The complete LNA schematic is illustrated in Fig. 3. The input matching is realized with a transformer as described in [2]. The inductors L_B and L_E are used to adjust the real part of the input impedance to $50\,\Omega$ while C_1 is used to resonate out the inductive imaginary part. Eq. (1) shows an approximation of the input impedance as proposed in [2].

$$\underline{Z}_{\mathrm{in}}^{*} = \frac{sL_B+s^2(1-k_{BE}^2)L_BL_E(g_m+sC_{BE})}{1+s(L_E+M_{BE})(g_m+sC_{BE})+s^2(L_B+M_{BE})C_{BE}} \quad (1)$$

A center tapped inductor L_C in parallel with the capacitance C_L acts as an integrated balun to perform single-enendt-to-differential conversion. The LC-tank shows a $-3\,\mathrm{dB}$-bandwidth of 230 MHz which is verified by measurements. The current consumption of the whole LNA is $750\,\mu A$ on a 2.5 V-supply. The switching feature is enabled with the MOS switches $M_{1,2}$ and M_4. A low level power down signal (PD) forces the switches $M_{1,4}$ to act as an open while M_2 is closed. A high level PD-signal causes the nMOS-switches $M_{1,4}$ to be

978-1-4799-4993-9/14 $31.00 © 2014 IEEE

closed, the base capacitances of $T_{1,3}$ are discharged via the on-resistances of the switches resulting in a cascode current close to zero. The LNA exhibits a noise figure of $NF_{LNA} = 4.1\,dB$ which was measured with a standalone chip.

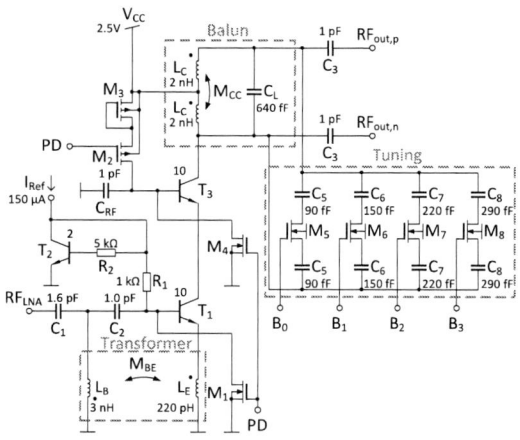

Fig. 3: LNA schematic

B. Multi-Tanh Gilbert Cell Mixer

The mixer consists of a multi-tangent hyperbolic (multi-tanh) transconductance (g_m) cell called hybrid doublet which is described in [3]. Input offsets are applied to superimpose the g_m of the pairs $T_{1,2}$ and $T_{3,4}$ resulting in a more constant g_m as illustrated in Fig. 4.

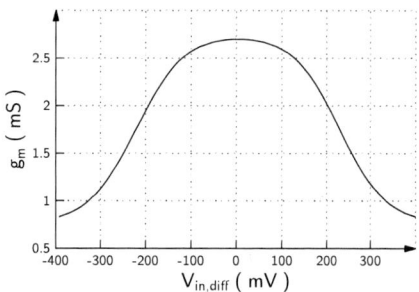

Fig. 4: PSS-simulated transconductance of hybrid doublet

The designed g_m-cell is linear within a range of $-100\,mV \leq V_{in} \leq 100\,mV$ for a 2.4 GHz input tone, which increases the linearity of the proposed frontend because the mixer is the limiting part. The transistors $T_{5,...,8}$ are the gilbert cell switches driven by the large LO-signal. To adjust the output common mode voltage applied to the Sallen-Key (SK) filter the common mode is sensed with the high ohmic resistors R_{CM} and fed back to the active loads $M_{1,2}$ via the error amplifier containing $M_{3,...,6}$. To enable a fast settling the bandwith of the feedback loop is optimized because C_{PC} and R_{PC} are used to compensate a right hand plane zero. A modified Sallen-Key topology as shown in Fig. 6 is added to reduce the voltage noise level at the mixer output. The unity gain amplifier of the SK-topology is realized with common collector stages. Additional gain is not necessary due to the high gain caused by the active loads. The filter is built pseudo

differentially and the real part of the mixer output impedance $R_M = Re\{\underline{Z}_{Mix,out}\}$ is exploited to build the Sallen-Key filter.

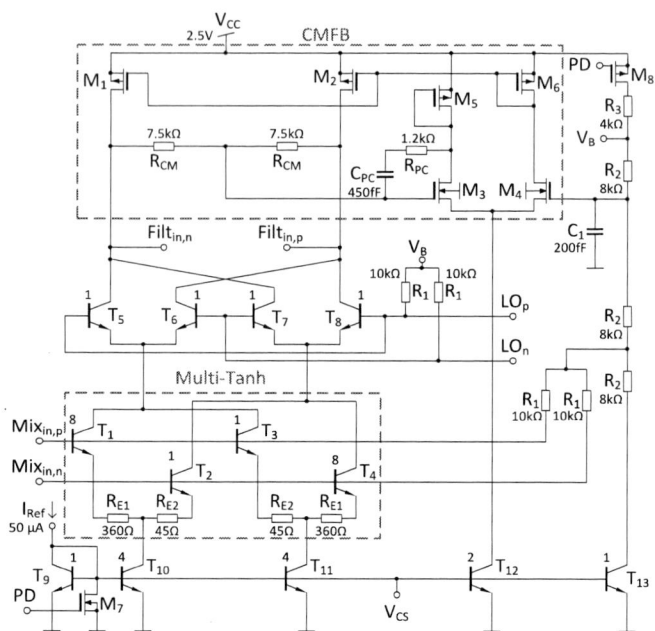

Fig. 5: Mixer core schematic

The -3 dB cutoff frequency $f_{SK,-3\,dB}$ is set to 70 MHz. The emitter followers $T_{20,21}$ are used because for high frequencies the RF signal passes directly through C_{SK2}, hence there is no second order behavior anymore. This is compensated with the $T_{20,21}$ that at least a first order pole is seen for high frequencies at the output. The mixers current consumption is $800\,\mu A$ including the filter with a simulated settling time $T_{S,Mix}$ of 50 ns and a conversion gain $G_{C,Mix}$ of 14 dB.

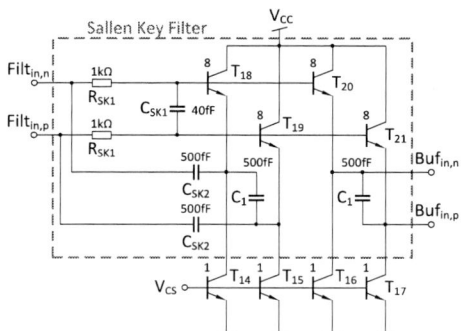

Fig. 6: Sallen-Key low pass filter

C. LC-Oscillator

A cross coupled pair is used to build the negative resistance r_n. Oscillation can only occur when the negative real part of $T_{1,2}$ at resonance frequency ω_0 overcompensates the tank losses and the system open loop phase is $n \times 360°$.

$$r_{CC}\big|_{\omega=\omega_0} = \frac{-r_n r_T}{r_T - r_n} \qquad (2)$$

978-1-4799-4993-9/14 $31.00 © 2014 IEEE   424

The resistance r_{CC} is negative under the condition $r_T > r_n$. A factor of 1.5 times higher r_T is chosen to optimize current and start-up time.

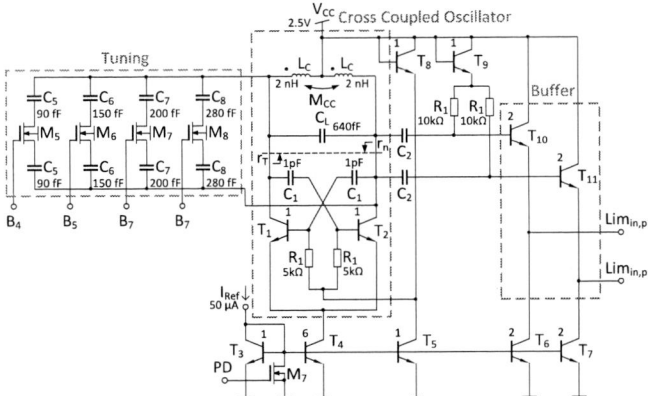

Fig. 7: Oscillator schematic

The common collector stage buffers are applied to ensure that the input impedance of the following stage is high enough to meet the oscillation conditions. The LC-oscillator has a current consumption of only $750\,\mu A$ with a $300\,mV$ peak-to-peak amplitude at the limiter load $\underline{Z}_{L,Lim}$.

D. Limiting Amplifier

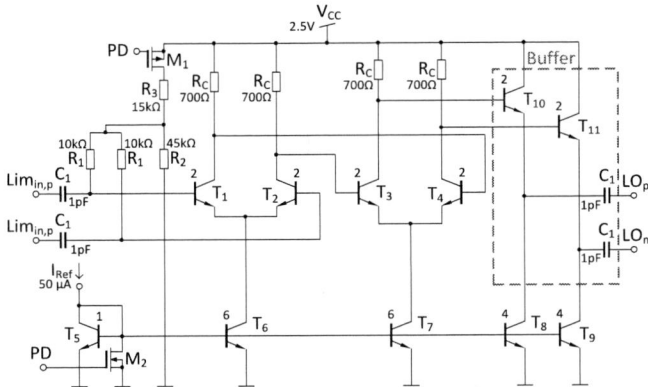

Fig. 8: Limiter schematic

The limiting amplifier performs shaping of the LO-sine wave to a rectangular wave which reduces the mixer noise caused by non-ideal transistor switches. There is only a headroom of $200\,mV$ until the signal swing is limited by the supply. The large LO-amplitude forces a voltage saturation in the differential pairs output which generates harmonics and leads to a $400\,mV$ peak-to-peak rectangular wave amplitude. To drive the mixer load and provide enough gain a relatively large current is needed that the current consumption of the limiter $I_{CC,Lim}$ is $1.1\,mA$.

E. Totem Pole Buffer

The totem pole buffer presented in [4] is used to provide a single-ended output which allows noise figure measurement. The current can be tuned to achieve proper matching.

The buffer is only needed to perform measurements with $50\,\Omega$-equipment, in the future it will not be part of the system.

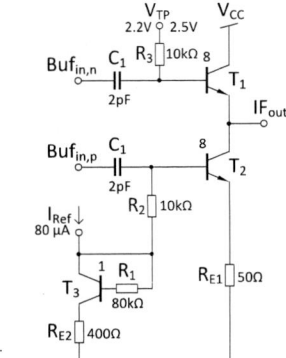

Fig. 9: Totem pole buffer schematic

III. MEASUREMENT AND SIMULATION RESULTS

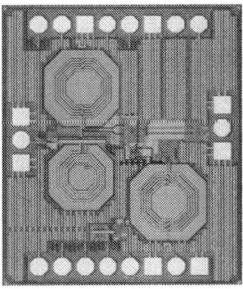

Fig. 10: Chip photo of proposed frontend ($1\,mm \times 1.2\,mm$)

In this section the measurement results of the proposed frontend are presented. Fig. 10 shows the microphotograph of the fabricated chip in IHP $130\,nm$-BiCMOS technology.

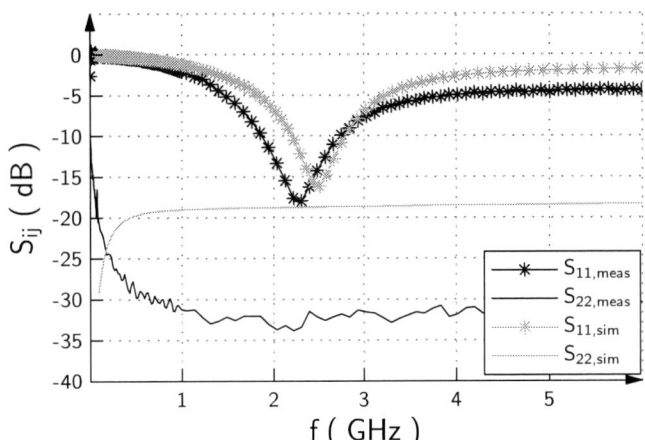

Fig. 11: Measured and simulated S-parameter

The chip was bonded on a PCB to supply all pads. The S-parameter measurement is illustrated in Fig. 11. The input return loss is -19 dB providing proper input matching at the 2.4 GHz RF-port, which confirms the functionality of the transformer. The output return loss S_{22} is better because a larger buffer current is applied for measurement. The measured conversion gain G_C is with 29 dB approximately 6 dB less than simulated and is caused by PCB losses. Also a lower LO-amplitude decreases the mixer conversion gain. The measured double-sideband noise figure NF_{DSB} is 7.6 dB for the LNA mixer combination at $f_{IF} = 51$ MHz (Fig. 12). Four on-chip inverters shape the used power-down signal (sine wave) internally to a rectangular one.

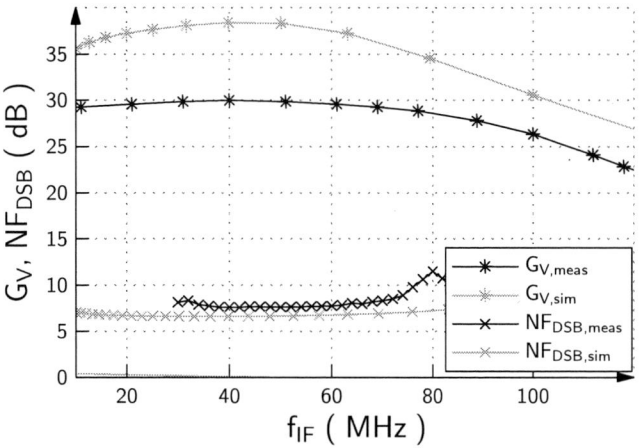

Fig. 12: Measured gain and noise figure

In Fig. 13 the settling behavior of the fabricated frontend is illustrated. The time between applying the power down (sine wave zero crossing) until the IF output signal reaches ± 1 dB of its steady state amplitude is defined as settling time T_S. A time of $T_S = 60$ ns was measured with the Rohde & Schwarz RTO1044 using an active differential probe. A 5 MHz-sine wave PD-signal is used , which is equal to on-times of 200 ns.

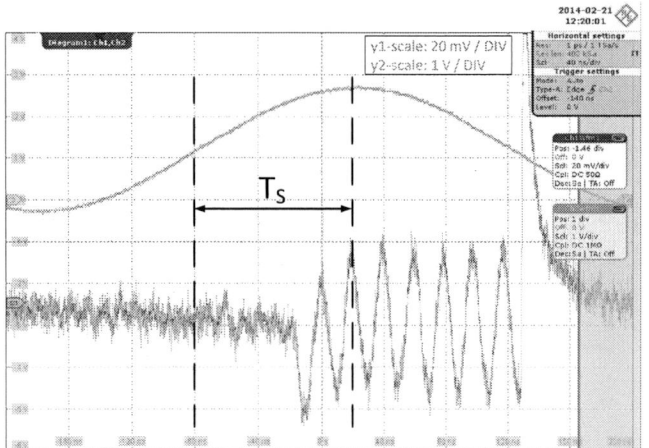

Fig. 13: Power down and IF-output signal

The complete frontend without buffer consumes a current I_{CC} of only 3.3 mA which is very low compared to other low power frontends in TABLE I whereby the conversion

gain G_C of the proposed chip is large with 29 dB. A double sideband noise figure NF_{DSB} of 7.6 dB was measured which is moderate but in relation to the used current very good. The input referred 1 dB-compression point of $P_{g,1\,dB} = -44$ dBm is moderate but linearity is no major design issue in an OOK-receiver frontend.

Ref.	**This Work**	[5] *	[6] ˣ	[7] +
Technology	130 nm BiCMOS	180 nm CMOS	130 nm CMOS	180 nm CMOS
I_{CC} (mA)	3.3	11.6	32	31.65
V_{CC} (V)	2.5	1.8	1.5	1.0
f_{RF} (GHz)	2.4	2.4	3.1-10.6	24
f_{IF} (MHz)	51	39	0	100
G_C (dB)	29	23	22.9-26.4	23.4
NF_{DSB} (dB)	7.6	8.1	4.8-7.7	5.3
S_{11} (dB)	-19	-14	<-10	-12.2
$P_{g,1\,dB}$ (dBm)	-44	-20	-21	-27

* only LNA and mixer
ˣ Quadrature-topology
+ with IF-VGA

TABLE I: Comparison with previous published frontends

IV. CONCLUSION

This work presented a 2.4 GHz analog frontend containing LNA, mixer, DCO and limiter and is fabricated in a SiGe 130 nm-technology. The overall current consumption without buffer is only 3.3 mA. All circuits are optimized to settle within times of several 10 ns which make them applicable in a 2.4 GHz pulsed wake-up receiver as proposed in [1]. This design flow differs from the conventional superheterodyne receiver design, because all circuit blocks are optimzed for low current consumption and are equipped with a fast on/off-switching ability. An integrated balun performs the single-ended-to-differential conversion on chip to enable differential signaling. Configuration of currents and the required DCO-frequency as well as the LC-tank center frequency can be done with the integrated serial peripheral interface (SPI).

REFERENCES

[1] H. Milosiu, F. Oehler, M. Eppel, D. Frühsorger, S. Lensing, G. Popken, T. Thönes, "A 3-μW 868-MHz Wake-Up Receiver with -83 dBm Sensitivity and Scalable Data Rate" European Solid State Circuits Conference (ESSCIRC), 2013

[2] C. Tzschoppe, R. Kostack, J. Wagner, R. Paulo, F. Ellinger, "A 2.4 GHz Fast Switchable LNA With Transformer Matching For Wireless Wake-Up Receivers" European Microwave Week Conference (EuMW), submitted 2014

[3] B. Gilbert, "The multi-tanh principle: a tutorial overview" IEEE Journal of Solid-State Circuits, page 10, 1998

[4] M. Wickert, R. Wolf, F. Ellinger, "Analysis of totem-pole drivers in SiGe for RF and wideband applications" International Journal of Microwave and Wireless Technologies, page 1 of 9, Cambridge University Press and the European Microwave Association, 2011

[5] Hee-Sauk John, Ichhyun. Song, In Man Kang, Hyungcheol Shin, "2.4 GHz ISM-Band Receiver Design in a 0.18 μm Mixed Signal CMOS Process" IEEE Microwave and Wireless Components Letters, VOL. 17, NO. 10, 2007

[6] Bo Shi, Michael Yan Wah Chia, "A CMOS Receiver Front-End for 3.1-10.6 GHz Ultra-Wideband Radio" Proceedings of 36th European Microwave Conference, pages 1829 - 1832, 2006

[7] Chen-Yuan Chu, Chien-Cheng Wei, Hui-Chen Hsu, Shu-Hau Feng, Wu-Shiung Feng, "A 24GHz low-power CMOS receiver design" IEEE International Symposium on Circuits and Systems (ISCAS), 2008

Design of a CMOS Image Sensor with a 10-bit Two-Step Single-Slope A/D Converter and a Hybrid Correlated Double Sampling

Yeonseong Hwang, Seongjoo Lee, and Minkyu Song

Dept of Semiconductor Science, Dongguk University, Seoul, Korea
E-mail : mksong@dongguk.edu

Abstract — In this paper, a low-noise CMOS Image Sensor (CIS) based on a 10-bit two-step Single Slope A/D Converter (SS-ADC) with Hybrid CDS is proposed. In order to reduce the pixel noise, a Hybrid Correlated Double Sampling (H-CDS) is discussed. With this technique, Column Fixed Pattern Noise (CFPN) is drastically reduced by about 55% or more, compared to that of analog CDS only. Furthermore, to overcome low conversion speed of SS-ADC , two-step SS-ADC is proposed. The conversion speed of proposed two-step SS-ADC is 5us, while that of the conventional SS-ADC is about 40us at 25MHz reference clock. The proposed CIS has been fabricated with 0.13um 1-poly 4-metal CIS process, and it has a QVGA (320x240) resolution. The fabricated chip size is 5mm x 3mm, and the power consumption is about 35mW at 3.3V supply voltage. The measured CFPN is 0.8LSB, and the frame rate is 220 frames/s.

Index Terms—Hybrid Correlated Double Sampling, Hybrid CDS, Digital CDS, Two-step Single Slope ADC, SS-ADC

I. INTRODUCTION

CMOS Image Sensor (CIS) has been widely used in various applications such as digital cameras, digital camcorder, CCTV, car security cameras, medical equipment, and so on. CIS is normally composed of a pixel, a Correlated Double Sampling (CDS), an Analog-to-Digital Converter (ADC), a memory, a digital control block, and so on. Among them, ADC is the most important block to convert the analog pixel voltage into the digital code. From the view point of ADC, there exists a single channel ADC, a column parallel ADC, and a pixel ADC. Recently, a column parallel ADC is normally used at the high resolution CIS, because it has many advantages in terms of frame rate, chip area, and power consumption. However, in a column-parallel CMOS image sensor, high Fixed Pattern Noise (FPN) generated by non-uniformity of pixels and readout circuits including ADC is a major factor causing noise. In general, an analog CDS circuit, which consists of capacitors and switches, has been widely used to eliminate the FPN. Although the analog CDS circuit is easy to design and operate, it is difficult to improve the accuracy when using this circuit

This research was supported by the Center for Integrated Smart Sensors funded by the Ministry of Science, ICT & Future Planning as Global Frontier Project.

because of a capacitance mismatch, a clock feed-through error at the switch, and so on. On the contrary, a digital CDS or H-CDS that uses both analog CDS and digital CDS enables the resolution of CDS to improve beyond 10-bit. Nevertheless, they also have drawbacks. Fundamentally, they have a low conversion speed because of digital double sampling for comparison of the reset and the signal. To overcome this drawback of H-CDS, two-step Single Slope ADC (SS-ADC) also described in this paper. The proposed CIS has been fabricated with 0.13um 1-poly 4-metal Samsung process, and it has a QVGA (320×240) resolution.

II. ARCHITECTURE

Fig. 1 shows the structure of the proposed CIS based on two-step SS-ADC with H-CDS. It has a 320 x 240 QVGA resolution with a 2.25um x 2.25um 4–Tr Active Pixel Sensor (APS). In order to increase the minimum layout column pitch, even columns and odd column are divided into two parts of pixel array [6]. And each columns consist of a comparator, sync block, and 5-bit, 6-bit SRAM array for reset signal and pixel signal A/D conversion. Further, a 6-bit fine ramp is used to improve the boundary error of two-step SS-ADC, and a digital correction logic is also adopted to compensate the errors. Therefore, this structure consist of a 6-bit global counter, not a 5-bit global counter. We discuss them in section III one by one.

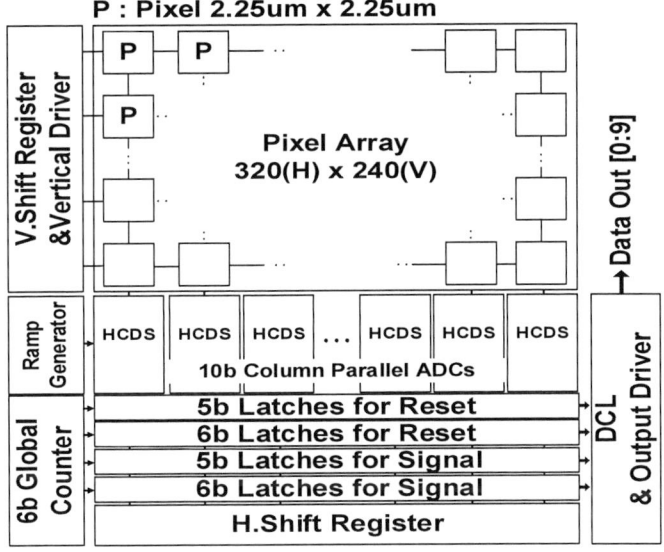

Fig. 1. The structure of the proposed CIS based on two-step SS-ADC with H-CDS

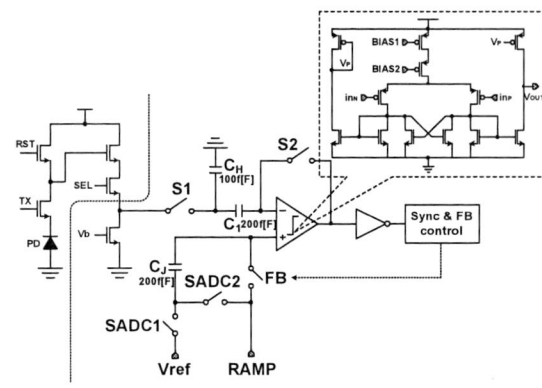

Fig.2. Circuit diagram of a two-step SS-ADC

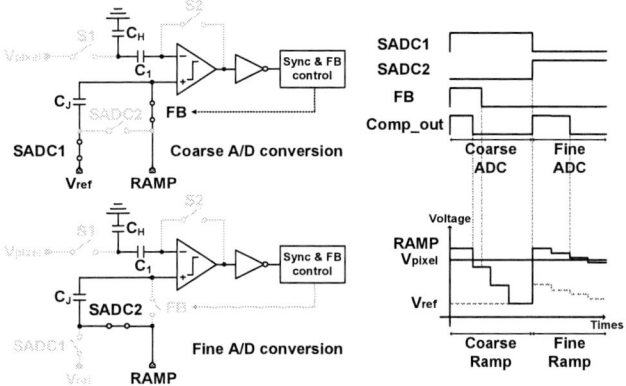

Fig. 3. Principle and timing diagram of two-step SS-ADC

III. CIRCUIT DESCRIPTION

A. The principle of two-step single-slope ADC

When we implement a conventional 10-bit SS-ADC, the maximum number of counting is 1024(2^{10}). Thus the conversion time of SS-ADC is very slow. However, in case of two-step SS-ADC, only 64 (2^5+2^5) counting is enough to satisfy the 10-bit resolution, because ADC is divided into 5-bit coarse block and 5-bit fine block. Therefore, the conversion time of two-step SS-ADC is theoretically 16 times faster than that of the conventional SS-ADC. Two-step SS-ADC is suitable for high speed CIS systems [3]-[8]. Fig.2 shows the circuit diagram of two-step SS-ADC. Two-step SS-ADC is normally divided into a coarse block and a fine block. A precise ramp generator is needed at each step. Fig 3 shows the principle and timing diagram of two-step SS-ADC. At the initial stage, a correlated double sampling between the pixel reset voltage and pixel signal voltage is starting with the switches S1, S2 and the capacitors C_H, C_1. The offset error of comparator and pixel FPN is almost removed with this CDS operation [7]. At the stage of coarse A/D conversion, then, a 5-bit resolution coarse ramp signal is driven to the comparator with the switch (SADC2) off, and the switches (SADC1, FB) on.

TABLE I
FPN ERROR BY MONTE CARLO SIMULATION

	Analog CDS only	Hybrid CDS
0 code error	50 %	77 %
+1 code error	19 %	10 %
-1 code error	20 %	13 %
+2 code error	6 %	0 %
-2 code error	5 %	0 %
Total average error	0.56 LSB	0.23 LSB

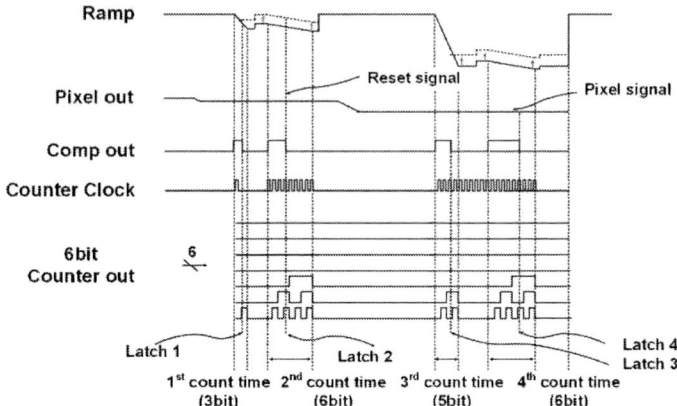

Fig. 4. Timing diagram of the proposed H-CDS

When the pixel signal voltage and the coarse ramp voltage are the same, the output of comparator is changed and the digital code of ramp generator is stored at the first memory. Simultaneously, the changed output of comparator turns off the switch FB. Thus the difference voltage (VH) between Vref and Vramp.coarse is stored at the capacitor CJ. Thus the coarse conversion is finished. At the stage of fine A/D conversion, a 5-bit resolution fine ramp signal is driven to the capacitor CJ with the switch (SADC2) on, and the switches (SADC1, FB) off. Since the fine ramp signal is driven to the capacitor CJ, the other side of CJ is VH + Vramp.fine. When the pixel signal voltage and the fine ramp voltage are the same, the output of comparator is changed and the digital code of fine ramp generator is stored at the second memory. Hence the fine conversion is finally finished

.

B. Hybrid Correlated Double Sampling (H-CDS)

Conventionally, an analog correlated double sampling (A-CDS) has been widely used to compensate the fixed pattern noise (FPN). However, the A-CDS has a lot of circuit noises generated from device mismatch, clock feed through error, and charge injection [9]. In this paper, a hybrid correlated double sampling (H-CDS) is discussed to reduce the circuit noises. Fig. 4 shows the timing diagram of the proposed H-CDS. Since the proposed technique includes the role of A-CDS and a new digital algorithm, we call it H-CDS. Thus H-CDS is different from a normal digital CDS. Generally, the working procedure of H-CDS is divided into two parts, the reset mode, and the signal-capture mode. When the H-CDS is on the reset mode, the reset voltage of a pixel is converted into digital code by two-step SS-ADC. Using a 3-bit coarse ramp and 6-bit fine ramp, the digital code of reset voltage is obtained. The digital code of reset voltage is stored at the additional memory.

978-1-4799-4993-9/14 $31.00 © 2014 IEEE   428

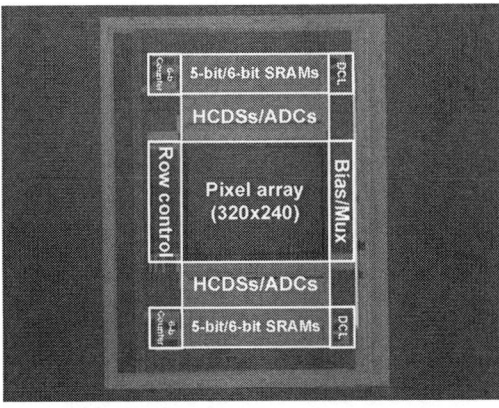

Fig. 5. Microphotograph of the fabricated CIS

Fig. 6. CIS measurement environment

(a)

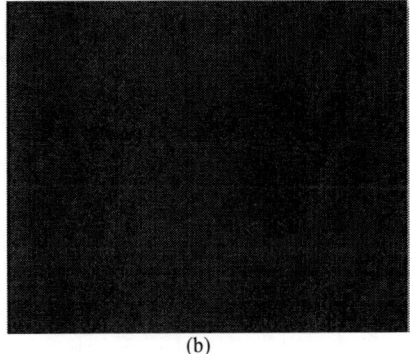
(b)

Fig. 7. The measured image at a strong dark condition
(a) only analog CDS (b) with H-CDS

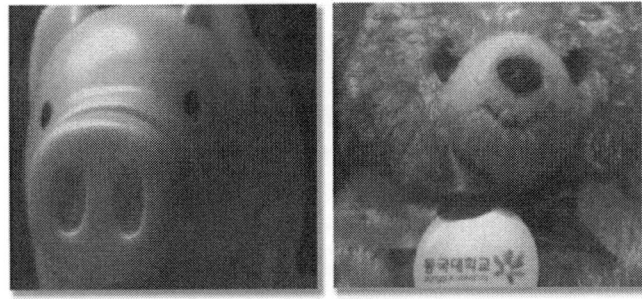

Fig. 8. Measured sample images (@25MHz, 450frames/s)

In case of the reset mode, only 3-bit coarse ramp is enough to obtain the desired value. When the H-CDS is on the signal-capture mode, the signal voltage of a pixel is converted into digital code by two-step SS-ADC. Using a 5-bit coarse ramp and 6-bit fine ramp, the digital code of signal-capture voltage is obtained. The digital code of signal-capture voltage is also stored at the additional memory. In case of the signal-capture mode, 5-bit coarse ramp is required. Finally, the digital code of reset voltage and the digital code of signal-capture voltage are subtracted at the digital subtractor. Thus the working procedure of H-CDS is finished. Tab.I shows the comparison data of FPN error between A-CDS only and H-CDS. The comparison data is based on the Monte Carlo simulation which makes a random mismatching at each column-parallel ADC. If all the columns are normal conditions, the output should be the same code. However, if there is a mismatch for all columns, code errors of up to ±2 codes are generated in the case of using only the analog CDS. However, when using the H- CDS, code error is reduced to ±1 code. As a result, the total average FPN error of H-CDS is smaller by 0.33 LSB than that of A-CDS only. Thus the FPN error of the proposed technique is smaller. Since the number of switches at the H-CDS is much smaller than that of A-CDS only, the clock feedthrough error and charge injection error are drastically reduced. Even though the chip area of H-CDS is bigger than that of A-CDS only, the proposed one has a low noise performance.

IV. MEASUREMENT RESULT

Fig. 5 shows the microphotograph of the proposed CIS with Samsung 0.13um technology. The size of the chip is 5mm x 3mm, and the pixel resolution is 320 x 240 QVGA level. In order to use the chip area efficiently, even column and odd columns are divided into two parts. The upper side of CIS is the part of even columns, and the lower side of CIS is the part of odd columns. At the end of ADC, 5-bit / 6-bit memory are placed to satisfy the specifications of two-step single-slope ADC. Fig. 6 shows the photos of the CIS test environments. The test environments are composed of a CIS measured board with chip on board (COB) technique and a CIS control board with Xilinx-XEM3050 FPGA. The role of FPGA is generating a control input digital signal for CIS, and capturing the output signal of CIS from ADC. Then, the captured digital signals of ADC are transferred to the computer by USB interface. Finally the transferred signals are displayed at the monitor of computer. Form the display of computer, the performance of CIS is measured.

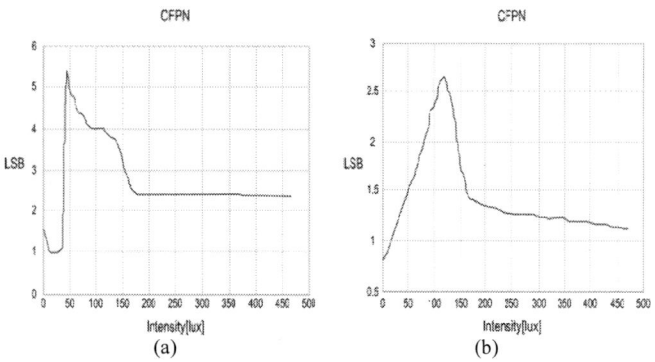

Fig. 9. Measured result of CFPN (a) only analog CDS (b) with H-CDS

TABLE II .
SUMMARY OF THE MEASURED PERFORMANCE AND COMPARISON

	[3]	[8]	This work
Pixel Resolution	400x330 pixels	320x240 pixels	320x240 pixels
Pixel size	7.4umx7.4um	5.6umx5.6um	2.25umx2.25um
Chip size	5mmx5mm	5mmx3mm	5mmx3mm
Dynamic range	N/A	64.8dB	62dB
Frame rate	142 fps	700 fps	450 fps
Column FPN	4.16LSB	3.2LSB	0.8LSB
CDS Type	A-CDS only	A-CDS only	H-CDS
Power Supply	2.5V	2.8V	2.8V(analog) 1.5V(digital)
Power Dissipation	52mW	36mW	35mW

Fig. 7 shows the measured results of the proposed CIS at a strong dark condition. Fig.9 (a) shows the measured results with A-CDS only, and (b) shows the results with H-CDS. In case of Fig.9 (a), there are many picture noises, but the noises are almost reduced at Fig.9 (b). Fig. 8 shows the measured images taken by the proposed CIS. The resolution of images satisfies the specifications of 10-bit and QVGA. The clock of CIS is 25 MHz and it has a high frame rate of 450frames/s. Fig. 9 shows the measured column fixed pattern noise (CFPN) versus light intensity. In case of A-CDS only, the measured CFPN at the strong dark light intensity is about 1.5 LSB, and the maximum CFPN is about 5.3 LSB. The average CFPN from 0 [lux] to 500 [lux] is about 3.2 LSB. In case of H-CDS, the measured CFPN at the strong dark light intensity is about 0.8LSB, and the maximum CFPN is about 2.7 LSB. The average CFPN from 0 [lux] to 500 [lux] is about 1.4LSB. When we compare the average CFPN between A-CDS only and H-CDS, the average CFPN of H-CDS is reduced by about 55% than that of A-CDS only. Therefore, the proposed CIS has a performance of a low noise. Further, the measured power consumption is about 35mW. With the 6-bit global counter to convert the reset voltage into digital code, the power consumption of proposed CIS is lower than the conventional two-step SS-ADC CIS.

V. CONCLUSION

Tab.II shows the performance summary and comparison results between the proposed one and conventional two-step single-slope ADCs. Most of the performances are almost same as the conventional ones except column fixed pattern noise(CFPN). The CFPN is drastically reduced by about 55% or more, compared to that of analog CDS only. This is because the proposed H-CDS is designed with a new digital algorithm. The conversion speed of CIS with the proposed two-step single-slope ADC is 5us, while that of the conventional single-slope ADC is about 40us at 25MHz reference clock. Thus the conversion speed of the proposed CIS is 8 times faster. Further, to improve the boundary error of two-step ADC, a novel error correction technique is also described. The technique is very useful to calibrate the errors.

ACKNOWLEDGMENTS

This research was supported by BK21 PLUS and the Center for Integrated Smart Sensors funded by the Ministry of Science, ICT & Future Planning as Global Frontier Project.

REFERENCES

[1] T. Sugiki et al., " A 60 mW 10 b CMOS image sensor with column-to-column FPN reduction," in Proc. IEEE ISSCC Dig. Tech. Papers, Feb. 2000,pp. 108-109,450.

[2] S. Lim, J. Cheon, S. Ham, and G. Han, "A new correlated double sampling and single slope ADC circuit for CMOS image sensors," in Proc. Int. SoC Des. Conf., Oct. 2004, pp. 129–131.

[3] M. F. Snoeij et al., "Multiple-ramp column-parallel ADC architectures for CMOS image sensors," IEEE J. Solid-State Circuits, vol. 42, no. 12, Dec. 2007, pp. 2968–2967.

[4] J. Lee, S. Lim, and G. Han, "A 10 b column-wise two-stage single-slope ADC for high-speed CMOS image sensor," in Proc. IEEE Int. Image Sensor Workshop, Jun. 2007, pp. 196-199..

[5] Y. Nitta et al., "High-Speed Digital Double Sampling with Analog CDS on Column Parallel ADC Architecture for Low-Noise Active Pixel Sensor," ISSCC Dig. Tech. Papers, Feb. 2006, pp. 500-501.

[6] Y. Yoshihara et al. "A 1/1.8-inch 6.4 MPixel 60 frames/s CMOS Image Sensor with seamless mode change," IEEE J. Solid-State Circuits, vol. 41, Dec. 2006, pp.2998-3006.

[7] M. F. Snoeij, et al, "A CMOS imager with column-level ADC using dynamic column fixed-pattern noise reduction," IEEE J. Solid-State Circuits, vol. 41, Dec. 2006, pp. 3007–3015.

[8] S. Lim, J. Lee, G. Han, "A high-Speed CMOS image sensor with Column-parallel Two-Step Single-Slope ADCs."in Proc. IEEE J. Solid-State Circuits, Vol. 56, no. 3, Mar. 2009, pp.393-398.

[9] Kyuik Cho, Daeyun Kim, and Minkyu Song, "A Low Power Dual CDS for a Column-Parallel CMOS Image Sensor", Journal of Semiconductor Technology and Science, Vol. 12, No.4, Dec., 2012, pp. 388-396.

978-1-4799-4993-9/14 $31.00 © 2014 IEEE

9781479949939